Parasitic Worms of Fish

Parasitic worms of fish 150 years ago, Plate II from the appendix (also referred to as an atlas) by de Blainville to: Bremser, Johann Gottfried. *1828 Traite zoologique et physiologique sur les vers intestinaux de l'homme*, 574 pp. Paris. Some of the worms figured in this plate are today most probably, known as *Caryophyllaeus* (fig. 1), trypanorhynchs (8, 8a, 8b, 9), *Triaenophorus* (10), *Ligula* (16) and *Anthobothrium* (14a).

Parasitic Worms of Fish

by

Harford Williams

and

Arlene Jones

Taylor & Francis
Publishers since 1798

UK	Taylor & Francis Ltd, 4 John St, London WC1N 2ET
USA	Taylor & Francis Inc., 1900 Frost Road, Suite 101, Bristol, PA 19007

British Library Cataloguing in Publication Data
A catalogue record for this book is available from the British Library

ISBN 0-85066 425X

Library of Congress Cataloging in Publication Data are available

Cover designed by John Leath based on original idea by Chris Meacham

Typeset by Santype International Limited, Salisbury, SP2 8PS

Printed in Great Britain by Burgess Science Press, Basingstoke on paper which has a specified pH value on final paper manufacture of not less than 7.5 and therefore 'acid free'.

Contents

Foreword

The Sidney Sussex College Annual for 1984 states: 'We expect to receive in the Lent and Easter terms of 1985, as the second Visiting Fellow of the year, Professor H. H. Williams. During his time here he proposes to complete a major work on the parasites of fishes'. The College's expectations were duly fulfilled, Haffi Williams arrived and ideas for this magnificent volume began to flow particularly during Friday evenings as we refreshed ourselves on the site where Franciscan scholars had wrestled with the theories and opinions of their day. Near to the Senior Combination Room of Sidney Sussex College, where many of Haffi's plans for this book were laid, is a plaque of Welsh slate with an inscription commemorating the residence of Duns Scotus in the convent on the site where the College now stands. How pleasing it is to reflect that there have remained for centuries safe havens for academics to pursue scholarship and share their knowledge and curiosity with others.

As Director of the Open University in Wales, Haffi found his project required help and he invited Dr Arlene Jones to join him in the research and writing of the book. The pages which follow are the fruits of their labours. Scholarship for its own sake, scholarship as a proper enterprise and scholarship giving pleasure to authors and readers alike are the hallmarks of this most imposing work. It is beautifully illustrated in a functional and helpful style, it is written with clarity and a passion for the subject and it offers readers an unsurpassed bibliography. Those of us who have been working for some years in our country's great universities hope we may be excused if we sometimes feel we have gone back into some new version of the dark ages. For me, a university is a community of scholars of all levels of learning, a place where research flourishes and a centre for the curation of knowledge. Harford Williams and Arlene Jones have written a book which will help to keep alive academic traditions, perhaps as the mediaeval friars did so many years ago. Their book proclaims that academic learning is a truly worthy activity which can still flourish despite the demands of accountability, audit and appraisal. We have now entered the era of biodiversity; *Parasitic Worms of Fish* should prove to be a significant contribution to our understanding of the fascinating complexity of the fauna of our planet.

D. W. T. Crompton
University of Glasgow
May 1993

Acknowledgements

We were fortunate to be introduced to the study of parasitology at the University College of Wales, Aberystwyth, by Professor Gwendolen Rees FRS. She has made outstanding contributions to the subject and stimulated in both of us a lasting interest in fish worms. We dedicate this book to Professor Rees.

We have also benefited from long-standing and close associations with both the CAB International Institute of Parasitology and the Natural History Museum, London, two institutions with a traditional and an internationally known interest in fish worms. One of us (H.W.) also gained from working at the Marine Laboratory, Aberdeen, in the early 1960s and from having since retained close collaborative links with the laboratory.

We would also like to thank Professor David Crompton, Department of Zoology, University of Glasgow, for suggesting the project and for his continuing interest and encouragement; Mr. Peter Morgan, Keeper of Zoology, National Museum of Wales for facilities in his department and so much encouragement and support and Professor M. Claridge, Chairman, School of Pure and Applied Biology, University of Wales College of Cardiff, for facilities at the School. The following kindly agreed to read and critize draft manuscripts: for Chapter 1, Drs. David Gibson, R. Bray and L. Khalil; Chapter 2, Professor J. Pearson, Drs. B. James and R.A. and B. Mathews; Chapter 3, Professors A. Dobson, O. Halvorsen and J. Holmes; Chapter 4, Professor M.M.D. Burt, Dr. R.A. Bray and Dr. B. James; Chapter 5, Professor Ron Roberts, Drs. J. Harris and M. Tatner; Chapter 6, Drs. A. McVicar, K. MacKenzie and J.W. Smith and Mrs. Judith Colin. We remain very much indebted to them all for sparing so much time to comment on the text and suggesting many improvements. We accept full responsibility for remaining errors, and omissions.

In the production of first typed drafts of each chapter and an initial reference list, no-one has been involved more closely than the senior author's wife, Margaret. His son, Richard, also volunteered much of his time in checking references against the text and figures in all drafts of various chapters. Mrs Rosemary Browne gave unstintingly of her time and energy towards completing all penultimate versions of each chapter at The Open University, and we are especially grateful for her help. Mrs Marion Williams, University of Wales College of Cardiff, placed the final list of references cited on a word-processor and dealt with all matters connected with copyright.

We wish to thank the following for financial support: The Research Committee of The Open University for bench fees to the National Museum of Wales throughout the period of the work and for financing a research assistant for 18 months; The Open University in Wales and the National Museum of Wales for drawing, photographic and stationery materials and postage costs. In this context the interest and support of Mr. Peter Morgan and Mr. Kevin Thomas (Chief Photographer) both of the National Museum of Wales have been outstanding.

The awards of generous research grants to one of us (H.W.) in 1990, one from the Natural Environment Research Council and another from the Department of the Environment to work on 'Tapeworms as biological tags' and 'Fish helminths as indicators of pollution' were a considerable stimulus and encouragement to complete Chapters 3 and 6 as a basis for proceeding with these two programmes.

One of us (A.J.) is indebted to The Open University and the other (H.W.) to Sidney Sussex College, University of Cambridge, for Visiting Fellowships to enhance progress in the production of this book. The senior author is also grateful to The Open University for a Senior Research Fellowship, from July 1991, and to the University of Wales College of Cardiff for an Honorary Professorship (from October 1991) to complete the work.

Dr. Rachel Bates, Dr. Andrew McCarthy, Dr. R. Bray, Dr. D. Gibson, Dr. K. MacKenzie, Judith Colin and Eileen Harries assisted, over a long period, in a variety of other ways. We are especially grateful to Drs. Gibson and Bray and Eileen Harries for solving so many of our literature and copyright queries and Dr. McCarthy for preparing figures 2.14, 3.1, 3.2, 4.2, 4.3, 4.4 and 4.6. Christopher Keeling, University of New Brunswick, during a visit to the National Museum of Wales, volunteered time in checking figure citations against the references cited. Professor Boris Lebedev, Institute of Soil Biology of the Far Eastern Scientific Centre, USSR Academy of Sciences, Vladivostok, USSR, during a three-day visit to The Open University in Wales in 1988, made a number of invaluable comments and constructive criticisms on the plan and contents of the book.

We are deeply indebted to Mr. John Kenyon, Librarian at the National Museum of Wales in whose care the senior author's reprint collection has now been placed and to the CAB International Institute of Parasitology for the use of all library and other facilities.

Many of the illustrations in the book have been reproduced directly from or are redrawn from published works. For this, we are grateful to the authors whose names, with date of publication, appear in the captions to the relevant figures; the full title and publisher of each source is cited in our reference list. We thank the following publishers/copyright holders for permission to reproduce these figures:

Academic Press
Figs 1.31–1.33, 3.23, 3.24, 4.2, 4.13, 5.13.1–5.13.4, 5.29, 6.5, 6.6
Acta Parasitologica Polonica,
Figs 2.22 (1–2)
American Fisheries Society,
Fig. 5.25
Anales del Instituto de Biologia,
University of Mexico.
Fig. 1.159
Annales de Parasitologie Humaine et Comparee,
Figs 1.10, 1.129, 1.130, 1.131, 4.10, 4.11
Australian Journal of Zoology,
Figs 2.15 (4–5), 2.16 (1–6)
Biology of the Antarctic Seas,
Antarctic Research Series, (American Geophysical Union)
Figs 1.6, 1.7, 1.9, 1.17
Biological Bulletin,
Fig. 2.19(5)
Blackwell Scientific Publications,
Figs 3.13–3.15, 3.18–3.20, 3.21, 3.22, 4.1, 5.1, 5.8.1–5.8.2, 5.14.1–5.14.3, 5.22.1–5.22.5, 5.23.1–5.23.4, 5.24.1–5.24.3, 5.28.

Wm. C. Brown Company Publishers,
Dubuque,
Figs 1.107, 1.108.
Bulletin de l'Institut Fondamental d'Afrique Noire,
Figs 1.23, 1.28.
Bulletin of the British Museum (Natural History),
Figs 1.135, 1.136, 1.138, 1.139, 1.140, 1.149, 1.150, 1.151, 1.152, 1.153, 1.154, 1.155
Bulletin du Museum National d'Histoire Naturelle, Paris
Figs 1.26, 1.217.
Bulletin of the Research Council of Israel,
Fig. 1.21
Bulletin de la Société Zoologique de France,
Figs 1.29, 1.163.
CAB International Information Services,
Figs 6.1, 6.2, 6.3, 6.4.
Cambridge University Press,
Figs 1.13, 1.30, 1.36, 1.37–1.54, 1.56–1.78, 1.82, 1.84, 1.85, 1.86, 1.87, 1.91, 1.93, 1.94, 1.113–1.119, 1.121, 1.126, 1.127, 1.132, 1.133, 1.160, 1.161, 1.162, 1.164, 1.169, 2.2(1), 2.2(2), 2.3(1), 2.3(2), 2.13, 2.19(1–2), 2.22(3), 2.39(9), 3.3, 4.14–4.16, 4.18, 4.19, 4.20–4.26, 4.27, 4.28, 4.29, 5.4.1–5.4.9 5.5.1–5.5.6, 5.7.1, 5.7.2, 5.11.1–5.11.2.
Canadian Government Publishing Centre, (Fisheries and Oceans Canada)
Figs 1.1, 1.2, 1.3, 1.4, 1.5, 1.19.
Canadian Journal of Zoology
Figs 1.186, 1.187, 1.195, 1.196, 1.197
Ceylon Journal of Science,
Fig. 1.141
Czechoslovak Academy of Sciences
Figs 1.173–1.185
Experimental Parasitology,
Figs 2.7, 2.8
Fish Pathology (Japan)
Figs 5.12.1–5.12.5
Folia Parasitologica
Figs 2.33(2–7) 2.36(2,3,6–10), 2.37(2 & 6)
Journal of the Fisheries Research Board of Canada,
Fig. 1.92
Journal of Helminthology,
Figs 1.110, 1.193, 1.194, 1.200–1.204, 2.4(1–15), 2.15(9), 2.39(1–8), 4.30, 5.18, 5.29
Journal of Parasitology,
Figs 1.8, 1.55, 1.123, 1.124, 1.125, 1.128, 1.146, 1.170, 1.214, 1.216, 1.220, 1.221, 1.222, 2.11(1–10), 2.12, 2.14, 2.15(1–2), 2.15(10), 2.15(12), 2.17(1–9), 2.19(3), 2.21(1), 2.21(2), 2.21(4), 2.34(2–5, 7–9 & 11–13) 2.38(1–7), 5.21.1–5.21.6, 5.32.1–5.32.4
Mémoires du Museum national d'Histoire naturelle,
Fig. 1.14
McGraw Hill
Figs 1.95, 1.98, 1.104, 1.105, 1.109, 1.120, 1.166, 1.218, 1.219
New Scientist,
Fig. 3.5
Ophelia,
Figs 2.19(4), 2.23(1–2), 2.23(3), 2.23(4), 2.24, 2.25, 2.27(2–4), 2.27(5), 2.27(6), 2.28

Oxford University Press,
Figs 1.35, 4.3
Parazit ologicheskii Sbornik
Fig. 4.7
Pergamon Press
Figs 1.11, 1.205–1.212, 3.4, 3.6–3.12, 3.16, 3.17, 4.5, 4.6, 5.3, 5.4.1, 5.4.2, 5.4.3, 5.4.4, 5.4.5.
Philosophical Transactions of the Royal Society of London, B.,
Fig. 1.81
Proceedings of the Helminthology Society of Washington,
Figs 1.122, 1.165, 1.213, 2.9(1–10), 2.21(3), 2.22(4–5), 4.4
Révue Suisse de Zoologie (Museum d'Histoire Naturelle)
Fig. 1.27
Sea Grant Quarterly (ISSN 0199–137X),
Figs 5.9.1, 5.9.2
Systematic Parasitology (Kluwer Academic Publishers)
Fig. 1.156
Taylor & Francis
Figs 1.16, 1.80, 1.157, 1.167, 1.168, 1.172, 2.26
Transactions of the American Microscopical Society,
Figs 1.144, 1.171
Tulane Studies in Zoology,
Fig. 1.12
University of Hawaii Press,
Figs 5.26.1–5.26.3
University of Mexico
Figs 1.143, 1.145
Universite Montpellier II,
Figs 1.79, 1.83, 1.88, 1.89, 1.90, 1.106, 1.111, 1.112, 2.29, 2.32, 4.8, 4.9, 5.6.1–5.6.2
University of South Carolina Press,
Figs 2.5, 4.12
The University of Wisconsin Press,
Figs 5.10, 5.17, 5.19, 5.20
Vestnik Ceskoslovenske Spolecnosti Zoologicke
Fig. 2.35(2–5)
Vie et Milieu,
Figs 1.24, 1.25
Zeitschrift fur ParasitenKunde
Figs 2.15(3), 2.20(1–3), 2.30(1–7), 2.31(1–9), 5.2.1–5.2.6, 5.16

Special thanks are due to Taylor and Francis Ltd. and the various editors allocated to this book, namely Dr. J. Cheney, Sara Waddell, Philippa McBain and Janie Curtis. Their patience, understanding and advice are all deeply appreciated.
The senior author wishes to thank his family for enormous patience, encouragement, help and understanding at all times of someone who allocated most of his spare-time, from 1970–1990, to the study of fish worm biology.

Introduction

Almost 400 of 1000 references cited by McGregor (1963) in 'Publications on fish parasites and diseases 330 B.C.-AD 1923' are on fish worms but few were described before 1800. Descriptions began to appear from around this date but many were brief and poorly figured. It is obvious from the plate used as frontispiece to this book that tetraphyllidean and trypanorhynch tapeworms were recognized and, most probably, *Ligula* and *Triaenophorus*, now very well known genera of fish tapeworms. All main groups of fish worms, monogeneans, cestodes, digeneans, aspidogastreans, nematodes and acanthocephalans, were described in the first half of the nineteenth century. During the second half of that century and at the beginning of the twentieth, some extraordinarily detailed accounts of fish parasitic worms, flatworms in particular, were published, some surpassing many present-day taxonomic descriptions. Among the most prominent pioneers were Braun, Cobbold, Diesing, Kolliker, Leidy, Leuckart, Linstow, Stossich, Beneden and Zschokke, whose publications are cited in the comprehensive two volumes on *Vermes* of Dr. H.G. Bronn's *Klassen und Ordnungen des Thier-Reichs*, published from 1879–1900 (See Braun 1879–1900). These remain an excellent source of early information on fish flatworms. In the same series, a two-part monograph on the Acanthocephala by Anton Meyer appeared in 1932 and 1933. In 1926, Yorke and Maplestone had published their book on the nematode parasites of vertebrates, many of them fish worms and this work was reprinted in 1962. Fuhrmann's excellent accounts of trematodes and cestodes appeared in 1928 and 1931, respectively, in Kukenthal and Krumbach's *Handbuch der Zoologie*. One of the first monographs to be devoted entirely to a group of fish worms was Southwell's work on tetraphyllidean tapeworms which appeared in 1925. In 1936 a comprehensive account of all tapeworms was published by Joyeux and Baer in the Faune de France series and this remains important for everyone interested in fish tapeworms. The same can be said of Sproston's monograph in 1946 on Monogenea, two volumes on the Trematoda by Dawes in 1946 and 1947, Wardle and McLeod's classic on the *Zoology of Tapeworms* in 1952, Hyman's comprehensive works on platyhelminths, nematodes and acanthocephalans in 1951, and Yamaguti's series of five volumes, from 1958 to 1963, on the Digenea, Cestoda, Nematoda, Monogenea and Acanthocephala. Other publications up to about 1960 include Baer and Euzet (1961) on monogeneans and Baer and Joyeux (1961) on platyhelminths and acanthocephalans in the *Traité de Zoologie* series edited by P. Grassé.

There followed an explosion in the literature on fish helminths. Of the 700 or so papers listed in the 1935 volume of Helminthological Abstracts about 30 were on fish worms, whereas in 1985 there were 400 from a total of about 5000. From 1985 to 1990 the same rate of publication appears to have been sustained. Concomitant with this increased number of research papers, many books, monographs, taxonomic works and checklists have been published, either on particular groups of fish parasites or on the parasites of particular groups of vertebrates including fish. Some of these are listed in the appendix to this introduction. In compiling this list we remained conscious of our lack of coverage of the Russian literature in particular. For example it is pointed out by Gusev, Roitman and Shul'man (1984) that in the five years following the 6th All-Union Conference on fish parasites and diseases held in 1976, 750 papers on fish parasites were published by Soviet scientists covering taxonomy, life-cycles, ecology, physiology, biochemistry and zoogeography. The list also excludes Skryabin's many invaluable volumes on cestodes, trematodes and nematodes.

The expansion of interest in fish helminths is also reflected in the increased number of papers on the topic at the International Congresses of Parasitology between 1961 and 1990 and the European Multicolloquia in Parasitology from 1974–1988. It is therefore surprising that there has been no previous attempt to cover all groups of fish helminths in one volume, given that almost 10,000 papers on fish worms have been published in the last twenty years.

A major aim, therefore, was to select important contributions from this literature and to bring it together in one volume under appropriate chapter headings. Difficulties arose in finalizing the contents of each chapter and our final choice will become evident in this introduction and in the contents list for each chapter. But it was not only our personal interests that influenced coverage. Shortly after commencing the work we wrote to the secretaries of 56 parasitological societies of the world to obtain some measure of the interest of members in fish worm biology. From 33 replies we estimated that of the 8000 or so active members of the various societies about 3500 were helminthologists and almost 500 of these were fish helminthologists. It is apparent from papers published by these fish helminthologists that their interests were varied and wide. We have attempted to prepare a reference work that will cater for all these interests.

Since more papers are published annually on the taxonomy and systematics of fish worms than on any other aspect of their biology, the volume begins with a chapter which indicates the diversity of body form amongst adult worms, and demonstrates the enormous scope for further systematic research in the areas of descriptive, revisionary and interpretative taxonomy. The chapter is intended to be a preliminary guide only to the literature on the taxonomy and systematics of the main groups of fish worms and an introduction to those mentioned in the following five chapters. We wish to encourage interest in the infinite variety of body form in adult worms in relation to the topography of the sites in which they are found because, as so clearly shown by Hayunga (1991), adaptation to the microenvironment, is a key to the success of parasitic worms, especially those which inhabit the alimentary canal and its associated organs. It is also clear from the figures in Chapter 1 that a wide range of reproductive systems have evolved amongst fish helminths. This variation and that shown by the eggs and larval forms mentioned in Chapter 2 indicate the scope for further research on the various reproductive strategies which must be involved, as discussed for instance by Nollen (1983) and Tinsley (1983).

Chapter 2 reviews the life-cycles of helminths which mature in fish, with emphasis on patterns of transmission for each group. The life-cycles reflect the close associations between the evolution, ecology and behaviour of host and parasite. Some groups have been more extensively studied than others and, where this is the case, examples have been selected to illustrate life-cycle patterns. It is also apparent that compared with the number of fish helminths now known, only relatively few have had their life-cycles partly or wholly elucidated.

Chapter 3 attempts to interpret and discuss the ecology of fish worms. In the last twenty years, ecology has become a fashionable and, not unexpectedly, a highly controversial area of research. Its importance is beyond doubt because progress will be made in further taxonomic and life-cycle work only when it draws upon the methodology and knowledge of ecology. It can also be argued that good ecological research is dependent upon sound taxonomic work. Consequently the scope for further research in ecology is enormous, as we hope the chapter indicates. Very few species have been the subject of intensive field and laboratory studies through to a mathematical interpretation of the results. As Table 3.1 shows, biotic and abiotic factors are referred to in host/parasite ecological relationships but the challenge of planning research programmes with due regard for more than one of these is formidable. Here, most probably, we have at least a partial explanation of why we have relatively little knowledge to explain seasonality and zoogeography. Nevertheless we believe that what has already been achieved in these areas is so interesting that our comments will encourage further progress. Research on the ecology of fish parasites usually depends on knowledge of host ecology and biology. The relationship is so close that parasites have often been used as indicators of host biology. This is considered in Chapter 3, and its potential practical applications in fish and fisheries research are covered in Chapter 6. The greatest challenge of all lies in the area of community ecology, and to encourage more research on this complicated area a comprehensive account of site selection is given in Chapter 3. This information on sites when used together with the two appendices giving more information on host/parasite systems and biogeographical studies is intended to facilitate choice of hosts and locality for further work in community ecology. It is also our aim that a fairly comprehensive section on sites will constantly remind everyone of the need for more accurate data on such aspects as prevalence, intensity and competition/negative interaction towards a better understanding of the concepts of niche, environment and habitat. Further research on sites should also consider the variety of helminth body forms found in these sites, especially variations in the reproductive systems, and strategies not only for invading sites but also those for liberating eggs or larvae to continue the life-cycle. In this context helminths of the body cavity, for example, present some intriguing problems for further investigation.

After considering the variety of adult worms which are to be found on or in fish, and their life-cycle patterns and ecology, it was appropriate to consider, in Chapter 4, some of the more precise host/parasite relationships that were implicit in the first three chapters. This chapter concentrates on host-specificity and its importance to the parallel evolution of hosts and parasites. Relationships between and amongst the phyla, and the morphological, physiological, ecological and zoogeographical factors influencing host-specificity are discussed.

Chapter 5 on host-parasite interactions discusses some of the consequences for the host of site-selection by a parasite, beginning with the pathology of various organ systems. An important consequence of parasitism is susceptibility of the host to stress, whether direct or indirect. The host's main defensive mechanism is an immune response, so the chapter concludes with a discussion of this response to monogeneans, cestodes, digeneans, nematodes and acanthocephalans.

Chapters 1–5 underline the importance of fundamental research in understanding economic aspects of fish worm biology and also show clearly that in the last twenty years or so an explosion of knowledge has occurred with regard to research on these parasites. Consequently, there is now greater acknowledgement of the socio-economic impact of fish diseases in nature, or under aquaculture conditions, their significance for other vertebrates, including man, and the need for control in fish farming conditions. These areas of particular concern to man are discussed in Chapter 6. Man has influenced the abundance and prevalence of fish worms by the inadvertent transfer of diseased fish, by man-made changes in water bodies and by pollution. The chapter ends with an account of the value of fish worms as biological indicators, as biological control agents and, because they are so abundant and relatively cheap to obtain, as ideal models for use in teaching and research.

Because of the need to limit the number of pages in this book we could not include a chapter on physiology and biochemistry. We have, therefore, included an appendix at the end of Chapter 6 to indicate sources of information on these aspects of the biology of fish worms.

Appendix: Sources of general information on fish worms*.

Diseases of fish	van Duijn (1956, 1973)
Parasitology of fish	Dogiel, Petrushevski and Polyanski (1958) Translated by Kabata (1961)
The biology of animal parasites	Cheng (1964)
Introduction to Animal Parasitology	Smyth (1976)
Parasites of North American freshwater fishes.	Hoffman (1967)
Bibliography of the monogenetic trematode literature of the World 1758–1969	Hargis, Lawler, Morales-Alamo and Zwemer (1969, 1971)
How to know the trematodes	Schell (1970)
A symposium on diseases of fishes and shellfishes	Sniezko (1970)
Aspects of Fish Parasitology	Taylor and Muller (1970)
The biology of trematodes	Erasmus (1972)
Diseases of Fish	Mawdesley-Thomas (1972)
Biochemistry of Parasites	von Brand (1973)
CIH Keys to the Nematode parasites of vertebrates Nos. 1, 2, 3, 4 & 6	Editors: Anderson, Chabaud and Willmott (1974–1983). See Hartwich (1974), Chabaud (1975a,b, 1978) and Petter and Quentin (1976).
Biology of the Turbellaria	Riser and Morse (1974)
Animal parasites: their life-cycles and ecology	Olson (1974)
Advances in the zoology of tapeworms	Wardle, McLeod and Radinovsky (1974)
Ecological Animal Parasitology	Kennedy (1975)
Pocket book of fish diseases for veterinarians and biologists	Amlacher (1976)
Ecological aspects of Parasitology	Kennedy (1976a)
Fish pathology	Roberts (1978a)
Methods of cultivating parasites *in vitro*	Taylor and Baker (1978)
Principal diseases of farm-raised catfish	Plumb (1979)
The Biology of Parasitism	Whitfield (1979)
Biochemistry of Parasitic Helminths	Barrett (1981)
Foundations of Parasitology	Schmidt and Roberts (1981)
Modern Parasitology—a textbook of Parasitology	Cox (1982)
Ecology of marine parasites	Rohde (1982a)
Biology of the Eucestoda Volumes I & II	Arme and Pappas (1983)
The physiology of trematodes	Smyth and Halton (1983)
Guide to the parasites of fishes of Canada, Parts I & III	Editors: Margolis and Kabata (1984, 1989) See Beverley-Burton (1984) and Arai (1989)

* Many major taxonomic works are omitted from this list because they are cited in Chapter 1.

Diseases of marine animals	Kinne (1980, 1984)
Diseases caused by Metazoans: Helminths	Rohde (1984c)
Biology of the Acanthocephala	Crompton and Nickol (1985)
Parasites and diseases of fish cultured in the tropics	Kabata (1985)
Trematodes of North America	Schell (1985)
Handbook of tapeworm identification	Schmidt (1986)
Diseases and parasites of marine fishes	Möller and Anders (1986)
General parasitology	Cheng (1986)
A guide to the freshwater fish parasites of the USSR, 2 volumes	Bauer (1984b, 1987)
Disease diagnosis and control in North American marine aquaculture	Sindermann and Lightner (1988)
Parasitology: the biology of animal parasites	Noble, Noble, Schad and MacInnes (1989)
The physiology and biochemistry of cestodes	Smyth and McManus (1989)
Marine fish parasitology	Grabda (1991)

To
Professor Gwendolen Rees FRS
for introducing us to the study of
Parasitology

1 The variety of fish worms

Introduction

This chapter presents a panoramic view of adult parasitic worms living on or in fish. It is not intended to be a detailed account of their taxonomy or evolution because the phylogenies of fish worms are discussed in Chapter 4. Our aim is to indicate the infinite variety of fish helminths, to emphasize that no other group of vertebrates has such a diversity of helminth species and that many of the groups referred to are unique to fish. In planning the chapter we used the following sources for the taxonomic groups mentioned and many references cited in these publications: Ehlers (1985a) for Turbellaria; Baer and Euzet (1961), Beverley-Burton (1984), Sproston (1946), Hargis and Thoney (1983), Bykhovskii (1937, 1957), Oliver (1982), Lambert (1982) and Lebedev (1988) for Monogenea; Williams, Colin and Halvorsen (1987) for Gyrocotylidea; Mackiewicz (1981) for Caryophyllidea; Schmidt (1986) for Cestoda; Gibson (1987) and Gibson and Chinabut (1984) for Digenea and Aspidogastrea; Durette-Desset (1983)[1] and Stone, Platt and Khalil (1983) for Nematoda; Amin (1982, 1985a), Crompton and Nickol (1985), Golvan (1958, 1959, 1960–1961, 1962) for Acanthocephala and Baer and Joyeux (1961) for platyhelminths generally. Although the term 'worm' covers all helminths, the Annelida, and to a certain extent the Coelenterata, its use in this book is restricted to helminths.

Fish are hosts to representatives of three major groups of helminths, the Platyhelminthes (flatworms), Nematoda (roundworms) and Acanthocephala (spiny-headed worms). When, why and how such a large variety of worms representing these three groups became established in fish is a fascinating and little-known topic. Kurochkin (1984, 1985) estimated that at least 30,000 species of helminths had already been described from marine animals. Since freshwater fish are also heavily infected with a variety of helminths, it can be assumed from Kurochkin's estimates, that at least 30,000 species are already known from fish. Between 20,000 and 30,000 species of fish are known and each is likely to harbour a few species of helminths. It follows that a very large number remains to be discovered and described, as was suggested by Rohde (1982a, 1986a) but further pursuit of estimating the total number of extant fish helminth species should be done only in the light of views expressed by May (1988).

The variety already found seems infinite and species are often difficult to identify, but within each of the three major groups, distinct classes, orders and families are recognisable. Adult representatives of these are illustrated and briefly described in this chapter with emphasis on distinctive body shapes and major morphological features as preliminary aids to their identification. The specificity of classes, orders and families to particular categories of fish is mentioned and key references will be given for those readers who may wish to identify to genus or species level. The examples chosen and the sequence of presentation are intended to reflect, if possible, views on the origin and evolution of the three groups.

We begin with the Platyhelminthes because it is generally agreed [although not by Malmberg (1986) and Ubelaker (1983)] that, within this phylum the 'Turbellaria', a largely free-living group, radiated as surface-creeping carnivorous forms and gave rise to parasitic worms (Clark, 1979). In contrast to the general agreement on the evolution of platyhelminths, Clark emphasized the great lack of consensus on the origin of parasitism in the Nematoda and Acanthocephala. Discussions on the origins and relationships of these two groups are, however, given by Anderson (1984), Chabaud (1982 a,b), Conway Morris and Crompton (1982a,b), Inglis (1985), Lorenzen (1985) and Stone, Platt and Khalil (1983). The evolution of helminths is further discussed in Chapter 4.

---

[1] Includes title for Nos. 1–9 of the CIH Keys to nematodes of vertebrates and index to whole series.

Platyhelminthes

There are at least eight different schemes of classification of the phylum Platyhelminthes (Brooks, O'Grady and Glen, 1985a, b). For convenience, we have avoided taxonomic controversy and adopted a division of the phylum into the 'Turbellaria', Digenea, Monogenea, Gyrocotylidea, Amphilinidea, Caryophyllidea and Cestoda. According to Ehlers (1984, 1985a, b) the 'Turbellaria' are a paraphyletic group which should not be retained as a valid taxon. He also grouped together the Trematoda, Monogenea and Cestoda as the Neodermata because the tegument of these well-known parasitic taxa is quite different from that of the Turbellaria which have a monolayered, normally ciliated and cellular body surface. In parasitic forms the tegument without exception is a peripheral syncytial dermis or neodermis with cytoplasmic connections to nucleated regions below the basement membrane. Ehlers included the Gyrocotylidea, Aphilinidea and Cestoidea within the Cestoda. The Neodermata of Ehlers is the equivalent of Cercomeridea of Brooks, O'Grady and Glen (1985a,b).

'Turbellaria'[2]

These are mostly free-living flatworms in terrestrial, freshwater and marine environments. The ancestral parasitic platyhelminth (i.e. the proto-neodermata) would have been a free-living dalyellioid-like 'turbellarian' whose ciliary locomotion was gradually replaced by an increased dependence upon muscular, leech-like movements. The worm then became associated with an ecto-commensal on other animals and developed anterior and posterior adhesive pads (Gibson, 1987). It is thought that from such a worm the ancestors of the Cercomeromorphae (the Monogenea-Cestoda line), i.e. the proto-cercomeromorphs and the ancestors of the Trematoda, the prototrematodes, diverged. The proto-cercomeromorphs became associated with vertebrates and the prototrematodes with bivalve molluscs and developed anterior and posterior muscular suckers. Present day descendants of a proto-cercomeromorph might be *Micropharynx parasitica* and *M. murmanica* (Figure 1.1).

More information on symbiotic and parasitic Turbellaria may be found in Ax (1963), Ehlers (1985a), Ball and Khan (1976), Beverley-Burton (1984), Cannon & Lester (1988), Jennings (1974), Kent (1981) and Kent and Olsen (1986). Some of these discuss affinities of 'Turbellaria' and Monogenea. For instance, Beverley-Burton (1984) discussed whether *Udonella caligorum* (Figure 1.2) is a turbellarian or monogenean, and accepted that it should be placed in the order Udonellida of the neoophoran 'Turbellaria'. The Neoophora have been regarded as a sub-group of 'Turbellaria' and they are so-called and characterized by the fact that the yolk glands are separate from a germarium. Recent work on sperm morphology by Justine, Lambert and Mattei (1985) suggests affinities between the neoophoran 'Turbellaria' and the Monogenea. In the other sub-group, the Archoophora, the yolk glands and germarium are combined to form an ovary. Although these two sub-groups have been prominent in discussions of the origin and evolution of Platyhelminthes from 'Turbellaria', Karling (1974) abandoned the division of the class into the Archoophora and Neoophora. The controversy over the relationships between the Turbellaria and Monogenea is demonstrated by two studies. *Udonella* was investigated by Nichols (1975) because of its similarities to the turbellarian, *Temnocephala*, whilst Williams (1981) concluded that the

[2] The name Turbellaria was abandoned by Ehlers (1976, 1985a).

Temnocephaloidea are not closely related to monogeneans. Such controversy influenced our choice of examples to illustrate adult monogeneans of fish.

Monogenea

Authorities on this group include Bykhovskii, Dawes, Euzet, Hargis, Kearn, Llewellyn, Price, Sproston and Yamaguti, as may be seen from references cited in Yamaguti (1963a), Schmidt and Roberts (1981), Hargis and Thoney (1983) and Lebedev (1988)[3]. There are five major monogenean groups (two families and three orders) according to Beverley-Burton (1984), Lambert (1982), and Llewellyn (1982).

1. Family Microbothriidae

These are parasites of the body surface of elasmobranchs and their affinities are uncertain. *Microbothrium apiculatum* (Figure 1.3) is a representative of this family and was once thought to be related to rhabdocoel turbellarians. Beverley-Burton (1984) concluded that it is a monogenean and suggested that microbothriideans may be degenerate monocotylids (see below). Two new microbothriids were described by Cheung and Nigrelli (1983) and Cheung and Ruggieri (1983).

2. Order Dactylogyrida

Family Acanthocotylidae

Within the Dactylogyrida the family Acanthocotylidae occupy a special position (Llewellyn. 1982). They are normally parasites of the body surface of elasmobranchs. A representative, *Pseudacanthocotyla* (Figure 1.4), has an unusual pseudohaptor and retains a 16-hooked larval haptor in the adult. Because of these features and the occurrence of one acanthocotylid species on a myxinid fish, the acanthocotylids may be similar to the monogenean ancestor. Malmberg and Fernholm (1989) described three new species of Acanthocotylidae from Myxinidae and these are of special interest because hitherto monogeneans belonging to this family had been found only on Rajidae (Beverley-Burton, 1984).

In addition to the Acanthocotylidae, five other well-known families of this large order are the Pseudomurraytrematidae, found on the gills of freshwater catastomid fish and of which *Anonchohaptor* is a member (Figure 1.5); the Ancyrocephalidae, found typically but not exclusively on Perciformes and of which, *Amphibdelloides* a member of the sub-family Ancyrocephalinae and a parasite of electric rays is shown (Figure 1.6); the Capsalidae, eg. *Megalocotyle* and *Nitzschia* (Figures 1.7 and 1.8) which are found on marine and brackish water teleosts; the Tetraonchidae which are often found on gills of salmonids and esocids, *Pavlovskioides* (Figure 1.9) from the gills and skin of a nototheniid being an interesting exception; and the Dactylogyridae, usually on Cypriniformes e.g. *Lamellodiscus* (Figure 1.10).

Recent useful publications on Dactylogyrida include those of Birgi and Euzet (1983), Dossou and Euzet (1984a, b), Euzet (1984), Euzet and Dossou (1975, 1976), Euzet and Oliver (1965a). Euzet and Vala (1977), Guegan, Lambert and Euzet (1988), Kritsky, Thatcher and Boeger (1986), Kritsky, Boeger and Thatcher (1988) and Young (1968).

---

[3] This important work and the other references it contains on Lebedev's contributions came to our attention after this manuscript had been completed.

3. Order Gyrodactylida

This is a large order, with one family, of viviparous monogeneans found on marine, fresh-water and brackish water teleosts according to Beverley-Burton (1984), Harris (1985a, b), Kritsky (1978), Malmberg (1970), Ogawa (1984a), Ogawa and Egusa (1978, 1979, 1985) and Paperna (1959, 1960a, b, 1964a, b, c). Amongst the best known examples are *Macro-gyrodactylus* and *Gyrodactylus* (Figures 1.11, 1.12).

4. Family Monocotylidae

These are monogeneans of holocephalans and elasmobranchs. *Calicotyle,* a typical monocotylid, is shown (Figure 1.13), together with the posterior adhesive organs of other forms belonging to four sub-families (Figure 1.14).

5. Order Polyopisthocotylida

Many interesting members of this order have recently been described by Dollfus and Euzet (1973), Euzet and Birgi (1975), Euzet and Cauwet (1967), Euzet and Ktari (1970a, b), Euzet and Maillard (1973, 1974), Euzet and Noisy (1978, 1979), Euzet and Wahl (1970a, 1977), Lebedev (1984a), and Ogawa and Egusa (1977, 1980, 1981). These papers cite a large number of references to the Polyopisthocotylea, a major group which has three sub-orders: the Hexabothriidea on elasmobranchs, Mazocraeidea on teleosts and Diclyobothriidea on Acipensiformes. *Squalonchocotyle* is an example of the Hexabothriidea whilst *Winkenthughesia*, *Microcotyle* and *Discocotyle* represent the Mazocraeidea (Figures 1.15, 1.16, 1.17 and 1.18). *Diclybothrium* is illustrated as an example of the Diclybothriidea (Figure 1.19).

To conclude this introduction to the variety of monogeneans of fish, attention is drawn to four particularly interesting forms. *Chimaericola* (Figure 1.20), a parasite of the Holocephali, may be amongst the most ancient members of the polyopisthocotyleans (Brinkmann, 1952; Williams, Colin and Halvorsen, 1987). *Enterogyrus* (Figure 1.21), a member of the Dactylogyridae and an endoparasite found in the gut of the cichlid fish, *Tilapia zilli* and *T. nilotica*, is regarded as important in discussions of the evolution of monogeneans (Paperna, 1963a: Llewellyn, 1965). *Diplozoon* (Figure 1.22), from the gills of salmonid fish, is curious because pairs of worms become attached in permanent copulation (Owen, 1959). A key to species of *Paradiplozoon* was given by Khotenovskii (1985) whilst speciation and specificity in *Diplozoon* was discussed by Le Brun, Renaud and Lambert (1988). Since the 1960s, Professor Louis Euzet and his co-workers in France have described, in over 30 publications, a remarkable range of body form in adult monogeneans (Figures 1.23–1.29).

Much additional information on the five groups of monogeneans already mentioned in this chapter and others which are particularly interesting from a biological or phylogenetic standpoint will be found in Baer and Euzet (1961), Bykhovskii (1957), Euzet and Combes (1980), Hargis (1959), Hargis and Thoney (1983), Hyman (1951a), Malmberg (1990), Lebedev (1978, 1979a, b, 1983, 1984, 1986, 1988), Lebedev and Mamaev (1976), Sproston (1946) and Yamaguti (1963a). Some of these references highlight relationships between monogeneans and cestodes and the importance of three monozoic sub-classes of the Platyhelminthes in these discussions, namely the Gyrocotylidea, Amphilinidea and Caryophyllidea which are briefly referred to in the following pages.

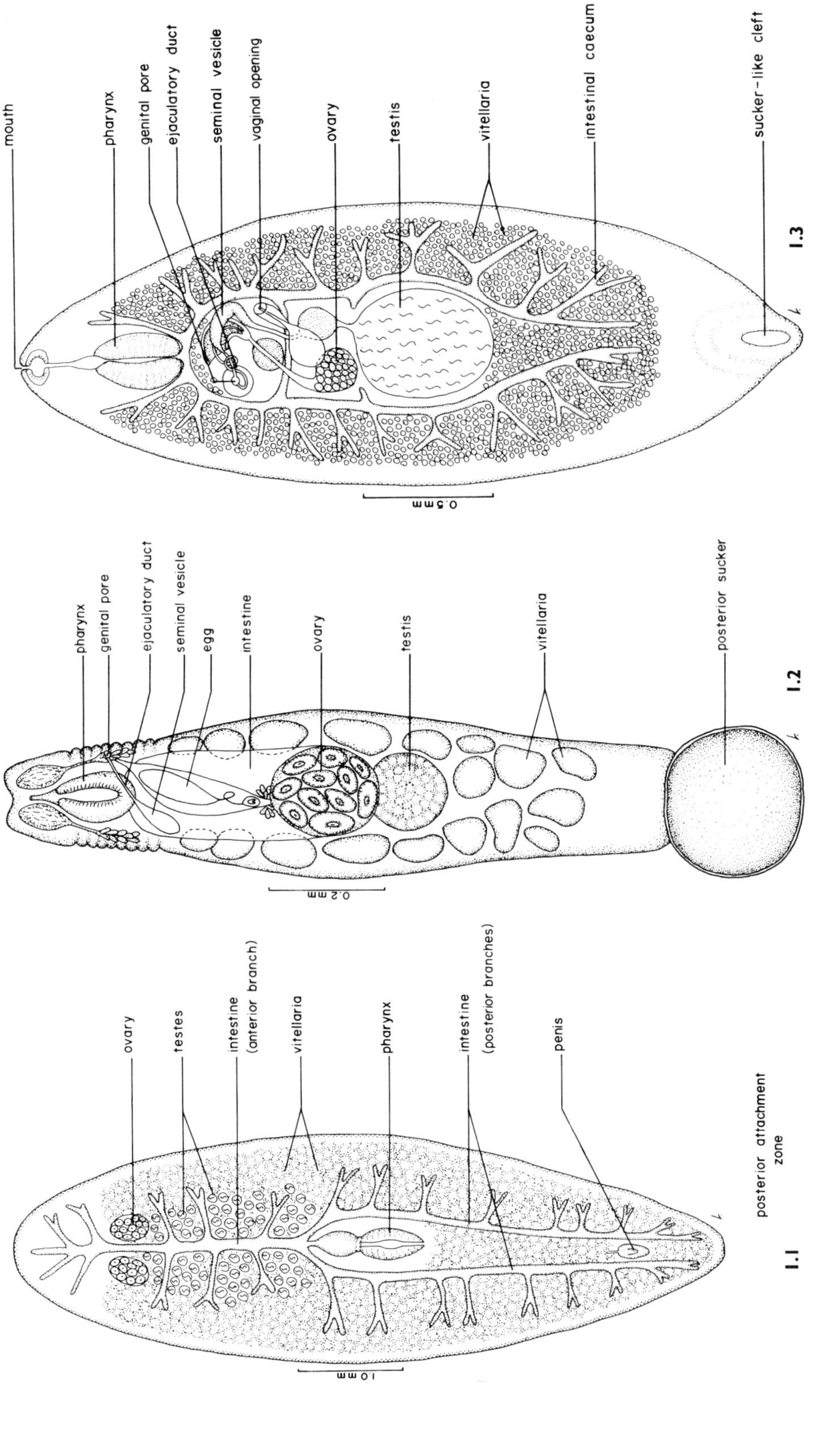

Figures 1.1–1.3 A fish turbellarian and two monogeneans of questionable affinities. 1.1 *Micropharynx* 1.2 *Udonella* 1.3 *Microbothrium* From Beverley-Burton (1984)

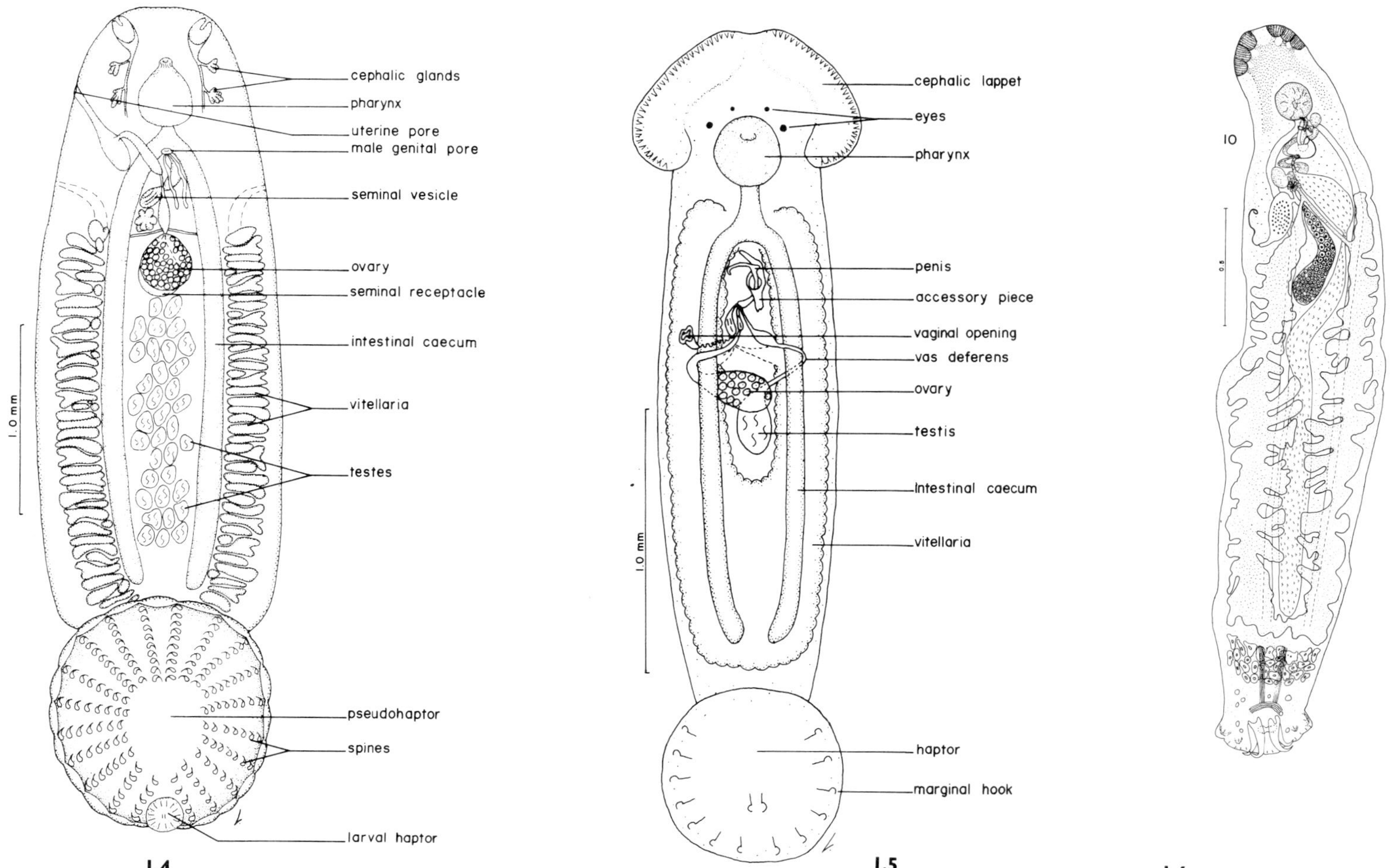

Figures 1.4–1.6 Three primitive monogeneans of the Dactylogyrida 1.4 *Pseudacanthocotyla* from Beverley-Burton (1984) 1.5 *Anonchohaptor* from Beverley-Burton (1984) 1.6 *Amphibdelloides* from Dillon and Hargis (1965a)

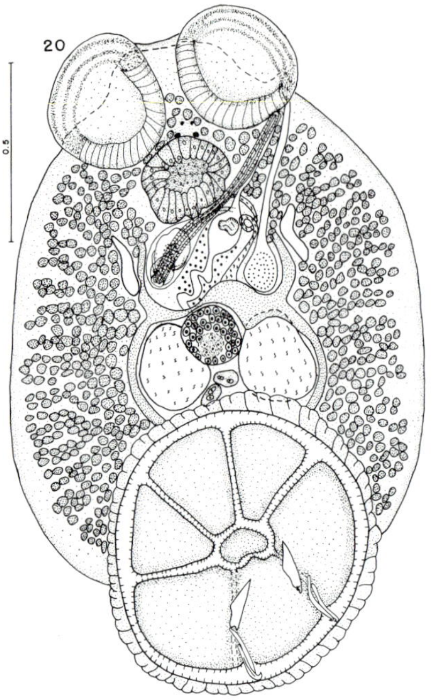

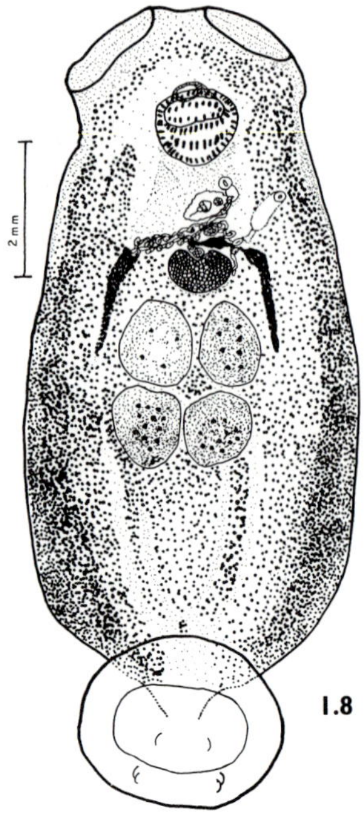

1.8

1.7

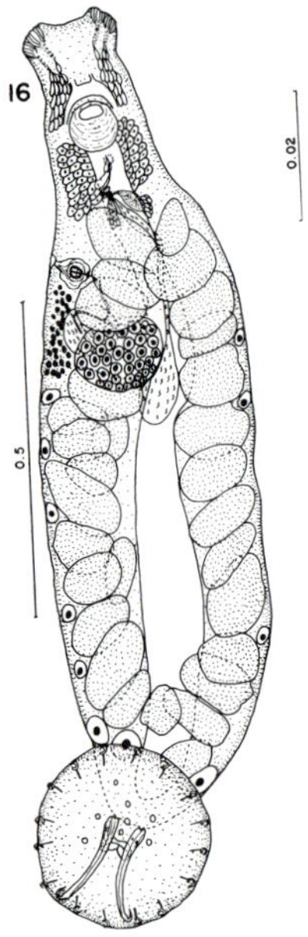

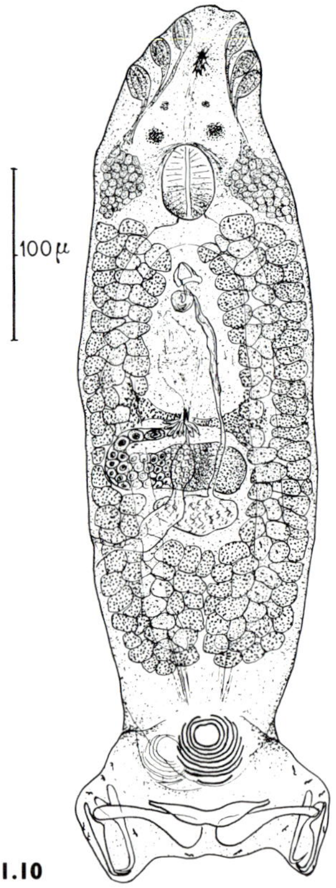

1.9

1.10

Figures 1.7–1.10 Other interesting Dactylogyridea representing four families 1.7 *Megalocotyle* from Dillon and Hargis (1965a) 1.8 *Nitzschia* from Pratt and Herrmann (1962) 1.9 *Pavlovskioides* from Dillon and Hargis (1968) 1.10 *Lamellodiscus* from Euzet and Oliver (1965a)

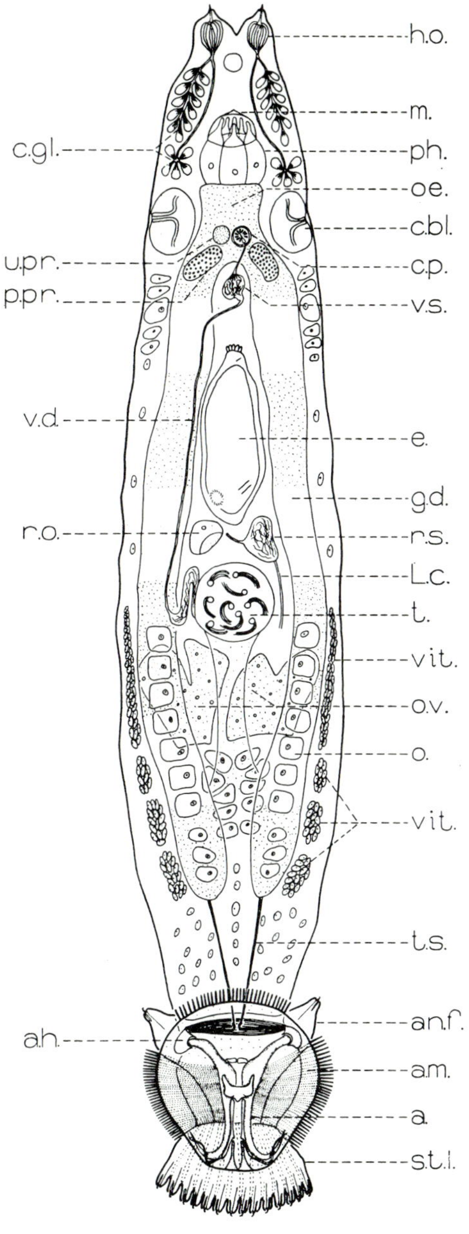

1.11

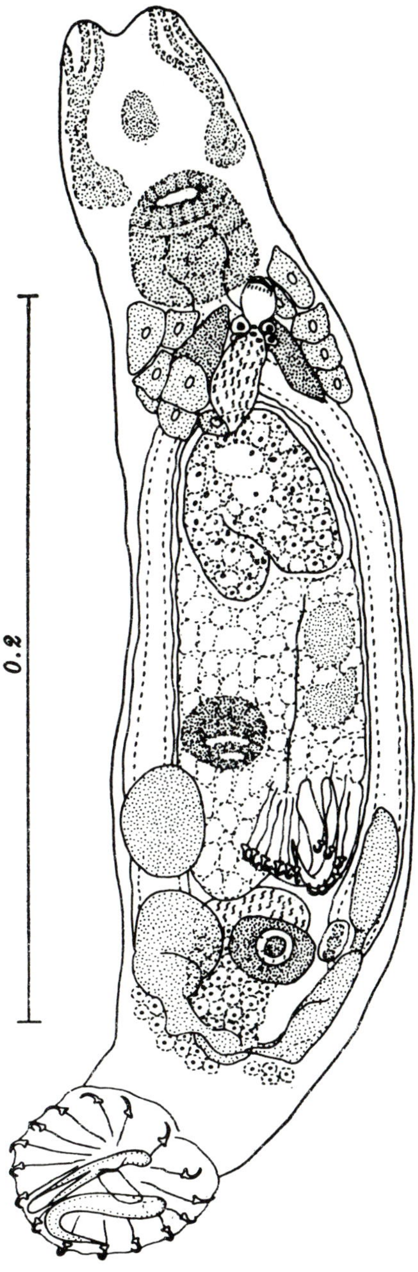

1.12

Figures 1.11–1.12 Viviparous monogeneans of the order Gyrodactylida 1.11 *Macrogyrodactylus* from Malmberg (1956) Key to lettering: a. anchor; a.h. hood-shaped structure of anchor root; a.m. anchor membrane; an.f. anteriorly pointing finger; c.bl. contractile bladder; c.gl. cephalic glands (some diagramatics); c.p. cirrus sac with cirrus; e. embryo in uterus; g.d. gut diverticulum; h.o. head organ; L.c. Laurer's canal; m. mouth; o. oocyte; oe. oesophagus; ov. ovary; ph. pharynx; p.pr. paired prostate gland; r.o. ripe ovum; r.s. receptaculum seminis; s.t.l. separate terminal lobe; t. testis; t.s. tendon-like structure; u.pr. unpaired prostate gland; v.d. vas deferens; vit. vitellarium; v.s. vesicula seminalis. 1.12 *Gyrodactylus* from Holliman (1963)

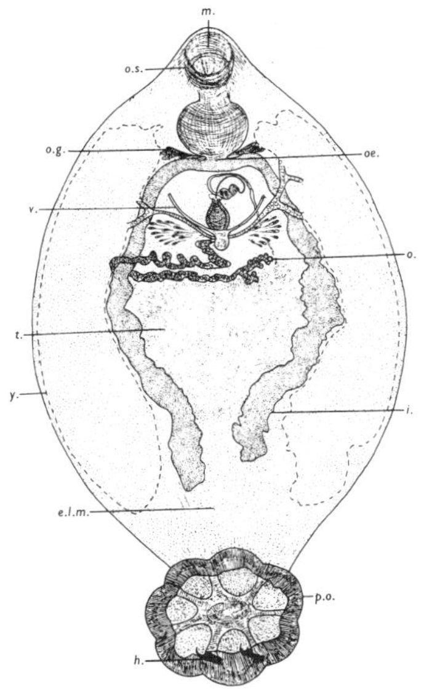

1.13

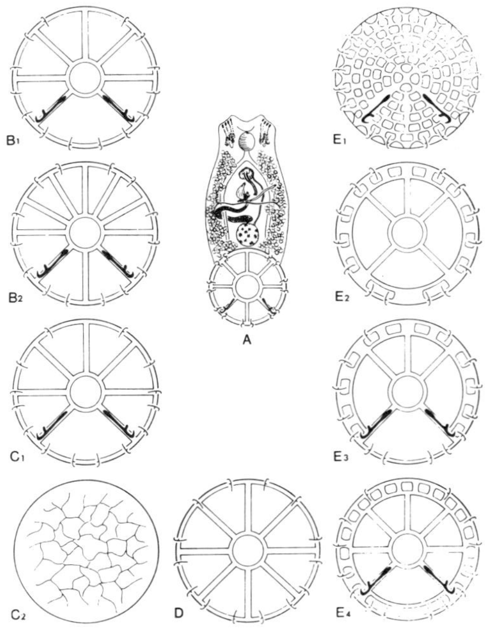

1.14

Figures 1.13–1.14 Examples of the variety in form of haptors in *Calicotyle* (a typical monocotylid) and other genera. 1.13 *Calicotyle* from Euzet and Williams (1960) Key to lettering: e.l.m. extrinsic longitudinal muscles; h. hook; i. intestine; m. mouth; o. ovary; oe. oesophagus; i.g. oesophageal glands; o.s. oral sucker; p.o. posterior adhesive organ; t. testis; v. vagina; y. yolk gland. 1.14 Haptors of a typical monocotylid and other genera from Lambert (1982) B Monocotylinae B_1 *Dasybatotrema*, *Horricauda* and *Troglocephalus* B_2 *Decacotyle*, *Papillocotyle*, *Allobeterocotyle* C Calicotylinae C_1 *Calicotyle* C_2 *Dictyocotyle* D Dendromonoctylinea E Merizocotylinae E_1 *Cathariotrema* E_2 *Empruthotrema* E_3 *Thaumatocotyle* E_4 *Merizocotyle*

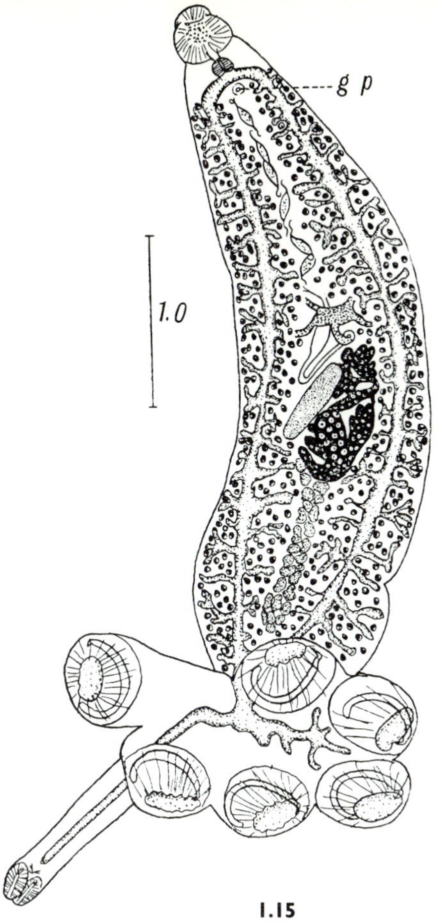

1.15

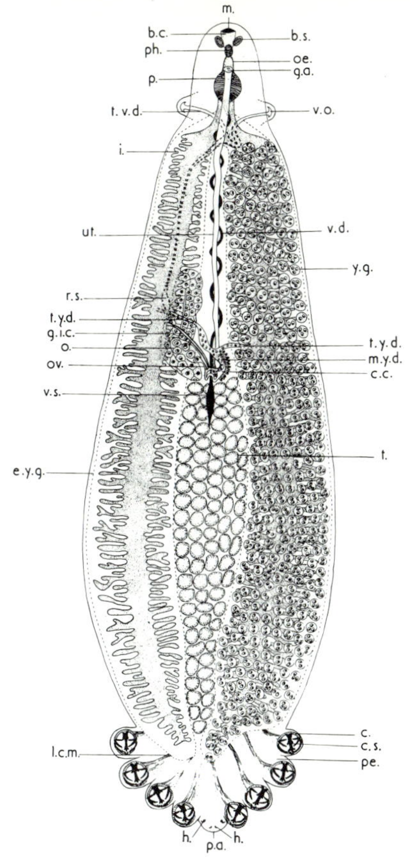

1.16

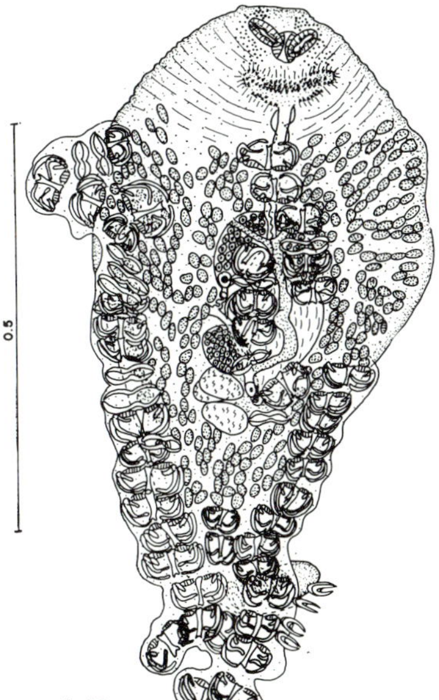

1.17

Figures 1.15–1.17 Polyopisthocotylidean monogeneans representing the super-families Hexabothriidea, Mazocraeidea and Hexabothriidea 1.15 *Squalonchocotyle* from Brooks (1934) Key to lettering g.p. genital pore 1.16 *Winkenthugesia* from Williams (1960b) Key to lettering: b.c. buccal cavity; b.s. buccal sucker; c. clamp; c.c. central chamber; c.s. clamp sclerite; e.y.g. extent of vitelline gland; g.a. genital atrium; g.i.c. genito-intestinal canal; h. hook; i. intestine; l.c.m. longitudinal clamp muscle; m. mouth; m.y.d. median vitelline duct; o. ovary; oe. oesophagus; ov. oviduct; p. penis; p.a. posterior appendage of haptor; pe. peduncle; ph. pharynx; r.s. receptaculum seminalis; t. testis; t.y.d. transverse vitelline duct; t.v.d. transverse vaginal duct; ut. uterus; v.d. vas deferens; v.o. vaginal opening; v.s. vesicula seminalis; v.g. vitelline gland. 1.17 *Microcotyle* from Dillon and Hargis (1965b)

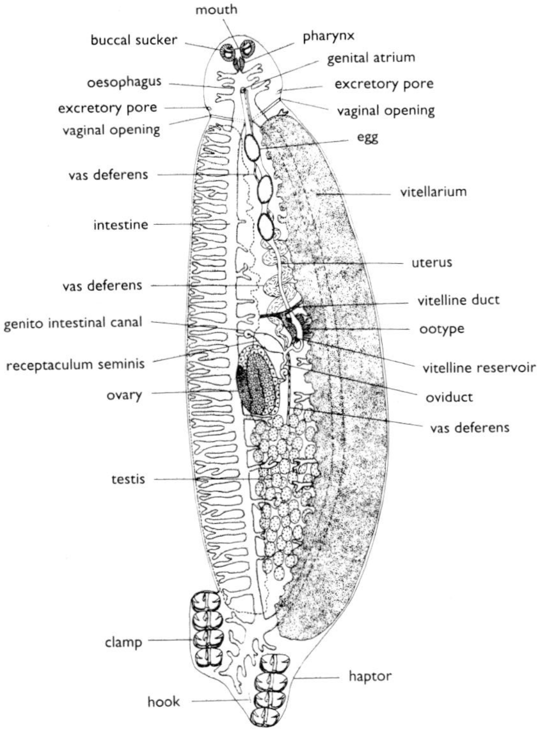

1.18

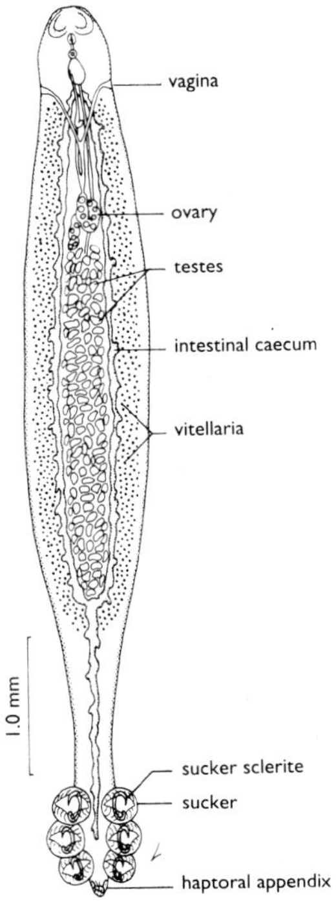

1.19

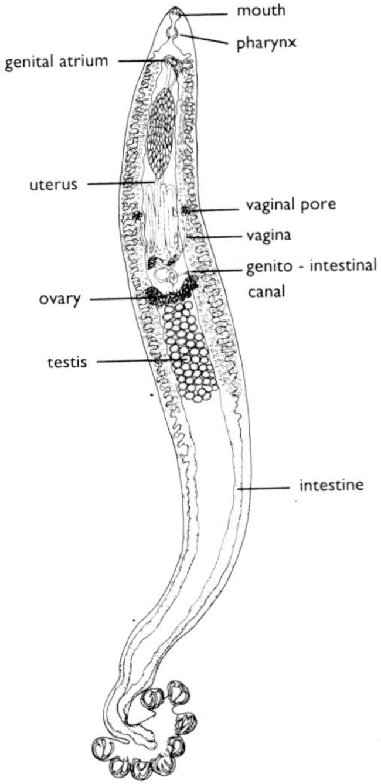

1.20

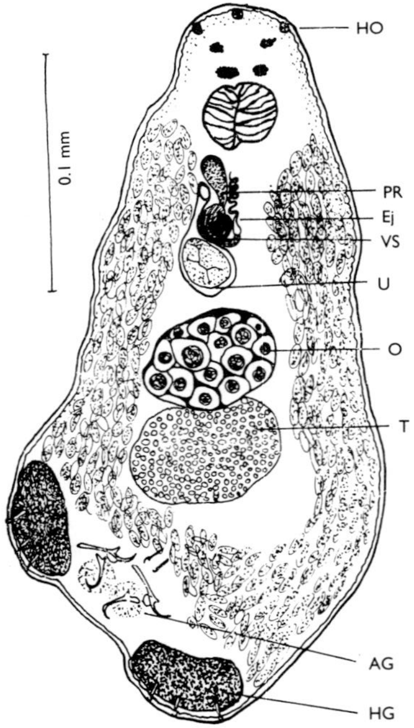

1.21

Figures 1.18–1.21 *Discocotyle* and three polyopisthocotyleans of importance in discussing the evolution of the Monogenea. 1.18 *Discocotyle* from Owen (1959) 1.19 *Diclybothrium* from Beverley-Burton (1984) 1.20 *Chimaericola leptogaster* 1.21 *Enterogyrus* from Paperna (1963a) Key to lettering: AG, anchoral gland; Ej, ejaculator (copulatory organ); HG, haptoral gland; HO, head organ; O, ovary; PR, prostate gland; T, testis; U, uterus; VS, seminal vesicle.

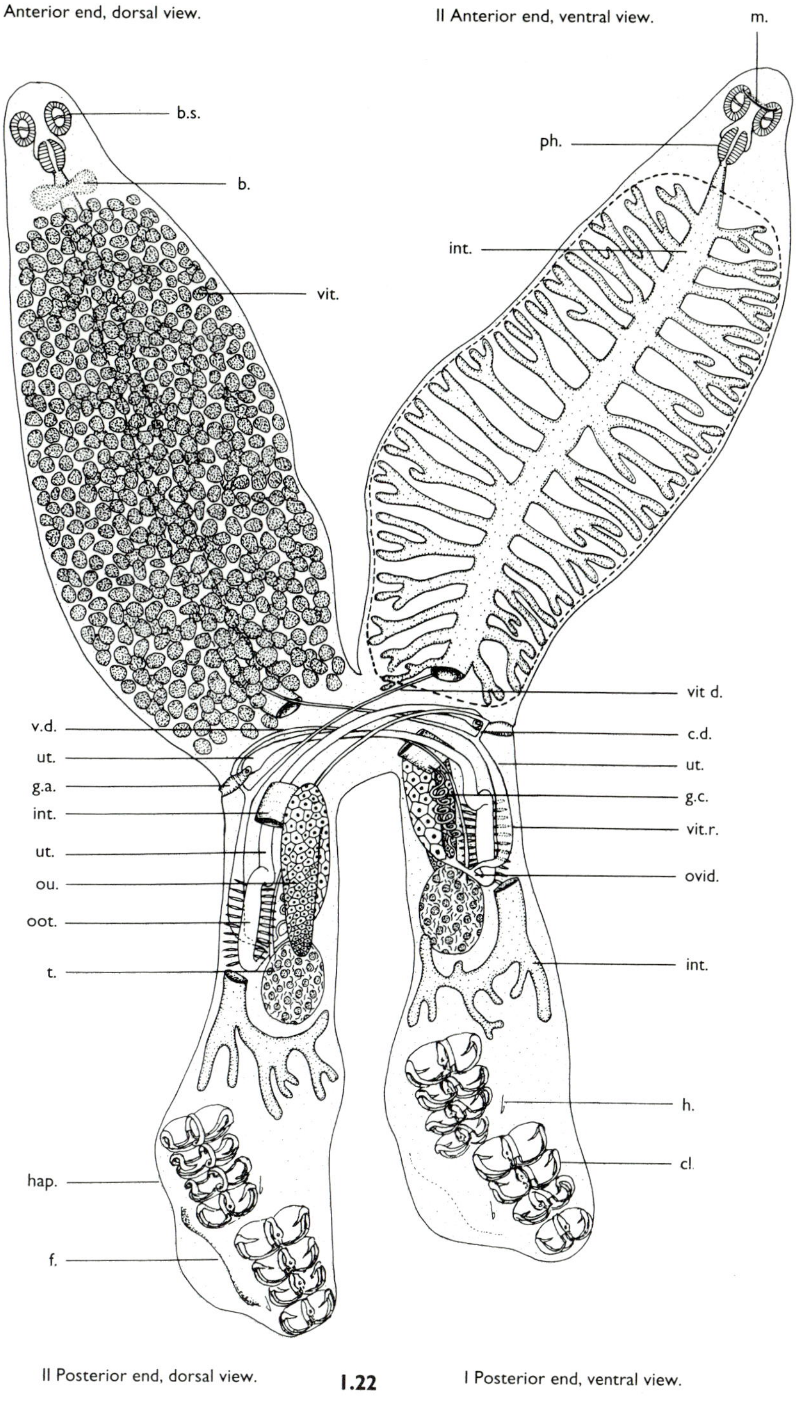

Figure 1.22 *Diplozoon*, a curious monogenean found fused into pairs in permanent copulation, from Owen (1959). Except for specimens from bream, most British material is now thought to be *Paradiplozoon homoion*.
Key to lettering: b. cerebral ganglion; b.s. buccal sucker; c.d. connecting duct; cl. clamp; f. dorsal lip-like fold; g.a. genital atrium; g.c. genito-intestinal canal; h. hook; hap. haptor; int. intestine; m. mouth; oot. ootype; ov. ovary; ovid. oviduct; ph. pharynx; t. testis; ut. uterus; v.d. vas deferens; vit. vitelline follicle; vit.d. vitelline duct; vit.r. vitelline reservoir.

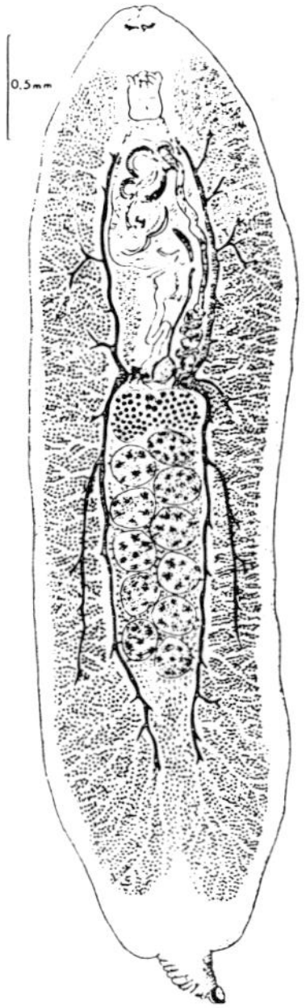

1.23

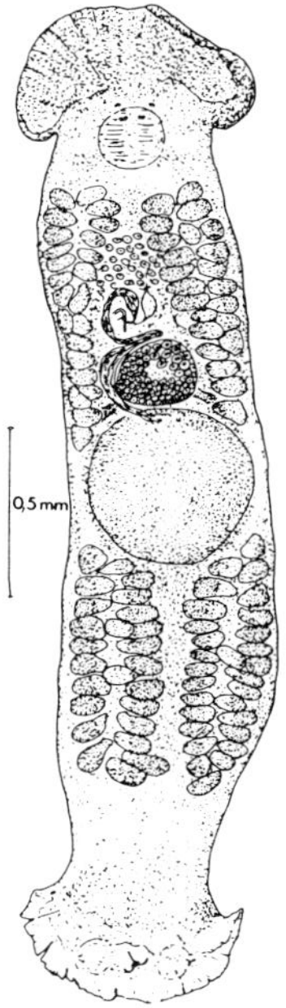

1.24

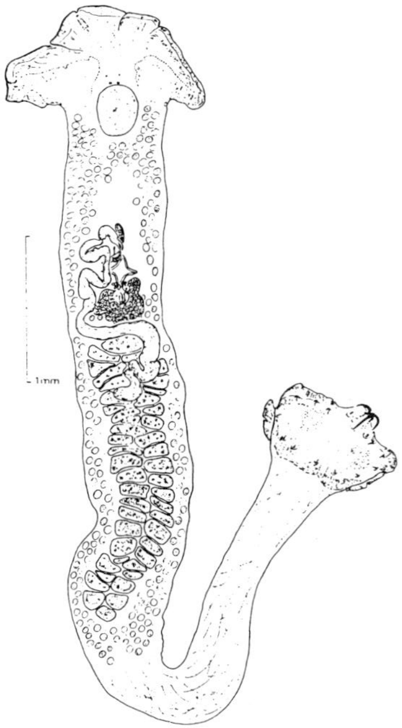

1.25

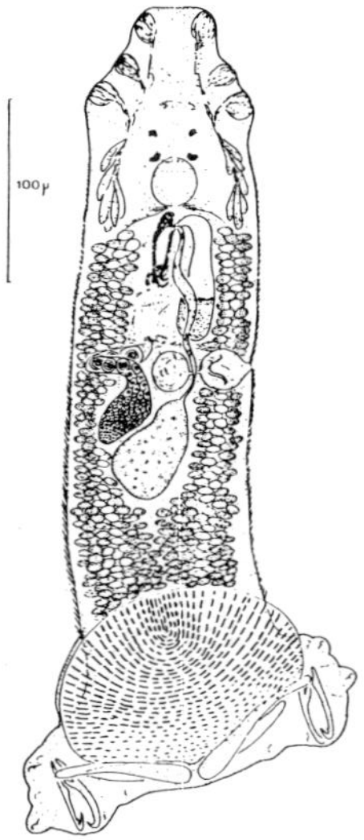

1.26

Figures 1.23–1.26 Four examples of the variety of body form in monogeneans 1.23 *Cadenatia* from Euzet and Maillard (1967) 1.24 *Calceostoma* from Euzet and Vala (1975) 1.25 *Ktariella* from Vala and Euzet (1977) 1.26 *Diplectanum* from Euzet and Durette-Desset (1973)

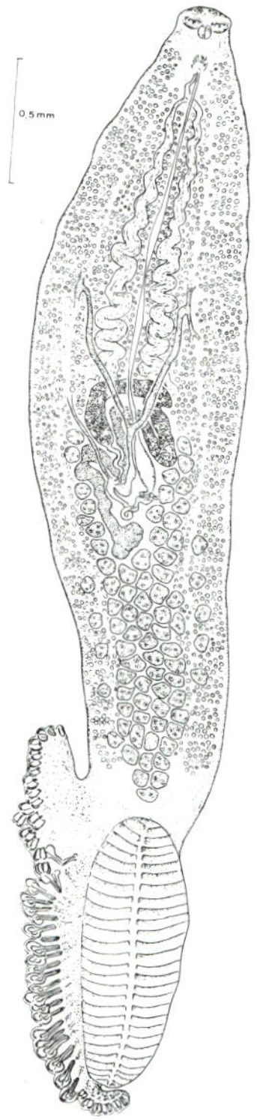

1.27

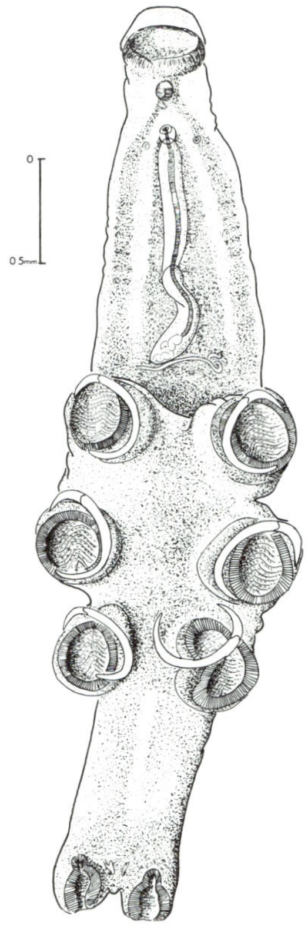

1.28

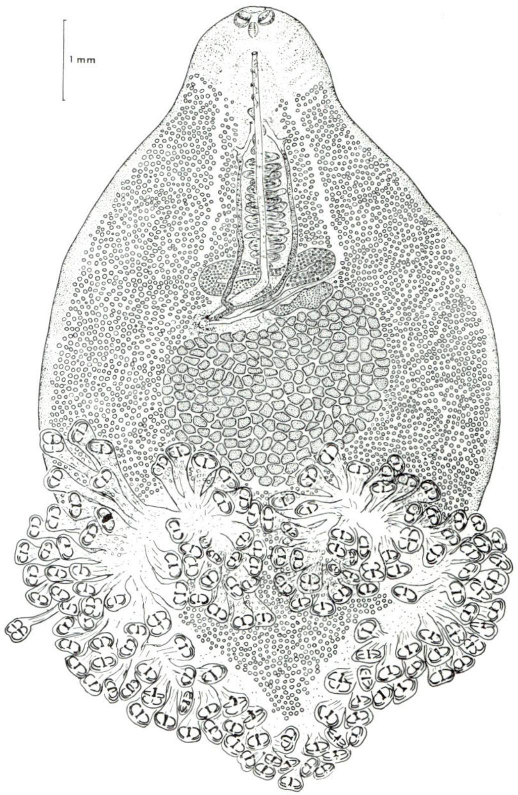

1.29

Figures 1.27–1.29 Further examples of the varied body form in monogeneans 1.27 *Rhinecotyle* from Euzet and Wahl (1970a) 1.28 *Erpocotyle* from Euzet and Maillard (1967) 1.29 *Aspinatrium* from Euzet and Ktari (1971)

Class Cestoidea

To avoid further confusion on a difficult taxonomic problem we have tentatively accepted Schmidt (1986) in this chapter in referring to two subclasses of the Cestoidea, namely the Cestodaria and Eucestoda, the former having two orders, the Gyrocotylidea and Amphilinidea, and the latter, eleven which are found in fish. It will, however, become apparent in the brief accounts of the Gyrocotylidea, Amphilinidea and Caryophyllidea that they may warrant class status within the Platyhelminthes.

Subclass Cestodaria

Order Gyrocotylidea

Gyrocotylideans are restricted to the intestine of holocephalans. These unusual Platyhelminthes are now generally given a status independent of monogeneans and cestodes, although they have affinities with both (Colin, Williams and Halvorsen, 1986). Papers by Xylander (1984–1989) are of particular interest with regard to the ultrastructure and affinities of the gyrocotylideans. Berland, Bristow and Grahl-Neilsen (1990) discuss the chemotaxonomy of *Gyrocotyle*. The features of an adult *Gyrocotyle* are shown (Figure 1.30).

Order Amphilinidea

Amphilinideans are secondarily monozoic cestodes found in the body cavity of teleost and chondrostean fishes and also in turtles. They have a status independent of gyrocotylideans and caryophyllideans (Dubinina, 1982). *Gyrometra* is illustrated as an example (Figures 1.31–1.33).

Subclass Eucestoda

Details of the classification of the Eucestoda are given by Wardle and McLeod (1952), Yamaguti (1959), Baer and Joyeux (1961), Schmidt (1970, 1986), Freeman (1973) and Stunkard (1983). These authorities recognised between nine and 13 orders of the Eucestoda, of which up to 12 are found in fish. We have avoided taxonomic controversies and accepted 12 orders of fish tapeworms belonging to the class Eucestoda. Nine of the 12 orders are restricted to fish. Eucestodes are usually elongate, ribbon-like and segmented (strobilate), but in the Spathebothriidea, external segmentation is absent, although there are many sets of genitalia (Figures 1.35, 1.36) whilst in some eucestodes of the order Pseudophyllidea an apparent prolific external segmentation far exceeds the actual number of internal sets of reproductive organs (Figures 1.37–1.52).

Order Caryophyllidea

Caryophyllideans are a large group of cestodes of fresh water siluriform and cypriniform fish. The monozoic body, one set of reproductive organs, and many features of their biology, set them apart from all other members of the subclass Eucestoda (Mackiewicz, 1972, 1981, 1982a) (Figure 1.34). Mackiewicz, however, does suggest that caryophyllideans are cestodes but that the amphilinids and gyrocotylideans are not.

Order Spathebothriidea

In this order, the scolex may be a funnel-shaped apical adhesive organ as in the Cyathocephalidae (Figure 1.35), two adhesive cups (Diplocotylidae, Figure 1.92), or a proboscis (Spathebothriidae, Figure 1.36). Some spathebothriidean tapeworms can mature and produce eggs in crustaceans and others in teleost fish. They may represent a fundamental ancestral type of the segmented tapeworms. On the one hand they may have affinities with Caryophyllidea and, on the other, with both Pseudophyllidea and Tetraphyllidea.

In ten other orders of eucestodes found in fish, the body is divided externally into few to many segments, each with one (or occasionally more than one) set of reproductive organs (Figure 1.54). Often in such tapeworms the posterior segment of the strobila becomes detached and continues to grow and live a separate existence in the fish intestine; the segments may develop anterior suckers for attachment to the gut mucosa. In one species, *Cathetocephalus thatcheri*, a multistrobilate condition (Figure 1.55) has been described by Dailey and Overstreet (1973). The anterior end in cestodes is generally expanded into a scolex with adhesive organs showing a wide range of variation and it is largely on the form of the scolex that various taxa are initially identified. The size, number of segments and shape of the adhesive organs of the scolex may vary considerably, not only amongst members of various orders of tapeworms, but also amongst species of related genera (Figures 1.56–1.65).

The following sequence of introducing the other ten orders of eucestodes which are found in fish is largely based on Euzet (1959, 1974), Freeman (1973), Gallegos (1982), Stunkard (1983) and Schmidt (1986). Two main lines of evolution are generally recognized. One comprises the Tetraphyllidea (Figures 1.79–1.91), Prosobothrioidea (Figure 1.106), Lecanicephalidea (Figure 1.107–1.109) and Proteocephalidea (Figure 1.110); and another the Trypanorhyncha (Figures 1.97–1.103), Haplobothridea (Figure 1.104) and Pseudophyllidea (Figures 1.93–1.96).

Order Tetraphyllidea

Adult tetraphyllideans are found only in the intestine of elasmobranchs. The scolex is very variable: the four bothridia may be sessile, pedunculate, simple in outline, folded or loculate: suckers may be present and the scolex may be armed with hooks and/or spines (Figures 1.79–1.91 and Figures 1.11–1.12), (Euzet, 1959; Williams, 1966, 1968c, 1969; Freeman, 1973; Schmidt, 1986, and Stunkard, 1983).

Order Prosobothrioidea

Prosobothrium armigerum is the only representative of the Prosobothrioidea. Adults are found only in the intestine of elasmobranchs. The four anteriorly-directed glandular suckers of the bothridia and a characteristic spiny neck (Figure 1.106) distinguish them from the Tetraphyllidea and Proteocephalidea (Baer and Euzet, 1956; Euzet, 1959; Freeman, 1973; Schmidt, 1970, 1986, and Stunkard, 1983).

Order Lecanicephalidea

Lecanicephalideans are tapeworms of the intestine of elasmobranchs in which the scolex is divided horizontally into anterior and posterior regions: the scolex may have four small suckers plus hooks and/or tentacles (Figures 1.107–1.109). There are four distinct families (Schmidt, 1970, 1986).

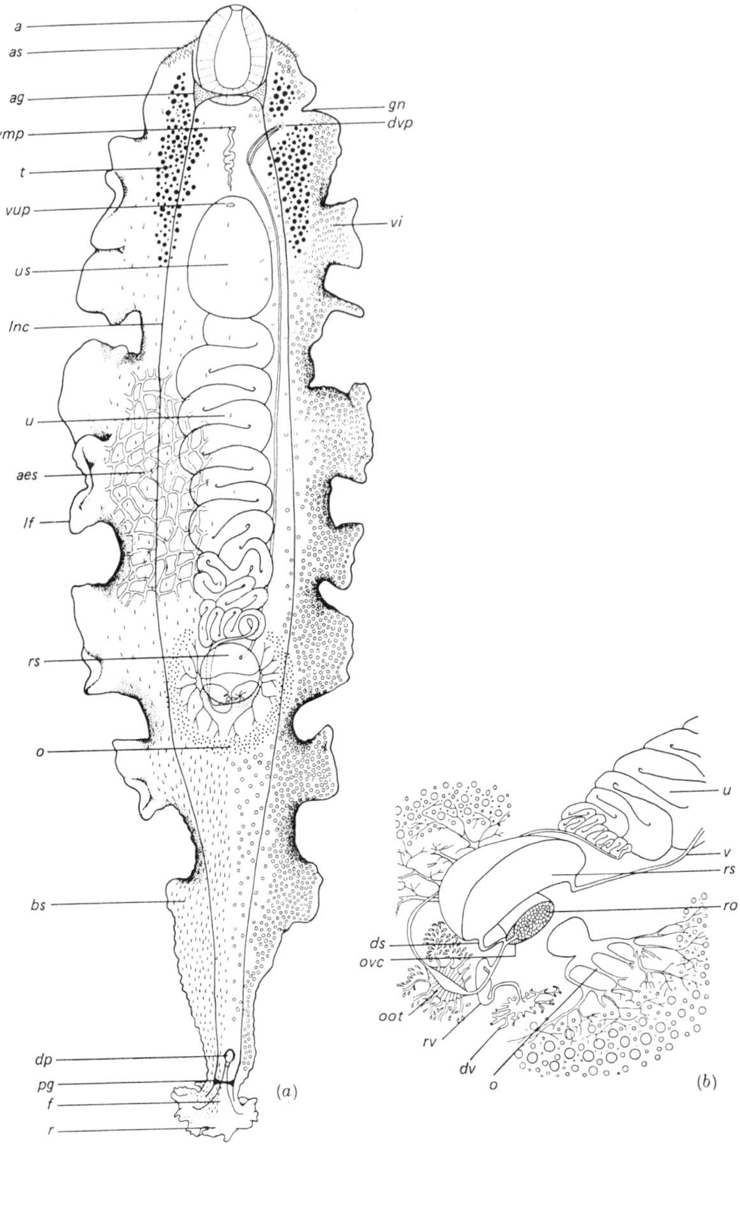

Figure 1.30 The distinctive features of *Gyrocotyle*, from Williams, Colin and Halvorsen (1987). Key to lettering: a. acetabulum; a.e.s. anastomozing excretory system; a.g. anterior ganglion; a.s. acetabular spines; b.s. body spines; d.p. dorsal pore; d.s. ductus seminalis; d.v. vitelline duct; d.v.p. dorsal vaginal pore; f. funnel; g.n. genital notch; l.f. lateral fold; l.n.c. longitudinal nerve cord; o. ovary; oot. ootype; o.v.c. ovicapt; p.g. posterior ganglion; r. rosette; r.o. receptaculum ovorum; r.s. receptaculum seminis; r.v. receptaculum vitellarium; t. testes; u. uterus; u.s. uterine sac; v. vagina; vi. vitelline glands; v.m.p. ventral male pore; v.u.p. ventral uterine pore.

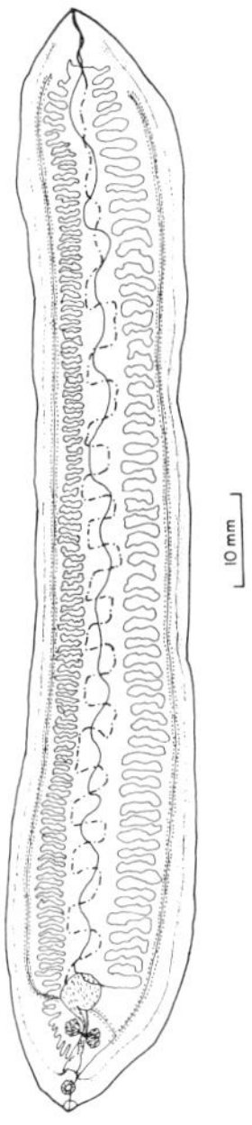

1.31

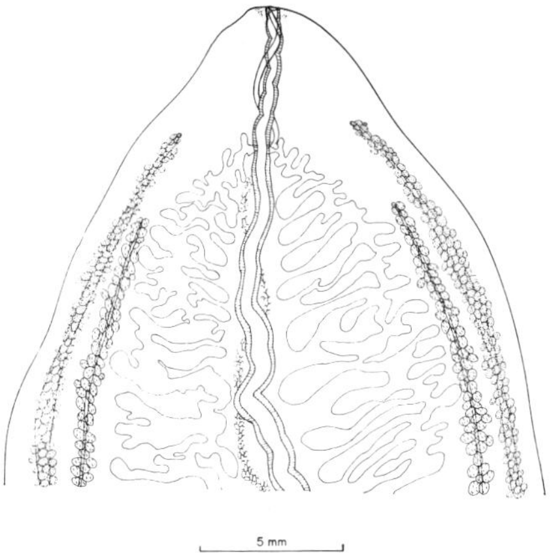

1.32

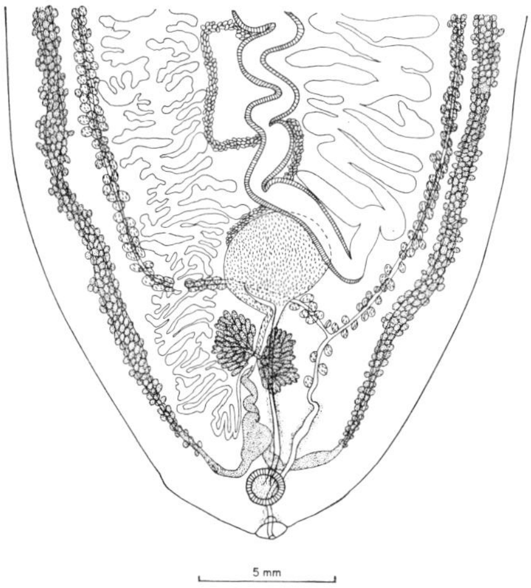

1.33

Figures 1.31–1.33 *Gyrometra*; a typical amphilinidean
1.31 Dorsal view of a complete worm 1.32 Dorsal view of the anterior end 1.33 Dorsal view of the posterior end
from Khalil (1977)

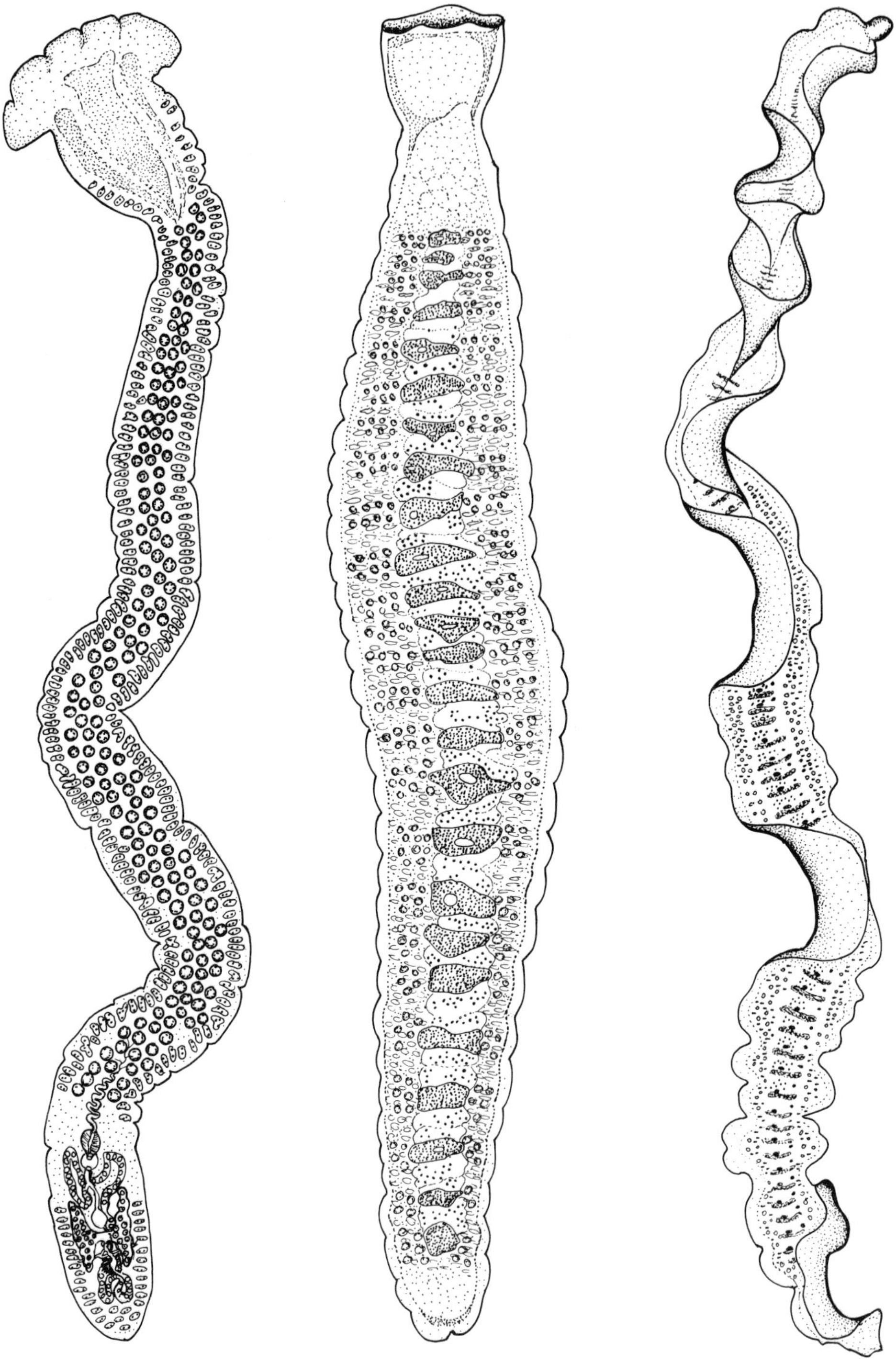

1.34 1.35 1.36

Figures 1.34–1.36 A Caryophyllidean, *Caryophyllaeus* (Figure 1.34), redrawn from Dogiel, Petrushevski and Polyanski (1961) and Nybelin (1922); and two spathebothriideans, *Cyathocephalus* and *Acompsocephalum* (Figures 1.35–1.36), redrawn from Prudhoe and Bray (1982) and Rees (1969) *Acomsocephalum* is now regarded as a synonym of *Anantrum*, described by Overstreet (1968). See Schmidt (1986).

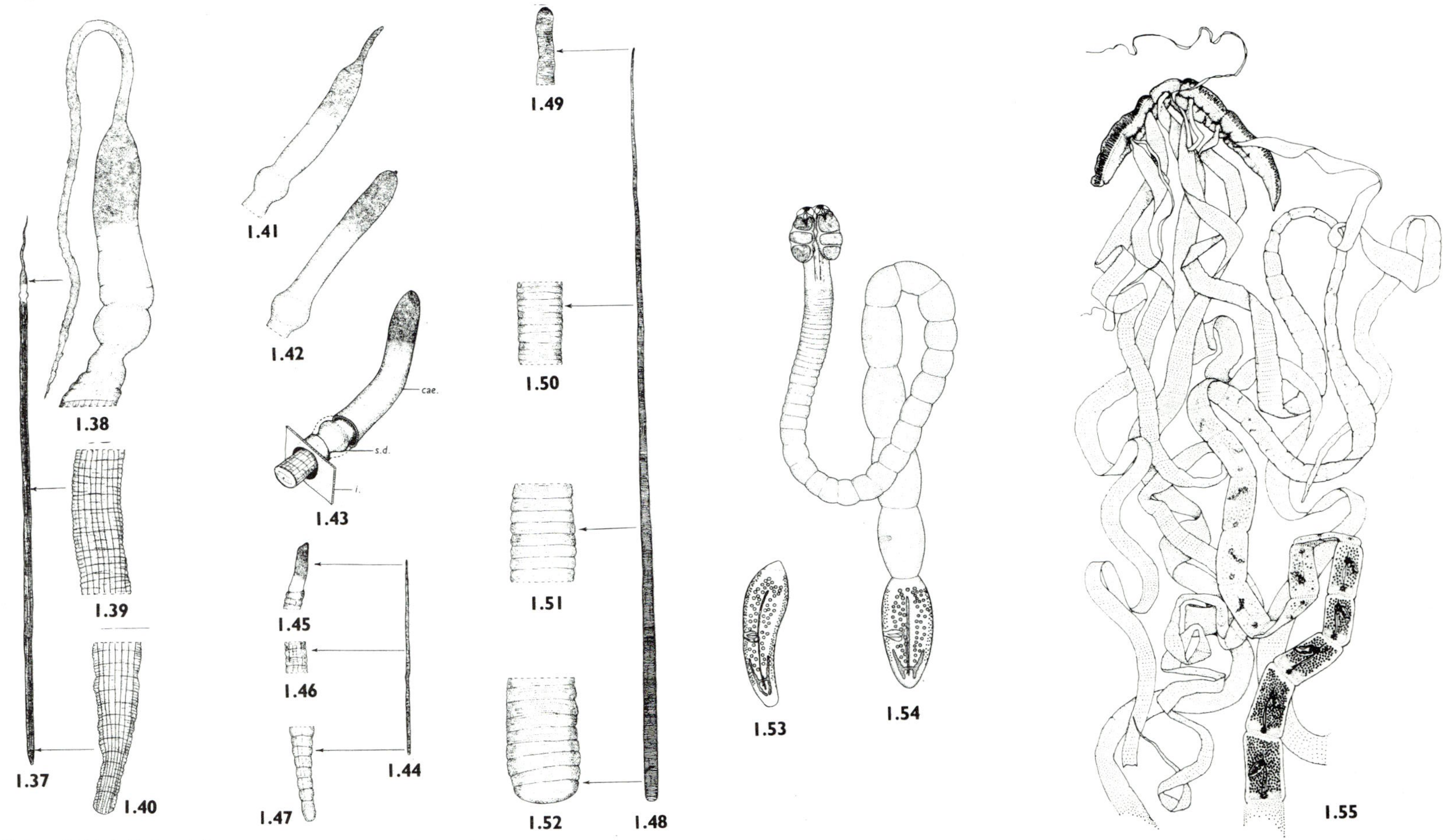

Figures 1.37–1.43 *Parabothrium gadi-pollachii* from Williams (1960c). 1.37 Entire strobila. 1.38 Scolex deformatus, first stage in degeneration. 1.39 Mid-strobila region. 1.40 Posterior tapering region of strobila. 1.41 Scolex deformatus, second stage in degeneration from Williams (1960c) 1.42 Scolex deformatus, third stage in degeneration. 1.43 Scolex deformatus, mode of attachment within pyloric caecum. 1.44–1.52 *Abothrium gadi* (from Williams (1960c)) 1.44–1.47 Specimen from the haddock showing various regions of the strobila 1.48–1.52 Specimen from the cod showing various regions of the strobila 1.53–1.54 *Acanthobothrium pearsoni* from Williams (1962b) 1.53 mature proglottid 1.54 Scolex and strobila 1.55 *Cathetocephalus* from Dailey and Overstreet (1973)

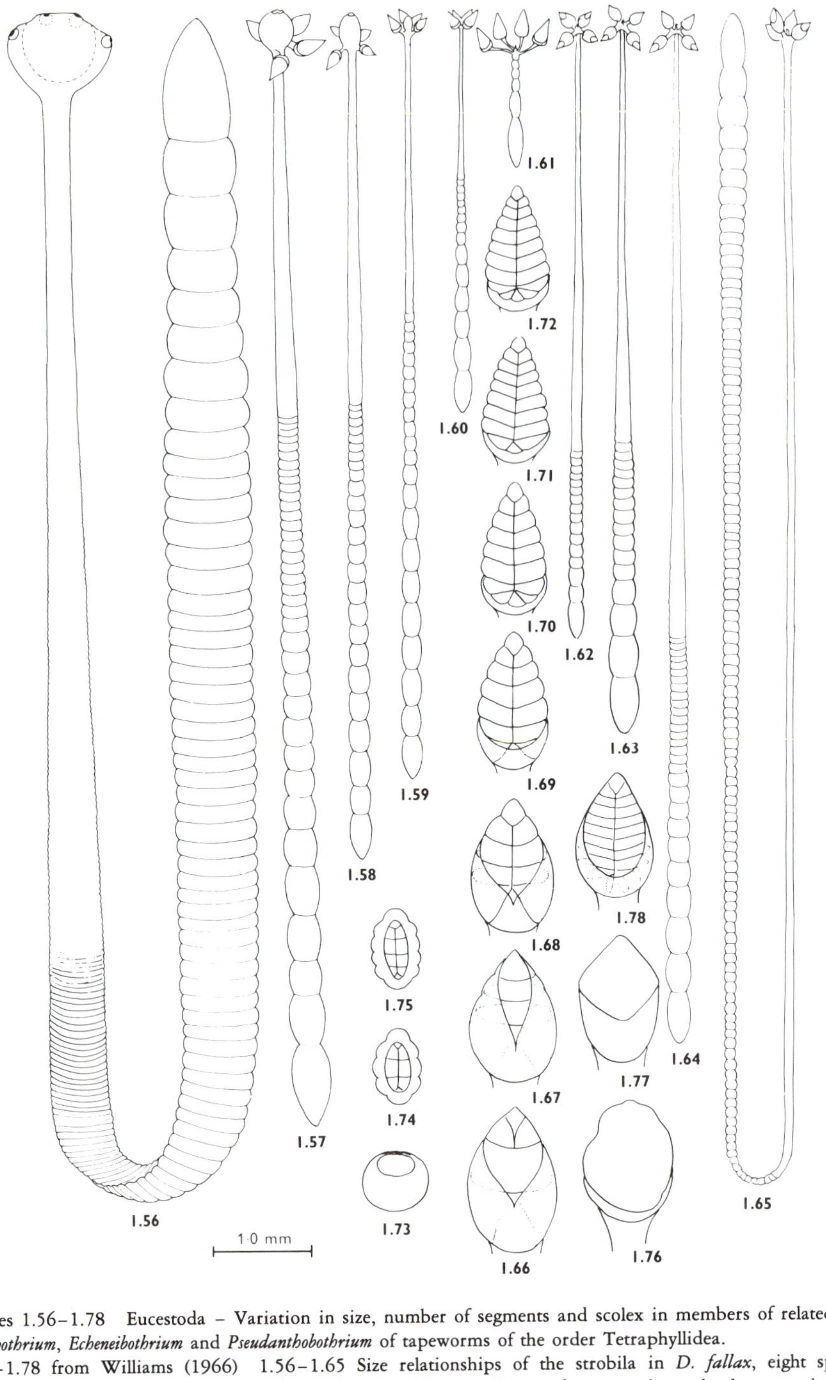

Figures 1.56–1.78 Eucestoda – Variation in size, number of segments and scolex in members of related genera, *Discobothrium*, *Echeneibothrium* and *Pseudanthobothrium* of tapeworms of the order Tetraphyllidea.

1.56–1.78 from Williams (1966) 1.56–1.65 Size relationships of the strobila in *D. fallax*, eight species of *Echeneibothrium* and one of *Pseudanthobothrium* from British rays. Figures drawn to the scale shown against Figure 1.56 1.56 *Discobothrium fallax* from *Raja clavata* 1.57 *Echeneibothrium* from *R. clavata* 1.58 *Echeneibothrium* sp. from *R. batis* 1.59 *E. maculatum* from *R. montagui* 1.60 *Echeneibothrium* sp. from *R. clavata* 1.61 *E. minutum* from *R. batis* 1.62 *E. dubium* from *R. batis* 1.63 *Pseudanthobothrium hanseni* from *R. radiata* 1.64 *Echeneibothrium.* sp. from *R. naevus* 1.65 *E. elongatum* from *R. circularis*

Figures 1.66–1.72 Increase in complexity of the bothridum in seven species of *Echeneibothrium*; possibly indicating stages in the evolution of this organ.

Figures 1.73–1.78 A comparison of the adhesive organs of *Discobothrium* (1.73), two species of *Echneibothrium* (1.74 & 1.75) *Anthobothrium* (1.76), *Pseudanthobothrium* (1.77) and *Rhinebothrium* (1.78)

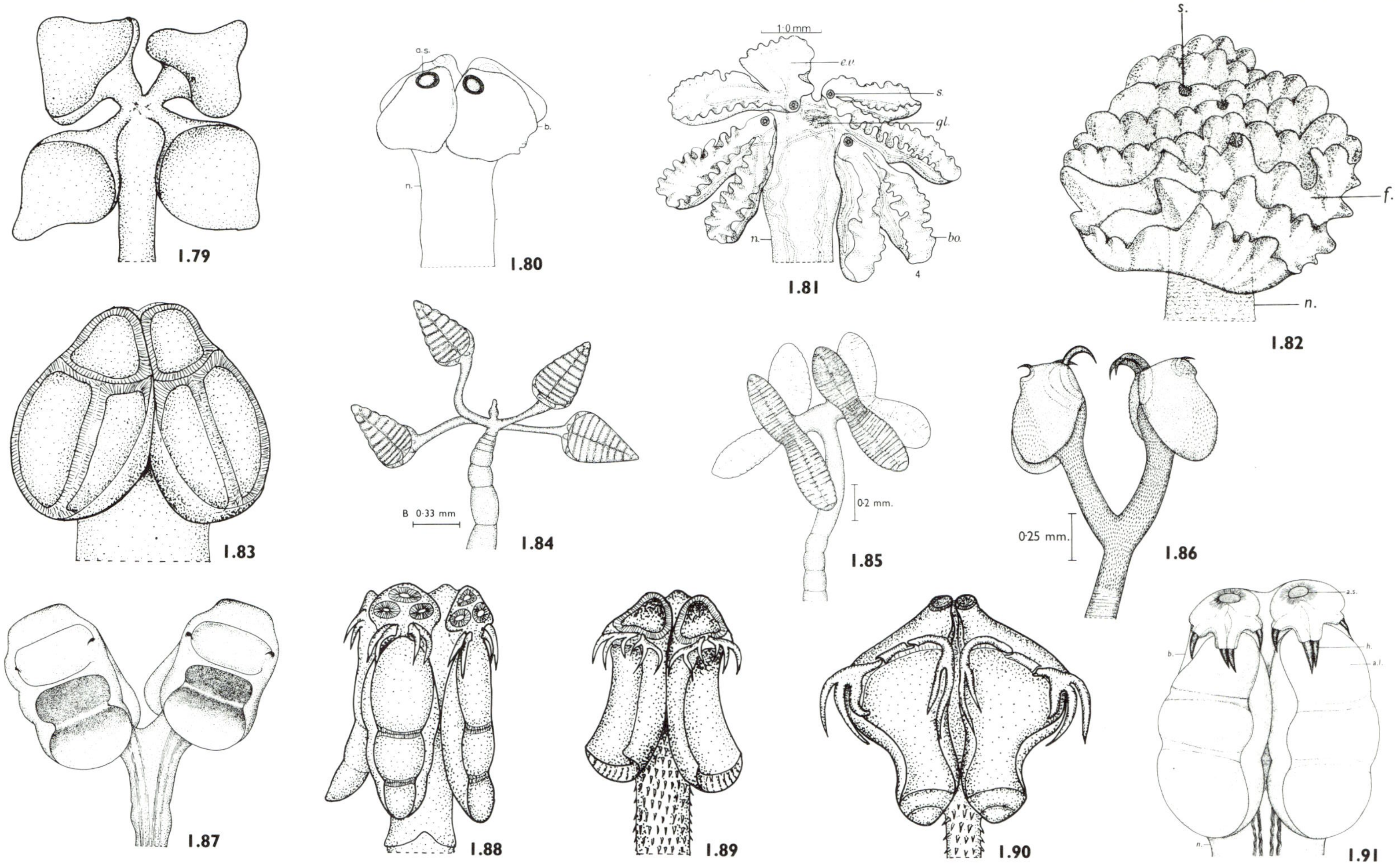

Figures 1.79–1.91 Variation in the scolex of representatives of the Tetraphyllidea. 1.79 *Anthobothrium* from Euzet (1959) 1.80–1.82 *Phyllobothrium* from Williams (1958c, 1959a, 1968c) 1.83 *Trilocularia* from Euzet (1959) 1.84 *Echeneibothrium* from Williams (1966) 1.85 *Rhinebothrium* from Williams (1964) 1.86 *Yorkeria* from Williams (1964) 1.87 *Spiniloculus* from Williams (1964) 1.88 *Calliobothrium* from Euzet (1959) 1.89 *Phoreiobothrium* from Euzet (1959) 1.90 *Platybothrium* from Euzet (1959) 1.91 *Acanthobothrium* from Williams (1962b) Key to lettering: a.l. anterior loculus; a.s. accessory sucker; b and bo. bothridium; e.v. excretory vessel; gl. glandular myzorhynchus; n. neck; s. sucker.

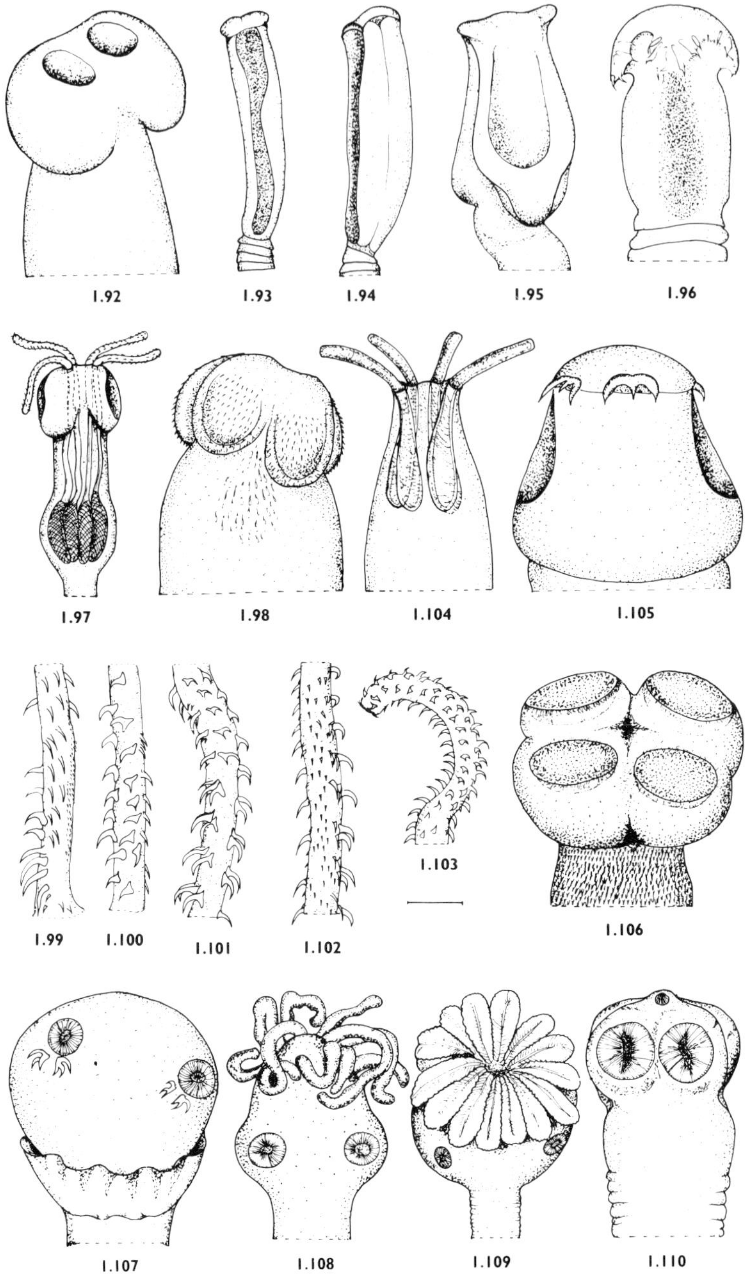

Figures 1.92–1.110 Variation in the scolex as seen from representatives of the Spathebothriidea, Trypanorhyncha, Haplobothriidea, Pseudophyllidea, Lecanicephalidea and Proteocephalidea. Redrawn from various sources.

1.92 *Bothriomonus* (spathebothriid) from Burt and Sandeman (1974) 1.93–1.96 pseudophyllideans: 1.93–1.94 *Bothriocephalus* from Rees (1958) 1.95 *Fistulicola* from Hyman (1951a). 1.96 *Polyonchobothrium* from Jones (1980). 1.97–1.103 trypanorhynchs: 1.97 *Grillotia* 1.98 *Aporhynchus* from Hyman (1951a) 1.99–1.103 tentacle of *Grillotia* showing arrangement of hooks from Bates (1987) on the bothridial face 1.99 antibothridial face 1.100 internal face 1.101 external face 1.102 tip 1.103 1.104 *Haplobothrium* from Hyman (1951a) 1.105 *Triaenophorus* from Hyman (1951a) 1.106 *Prosobothrium* from Baer and Euzet (1956) 1.107 *Balanobothrium* from Schmidt (1970) 1.108 *Polypocephalus* from Schmidt (1970) 1.109 *Anthemobothrium* from Hyman (1951a) 1.110 *Proteocephalus* from Jones (1980)

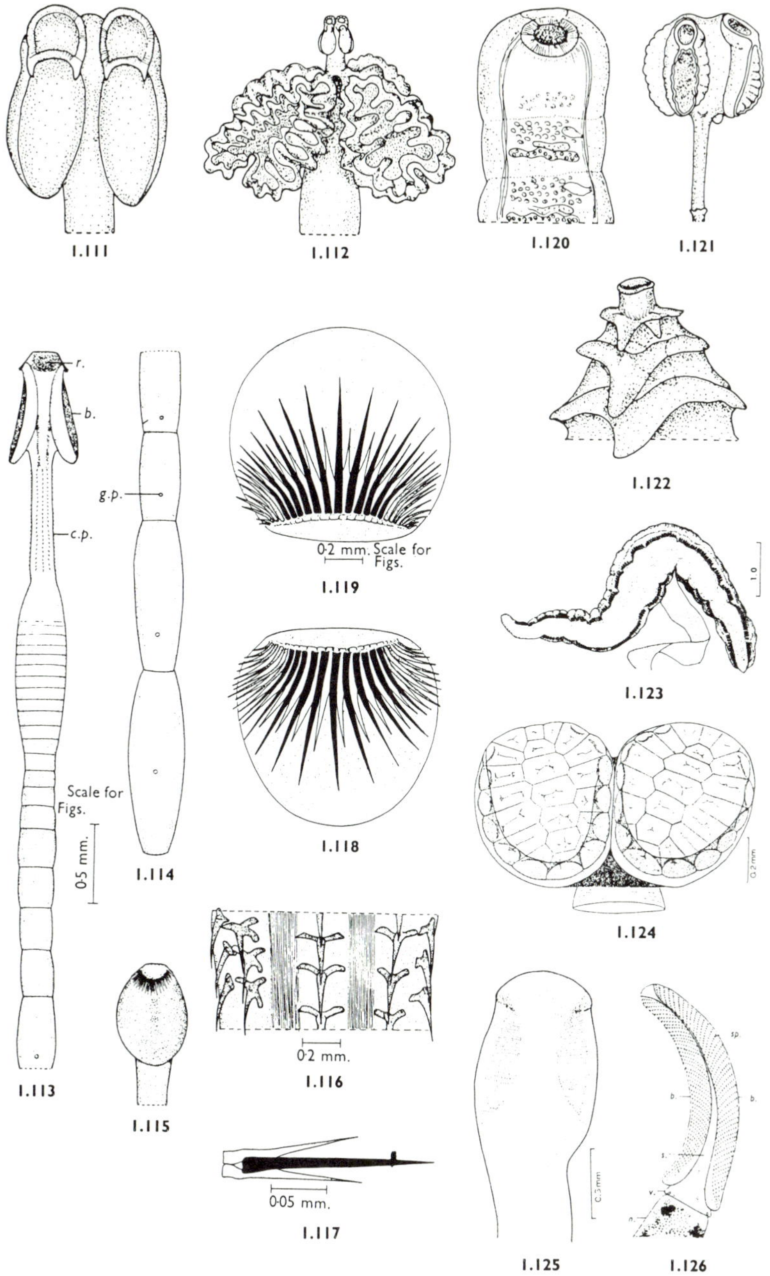

Figures 1.111–1.126 1.111 *Ceratobothrium* (tetraphyllidean) from Euzet (1959) 1.112 *Thysanocephalum* (tetraphyllidean) from Euzet (1959)

1.113–1.119 *Echinobothrium heroniensis* 1.113–1.114 Scolex and strobila; 1.115 Scolex (head proper) dorsal or ventral view, 1.116 Region of peduncle to show hooks, 1.117 Group of three apical hooks, one from outer row in black, 1.118 and 1.119 The two groups of apical hooks in anterior view. From Williams (1964) 1.120 *Nippotaenia*, from Hyman (1951a) 1.121 *Chimaerocestus* from Williams and Bray (1984) 1.122 *Litobothrium* from Dailey (1969) 1.123 *Cathetocephalus* 1.124 *Dioecotaenia* 1.125 *Phyllobythos* 1.126 *Ditrachybothridium* Key to lettering: b. bothridium; c.p. cephalic peduncle; r. rostellum 1.113–1.119 original from Williams (1964a) 1.111, 1.112, 1.120, 1.121, 1.122—redrawn from various sources 1.123 from Dailey and Overstreet (1973) 1.124 from Schmidt (1969) 1.125 from Campbell (1977b) 1.126 from Rees (1959).

Order Proteocephalidea

Proteocephalideans are very common in freshwater teleost fish but may also be found in amphibians and reptiles. The scolex has four suckers and occasionally an apical sucker or armed rostellum (Figure 1.110), (Jones, 1980; Andersen, 1979a; Brooks and Deardorff, 1980).

Order Trypanorhyncha (Tetrarhynchidea)

This order is found in elasmobranchs only. Except for *Aporhynchus norvegicus* (Figure 1.98), the scolex has four eversible armed tentacles and two or four bothridia (Figures 1.97, 1.99–1.103) (Bates 1990; Beveridge 1990a, b; Beveridge and Campbell 1987, 1988a,b, 1989; Schmidt, 1986; and Sakanari, 1989).

Order Haplobothriidea

Haplobothrium globuliforme and *H.bistrobilae* are the only representatives of this order. They are found in the intestine of thc holostean *Amia calva*, described by Baer (1952) as the most archaic bony fish host known. The scolex has four unarmed tentacles (Figure 1.104), (MacKinnon and Burt, 1985a, b).

Order Pseudophyllidea

This is a large order of nine families of which the following are found in teleost fish: Ptychobothriidae, Bothriocephalidae, Echinophallidae, Triaenophoridae, Amphicotylidae and Parabothriocephalidae. The scolex has two bothria, with or without hooks (Figures 1.93–1.96, 1.105) (Schmidt, 1970; Khalil, 1971; Andersen, 1979b; Kuperman, 1981b).

In addition to the two main categories of eucestodes already mentioned, representing two major lines of evolution, a third category of the Eucestoda comprises some orders, families and genera of intriguing phylogenetic relationships, the systematic position of which is difficult to determine. Examples are the Diphyllidea (see Rees, 1961a, b; Khalil and Abdul-Salam, 1989), Nippotaenidea (see Schmidt, 1970; Hine, 1977a), Litobothridea (see Dailey, 1969), Dioecotaeniidae (see Schmidt, 1969), Chimaerocestidae (see Williams and Bray, 1984), Philobythiidae (see Campbell, 1979), Cathetocephalidae (see Dailey and Overstreet, 1973, and Schmidt and Beveridge, 1990) and *Ditrachybothridium* (see Rees, 1959) (Figures 1.113–1.126).

Class Trematoda

Subclass Aspidogastrea

Gibson and Chinabut (1984) recognized two subclasses of trematodes, the Aspidogastrea and Digenea and accepted two orders of the former, the Aspidogastrida with one family, Aspidogastridae, and the Stichocotylida with three, Stichocotylidae, Multicalycidae and Rugogastridae. Members of the three sub-families of the Aspidogastridae which occur in molluscs, teleost fish and turtles, have as a holdfast a ventral disc divided into a marginal ring of alveoli surrounding medial alveoli (Figures 1.127, 1.130–1.131).

In the Stichocotylida, one family (Stichocotylidae) has a holdfast composed of isolated suckers (Figure. 1.129); in another (Multicalycidae) it is composed of fused suckers; and in the third (Rugogastridae), it is a series of raised transverse septa (ruga) (Figure. 1.128). Gibson and Chinabut (1984) comment upon interesting similarities between the Rugogastridae and endoparasitic monogeneans.

Subclass Digenea

Yamaguti (1971) recognised about 70 families of fish digeneans. Except for some species of Fellodistomidae, Azygiidae, Gorgoderidae, Sanguinicolidae, Syncoeliidae, Zoogonidae and Didymozoidae, and all species of Ptychogonimidae and Aphanhysteridae, they all occur exclusively in teleosts. Disagreements on the relationships of families have occurred because some classifications are largely based on larval stages and life-histories and others mainly on adult morphology (Gibson and Bray, 1979; Cable, 1982; Gibson, 1987).

Five of about 40 characters used by Yamaguti (1958) to distinguish between families have been used in this chapter to introduce the variety of adult body form in 26 families of fish digeneans. They are: (1) position of mouth, (2) position of testes, preovarian or postovarian, (3) position and size of suckers, (4) absence of one or both suckers, and (5) presence of one testis only. The following sequence of presentation is not intended to suggest phylogenetic relationships. We have attempted to present the other platyhelminth groups in a more or less phylogenetic order but within the digeneans, by far the largest and most complex group of parasitic worms, it is necessary to rely on superficial phenetic characters. Gibson and Bray (1979) give an excellent account of a functional morphology approach to understanding the phylogenetic relationships of the superfamily Hemiuroidea and its 14 families of fish digeneans, whilst Cable (1974, 1982), La Rue (1957), Pearson (1972) and Stunkard (1963, 1975) argue for a life-cycle approach to classification.

Order Gasterostomata

For convenience of presentation only, we have retained in this chapter a traditional division of Digenea into the Gasterostomata (Bucephalidae) and Prosostomata (other families). As Gibson (1987), however, has emphasized, the Bucephalidae are now associated with more conventional digenean groups and are included amongst six other families as possible candidates for the most primitive of extant digeneans.

Members of the Gasterostomata are distinguished from all other Digenea because the mouth is in the middle of the body and leads, by way of a pharynx, into a small sacciform intestine (Figure. 1.132). This feature is reminiscent of the gut of rhabdocoel turbellarians and has, therefore, been regarded by some authors as primitive. A ventral sucker is lacking. Adults are restricted to fish and usually are found only in carnivorous fish; metacercariae are common. (Nagaty, 1937; Hyman, 1951a; Velasquez, 1959; Yamaguti, 1958; Rees, 1970, and Stunkard, 1974 a,b,c).

Order Prosostomata

The Prosostomata are Digenea with a terminal or subterminal mouth (Figures. 1.133–1.134). As already stated, five morphological characters have been used to introduce 26 digenean families:

1. Digeneans with postovarian testes

Species showing this feature may often be found amongst the Allocreadiidae. (Figures 1.133, 1.134 and Figures. 1.142–1.147), Opecoelidae, Ptychogonimidae, Azygiidae and

Acanthocolpidae (Figures 1.135–1.137), Fellodistomidae (Figures 1.138–1.140) and Gorgoderidae (Figure 1.141). The range of body form in the Allocreadiidae and related families can be extensive (Figures 1.142–1.147).

For more information on the digenean families figured and others with postovarian testes, see Argumedo (1963), Bakke (1984), Bakke and Bailey (1987), Bray (1982, 1983, 1985a,b), Bray and Gibson (1977, 1980), Brooks and Mayes (1975), Caballero (1952), Caira (1989), Campbell (1975b, 1977a), Crusz (1957), Gibson and Bray (1982), Goto and Ozaki (1929), Lasee *et al.* (1988), Manter (1945), Martin (1958a), Mueller and van Cleave (1932), Peters (1960), Razarihelisoa (1959), Rees (1968), Shimazu (1990a,b,c,d), Sparks (1954), Sparks and Thatcher (1958) and Yamaguti (1971).

2. Digeneans with preovarian testes

Species showing this feature may often be found amongst the Hemiuridae, Accacoeliidae, Syncoeliidae, Hirudinellidae, Sclerodistomidae, Lampritrematidae and Zoogonidae (Figures 1.148–1.150 and Figures 1.151–1.155). All but the Zoogonidae are members of the superfamily Hemiuroidea (see Gibson and Bray, 1979). For more information on the digeneans figured with preovarian testes, see Bray (1987a, b), Bray and Gibson (1977, 1986), Gibson and Bray (1977, 1979, 1986), Hyman (1951a) and Margolis (1962). From these publications it will be noted that in some Zoogonidae, the testes are postovarian.

3. Digeneans with unusual oral or ventral suckers

In the families Paramphistomidae and Cephaloporidae, the ventral suckers are posterior or subterminal and often large, whilst in some members of the Megaperidae and Cryptogonimidae there are relatively large oral suckers (Figures 1.156–1.159). For more information on those figured and others, see Arai (1962), Beverley-Burton and Margolis (1982), Bray (1985b), Caballero and Bravo-Hollis (1952), Jones and Seng (1986), Khalil (1963, 1981), Manter (1933), McClelland (1957), Velasquez (1961) and Vaz (1932).

4. Digeneans without oral and/or ventral suckers

The absence of one or both suckers is often a feature in members of the unusually shaped Didymozoidae (Figures 1.160–1.162), Angiodictyidae, Transversotrematidae, Bivesiculidae and Sanguinicolidae (Figures 1.163–1.166) and the Aporocotylidae and Mesometridae (Figures 1.167–1.169). For more information on those figured and other members of these families, see Angel (1969); Gibson, MacKenzie and Cottle (1981), Gladunko (1981), Bray (1984), Dawes (1947), Madhavi (1982), Manter (1961, 1970), Nikolaeva (1985) Pearson (1968), Razarihelisoa (1959), Smith (1972), Thulin (1980a, b, c, 1982), Wierzbicka (1980) and Williams (1958a).

5. Digeneans with one testis

Members of the Monorchiidae, Haploporidae and Haplosplanchnidae have only one testis (Figures 1.170–1.172); see Bray (1984), Brooks (1977), Manter and Pritchard (1961), Overstreet (1969) and Thomas (1959).

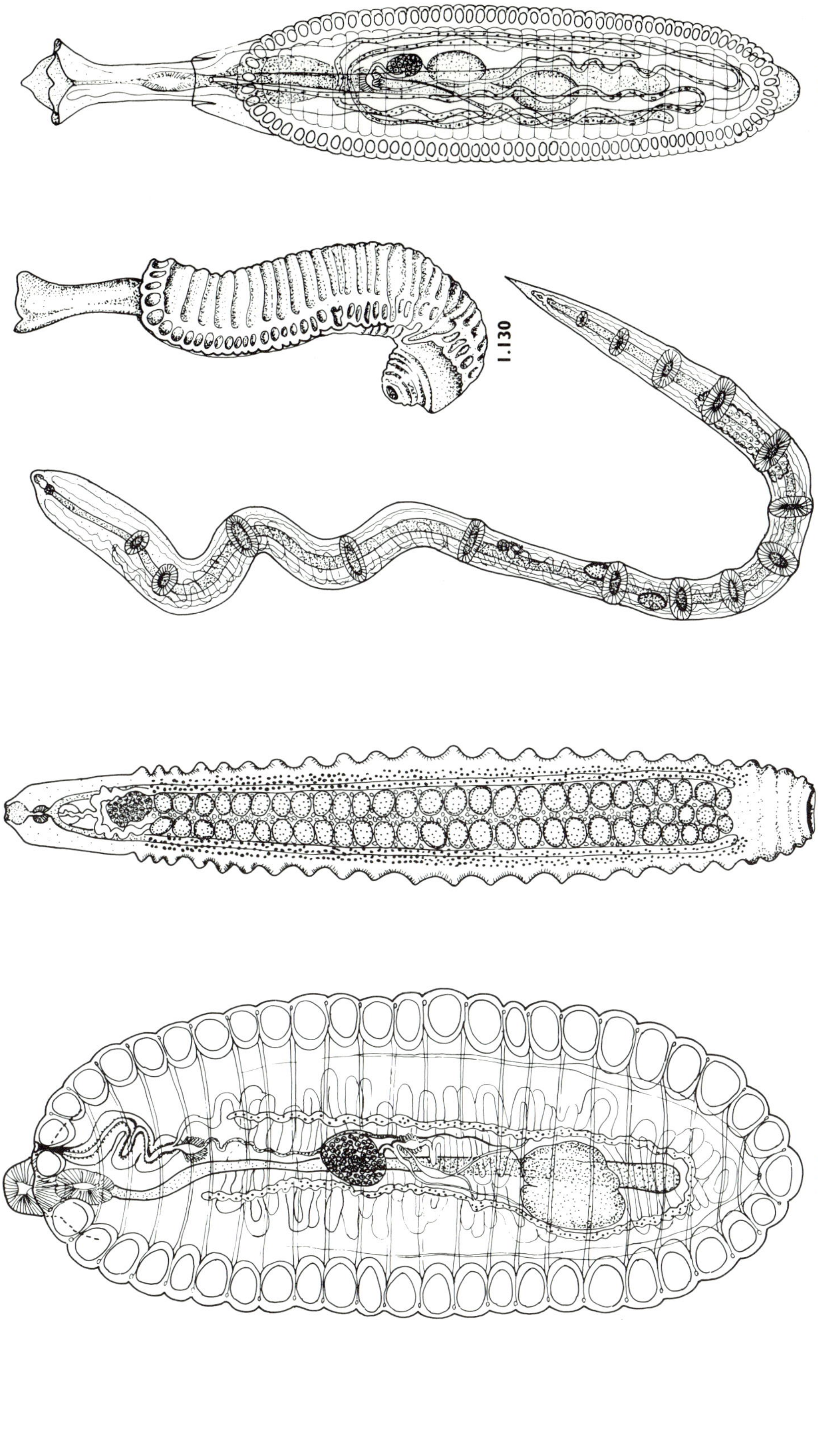

Figures 1.127–1.131 Some representatives of the Aspidogastrea 1.127 *Rohdella siamensis* (family Aspidogastridae) redrawn from Gibson and Chinabut (1984) 1.128 *Rugogaster hydrolagi* (Stichocotylidae) redrawn from Schell (1973) 1.129 *Stichocotyle nephropis* (Stichocotylidae) redrawn from Dollfus (1958) 1.130–1.131 *Cotylogaster occidentalis* (Aspidogastridae) redrawn from Dollfus (1958)

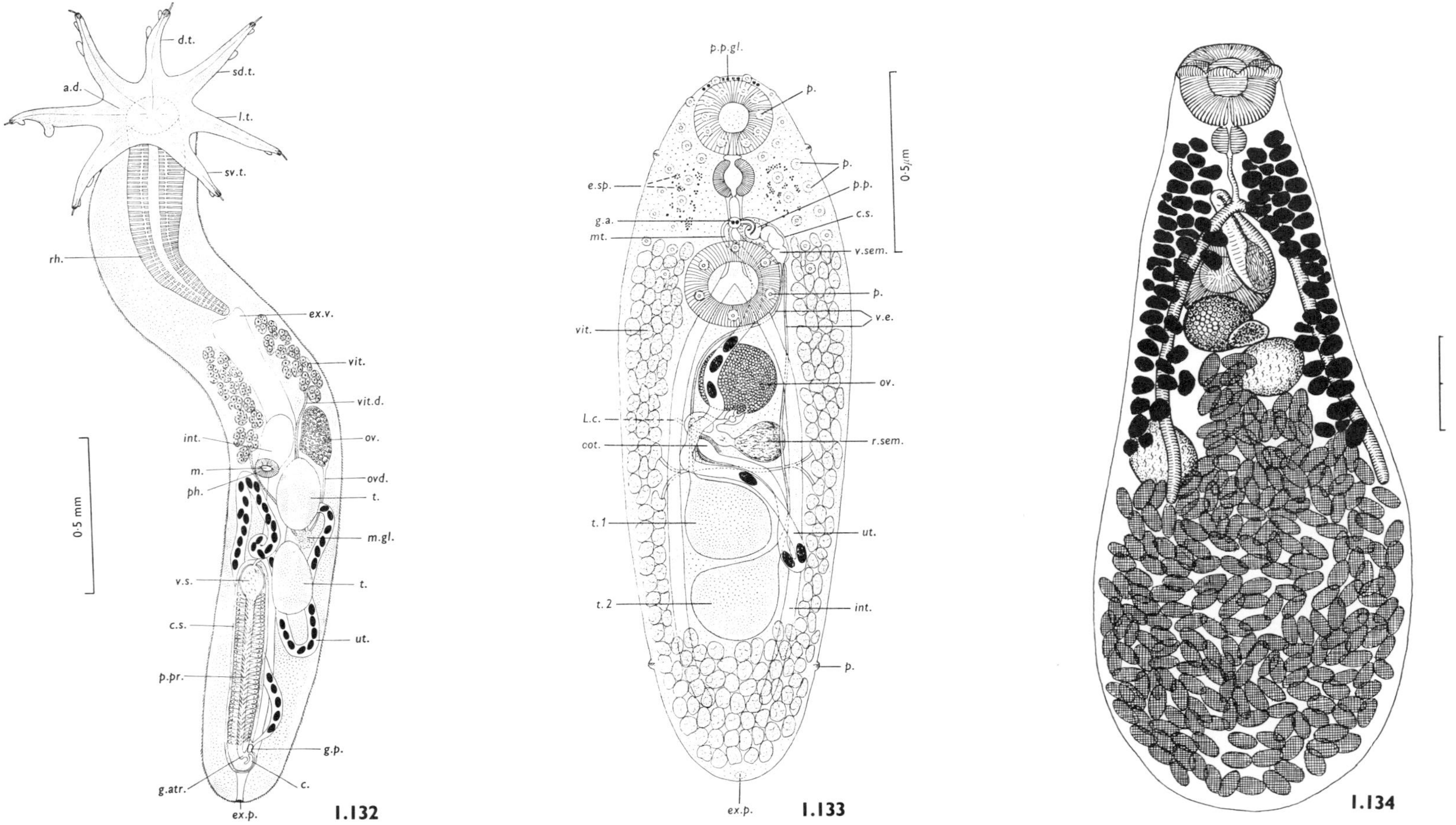

Figures 1.132–1.134 Digeneans belonging to the Bucephalidae and Allocreadiidae 1.132 Bucephalidae: *Alicornis* from Rees (1970) Key to lettering: a.d. apical disc; c. cirrus; c.s. cirrus sac; d.t. dorsal tentacle; ex.p. excretory pore; ex.v. excretory vesicle; g.atr. genital atrium; g.p. genital pore; int. intestine; l.t. lateral tentacle; m. mouth; m.gl. Mehlis' gland; ov. ovary; ovd. oviduct; p.pr. pars prostatica; ph. pharynx; rh. rhynchus; sd.t. subdorsal tentacle; t. testis; ut. uterus; v.s. vesicula seminalis; vt. vitelline glands; vit.d. vitelline duct. 1.133 Allocreadiidae: *Macrolecithus* from Rees (1968) Key to lettering: c.s. cirrus sac; e.sp. eyespot; ex.p. excretory pore; g.a. genital atrium; int. intestinal caecum; L.c. Laurer's canal; mt. metraterm; oot. ootype; ov. ovary; p. papilla; p.p. pars prostatica; p.p.gl. pore of penetration gland; r.sem. receptaculum seminis; t.1. anterior testis; t.2. posterior testis; ut. uterus; v.e. vas efferens; v.sem. vesicula seminalis; vit. vitelline glands 1.134 Allocreadiidae: *Bunodera* from Mueller and Van Cleave (1932)

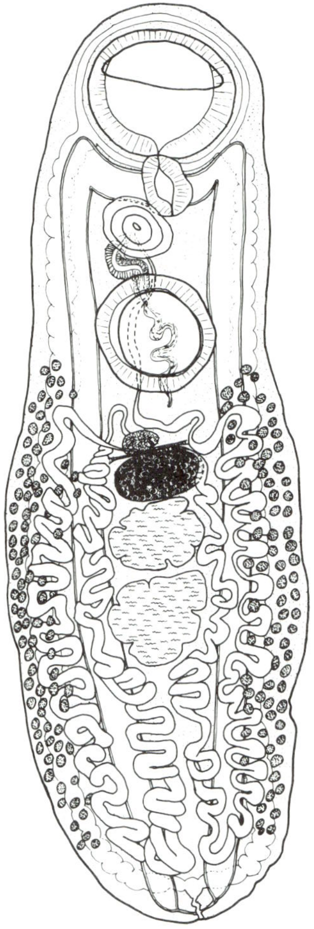

1.135

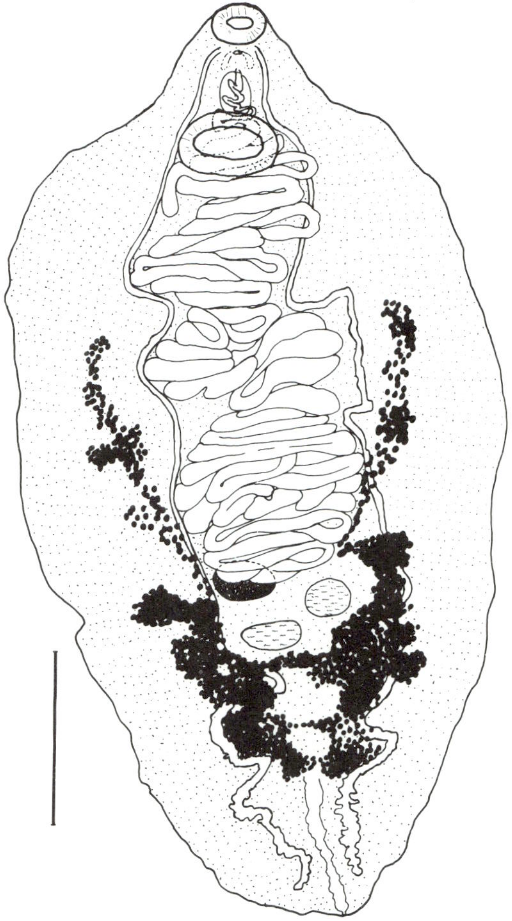

1.136

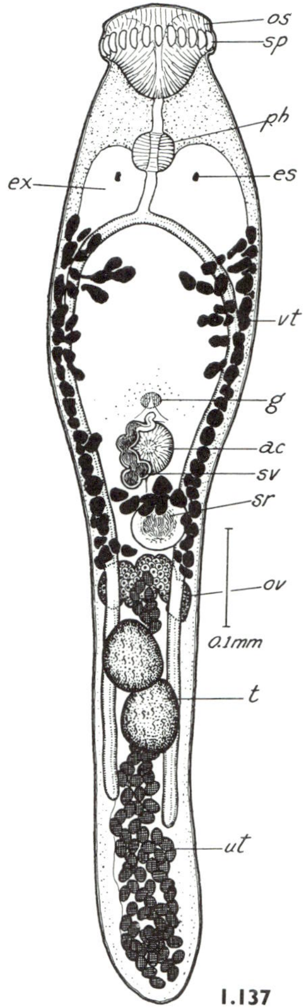

1.137

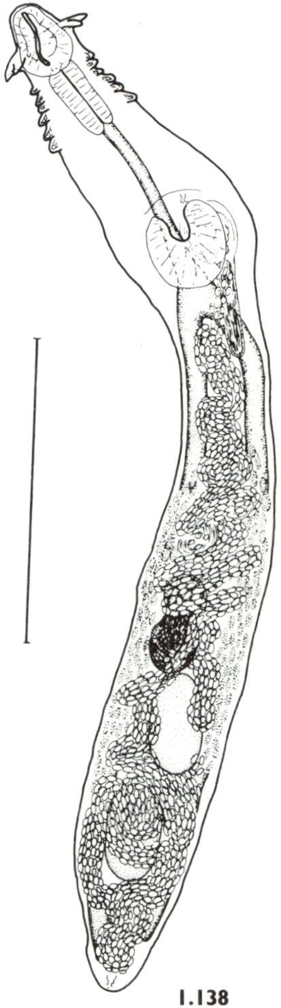

1.138

Figures 1.135–1.138 Digenea, usually with post-ovarian testes, belonging to the Ptychogonimidae, Azygiidae, Acanthocolpidae and Fellodistomidae. 1.135 Ptychogonimidae. *Ptychogonimus* from Gibson and Bray (1977). Bar: 1 mm 1.136 Azygiidae. *Otodistomum* from Gibson and Bray (1977) Bar: 5 mm 1.137 Acanthocolpidae. *Allacanthochasmus* from Mueller and Van Cleave (1932) Key to lettering ac. acetabulum; e.s. eye spots; ex. excretory bladder; g. gonotyle; o.s. oral sucker; ov. ovary; ph. pharynx; sp. spines; s.r. seminal receptacle; s.v. seminal vesicle; t. testis; ut. uterus; vt. vitelline glands. 1.138 Fellodistomidae. *Tergestia laticollis* from Bray and Gibson (1980)

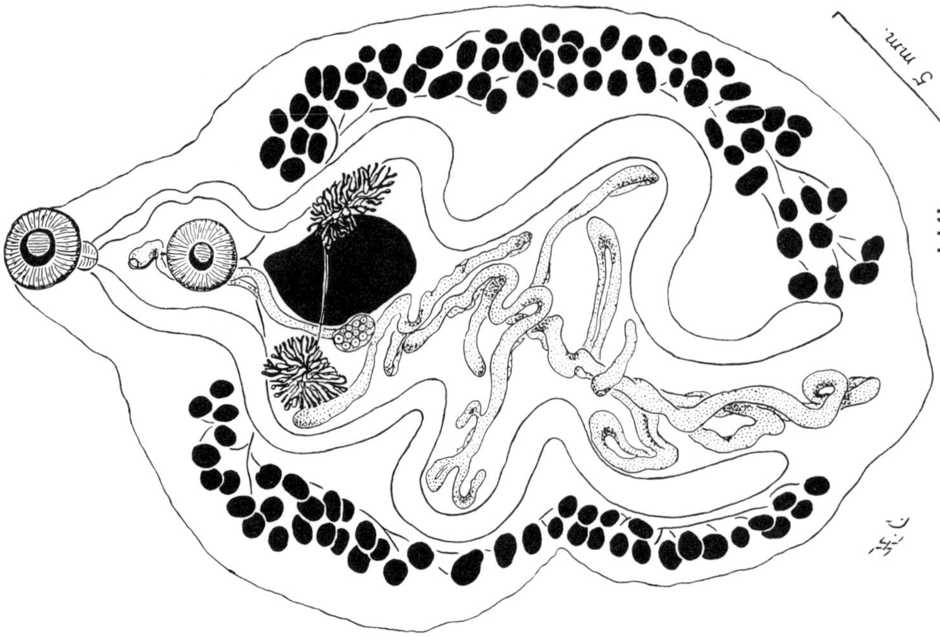

1.141

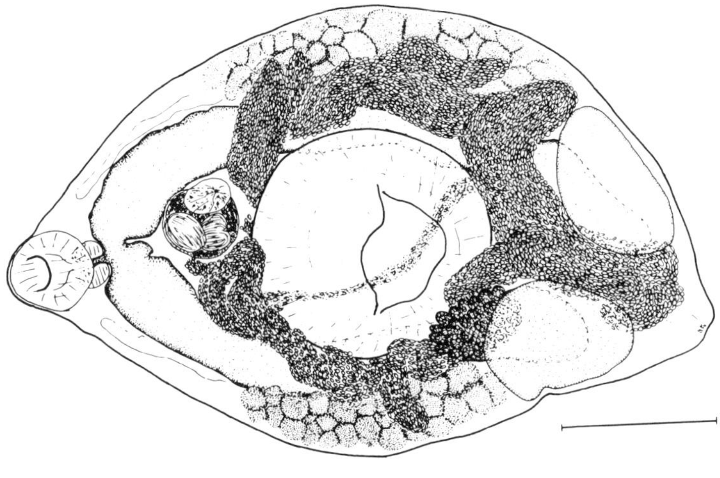

1.140

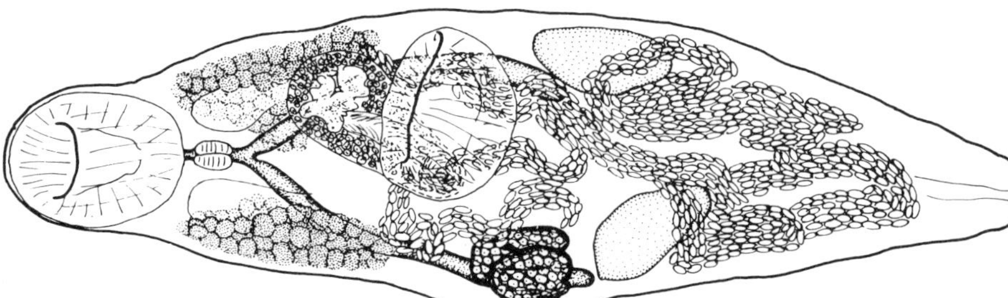

1.139

Figures 1.139–1.141 Digeneans with post-ovarian testes belonging to the Fellodistomidae and Gorgoderidae 1.139, 1.140 Fellodistomidae. *Olssonium* and *Fellodistomum* from Bray and Gibson (1980) 1.141 Gorgoderidae. *Staphylorchis* from Crusz (1957)

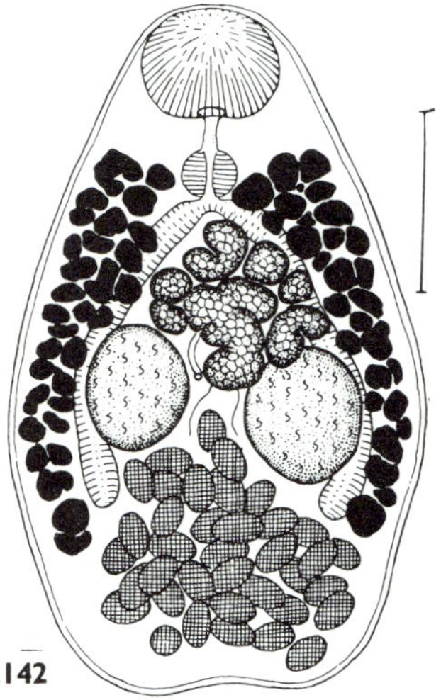

1.142

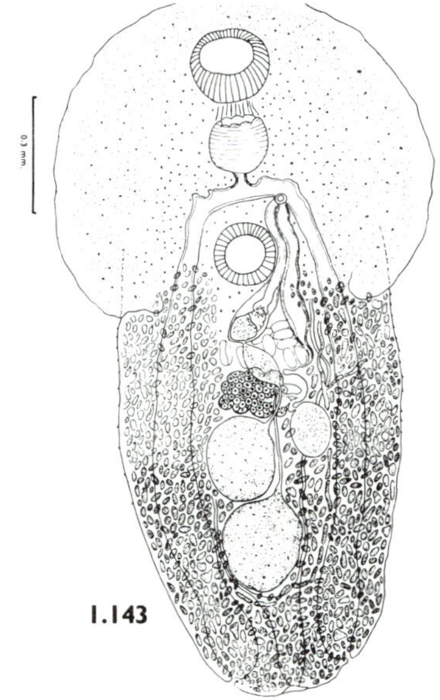

1.143

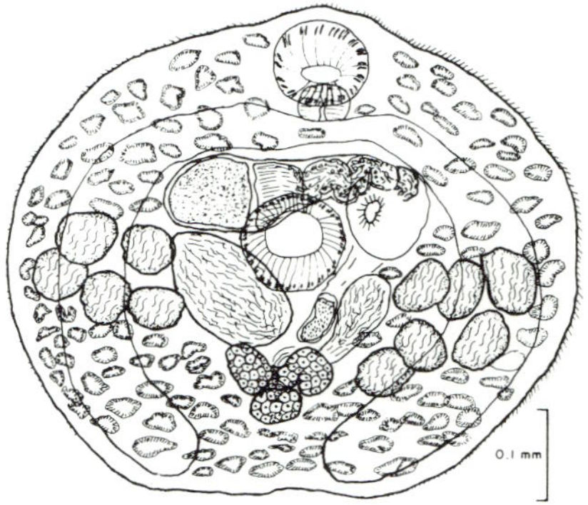

1.144

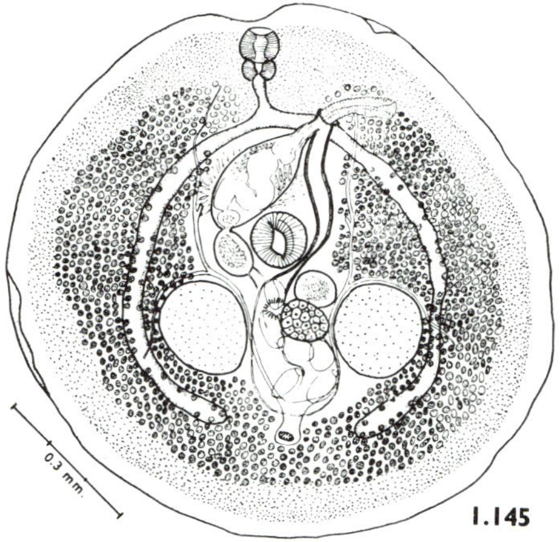

1.145

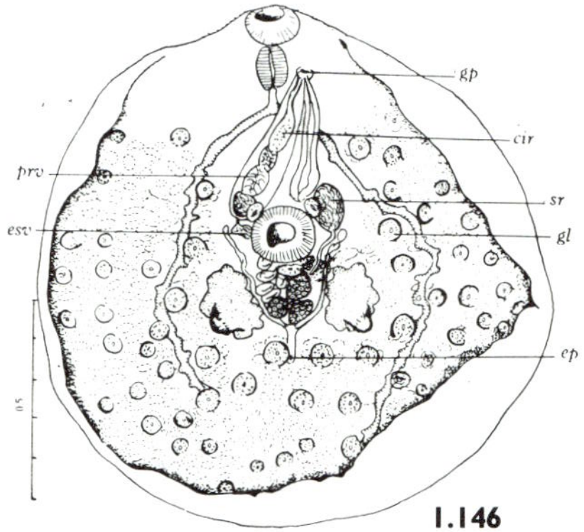

1.146

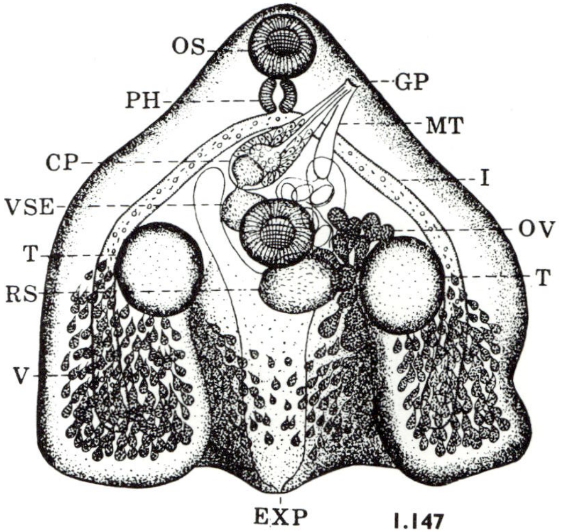

1.147

Figures 1.142–1.147 A range of body form in the Allocreadiidae, usually with post-ovarian testes 1.142 *Vietosoma* from Mueller and Van Cleave (1932) Bar: 0.1 mm 1.143 *Bianium* from Argumedo (1963) 1.144 *Multitestis* from Sparks (1954) 1.145 *Pseudocreadium* from Argumedo (1963) 1.146 *Dermadena* from Manter (1945) Key to lettering cir. cirrus; ep. excretory pore; esv. external seminal vesicle; gl. ventral gland; gp. genital pore; prv. prostatic vesicle; sr. seminal receptacle 1.147 *Trigonotrema* from Goto and Ozaki (1929) Key to lettering CP. cirrus pouch; GP. genital pore; I. intestinal caecum; MT. metraterm; OS. oral sucker; OV. ovary; PH. pharynx; RS. receptaculum seminis; T. testis; V. vitelline glands; VSE. vesicula seminalis externa.

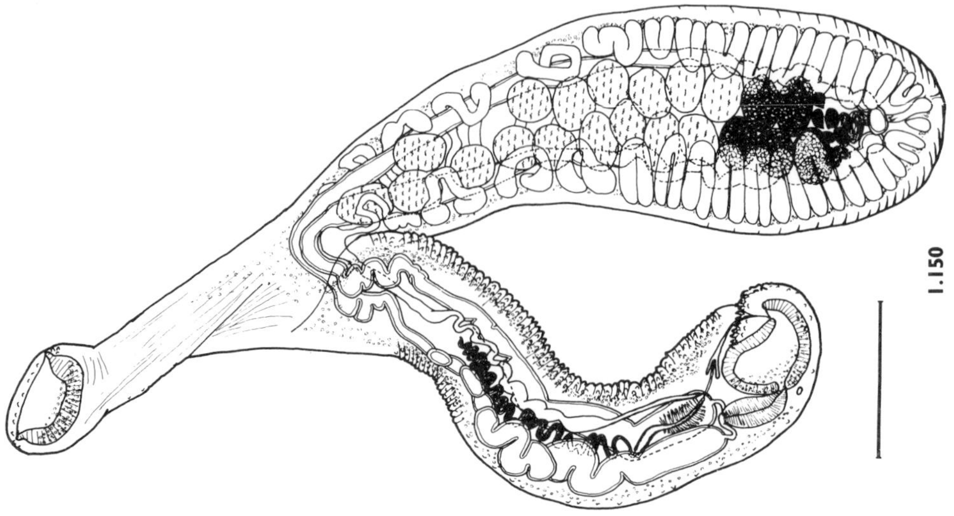

1.150

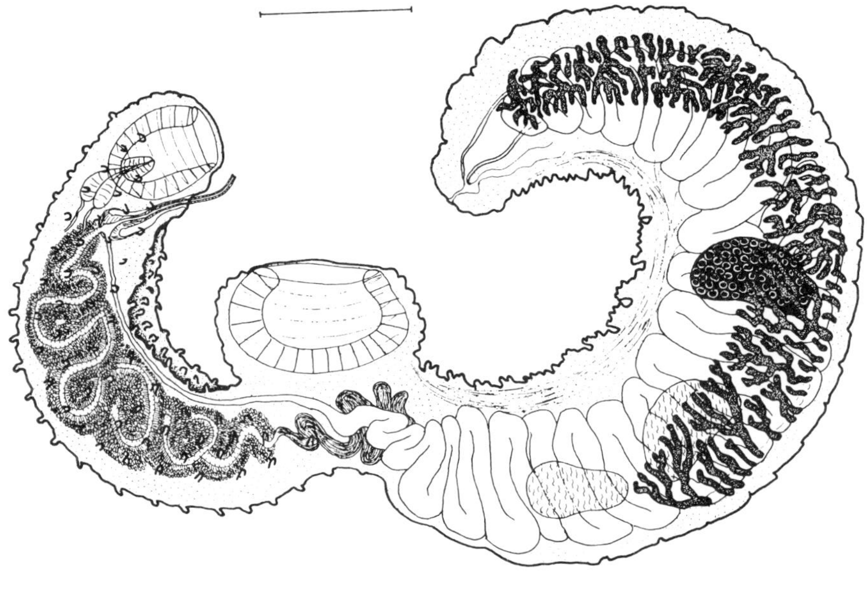

1.149

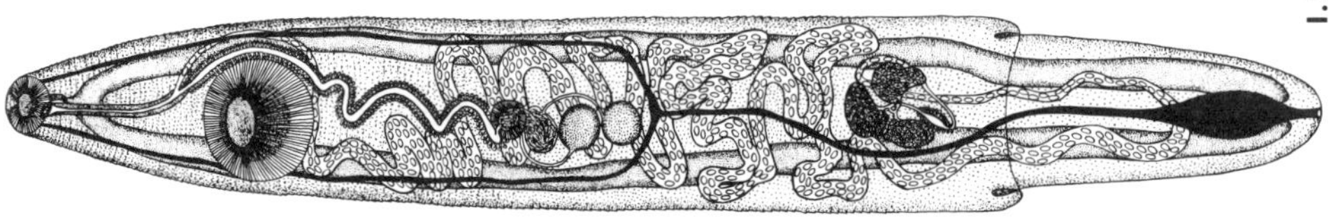

1.148

Figures 1.148–1.150 Digeneans with pre-ovarian testes, belonging to the Hemiuridae, Accacoeliidae and Syncoeliidae. 1.148 Hemiuridae. *Hemiurus* from Williams (1961b) 1.149 Accacoeliidae. *Accacoelium* from Bray and Gibson (1977) Bar: 1 mm 1.150 Syncoeliidae. *Copiatestes* from Gibson and Bray (1977) Bar: 1 mm

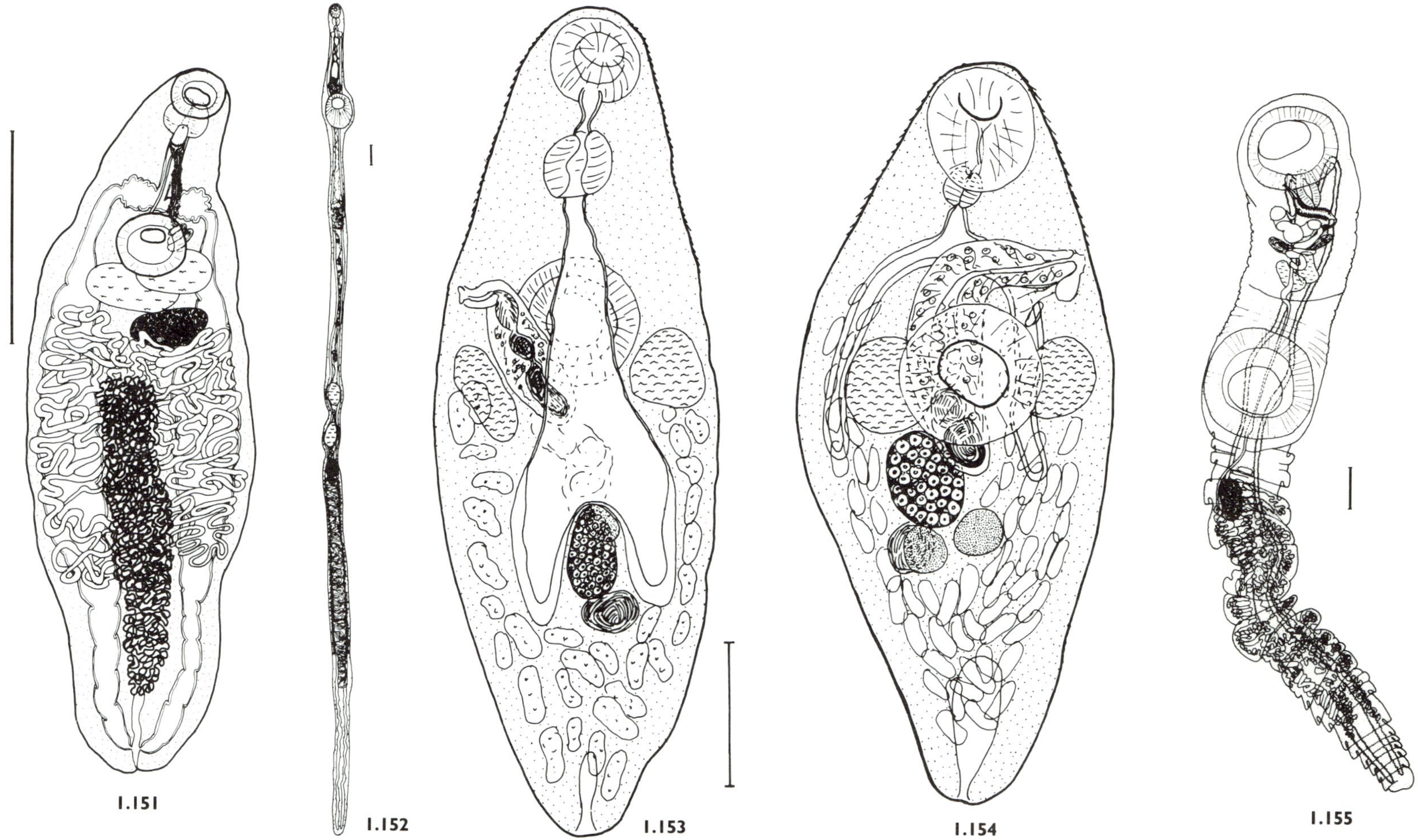

Figures 1.151–1.155 Digeneans with pre-ovarian testes belonging to the Hirudinellidae, Zoogonidae and Sclerodistomidae. 1.151 Hirudinellidae. *Botulus* from Gibson and Bray (1977) Bar: 10 mm 1.152 Hirudinellidae. *Lampritrema* from Gibson and Bray (1977) Bar: 1 mm 1.153–1.154 Zoogonidae. *Zoogonus* and *Pseudozoogonoides* from Bray and Gibson (1986) 1.155 Sclerodistomidae. *Prosorchiopsis* from Gibson and Bray (1977) Bar: 1 mm

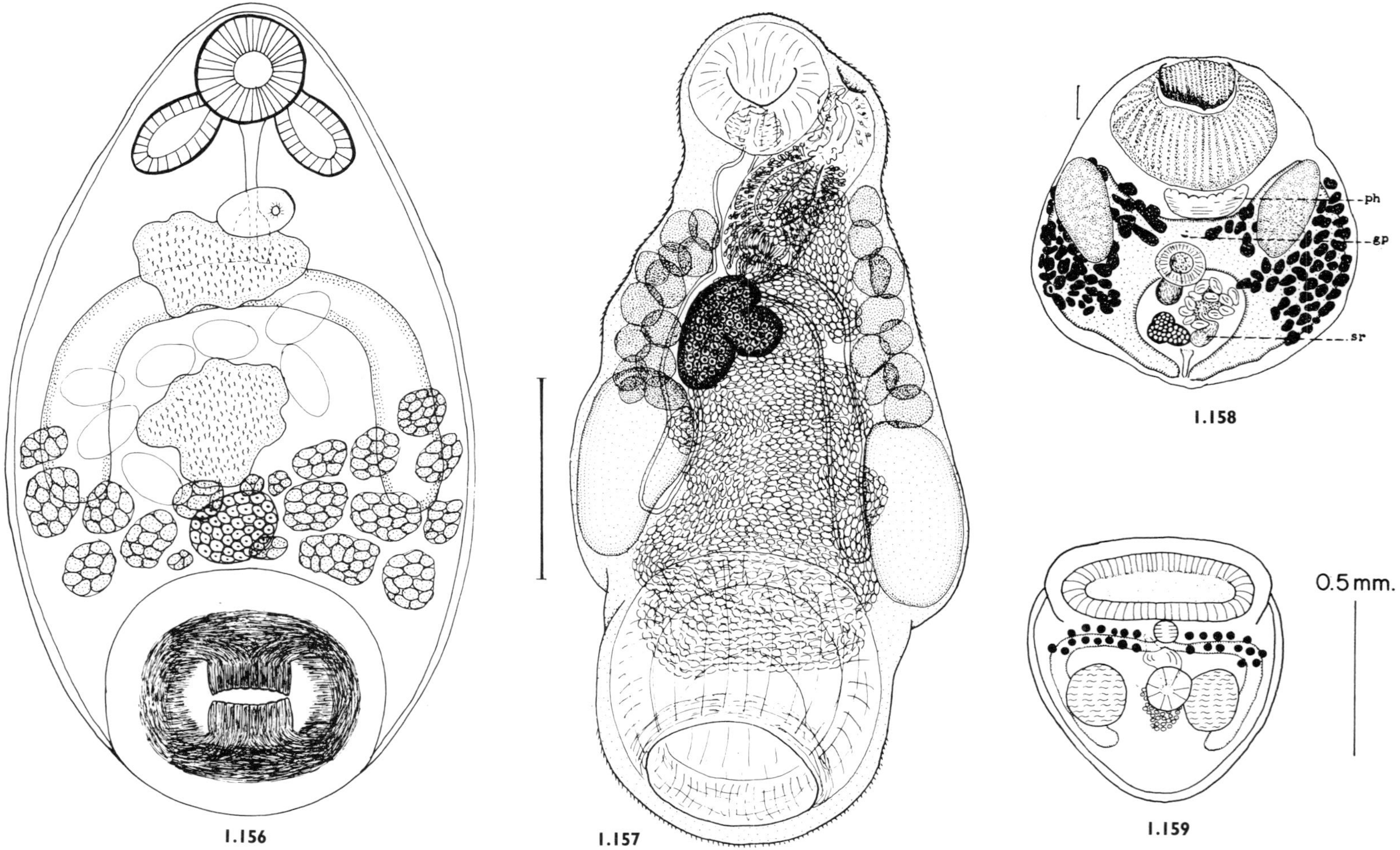

Figures 1.156–1.159 Digeneans with unusual oral or ventral suckers 1.156 Paramphistomidae. *Australotrema* from Khalil (1981) 1.157 Cephaloporidae. *Cephaloporus* from Bray (1985b) Bar: 0.5 mm 1.158 Megaperidae. *Eurypera* from Manter (1933). Bar: 0.1 mm Key to lettering: ph. pharynx; gp. genital pore; sr. seminal receptacle 1.159 Cryptogonimidae. *Metadena* from Arai (1962)

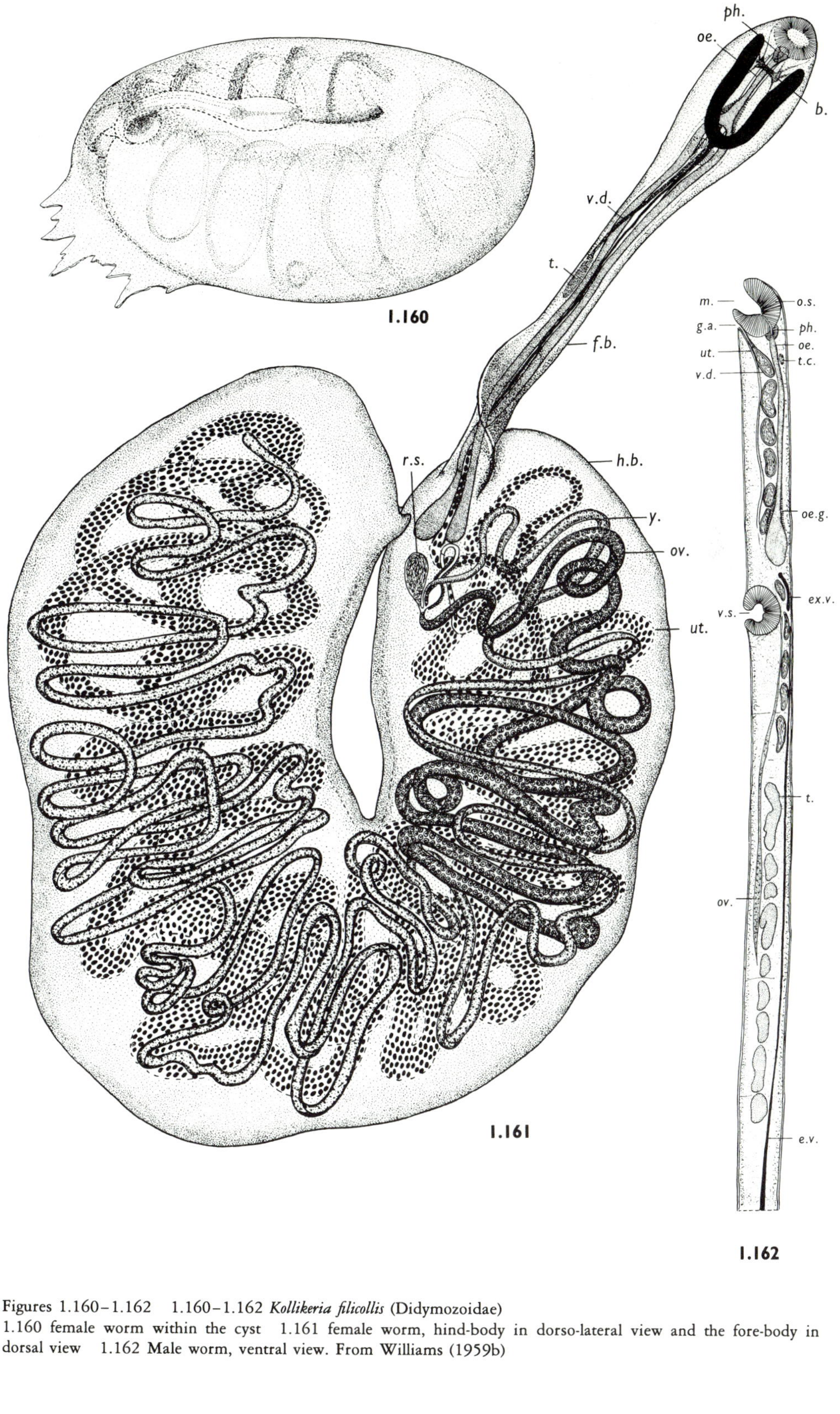

1.160

1.161

1.162

Figures 1.160–1.162 1.160–1.162 *Kollikeria filicollis* (Didymozoidae)
1.160 female worm within the cyst 1.161 female worm, hind-body in dorso-lateral view and the fore-body in dorsal view 1.162 Male worm, ventral view. From Williams (1959b)

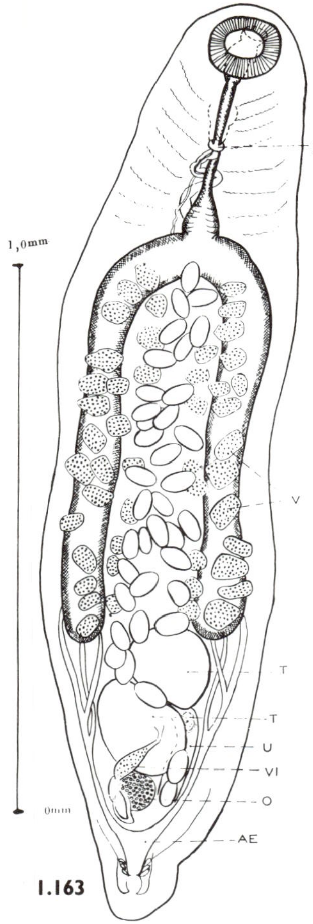

1.163

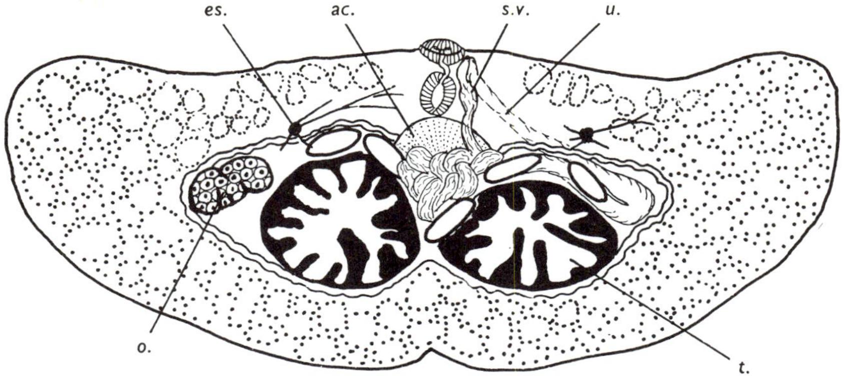

1.164

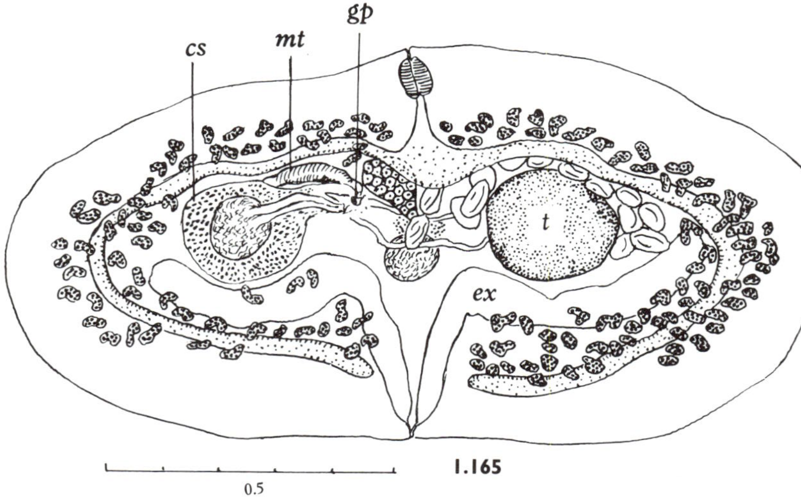

1.165

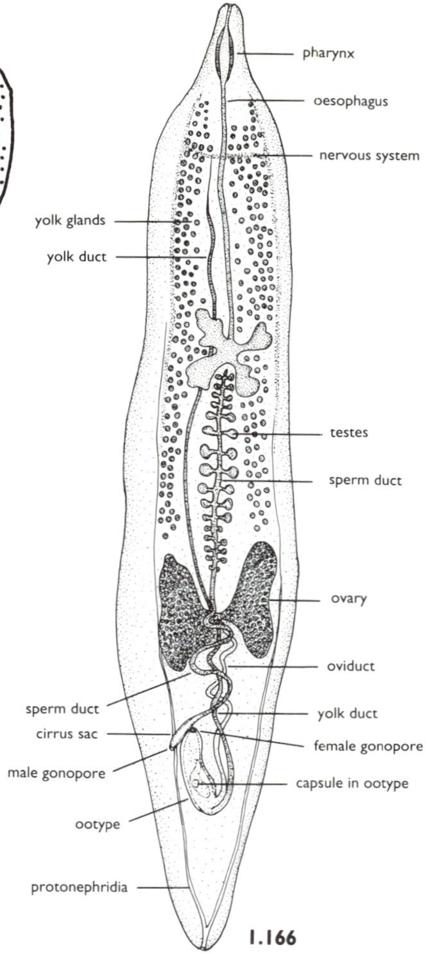

1.166

Figures 1.163–1.166 Digeneans lacking oral, ventral or both suckers

1.163 Angiodictyidae. *Hexangium* from Razarihelisoa (1959) Key to lettering: AE. excretory vesicle; O. ovary; PG. genital pore; T. testis; U. uterus; V. vitelline glands; V.I. single vitelline duct. 1.164 Transversotrematidae. *Prototransversotrema* from Angel (1969) Key to lettering: ac. acetabulum; es. eyespot; o. ovary; s.v. seminal vesicle; t. testis; u. uterus. 1.165 Bivesiculidae. *Treptodemus* from Manter (1961). In bivesiculids an oral sucker is present but the pharynx is missing. Key to lettering: cs. cirrus sac; gp. genital pore; mt. metraterm. 1.166 Sanguinicolidae. *Sanguinicola* redrawn from Hyman (1951a).

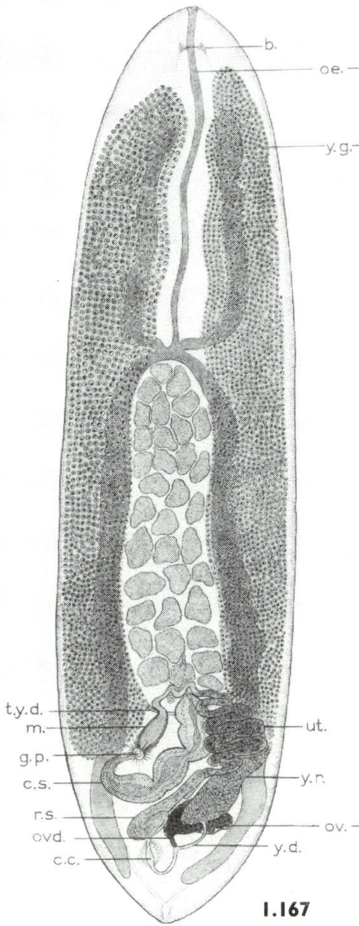

1.167

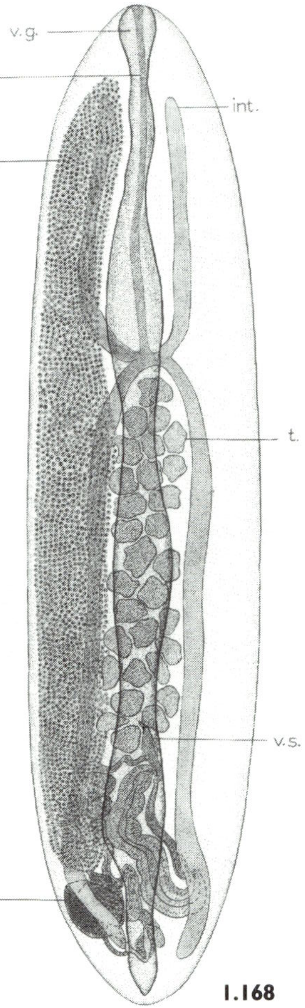

1.168

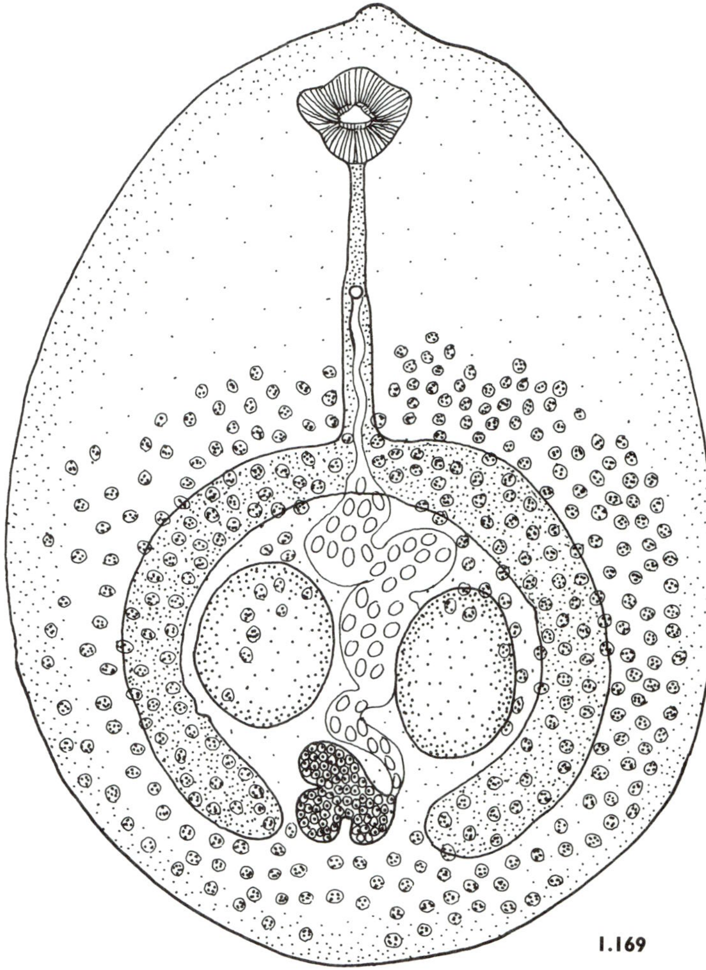

1.169

Figures 1.167–1.169 Digenea belonging to the families Aporocotylidae and Mesometridae. 1.167 Aporocotylidae. *Aporocotyle*, dorsal view showing general features, from Williams (1958a) 1.168 Aporocotylidae. *Aporocotyle*, ventral view, showing median longitudinal spinous canal, from Williams (1958a) Key to lettering. b. cerebral ganglion; c.c. central chamber; c.s. cirrus sac; g.p. genital pore; int. intestine; m. metraterm; oe. oesophagus; ov. ovary; ovd. oviduct; r.s. receptaculum seminis; t. testis; t.y.d. transverse vitelline duct; ut. uterus; v.g. ventral groove; v.s. vesicula seminalis; y.d. vitelline duct; y.g. yolk gland; y.r. vitelline reservoir. 1.169 Mesometridae. *Mesometra* from Dawes (1946).

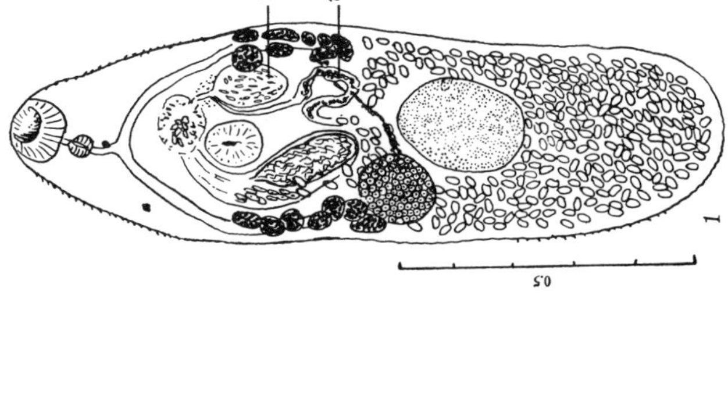

1.170

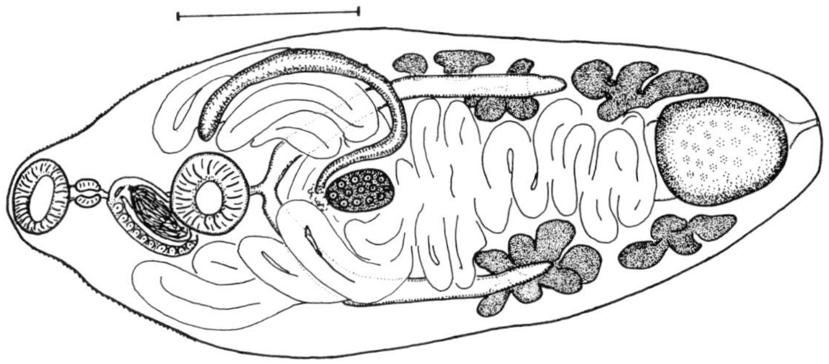

1.171

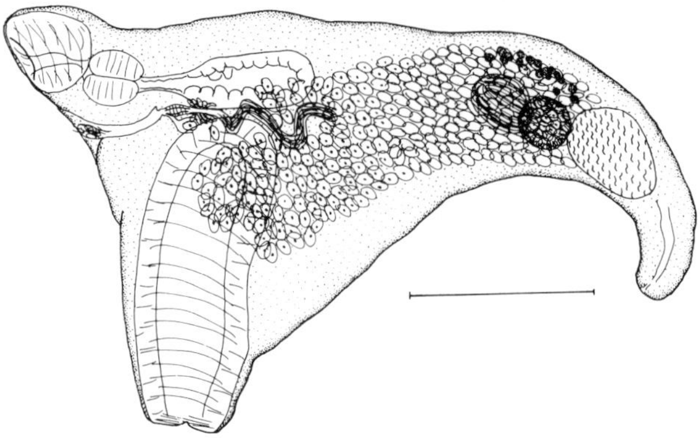

1.172

Figures 1.170–1.172 Digenea with only one testis 1.170 Monorchiidae. *Genolopa* from Manter and Pritchard (1961). Some monorchiids have two testes Key to lettering: tov. terminal organ vesicle; ut. uterus. 1.171 Haploporidae. *Chalcinotrema* from Brooks (1977) Bar : 1 mm 1.172 Haplosplanchnidae. *Haplosplanchnus* from Bray (1984) Bar : 0.5 mm

Phylum Nematoda

The nematodes are a large, extremely diverse and successful group, consisting of 256 families (Anderson, 1984). The majority are free-living and occupy environments ranging from the sea depths to deserts and icecaps, while the parasitic nematodes exploit a great variety of hosts in both plant and animal Kingdoms. The 125 families of zooparasitic nematodes are thought to have originated from freeliving soil-dwelling forms on land rather than in the sea. Fish nematodes are suggested by Anderson (1984) to have been derived from terrestrial vertebrate nematodes, as nearly three quarters of the 17 nematode families found in fish are also found in land vertebrates. This hypothesis is supported by the absence of any superfamily, and the presence of only five families of nematodes unique to fish, suggesting that they are an acquired fauna.

The following classification, references and seven figures show the variety of form amongst selected adult nematodes of fish. Although the C.I.H. Keys refer to the class Nematoda they are accorded phylum status in this chapter. Anderson, Chabaud and Willmott (1974) introduce these keys, and an index to the whole series is included in Durette-Desset (1983). Other comprehensive sources of information are Berland (1961), Chabaud (1965), Fagerholm (1982), Hooper (1983), Inglis (1983), Stone, Platt and Khalil (1983) and Yamaguti (1961).

Nematodes are usually divided into two groups according to the presence or absence of: 1. special gland cells associated with the oesophagus known as stichocytes which, if arranged in a longitudinal row, are known as a stichosome, 2. caudal papillae, 3. a reserve organ, called a trophosome, apparently syncytial, formed by the transformation of the oesophagus, 4. a pair of glandular sensory organs, called phasmids, situated laterally in the caudal region and opening to the surface by a slit or pore, 5. polar plugs to the eggs.

In the Adenophorea, there are no phasmids, caudal papillae are absent or few in number, the eggs usually have polar plugs and, in one major superfamily, the Trichuroidea, a stichosome or trophosome is present. In the Secernentea, there is no stichosome or trophosome, caudal papillae are almost always numerous (the basic number being 21), phasmids are present and eggs are without polar plugs. A comprehensive study of some members of the Adenophorea and Secernentea found in fish is that of Fagerholm (1982), and the work of Moravec and his co-authors, as will be seen from the references cited throughout this book, is indispensable to fish nematologists.

I. Adenophorea

Two families of Adenophorea, the Trichuridae and Cystoopsidae contain, respectively, the genera *Capillaria* and *Cystoopsis* which are important fish parasites. *Capillaria* (Figures 1.173–1.187) may occur in the skin, viscera, spleen, respiratory and excretory systems. The genus has been well described by Anderson and Bain (1982), Bell and Beverley-Burton (1981), Moravec (1980c, 1982b, 1983b, 1984a, 1987), Moravec, Margolis and McDonald (1981), and Moravec and McDonald (1981).

Capillariids are easily distinguished by their filiform bodies (Figure 1.188) and an explanation of why such a slim and elongated body has evolved may be found by further work on the site which *Capillaria* occupies amongst the villi of the host's intestine (Figure 1.191 and Figure 1.192). Other characteristics include an oesophagus consisting of a muscular anterior portion followed by 20–60 stichocytes in 1–3 rows. Moravec (1982b, 1983a,b) regards the structure of the posterior end of the male as the most important feature for separating capillariids.

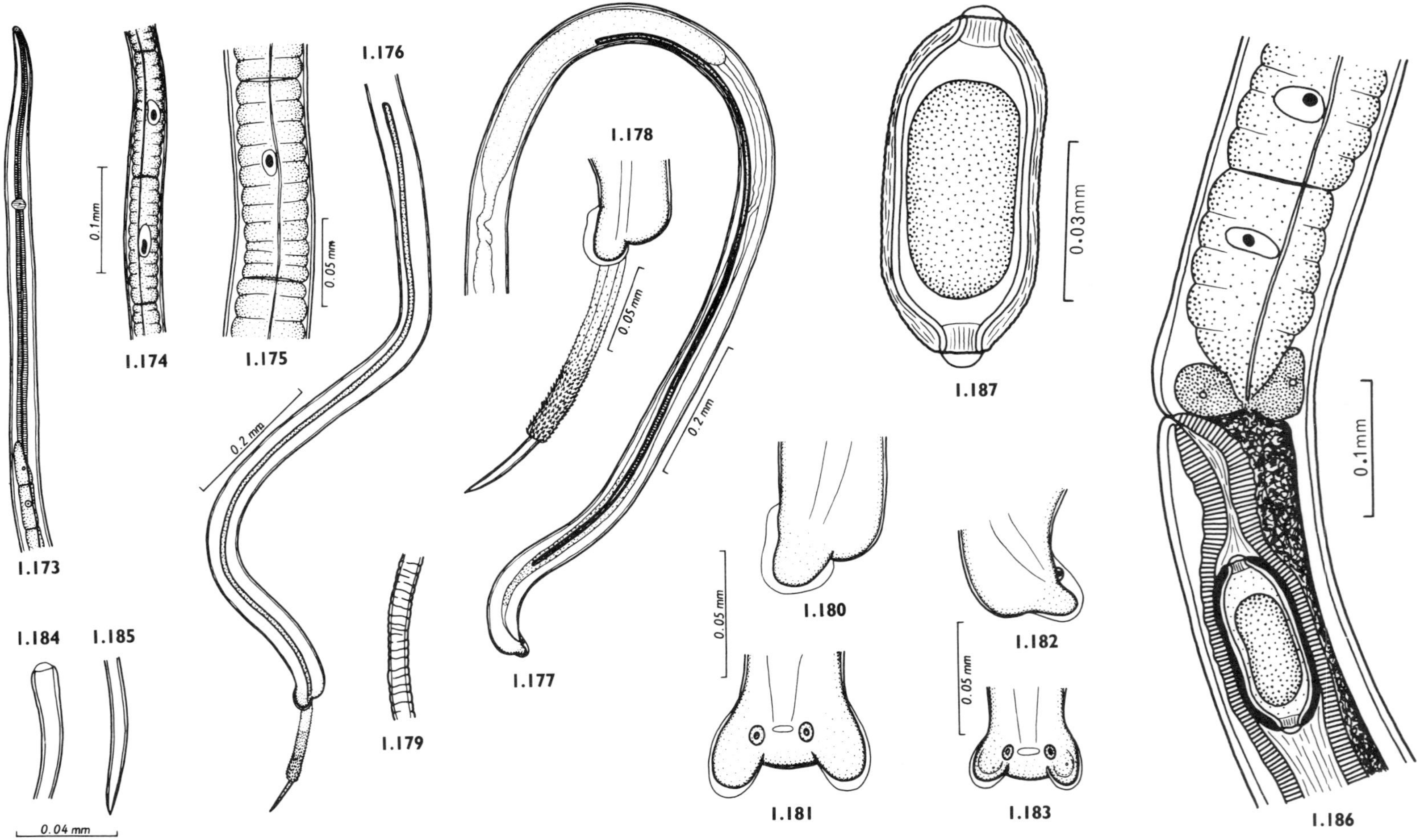

Figures 1.173–1.187 1.173–1.185 *Capillaria gracilis* (Bellingham, 1840) from gadiform fishes from Moravec (1987) 1.173 head end of female 1.174,1.175 stichosome region 1.176,1.177 posterior end of male 1.178 Male tail with evaginated spicular sheath and protruding spicule 1.179 grooves on middle part of spicule 1.180–1.183 tail of male (lateral and ventral views) 1.184,1.185 proximal and distal ends of spicule 1.173, 1.174, 1.177, 1.179, 1.182, 1.183 from *Gadus morhua* 1.175, 1.176, 1.178, 1.180, 1.181, 1.184, 1.185 from *Trisopterus luscus* 1.186 Vulvar region in female 1.187 Mature egg 1.186–1.187*Capillaria parophrysi* from Moravec, Margolis and McDonald (1981)

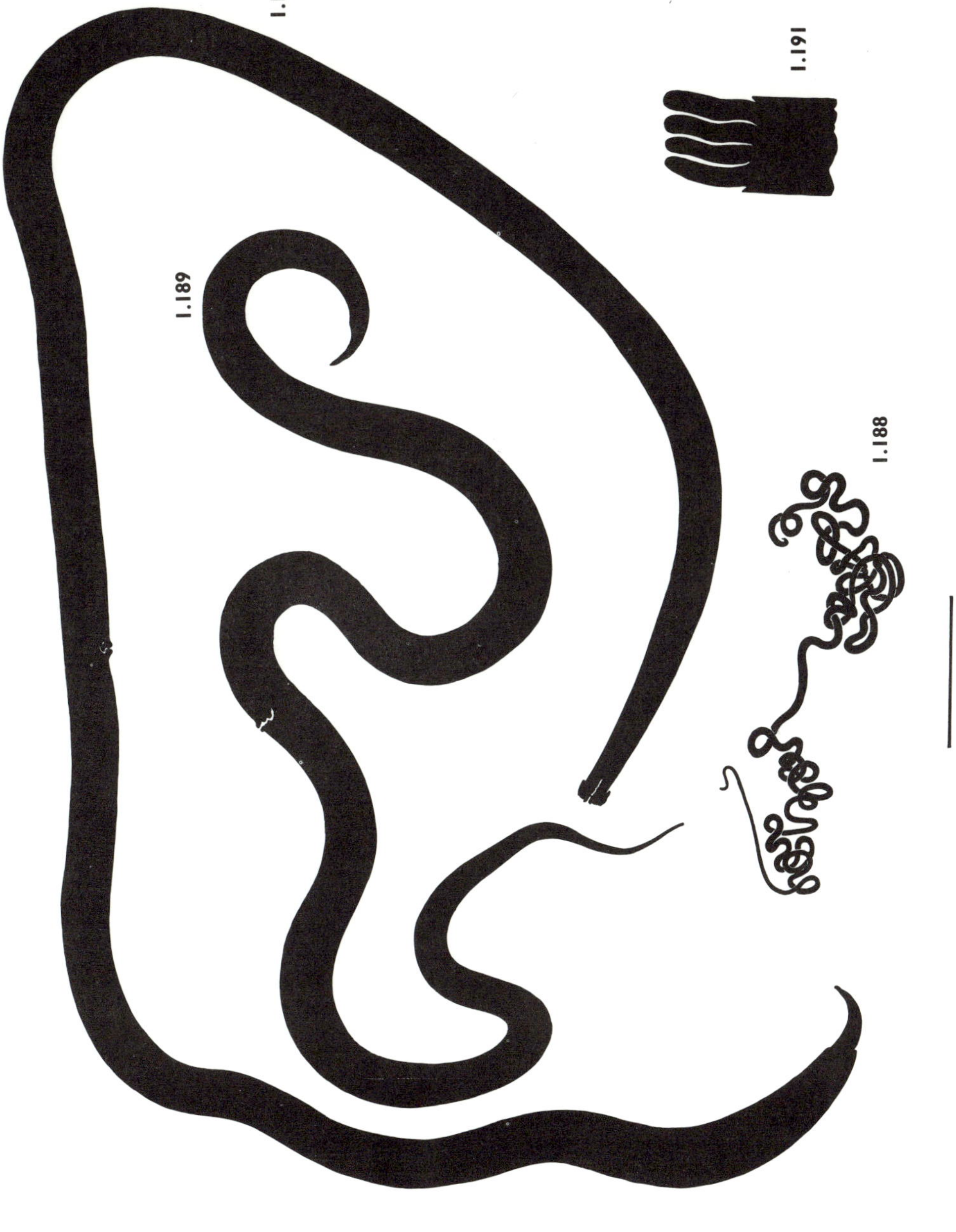

1.190 1.191 1.189 1.188

Figures 1.188–1.191 The distinctive silhouettes of three well-known genera of fish nematodes from elasmobranchs 1.188 *Capillaria* 1.189 *Pseudanisakis* 1.190 *Proleptus* 1.191 Villi of the gut mucosa of a species of *Raja* drawn to the same scale as Figure 1.188.

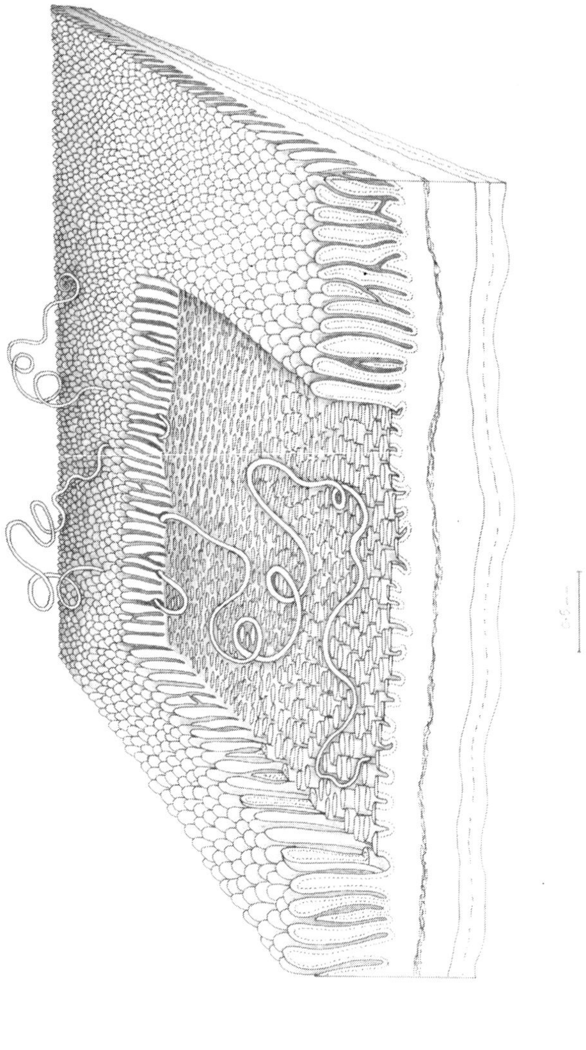

Figure 1.192 The relationships berween the capillariid figured (1.188) and the gut mucosa of a species of *Raja*

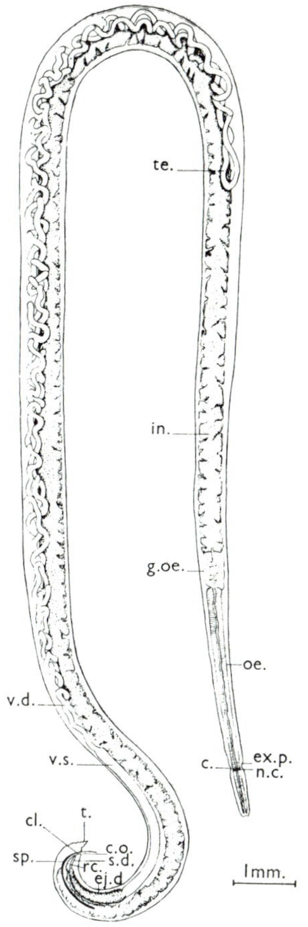

1.193

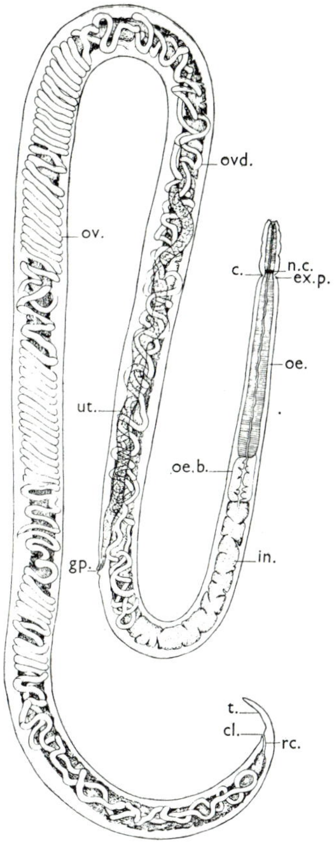

1.194

Figures 1.193–1.194 Nematoda: Ascaridoidea; Acanthocheilidae; *Pseudanisakis*
1.193 *Pseudanisakis* male 1.194 *Pseudanisakis* female
Figures from Williams and Richards (1968)
Key to lettering: c. constriction; cl. cloaca; e.j.d. ejaculatory duct; ex.p. excretory pore; gp. gonopore; in. intestine; n.c. nerve collar; oe. oesophagus; oe.b. ventriculus; ov. ovary; ovd. oviduct; rc. rectum; s.d. sperm duct; sp. spicule; t. tail; te. testis; ut. uterus; v.d. vas deferens; v.s. vesicula seminalis.

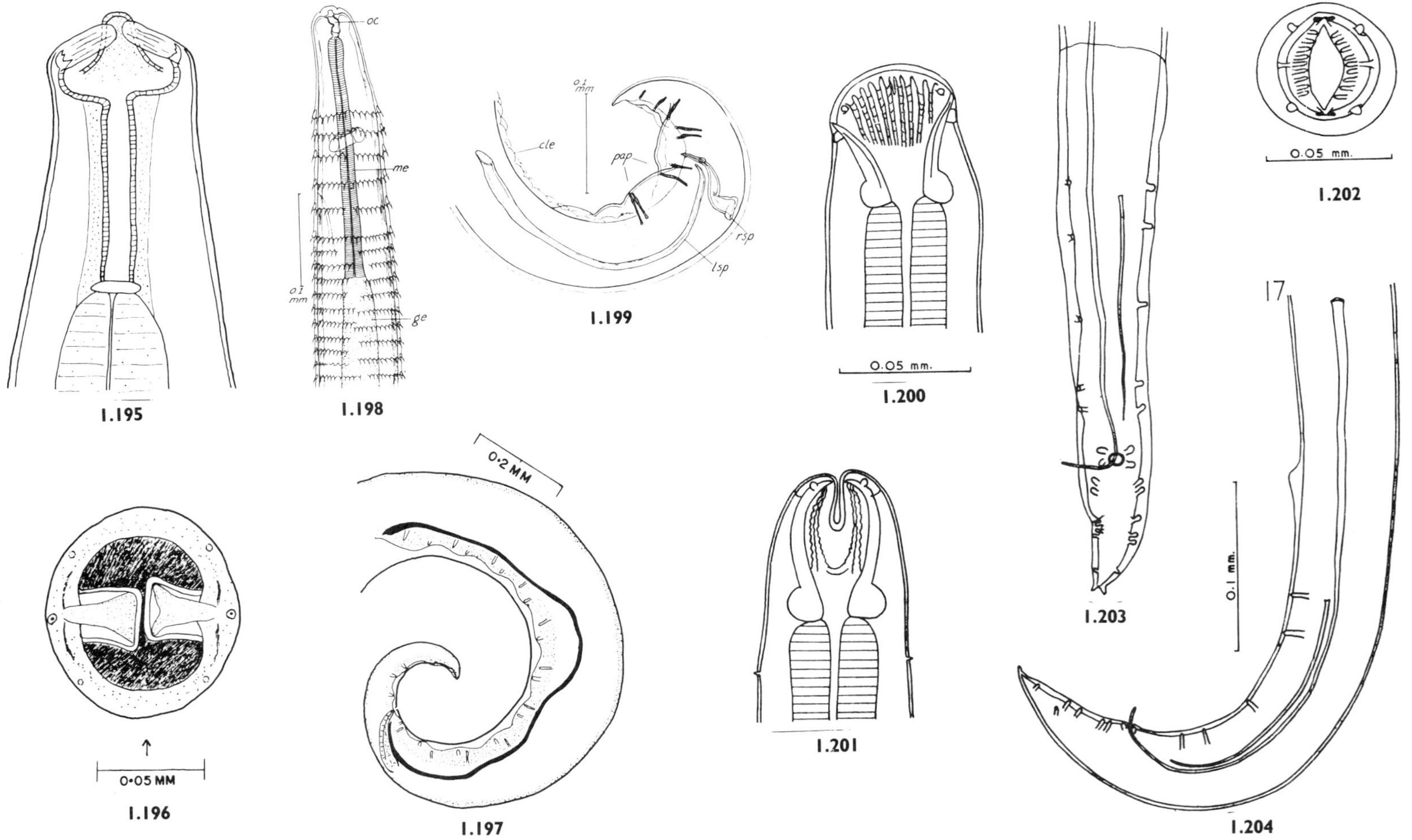

Figures 1.195–1.204 Nematoda: Habronematoidea and Camallanoidea 1.195–1.199 Habronematoidea 1.195 *Salvelinema*, buccal cavity of male, from Margolis (1967b) 1.196 *Salvelinema*, anterior end of female, *en face* view from Margolis (1967) 1.197 *Salvelinema*, posterior region of male, from Margolis (1967b) 1.198 *Spinitectus* from Mueller and Van Cleave (1932) 1.199 *Spinitectus*, tail of male with a crooked short spicule, from Mueller and Van Cleave (1932) Key to lettering: cle. cleats on cuticle; g.e. glandular oesophagus; m.e. muscular oesophagus; l.sp. left spicule; o.c. oral capsule; pap. papillae; r.sp. right spicule 1.200–1.204 Camallanoidea, from Yeh (1960) 1.200 *Camallanus*, lateral view of head 1.201 *Camallanus*, ventral view of head 1.202 *Camallanus*, *en face* view 1.203 *Camallanus*, ventral view of male tail 1.204 *Camallanus*, lateral view of male tail

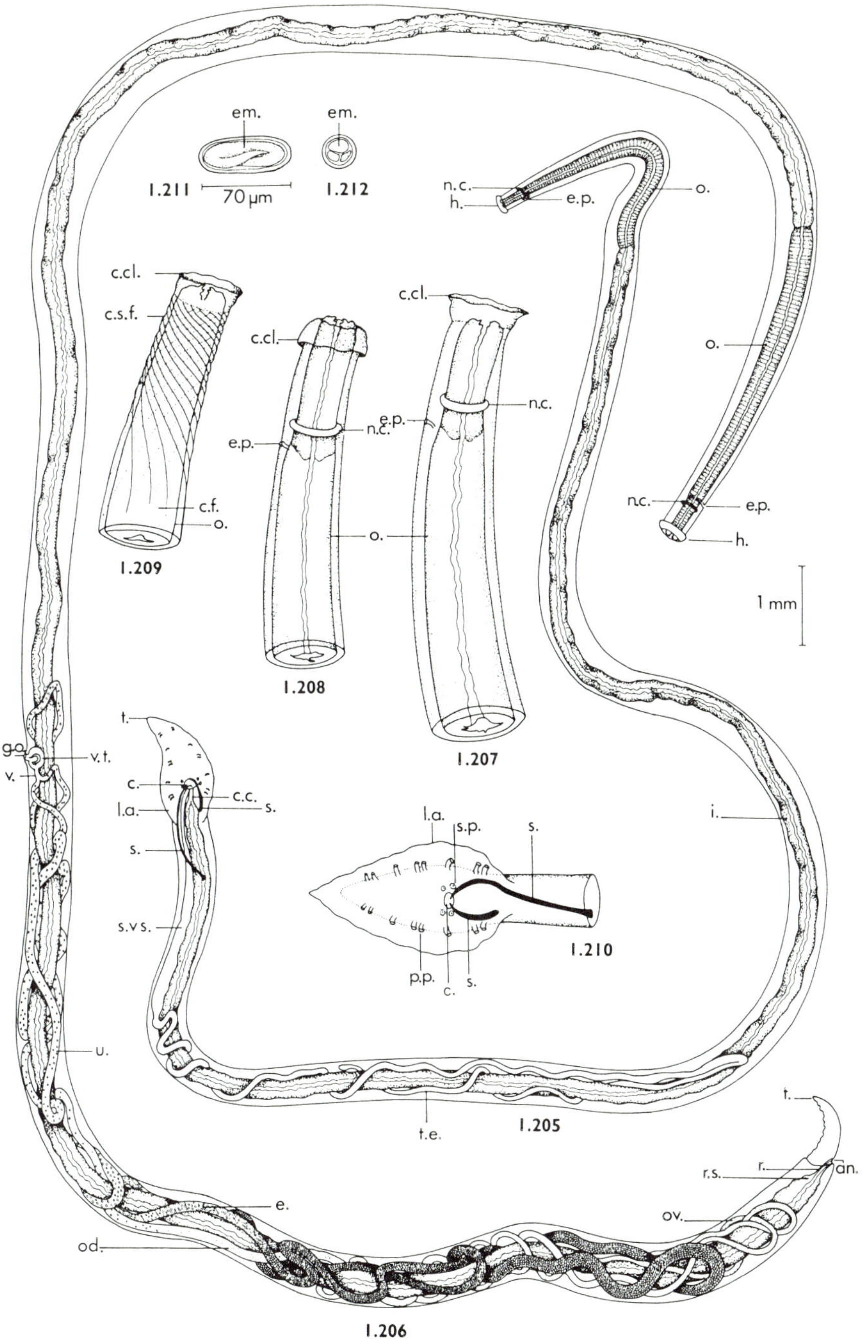

Figures 1.205–1.212 Nematoda: Physalopteroidea

1.205–1.212 ***Proleptus*** from Williams and Richards (1978) 1.205 Male, entire 1.206 Female, entire 1.207 Anterior end showing everted collar 1.208 Anterior end showing collar folded backwards 1.209 Anterior end showing spiral foldings of cuticle 1.210 Male tail, ventral view 1.211 Egg, lateral view 1.212 Egg, end view Key to lettering: a. anus; c. cloaca; c.c. cloacal canal; c.cl. cuticular collar; c.f. cuticular folds; c.s.f. cuticular spiral folds; e. eggs; em. embryo; e.p. excretory pore; g.o. genital opening; h. head; i. intestine; l.a. lateral alae; n.c. nerve collar; o. oesophagus; ov. ovary; od. oviduct; p.p. pedunculate papilla; r. rectum; r.s. rectal sphincter; s. spicule; s.p. sessile papilla; s.vs. seminal vesicle; t. tail; te. testis; u. uterus; v. vagina; vt. vestibule

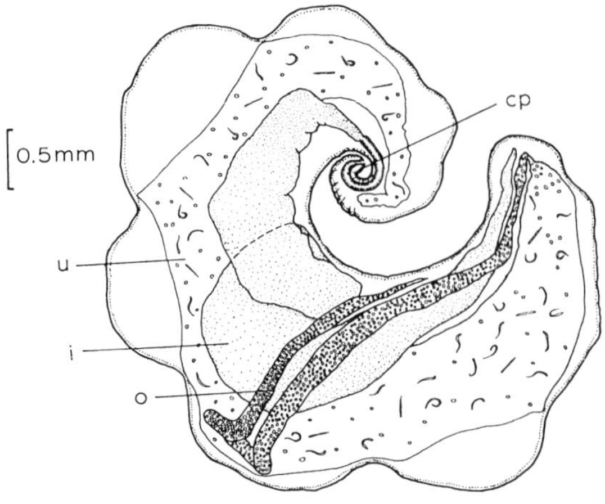

1.213

1.214

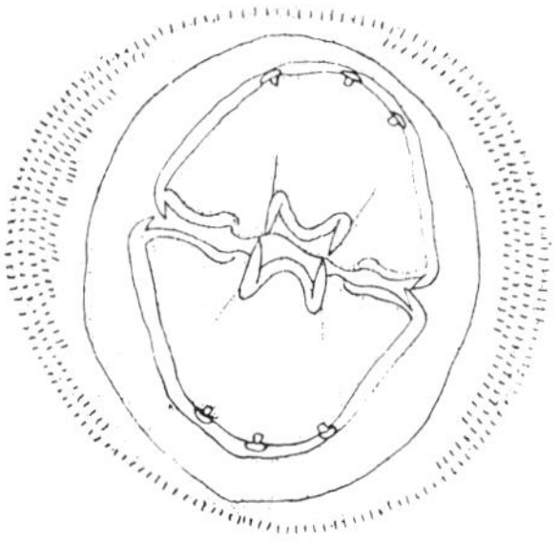

1.216

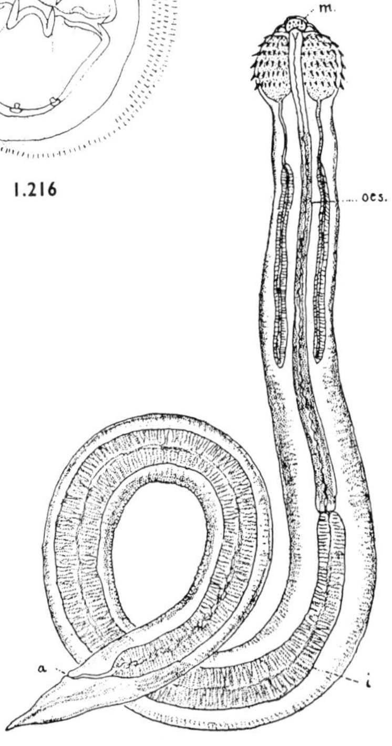

1.215

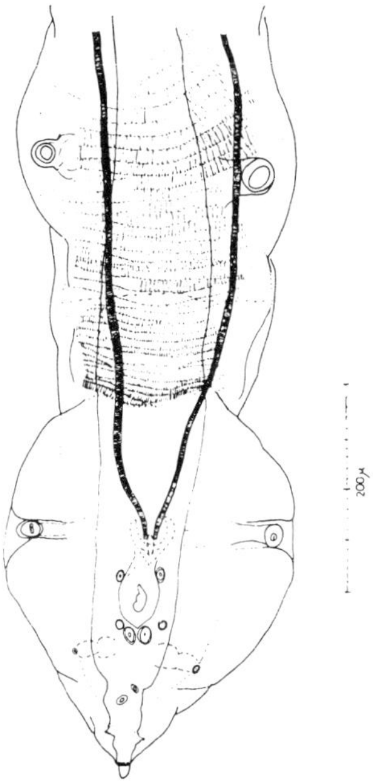

1.217

Figures 1.213–1.217 Nematoda: Dracunculoidea, Thelazioidea and Gnathostomatoidea
1.213 Dracunculoidea: *Phlyctainophora* from Mudry and Dailey (1969) Key to lettering: cp. cephalic protrusion; i. intestine; o. ovary; u. uterus 1.214 Thelazioidea: *Vasorhabdochona* from Martin and Zam (1967) Key to lettering: E1. buccal capsule; E2. oesophagus; s. spicules; sp. sperm; t. testes 1.215 Gnathostomatoidea: *Echinocephalus* from MacCallum (1921) Key to lettering: oes. oesophagus 1.216 Anterior end of *Echinocephalus* from Millemann (1963) 1.217 Tail end of *Echinocephalus*, from Troncy (1969)

The genus *Cystoopsis*, a parasite of the conjunctiva of chondrostean fish (*Acipenser*) is described by Anderson and Bain (1982), Hyman (1951b) and Yorke and Maplestone (1962).

II. Secernentea

Three orders of this subclass, the Oxyurida, Ascaridida and Spirurida, contain a number of important genera found in fish.

1. Order Oxyurida

Five genera, *Travnema*, *Icthyouris*, *Synodentisia*, *Laurotravassoxyuris* and *Cithariniella* are found in fish, the first of these having an oesophagus divided into two equal parts, each dilated into a bulb at its posterior extremity. Petter and Quentin (1976) provide a key to these genera.

2. Order Ascaridida

There are two superfamilies of this order, the Ascaridoidea and Seuratoidea, each with many genera of fish nematodes. Excellent accounts of the former are given by Fagerholm (1990), Hartwich (1974), Gibson (1983), Soleim and Berland (1981) and Sprent (1983), and of the latter by Berland (1970, 1983), Chabaud (1978), Arthur and Margolis (1975) and Gibson (1972b). *Goezia*, *Cucullanus* and *Pseudanisakis* (Figures 1.193, 1.194 and 1.189) are representatives of these two superfamilies. *Goezia* is discussed by Deardorff and Overstreet (1980a); *Cucullanus* by Gibson (1972b) and *Pseudanisakis* by Williams and Richards (1968) and Gibson (1973a). Other well-known genera include *Hysterothylacium* (= *Thynnascaris)* discussed by Deardorff and Overstreet (1981) and Petter (1969a,b); *Raphidascaris* discussed by Moravec (1971e); *Anisakis* by Hartwich (1974) and Smith (1983a,b); *Truttaedacnitis* by Berland (1970); *Paraquimperia* by Chabaud (1978); and *Haplonema* by Arthur and Margolis (1975).

3. Order Spirurida

Six superfamilies and about 20 genera of the Spirurida are important: (i) the superfamily Habronematoidea contains one large family, the Cystidicolidae, with twelve genera of which *Salvelinema* (Figures 1.195–1.197) and *Spinitectus* (Figures 1.198–1.199) are shown (see Petter, 1969b, and Moravec, 1979b). Other genera mentioned by Chabaud (1975b) are *Cristitectus*, *Cystidicoloides*, *Cyclozone*, *Pseudoproleptus*, *Ascarophis* (see Ko, 1986), *Pseudascarophis* (see Ko, Margolis and Machida, 1985), *Parascarophis*, *Spinitectoides* and *Cystidicola* (see Margolis, 1968) (ii) the superfamily Camallanoidea contains *Camallanus*, a member of the Camallanidae, (Figures 1.200–1.204) (see Petter, 1969b and Chabaud, 1975a), (iii) the superfamily Physalopteroidea has four genera of the family Physalopteridae, *Proleptus*, *Paraleptus*, *Heliconema* and *Bulbocephalus*, which are listed by Chabaud (1975a) as nematodes of fish. The genus *Proleptus* (Figure 1.190 and Figures 1.205–1.212) was reviewed by Specian, Ubelaker and Dailey (1975), and a detailed description of *P. mackenziei* given by Williams and Richards (1978); (iv) the superfamily Dracunculoidea has three families, the Anguillicolidae, Guyanemidae and Philometridae, all of which are found in fish according to Chabaud (1975a). According to Moravec and Taraschewski (1988) a general interest in the Anguillicolidae has increased considerably since *Anguillicola* appeared in Europe in the

1980s. These authors briefly described and illustrated all species of the genus and provided a key to their identification. The Guyanemidae has one genus, *Guyanema*, which is found in the body cavity of freshwater fish, and the Philometridae have ten genera including *Philometroides* and *Phlyctainophora* (Figure 1.212) (see Moravec, 1971e, 1978a, b and 1984a, and Rasheed, 1963); (v) members of the superfamily Thelazoidea family Rhabdochonidae are parasites of the intestine or of various other organs of fish. This family has at least six genera, including *Rhabdochona* and *Vasorhabdochona* (Figure 1.214) and *Pancreatonema* (McVicar and Gibson, 1975). Moravec, Margolis and Boyce (1981) redescribed three species of *Rhabdochona*: (vi) the superfamily Gnathostomatoidea of which *Ancyracanthus* and *Echinocephalus* are members of the Gnathostomatidae. *Ancyracanthus* larvae occur in teleosts whilst *Echinocephalus* adults occur in elasmobranchs. The head of *Ancyracanthus* is unmistakable by virtue of its cephalic appendages (see Chabaud, 1975a), and of *Echinocephalus* by virtue of its pronounced cephalic bulb which is armed with transverse rows of recurved hooks (Figures 1.215–1.217). Brooks and Deardorff (1988), Deardorff and Ko (1983) and Beveridge (1987) discuss *Echinocephalus*. Daengsvang (1980) gives a detailed account of *Gnathostoma*, the third-stage larvae of which occur in freshwater fish.

Phylum Acanthocephala

Acanthocephalans, or spiny-headed worms, are a phylum of about 800 species and, as adults, are exclusively parasites of the vertebrate intestine. Most genera and species occur in fish, particularly freshwater fish. A classification and list of all species is given by Amin (1985a). The sexes are separate with females usually larger than males (Figures 1.218–1.219). The body consists of a proboscis, neck and trunk. Except in one order, the Apororhynchida, the proboscis is armed with recurved sclerotized hooks and is retractable into a receptacle which houses the cerebral ganglion. A pair of sacs, connected to a lacunar system in the proboscis wall, the lemnisci, originates from the base of the neck. One or two ligament sacs, extending from the base of the proboscis receptacle to near the distal genital pore, suspend the gonads and accessory reproductive organs of both sexes in the body cavity, which is a pseudocoelom. Among parasitic adaptations are a reduction of the muscular, nervous, circulatory and excretory systems and a complete loss of the digestive system (Amin, 1982). The phylum includes 3 classes: Archiacanthocephala, Eoacanthocephala and Palaeacanthocephala. Most archiacanthocephalans are parasites of predaceous birds and mammals.

The Eoacanthocephala are mainly parasites of fish but also occur in amphibians and reptiles. The proboscis is retractable and has radially arranged hooks. The proboscis receptacle is single-walled with the cerebral ganglion near its middle or anterior end. The nuclei of the lemnisci are few in number and giant. Males have a single syncytial cement gland with several giant nuclei and a cement reservoir. The class consists of two orders: Gyracanthocephala with one family, and Neoechinorhynchida with three families. There are about 25 genera including *Quadrigyrus*, *Pallisentis*, *Acanthogyrus*, *Neoechinorhynchus*, *Octospinifer*, *Microsentis*, *Tenuisentis* and *Paulisentis* (Figures 1.220–1.223).

In the Palaeacanthocephala, the proboscis is retractable with hooks arranged in alternating radial rows. The proboscis receptacle is double-walled with the cerebral ganglion near its middle or posterior end. The nuclei of the lemnisci and the cement glands are fragmented. Males usually have six separate tubular-to-spheroid cement glands. The class is the largest and most diversified of acanthocephalans, parasitizing fish, amphibia, reptiles, birds and mammals. There are two orders but members of only one, the Echinorhynchida with 10 familes, are found in fish. There are about 50 genera including *Diplosentis*, *Heter-*

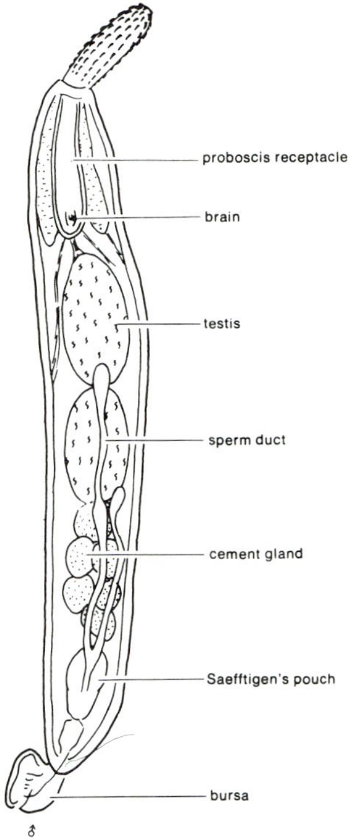

1.218

Acanthocephalus parksidei
(Echinorhynchidae)

♀ **1.219**

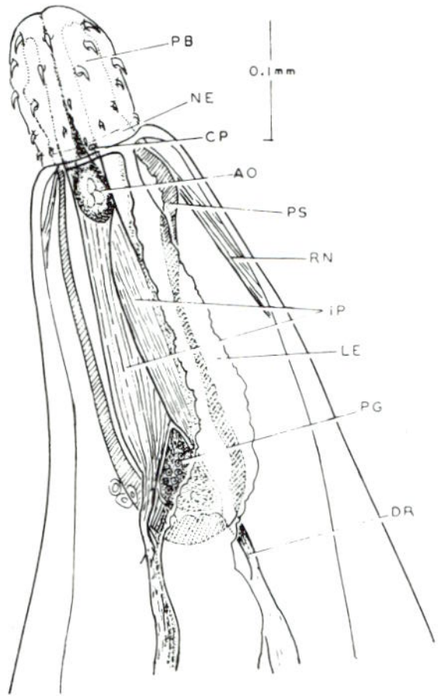

1.221

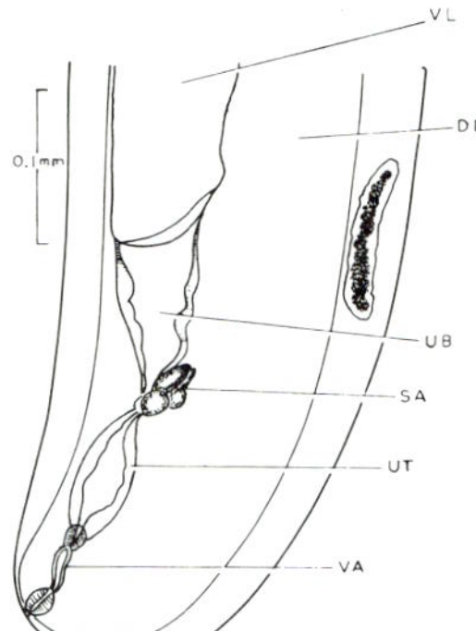

1.222

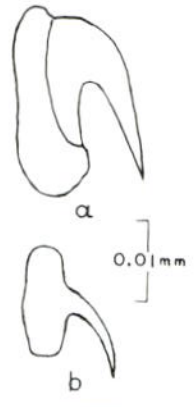

1.223

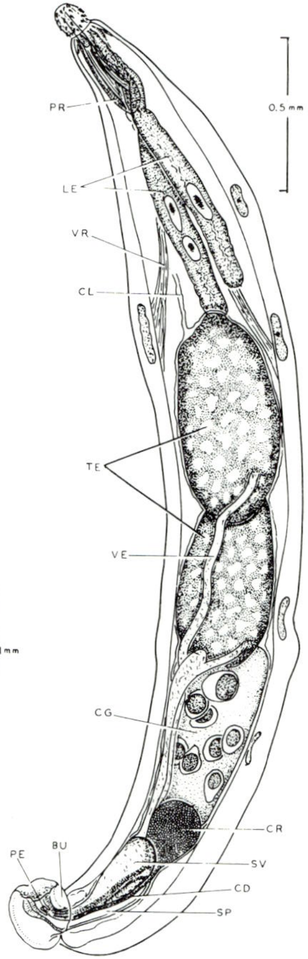

1.220

Figures 1.218–1.219 *Acanthocephalus parksidei* (Palaeacanthocephala: Echinorhynchidae) from Amin (1982) Figures 1.220–1.222 Adults of *Paulisentis fractus* (Eoacnthocephala: Neoechinorhynchidae) from Cable and Dill (1967). Fig.1.220, Mature male, lateral view. Fig.1.221. Same, proboscis and arterior trunk region enlarged. Fig.1.222. Posterior portion of female, lateral view. Fig.1.223. Proboscis looks enlarged: a, second look in a spiral; b, basal look. Key to lettering: AO, apical organ; BU, bursa; CD, cement duct; CG, cement gland; CL, central ligament; CP, cervical papilla; CR, cement reservoir; DL, dorsal ligament sac; DR, dorsal retractor of proboscis receptacle; IP, inverter of proboscis; LE, lemniscus; NE, neck; PB, proboscis; PE, penis; PG, proboscis ganglion; PR, proboscis receptacle; PS, muscular sheath of proboscis receptacle; RN, retractor of the neck; SA, selector apparatus; SP, Saefftigen's pouch; SV, seminal vesicle; TE, testes; UB, uterine bell; UT, uterus; VA, vagina; VE, vas efferens; VL, ventral ligament sac; VR, ventral retractor of proboscis receptacle.

acanthocephalus, *Illiosentis*, *Metarhadinorhynchus*, *Telosentis*, *Rhadinorhynchus*, *Leptorhynchoides*, *Arhythmacanthus*, *Hypoechinorhynchus*, *Bolborhynchoides*, *Pomphorhynchus*, *Echinorhynchus*, *Fessisentis*, *Echinorhynchoides*, *Neorhadinorhynchus* and *Acanthocephalus* (Figures 1.218–1.219).

Authorities on the phylum include Amin, Arai, Crompton, Golvan, Kennedy, Nickol, Petrochenko, van Cleave and Yamaguti, as may be seen from Amin (1982, 1985a, 1987a), Arai (1989), Crompton and Nickol (1985) and Valtonen and Crompton (1990). Host-lists for acanthocephalans are given by Buron and Golvan (1986) and Golvan and Buron (1988).

2 Life-cycles

General Introduction

The importance of understanding the life-cycle of an animal was discussed by Bonner (1965). We have interpreted his views diagrammatically in Figure 2.1. This is intended to show that a thoroughly elucidated life-cycle embraces the whole of biology. The life-cycle of every sexually reproducing animal species involves the production of zygotes and their subsequent growth, through an embryo, larval or juvenile stage, to adults which reproduce inexact copies of their own kind (1 and 2). This sequence of events provides a basis for investigating biology at molecular, developmental and evolutionary levels (3, 4 and 5). Investigations at these levels lead to pursuit of knowledge in seven major disciplines of biology (6–11) and also to research on the seven functions of life (12–18). The above comments summarize the fundamental aspects of the biology (= life-cycle) of a free-living animal. To elucidate the life-cycle of a parasite challenges a biologist to consider ultimately all these aspects at the same time for at least two species of animal. A major consequence is the need to consider the additional dimensions of host-parasite reactions and host-parasite interrelationships and these have important implications in the areas of transmission, epizootiology, pathology, immunology, host-specificity and co-evolution.

Having introduced the variety of sexually reproducing adult helminths of fish in Chapter 1, our main aim in this chapter is to indicate how the further development of their fertilized eggs and larvae fit into well-defined life-cycle patterns. The scope of the chapter is almost entirely limited to species which mature in fish. Many of the species which use fish as intermediate hosts have been omitted because their life-cycles are well-known and have been described elsewhere. Those which are of economic and medical significance are considered in Chapters 5 and 6.

Monogenea

Introduction

Monogeneans have a direct life-cycle, i.e. have only one host. Suggestions that pelagic plankton-feeding fish may act as intermediate hosts for gastrocotylids have been discounted by Popova and Gichenok (1978). Most monogeneans are oviparous but a minority, the gyrodactylids, are viviparous and will be considered in more detail later. Ovoviviparity has been described in the gill parasites *Diplectanum aequans* and *D. sciaenae* by Oliver (1978).

In oviparous species, the eggs hatch to release a larva, the oncomiracidium, which is usually provided with cilia borne on epidermal plates; it also has gland cells which open in the anterior part of the body and a posterior armed haptor. Many species have one or two pairs of eyespots. On contact with the host, the ciliated plates are shed and the parasite becomes attached and develops to maturity. It may migrate from the original point of contact to another site in which it becomes established.

The following general account of features of monogenean life-cycles is based mainly on reviews by Bykhovskii (1957), Kearn (1971) and Llewellyn (1963, 1968, 1972). Kearn (1986) has provided a comprehensive account of current knowledge of monogenean eggs from their assembly and structure to hatching, including the factors which affect egg output, egg laying, development and hatching.

Eggs

All fish monogeneans have operculate eggs which show a great variation in shape, determined by the internal structure of the ootype. Many have elongated appendages which attach the eggs together in groups or to the substrate by means of sticky droplets along or at the expanded ends of the appendages. Early opinion was that these outgrowths attached the egg to the body of the host of the parent parasite and a frequently quoted example of

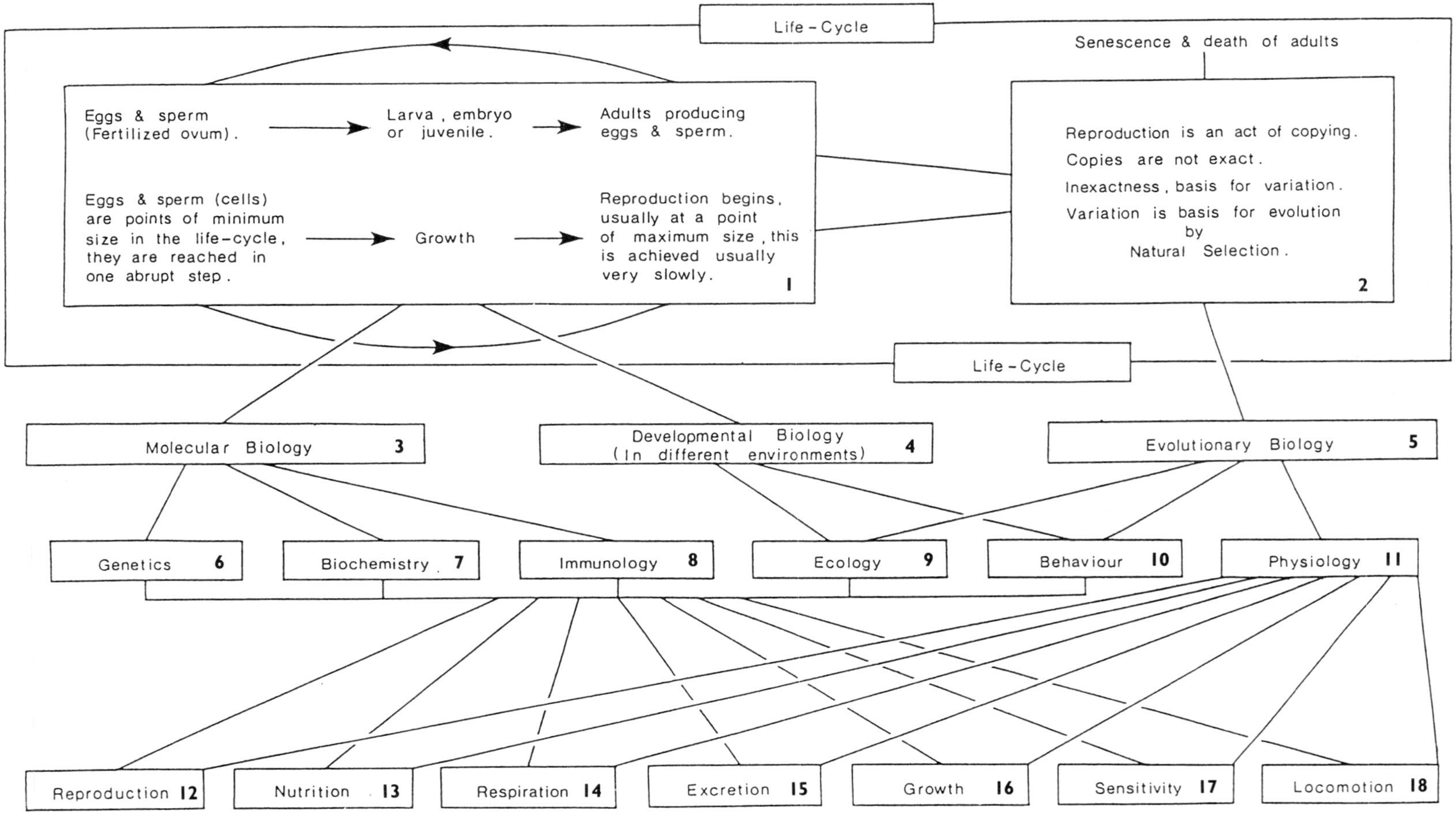

Figure 2.1 (1–18) The significance of a life-cycle in biology—a diagrammatic interpretation of Bonner (1965).

this is the attachment of eggs of *Nitzschia sturionis* to the buccal and branchial cavities of sturgeon (Bykhovskii, 1957). The eggs of *Concinnocotyla australensis*, an unusual polystome from the gills, oral cavity and skin of the Australian lungfish, *Neoceratodus forsteri*, are found attached to the surface of the tooth plates of the host; it is not clear how they escape damage when the host feeds (Whittington and Pichelin, 1991, and Pichelin, Whittington and Pearson, 1991). In two species of *Dionchus, D. agassizi* and *D. remorae*, parasites of the shark-sucker *Echeneis naucrates*, the eggs remain attached in bunches to the gills of the host (Ktari, 1977). The egg appendages are entangled together and the egg bunches are themselves attached, by their appendages, to a loop of shell-like material which encircles the gill filament. This strategy presumably enhances the chances that the non-ciliated, and therefore non-swimming, larvae of these species will infest other shark-suckers.

In most species the eggs fall to the bottom where they may become attached to the substrate, e.g. *Entobdella soleae, Acanthocotyle lobianchi*. Egg bunches of *Diclidophora luscae* sink more rapidly than isolated eggs to the bottom where the host, *Trisopterus luscus*, spends most of its time; the egg appendages may anchor the bundles to the substrate (Whittington and Kearn, 1988) which helps to retain the eggs in the environment of the host. Another advantage of depositing eggs in bunches is mass infection of a single host by larvae from eggs from a single parasite, when the eggs hatch in response to mechanical disturbance caused by the host. In some species, appendages are a means of retaining bunches of eggs in the terminal region of the uterus. This occurs in *A. greeni* where the bases of the appendages fuse to form a pedal disc and up to 80 eggs occur in a bunch (Macdonald and Llewellyn, 1980), which protrudes from the uterus and is exposed to seawater, although protected from water currents by an expansion of the anterior part of the right side of the body. Egg appendages of *A. lobianchi* and *A. elegans* are held together by a sticky droplet and are protected in *A. lobianchi* by a shield similar to that of *A. greeni* but smaller. A shield does not occur in *A. elegans*. The reason for this type of egg retention is not clear. Macdonald and Llewellyn (1980) suggested that if egg bunches of *A. greeni* were shed, the younger eggs in the bunch would continue their development on the substrate and might later infect a new host. *A. greeni* eggs hatch in response to host skin mucus; their larvae lack cilia and therefore cannot swim.

Incubation period

The incubation period of the eggs is temperature-dependent. The rate of development is retarded or stopped at low temperatures but increases as the temperature rises until it reaches a point above which development will not take place. Llewellyn (1957a) found little or no development in any of 13 marine species at temperatures less than 8°C. Eggs of *Anthocotyle merluccii* failed to hatch in 28 days at 13°C but did so in 21 days at 18°C. *Plectanocotyle gurnardi* eggs hatched in 13–16 days at 18°C and 8–11 days at 20°C, and those of *Urocleidus adspectus* in 5–6 days at 20°C (Cone, 1979).

Life-span and behaviour of oncomiracidia

Research during the last twenty years has focussed on the various behavioural and physiological adaptations which maximize opportunities for successful invasion of the host during the limited free-living life of the oncomiracidium. Several species have been shown to have a circadian hatching rhythm related to the natural cycle of illumination and also correlated to host behavioural patterns, so that most larvae hatch when the host is likely to be nearby and to be sedentary. Hatching may also be chemically stimulated by a component of host skin mucus or by water turbulence, both indicative of the actual presence of a host fish.

Conversely, no specific stimuli were required to initiate hatching of *Urocleidus adspectus* oncomiracidia (Cone, 1979).

The life span of oncomiracidia is short, up to 10 hours in *Diplozoon paradoxum* in natural conditions (Bovet, 1967), 4–6 hours in *Discocotyle sagittata* (Paling, 1969) and 20–30 hours at 7°C in *Entobdella soleae*, reducing to 9–14 hours at 17°C (Kearn, 1967b). Larvae of *Diclidophora merlangi* are active for 12–15 hours at 13°C and for 9 hours at 18°C, but lose the cilia at 24 hours and die within 30 hours (Macdonald, 1974). Information on the speed at which oncomiracidia swim and the distance they may travel indicates that they probably infect the hosts close to the site where the eggs hatch (Llewellyn, 1972).

Responses to light, chemical stimuli and water currents have been described in various species. Bykhovskii's (1957) observation that many oncomiracidia are initially photopositive but lose this response as they age is supported by Bovet (1967) for *Diplozoon paradoxum* and Paling (1969) for *Discocotyle sagittata*. Oncomiracidia of *Entobdella soleae*, initially photopositive, exhibit increasingly long photonegative periods as they age (Kearn, 1980). Most polyopisthocotylinean oncomiracidia have a single pair of eyes (some have two pairs) but most monopisthocotylinean oncomiracidia have two pairs of eyes with lenses, the anterior pair orientated posterolaterally and the posterior pair anterolaterally. The two pairs of eyes of monopisthocotylineans are provided with permanent crystalline lenses. However, Llewellyn (1972) pointed out that although most polyopisthocotylineans have only a single pair of pigment cups without special lenses, two of them, *D. sagittata* and *Diplozoon paradoxum*, are nevertheless positively phototactic. Additionally, species of *Diclidophora*, which apparently lack eyes entirely, have been shown to exhibit circadian hatching rhythms (Macdonald, 1975) and perhaps may have unpigmented photoreceptors. Llewellyn suggested that oil droplets associated with the eyes of diclidophoroid polyopisthocotylineans might act as temporary lenses. However, oncomiracidia of *Diplozoon paradoxum* have since been shown to have, as well as a median pair of pigment-shielded eyes, two additional eyes, lateral in position, which are rhabdomeric in structure but lack pigment shields and are, therefore, inconspicuous when viewed by light microscopy (Kearn, 1978b). *D. paradoxum* lacks both solid lenses and oil droplets which might function as temporary lenses, but Kearn suggested that the median pigment-shielded eyes would have directional sensitivity to light, thus helping to orientate the free-swimming larva. The unshielded eyes presumably lack directional sensitivity but might respond to changes in illumination such as shadowing by a passing fish. The swimming pattern of the oncomiracidium, which rotates around its longitudinal axis and also follows a spiral path, would allow it to scan in all directions as it swims, although the pigment-shielded eyes are laterally directed. This study suggests that polyopisthocotylinean larvae previously reported to have a single pair of eyes or none should be re-examined. Responses of oncomiracidia pre- and post-hatching to chemical stimuli are discussed in more detail below in relation to the life-cycles of *E. soleae* and *A. lobianchi*. An example of a rheotactic response is shown by *Diplozoon paradoxum* oncomiracidia, which move at random until they arrive in front of the mouth of the fish, at which point they stop swimming if water is entering the mouth but not otherwise, thus passing into the buccal cavity passively with the water current (Bovet, 1967).

Oncomiracidia of some species migrate from the point of contact with the host to the final location (Llewellyn, 1972). Some shed their cilia on contact with the host and attain the gill chamber by crawling over the body surface (and entering via the opercular chamber in the pause between exhalations) or that of the buccal cavity, e.g. *Neogyrodactylus crucifer, Tetraonchus monenteron* and *Diplectanum aequans* (Kearn, 1968b). Others retain the cilia and are borne to the gills in the ventilating current, e.g. *Diplozoon paradoxum* and *Discocotyle sagittata*. The latter species, having arrived on the gills, migrates to reach its preferred site (Paling, 1969). Migration over the body surface may be assisted by the anterior adhesive apparatus described in some monogeneans, e.g. *Gyrodactylus eucaliae, Dactylogyrus amphibothrium* and *D. hemiamphibothrium, Entobdella soleae* and *Acanthocotyle lobianchi* (Kritsky, 1978; El Naggar and Kearn, 1980, 1983a; and Rees and Kearn, 1984).

The majority of skin- and gill-dwelling monogeneans inhabit the same site as larvae and as adults and perform only minor relocations rather than migrations. Other methods of invasion occur in non-ciliated, therefore non-swimming, oncomiracidia of some oviparous species and in the viviparous gyrodactylids. Some well-elucidated monogenean life-cycles are described below.

Some Monogenean Life-Cycles

The Life-Cycle Of *Entobdella soleae* (Fig. 2.2.1)

Entobdella soleae is a skin parasite of the common sole, *Solea solea.* Mature adults occur on the ventral surface of the fish, orientated with the haptor anterior with respect to the longitudinal axis of the fish. The sole is active at night and during the day is buried in sand with only the eye region exposed (Kearn, 1963a, b).

Egg structure and factors affecting hatching

Eggs are produced at an average of two per hour at 15°C. They are tetrahedral (Figure 2.2.2), with a long appendage at one apex which bears groups of sticky droplets along its length. One of the other three apices serves as the operculum. The opercular joint lies about 20 μm below the apex, will withstand pressures which crush the eggshell and is resistant to external application of proteolytic enzymes (Kearn, 1975a). There is no viscous cushion. The droplets on the appendages anchor the eggs to sand particles and may prevent them from being swept by currents to hard substrates where soles are not found, or lifted by turbulence into suspension above the zone inhabited by soles. *E. soleae* eggs hatch in 27 days at 13–17°C and hatching follows an endogenous circadian rhythm related to the behaviour pattern of the host. Kearn (1973) has shown that if eggs are subjected to alternating periods of light and dark (L:D 12:12, L:D 6:18 or L:D 18:6) at constant temperatures (12–16°C), the majority hatch during the first four hours of illumination on each successive day. In nature, hatching just after dawn would have survival value because soles, active at night, bury themselves at dawn in the sediment where the eggs are to be found and they are sedentary targets during the day. Experiments showed that the rhythmicity established in response to L:D 12:12 persists in a wide range of lighting conditions, including natural daylight and dim blue light similar in intensity and spectral quality to that at depths where soles live. Exposure of eggs to L:D 12:12 until the start of hatching produces a hatching rhythm which persists with a periodicity of 24 hours in either constant light or darkness, indicating a strong endogenous component in the rhythm. Eggs laid and kept in total darkness, however, will hatch at any time in each 24-hour period and show little evidence of periodicity. If alternating periods of 12 hours each of light and darkness are reversed after hatching begins, the hatching pattern will reverse after 24 hours so that oncomiracidia emerge at the beginning of the new light period. Mechanical disturbance of the eggs has little or no effect in stimulating hatching (Kearn, 1973) but hatching is affected by a component (host hatching factor) of fish skin mucus which enhances normal hatching during the first hours of a period of illumination and also stimulates hatching of mature eggs during the second half of an illuminated period or in darkness. *E. soleae* eggs hatched in response to mucus from fish other than sole (plaice, dab, halibut, whiting, ray), but the oncomiracidia, when offered scales from dab or sole, usually selected sole scales regardless of which host had produced the hatching factor. Kearn (1974b) suggests that the effect of host hatching factor is to maximize the opportunity for larvae to infect the fish; a sole might rest on the substrate long enough to stimulate any eggs present to hatch and would then become infected. Hatching in response to mucus from non-host fish would be disadvantageous but such species might be uncommon in the habitat of the sole. The host hatching factor is produced in the mucus of large and small fish and resists boiling or freezing (Kearn, 1974b). *E. soleae* eggs will hatch in response to dilute solutions of urea or ammonium chloride in sea water and sole skin mucus contains enough urea, but not ammonia, to stimulate hatching (Kearn and Macdonald, 1976).

In contrast, eggs of *Entobdella hippoglossi*, a parasite of halibut (*Hippoglossus hippoglossus*), do not hatch in response to fish skin mucus but do show a hatching rhythm with a 24-hour periodicity. In this species, however, the larvae emerge at the end of the first two hours of each period of darkness, suggesting that the halibut, the behaviour of which is little known, may rest on the substrate at night and be active during the day (Kearn, 1974a).

Oncomiracidial morphology

The oncomiracidium of *E. soleae* (Figure 2.3.1,2) has three transverse bands of cilia borne on three tiers of epidermal plates, two pairs of pigmented eyespots and various sets of gland cells, opening anteriorly, arranged in anterior median, posterior median and lateral groups. Their secretions may assist attachment of the oncomiracidium to the host. Inactive larvae removed from the egg before hatching also have two pairs of gland cells with granular contents (ventral gland cells) which open ventrolaterally. These are not found in freshly hatched larvae and may secrete a proteolytic hatching fluid which acts on the inner surface of the opercular joint. The haptor bears ten pairs of sclerites, seven marginal and three median (Kearn, 1963b, 1974c, 1975a).

The hatching process and invasion of the host

Before hatching begins, the anterior end of the oncomiracidium lies in the opercular corner of the egg. Larvae hatching in response to light or host hatching factor rotate anticlockwise about the longitudinal axis by means of ciliary activity until the operculum opens. Rotation presumably spreads the secretion of the ventral glands over the opercular joint. At this stage, the larva uses muscle activity for the first time to dislodge the operculum and squeeze out of the egg, then swims away. Escape from the egg is easier if the egg is firmly anchored to the substrate. In contrast, *Urocleidus adspectus* oncomiracidia appeared not to use glandular secretions to dislodge the operculum but moved it mechanically; they hatched in total darkness and did not respond to perch mucus (Cone, 1979).

Hatching of *E. soleae* can occur as quickly as four or five minutes after the addition of fish washings to eggs but may take up to 18.5 minutes. Freshly-hatched larvae are photo-positive but exhibit some photonegative periods, in which they swim downwards, which increase in frequency and duration as they age (Kearn, 1980). They perform a series of ascents and descents which, with horizontal movements due to currents, would amount to a search strategy for the host. Oncomiracidia stimulated with urea to hatch in darkness are geonegative, can extricate themselves from sand 1 cm deep, and in aquarium conditions can invade the host from below (Kearn, 1980). Invasion usually takes place from above, in the anterior part of the upper surface, which is the only part of the fish exposed above the substrate. The larvae are chemically attracted to sole skin and attach to it in preference to that of other soleids, pleuronectids and elasmobranchs. The parasites migrate from the

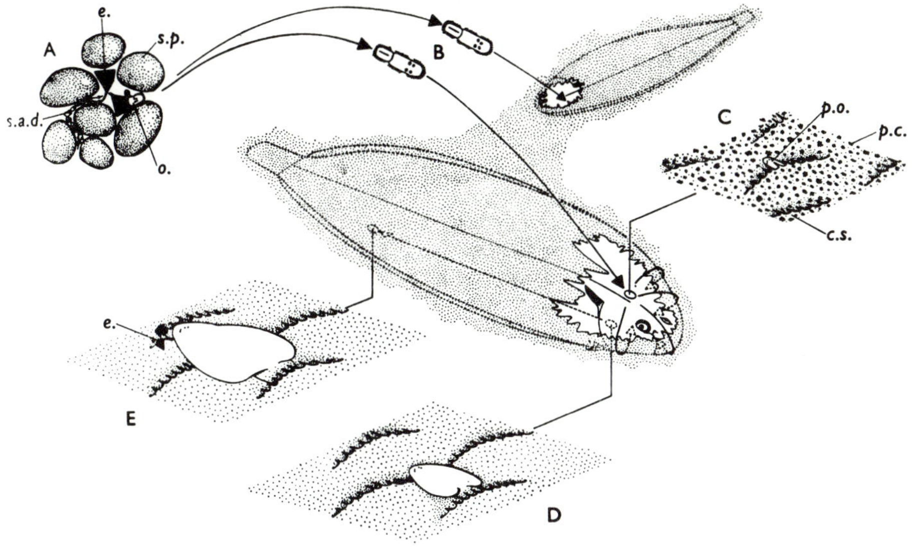

2.2.1

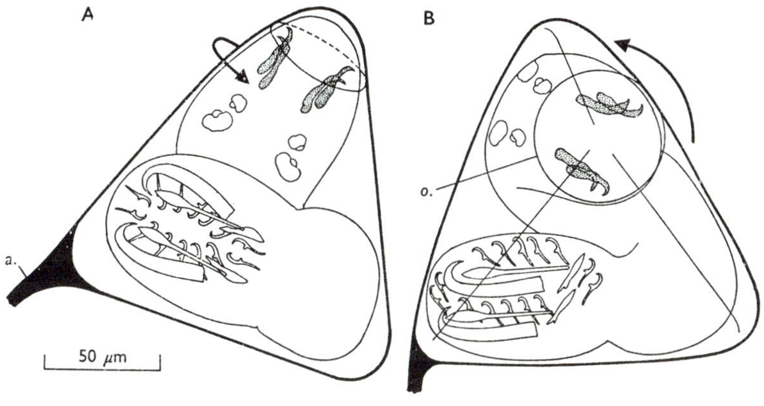

2.2.2

Figure 2.2.1 The life-cycle of *Entobdella soleae*. A. Two eggs attached to sand particles at the sea bottom; B. free-swimming oncomiracidia invade the upper surface of the heads of soles buried in the sand; C. the post-oncomiracidium attached to the skin of the upper surface of the fish; D. parasites emigrate to the lower surface and E. become mature. c.s. ctenoid scale; e. egg; o. oncomiracidium; p.c. pigment cell; p.o. post-oncomiracidium; s.a.d. stalk and adhesive globules of the egg; s.p. sand particle. Kearn (1963a).

Figure 2.2.2 Two views of the egg of *Entobdella soleae* showing the position of the fully-developed oncomiracidium. The ventral head glands are stippled and the arrows show the direction of rotation of the hatching larva. a. appendage; o. opercular joint. Kearn (1975a).

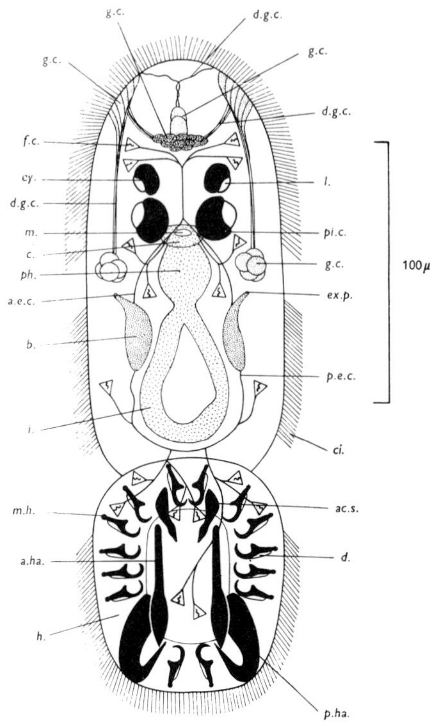

2.3.1

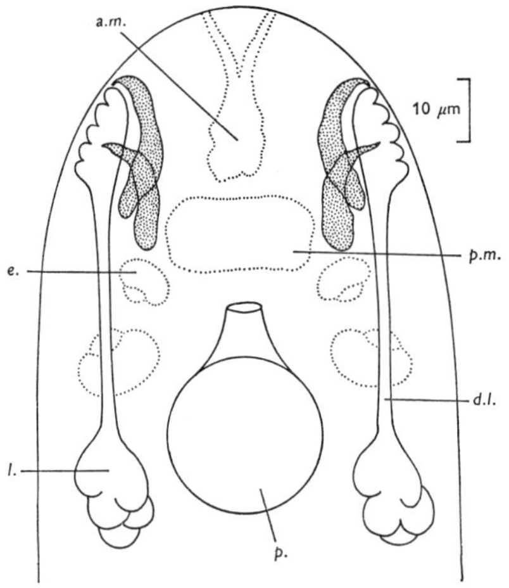

2.3.2

Figure 2.3.1 The oncomiracidium of *Entobdella soleae*, ventral view. Kearn (1963b).

Figure 2.3.2 Anterior end of the oncomiracidium of *Entobdella soleae* before hatching, showing the position of the ventral head glands (stippled), a.m. anterior median head glands; d.l. ducts of the lateral head glands; e. eye; l. lateral head glands; p. pharynx; p.m. posterior median head glands. Kearn (1975a).

upper to the lower surface of the fish where they develop to maturity and commence egg production (Kearn, 1963a, b).

The Life-Cycle Of *Acanthocotyle Lobianchi*

A. lobianchi is an acanthocotylid parasitic on the surface of various species of *Raja*, hosts which spend much time resting on the sea bottom. The life-cycle shows some interesting parallels with that of *E. soleae* but also some significant differences.

The eggs, released singly or in groups of up to eight, are sausage-shaped with a short appendage with a sticky end which adheres the eggs together or to the substrate. Eggs laid by adults attached to the host fall to the bottom and become attached to the substrate. The oncomiracidia develop in 15–20 days at 15°C, with the head at the opercular end of the egg; unlike those of *E. soleae* they have no cilia, and therefore cannot swim, and lack pigmented eyespots. They have two groups of five gland cells, the secretions of which are probably involved in attaching larvae to the substrate, but have no hamuli and no trace of the adult pseudohaptor (Kearn, 1967a). Unlike *E. soleae*, hatching does not occur in response to cyclical changes in illumination and is not stimulated by shadowing designed to simulate the effect of a descending ray. It is, however, stimulated by host skin mucus, taking place within 2–4 seconds. The larva forces off the operculum by vigorous movement but often fails to leave the shell; instead it performs exploratory movements with the anterior end, then withdraws (Macdonald, 1974). Eggs will hatch in response to mucus from a *Raja* species other than that of the adult's original host. The active component of host mucus is urea, but mucus from sole will not prompt *A. lobianchi* eggs to hatch (Kearn and Macdonald, 1976). The response to mucus increases with age of the egg, and larvae can stay alive and active within the egg for an average of 60 days (maximum recorded 83 days). The larvae probably require a host in the immediate vicinity before they hatch and the longer they can survive in their shells, the better are their chances of finding one (Macdonald, 1974). They contain lipid droplets which may provide a food source. As they cannot swim, they contact a host only if a ray settles on the eggs.

Other Oviparous Monogeneans

Further examples of a correlation between hatching rhythms in marine monogeneans and host behaviour are provided by three species of the genus *Diclidophora* studied by MacDonald (1975). Eggs of *D. merlangi* subjected to alternating 12-hour periods of light and darkness hatched before or just after 'dawn', the period when the host, whiting (*Merlangius merlangus*), would descend to the sea bottom where it remains during the day. Eggs of *D. denticulata* hatched after 'dark', when the host, coalfish (*Pollachius virens*), would have moved to the sea bottom for the night. However, pouting (*Trisopterus luscus*), the host of *D. luscae*, always remain close to the sea bed but tend to be less concentrated in shoals during the night. *D. luscae* eggs hatch at 'dusk' and, if they are wide-spread over the sea bottom, it may be to the advantage of the larvae to hatch when the fish are also scattered over a wide area. It is of interest that the larvae of the three *Diclidophora* species lack pigmented eyespots, which therefore cannot be essential for detecting changes in lighting conditions, in contrast to *E. soleae* oncomiracidia which have pigmented eyespots (Kearn, 1963b) and may also have ciliary photoreceptors (Lyons, 1972). Macdonald (1975) suggested that *Diclidophora* larvae may have unpigmented photoreceptors. She found that neither mechanical disturbance nor application of host mucus stimulated hatching in *D. merlangi* or *D. luscae*, but *D. luscae* eggs have subsequently been found to hatch, in light or darkness, in response to more sustained and vigorous disturbance than that employed by Macdonald (Whittington and Kearn, 1988).

Similar observations on hatching stimuli and rhythms have been reported for other species. Hatching in response to host skin mucus occurs in *Squalonchocotyle torpedinis*, a gill parasite of *Torpedo marmorata*, and *Microcotyle salpae* from the gills of *Box salpa* (see Euzet and Raibaut (1960) and Ktari (1969) respectively). Eggs of *M. salpae* also hatched in response to water turbulence. The oncomiracidia are non-ciliated and lack eyespots. Ktari (1969) suggests that the eggs enter the buccal cavity with the algae on which the host browses or with the respiratory current, to hatch there liberating the larvae which are carried into the gill chamber and attach to the gills. Eggs of *Acanthocotyle greeni* develop at different rates within the same egg bunch. They hatch in response to host mucus and the oldest eggs probably hatch while the egg bunch is still retained by the parent, re-infecting the same host fish. Others hatch only after the bunch has been deposited and infect new hosts. This strategy results in a reservoir population on the original host and the infection of new hosts following dispersal of the eggs (Macdonald and Llewellyn, 1980).

Most investigations have been carried out on marine monogeneans but there is evidence to suggest that freshwater species show similar behaviour patterns. *Diplozoon homoion gracile*, a gill parasite of *Barbus meridionalis*, exhibits a rhythm both in egg production and in hatching, the latter being much more clearly defined (Macdonald and Jones, 1978). The parasite laid eggs both day and night, but a comparison of the rate of egg laying in light and darkness (natural illumination, dawn 0600 h and dusk 2200 h) over 14 days at different temperatures showed that more eggs were laid during the night and that the difference was statistically significant at the 5 percent probability level. The eggs hatched after 5–6 days under conditions of natural illumination and temperature, the majority of larvae emerging around dusk. Hatching ceased completely around dawn. What is known of the behaviour of the southern barbel suggests that it is active during the day and probably rests under banks and ledges at night. The eggs of the parasite would, therefore, be deposited and would later hatch in places where opportunities for locating and invading the host are maximized.

The Life-Cycle of *Macrogyrodactylus polypteri* (Figure 2.4.1–15)

The life-cycle of this viviparous species, a parasite of the skin and fins of an African freshwater fish, has been described in detail by Khalil (1970) (Figure 2.4.1–15). An adult worm has within its uterus a fully developed embryo which may have a second embryo in its uterus, a third embryo within the second and even a fourth embryo inside the third. The oocyte enlarges within the ootype by fusion with other oocytes, the nuclei of which degenerate. The fertilized egg passes into the uterus and develops into the first embryo, within which a group of large cells surrounded by small ones appears, representing the second embryo forming inside the future uterus of the first. Meanwhile, a new oocyte appears in the ootype of the parent. When the development of the first embryo is complete, it usually lies in the uterus in an inverted U-shape and is born through a temporary birth pore on the ventral surface of the parent. The newly born worm is protogynous with a well-developed female reproductive system; the male system develops completely only after birth. At birth the worm already contains an embryo in an advanced state of development and will probably give birth at least twice before it copulates and before its own eggs are fertilized. The first birth is that of the second embryo inside the first, now a newborn parent, which derives from the same egg as the parent. The second is of the embryo formed from the oocyte in the ootype of the new parent at the time of its birth, also presumed to originate from the same egg as the parent. Later, the female system ceases to function and the worm becomes functionally male. Birth can be induced by an increase in temperature and by a variety of chemicals in solution (sodium chloride, formalin, potassium permanganate). Newborn worms attach to the skin of the host and very heavy infestations develop within a short space of time. Mature worms, with the sclerites of the opisthaptor fully developed, are capable of transferring from one host to another under aquarium conditions, ie. infestation of the new host occurs by direct contact.

Cestoidea

Introduction

The overwhelming majority of cestodes have indirect life-cycles incorporating one or two intermediaries and not uncommonly paratenic hosts as well. Direct cycles are known (see *Archigetes* below) but are extremely rare. Transmission of cestodes from host to host is always passive (with the exception of the amphilinids), achieved when the egg, oncosphere or metacestode is eaten as or with food, and is therefore highly correlated with host diet at each stage of the life-cycle. Successive hosts occupy progressively higher levels in the trophic pyramid. Mackiewicz (1981) attached more importance to host feeding habits in determining the evolutionary relationships of hosts to parasites than to the phylogenetic relationships between hosts and parasites. The following introduction is drawn largely from reviews of cestode life-cycles by Freeman (1973, 1982a,b) and Mackiewicz (1988).

The basic cestode cycle incorporates an adult in the gut lumen of a vertebrate which produces eggs containing a six-hooked or hexacanth oncosphere in the eucestodes and a 10-hooked larva in the gyrocotylids and amphilinids. In the first intermediate host, the larva migrates to the parenteron where it metamorphoses and grows into the metacestode stage. The first intermediary is most commonly an arthropod. An exception is the caryophyllaeid genus *Archigetes*, members of which can complete the cycle in the coelom of tubificid annelids without using a vertebrate host. The oncosphere stage always invades a parenteral site in the invertebrate host, where metamorphosis takes place. Metacestodes of species which mature in fish utilize a range of invertebrate hosts as intermediaries but some species utilize fish as both intermediary and definitive hosts. The use of paratenic hosts also occurs in many cestode cycles, although predominantly in species which mature in birds and mammals. The metacestode stage becomes a sexually mature adult in the gut lumen of a vertebrate, with the exception of *Archigetes* in the coelom of annelids and of the spathebothriid *Bothrimonus* in that of gammarids.

In the majority of cestode groups, the adults have increased their reproductive potential by serial repetition of the reproductive organs, usually accompanied by segmentation of the body so that each segment corresponds to a single set of genitalia. A minority of groups are monozoic (gyrocotylids, amphilinids and caryophyllaeids), having a non-segmented body and a single set of reproductive organs while others (spathebothriideans and one pseudophyllidean genus) have serially repeated genitalia but lack external segmentation.

In polyzoic cestodes, gravid segments are shed by a process of apolysis. Some fish cestodes, the tetraphyllideans, trypanorhynchs and lecanicephalideans, are hyperapolytic, ie. their segments are shed before they are fully gravid and they complete their development in the host gut. Mackiewicz (1988) suggests that this enables the worm to invest energy in proglottis formation rather than in the growth and maintenance of older proglottides. Cestodes which have retained the uterine pore, eg. pseudophyllideans, discharge their eggs through the pores and shed spent proglottides in groups by a process of pseudoapolysis.

A feature of the life-cycles of some cestodes and digeneans is paedomorphosis, defined by

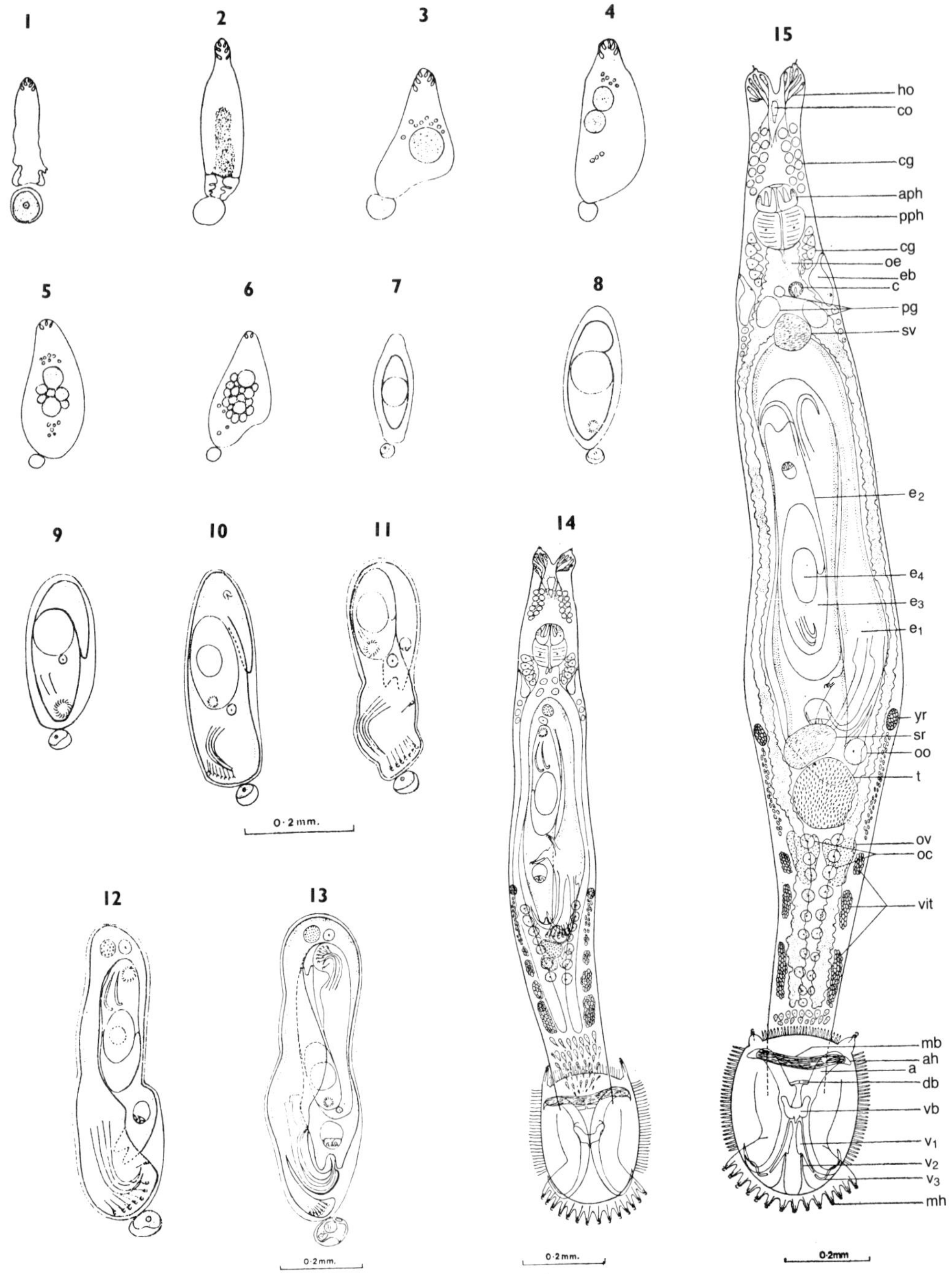
1
2
3
4
5
6
7
8
9
10
11
12
13
14
15
ho
co
cg
aph
pph
cg
oe
eb
c
pg
sv
e2
e4
e3
e1
yr
sr
oo
t
ov
oc
vit
mb
ah
a
db
vb
v1
v2
v3
mh
0·2 mm.
0·2mm.
0·2mm.
0·2mm

Fig 2.4

Figure 2.4. (1–15) Developmental stages of *Macrogyrodactylus polypteri* after Khalil (1970).
1. Uterus and ootype of a specimen that has just given birth; note small cells in the inner anterior wall of the uterus, the posterior fold of the uterus wall and the mature oocyte in the ootype. 2 The passage of the egg from the ootype into the lumen of the uterus in a longitudinal form. 3. Undivided spherical egg in the uterus; note vitelline-like cells anterior to the egg. 4. Two-cell stage. 5. Seven-cell stage. 6. Epibolic gastrula. 7. First embryo (oval in outline) with the second embryo as a round mass of cells in its middle; note the new oocyte in the ootype. 8. First embryo with a ring of hooklets at the posterior end. 9. First embryo with the anterior end bending backwards, anchors anterior to ring of hooklets and a large cell in the future ootype. 10. First embryo with its pharynx just appearing; second embryo with a ring of hooklets at its posterior end and the third embryo as a round mass of cells in the middle of the second embryo. 11. First embryo with the marginal hooks well developed; the second embryo with its anterior end bending backwards and anchors anterior to the ring of hooklets. 12. First embryo with the testis just appearing; second embryo moves into the anterior half of the first embryo; the third embryo as an oval mass of cells with the fourth embryo as a round mass of cells in its middle. 13. First embryo developing a skin fold of the opisthaptor with 14 marginal hooks; the second embryo in an advanced stage of development and the third embryo with a ring of hooklets; note the developing oocyte in the ootype acquiring another egg cell. 14. Newly born specimen, ventral view; note the absence of cirrus, prostate glands, seminal vesicle and seminal receptacle. 15. Mature specimen, ventral view. Key to lettering, 15 a.—anchor; ah.—anchor hood; aph.—anterior section of pharynx; c.—cirrus; cg.—cephalic glands; co.—cerebral organ; db.—dorsal bar; e_1, e_2, e_3, e_4, first to fourth embryos; eb.—excretory bladder; ho.—head organs; mb.—muscle bar; mh.—marginal hooks; oc.—oocytes; oe.—oesophagus; oo.—ootype with a maturing oocyte; ov.—ovary; pg.—prostate glands; pph.—posterior section of pharynx; r_1, r_2, r_3,.—rods of the opisthaptor; sr.—seminal receptacle; sv.—seminal vesicle; t.—testis; vb.—ventral bar; vit.—vitellaria; vr.—vitelline reservoir.

Gould (1977) as the retention of ancestral juvenile characters by later ontogenetic stages of the descendants. Paedomorphosis can result from either of two separate phenomena, progenesis and neoteny, and Gould uses the term paedomorphosis to embrace both. He defines progenesis as paedomorphosis produced by precocious sexual maturation of an organism still in a morphologically juvenile stage. Neoteny is the retention of juvenile characters in the sexually mature adult stages of an organism. Mackiewicz (1982a) has emphasized that it is important to distinguish between them because the evolutionary implications differ for each. A progenetic organism must have lost an adult stage from its life-cycle but, if an organism is neotenic, all stages of the original cycle, including the adult, are still present. To decide which condition pertains, the adult must be defined, easily done for strobilate cestodes but not obvious for the monozoic cestodes. Progenesis is particularly marked in the caryophyllideans and also occurs in some spathobethriideans and, to a degree, in some pseudophyllideans which use fish as second intermediaries. It should be noted that different meanings have been attached to the term progenesis by different workers, and that many helminthologists have used it to signify the achievement of sexual maturity in an intermediate host.

Freeman's (1973) comprehensive review of cestode ontogeny includes accounts of their eggs, oncospheres and metacestodes. Eucestode oncospheres may complete development *in utero* or undergo delayed embryogenesis outside the worm and its host. Delayed embryogenesis may occur in operculate or non-operculate eggs. Operculate eggs, which liberate ciliated coracidia, have been assumed to be the most common type in cestodes with aquatic cycles, a view with which Freeman disagrees. Such eggs are comparatively rare, occurring in some but not all of the pseudophyllideans and some trypanorhynchs and are not known from tetraphyllideans or proteocephalideans. Most eucestodes, whether aquatic or not, lack operculate eggs and not all those with operculate eggs have coracidia (Freeman, 1982b).

The metacestode most often completes most of its development in the parenteral site invaded by the oncosphere, undergoing a metamorphosis in the course of which the scolex develops on the end opposite to the oncospheral hooks. The region around the oncospheral hooks forms, in many but not in all cestodes, an appendage-like cercomer which is later shed. Because the adult scolex is anterior but the oncosphere had moved with the hooks anteriorly, development of the scolex on the end opposite to the oncospheral hooks is accepted by some workers as an indication of a reversal of polarity. The scolex usually develops to a recognizable stage while in the invertebrate parenteron but often completes its development to adult size and shape in a vertebrate, whether in the gut or coelom. A degree of sexual maturation may occur in some species in metacestodes while they still occupy parenteral sites, but in most it takes place after the metacestode has entered the vertebrate gut. Freeman (1973) has provided a very detailed review of the metacestode stage.

Mackiewicz (1988) has reviewed transmission strategies adopted by the cestodes. He argued that, apart from the short-lived free-living coracidium which occurs in comparatively few cestodes, cestodes lack a free-living stage. This implies that eggs and coracidia are undirected and passive with respect to their hosts and therefore to transmission.

The three basic transmission strategies developed by cestodes are: (i) evolution of life-cycles interpolated into host biology; (ii) presentation of infective stages that increase the probability of contact between host and parasite; and (iii) increased reproductive potential. Some examples which illustrate these strategies are those of cestodes which use fish as intermediate or definitive hosts.

With regard to life-cycles, the most common aquatic cestode cycle involves three hosts, which is true of many fish cestodes; two-host cycles are also common. Paratenic hosts are used to bridge trophic levels between the host in which the metacestode develops and the definitive host. Fish are used as paratenic hosts by cestodes which mature in fish, birds or mammals.

Cestodes have evolved a number of ways to increase contact between host and parasite. These include the use of a diverse range of intermediaries which is the case in some trypanorhynchs and tetraphyllideans. Another method is by adapting the eggs, eg. the large floating eggs of *Proteocephalus percae* which are eaten by copepods. A third method is to induce behavioural changes which render the host more vulnerable to predation, eg. *Schistocephalus* and *Ligula* plerocercoids in the body cavity of sticklebacks and minnows respectively (see Chapter 5).

Reproductive potential in cestodes may be increased by serial repetition of the reproductive organs, which has been considered above, or by the development of proliferative larval stages.

Life-Cycles In Some Cestode Orders

Order Gyrocotylidea

Comparatively little is known about the life-cycles of gyrocotylideans. Reviews by Simmons (1974), Simmons and Laurie (1972) and Williams, Colin and Halvorsen (1987) suggest that the life-cycle may be direct and that a tissue-dwelling post-larval stage may be a normal part of the life-cycle. Xylander (1988b), however, speculated that the life-cycle is indirect and incorporates a crustacean intermediary.

Gyrocotylideans inhabit the spiral valve of chimaeroid fish and usually occur in pairs. It has been suggested that two worms comprise a functional sexual unit; two is the minimum number which would permit cross-fertilization, a necessary step since gyrocotylideans are protandrous with a ventral male pore, a dorsal female pore and no permanent protrusible intromittent organ. The eggs are probably released with the faeces of the host but there is evidence, in at least some species, that they are retained *in utero* and that dissemination may be facilitated by escape of gravid adults from live or dead hosts or by passage unharmed through the gut of a predator. Adult worms will leave a dead or moribund host very quickly, usually via the anus but also by the gills or mouth; some species survive for long periods in seawater and might be eaten by a benthophagous fish.

The eggs are thin-shelled and untanned when laid. Those of *Gyrocotyle rugosa* are fully embryonated *in utero* and hatch within seconds of exposure to seawater, but those of *G. fimbriata* and *G. parvispinosa* require 3 weeks at about 12°C in seawater to develop and *G. urna* eggs take longer. Gyrocotylidean larvae, known as decacanths, are ciliated with ten morphologically similar hooks at the posterior end and are provided with gland cells which open anteriorly (Figure 2.5). There is so far no conclusive evidence that an intermediate host is a necessary part of the life-cycle although larvae of *G. fimbriata* are known to be ingested, but also digested by serpulid annelids (*Mercierella enigmatica*) and harpacticoid copepods (*Tigriopus californicus*). Similarly, *Gyrocotyle* larvae have been found in the pharynx of cumaceans but are dead by three days after ingestion. Cumaceans and other invertebrates might concentrate the larvae for transport to the final host. There are indications, for example, that infection of *Chimaera* with *Gyrocotyle* first occurs when the fish ceases to rely on its yolk sac and begins to feed on polychaetes and cumaceans.

An intriguing feature is the presence of postlarval stages, which lack the cilia and one pair of glands of free-swimming larvae, in the parenchyma of adult gyrocotylideans. These are not derived from eggs hatched *in utero*. They are capable of tissue penetration and will invade gyrocotylideans of all ages from juvenile to adult and will also penetrate cut surfaces

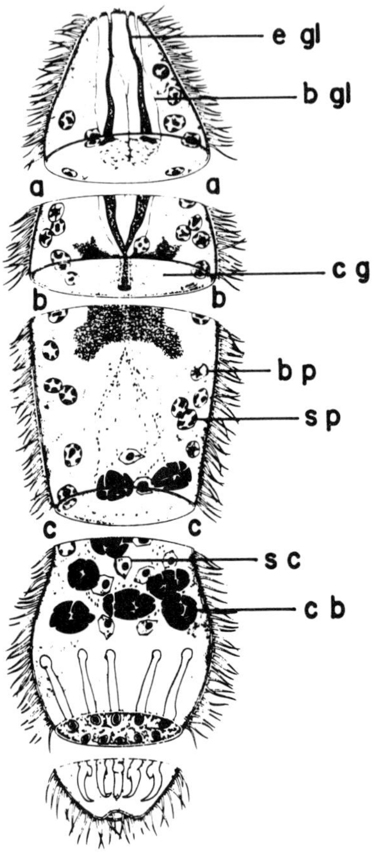

2.5

Figure 2.5. Diagrammatic representation of the larva of *Gyrocotyle fimbriata* a-a, transverse section anterior to the cerebral ganglion b-b, transverse section through the cerebral ganglion c-c, transverse section through the region of presumptive secretory cells and clustered bodies associated with them e gl, ducts of eosinophilic glands b gl, ducts of basophilic glands c g, cerebral ganglion b p, nucleus of body parenchymal cell s p, nucleus of subepithelial parenchymal cell s c, presumptive secretory cell c b, clustered body, one of a group associated with the secretory cells (from Simmons, 1974).

of the gut of *Callorhynchus milii* and enter the blood vessels. Larvae may enter a fish by invading the surface or gills, eventually reaching the intestine and emerging into its lumen. The occasional massive infection of a fish with *Gyrocotyle* may have resulted from the ingestion of a free adult with many postlarvae in its parenchyma. Usually, only a few postlarvae are found in each adult. A phase of tissue development may precede entry into the fish gut. It is as yet unclear whether the pair of worms commonly found in each host are the survivors of a much larger infection or not, or by what mechanism their numbers might be restricted.

Order Amphilinidea

Amphilinideans resemble gyrocotylideans in that the adults are monozoic and the larvae have ten hooks, a ciliated epidermis and anteriorly-opening gland cells. In amphilinideans, however, the hooks are not uniform in morphology but consist of two dissimilar groups of six and four hooks. The adults occupy the body-cavity of their hosts (chondrosteans, teleosts and chelonians) and the life-cycle differs from that of gyrocotylideans in the presence of an intermediate host. The phylogeny of the two groups is entirely different.

Amphilina foliacea is a parasite of sturgeons (*Acipenser* spp.) in the Palaearctic region. The eggs are thin-shelled, non-operculate, with a tubular filament at one end; according to Janicki (1928) they do not hatch until eaten, usually by an amphipod (but see Rohde and Georgi (1983) below). In the gut, the cilia are lost and the larva penetrates into the coelom, where it becomes attached by its hooks and develops to form a larval stage with a cercomer. The hooks do not pass into the cercomer, which is subsequently lost while the hooks remain embedded in the parenchyma. The larva elongates and develops to maturity when eaten by a sturgeon (Janicki, 1928). It has been suggested (Janicki, 1930) that the original definitive host, subsequently lost from the cycle, was a predatory marine reptile. The intermediaries implicated by Janicki (1928), Rasin (1931) and Dubinina (1974), include *Dikerogammarus haemobaphes, Pandorites platycheir, Corophium curvispinus, Carinogammarus roeseli, Rivulogammarus pulex, D. caspius, Pontogammarus crassus, P. robustoides* and *P. obesus*.

A freshwater prawn, *Desmocaris trispinosa*, is an intermediary for *Nesolecithus africanus*, an amphilinid from the body cavity of a freshwater teleost, *Gymnarchus niloticus*, in West Africa. It inhabits the haemocoel of female prawns and restricts development of the ovary. This report extends the intermediate host range of the amphilinids to the palaemonoid decapods (Gibson, Bray and Powell, 1987).

Rohde and Georgi (1983) have shown that the eggs of *Austramphilina elongata*, a parasite of turtles, hatch in water to liberate a ciliated larva which penetrates through the cuticle of crayfish and shrimps. They therefore question Janicki's (1928, 1930) statement that amphilinid eggs do not hatch until eaten by crustaceans. Furthermore, they point out that penetration of the intermediate host by the larva constitutes the first known case of active invasion by a cestode-cestodarian larva through the surface of an intermediate host. Janicki's account of *Amphilina foliacea* was not supported by observations of ingestion of the egg. In addition, the larva of *Austramphilina elongata* is ciliated, which implies a free-swimming existence (Rohde and Georgi, 1983). Shedding of a cercomer, reported by Janicki (1928) for *Amphilina foliacea*, was not seen in *Austramphilina elongata*.

Order Caryophyllidea

The life-cycle of the majority of caryophyllideans involves an oligochaete of the families Tubificidae or Naididae as the intermediate host and a freshwater fish as the definitive host. Use of an oligochaete intermediary is highly unusual in cestodes with aquatic cycles, the majority of which use arthropods, and in helminths as a whole. Other than caryophyllideans, only ten helminth species are known to have oligochaete intermediaries (Mackiewicz, 1972, 1981, 1982a). Mackiewicz considers caryophyllideans to be progenetic and consequently that term has been adopted here.

The fish hosts are predominantly ostariophysans of the families Cyprinidae and Catostomidae, with six siluriform families next in importance. Caryophyllids also parasitize fish with a migratory phase in the life-cycle but not true marine species. A common factor is that the hosts are benthic feeders, and host specificity is maintained largely by this habit and also, perhaps, by a chemical factor—the composition of bile in different host groups (see Chapter 4).

There is no free-swimming larval stage. Eggs are operculate but hatch only in the oligochaete's gut, releasing a non-ciliated hexacanth which penetrates the intestinal wall and develops to the procercoid stage in the coelom or seminal vesicle. From this stage, three types of life-cycle can be distinguished, the first of which, in the genus *Archigetes*, is characterized by progenesis in the oligochaete and the absence of a fish host in the life-cycle of some species. *A. limnodrili* completes its cycle in *Limnodrilus* (Figure 2.6) but *A. sieboldi* and *A. iowensis* (Figure 2.7) may also mature in fish (Mackiewicz, 1972, 1982a). The egg-producing stage in tubificids is considered a progenetic procercoid (Mackiewicz, 1982a) because it retains juvenile characters—the cercomer and non-functional gonopores. In species (e.g. *A. iowensis*) with a facultative cycle, a procercoid eaten by a fish loses the cercomer and has functional gonopores (Figure 2.7).

Archigetes provides the only example of a cestode attaining maturity in an invertebrate and this type of cycle represents an end-point in the evolution of progenesis in the caryophyllids. Mackiewicz (1982a) considered this type of life-cycle to have been achieved by reduction, through progenesis, from a two-host cycle involving vertebrates and tubificids (see Chapter 4). The other two types of cycle differ according to the length of time spent in the oligochaete and the degree of development of the genitalia in this host. For example, *Caryophyllaeus* species may remain in the oligochaete host for a prolonged period during which time the genitalia can develop to an advanced state, but it does not produce eggs until it enters a fish host (Kulakovskaya, 1962). Procercoids of *Glaridacris confusa* in tubificids and naidids have very advanced reproductive organs, so that egg production would begin very shortly after establishment in a fish (Calentine and Williams, 1967). Williams (1978) showed that procercoids of *G. vogei* would develop in tubificid oligochaetes but not in naidids. The adults occur in *Catostomus macrocheilus*. Species which exemplify the third type of cycle spend a relatively short time in the oligochaete and development of the genitalia occurs only after invasion of the definitive host. *Khawia japonensis*, for example, develops in *Limnodrilus udekemianus* to the point of forming the genital anlagen (Demshin, 1978). *K. sinensis*, from carp fed with infected tubificids, had lost the cercomer and developed rudimentary genitalia after five days in the fish (Demshin and Dvoryadkin, 1980). *Markevitschia sagittata*, like *K. japonensis*, forms a genital anlagen during its development in *Limnodrilus* spp. (Demshin and Dvoryadkin, 1981). *Hunterella nodulosa* procercoids developed in *L. udekemianus* in 45–50 days at 20°C but still required 50 days in experimentally infected *Catostomus commersoni* to produce eggs (Mudry and Arai, 1973). The life-cycle of *Biacetabulum macrocephalum*, parasitic in tubificids and *Catostomus commersoni* in North America, is illustrated in Figure 2.8.

Order Spathebothriidea

The best-known life-cycles in this order are those of *Cyathocephalus truncatus*, a member of the Cyathocephalidae, and *Bothrimonus sturionis*, the taxonomic position of which is controversial. Most recently, it has been allocated its own family, Bothrimonidae, within the Spathebothriidea, by Schmidt (1986).

Neither species has a free-swimming stage in the life-cycle and both are capable of advanced development in the intermediate hosts, usually gammarids, so that the two-host cycle may be telescoped to one involving only a single, invertebrate host. The eggs are operculate. Those of *B. sturionis* have filaments radiating from an opercular pole by which they become attached to seaweed before ingestion by gammarids. The eggs hatch only after ingestion, releasing non-ciliated larvae. Those of *B. sturionis* are hexacanths but those of *C. truncatus* lack hooks (but see below). They penetrate the gut wall and develop in the coelom. Wisniewski (1932b, c), who elucidated the life-cycle of *C. truncatus*, did not observe egg production in the invertebrate host but this was described for *C. truncatus* in *Pontoporeia affinis* in North America by Amin (1978a). The fish hosts include trout, other salmonids, pike, perch and burbot.

Paedomorphic plerocercoids of *B. sturionis* have been reported in *Marinogammarus finmarchicus*, *M. pirloti*, *Gammarus oceanicus* and *Anonyx* (Burt and Sandeman, 1969; Sandeman and Burt, 1972). Adults occur in many fish, including salmonids, gadids, sturgeons and flatfish. Only about 10 per cent of eggs released from fish, and fewer of those from invertebrates, had hooks, suggesting that embryonation and development of infectivity takes a long time. This may account for the absence of hooks in many *Bothrimonus* eggs and in those of *C. truncatus*. The life-cycle suggests that these genera have a closer relationship with the caryophyllaeids than with the Pseudophyllidea, in which they were once placed (Sandeman and Burt, 1972).

Diplocotyle plerocercoids from the haemocoel of *Gammarus zaddachi* in the Yorkshire Esk had eggs in the uterus. They were present only in December–April, coinciding with the migration of *Salmo salar*, a possible fish host, to the sea in March–May, after which they disappeared (Stark, 1965). This genus is considered synonymous with *Bothrimonus* by Schmidt (1986).

Order Trypanorhyncha

The life-cycle is similar in several respects to that of the tetraphyllideans but the eggs hatch to release ciliated coracidia in some species but not in others, suggesting two different types of life-cycle involving two or three hosts (Mudry and Dailey, 1971). *Lacistorhynchus tenuis* and *Grillotia erinaceus* have operculate eggs which tan in seawater and hatch to release free-swimming coracidia (Mudry and Dailey, 1971). The eggs hatched in salinities greater than 25 per cent seawater. They hatched in 5–11 days at 11–19°C, developing more rapidly at higher temperatures. Coracidia remained viable for 14 days and tolerated some dilution of seawater. They were ingested by copepods, *Tigriopus californicus*, with a higher prevalence of infection at 11°C than at 10°C, and procercoids grew larger at the higher temperature. Mudry and Dailey (1971) found no further development of the procercoid after three weeks. In experimentally-infected second intermediaries, *Gambusia affinis*, plerocercoids developed immature scolices by the fourth week after infection and by the twelfth week, their bothridia were active and their tentacles protrusible. They successfully infected leopard shark, *Triakis semifasciata*, force-fed with experimentally infected *G. affinis*. This species has a plerocercoid stage in over 60 species of teleosts. Plerocercoids from *Cymatogaster aggregata* successfully infected leopard shark (*Triakis semifasciata*) (Young, 1954).

Sakanari and Moser (1985a, b) completed the life-cycle of *L. dollfusi* (syn. *L. tenuis*) with *Tigriopus californicus and Gambusia affinis* as experimental first and second intermediaries, respectively; plerocercoids occurred naturally in *Genyonemus lineatus*. Worms were recovered from the spiral valve of leopard shark but were not sexually mature. Subsequently, Sakanari

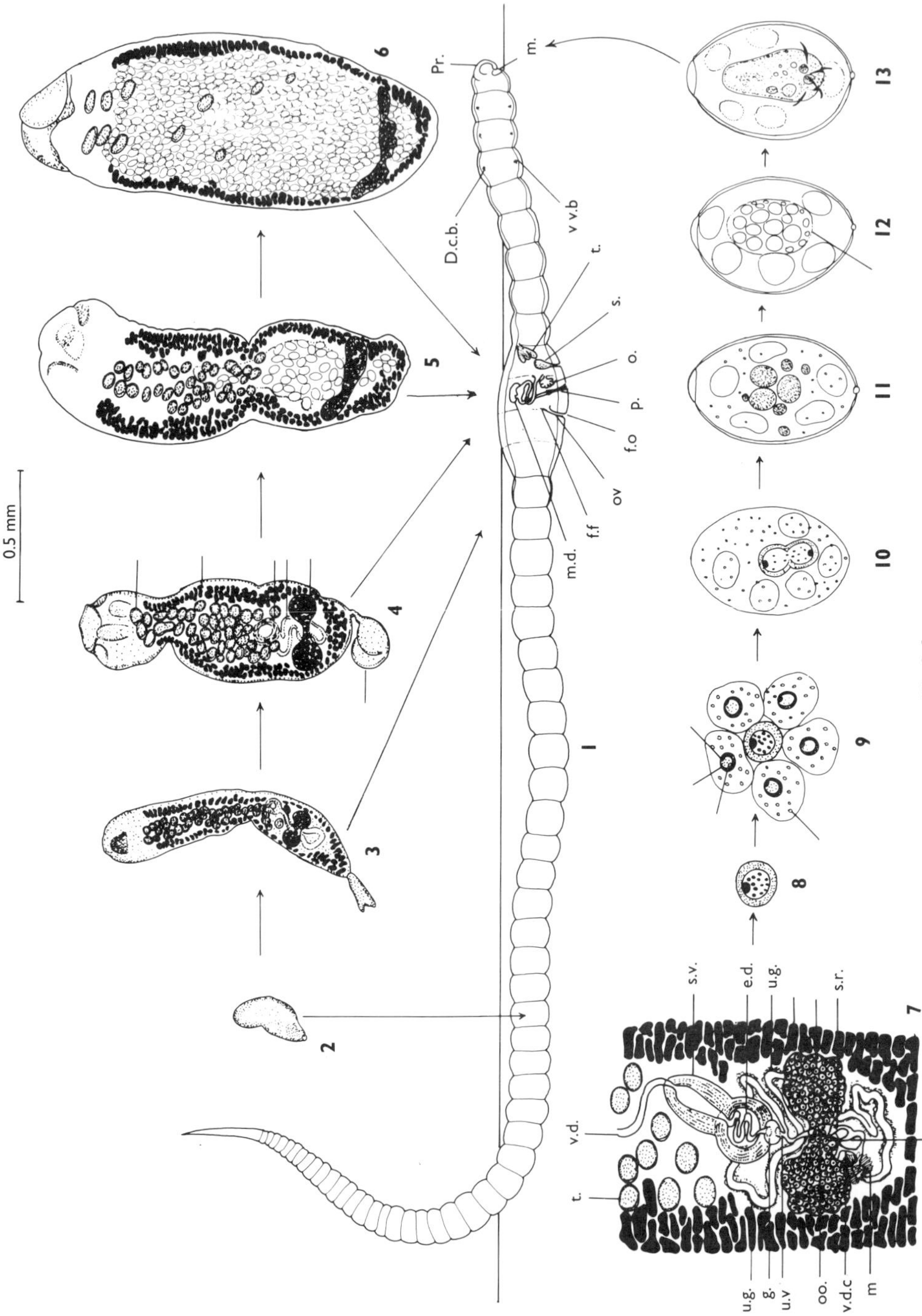
0.5 mm
Pr.
m.
D.c.b.
v v.b
t.
s.
o.
p.
f.o
ov
f.f
m.d.
s.v.
e.d.
u.g.
s.r.
v.d.
t.
od.
u.g
g.
u.v
oo.
v.d.c
m
1
2
3
4
5
6
7
8
9
10
11
12
13

Fig 2.6

Figure 2.6. 1–13 *Archigetes limnodrili*. Diagrammatic representation of the life-cycle.
1. A tubificid worm partially embedded in mud. D.c.b. dorsal chaetal bundle; f.f. female funnel; f.o. female opening; m. mouth; m.d. male ducts o. ovary; ov. oviduct; p. penis; Pr. prostomum; s. spermatheca; t. testis; v.c.b. ventral chaetal bundle. 2–6. Stages in the development of the cestode in a tubificid worm. As it grows the cestode migrates forwards. 7. Reproductive system. e.d. ejaculatory duct; g. genital atrium; m. Mehlis' gland; o. ovary; oo. oocyte; ov. ovicapt; s.r. seminal receptacle; s.v. seminal vesicle; u.v. utero vaginal canal; v. vitellarium; v.d. vas deferens; v.d.c. vitelline duct canal 8–13. Stages in the development of the fertilized ovum to an oncosphere in the egg. 8. Ovum with large nucleus; 9. Ovum surrounded by 5 yolk cells each having a vacuolated nucleus; 10. First division of ovum into two blastomeres and formation of egg-shell; 11. Subsequent division of ovum into macro, meso and micromeres and thickening of egg shell which has operculum; 12. Formation of embryophore around embryo; 13. Fully developed embryo or oncosphere in the egg which is eaten by a tubificid worm.

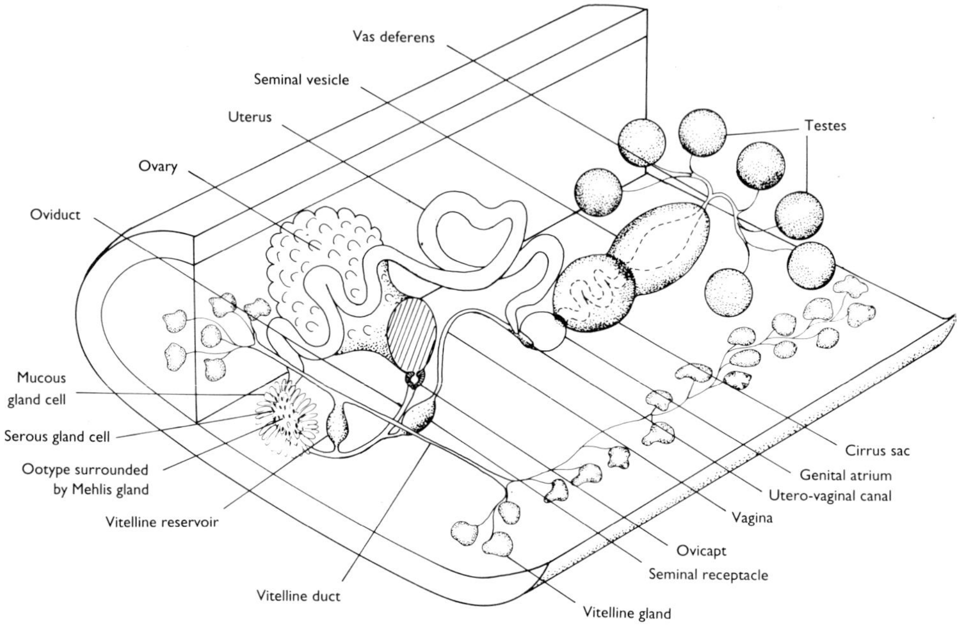

Figure 2.6.–14. The reproductive complex of a mature *Archigetes*
c.s.—Cirrus sac g.a.—genital atrium m.g.c.—mucous gland cell o.—ovary oo.—Ootype surrounded by Mehlis' gland od.—oviduct ov.—ovicapt s.g.c.—serous gland cell s.r.—seminal receptacle s.v.—seminal vesicle t.—testes u.—uterus u.v.c.—utero-vaginal canal v.—vagina v.d.—vas deferens v.g.—vitelline gland vi.d.—vitelline duct v.s.—vitelline reservoir

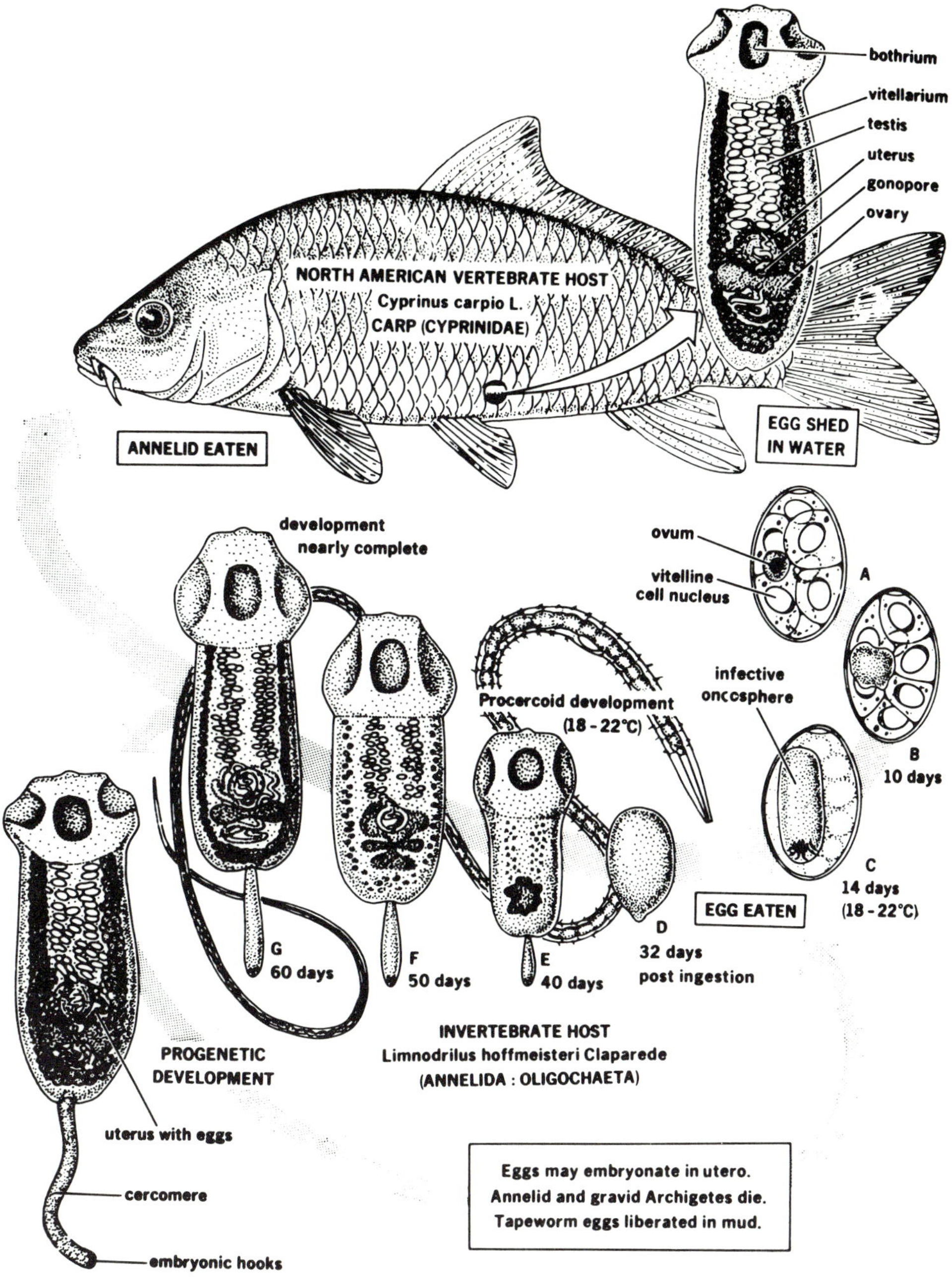

2.7

Figure. 2.7 The life-cycle of *Archigetes iowensis* based on data from Calentine (1964). Cestode stages from Calentine (1964). From Mackiewicz (1972)

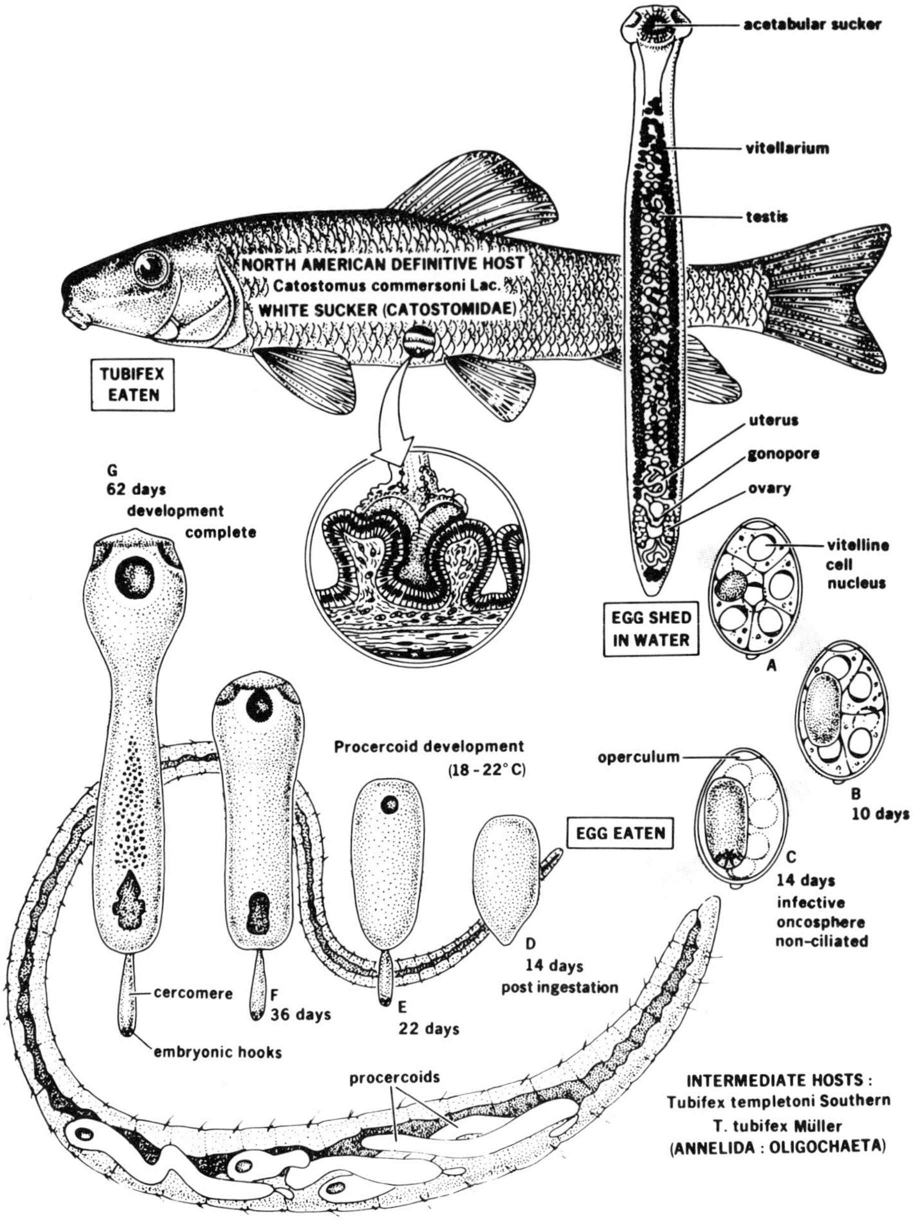

2.8

Figure 2.8 The life-cycle of *Biacetabulum macrocephalum* based on data from Calentine (1965). From Mackiewicz (1972).

and Moser (1989), using the same species of first and secondary intermediaries and definitive hosts, succeeded in obtaining gravid worms after 21 months in experimentally infected leopard shark and completed the life-cycle by infecting *Tigriopus californicus* with coracidia hatched from their eggs. Beveridge and Sakanari (1987) had shown that the species of *Lacistorhynchus* in Californian and Australian elasmobranchs was a new species, *L. dollfusi*, but that in Atlantic hosts was *L. tenuis*. Previous reports of *L. tenuis* in Californian and Australian waters should be referred to as *L. dollfusi*. Following the same protocol as for *L. dollfusi*, Sakanari and Moser (1989) completed the life-cycle of *L. tenuis*, originally obtained from *Mustelus canis* in the Atlantic, recovering adult worms from experimentally infected leopard shark 18 months after infection.

Parachristianella monomegacantha lacks a free-swimming coracidium and its development in the copepod culminates not in a procercoid but in an immature plerocercus with developing tentacles and bothridia, thought to be directly infective to elasmobranchs, an advantage to species which parasitize non-piscivorous hosts (Mudry and Dailey, 1971). A three-host life-cycle occurs in *Callotetrarhynchus nipponica*, plerocerci of which, from the body-cavity of *Seriola quinqueradiata*, successfully invaded *Triakis scyllia* and had a life-span of about 25 days. However, a 'procercus' of *Callotetrarhynchus* from anchovy, *Engraulis japonica*, failed to infect *T. scyllia* and was digested or passed out of the gut (Nakajima and Egusa, 1972a, b; 1973). It is not known how anchovy become infected. Attempts at experimental infection have failed. Metacestodes of *Eutetrarhynchus* and *Parachristianella* have been found in a wide range of gastropods and pelecypods (Cake, 1976, 1977) (Figure 2.9.1,2) and *Eutetrarhynchus* blastocysts have been found free in plankton samples (Reimer, 1977).

Order Haplobothriidea

Haplobothrium globuliforme is the representative of this recently erected order which occupies a position intermediate between the Trypanorhyncha and the Pseudophyllidea (MacKinnon, Jarecka and Burt, 1985). Adults are found in the bowfin, *Amia calva*, in North America and resemble the trypanorhynchs in that the scolex has four armed proboscides. However, other morphological features (presence of bothria, arrangement of the reproductive organs, structure of the sperm) are close to those of Pseudophyllidea and the life-cycle, which involves two intermediaries of which the second is a fish, resembles the pseudophyllidean pattern. The eggs, laid in summer, are operculate and hatch on contact with water to liberate fully-developed coracidia which are eaten by cyclopoid copepods (*Cyclops* sp., *C. brevispinosus, Macrocyclops albidus*). Procercoids develop in the haemocoel in 12–20 days at about 20°C and plerocercoids, with the characteristic scolex, are found in cysts in the liver of small fish which serve as second intermediaries. Examples of fish intermediaries are *Ictalurus nebulosus, Ameiurus nebulosus, Lebistes reticulatus* and sunfish (Essex, 1929c; Thomas, 1930; Meinkoth, 1947; MacKinnon and Burt, 1985b,c, and Mackinnon, Jarecka and Burt, 1985).

Order Pseudophyllidea

The eggs of pseudophyllideans develop in water rather than *in utero*, unlike some other cestode groups and they have numerous vitelline cells rather than a single one (Jarecka, 1975). They usually, but not invariably, hatch in water liberating a coracidium, a hexacanth enclosed in a ciliated envelope, which is eaten by a crustacean, generally a copepod, after which the hexacanth penetrates into the haemocoel where it develops to the procercoid stage. Pseudophyllideans that are adult in fish have life-cycles incorporating either one or two intermediaries. If two, the second is also a fish, in which the plerocercoid

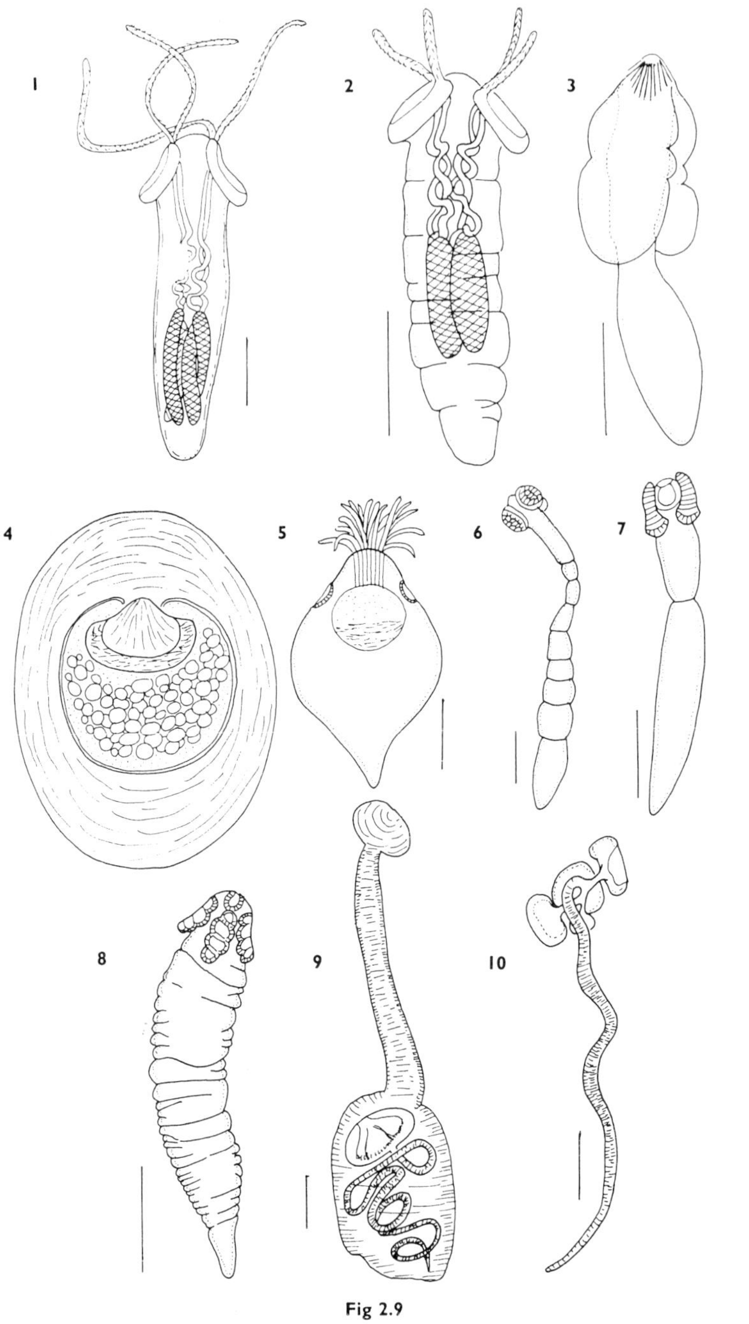

Fig 2.9

Figure 2.9 Metacestodes of the orders Trypanorhyncha, Diphyllidea, Lecanicephalidea, Dioecotaeniidea and Tetraphyllidea from molluscs (from Cake, 1976).
1. *Eutetrarhynchus* sp. (Trypanorhyncha) from *Pleuroploca gigantea* (scale bar 0.2 mm) 2. *Parachristianella* sp. (Trypanorhyncha) from *Macrocallista nimbosa* (scale bar 0.5 mm) 3. *Echinobothrium* sp. (Diphyllidea) from *Nassarius vibex* (scale bar 0.1 mm) 4. Encysted metacestode of *Tylocephalum* sp. (Lecanicephalidea) from *Argopecten irradians concentricus* 5. *Polypocephalus* sp. (Lecanicephalidea) from *A. i. concentricus* (scale bar 0.1 mm) 6. *Dioecotaenia cancellata* (Dioecotaeniidea) from *Chione cancellata* (scale bar 1.0 mm) 7–10 Tetraphyllidea 7. *Rhinebothrium* sp. from *A. i. concentricus* (scale bar 0.5 mm) 8. *Acanthobothrium* sp. from *Oliva sayana* (scale bar 0.5 mm) 9–10 *Anthobothrium* sp. from *Anadara transversa* enclosed in capsule (9) and free (10) (scale bars 1 mm (9) and 5 mm (10))

stage develops in the musculature or body cavity. Fish also serve as second intermediaries for cestodes *(Diphyllobothrium, Ligula, Schistocephalus, Valipora, Digramma)* whose definitive hosts are piscivorous birds and mammals.

Development of eggs and coracidia

The development of pseudophyllidean eggs and coracidia is affected by a variety of environmental factors of which water temperature is the most important. Table 2.1 gives examples of the effect of temperature on the development of procercoids to an infective stage. Above the minimum, development accelerates with rising temperature to a maximum beyond which the eggs are killed. Eggs of *Bothriocephalus scorpii*, for example, will develop in 16 days at 8–9°C and in 4–5 days at 20–24°C, but are killed at 30°C (Solonchenko, 1979). The maximum for eggs of *T. crassus* and *T. nodulosus* is 26°C (Kuperman, 1973). Eggs of some species, eg. *T. nodulosus*, survive freezing for short periods (15 minutes) but others do not. *Valipora campylancristota* eggs die when frozen but survive for 40 days at 4°C (Skvortsova, 1979). *Triaenophorus* eggs will develop at temperatures as low as 2–7°C but only over very long periods. Species of this genus have different optimal temperatures for development, eg. 22–24°C for *T. meridionalis*, 20°C for *T. nodulosus* and 17–20°C for *T. crassus*, and they also develop at different rates even at the same temperature (Kuperman, 1973). Ligulids, however, develop at a higher temperature range (10–32°C) than *Triaenophorus*. Below 10°C, egg development is retarded or ceases and above 32°C they die (Dubinina, 1966). Kuperman (1973) has emphasized that eggs of *Triaenophorus* species, e.g. *T. amurensis*, *T. crassus* and *T. orientalis*, will develop to the coracidial stage in the uterus of the parent worm if the water temperature is sufficiently high (15°C) and hatch about 3 hours after deposition. Water pressure and light have no effect on egg development in *Triaenophorus*, ligulids and *Diphyllobothrium latum*, although hatching in all of them appears to require the stimulus of light (Kuperman, 1973; Dubinina, 1966). Investigations of salinity tolerance show that the eggs of many species will develop in water of salinity greater than that in which they can hatch and in which the coracidia can survive. *T. nodulosus* larvae develop normally in water of salinity up to 22.5‰ but do not hatch well at 7‰ or above and die at 25‰. *D. latum* corcidia develop in water of 12.05‰ but emergence is poor above 9‰; emergence and survival are good at 3.01‰ and 6.03‰. *Schistocephalus solidus* larvae will develop but not hatch at 18‰ although coracidia will emerge if the eggs are subsequently transferred to fresh water. At 35‰, they do not develop and

Table 2.1. Effect of temperature on procercoid development

Species	Procercoid infective (days)	Temperature °C	Reference
Bothriocephalus acheilognathi	7–10	25–27	Urazbaev and Allaniyazova (1977)
Eubothrium rugosum	10–12	18–20	Kuperman (1978a)
Triaenophorus nodulosus	30	4–5	Kuperman (1973)
T. crassus	11–12	9–10	
	7–9	16–20	
Diphyllobothrium latum	15–16	14–20	Dubinina (1966)
Digramma interrupta (from goldfish)	13–14	15–20	Dubinina (1966)
	10–12	20–22	”
D. interrupta (from bream)	9–10	24–28	”
Ligula intestinalis	9–10	24–28	”
Schistocephalus solidus	7–8	22–25	”
S. pungitii	13–14	16–18	”

most die even after removal to fresh water (Dubinina, 1966; Kuperman, 1973). Eggs of *Eubothrium crassum* and *E. salvelini* are sensitive to changes in salinity and are adapted to ranges of 5–20‰ and less than 5‰ respectively (Kuperman, 1978b).

Two-host life-cycles

Bothriocephalus is an example of a genus with a single intermediate host. Its best-known representative, *B. acheilognathi* (syn. *B. gowkongensis*) parasitizes many genera of freshwater fish including fish which are commercially farmed. In this context, its requirement for only a copepod intermediary has allowed it to become a very significant problem (Mitchell and Hoffman, 1980; Hoffman, 1976a). Of Asian origin, it has been disseminated throughout Europe and parts of North America through importation of infected fish (Hoffman and Schubert, 1984). Species of several genera of copepods have been confirmed as first intermediaries. Several species of North American cyclopoid, but not calanoid, copepods were found to be susceptible to experimental infection by Marcogliese and Esch (1989). Both planktonic cyclopoids *(Ciacyclops thomasi, Mesocyclops edax, Tropocyclops prasinus)* and benthic cyclopoids *(Eucyclops agilis, Paracyclops fimbriatus poppei)* became infected under laboratory conditions. Planktonivorous fish hosts (*Gambusia affinis, Notropis lutrensis*) would acquire infection by ingesting the former species, while benthic feeders *(Pimephales promelas)* would ingest benthic copepods. Eggs of *B. acheilognathi* sink and stick to the substrate and the susceptibility of benthic copepods to infection would facilitate transmission to detritus-feeding fish; although natural infections of these copepods were not found during summer, they may well occur in the spring and autumn periods of recruitment. The susceptibility of a large number of invertebrate and fish hosts to *B. acheilognathi* suggests that it will extend its range still further.

A single copepod may harbour 9–15 procercoids, occasionally up to 30, and development of the procercoid is temperature-dependent: at 25–27°C, the cercomer forms in 3–4 days and development is completed in 7–10 days (Urazbaev and Allaniyazova, 1977; Allaniyazova, 1975). The life-cycle takes about a year to complete (Mitchell and Hoffman, 1980). The two-host life-cycles of other species of the genus in fish have been described by Essex (1928), Jarecka (1964) and Solonchenko (1979).

Eubothrium is a genus of interest because it includes marine and freshwater species and others which can live in both environments. The biology of the genus has been reviewed by Kennedy (1978b), who outlined the life-histories of *E. crassum* (illustrated in Figure 2.10) and *E. salvelini* (summarized in Table 2.2). The distribution of the freshwater race of *E. crassum* coincides with that of *Salmo trutta*; its preferred host is the non-migratory *S. t. fario*, but it also occurs in other salmonids. The life-cycle takes place in fresh water and this race may survive in *Salmo trutta* but not in *S. salar*, if taken to sea. The distribution of the marine Atlantic and Pacific races coincides with that of the preferred host species. The life-cycle of neither race is known but transmission of the Atlantic race takes place in marine coastal waters. Both may be carried into fresh water by migrating hosts, in which case the incidence and intensity of infection of both decreases. Marine and freshwater races of *E. crassum* replace each other in *S. salar* migrating between the two environments (Kennedy, 1978a) but complete replacement does not occur in *S. trutta trutta* which is at sea for a shorter period.

Of the two races of *E. salvelini* the European race is co-extensive with *Salvelinus alpinus*, but the American race has a much wider specificity that may result from the absence of *E. crassum* from this continent. The European race survives migration of *S. alpinus* to sea but the American race has not yet been shown to do so.

E. fragile, adult in twaite shad, is regarded as a marine species which may be carried into fresh water during the spawning migration of the host (Kennedy, 1981a).

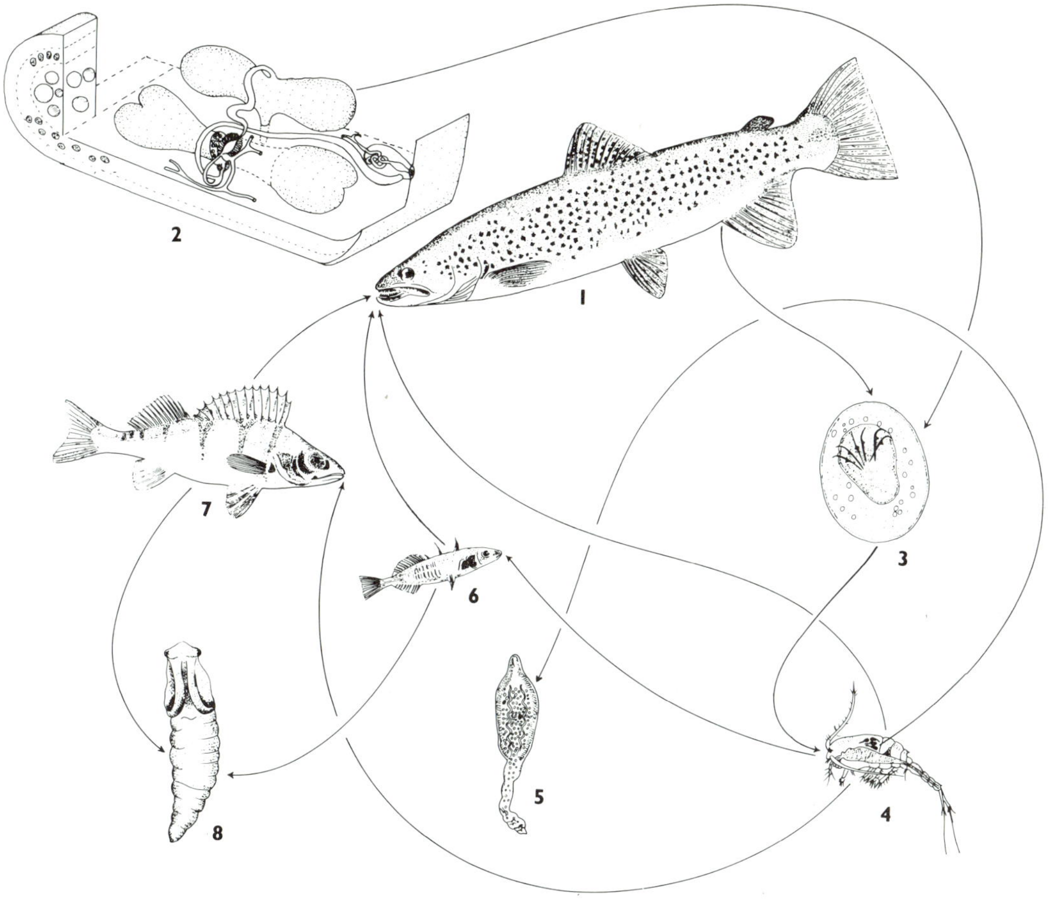

Fig 2.10

Figure 2.10. 1–8. *Eubothrium crassum*. Diagrammatic representation of the life-cycle (figures not drawn to scale) Based on information in Nybelin (1922), Vik (1963), Rawson (1957), Kennedy (1969a),
1. *Salmo* sp. definitive host becomes infected by eating the second intermediate host. 2. Reproductive organs of the adult tapeworms from the intestine of the definitive host. 3. Egg of *E. crassum* containing hexacanth embryo. 4. A copepod, the first intermediate host. 5. Procercoid from the body-cavity of a copepod having a long slender cercomer 6 & 7 *Gasterosteus aculeatus* and *Perca fluviatilis* second intermediate hosts. 8. A plerocercoid from the body cavity of the second intermediate host develops into the adult tapeworm on being eaten by a definitive host

Dr. Kennedy (personal communication) has drawn attention to the lack of supporting evidence for *Gasterosteus aculeatus* and *Perca fluviatilis* as second intermediate hosts. See Kennedy (1978a) and Kennedy and Burrough (1978).

Table 2.2. The life-histories of races of two species of *Eubothrium* (from Kennedy, 1978b)

	Eubothrium crassum			*Eubothrium salvelini*	
Races	Freshwater	Marine		European	American
Geographical distribution	Europe Eurasia	Atlantic,	Pacific	Europe	North American East Asia
Preferred host	*Salmo trutta fario* (non-migratory)	*Salmo salar*	*Oncorhynchus* spp.	*Salvelinus alpinus*	*Salvelinus, Salmo, Oncorhynchus, Cristivomer, Mytocheilus Ptychocheilus*
Other hosts	parr and smolt of *S. salar;* land-locked *S. salar*; *S. gairdneri*	*S. trutta trutta*	–	*Coregonus*	–
Accidental hosts	*Coregonus, Perca*				
Intermediate host	*Cyclops* spp.			*Cyclops spp.*	*Cyclops* spp.
Life-cycle	freshwater	marine			

Kuperman (1978a) has described for *E. rugosum* a life-cycle with features unusual for a pseudophyllidean and for this genus. The eggs develop *in utero* and development is prolonged, taking 5–6 months (from December to May). The larvae lack a ciliated envelope and, although the eggs survive in water for 300 days at 5–7°C, will die within half an hour should they hatch in water. Procercoids develop in species of *Cyclops* and *Microcyclops*, which ingest the eggs, and the plerocercoids in the intestine, mainly the pyloric caeca, of *Acerina cernua*. Plerocercoids have a fully-formed scolex and three or four proglottides and are present in *A. cernua* throughout the year. However, *E. rugosum* infests the definitive host, *Lota lota*, in September to October when the host feeds voraciously due to cold conditions, with a peak in February to March as a result of intensive feeding on *A. cernua*. Although the life-cycle is completed in a year and the worms destrobilate in August, many will regenerate and reproduce again the next year.

Three-host life-cycles

The genera *Senga* and *Triaenophorus* have three-host life-cycles with fish as both second intermediaries and final hosts. *S. visakhapatnamensis*, adult in *Ophicephalus punctatus* in India, develops in cyclopoid copepods attaining the procercoid stage in 15 days but requires a second intermediary in which development is very slow—four months in topminnows, *Panchax panchax* and *P. melastigma*. By this time, the plerocercoid has encysted in the abdominal muscles and has a scolex with the same features as that of the adult (Ramadevi, 1976). *Ptychobothrium belones* has a similar cycle involving garfish, *Tylosurus leiurus*, as final hosts and a planktonivorous fish, *Hyporhamphus knysnaensis*, as a second intermediary (Whitfield and Heeg, 1977). The metacestodes localize in the epaxial musculature; the first intermediary is as yet unknown.

Order Diphyllidea

The intermediaries of diphyllideans, adults of which parasitize elasmobranchs, include molluscs, gammarids and crabs. Dollfus (1964a) reported metacestodes of *Echinobothrium affine* in *Carcinus maenas* and summarized accounts of metacestodes of *E. typus* in *Gammarus locusta, Perioculoides (Oediceros) longimanus, Pagurus* and *Crangon*, and of *E. lateriporum* in *Bullia melanoides, Murex tropa* and *Matuta victor*. He reported a form of asexual multiplication for *E. affine* in *Carcinus maenas*. Cake (1976, 1977) recorded *Echinobothrium* larvae in the gastropods *Cantharus cancellarius* and *Nassarius vibex* in the Gulf of Mexico (Figure. 2.9.3).

Order Lecanicephalidea

Metacestodes of *Tylocephalum* and *Polypocephalus* have been found in edible molluscs from the Gulf of Mexico (Cake, 1976, 1977) (Figure. 2.9.4,5). Cake and Menzel (1980) also found *Tylocephalum* in the American oyster, *Crassostrea virginica*, and in three molluscivorous gastropods, *Busycon contrarium, Murex pomum* and *Thais haemastoma canaliculata. Polypocephalus* metacestodes occurred in the digestive gland of bay scallops, *Agropecten irradians concentricus*, which showed a high incidence and intensity of infection and were believed to act as intermediate hosts (Cake, 1976, 1979).

Order Dioecotaeniidea

Metacestodes of *Dioecotaenia cancellata* (Figure. 2.9.6) have been found in edible molluscs by Cake (1976, 1977) in the Gulf of Mexico. Adults parasitize the spiral intestine of the cow-nosed ray, *Rhinoptera bonasus*. Gastropods *(Melongena corona)* and pelycypods *(Anadara ovalis* and *Chione cancellata)*, shallow-water benthic molluscs, serve as intermediaries or paratenic hosts.

Order Tetraphyllidea

Very little information is available on the life-cycles of this order, the adults of which are parasites of elasmobranch rather than teleost fish. Tetraphyllideans have a three-host life-cycle comprising a procercoid stage in copepods, a plerocercoid in teleosts and cephalopods, and adults in elasmobranchs (Mudry and Dailey, 1971). Experimental infection of copepods with eggs, which they ingest, has been achieved for *Acanthobothrium hispidum* in *Tigriopus fulvus* and *A. olseni* in *T. californicus* (see Riser (1956) and Mudry and Dailey (1971), respectively). There is no ciliated oncosphere. *A. olseni* procercoids had hooks posteriorly but lacked a cercomer. Plerocercoids of tetraphyllideans *(Pelichnibothrium, Phyllobothrium)* have been found in squid, *Illex argentinus*, by Threlfall (1970).

Cake (1977) found metacestodes of *Anthobothrium, Acanthobothrium* and *Rhinebothrium* (Figure. 2.9.7–10) in edible molluscs in the Gulf of Mexico and listed (Cake, 1976) numerous gastropods and pelycypods as hosts of these genera. Phyllobothriid metacestodes (probably *Phyllobothrium*) have been found in chaetognaths *(Sagitta friderici, S. decipiens, Spadella cephaloptera* and *Pterosagitta draco)* from plankton samples off northwest Africa (Reimer, 1977). Other tetraphyllidean metacestodes have been found in *S. cephaloptera, P. draco, Krohnitta subtilis, Tomopteris* sp. and five species of *Sagitta*.

Order Proteocephalidea

Proteocephalid eggs pass into water in the faeces of the host or in detached segments which swell on contact with water, expelling the eggs. Increased activity of the oncospheres after exposure to water has been reported in *Proteocephalus tumidocollis* by Wagner (1954) and in *P. parallacticus* by Freeman (1964a). The eggs hatch only after ingestion by a copepod; the hatching mechanism is unknown but may result from mechanical damage by the mouthparts or the activity of the host's digestive enzymes. The larva escapes from the surrounding membranes by using its hooks and penetrates the intestinal wall of the copepod by using the hooks and possibly secretions from glands thought to be penetration glands. These have been reported in *P. percae* by Wootten (1974), *P. fluviatilis* by Fischer (1968), *Corallobothrium parafimbriatum* and *Corallotaenia minutia* by Befus and Freeman (1973a,b). Those of *P. percae* had disappeared in larvae removed from the copepod's haemocoel and were assumed to have discharged their contents during penetration (Wootten, 1974). In this species and *P. parallacticus*, the hooks seem to be used to displace the intestinal cells rather than rupture them (Freeman, 1964a; Wootten, 1974).

Development in the copepod is similar in those species of *Proteocephalus* whose life-cycles are known. In *P. tumidocollis, P. parallacticus, P. fluviatilis, P. percae* and *P. neglectus*, development is slow for the first 8–10 days and then growth is more rapid and a cercomer is formed between 9–20 days. The larvae are fully developed after 18–35 days, depending on the species (Wootten, 1974). The suckers usually appear at about the same time or slightly after the cercomer, although Wagner (1954) reported that, in *P. tumidocollis*, sucker development began before the cercomer was formed. The larval development of *P. tumidocollis* is illustrated in Figure 2.11. 1–10. This stage, with the suckers characteristic of the adult, has been termed the plerocercoid, plerocercoid I or cercoscolex by various authors (Fischer, 1968; Freeman, 1964a; Priemer, 1980).

The rate of development in the copepod is influenced by the water temperature, species of host and, occasionally, its sex, and sometimes by the number of metacestodes present. *P. neglectus* plerocercoids are infective to trout after 21 days in *Cyclops strenuus* at 14°C (Priemer, 1980). The optimum temperature for development of *P. tumidocollis* in *C. vernalis* was 20°C but it did not occur below 10°C or over 26°C (Wagner, 1954). Similarly, *P. exiguus* develops optimally in *Eucyclops gracilis* at 18–20°C but is arrested at 26°C (Anikieva, 1982c). That the factors influencing development are complex was illustrated by Wootten (1974), who found that *P. percae* developed normally in *C. viridis* only at 14°C but in *C. agilis* and *C. leuckarti* would develop normally at 20°C in about the time taken in *C. viridis*. The optimum temperature for *P. parallacticus* in mature *C. bicuspidatus* is 16°C but, in immature specimens, it is slightly less (Freeman, 1964a). Only mature female *C. bicuspidatus, C. vernalis* and *Tropocyclops prasinus* and immature females of the first species contained fully developed plerocercoids of *P. fluviatilis*. They were not found in males or copepodids of any species or in a copepod containing more than 13 metacestodes (Fischer, 1968). Up to five mature plerocercoids were found in a mature *C. bicuspidatus* but two to four were more usual; infection with more than 15 killed the host. Wootten (1974) suggested that the number of *P. percae* completing development in *Cyclops* spp. is determined by the amount of nutrients available or by a physical crowding effect.

The above species have a two-host life-cycle but other proteocephalids have a three-host cycle, involving two fish hosts, which may be facultative in one species. *Corallobothrium parvum* plerocercoids became established in *Glaridichthys talcatus* under experimental conditions and this host was considered to be a necessary part of the life-cycle (Larsh, 1941); the definitive hosts are bullheads (*Ameiurus nebulosus*). *Proteocephalus ambloplitis*, however, parasitic as adults in smallmouth bass (*Micropterus dolomieui*), yellowfin perch (*Perca flavescens*) and other fish, use fry of bass or pumpkin seed (*Lepomis gibbosus*) as second intermediaries, but the plerocercoids must undergo development before they are infective to larger fish of a suitable host species. If eaten by such a fish at too early a stage, they migrate to the body cavity; they become sexually mature if subsequently ingested by a suitable host (Hunter and Hunter, 1929). Subsequently, Fischer and Freeman (1969) showed that parenteral plerocercoids in bass over 7.6 cm fork length migrated into the gut when the temperature was raised from 4 to 7°C under experimental conditions. The seasonal incidence in the gut was highest in spring, declined in summer and none were found in late autumn, and the results suggest that parenteral plerocercoids overwintered in the viscera then migrated in May and

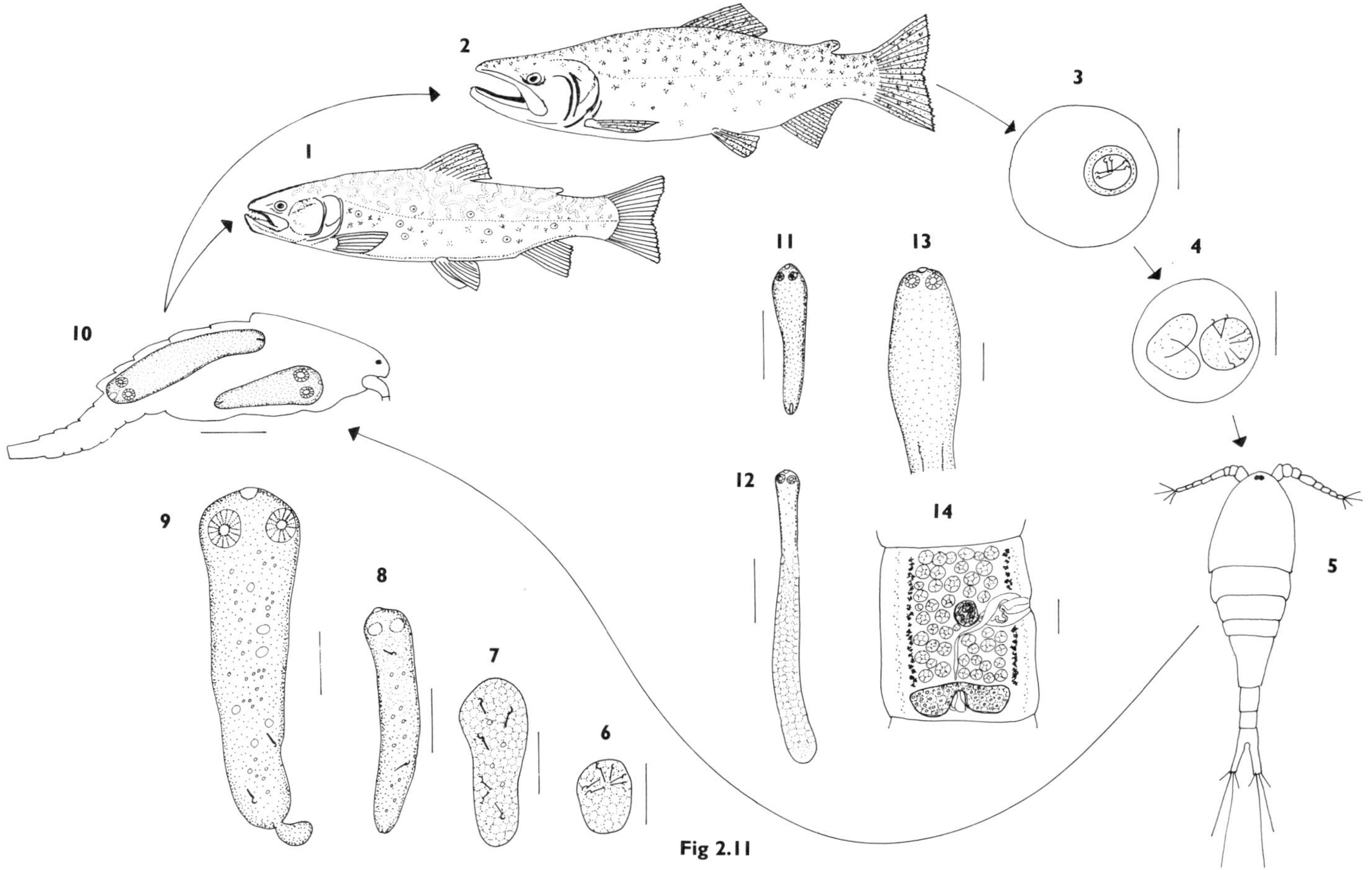

Fig 2.11

Figure 2.11. 1–14 The life-cycle of *Proteocephalus tumidocollus*. Wagner (1954) 1. *Salvelinus fontinalis* and 2. *Salmo gairdneri*, naturally infected definitive hosts. 3. Mature egg. 4. Egg with oncosphere escaped from middle membrane. 5. *Cyclops*, first intermediate host. 6–8. Larvae at 7, 10 and 13 days respectively. 9. Larva at 16 days, with cercomer. 10. *Cyclops* with larvae which have developed beyond the cercomer stage. 11. Young worm from fish intestine, 11 days after ingestion of infected *Cyclops*. 12. Six-week old tapeworm. 13. Adult scolex. 14. Mature proglottid. Scale bars: 3, 0.05 mm; 4, 6 0.04 mm; 7,8,9 0.06 mm; 10,13,14 0.12 mm; 11, 0.25 mm; 12, 0.12 mm. Redrawn after Wagner (1954) (Figures 1, 2, 5, 7, 9, 10, 11, 12, 13, 15, 16).

early June when lake temperatures rose. Freeman (1964a) described both a direct and an indirect cycle in *P. parallacticus*. In the direct cycle, the infected copepod is eaten by a trout and development to maturity proceeds if a minimum temperature between 10 and 14°C is maintained. If, however, the copepod is eaten by a fish other than trout, the plerocercoid remains in the gut without developing until ingested by a trout.

Trematoda

Aspidogastrea

The aspidogastreans are a small group inhabiting molluscs, marine and freshwater fish and chelonians. Some can complete the life-cycle in molluscs alone but others mature in both molluscs and vertebrates and the majority achieve maturity only in a vertebrate. Only one instance is known of their occurrence in groups other than molluscs or vertebrates. *Stichocotyle nephropsis*, adult in rays, occurs as a pre-adult encysted in crustaceans, which may be either accidental or transport hosts. If the latter, the cycle would be intermediate between the majority of the group and the digeneans (Rohde, 1972). However, Gibson and Chinabut (1984) have argued that the stichocotylids, which have no known associations with molluscs, represent a separate order within the aspidogastreans from the aspidogastrids. The two groups probably diverged well before the appearance of the digeneans and their relationship may not be as close as previously believed. Much of the following account is drawn from Rohde's (1972) review of aspidogastrean biology.

In molluscs, aspidogastreans occupy the gut and the mantle, pericardial and renal cavities, but live on the surfaces of organs rather than in the tissues. In fish, they inhabit the intestine, gall-bladder, bile-ducts and, rarely, the rectal glands (Rohde, 1972; Schell, 1973). The eggs are operculate (*Aspidogaster conchicola, Macraspis elegans, Rugogaster hydrolagi*) or non-operculate (*Cotylogaster occidentalis*) and are either embryonated when laid and hatch rapidly (some *Aspidogaster* spp., *R. hydrolagi, C. occidentalis*), or require a more prolonged period of development (*Cotylaspis insignis, Multicalyx cristatus*). *Cotylogaster occidentalis* eggs may hatch *in utero*. In this species, hatching is preceded by the production of mucus from the cephalic glands of the larva and the non-operculate egg ruptures antero-laterally within a few minutes of deposition (Fredericksen, 1978). In contrast, eggs of *A. indica* hatch after 1 hour at 30–32.5°C, the larva pushing off the operculum by muscular activity (Rai, 1964a).

The cotylocidium larva is characteristic of the group; it has a digestive system (oral cavity, prepharynx, pharynx and simple caecum) and a postero-ventral sucker which is replaced by the multiloculate ventral disc of the adult. Some cotylocidia are aciliate (*Aspidogaster conchicola, A. indica, Cotylaspis insignis, Lobatostoma manteri, Rugogaster hydrolagi*) but others, e.g. *Cotylogaster occidentalis*, have tufts of cilia at the posterior end of the body and in the mid-region. *C. occidentalis* cotylocidia have six tufts at the posterior end and eight in the mid-region (Figure 2.12) and Fredericksen (1978) has argued that, ciliation being a primitive feature, it is the most primitive known type in the group as it has the greatest number of ciliary tufts. Emergence from the egg is rapid in ciliated cotylocidia but aciliate types take much longer, e.g. from 1–3 hours for *A. indica* (Rai, 1964a).

Little information is available on the route of invasion of molluscs but it is probably through the gut as there is evidence that visceral infections first establish in organs associated with the gut (Rohde, 1972). Several species (*A. conchicola, C. occidentalis, L. manteri*) hatch only after ingestion by the mollusc and *L. manteri* cotylocidia are known to migrate into the ducts of the digestive gland (Rohde, 1973). Under experimental conditions, pre-adult *L. manteri* were inhaled by *Planaxis sulcatus* and stayed in the mantle cavity for 2–6 hours but permanent infection was not achieved. Cotylocidia are not directly infective to

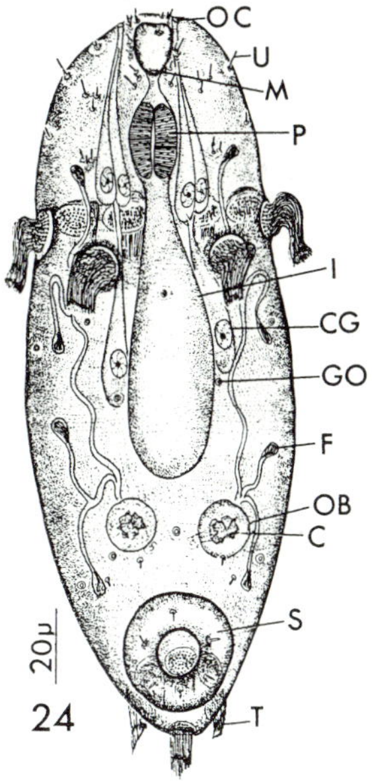

2.12

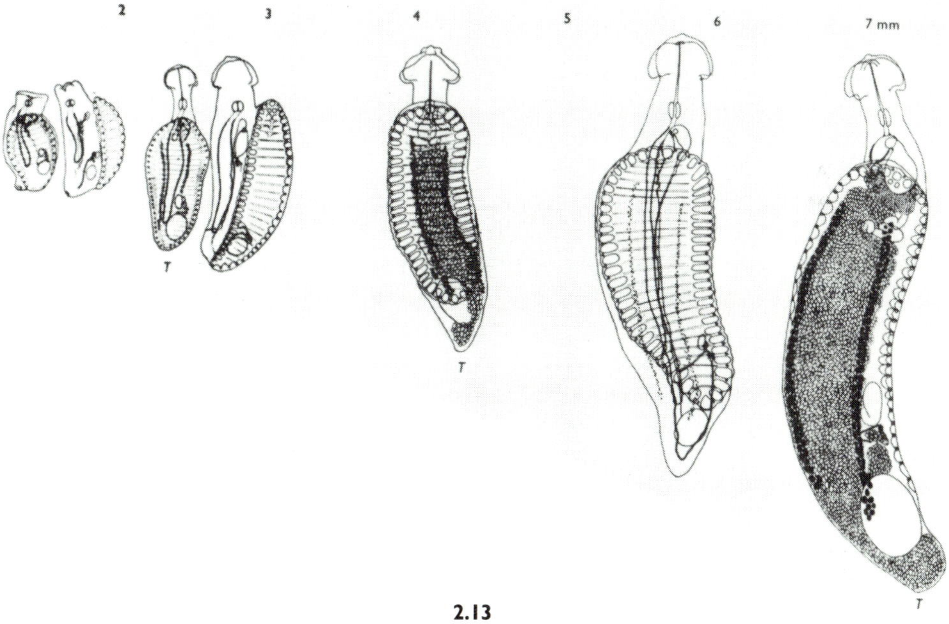

2.13

Figure 2.12 Composite drawing of cotylocidium of *Cotylogaster occidentalis*. C. concretion; CG, cephalic gland; GO, opening of gobletlike gland cells; F, flame cell; I, intestine; M mouth; OB, osmoregulatory bladder; OC, opening of cephalic gland; P, pharynx; S, sucker; T, tuft of cilia; U, uniciliate sensory structure. (From Fredericksen, 1978.)

Figure 2.13 Developmental stages of *Lobatostoma manteri* (*T* from the fish *Trachinotus blochi*, others from the snails *Cerithium moniliferum* and *Peristernia australiensis*). Numbers indicate length of worms. (From Rohde, 1973).

vertebrates, which acquire infection by eating molluscs. Rohde (1972) noted that some aspidogastreans parasitize a wide variety of molluscs, indicating a lack of host specificity.

Maturation of the cotylocidium has rarely been described. Rohde (1972, 1973, 1975) has reported it for *L. manteri* (Figure 2.13). The larvae first increase three or four times in size but otherwise show little change. At 0.5–0.6 mm length, the anterior part of the ventral sucker begins to divide into alveoli and rapidly grows forward as new alveoli are formed in the more posterior, undivided, area. The alveoli then grow larger. The rudiments of the reproductive system appear at 0.3–0.4 mm long, and at 0.6 mm the testis, sperm duct and cirrus pouch and the ovary, uterus and metraterm are represented by solid cords of cells (Rohde, 1973, 1975b).

Although some species achieve maturity in molluscs, most mature in vertebrates; species capable of maturing in both groups may do so more successfully in vertebrates. *Cotylogaster occidentalis* shows greater egg production and a greater percentage hatch in fish (*Aplodinotus grunniens*) than in molluscs (*Ligumia nasuta, Lampsilis radiata siliquoidea*) (Fredericksen, 1978) and probably achieves its maximum reproductive potential only in a fish. Cable (1974) found very few gravid *C. occidentalis* in molluscs over a 30-year period. At Heron Island on the Great Barrier Reef, *Lobatostoma manteri* matures only in *Trachinotus blochi*. Infections were not found in other fish in the same area and attempts to infect various fish and chelonians failed (Rohde, 1973). The main mollusc host, *Cerithium moniliferum*, is a major component of the diet of *T. blochi*. Hendrix and Overstreet (1977) regarded a fish host as certainly obligatory for *L. ringens*, reported from 14 species of teleosts, and probably obligatory for *Cotylogaster basiri*.

Digenea

Introduction

Of helminth life-cycles, those of digeneans are the most diverse and complex. They are remarkable in having a sequence of stages, in their molluscan primary intermediate hosts, which are usually morphologically distinct from each other and which reproduce nonsexually. The sexually reproducing adult stage normally occurs in the definitive, usually vertebrate, host. The interpretation of the nature of the intramolluscan stages and their mode of reproduction has been controversial and has been discussed in detail by Cable (1965, 1974, 1982), James and Bowers (1967), Ginetsinskaya (1968), Pearson (1972), Rohde (1972), Clark (1974), Gibson (1981, 1987) and many others. Most regard the cycle as an alternation of homologous, polymorphic generations which reproduce by parthenogenesis in the molluscan host and sexually as a single generation in the vertebrate definitive host (Cable, 1982). The basic sequence of generations, i.e. miracidium—mother sporocyst—daughter sporocyst or redia, and cercaria—metacercaria—adult, is highly variable. James and Bowers (1967) and Pearson (1972) have proposed a phylogeny of life-cycle patterns, bringing into consideration the origin of alternation of generations and the routes by which cycles involving two or more hosts may have developed from a primitive one-host cycle, and how some later became secondarily reduced.

The origin of the digenean life-cycle

There is consensus that the original hosts of the digeneans were the molluscs (e.g. Pearson, 1972; Cable, 1965, 1974; Rohde, 1972; Gibson, 1987; Shoop, 1988), which continue to be associated with digeneans as their primary hosts, to which they have a high degree of host specificity in virtually all known life-cycles.

The life-cycles of digeneans have been used as a basis for developing schemes of the evolution, phylogeny and systematics of the group, superseding adult morphology on which most early schemes were based. These subjects will be discussed in greater detail in Chapter 4 on host-parasite relationships, but an account of different ideas on the origin of the various stages in the life-cycle of digeneans is presented here.

Digeneans are generally agreed to have evolved in the sea from free-living dalyellioid-like rhabdocoel turbellarians which became parasitic in molluscs. Gibson (1987) has suggested that the original mollusc hosts were bivalves but prosobranch gastropods are favoured by the majority of authors. These had evolved and diversified by the Cambrian period, whereas freshwater prosobranchs did not appear until the Permian, and pulmonates and freshwater lamellibranchs not until the Cretaceous (Cable, 1974). Beyond this point, opinions conflict as to how and in what order the various life-cycle stages arose and which stage should be considered of most phylogenetic significance. Some authors (eg. Sinitsin, 1931; Cable, 1965, 1974; Pearson, 1972, and others) believe that the present adult developed from a mature free-swimming cercaria-like adult with a tail, which left the mollusc to lay eggs and, later in evolution, was ingested by fish, the earliest vertebrates, eventually colonizing them as hosts. In this case, the present cercaria would be a more reliable phylogenetic indicator than the present adult. Conversely, others (e.g. James and Bowers, 1967; Rohde, 1971, 1972; Gibson, 1981, 1987) reject the idea that the present cercaria is primitive, i.e. was once a sexually-reproducing adult, in favour of the hypothesis that it is a derived stage developed for transmission after the acquisition of the fish host. Rohde (1971, 1972) and Gibson (1981, 1987) believe that the vertebrate host acquired parasites by feeding on molluscs which contained adults already adapted for parasitism. In this case, the present adult is homologous with the original prototrematode adult and is, therefore, the most reliable source of phylogenetic information. In contrast, James and Bowers (1967) believed that the primitive life-cycle involved an alternation between one or more paedomorphic parthenogenetic and ovoviviparous generations in the mollusc with a single free-swimming ciliated, sexual and oviparous generation. The latter was later swallowed by and became a gut parasite of fish. In this case, the intramolluscan stages, the sporocysts and rediae, would be the most reliable source of phylogenetic information.

The views of Pearson (1972) and Gibson (1987) are outlined below. Both support the theory of an alternation of generations in the cycle, but differ on their origin and order of appearance.

Pearson (1972) postulates that digeneans arose from a dalyellioid-like rhabdocoel ancestor which became a visceral parasite of molluscs. A swimming larva penetrated the mollusc and grew to the adult stage; this protodigenean escaped from the mollusc, presumably as a free-swimming tailed adult, to lay and disperse its eggs. Pearson supports James and Bowers' (1967) view that alternation of generations then arose by the evolution of a viviparous adult generation (which developed into the proto-mother sporocyst), by regression of the somatic tissue together with the reproductive system, and the development of parthenogenesis and viviparity. The proto-miracidium was the larval stage of this generation. Pearson (1972) differs from James and Bowers (1967) in suggesting that the developing oviparous adults, retained within the body of the viviparous generation, came to represent protocercariae which, when mature, left the mollusc to lay eggs. He considers retention of the developing oviparous generation within the viviparous generation to be a late development. The oviparous adult was presumed to be eaten by a fish in which it persisted and continued to lay eggs. Subsequently, sexual maturation occurred only after the invasion of the vertebrate host, which became obligatory. The miracidium and mother sporocyst lost the gut. The redial generation, which has a gut, was derived from the oviparous adult, i.e. the protocercaria and later, in some advanced forms, gave rise to daughter sporocysts by regression of the gut. Both fed through the tegument. Finally, an encysted metacercarial stage, an adaptation to prolong the infective life of the cercaria, was achieved, perhaps by secretion by the cercaria of a protective covering like that produced by free-living rhabdocoels in adverse conditions (Pearson, 1972).

The site of encystment, whether in the open or in another animal, was related to the feeding habits of the definitive host and when cercariae penetrated and encysted in another animal, a second intermediary was incorporated into the life-cycle. Using echinostomes as an example, Pearson (1972) suggested a sequence beginning with encystment in the open, then on an animal's shell or carapace, progressing through invasion of a natural orifice followed by encystment to penetration of and encystment in the tissues. Ginetsinskaya (1968) had made a similar suggestion. A tendency towards growth and development of metacercariae in the second intermediaries accelerated a switch in favour of production of a greater number of smaller, less advanced cercariae and smaller adults.

Precocious sexual development in intermediate hosts, discussed in more detail below, is a feature of many digenean life-cycles which often results in cycles which are telescoped by comparison with the classical or typical cycle.

In Gibson's (1987) view, a proto-digenean parasitic as an adult in a mollusc, probably a bivalve, was ingested within its host by a fish and survived in the gut. Later, it laid eggs in the fish and ultimately became incapable of egg production in the mollusc, at which point the vertebrate host became obligatory. A viviparous generation next developed in the mollusc to compensate for losses in transmission incurred by natural wastage, i.e. those infected molluscs which were not eaten by fish. The viviparous stage moved into the tissues of the mollusc, lost its adult characters by paedomorphosis and became parthenogenetic, i.e. became the precursor of the mother sporocyst with the miracidium as its larval stage. The offspring of the proto-mother sporocyst remained in the mollusc as a proto-metacercaria, presumably encysted. Immature proto-metacercariae which left the mollusc before encysting developed into free-living proto-cercariae with a tail to aid transmission. Proto-cercariae developed associations with other invertebrates, thus incorporating second intermediaries into the cycle, or were eaten directly by the vertebrate host. A second parthenogenetic generation evolved, again to combat wastage during transmission, by the repetition of the proto-mother sporocyst stage, probably before the latter lost the gut. This immensely increased the numbers of proto-cercariae produced. In Gibson's opinion, this stage, the redia, is the least paedomorphic and most recently evolved generation, bearing a close resemblance to the dalyellioid ancestors. Daughter sporocysts and rediae are homologous, the former being derived from rediae by the loss of the gut. This had been suggested by, among others, Popiel and James (1978) and is supported by the work of Matthews (1980) on the hemiurid *Lecithochirium furcolabiatum*. This species is unusual in the hemiurids in that cercariae are produced not in rediae but in daughter sporocysts. However, the daughter sporocysts have a birth canal considered homologous to the gut of rediae, suggesting that the daughter sporocysts of this species correspond to the rediae of other hemiurids and supporting the view that daughter sporocysts originated from rediae.

The two schemes have several points of similarity but differ in that Pearson (1972) postulates that the proto-digenean adult was cercaria-like, emerging from the mollusc to lay eggs, that alternation of generations preceded acquisition of a vertebrate host and that this host was acquired when it accidentally ingested free-living oviparous adults for which it eventually became an obligate host. Gibson (1987) holds that the original adult was not cercaria-like, that it was a parasite of molluscs, and that the vertebrate host was invaded because it consumed infected molluscs. Alternation of generations arose subsequently. In both schemes, the proto-mother sporocyst is seen as the first intramolluscan generation with the miracidium as its larval stage. In Gibson's scheme, proto-cercariae were derived from proto-metacercariae which left the mollusc before encysting and became associated with other molluscs or other invertebrates, i.e. achieving a three-host life-cycle. Pearson (1972) regarded the metacercaria as an adaptation to prolong the infective life of the cercaria in a two-host cycle, with the subsequent interpolation of a second intermediary host. The redia-daughter sporocyst generation was derived from the cercaria in Pearson's scheme and by repetition of the proto-mother sporocyst in that of Gibson. The redia and daughter sporocyst appear to be mutually exclusive, as they do not usually occur together in known life-cycles. Pearson (1972), however, regarded daughter sporocysts as simplified rediae and has pointed out (personal communication) that, in two lepocreadiid life-cycles described by Watson (1984), the mother sporocyst is suppressed and daughter sporocysts give rise to rediae.

Other recent views include those of Clark (1974), who argued against the concepts of alternation of generations and polyembryony and in favour of derivation of the larval stages by budding rather than by parthenogenesis, resulting in a sequence of larval forms linked by metamorphosis and regenerative multiplication.

Brooks, O'Grady and Glenn (1985a, b) and O'Grady (1985) applied cladistic methods and concluded that digenean life-cycles are derived by the intercalation or non-terminal addition of ontogenetic stages and the terminal addition of a vertebrate host to the ancestral life-cycle.

Suggestions very similar to those of Gibson (1987) have been advanced by Shoop (1988) in a review of trematode transmission patterns. He derived digenean life-cycles from those of aspidogastreans by means of a vertebrate host entering the cycle by ingesting infected molluscs. He considered each digenean larval stage to represent a viviparous generation equivalent to the single generation in the aspidogastrean cycle. Each generation was initially capable of producing free-living transmission stages that left the mollusc. After the adoption of a vertebrate host, with consequent improved dissemination of the trematode, viviparity replaced oviparity in all but the last generation (alternation of viviparous and oviparous generations had already been advanced by James and Bowers, 1967). The proto-mother sporocyst simplified to become an incubator for cercariae and transmission to the vertebrate host was centred in the cercaria-adult generation. A redia-daughter sporocyst generation was added to increase cercarial production and thus offset wastage among free-living cercariae. Shoop derived the metacercarial stage from the cercaria and derived three- and four-host cycles by encystment on and in invertebrates in a manner similar to that of Pearson (1972).

Transmission at various stages of the life-cycle

Transmission strategies in the digeneans are illustrated in Figure. 2.14 based on Shoop (1988).

In contrast to other helminth groups, the digeneans invade their hosts by active as well as passive means. The miracidium actively penetrates the body surface or gut wall of the mollusc in all cases (except notocotylids) and cercarial invasion of the second intermediary in three-host cycles or the definitive host in some two-host cycles is often by active penetration of the body surface.

The mollusc hosts of most digeneans are prosobranch gastropods but bivalves are used by a minority of families, including the Gorgoderidae, most Fellodistomidae, some Monorchiidae, most Allocreadiidae, the Bucephalidae and some Sanguinicolidae, both marine and freshwater.

As detailed descriptions of the morphology of digenean larvae can be located by reference to papers cited by Yamaguti (1975), this account is limited to the features of the miracidia and cercariae significant in transmission.

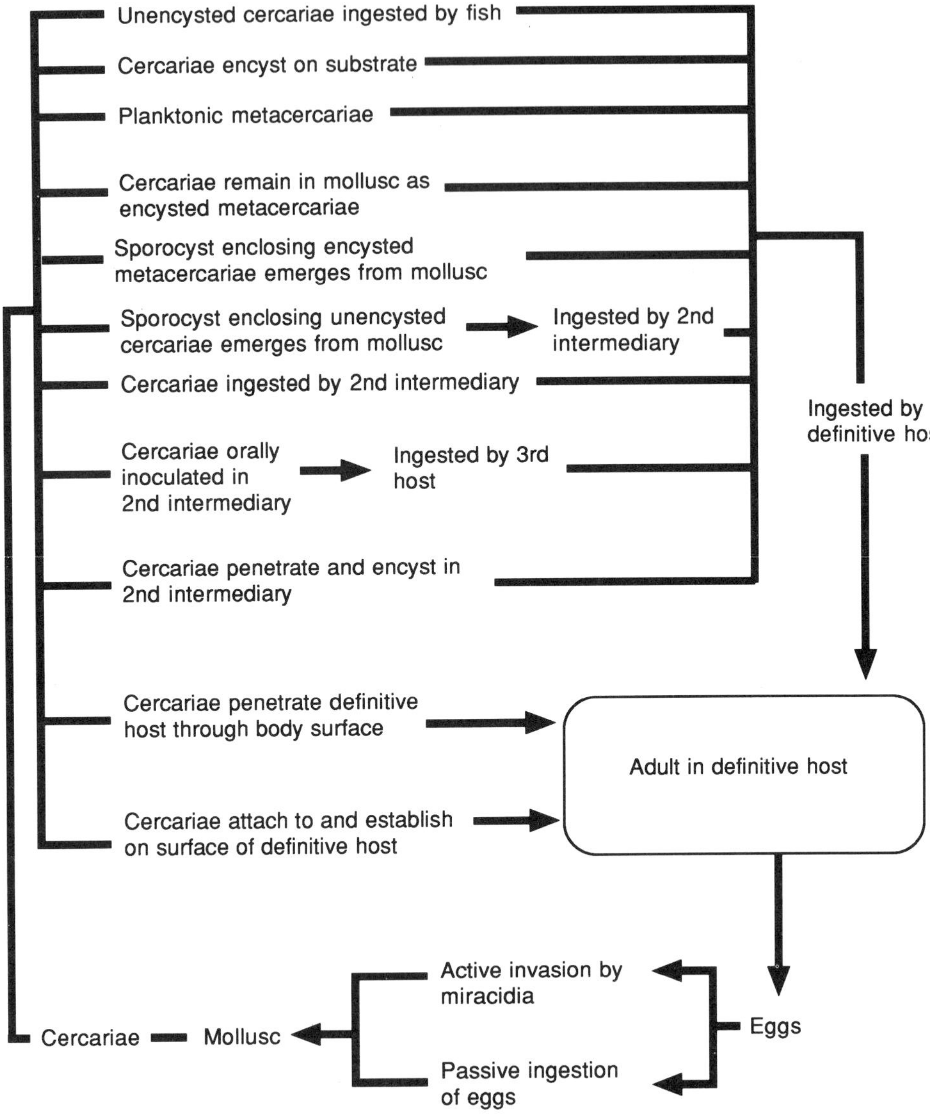

2.14

Figure 2.14. The life cycle patterns of digeneans maturing in fishes, with particular relevance to cercarial behaviour (based on Shoop, 1988).

Morphology of the miracidium (Figure 2.15.1–12)

The morphology of the miracidia of representatives of various families is illustrated in Figure 2.15.1–12. The majority are covered in ciliated epidermal plates arranged in tiers but some, the azygiids, ptychogonimids, some hemiuroids, didymozoids and fellodistomes, have bristles arranged in groups or rows. As the miracidium is considered a morphologically stable stage, this is taken to indicate relationships between the families in which they occur. Some hemiuroid miracidia are ciliated, however, e.g. *Lecithaster salmonis* (Lecithasteridae) (see Schell, 1975) and *Genarchopsis goppo* (Derogenidae) which also has an apical crown of spines (Madhavi, 1978b). The first description of a hemiurid miracidium, *Lecithochirium furcolabiatum*, has shown that it has no bristles but groups of long cilia (Matthews and Matthews, 1988). Some bucephalid miracidia have cilia borne on cephalic plates and caudal appendages. Members of the Bucephalinae, Paurorhynchinae and Sanguinicolidae have stylets.

Ocellate miracidia occur in various families including bivesiculids, sanguinicolids, zoogonids, allocreadiids and lepocreadiids. Most miracidia contain a granular apical gland and at least one pair of penetration glands, all of which open apically, a pair of flame cells (two pairs in strigeoids, clinostomes and schistosomatoids) and germinal cells. After the miracidium has entered the mollusc and metamorphoses into the mother sporocyst, the germinal cells give rise to the next generation, the redia or daughter sporocyst. Some departures from this pattern have been reported. Watson (1984) found that the mother sporocyst in the lepocreadiids *Tetracerasta* and *Stegodexamene* was suppressed and that the larval stage delivered to the snail was a daughter sporocyst which gave rise to rediae. Madhavi (1976, 1978a) has reported that the miracidium of *Allocreadium fasciatusi* contains a redia.

Hatching of the egg and invasion of the mollusc host

Miracidia invade molluscs by penetrating the integument after the eggs have hatched in water to liberate free-swimming miracidia, or by penetrating the gut following ingestion of eggs by the mollusc. In the first case, eggs may either be non-embryonated when laid and hatch only after a period of development, the length of which is temperature-dependent, or they may have embryonated *in utero* and hatch rapidly, within minutes or even seconds, on contact with water. Most digeneans are in the first category. Examples of the second category, with eggs embryonated when laid, are the hemiurids, cryptogonimids, acanthostomids, plagiorchiids, renicolids (which use fish as second intermediate hosts), the gorgoderids and various members of other families, e.g. Bivesiculidae (*Paucivitellosus fragilis*; Pearson, 1968), Bucephalidae (*Rhipidocotyle illense*; Woodhead, 1929) and some zoogonids, haploporids, waretrematids and haplosplanchnids. Pearson (1968) has suggested that rapid hatching is an adaptation to transmission in the intertidal zone but it has also developed in freshwater forms in a more constant environment.

Ingestion of embryonated eggs by molluscs occurs in representatives of several families, e.g. the azygiids (*Azygia acuminata, A. longa*), monorchiids (*Asymphylodora amnicola, Monorcheides cumingiae, Telolecithus pugetensis*), some acanthostomids (*Acanthostomum imbutiforme, Timoniella praeteritum* (see Maillard, 1976)), some hemiuroids (Derogenidae—*Genarchopsis goppo* (see Madhavi, 1978b) and *Genarchella genarchella*; Lecithasteridae—*Pseudodichadena lobata*), some cryptogonimids (*Aphalloides coelomicola*; see Maillard, 1976; *Caecincola parvulus*), homalometrids (*Homalometron pallidum*), opecoelids (*Cainocreadium labracis*; see Maillard, 1976) and some macroderoidids (*Macroderoides spinifer* and *Paramacroderoides echinus*). All references to the above examples are cited by Yamaguti (1975) unless otherwise stated.

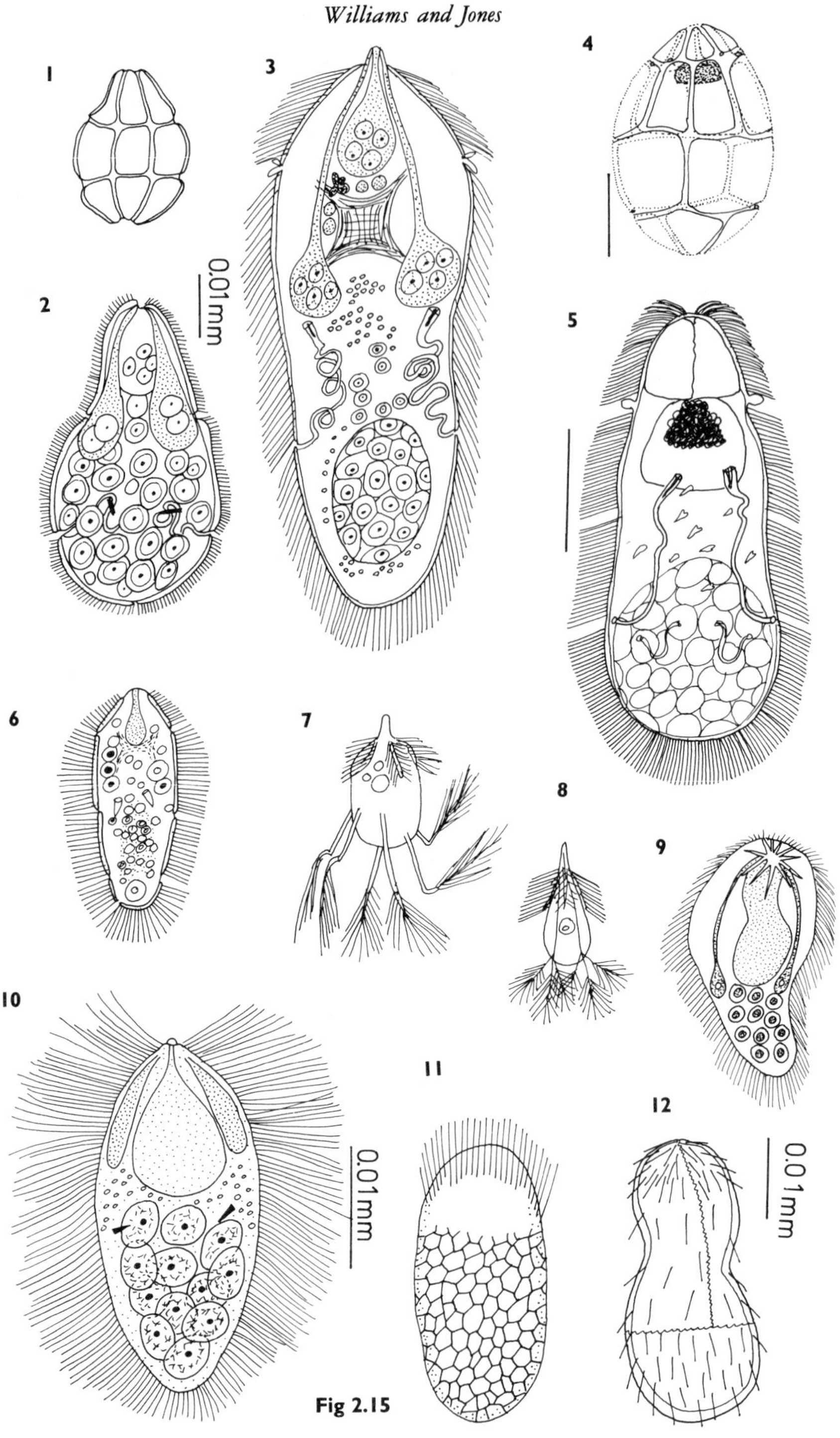

Figure 2.15. Digenea: miracidia of various families.
1 and 2. *Phyllodistomum staffordi* (Gorgoderidae), epidermal plates and miracidium; after Schell (1967). 3. *Allocreadium handiai* (Allocreadiidae); after Madhavi (1980). 4 and 5. *Tetracerasta blepta* (Lepocreadiidae); epidermal plates and miracidium; after Watson (1984). 6. *Plagioporus sinitzini* (Opecoelidae); after Dobrovolny (1939). 7. *Rhipidocotyle papillosa* (Bucephalidae); after Woodhead (1929). 8. *Bucephalus elegans* (Bucephalidae); after Woodhead (1930). 9. *Genarchopsis goppo* (Derogenidae); after Madhavi (1978b). 10. *Lecithaster salmonis* (Lecithasteridae); after Schell (1975). 11. *Ptychogonimus megastoma* (Ptychogonimidae); after Palombi (1955). 12. *Nematobothrium texomensis* (Didymozoidae); after Self *et al.* (1963).

Examples of eggs hatching in unusual sites are provided by the bloodflukes and by many digeneans with telescoped life-cycles. Sanguinicolid eggs lodge in the gill filaments of the host where they embryonate and hatch, the miracidia escaping through the host tissue into the water (Smith, 1972). In species with markedly precocious cercariae or metacercariae, miracidia may hatch out in the germinal sacs or the cyst, e.g. miracidia of the azygiid *Proterometra dickermani* hatch in the redia, telescoping the life-cycle to a single mollusc host (Anderson and Anderson, 1963); some miracidia reinfect the same individual snail, others escape to infect other snails. Occasionally, rediae are extruded from the genital aperture of the snail and disintegrate, releasing their contents. Possibly, a cercaria is occasionally eaten by a fish but this would not be an essential part of the life-cycle. Miracidia of *Coitocaecum anaspidis*, an opecoelid, hatch in metacercarial cysts within the amphipod second intermediary (Hickman, 1934).

The hatching process

There are very few descriptions, for fish helminths, of the processes of hatching and penetration of the mollusc host, but Watson (1984) has recently described both for miracidia of the lepocreadiids *Tetracerasta blepta* (Figure 2.16.1–6) and *Stegodexamene callista*. Slow stretching and contracting movements during the hours preceding hatching are succeeded in the last hour by exertion of pressure at about 30-second intervals on the operculum, accompanied by ciliary activity and phases of rapid expansion and contraction. Flame cell activity also increases before hatching. Eventually the operculum opens and the miracidium is partly ejected, temporarily restrained within the vitelline membrane which does not pass completely out of the shell. It soon pierces this and escapes. Whilst probing the operculum, the miracidium may be releasing a secretion to weaken the opercular seal, or damaging the vitelline membrane to allow water in, or both. Increased flame cell activity suggests that osmosis is taking place and that an increase in osmotic pressure is responsible for ejection of the miracidium. At the same time, vitelline remains in the shell swell and burst, also indicative of osmosis. These species lack the viscous cushion observed in some other trematodes. The miracidia are positively phototactic and appear to respond at close quarters to a chemical attractant in snail mucus. On touching the snail, the anterior end of the miracidium becomes cone-like and penetrates slightly into the tissue. The sporocyst becomes active and eventually passes into the snail's tissues. Cannon (1971) has described a similar process in *Bunodera* miracidia, in which entry of the sporocyst into the snail is preceded by secretion of adhesive droplets by the apical gland and extrusion of the contents of the lateral glands into the snail's tissue. The miracidium then burrows into this secretion. Miracidia of *Bunodera mediovitellata* show a negative photoresponse and a positive georesponse, which would help in location of the snail host, the bottom-dwelling fingernail clam, *Pisidium casertanum* (see Kennedy, 1979).

Development within the mollusc

The miracidium transforms into the mother sporocyst which produces rediae or daughter sporocysts. One or more such generations may occur before the production of cercariae.

Escape from the mollusc host

In most instances, cercariae escape from the redia (through the birth pore) or daughter sporocyst into the tissues or haemocoel of the mollusc but little is known about how they escape from the mollusc. The process must be an active one. Forms with slender locomo-

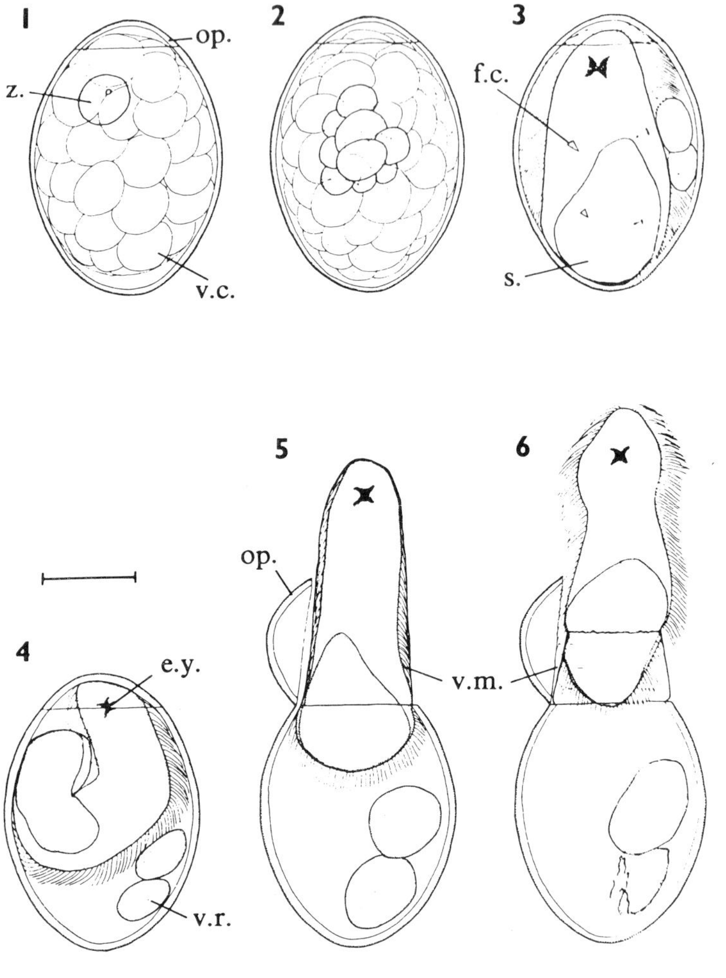

Figure. 2.16. Development and hatching of the eggs of *Tetracerasta blepta*.
1. Miracidium soon after touching snail; 2. Egg with developing miracidial embryo; 3. Egg with fully developed, active miracidium; 4. Miracidium pressing on egg operculum; 5. Miracidium ejected within vitelline membrane; 6. Miracidium piercing vitelline membrane.
Scale lines, 25 µm. From Watson (1984)

tory tails are known to swim or crawl through tissues or blood vessels to the mantle cavity (various authors in Matthews, 1982). Gorgoderine cercariae use up the contents of a pair of gland cells in the process (Fischthal, 1951).

The escape of large cercariae, e.g. those with cystophorous or cystocercous tails, is more difficult as they could not easily migrate to the exterior. The site and method of release of cercariae from the parental generation are probably related to the size and motility of the cercariae and facilitate their escape from the mollusc host (e.g. Pearson, 1972; Matthews, 1980, 1982). Forms which do not retract into their tails or do not do so until in water (bivesiculids, some gorgoderids, most azygiids) escape by active body movements. Others, e.g. the hemiuroid *Lecithochirium furcolabiatum* (described by Matthews, 1980, 1982) are helped out by the daughter sporocyst which projects its anterior end, and birth pore, into the mantle cavity. The filamentous unbranched daughter sporocysts are located in the haemocoel of the digestive gland of *Gibbula umbilicalis* but have a specialized anterior end which extends through the afferent renal vein, right kidney haemocoel and transverse pallial vein into a blood channel inside a gill filament. Cercariae progress towards the birth pore by a combination of body movement and peristaltic activity of the sporocyst wall, dragging their tails. The anterior end penetrates through the epithelium of the gill filament and cercariae emerge through the birth pore onto the gill surface. Within seconds, the body retracts into the cystophorous tail and the cercaria leaves the mantle cavity with the exhalent water current.

A direct escape route occurs in the bucephalid *Bucephaloides gracilescens*, the cercaria of which has a large tail complex (Matthews, 1974). The wall of the daughter sporocyst ruptures into the exhalent chamber of the host, *Abra alba*, allowing the cercariae to escape. In this case, there is no specialization of the sporocyst.

Daughter sporocysts of gorgoderids are almost ectoparasitic outside the tissues of the mollusc, which may be an adaptation to aid the escape of large cercariae (Pearson, 1972). Mother sporocysts of *Phyllodistomum staffordi* (Figure 2.17.1–9) between the gill bars of *Musculium ryckholti*, are at first enveloped by host tissue but later, as they grow, break through it and establish on the inner surface of the gill lamellae, held in position by strands of connective tissue. They rupture, releasing daughter sporocysts which are only loosely attached to the gill lamellae, not enclosed in a cellular envelope (Schell, 1967).

The daughter sporocyst of *Phyllodistomum folium*, enclosing encysted cercariae, is itself discharged from the gill chamber, floats by reason of its high lipid content and is eaten by the fish host (Sinitsin, 1901). This also happens in the ptychogonimid *Ptychogonimus megastomus* (Figure 2.18) but, in this case, the daughter sporocyst contains mature but unencysted cercariae and is eaten by crustacean intermediaries in which the metacercariae develop (Palombi, 1941, 1942a, b). Daughter sporocysts of the opecoeliid *Plagioporus sinitzini*, containing encysted cercariae, emerge from the rectum of the snail to be eaten by fish (Dobrovolny, 1939).

In representatives of several families, the cercariae leave the germinal sac (usually the redia) while still immature to complete their development in the tissues of the snail. Pearson (1972) suggests that this primitive habit may have been retained to facilitate the escape of large-tailed cercariae from daughter sporocysts or rediae. The birth of cercarial embryos into the haemocoel of the snail occurs in some homalometrids, azygiids, transversotrematids, waretrematids, zoogonids, acanthocolpids, allocreadiids, lepocreadiids and acanthostomids (Table 2.3) and also in paramphistomids, notocotyloids and opisthorchioids.

Matthews (1982) described the first instance of a redia involved in transporting cercariae to the point of emergence from the snail. Rediae of the hemiuroid *Cercaria calliostomae* Dollfus, 1923 (probably a hemiurid) migrate from the gonad haemocoel of *Calliostoma zizi-*

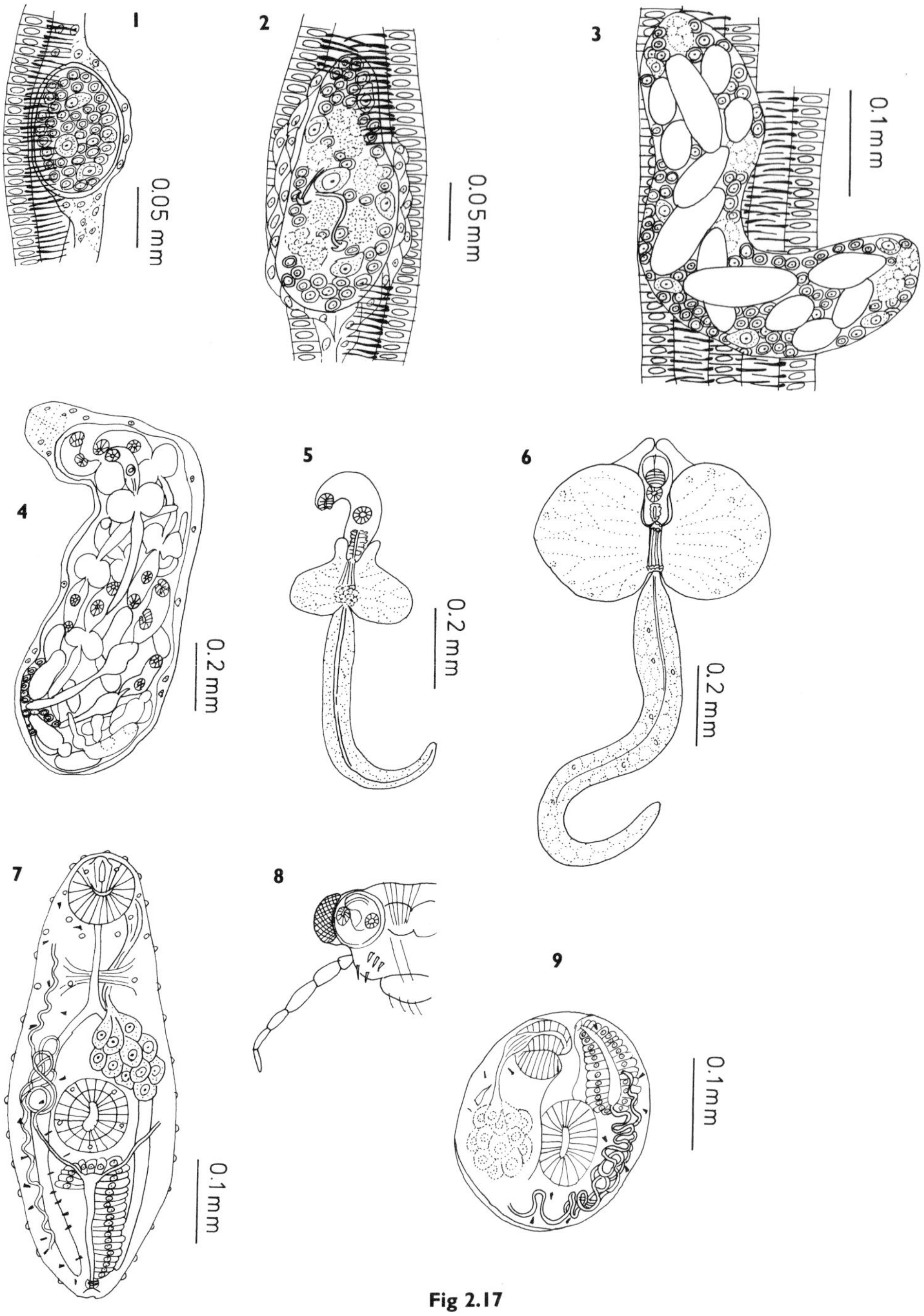

Fig 2.17

Figure. 2.17. Stages in the life-history of ***Phyllodistomum staffordi*** (Gorgoderidae) (after Schell, 1967)
1. Mother sporocyst, 4 days old. 2. Mother sporocyst, 7 days old. 3. Mother sporocyst, 12 days old. 4. Daughter sporocyst, 42 days old. 5. Late developmental stage of cercaria. 6. Fully-developed cystocercous cercaria, ventral view. 7. Ventral view of body of cercaria. 8. Encysted metacercaria in head of damsel fly naiad. 9. Encysted metacercaria.

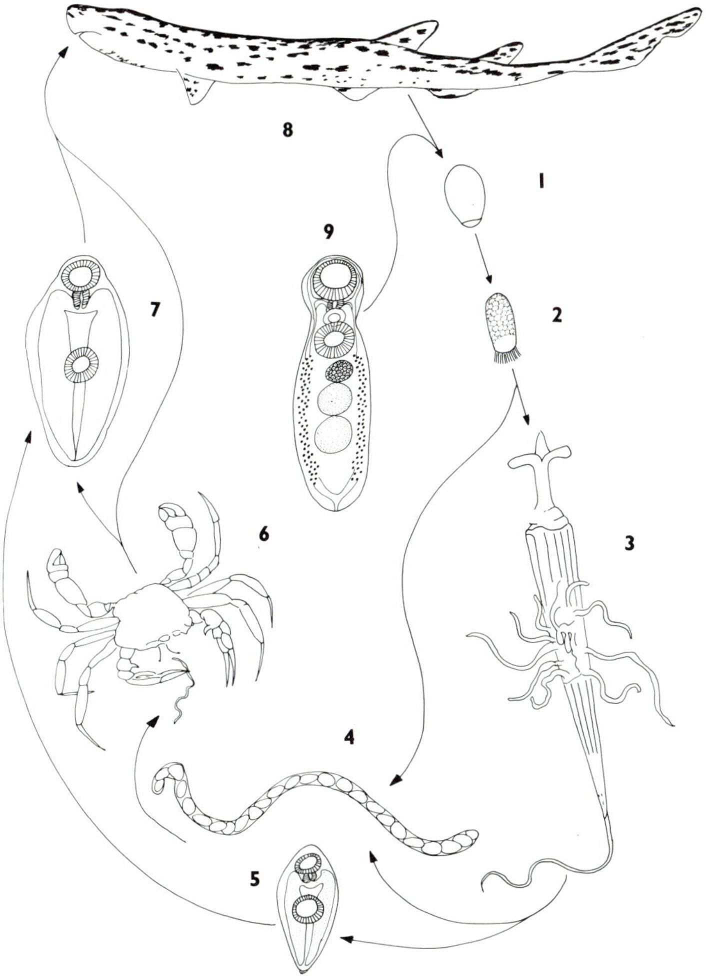

2.18

Figure. 2.18 Life-cycle stages of *Ptychogonimus megastomus*. From Palombi (1942b).

phinum through the afferent renal vein to the haemocoel of the right kidney. There the pharynx, probably aided by pharyngeal gland secretions, disrupts the kidney tubule wall. The anterior end, with the birth pore, protrudes through the wall and the cystophorous cercariae emerge through the kidney sac into the mantle cavity. No free cercariae were found in the tissues of the snail.

Invasion of the second intermediary and the definitive host

The morphology of cercariae of various families is illustrated in Figures 2.19–2.22.

Having emerged from the mollusc, cercariae may infest the final host directly, by being ingested into the gut lumen or by direct penetration of the body surface, or by means of a metacercarial stage which may be encysted on the substrate or occurs, encysted or not, within the body of a second intermediate host which will be ingested by the fish host. Invasions of the fish by means of a second intermediary constitutes an indirect invasion. Molluscs, annelids, arthropods, fish and even occasionally amphibian tadpoles, serve as second intermediaries in the life-cycles of various fish digeneans. Entrance into the gut lumen of a second intermediary is accomplished passively with ingested food but entry into its tissues is by active penetration. Patterns of cercarial behaviour related to the infection of intermediary and definitive hosts are illustrated in Figure 2.14.

Pearson (personal communication) believes that the metacercaria is essentially the juvenile of the adult stage, infective for the definitive host in which it reaches sexual maturity, and is basically a resting stage, which prolongs its infective life. In his opinion, metacercariae are distinguished from cercariae by their ability to secrete a cyst, or the acquisition of a second intermediary in the life-cycle without prior acquisition of a cyst, e.g. fellodistomes. The ability to form a cyst is believed primitively absent in azygioids and transversotrematids (Pearson, 1968) and to have been secondarily lost in some groups. The metacercarial stage is primitively lacking in bivesiculids, azygiids and transversotrematids and may be secondarily lost in schistosomes and some hemiuroids.

Precocious sexual maturity achieved in what was originally an intermediate host is a feature of many digenean life-cycles. The term progenesis has usually been applied to this phenomenon in digeneans but its use is controversial. Pearson (personal communication) prefers the term precocious, and believes that a sexually mature form, in whichever host it occurs, is the adult and is only progenetic (= precocious) in the sense that maturity has been achieved earlier than in classical or typical life-cycles. Buttner (1951a,b,c) has pointed out that progenetic digeneans are mature in both morphological and physiological terms. The term precocious has been adopted here for digeneans on the grounds that its general meaning is well-known, i.e. the early development of a feature or faculty (Pearson, personal communication).

Cercariae of bivesiculids, some fellodistomes and azygiids are eaten by fishes, which may reflect the means by which vertebrates originally acquired digenean infections; the cercariae are often conspicuous to fish by reason of their large size, clumsy swimming habits or pigmentation or all three. Azygiid cercariae, among the largest cercariae known, are usually pigmented and swim very slowly. The habits of cercariae are related to the feeding habits of the host, e.g. some bivesiculid cercariae (*Bivesicula* spp.) are planktonic and are ingested by planktonivorous fish. Another bivesiculid cercaria, *Paucivitellosus fragilis* (Figure 2.19.2), attaches to the substrate by means of glandular secretions. It is eaten by a fish which grazes on rocky surfaces (Pearson, 1968). Attachment to the substrate occurs in *Azygia sebago* (Figure 2.19.5), and in various cotylomicrocercous cercariae which penetrate bottom-dwelling intermediaries through the exoskeleton. Planktonic cercariae also occur in tergestine fellodistomes and in hemiuroids which are swallowed by intermediate hosts.

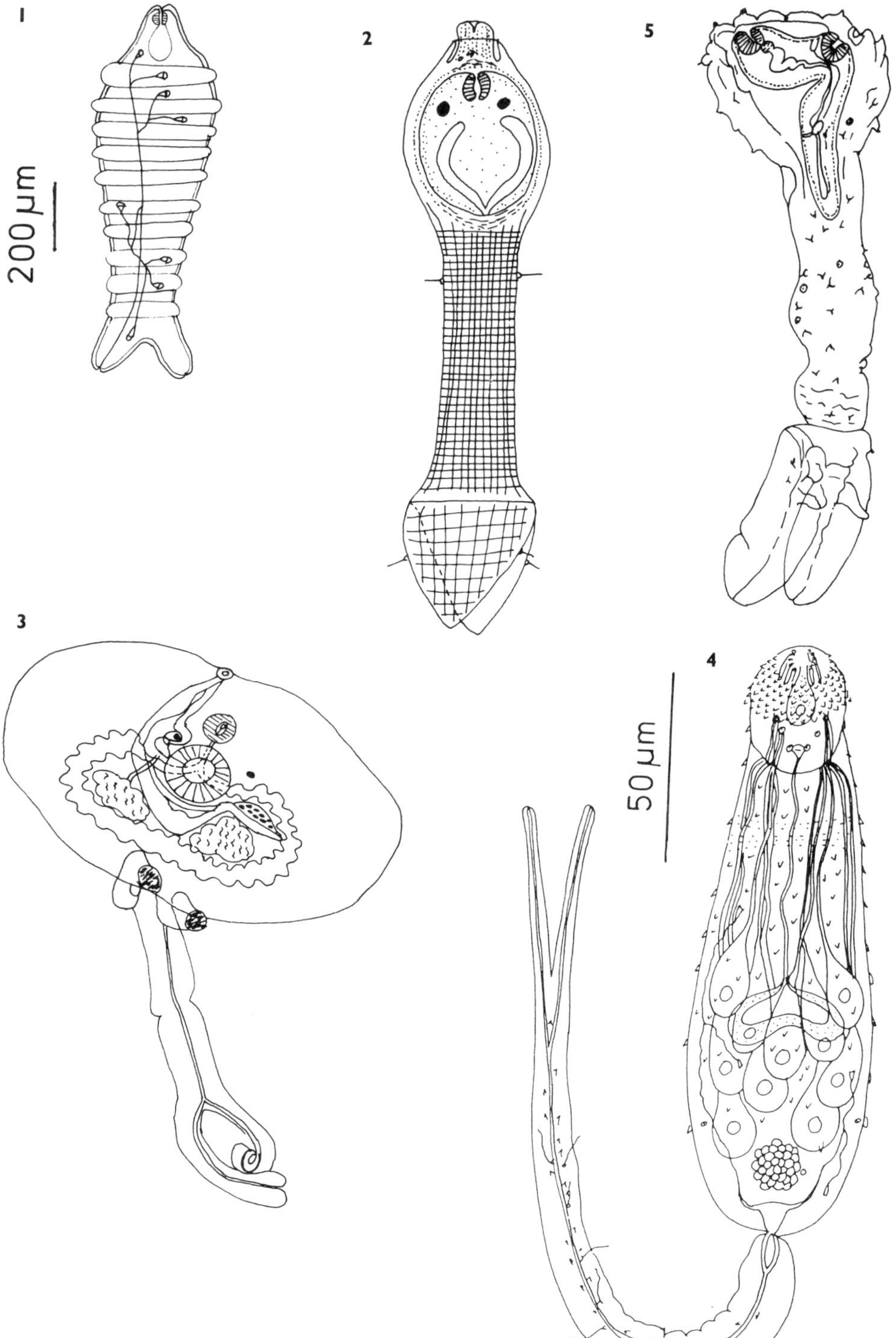

Fig 2.19

Figure 2.19.1–5 Digenean cercariae
1, 2 *Paucivitellosus fragilis* (Bivesiculidae) redia and cercaria; after Pearson (1968). 3 *Transversotrema laruei* cercaria (Transversotrematidae); after Velasquez (1961). 4 *Aporocotyle simplex* cercaria (Sanguinicolidae); after Køie (1982). 5 *Azygia sebago* cercaria (Azygiidae); after Stunkard (1956).

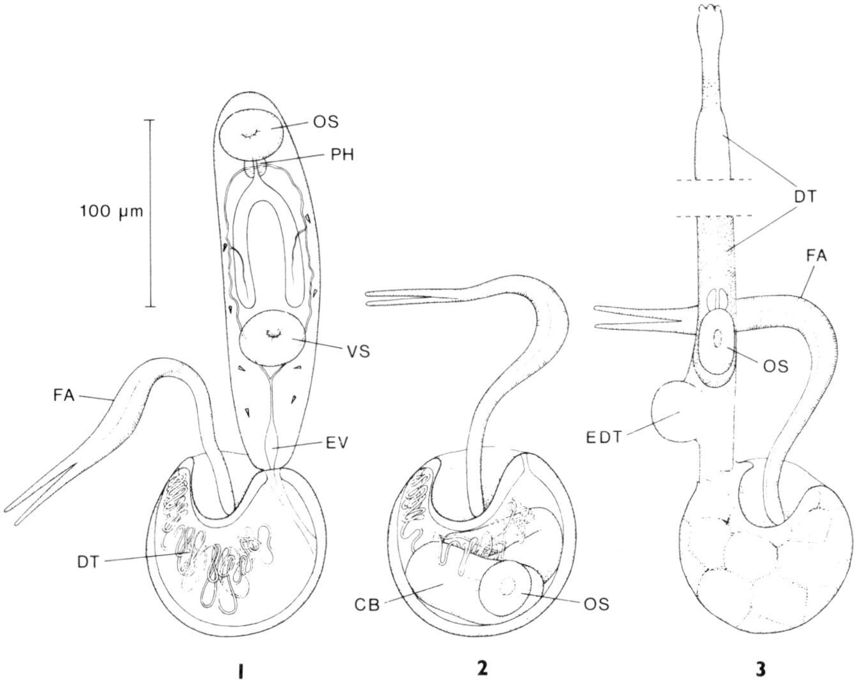

Fig 2.20

Figure. 2.20 Digenean cercariae. *Derogenes varicus* (Derogenidae).
1. Intraredial cercaria of *Derogenes varicus* with withdrawn delivery tube. 2. Fully developed free-swimming cercaria. 3. Cercaria with evaginated delivery tube. CB, cercarial body; DT, delivery tube; EDT, extension on delivery tube; EV, excretory vesicle; FA, furcate appendage; OS, oral sucker; PH, pharynx; VS, ventral sucker. After Køie (1979b).

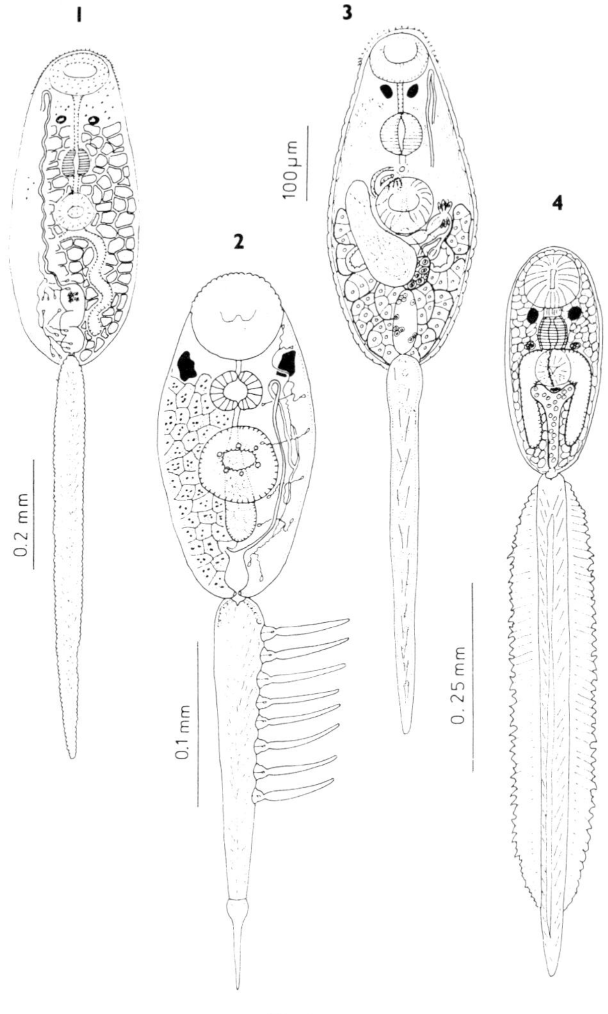

Fig 2.21

Figure 2.21 Digenean cercariae

1. *Hapladena varia* cercaria (Waretrematidae); after Cable (1962) 2. *Schikhobalotrema acutum* cercaria (Haplosplanchnidae); after Cable (1954a) 3. *Saccocoelioides sogandaresi* cercaria (Haploporidae); after Cable and Isseroff (1969) 4. *Megapera gyrina* cercaria (Megaperidae); After Cable (1954b).

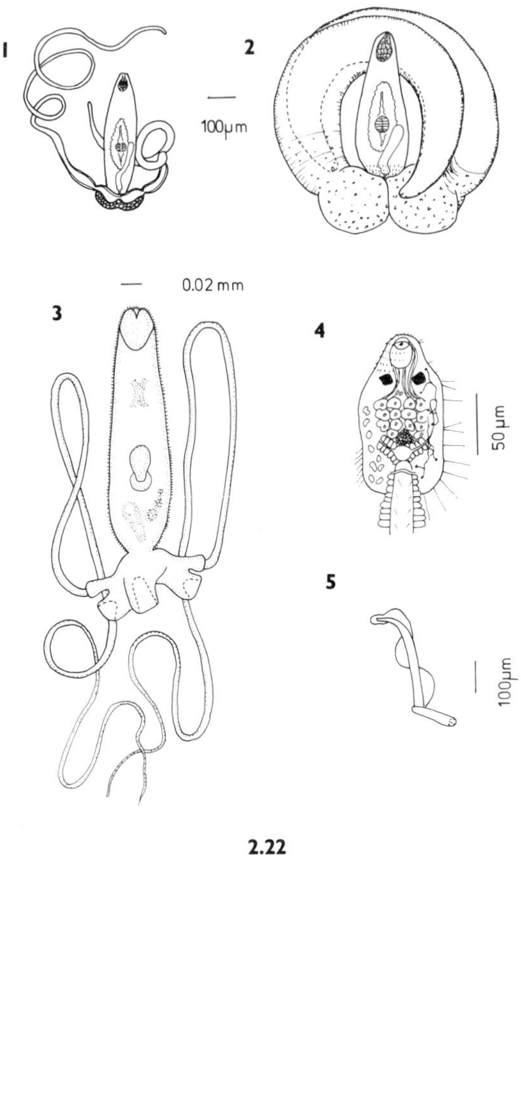

Figure 2.22 Cercariae of the Bucephalidae and Cryptogonimidae.
1–3 Cercariae of the family Bucephalidae (gasterostome type). 1 *Bucephalus polymorphus* (after Baturo, 1977) 2 *Rhipidocotyle illense* (after Baturo, 1977) 3 *Prosorhynchus squamatus* (after Matthews, 1973) 4, 5 Cercaria of the family Cryptogonimidae. 4 *Caecincola latostoma*, ventral view of body 5 Side view of cercaria (after Greer and Corkum, 1979).

Cercariae ingested by the definitive or second intermediary host have developed morphological adaptations to protect the body of the cercaria during ingestion. It withdraws into the expanded base of the tail. Two conditions are known: cystocercous and cystophorous. The body of a cystocercous cercaria retracts into a simple cavity in the tail from which it escapes actively following ingestion by the next host in the life-cycle, e.g. the gorgoderids (ingested by the second intermediate host) and the bivesiculids and azygiids (ingested by the definitive host). Cystophorous cercariae occur in the hemiuroids: the cercarial body retracts into a cavity in the tail containing a delivery tube through which the body is forcibly expelled when the cercaria is ingested by the second intermediate host. The cercarial body is inactive during this process. At this stage, the cercarial body is at a relatively early stage in its development.

In hemiuroids, e.g. *Derogenes varicus* (Figures 2.20.1–3), eaten by crustacean second intermediaries, the cercarial cyst contains a long, coiled delivery tube which is everted in response to pressure exerted on the cyst by the copepod's mouthparts. It pierces the gut wall and the cercaria, usually adopting a filiform shape, is shot through it into the haemocoel (Køie, 1979b). A similar mechanism has been described in a species of *Bunocotyle* by Chabaud and Biguet (1959) and in *Lecithochirium furcolabiatum* by Matthews (1981a, b). In contrast to other families which have crustacean intermediaries, hemiuroid cercariae lack penetration and cystogenous glands, and the delivery tube is an alternative means of entry to the haemocoel. These glands are also lacking in cercariae eaten by the definitive host; cystogenous glands make their appearance with the inclusion of a metacercarial stage and penetration glands with the acquisition and invasion of a second intermediary. Pearson (1972) has speculated on their origin.

Two-host cycles with an encysted metacercaria occur in the families Waretrematidae, Haploporidae, Haplosplanchnidae, Megaperidae, Mesometridae and Paramphistomidae (some of which occur in fish). Waretrematid metacercariae are planktonic and those of the other families encyst on vegetation eaten by their herbivorous hosts. The morphology of cercariae of these families is illustrated in Figure 2.21. The mode of encystment in *Mesometra orbicularis* is of interest because its long glandular tail encircles the body and secretes cystogenous products (Palombi, 1937). The cercariae of species which infect fish directly or by a metacercaria encysted in the open are sometimes sexually advanced.

The two-host cycles of transversotrematids and sanguinicolids also involve direct invasion by cercariae. Transversotrematids are a phylogenetically isolated group (Cable, 1974), ectoparasitic beneath fish scales but not within the epidermis, e.g. *Transversotrema laruei* (Figure 2.19.3), with furcocercous, markedly precocious cercariae which invade the host possibly by utilizing the respiratory current to reach the gill chamber, attach by means of paired adhesive appendages, shed the tail and migrate to the final location (Velasquez, 1961). Cercariae of *T. patialense* contact the host by chance but show a strong positive response on making contact (Whitfield, Anderson and Moloney, 1975). Transversotrematid cercariae have, arising at the anterior end of the tail, a pair of processes (illustrated for *T. laruei* in Figure 2.19.3), the ends of which are modified for adhesion. In *T. patialense*, these always project forward beyond the cercarial body, which is flexed ventrally around the body/tail junction. Free cercariae alternate active bursts of tail-first swimming with periods when they drop through the water, body-down, with the processes in advance. On contact with a fish, the adhesive pads rapidly attach and the body unrolls. Because the processes originate on the dorsal surface of the tail, the dorsal surface of the body is nearest the fish and the cercaria therefore rolls over to bring the ventral sucker into contact with the fish. The cercaria migrates until its anterior edge enters a recess beneath a scale; muscular activity of body and tail then forces the body completely into the recess, after which the tail is shed.

Sanguinicola sp. cercariae shed by *Ancylus fluviatilis* in a river in Spain were not attracted to the host, *Rutilus arcasii*, under experimental conditions but appeared to contact it only when drawn into the mouth with the inhalent current, so that they penetrated the skin lining the buccal cavity (Martin and Vasquez, 1984). The cercariae were shed from 1500 to 2100 h and survived for three hours at 15°C. Experimental infection of fish was achieved only in darkness when their swimming activity ceased. In natural conditions, the fish are concentrated in pools, and the chances of infection must be maximal in pools in the evening and early part of the night. Sanguinicolid cercariae invade through the skin by an apical papilla or cephalic organ bearing rows of spines and penetration glands. Having penetrated the skin, the cercariae migrate to the blood system, a process described in detail for *Aporocotyle simplex* (Figure 2.19.4) by Køie (1982). The cercariae travel from beneath the skin of dabs between the muscles to the lymphatic system, reaching the interspinal and neural lymph ducts. Most locate in the lymph spaces of the head from which they reach the branchial vessels through the ducts of Cuvier.

An interesting cercarial type is the gasterostome cercaria of the Bucephalidae (Figure 2.22.1–3) which has a pair of highly extensile tail furcae arising from a short tail stem, with glands which secrete an adhesive substance. The cercaria alternates periods of swimming activity with floating or gliding with the furcae fully extended. On contacting the surface of a fish, the furcae contract, drawing the cercarial body towards the host to which it adheres by means of the secretion of the tail stem glands before penetrating the skin.

Little is known about the sense organs of cercariae of fish digeneans other than the eyespots which have been described in many families (see Table 2.3). In general, cercariae have been shown to have uniciliate sensory structures on the body and tail (Lyons, 1973; Smyth and Halton, 1983). Whitfield, Anderson and Maloney (1975) demonstrated that the arm processes of *Transversotrema patialensis* cercariae bear sensory structures, with long, protruding cilia, near their proximal ends and distally, above the adhesive pads, they have a group of mammiform sensory receptors, believed to be contact chemoreceptors.

Cercariae of families which invade arthropods often have stylets to penetrate the weakest areas of the hard body surface. Cercariae of *Bunodera mediovitellata* showed a positive response to light but expressed it by horizontal rather than vertical movement, and they remained on or near the bottom of containers under test conditions for response to light and gravity (Kennedy, 1979). Their behaviour patterns would tend to keep them at the bottom of ponds but in the more illuminated areas on top of debris. This would bring them into contact with their second intermediaries, larvae of a caddis-fly, *Limnephilus* sp., which are bottom-dwellers.

Life-cycle patterns

Two-host life-cycles, other than those achieved by secondary reduction, are considered to be primitive and three-host cycles, the most common type, to be advanced. The three-host cycle may be reduced by loss of a host, sometimes the original definitive host, following achievement of sexual maturity in the second intermediary, or loss of the second intermediary when the cercariae fail to leave the mollusc host, the definitive host becoming infected by eating the mollusc. A further reduction to a one-host life-cycle in which maturity is reached, facultatively or habitually, in the mollusc host is known in a few families, the azygiids (*Proterometra dickermani*), hemiurids (*Parahemiurus bennettae*), bunocotylids (*Bunocotyle progenetica*) and derogenids (*Genarchella genarchella*). At the other extreme from such telescoped cycles are those which require four hosts for completion, a situation which has occurred in some hemiuroids and didymozoids.

Families Bivesiculidae, Azygiidae, Transversotrematidae, Sanguinicolidae

Among the fish helminths, the most simple life-cycle patterns are those of the bivesiculids, azygiids and transversotrematids, incorporating two hosts but no metacercarial stage and achieving direct invasion of the fish host. Bivesiculid and azygiid cercariae are ingested by fish and show consequent morphological and behavioural modifications as outlined above. The bivesiculids are a primitive group. Cable (1974) summarized their primitive features as the lack of suckers, non-fusion of the left and right excretory ducts, a posteriorly directed male copulatory organ, furcocercous rediae very similar to young cercariae, ie. reduction of the polymorphism common in digenean life-cycles, and marked host specificity. The azygiids, although similar in the pattern of their life-cycles, are more advanced and have cystocercous cercariae. The transversotrematids, cercariae of which invade the surface of the host directly to become established beneath its scales as described above, are a phylogenetically isolated group (Cable, 1974). The two-host cycle of sanguinicolids is not originally primitive but achieved by reduction from a three-host cycle.

Family Fellodistomidae

Some fellodistomes also infect the fish directly when it ingests cercariae (eg. *Burnellus trichofurcatus*, see Angel, 1971) or indirectly when it ingests a smaller fish which serves as a paratenic host (*Fellodistomum furcigerum*, *Steringotrema pagelli* and *Monascus filiformis* (Figure 2.23) (Køie, 1979a, 1980, 1983). The cercariae of these species are not, however, cystocercous. The appearance in the two-host cycle of a metacercarial stage encysted in the open or on substrates or vegetation eaten by the fish hosts has been outlined above. Most digeneans have three-host cycles with a metacercarial stage in second intermediaries but different life-cycle patterns often occur within the same family, eg. Fello-distomidae (Figure 2.24 and 2.25) with cycles incorporating from one to three hosts, including bivalves rather than gastropods as mollusc hosts. The cercariae, produced in daughter sporocysts, are basically furcocercous but modified by loss or reduction of the tail stem or furcae, and the metacercariae, when present, are usually unencysted and often precocious. *Fellodistomum fellis* has an unencysted metacercaria attached to the stomach of brittle stars; the behaviour of the cercaria, which swims sluggishly just above the substrate, accords with the bottom-dwelling habits of the ophiuroid hosts (Køie, 1980). Køie (1980) regards the fellodistomines and monascines as the most primitive of four groups in the family, defined according to cercarial morphology and the first intermediaries employed. This and the second group, the tergestines, use protobranch and filibranch bivalves respectively and the cercariae lack spines, penetration and cystogenous glands and have non-trichocercous furcate tails. Tergestine cercariae are planktonic (Angel, 1960) and their metacercariae have been found unencysted in ctenophores (Stunkard, 1978). *Proctoeces maculatus*, sole representative of Køie's third group, has a cycle (Figures 2.25, 2.26) telescoped in the cold part of its range so that sporocysts, metacercariae and adults all occur in the first intermediary (Bray, 1983). In warmer latitudes, adults occur mainly in labrid and sparid fishes and metacercariae in polychaete annelids, echinoid echinoderms and several groups of molluscs. Machkevskii (1982) has shown that sporocysts in *Mytilus galloprovincialis* in the Black Sea will switch from production of cercariae to that of sporocysts when water temperature falls in autumn and potential fish hosts migrate offshore. Køie's fourth group includes only *Bacciger bacciger*, which has a trichofurcocercous cercaria with a spinous body, and an encysted metacercaria in amphipods.

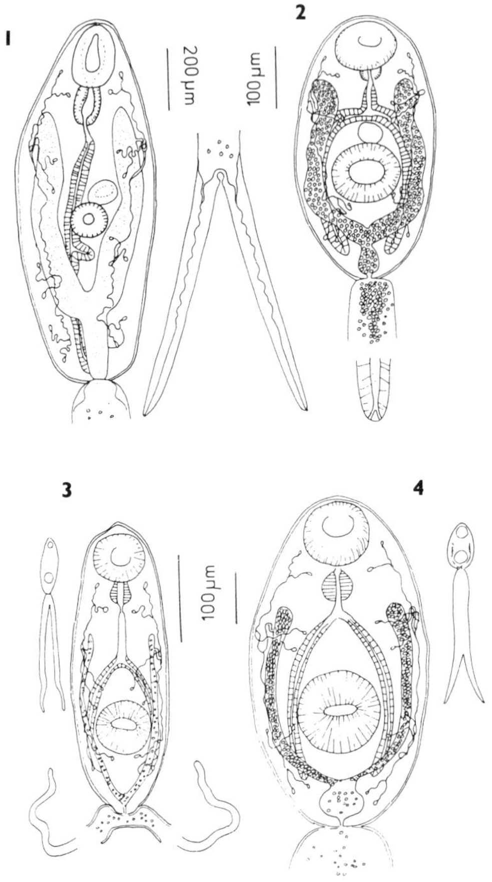

2.23

Figure 2.23 Cercariae of the family Fellodistomidae
1 *Monascus filiformis* (after Køie, 1979a). 2 *Steringophorus furciger* (after Køie, 1979a). 3 *Steringotrema pagelli* (after Køie, 1980). 4 *Fellodistomum fellis* (after Køie, 1980).

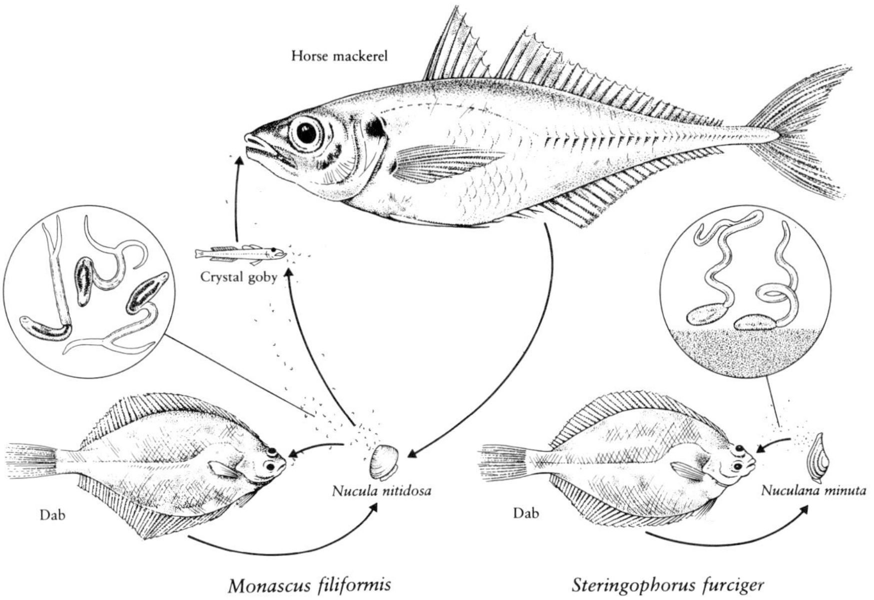

2.24

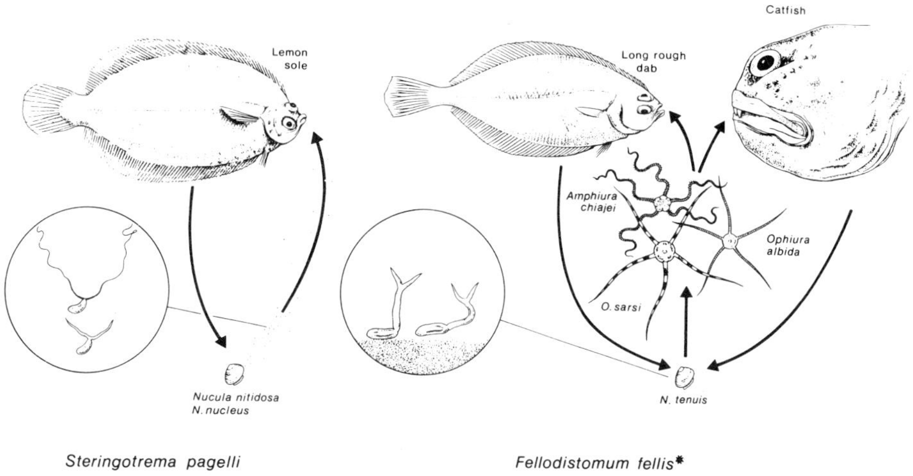

2.25

Figure 2.24 and 2.25 Life-cycles of four members of the Fellodistomidae
2.24 Life-cycles of *Monascus filiformis* (left) and *Steringophorus furciger* (right). From Køie (1979a). 2.25 Life-cycles of *Steringotrema pagelli* (left) and *Fellodistomum fellis* (right). *This is the life-cycle of *Steringotrema Ovacutum* which is not a synonym of *Fellodistomum fellis* (Køie, pers. comm.). From Køie (1980)

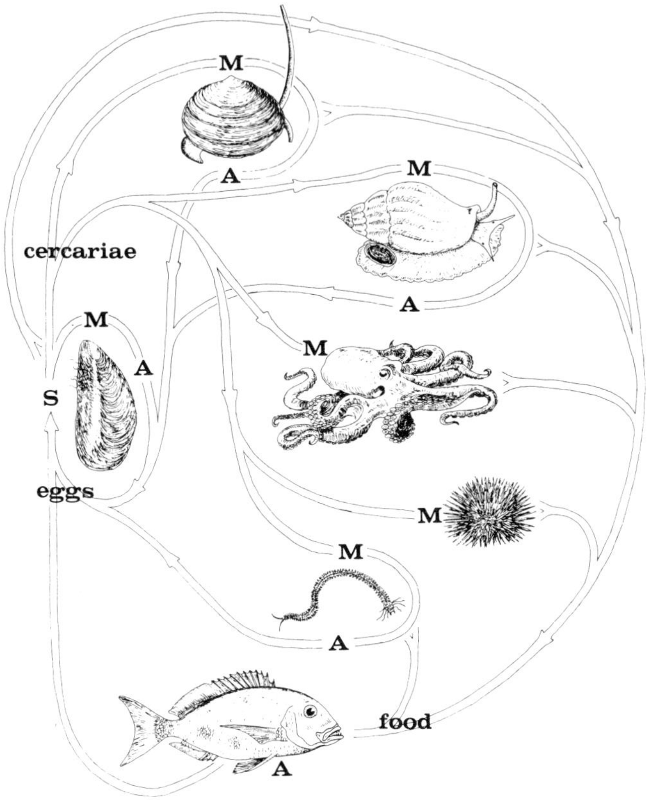

2.26

Figure 2.26. Life-cycle of *Proctoeces maculatus*. From Bray (1983).
Diagram showing the complexities possible in the life-cycle of *Proctoeces maculatus*. The bivalve mollusc which harbours the sporocyst (mid-left) may also harbour metacercariae and adults. Other bivalves (top), gastropods (top right) and annelids (bottom right) may harbour metacercariae and adults. Cephalopods (mid-right) and echinoderms (right) are only known to harbour metacercariae. Fish (bottom) are infested by ingestion of metacercariae and possibly adults in prey organisms. Abbreviations: S sporocyst; M metacercaria; A adult.

Family Acanthocolpidae

Cable (1974) places the Acanthocolpidae near the beginning of an evolutionary line which incorporates, of the families under consideration here, the lepocreadiids, monorchiids, zoogonids, opecoelids and allocreadiids and culminates in the gorgoderids. Cercariae (Figure 2.27.6) of this marine family develop in rediae in marine prosobranchs and are biocellate xiphidiocercariae, encysting in second intermediaries which include fish and polychaete annelids. *Neophasis lageniformis* has a telescoped life-cycle in which cercariae lose their tails within the redia and are considered to be metacercariae; infected molluscs are eaten by the fish host (Køie, 1973).

Family Homalometridae

The family Homalometridae has marine, brackish and freshwater members and represents a line of evolution from marine ancestors to freshwater parasites (Manter, 1963; Stunkard, 1964). Cercariae formed in rediae are biocellate with a setose body and straight tail, with dorsal and ventral finfolds and three pairs of setae borne on papillae—the anallocreadiine type of Hopkins (1937) (Figure 2.27.1). They complete their development in the lymph spaces. Encysted metacercariae occur in molluscs or annelids and are markedly precocious.

Family Lepocreadiidae

The Lepocreadiidae are predominantly marine with a three-host life-cycle using prosobranch gastropods in which several generations of rediae produce ophthalmotrichocercous cercariae (Figure 2.27.2–4). Lepocreadiinae metacercariae, eg. *Lepocreadium pegorchis*, *Opechona bacillaris*, (Figure 2.28) occur unencysted in a wide variety of invertebrates including turbellarians, lamellibranchs, prosobranchs, spionid and polychaete worms, ctenophores and medusae of hydrozoans and scyphozoans. Metacercariae of stegodexamenines encyst in small fish and those of lepidapedines in polychaete annelids. *Tetracerasta blepta*, which resembles the Stegodexameninae, encysts in fish but also in tree frog tadpoles (Watson, 1984) while *Holorchis pycnoporus* placed in the Aephidniogeninae by Bartoli and Prévot (1978) encysts in its prosobranch first intermediary or in a cardiid bivalve. Some lepocreadiids have precocious metacercariae, eg. those of *Stegodexamene anguillae*, in the gobiid secondary intermediary (Macfarlane, 1951).

Family Gorgoderidae

Several significant features of the gorgoderid life-cycle are the rapidly hatching eggs, bivalve hosts, 'ectoparasitic' sporocysts and cystocercous cercariae. In members of this group with three-host cycles, insects and crustaceans are used as second intermediaries and in many species the cercariae have stylets to facilitate penetration into the haemocoel where the cercariae encyst. Some species, eg. *Phyllodistomum macrobrachicola*, *P. lesteri* and *P. srivastavai*, have shrimps as second intermediaries rather than the more usual insect larvae (Yamaguti, 1975; Wu, 1938; Rai, 1964b). Abbreviated cycles, in which cercariae can encyst in the sporocyst, are known for *P. folium*, *P. simile*, *P. lohrenzi*, *P. nocomis* and *P. caudatum* (see Yamaguti, 1975); in at least *P. lohrenzi*, the cercariae may also leave the host to encyst in insect larvae. In this species, miracidia are actually passed in the urine of the fish host (Beilfuss, 1954). *P. elongatum* is unusual in the family in that the rhopalocercous cercaria encysts within its expanded tail in water (Orecchia *et al.*, 1975).

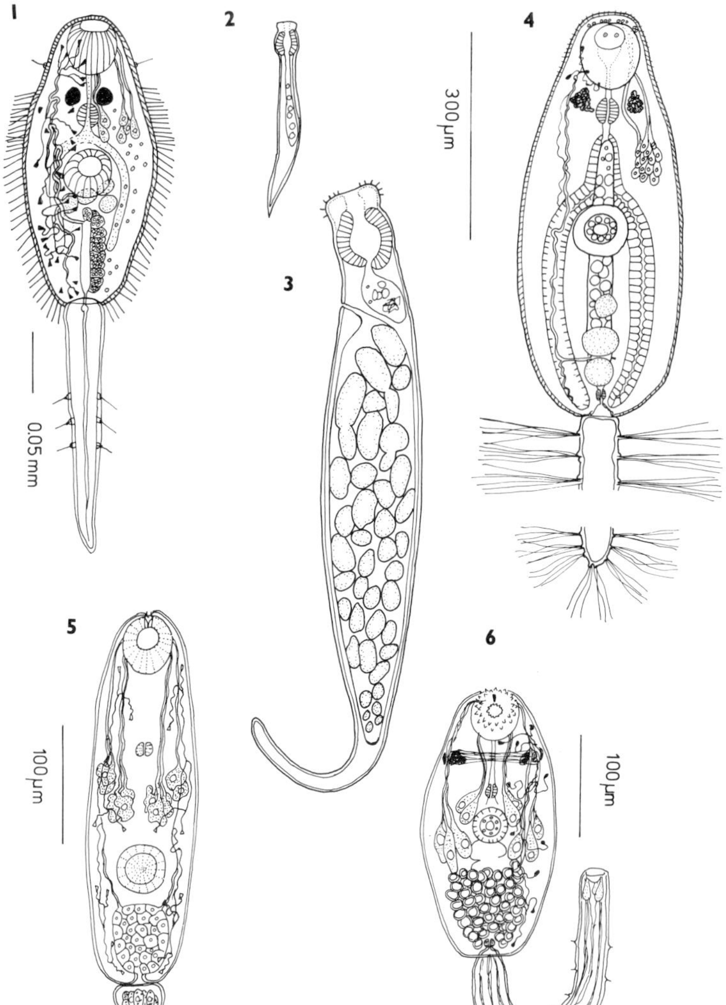

Fig 2.27

Figure 2.27.1–6 Larval stages of Homalometridae, Lepocreadiidae, Opecoelidae and Acanthocolpidae.
1 Cercaria of *Microcreadium parvum* (Homalometridae), anallocreadiine type (after Hopkins, 1937) 2–4 *Opechona bacillaris* (Lepocreadiidae) 2 Small redia 3 Mature redia 4 Cercaria, ventral view (after Køie, 1975) 5 Cercaria of *Podocotyle reflexa* (Opecoelidae), cotylomicrocercous type (after Køie, 1981) 6 Cercaria of *Stephanostomum caducum* (Acanthocolpidae), ophthalmoxiphidocercous type (after Køie, 1978).

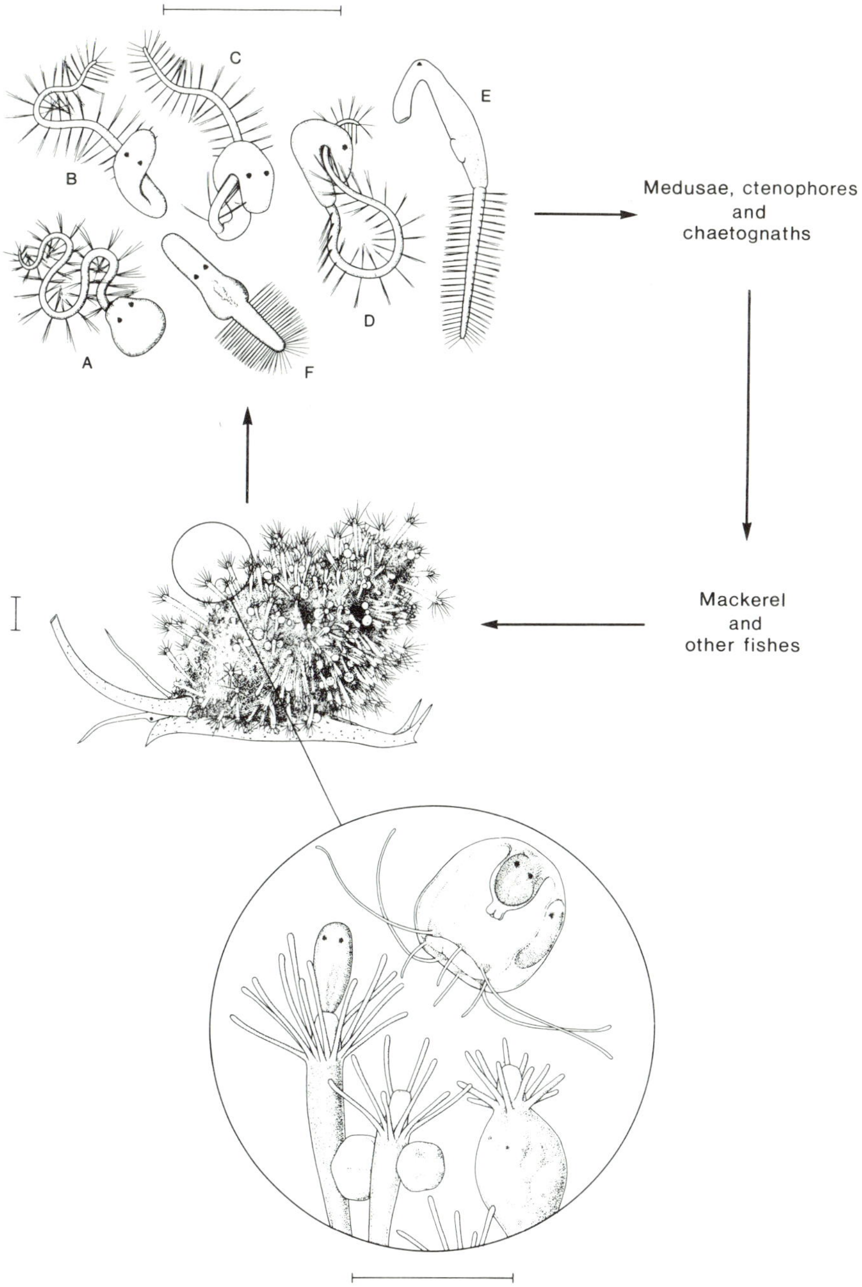

2.28

Figure 2.28. The life-history of *Opechona bacillaris*. The shell of the first host *Nassarius pygmaeus* is covered with the hydroid *Podocoryne carnea*. Both the hydroid generation and the free-swimming medusae may feed on the emerged cercariae. Different swimming and behaviour positions of the cercaria are shown. Scale bars 1 mm. (From Køie, 1975).

Family Lissorchiidae

The lissorchiid life-cycle is similar to that of gorgoderids in that the cercariae are xiphidiocercous and use insect larvae as intermediaries but the cercaria is simple-tailed and the first intermediary is a prosobranch gastropod (Magath, 1918).

Family Monorchiidae

Monorchiid eggs hatch only after ingestion by the mollusc (see above) and the cercariae usually encyst in the tissues of the same snail or in another of the same or a different species, in which case the fish hosts eat the snails. Cercariae of several species lack tails (cercariaeum type) and cannot swim, e.g. *Asymphylodora tincae*, *Triganodistomum mutabile* (see Table 2.3 for references) and *Parasymphylodora markevitschii* (Lambert, 1976). They crawl on the substrate and are eaten by the second intermediary (e.g. *T. mutabile*) or attach to and penetrate the tentacles of a snail (e.g. *P. markevitschii*). Two marine species, *Paratimonia gobii* (Figure 2.29) and *Monorcheides cumingiae*, encyst in the inhalent siphon of the mollusc, which cannot be withdrawn and is eaten by the definitive host (Maillard, 1976; Martin, 1940). Monorchiid metacercariae are markedly precocious.

Family Zoogonidae

The Zoogonidae are predominantly marine with prosobranch gastropod first intermediaries and tailless xiphidiocercariae, lacking eyespots, which encyst in polychaetes, ophiuroids and molluscs (Zoogoninae), prosobranchs (*Diphtherostomum brusinae*) and mysids (*Deretrema minutum*).

Family Opecoelidae

The Opecoelidae has both marine and freshwater members usually with a three-host cycle, reduced in some instances to two, involving prosobranch gastropods. The second intermediaries include a wide range of invertebrates, for example amphipods and mysids for the Opecoelinae, snails and leeches for the Sphaerostominae, and crayfish, molluscs, insects, mysids and fishes for the Plagioporinae. The cercariae are cotylomicrocercous xiphidiocercariae, developing in daughter sporocysts. They cannot swim but move on the substrate by means of their suckers, and their behaviour is well-adapted to that of their benthic second intermediaries. Cercariae of some species encyst within the daughter sporocyst (e.g. *Sphaerostostoma bramae, Plagioporus sinitzini* and *Podocotyle virens* (Sinitsin, 1931). The daughter sporocyst of *Plagioporus sinitzini*, like that of *Phyllodistomum folium* (Gorgoderidae) emerges from the snail to be eaten by fish. Species encysting in daughter sporocysts lack a stylet while others of the same genus which penetrate another host have them (Pearson, 1972). However, *Podocotyle virens*, which encysts in the daughter sporocyst or in the liver of the snail, has a stylet (Sinitsin, 1931). Metacercariae often tend towards precocity.

Family Allocreadiidae

The Allocreadiidae is an almost entirely freshwater family with three-host or abbreviated life-cycles. The first intermediaries are usually bivalves and the second intermediaries are bivalves, gastropods, crustaceans and insect larvae. Rediae produce ophthalmoxiphidiocercariae with morphologically diverse tails. Abbreviated cycles occur in *Allocreadium isoporum*, cercariae of which can encyst in the molluscan host (Dollfus, 1949), and

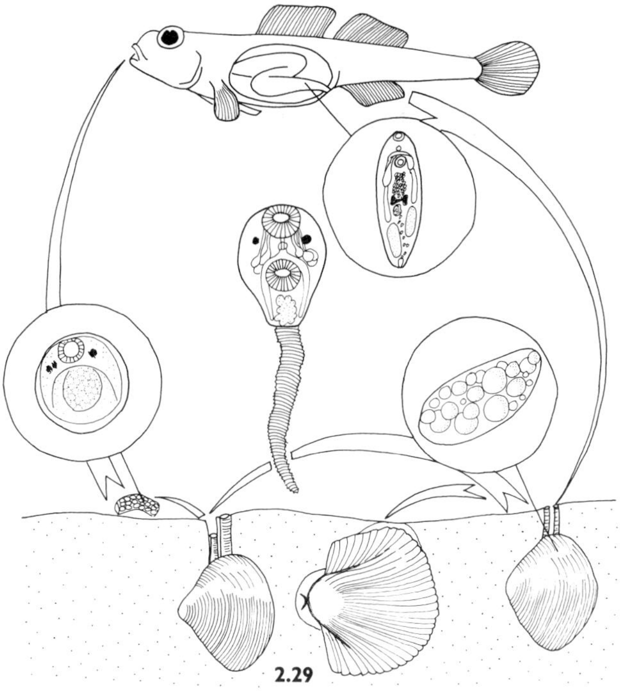

Figure 2.29 The life-cycle of *Paratimonia gobii* (Monorchiidae), after Maillard (1976)

A. lobatum, which can produce viable eggs within the second intermediaries (DeGiusti, 1962). The ophthalmoxiphidiocercariae of *Pseudoallocreadium alloneotenicum* lack cystogenous glands; they penetrate but do not encyst in caddis fly larvae, in which they mature. The original final host has been eliminated, to the extent that attempts to infect fish have not succeeded (Wootten, 1957a). Two species of *Allocreadium* illustrate the point that the choice of second intermediary is related to the eating habits of the final host. *A. handiai* (Figure 2.30), which encysts in gastropods, is adult in benthophagous fish while *A. fasciatusi* encysts in amphipods consumed by planktonivorous fish (Madhavi, 1978a, 1980). They differ from almost all other members of the family in using a gastropod first intermediary.

Family Macroderoididae

The Macroderoididae is another freshwater family with xiphidiocercariae encysting in insect larvae and small fish. They are, however, non-oculate and the family is one in which the mollusc hosts, pulmonate gastropods, must ingest the eggs. This is also the case in the marine acanthostomids, the molluscan hosts of which are prosobranch gastropods. Their monostomatous biocellate pleurolophocercous cercariae leave the rediae to complete development in the snail's tissue, after which they penetrate and encyst in small fish.

Family Bucephalidae

The Bucephalidae is a predominantly marine family considered by Cable (1974, 1982) to be close to the fellodistomes and brachylaimids and to the azygiids, all of which have miracidia with cilia or bristles in patches or on appendages. As with the fellodistomes, the molluscan hosts are bivalves. The sporocysts, in contrast to those of most digeneans except

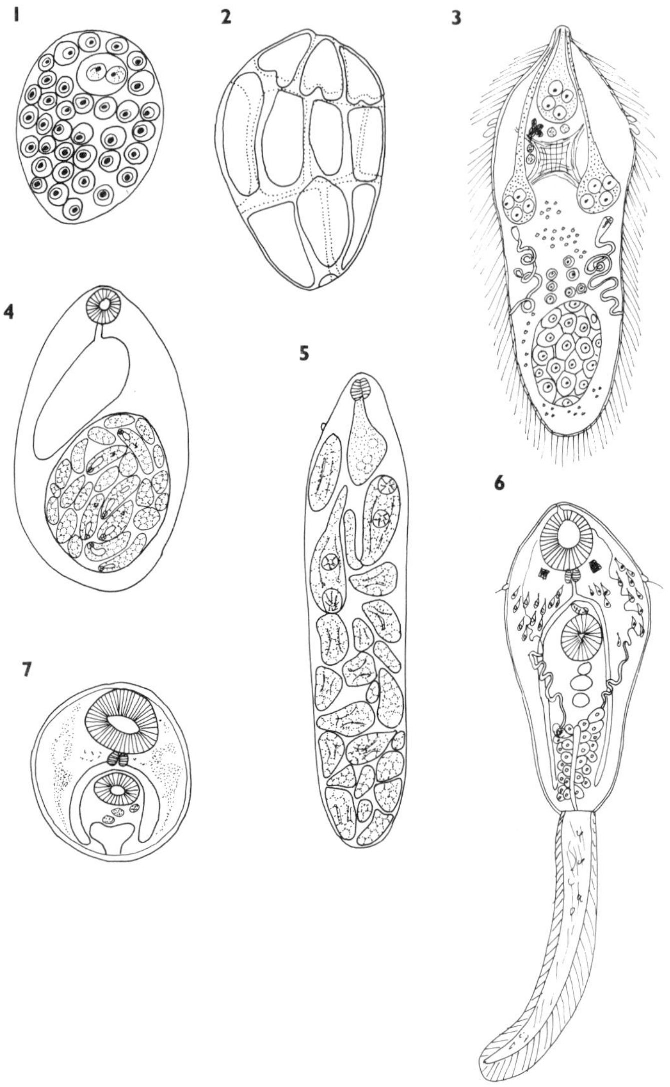

2.30

Figure 2.30.1–7. Stages in the life-history of *Allocreadium handiai* (Allocreadiidae). After Madhavi (1980).
1. Egg. 2. Epidermal cells of miracidium. 3. Miracidium. 4. Mother redia. 5. Daughter redia. 6. Cercaria.
7. Metacercaria.

the brachylaimids, are branched. They produce gasterostomatous cercariae, with forked tails composed of long furcae and a short stem, which penetrate and encyst in small fish, usually in the skin, subcutaneous muscles or extrinsic fin muscles. *Bucephaloides gracilescens* encysts in the cranial nerves (Matthews, 1974). The final hosts are piscivorous fish and, in some instances, their fry may serve as second intermediaries.

Superfamily Hemiuroidea: Families Derogenidae, Hemiuridae, Lecithasteridae, Ptychogonimidae

The Hemiuroidea is a mainly marine group with life-cycles incorporating from one to as many as four hosts, although the most common is a three-host cycle incorporating a prosobranch gastropod with copepod or ostracod second intermediaries. The life-cycle of *Derogenes varicus* (Derogenidae) is illustrated in Figure 2.31. Fish often serve as transport or paratenic hosts. *D. varicus* will mature in crustaceans and chaetognaths (Køie, 1979b). In the majority of hemiuroids, the cercariae develop within rediae rather than daughter sporocysts, but *Lecithochirium furcolabiatum* (described above) and *Cercaria 'A'* Miller, 1925 are exceptions (Matthews, 1980). Both are hemiurids. The cercariae are cystophorous with a specialized delivery mechanism, already described, for the cercarial body, which lacks penetration and cystogenous glands. The appendages of hemiuroid cercariae are very varied (Cable and Nahhas, 1963); all have a delivery tube but other appendages include the excretory appendage, described for *Pseudodichadena lobata* (= *Dichadena acuta*) (Lecithasteridae) by Cable and Nahhas (1963) and for *Genarchopsis goppo* (Derogenidae) by Madhavi (1978b), paired ear-like appendages in *P. lobata*, and a bifurcate tail like the tail of furcocercariae in, for example *Derogenes varicus* (Køie, 1979b). Cable and Nahhas (1963) considered that the homologies of the appendages could be resolved only by embryological studies and that the bifid appendages are secondarily developed rather than indicative of a close relationship with other groups with furcocercous cercariae. Matthews (1981a, 1982) has described the embryology of *Lecithochirium furcolabiatum* cercariae and of *Cercaria calliostomae*. The metacercariae do not encyst and are markedly precocious in some species. An unusual cycle in the group is that of *Genarchella genarchella* (Derogenidae) which has an elongated cercariaeum able to mature and produce eggs in the redia (Szidat, 1956). Infected snails die in winter, releasing rediae and cercariae, eggs from which may eventually be eaten by snails, establishing a one-host cycle. Paratenic hosts have been incorporated into the cycles of several species (see Table 2.3). Chabaud and Campana-Rouget (1959) considered blennid, sparid and labrid fish to be essential in the life-cycle of *Lecithochirium fusiforme* but their interpretation of this life-cycle has been questioned by Gibson and Bray (1986); the metacercariae occurred encapsulated, not encysted, in the liver. *Lecithochirium furcolabiatum* has a four-host life-cycle incorporating a prosobranch mollusc, a harpacticoid copepod, species of rock pool fish in which metacercariae become encapsulated in the body-cavity (*Blennius pholis, Gobius paganellus*) and a piscivorous definitive host, *Ciliata mustela*. The adult occupies the body-cavity of its host. This life-cycle has been completed experimentally and the processes of escape of cercariae from the mollusc (described above) and of infection of the copepod host have been described in detail (Matthews, 1980, 1981a, b; Matthews and Matthews, 1988; Gibson, Rollinson and Matthews, 1985). The stage in copepods has been termed a mesocercaria. In contrast, abbreviated life-cycles have been described in *Bunocotyle progenetica* (one-host, precocious adults in *Hydrobia ventrosa*) and *B. meridionalis* (two-host, precocious adults in copepods) by Chabaud and Buttner (1959) and in *Parahemiurus bennettae* (sexually mature in the gastropod *Salinator fragilis*) by Jamieson (1966a,b).

Most hemiuroids so far studied (but not all, see Matthews and Matthews (1988) on *Lecithochirium furcolabiatum*) resemble the azygiids and didymozoids in having miracidia with

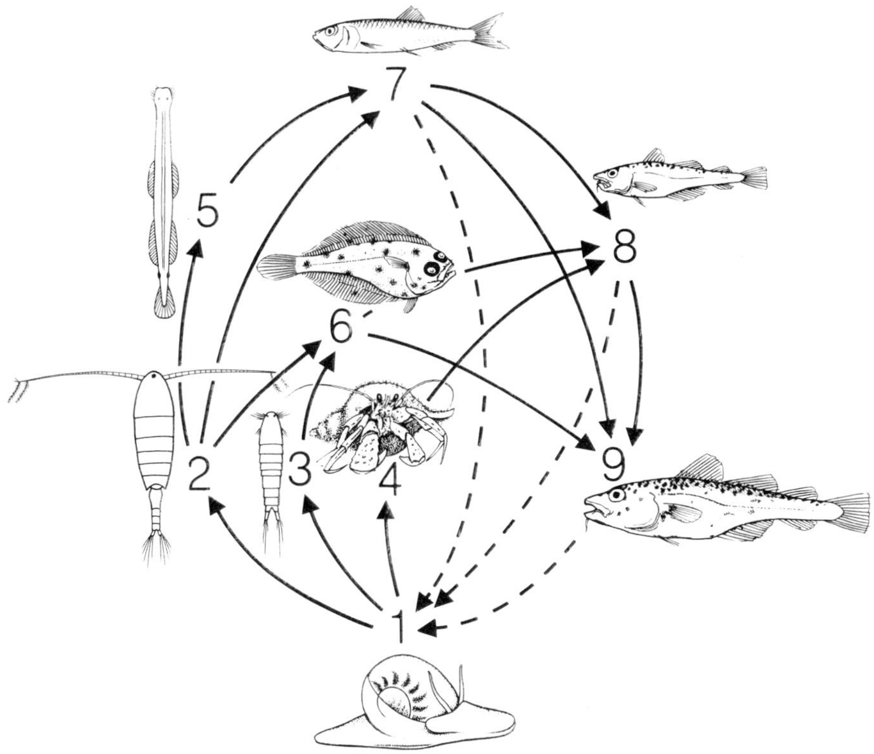

Figure 2.31.1–9 The life-cycle of *Derogenes varicus*
1. *Natica* spp., 2 calanoid copepod, 3 harpactacoid copepod, 4 hermit crab and probably other decapods and barnacles, 5 *Sagitta* spp., 6 small fishes, eg. gobies and fry of other fishes, 7 planktophagous fishes, eg. salmon, herring, and juvenile gadid fishes, 8 benthophagus and piscivorous fishes, eg. gadid fish and flatfishes, and *Sepia officinalis*, 9 piscivorous fishes, eg. large cod. From Køie (1979b)

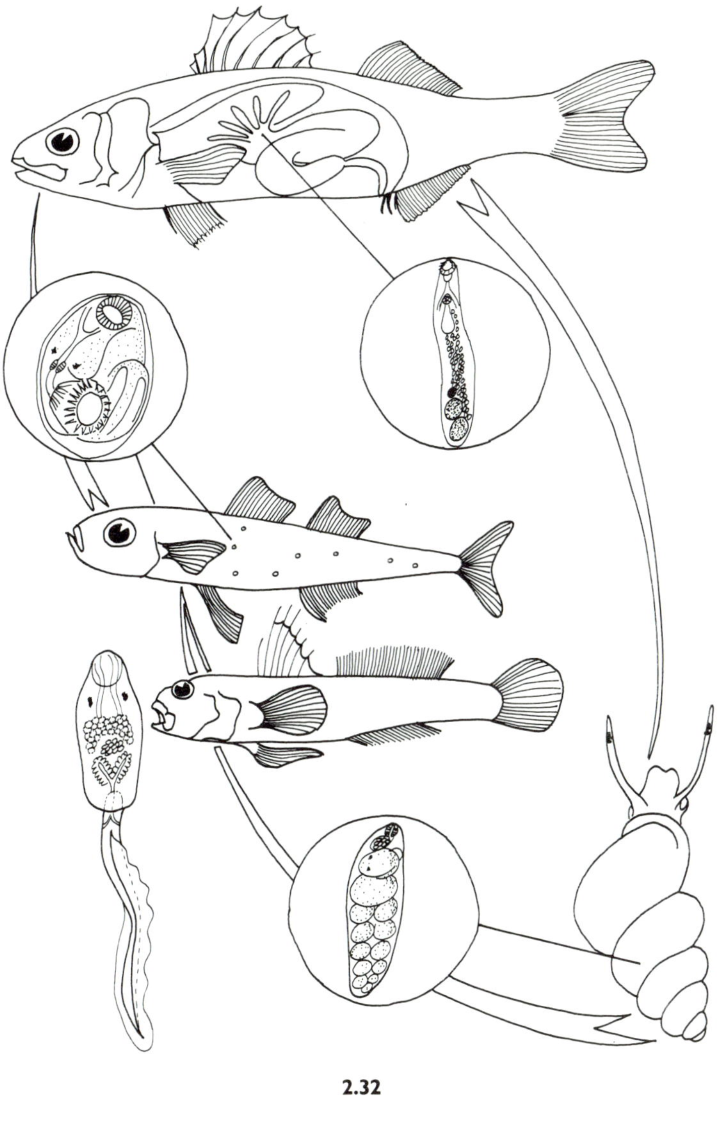

2.32

Figure 2.32. The life-cycle of *Timoniella praeteritum* (Acanthostomidae) after Maillard (1976).

bristles rather than cilia, and therefore they, together with the didymozoids which arose from hemiuroid stock, are considered to represent one of the major lines of evolution in the digeneans (Cable, 1974, 1982).

The Ptychogonimidae, e.g. *Ptychogonimus megastomus*, are another hemiuroid group which has miracidia with bristles rather than cilia; transmission to the second intermediary is by means of the daughter sporocysts which, containing a row of mature cercariae, are discharged from the mollusc into sea water and are subsequently eaten by a wide variety of crustaceans.

Superfamily Didymozoidea

Information on didymozoid life-cycles is fragmentary and is summarized in Table 2.3. The life-cycle involves three or four hosts, and fish serve both as paratenic and definitive hosts.

Because of the great volume of information available on digenean life-cycles, the species included in Table 2.3 have been selected to illustrate representative, and sometimes unusual, cycles within each family.

Table 2.3. Definitive and intermediate hosts of some fish digeneans[1]

Family Bivesiculidae	
Paucivitellosus fragilis 1. *Cerithidium penthusarus* D. *Salarias meleagris* Coil, Reid and Kuntz (1965); Pearson (1968) Cable and Nahhas (1962b)	Eggs contain fully developed miracidia. Redia furcocercous. Cercariae furcocystocercous, attach to surfaces grazed by fish host and are ingested.
Family Waretrematidae	
Hapladena varia 1. *Zebrina browniana* D. *Acanthurus* spp. *Sparisoma, Pseudocanes* Cable (1962)	Rediae → biocellate gymnocephalous cercariae which complete development outside redia. Metacercarial cysts float on water surface.
Family Haploporidae	
Saccocoelioides sogandaresi *Amnicola comalensis* D. *Mollienesia latipinna* Cable and Isseroff (1969)	Rediae → biocellate distomatous cercariae, precocious, encyst in open on vegetation.
Family Haplosplanchnidae	
Schikhobalotrema acutum 1. *Cerithium variabile* D. *Hyporhampus unifasciatus Strongylura* sp. Cable (1954a)	Daughter sporocysts → biocellate, distomatous cercariae, metacercariae encysted on vegetation, final hosts are herbivorous fish.
Family Mesometridae	
Mesometra orbicularis 1. gastropod D. *Sparus salpa* Palombi (1937)	Cercaria has long, glandular tail which encircles body and secretes cystogenous products. Metacercariae encyst on *Posidonia caulini*, on which the gastropod host is found and on which the fish host browses.

[1] Footnote for Table 2.3
1. Intermediate host, or first intermediary in cycles with more than one intermediate host
2. Second intermediate host
D. Definitive host.

Table 2.3. *Continued*

Family Megaperidae	
Megapera gyrina 1. *Crepidura convexa Lactophrys, Monacanthus* Cable (1954b)	Redia → biocellate cercaria with tail with ventral fin and pair of lateral fins, encysts in open, fish host feeds on seaweeds and molluscs.
Family Transversotrematidae	
Prototransversotrema steeri 1. *Posticobia brazieri* D. *Gambusia affinis, Pseudomugil signifer, Mugil cephalus, Aldrichetta forsteri, Arripis trutta, Taeniomembras microstoma, Pomatomus saltatrix, G. affinis, Xiphophorus maculatus, X. helleri, M. Cephalou Craterocephalus marjoriae* (experimental) Cribb (1988)	Miracidium transforms into mother sporocyst which produces a single mother redia. It produces daughter rediae which in turn produce other daughter rediae, or cercariae. Cercariae emerge from redia while still embryonic and continue development in snail tissues; biocellate, furcocercous. Produces eggs within six days of infecting definitive (natural) host.
Transversotrema laruei 1. *Thiara riguettii* D. *Lates calcarifer, Mollienesia latipinna* Velasquez (1961)	Biocellate furcocercous cercariae produced in rediae which they leave before completing their development, sexually mature and protandrous. Shed tail on contact with fish and migrate to final site under scales, in opercular cavity or on gills. May reach fish passively with inhalent current.
Family Sanguinicolidae	
Sanguinicola idahoensis 1. *Liphoglyphus virens D. Salmo gairdneri* Schell (1974)	Adults in blood system. Eggs embryonate and hatch in gill tissue; miracidia emerge from gills. Furcocercous apharyngeate cercariae produced in daughter sporocysts, penetrate skin of fingerling trout.
Aporocotyle simplex 1. *Artacamia proboscidea* D. Flatfishes Køie (1982)	Miracidium contains fully developed mother sporocyst. Two or more redial generations. Cercariae apharyngeate, furcocercous, develop in serpulid, ampharetid or terebellid polychaetes, invade definitive host through skin and migrate to blood vessels.
Family Azygiidae	
Azygia sebago 1. *Amnicola limosa* D. *Anguilla rostrata* and freshwater fish Stunkard (1956)	Eggs *in utero* contain fully developed bristled miracidia. Mollusc ingests eggs. Cercariae produced in rediae; complete development in branchial cavity, furcocystocercous, swim with a flapping movement and may become attached to substrate by tail tip. Eaten by fish.
Proterometra sagittaria 1. *Goniobasis* sp. *Pleurocera* sp. D. *Eupomotis gibbosus* Dickerman (1946)	Eggs contain bristled miracidia. Rediae produce furcocystocercous cercariae which are functionally adult. Eggs with mature miracidia often found in rediae from which cercariae have been discharged. Cercaria 20–22mm long, largest cercaria known. Swims slowly and is ingested by fish.
Proterometra dickermani 1. *Goniobasis livescens* Anderson and Anderson (1963)	Cercariae mature within rediae in the snail and deposit embryonated eggs. Bristled miracidia hatch within rediae and infect the same or a different snail. Rediae sometimes extruded from snail, disintegrate to release contents. Life-cycle completed within mollusc host. No naturally-infected fish found and attempts to infect fish (*Lepomis gibbosus, L. macrochirus, Micropterus salmoides*) succeeded only rarely.
Family Fellodistomidae	
Burnellus trichofurcatus 1. *Corbiculina angasi* D. *Tandanus tandanus* Angel (1971)	Eggs contain well-developed embryos. Cercariae trichofurcocercous, probably ingested by fish. No metacercarial stage
Steringophorus furciger 1. *Nuculana pernula N. minuta* D. pleuronectid, cottid and anarrichadid fishes Køie (1979a, 1983)	Cercariae leptocercous, ingested by and directly infective to 0-group dab. Larger fish ingest cercariae and also infected smaller fish. Whelks (*Buccinum undatum*) acquire infection by scavenging dead fish, as do fish which serve as paratenic hosts.

Table 2.3. *Continued*

*Fellodistomum fellis** 1. *Nucula tenuis Ophiura* spp., *Amphiura chiajei* D. *Anarrhichas lupus* Køie (1980)	Cercariae non-oculate, furcocercous, swim sluggishly just above substrate. Unencysted metacercariae in brittle stars, in lumen of stomach attached to wall.
Steringotrema pagelli 1. *Nucula nitidosa, N. nucleus, Microstomus kitt, Limanda limanda*, various flatfish and sparids Køie (1980)	Cercariae have bifurcate tail without common stem. May be drawn into mouth of fish with respiratory current, get entangled in gills by furcae then enter stomach. No secondary intermediary.
Monascus filiformis 1. *Nucula nitidosa* D. *Limanda limanda, Trachurus trachurus* Køie (1979a, 1983)	Cercariae furcocercous, eaten by and mature in small dabs. Intermediate-sized dabs acquire infection by eating smaller fish and a few cercariae but large fish eat cercariae by chance when feeding on benthic invertebrates.
Tergestia haswelli 1. *Mytilus latus* Angel (1960)	Cercaria furcocercous with tail base expanded into prominent crest with lobed margin, thought to be a pneumatophore. Tergestine cercariae planktonic, complete life-cycles unknown.
Proctoeces maculatus 1. *Mytilus, Ischadium, Mytilopsis, Nereis, Patella, Hydroides, Acanthochites* and others D. labrid, sparid, blennid and gobiid fish, molluscs including *Mytilus edulis, Scrobicularia plana, Rissoa splendida* Bray and Gibson (1980) Bray (1983), Machkevskii (1982), Køie (1980)	Adult in fish in the warmer waters of its range but precocious forms occur in molluscs in temperate waters. Metacercariae also occur unencysted in molluscs and polychaetes. Life-cycles include three or more generations of sporocysts which produce cercariae in spring-summer when water temperature is high and labrid fish present but sporocysts in autumn when temperature falls and fish move offshore.
Bacciger bacciger 1. *Laevicardium mortoni, Tapes philippinarum, T. aureus, T. decussatus, Donax vitattus, Pholas candida, Chione gallina* 2. *Erichthonius difformis, Mnemiopsis leidyi* D. *Atherina presbiter, A. hepsetus, A. boyeri* Palombi (1933, 1934, 1940) Martin (1945) Dolgikh (1968)	Daughter sporocyst → gymnocephalous cercariae with setiferous tail, encysts in amphipods. May be carried into gut of ctenophore with ciliary current but does not encyst. *M. leidyi* may be an accidental host.
Family Gorgoderidae	
Phyllodistomum staffordi 1. *Musculium ryckholti* 2. naiads of dragonflies, damsel flies, trichopteran larvae, chironomid larvae, larvae of diving beetles D. *Ameiurus nebulosus* Schell (1967)	Eggs hatch rapidly, miracidia enter clam via inhalent syphon or swim between valves. Mother sporocyst between gill bars or on lamellae. Daughter sporocysts produce cystocercous xiphidiocercariae which leave clam via exhalent siphon to be eaten by insect larvae and encyst. May also encyst in gill lamellae of clam.
Phyllodistomum folium 1. *Dreissena polymorpha* D. "Karausche", "Brachse" Sinitsin (1901)	Miracidia penetrate gill tissue of clam; mother sporocyst produces 12–14 daughter sporocysts which migrate to branchial cavity and produce more generations. Cercariae lack stylets, lose tails and encyst within sporocyst which is discharged from clam, floats due to high lipid content, eaten by fish.
Phyllodistomum nocomis 1. *Sphaerium striatinum* D. *Hybopsis biguttata* Wanson and Larson (1972)	Cercariae with immobile tails produced in and encyst in daughter sporocysts. Lack stylets. Fish eats infected clams.
Family Lissorchiidae	
1. *Lissorchis fairporti Planorbis trivolvis* 2. *Chironomus lobiferus, Tanypus decoloratus* D. *Ictiobus cyprinella, I. bulbulus* Magath (1918)	Daughter sporocysts produce simple-tailed xiphidiocercariae which penetrate and encyst in chironomid larvae.

*See footnote for fig. 2.25.

Table 2.3. *Continued*

Family Monorchiidae	
Asymphylodora tincae 2. *Lymnaea limosa, L. stagnalis, Planorbis carinatus, Bithynia tentaculata, Radix auricularia* D. *Tinca tinca* Zietse *et al.* (1981); Broek and Jong (1979); various authors cited by Yamaguti (1975)	Tailless cercariae (cercariaea) develop in rediae and encyst in the snail. The metacercariae are markedly precocious. Fish infected by eating molluscs.
Triganodistomum mutabile 1. *Helisomum campanulata, H. trivolvis* 2. *Chaetogaster, Planaria* D. *Erimyzon sucetta kennerlii* Wallace (1941)	Cercariae (cercariaeum type) develop in redia, lack tail throughout development, cannot swim, crawl on substrate. Eaten by *Chaetogaster*, penetrate gut wall and encyst in body cavity. Second intermediate host required.
Paratimonia gobii 1. *Abra ovata* 2. *A. ovata, Cardium glaucum*, other lamellibranchs D. *Pomatoschistus microps* Maillard (1976)	Daughter sporocysts produce biocellate, gymnocephalous cercariae with a tail ornamented by folds. Encyst in inhalent siphon of lamellibranchs, which cease to be capable of withdrawal and are eaten by fish. Only metacercariae in this site will complete the life-cycle.
Family Zoogonidae	
Zoogonoides viviparus 1. *Buccinum undatum* 2. Brittle stars, polychaetes, lamellibranchs, prosobranchs D. plaice, flounder, dab, long rough dab Køie (1976, 1983)	Tailless xiphidiocercariae produced in daughter sporocysts; unable to swim, creep on substrate using suckers. Penetrate and encyst in a wide variety of of second intermediaries, the most important of which in Danish waters is *Ophiura* albida.
Diphtherostomum brusinae 1. *Nassa mutabilis, N. reticulata, Natica poliana* D. *Blennius gattorugine, Crenilabrus pavo, Sargus vulgaris* Palombi (1930); Prévot (1966)	Eggs embryonated in uterus. Xiphidiocercous cercariae produced in daughter sporocysts within which they encyst. No second intermediary.
Deretrema minutum 1. Unknown? 2. *Tenagomysis chiltoni* D. *Retropinna retropinna, Galaxias maculatus* Holton (1983)	Metacercariae in mysids, encysted beneath cuticle of abdominal segments, markedly precocious.
Family Acanthocolpidae	
Deropristis inflata 1. *Bittium alternatum, B. reticulum, Hydrobia acuta* 2. *Nereis dumerilii, N. diversicolor* D. *Anguilla rostrata* Cable and Hunninen (1942); Maillard (1976)	Biocellate cercariae produced in rediae, have ornate tail with ventral fin fold. Penetrate nereids and encyst along nerve cord and ventral blood vessel.
Stephanostomum caducum 1. *Natica alderi* 2. *Pomatoschistus pictus, P. microps, Gobiusculus flavescens* D. *Gadus morhua* Køie (1978)	Rediae → biocellate ophthalmoxiphidiocercariae, weakly photopositive. Encyst in gobies, usually just under epithelial lining of mouth.
Neophasis lageniformis 1. *Buccinum undatum* D. *Anarrhichas lupus* Køie (1973)	Rediae produce biocellate cercariae which have tails but lose them during development and are considered metacercariae although they do not encyst. Both stages are biocellate. Cercaria production ceases when the redia is full. The fish host eats infected whelks.
Family Homalometridae	
Microcreadium parvum 1. *Amnicola peracuta* 2. *Musculium ferrissi* D. *Aplodinotus grunniens* Hopkins (1937)	Cercariae develop in rediae but leave to complete development in lymph spaces. Cercaria biocellate, of the anallocreadiine type. Encyst in *M. ferrissi* and when mature differ from adult only in absence of eggs.

Table 2.3. *Continued*

Family Allocreadiidae	
Allocreadium isoporum 1. & 2. *Sphaerium rivicola* D. many species of freshwater fish. Dollfus (1949)	Rediae produce ophthalmoxiphidiocercariae which encyst in the same snail or another of the same species.
Allocreadium handiai 1. *Alocinma travancorica* 2. *A. travancorica, Lymnaea luteola, Thiara (Melanoides) tuberculata* *Channa orientalis, Clarias batrachus* Madhavi (1980)	No mother sporocyst stage observed but two redial generations present. Cercariae ophthalmoxiphidiocercous, escape to penetrate and encyst in the same or other species of mollusc. Fish hosts are benthophagous. D.
Crepidostomum cooperi 1. *Musculium transversum* 2. *Hexagenia* nymphs *H. limbata* adults, *H. recurvatus* nymphs, *Polymitarcys* nymphs D. Many genera of fresh water fish Choquette (1954); Hopkins (1933)	Slender-tailed ophthalmoxiphidiocercariae develop in rediae, penetrate and encyst in mayfly nymphs.
Bunodera luciopercae 1. *Sphaerium* spp., *Pisidium* spp. 2. *Mesocyclops oithonoides, M. crassus, Hyalella azteca, Crangonyx gracilis, Daphnia pulex, Simocephalus exspinosus, Eurycercus lamellatus, Notodromus monacha* D. Many freshwater fish Wisniewski (1958); Moravec (1969b); Cannon (1971)	Miracidium ocellate. Mother sporocyst and two redial generations present. Lophocercous, ophthalmoxiphidiocercariae leave redia to complete development in host tissue, penetrate and encyst in copepods, phyllopods and ostracods.
Family Lepocreadiidae	
Lepocreadium pegorchis 1. *Sphaeronassa mutabilis, Amyclina corniculum* 2. Lamellibranchs (approx. 12 spp.) D. *Sparus auratus, Pagellus erythrinus, P. mormyrus, Pomatoschistus microps* Bartoli (1967, 1983)	Ophthalmotrichocercariae produced in rediae but leave before completing their development, penetrate lamellibranchs but do not encyst.
Opechona bacillaris 1. *Nassarius pygmaeus* 2. *Pleurobrachia pileus, Sagitta elegans, Podocoryne carnea, Eutonina indicans* D. *Cyclopterus lumpus, Scomber scombrus* Køie (1975)	Rediae produce ophthalmotrichocercariae which complete their development outside the redia. They penetrate ctenophores, chaetognaths and medusae but do not encyst. They have been reported from a wide variety of planktonic invertebrates. In molluscs infection castrates both sexes. Hydroids on snail shells often eat emerging cercariae.
Stegodexamene callista 1. *Posticobia brazieri* 2. *Retropinna semoni, Pseudomugil signifer, Ambassis* sp., *Bufo marinus, Craterocephalus stercusmuscarum, Gobiomorphus* sp., *G. australis, Hypseleotris compressus, H. galii, Litoria* tadpoles, *Mixophyes* tadpoles D. *Anguilla reinhardtii, A. australis, Leiopotherapon unicolor, Macquaria novemaculeata, Glossamia aprion* Watson (1984)	Sporocyst and several redial generations present. Cercariae ophthalmotrichocercous, attractive to and eaten by fishes, encyst in pharyngeal muscles, viscera, muscles at base of fins.
Tetracerasta blepta 1. *Posticobia brazieri* 2. *Litoria leseuri* tadpoles, *Mixophyes* tadpoles, *Bufo* marinus, *Hypseleotris galii, H. compressus, Gobiomorphus* sp., *G. australis, Ambassis* sp., *Craterocephalus stercusmuscarum, Nematocentris* spp., *Pseudomugil signifer, Retropinna semoni* D. *Anguilla reinhardtii, A. australis, Leiopotherapon unicolor, Glossamia aprion, Macquaria novemaculeata* Watson (1984)	Sporocyst and at least two redial generations present. Cercariae ophthalmotrichocercous with penetration and cystogenous glands. Positively phototactic. Usually contact and encyst in tree frog tadpoles at junction of body and tail. If eaten or inhaled by small fish or tadpoles, they encyst in the pharyngeal muscles and viscera.

Table 2.3. *Continued*

Family Opecoelidae	
Coitocaecum anaspidis 1. *Potamopyrgus badia, P. antipodum Anaspides tasmaniae, Gammarus* sp., *Paracalliope fluviatilis* D. *Salmo trutto* m. *fario, Gobiomorphus gobioides, Galaxias* spp., *Anguilla* spp. Hickman (1934); Macfarlane (1939)	Daughter sporocysts produce non-ocellate, cotylocercous xiphidiocercariae which penetrate and encyst in amphipods. Thigmotropic behaviour of *P. fluviatilis* fits well with the habits of the ground-living stumpy-tailed cercariae. Precocious metacercariae in second intermediaries.
Nicolla skrjabini 1. *Lithoglyphus naticoides* 2. *Gammarus balcanicus, Pontogammarus crassus, Dikerogammarus haemobaphes* D. *Acerina cernua, Carassius auratus gibelio* Sten'ko (1976)	Cotylomicroxiphidiocercariae produced in daughter sporocysts, penetrate and encyst in gammarids.
Sphaerostoma bramae 1. *Bythinia tentaculata, B. leachi* 2. Some spp. of snail and rarely *Herpobdella* sp. D. Many freshwater fish Wikgren (1956); Chernogorenko-Bidulina and Bliznyuk (1960)	Cotylomicrocercous xiphidiocercariae produced in daughter sporocysts. Encyst in daughter sporocysts or in other snails of the same species. Metacercariae precocious.
Plagioporus hypentelii 1. *Leptoxis (Mudalia) carinata* 2. *Sialis infumata, Culex pipiens* larvae D. *Hypentelium nigricans, Xiphophorus helleri* Hendrix (1978)	Daughter sporocysts → cotylomicrocercous xiphidiocercariae which emerge in darkness and attach to substrate. Penetrate insects and larvae through weak spots in cuticle and most encyst in abdominal haemocoel. Stylet resorbed within four days.
Plagioporus sinitzini 1. & 2. *Goniobasis livescens* D. *Nocomis biguttatus, Notropis cornutus chrysocephalus, N.c. frontalis* Dobrovolny (1939)	Daughter sporocysts in rectum of snail → cotylomicrocercous cercariae lacking stylet, encyst in sporocysts which emerge from rectum of snail to be eaten by fishes. Cercarial tail lost within cyst or before encystment. Gonads of metacercaria well defined.
Plagioporus skrjabini 1. *Theodoxus fluviatilis* 2. *Pontogammarus* spp., *Dikerogammarus* spp., *Chaetogammarus ischnus, Amathillina cristata* D. *Neogobius kessleri, N. fluviatilis, Blicca bjoerkna* Chernogorenko, Komarova and Kurandina (1978).	Daughter sporocysts → cotylocercous xiphidiocercariae shed between 1800–2400h, survival time temperature-dependent; invade gammarids. Metacercariae precocious.
Podocotyle enophrysi 1. *Lacuna marmorata* 2. *Hyale plumulosa* D. *Oligocottus maculosus* Ching (1979)	Daughter sporocysts → cotylomicrocercous xiphidiocercariae. Metacercariae not precocious.
Podocotyle reflexa 1. *Buccinum undatum* 2. *Crangon crangon, Haploops* sp., *Pandalus* spp., *Spirontocaris* spp., *Eualus hetairus*, mysids. D. *Gadus morhua, Pholis gunnellus, Rhinonemus cimbrius, Cyclopterus lumpus, Zoarces viviparus* Køie (1981)	Cotylomicrocercous xiphidiocercariae formed in daughter sporocysts, invade and encyst in amphipods, decapods and mysids.
Cainocreadium labracis 1. *Gibbula adansoni* 2. gobiid fish and other teleosts, eg. *Syngnathus* D. *Dicentrarchus labrax Maillard (1976)* Maillard (1976)	Daughter sporocysts → cotylomicrocercous xiphidiocercariae, penetrate and encyst in muscles of many teleosts, especially Gobiidae. Other members of the family use amphipod and decapod crustaceans.
Family Macroderoididae	
Macroderoides typicus 1. *Helisoma trivolvis, H. campanulata* 2. tadpoles, salamander larvae, bullheads, *Physella parkerii, P. magna lacustris.* D. *Amia calva* McMullen (1935); Cort and Ameel (1944)	Daughter sporocysts → non-oculate xiphidiocercariae with ventral tailfin which penetrate and encyst in various small fish, tadpoles and salamander larvae.

Table 2.3. *Continued*

Family Acanthostomidae	
Timoniella praeteritum 1. *Hydrobia acuta, H. ventrosa* 2. *Pomatoschistus microps Atherina mochon* D. *Dicentrarchus labrax* Maillard (1976) Maillard (1976)	Eggs presumed eaten by molluscs. Mother sporocysts → rediae → monostome, biocellate, pleurolophocercous cercariae which penetrate small teleosts and encyst in the musculature (Figure 2.32).
Family Cryptogonimidae	
Aphalloides coelomicola 1. *Hydrobia ventrosa* 2 & D. *Pomatoschistus microps* Maillard (1976)	Eggs hatch after ingestion by mollusc. Two redial generations. Cercariae biocellate, parapleurolophocercous, penetrate gobies and encyst in body cavity or caudal subcutaneous musculature. Those in body cavity encyst after 8–10 days and reach maturity. Those in muscles develop *in situ* but some die. Adults regarded as precocious metacercariae by Maillard (1976).
Caecincola latostoma 1. *Cincinnatia* (= *Amnicola*) *peracuta* 2. *Micropterus salmoides, Elassoma zonatum, Lepomis Lepomis megalotis, L. macrochirus, L. gulosus, L. symmetricus* D. *M. salmoides, M. punctulatus* Greer and Corkum (1979, 1980)	Miracidia ocellate, anterior two-thirds ciliated. Probably more than one redial generation. Cercaria biocellate, pleurolophocercous, penetrate and encyst in subcutaneous musculature and fins of fish.
Family Bucephalidae	
Bucephalus polymorphus 1. *Anodonta mutabilis, A. cyanea, Dreissena polymorpha, Unio pictorum* 2. *Rutilus rutilus, Scardinius erythrophthalmus, Blicca bjoerkna, Tinca tinca, Alburnus alburnus, Gobio gobio, Abramis brama.* D. Many freshwater fish Müller (1975); Baturo (1977); Tuffery (1978)	Branched daughter sporocysts → gasterostome cercariae which float with the furcae extended and swim upwards in response to water movement. According to Baturo (1977), cercaria contacts fish by rapidly extending a furca in its direction and attaching it to the skin. It then contracts drawing the cercarial body to the fish surface to attach itself with the adhesive glands of the tail stem. Cercaria penetrates skin using body spines and secretion of gland cells.
Rhipidocotyle illense 1. *Anodonta piscinalis, Unio pictorum* 2. *Rutilus rutilus, Blicca bjoerkna, Scardinius erythrophthalmus, Abramis brama, Leucaspis delineatus* D. *Perca fluviatilis* and other freshwater fish. Baturo (1977); Chernogorenko and Ivantsiv (1980); Ivantsiv and Chernogorenko (1984)	Cercariae produced in branched daughter sporocysts, shed between 0600 and 1400 h at 22–24°C. Activity alternates between active swimming and passive gliding. Contact fish at the gills and fins, penetrate and encyst mainly in the cephalic or caudal fin area. Adults recovered from perch 12 days after infection. According to Baturo (1977) cercariae are planktonic and enter the mouth of the fish with food or the inhalent current.
Prosorhynchus squamatus 1. *Mytilus edulis* 2. *Caprella septentrionalis Liparis liparis Myoxocephalus scorpius* D. *M. scorpius, L. liparis, Cottus bubalis, Gadus callarias, Eleginus nava, Onchocottus quadricornis, Zoarces viviparus* Chubrik (1952, 1966) Matthews (1973) Valter (1977)	Gasterostome cercariae produced in branched daughter sporocyst, contact fish by long tail furcae which contract. Lobes of tail stem produce adhesive secretion which fixes cercaria to host skin, ends of lobes invaginate forming effective attachment mechanism. Penetrate and encyst in fish. Precocious metacercariae described by Chubrik (1966) in *M. edulis*.
Bucephaloides gracilescens 1. *Abra alba* 2. *Ciliata mustela, Merlangius merlangus, Phycis blennoides, Merluccius merluccius* D. *Lophius piscatorius* Crofton and Fraser (1954) Matthews (1974)	Branched daughter sporocysts → gasterostome cercariae, penetrate and encyst in gadoid fish to which the species shows a high degree of specificity. Localize mainly in cranial nerves, found both encysted and free. If freed by breakdown of cyst wall, cercariae do not become encapsulated by host tissue as the nervous system is immunologically inactive.

Table 2.3. *Continued*

Family Derogenidae	
Genarchopsis goppo 1. *Amnicola travancorica* 2. *Stenocypris malcomsoni* Paratenic host: *Aplocheilus pancha* D. *Channa punctata* and other freshwater fish Madhavi (1978b)	Eggs contain fully developed miracidia, hatch after ingestion by mollusc. Mother sporocyst → rediae → cystophorous cercariae (syn. *Cercaria indicae* XXXV Sewell). Body and delivery tube remain outside caudal cyst while cercaria is in the redia but withdraw into it before the cercaria leaves the snail. Cercariae do not swim, lie on substrate with excretory appendage extended upwards, eaten by ostracods. Gonads well developed in metacercaria but no eggs produced.
Derogenes varicus 1. *Natica catena, N. alderi, N. pallida* 2. *Paracalanus parvus, Pseudocalanus elongatus, Temora longicornis, Acartia* sp., *Centropages hamatus, Calanus finmarchicus* D. About 100 spp. of teleosts. Køie (1979b)	Rediae → cercaria, cystophorous with a furcate appendage used for swimming (syn. *Cercaria appendiculata* Pelseneer, 1906). Body enclosed in caudal vesicle. Cercaria eaten by calanoid or harpacticoid copepods, pressure of mouth parts ruptures vesicle, delivery tube evaginates through hole and cercarial body is discharged through tube into haemocoel of copepod. Metacercaria unencysted, precocious, has been reported from body-cavity of *Sagitta elegans* and *Pagurus* pubescens and from digestive tract of other hosts. Can mature in crustaceans and chaetognaths. Gobies infected experimentally but probably act only as paratenic hosts.
Family Lecithasteridae	
Lecithaster confusus 1. *Odostomia trifida* 2. *Acartia tonsa Clupea harengus Scomber scombrus, Tautogolabrus adspersus, Fundulus* spp., *Apeltes quadracus* Hunninen and Cable (1943)	Rediae → cystophorous cercariae with delicate filaments as only external appendage. Delivery tube forcibly everted, probably penetrates gut of copepod *in vivo*, cercaria projected through it. Metacercaria unencysted.
Lecithaster gibbosus 1. *Odostomia eulimoides* 2. *Acartia* sp. D. Many marine teleosts; *Gasterosteus aculeatus* (experimental) Køie (1989)	Germinal sacs (nature not established) produce cystophorous cercariae. Copepods ingest cercariae; pressure from mouthparts causes eversion of delivery tube and injection of cercarial body into haemocoel. Metacercariae unencysted; become infective only in *Acartia* sp.
Family Hemiuridae	
Lecithochirium furcolabiatum 1. *Gibbula umbilicalis* 2. *Tigriopus brevicornis* 3. *Blennius pholis, Gobius paganellus* D. *Ciliata mustela* Matthews (1980, 1981a, b); Matthews and Matthews (1988); Gibson, Rollinson and Matthews (1985)	Filamentous unbranched daughter sporocyst, in digestive gland of prosobranch, specialized to facilitate escape of cystophorous cercariae. Harpacticoid copepods ingest cercariae, cercarial body injected into haemocoel (mesocercaria). Infected copepods eaten by rockpool fish; metacercariae become encapsulated in body-cavity. Adults in body-cavity of piscivorous fish (five-bearded rockling).
Tubulovesicula pinguis 1. *Nassarius trivittatus* 2. *Acartia tonsa* D. *Menidia menidia* and other marine fish. Stunkard (1980c)	Snails ingested eggs under experimental conditions and sporocysts were found a month later. Cercariae cystophorous. Unencysted metacercariae occur in haemocoel of copepod. The largest are well-developed juveniles with conspicuous ecsomas.
Lecithaster gibbosus 1. *Odostomia eulimoides* 2. *Acartia* sp. D. Many marine teleosts; *Gasterosteus aculeatus* (experimental). Køie (1989)	Germinal sacs (nature not established) produce cystophorous cercariae. Copepods ingest cercariae; pressure from mouthparts causes eversion of delivery tube and injection of cercarial body into haemocoel. Metacercariae unencysted; become infective only in *Acartia* sp.
Family Ptychogonimidae	
Ptychogonimus megastomus 1. *Antalis vulgaris* (= *Dentalium tarentinum*), *A. inaequicostatum* (= *D. alternans*) 2. Crustaceans (Oxystomata, Oxyrhyncha, Brachyrhyncha)	Miracidia have bristles at anterior end; penetrate scaphopod molluscs. Daughter sporocysts containing 20–30 cercariae emerge from mollusc and are eaten by crabs. Cercariae penetrate gut wall and enter body cavity, where they become unencysted metacercariae which may develop

Table 2.3. *Continued*

D. Carchariniform sharks (*Prionace, Mustelus, Galeus, Dasyatis, Scyliorhinus*) Palombi (1941, 1942a, b); Gibson and Bray (1977)	precociously but usually mature when crab eaten by shark. Reports of adults from teleosts (eg. *Dentex*) may refer to accidental infestations.

Superfamily Didymozoidea

No complete life-cycles known. Adults occur mainly in the viscera and musculature of marine and, more rarely, freshwater fish, sometimes surrounded by connective tissue capsule of host origin, eg. *Nematobothrium spinneri* in muscles of *Acanthocybium solandri*. Only *Kollikeria filicollis* secretes a cyst. Eggs escape by ulceration of nodules, causing discharge of worms (e.g. *Neometadidymozoon helicis* in buccal cavity and gill arches of *Platycephalus fuscus*), by discharge with fish spawn (*Nematobothrium texomensis* in ovaries of *Ictiobus bubalis*) or by predation. Eggs of *Neometadidymozoon helicis* and *Didymozoon brevicolle* pass unharmed through gut of *P. fuscus*, suggesting predation as a means of dispersal. Miracidium of *Nematobothrium texomensis* has spinelets scattered over the body and in three groups apically. It resembles some hemiuroid miracidia (halipegids). Cystophorous marine cercariae thought to be didymozoids have been reported and metacercariae have been found in copepods and cirripedes. Various fishes and chaetognaths act as paratenic hosts. The life-cycle and morphology of the larvae indicates that didymozoids are digeneans closely related to the hemiuroids (Cable and Nahhas, 1962a; Madhavi, 1968; Lester, 1980; Stunkard, 1963; Self *et al.*, 1963; Williams, 1959).

Nematoda

Introduction

The life-cycles of most fish nematodes require an intermediate host for their completion but there is evidence that an intermediary is not obligate for some species, e.g. some cucullanids. In some families, use of paratenic or transport hosts is common: this is particularly so in the Camallanidae and Anisakidae. Transport hosts have the potential to increase the host range of the species by means of predator-prey relationships and, in some cycles, they are essential for transmission to a final host which may not feed directly on the infected intermediary. Some life-cycles of fish nematodes are presented in Table 2.4. The life-cycles of freshwater fish nematodes have been comprehensively discussed by Moravec (1984a), who formulated general principles of their bionomics and analyzed the roles of the different types of host which may participate.

Anderson (1984, 1988) has reviewed the transmission patterns of the nematodes, emphasizing the importance of monoxeny, heteroxeny, paratenesis and precocity. He believed nematode parasites of fish to be derived from those of early terrestrial vertebrates which had developed transmission strategies that enabled them to colonize fish. These strategies were principally the development of a heteroxenous life-cycle, i.e. the acquisition of intermediate hosts, and paratenesis, i.e. the use of paratenic or transport hosts. Both these strategies are highly significant in the transmission of fish nematodes.

Nematodes which mature in fish belong mainly to the secernentean orders Ascaridida (superfamilies Seuratoidea and Ascaridoidea) and Spirurida (superfamilies Camallanoidea, Habronematoidea, Thelazioidea, Gnathostomatoidea, Dracunculoidea, Physalopteroidea) and the adenophorean superfamily Trichinelloidea.

Anderson (1988) has emphasized that in the secernenteans it is the early third-stage larva which is the infective or invasive stage.

The Spirurida generally use arthropod intermediate hosts. The Ascaridoidea utilize both vertebrates and invertebrates as intermediaries, the former being the more primitive host group (Anderson, 1984). In both the Spirurida and Ascaridoidea, precocity, which Anderson (1988) defines as growth and/or development beyond that expected in an intermediate host, is a feature of many life-cycles and is probably aimed at accelerating the onset of reproduction after the nematodes reach the definitive host.

Table 2.4. Definitive, intermediate and paratenic hosts of some fish nematodes

Species	Definitive host	Intermediate host	Paratenic host	Reference
		Superfamily Seuratoidea		
Family Cucullanidae				
Cucullanus truttae	Salmonid fish, *Lampetra planeri* adults	*L. planeri* larvae		Moravec (1976a, 1979a, 1980b); Moravec and Malmqvist (1977)
(as *Truttaedacnitis stelmioides*)	*L. lamottenei* adults	*L. lamottenei* ammocoetes		Pybus *et al.* (1978)
Cucullanus minutus	*Platichthys flesus*			Gibson (1972b)
C. heterochrous	*P. flesus*			Gibson (1972b)
Dacnitis sphaerocephalus caspicus	*Acipenser guldenstadti*	?*Nereis diversicolor*		Khromova (1975)
Family Quimperiidae				
Paraquimperia tenerrima	*Anquilla anquilla*		Some invertebrates and forage fish assumed to act as transport hosts for Lsm.inf.	Moravec (1974b)
		Superfamily Ascaridoidea		
Family Acanthocheilidae				
Pseudanisakis rotundata Uspenskaya (1955)	*Raja radiata*, other elasmobranchs	*Lithodes* sp. (decapod)	Pleuronectid and gadid fish	
Family Anisakidae				
Raphidascaris acus	*Salmo trutta* m. *fario*, *Parasalmo gairdnerii* and other fish	*Cottus gobio*, *C. poecilopus*, *Lampetra planeri*, *S. trutta* m. *fario*	Chironomid larvae, oligochaetes, gastropods, copepods and malacostracans	Moravec (1970a, b, 1980c); Alvarez-Pellitero (1979b)
Goezia ascaroides	*Silurus glanis*	*Diaptomus castor*	Young bream, white bream, bleak, rudd, catfish	Mozgovoi *et al.* (1971)
Thynnascaris aduncum	Many genera of marine fish	*Acartia bifilosia*, *Eurytemora affinis*, *Jaera albifrons*, *Lepidonotus*, *Nereis*, *Harmothoe*, *Gattiana*, many other invertebrates	Many species of invertebrates and marine fish	Markowski (1937); Val'ter (1980); Popova *et al.* (1964); Norris and Overstreet (1976)[1]
		Superfamily Camallanoidea		
Family Camallanidae				
Camallanus sweeti	*Ophicephalus gachua*	*Mesocyclops leuckarti*, *M. hyalinus*	*O. gachua*, *Barbus puckelli*, *B. (Puntius) ticto*, *Lepidocephalichthys thermalis*, *Gambusia* sp.	Moorthy (1938)
Camalanus oxycephalus	*Morone chrysops* and other freshwater fish	*Cyclops vernalis*, *C. bicuspidatus*, *Diaptomus* sp.	*Dorosoma cepedianum*	Stromberg and Crites (1974a)
Camallanus truncatus	*Lucioperca lucioperca* x *L. volgensis*	*Macrocyclops albidus*, *Cyclops strenuus*, *Megacyclops viridis*		Moravec (1971b)
Camallanus lacustris	*Perca fluviatilis* and other freshwater fish	*Megacyclops viridis*, *Macrocycops albidus*, *Acanthocyclops vernalis*, *Eucyclops serrulatus*,	Non-predatory cyprinids and cobitids	Moravec (1969a, 1971b); Campana-Rouget (1961a)
		Cyclops strenuus, *A. viridis*	cobitids	

Table 2.4. *Continued*

Camallanus adamsi	*Channa (Ophicephalus) punctatus*	*Mesocyclops leuckarti*		Bashirulla and Ahmed (1976a)
Camallanus fotedari	*Lebistes reticulatus, Danio rerio*	*Cyclops*		Campana-Rouget *et al.* (1976)
Spirocamallanus cricotus	*Micropogonias undulatus*	*Tigriopus californicus, Penaeus setiferus*		Fusco (1980)
Spirocamalllanus intestinecolas	*Mystus vittatus*	*Mesocyclops leuckarti, Thermocyclops crassus*		Bashirullah and Ahmed (1976b)
Spirocamallanus pereirai	*Leiostomus xanthurus, Micropogon undulatus*	*Penaeus setiferus*		Overstreet (1973)
Procamallanus laeviconchus	Siluroid fish, eg. *Clarias lazera*	*Mesocyclops leuckarti*	*Gambusia affinis*	Moravec (1975)
Procamallanus cearensis	*Astyanax bimaculatus vittatus*	*Diaptomus cearensis, D. azevedoi*	*Curimatus elegans* fry	Pereira *et al.* (1936)
Paracamallanus cyathopharynx	*Clarias lazera* and other clariid fish	*Mesocyclops leuckarti*	Present but not named	Moravec (1974a)

Superfamily Habronematoidea

Family Cystidicolidae

Cystidicola cristivomeri	*Salvelinus namaycush, S. alpinus*	*Mysis relicta*		Black and Lankester (1980, 1981); Smith and Lankester (1979)
C. farionis	*Salmo trutta* m. *fario, S. gairdneri*	*Gammarus pulex, G. fasciatus, **Hyalella azteca, Pontoporeia affinis***	*Coregonus clupeaformis*	Awachie (1973); Black and Lankester (1980); Smith and Lankester (1979)
Cystidicoloides ephemeridarum (syn. ***C. tenuissima, Metabronema salvelini***)	***Salmo trutta*** m. ***fario, Salvelinus fontinalis***	***Ephemera*** sp. ***E. danica, Habroleptoides modesta, Habrophlebia lauta, Hexagenia recurvata, Polymitarcys*** sp.	*Cottus gobio,* ***Noemacheilus barbatulus***	Moravec and De (1982); Moravec (1971c, d); De and Moravec (1979); Choquette (1955)
Spinitectus gracilis	***Lepomis cyanellus, Ambloplites rupestris*** and many other fresh-water fish	Mayfly naiads, dragonfly nymphs, stonefly larvae, collembolan larvae	Lsm.inf. and Lsm.inf. could be transferred from fish to fish by gavage	Jilek and Crites (1982a)
S. carolini	***Lepomis macrochirus***	Mayfly naiads, dragonfly nymphs, stonefly larvae, midge larvae	Lsm.inf., Lsm.inf. and adults could be transferred from fish to fish by gavage	Jilek and Crites (1982b)
S. micracanthus	***Lepomis macrochirus***	***Hexagenia*** sp. naiads		Keppner (1975)
Salvelinema walkeri	Salmonid fish	***Ramellogammarus vancouverensis***		Margolis and Moravec (1982)
Ascarophis pseudoargumentosa	***Acipenser brevirostrum***	***Gammarus tigrinus, G. fasciatus***		Appy and Dadswell (1983)
Ascarophis morhua	***Gadus morhua, Melanogrammus aeglefinus***	***Pagurus*** sp., ***Carcinus maenas***		Uspenskaya (1953); Poinar and Thomas (1976)
A. filiformis	***G. morhua, M. aeglefinus***	***Sclerocrangon*** sp.		Uspenskaya (1953)

Superfamily Thelazioidea

Family Rhabdochonidae

Rhabdochona phoxini	***Phoxinus phoxinus***	Mayfly nymphs, (***Ephemera, Habrophlebia, Ecdyonurus***)		Moravec (1976b, 1977a)
R. ergensi	***Noemacheilus barbatulus***	Mayfly nymphs (***Habroleptoides modesta***)		Moravec (1972)

Table 2.4. *Continued*

		Superfamily Gnathostomatoidea		
Family Gnathostomatidae				
Echinocephalus overstreeti	***Heterodontus portusjacksoni*** (main host)	***Chlamys bifrons, Pecten albus***		Beveridge (1987); Andrews, Beveridge, Adams and Baverstock (1988)
E. pseudouncinatus	*Heterodontus francisci, Myliobatis californicus*	*Haliotis corrugata*		Milleman (1963)
E. sinensis	*Aetobatus flagellum*	*Crassostrea gigas*		Ko *et al.* (1975)
E. uncinatus	*Trygon, Myliobatis, Balistes* and other fish	*Hemifuscus pugilinus*		Anantaraman (1964)
		Superfamily Dracunculoidea		
Family Philometridae				
Philometra abdominalis	*Phoxinus phoxinus, Gobio gobio, Leuciscus cephalus, L. leuciscus, Barbus meridionalis*	*Macrocyclops albidus, M. fuscus, Megacyclops viridis, Acanthocyclops vernalis, Diacyclops bisetosus*		Molnár (1967); Moravec (1977b, c)
P. cylindracea	*Perca flavescens*	*Cyclops vernalis*		Molnár and Fernando (1975)
P. obturans	*Esox lucius*	*Cyclops strenuus, Eucyclops serrulatus, Acanthocyclops vernalis, A. viridis, Macrocyclops albidus, M. fuscus*	reservoir hosts: *Perca fluviatilis, Scardinius erythrophthalmus*	Moravec and Dyková (1978); Moravec (1978a, b)
P. ovata	*Abramis brama, A. ballerus, Rutilus rutilus* and others	*Acanthocyclops vernalis, Cyclops strenuus, Macrocyclops albidus, Megacyclops gigas*		Moravec (1980a)
P. lusiana	*Cyprinus carpio*	*Acanthocyclops viridis, A. vernalis, Eucyclops macruroides, Cyclops strenuus, Mesocyclops leuckarti*		Vasil'Kov (1968)
Philometroides sanguinea	*Carassius carassius, C. auratus gibelio*	*Cyclops strenuus, Acanthocyclops viridis, Mesocyclops leuckarti*		Yashchuk (1974, 1975)
P. huronensis	*Catostomus commersoni, C. catostomus, Moxostoma macrolepidotum*	*Cyclops bicuspidatus thomasi, C. vernalis*		Uhazy (1977a, b)
P. nodulosa	*Catostomus commersoni*	*Cyclops vernalis*		Dailey (1966)
P. carassii	*Carassius carassius*	*Cyclops strenuus, C. vicinus, Tropocyclops prasinus*		Nakajima and Egusa (1977a)
Philonema oncorhynchi	*Oncorhynchus nerka* and other salmonids	*Cyclops bicuspidatus, C. b. thomasi*		Ko and Adams (1969); Platzer and Adams (1967)
P. agubernaculum	*Salmo gairdneri* and other salmonids	*Cyclops bicuspidatus*		Ko and Adams (1969)
		Superfamily Physalopteroidea		
Family Physalopteridae				
Proleptus acutus	*Raja clavata* and other elasmobranchs	*Carcinus maenas, Pagurus* sp.		Wülker (1929)
P. obtusus	Several genera of elasmobranchs	*C. maenas, Eupagurus bernhardus*		Lloyd (1928)

Table 2.4. *Continued*

		Superfamily Trichinelloidea	
Family Cystoopsidae			
Cystoopsis acipenseris	*Acipenser ruthenus*	*Dikerogammarus haemobaphes* *Gammarus platycheir*	Janicki and Rasin (1930)
Family Trichuridae			
Capillaria tuberculata	*Acipenser ruthenus*	*Gammarus* sp.	Markevich (1951)
C. tomentosa	*Idus melanotus*, *Leuciscus idus*, *Scardinus erythrophthalmus*, other cyprinids	*Cyclops* sp.	Markevich (1951)
C. brevispicula	*Blicca bjoerkna*, *Lota vulgaris*, *Leuciscus idus*, *Rutilus rutilus*	*Gammarus* sp.	Markevich (1951)

[1] Norris and Overstreet (1976) listed as invertebrate hosts (some experimental) three named species of gastropods, four of annelids, and seven of copepods, one isopod, one amphipod and one decapod, but noted that many serve as transfer hosts rather than as true intermediaries. Copepods, especially, transmit infection to planktonivorous fish.

Early development

Development in the nematodes follows a basic pattern of five stages, each separated from its predecessor by a moult, the last of which is a juvenile adult. Gravid females in the definitive host produce eggs which, when released into water, may be non-embryonated (e.g. anisakids, some cucullanids) or partly or fully embryonated (e.g. rhabdochonids, cystidicolids). Philometrids and camallanids, however, are viviparous, releasing fully-developed first-stage larvae which are ingested by the intermediaries. Anisakid and cucullanid eggs develop in water; the first moult takes place within the egg in anisakids and in some cucullanids and second-stage larvae usually hatch out. They are infective to the intermediate host which may consume either eggs containing the Lsm.inf., or free Lsm.inf.. Rhabdochonid and cystidicolid eggs hatch after ingestion by the intermediate host.

The rate of development of eggs in water and the survival times of eggs and first- or second-stage larvae is temperature-dependent (Tables 2.5, 2.6). Survival declines with increasing temperature and, in some species e.g. *Camallanus oxycephalus*, the ability of larvae to infect the intermediate host declines with age (Stromberg and Crites, 1974b). The efficiency with which larvae of *C. oxycephalus* penetrated the gut wall of the copepod intermediary decreased logarithmically with age, the deterioration being rapid between 20° and 25°C. None succeeded after 18 days at 25°C but some did so after 28 days at 20°C. Larval activity was the most important factor in penetrating the gut and it declined in older larvae. Moravec (1969a) noted that first; second- and third-stage larvae of *C. lacustris* could escape from the body of a dead copepod but only young Lsm.inf. could reinfect a new host.

The behaviour of larvae of some camallanids, philometrids and cucullanids maximizes their chances of infecting an intermediate host, usually by attracting its attention by active movement. First-stage larvae of *Spirocamallanus intestinecolas, Camallanus adamsi, C. oxycephalus* and *Paracamallanus laeviconchus* attach themselves to the substrate by their tails and undulate vigorously. In *C. adamsi* this behaviour declines with age which suggests that young larvae, which undulate most energetically and are the most efficient at penetrating the intermediate host's gut wall, are those most likely to attract an intermediary (Stromberg and Crites,

Table 2.5. Egg hatching times

Species	Temperature °C	Time (days)	Reference
Cucullanus truttae	22–24	7–8	Moravec (1979a)
C. minutus	19	7	Gibson (1972b)
C. heterochrous	19	7	Gibson (1972b)
Paraquimperia tenerrima	20–25	5–6	Moravec (1974b)
Raphidascaris acus	22	7–30	Moravec (1970a)

Table 2.6 Survival of eggs or larvae in water

Species	Temperature °C	Stage	Survival time (days)	Reference
Camallanus oxycephalus	20	L1	39	Stromberg and Crites (1974b)
	25	L1	24	
C. lacustris	22	L1	12	Moravec (1969a)
	7	L1	80	
C. adamsi	27	L1	17	Bashirullah and Ahmed (1976a)
	24	L1	28	
Spirocamallanus intestinecolas	26	L1	17	Bashirullah and Ahmed (1976b)
Cucullanus truttae	13	L2	75	Pybus *et al.* (1978)
Raphidascaris acus	22	L2	1–2	Moravec (1970a)
	7	L2	3–10	
Philometra abdominalis	7	L1	25	Moravec (1977c)
	15–18	L1	16	
	20–24	L1	7	
	30–34	L1	2	
P. obturans	5	L1	26	Moravec (1978b)
	20–22	L1	7	
Philonema oncorhynchi	10	L1	25	Ko and Adams (1969)
Goezia ascaroides	?	eggs	21	Mozgovoi *et al.* (1971)

Table 2.7. Penetration of larvae into the haemocoel of the intermediate host (in hours (h) or days (d) after ingestion of eggs or larvae)

Species	Temperature °C	Time post ingestion	Reference
Camallanus sweeti	–	4– 8 h	Moorthy (1938)
C. oxycephalus	–	2h	Stromberg and Crites (1974b)
C. lacustris	–	1h	Moravec (1969a)
C. lacustris	20–24	3–8 h	Campana-Rouget (1961a)
C. adamsi	–	1h	Bashirullah and Ahmed (1976a)
Spirocamallanus intestinecolas	25–26	1h	Bashirullah and Ahmed (1976b)
Paracamallanus cyathopharynx	–	several hours	Moravec (1974a)
Cystidicola farionis	19	24h	Smith and Lankester (1979)
	5–7	10–20 d	
Cystidicoloides ephemeridarum	13–15	by 24h	Moravec (1971c)
Spinitectus gracilis	20	by 6h	Jilek and Crites (1982a)
S. carolini	20	by 6h	Jilek and Crites (1982b)
S. micracanthus	21–23	by 12h	Keppner (1975)
Rhabdochona ergensi	13–15	by 24h	Moravec (1972)
Philometra ovata	25	1h	Moravec (1980a)
Philonema oncorhynchi	10	5–5.5h	Ko and Adams (1969)

1974b). Because *Camallanus* larvae are released into the plankton and sink to the bottom, their exposure to copepods is limited. Although they can survive for relatively long periods, this is of limited value in, e.g. *C. oxycephalus* and other species in temperate zones whose intermediaries are abundant for only a few weeks each year. Constant expenditure of energy by the Lsm.inf. is, therefore, probably a worthwhile investment in attracting a copepod. *Philometra abdominalis* Lsm.inf. similarly attract copepods by active movement (Moravec, 1977c) but those of *Philonema oncorhynchi* tend to remain in suspension after release from gravid females, thus prolonging their availability to their planktonic intermediaries (Platzer and Adams, 1967). In the Cucullanidae, *Cucullanus truttae* larvae are not very mobile, tending to remain spirally coiled and sometimes straightening and contracting vigorously, but those of *C. minutus* and *C. heterochrous* are very active and are capable of swimming (Moravec, 1979a; Pybus *et al.*, 1978; Gibson, 1972b). Penetration of the first-stage larvae into the haemocoel of the intermediate host is achieved in times varying from less than an hour in some camallanids (Table 2.7) to several hours or even days. The process is affected by temperature since larvae of *Cystidicola farionis* require 24 hours at 19°C but 10–20 days at 5°–7°C.

Of the Spirurida of fishes, larval development is best known in the families Camallanidae, Cystidicolidae and Philometridae, in which the first two moults take place within the intermediary and the third and fourth moults in the definitive host. Developmental times are presented in Table 2.8. Development is not only temperature-dependent but may also differ in different species of intermediaries: *Cystidicola farionis* developed to the third stage in five weeks in *Gammarus fasciatus* at 12–14°C as compared with six weeks in *Hyalella azteca* at the same temperature (Smith and Lankester, 1979). Maturation times of some species in their definitive hosts are presented in Table 2.9.

Life-Cycle Patterns

Secernentea
Order Ascaridida
Superfamily Seuratoidea
Families Cucullanidae and Quimperiidae

Cucullanid eggs may or may not be partly embryonated when deposited and develop in water at a temperature-dependent rate. The first moult occurs inside the egg in *Cucullanus truttae* and may do so in *C. minutus* (Moravec, 1979a; Gibson, 1972b). In *C. heterochrous* and *Paraguimperia tenerrima*, it takes place in water (Gibson, 1972b; Moravec, 1974b). Moravec (1976a, 1979a, 1980b) has shown that the life-cycle of *C. truttae* in salmonids in the Elbe Basin, Czechoslovakia, shows great inherent variability (Figure 2.33). Encapsulated Lsm.inf. have been found in larval *Lampetra planeri*, detritus feeders which ingest second-stage larvae, free or still within the egg. Second-stage larvae failed to infect trout directly when given by stomach tube, suggesting that an intermediate host is obligatory. Third-stage larvae eaten by trout undergo the third moult in the intestine and migrate to the pyloric caeca where they complete their development. Lampreys may also become definitive hosts since third-stage larvae not encapsulated at metamorphosis survive to mature in adult lampreys. Trout acquiring such parasites from adult lamprey have the status of postcyclic hosts. Since larval and adult lampreys are rather large prey for trout, only fish over two years old become infected with *C. truttae*, but lampreys are the main and obligatory intermediaries. Similarly, an obligate intermediary, possibly *Nereis diversicolor*, is thought to exist in the life-cycle of *Dacnitis sphaerocephalus caspicus*, larvae of which failed to infect sturgeons directly (Khromova, 1975). In contrast, Gibson (1972b) thought it possible that *C. heterochrous*

Table 2.8. Developmental times of larval stages in the intermediate and definitive hosts

	Temperature °C	First moult (M1)	Second moult (M2)	Third moult (M3)	Fourth moult (M4)	Reference
Family Camallanidae						
Camallanus sweeti	32–38	24–36h	5–7d	–	–	Moorthy (1938)
O. oxycephalus	25	3d	6d	–	–	Stromberg and Crites (1974a)
	26	–	–	9–10d	♂ 17–18d ♀ 24d	
C. truncatus	?	7d	20d	–	–	Moravec (1971b)
C. lacustris	20–25d	4d	11–12d	13–15d	♂ 35–69d ♀ 65–69d	Moravec (1969a)
	20–24	4–5d	8d	–	–	Campana-Rouget (1961a)
C. adamsi	24	117–144h	190–222h	–	–	Bashirullah and Ahmed (1976a)
	27	72–87h	135–150h	–	–	
Spirocamallanus intestinecolas	25	39h	72h	–	–	Bashirullah and Ahmed (1976b)
Procamallanus laeviconchus	23–24	1–2d	5–6d	–	–	Moravec (1975)
Paracamallanus cyathopharynx	23–24	1–2d	7–9d	–	–	Moravec (1974a)
Family Cucullanidae						
Cucullanus truttae	22–24	6–7d[1]	?	20d	30–40d	Moravec (1979a)
Family Cystidicolidae						
Cystidicola cristivomeri	4–5	10 wks	17 wks	–	–	Smith and Lankester (1979);
	9–11	3 wks	8 wks	–	–	Black and Lankester (1980)
	4–10	–	–	20d	♂ 84–104d ♀ 85–203d	
Cystidicola farionis	12–14	3wks[2]	5 wks[2]	–	–	Smith and Lankester (1979)
	16–18	2.5 wks	3 wks	–	–	
	4–10	–	–	♂ 1d ♀ 12d	♂ 74d ♀ 111d	Black and Lankester (1980)
Cystidicoloides ephemeridarum	13–15	5–18d	23–26d	?	12–20d	Moravec (1971c);
	7–15	–	–	15d	♂ ? ♀ 32d	De and Moravec (1979)
Spinitectus gracilis	20	18–36h	8d	2d	? 15d	Jilek and Crites (1982a)
S. carolini	20	18–36h	8d	2d	14d	Jilek and Crites (1982b)
S. micracanthus	21–23	6–10d	19–20d	–	16d	Keppner (1975)
Ascarophis pseudo-argumentosa	21–25	–	10–15d	–	–	Appy and Dadswell (1983)
	10–14	–	28–40d	–	–	
	15–16	–	–	15d	–	
Family Rhabdochonidae						
Rhabdochona phoxini	13–15	12–16d	20–36d	3d	22–38d	Moravec (1976b, 1977a)
R. ergensi	13–15	13d	16d	from 22d	♂ 6–20d ♀ 33–43d	Moravec (1972)
Family Anisakidae						
Raphidascaris acus	22	4d	15–55d	3–4d	21d	Moravec (1970a)
Family Philometridae						
Philometra abdominalis	20–24	5–6d	7–9d	–	–	Moravec (1977c)
P. obturans	20–22	5–7d	9–11d	–	–	Moravec (1978b)
	10	30–35d	–	–	–	
P. ovata	25	3–4d	5–7d	–	–	Moravec (1980a)

Table 2.8. *Continued*

P. lusiana	20–22	?3–4d	7–8d	14–17d	20–21d	Vasil'kov (1968)
Philometroides huronensis	10	14–18d[3]	30d[3]	–	–	Uhazy (1977a)
	23	6–9d[4]	14–20d[4]	–	–	
P. nodulosa	room temp.	11–14d	18–20d	–	–	Dailey (1966)
Philonema	4	23–25d	74–78d	–	–	Ko and Adams (1969);
oncorhynchi	10	13d	30–31d	–	–	Bashirullah (1973)
	15	6d	17–19d	–	–	
	8–10	–	–	240–360d	–	
P. agubernaculum	10	12–15d	30–34d	–	–	Ko and Adams (1969);
	5–14	–	–	from 45d	from 105d	Adams (1974);
	8–10	–	–	58d	–	Bashirullah (1973)

1. The first moult in *C. truttae* occurs within the egg.
2. In *Gammarus fasciatus*: developmental times differ in other intermediaries.
3. In *Cyclops bicuspidatus thomasi*
4. In *C. vernalis*

Table 2.9. Maturation in the definitive host

Species	Time postinfection		Reference
Camallanus lacustris	3 months.		Moravec (1969a)
Cucullanus truttae	3 months		Moravec (1979a)
Cystidicola cristivomeri	mature ♂	67d	Black and Lankester (1980)
	mature ♀	210d	
C. farionis	mature ♂	112d	Black and Lankester (1980)
	mature ♀	235d	
Cystidicoloides ephemeridarum	adult ♂	12d	Moravec (1971c)
	young ♀	20d	
Spinitectus gracilis	ovigerous ♀	24d	Jilek and Crites (1982a)
S. carolini	gravid ♀	21d	Jilek and Crites (1982b)
S. micracanthus	mature ♂	26d	Keppner (1975)
	mature ♀	36d	
Rhabdochona phoxini	2 months		Moravec (1976b)
R. ergensi	adult ♂	20d	Moravec (1972)
	juvenile ♀	33d	
	gravid ♀	43d	
Raphidascaris acus	mature ♂	33d	Moravec (1970a)
	gravid ♀	2 months	

Lsm.inf. infected flounder directly. Third-stage larvae occur encapsulated in the gut wall, moulting inside the capsules and subsequently emerging to wander through the gut tissue. Immature adults inhabit the lumen but the site of the fourth moult is not known. The life-cycle of *C. minutus* may involve an intermediary, but attempts to infect a variety of invertebrates failed. Flounders have not been found to harbour second-stage larvae of either species. Again, Moravec (1974b) failed to establish *Paraquimperia tenerrima* in *Tubifex tubifex* and *Lymnaea peregra* and recovered only one Lsm.inf. from eels, the natural definitive hosts, given Lsm.inf. by stomach tube. Invertebrates and small forage fish are presumed to act as transport hosts for this species.

Superfamily Ascaridoidea: Family Anisakidae

Of the fresh water anisakids, the life-cycle of *Raphidascaris acus* is the best known. Studies in Czechoslovakia have shown that the first moult occurs within the egg and that second-stage larvae, free or still unhatched, are eaten by a variety of invertebrates, predominantly chironomid larvae (*Prodiamesa olivacea*), by lampreys or by small fish. Development does not continue in the invertebrates, which serve as preintermediate paratenic hosts, but in fish the larvae penetrate the gut wall to enter the body cavity or liver, eventually becoming encapsulated. The third moult takes place in this host and development continues following ingestion of the fish by trout or another predator (Moravec, 1970a, b, 1980d). Trout have been found to serve as both the definitive host and the main intermediary in Leon, Spain (Alvarez-Pellitero, 1979b). Precocity is a marked feature of the life-cycle of *R. acus* (Anderson, 1988). The use of paratenic hosts is a frequent characteristic of anisakid life-cycles and many species employ fish in this role.

Two marine species, *Thynnascaris aduncum* and *Anisakis simplex* (the latter adult in cetaceans) have invertebrate intermediaries in which the second moult occurs (see Table 2.4 and Smith, 1983a,b). Third-stage larvae in copepods and isopods (*T. aduncum*) and in euphausiids (*A. simplex*) are directly infective to the final hosts but, in both species, a wide range of transport hosts, usually fishes, are interpolated as a result of predator-prey relationships and serve to distribute the parasite larvae both in space and in time. Anderson (1988) states that precocity occurs in the cycles of marine ascaridoids, including those maturing in carnivorous fish, in both the vertebrate intermediary and in what was primitively the invertebrate paratenic host.

Order Spirurida

Spirurids produce eggs containing first-stage larvae and use arthropod intermediaries.

Superfamily Camallanoidea:
Family Camallanidae

Camallanids are ovoviviparous and first-stage larvae are released when the female's body ruptures in water, eg. by protruding from the anus of the host as in *Camallanus oxycephalus* (Stromberg and Crites, 1974a); in *Paracamallanus cyathopharynx*, the larvae are liberated into the gut of the host and pass out with the faeces (Moravec, 1974a). The first two moults take place in the intermediate host (usually cyclopoid copepods) and the last two in the fish. The growth pattern of *C. oxycephalus* in white bass is temperature-related but may also be regulated by the hormones of the host. The dispersal period coincides with the maximum density of the intermediate host, favouring successful transmission (Stromberg and Crites, 1975a). Camallanids commonly incorporate transport hosts into the life-cycle. Adults and L4 of *C. lacustris* have been successfully transferred between pike by stomach tube, suggesting that predation is a common mode of transmission (Moravec, 1971a).

The life-cycle of *Camallanus oxycephalus* is illustrated in Figure 2.34. First-stage larvae ingested by *Cyclops vernalis* and *C. bicuspidatus* penetrate the gut wall to enter the haemocoel where they moult twice (Stromberg and Crites, 1974a). When the copepod is eaten by a fish, the third-stage larvae are stimulated to activity by the presence of bile and escape from the body of the copepod. In some fish, eg. white bass (*Morone chrysops*), the larvae develop through a further two moults to maturity but, in others, development does not proceed beyond the fourth stage. When infected copepods are eaten by small planktonivorous fish, eg. gizzard shad (*Dorosoma cepedianum*), the final moult is inhibited and takes place only after the host is eaten by a larger predatory species. The life-cycle may be completed directly by ingestion of copepods by the definitive host or indirectly through the incorporation of a forage fish transport host.

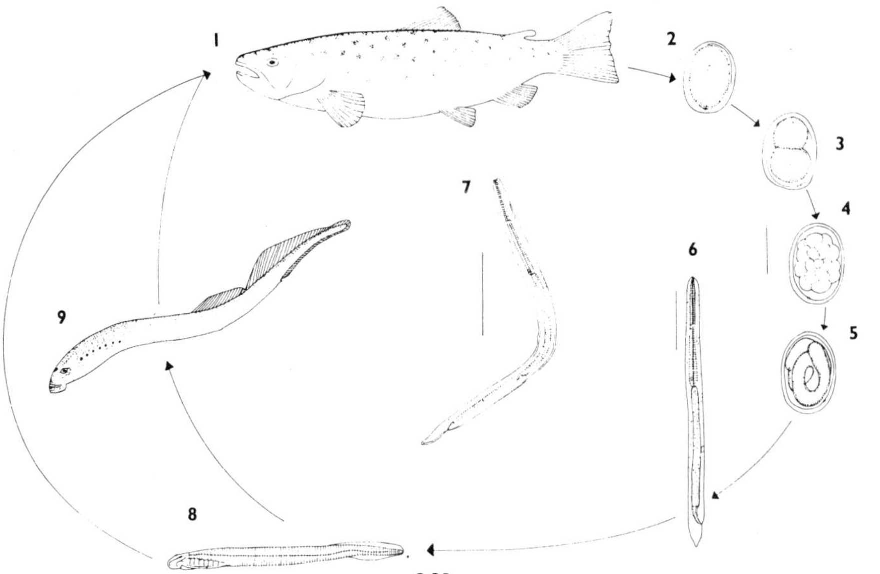

2.33

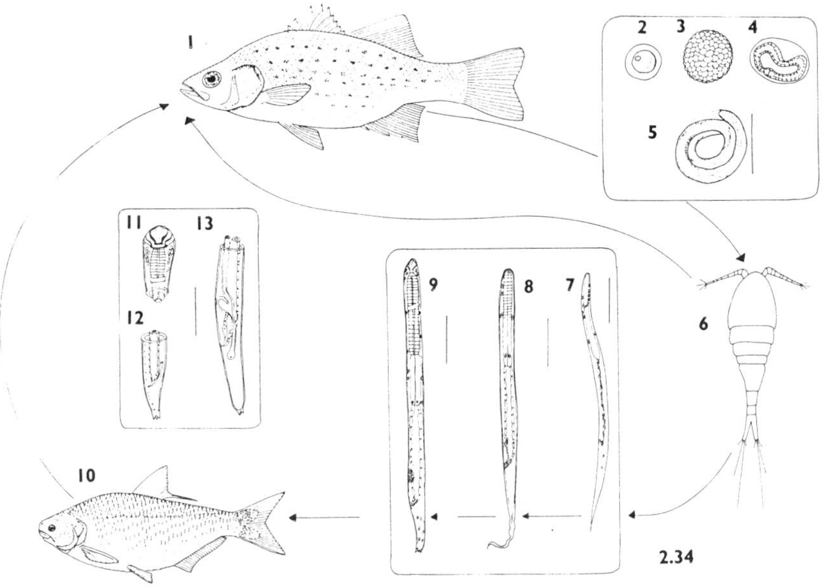

2.34

Figure 2.33.1–9 The life-cycle of *Cucullanus truttae*.
1. Definitive host, *Salmo trutta* m. *fario*. 2–5. Stages in the development of the first-stage larva. 6. Newly-hatched second-stage larva. 7. Third-stage larva from larval lampreys. 8. *Lampetra planeri*, larval stage. 9. *L. planeri* adult.
Scale bars: 2–5, 0.05 mm; 6, 0.1 mm; 7, 0.2 mm.
Figures 2–7 after Moravec (1979a).

Figure 2.34.1–13 The life-cycle of *Camallanus oxycephalus*.
1. Definitive host, *Morone chrysops*. 2–5. Developmental stages of *C. oxycephalus*. 2. Zygote. 3. Blastula. 4. Developing larva. 5. Developing first-stage larva enclosed in egg membrane. 6. Cyclopoid copepod intermediary. 7. First-stage larva. 8. Second-stage larva. 9. Third-stage larva. 10. Transport host, *Dorosoma cepedianum*. 11–13. Fourth-stage larva. 11. Anterior end. 12. Male tail just before final moult. 13. Female tail just before final moult.
Scale bars: 2–5, 0.06 mm; 7–9, 11–13, 0.15 mm.
Figures 2–5, 7–9, 11–13 after Stromberg and Crites (1974a).

Superfamily Habronematoidea:
Family Cystidicolidae

Cystidicolid eggs are fully embryonated when deposited and hatch after ingestion by benthic intermediaries in which the larvae moult twice. The intermediaries of most species for which the life-cycles are known, which are mainly freshwater species, are insect nymphs and naiads, although amphipods have a role in some freshwater and marine cycles and crabs and other large crustaceans have been implicated in the cycle of species of *Ascarophis*. Within the intermediary, *Spinitectus* species become encapsulated within the abdominal muscles and undergo the second moult within the capsule. Third-stage larvae of *Cystidicola cristivomeri* in *Mysis relicta* also become encapsulated but those of *C. farionis* and *Cystidicoloides ephemeridarum* (synonyms *C. tenuissima, Metabronema salvelini*) remain free in the body cavity (see Table 2.4 for references).

Transport hosts have been reported for several species, e.g. *Cystidicola farionis* and *Cystidicoloides ephemeridarum*, and the latter was capable of developing to the fourth stage in experimentally-infected *Noemacheilus barbatulus* (De and Moravec, 1979), which is therefore considered to be a metaparatenic host. In the natural definitive host, trout, the two final moults take place in the stomach. Third- and fourth-stage larvae of *Spinitectus gracilis* and *S. carolini* have been successfully transferred from fish to fish (*Lepomis* spp.) by gavage, indicating that predation increases the number of potential host species and establishes a transmission route to fish that do not feed on insect larvae.

The life-cycle of *Cystidicoloides ephemeridarum* is illustrated in Figure 2.35. The eggs contain first-stage larvae when deposited and pass into water with the faeces of the host, salmonid fish, to be ingested by detritus-feeding mayfly nymphs. The larvae hatch, penetrate into the haemocoel and moult twice to the infective third stage but do not become encapsulated. They remain free, coiled in the abdominal haemocoel or in the muscles and may even develop to the fourth stage (Moravec, 1971c, d). *Habrophlebia lauta*, *Habroleptoides modesta* and *Ephemera danica* all became infected under experimental conditions. Nymphs are killed by massive infections but smaller numbers do not prevent metamorphosis of the host. Moravec (1971c, d) has emphasized the significance of this for the infection of trout during spring and summer, when it feeds mostly on winged insects. On ingestion by a salmonid fish, the nematode completes its development to maturity but, in some non-salmonid hosts, development does not proceed beyond the third or fourth stage. Loach, *Noemacheilus barbatulus*, and bullhead, *Cottus gobio*, small fish which fall prey to trout, serve in this way as paratenic hosts (De and Moravec, 1979; Moravec and De, 1982). Information on the life-cycle of this nematode has been published under several different names: Moravec (1971c) and Moravec and De (1982) list *Metabronema salvelini* and *Cystidicoloides tenuissima* as synonyms of *C. ephemeridarum*.

Superfamily Thelazioidea:
Family Rhabdochonidae

The best known rhabdochonid life-cycles are those of *Rhabdochona phoxini* and *R. ergensi*. The life-cycle of *R. phoxini* is illustrated in Figure 2.36. When deposited, the eggs contain a first-stage larva which hatches after the egg has been ingested by a mayfly nymph (*Habrophlebia lauta*, *H. fusca*, *Ecdyonurus dispar*). The toothed larva penetrates into the haemocoel and moults twice to reach the infective stage. Third-stage larvae become encapsulated singly or in groups, usually in the dorsal abdominal region of the host. The third moult may occur at this stage. Subsequently, the definitive host, *Phoxinus phoxinus*, ingests infected nymphs and one or two moults occur within the fish, depending on whether the larvae were at the third

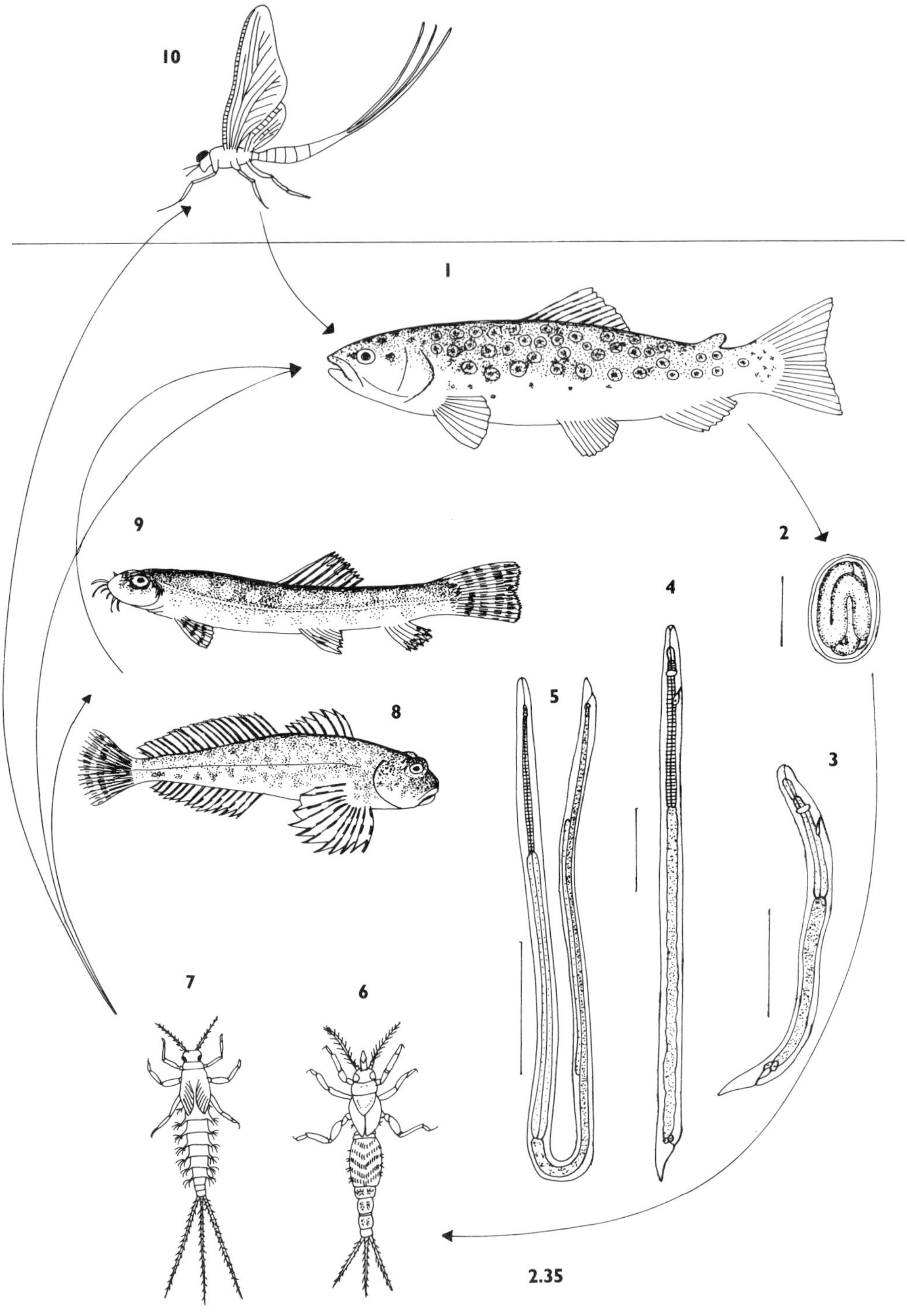

Figure 2.35.1–10 The life-cycle of *Cystidicoloides ephemeridarum*.
1. Definitive host, *Salmo trutta*. 2. Egg containing first-stage larva. 3. First-stage larva. 4. Second-stage larva. 5. Third-stage larva. 6,7. Intermediate hosts, *Ephemera* sp. nymph (6), *Habrophlebia fusca nymph* (7). 8,9. Transport hosts, *Cottus gobio* (8) and *Noemacheilus barbatulus*. 10. Adult mayfly.
Scale bars: 2–4, 0.02 mm; 5, 0.5 mm.
Figures 2–5 after Moravec (1971c).

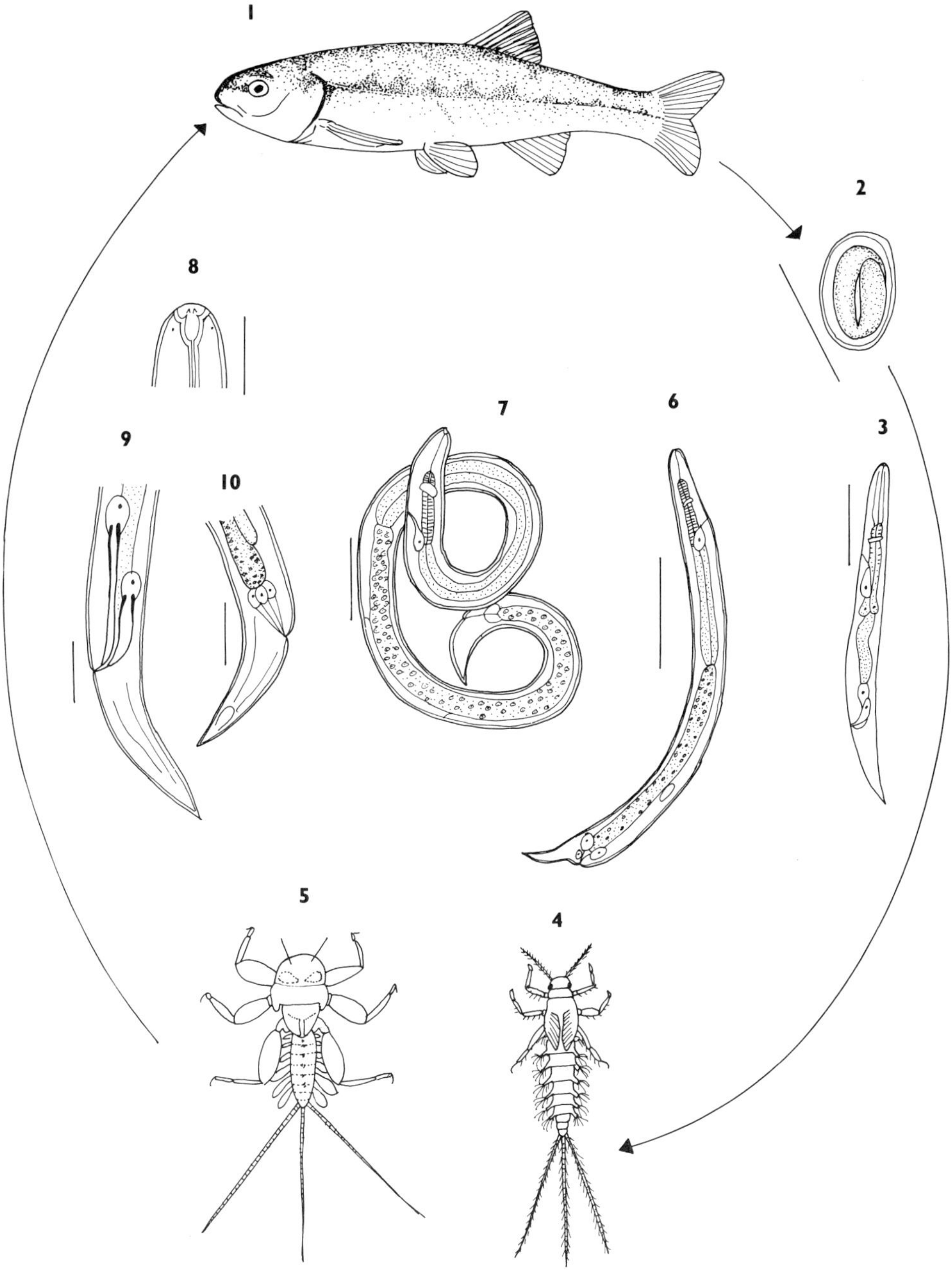

2.36

Figure 2.36. 1–10 The life-cycle of *Rhabdochona phoxini*.
1. Definitive host, *Phoxinus phoxinus*. 2. Egg containing first-stage larva. 3. First-stage larva. 4,5. Intermediate hosts, *Habrophlebia fusca* nymph (4) and *Ecdyonurus* nymph (5). 6. Second-stage larva. 7. Third-stage larva. 8–10. Fourth-stage larva: Anterior end (8); Male tail (9); Female tail (10).
Scale bars: 2, 3, 9, 10, 0.05 mm; 6,8, 0.02 mm; 7, 0.1 mm. Figures. 2,3, 6–10 after Moravec (1976b).

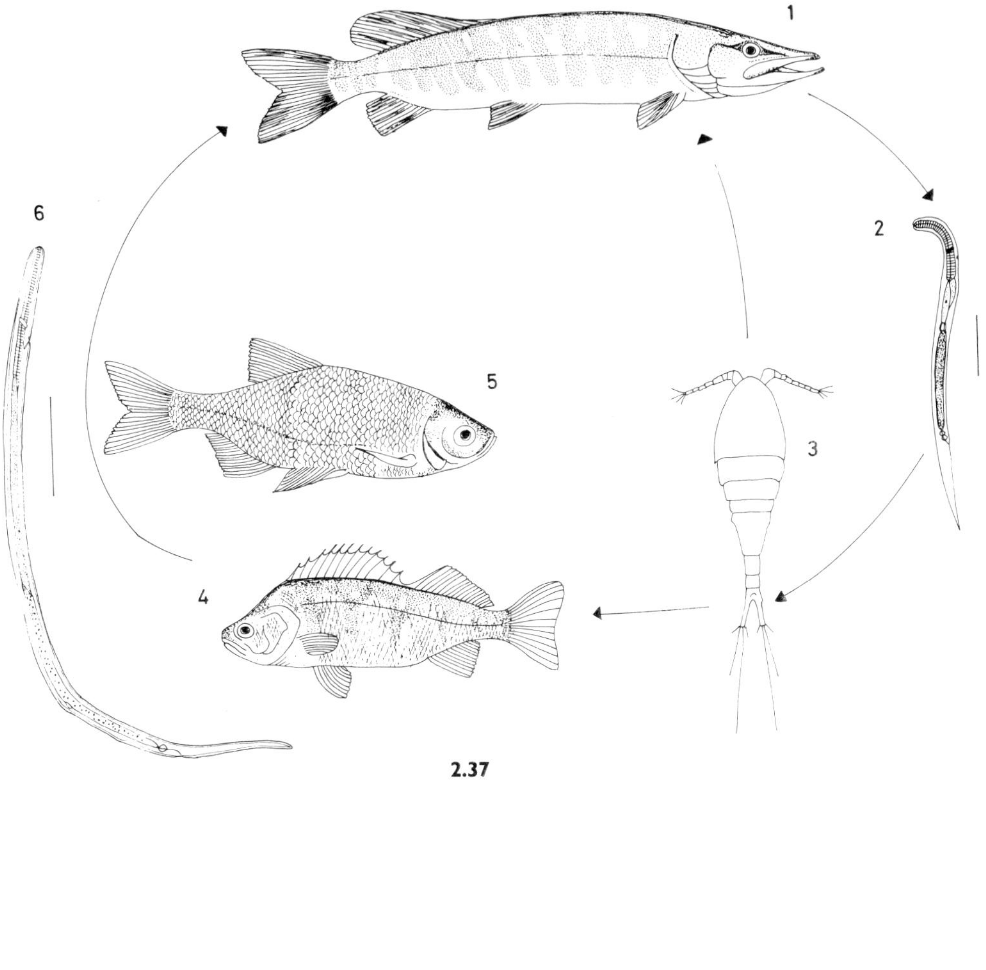

2.37

Figure 2.37.1–6 The life-cycle of *Philometra obturans*. 1. Definitive host, *Esox lucius*. 2. First-stage larva. 3. Intermediate host, a cyclopoid copepod. 4,5. Reservoir hosts, *Perca fluviatilis* (4) and *Scardinius erythrophthalmus* (5). 6. Third-stage larva from the eye of *P. fluviatilis*.
Scale bars: 0.1 mm. Figures 2 and 6 after Moravec (1978b).

or fourth stage. The presence of fourth-stage larvae in mayfly nymphs may explain the occurrence of *Rhabdochona* in atypical host species in which they cannot complete their development. Mayfly nymphs infected with small numbers of *R. phoxini* larvae were not prevented from metamorphosing and the imagos were capable of flight (Moravec, 1972, 1976b, 1977a).

Superfamily Gnathostomatoidea:
Family Gnathostomatidae

Several species of *Echinocephalus*, parasitic as adults in the spiral valve of elasmobranchs, are known to have molluscan intermediate hosts. These are bivalves (abalone and oysters) in the cases of *E. pseudouncinatus* and *E. sinensis*, but gastropods in that of *E. uncinatus*. *E. pseudouncinatus* larvae have also been found in sea urchins (Pearse and Timm, 1971).

An unusual feature of *E. sinensis* is that its intermediate host, *Crassostrea gigas*, is essentially a brackish water species, in contrast to the marine intermediaries of other members of the genus. Ko *et al.* (1975) speculated that *E. sinensis* represents an intermediate stage linking marine members of the family with the freshwater members in gnathostomatid evolution.

E. overstreeti larvae have been found in *Chlamys bifrons* and *Pecten albus*; the adult is found in *Heterodontus portusjacksoni* and other elasmobranchs (Beveridge, 1987; Andrews, Beveridge, Adams and Baverstock, 1988). Anderson (1988) suggested that oysters were probably paratenic hosts for this genus.

Superfamily Dracunculoidea:
Family Philometridae

The first two moults in the philometrid life-cycle take place in the cyclopoid copepods which ingest first-stage larvae freed by the bursting of the gravid female on exposure to water. The fish definitive hosts ingest infected copepods. The third and fourth moults occur in the definitive host, in the liver and air-bladder respectively in the case of *Philometra lusiana* (Vasil'kov, 1968). Most philometrid adults occupy sites in the serosa of the swim-bladder or other organs or in the skin. Invasive larvae must penetrate the gut and then perform a migration to this preferred habitat. *Philonema oncorhynchi* third-stage larvae penetrate the gut near the pyloric caeca and migrate to the swim bladder via the coelom and, to a lesser extent, across the mesentery and associated tissues (Adams, 1969). Males and unfertilized females of *Philometra abdominalis*, *P. lusiana* and *Philometroides huronensis* are found in the serosa of the swim bladder but fertilized females migrate to the skin or another site from which larvae can be discharged to the exterior. Those of *Philometra abdominalis* migrate to the abdomen, where they remain for some time, but they eventually penetrate the rectum, protrude from the anus and burst as a result of an increase in osmotic pressure. In the Rokytka brook, Czechoslovakia, this coincides with a high incidence of *Acanthocyclops vernalis*, the main intermediate host (Moravec, 1977b, c). Gravid females of *Philometra* sp. of Bier, Payne and Jackson (1974) and of *Philometroides huronensis* penetrate the skin to discharge their larvae and those of *Philonema oncorhynchi* are expelled from the body cavity of the fish with the roe during spawning (Uhazy, 1977a, b).

Most philometrids complete the life-cycle within a year but that of *Philonema oncorhynchi* is correlated with the four-year spawning cycle of its anadromous migratory host, *Oncorhynchus nerka*. The cycles of maturation of both host and parasite are such that both are gravid on reaching the spawning grounds, suggesting that production of parasite larvae is mediated by the hormones of the host (Platzer and Adams, 1967). The larvae are released at the optimum time for transmission to the copepod host and subsequently to other salmon. Similarly, Bier, Payne and Jackson (1974) found that embryogenesis of *Philometra* sp. in *Morone saxatilis* is correlated with patency of the host.

Philometra obturans is unusual in the family by virtue of the site occupied in the host, the small size of adults of both sexes, its great pathogenicity to its final host, *Esox lucius*, and its use of paratenic hosts. Gravid and subgravid females locate in the ventral aorta, often with the mid-region of the body in the bulbus arteriosus and the free ends in two different gill arches. To discharge the larvae, the females enter a gill artery, penetrate its wall in the middle of a gill and extrude into the water. The species is also unusual in philometrids in that gravid females are present throughout the year. In other species, they are present only in late spring and early summer, e.g. *P. abdominalis*, *P. ovata*, *P. rischta*, *P. kotlani*, *P. cylindracea*, *Philometroides lusiana*, *P. sanguinea* and *P. huronensis* (Moravec and Dyková, 1978). The reservoir hosts of *Philometra obturans* are mainly perch and, less frequently, rudd, in both of which the larvae localize in the vitreous humour. Perch is particularly important in transmitting larvae to pike because it consumes very large quantities of plankton. As adults, this species is highly specific to pike. The life-cycle of *P. obturans* is illustrated in Figure 2.37.

Superfamily Physalopteroidea:
Family Physalopteridae

The two species of *Proleptus* for which the cycles are known (*P. acutus* and *P. obtusus*) use crabs as intermediaries. The final hosts are elasmobranchs (see Table 2.4).

Adenophorea:
Order Enoplida
Superfamily Trichinelloidea: Families Cystoopsidae and Trichuridae

Members of both families use copepods and amphipods as intermediaries. *Cystoopsis acipenseri* inhabits small subcutaneous tumours, each containing one male and one female worm, on the abdominal surface of sturgeons. Fully developed larvae are released into water when the tumour bursts.

Acanthocephala

Introduction

The life-cycles of acanthocephalans involve a vertebrate definitive host and an arthropod intermediary which, in the case of fish parasites, is an amphipod, ostracod or copepod. A few cycles also incorporate paratenic or transport hosts which, as in nematode life-cycles, bridge gaps in the food chain between intermediate and definitive hosts if no direct predator-prey relationship exists. The life-cycles of some fish acanthocephalans are summarized in Table 2.10. The most recent reviews of acanthocephalan life-cycles and reproductive biology are those by Nicholas (1967, 1973); Crompton (1975, 1985); Parshad and Crompton (1981) and Schmidt (1985).

Development of the egg, acanthor and acanthella

Acanthocephalan eggs develop in the fluid-filled body-cavity of the female by a process of spiral cleavage which produces an elongated embryo, the acanthor, surrounded by

Table 2.10. Definitive, intermediate and paratenic hosts of some fish acanthocephalans

Species	Intermediate host (and paratenic host if present)	Final host	References
		Class Palaeacanthocephala	
Family Echinorhynchidae			
Acanthocephalus anguillae	*Asellus aquaticus*	*Percottus glehni*, *Esox lucius*, *Leuciscus idus*, *Abramis brama*	Andryuk (1979a, b)
Acanthocephalus lucii	*Asellus aquaticus*	*Perca fluviatilis*, *E. lucius*, *L. idus*, *A. brama*	Andryuk (1979b, c)
A. parksidei	*Caecidotea militaris*	*Catostomus commersoni* and 10 other fish species	Amin (1975); Amin, Burns and Redlin (1980)
Leptorhynchoides thecatus	*Hyalella azteca*	*Lepomis cyanellus*, *L. gibbosus*, *Ambloplites rupestris*	De Giusti (1949); Uznanski and Nickol (1976, 1980a, b)
Echinorhynchus gadi	*Gammarus duebeni*	*Gadus morhua kildinensis*	Kulachkova and Timofeeva (1977)
E. truttae	*Gammarus lacustris*, *G. fossarum*	*Salmo trutta*, other salmonids	Awachie (1966b); Maren (1979)
E. lageniformis (Figure 2.40.1–14)	*Corophium spinicorne*	*Platichthys stellatus*	Olson and Pratt (1971)
Family Pomphorhynchidae			
Pomphorhynchus laevis	*Gammarus pulex*, *G. fossarum* Paratenic hosts: cyprinids, *Gasterosteus aculeatus*	Several families of freshwater fish including cyprinids, salmonids and percids	Schäperclaus (1954); Maren (1979)
Family Iliosentidae			
Tegorhynchus furcatus	*Lepidactylus* sp., *Haustorius*, sp.	*Menticirrhus*, *Fundulus similis*, *Dasyatis* spp.	Buckner, Overstreet and Heard (1978)
Dollfusentis chandleri	*Lepidactylus* sp., *Grandidierella bonnieroides*, *Corophium lacustre*	*Micropogonias undulatus*, *Leiostomus*, *xanthurus Bairdiella chrysura*, *Orthopristis chrysoptera*, *Archosargus probatocephalus*, *Dasyatis sabina*	Buckner, Overstreet and Heard (1978)
		Class Eoacanthocephala	
Family Tenuisentidae			
Paratenuisentis ambiguus	*Gammarus tigrinus*	*Anguilla rostrata*	Samuel and Bullock (1981)
Family Neoechinorhynchidae			
Neoechinorhynchus cylindratus	*Cypria globula* Paratenic host: *Lepomis pallidus*	*Huro salmoides*	Ward (1940); Harms (1965)
N. saginatus	*Cypridopsis vidua*	*Semotilus atromaculatus*	Uglem and Larson (1969)
N. rutili	*Cypria turneri*	freshwater fish (14 families)	Merritt and Pratt (1964)
N. cristatus	*Cypridopsis helvetica*	*Catostomus macrocheilus* and other catostomids	Uglem (1972a)
Paulisentis fractus	*Tropocyclops prasinus*	*Semotilus atromaculatus*	Cable and Dill (1967)
Octospiniferoides chandleri	*Cypridopsis vidua*, *Physocypria pustulosa*	*Gambusia affinis*	DeMont and Corkum (1982); Corkum (1982)
Octospinifer macilentis	*Cyclocypris serena*	*Catostomus commersoni*	Harms (1965)
Family Quadrigyridae			
Pallisentis nagpurensis	*Cyclops strenuus* Paratenic hosts: *Macropodus cupanus*, *Aplocheilus melastigma*, *Barbus* sp., *Ophicephalus gachua*	*Ophicephalus striatus*	George and Nadakal (1973)

three or four envelopes, one of which forms a thick shell (Schmidt, 1973, 1985; Parshad and Crompton, 1981). The acanthor bears hooks or spines arranged in circles or spirals at the anterior end; the hooks and their associated contractile muscle fibres constitute the aclid organ. Some species also have body spines and some lack anterior hooks (e.g. *Neoechinorhynchus cylindratus* and *N. rutili*, the anterior ends of which show small invaginations (Ward, 1940). Neoechinorhynchid acanthors (order Eoacanthocephala) have an organ containing three apical nuclei, not so far reported in palaeacanthocephalans and archiacanthocephalans (Uglem, 1972a). Acanthors contain a group of germinal cells, the entoblast, which gives rise to most of the organs of the adult. The subsequent developmental stage, the acanthella, develops into the infective juvenile in the haemocoel of the intermediate host and more detailed descriptions of these stages will be given below.

When the egg leaves the body of the female and reaches the exterior in the faeces of the host, it already contains a fully-developed acanthor. In some palaeacanthocephalans, the outermost egg envelope breaks down into fibrils which anchor the eggs to each other or to vegetation, increasing the probability of ingestion by intermediaries, eg. *Acanthocephalus jacksoni* (Oetinger and Nickol, 1974), *Leptorhynchoides thecatus* (Uznanski and Nickol, 1976) and several other fish parasites listed by Whitfield (1973). Fibrils of eggs of *L. thecatus* tangle with filamentous algae, attaching them to the feeding site of the amphipod host; amphipods fed in dishes containing algae and eggs subsequently showed a greater incidence and intensity of infection than those fed in dishes containing eggs alone. Embryonated eggs of *Pallisentis nagpurensis* are adapted to float in water (George and Nadakal, 1973).

The egg is adapted for survival in several different environments that include the gut of the definitive host, the environment of the host and the gut of the intermediary; it is resistant to environmental conditions and is probably the stage by which many species overwinter (De Giusti, 1949; Crompton, 1970a). At 4°C, eggs of *Leptorhynchoides thecatus, Pomphorhynchus bulbocolli, Octospinifer macilentis, Neoechinorhynchus rutili* and *Paulisentis fractus* will survive in water without loss of infectivity for 9, 6, 9, 6 and many months, respectively (various authors in Crompton, 1970a). Eggs hatch only after ingestion by the intermediary but the hatching mechanism is incompletely known. It was attributed in *L. thecatus* to the combined effects of digestive enzymes and active movements of the acanthor which ruptured the membranes at the anterior end of the egg (De Giusti, 1949). Harms (1965), however, failed to induce hatching in *Octospinifer macilentis* eggs with pepsin, trypsin or pancreatin. Eggs of *Echinorhynchus truttae* hatch in response to the action of the gastric mill and digestive fluid of the host and the activity of the acanthor (Awachie, 1966b). Uglem (1972a) found that when eggs of *Neoechinorhynchus cristatus* hatch the acanthors become active, the two outer egg envelopes split transversely near the anterior end and the acanthor, still enclosed in the two inner envelopes, leaves the egg. It frees itself using its hooks and rapidly increases in size. Hatching is completed within two hours. In *Paratenuisentis ambiguus* eggs, the split is in the posterior half of the egg (Samuel and Bullock, 1981). Increase in size of the acanthor immediately after hatching has been described in several neoechinorhynchids and has been attributed by Uglem (1972a) to muscular activity rather than uptake of water as previously believed. The time taken to hatch varies between species as does the time taken to reach the haemocoel of the intermediate host (Table 2.11).

Penetration of the intestinal wall begins immediately after hatching in many species but some have an initial feeding period in the intestine, e.g. *N. cristatus*, which remains in the gut for up to 1.5 days (Uglem, 1972a). The acanthor penetrates the gut wall using its hooks and enters the haemocoel. Early and late acanthor stages, in the gut and haemocoel respectively, are distinguished by some authors. Similarly, early and late acanthellae are distinguished depending on the degree of differentiation and development of the organs. Details of development are described in more detail for an eoacanthocephalan, *Paratenuisentis ambiguus*, and a palaeacanthocephalan, *Echinorhynchus truttae*, below. Development culminates in juveniles which resemble the adults, and those that become enveloped in a capsule are known as cystacanths; the capsule is thought to be mainly of host origin. Wanson and Nickol (1973) considered that the capsule was the stretched outer covering of the acanthor, supplemented in some species by host cells. Although a cystacanth occurs in many acanthocephalan life-cycles, it is absent in, for example, *Paratenuisentis ambiguus, Paulisentis fractus* and *Echinorhynchus lageniformis* (Table 2.10).

Effect on intermediate host behaviour

Some species modify the behaviour of the intermediary so as to maximize their chances of being eaten by the next host. Amphipods infected with *E. truttae* are unable to swim against the water current and are more easily caught by trout (Crompton, 1970a), while ostracods infected with *Octospiniferoides chandleri* and *Neoechinorhynchus cylindratus* seek the surface of the water, possibly in a photophilic response (DeMont and Corkum, 1982), making them more vulnerable to their definitive and second intermediate hosts respectively.

Use of transport or paratenic hosts

Some species incorporate a transport or paratenic host into the life-cycle. In some cases, paratenic hosts acquire infection accidentally by predation on fish with immature worms which will develop further when the paratenic host is eaten by the usual definitive host. In this way the occurrence of *Paratenuisentis ambiguus* in *Microgadus tomcod, Fundulus heteroclitus* and other fish may contribute to heavy infections in *Anguilla rostrata* (Samuel and Bullock, 1981), but these hosts are not essential to completion of the life-cycle. In contrast, *Lepomis pallidus*, in the liver of which encapsulated *Neoechinorhynchus cylindratus* occur, is essential because the final host, *Huro salmoides*, is a predator which does not eat the ostracod host *Cypria globula* in natural conditions (Ward, 1940; Crompton, 1970a). Ingestion of immature, i.e. not yet infective, worms by the normal definitive host may result in their migration from the gut to the mesenteries where they become encapsulated. De Giusti (1949) found that 30-day-old juveniles of *Leptorhynchoides thecatus* could establish in the fish gut but those 26–28 days old could not do so and migrated to the mesenteries. *Lepomis cyanellus* was a poor transport host for this species but would support 10–15 worms in the pyloric caeca (Uznanski and Nickol, 1982). Awachie (1966b) noted that *Echinorhynchus truttae* acanthellae up to 75 days old could not evert their proboscides and speculated that this accounted for their failure to penetrate the gut wall and hence for the probable absence of paratenic hosts in the life-cycle. Cannibalism has been suggested as a means of transmission of *Octospiniferoides chandleri* in mosquito fish (*Gambusia affinis*) by DeMont and Corkum (1982), who drew attention to a similar mechanism in the life-cycle of *E. salmonis* (Hnath, 1969) and to transfer of *Neoechinorhynchus saginatus* from prey to predator fish (Muzzall and Bullock, 1978).

Development in two acanthocephalan models

Details of development have been studied in most of the species listed in Tables 2.11 and 2.12. The developmental stages of *Paratenuisentis ambiguus* and *Echinorhynchus truttae* are illustrated in Figures 2.38.1–7 and 2.39.1–9 as examples of an eoacanthocephalan and a palaeacanthocephalan respectively.

Table 2.11. Hatching and penetration of the gut wall of the intermediate host

Species	Temp °C	Hatching time	Penetration time	Reference
Paratenuisentis ambiguus	22–25	1–6h	20–96h	Samuel and Bullock (1981)
Paulisentis fractus	21–23	?	2–4h	Cable and Dill (1967)
Octospinifer macilentis	21	1–4h	24h	Harms (1965)
Neoechinorhynchus saginatus	25	1–2h	by 36h	Uglem and Larsen (1969)
N. rutili	15	6–12h	?	Merritt and Pratt (1964)
N. cristatus	?	1–2h	1–5 days	Uglem (1972a)
Pallisentis nagpurensis	?	8–12h	30–48h	George and Nadakal (1973)
Echinorhynchus truttae	17	1–20h	11–20h	Awachie (1966b)
E. lageniformis	23	?	by 48h	Olson and Pratt (1971)
Leptorhynchoides thecatus	25	45 min	?	De Giusti (1949)

Table 2.12. Development times (in days) for stages in the life-cycle of some fish acanthocephalans

Species	Temp. °C	Acanthor	Acanthella	Infective Juveniles	Cystacanth (present +, absent −)	Reference
Paratenuisentis ambiguus	22–25	1–14	13–32	27–32	—	Samuel and Bullock (1981)
Paulisentis fractus	21–23	1–7	7–11	13	—	Cable and Dill (1967)
Octospinifer macilentis	21	1–4	5–30	?	+	Harms (1965)
Octospiniferoides chandleri	23	—	by 14	by 20	+	DeMont and Corkum (1982)
Neoechinorhynchus saginatus	25	1–3.5	3–14	16	—	Uglem and Larson (1969)
N. rutili	15	1–6	6–48	48–57		Merritt and Pratt (1964)
Acanthocephalus anguillae	24	2–3	15			Andryuk (1979a, b)
	19		43			
A. lucii	25		19			Andryuk (1979c)
	22		32			
	19		51			
	18		60			
	16		72			
	15		89			
Echinorhynchus truttae	17	1–25	18–56	57–82	+	Awachie (1966b)
E. lageniformis	23	1–4	5–25	30	—	Olson and Pratt (1971)
Leptorhynchoides thecatus	25	1–2	3–28	30	—	De Giusti (1949)

The life-cycle of *Paratenuisentis ambiguus* (Figure 2.38.1–7)

The late acanthor in the haemocoel of *Gammarus tigrinus* becomes spherical with 6 to 19 or more conspicuous giant nuclei, two or three of which are anterior in position and probably give rise to the apical nuclei of the acanthella. The hooks of the acanthor are lost by day 7 and the acanthella grows in size and the organs become differentiated. The primordia of the brain, reproductive system and presoma appear on day 13 and three apical nuclei occur anteriorly with several giant nuclei, those of the future lemnisci, behind them. The nuclei of the proboscis ring and of the apical organ lie between the lemniscal giant nuclei and the brain primordium. The uncinogenous band which gives rise to the hooks is anterior to the proboscis ring. By 15–18 days the acanthella has elongated and its sex can be determined. The proboscis and its receptacle are well-developed and the muscle layers of the body wall, the proboscis retractors and the ligament in the female are visible. The proboscis develops in the inverted position but everts as the hooks appear. It inverts again by day 27–32, at which point the lemnisci have developed and the juvenile stage has been

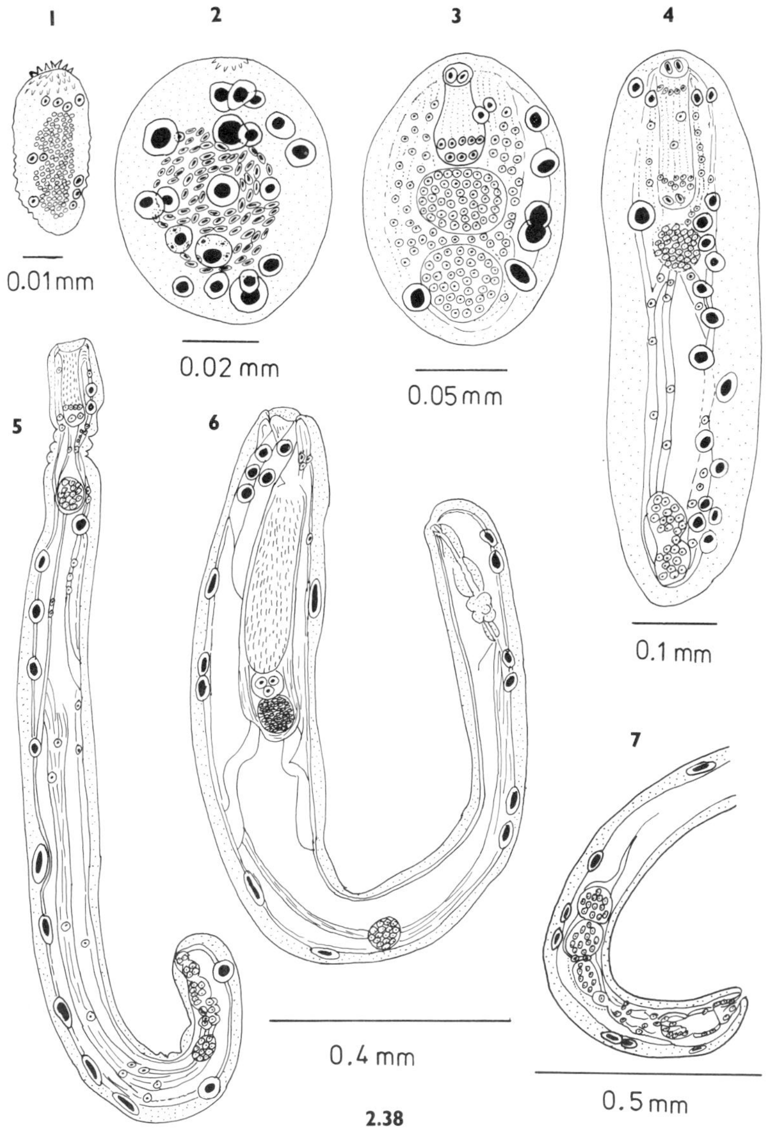

Figure 2.38. 1–7 Larval stages of *Paratenuisentis ambiguus* from the intermediate host, *Gammarus tigrinus*. After Samuel and Bullock (1981).

1 Early acanthor from intestine of amphipod 2 Late acanthor 3 Thirteen-day acanthella 4 Eighteen-day female acanthella 5 Twenty-day female acanthella 6 Twenty-seven day juvenile female 7 Twenty-seven day juvenile male, posterior end.

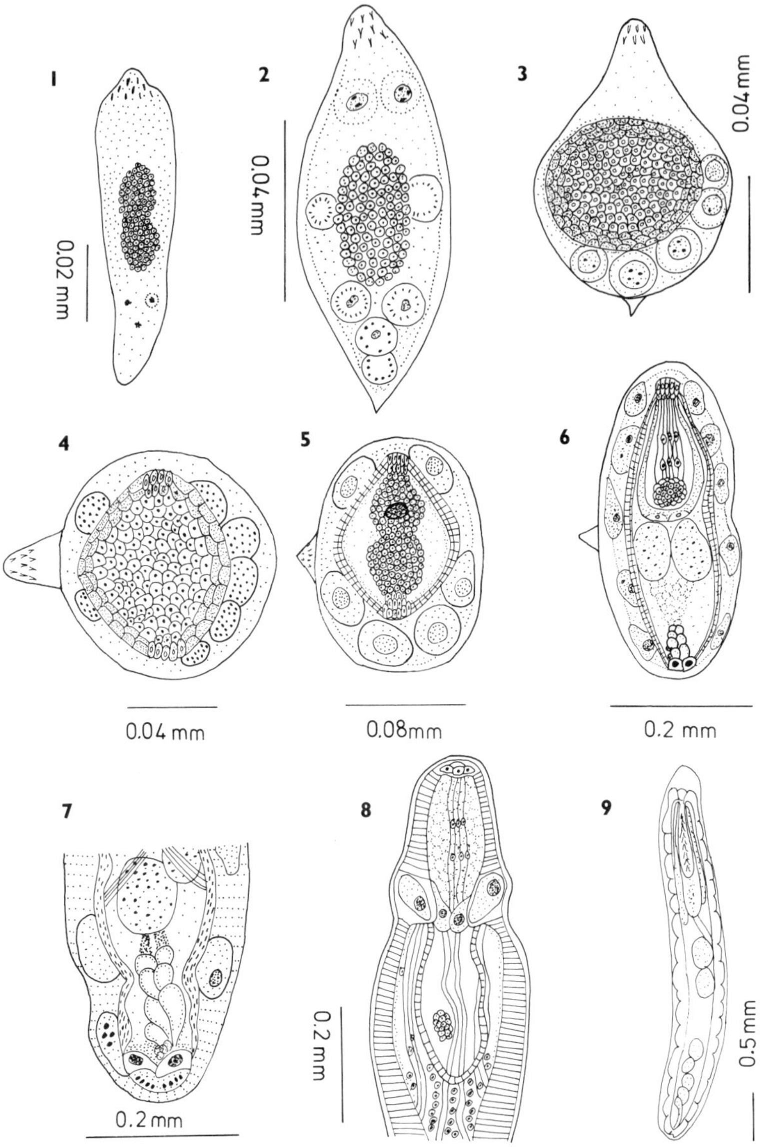

2.39

Fig. 2.39. 1–9. Stages in the life-cycle of *Echinorhynchus truttae*. After Awachie (1966b) (Figures 1–8) and Crompton (1970a) Figure 9. 1. Acanthor from haemocoel of *Gammarus pulex* 11.5 hours after infection. 2. 12-day acanthor. 3. 18-day acanthella. 4. 24-day acanthella. 5. 32-day acanthella. 6. 37-day acanthella. 7. Posterior half of male acanthella, 39 days post-infection. 8. Anterior half of male acanthella, 49 days post-infection. 9. Cystacanth.

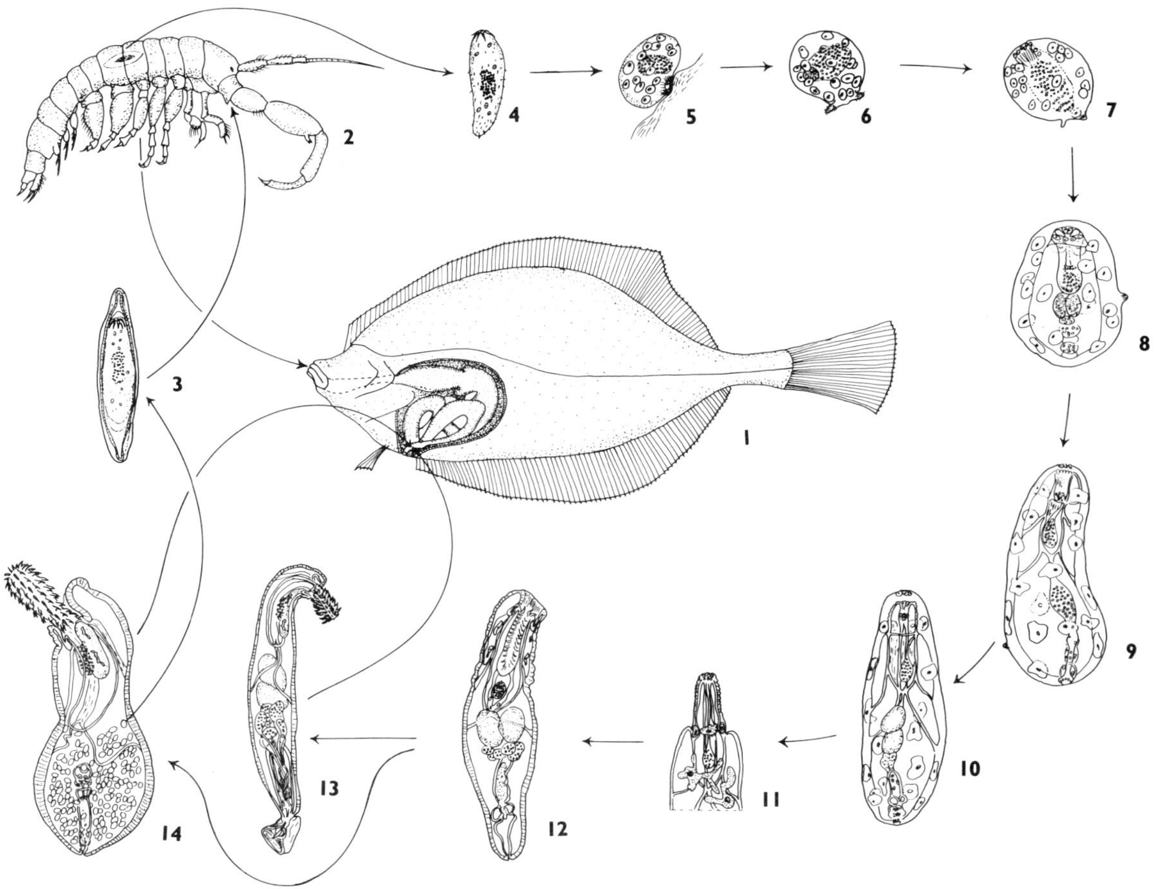

Figure 2.40. 1–14 *Echinorhynchus lageniformis* Diagrammatic representation of the life-cycle (figures not drawn to scale) based on Olson and Pratt (1971). 1. *Platichthys* sp. definitive host becomes infected on eating *Corophium spinicorne* which contain infective larvae. 2. The amphipod intermediate host *C. spinicorne.* 3. Egg of the acanthocephalan which is eaten by an amphipod. 4. Acanthor removed from amphipod serosa 2 days after infection. 5. Young acanthella attached to host serosa 5 days post-infection. 6. 10-day acanthella. 7. 13-day acanthella. 8. 15-day male acanthella. 9. 18-day female acanthella. 10. Male acanthella 20 days post-infection. 11. Acanthella after 22 days development showing everted proboscis. 12. Infective juvenile after 30 days development in amphipod. 13. Mature adult male acanthocephalan from the fish final host. 14. Adult female (immature) from the final host.

attained. Development of the male system begins with the appearance of the testes anterior to the genital primordium by day 16; most other organs have differentiated by day 20 and development is completed when Saefftigen's pouch and the genital opening have formed. In females, the ovarian mass separates from the genital primordium and the remainder of the reproductive organs have differentiated by day 24. Juveniles are not encapsulated and both sexes have everted proboscides. Having penetrated into the haemocoel of the intermediate host, the acanthors attach by their spines to the outer surface of the intestine or caeca. This also occurs in *E. lageniformis* (the life-cycle of which is illustrated in Figure 2.40) (Olson and Pratt, 1971) and in *Leptorhynchoides thecatus* (De Giusti, 1949), but not in *P. ambiguus* or the neoechinorhynchids which have been studied. The acanthor grows and its original width becomes the future longitudinal axis of the acanthella. By 20–25 days it is almost spherical with or without the visible remnants of the original poles of the acanthor. The entoblast elongates along the new axis and the primordia of the proboscis apparatus and reproductive organs are visible at the anterior and posterior end, respectively. By day 32, a cavity separates the body-wall primordia from the central mass of nuclei which is constricted in the middle, the anterior portion of which will give rise to the proboscis retractors and sac and the posterior part to the proboscis sac retractors, gonads and ligament proximally and the genital organs distally.

During days 32–56 the acanthella elongates rapidly and the organs appear. At 40–41 days the cortical giant nuclei are arranged as: an anterior apical nucleus which later divides into four and helps form the proboscis; a group of four neck nuclei (lemniscal ring); and three groups of four nuclei, one of which will form the copulatory bursa in the male. The hooks appear about day 49 and the proboscis retracts about day 56, the four neck nuclei migrating to form the lemnisci. The neck and forebody retract later. Slow differentiation and maturation continue through days 57–82, the acanthella grows in size and becomes coiled or folded within a capsule. The ovary is formed about day 68 from 2–5 groups of nuclei within the ligament which coalesce; it subsequently breaks down into ovarian masses. Acanthellae cannot evert the proboscis by day 75 but can do so by day 81. Cystacanths are orange in colour as a result of pigment absorbed from the host. When eaten by the final hosts, they are capable of sexual activity immediately they reach the pyloric region of the intestine (Awachie, 1966b) in contrast to, e.g. *P. ambiguus* and *Octospiniferoides chandleri* which require about a month to mature in the definitive host (Samuel and Bullock, 1981; DeMont and Corkum, 1982).

Development in *E. truttae* is very similar to that in other palaeacanthocephalans but shows some differences from *E. lageniformis* (Figure 2.40.1–14) and *L. thecatus*. In the former, the acanthor penetrates the gut wall and attaches to the serosa for about 10 days; in the latter, early development takes place within the gut wall which eventually ruptures at about day 10 due to the increasing size of the acanthella. The acanthella remains attached to the gut wall by the remnant of the original acanthor body which breaks at about day 14, allowing the acanthella to float freely in the haemocoel. Other differences from *E. truttae* include the absence of a cavity formed in the entoblast, the origin of the apical nuclei from the entoblast rather than by division of one giant nucleus and the absence of an encapsulated juvenile (Olson and Pratt, 1971).

3 Ecology

Introduction

This has been a difficult chapter to plan and write because, as stated by Begon, Harper and Townsend (1986), ecology is a complex science and must deal explicitly with three levels of biological hierarchy, the organism (species) under investigation, populations of these species and communities of species. Ecology has been defined as the total relation of an animal species to its abiotic and biotic environments. This being so, there is very little in biology that is not ecology; Krebs (1972) demonstrated this schematically as shown in Figure 3.1. Begon, Harper and Townsend (1986) support this view and extend the definition in saying that ecology feeds, in a peripheral way, on advances in biochemistry, behaviour, climatology, plate tectonics and so on. But it also feeds back to our understanding of vast areas of biology. They paraphrase Dobzhansky (1937) and emphasize that very little in evolution makes sense, except in the light of ecology, i.e. in terms of the very large number

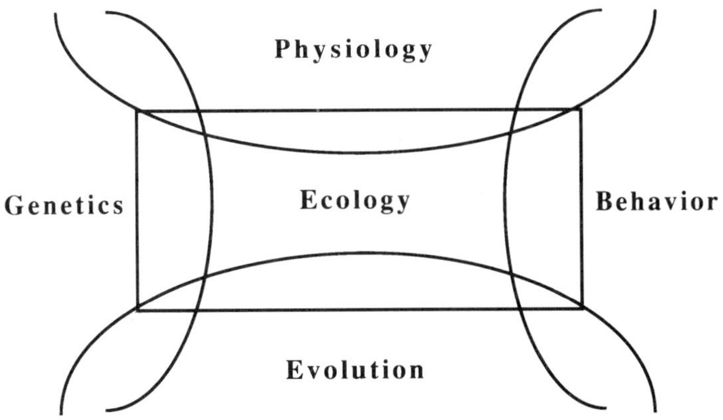

Figure 3.1 A schematic definition of ecology. After Krebs (1972).

of interactions between organisms and their physical, chemical and biological environments. Most textbooks of animal ecology exclude discussions of parasitology, although Begon, Harper and Townsend (1986) is a notable exception. The reasons for this are, most probably, those suggested by Holmes and Price (1986), ie. the daunting task of having to consider two animals (host and parasite) at the same time and at the three levels mentioned above and with consideration for the overlap between ecology and other disciplines of biology. The complexities of the life-cycles of most parasites and those of the innumerable host-parasite interactions involved add to the complexity of ecological parasitology.

There is, however, an additional simple explanation for the omission of parasites from ecological texts. Generally speaking, ecologists are loathe to dissect and, bearing in mind the occasionally unreliable variety of dissection and collection techniques used by parasitologists, ecologists are probably wary of using data which have been accumulated in this way unless it can easily be confirmed by further sampling. Elton was an exceptional ecologist because, about sixty years ago, he was advocating ecological surveys of parasites on the same principles, though not by the same techniques, as those of free-living animals. His survey, jointly with others, on the parasite fauna of wood-mice (*Apodemus sylvaticus*) in about 450 acres of woodland, disclosed 47 species of parasites and highlighted important principles in both ecological parasitology and community ecology (Elton *et al.*, 1931). Regrettably most ecologists did not follow Elton's lead in expressing an interest in parasites and conversely, until the 1970s, few parasitologists had realized the importance of adopting an ecological approach to the study of parasites.

From the 1970s, many parasitologists have faced these formidable problems and have focussed attention on several aspects of parasite ecology which until then had received little or no attention. This can be seen from Kennedy (1976a), and its treatment by various authors of parasite strategies in adapting to hosts, in adapting to habitats within hosts and of population ecology, i.e. the dynamic relationships between populations of two species (parasite and host).

The realization that parasitology could be approached from an ecological standpoint has actually grown very slowly since the pioneer studies of V.A. Dogiel and his colleagues in Russia during the 1930s (Kennedy, 1976a), Petrushevskii and Petrushevskaya (1960) referred to Dogiel's work in 1933, and his belief that an examination of 15 fish in a particular locality gives a reasonably complete picture of the parasite fauna of a particular host species. For investigations of seasonal and age dynamics, of helminth faunas associated with artificial waters and the zoogeography of a particular species, Dogiel suggested that perhaps up to 100 fish would have to be examined. The statistical significance of these predictions was tested and discussed with reference to particular species of fish helminths, and the authors felt that they had completely substantiated Dogiel's view by examining the results of numerous subsequent researches and subjecting them to an experimental mathematical analysis. Assuming that this is so, then a considerable amount of field data already published are worthy of consideration, as has been done by Price and Clancy (1983), and if deemed to be inadequate for a mathematical analysis they will at least point to areas of fruitful ecological research in the future. The study of fish parasites occupied a special place in the work of V.A. Dogiel's school. This is evident from various chapters in Dogiel, Petrushevskii and Polyanskii (1961) on the ecology of parasites (freshwater and marine), on relationships between host fish and their parasites, and also on specificity and zoogeography. Since the appearance of the first English translation of this volume in 1961 there has been an explosive growth of interest in the ecology of fish parasites, helminths in particular, as may be seen, for instance, from Anderson (1982a,b,c,d), Anderson and Kikkawa (1986), Chubb (1970), Kennedy (1981b, 1985b, 1990), MacKenzie and Gibson (1970), Rohde (1982a) and Williams, McVicar and Ralph (1970). Recent studies on fish parasite ecology, with particular emphasis on the work of Soviet researchers, were reviewed by Bauer (1984a, 1986). A large number of other publications will become evident throughout this chapter. From this literature it can be seen that all three standard ecological methods of approach (ie. mathematical, field and experimental) have been used in researches into the ecology of fish helminths. These three approaches differ in their strengths and weaknesses and essentially complement each other (Diamond and Case, 1986). Thus, when field and mathematical approaches have been combined with laboratory experiments, considerable progress has been made in understanding the ecology of fish helminths.

Before this combined approach had taken place, however, a wealth of published ecological information, largely from field studies, had accumulated as (i) checklists or surveys of major groups or families of fish for their entire helminth faunas or for particular groups of helminths and (ii) helminth surveys of freshwater, marine or brackish-water fish. These surveys have formed important bases for investigating seasonality, zoogeography and the co-evolution of hosts and parasites. From the 1970s they have also led to very many intensive ecological researches by field workers, theoreticians and/or experimentalists on a single species of helminth. Occasionally, more than one species has been chosen for concomitant studies and, more rarely, some parasitologists have aimed towards understanding the community ecology of fish helminths (Holmes, 1990 and Kennedy, 1990). A vast literature has thus accumulated in recent years in which the effects of a number of biotic and/or abiotic factors on the abundance and prevalence of fish helminths are discussed. Consequently a main objective of research on ecology of fish helminths is to understand which combinations of these two groups of factors account for the success of a species in a particular 'niche'. Before examining these groups of factors, however, we need to define some ecological terms and give some examples demonstrating the usefulness of combining theoretical with field and experimental work.

Ecological terms

Succinct definitions of co-evolution, competition, frequency-dependent, intensity, mortality, multiple infection, prevalence, reproductive rate, resistance, and of many other terms used in the study of parasites are given in a glossary to the book edited by Anderson and May (1982b). Margolis *et al.* (1982) established working definitions of a few terms (mostly numerical) used and misused in ecological parasitology and until their recommendations, or emended recommendations, are universally accepted, much confusion and misunderstanding will bedevil research into fish helminths relative to their environments. Because parasitologists are inconsistent even in the use of the word 'environment' it is not surprising that the confusion is enhanced when attempting to explain the concepts of habitats, microhabitat sites and niche, to name only a few examples. This major problem cannot be resolved without further debate of these and many other terms such as invasion, infestation, infection, extensity and intensity. Zmoray (1980) argued that 'invasion' is the penetration of helminths, or their eggs and larvae, into animals and he provides a useful distinction between parasitic and infectious disease; that 'infestation' has been regarded as a synonym of 'infection' and should be discarded, that 'prevalence' should be replaced by 'extensity', which pairs well with 'intensity'. 'Extensity' designates the occurrence and spread of infection in a population or geographical region and 'intensity' expresses the number of parasites per host. Towards clarifying this confusion we have, for convenience divided ecological terms into those which deal with (a) niche, environment, habitat and site (b) numbers and distribution. Other terms used in community ecology studies will be mentioned in our discussion of this topic (p. 234).* We support Margolis *et al.* (1982) who suggested that all publications in future should include definitions of terms used until uniformity is reached.

Niche, environment, habitat and site

The term 'niche' is used and often misused by parasitologists, as has already been pointed out by, for example, Crompton (1973) and Margolis *et al.* (1982) as a very broad and complicated ecological concept which is difficult to define. Not surprisingly, therefore, the literature on the niche theory is vast and complex. Amongst the best accounts are those of Esch, Bush and Aho (1990), Hutchinson (1958), Begon, Harper and Townsend (1986) and Pianka (1981). For a discussion of a mathematical approach in defining the concept, readers are referred to Levin (1970). Some of the salient features of modern niche theory were summarized by Whitfield (1979) before discussing the crucial niche dimensions of a parasite, these being time, nutrient resources, microhabitat (site?) and total pool of potential host species. The niche of a parasite is the sum total of its requirements (within two environments) in terms of food, oxygen and other factors, including competition with other species. The 'requirements' encompass all of the abiotic factors in the larger environment as well as the biotic factors associated with the host itself. According to Rohde (1982a) the fundamental niche of a species may be compressed by a competitor species to a reduced niche. The 'fundamental niche' according to Professor J. Holmes (personal communication) is that range of conditions in which a parasite can live and reproduce; the 'realized niche' is that part of the fundamental niche in which it actually does live. The realized niche is always smaller than the fundamental niche, but only in part because of the actions of competitors. A parasite's population within a host may not be large enough to force some individual parasites to use livable, but less favourable, parts of the environment. It has often been implied in the parasitological literature that important aspects of a parasite's niche are the presence of competitors or hyperparasites, host specificity and the geographical range of susceptible hosts. Competitors, predators, or hyperparasites, however, are not part of the niche but factors which restrict occupation of parts of the niche.

Similarly, such features as host specificity are not aspects of the niche, but are consequences of the fact that only some hosts meet the needs of a parasite, together with the fact that some of those which do provide the requirements cannot be invaded for geographical or other reasons which prevent exposure to the parasite. Host range, however, can be used as clues to the requirements of a parasite. Thus, Whitfield (1979) used host specificity or geographical ranges of susceptible hosts as 'niche axes' which are measurable attributes correlated with the requirements or which reflect the requirements of the parasite in question. Lebedev (1981) considered various definitions of the concept of 'niche' and argued that host specificity in parasites may be considered as roughly equivalent to the niche of free-living animals.

The 'niche', then, is the sum total of requirements; the 'habitat' is the location where those requirements are met plus the conditions which allow them to be met; the 'site' is the precise location within the host in which the parasite is found, and which thus must provide these requirements and can be occupied given other constraints on the parasite (J. Holmes, personal communication.)

According to Allee *et al.* (1949), plants and animals form one of three major environments comparable to the aquatic and terrestrial environments in which they themselves live. In this context a host animal is regarded as an environment. It follows, as was emphasized by Polyanski (1957), that a parasite has two environments, the host itself and its external environment. If this concept or definition of a parasite's two environments is acceptable, then it should be noted that many researchers in the past, including Williams,

McVicar and Ralph (1970) may have used the term environment carelessly when referring to one organ system of the vertebrate body, eg. the alimentary canal, as a parasite's environment. In contrast others have referred to the alimentary tract and, by implication all other systems, as a parasite's habitat. More confusing, perhaps, is the interchangeable use of environment and habitat for one organ system. If referring to any organ system as the habitat of a parasite is acceptable then use of the word 'site' for any tissue, organ or region of that system which is occupied by a parasite is meaningful. This is what has been suggested, for example, by Crompton (1973) and Margolis *et al.* (1982), in referring to the word 'site'. According to Margolis *et al.* (1982) the word 'habitat' refers not only to the spatial location of a parasite but to its physical and chemical environments as well and, therefore, should not be used synonymously with 'site' and 'location'. Crompton (1973) advocated against the use of 'location'.

Because of the emphasis placed on sites and site-selection in ecological parasitology the topic is treated separately in this chapter following a discussion of abiotic and biotic factors, seasonality and zoogeography. A knowledge of these three areas is invaluable in explaining host and site-selection. Two aspects of site-selection, however, involve the use of the terms emigration and migration and, for this reason, they are mentioned here. **Emigration** is used to describe a change of site without a return journey and **migration** is reserved for movements involving a regular phase of coming and going between sites (Crompton, 1973).

Numbers and distribution

(a) Numbers

Of all ecological terms in parasitology the following are the most ambiguous, as may be seen from Margolis *et al.* (1982), Rohde (1982a), Kennedy (1981b), Pennycuick (1971a,b) and Zmoray (1980).

Prevalence is the percentage of host individuals infected with a particular species or the number of host species infected divided by the number examined. The term is frequently misused for 'incidence' and, more frequently, 'incidence' is used for prevalence.

Incidence should be retained for the number of new cases of infected individuals appearing in a population within a given period of time divided by the number of uninfected individuals at the beginning of the time period. The term is unlikely to be applicable to populations of wild animals because the number of uninfected individuals at the beginning of the time period is rarely known.

Intensity is the number of individuals of a particular parasite species in each host, i.e. in an infrapopulation.

Mean intensity refers to the total number of individuals of a particular parasite species in a sample of host species divided by the number of infected individuals of the host species in the sample. It is the mean number of individuals of a particular parasite species per infected host in a sample.

Density is the number of individuals of a particular parasite species per unit area, volume or weight of infected host tissue or organ.

Relative density or **abundance** is the total number of individuals of a particular parasite species in a sample of hosts divided by the total number of individuals of the host species (infected and uninfected) in the sample. It is the mean number of a particular parasite species per host examined. It is equivalent to mean intensity × prevalence. This concept seems to cause the greatest problem in ecological parasitology. Mean intensity has been misused for relative density. Density has also been misused for relative density. If relative density or abundance is used, Margolis *et al.* (1982) recommend that authors include a definition in their papers to avoid further misunderstanding. In the British literature a preferable term, 'worm burden', and variations on this, is used in the same way as 'relative density' and 'abundance'.

Diversity has been used to denote the mean number of parasite species per host animal by Kennedy, Bush and Aho (1986) and is generally used in discussions of parasite and host populations. The term has a well-defined meaning which encompasses two features, the number of species present **(species richness)** and the degree to which individuals are evenly distributed among the species present **(evenness)**. The two are combined in various ways in most indexes of diversity, e.g. Simpson's index as used in Holmes (1973, 1976, 1979, 1982, 1983), Brillouin's as used in Kennedy, Bush and Aho (1986) or the Shannon H' diversity index as used in Janovy and Hardin (1988).

(b) Distribution

The distribution of parasites within a host population can usually be described by a frequency distribution. In general three fundamental distributions exist and theses are: regular, random and contagious (the latter may also be termed 'clumped' or 'aggregated'). Each one of these distributions may be modelled using mathematical models. For example the negative binomial model is used for contagious distributions and the Poisson model is widely used to describe a random distribution.

In nature, regular and random distributions of parasites within host populations rarely occur since heterogeneity (or uneveness) almost always exists with respect to the probability that an individual host will be infected by a parasite. This heterogeneity may be generated in a number of different ways, but some of the most important relate to (i) behavioural, genetic or immunological differences in host susceptibility to infection (Wakelin, 1985), and (ii) the spatial distribution of infective stages in the habitat of the host (Anderson, 1974a, 1978, 1979, 1982a, b, c, d, 1986; Anderson and Gordon, 1982; Kennedy, 1975). Heterogeneity in host susceptibility to infection, in most cases, results in a contagious distribution in which the variance to mean ratio ($S^2/\bar{x}$) of the parasite counts is significantly greater than unity, and the parasites are said to be overdispersed within the host population. This type of distribution is usually well-described by the negative binomial model, a theoretical distribution defined by two parameters, the arithmetic mean ($\bar{x}$) and a positive value (k) which measures inversely the extent to which parasites are clumped within the host population. Among the helminth parasites that have been shown to be contagiously distributed within populations of host fish are the adults of the acanthocephalan *Pomphorhynchus laevis* in the flounder, *Platichthys flesus*. (Munro, Whitfield and Diffley, 1989), and the metacercariae of the digenean *Apatemon gracilis pellucidus* in the brook stickleback, *Culaea inconstans* (Rau and Gordon, 1978).

A further example of a contagious distribution involving a fish parasitic helminth is illustrated (Figure 3.2). The diagram demonstrates the goodness-of-fit of the theoretical negative binomial distribution to observed data on the distribution of *Apatemon gracilis gracilis* metacercarial cysts among a sample of 87 fish from a British population of three-spined sticklebacks, *Gasterosteus aculeatus*.

Crofton (1971 a, b) suggested that it was characteristic of parasitism to produce a contagious distribution of parasites within host populations, yet despite this there are docu-

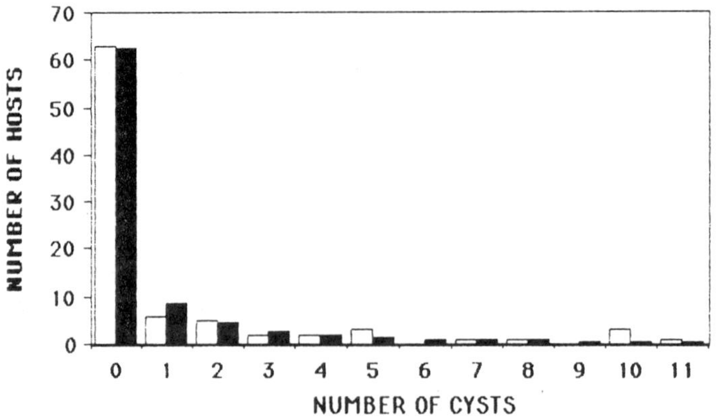

Figure 3.2 The frequency distribution of *Apatemon gracilis gracilis* metacercarial cysts in *Gasterosteus aculeatus*. White columns represent observed frequencies and black columns indicate frequencies predicted by the negative binomial model. (A. McCarthy, unpublished)

mented cases of parasite distributions which depart from this general rule. For example, Anderson (1974b) described a host-parasite system involving the fish-parasitic monogenean *Diplozoon paradoxum* in which a regular distribution existed. However, as Whitfield (1979) states, the vast majority of host-parasite relationships which have been studied quantitatively demonstrate a contagious distribution.

Infrapopulation and suprapopulation

Esch *et al.* (1975, 1977) referred to all the individuals of a single species in a single host individual as an 'infrapopulation' and all the individuals of all stages of a single species in an ecosystem, whether they are free-living or in their intermediate or definitive hosts, as a 'suprapopulation'. Factors which control an infrapopulation size include host susceptibility, host physiology (diet, age, sex and state of sexual maturity) and host genetics and behaviour. Inter- and intra-specific interactions and environmental factors extrinsic to the host which affect rates of parasite transmission (temperature and dispersion) also determine observed infrapopulation sizes. Regulation of suprapopulations occurs through density-dependent reductions in parasite survival and establishment and by parasite-induced host mortality (Anderson and May, 1978; May and Anderson, 1978). These two levels of population in parasitology have also been discussed by Holmes, Hobbs and Leong (1977).

Theoretical, mathematical, experimental and field studies.

Introduction

In the late 1950s it was realized that a manual analysis in the field and laboratory of the large number of abiotic and biotic factors involved in understanding host/parasite relationships was likely to remain tedious and complex unless supported by an alternative approach. Thus a mathematical approach was pioneered by Macdonald (1957) building on the much earlier work of Ross, who had used *Plasmodium* as a model, and by Ratcliffe *et al.*, (1969) who used the nematode *Haemonchus contortus*. Crofton (1971a,b) generalized Ratcliffe's work by constructing a quantitative simulation model of host-parasite systems that highlighted the essential parameters of parasite-host population dynamics, thus paving the way to a better understanding of the population dynamics of these relationships. More recently Anderson and May (1978, 1979, 1981) and May and Anderson (1978, 1979) have developed analytical models for both helminth and protozoan parasites, which have provided a general mathematical framework for understanding the population dynamics of interactions between host and parasite. Models have proved invaluable in stimulating ideas for further research and in studying the many complicated variables of host/parasite relationships.

Hybrid mathematical models combining those used for host-parasite relationships and those used by fisheries biologists were developed by Dobson and May (1986) to determine how pathogens can affect the dynamics of exploited fish populations. They concluded that the ability of a pathogen to establish itself is dependent upon the relative magnitudes of the threshold host density for parasite establishment, H_T, and the level of the host population density at its current level of exploitation $H(E)$. If H_T is less than $H(E)$, the parasite will always be able to establish itself.

Disease prevalence would also seem to be roughly independent of the level of exploitation of the fishery, providing the exploited population density is significantly higher than the threshold density for disease establishment. Dobson and May suggested extending the models they used to include such factors as age-structure, more realistic density-dependent recruitment functions in the host, immune response of the host and environmental stochasticity.

Some of the disagreements which have arisen between theoretical ecologists and those who have concentrated on a fieldwork approach may be seen, for instance from Holmes (1982) who agreed with Anderson and May (1978, 1979) on (a) parasites causing host-mortality, (b) patterns of epizootic diseases, (c) the relation of epizootics to stress in the host population and (d) that the population dynamics of the parasite may be controlled by selective mortality of the more heavily infected hosts (although there was little hard data from field studies to support this). He disagreed with the assumption that parasite-induced mortality is additive to mortality from other factors and, therefore, whether or not parasites regulate host populations.

The major contributions by Anderson and May to theoretical and mathematical ecology of parasites and criticisms of them, however, have considerably narrowed the gap between the theoreticians and field workers. Anderson (1978) reviewed previous work on theoretical, experimental and field study aspects of parasite distributions and focused on three topics: (i) processes that generate various dispersion patterns, (ii) the dynamic nature of parasite distributions and (iii) the relevance of (i) and (ii) to studies of the population dynamics of host-parasite interactions. He concluded that much of our understanding of parasite aggregation and the dynamics of host parasite population interactions has been created by theoretical studies of population dynamics although there was a need for an integrated approach to parasite ecology by concomitant use of theoretical, experimental and field studies. He appealed for greater collaboration between the theoreticians and field ecologists. Anderson (1982 a, b, c, d) referred to marked progress in theoretical ecology and the parallel development of experimental laboratory work to ascertain, for example, the dependance of parasite birth, death and establishment rates on population density, the factors which generate aggregation in the distribution of parasite numbers per host and the processes influencing epidemic or endemic patterns of infection. He emphasized that, in contrast, progress in most field studies in ecology had been less extensive and was largely based on survey as opposed to experiment. He reviewed recent theoretical developments on two-species interactions (host and parasite) and again appealed for a greater degree of collaboration between theoreticians and field workers, on the grounds that the latter interpreted population trends largely on intuitive and qualitative grounds, as opposed to quantitative judgements of the numerous processes that determine dynamic changes in parasite and host abundance. Good progress in this direction over the last 15 years, led mainly by the theoreticians, has been achieved through intensive research on a single species (some using a combination of field and laboratory work), on competition between species, on community ecology and on the use of published checklists for analyzing independent variables which account for the distribution and richness of helminth species on various groups of fish. In this chapter emphasis is placed on the biological significance of the results. For the methodology of mathematical modelling it is essential to consult the original Anderson and May, and May and Anderson (1978) papers and Anderson (1982a), which is one of 10 chapters by 11 authors, mathematicians and biologists, actively engaged in making models of the dynamics of infectious disease transmission. Complicated mathematical equations are included in this book but most chapters also include a non-mathematical summary of the principal conclusions. The principles involved are also illustrated in many graphs and tables. The approach enables complete life-cycles to be broken down into compartmentalized flow diagrams suitable for algebraic or calculus methods of description. The resultant dynamic models can simulate the effects of changing variables and the results can be compared with actual field data. The use of mathematical models in parasite population studies was reviewed by Fedorov (1981).

The importance of a theoretical, mathematical, experimental and/or a field studies approach to understanding the ecology of fish worms can be seen from the results of intensive studies on *Gyrodactylus turnbulli*, *Diplozoon paradoxum*, *Caryophyllaeus laticeps*, *Bothriocephalus acheilognathi*, *Pomphorhynchus laevis*, *Transversotrema patialense* and on some larval digeneans. Information on these is arranged below in order of increasing life-cycle complexity rather than on the taxonomic lines of Chapter 1. Such an approach highlights the increasing complexity of ecological problems with regard to parasite birth and death rates, seasonal infection dynamics, the effects of host age and feeding habits, parasites as regulators of host populations in nature, as regulators of host populations in the laboratory and competition between species. Furthermore, researchers on these problems focus attention on the most influential biotic and abiotic factors in ecological parasitology.

Birth and death rates of *Gyrodactylus turnbulli* (= *G. bullatarudis*).

At 25°C the viviparous monogenean *Gyrodactylus bullatarudis* on guppies, *Poecilia reticulata*, has an average fecundity of 1.68 offspring during its expected life span of 4.20 days (Scott, 1982). The average instantaneous birth rate is 0.63/parasite/day. The first offspring is born about one day after the birth of the parent and subsequent offspring are born at 2-25-day intervals. The death rate increases exponentially with the age of the parasite and has an average value of 0.24/parasite/day. Scott and Nokes (1984) investigated the effect of water temperature on the survival and reproduction of *G. turnbulli* as an extension of the previous study of the dynamics of the birth and death processes of the parasite on isolated guppies maintained at 25°C. The intrinsic rate of increase of the parasite population was maximum (0.230/parasite/day), at 27.5°C *G. bullatarudis* was not able to survive at 30°C.

Scott and Anderson (1984) and Scott (1985) further extended these studies and combined a long-term experimental epidemiological approach in combination with short-term directed experiments and theoretical studies to illuminate the population dynamics of *G. turnbulli* on *P. reticulata*. On some guppies the parasite never establishes, some fish recover from an infection and some fish die during the exponential growth of the parasite population. Parasite-induced host mortality is thought to be an important regulatory factor. These three different observed patterns reflect genetic differences in the host. Aggregation is a dominant factor in the association and varies both temporally and spatially. Continual additions of susceptible fish to an infected population resulted in recurrent epidemics of *G. turnbulli* with increasing amplitude at medium and high levels of immigration and of decreasing amplitude at low levels of immigration. In the absence of immigration the parasite becomes extinct after the first epidemic. The driving force behind the cycles may be associated with a temporary, partial refractory period to infection.

Scott (1987) pursued the topic of temporal changes in aggregation by examining *G. turnbulli* in laboratory populations of *P. reticulata*. She also investigated aggregation and two common measures of aggregation, the variance to mean ratio and the parameter *k* of the negative binomial distribution. As peak prevalence and abundance was approached, *G. turnbulli* became less aggregated with lowest clumping occurring during the declining phase of the cycle. This, perhaps, was a function of density-dependent death of infected hosts and density-dependent reduction in parasite survival and reproduction on hosts that recover from infection (Anderson and Gordon, 1982). The variance to mean ratio and parameter *k* of the negative binomial distribution did not consistently quantify the same aspect of the observed degree of aggregation. Scott (1987) suggests that the variance to mean ratio may be a better measure when the prevalence and/or mean burden are changing rapidly and when the tail of the distribution is of particular interest. The parameter *k* may be a preferred measure when the zero class on the lightly infected hosts is of primary interest.

Madhavi and Anderson (1985) compared the establishment, reproduction and survival of *G. turnbulli* on four inbred strains of guppies following low and high levels of exposure to infection. The results indicated two broad categories of innately 'resistant' and 'susceptible' amongst the four strains of fish. Experiments with hybrids indicated dominance of resistance over susceptibility. The guppy shows promise in investigating the role of genetic factors in the study of host/parasite interactions.

With regard to susceptibility, Scott and Robinson (1984) investigated the establishment success, mean parasite population size, peak parasite burden, time to peak burden and duration of the infection following challenge infections of *P. reticulata* with *G. turnbulli*. Initial infections were removed by formaldehyde treatment. The challenge infections were significantly lower in all the aspects investigated than in the respective initial or control (non-treated unchallenged) infections.

Seasonal dynamics of *Diplozoon paradoxum*

Anderson (1974c) studied the population dynamics of *Diplozoon paradoxum* on *Abramis brama* and observed that the number of parasites per host increased with the age of the fish, and also that the parasite counts per fish were under-dispersed in the younger age groups of hosts. He therefore examined the relationship between some easily-measurable morphometric characters of the host, relating to host size and hence indirectly to the physical carrying capacity, and the number of parasites a host harboured. The aims were (i) to develop a predictive equation relating the dependent parasite variable (numbers per host) to a set of supposedly independent morphometric characters of the host and, more importantly, (ii) to investigate the relative importance of each morphometric character in influencing the size of the parasite population. Results from the use of principle components in multiple regression analysis demonstrated a close positive correlation between the host's age and thus size as measured by various morphometric features and the number of *D. paradoxum* harboured. This was associated with (i) space available in the gill chamber for establishment of infective stages, (ii) volume of water carrying infective stages over the gill filaments and (iii) the accumulation of parasites in time due to continuous exposure to infection. There was a significant relationship between large parasite burdens and a reduction in host weight when specific age-groups of fish were competing for limited food resources.

Seasonal dynamics of *Caryophyllaeus laticeps*

Anderson (1974b) found the statistical distribution of *C. laticeps* counts per host to be aggregated, with the negative binomial providing a good empirical model for their distribution. This system was also a good model for examining the spatial distribution of infected and uninfected tubificid intermediate hosts. These were used to consider possible processes generating the observed negative binomial distribution. Anderson (1986) examined in detail the biological parameters which cause general seasonal fluctuations in *C. laticeps* populations by comparing the prediction of a model with observed population dynamics data. He concluded that seasonal incidence was related to host feeding habits and a temperature-dependent immune response may have accounted for the death rate of adult parasites. A deterministic immigration death model was then explored which contained non-homogeneous immigration and death rates to describe the adult population size in a single host, while a temperature-dependent mortality rate and fluctuations in host feeding controlled the immigration rate of larval parasites. This suggested that observed aggregation of adult tapeworm numbers within the host population was predominantly generated by variability in feeding habits of individual fish within a population. The theoretical population model was also used to investigate the comparative influences of the immigration and death rates on the dynamics of the adult parasite, the future behaviour of the system under altered environmental conditions and the importance of chance effects in the dynamics of individual parasite populations within a single host. The stability of the dynamics of the complete life-cycle of *C. laticeps* was discussed relative to the inherent biological complexity of host/helminth parasite interactions.

Natural regulation of host populations by *Bothriocephalus acheilognathi*

May and Anderson (1979) have suggested that 'parasites (broadly defined) are probably at least as important as the more usually studied predators and insect parasitoids in regulating natural populations'; unfortunately there are woefully few laboratory and field studies on the population dynamics of a single host-parasite system to adequately support the contention. Consequently, Granath and Esch (1983c) used a combination of laboratory and field experiments to investigate the population biology of *Bothriocephalus acheilognathi* in the mosquito fish, *Gambusia affinis*, to determine the potential for mutual regulatory interaction. Specifically they set out to assess the effect of *B. acheilognathi* on the mortality and survivorship of mosquito fish and to evaluate the impact of the tapeworm on certain demographic features of the mosquito fish population, although temperature was shown to be important in determining survivorship of infected mosquito fish. Survivorship was also a function of parasite density and host size. Since parasite-induced mortality is a function of host size it was assumed that male fish are more vulnerable to parasite-induced mortality than females. Granath and Esch suggested that temperature-dependent growth and development, intraspecific competition, a contagiously distributed parasite population and size-related parasite-induced host mortality all combined to effectively regulate parasite population density and composition, and at the same time manipulated some of the demographic features of the mosquito fish population in Belewes Lake.

Riggs, Lemly and Esch (1987) demonstrated that host suitability for *B. acheilognathi* in *Gambusia affinis* and *Pimephales promelas* is determined by morphological, physiological and behavioural differences in the host species. These factors affect transmission dynamics and the quality and stability of the enteric environment. Riggs and Esch (1987) investigated the tapeworm from the two host species named above and also from *Nototropis lutrensis*. They found that mean infrapopulation density and prevalence differed by site, season, species and size of hosts. Degree of aggregation and abundance and prevalence of gravid worms differed by species of host. Abundance of gravid worms was significantly lower in metapopulations from localities that received power plant effluents. The differences in infrapopulation density, prevalence, and aggregation appeared to be related to predator-prey interactions, which varied with season and local community structure. Difference in abundance of gravid worms, however, was probably caused by differential suitability of hosts and by local variation in selenium concentration in the water column. Biotic and abiotic components of the host community may, therefore, determine the suprapopulation dynamics of *B. acheilognathi*.

Infection dynamics of *Pomphorhynchus laevis* and other acanthocephalans.

Brown (1986) refers to evidence (Kennedy and Rumpus, 1977) showing that the occurrence of *Pomphorhynchus laevis* in the fish of the River Avon in England is a stable parasite population and is, therefore, regulated by some density-dependent influence (see Anderson and May, 1978: Anderson, 1978). He also referred to work which demonstrated that the establishment and survival of *P. laevis* in laboratory-infected goldfish, *Carassius auratus*, are independent of the initial infection density (Kennedy, 1972b, 1974a; Kennedy and Rumpus, 1977). Brown suggested that since *P. laevis* only grows, matures and reproduces regularly within chub (*Leuciscus cephalus*), barbel (*Barbus barbus*) and rainbow trout (*Salmo gairdneri*), and not at all in *C. auratus* (Kennedy *et al.*, 1978), a density-dependent mechanism may operate in any or all of these preferred hosts. Brown investigated the possibility that such a mechanism operates in populations of *P. laevis* in laboratory-infected *S. gairdneri*. He also discussed the significance of the relationship between the size of *L. cephalus* and its worm burden, in a natural population in the River Severn in England, in the light of his experimental work. The experimental primary and superimposed infections of *Salmo gairdneri* with various dosages of *P. laevis* indicated that density-dependent establishment and survival of the acanthocephalan reaches a high level of survival and reproductive capacity in the individual host. The level is partly dependent on host size although the size of the parasites themselves appears to have an important influence, superimposed infections being possible where pre-established parasites are small (less than seven days old) and in numbers below the carrying capacity of the host. The number of parasites harboured with increasing infection dosage reached a plateau, proportional to the length of the fish host. The idea that the processes leading to predictable worm burdens in relation to fish size are density-dependent is an alternative to that commonly suggested, that older and larger fish consume more food and therefore more intermediate hosts than smaller, conspecific hosts, and thus harbour more parasites. Brown emphasized that no other studies have considered these two alternatives. Whilst acknowledging the action advocated by Keymer *et al.* (1983) in extrapolating the results of controlled laboratory experiments to field studies where population dynamics and host nutrition are more uncertain, Brown concluded that the identification of a possible regulatory mechanism in a laboratory population raised the possibility that it may operate in the natural population of *P. laevis* in *Leuciscus cephalus*.

Other acanthocephalan genera which have attracted most attention in recent years are *Acanthocephalus, Echinorhynchus, Neoechinorhynchus* and *Leptorhynchoides*. Three species of *Acanthocephalus* have been studied, *A. clavula* by Kennedy (1984b) and Kennedy and Lord (1982); *A lucii* by Lee (1981) and *A. parksidei* by Amin (1975a,c) and Amin, Burns and Redlin (1980).

Kennedy and Lord (1982) investigated the site-specificity of *A. clavula* in *Anguilla anguilla* both in natural infections and in fish maintained experimentally under different regimes. The fish alimentary canals from the oesophagus (0 per cent) to the anus (100 per cent) were removed, slit longitudinally and searched. The acanthocephalan showed a preference for the 55–65 per cent area of the alimentary canal and thus did not differ in different species of host. This site-specificity, however, was not very precise; there was considerable variation between individual hosts and *A. clavula* was capable of surviving and maturing in all regions of the intestine. The mean position of male parasites was slightly anterior to that of females and the overall sex ratio was in favour of females. At high levels of infection the range of the parasites was extended and the mean position was significantly more anterior when compared with single worm infections. The mean position did not change when the host was starved or maintained in 100 per cent sea water, or with reduced aeration or at high temperatures. The results were compared with site-specificity shown by other species of freshwater fish acanthocephalans. Kennedy (1984b) investigated changes in the size and composition of a population of *A. clavula* in *Anguilla anguilla* and *Platichthys flesus* from 1979–1983, in the course of which the population underwent a steady and continuous decline. This was thought to be due to the decrease in abundance of the intermediate host, *Asellus meridianus*, as a result of interspecific competition with a congeneric species.

Valtonen (1983b) found no pronounced seasonal changes in prevalence and intensity of infection of *Coregonus lavaretus* with *Echinorhynchus salmonis*. Valtonen also reported briefly on the sex ratio of the worms, the developmental stage of the females and the site of infection in the fish intestine relative to the results of other workers on various fish acanthocephalans. The overdispersion of *E. salmonis* in whitefish was thought to be due to a rather homogeneous infection mechanism. Some local areas of deeper waters in the Gulf of Bothnia are known to be favoured by whitefish in spring and summer, and, therefore, infection foci may occur in these areas. The high levels of infection of *E. salmonis* in some whitefish was assumed to be due to aggregations of infected copepods. Of 6300 *Pontoporeia affinis* collected in one area and 545 in another, only three in the second small sample were infected, each with only one specimen of *E. salmonis*. Thus it was suggested that, in order to acquire a burden of 50 worms, over 9000 *P. affinis* would have to be consumed. Every nineteenth fish had in excess of 50 *E. salmonis*.

Leptorhynchoides thecatus is a common, widely-distributed acanthocephalan of centrarchid fish in North America (Uznanski and Nickol, 1982). These authors provided new information on growth, maturation, site-selection and parenteric infections with the parasite. *Lepomis cyanellus* were fed 10, 20 or 40 *L. thecatus* cystacanths and examined after 1, 3, 7, 14, 28 or 56 days. Some worms remained in the stomach or intestine for up to 7 days but by 14 days they were restricted to the pyloric caeca and mesenteries. The number of worms recovered from fish fed 40 cystacanths declined between days 3–14, to levels similar to those observed in fish fed 20 cystacanths. No decline was detected at either of the lower dosages. Parasite activation and site selection are density-independent processes but establishment and survival are apparently density-dependent. *Lepomis cyanellus* provides sufficient resources for the establishment of 10 to 15 *L. thecatus* or approximately two per caecum. Excess parasites are lost within 14 days. Uznanski and Nickol's results are not consistent with a proportional maturation mode, preferred by Holmes *et al.* (1977) for *Echinorhynchus salmonis* in *Coregonus clupeiformis*. Uznanski and Nickol conclude that more detailed field and experimental data are needed before the population processes of any fish parasite is fully understood. Because of its distribution, abundance and suitability as an experimental animal they suggest that *Leptorhynchoides thecatus* is an ideal parasite for studies of population ecology.

Experimental regulation of host populations by *Transversotrema patialense*

Anderson, Whitfield and Mills (1977) referred to the difficulties created by the numerous population variables and rate parameters involved in host-parasite interactions in field and experimental ecology. They investigated the population dynamics of the ectoparasitic digenean, *Transversotrema patialense*, under laboratory conditions of constant temperature and dark-light regimes. The system could then be used to estimate the form of functional rate parameters such as birth, death and infection rates which control the population dynamics of a helminth with a complex life-cycle. The results suggested that the life-cycle of *T. patialense* contains many potential density-dependent population processes and that a series of stable equilibrium states may exist for both host and parasite populations in a given habitat. Density-independent factors such as water temperature also play a major role in determining the dynamics of a parasite that has two poikilothermic hosts and two different larval stages free-living in the aquatic habitat. The authors emphasized that a great deal of experimental work remained to be done before even a superficial understanding was achieved of the many population processes outlined in Figure 3.3.

Mills (1980a, b) examined growth, development and senescence in adult *T. patialense* to investigate generative mechanisms behind age-dependent egg production in the species. The density-dependent nature of parasite growth was also investigated and found to be age-dependent, ceasing 15–20 days post-infection. The vitelline glands expanded greatly in size after infection from 1.25 per cent of cercarial area to 20.7 per cent of that of the mature adult. There was an increase in the occurrence of reproductive abnormalities in old parasites, but this alone did not account for the decline in egg production as populations of *T. patialense* grew older. Growth of adult *T. patialense* is density-dependent with reduced growth at high initial parasite densities per host.

Bundy (1981a) carried out laboratory experiments on the miracidia of *T. patialense* and found that they had a maximum longevity of 8 hours with 50 per cent surviving to 5.5 hours after hatching at 25°C. Miracidial survival was age-dependent, the instantaneous mortality rate increasing exponentially with miracidial age. A model was used to estimate the mean expected life-span.

Anderson and Whitfield (1975) showed the survival characteristics of free-living cercarial populations of *T. patialense* to be age-dependent. The maximum life-span was 44 hours with a 50 per cent survival at 26 hours. An attempt was made to interrelate activity and infectivity, in a theoretical manner, with the availability of energy reserves. A simple mathematical model was formulated to aid the conceptual understanding of the biological processes involved.

The number of cercariae of *T. patialense* which attached to *Brachydanio rerio* during a fixed exposure period was directly proportional to cercarial density within an experimental infection arena (Anderson, Whitfield and Dobson, 1978). The distribution of successful infections/host was shown to change from a random pattern to an increasingly aggregated distribution as cercarial exposure density or duration of host exposure increased. Differences in host behaviour during exposure to infection appeared to generate variability in host susceptibility to infection. The functional response of *Brachydanio rerio* to changes in prey density (*Daphnia magna*) and to changes in density of *T. patialense* cercariae, and how the latter response influenced infection of the host were investigated by Anderson *et al.* (1978). The rate of infection appeared to be approximately directly proportional to cercarial exposure density, the number of parasites establishing themselves varying almost linearly with exposure density. A model of predation and infection dynamics was presented and the relevance of concomitant predation and the infection process to the dynamics of digenean life-cycles discussed.

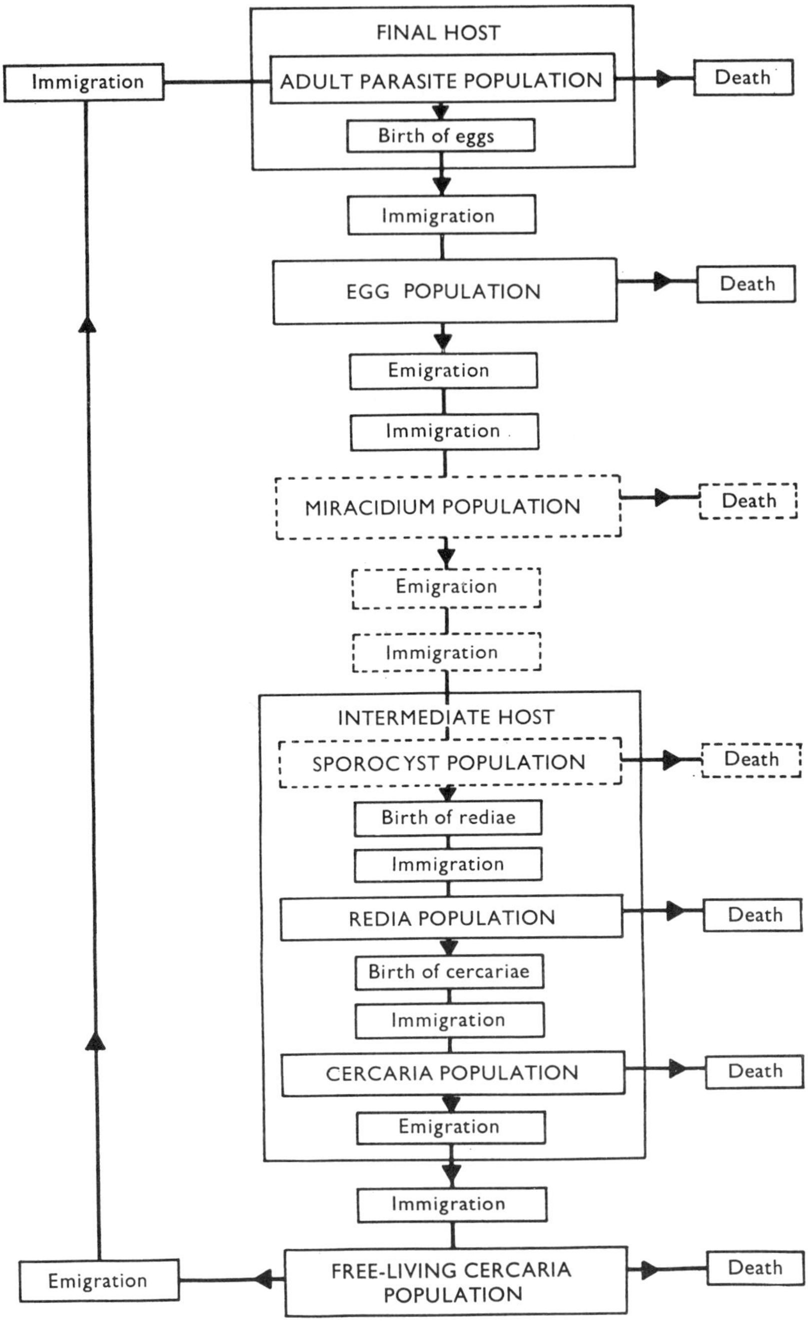

Figure 3.3 Diagrammatic flow chart of the population processes involved in the life cycle of a typical transversotrematid.
From Anderson and Whitfield (1975)

Infection dynamics of *Uvulifer ambloplitis* and other larval digeneans

Lemly and Esch (1983) investigated the survivability of metacercariae of *Uvulifer ambloplitis* in juveniles of several species of centrarchid, up to the point at which individual fish were too large to be ingested by a kingfisher, and they also assessed the impact of over-wintering on metacercarial infrapopulations. Significantly greater survival was found in the smallest size-class of all species of hosts examined ($p < 0.05$). This may maximize the probability of successful transmission to the definitive host because of selective predation pressure on this size-class of fish. The percentage of live metacercariae dropped significantly over winter although the empty cysts remained pigmented and visible. Parasite surveys based on cyst counts may, therefore, contain a significant amount of error. This study was extended by Lemly and Esch (1984a,b) to investigate the population biology of *U. ambloplitis* in the snail intermediate host, *Heliosoma trivolvis*, to study the effects of the parasite at the fish host, *Lepomis macrochirus*, and to assess the impact of the parasite on the level of the host population. The snail host was most abundant in July of each year, with densities of up to $95/m^2$. Fewer than 4 per cent of the population shed cercariae in any month except July when 13 per cent were shedding, coinciding with peak abundance of *H. trivolvis*. No difference was found in the prevalence of *U. ambloplitis* among snails collected at different locations within the pond. The high prevalence of infection in July may be related to the breeding behaviour of the bird definitive host during early summer.

With regard to the effects of the parasite on the fish host, Lemly and Esch emphasized that, although the possibility of parasite-induced host mortality had been considered in great detail by many theoreticians, almost no direct evidence was available to show that such mortality occurs in natural populations of fish. Their data, however, for one pond provide clear evidence of mortality in natural populations of fish due to the direct effects of parasitism. This was demonstrated through both laboratory and field evidence. They also established simple cause-and-effect relationships between mortality and parasitism by physiological and biochemical data. Lemly and Esch (1985), however, point out that they may well have been dealing with a unique situation in a particular pond because a survey of 43 other ponds, lakes and streams in the same area indicated that prevalence and intensity of parasitism in juvenile centrarchid fish were quite low. They concluded that the observed elimination of heavily parasitized juvenile *L. macrochirus* (50 cysts/fish) among overwintering hosts in one pond is exceptional, and that such mutual regulatory interaction involving *U. ambloplitis* and juvenile *L. macrochirus* is the exception and not the rule.

Aspects of the ecology of *Posthodiplostomum cuticola* were studied by Dönges (1964) who found that the species survived the cold season in the metacercarial, sporocyst and probably in the egg stage. Swennen, Heessen and Hocker (1979) investigated three species of *Cotylurus* and found *C. erraticus* in Salmoniformes, especially in the pericardial cavity, *C. variegatus* in Percidae, mainly on the wall of the swimbladder, and *C. platycephalus* in Cyprinidae and in Percidae, mainly in the pericardial cavity and on the mesentery. Since cercariae of *Cotylurus* spp. released by infected snails can live for only one day, fish acquire infections only in waters where infected snails occur. This accounts for the absence of infected fish in some areas. Although encysted metacercariae were found in many species of fish, the highest infections occurred in *Osmerus eperlanus* and *Gymnocephalus cernua*. Both species are relatively small and extraordinarily numerous and, therefore, are of great importance in continuing the life-cycle. Species which grow to sizes too large to be swallowed by gulls were considered as ecological traps rather than efficient intermediate hosts of *Cotylurus* spp. The occasional death of large numbers of *O. eperlanus* and *G. cernua* in nature were attributed to *Cotylurus* spp. cercariae.

Kennedy (1987) studied annual changes in the population size of *Tylodelphys podicipina* in the eyes of *Perca fluviatilis* in a small eutrophic lake over ten years, by following changes in prevalence, abundance and over-dispersion of parasites throughout the life of each year class of fish. The population increased rapidly in the first two years but for the next six years fluctuated within very narrow constrained limits before declining as a result of a catastrophic decline in the perch population.

The prevalence and mean infrapopulation densities of *Diplostomulum scheuringi* metacercariae were monitored for 4.5 years in populations of *Gambusia affinis* in ambient and thermally-altered locations of a pond in South Carolina by Aho, Camp and Esch (1982). The various seasonal changes in parasite population dynamics in the two locations were attributed to a combination of seasonal variation in cercarial shedding from infected snails, recruitment of metacercariae, and to seasonal patterns of birth and mortality of the short-lived mosquito fish. At the heated location, periods of cercarial shedding and metacercarial recruitment were altered by thermal effluent that also reduced the normal life expectancy of mosquito fish, thereby further modifying the population biology of the parasite.

Poole and Dick (1983) investigated the distribution of *Apophallus brevis* in *Perca flavescens*, and whether the population size of fish predators of perch influenced the recruitment of the parasite into the perch population. Prevalence levels remained high in most age-classes of perch and mean intensity differences were not significant for all ages beyond one year old. Prevalence levels remained more or less constant during the open-water seasons. Mean intensities dropped significantly from June to July and increased in August and October. Monthly fluctuations are probably the results of recruitment of the parasite and a loss of heavily infected hosts. The variance to mean ratios of *A. brevis* showed over-dispersed parasite populations and the distribution of each in two lakes was found to fit the negative binomial. The authors questioned the use of variance to mean ratios or other statistics dealing with overdispersion in predicting loss or recruitment for all parasite systems.

Hazen and Esch (1978) investigated *Clinostomum marginatum* in various species of Centrarchidae (*Lepomis macrochirus*, *L. gulosus*, *L. auritus*, *Pomoxis nigromaculatus* and *Micropterus salmoides*) from a reservoir receiving thermal effluent from a nuclear reactor. Except for *M. salmoides* infection percentages among the five species were less than 1 per cent. Attention was given to temperature, the condition of *M. salmoides*, host size, seasonal changes of infection in *M. salmoides* and to differences in infection percentages in various localities of the reservoir. Among *M. salmoides*, infection varied seasonally, being highest from January to June. From the spring highs of approximately 25 per cent the percentages dropped to lows of 10 per cent in July and August, but another peak of 30 per cent was reached in September–October, leading to a steady decline of less than 10 per cent in December. Neither body-condition nor size of *M. salmoides* were related to infection percentage or metacercarial density. Infection percentages were different in various localities but they could not be related to different temperatures. Locality differences in infection percentages were related to the amount of littoral zone present, an extensive zone leading to a greater aquatic/terrestrial interaction and in turn to more predator/prey relationships with consequent increases in infection percentages.

Biotic and Abiotic Factors

Introduction

In the previous section we discussed intensive field and laboratory studies on various species of fish helminths (according to their life-cycle complexity) in order to introduce and highlight a number of influential biotic and abiotic factors in ecological parasitology. In this section our aim is to extend a discussion of these factors by bringing together information on a number of other species which have been the subject of ecological studies.

A significant early paper on biotic and abiotic factors is that of Bauer (1959a), where he drew attention to Soviet investigations which showed that age, physiological condition of the host, migration and season of the year were important. Bauer discussed the effects of temperature, salinity, oxygen content, pH, light and pressure. *Dactylogyrus vastator*, *Triaenophorus nodulosus*, *Ligula, Contracaecum*, *Camallanus* and *Cystoopsis acipenseris* had been the subject of much attention with regard to temperature. Salinity was said to have important effects on *Dactylogyrus*, *Gyrodactylus*, *Tetraonchus*, *Amphilina* and *Triaenophorus*, and some of these parasites had also been the focus of attention in investigations on the influence of oxygen and light on the development of helminths. No studies were cited by Bauer in which it was known that environmental pH directly influenced the ecology of fish worms, but some work was already known on the effects of pressure, namely that of Thomas (1953) on *Bothriocephalus abyssus* and of Michajłow (1951) on *Triaenophorus*. Bauer also discussed, in general terms, the importance of the feeding habits of fish, the physiological condition of the host as well as the immunity it develops, and the abundance of parasites and their reproduction and development.

It is now accepted that a large number of biotic factors affect the abundance and prevalence of parasites but, amongst the more important group of factors, are host-parasite interactions (in particular the host-immune response) and the effects of parasites on their hosts (Kennedy, 1982). Intra- and inter-specific interactions are also important, although it is still not possible to identify the many variables involved in competition and niche-selection. For each of these two broad concepts, which involve many biotic factors, there are important population biology consequences resulting from the regulation of host and parasite abundance (Anderson, 1982 a,b,c,d.). The association is under a continuous dynamic evolutionary process and this underlines the importance of considering co-evolution in population ecology studies (Anderson and May, 1982a).

Although a large number of chemical and physical factors are known to affect a wide range of animal life-cycle processes, to compile such a list, according to Esch (1982), would add little to our understanding of ecological relationships between hosts and parasites. He therefore divided abiotic factors into two groups; those which are determinate and predictable in character and those which are not. In nature, temperature is a determinate predictable character (eg. according to season) but through man's influence in artificially heating water by thermal effluent from coal-fired and nuclear electricity power stations, it is most often unpredictable and indeterminate. Most indeterminate factors are man-made and include fire, flood, artificially-elevated temperature, radical temperature change on a short-term basis and nutrient enrichment of water. Esch discussed determinate abiotic factors in the context of a parasite's life-cycle process which involves a pool of infective stages (eg. eggs, miracidia, cercariae and coracidia) and encysted stages within intermediate hosts and their transmission to a host where growth, reproduction and mortality might occur. Determinate abiotic factors are important in host-finding and comprise (i) light and gravity stimuli which enhance spatial distribution and thus enhance contact between parasite and host, (ii) random swimming, and (iii) host-chemistry which influences location and penetration by a parasite.

With regard to development and reproduction in a host an extensive literature has shown that they are directly related to temperature. Experimental work on *Transversotrema patialense* has demonstrated this unequivocally. Without doubt, temperature is also one of the most significant factors in determining the rates and seasonal periodicity of mortality. This has been demonstrated for *Caryophyllaeus laticeps*, but in *Bothriocephalus acheilognathi*, mortality may be due to competition, although this result of a negative interaction process is initially stimulated by temperature.

Two areas involving indeterminate factors have received much attention. One concerns unpredictable habitat quality and the other ecological succession from an oligotrophic to eutrophic state, a process which involves many variables such as lake morphometry, depth, nature of the drainage basin, altitude and latitude and man's activities in the vicinity. Esch (1982) aimed at discussing abiotic factors only, although well aware of the fact that their interactions with biotic factors are important in studying parasite ecology. While agreeing with Esch that a mere list of biotic and abiotic factors is unlikely to clarify our knowledge of host/parasite ecological relationships we have, nevertheless, compiled such a list (Table 3.1) to enable us to discuss those which have received most attention and thus to highlight many which are in need of further work.

Table 3.1. Biotic and abiotic factors mentioned in ecological studies of fish worms

Biotic	Abiotic
1. Host species	Temperature
2. Schooling behaviour	Season
3. Age and length of host	Latitude/Longitude
4. Condition of host	Salinity
5. Pathogenic effects of parasite	Oxygen
6. Post-spawning migrations of host	Water mineral content
7. Host sex and reproductive behaviour	Ammonia
8. Host hormone levels/state of maturity	pH
9. Longevity of host species	Light
10. Immunological response of host	Depth/water pressure
11. Host life-history	Pollution
12. Host diet/feeding behaviour	
13. Availability of infected intermediate hosts as food	
14. Site of infection	
15. Prevalence and intensity of infection	
16. Interspecific competition/negative interaction between parasites	

In preparing this table our aim was to identify the best-known factors, those which actually represent a complexity of factors and those which require further investigation. Such an arrangement of available information, wherever possible, in order of increasing complexity, will hopefully achieve the objective of encouraging further research, the results of which will lead to more precise definitions of biotic factors in particular and thus, eventually, to a better understanding of community ecology. Factors 1–13 in the left hand column of Table 3.1 relate directly to what collectively might be termed host biology and its influence on parasite numbers, whilst 14–16 comprise those which concern the effects of parasites themselves, be it the same or a different species, on their ecology. Of the abiotic factors, temperature is recognized as the most important in influencing the abundance and distribution of fish worms (Overstreet, 1982). Since temperature will vary according to the season of the year at various latitudes and longitudes and also at varying depths, it is not surprising that much has been written on the seasonality and zoogeography of parasitic worms in fish.

In the light of this analysis of Table 3.1 it is appropriate to present the remainder of this chapter around discussions of ten main topics: (i) host age, size, weight and/or maturity, (ii) host food and feeding behaviour, (iii) host sex, (iv) mixed biotic and abiotic factors, (v) competition/negative interaction, (vi) temperature, (vii) seasonality, (viii) zoogeography, (ix) sites and site selection and community ecology.

A large number of papers could not be incorporated under these headings although they include valuable information in planning further work on fish helminth ecology. Two appendices have, therefore, been compiled to conclude the chapter. The first lists fish genera which have been the subject of previous work and the second refers to various countries which have a particular interest in fisheries helminthology. It is hoped that these appendices will draw attention to gaps in knowledge and therefore encourage further work, in particular, on community ecology and zoogeography.

Host age, size, weight and/or maturity

This complexity of factors is in need of further research based on better accounts of methodology in determining host age and its relationships to size, weight and/or state of host maturity. The papers referred to below also reveal an additional problem, that of host age being investigated alongside other subsidiary factors but often without explanation of the relative importance of each factor investigated. This is further compounded because some factors e.g. feeding and sexual behaviour, are intricately bound up with age. According to Frankland (1955), for instance, an age resistance occurs with *Diclidophora denticulata* on the gills of *Gadus virens*. The average number of flukes per host was 2.24, the first and second gill arches carrying three times as many flukes as the third and fourth. She found that larvae were unable to establish on older fish, perhaps due to changed host behaviour. Another study, by Llewellyn (1962), showed that *Gastrocotyle trachuri* infects young three- or four-month-old adolescent *Trachurus trachurus* as soon as they descend to the sea bottom in October. There are about eight parasites per fish and they mature in 3–4 months. The life span is normally no longer than a year. The species is much less frequent on two- and three-year-old fish and occurs rarely on older fish. The limiting factor is not an age immunity but a post-spawning migration of the host away from the concentration of free-living infective stages in the coastal waters.

Discocotyle sagittata invades *Salmo trutta* from April to November but young trout in nursery streams do not harbour the parasite (Paling, 1965). Invasion takes place in deeper waters of the lake from April to November with a peak in July and August. Sexual maturity was reached, on average, in two months and the parasites lived for 3–4 years. Older trout bear, on average, more parasites per fish.

Noisi and Maillard (1980) found that *Microcotyle chrysophryi* occurred on 85 percent of *Sparus aurata* with fish less than 25cm long having 12.38 parasites per host and those of 25–47cm in length having 6.42 per fish. Differences in the distribution of the parasites on the hemibranchs of the gill arches of these two size groups of fish were thought to be related to the sexual maturity of *S. aurata*. Oliver (1984) described the prevalence and intensity of *Microcotyle chrysophrii* on 0 + *Sparus aurata* at different seasons over a four-year period (1975–78) in three littoral pools in the Mediterranean.

Winch (1983) investigated the occurrence of *Atrispinum labracis* on *Dicentrarchus labrax* in relation to the host's life-history and habitat. *A. labracis* does not begin to infect its host until fish are at least three years of age. Prevalence is much higher on open-sea than on estuarine bass although the intensity is low from both locations. On open-sea bass it is rarely more than two per fish but the strict localization of the habitat to the outer hemibranch of the ventral end of the first gill arch is thought to enhance the mating opportunities. Unlike the gill monogeneans of the grey mullet, *Chelon labrosus*, salinity is probably not responsible for the change in prevalence. No satisfactory explanation of the difference could be offered. A parallel situation may be *Ancylodiscoides siluri* which infects 100 per cent of *Siluris glanis* in the lower part of the Volga but is found only rarely on fish in the upper part. Likewise the absence of *A. labracis* from fish under three years of age is similar to

Microcotyle spinicirrus on the freshwater fish, *Aplodinotus grunniens*, which is absent from hosts under one year and present in most which are over two.

Cone and Burt (1981, 1985) investigated *Urocleidus adspectus* on *Perca flavescens* and found a prevalence ranging from 47–100 per cent over a two-year period. Intensity peaked in July and generally increased with fish age. The largest *U. adspectus* were found on the largest hosts and spring growth of this overwintering species, as well as spring invasion, commenced earlier on mature perch than on immature fish.

As for monogeneans, some researchers on fish tapeworms have emphasized the importance of age but also with reference to a number of other factors which influence their abundance and prevalence. For example, aspects of the ecology of various caryophyllidean species were reviewed by Mackiewicz (1972) with regard to location of the worms in the gut and worm burden, their growth, maturation and longevity, seasonal prevalence, immunity and age resistance, and their association, including possible competitive antagonism, with other species of helminths. The population structure of *Hunterella nodulosa* in 646 *Catastomus commersoni* was observed by Mudry and Arai (1973). There was a mean prevalence of 52.8 per cent and mean intensity of 8.7 worms per fish. Prevalence increased with fish weight up to about 50g but in fish above this weight it was found to be fairly constant at about 70 per cent. Intensity of infection was directly proportional to fish weight. No seasonal changes in prevalence and intensity were observed.

According to Anikieva and Malakhova (1982), *Coregonus albula* aged 0–6+ years in the USSR were infected with *Proteocephalus exiguus* at a prevalence rate of 96.5–99.0 per cent with highest intensities occurring in fish aged 1+ and 3+ years. Fischer and Freeman (1969) also mention fish length (age?) in their study of *P. ambloplitis* in *Micropterus dolomieui* and found that all fish more than 7.6cm long had plerocercoids. Since their paper refers also to temperature and seasonality it will be mentioned again in a discussion of these factors.

With regard to digeneans, Halvorsen (1968) investigated the occurrence of *Azygia lucii* in *Esox lucius* and found an increase in frequency and intensity with the age of the fish. Halvorsen's speculation on aspects of the life-cycle of *Azygia lucii* is of particular interest relative to the observation of Odening and Bockhardt (1976) who found that *Esox lucius* became infected either directly as fingerlings up to 3cm long, penetrated by free-swimming cercariae, or as older fish, by eating infected fish of the same or different species. The second mode is not seasonally limited as was speculated by Halvorsen. Cercarial penetration is possible only in fingerling pike and the older these fingerlings are on infection the longer the transmission time to older pike. *A. lucii* in older pike were small ($<$4mm) and young pike which had survived one summer harboured large specimens (20mm) throughout the year.

The age and seasonal dynamics of *B. luciopercae* in *Perca fluviatilis* were studied by Ieshko and Golitsina (1984) for three groups of fish; those less than 7.1cm long, those ranging from 7.2–11.3cm and fish longer than 11.3cm. It was found that the population of *B. luciopercae* is maintained by the interaction of sub-populations in fish of different age groups. The core of the population is provided by the digeneans in the medium-sized groups of fish which inhabit littoral waters and have both stable and high numbers of parasites. The smallest group of fish maintained infection in the early season and served as a source of infection to medium-sized and larger fish. The largest fish are probably not important in maintaining local populations of *B. luciopercae* but since they make extensive feeding and spawning migrations and harbour adult digeneans they have a role in spreading infection and thus ensuring exchanges with other populations of the parasite in different locations.

The nematode *Eustrongylides* sp. is almost entirely confined to *Fundulus heteroclitus* over

70mm long, according to Hirshfield, Morin and Hepner (1983), while the rate of infection of *Salmo trutta m. fario* in Czechoslovakia with *Cystidicoloides ephemeridarum* was found to be related to host size by Moravec and De (1982), 73 per cent of trout less than 10cm and 100 per cent of those over 10cm long being infected. Two previous hypotheses on aspects of the ecology of *C. stigmatura* were tested by Black (1985). The first hypothesis is that within age-classes of fish, length-frequency distributions of gravid nematodes shift towards smaller worms as the intensity of infection increases. The second is that the number of eggs released by gravid female nematodes in nature is positively correlated with their length. Black (1985) also estimated the potential egg output of parasite populations within individual fish, in order to relate this to the nematode egg production within the whole fish population. The size and number of gravid female nematodes did increase with fish age, thus supporting the suggestion that worms are long-lived and grow throughout life. Fecundity was linearly related to the length of gravid female nematodes. Reproductive output of parasites in individual fish (estimated by the number of free eggs in swim-bladders) was overdispersed in the host population and was concentrated in the oldest age-classes of char. The latter feature makes populations of the parasite susceptible to exploitation of their fish hosts. It was estimated that in one lake exploitation of larger fish decreased nematode egg production by about one-half over a seven-year period. Decline of fish stocks has probably resulted in extinction of populations of *C. stigmatura* in at least nine lakes in southern Ontario and northern New York State.

The incidence and intensity of *Spirocamallanus pereirai* in *Gillichthys mirabilis* was investigated by Noble and King (1960). As the host increases in size it acquires more parasites. Hosts ranging from 91–100mm long harboured an average of 3.43 worms whilst those measuring 141–150mm had 13.62. As the worm grows it tends to migrate posteriorly in the gut. Eight hundred of 810 hosts were infected with the worm.

Amin (1985b) studied the intensity of infection of salmonid fish with *Echinorhynchus salmonis* in relation to host age. From his own and other results he concluded that the progressive increase in abundance of *Echinorhynchus* spp. in *Coregonus artedii* (cisco) and *Salmo truttae* as they grow older was associated with feeding on invertebrates, including the infected intermediate host, in larger volumes by older fish. Loss of this pattern in *Onchorhynchus kisutch, O. tshawytscha* and *Salvelinus namaycush* and in the oldest cisco and brown trout was due to the partial replacement of the invertebrate diet with a fish diet by these fish. A similar observation was made by Valtonen (1983b) on *E. salmonis* in *Coregonus lavaretus*. The acanthocephalan was found only occasionally in small fish (<174mm). In fish over three years a considerable increase in the numbers of *E. salmonis* was found. This was explained by a change in feeding habits, because *C. lavaretus* of this age revert to feeding on amphipods, including *Pontoporeia affinis*, the intermediate host.

Host food and feeding behaviour

Price and Clancy (1983) used the key and checklist by Maitland (1972) and Kennedy (1974b) respectively to test predictions on host feeding preferences influencing the accumulation of parasites, i.e. whether top predators such as *Esox lucius* were likely to support a much richer parasite fauna than say, *Phoxinus phoxinus*, which feeds on algae and invertebrates and in turn is likely to be fed upon by pike. Two other independent variables, geographic range of host and host size were also used in an attempt to explain richness in parasite fauna. Of the five major groups of helminths investigated, only in the Monogenea, in which feeding plays no role in parasite colonization, were host feeding habits an insignificant factor. Together, the effects of geographic range of host and feeding habits accounted for 73 per cent of the variation in total parasite number per host species and from 27–70 per cent of the variation when the five parasite groups were considered individually. Price and Clancy advocated using other host-parasite lists for similar kinds of analysis providing they included variables such as number of habitat types inhabited by the host, food available in each habitat and geographic area, and parasites acquired in each habitat, as well as effects of age and size of fish.

A relationship between feeding behaviour of the host and infection with cestodes, digeneans, nematodes and acanthocephalans is well-documented. For example, *Cyathocephalus truncatus* had a definite annual cycle, *Salmo trutta* becoming infected in late autumn, the worms maturing in late winter and early spring and disappearing by late summer. The intensity of infection was correlated with feeding habits of the trout and decreased with age. Only 0.1 per cent of 14,260 *Gammarus pulex*, the intermediate host, were infected (Awachie, 1966a). Hazen and Esch (1978) investigated *Clinostomum marginatum* in various species of Centrarchidae (*Lepomis macrochirus*, *L. gulosus*, *L. auritus*, *Pomoxis nigromaculatus* and *Micropterus salmoides*) from a reservoir receiving thermal effluent from a nuclear reactor. Locality differences in infection percentages were related to the amount of littoral zone present, an extensive zone leading to a greater aquatic/terrestrial interaction and, in turn, to a greater predation pressure from fish and birds with consequent increases in infection with *C. marginatum*.

Host feeding behaviour, the availability of infective larvae and host diet control the rate at which *Cystidicoloides ephemeridarum* adds to its own population in *Salmo trutta* and junior *S. salar*, according to Aho and Kennedy (1984), but as will be discussed later, the pattern of gains and losses was determined by a temperature-dependent rejection. Valtonen and Valtonen (1978) investigated *Cystidicola farionis* in *Coregonus nasus* in the Bothnian Bay and occasionally found the species in immature fish (<175mm). In larger fish an increased intensity of infection with age was explained by the more frequent feeding on *Pontoporeia affinis*, the intermediate host.

Kennedy (1984b) investigated changes in the size and composition of a population of *Acanthocephalus clavula* in *Anguilla anguilla* and *Platichthys flesus* from 1979–1983, in the course of which the population underwent a steady and continuous decline. This was thought to be due to the decrease in abundance of the intermediate host, *Asellus meridianus*, as a result of interspecific competition with a congeneric species. Amin (1975a) reported *A. parksidei* from 11 fish species, and showed (Amin, 1975b) that growth of adult worms in their fish hosts parallels that of cystacanth seasonal increase in size in the isopod intermediate host, *Caecidotea militaris*. In other words, cystacanths ingested during the winter are larger and more developed than those ingested earlier in the year and will require less time to mature and reproduce in the fish host. The cystacanth developmental cycle was closely related to that of its isopod host, each with one generation per year. The cystacanth cycle was also closely synchronized with that of the adults in the fish hosts (Amin, Burns and Redlin, 1980).

Chubb (1964) described the occurrence of *Echinorhynchus clavula* in *Thymallus thymallus, Esox lucius, Rutilus rutilus* and *Anguilla anguilla* in a mesotrophic lake in Britain. The prevalence in these fish was 46 per cent, 11.5 per cent, 16.1 per cent and 27.7 per cent respectively. The degree of infection was correlated with feeding of the fish on *Asellus aquaticus*, the only known intermediate host in the lake. There was no seasonal periodicity of occurrence and maturation of *E. clavula* between ostracods in the diet and infection of *Gasterosteus aculeatus* with *N. rutili*. He further observed that the worms mature rapidly in spring, immature individuals develop during the summer and autumn, and a stable condition was reached in winter. Valtonen (1979) confirmed from observations on *N. rutili* in *Coregonus nasus* in the Bay of Bothnia that the pattern of infection could be explained by fish feeding on ostracods. In the Bay of Bothnia, however, the infection of whitefish with *N. rutili* is accidental and in this area the most important final host of the species is not known.

As was mentioned earlier, an increase in the number of *Echinorhynchus salmonis* in *Coregonus lavaretus* over three years can be correlated with feeding habits (Valtonen, 1983b). The over-dispersion of *E. salmonis* in whitefish was thought to be due to a rather homogeneous infection mechanism. Some local areas of deeper waters in the Gulf of Bothnia are known to be favoured by whitefish in spring and summer and, therefore, infection foci may occur in these areas. The high level of infection of *E. salmonis* in some whitefish was assumed to be due to aggregations of infected copepods. Of 6,300 *Pontoporeia affinis* collected in one area and 545 in another, only three in the second small sample were infected each with only one specimen of *E. salmonis*. Aspects of the ecology of *Neoechinorhynchus rutili* were discussed by Walkey (1967) and Valtonen (1979). Walkey observed a distinct correlation between ostracods in the diet and infection of *Gasterosteus aculeatus* with *N. rutili*.

Host sex

A number of researchers have remarked on the differences in the prevalence of infection with helminths in male and female hosts, the published papers indicating that different factors may be responsible for the unequal distribution and that the problem requires further investigation. For example, Hickman (1960) examined 1,428 Tasmanian anurans for the cestode *Nematotaenia hylae* and found that male frogs were more frequently, and slightly more heavily, infected than females, while Lees and Bass (1960) found that the level of parasitism in male frogs during and immediately before the breeding season was considerably higher than in females. Lees and Bass suggested that the hormone oestradiol in the host tissues and fluids leads to a reduction in the number of helminths. With regard to fish, Williams (1965) found that 34 of 46 male rays examined harboured *Calicotyle*, but only 20 of the 46 females did so. On further analysis it was found that, of the 26 female rays without *Calicotyle*, 12 were mature female *R. radiata* with eggs in their uteri and 12 were either *R. batis* or *R. oxyrhynchus* species which do not normally harbour *C. kroyeri* in large numbers. Of the 37 male *R. radiata* examined, only four were free from *C. kroyeri* but 13 of the females were uninfected. This difference in the prevalence of infection in male and female *R. radiata* was statistically significant (χ^2 test $p < 0.01$) Paling (1965) found that male trout between 5–7 years bear significantly more parasites than do females of comparable age.

Kennedy (1968) described the pattern of infection attained by *Caryophyllaeus laticeps* over a year in a population of dace, *Leuciscus leuciscus*, in the River Avon, Britain. The degree of infection was independent of the size of fish but was generally higher in females. The parasite was present only from December to June and became gravid after March. These seasonal cycles in prevalence and maturation may have been due to differences in fish feeding habits and resistance. The maturation cycle did not appear to be initiated by annual temperature changes. It was postulated that the host hormone levels may play a significant role in the seasonal maturation of caryophyllaeids.

A striking feature of *Caryophyllaeides fennica* in *Rutilus rutilus* in Norway was the low prevalence and intensity of infection, only 12 of 176 roach being infected and the maximum number of worms found in one fish was three (Borgstrom and Halvorsen, 1968). The cestodes were restricted to female roach. Halvorsen and Andersen (1984) found that 466 (77.4 per cent) of 602 *Salvelinus alpinus* harboured *Diphyllobothrium ditremum* plerocercoids in an overdispersed infection. Prevalence increased with age, reaching 100 per cent at 7+ years for males and 8+ years for females. Borgstrom (1970) found that a greater number of female than male *Esox lucius* were infected with *Triaenophorus nodulosus*. This was thought to be due to physiological conditions or different feeding habits of the host. As a final example we draw attention to the work of Hirschfield, Morin and Hepner (1983) who investigated the prevalence of *Eustrongylides* sp. in two populations of *Fundulus heteroclitus*, one from the intake and another from the discharge canals of an electricity plant in Maryland, USA. The nematode was four times as prevalent in female fish caught in the outlet canal and was almost entirely confined to fish over 70mm long. It was suggested that this was due to the greater abundance of oligochaetes, the suspected first intermediate hosts, in the outlet canal, presumably as a consequence of elevated temperatures and organic enrichment from the power station. This last example is a good illustration of the difficulties in ranking the importance of such factors as age, feeding and sex with regard to their influence on the prevalence and abundance of parasites.

Mixed biotic and abiotic factors

A number of publications draw attention to difficulties in assessing the precise effect of a particular biotic or abiotic factor on the population dynamics of various helminth species and state that a number of factors are involved. The following selected examples from amongst the monogeneans, gyrocotylideans, cestodes, aspidogastreans and nematodes illustrate this point. The same applies to acanthocephalans which have already been discussed on page 245.

Gelnar (1987b,c) extended previous studies on the effects of temperature and other factors on the growth of *Gyrodactylus* spp. by investigating the effect of host condition on the development of *G. gobiensis* on *Gobio gobio* and of *Gyrodactylus katharineri* on *Cyprinus carpio*. Depriving *Gobio gobio* of food and oxygen resulted in a greater number of parasites per fish than those deprived of oxygen only. Oxygen deficiency did not affect the parasites directly. On both scaly and scaleless *C. carpio*, and at 12°C, *Gyrodactylus katharineri* increased on the two host forms at about the same intensity. At 18°C however the species reproduced intensively only on the scaly form of the host and no parasites were found on the scaleless variety.

Paperna (1963c) described the following factors which determine the population dynamics of *D. vastator*: (i) the extent of initial infection, (ii) growth rate of fingerlings in the pond, and (iii) time-interval between hatching of the fingerlings and contamination of the pond. The parasite induces hyperplasia of the gills and these pathological changes constitute an important factor in regulating its propagation, ie. there seem to be host and parasite regulatory mechanisms involved in controlling their abundance. The prevalence and intensity of *Dactylogyrus legionensis* on *Barbus barbus bocagei* rose significantly with the age and length of host (Gonzalez-Lanza and Alvarez-Pellitero (1982). It was higher in females than in males and there was a site preference for second and first gill arches. Whilst most authors consider temperature to be the major factor determining seasonal variations in dactylogyrids, this was not clear with *D. legionensis* because maximum infections occurred in autumn-winter. This may be due to other abiotic factors such as light, pH, O_2 and salinity, or biotic ones, e.g. the shoaling of fish in winter. Gravid specimens were found from September to January and, except for May, from March to June.

A statistical analysis of the abundance of *Cleidodiscus pricei* on *Ictalurus platycephalus* over 12 months showed that the frequency distribution per month was overdispersed towards the lower numbers of parasites per host. One- and 4-year-old *I. platycephalus* had fewer parasites than 2- and 3-year-old hosts. Seasonal abundance peaked in May and remained high in July. Seasonal variations were correlated with water temperature and the immune response of the host (Cloutman 1978).

Diplozoon paradoxum on roach and bream and a hybrid of roach and bream and white-bream was investigated by Halvorsen (1969) who found an annual reproductive cycle with main egg production in June. Most *D. paradoxum* live for about a year in the area studied. Prevalence was greater on bream and hybrids than on roach. This may be due to differences in host ecology. There was an average of 2–4 worms per fish.

Regarding *Diclidophora* spp., Munroe, Campbell and Zwerner (1981) found an average of about 3.2 *D. nezumiae* on the gill filaments of the first gill arch of each of the 106 *Nezumia bairdii* examined. The prevalence and intensity was greater for fish collected at depths of 700–1000 metres. With *Diclidophora denticulata* on the gills of *Gadus virens* reproduction occurs throughout the year and an average number of flukes per host is 2.24. The first and second gill arches carry three times as many flukes as third and fourth. An age resistance occurs—larvae being unable to establish on older fish—perhaps due to changed host habits (Frankland, 1955). *D. merlangi* also, like most monogeneans, has a preferred site of attachment, but the factors influencing its distribution may be complex, possibly involving parasite-parasite interactions and/or the responses of flukes to other stimuli, e.g. where maximum oxygen tension occurs (Arme and Halton, 1972).

Several aspects of the population biology of gyrocotylideans are intriguing. These include the rarity of some species, the niche segregation between species pairs and infrapopulation size. Nearly every holocephalan species so far examined is said to be parasitized by one very prevalent and one rare species of *Gyrocotyle*. Most authors who have recognized two sympatric species have noted that mixed infections never, or very rarely, occur. Except for Simmons and Laurie (1972), who found that the two species differed statistically in their distribution along the spiral valve, there are no indications of any niche difference between pairs of species. *Gyrocotyle urna* in *Chimaera monstrosa* shows high prevalence and a low intensity with a strong tendency towards underdispersion with two worms as the dominant infrapopulation size (Williams, Colin and Halvorsen, 1987). The smallest *C. monstrosa* have a lower prevalence but this increases rapidly to 90–100 per cent in nearly all size groups of *Chimaera monstrosa* as a result of a population regulation mechanism by *Gyrocotyle* rather than by the host fish.

Sixty-two of 289 *Gadus morhua* were found to harbour *Abothrium gadi* by Williams and Halvorsen (1971). The prevalence and degree of infection increased with fish length. In most fish only one tapeworm was found and this showed a preference for pyloric openings adjacent to the bile duct opening (Figure 3.4). The results were discussed relative to 'the crowding effect' and 'premunition' in the Cestoda, but it was concluded that *A. gadi* may be able to limit its population density relative to the available resources of food and/or other chemicals. This work on *A. gadi* may be worthy of comparison with that of McKinnon and Featherston (1982) on *Bothriocephalus scorpii* and its site of attachment in the intestine of *Pseudophycis bacchus*. They found that large worms of *B. scorpii* from the posterior regions of the intestine were looped forwards towards the anterior region.

The ecology of the larval stage of another pseudophyllidean, namely *Diphyllobothrium*, is obviously influenced by a number of complex factors. For example, in Norway 466 (77.4 per cent) of 602 *Salvelinus alpinus* harboured *Diphyllobothrium ditremum* plerocercoids in an overdispersed infection (Halvorsen and Andersen, 1984). Prevalence increased with age, reaching 100 per cent at 7+ years for males and 8+ years for females. The rate of infection appeared to be related to feeding habits and the intensity decreased with age as did the degree of overdispersion. According to Halvorsen and Andersen the only detailed analysis of the occurrence of diphyllobothriid plerocercoids in a fish population are those of Henricson (1977, 1978). In an unpublished report of a roundtable conference held at the Sixth International Congress of Parasitology there is a reference to work in progress on the dynamics of *Diphyllobothrium* populations in various countries (Bylund and Fagerholm, personal communication). Further information on *Diphyllobothrium* is given in Chapter 6.

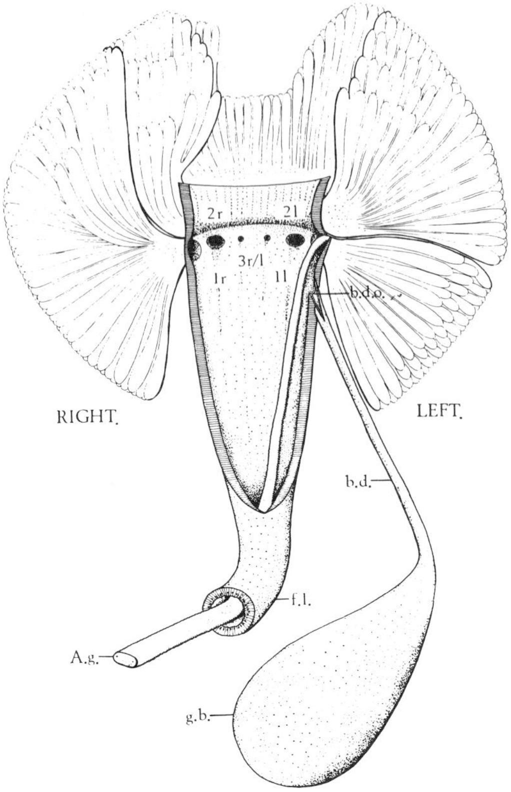

Figure 3.4 Pyloric region of intestine showing pyloric openings and *A. gadi, in situ.*

Three trypanorhynch species have been the subject of intensive investigations with regard to their distributions in the flesh of fish of economic importance and, as with *Diphyllobothrium*, many factors are involved. They are worthy of intensive further research for which the following publications will be essential: Collins, Marshall and Lanciani (1984) and Overstreet (1977, 1978a, b) for *Poecilancistrium caryophyllum*; MacKenzie (1975b) for *Gilquinia squali*; and Rae (1958) for *Grillotia erinaceus*.

Little is known of the ecology of aspidogastreans in fish with the exception of *Multicalyx cristata* investigated by Thoney and Burreson (1986). Sixteen species of elasmobranchs were collected between Cape Fear, North Carolina and Long Island, New York mainly in March and October (1982–1984). Specimens of *M. cristata* were found only in *Myliobatis freminvillei* (prevalence 10.7 per cent, 15 of 140) and *Rhinoptera bonasus* (prevalence 1.3 per cent, 1 of 76). Only *M. freminvillei* wider than 68 cm were infected. No differences in prevalence were found with regard to host sex or season. The intensity of infection was one or two worms per fish with 38 per cent of the hosts having two worms. The large size of *M. cristata* relative to that of the gall bladder in which it is found is thought to lead to intraspecific competition, which restricts infection intensity to one or two worms as has been suggested for a few other species of fish helminths.

Many complicated factors influence the ecology of nematodes, especially anisakid larvae in the flesh of marine fish as discussed, for example, by Khalil (1969), Novotny and Uzmann (1960), Linnik (1980a) and Smith and Wootten (1978a). Their papers are discussed elsewhere in this book especially in Chapter Six. The prevalence, mean worm burdens and range of worm burdens of *Eustrongylides tubifex* in *Ictalurus punctatus. Aplodinotus grunniens* and *Perca flavescens* in Lake Erie, North America were investigated by Cooper, Crites and Sprinkle-Fastkie (1978). *P. flavescens* were the most frequently infected with young fish and fish aged 1+ year showing increasing monthly larval burdens during parts of the sampling period but no significant increases in worm burden were noted in older (2+) fish. Most larvae occurred in mesenteric cysts and they migrated from cysts through the body wall of dead fish when the environmental temperature was raised to 40°C or higher. Fourth stage larvae from perch became established, usually in the body cavity, in six species of experimentally infected fish, one amphibian species and three reptile species.

Moravec and De (1982) found that the rate of infection of *Salmo trutta* m. *fario* in Czechoslovakia with *Cystidicoloides tenuissima (C. ephemeridarum)* was related to the size of the host, 73 per cent of trout smaller than 10 cm and 100 per cent of those over 10 cm long being infected. He observed two peaks, February and July, for intensity of infection, followed by sudden drops. Gravid females with mature eggs were present mainly from May to December. Third- and fourth-stage larvae were found throughout the year. Moravec agreed with Aho and Kennedy (1984, 1987) in suggesting that the seasonal maturation cycle is influenced by water temperature and mayfly populations in particular. *Ephemera danica* was the only intermediate host but *Cottus gobio* is a paratenic host. Alvarez-Pellitero (1976a) showed that two generations of *C. ephemeridarum* occurred each year in *Salmo trutta* m. *fario* in Spain, one in Spring and one in Autumn, and that temperature, directly and indirectly was the most important single ecological factor involved. Fagerholm, Kuusela and Valtonen (1982) briefly discussed the taxonomy of *C. ephemeridarum*, its life-cycle in *Thymallus thymallus* and previous observations on its ecology. Muzzall (1986) has reported on the ecology of *C. ephemeridarum* in *Salvelinus fontinalis Salmo trutta* and with *Ephemera simulans* as the intermediate host. He discussed the seasonal occurrence and distribution of this nematode as well as *Spinitectus gracilis* and *Crepidostomum cooperi* in fresh and frozen trout.

Cystidicola cristivomeri has a life span of at least 10 years. Growth of female worms and the rate at which they release eggs is regulated in a density-dependent way. Density-dependent regulatory mechanisms minimize fluctuations in the size of parasite populations within a system, and had previously been identified in only a few parasite populations (Black and Lankester, 1984).

Uhazy (1977a, b) reported on some previously unknown details of the biology and population dynamics of *Philometroides huronensis* in *Cyclops bicuspidatus* and *C. vernalis* as intermediate hosts and *Catostomus commersoni* as final hosts. Prevalence in *C. commersoni* was high, 83–100 per cent, throughout the year, and showed no significant seasonal trend. Numbers of worms per fish ranged from 1–32, but 77 per cent of fish were infected with from 1–7 worms. Intensities of infection varied significantly with season, being highest during autumn, winter and early spring. All age groups of fish were infected but intensity did not vary significantly with host age. Although the life-cycle is completed in one year prevalence was high throughout the year and from year to year. *P. huronensis* was less prevalent in *Catostomus catostomus* and *Morostoma marcolepidotum*. Uhazy suggests that three valid species of Philometrinae have evolved from *C. commersoni* in North America on the basis that before the Pleistocene glaciation there were three separate forms of the fish species.

Competition/negative interaction

Inter- and intra-specific relationships of parasites within a host are well discussed by Kennedy (1975). He cited the work of MacKenzie and Gibson (1970) on *Cucullanus heterochrous* and *C. minutus* in *Platichthys flesus*. When both species are present they are found in different regions of the gut. Other examples and evidence for negative interaction amongst fish helminths are discussed by Halvorsen (1976) with reference to *Dactylogyrus extensus* and *D. vastator* on reared carp, *Proteocephalus exigus* and *Neoechinorhynchus* sp. in ciscoes, *Crepidostomum metoecus*, *C. farionis*, *Dacnitis truttae* and *Neoechinorhynchus rutili* in brown trout, and intensive infestations (80–200) of *Pomphorhynchus laevis* appearing to exclude other helminths in a number of fish species. Lee (1981) presented data and evidence to show antagonism between *Acanthocephalus lucii*, *Proteocephalus percae* and *Camallanus lacustris* in the intestine of *Perca fluviatilis* from Lake Serpentine, central London. *A. lucii* was rare in this lake compared with other locations in Britain and this was attributed to the rarity of the intermediate host, *Asellus aquaticus*. The restriction of *Acanthocephalus lucii* to the third quarter of the intestine was due either to the worm actively seeking this part of the intestine or to the large numbers of *Proteocephalus percae* and *Camallanus lacustris* excluding it from more anterior sites which are said to be more favourable to *Acanthocephalus* spp. The mutual exclusion theory was favoured.

Dobson (1985) used theoretical models of host-parasite associations for two classes of competition ('exploitation' and 'interference') to examine the expected pattern of population dynamics shown by simple two-species communities of parasites using the same host population. 'Exploitation' was defined as the joint use of a host species by two or more parasite species, and 'interference' as antagonism used by one species to reduce the survival or fecundity of a second or to displace it from a preferred site of attachment. The analysis suggested that the most important factor allowing competing species of parasites to co-exist is the statistical distribution of the parasites within the host population. A joint stable equilibrium should be possible if both species are independently aggregated in distribution. The potential application in biological control was discussed. Competition in parasites was compared with that in free-living animals and plants and it is suggested that further experimental tests may help to assess the importance of competition in determining the structure of more complex parasite-host communities. The population data used for 'exploitation' included those on *Neoechinorhynchus* and *Proteocephalus* in ciscoes, *Neoechinorhynchus, Dacnitis, Crepidostomum* spp. in *Salmo trutta, Neoechinorhynchus rutili* and *Proteocephalus filicollis* plerocercoids in *Gasterosteus aculeatus, Cyathocephalus truncatus and Crepidostomum* spp. in *Salmo trutta, Neoechinorhynchus cristatus* and *N. crassus* in *Catostomus macrocheilus, Diplostomum spathaceum* and *Tylodelphys clavata* in *Scardinius erythrophthalmus*, and *Asymphylodora kubanicum* and *Sphaerostoma bramae* in *Rutilus rutilus*. For 'interference' data the following were used: *Dactylogyrus vastator* and *D. extensus* in *Cyprinus carpio*, and *Pomphorhynchus bulbocolli* and *Glaridacris laruei* in *Catostomus commersoni*. As mentioned earlier in this chapter, a mathematical model to test the concepts of niche and competition was developed by Levin (1970). A number of other references in this chapter especially those cited under sites and site-selection also provide evidence for competition and negative interactions amongst fish helminths.

Temperature

Already in this chapter we have made a few references to the importance of temperature on the population dynamics of helminths. Here we give further examples to underline its importance. Kirby (1981) showed that the highest densities of *Gyrodactylus atratuli* on *Notoropis spilopterus* occurred in early spring at 15°C when the fish schooled; and was lowest in summer when water temperatures were higher and fish scarce. Laboratory studies confirmed that infected fish retained more *G. atratuli* at 8°C than at 15 or 24°C. Most parasites attached to fins, especially caudal and pectoral, but in summer the proportion attached to gills increased. According to Hanzelova and Zitnan (1982, 1983) the prevalence of *G. katharineri* on *Cyprinus carpio* in three ponds on a fish farm was 95 per cent in April and 100 per cent in May, and the intensity for these two months increased from 15.2 to 137.1 per fish. This was thought to be influenced by an increase in temperature. The prevalence and intensity of *D. vastator*, unlike that of *G. katharineri* as described above, peaked in July or August, the species having an optimum temperatue of 20–26°C for its development in the USSR. Interesting experimental observations were made by Ergens and Gelnar (1985) on the effect of temperature on the anchors, ventral bars and marginal hooks of the opisthaptor of *G. katharineri*. Their growth responded to much smaller changes in temperature than those which occur naturally from one season to another. Specimens which are born and grow in the warm season have significantly smaller anchors, bars and hooks than those which develop in colder weather. The taxonomic implications of this study are considerable and they will have to be borne in mind when identifying those species of the genus which are thought to be dangerous pathogens, eg. *G. salaris*. Gelnar (1987a) showed that reproduction in *G. katharineri* was related to temperature. At 12°C a mean number of 64 parasites per host was attained on day 27, at 14°C the mean was 96 per host on day 16, but at 18°C it was 154 on day 15.

Paperna (1963b) described the rate of development of *D. vastator* on carp fry from egg to sexually mature monogeneans and the rate of oviposition at 24–28°C. The egg ripens in 2–3 days and the larva survives for up to 24 hours. After attaching itself to the fish it matures in five days and lives for another five days. *D. vastator* lays an average of 1.68 eggs per hour, with salinity and temperature being important factors in this process. The ammonia and oxygen content of the water however are not important so long as the fish can survive. *D. vastator* can tolerate a salinity of up to 6000 mg Cl/L. The optimal temperature for egg laying is 22–24°C and no egg-laying takes place below 12°C or above 37°C. The eggs will develop at 12°C but cease to do so at 5°C, although they remain viable. In contrast Belova (1977) found that the ova of *Dactylogyrus hypophthalmichthys* developed at 6–35°C and that the optimum temperature was 30°C.

A population of *D. extensus* has established itself in Israel which is capable of thriving at 24–28°C. It has also become resistant to oxygen deficiency. The form from which it is descended has an optimal temperature of 17°C. Paperna (1964a,b) described competition/negative interaction between two *Dactylogyrus* spp. When there are more than 50 *D. vastator* per fish, infection with *D. extensus* is only one or two per fish. In fish infected with *D. extensus* and then exposed to *D. vastator* there was a gradual disappearance of the former. If they became free of *D. vastator* they were again susceptible to *D. extensus*. A similar situation occurs between *D. vastator* and *D. anchoratus* (Paperna and Kohn, 1964).

The abundance of *Dactylogyrus amphibothrium* on *Acerina cernua* depended on various ecological and biological factors, primarily temperature (Kashkovskii, 1982). Two generations are produced each year—one in early summer which dies off in August/September and a second in late summer which dies off the following May/June. One worm of the early summer generation produced 853 eggs in its lifetime and a worm of the late summer generation, 411 eggs.

Kennedy (1971) experimentally infected *Leuciscus idus* with *Caryophyllaeus laticeps* and maintained the hosts at different temperatures. At 12°C and below there was an initial loss of parasites after infection but survivors persisted for up to one month. At 18°C only temporary survival occurred and worms were killed and rejected after three days. It was thought that the rejection of *C. laticeps* did not involve antibodies, although a comparison was made with the work of McVicar and Fletcher (1970) who suggested that non-specific antibodies were responsible for the rejection of a tapeworm of *Raja radiata*. The results however were thought to provide further evidence for the importance of temperature as a major controlling factor in the seasonal dynamics of many fish parasites.

Fischer and Freeman (1969) showed that plerocercoids of *Proteocephalus ambloplitis* may penetrate into the gut of the same bass, *Micropterus dolomieui*, following a suitable stimulus. The incidence of this species in the gut was seasonal, being highest in the spring, declining in the summer and disappearing in late autumn. When bass, kept over winter at 4°C, were raised to 7°C or higher some plerocercoids left the viscera and penetrated into the gut lumen. Similar penetration, observed only in May and early June at comparable lake temperatures, apparently accounts for the seasonal incidence of *P. ambloplitis* in the gut. All 877 bass more than 7.6 cm long had plerocercoids of the tapeworm in the viscera. These observations were confirmed experimentally by Fischer and Freeman (1973) thus further emphasizing the importance in the biology of *P. ambloplitis* of the re-entry of plerocercoids into the gut of smallmouth bass.

Eure (1976) working on *P. ambloplitis* in more southern latitudes found that the appearance of adult tapeworms in the gut coincided with a decrease in temperature. It was postulated that the decline in temperature in southern latitudes and an increase in northern latitudes initiated the same response. Esch, Johnson and Coggins (1975) suggested that hormones may be involved in stimulating parenteric plerocercoids to migrate into the gut, and that the density of *P. ambloplitis* in the gut was independent of acanthocephalan infections. The potential space in the gut was not being fully exploited by the enteric helminths.

Because temperature and light are important factors influencing the liberation of cercaria from snail hosts, Halvorsen (1968) investigated the occurrence of *Azygia lucii* in *Esox lucius* in a locality in South Norway with pronounced seasonal fluctuations of these factors. The high frequency of pike with *A. lucii* in Bogstad lake at all times of the year and the rather uniform size composition of the *A. lucii* 'gross population' suggested a well-balanced relationship between host and parasite in the system. There was an increase of frequency and intensity with the age of the fish. Small, immature *A. lucii* were found in pike at all seasons and, as no crowding effect could be detected, they were interpreted as representing newly acquired infection. Such observations could be explained only if there is a second intermediate host in the life-cycle of *A. lucii* because, at Bogstad lake, there is a pronounced seasonal fluctuation in temperature and light. When Bogstad lake was compared with other localities it was found that *A. lucii* was less frequent where stream velocity is greater.

Andrews and Chubb (1980) investigated the effect of temperature, at four thermal régimes, on the development of *Bunodera luciopercae* in *Perca fluviatilis*. At 16–20°C development did not proceed normally and the infection had disappeared after 130 days. At 4–8°C and at natural environmental temperatures (0–15°C air temperatures) gravid flukes were produced after 158–168 days. The failure of parasites to mature at a temperature of 16–20°C even after exposure to 4–8°C for 43 days suggested that *B. luciopercae* in its definitive host requires an extended period of vernalization in order to develop and mature normally. Whilst temperature has a major influence on the life-history of *B. luciopercae*, maturation in this species is probably not dependent upon normal seasonal changes in gonadotrophic hormone levels in adult perch. Skörping (1981) found that *B. luciopercae* in *P. fluviatilis* showed marked seasonal variations in prevalence and intensity of infection, and he recognized five developmental stages of the fluke, including immature stages resembling fully developed metacercariae found in the second intermediate host occurring from August to March, with a peak in January and gravid flukes being present mainly in May–June.

In *Semotilus corporalis*, infection with *Allocreadium lobatum* in a Canadian lake ranged from 1–86 worms per fish but the mean intensity was 6.0 (Rand and Burt, 1985). There was also a low prevalence from July to September but this increased to high levels from November through May, with high parasite recruitment from August to November. From February through July over 60 per cent of the digenean population were gravid and these died after oviposition. This annual cycle and seasonality are related to temperature changes in the water, host diet and trends in the occurrence of amphipods, the second intermediate hosts.

Crepidostomum metoecus is an abundant and widespread intestinal parasite of salmonid fish of the Holarctic (Moravec, 1982a). Moravec discussed previous work on its ecology. In contrast to results by some other workers he found that in the Elbe river basin *C. metoecus* has only one annual maturation cycle when all the adults lay eggs until May–June and then die. New infections of *Salmo trutta fario* may be acquired throughout the year but mostly in spring and autumn. The occurrence of spring and autumn generations of the adults in some regions, e.g. North-Eastern Spain, is attributed by Moravec to differing water temperatures and seasonal changes in the availability of mayflies, the intermediate hosts.

An *Eustrongylides* sp. in *Salmo trutta* and *S. gairdneri* from two stations of a river in Wyoming, USA, each having different temperatures, was studied by Kaeding (1981). At the colder station the overall prevalence was 1.2 per cent for *S. trutta* and at the downstream warmer station it was 17.5 per cent and 17.6 per cent respectively for *S. trutta* and *S. gairdneri*. Whilst at this station prevalence and intensity of infection increased with the age of *S. gairdneri* it did not do so in *S. trutta*. Seventy-eight per cent of infected fish contained only one worm and the maximum worm burden was eight. The differences between infection at the two stations may be due to the absence of an oligochaete first intermediate host at the colder upstream station.

Already in this chapter we have highlighted temperature as an important factor in influencing the abundance and distribution of fish worms, especially in our discussion of abiotic factors and the reference to the effect of temperature on the development and reproduction of *Transversotrema patialense*. Since temperature will vary according to the season of the year at various latitudes and longitudes, it is not surprising that much has been written on the seasonality and zoogeography of parasitic worms in fish. It is logical, therefore, to proceed to a discussion of these two topics.

Seasonality

Linton (1914) investigated seasonality of worms in the flesh and other organ systems of various named marine species of food fish. He concluded that 'there does not appear to be evidence of any marked periodicity in the occurrence of helminth parasites of marine fishes, either adult in the alimentary canal, or immature encysted in the tissues of their hosts, beyond what may be expected where fishes are exposed to varying sources of infection in the course of their migrations.' Since 1914, however, seasonal variations in the incidence and intensity of infection of fish with helminths have become well-known, although the causes of these have not. Not only do the numbers of parasites recovered at given times of the year vary but also the composition by age of the parasite population, i.e. there is also a seasonal cycle of maturation. A great deal of information on this topic has been amassed from surveys and life-cycle studies and, in most instances, attempts have been made to identify the factor or factors responsible. One of the most important factors to emerge is environmental temperature, which affects the life-cycles of intermediate and definitive hosts and parasites, exerting a direct effect on ectoparasites and an indirect effect, through their poikilothermic hosts, on adult parasites in fish and larval stages in the intermediate hosts. Temperature affects the abundance of intermediate hosts and is correlated with the population dynamics, feeding behaviour and breeding cycle of the fishes. Maturation of some parasites has been shown to be closely linked to fish spawning. In addition, temperature affects the immune response of the fish hosts, which has itself emerged as an important factor in seasonal variations, producing seasonal changes in resistance to infection. It has also been established that the same parasite may have a seasonal cycle of incidence and/or reproduction in one locality but not in another and that the timing of the cycle may be quite different.

Many of these aspects were discussed by Dogiel (1961) and Polyanski (1958) in reviews of the ecology of the parasites of freshwater and marine fish which summarized work done, especially in the USSR, up until about 1960. Kennedy (1970) reviewed the population biology of fish helminths and referred to basic principles controlling population changes in parasites for which a fish is the definitive host. Changes in structure and size of the population depend on changes in input and output, the former being affected by the availability of infective larvae and the feeding behaviour of the fish and the latter by the failure of parasites to establish themselves, by rejection of established parasites and by natural parasite mortality. Biotic factors such as overcrowding or interaction with other parasites, and abiotic factors such as temperature, may affect the factors which influence input and/or output and all factors will vary with time. They all contribute to the pattern of seasonal variation. Kennedy presented models of definitive host-parasite systems for endoparasites present throughout or for less than a year.

In one model, incidence shows seasonal peaks or troughs, and recruitment occurs only in autumn or early winter at the end of a period of decline in incidence and intensity of infection which follows a summer maturation peak; the annual cycle is usually correlated with the annual cycle of temperature changes. Examples of species showing this pattern are *Cyathocephalus truncatus, Crepidostomum metoecus, Allocreadium isoporum, Crepidostomum* sp. and *Sphaerostoma bramae* (Awachie, 1968; Chubb, 1979). Recruitment of *Cyathocephalus truncatus* in trout was not restricted to a short period because of changes in the food of the fish host; the intermediary, *Gammarus pulex*, was part of the diet of trout all year round but the availability of infective larvae showed seasonal periodicity and was probably the limiting factor. Water temperature controlled the establishment of new infections of *C. truncatus* and *Crepidostomum metoecus* since their larvae were unable to establish at temperatures greater than 10°C and, therefore, recruitment became possible only in autumn when water temperatures fell. Worm loss by mortality begins soon after infection. In *A. isoporum, S. bramae* and *Crepidostomum* infections, there was an initial heavy loss which levelled off, the survivors maturing, breeding and then dying at the end of their life-span. This was suggested as a major cause of seasonal mortality but the proportion surviving to breed was not known and no seasonal cause of mortality has yet been established.

In another model, parasites are present for only part of the year, as for many monogenean species in which parasites are recruited for a short life-span in spring; most survive only to breed and the cycle is repeated until high summer water temperatures inhibit egg production and recruitment decreases (Kennedy, 1975). Endoparasites exhibiting similar cycles are *Raphidascaris acus* in pike and *Caryophyllaeus laticeps* and *Proteocephalus torulosus* in dace as discussed by Davies (1967), Kennedy (1968), Kennedy (1969b) and Kennedy and Hine (1969). *R. acus* infective larvae are available throughout the year and recruitment is, therefore, not governed by their availability. *C. laticeps* and *P. torulosus* are both recruited in winter although infective larvae are available at other times. They grow during winter and mature in late spring/early summer. Only part of the total loss is attributable to death of spent adults and it appears that a temperature-dependent rejection mechanism is responsible. This has also been suggested as an explanation for the seasonal periodicity of incidence and intensity of infection and maturation of *Monobothrium ulmeri* in creek chubsucker in North Carolina (Grimes and Miller, 1976). *M. ulmeri* is present for only eight months of the year; its maturation corresponds with the host's reproductive cycle and may be influenced by the host's hormonal changes in the spring, as was also suggested for *C.*

laticeps in dace (Kennedy, 1969b). In this cestode, intensity of infection is related to host sex, which is not the case for *Biacetabulum meridianum* and *Penarchigetes* sp., both also present in chubsucker. The former is present throughout the year and shows a seasonal cycle in incidence, maturation and length distribution, but its establishment appears to be inhibited by previous heavy infections of *M. ulmeri*, an example of interspecific interaction mentioned earlier. *Penarchigetes* shows no seasonality in incidence, intensity or maturation, possibly because its procercoids are not pathogenic to the intermediary and are, therefore, available all year round, or because in the fish, it occurs only in small numbers and is too loosely attached to provoke an immune response.

The first comprehensive review of seasonal occurrence of fish parasites was given by Chubb (1977b) who concentrated on the monogeneans of freshwater fish, and this was followed by equally extensive studies of the same topic for Digenea by Chubb (1979), larval Cestoda and Nematoda by Chubb (1980b) and Cestoda, Nematoda and Acanthocephala by Chubb (1982). Chubb (1977b) suggested that, for monogeneans, water temperature is the primary factor in explaining seasonal incidence and intensity of infection could also be related in a fairly precise manner to the effect of another abiotic factor, ie. oxygen concentration, and to many biotic factors, eg. behaviour, migrations and immunity of host species, or to the interaction of a range of both abiotic and biotic factors. Chubb related a number of important aspects of the biology of monogeneans to seasonal effects, eg. their reproduction, longevity, mortality and even different morphological forms of the same species. Seasonal changes in infection were also noted according to principal and auxiliary hosts, host microhabitats, host macrohabitats and other biotic and abiotic factors. More detailed accounts of the influence of these factors on monogeneans in relation to seasonality have already been cited in this book, eg. Frankland (1955), Llewellyn (1962), Paling (1965), Noisi and Maillard (1980), Winch (1983) and Cone and Burt (1981, 1985)

The precise effect of seasonality on the ecology of Cestoda, Digenea, Nematoda and Acanthocephala is more difficult to demonstrate than it is for monogeneans and the problem becomes more complicated relative to the increasing number of hosts involved in the life-cycle. Chubb (1982) emphasized that for adult Cestoda, Nematoda and Acanthocephala of freshwater fish, water temperature acting directly on the helminth or indirectly through the host behaviour, especially feeding behaviour and metabolism, seems to be the most important factor determining seasonal biology in the sub-tropical and mid-latitude climatic zones of the world. He advocated further investigations on each species of helminth on the relevant factors which have been demonstrated or postulated to control their seasonal patterns of occurrence. They will inevitably vary, if only in a subtle way, due to differences in the genetic composition of each species.

Caryophyllaeid tapeworms have been a focus of attention with regard to seasonality probably because their life-cycles are relatively less complicated than those of other cestodes. Kennedy (1969b) followed up a previous study suggesting that one of the factors that could contribute to a well-defined seasonal incidence and maturation cycle of *Caryophyllaeus laticeps* was seasonal variation in the availability of infective larvae. Infective larvae occurred in the tubificid intermediate hosts in all months except August. Fish, however, acquired infections between December and March, the parasite maturing in April and May and disappearing in July. Maturation appeared to be governed by fish hormone levels. The most important factor controlling the dynamics of the *C. laticeps*-dace system appeared to be a temperature-dependent rejection response by the host. Milbrink (1975) observed two peaks of infection for *Caryophyllaeus laticeps* in *Abramis brama* in Sweden, one in spring and the other in autumn. During the peak periods 90–100 per cent of the bream had consumed oligochaetes. When *Ligula intestinalis* was also present the number of *C. laticeps* was lower than when it was the only cestode present. Begoyan (1977) found an annual cycle in *Khawia armeniaca*, a parasite of *Varicorhinus capoeta*, procercoids being found in fish from mid-May and mature parasites in the autumn. Young specimens, probably newly acquired, were occasionally found amongst adults in November and December. Old, presumably overwintered specimens, were also found in the spring. Williams (1979b) found increased peaks of *Isoglaridacris wisconsinensis* in *Hypentelium nigricans* in the USA in the spring and autumn, and decreases in July and February. This result supports the work of Milbrink (1975) on *Caryophyllaeus laticeps*.

The work of Anderson (1976a,b) on a caryophyllaeid is of particular interest and value in relation to the above, because he used a mathematical model of *Caryophyllaeus laticeps* infections in bream and showed that seasonal variations in the size of the parasite population are caused by the combined effect of a temperature-dependent mortality rate and of fluctuations in the feeding activity of the host which affect the recruitment rate of larvae. In the life-cycle, evolutionary and other aspects of their biology, cyathocephalid tapeworms are generally regarded as being more advanced than caryophyllaeids, but little is known of their seasonality and they are, therefore, worthy of further work from this standpoint. Reimchen (1982) found that the highest incidence of *Cyathocephalus truncatus* infection of *Gasterosteus aculeatus* was from February to May (80 per cent) and of the same host with *Schistocephalus solidus* from April to September (50 per cent). Infection rates for *C. truncatus* during different seasons and among different length classes of fish were directly correlated with the relative abundance of amphipods in the fish stomachs. Possible modifications of host feeding behaviour were indicated by a relative increase in the consumption of amphipods by infected fish.

Amongst the most well-known and complicated of fish tapeworm life-cycles are those of the proteocephalideans and pseudophyllideans. The following information on these worms is of particular value in understanding aspects of the seasonality and planning further research on the topic. Hopkins (1959) investigated *Proteocephalus filicollis* including its rate of growth in *Gasterosteus aculeatus* and the correlation between size of worm, period of year and stage of development reached because such a correlation had been observed for other tapeworm species. An annual cycle was confirmed for *P. filicollis*. The most advanced stages commonly occurring were: plerocercoids (0.25–5 mm) July–November; strobilate worms with genital primordia (5–8 mm) December–April; mature worms (1 cm) April–May; gravid (2 cm) June–July. From a consideration of the seasonal variation in incidence it was deduced that the parasite population is in dynamic balance and that approximately 1 per cent of worms present are lost daily.

Kennedy and Hine (1969) investigated the abundance and prevalence of *Proteocephalus torulosus* in *Leuciscus leuciscus* over two years in Britain. The tapeworm is present only from November or December through to June and July and becomes gravid in April. The population biology of this species was found to be similar to *Caryophyllaeus laticeps* in the same fish population and differs from that of other fish tapeworms. It was postulated that seasonal differences in resistance of dace to infection, which is directly dependent on river temperature, were responsible for the observed seasonal rhythms in abundance and prevalence in the fish population. Observations by Sysoev (1983) showed that copepods infected with *P. torulosus* were first found at the end of May, the prevalence peaking in July before dropping to zero in August, thus supporting the findings of Kennedy and Hine on development of the adult worm.

A seasonal cycle of intensity of infection has been established for a species of *Proteocephalus* in *Salmo trutta* from a lake in South Norway, although the incidence of infection remained fairly stable (Lien and Borgström, 1973). Recruitment occurred in October–November and intensity fluctuated considerably during winter and spring; incidence decreased to almost nil in August–September as water temperature increased. The feeding habits of the host might have been significant since larger fish, which ate more of the copepod intermediaries, were most frequently infected and harboured heavier infections. Variations in the prevalence and intensity of *Proteocephalus exiguus* infection in *Coregonus albula* were studied in the summer of 1978 and 1979 in two mesotrophic lakes of Karelia, USSR, by Yakushev (1985). Mathematical analysis of the data showed considerable variation in the rate of infection in the two water bodies and these may be represented by various polynomial models. These differences depend mainly on the times when the prevalences reach maximal and minimal values in the two lakes, and these are determined by the ecological peculiarities of the water bodies.

Regarding pseudophyllideans, Borgström (1970) discussed the seasonal occurrence of *Triaenophorus nodulosus* in *Esox lucius* in Norway and found distinct seasonal differences in numbers of worms present, the maximum occurring in April–May, and in the sexual maturity of the worms which seemed to take place in May–June to coincide with the fish spawning. Subsequently, temperature and/or host physiological conditions made the pike more resistant to infection although a greater number of females than males were infected. This was also thought to be due to physiological conditions or to different feeding habits of the host. In contrast to these results, Chubb (1963a) had found a constant infection of pike with the tapeworm throughout the year in Britain and he attributed the result to a dynamic equilibrium between gain and loss of worms in the pike intestine. The recovery of procercoids of *T. nodulosus* from seven species of copepod found to be the intermediate hosts of this tapeworm by Sysoev (1982) corroborates some of the results of Borgström. About 1,035 procercoids were found from June to August.

The seasonal changes in prevalence, density and recruitment of *Bothriocephalus acheilognathi* in *Gambusia affinis* coincided with temperature changes. Granath and Esch (1983b) used the same model to examine the possible influence of biotic and abiotic factors on various aspects of the seasonal dynamics and composition of *B. acheilognathi* infrapopulations. As previously stated (p. 244) Granath and Esch discussed temperature-dependent rejection, immunity and selective host mortality, as well as intraspecific exploitative competition, as the causes of the decline of mean infrapopulation densities of *B. acheilognathi*, but they also considered density-independent mechanisms, primarily temperature and host behaviour, both dietary and social, as causative factors. They concluded that it is most likely that a combination of both sets of factors regulate infrapopulation densities of the species.

Hitherto we have suggested that one of the best ways of understanding the complicated aspect of seasonality is to investigate species with life-cycles of increasing complexity. It can be argued that acanthocephalan life-cycles, insofar as number and type of intermediate hosts are concerned, as well as the location of the adult in the alimentary canal of the fish, are no more complicated than those of tapeworms. Using the same criteria of sites occupied by the adults and the number and range of intermediate hosts involved, it is clear that the Nematoda and Digenea respectively have the most complicated life-cycles of the groups of worms discussed in this book. For these reasons we now propose to highlight some previous work on the seasonality of Acanthocephala, Nematoda and Digenea.

Almost all adult acanthocephalans have, at longest, an annual turnover of occurrence in their hosts, so that invasion, establishment, growth, maturation of genitalia, egg accumulation and loss of gravid worms are all achieved within 12 months (Chubb, 1982). Chubb further discussed the seasonal dynamics of acanthocephalans according to their (i) incidence and intensity of occurrence, (ii) principal and auxiliary hosts, (iii) invasion of fish by larvae, (iv) growth and maturation, (v) abiotic and biotic factors, and (vi) long-term population studies and experimental work. Of particular importance in understanding the seasonality of acanthocephalans is the work of Holmes, Hobbs and Leong (1977) on the flow rates of *Echinorhynchus salmonis* in ten species of fish, of Awachie (1965) on *E. truttae*, of Chubb (1963b, 1964a) on *E. clavula*, of Lee (1981) on *Acanthocephalus lucii*, of Hine and Kennedy (1974a,b), Kennedy (1972b) and Rumpus (1975) on *Pomphorhynchus laevis*, of Valtonen (1980a,b,c) on *Metechinorhynchus salmonis*, of Walkey (1967) on *Neoechinorhynchus rutili* and of Bromage and Whitehead (1980) on the potential for experimental studies on *E. salmonis* in *Salmo gairdneri*.

Seasonal variations in incidence of acanthocephalans have been linked with many ecological factors (including water temperature) and the feeding behaviour and diet of the host. No relationship was found between feeding and breeding of the host and incidence and maturation of *Echinorhynchus salmonis* in yellow perch in the Bay of Quinte, Lake Ontario, but a marked seasonal cycle related to water temperature did occur (Tedla and Fernando, 1969). Fish became infected in autumn and incidence peaked late in winter, began to decrease in March and dropped to nil in August–September. Such seasonality had already been described for species inhabiting water bodies which freeze in winter, as does the Bay of Quinte, e.g. *Acanthocephalus lucii* in fish in the River Dnieper and *E. gadi* in those of the White Sea. A single cycle of egg production occurred, mature acanthors being produced in spring and early summer, and the ratio of males and females gradually decreased as the proportion of females with acanthors increased. The worms died when the cycle was completed. The intermediaries are infected early in spring when their population levels are high and the ensuing warm spring and summer period provides a suitable environment for development of the parasite. In autumn, the cystacanths infect the final hosts, in which they are protected from the worst effects of low winter temperatures.

Another example where temperature was thought to be the main cause of the seasonal cycle is that given by Moravec (1984b) who found that *Neoechinorhynchus rutili* in carp, *Cyprinus carpio*, showed an obvious annual maturation cycle because gravid egg-producing females are present only in May. Carp acquire new infections from June until March of the next year, but mainly in autumn and early spring.

A study which relates the effect of temperature to increased predation is that of Shotter (1976). He showed that the intensity of infection of *Tilapia zillii* from a river and lake in Zaria, Nigeria, with *Acanthogyrus (Acanthosentis) tilapiae* was highest in November to February, i.e. in the middle of the dry season when the water temperature was at its coldest and the volume of water was reduced. This may result from a reduction in water volume leading to increased density of the animals present, including the intermediaries, and increased predation on them. Feeding activity is also emphasized by Pennycuick (1971a,b). Studying *Echinorhynchus clavula* in stickleback in a pool in Somerset, UK, he found that infections were acquired in August–September and decreased during the spring, reaching a very low level in April. This was correlated with feeding activity which was low in winter, increased in spring, was intermittent during the breeding season and reached a peak in July–August.

The observations of Hubschman (1985) are less clear with regard to the main factors involved in seasonality but are nevertheless interesting. Of 250 *Dorosoma cepedianum* collected by Hubschman from Caesar Creek Lake, Ohio, from March–October 1982, 55 were infected with *Tanaorhamphus longirostris*. Most infected fish (52) were found in April, May and June, none was found to be infected in March, August or October and only three were found in July. In June and July, infection was dominated by juvenile worms, and the failure of these late-appearing juveniles to mature may account for the earlier short period of infection.

Whilst in the above studies seasonality was pronounced, no seasonal cycle was observed by Kennedy (1972b) for *Pomphorhynchus laevis* in goldfish in Britain, which is present in nature throughout the year. *P. laevis* does not provoke a strong temperature-dependent rejection response and such fluctuations as do occur probably result from host feeding behaviour. The fish feed intensively in summer and ingest more cystacanths, compensating for those which fail to establish. In winter, fewer are ingested although more will establish; because of the decline in feeding intensity and of periods of starvation, the survival rate of established worms decreases during subsequent starvation. During a complete year, the effect of temperature on host feeding, and hence on the intake of parasites, is counter-balanced by its effect on the establishment of ingested worms. The overall effect is to even out seasonal variations in parasite input and to induce a more-or-less steady rate of recruitment.

According to Amin (1981) *Echinorhynchus salmonis* showed no pronounced seasonal periodicity in rainbow smelt, *Osmerus mordax*. The observed monthly distributions were determined by climatic, especially temperature and other ecological conditions, eg. those affecting the infection sources, that is, the amphipod *Pontoporeia affinis* and *Alosa pseudoharengus* and the feeding behaviour of rainbow smelt.

Chubb (1980b, 1982) discussed the seasonal dynamics of larval and adult nematodes. He suggested that for larval nematodes the calendar of events is determined by the interactions of the annual life-cycle pattern of the parasite with those of all its hosts, principal and auxiliary, intermediate or definitive. Water temperature, as determined by climate, the sum of weather conditions over a long period of time and place, will affected the rate and time of these interactions of the parasite with potential hosts. Exceptions include adult worms in warm-blooded definitive hosts where rate of development to maturity is determined by the body temperature of the host. That an understanding of seasonality amongst fish nematodes is intricately bound up with unravelling the complexity of life-cycles and the important influence of temperature on these can be demonstrated by taking two examples *Camallanus lacustris* and *Cystidicoloides ephemeridarum*. From 73.3 to 100 per cent of *Perca fluviatilis* were infected with *Camallanus lacustris* in a Norwegian lake near Oslo in 1975/76 and had mean worm burdens of 3.1 to 7.3 per infected fish (Skörping, 1980). All age groups of worms were present throughout the sampling period. The mean numbers of larvae, however, were higher for September–February than at other times of the year, indicating that recruitment is greater in autumn–winter at a time when microcrustaceans in the diet were also increasing. An increase in larval recruitment appeared to be compensated by an increase in mortality of older worms as indicated by a fall in the mean number of gravid females. The mean intensity of infection increased with fish length. In May–October most worms were located in the pylorus region of the gut, whereas in December–March most were in the posterior region.

Amongst the most intensive studies of the seasonal dynamics of larval and adult '*Cystidicoloides tenuissima*' = *C. ephemeridarum* (see Moravec, 1981) is that of Aho and Kennedy (1984). The principal intermediate host is *Leptophlebia marginata* although 17 other species are involved in which development occurs only to the first stage larval forms. The definitive hosts are *Salmo trutta* and *S. salar*. Prevalence and mean intensity indicated that *C. ephemeridarum* is an annual parasite exhibiting seasonal periodicity and systematic variation with demographic characteristics of its invertebrate and vertebrate hosts. Maturation of the worms was correlated with river water temperature. Adult *L. marginata* emerge and lay eggs in May and June. Small nymphs appear in June and July and grow rapidly in summer and early autumn. Growth slows in winter and accelerates in spring. The larval nematodes were generally randomly distributed in *L. marginata*, and *S. trutta* was consistently more heavily infected than *S. salar*. In *S. trutta*, prevalence was low during the summer months but it rose during winter to reach a peak in spring. There was a similar pattern of mean intensity. The salmon infection parameters varied in a similar manner but were less pronounced. A large proportion of the trout fry were infected in April and May but there was a subsequent summer maturation with third-stage larvae being predominant over winter. A large proportion of adult worms was found in June, and mature gravid females in late June and early July. It was concluded that the seasonal infections of both mayfly and fish were interdependent and controlled by the availability of infective larvae.

Cystidicoloides ephemeridarum shows a seasonal periodicity and a variation with the age of *Salmo trutta* and juvenile *S. salar* (Aho and Kennedy, 1984). Infective larvae are present in each generation of the mayfly, *Leptophlebia marginata* for almost its entire generation in the stream benthos. Variations in the infection of fish regarding time, site and host species appeared to be controlled by transmission-related events such as availability of infective larvae, host feeding behaviour and water temperature. The availability of infective larvae and host diet controlled the rate at which parasites were added to the parasite population but the pattern of gains and losses was determined by a temperature-dependent rejection. Moravec and De (1982) found that the rate of infection of *Salmo trutta fario* in Czechoslovakia with *C. ephemeridarum* was related to the size of the hosts, 73 per cent of trout smaller than 10 cm and 100 per cent of those over 10 cm long being infected. He observed two peaks, February and July, for intensity of infection, followed by sudden drops. Gravid females with mature eggs were present mainly from May to December. Third- and fourth-stage larvae were found throughout the year. Moravec agreed with Aho and Kennedy in suggesting that the seasonal maturation cycle is influenced by water temperature and mayfly populations in particular. *Ephemera danica* was the only intermediate host but *Cottus gobio* is a paratenic host. Alvarez-Pellitero (1976) showed that two generations of *C. ephemeridarum* occurred each year in *Salmo trutta fario* in Spain, one in spring and one in autumn and that temperature, directly and indirectly, was the most important single ecological factor involved. Fagerholm, Kuusela and Valtonen (1982) briefly discussed the taxonomy of *C. ephemeridarum*, its life-cycle in *Thymallus thymallus*, and previous observations on its ecology. Muzzall (1986) has reported on the ecology of *C. ephemeridarum* in *Salvelinus fontinalis* and *Salmo trutta* and with *Ephemera simulans* as the intermediate host. He discussed the seasonal occurrence and distribution of this nematode as well as *Spinitectus gracilis and Crepidostomum cooperi* in fresh and frozen trout.

It has been found that the population dynamics of *C. ephemeridarum* differs in various European localities. This was confirmed by Aho and Kennedy (1987) who studied and quantified the circulation pattern and transmission dynamics of larval and adult stages of the species in all its intermediate and definitive hosts at three sites in a small upland stream over a period of one year. The brown trout, *Salmo trutta*, was the only required definitive host and was alone responsible for the perpetuation of the parasite suprapopulation in the river, but the parr of *S. salar* are also suitable hosts. Within the intermediate host complex, the mayfly *Leptophlebia marginata* is the required host, and all other species of invertebrate harbouring the species are unsuitable hosts. The greater efficiency of transfer to trout appears to be related primarily to the food web in the river and particularly to the extent to which *L. marginata* was important as a dietary item to each species of fish. Aho and Kennedy discussed the relative importance of ecological factors, host community structure and parasite specificity in determining circulation routes and transmissions efficiences of *C. ephemeridarum*, and suggested that the parasite's suprapopulation dynamics in the area they studied may not be typical of the species in general. Further comments on these results will be made later in this chapter under community ecology.

The importance of temperature in influencing seasonal distribution was pursued by Chubb (1980b) after reviewing the seasonal distributions of digeneans. By taking a comprehensive study of one species as an example, that of Awachie (1963, 1968) on *Crepidostomum metoecus*, he concluded that temperature operated in different ways throughout the life-cycle. The overall effect of environmental temperature was that of synchronization of the life-cycles of the intermediate and definitive hosts with that of the parasite. Thus, a cold spring will delay, or a warm spring speed, the progress of all cycles. Chubb referred to three examples where temperature optima could be used to explain the phenomena of seasonal occurrence of adult digeneans in mid-latitude climates. *Bunodera luciopercae* has a low temperature optimum, *Asymphylodora kubanicum* a high temperature optimum and *A. tincae* will develop over a wide range of temperatures. There are, however, exceptions, e.g. with *Allocreadium fasciatusi* it is not temperature but an interaction of the life-cycles of the mollusc, copepod and fish hosts that determines the major seasonal pattern of occurrence of the adult worms. Chubb called for more experimental work of the kind described elsewhere in this chapter on *Transversotrema patialense* and *Bunodera luciopercae* to confirm the factors involved in the seasonal patterns of occurrence of digeneans. Temperature was thought by Halvorsen (1972) to affect the maturation cycles of *Allocreadium isoporum, Bunodera luciopercae* and *Azygia lucii* in fish in the River Glomma, Norway. Ectoparasites on the same fish were affected directly by water temperature, and maturation and egg production of the above gut parasites were similarly influenced, allowing for the seasonality of the intermediaries which might modify the basic temperature-dependent cycle. The maturation cycles of these digeneans were not correlated with the breeding cycle of their hosts but the cestode *Triaenophorus nodulosus*, present in pike all year round, was an exception. It produced gravid proglottides in spring when temperature was not likely to be the controlling factor, maturation probably depending on the hormone level of the host. A study of *B. luciopercae* in perch in a British reservoir revealed an annual cycle of incidence and maturation (Wootten, 1973). Juveniles were recovered in July and incidence rose to a peak in December although intensity of infection was fairly constant until December, when it increased and remained constant until February. Maturation occurred in autumn and early winter, mature worms were found in December and gravid ones in January, coinciding with high levels of incidence and intensity. By May, nearly all the worms found were gravid; worms were then lost, the process possibly being linked to an increase in water temperature and, by June, only a few worms, all gravid, remained. Rahkonen, Valtonen and Gibson (1984) confirmed this result by working on *B. luciopercae* in Finland. Others who have reported on recruitment and loss of worms from a subpopulation linked with temperature are Pojmanska (1984) and Burreson and Olson (1974).

Pojmanska investigated the seasonal dynamics of *B. luciopercae* and compared the results with data in the literature. It was confirmed that seasonal trends occur earlier in southern than in northern waters and that temperature was the important factor determining times of infection and death of the subpopulations. Pojmanska (1985a, b) continued her researches on temperature as an important factor in the seasonal dynamics of fish digeneans. Burreson and Olson (1974) studied the life-cycles of *Genolinea laticauda* and *Tubulovesicula lindbergi* (from *Leptocottus armatus* in Yaquina Bay, Oregon, USA. The incidence and intensity of infection with both species increased in late spring, peaked in summer and decreased in winter, the decrease resulting (at least in *G. laticauda* infections) from the loss of mature specimens at the end of their life-span.

Whilst discussing recruitment and loss of worms from a host in relation to seasonality, the work of Poole and Dick (1983) is important. They investigated the distribution of *Apophallus brevis* in *Perca flavescens* and whether the population size of fish predators of perch influenced the recruitment of the parasite into the perch population. Prevalence levels remained high in most age-classes of perch and mean intensity differences were not significant for all ages beyond 1-year-old. Prevalence levels remained more-or-less constant during the open-water seasons. Mean intensities dropped significantly from June to July and increased in August and October. Monthly fluctuations are probably the result of recruitment of the parasite and a loss of heavily infected hosts. The variance to mean ratios of *A. brevis* showed overdispersed parasite populations and the distribution of each in two lakes was found to fit the negative binomial distribution. The authors questioned the use of variance to mean ratios or other statistics dealing with overdispersion in predicting loss or recruitment for all parasite systems.

Kuntz and Font (1984) reported on seasonal changes in relation to host-specificity in the population dynamics of *Allopodocotyle boleosomi* in the fantail darter, *Etheostoma nigrum*, in Wisconsin over 14 consecutive months. Most recruitment occurred in May and June and peak occurrence of gravid specimens was reached in July. Little growth or development of *A. boleosomi*, took place during autumn, winter or early spring. Each of five other species of darter were less suitable hosts for *A. boleosomi*, some because of ecological differences and others because of physiological factors. The differing adaptations according to host species suggest the possibility of using darters and *A. boleosomi* for evolutionary and host-specificity studies but, as with understanding the mechanisms of seasonal population dynamics of this digenean, these would be dependent on elucidating the life-cycle of the species and of five other species of helminths which were recorded from *E. nigrum*.

The foregoing information has highlighted recent work on the seasonality of particular species or groups of helminths. Little seems to have been done on seasonality with regard to possible interactions between different species and different groups of helminths. We therefore conclude this account by citing other attempts to unravel one of the most complicated of influencing factors on seasonality, i.e. the presence of more than one species and of mixed groups of helminths in a host as was indicated above in our reference to Kuntz and Font (1984) and their discovery of five species of helminths in *Etheostoma nigrum*. Work by Halvorsen (1972), Font (1983) and Scholz (1986) also highlights some of the interesting features of such studies. Halvorsen (1972) chose the River Glomma in Norway as a study area towards understanding mechanisms regulating seasonality in fish helminths and, for this purpose, he obtained data on water temperature as well as those on fish activity and spawning. Water temperature acted directly in governing maturation and egg production in 15 parasite species on five host species. But the seasonal cycles of incidence and/or reproduction were different in different localities. In one case, i.e. *Triaenophorus nodulosus* in pike, the fish hormone level initiated maturation and egg production.

Font (1983) described seasonal variations in abundance and maturation which occurred in populations of intestinal helminths of the brook stickleback, *Culea inconstans*, in Wisconsin, USA. *Bunodera luciopercae* exhibited a pattern of development that was strikingly different from that of *B. eucaliae*, the former becoming gravid at the coldest time of the year whereas the latter produced eggs during the warmest months. This was regarded as an example of resource partitioning, i.e. specific adaptations that avoid interspecific competition within the intestine. Each of these species, however, had a different principal host. Font also found several differences between the population dynamics of *Neoechinorhynchus pungitius* in the brook stickleback and those reported for *N. rutili* in the three spined stickleback, the former displaying much more distinct seasonal changes of prevalence (2 per cent in July to 80 per cent in December), relative density (0.02 worms per fish in July to 3.00 in March), and maturation. The largest and most mature worms were collected in late spring, just prior to the total expulsion of the Acanthocephala from fish in summer. The *N. pungitius* population was at its highest in April at a time when that of *B. eucaliae* was at its lowest; *B. eucaliae* peaked in August. *Halipegus* sp. metacercariae and the plerocercoids of *Proteocephalus* sp. were also investigated and it was concluded that this parasite community of five species was structured in such a manner that most species reached peaks of abundance or development during different times of the year. Thus, stresses imposed upon the host species and potential interspecific competition for host resources were minimized. The brook stickleback was recommended as an excellent model for further research in this area.

Scholtz (1986) investigated the seasonal dynamics of five species of intestinal helminths in *Perca fluviatilis* in Czechoslovakia. *Bunodera luciopercae*, *Camallanus lacustris* and *Acanthocephalus lucii* occurred throughout the year but *Bothriocephalus claviceps* and *Proteocephalus percae* were found from June to October and from October to May respectively. *B. luciopercae* and *P. percae* showed prominent seasonal cycles of maturation whereas *C. lacustris* and *A. lucii* did not. Distinct seasonal changes in the location of *B. luciopercae* within the gut, were also found and this was related to their state of maturation, a greater number of mature egg-laying parasites being found in the middle and posterior regions than in other regions of the gut.

Biogeography

Introduction

An interest has occurred in the biogeography of parasites since Von Ihering, in 1891, studied the parasites of vertebrates of North and South America and concluded that the land bridge between the two countries was established after the Pliocene era, i.e. within the last 10 million years or so. Metcalf's work, in 1930, on parasites of frogs in South America and Australia supported a view that there was once a land link between the two countries and this work, as well as research on the hookworms of man, has also fostered interest in biogeography. We know, for instance, that one genus of hookworm, *Necator*, occurs in Africa south of the Sahara and also in south-east Asia. The other, *Ancylostoma*, occurs in Africa north of the Sahara, in parts of southern Europe and in Asia north of the Himalayas. Both genera occur in certain parts of South America, eg. Ecuador. From anthropological and parasitological evidence, considered together, it has been suggested that Japanese fishermen first introduced *Ancylostoma* into Ecuador around 3000 BC and that *Necator* was brought by south-east Asians almost 3,000 years later, in about 200 BC. Regrettably, however, these fascinating aspects of the biogeography of parasites in general were not taken up by fish parasitologists until the 1950s. Actually, in the 1950s, it was already clear from many papers, including Rees (1953a, b, c), that the helminth parasites of fish had been studied mainly in the northern hemisphere and that very little was known of forms occurring in fish caught off the coasts of South America and Africa.

The extensive efforts made to study the biogeography of fish parasites in all major freshwater systems of the USSR were described by Shulman (1961a). He listed all the species known at the time and indicated their biogeographical groupings. At the same time, Polyanski (1961a) drew attention to the lack of information on the biogeography of marine fish parasites and the importance of increasing work in the area.

Lebedev (1969) distinguished the following 10 biogeographical regions from an examination of 216 monogenean and 420 digenean genera: Western Tropical Atlantic, Northern Atlantic, Mediterranean-Atlantic, Red Sea, Indian Ocean, Sunda-Malayan Seas, Japan, New Zealand-Australia, Eastern Tropical Pacific and the North Pacific. According to Rohde (1982a), however, the parasites of many seas have not been examined sufficiently and large areas are, therefore, not included in Lebedev's scheme. Rohde (1982a) pointed out that the Indo-Pacific has generally more species of parasites than the Atlantic Ocean. The species composition in the two areas is also different. Whereas one study had shown that 56 of 96 genera of monogeneans and 136 of 224 genera of digeneans are common to both oceans, another investigation on species revealed that of 147 found off Tortugas, Florida, only 26 occurred in the Pacific.

Rohde also referred to latitudinal gradients in species diversity of hosts and parasites, emphasizing that, for monogeneans of teleosts, there is not only an increase in absolute numbers of species towards the equator (expected because of the greater number of host species) but also a relative increase. He also compared frequencies (percentage of fish individuals infected) and intensities of infection (number of parasite individuals per host individual) of Monogenea at different latitudes and found that frequencies are greater in tropical areas whereas intensities vary greatly and show no definite trends.

Rohde was of the view that, corresponding to the increase in species numbers towards the tropics, more species can co-exist because their niches are smaller, a niche being defined as the total of the relationship of an organism to its environment. He discussed this in relation to three important niche dimensions, host range, microhabitat and food, of the monogeneans, that is, whether they are blood or mucus and epithelial cell feeders. He did not think the number of species in a community, with consequential interspecific effects, competition and reinforcement of reproductive barriers, affected the fundamental niche. Other aspects of zoogeography discussed by Rohde (1982a) are: (i) fluctuations of parasite infections in cold and warm seas, (ii) differences between parasite faunas in shallow and deep water, (iii) relict parasite faunas in isolated seas and hosts—a relict species being one which has been isolated from its parent group for a long time, as in some of the marine helminths of the freshwater fish *Lota lota*, (iv) parasite endemicity of remote oceanic islands, (v) importance of temperature for parasite distribution, and (vi) importance of age of oceans for parasite diversity, the Pacific Ocean, for example, being much older than the Atlantic which began to form approximately 150 million years ago according to the theory of continental drift. It is, therefore, assumed that more species are found in the Pacific because there has been more time for species to originate and accumulate. Rohde emphasized the importance of biogeography in studying the effects of host migrations on parasites and in using parasites as biological tags. Rohde (1984) further emphasized that a study of the biogeography of marine parasites is in its infancy and that comprehensive treatises of marine biogeography ignore parasites completely. Nevertheless the information that was already available in the 1950s and 1960s enabled Manter (1967) to introduce an important general rule or principle, namely, a host is likely to have the greatest variety of highly specific parasites in the region where it has lived the longest, usually its place of origin. The migration, or the introduction, of this host into another area is likely to result in the loss of many, or most, of its parasites (especially those with complicated life-histories) and, perhaps, the acquisition of new ones. Following Manter's observations in 1967 many investigations of his general rule have contributed to our knowledge of biogeography and emphasized its importance in the following areas:

(i) the use of parasites as indicators of fish movements and the existence of separate breeding stocks of the same host species;
(ii) host specificity;
(iii) origin and evolution and parallel evolution of parasites and hosts;
(iv) the origin of various fish genera, e.g. *Salmo*, *Anguilla*, *Merluccius* and *Potamotrygon*.

Our aims, therefore, will be to comment on these four aspects and also to draw attention, in an appendix, to selected checklists and surveys of fish worms which may enable further research in biogeography.

Helminths as indicators of fish population biology

Success in the use of worms to enhance our knowledge of fish biology is dependent upon reliable data on the biogeography of both host and parasite. Because of its applied nature, this topic is discussed in detail in Chapter 6, but is briefly introduced here to indicate the links with biogeography.

Shchupakov (1954), for example, found that when fish were transferred from one region to another to stock fish farms or rivers in the USSR, they often lost at least some of their original helminth fauna and acquired other species characteristic of the new habitat. *Coregonus lavaretus maraenoides* transferred from Lakes Cheedskoye and Ladoga to lakes in the Urals lost all its original parasites and acquired *Tylodelphys clavata*, *Tetracotyle variegata* and *Agamonema* sp., while *C. albula ladogensis* acquired *T. variegata*, *Camallanus lacustris*, *Diphyllobothrium* sp. and *Agamonema* sp. Analysis of the helminth fauna of various fish species introduced into other regions of the USSR led Petrushevskii (1954) to conclude that the original parasites were reduced or completely lost and some new ones were acquired; parasites with a direct life-cycle in fish acclimatized as adults were likely to be retained. Fish transferred as roe were parasite-free and this has been accepted as a highly desirable feature of fish-farming.

The same, or closely-related, species of fish may have a different helminth fauna in different areas of its geographical range. Analysis of the fauna, therefore, provides a guide to the geographical origin of different fish stocks and to their subsequent migrations. The variations in parasite fauna depend on the life-cycle of the parasite, including availability of suitable intermediate hosts in different environments. Thus, the helminth fauna of *Osmerus eperlanus* from the Baltic Sea, White Sea, Siberian rivers and Kamchatka has been found to differ, fish from saline habitats harbouring more species than those from freshwaters (Kazakov, 1967).

From an examination of the parasites of coregonids throughout the world, Lawler (1970) has shown that some groups have a similar fauna while others do not, for example, in North America, the fauna of *Prospium* was different from that of other species. Parasites could also be used to follow movements of coregonids to identify fish stocks, in food studies and in studies on coregonid taxonomy or systematics. From records of helminths, particularly monogeneans of fishes (*Acanthocybium solandri, Makaira ampla, Neothunnus macropterus, Parathunnus sibi*) from the Central Equatorial Pacific, Iversen and Hoven (1958) suggest that the fish stocks from various areas in the Pacific and Atlantic are not, or were not, completely independent.

A survey of *Parona signata* in Argentine waters by Szidat (1969) showed all to be infected with *Lecithochirium microstomum*, the range of which includes the Galapagos Islands in the Pacific, Galveston Bay (Texas, USA) in the Atlantic, and Jamaica, i.e. on both sides of the central American isthmus, corresponding to the tropical Tethys Sea of the Cenozoic. The presence of *L. microstomum* suggests that *P. signata* originated in this area and subsequently migrated to the south Atlantic, which coincides with the present-day distribution of the coregonids.

A comparison of the fauna of mullet, *Mugil capito* and *M. cephalus*, from Lake Borullus, a brackish lake in the Nile Delta permanently connected with the Mediterranean, with that of fish from the sea has shown that they lose their intestinal digeneans (*Haplosplanchnus pachystoma* and *Saccocoelium obesum*) after migrating from the sea to the lake (Moravec and Libosvarsky, 1975). In Britain, dissimilar helminth faunas found by Gibson (1972a) in flounder from the Ythan and Dee estuaries and from the open sea off Aberdeen, made it possible to characterize the fauna of each group and to recognize individuals from each locality. 'Foreign' flounders migrating into any particular group could also be identified by their different parasites. The most useful indicator species in this study were found to be *Podocotyle* sp. and *Zoogonoides viviparus*. Plaice from three localities in the southern North Sea were also found to harbour different parasite species by Wickins and MacFarlane (1973). In this case, *Z. viviparus* was not a useful indicator but fish from the Southern Bight were more likely to harbour the nematode *Capillaria wickinsi* than those from Flamborough or the German Bight, and those from Flamborough were more likely to harbour anisakid larvae than were fish from the other two areas. *Cucullanus heterochrous* was present in all three areas but comparatively fewer were found from fish from Flamborough. Small numbers of fish from all three areas were infected with *Derogenes varicus*. Similarly, American plaice, *Hippoglossoides platessoides*, from the southern Gulf of St. Lawrence, Scotian Shelf and northeast Gulf of Maine harboured different faunas, indicating differences in the origin of each stock (Scott, 1975a, b). Fish from the Gulf of St. Lawrence had a low incidence of *Z. viviparus* and a high incidence of *Fellodistomum furcigerum*, suggesting that they are distinct from other stocks. No *Steringotrema ovacutum* were found in plaice from Banquereau. Olson and Pratt (1973) found that the nursery ground of English sole, *Parophrys vetulus*, in Yaquina Bay, Oregon, was the Yaquina Bay estuary by comparison of the helminths of bay fish before emigration offshore and of 0-group fish after migration. The incidence of *Echinorhynchus lageniformis*, acquired only in estuaries, was similar in both groups indicating that few or no fish other than those from estuaries joined the offshore population. *Philometra americana* infected sole only in the upper estuary and its presence indicated that fish had spent some time in that area. In contrast, infection with *Zoogonus destrocirrus* and with metacercariae of *Otodistomum veliporum* occurred only offshore and these parasites were not often found in fish collected in the estuary, indicating little or no migration between the open sea and the estuary.

Host-specificity

Manter (1967) discussed how host-specificity is related to biogeography because it limits the distribution of a parasite species to certain host species. Theoretically, the less restricted as to the kind of host a parasite can invade, the greater the chances of dispersal. The effects of host specificity, however, in restricting the dispersal of parasites is offset to some extent in two ways: (1) the attainment of sexual maturity in an intermediate host, i.e. progenesis, and (2) survival in a submature form in a number of paratenic or auxiliary hosts. Host-specificity is the subject of more extensive treatment in the next chapter.

Origin and evolution of parasites and hosts

An extensive discussion of the evolution of fish worms occurs in the next chapter and here we are concerned only in introducing its relationship to biogeography.

Rohde (1982a) has indicated how detailed studies of various groups of parasites in different seas may yield important information on absolute speciation rates at different latitudes, as well as speciation rates relative to those of hosts, interchange of parasite faunas between different seas and origin of host groups. In pursuing such information, more knowledge is required on latitudinal gradients in species diversity, and diversity gradients from shallow to deep water, in order to test the hypothesis that temperature is the predominant factor in determining evolutionary speed, and thus species diversity.

With regard to some of these factors mentioned by Rohde, especially gradients and species diversity in relation to temperature, the observations of Strelkov and Shul'man (1971), Bykhowskii and Mikhailov (1975) and Rumyantzev (1975) are interesting. Shul'man found that the helminth fauna of fish in the Amur basin, USSR, has been shown to consist of two zoogeographical groups: Sino-Indian species and Holarctic species (Strelkov and Shul'man, 1971). Of the latter, some originate from the cold waters of the north and the mountain regions, some from the warmer waters of the middle and lower reaches of rivers, and some have a discontinuous geographical distribution representing a heterogeneous group composed of species remaining from the Tertiary period.

Fish in the Kura and Volga rivers and the Caspian Sea harbour parasites of various origins (Bykhovskii and Mikhailov, 1975). Fish in the Kura river basin have many parasites of southern origin, the Ponto-Arolo-Caspian species predominating, and relatively many middle western species, but few from the colder northern regions. Volga fish harbour mainly northern species with some southern forms but none from the Middle East fauna and no endemic species of the mountain type. Caspian fish parasites are mainly euryhaline freshwater species, with numerous marine and freshwater representatives, some southern types and a few northern species.

The parasite fauna of *Coregonus albula* from various water systems in Northern Karelia, USSR, zoogeographically the European part of the Arctic Ocean, has been examined by Rumyantzev (1975). Fish from Lake Kuyto in the north (White Sea system) harboured 23 parasite species but those from southern Lakes Onega and Ladoga (Baltic system) harboured 11–12. *Philonema sibirica* and *Capillaria coregoni*, considered to be northern species, were absent from the southern region. Fish in the Syamozerskoy lakes which are considered to be in the same region as Lakes Onega and Ladoga, although their locality is described as the Baltic province of the Mediterranean subregion, had only 6–9 parasite species. Fish from the most southern fringe of the area investigated had only 5–6 parasite species.

Michajłow (1962) drew attention to the parallel evolution of the genus *Esox* and its tapeworm, *Triaenophorus*. All known species of *Esox* harbour this tapeworm genus. Accepting that parallel evolution has occurred, the problem of the origin of both host and tapeworm was also addressed by Michajłow and he concluded that *T. nodulosus* and *T. crassus* first occurred in Eurasia and then spread to North America. There is no doubt in his view, however, that *T. stizostedionis* originated in North America following the establishment of pike in the region. Analysis of the helminth fauna, particularly *Dactylogyrus extensus* and *Caryophyllaeus fimbriceps* of carp, together with results from electrophoresis of host blood serum, has shown that carp had a continuous Eurasian distribution in the Oligocene and split into European and Far-Eastern populations in the cold Quaternary period (Iwasik and Swirepo, 1967).

The work of Manter (1963) on the distribution patterns of the digeneans of fish provides evidence of present and ancient continuity of oceans. Further evidence by Manter (1967) suggests that the ancient Tethys Sea once formed a connection between Australia and the Caribbean, with the present Indian Ocean being connected with the present Mediterranean and the Atlantic and Pacific Oceans being continuous across Panama. Manter also provided parasitological evidence to support the continental drift theory, a topic also discussed by Dönges (1967).

According to Donges, his work on *Nesolecithus africanus* supported a view that South America and Africa were connected in former times, as assumed by the continental drift theory. *N. africanus* occurs in the intestine of *Gymnarchus niloticus* (the 'electric pike') but while the host is found in all North African rivers, the parasite occurs only in fish living in the River Yewa in Nigeria. The nearest relative of *N. africanus* is *N. janickii*, found in the intestines of *Arapaima gigas*, the biggest South American freshwater fish populating the Amazon. Although the two parasites are different species they are very similar and must be assumed to have common ancestors in the not-so-distant past. Both use freshwater fish and amphipods as hosts. According to Donges, it seems very unlikely indeed that some of their common ancestors crossed the thousands of miles of present-day Atlantic. If, however,

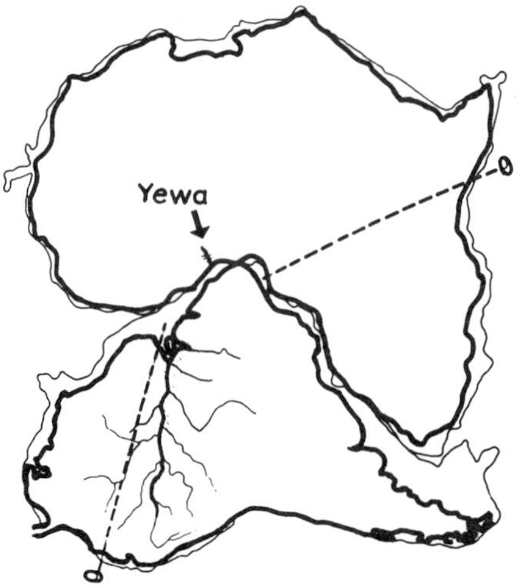

Figure 3.5 Parasites support continental drift. From Dönges (1967).

south America and Africa were connected up to the Jurassic or Cretaceous times, as may be assumed from geological evidence, the rivers Yewa and Amazon were once near each other (Figure 3.5). The common ancestors of the two *Nesolecithus* species then lived originally in one closed area which was later divided by continental drift.

The continental drift theory was also used by Rohde (1986b) as a possible explanation of the distribution and abundance of monogeneans in two zoogeographical areas. He researched on species diversity of marine Gyrodactylidae, an exclusively cold-water family of Monogenea, and showed that it was much greater in the northern Pacific than the northern Atlantic. He plotted relative species diversities against mean annual surface temperatures at the various localities and not against latitude for two reasons, namely, the assumption that temperature is the predominant factor determining species richness, and the lack of data to distinguish gradients in the southern and northern hemispheres. Rohde concluded that the only explanations for the difference between the northern parts of the two oceans appear to be evolutionary and/or the ecological time hypothesis. According to the continental drift theory, the Pacific has existed for considerable geological time whereas the Atlantic began to form only about 150 million years ago. Hence, more time has been available in the Pacific for species accumulation. In earlier papers, Rohde had concluded that only a time hypothesis based on the assumption of greater evolutionary speed in the tropics could explain latitudinal gradients in general and the gradient of relative species diversity of Monogenea in particular.

Studies on the caryophyllaeids suggest that they originated in eastern Asia, giving rise to one branch associated with Eurasian cyprinids and another radiating with the catostomids in North America (Mackiewicz, 1965). Neoarctic and palaearctic caryophyllaeids appear to be distinct from each other and the presence of *Glaridacris catostomi* in *Catostomum catostomus*, the only catostomid common to both regions, is of zoogeographical importance. Mackiewicz (1972), in the absence of a comprehensive analysis of the biogeographical distribution of caryophyllaeid cestodes, relied on host parasite checklists for general distribution data. From these lists and other data, he concluded that caryophyllaeids form almost one-quarter of the cestode fauna of the freshwater fish of North America and Russia. He showed on a map that, with few exceptions, the caryophyllaeid fauna of each region appears to be distinct with an apparent high degree of endemism reflected in 23 of 37 genera being monotypic. Furthermore, there were no cosmopolitan species and few genera were found in more than one region. Their general absence from the neotropical region may have reflected our lack of knowledge of the parasites of the freshwater fishes of that region; on the other hand, caryophyllaeds may not yet have radiated to South America in the absence of cyprinid and catostomid hosts.

From amongst more advanced tapeworms than caryophyllideans, we have selected the comprehensive studies by Kennedy (1969a, 1978b) on the pseudophyllidean *Eubothrium* and by various workers on *Ligula intestinalis*. Kennedy (1969a) found adult *Eubothrium crassum* only in *Salmo salar* and *S. trutta* (sea trout) migrating up-river to spawn, and they ranged in size from small and immature to large and mature, though rarely gravid, worms. It was thought that the infection was acquired in the sea and that gravid worms had been lost on or before arrival in the rivers. Most of the worms in salmon die with the fish in fresh water but, in sea trout, the worms continue to grow and probably mature on return to the sea. It was suggested that *E. crassum* found in adult Atlantic and Pacific salmon in rivers are rarely, if ever, the same worms that are carried to sea by smolts but are of a marine origin and are residues of a marine life-cycle; and that two biological races of *E. crassum* exist, a marine and a freshwater one. Kennedy (1978a,b) found *Eubothrium salvelini* specific to *Salvelinus alpinus* and *E. crassum* to *Salmo trutta* and *S. salar*. Distributions of both species are co-extensive with their hosts but levels of infection in any locality in North Norway and the islands of Spitzbergen and Jan Mayer relate to the abundance of zooplankton. *E. salvelini* is a freshwater parasite but there is a distinct marine race of *E. crassum* which completes its life-cycle at sea. In salmon moving from one medium to the other, the marine and freshwater races replace each other, but in sea trout, which spend shorter periods at sea, complete replacement may not occur and at any one time fish may harbour individuals of both races. *E. fragile* found in *Alosa fallax* is also thought to be a marine species and eggs released in fresh water during the spawning migration are part of the parasite's natural reproductive wastage (Kennedy, 1981a).

Invaluable background publications on the distribution of *Ligula intestinalis* in various hosts and geographically in Northern and Southern Hemispheres are those of Orr (1967, 1968a) and Arme and Owen (1968). The introduction, establishment and subsequent history of a population of *L. intestinalis* in roach *Rutilus rutilus* over a period of seven years was described by Kennedy and Burrough (1981). Following a chance introduction by the return of great crested grebes (*Podiceps cristatus*) the tapeworm had within two years infected 33 per cent of the roach between 60 and 120 mm long. Infection levels and parasite growth rates varied from year to year but these annual fluctuations were irregular and showed no consistent pattern. The tapeworm population exhibited both regulatory and destabilizing features. Black and Fraser (1984) reported annual changes in *L. intestinalis* in *Catostomus commersoni* from small lakes in Canada from 1963 to 1981. Prevalences were low (50 per cent) in the four lakes compared with a frequent 100 per cent recorded by Kennedy and Burrough. Differences in the levels of prevalences among the lakes were probably most dependent on differences in the morphology of the lake basins.

An interesting example from amongst the acanthocephalans is that of Amin (1985c) on *Acanthocephalus dirus*. Geological and morphological evidence supports the proposition that the Wisconsin–Lake Michigan population of *A. dirus* was geographically isolated from Mississippi River *A. dirus* less than 15,000 years ago. Amin is of the view that, given sufficient time and continued isolation, this situation would be a classical example of Van Cleave's (1952) proposition that, in the Palaeacanthocephala, 'isolation might allow the normal extremes in a highly variable species to become segregated as distinct species'. The New England population of *A. dirus* may have originated in a like manner. The discovery of additional species of *Acanthocephalus* in new hosts from the Mississippi River drainage system which might have evolved through isolation is not unlikely.

With regard to the Nematoda, species of *Cystidicola* have been intensively investigated from an ecological and zoogeographical standpoint, especially by Black and his co-workers in Canada. Their work has a sound taxonomic basis in Black (1983a), who considered *C. stigmatura* a valid species which could be distinguished from the only other species in the genus, namely *C. farionis*, by presence of lateral lobes on its eggs rather than filaments. He listed the North American hosts of both species, including *Salvelinus alpinus*, *S. fontinalis* and *S. namaycush* for *C. stigmatura* which has apparently been absent from lake trout collected from large areas of the Great Lakes over the last 50 years. It may have disappeared with the decline of the lake trout populations during the last century. Some local populations of native lake trout still exist and *C. stigmatura* may have survived in these remnant populations.

The foregoing observations prompted Black (1983b) to investigate the distribution of *C. stigmatura* in the North America so that he could provide an explanation for the observed distributional pattern. The results indicated that *C. stigmatura* occurs in lake trout (*Salvelinus namaycush*) and Arctic char (*S. alpinus*) in the Great Lakes and St. Lawrence River, drainage basins in lakes along an arc extending from north-western Ontario to Great Bear Lake and into the Canadian Arctic Archipelago, but not elsewhere. This distribution was explained by the nematode's probably post-glacial history. It was hypothesized that *C. stigmatura* survived in only one glacial refugium and on deglaciation probably dispersed with lake trout from its refugium in the upper Mississippi River region into the Lake Michigan basin. Its northward dispersal via glacial waters resulted in Arctic char acquiring the parasite. The apparently restricted preglacial distribution of the nematode and recent history of its intermediate host (*Mysis relicta*) in fresh water suggests that *C. stigmatura* may have arisen as a species during the last glaciation. Black (1984a) tested the hypothesis by examining the morphology of *C. stigmatura* across its range and concluded that the nematode survived in one glacial refugium. Black and Lankester (1984) reviewed the distribution of both *C. cristivomeri* and *C. farionis* in North America and Eurasia. They suggested that *C. cristivomeri* in *S. alpinus* in certain lakes and in *S. namaycush* in others may have a life span of at least 10 years. There were considerable differences in the numbers of worms in fishes between the lakes. Mature female worms were not abundant until fish were five years old, suggesting that the development of sexual maturity in *C. cristivomeri* may be retarded when large numbers of worms are present in the swimbladder. They also suggested that the maximum length attained by female worms might be determined by the total number of worms present and perhaps by the space available in the swimbladder. *In vitro* studies indicated that the number of eggs released was positively correlated with worm length.

As for our account of seasonality, we have until now deferred commenting on the origin and spread of digeneans and for the same main reason, i.e. the complicated nature and little-known aspects of their life-cycles, especially of marine forms. We have, however, already emphasized the importance of Manter's contributions, especially Manter (1967), in this context. We shall restrict our comments to a mention of work by Fischthal and by Bakke. A study of the zoogeography of the 107 digenean species parasitic in West African marine fishes has shown that only five genera, including 55 species, are endemic, many of the others occurring in all of the seas of the world except the polar seas (Fischthal, 1972). The endemic species, and those most similar to them, are widely distributed, particularly in the west and north Atlantic and in the Mediterranean. The zoogeography of the parasites closely parallels the distribution and migrations of their fish hosts. Bakke (1985) referred to the geographical distribution and host records of *Phyllodistomum umblae* so far known from Norway and showed that it has a wide distribution in the northern parts of the Palaeartic region. Other papers on the zoogeography of digeneans are cited in Appendices (pp. 322 to 331).

Origin of *Anguilla*, *Merluccius*, Salmonidae and Potamotrygonidae

Manter (1955) discussed the parasites of *Anguilla* in a zoogeographical sense and concluded that, although a smaller number of species are known from Pacific eels, perhaps due to less study, this is not so significant as the greater uniqueness of the Pacific parasites. North Atlantic eels have acquired a large number of rather non-specific species of trematodes common in other fishes, but have only one species unique to them and it occurs in both Europe and America. The three genera and eight species known only from Pacific eels stand in considerable contrast and suggest a Pacific origin of *Anguilla*. The large number of marine species of trematodes found in North Atlantic eels suggests longer contact with the sea than occur in Pacific eels where the trematodes have greater freshwater trends.

According to Szidat (1960), the parasites of *Merluccius hubbsi* from coastal waters of Argentina are similar to those found in other species of hake in the Pacific and not to those of hake of the North Atlantic. He concluded, therefore, that *M. hubbsi* migrated to the South Atlantic from the Pacific via Cape Horn in the Pleistocene, and the genus *Merluccius* may have originated in the North Pacific. From observations on the parasites of other fish species, Szidat concluded that the fish fauna of the South Atlantic had a dual origin, from the Mid- and North-Atlantic and from the Pacific, and that the migration took place fairly recently. The failure of parasites of fish of Atlantic origin to infect fish of Pacific origin and *vice versa*, even though these fish are thought to be related, indicated to Szidat that there are profound differences in the protein composition of the two groups of fish and that this should be investigated biochemically. According to Rohde (1984b), however, who cited investigations by Kabata and Ho on the copepods of *Merluccius*, and by Reimer (1981) on its digenean parasites, Szidat's views are not supported by facts. Rohde felt that palaeontological and other ichthyological evidence, together with parasite data, now firmly confirm the view that *Merluccius* originated in the northern Atlantic, as may be seen from references cited by Kabata and Ho. There is, however, still some doubt on how *M. hubbsi* reached the South Atlantic coast.

Ichthyologists have often discussed the question of a freshwater versus a marine origin of salmonids (Margolis, 1965). After discussing some of the parasites of salmon and drawing attention to the views of Soviet parasitologists who think that evidence from parasitology supports the theory of a freshwater origin of salmonids, Margolis also concluded that the facts indicate that the freshwater environment must be an older habitat of salmon than is the marine environment, and that anadromy is a comparatively recent development in the life of the salmon. Other ecological and zoogeographical considerations on the parasites of Pacific salmon are given by Margolis (1982a, b).

As our last chosen example of helminths as possible indicators of the zoogeographical origin of various groups of fish, we refer to Brooks, Thorson and Mayes (1981) who examined four hypotheses concerning the origins and evolution of Potamotrygonidae (freshwater stingrays) in terms of the tapeworms found. These were concerned with the monophyletic or polyphyletic origin of the family, its freshwater or marine origin, a Pacific or Atlantic origin, designated or non-designated ancestry and the zoogeography being consistent with a vicariance hypothesis. Of the four hypotheses proposed the most promising for further research seems to be the one which suggests that the potamotrygonids represent a monophyletic group whose ancestor was a non-designated marine stingray species trapped in South America by Andean orogeny.

Sites and site-selection

Introduction

It has already been emphasized in this chapter that the term 'niche' is used for a very broad and controversial concept in ecological parasitology. Although the site-dimension of a 'niche' has received some attention it needs further research towards clarifying other aspects of the niche concept. For this reason we shall now attempt to bring together and summarize a large volume of information on sites occupied by larvae and adults of the nine groups of fish helminths discussed in Chapter One. Our aims are to discuss previous work on site-selection by fish helminths and refer to those which attack the ten main body systems of fish, as described by Brown (1957). With regard to the skin, gills and alimentary canal, in particular, we have been highly selective in our examples.

Sites and site-selection are challenging areas for further research in helminthology because the same helminth species may occupy different sites during various stages in its development, from a larva to an egg-producing adult. Furthermore, a parasitic species might occur in a tissue, organ or body system which performs more than one physiological function. For instance, the skin (integument) of a fish, according to Brown (1957), has many functions, including protection against parasites, whilst the swim bladder may operate as a hydrostatic organ, sense organ, respiratory system, digestive organ or serve in sound production. Currently, for most groups of helminths, there is insufficient information to determine whether the preference of a helminth species for a particular tissue, organ or body system is directly related to the various aspects of host physiology, namely, locomotion, sensitivity, respiration, nutrition, reproduction or excretion. It is thought that almost all adult acanthocephalans and cestodes live in the alimentary canal because they are gutless and have to live in a bath of absorbable nutrients. The presence of other helminths, with well-developed guts, in the alimentary canal of their host is, therefore, interesting, and does emphasize the need for further research to determine the precise nutritional factors which may influence site-specificity. The challenging nature of the topic may also be illustrated by reference to the muscle system and movement in the Scombridae. The dominant characteristic of this family is adaptation for rapid locomotion—feeding, courtship, spawning, and even resting, are all done on the move. Their swimming capacity, is regulated by the metabolic capacity to convert chemical energy into propulsive thrust by muscular contraction, which is influenced by environmental temperature, oxygen and carbon dioxide concentrations of the water and also salinity. Since parasites of muscles are known to affect locomotion in fish (see Beamish, 1978, p. 144) one of the consequences of selecting a muscle system as a site in which to live would be to make the host more vulnerable to a predator and thus ensure continuation of the parasite's life cycle.

For these and other interesting reasons we felt that a brief review of previous work on site-selection followed by an account of which helminths may be found in ten fish body systems would encourage further research on (i) the implications of successful site-selection for life history strategies, ecology and helminth physiology, and (ii) the little-known effects of helminths on host physiology.

Previous work on site-selection

There is much information on sites occupied by fish helminths, but it is widely scattered, as will become evident from the following account. Dawes (1947) and Erasmus (1972) emphasized the importance of site-selection amongst monogeneans and digeneans. Dawes referred to monogeneans, usually skin- or gill-dwelling forms, from the following sites:

Thaumatocotyle concinna in the nasal fossae of *Trygon pastinaca*; *Microbothrium centrophori* on the caudal fin of *Centrophorus squamosus*; *Leptocotyle minor* on the skin of the head just behind the eye of *Scyliorhinus caniculus*; *Udonella caligorum* parasitic on *Caligus* spp. recovered from various teleosts and from *Anchorella uncinata* parasitic on *Gadus morhua*; *Acanthocotyle monticelli* in the gill chambers of rays; *A. borealis* on the ventral surface and *A. lobianchi* on the dorsal surface of the thornback ray, *Raja clavata*; *Encotyllabe pagelli* and *Cyclocotyle charcoti* (hyperparasitic on *Cymothoa oestroides*) in the buccal cavity of *Trachurus trachurus*; and *Diclidophoropsis tissieri* on the skin near the gills of *Macruris laevis*. According to Kearn (1968a) no other group of monogeneans exploits such a wide range of habitats as the monocotylids, which may inhabit the skin, gills, nasal fossae, cloaca, or body cavity of their hosts. Dawes (1947) also referred to different digenean species from many specific sites in the alimentary canal of fish and also from the nervous system, skin, gills, urinary bladder, coelom and blood vascular system.

Erasmus (1972) stressed the unusual locations of some monogeneans in the cloaca, ureter, oviducts, body cavities, alimentary canal and blood system of their hosts; and the precise locations of digeneans, especially metacercariae in fish hosts, e.g. *Bolbophorus* in the muscles and fins of brown trout *Salmo trutta*, *Crassiphialia bulboglossa* beneath the epidermis of freshwater fish, *Posthodiplostomum minimum* in the liver, kidney, pericardium and spleen of fish and *Diplostomum* spp. in the lens, humour of the eye or ventricles of the brain.

Crompton (1973), Kennedy (1976b), Fallis (1971) and Rohde (1982a) are valuable sources of information on site-selection, with chapters in Kennedy by various authors, on body surfaces, gills, body cavities, the eye and brain, and blood systems, as well as a general chapter on entry into the host and site-selection. A classic account by Ulmer (1971) describes site-finding behaviour.

Crompton (1973) appealed for more accurate descriptions of a helminth's site because it is fundamental in our understanding of ecology. He gave the distribution of sites occupied by the adults of 252 species of helminths in the alimentary tract of vertebrates including 24 acanthocephalans, 13 tapeworm and 6 nematode species from fish. Crompton's important contribution to the study of sites is further discussed in our treatment of the alimentary canal as a habitat for parasites.

Crompton (1976) considered entry into vertebrate hosts as processes of oral, cutaneous and diaplacental infection. The cloacal opening is an additional route for monogeneans of Amphibia (and possibly ureter and urinary bladder parasites) and the abdominal pores are the possible point of entry for abdominal cavity parasites of elasmobranchs. Of the fish helminths mentioned, *Plagioporus sinitsini* shows a *preference* for the gall bladder of *Nocomis biguttatus* and *Crepidostomum cooperi* for the pyloric caeca of *Perca flavescens*; *Aporocotyle macfarlani* for the afferent branchial arteries, ventral aorta and, very occasionally chambers of the heart of *Sebastes caurinus*, and *Sanguinicola klamathensis* for the efferent renal vein of *Salmo clarkii*. He referred to changes of sites if other parasites are present, eg. populations of the blood flukes *A. macfarlani* and *Psettarium sebastodorum* changed during concurrent infections in the vascular system of the rockfish; and populations of the cestode *Proteocephalus filicollis* and of immature forms of the acanthocephalan *Neoechinorhynchus rutili* appeared to move apart in the small intestine of *Gasterosteus aculeatus*. Crompton pointed out that most plerocercoids of *Triaenophorus crassus* were located on the right side of *Leucichthys* spp. because the stomach lies against the right body wall of the fish.

Ulmer (1971) summarized interesting work on the preference of strigeoid cercariae for the brain of fish, those of *Diplostomum baeri* localizing particularly in the choroid plexus and optic lobes of fish, apparently reaching these sites via the bloodstream. Stimuli which attract cercariae to such sites, were briefly discussed by Ulmer, who doubted Szidat's hypothesis that 'memory' is involved. Of the adult digeneans selected by Ulmer, *Phyllodisto-*

mum staffordi matures in the urinary bladder of bullheads, but requires a sojourn in the ureter before migrating to the bladder. He drew attention to the restricted number of sites occupied by larval and adult cestodes in intermediate and definitive hosts, in marked contrast to the diversity of sites occupied by digeneans. In some instances, the preference of a species of cestode larva for a particular site appears to vary with the host used. Different tapeworm species may be involved, eg. plerocercoids of *Triaenophorus crassus* are almost exclusively limited to epaxial muscles between the skull and dorsal fin of *Leucichthys* (= *Coregonus*) but those of *T. nodulosus* in *Lota* are restricted almost entirely to the liver.

Amongst other aspects of site-selection discussed by Ulmer are (i) the hormonal condition of the host and effects of temperature in influencing the migratory behaviour of *Proteocephalus ambloplitis* plerocercoids, (ii) the physicochemical and morphological features of the alimentary canal which influence the establishment of adult cestodes at particular sites, and (iii) the sensory mechanisms in site-finding behaviour.

Needham and Wootten (1978) refer to *Dactylogyrus vastator, Benedenia monticellii, Axine heterocerca, Discocotyle sagittata, Posthodiplostomum cuticola, Cryptocotyle lingua* and *Philometra lusiana* as examples of helminths of the integument of fish; *Sanguinicola* as a parasite of the blood system; *Diplostomum, Tylodelphys* and *Apatemon* as eye parasites; *Diphyllobothrium spp.*, *Triaenophorus* spp., *Proteocephalus ambloplitis, Posthodiplostomum minimum*, *Clinostonum marginatum, Nanophyetus salminicola. Contracaecum, Thynnascaris, Anisakis, Phocanema, Eustrongylides, Philonema, Philometra* and *Cystidicola* as helminths of the viscera and muscles; and *Caryophyllaeus fimbriceps, Khawia sinesis, Bothriocephalus acheilognathi, Eubothrium, Crepidostomum* spp, *Capillaria, Pomphorhynchus laevis* and *Acanthocephalus jacksoni* as parasites of the alimentary canal. The emphasis is placed on the pathogenic nature of the helminths at the sites named.

Much information on sites occupied by fish helminths is available in a wide range of publications but this is not apparent from their titles. Among the useful checklists and monographs are: Beumer, Ashburner, Burbury, Jette and Latham (1983): Bravo-Hollis and Deloya (1973); Crane (1972); Hewitt and Hine (1971); Hooper (1983); Khalil (1969); Margolis and Arthur (1979); Nahhas and Cable (1964); Overstreet (1969, 1978a); and Varela (1975). Others are cited in the zoogeography section of this chapter. Survey work on particular fish species or groups also gives information on sites, e.g. Appy and Anderson (1982) for lampreys (Petromyzontidae); Appy and Burt (1982) for cod, *Gadus morhua*; Davis and Huffman (1978) for the mosquitofish; *Gambusia affinis*, Fischthal, Carson and Vaught (1982a) for *Seriola dumerili*; Hine (1978) for 30 species of helminths in *Anguilla* spp; Margolis (1982a, b) for salmon (*Oncorhynchus* spp.); Nikolaeva and Dubina (1978, 1985) for various sites of didymozoids in Scombroidei; Paperna and Overstreet (1981) for helminths of Mugilidae; Sekerak and Arai (1973) for *Sebastes*; Seyda (1973) for eels *Anguilla anguilla*; Shotter (1976) for whiting *Merlangus merlangus*: Skinner (1975, 1978) for a number of hosts, especially mullet, *Mugil cephalus*: Srivastava (1966a, b, c, d) for five-bearded rockling, *Ciliata* (= *Onos*) *mustelus*, and Wootten and Waddell (1977) for larval roundworms in *Gadus morhua* and *Merlangus merlangus*: and very many others as can be seen from the community ecology section and appendices of this chapter. The need to search these references for useful information on sites is well illustrated in Gibson and Bray (1977, p. 195). They

> 'suggest that the reason why adult helminths are fairly common in the coelom of elasmobranchs may have something to do with the fact that these fish possess an abdominal pore which permits the escape of eggs. Forms associated with this habitat tend to have a broad or spatulate body, sometimes in contrast to their near relatives from the gut. This may be an adaptation which has developed to prevent individual specimens being lost through the abdominal pore. Helminths from similar habitats such as the urinary and

swim bladders, where the relatively static conditions permit a reduction in sucker-size and where there is an outlet through which the parasite might be lost, tend to have similar adaptations.'

Main Body Systems

Skin

According to Brown (1957) the skin of all fish, like that of any vertebrate consists of two basic layers; an outer epidermis and an inner dermis (also called the corium). The lowest layers of the epidermis are composed of columnar cells known as the stratum germinativum which gives rise to new epidermal cells. Except for cells that are in the process of being shed, the entire epidermis is alive. In cyclostomes, the epidermis secretes a thin film of noncellular dead cuticle that covers the body. No such specialization of the epithelium occurs in other fish, except for the peal organs (horny, nuptial tubercles, warts or granules) commonly found on the heads, fins and scales, of the breeding males of many teleosts, especially members of the Catostomidae and Cyprinidae. It is not known whether helminths attack such a secretion of the epidermis.

The fish dermis is a fibroelastic connective tissue with relatively few cells and consists of two layers, a relatively thin stratum vasculorum or spongiosum and a thick dense lower one, the stratum compactum. Loosely organized connective tissue, the subcutis, binds the dermis to the underlying muscle. It is important to consider the structure and function of fish skin in parasitological work since Kearn (1963c, 1976) suggested that most monopisthocotylean monogeneans are epidermis feeders resulting in minimal damage to the host because of the rapidity with which the epidermis is regenerated. This therefore prevents exposure of subepidermal tissues to infective micro-organisms.

There is probably a relationship between the nature of the skin of fish and the helminths they attract. Kearn's (1976) classic work on the body surface of fish draws attention to the need to relate variation in fish teguments in different species to variations in helminth groups or species they might attract. In most fish, the connective tissue of the dermis is replaced by a dermal skeleton in the form of scales, but there are a few studies on the chemical composition of the latter (Brown, 1957). There are, however, scaleless species such as the North American catfish (Ameiuridae) and their skin is so tough that it can be stripped off like the hide of a mammal. The skin of some Salmonidae, Gadidae and *Hippoglossus* spp. is also very thick. There is a need, therefore, to study their parasites in relation to the structure and chemical composition of the skin. An unusual situation is found in *Mola* where the skin is reinforced underneath by a hard cartilaginous layer which may be as much as two or three inches (5–8 cm) thick; and in the sea snails (*Liparis*) the skin is separated from the body by a layer of translucent, jelly-like, substance as much as 25μm in thickness. In the electric catfish *Melapterinus* this gelatinous layer contains a scattered number of electric plates derived from epidermal cells and not from muscles as is the case with all other types of electric organs in fish. If there are records of monogeneans or of other helminths having invaded the teguments of the above named hosts, the species are worthy of further study relative to the structure and function of the skin. The most obvious function is protection, both against injury and against infections with parasites. Other functions are: osmoregulation, thermal regulation, respiration, food absorption and as a locus for many modified structures such as mucous cells, poison glands, photophores, chromatophores, trichocytes, scales, plates, denticles and numerous sensory organs, including the lateral line, taste buds, pil organs and free nerve endings.

Kearn (1976) emphasized the important role of the skin in immunology with particular reference to host specificity and host protection. Serum antibiotics have been found in skin mucus and may be synthesized in the skin. A skin secretion from *Lymphysodon* may contain antibodies against important pathogens and when eaten by its fry, gives immunity. Thus, the secretion may be analogous to the colostrum of mammals. Nine papers by Rhee and his co-authors (1980–1984) show that the epidermal mucus of *Cyprinus carpio nudus* showed anthelminthic activity against *Clonorchis sinensis* and was more effective than the mucus of *Carassius carassius*. These observations point to promising lines of future research.

Consequently many monogeneans and other helminths parasitic in or on the skin of fish are worthy of intensive research (see Kearn, 1976), bearing in mind the variety in structure and function of the teguments in the hosts they prefer. Some examples, other than Monogenea are given below (Tables 3.2–3.4).

Alimentary canal and associated organs

Some groups of helminths in the gastrointestinal tract of fish indicate preferences for certain sites, e.g. monogeneans for the buccal cavity or the cloaca, and cestodes for sites where there is a direct supply of bile. Others are absent from particular hosts and sites, e.g.

Table 3.2. Some Adult Digenea of the Skin

Parasite	Host	Site	Reference
Prototransversotrema steeri	*Aldrichetta forsteri*	Under the scales	Angel (1969)
Transversotrema patialense	*Brachydanio rerio*	Recesses under the scales	Anderson, Whitfield and Mills (1977), Mills (1980a,b)
T. patialense	*Ophicephalus punctatus Macropodus cupanus*	Under the scales	Crusz, Ratnayake and Sathananthan (1964)
T. chakai	*Macropodus cupanus* and *Barbus puntius*	Under the scales	Mohandas (1983)
T. laruei	*Molienesia latipinna*	Under scales, opercular cavity and gills	Velasquez (1961)
T. licinum	*Microcanthus* sp.		Manter (1970)
Didymozoon tetragyne	*Sphyraena picuda* and *S. jello*	Encysted on paired fins	Job (1961a)
Platocystis polyastra	*Sphyraena obtusata*	Pairs encysted under scales	Job (1962)
Didymocystis apharyngi	*Sphyraena picuda S. jello*	Gill arch epithelium and buccal cavity	Job (1961b)
D. pseudobranchialis	*Sphyraena picuda*	Pairs in cysts on pseudobranch	Job (1964)
Nematobothrium spinneri	*Acanthocybium solandra*	Cysts in body muscles	Lester (1979)
Neometadidymozoon helicis	*Platycephalus fuscus*	Cysts in walls, buccal cavity	Lester (1979)
Allodidymozoon cylindricum	i.e. *Sphyraena obtusata* and *S. picuda*	Stomach wall, buccal epithelium and maxilla	Madhavi (1982)
A. operculare	*Sphyraena obtusata* and *S. picuda*	Opercular muscles	Madhavi (1982)
Didymozoon wedli	*Auxis thazard*	Buccal epithelium	Madhavi (1982)
Platocystoides polyaster	*Sphyraena obtusata* and *S. picuda*	Scales	Madhavi (1982)
Neometadidymozoon polymorphis	*Priacanthus harmur*	Inner side of operculum	Madhavi (1982)
Indodidymozoon platycephali	*Platycephalus scaber*	Opercular muscles	Madhavi (1982)
Lobatocystis vaito	*Euthynnus affinis*	Gills	Madhavi (1982)
Metadidymozoon branchiale	*Xiphias gladius*	Gills	Madhavi (1982)
Allonematobothrium epinepheli	*Epinephelus tauvina*	Opercular muscles	Madhavi (1982)
Neolampridіdymozoon tenuicole	*Lampris guttatus*	Gill, operculum and muscles	Wierzbicka (1980)
Metadidymocystis cymbiformis	*Lampris guttatus*	Subcutaneous muscles	Wierzbicka (1980)

Table 3.3. Some Larval Digenea of the Skin.
Knowledge of larval digeneans of the body surface of fish is scanty. Often the degree of host- and site-specificity involved is far from clear, largely because of the confusion about the taxonomy of various species. Those examples listed in the following table are amongst those worthy of further research and some are of particular interest because they invade more deeply situated tissues than the skin.

Parasite	Host	Site	Reference
Neodiplostomum	*Atherina presbyter* and *Trisopterus luscus*	Skin	Bamber, Glover, Henderson and Turnpenny (1983); Bamber and Henderson (1985)
Posthodiplostomum	*Phoxinus phoxinus* and *Squalius cephalus*	Skin	Dönges (1964)
Allonematobothrium ghanensis	*Epinephelus aeneus*	Encysted in sub-epidermal tissue	Fischthal and Thomas (1968)
Uvulifer ambloplitis	*Lepomis* spp. (6 species) *Micropterus salmoides Pomoxis annularis P. nigromaculatus* and *Perca flavescens*	Skin	Lemly and Esch (1985)
Stephanostomum baccatum	*Platichthys flesus, Solea solea*	Encysted in fins and muscles	Mackenzie and Liversidge (1975)
S. baccatum	*Pseudopleuronectes americanus* and other flatfish	Fins and skin and subcutaneous in muscles	Wolfgang (1954)
S. baccatum	*Pleuronectes platessa Limanda*	Encysted in fins and muscles	Manter and Pritchard (1969)
Clinostomum spp.	*Tilapia nilotica, Auchenoglanis* sp. *Chrysichthys walkeri, C. nigrodigitatus, Clarias* sp.	Skin, outside wall of intestine and gills	Manter and Pritchard (1969)
Cryptocotyle lingua	*Tautogolabrus adspersus*	Skin and various tissues	Sekhar and Threlfall (1970b)
Stephanochasmus baccatus	*Pleuronectes platessa, Scophthalmus maximus, Limanda limanda, Glyptocephalus cynoglossus, Microstomus kitt Hippoglossoides platessoides* and (experimental) *Solea solea*	Fins, skin muscles	Sommerville (1981a)
Haplorchis pumilio	*Tilapia* sp. and *Sarotherodon spilurus*	Skin below scales, muscles and gills	Sommerville (1982a,b)
Rhipidocotyle transversale and *R. lintoni*	*Menidia menidia*	Gills, buccal wall, around the eyes and especially bases of fins	Stunkard (1974c, 1976)

Table 3.4. Some Adult Nematoda of the Skin

Parasite	Host	Site	Reference
Philometra americana	*Parophrys vetulus*	Under skin at base of fins	Di Conza and Cooper (1969)
P. rischta	*Alburnus alburnus, Aspius aspius Scardinius erythrophthalmus, Blicca bjoerkna* and *Abramis brama*	Gill surface and head skin	Molnár (1966b)
P. sanguinea	*Carassius carassius*	Fins and body cavity	Molnár (1966b)
Filaria carassius	*Carassius auratus*	Embedded in caudal fin	Ishii (1933)
Philometroides sp.	*Labeo altivelis*	Under skin of opercular region	Khalil (1965)
Thwaitia bagri	*Bagrus bayad*	Under skin near mouth	Khalil (1965)
Phlyctainophora lamnae	*Mustelus mustelus*	In tumours at the base of the fins	de Ruyck and Chabaud (1960)
Philometroides huronensis	*Catostomus commersoni*	Base of fins	Uhazy (1977a,b)

acanthocephalans from elasmobranchs and nematodes from the gall-bladder. Increasing attention has been given to the gastrointestinal tract of vertebrates, especially of birds and mammals (Crompton, 1973; Crompton and Nesheim, 1976; Read, 1950; 1971; Mettrick, 1970; Mettrick and Podesta, 1974), as a habitat for parasites. Their results and reviews are relevant to research on the fish gut. Over 600 references were cited by Mettrick and Podesta in a convincing demonstration that the gastrointestinal tract is the most favoured site for adult digeneans, cestodes, nematodes and acanthocephalans. An understanding of the environment in which these worms live is a prerequisite for understanding their behaviour and their interactions with their hosts.

The establishment of a species at a site includes direct arrival, emigrations in and against the direction of gastrointestinal flow and site-selection (Crompton, 1973). Crompton suggested that accurate descriptions of helminth sites in the gut should include information about '(1) the helminth's linear distribution, (2) the helminth's radial distribution, (3) the length of time which passed between the death of the host and finding the parasites, (4) the time of day when the search was made, (5) the stage of digestion in progress on the death of the host, (6) the season of the year when the parasites were observed, (7) the parasite's reproductive state, and (8) the worm burden and the other species of parasite present in the host's alimentary tract.' Amongst many other important aspects discussed by Crompton were the co-existence of various species of nematodes in the same site, extensions of site in response to intra- and inter-specific reactions and both the emigration and migration of helminths. Further research on sites must consider all these features as well as the response of the host to a parasite. Any response by the parasite to digestive and absorptive functions is known to induce a counter-response by the host of one form or another. The whole interaction is in a continual state of dynamic equilibrium. When this is irrevocably upset, death of the parasite, host or both results (Mettrick and Podesta, 1974).

The distribution and survival of some species of helminths in the gut of vertebrates may be related to phases of digestive activity, or inactivity in a manner analogous to the dependence of inhabitants of the littoral zone on the tides (Crompton, 1973). It is also known that the process of digestion and the nature of the diet produce different conditions in different parts of the tract and that these may cause migrations, ie. movement from one site to another with return to the original location (Read, 1950, 1971; Smyth, 1969; Crompton, 1969, 1970a, 1973; Crompton and Nesheim, 1976; MacKenzie and Gibson, 1970; Mettrick and Podesta 1974; Williams, McVicar and Ralph, 1970).

The major divisions of the alimentary canal of fish are the mouth, buccal cavity, pharynx, oesophagus, stomach, intestine, rectum and related organs. In terms of number of species, fish are a dominant vertebrate group and have adopted many nutritional styles. Piscivores, insectivores, molluscivores, large plant feeders (herbivores), phyto- and zoo-planktivores, mud feeders (detrivores), cleaner fish and, especially in the specialized (primitive?) Cyclostomata, parasites and feeders on carcasses, can be distinguished amongst fish. Some species are extremely specialized in their feeding habits, while others are omnivores. Like other animals fish have specific amino acid, lipid, carbohydrate, vitamin and inorganic ion requirements and are structurally adapted to obtain these requirements from their food. Their structural adaptations include different elaborations (or losses) of parts of the alimentary canal and the consequent variety of these morphological adaptations might explain why they harbour such a wide variety of worm parasites, especially in their gastrointestinal tracts. It should be possible to test this speculation by comparing current knowledge of the alimentary canal in well-known genera, e.g. *Esox, Poecilia, Gasterosteus, Gambusia, Gadus, Polyodon, Raja*, and in various salmonid and cyprinid fish as described, for instance, by Brett (1979), with what is known about the variety of worms already described from these fish.

Living agnathans, lampreys and hagfishes, have a mouth surrounded by a sucker which leads into a large buccal cavity, communicating through a dorsal opening with the oesophagus which leads directly into a midgut, no true stomach being present. The surface area of the gut is greatly increased by a spiral typhlosole and the gut is associated with the liver, gall-bladder and bile-duct. The walls of the midgut have cells resembling both the digestive and the endocrine components of the typical vertebrate pancreas. The seven species of lamprey harbour 59 species of parasites most of which live in the gut, including 15 species of cestodes, 20 digeneans, 9 nematodes and 5 acanthocephalans (Wilson and Ronald, 1967), but the precise locations have not been studied.

Living holocephali (chimaeroids) have a muscular oesophagus, bile-duct and gall-bladder, distinct pancreas and spiral valve within a short gut leading into a cloaca, but have no stomach. Although there are only about 25 known species of this fish group, little is known of their helminths. *Chimaera monstrosa* harbours *Gyrocotyle urna* or perhaps a complex of *Gyrocotyle* spp. in the intestine, *Calicotyle affinis* in the cloaca, *Plagioporus minutus* in the intestine and *Taeniocotyle elegans* in the gall bladder, (Williams, Colin and Halvorsen, 1987; Dienske, 1968). The gastrointestinal tract of *C. monstrosa* is shown in Figures 3.6–3.12.

Of the sharks and rays which belong to the class Selachii, the gut of *Raja* spp. is typical of the group (Figures 3.13–3.17) The helminth fauna of the gut *Raja* spp. is also typical of the group and has been described by McVicar (1972; 1976; 1977a) and Williams, McVicar and Ralph (1970).

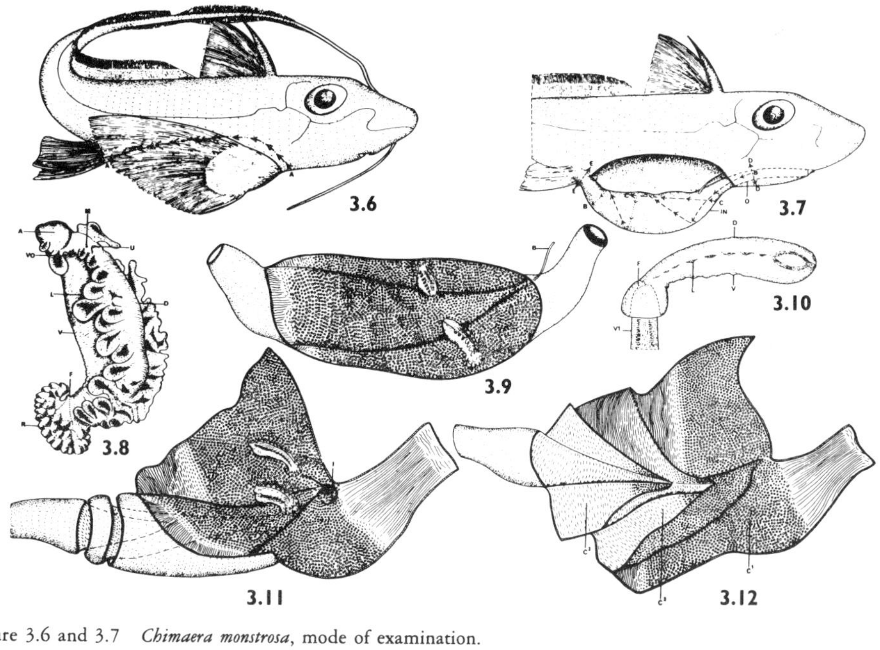

Figure 3.6 and 3.7 *Chimaera monstrosa*, mode of examination.
Figure 3.8 Result of incision BB and orientation of *Gyrocotyle*.
Figure 3.9 Result of incision CC and orientation of *Gyrocotyle*.
Figure 3.10 Result of incision FF, distribution of ridges (broken and continuous lines) and villi (circles) in mucosa.
Figure 3.11 Adult *Gyrocotyle*, external features and orientation, i.e. funnel opening is nearest to mucosa.
Figure 3.12 Young *Gyrocotyle*, external features and orientation, i.e. funnel opening is away from mucosa.

Key to lettering

A. acetabulum
B. bile duct
C^1-C^3. first, second and third tiers of spiral valve intestine
E. dorsal
F. funnel opening
I. imprint of rosette
IN. intestine
L. lateral frills
M. male genital opening
O. oesophagus
R. rosette
U. uterine pore
V. ventral
VI. villus
VO. vaginal opening

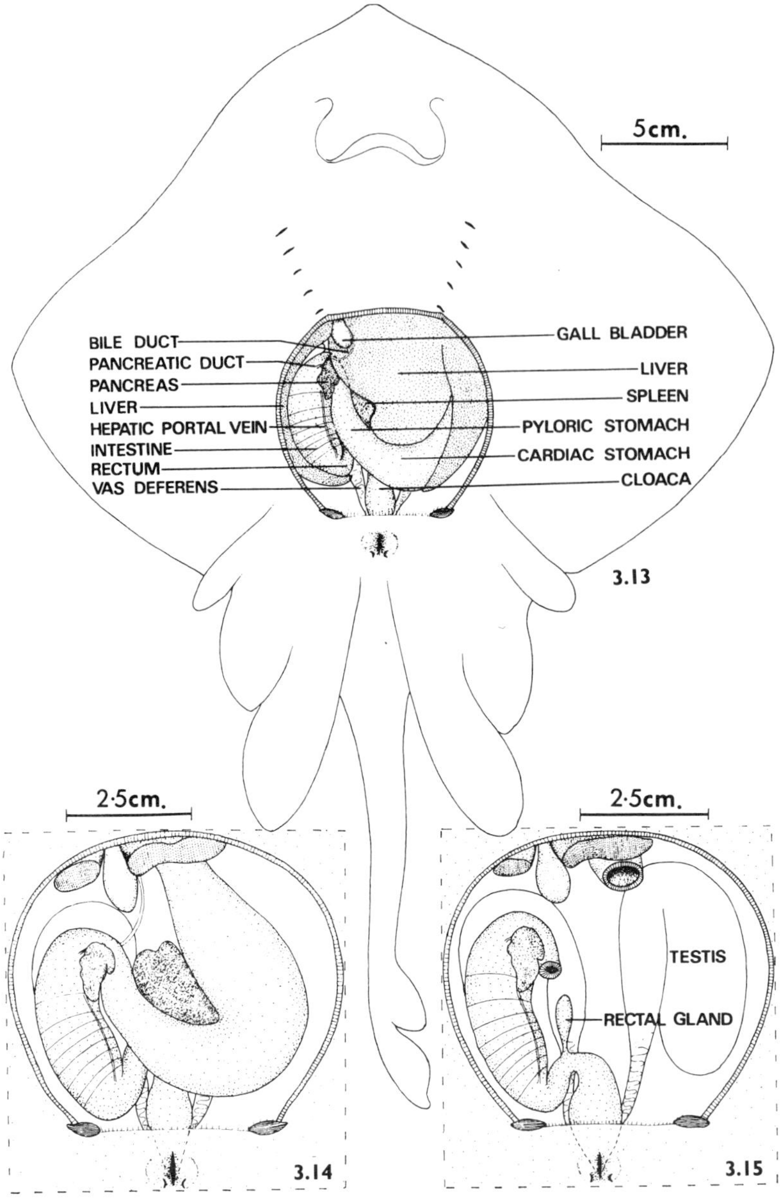

Figures 3.13, 14, 15 *Raja radiata*: the orientation of the alimentary canal and its associated glands in the body cavity: 13 on opening the body cavity; 14, after the lobes of the liver have been removed; 15 after the stomach and spleen have been removed.

More variation is seen in the alimentary canal of teleosts than in any other group of fish (Crompton, 1973). Some, e.g. *Catostomus catostomus*, have no stomach and in others such as *Rutilus rutilus* the first limb of the intestine becomes distended and acts as a stomach. Kapoor, Smith and Verighina (1975) review the alimentary canal and digestion in teleosts and cite about 600 references on food and feeding, morphology, histology and cytology, digestive enzymes, regulation of gastric secretion and absorption and conversion of food.

The alimentary canal of the cod illustrates some general features of the teleost gut (Figures 3.18–3.20). The short wide oesophagus is not well-differentiated externally from the poster-

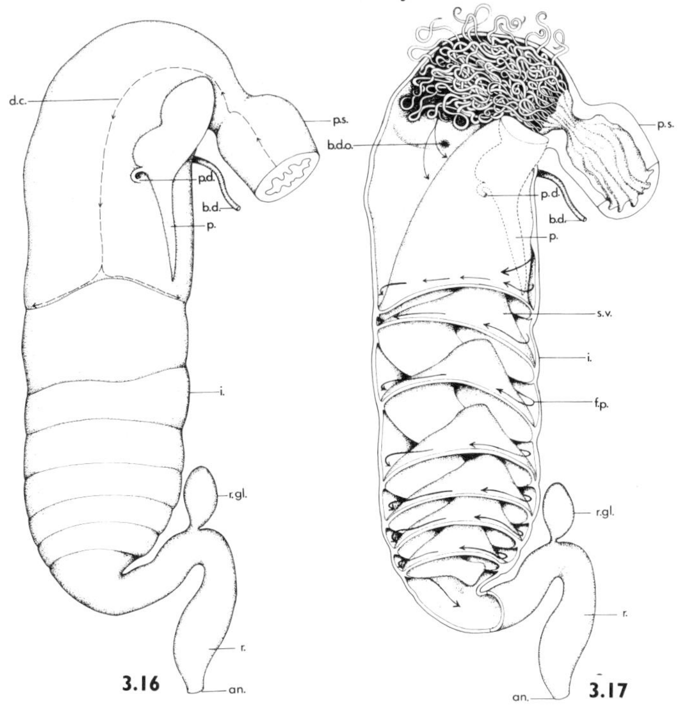

Figures 3.16, 17 *Raja fyllae.*—16. Intestine, external view.—17. Intestine, dissected to show attachment of *Proleptus mackenziei.*

iorly-directed cardiac stomach which is usually dilated with food. The cardiac limb curves into the forwardly-directed pyloric limb which narrows to join the intestine. About 300 pyloric caeca enter the anterior end of the intestine by either five or six openings. The bile duct opens into the intestine dorsally, to the left of and slightly behind the left dorsolateral group of pyloric caeca. Bishop and Odense (1966) state that the pancreatic ducts open into the intestine near the bile-duct; posterior to the opening of the bile-duct, the intestine loops as illustrated in Figures 3.18, 3.19 and finally passes downwards and backward into the rectum (Williams and Halvorsen, 1971; Williams, McVicar and Ralph, 1970). The sites occupied by some common helminths of the gut of cod are given below (Table 3.5).

Table 3.5 Some Common Helminths of the Gut of *Gadus Morhua* in Scottish Waters

	Species	Site
DIGENEA	*Podocotyle atomon*	Pyloric caeca to rectum
	Lepidapedon rachion	Mid-gut to rectum
	L. microcotyleum	Pyloric caeca and foregut
	L. elongatum	Pyloric caeca
	Stephanostomum pristis	Pyloric caeca
	Hemiurus communis	Stomach
	Lecithaster gibbosus	Fore-gut and mid-gut
	Derogenes varicus	Oesophagus and stomach
CESTODA	*Abothrium gadi*	Pyloric caeca and foregut
NEMATODA	*Contracaecum aduncum*	Mostly in stomach pyloric caeca and fore-gut
	Ascarophis morrhase	Gills, stomach and pyloric caeca and fore-gut
	Cucullanus cirratus	Mostly in fore-gut and mid-gut
	Capillaria sp.	Hind-gut and rectum
ACANTHOCEPHALA	*Echinorhynchus gadi*	Mostly in fore-gut and mid-gut

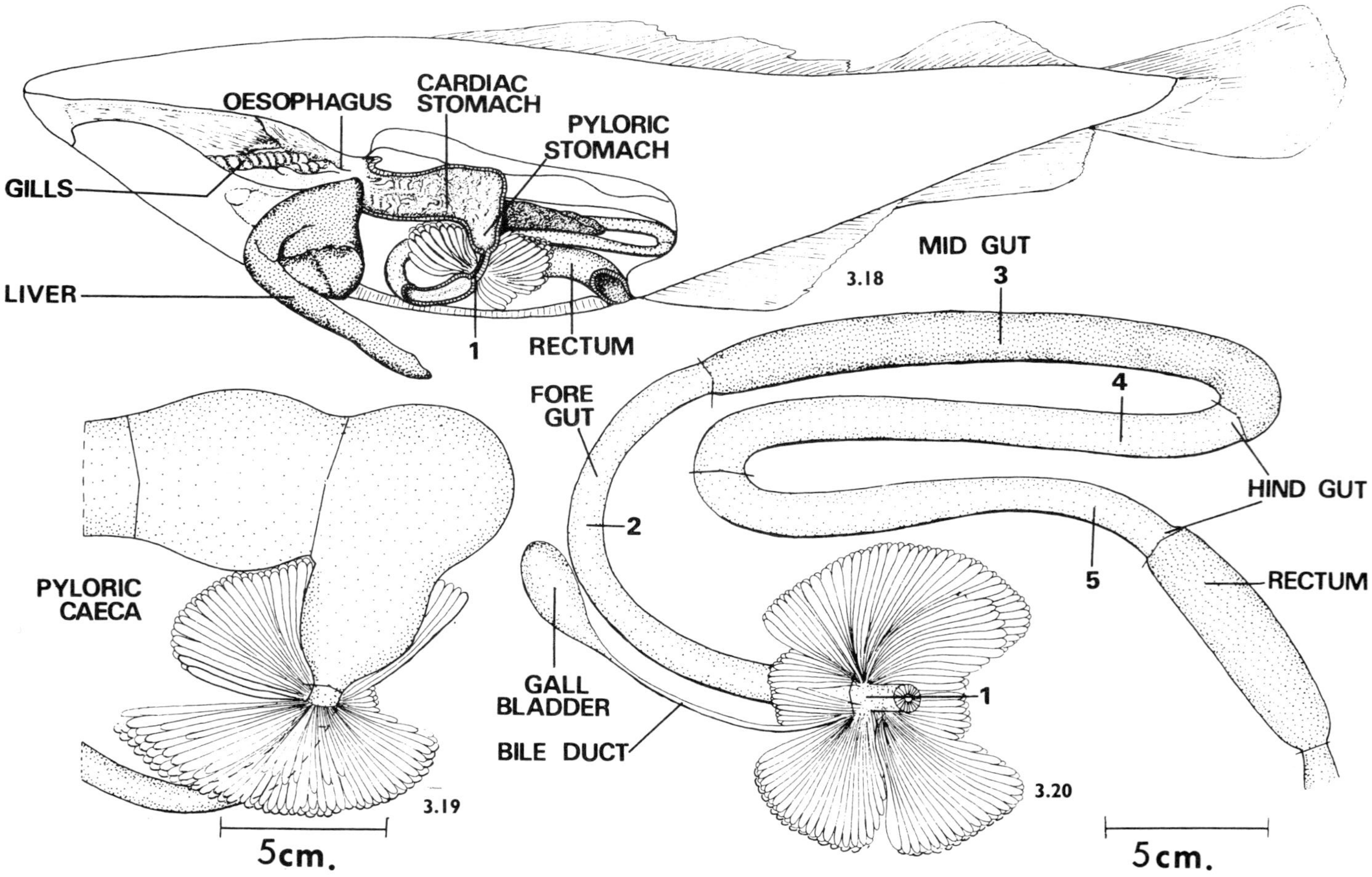

Figures 3.18–3.20 The gastrointestinal tract in the teleost *Gadus morhua*; 3.18 Diagrammatic representation of the gastrointestinal tract as interpreted from median vertical sections of the whole fish; 3.19 The stomach; 3.20 The intestine showing the regions referred to in the physiological study by Williams, McVicar and Ralph (1970).

Table 3.6. Some Monogenea Of The Alimentary Canal

Species	Host	Site	Reference
Austrocalicotyle spp.	*Raja* spp.	Cloaca	Suriano (1977)
Clemacotyle australis	*Aetobatis narinari*	Skin of branchial cavity	Young (1967)
Calicotyle kroyeri	*Raja naevus*, *Raja* spp.	Cloaca	Kearn (1970a, 1987)
*Calicotyle palombi**	*Mustelus mustelus*	Rectum and cloaca	Euzet and Williams (1960)
*Calicotyle stossichi**	*Mustelus mustelus* and *M. canis Raja* spp	Rectal gland and cloaca	Euzet and Williams (1960) Kearn (1987)

* Kearn (1987) suggested that these are morphological forms of the same species.

Table 3.7. Some Interesting or Unusual Cestoda of the Alimentary Canal

Species	Host	Site	Reference
Abothrium gadi	*Gadus morrhua*	Pyloric caeca	Williams and Halvorsen (1971)
Bothriocephalus ganapatti	*Saurida tumbil*	Strobila in intestine but scolex embedded in liver	Rao (1954)
B. scorpii	*Pseudophycis baccus*	Anterior intestine	McKinnon and Featherston (1982)
Capingens singularis	Catostomid fish	Stomach	Yamaguti (1959)
Cyathocephalus truncatus	*Salmo trutta*	Anterior caeca	Halvorsen and Macdonald (1972)
Cyathocephalus truncatus	*Gasterosteus aculeatus*	Anterior intestine	Reimchen (1982)
Diphyllobothrium vogeli (plerocercoid)	*Gasterosteus aculeatus*	Preference for, but not restricted to, liver	Vik, Halvorsen and Andersen (1969)
Dilepis unilateralis (cysticercoids)	*Cyprinus carpio*	Gall-bladder	Safonov and Vasil'kov (1971)
Echinobothrium harfordi	*Raja naevus*	First two tiers of spiral intestine	McVicar (1976, 1977a)
Glaridacris laruei	*Catostomus commersoni*	Anterior or posterior intestine depending on presence of other helminths	Grey and Hayunga (1980)
Gryporhynchus cheilanocristrotus (larval stage)	*Cyprinus carpio*	Gall-bladder	Andakulova (1976)
Lytocestus indicus	*Clarias batrachus*	Anterior intestine	Bose and Sinha (1980)
Mastacembellophyllaeus nandedensis	*Mastacembelus armatus*	Intestine	Shinde and Chincholikar (1977)
Parabothrium gadi-pollachii and *Abothrium gadi*	*Gadus* spp.	Strobila in intestine and scolex embedded in gut wall	Williams (1960c)
Paradilepis simoni (larvae)	*Oncorhynchus nerka*, *Salmo gairdneri*, *Ptychocheilus oregonensis*, *Prosopium williamsoni* and *Cottus asper*	Encysted in liver	Ching (1982)
Polyonchobothrium clarias (larvae)	*Clarias mossambicus*	Gall-bladder	Wabuke-Bunoti (1980); Amin (1978)
Proteocephalus filicollis	*Gasterosteus aculeatus*, *Triaenophorus* sp. *Silurus triostegus*	Anterior intestine	Chappell (1969a,b) Shamsuddin *et al.* (1971)
Triaenophorus crassus (plerocercoids)	*Salmo gairdneri*	Liver	Bauer and Solomatova (1984)

Comprehensive illustrated accounts of the gut helminths of *Platichthys flesus* and of *Anarhichas lupus* are given by Mackenzie and Gibson (1970) and Bray (1987) respectively (Figures 3.21–3.24). All major groups of helminths have representatives in the alimentary canal of fish, and adult tapeworms and acanthocephalans are not found in any other body system. The number of species recorded from the alimentary canal are too numerous to mention. Tables 3.5–3.12 therefore, list only a few, selected either because useful informa-

Table 3.8. Some Adult Digenea of the Alimentary Canal

Species	*Host*	*Site*	*Reference*
Asymphylodora kubanicum	*Rutilus rutilus*	Anterior intestine	Evans (1977)
Crepidostomum metoecus	*Salmo trutta*	Anterior pyloric caeca	Halvorsen and McDonald (1972)
C. Metoecus	*Salmo salar* and *S. trutta*	Pyloric caeca	Conneely and McCarthy (1984)
Deretrema ovale	*Myripristis violaceus*	Gall-bladder	Machida (1984a)
D. philippe	*Galaxias divergens*	Gall-bladder	Hine (1977b)
D. philippinensis	*Anomalops katoptron*	Gall-bladder	Beverley-Burton and Early (1982)
Derogenes varicus	*Merlangius merlangus*	Stomach	Shotter (1976)
Dimerosaccus onchorhynchi	*Oncorhynchus rhodurus*	Pyloric caeca	Shimazu (1980)
Hemiurus communis	*Merlangius merlangus*	Stomach	Shotter (1976)
Lepocreadium oyabitcha	*Abudefduf vaigiensis*	Gall-bladder	Machida (1984a)
Lecithaster gibbosus	*Merlangius merlangus*	Anterior intestine and caeca	Shotter (1976)
Nematobothrium spinneri	*Platycephalus fuscus*	Buccal cavity	Lester (1979)
Opisthorchis piscicola	*Gymnarchus niloticus*	Gall-bladder	Fischthal and Thomas (1972)
Podocotyle atomon	*Merlangius merlangus*	Anterior intestine and caeca	Shotter (1976)
Pseudochetosoma salmonicola salmonicola	*Acanthobrama marmid*	Gall-bladder	Kassim *et al.* (1977)
Sphaerostoma bramae	*Rutilus rutilus*	Different sites in intestine	Evans (1977)
Stephanostomum pristis	*Merlangus merlangus*	Anterior intestine and caeca	Shotter (1976)
Steringophorus pritchardae	*Alepocephalus agassizi*	Pyloric caeca	Campbell (1975a)

Table 3.9. Some Larval Digenea of the Alimentary Canal

Species	*Host*	*Site*	*Reference*
Ascocotyle gemina	*Cyprinodon varoegatis*	Visceral organs, mesentery, nervous tissue, cranial cavity, gills, pseudobranch and musculature	Font, Heard and Overstreet (1984)
Bucephalus haimeanus	*Pomatoschistus microps*	Liver	Higgins, Wright and Matthews (1977)
Euclinostomum sp.	*Macropleurodus* and *Haplochromis* sp.	Peritoneal wall	Khalil and Thurston (1973)
Phagicola nana	*Micropterus salmonoides, Lepomis microlophus, L. macrochirus* and *L. humilis*	Pyloric caeca, intestine, pancreatic tissue, fat bodies, mesentery, visceral and parietal peritoneum, liver, spleen, bulbus anteriosus, ventricle, somatic musculature, gills	Font, Overstreet and Heard (1984)
Pseudascocotyle mollieniesicola	*Mollieniesia latipinna*	Encysted on wall and surface of intestine, body musculature and gill branchiae	Sogandares-Bernal and Bridgman (1960)

Table 3.10 Some Aspidogastrea of the Alimentary Canal

Species	*Host*	*Site*	*Reference*
Multicalyx cristata	*Rhinoptera bonasus Myliobatis freminvillei*	Gall bladder	Thoney and Burreson (1987)
Rohdella siamensis	*Osteochilus melanopleurus, Barbus daruphani*	Intestine	Gibson and Chinabut (1984)
Rugogaster hydrolagi	*Hydrolagus colliei*	Rectal gland	Schell (1973)
Taeniocotyle elegans	*Chimaera monstrosa*	Gall-bladder	Dienske (1968)

Table 3.11. Some Nematoda Of The Alimentary Canal.

Species	*Host*	*Site*	*Reference*
Contracaecum clavatum	*Merlangius merlangus*	Unencapsulated forms in stomach and anterior intestine. Encapsulated forms in liver	Shotter (1976)
Thwaitia macronesi	*Macrones seenghala*	Buccal cavity	Shendge and Deshmukh (1977)
Raphidascaris acus (larvae)	*Perca flavescens*	Liver	Poole and Dick (1984)
Pancreatonema torriensis	*Raja naevus*	Pancreatic duct	McVicar and Gibson (1975)

Table 3.12. Some Acanthocephala of the Alimentary Canal

Species	*Host*	*Site*	*Reference*
Echinorhynchus salmonis	*Coregonus lavaretus*	Various sites in intestine	Valtonen (1983b)
Neochinorhynchus cylindratus	*Huro salmoides*	Small intestine	Venard and Warfel (1947)
Neoechinorhynchus rutili	*Gasterosteus aculeatus*	Rectum	Chappell (1969a,b)
Pomphorhynchus laevis	*Leuciscus cephalus, Barbus vulgaris, Salmo gairdneri, L. vulgaris, Carassius auratus*	Survives in all regions of intestine but selects different regions according to host species	Kennedy, Broughton and Hine (1976)
P. bulbocolli	*Catostomus commersoni*	Posterior half of intestine	Grey and Hayunga (1980)
Leptorhynchoides thecatus	*Huro salmoides*	Pyloric caeca	Venard and Warfel (1947)
Acanthocephalus lucii	*Perca fluviatilis*	Pyloric caeca	Lee (1981)
Acanthocephalus clavula	*Anguilla anguilla*	All regions of intestine	Kennedy and Lord (1982)

tion is already available on their preferred sites, or because they belong to helminth groups in which most species locate in other systems. Some Acanthocephala of the alimentary canal which have been the subject of intensive ecological studies are discussed on pages 167–168, 177, 179 and 192.

With reference to the last-named species in Table 3.12, Kennedy and Lord (1982) investigated the site-specificity of *A. clavula* in *Anguilla anguilla* both in natural infections and in fish maintained experimentally under different regimes. The fish alimentary canals from the oesophagus (0 per cent) to the anus (100 per cent) were removed, slit longitudinally and searched. The acanthocephalan showed a preference for the 55–65 per cent region of the alimentary canal and this did not differ in different species of host. This site-specificity, however, was not very precise and there was considerable variation between individual hosts, and *A. clavula* was capable of surviving and maturing in all regions of the intestine. The mean position of male parasites was slightly anterior to that of females and the overall sex ratio was in favour of females. At high levels of infection the range of the parasite was extended and the mean position was significantly more anterior when compared with single-worm infections. The mean position did not change when the host was starved or maintained in 100 per cent sea water, or with reduced aeration or at high temperatures. The results were compared with site-specificity shown by other species of freshwater fish acanthocephalans.

Lee (1981) presented data and evidence to show antagonism between *Acanthocephalus lucii*, *Proteocephalus percae* and *Camallanus lacustris* in the intestine of *Perca fluviatilis* from the Serpentine, central London. *A. lucii* was rare in this lake compared with other locations in Britain, and this was attributed to the rarity of the intermediate host, *Asellus aquaticus*. The restriction of *Acanthocephalus lucii* to the third quarter of the intestine was due either to the worm actively seeking this part of the intestine or to the large numbers of *Proteocephalus percae* and *Camallanus lacustris* excluding it from the more anterior sites that are said to be more favourable to *Acanthocephalus* spp. The mutual exclusion theory was favoured.

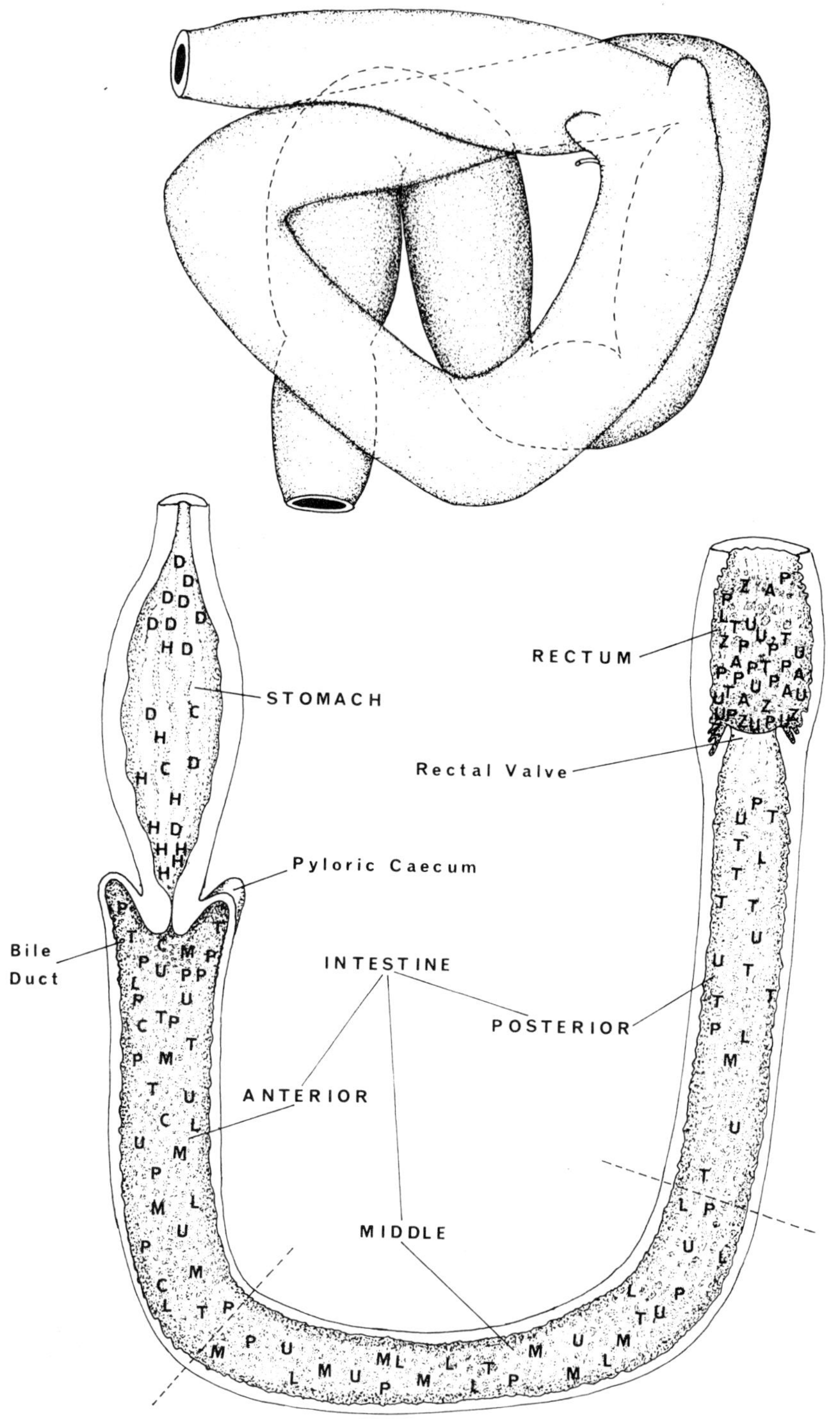

Figure 3.21 The flounder gut, as it appears *in situ*, and opened to show the regional divisions and parasite distributions: A, *Pomphorhynchus* sp.; C, *Contracaecum aduncum*; D, *Derogenes varicus*; H, *Hemiurus communis*; L, *Lecithaster gibbosus*; M, *Cucullanus minutus; P, Podocotyle* sp.; T, Tetraphyllidean larvae; U, *Cucullanus heterochrous*; Z, *Zoogonoides viviparus*.

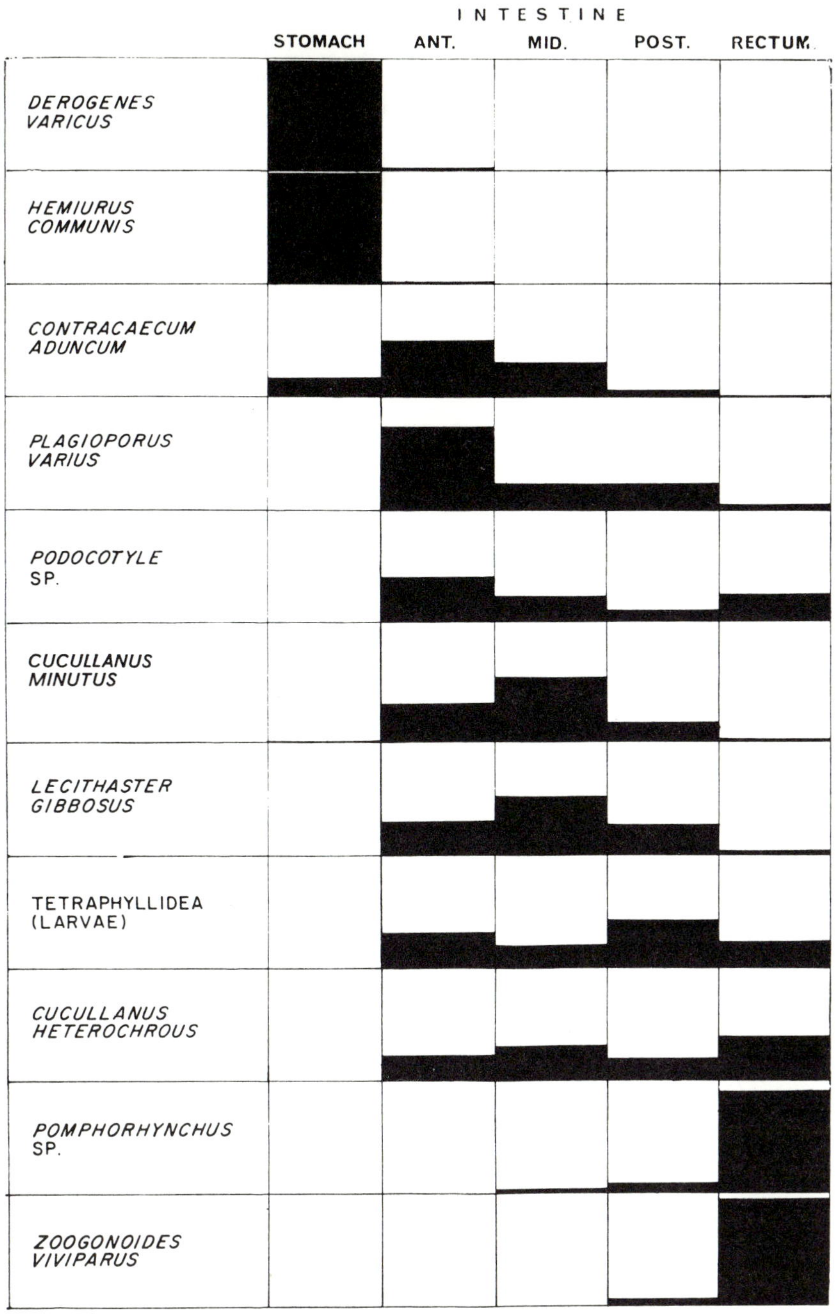

Figure 3.22 Percentage distribution of parasites in the flounder gut. Figures for *Plagioporus varius* are from plaice.

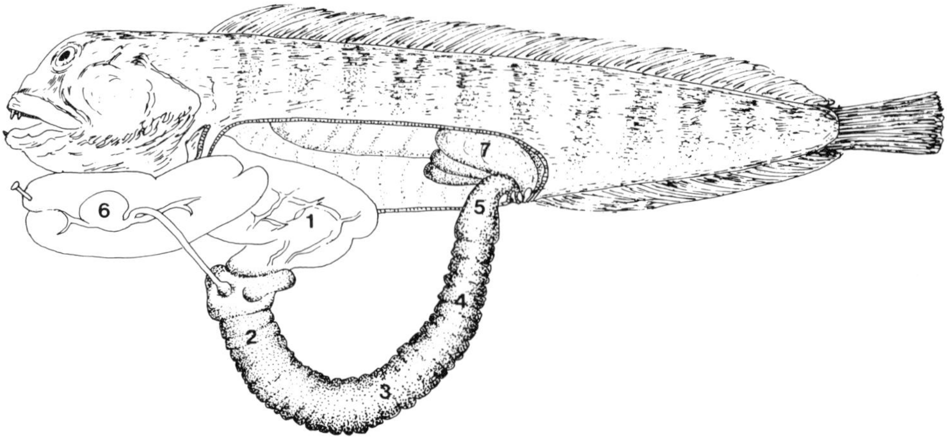

Figure 3.23 Drawing of *Anarrhichas lupus* with gastrointestinal tract etc. exposed, and regions sampled for helminth-parasites number 1–7.

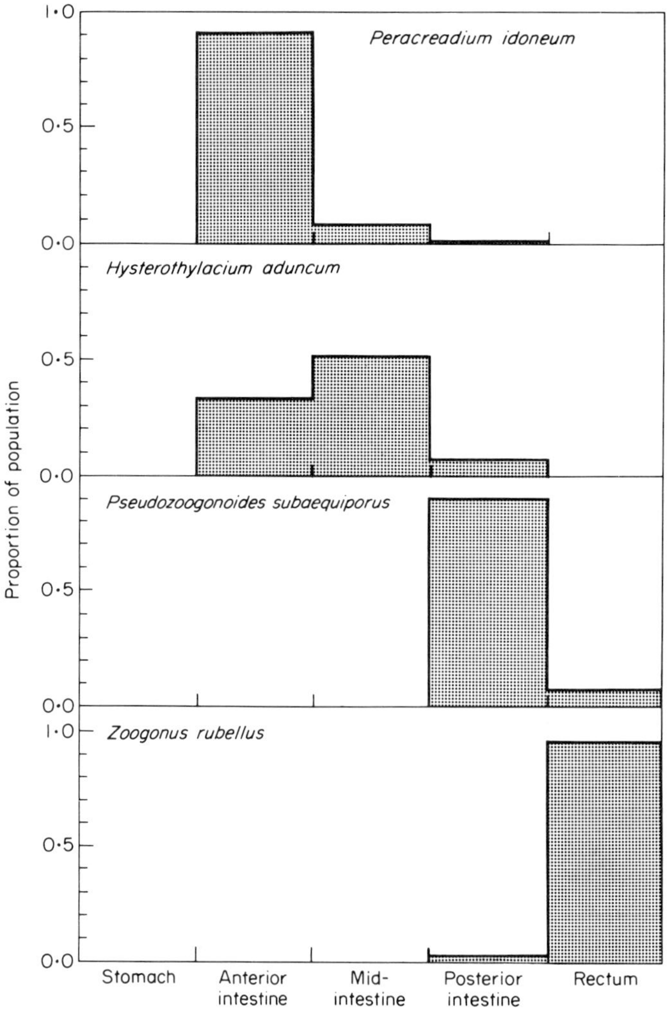

Figure 3.24 Distribution of the four common helminths species from the gastrointestinal tract of *Anarrhichas lupus*.

Muscles

Despommier (1976) discussed the suitability of mammalian muscle as a site for parasites. He thought muscle was an unsuitable site for most parasites because of its variable physical, chemical and physiological environment. He suggested that in a balanced relationship between host and parasite both specific (e.g. immunological) and non-specific (e.g. inflammatory) host responses serve to control the amount of damage the host incurs through infection. Similarly, the parasite must find new ways of avoiding host responses. The following account will indicate that an abundance of infected fish muscle should be easily available to research Despommier's views.

Well over half of the total body mass in most species of fish is composed of locomotor muscles. Other muscles in fish include those connected with the gills, alimentary canal, eyes, swim bladder, heart, electric organs and barbels, which are concerned with coordination and control. The larval stages of many cestodes, digeneans and nematodes localize in fish muscles, but little is known of their preference for and effects on different types of muscle. Beamish (1978, p. 144) states that *Bolbophorus confusu, Cotylurus erraticus, Crepidostomum farionis, Nanophyetus salmincola, Proteocephalus* sp. and *Eubothrium salvelini* have been studied with regard to their effect on reducing swimming performance. Metabolic changes associated with fish movement, however, are with few exceptions, little understood. Tunas and lamnid sharks maintain body temperatures well above that of the water and, like the Atlantic salmon, *Salmo salar*, are capable of very long-distance swims. Swimming capacity is directly linked to particular metabolic capacities to convert chemical energy into propulsive movement through muscular contraction. In this context tunas are unique in being the fastest swimmers, amongst the largest of fish, having the warmest of bodies, the highest metabolic rates and are exceptional in many biochemical respects, for example, having unique haemoglobins and very high lactate, glycogen and actinogen concentrations. They are so extraordinary that the tuna researcher often divides fishes into 'tunas' and 'non-tunas'. It is not known whether such a view is reflected in the helminth fauna of tunas.

Bearing in mind these interesting aspects of fish movement, the extent to which larval worms migrate to and from fish muscles is of particular interest. Smith (1984a,b) has shown that larval *Anisakis simplex* undertake a post-mortem migration from the body cavity into the flesh of some teleost fish (herring, mackerel) but not of others (blue whiting, *Micromesistius poutassou*, walleye pollack, *Theragra chalcogramma)*. Although the reasons for this difference are not clear it may be related to the fat content of the fish species. In freshly-caught teleost fish, larval *Anisakis* occupy either, or both, of two broad microhabitats, that is, the body cavity and the muscles. The relative abundance of larval *Anisakis* in these microhabitats varies with host species (Smith and Wootten, 1978a). In one group of teleosts (*Clupea harengus, Micromesistius poutassou, Stizostedion vitreum, Pollachius pollachius* and *Scomber scombrus*) the proportion of the total worm burden found in the flesh at capture is relatively low. In contrast, in a second group of teleosts, including whiting and cod, a high or low proportion of the total worm burden may occur in the muscles, depending on the species and age of the individual host examined.

The possible advantages to larval strigeids of attacking vulnerable muscle sites and weakening or killing their hosts in order to facilitate completion of the life-cycle is implied in the work of Swennen, Heessen and Hocker (1979). They investigated three species of *Cotylurus* and found *C. erraticus* in Salmoniformes, especially in the pericardial cavity, *C. variegatus* in Percidae, mainly on the wall of the swim bladder, and *C. platycephalus* in Cyprinidae and in Percidae, mainly in the pericardial cavity and on the mesentery. Since cercariae of *Cotylurus* spp. released by infected snails can live for only a day, fish acquire infections only in waters where infected snails occur. This accounts for the absence of infected fish in some areas. Although encysted metacercariae were found in many species of fish, the highest infections occurred in *Osmerus eperlanus* and *Gymnocephalus cernua*. Both species are relatively small and extraordinarily numerous and, therefore, are of great importance in continuing the life-cycle. Species which grow to sizes too large to be swallowed by gulls were considered as ecological traps rather than efficient intermediate hosts of *Cotylurus* spp. The occasional death of large numbers of *O. eperlanus* and *G. cernua* in nature was attributed to *Cotylurus* spp. cercariae. As will be seen later similar advantages may be conferred by strigeids which attack the central nervous system and sense organs of fish.

Table 3.13. Some Larval Tapeworms in Muscle

Species	*Host*	*Site*	*Reference*
Dilepidid larvae	*Tilapia nilotica, T. zilli, Hemihaplochromis multicolor* and *Haplochromis* (6 spp.)	Intestinal wall	Khalil and Thurston (1973)
Diphyllobothrium latum	*Stizostedion vitreum* and *Esox lucius*	Muscle	Dick and Poole (1985)
Nybelinia surmenicola	*Theragra chalcogramma*	Muscles	Arthur, Margolis, Whitaker and McDonald (1982)
Poecilancistrium caryophyllum	*Cynoscion nebulosus*	Muscles	Overstreet (1977)
Pseudophyllidean plerocercoids	*Theragra chalcogramma*	Muscles	Arthur, Margolis. Whitaker and McDonald (1982)
Pseudogrillotia sp. *Tentacularia* sp. and other genera of trypanorhynch plerocerci	*Seriola dumerili, Katsuwonus pelamis,* and other commercially important fish	Muscles of the head and body muscles adjacent to the body cavity	Deardorff, Raybourne and Mattis (1984)
Triaenophorus crassus	*Coregonus clupeaformis*	Muscles	Rosen and Dick (1983)
**T. crassus*	*Salmo gairdneri* and *Oncorhynchus kisutch*	Body cavity and muscles	
Trypanorhynch plerocercoids	*Xiphias gladius*		Muzykovskii (1972)

* *T. crassus* always use muscles and reports of *Triaenophorus* from viscera are *T. nodulosus* or *T. stizostedionis* (J. Holmes, pers. comm.)

Table 3.14. Some Larval Digenea of Muscle

Species	*Host*	*Site*	*Reference*
Apophallus brevis	*Perca flavescens*	Eyes, back muscles close to dorsal fins	Poole and Dick (1983)
Clinostomum sp.	*Haplochromis obliquidens*	Jaw muscles	Khalil and Thurston (1973)
Diplostomulum sp.	*Gasterosteus aculeatus*	Connective tissue muscle and eye	Erasmus (1958, 1959)
Heterophyes heterophyes	*Mugil cephalus*	Muscles beneath pectoral fin	Seo, Cho, Chai and Hong (1980)
Metanematobothrioides branchialis	*Pristipomoides typus*	Muscles of branchial region	Madhavi (1982)
Neodiplostomum sp.	*Heteropneustes fossilis*	Muscles	Herzog (1969)
Otodistomum veliporum	*Chimaera monstrosa*	Muscles	Dienske (1968)
Renicola spp.	*Clupea harengus, Engraulis encrasicholus*	Muscles	MacKenzie (1974, 1975a, 1985)
Stephanostomum spp.	*Antigonia steindachneri, Caprodon schlegeli* and *Xanthichthys lineo-punctatus*	Muscles	Korotaeva (1975)
Strigeids	*Mugil cephalus*	Muscles	Hutton (1957)
Cotylurus spp	Salmoniformes Percidae and Cyprinidae	Various muscles and body cavities	Swennen, Heessen and Hocker (1979)

Table 3.15. Some Larval Nematoda in Muscle

Species	*Host*	*Habitat*	*Reference*
**Anisakis* sp. and *Pseudoterranaova* sp.	*Merluccius gayi*	Viscera and muscles	Carvajal Cattan, Castillo and Schatte (1979)
**Pseudoterranova decipiens* *Anisakis simplex*	*Theragra chalcogramma*	Muscles	Arthur, Margolis, Whittaker and McDonald (1982)
Anisakis sp.	*Oncorhynchus keta*	Muscles	Novotny and Uzman (1960)

* See Chapter 6 for a number of additional references on *Anisakis* and *Pseudoterranova* spp.

It is against this background of scant knowledge on the preference of worms for the particular kinds of muscle, variation in the preference according to species and age of the individual host, that the following examples in Tables 3.13–3.15 were chosen to indicate potential for further research.

Peritoneal Cavities

The peritoneal cavity is an important habitat for fish worms. Howell (1976) reviewed the peritoneal cavities of vertebrates as a habitat for parasites under the general headings 'physio-chemical properties', 'parasites of the peritoneal cavity' and 'parasitological problems posed by this site'. The suitability of the cavity as a habitat lies in its proximity to the gut and the security from dislodgement that it offers. Conversely, the problems it poses include establishment of infection, transmission to subsequent hosts and/or release of genital products. Developing parasites also have to contend with the difference between this habitat and that of the preceding stage in the life-cycle, together with physico-chemical changes it may induce in the cavity, the immunological hostility of the habitat and the limited availability of space and nutrients. Nevertheless, as can be seen from the following selected examples, a relatively large number of fish parasites have successfully invaded and established themselves in body cavities (Tables 3.16–3.22).

Table 3.16. A monogenean of the Peritoneal Cavity

Species	*Host*	*Reference*
Dictyocotyle coeliaca	*Raja radiata* and *R. naevus*	Hunter and Kille (1950) Kearn (1970a) Williams (1965a)

Table 3.17. Amphilinidea in the Peritoneal Cavity

Species	*Host*	*References*
Amphilina foliacea	Acipenseridae	Yamaguti (1959)
A. japonica	*Acipenser mikadoi*	Goto and Ishii (1936)
A. tengaria	*Mystus tengara*	Rehana and Bilquees (1979)
Gephyrolina paragonopora	Siluroid fish	Yamaguti (1959)
Gyrometra albotaenia	Pristipomid fish	Yamaguti (1959)
Nesolecithus janickii	Arapaimid fish	Yamaguti (1959)

Table 3.18. Some Larval Eucestoda of the Peritoneal Cavity

Species	*Host*	*Site*	*Reference*
Diphyllobothrium dendriticum	*Coregonus lavaretus Salmo trutta*	Exclusively encysted on walls of stomach and oesophagus	Bylund (1972)
D. dendriticum	*Coregonus artedii* and *C. clupeaformis*	Encapsulated on the viscera	Dick and Poole (1985)
D. dendriticum	*Salmo gairdneri*	Alimentary canal wall, liver, gonads, swim-bladder, spleen and mesenteries	Torres *et al.* (1981)
D. dendriticum[1]	*Gasterosteus aculeatus*	Free in peritoneal cavity and encysted in other tissues	Vik, Halvorsen and Andersen (1969) Halvorsen (1970)
D. dendriticum	*Salmo trutta*	Encysted among pyloric caeca	Conneely and McCarthy (1984)
D. ditremum	*Osmerus eperlanus*	Encysted on outside wall of stomach	Bylund (1973)
D. ditremum	*Salmo trutta Salvelinus alpinus*	Encysted among pyloric caeca	Conneely and McCarthy (1984)
D. ditremum	*Esox lucius Gasterosteus aculeatus* and *Pungitius pungitius*	Body cavity	
D. vogeli[2]	*Gasterosteus aculeatus*	Preference for but not restricted to liver	Vik, Halvorsen and Andersen (1969) Halvorsen (1970)
	G. aculeatus and *Pungitius pungitius*	Free in abdomen but preference for liver	Bylund (1975c)
Diphyllobothrium sp.	*Micromesistius poutassou*	Encysted in or outside stomach wall	Andersen (1977)
Grillotia dollfusi[3]	*Merluccius gayi*	On gonads and serosal surface of instestine	Carvajal and Cattan (1978)
G. erinaceus	*Hippoglossus hippoglossus*	Peritoneal cavity on viscera and muscles	Rae (1958)
Ligula intestinalis	*Catostomus commersoni*	Body cavity	Black and Fraser (1984)
"	Many host species listed	"	Orr (1967, 1968a, b)
"	Several species of cyprinid fish	"	Arme (1968) Arme and Owen (1968, 1970)
"	*Engraulicypris argenteus* and *Haplochromis* sp	"	Khalil and Thurston (1973)
"	*Abramis brama*	"	Conneely and McCarthy (1984)
Proteocephalus ambloplitis	*Micropterus dolomieui*	Viscera	Fisher and Freeman (1969, 1973)
Schistocephalus solidus	*Gasterosteus aculeatus*	Body cavity	Conneely and McCarthy (1984); Hopkins and McCaig (1963)
S. solidus	*Gasterosteus aculeatus Pungitius pungitius*	Body cavity	Bråten (1966)

[1] Halvorsen (1970) should be consulted for details of the distribution of *Diphyllobothrium* spp. in the body cavity.
[2] Most authors except for Bylund now regard *D. vogeli* as a synonym of *D. ditremum*.
[3] Bates (1990) has reviewed the literature for the last fifty years on trypanorhynchs and includes a number of references to species which have invaded muscles and body cavities

Table 3.19. Some Adult Digenea of the Peritonel Cavity

Species	*Host*	*Site*	*Reference*
Aphalloides coelomicola	*Gobius microps*	Body cavity	Dollfus, Chabaud and Golvan (1957)
A. coelomicola	*Pomatoschistus microps*	"	Vaes (1978)
A. timmi	*P. microps*	"	Bakke (1980)
Crocodilicola pseudostoma	*Rhamdia hilari*	"	Conroy (1986)
Gonapodasmius microavatus	*Megalaspis cordyla*	"	Reimer (1980)

Table 3.19 *Continued*

Species	*Host*	*Site*	*Reference*
G. spilonotopteri	*Cypselurus comatus*	Connective tissue over swim bladder	Madhavi (1982)
Nagmia floridensis	*Dasyatis sabina* *D. americana*	Body cavity	Brooks and Mattis (1978).
N. pacifica	*Carcharhinus natator*	Peritoneal cavity intestinal mesenteries	Sogandares-Bernal (1959b)
Nagmioides trygonis	*Trygon marmorata*	Body cavity	Dollfus (1958)
Otodistomum plunketi	*Scymnodon plunketi*	Body cavity	Gibson and Bray (1977)
O. hydrolagi	*Hydrolagus colliei*	"	"
Paranematobothrium triplovitellatum	*Scomber scombrus*	"	Nikolaeva and Tkachuk (1979, 1982)
Paurorhynchus hiodontis	*Hiodon tergisus*	"	Dickerman (1954)
P. hiodontis	*Hiodon alosoides*	"	Margolis (1964)
Phyllodistomum symmetrorchis	*Auchenoglanis occidentalis and Bagrus docmac*	"	Thomas (1958), Fischthal and Thomas (1972)
Probolitrema richiardii	*Raja marginata*	"	Gibson (1976)
P. richiardii var. *pteromylaei*	*Pteromylaeus bovina*	Peritoneal cavity on liver	Dollfus (1958)
Pseudocolocyntotrema yaito	*Euthynnus affinis*	Intestinal wall and hepatic caeca	Madhavi (1982)
Staphylorchis scoliodonii	*Scoliodon sorrakowah*	Peritoneal cavity	Mehra (1960)

Table 3.20. Some Larval Digenea of the Peritoneal Cavity

Species	*Host*	*Site*	*Reference*
Euclinostomum sp.	*Macropleurodus bicolor* and *Haplochromis* sp.	Peritoneal wall	Khalil and Thurston (1973)
Otodistomum veliporum	*Chimaera monstrosa*	Encysted on walls of the bile duct and oesophagus	Dienske (1968)
Renicola spp.	*Clupea harengus* and *Engraulis encrasicholus*	Visceral cavity between pyloric caeca	MacKenzie (1974, 1985)

Table 3.21. Some Adult Nematoda of the Peritoneal Cavity

Species	*Host*	*Site*	*Reference*
Camallanus trichiurusi	*T. muticus* and *Arius venosus*	Body cavity	Gupta and Srivastava (1975a)
Hysterothylacium haze	*Acanthogobius flavimanus*	Body cavity, occasionally liver muscle, sub-cutaneous tissue and orbit	Machida, Takahashi and Masuuchi (1978)
Indocucullanus puriensis	*Arius venosus*	Body cavity	Gupta and Srivastava (1975b)
Philometra sp.	*Canthigaster jactator* *Barbus luteus*	"	Deardorff and Stanton (1983)
Philometra ovata	*Abramis brama* and *Rutilus rutilus*	"	Molnár (1966a,b)
Philonema agubernaculum	*Salmo salar* and *Salvelinus fontinalis*	"	Vik (1964)
Skrjabillanus amuri	*Scardinius erythrophthalmus*	Mesentery, intestinal serosa	Lomakin and Chernova (1980)
Skrjabillanus erythrophthalmi	*Ctenopharyngodon idella*		(Molnár (1966b)

Table 3.22. Some Larval Nematoda of the Peritoneal Cavity

Species	*Host*	*Site*	*Reference*
Contracaecum sp.	*Eutropius depressirostris, Hydrocynus vittatus* and *Mormyrops deliciosus*	Body cavity	Khalil (1974)
Contracaecum multipapillatum	*Mugil cephalus, M. curema* and probably *Liza ramada*	Liver, kidneys and mesentery of mullets	Deardorff and Overstreet (1980b)
Cucullanus truttae	*Lampetra planeri*	Body cavity	Moravec and Malmqvist (1977)
Eustrongylides sp.	*Protopterus aethiopicus Engraulicypris argenteus, Bagrus docmac, Haplochromis* sp. *H. angustifrons*	Peritoneal cavity and intestinal wall	Khalil and Thurston (1973)
Thynnascaris sp.	*Ammodytes personatus Sebasticus marmoratus Rudarius ercodes Scomber japonicus Acanthopagrus schlegelii*	Peritoneal cavities and adjacent tissues	Sakaguchi Kuniyuki and Ueda (1980)

Table 3.23. A Larval Cestode and Some Digeneans of fish Gills

Parasite	*Host*	*Site*	*References*
Grillotia branchi metacestode	*Scomberomorus commersoni*	Encysted in branchial arteries	Shaharom and Lester (1982)
Didymozoon lobatum	*Euthymus affinis*	Cysts on gills	Madhavi (1982)
Clinostomum tilapiae metacercaria	*Oreochromis mossambicus*	Gills	Britz, van As and Saayman (1985)
Euclinostomum heterostomum metacercaria	*Oreochromis mossambicus*	Gills, gill chambers and muscles	Britz, Saayman and van As (1984) Britz, van As and Saayman (1985)
Parascocotyle diminuta metacercaria	*Cyprinodon variegatus, Fundulus grandis, F. jenkinsi, Lucania parva* and *Mollienesia latipinna*	Gill filaments of all hosts	Sogandares-Bernal and Bridgman (1960)
Syncoelium spathulatum	*Prognichthys*	Gills	Coil and Kuntz (1963)

Table 3.24. Some Parasitic Worms of the Swim Bladder

Parasite	*Host*	*Site*	*References*
Phyllodistimoides duncani	*Astyanax* sp.	Swim-bladder	Brooks (1977)
Cotylurus variegatus metacercaria	Percidae	Mainly on wall of swim-bladder	Swennen, Heessen and Hocker (1979)
Anguillicola australensis	*Anguilla anguilla*	Swim-bladder	Paggi, Orecchia, Minervini and Mattiucci (1982)
Anguillicola globiceps	*A. japonica*	Swim-bladder	Yamaguti (1963) Huang (1981)
Comephoronema johnsoni	*Cybium guttatum*	Swim-bladder	Arya (1979)
*Cystidicola cristivomeri***	*Salvelinus namaycush S. alpinus Salmo gairdneri*		Black and Lankester (1980, 1981)
C. farionis	*Salvelinus namaycush*		Black (1983c) Black and Lankester (1980)
C. farionis	*Salmo salar S. trutta* and *Salvelinus alpinus*		Conneely and McCarthy (1984)
C. farionis	*Coregonus nasus*		Valtonen and Valtonen (1978)
C. farionis	Salmonid fish		Lankester and Smith (1980)

Table 3.24 *Continued*

Parasite	*Host*	*Site*	*References*
C. stigmatura	*Salvelinus alpinus*		Black (1983a, b) 1984a, b, 1985)
C. stigmatura	*Salvelinus namaycush* and *S. alpinus*		Black (1983a, b).
Philometroides huronensis	*Catastomus commersoni*	Peritoneum Around swim-bladder	Uhazy (1977a, b, 1978)
Philonema oncorhynchi	*Oncorhynchus nerka*	Wall of swim-bladder	Adams (1969)
Salvelinema walkeri	*Oncorhynchus kisutch*		Margolis (1967a, b)
Skrjaillanus scardinii	*Scardinius erythrophthalmus*	Serosa of the posterior sac of the air bladder, renal serosa	Molnár (1966b)

** See Black & Lankester (1984) for taxonomy of *C. cristivomeri*.

Respiratory system

Fish breathe sea water, fresh water, or air—or some combination of these media (Randall, 1970 p. 253). Although the skin, mouth, pharynx and portions of the gastrointestinal tract of many species are adapted for air-breathing, the gills are the primary surfaces for respiration in most species. It is now generally conceded that the air or swim-bladder in fish originally functioned as an accessory respiratory organ. Members of the primitive orders Chondrostei (*Polypterus*), Holostei (*Amia*) and Dipnoi (*Protopterus* and *Lepidosiren*) have retained a respiratory function of the air bladder (Johansen, 1970). Tables 3.23–3.24 list helminths which occupy the gills and swim bladders as sites of infection. Monogeneans have been omitted from these tables because such a large number of species are gill parasites, as can be seen from Dawes (1947), Sproston (1946), Yamaguti (1963a) and a number of other sources cited in Chapter One. According to Fernando and Hanek (1976), more monogeneans infect gills than any other group of true ectoparasites. Llewellyn (1957b) was the first to investigate the microhabitats of monogeneans on gills but, subsequently, there have been many investigations of this topic, e.g. by Euzet (1984), Kearn (1976), Rawson (1977) and Rohde (1982a). Rohde refers to the universal preference of helminths for certain microhabitats, e.g. gill monogeneans for (1) certain gill arches, (2) certain microhabitats along the longitudinal axis of the gills, (3) microhabitats along the gill filament, (4) external or internal gill filament, (5) anterior or posterior surface of the gill filament, and (6) gills on the left or right side of the body. The advantages and disadvantages of the gills as an ideal site for parasites are well summarized by Fernando and Hanek (1976). Gill sites offer a large surface and a good supply of blood and oxygen. Ventilating currents passing over the gills carry infective stages to the site and eggs away from it. The currents, however, together with the abrasive action of the moving gill arches and the host tissue responses, do require the development of strategies for protection by the parasite, including efficient organs of attachment. In view of the large number of gill helminths already described, mainly monogeneans, there is enormous scope for further research on their preference for particular sites on the gills in relation to the total surface area, differing capacity, resistance to water flow, primary and secondary lamellae, whether the fish are operculate or have gill slits, and the behaviour of the host generally. Arme and Halton (1972) indicated the challenging nature of the problem in their work on *Diclidophora merlangi*. This species, like most monogeneans, has a preferred site of attachment but the factors influencing its distribution may be complex, possibly involving parasite-parasite interactions and/or the responses of flukes to other stimuli, e.g. where maximum oxygen tension occurs.

Swim Bladder

In teleosts the swim-bladder acts as a hydrostatic organ by replacing part of the fish body, which is heavier than water, with gas. While most natural waters, and consequently the arterial blood of fish, have a pO_2 and pN_2 of about 0.2 and 0.8 atm respectively, the corresponding partial pressures of O_2 and N_2 in the swim-bladder may be 100 and 20 atm. The ability of the bladder to concentrate O_2 500 times and N_2 some 30 times respectively, is the unique property of this organ. A swim-bladder is found most frequently in fish which live in the upper 200 metres of the water body. Except for typical bottom-dwelling fish, most freshwater teleosts possess a swim-bladder. It is absent in elasmobranchs (Steen, 1970).

Teleosts have two main types of swim-bladder, physostome and physoclist. In physostome fish a connection between the bladder and digestive tract is retained in adult life. The gas content of this kind of bladder is potentially in direct communication with the exterior, although the avenue, the pneumatic duct, is always guarded by a muscular valve. Fishes with a closed swim-bladder are called physoclists. The anatomy of a swim-bladder is well seen in the eel, *Anguilla vulgaris*, but although it is a physostome, its swim-bladder has the same organization as that of a physoclist. The perch, *Perca fluviatilis*, is a physoclist but the variability in architecture of a physoclist swim-bladder is considerable (Steen, 1970, p. 418). Further research on the parasitic worms listed in Table 3.24 and carried out with due regard to the structure and function of the swim-bladder should prove rewarding in elucidating the advantages and disadvantages of this site to its inhabitants.

Cardiovascular system and endocrine organs

The cardiovascular system, as an environment for helminths, was well described by Smithers and Worms (1976). The blood flukes of cold-blooded vertebrates were extensively reviewed by Smith (1972), who listed 12 genera containing about 50 species of sanguinicolids from 91 host species, comprising 40 in fresh water and 51 in the sea representing 35 fish families. Smithers and Worms discussed problems of life in the vascular habitat under the headings: entrance to the vertebrate host and migration to the final habitat; life in the vascular environment; nutrition and metabolism; exit from the vertebrate host and transmission; and evading the immune response. As a site the cardiovascular system may differ both physiologically and chemically at different points along its length. In fish, as in other vertebrates, various parasite groups use the cardiovascular system as a migration route to preferred sites and also as a habitat. The system supplies nutrients to the parasites and is a means of transmission for eggs and larvae. Smithers and Worms (1976) emphasized how filarial worms and blood flukes exhibit major differences in the manner in which the problems of life in the blood system have been overcome. The problem of transmission in filarial worms has been overcome by the development of viviparity, the movement of larvae to peripheral sites and the use of arthropods as intermediate hosts. The problems particularly associated with parasitism of the system are those of entrance and exit, maintenance at a particular site in a flowing medium and the avoidance of damage to the blood vessels. Advantages to the parasite include constancy of the environment, protection, facilitated respiratory and excretory exchange and, in particular, availability of nutrients such as glucose and amino acids.

Other helminths which invade the cardiovascular system of fish include the monogenean *Amphibdella flavolineata*, juveniles of which are found in the heart of *Torpedo nobiliana* according to Llewellyn (1960), whilst the adults locate in the parietal mucosa of the gills; the nematode *Vasorhabdochona cablei* in the mesenteric bloodvessels of *Gillichthys mirabilis*

(Martin and Zam, 1967); *Goezia* sp. in the heart of *Barbus* sp. (Herzog, 1969); and various larval digeneans of the genera *Ascocotyle, Cotylurus, Holostephanus* and *Phagicola* in *Molienesia, Cyprinodon, Gasterosteus, Mugil* and various Cyprinidae, Percidae and Salmonidae (Hutton and Sogandares-Bernal, 1958; Erasmus, 1962; Sogandares-Bernal and Bridgman, 1960; Sogandares-Bernal and Lumsden, 1963, and Swennen *et al.*, 1979).

Although new species are occasionally being added to the long list given by Smith (1972), e.g. *Sanguinicola lungensis* found by Tang and Ling (1975) in the branchial artery and hepatic blood vessels of *Hypophthalmichthys molitrix, Aristichthys nobilis* and *Carassius auratus*, the study of fish blood flukes is still in its infancy. Further research on problems of site-location, mate finding, nutrition, metabolism and morphology of these worms in relation to the physical and physiological characteristics of this system should prove rewarding. Since the endocrine organs produce secretions which pass into the blood stream, it is appropriate to mention here that no worms have yet been recorded from the endocrine organ system of fish.

Reproductive system

Amongst the few worms recorded from this system are: *Triaenophorus* larvae from the ovaries of *Tinca tinca* by Aisa (1976); *Philometra* sp. from the testes and ovaries of *Chrysophrys major* by Nakajima and Egusa (1979); philometrid larvae from the ovaries of *Carcharhinus limbatus* by Rosa-Molinar and Williams (1983); *P. (P.) macroandri* in the ovary of *Thunnus albacares* and *P. (P.) katsuwoni* in the ovary of *Katsuwonus pelamis* by Petter and Baudin-Laurencin (1986); *Ovarionematabothrium texomensis* from the ovaries of *Ictiobus bubalus* by Self *et al.* (1963) and *O. saba* from the ovary of *Pneumatophorus japonicus* by Kamegai and Shimazu (1982).

Further research is necessary on the parasitic worms of the reproductive system of fish with due regard to the amazing array of curious modifications for reproduction that occur in these vertebrates as described, for instance, by Hoar and Randall (1969). Of interest also is the fact that the number of parasites of a particular species is often significantly different in male and female hosts, regardless of the site of infection, and it must be assumed that this is partly due to the influences of different hormones. Such differences in the infection of male and female fish have been referred to by Hine and Anderson (1982) for *Philometra* in *Chrysophrys auratus* and very many others whose publications are cited elsewhere in this chapter (pp. 257–258).

In view of the ease of entry and exit for helminth larvae and eggs through the cloaca and genital pore, and the rich nutrient source which must exist in the ovaries and testes to support the developing gametes of fish and, consequently, any parasites that may be present, it is surprising that so few helminths have been recorded in the reproductive system. It may be that fish helminthologists have neglected this complicated system in their search for parasites or that it has been examined when physiological and behavioural changes connected with the onset of maturity and reproduction provide a hostile environment to the parasite. On the other hand, one would expect the reproductive behaviour of parasites in this system to be nicely synchronized with the maturation of the gonads and external environmental conditions appropriate to breeding of the host.

Excretory and osmoregulatory systems

The foregoing comments on the paucity of helminths in the reproductive system of fish is also difficult to understand, bearing in mind the relatively large number of species, especially digeneans, recorded from the excretory system. The closeness of the two systems (they are generally referred to as the urinogenital system) and also their proximity to the peritoneal cavity, also an attractive site for helminths, does support the suggestion that the reproductive systems of fish should be examined more closely for helminths and, if possible, throughout the host's life-history. Should the results be unrewarding, then further research should focus not only on its disadvantages as a site but also on the advantages of adjacent body systems.

The function of the kidney and nature of excretory products vary considerably in different groups of fish (Hickman and Trump, 1969; Forster and Goldstein, 1969). In freshwater forms the kidney functions as an excretory device, filtration at the renal glomerulus resulting in the conservation of filtered ions and the excretion of dilute urine. In marine teleosts, the kidney functions chiefly as an excretory device for magnesium and sulphate ions. Sharks, skates and rays, which are hyperosmotic in their marine environment, have kidneys which combine the principal functions of those found in both freshwater and marine bony fishes. Excretory products include ammonia, urea and trimethylamine oxide and it is important to realize that not only kidneys, but also the liver, gills, intestine and other tissues of fish are involved in the synthesis and selective handling of excretory products. Examples of some digenea of the excretory system of fish are given in Table 3.25.

Nervous system and sense organs

The nervous system and sense organs of fish have been extensively studied in conjunction with their behaviour and sensory physiology and also because fish have certain receptors not present in other vertebrate groups (Hoar and Randall, 1971). Surprisingly, however, relatively little attention has been given to the presence of parasites in the nervous system and how they might bring about behavioural responses differing from those found in non-

Table 3.25. Some Digenea of the Excretory System

Parasite	*Host*	*Site*	*Reference*
Phyllodistomum carangis	*Caranx ruber*	Urinary bladder Body cavity	Bravo-Hollis and Manter (1957)
Cetiotrema crassum	*Thunnus thynnus maccoyii*	Ureter	Manter (1970)
P. conostomum	*Coregonus lavaretus*	Ureter and bladder	Bakke (1985)
P. conostomum	*Salvelinus alpinus Salmo trutta*	Ureters	Conneely and McCarthy (1984)
P. conostomum	*Coregonus albula C. lavaretus lavaretoides*	Urinary bladder	Kudinova (1979)
P. ghanense	*Mastacembelus nigromarginatus*	Urinary bladder	Thomas (1958)
P. marinae	*Mycteroperca pardalis*	Urinary bladder	Bravo-Hollis and Manter (1957)
P. marinae	*Alburnus alburnus*	Urinary bladder	Kudinova (1979)
P. pseudofolium	*Acerina cernua*	Urinary bladder	Kudinova (1979)
P. scrippsi	*Pimelometopon pulchrum*	Urinary bladder	Brooks and Mayers (1975)
P. simile	*Mesocottus* sp.	Urinary bladder	Kudinova (1979)
P. umblae	*Salvelinus alpinus Salmo trutta Coregonus lavaretus C. albula Thymallus thymallus*	Ureter and bladder	Bakke (1984, 1985), Rahkonen and Valtonen (1987)
P. undulans	*Cottus bairdi*	Urinary bladder	Fallon and Wallace (1977)
Renodidymocystis yamagutii	*Rastrelliger kanagurta*	Kidney	Madhavi (1982)

parasitized fish. The influences of parasites on host behaviour are discussed elsewhere in this book. According to O'Connor (1976), helminths of vertebrates other than fish, with few exceptions, appear to have no predilection for the eye and brain, but they usually become trapped there fortuitously during migration. Williams (1966) noted, however, that it was easier to distinguish between two species of *Diplostomum* in *Gasterosteus aculeatus* by their extremely restricted sites (*D. spathaceum* in the eye lens and *D. gasterostei* in the humour) than by morphological comparison. This suggests specialized site-selection by the two physiologically distinct species. Some monogeneans, larval cestodes, larval digeneans and nematodes of fish that appear to have a preference for the nervous system are listed in Table 3.26. Larval digeneans appear to be the most common helminths of the eye and

Table 3.26. Some Helminths of the Brain And Sense Organs

Parasite	*Host*	*Site*	*Reference*
Bucephaloides gracilescens	*Melanogrammus aeglifinus*, *Merlangus merlangius*, *Merluccius merluccius*	Cranial fluid surrounding brain	Johnston and Halton (1981)
Caballerocotyla klawei	*Neothunnus macropterus*	Nasal capsule	Stunkard (1962a)
Dactylogyrus editus	*Schizophygopsis stoliczkai*	Nasal cavity	Dzhalilov (1976a)
Paraquadriacanthus nasalis	*Clarias lazera*	Nasal cavity	Ergens (1988a)
Diplostomum gasterostei (larvae)	*Perca fluviatilis*	Eye	Kennedy and Burrough (1977)
D. gasterostei (larvae)	*Anguilla anguilla* *Perca fluviatilis*	Eye humour	Conneely and McCarthy (1984)
D. gasterostei (larvae)	*Gasterosteus aculeatus*	Retina	Williams (1966)
D. phoxini (larvae)	*Phoxinus phoxinus*	Brain	Bell and Hopkins (1956)
D. spathaceum indistinctum and *D. baeri bucculentum* (larvae)	*Coregonus clupeaformis*	Lens, and vitreous humour and retina, respectively	Dick and Rosen (1981)
D. spathaceum (larvae)	*Abramis brama* and *Anguilla anguilla*	Eye lens	Conneely and McCarthy (1984)
Diplostomulum stahli (larvae)	*Diplodus* (= *Sargus*) *annularis*	Optic lobes of brain	Rebecq and Leray (1960)
Diplostomum sp. (larvae)	*Heteropneustes fossilis*	Eye	Dubey, Dubey and Pandey (1981)
Diplostomum sp. (larvae)	*Salmo trutta Salvelinus alpinus Anguilla anguilla Perca fluviatilis*	Eye	Conneely and McCarthy (1984)
Diplostomum spp.	*Salmo gairdneri*	Eye lens and retina	Bortz, Kenny, Pauley and Bunt-Milam (1988)
Empruthotrema raiae	*Raja* spp.	Nasal fossae	Kearn (1976)
Gilquinia squali (larvae)	*Merlangus merlangus*	Eye	MacKenzie (1965, 1975b).
Larval trypanorhynch	*Seriola dumerili*	Muscle near brain	Deardorff Raybourne and Mattis (1984)
Merizocotyle	*Raja undulata*	Nasal fossae	Kearn (1976)
Ornithodiplostomum ptychocheilus (larvae)	*Pimephales promelas*	Brain	Hendrickson (1979)
Paronatrema sp.	*Heteropneustes fossilis*	Ventral side of brain	Toteja, Sood and Saxena (1982)
Pellucidohaptor catostomi	*Catostomus catostomus*	Nasal cavity	Dechtiar (1969)
Pellucidohaptor nasalis	*Catostomus commersoni*	Nasal cavity	Dechtiar (1969)
Philometra sp.	*Lepomis macrochirus* and *L. gulosus*	Eye orbit	Benz and Pohley (1980)
Rhinoxenus spp.	*Serrasalmus nattereri* *Schizodon fasciatum* *Rhytiodus argenteofuscus*	Nasal cavities	Kritsky, Boeger and Thatcher (1988)
Squalotrema llewellyni	*Squalus acanthias*	Nasal fossae	Kearn and Green (1983)
Trematode larvae	*Channa punctatus*	Brain ventricles close to pituitary	Joshi and Sathyanesan (1981)
Tylodelphys clavata (larvae)	*Perca fluviatilis*	Eye	Kennedy and Burrough (1977), Coneely and McCarthy (1984)

brain and these appear to affect host behaviour enough to increase predation rates by definitive hosts. This is clearly an advantage with regard to continuation of the life-cycle. It is reasonable to speculate that this is also the advantage for *Gilquina squali* in the eye of *Merlangius merlangus*. Further studies on these should prove rewarding since neurophysiology remains a very active area of research in biology. The fascinating and detailed studies by La Rue, Butler and Berkhout (1926), Ferguson (1943) and Erasmus (1958, 1959, 1962) will remain primary sources of information in planning further research including the advantages and disadvantages of particular sites within the central nervous system and associated sense organs. Points of entry and escape to the system are little known.

Community ecology

Introduction

According to Anderson and Kikkawa (1986), community ecology may be briefly defined as the study of assemblages of different species which interact with one another whilst growing, reproducing and surviving in differing environments including, in the case of parasites, the hosts themselves and the environment in which they dwell. As Aho and Kennedy (1987) have stressed, most ecological studies have 'concentrated on infrapopulation dynamics in one host-one parasite systems with little regard for integrating ecological factors for the other life-history states within the host community complex'. From one point of view this is surprising since it is often said, e.g. by May (1988) that there are far more species of parasites than of hosts, each host species may harbour several species of parasite and the commonest and most widespread method of interspecific coexistence is by resource partitioning in space. Most species exhibit resource partitioning in space by means of selective site segregation, a well-documented aspect of parasite behaviour, as already seen in our discussion of sites and site-selection.

From another point of view, however, it is easy to understand why so little has been done on the community ecology of parasites because, as Bradshaw and Mortimer (1986) have stated, there are a myriad interactions, ecological and evolutionary, between hosts and predators, parasites and pathogens. Toft (1986) recognized 11 categories of parasites that may potentially exhibit different population dynamics and therefore, different community characteristics. The ecological and evolutionary aspects of parasitism are inextricably bound (Whitfield, 1979), and ecology is also closely linked with physiology, behaviour and genetics (Krebs, 1972); the genetic link was emphasized by Rollinson and Anderson (1985). A few ecological parasitologists, however, amongst them Holmes (1990), Holmes and Price (1980) Price (1986) and Kennedy (1990) have faced the challenge of such a complicated situation. Our aim in this section, therefore, is first to comment on how these internationally-recognized community ecologists see the situation and the ways of resolving the many problems involved, and then to select a few papers where approaches have been made towards researching aspects of the community ecology of fish parasites.

Holmes and Price (1986) addressed themselves to two main topics: the nature of parasite communities and how these parasite communities influence host communities. The main features of parasite communities were related to (i) resources available to the parasite at the level of the host species or the particular part of the host that is exploited, (ii) host species representing a habitat for its parasites and thus that there are replicated habitats, (iii) the specialized natures of most parasites which restrict themselves to relatively few host species, and (iv) host individuals, host populations and communities of hosts providing resources for parasites also provide a hierarchy of organizational levels. Holmes and Price point out that it was on this basis that Esch, Gibbons and Bourque (1975) recognized infrapopulations and suprapopulations of parasites.

Holmes and Price summarized the assumptions and predictions of mechanisms involved in the organization of parasite infrapopulations as well as those involved in the organization of communities of parasites. With regard to infracommunities, three hypotheses were proposed: (i) the competition hypothesis which states that competition (of three differing kinds) has been important in organizing forces in parasite communities; (ii) the population concentration hypothesis which relates to high host-specificity and site-specificity as important reproductive isolating mechanisms to prevent hybridization, and (iii) the individualistic response hypothesis which suggests that species of parasites are individually and independently adapted to narrow niches, a niche being regarded as the complex of environmental characteristics permitting species to persist and reproduce indefinitely.

With regard to concepts at the suprapopulation level discussed by Holmes and Price, they focused on what determines species richness of the parasites of a host species and of a host population. They addressed themselves to four hypotheses: (i) the cospeciation hypothesis which relates to the coevolution of hosts and parasites; (ii) the island size hypothesis, the species richness on an island (the host population being related to its size); (iii) the island distance hypothesis which regards the difficulty of invasion of a host population by a parasite as a prime determinant of parasite communities, and (iv) the time hypothesis—younger communities have fewer species than older communities because evolutionary time is required for the increase in species number.

Finally, Holmes and Price discussed concepts at the compound community level, i.e. which mechanisms determine the distribution and abundance of parasites within and between lakes and other habitat units. Few studies have been made in this area, those of Wisniewski (1958), Halvorsen (1971), Wootten (1973) and Kennedy (1978c) being exceptions. Despite the variety of hypotheses applied to communities of parasites, the assumptions and predictions of these hypotheses led Holmes and Price to recognize initially, and on a crude basis, only two types of communities which they designated interactive and isolationist. The differences between interactive and isolationist infracommunities appear to be consequences of different assumptions on colonization possibilities, i.e. transmission rates which limit parasite populations. Many features can limit transmission rates, thereby producing isolationist communities, e.g. time (for introduced species, for example), physical distance (for habitat islands), a small host population size, food habits or other aspects of the way of life of the host which make colonization difficult or, by implication, severe physical or chemical stresses in a particular microhabitat. Holmes and Price also discussed the influence of parasites on host populations, on host species coexistence, on predator-prey interactions and on environmental patchiness. They concluded that existing data are totally insufficient to estimate the extent to which parasites do influence host communities.

Holmes (1986) drew attention to two major questions of interest to community ecologists over the previous decade: (i) do communities show predictable structure, i.e. patterns of species occurrence, relative abundances and resource use and, if so, (ii) what processes produce the structure? Debates of these two questions had led to the acceptance of two types of communities, interactive and isolationist, the former involving interaction between species and the latter without interaction. Holmes referred to three hierarchical scales of helminths: (i) interactions in infracommunities, i.e. at the level of the host individual. Such interactions are easier to detect where there are dominant species (core species) or within groups of species (guilds of species) that use the same set of resources in the same way; (ii) larger-scale interactions between core species, largely specialists, in the host population, and (iii) other larger-scale interactions in the host community involving generalist species and satellite species. The richness of each of the two large-scale communities is affected by various ecological, historical and evolutionary factors. To further understand these complex situations, Holmes appealed to parasitologists to familiarize themselves with the language and developments in ecology generally.

With regards to the evolution of parasite communities, Price (1986) asked five important questions. These were: (i) Does competition occur in parasite communities at present? (ii) Do vacant niches exist, that is, 'Could additional parasite species be introduced into the community so that they persist and reproduce, without causing extinction of other species?' (iii) Can colonists successfully invade an existing community whether or not they cause extinction of resident species? (iv) Does extinction of species in a community occur frequently, suggesting maintenance of a dynamic equilibrium? (v) Do phylogenies of hosts and parasite communities show cospeciation? Price believed that methods are available for rigorous testing of each of the five questions. The mechanisms involved in parasite community organization could, therefore, be evaluated. He thought that experimental testing of the questions, especially using field tests, would provide the most cogent results; the main impediment to progress is the availability of experimental parasite-host systems. He further believed that we had answers to questions (i) and (ii) on competition and vacant niches through the work of Holmes (1961, 1962) and Rohde (1978a,b, 1979a, 1981a) and that Brooks had addressed himself to question (v) on coevolution in Brooks 1979a,b, 1980; 1986; Brooks and Glen, 1982; Mitter and Brooks, 1983. Questions (iii) and (iv) however, had not been addressed by parasite ecologists. Price advocated further research on where communities are rich and 'depauperate' (= few species), where competition is strong or weak and where vacant niches are common.

A large number of papers that are mostly concerned with surveys of fish for their helminth parasites will undoubtedly form a basis for more intensive investigations of the community ecology of fish parasites. A list is given on pp. 245–254. Some of these have concentrated on families of fish, e.g. Zubchenko (1980) on the Anarhichadidae and Pleuronectidae, and Zubchenko (1981) on Macrouridae. Others have investigated various fish genera and species in the same locality, e.g. Izyumova *et al.* (1982a,b) on *Abramis brama, Stizostedion (Lucioperca) lucioperca* and *Acipenser ruthenus*, and Pojmanska *et al.* (1980) on *Abramis brama, Perca fluviatilis, Esox lucius* and *Lucioperca lucioperca*. Another approach has been to concentrate on a particular locality, e.g. Rybak (1982) on the Vygozero lake reservoir in Kareliya, and Lee (1977) on the fish population of the Serpentine in central London. A few researchers, including Price and Clancy (1983) and Kennedy, Bush and Aho (1986) have carefully and successfully used such survey work and lists in advancing our knowledge of ecology. Price and Clancy's work has already been referred to earlier in this chapter. Kennedy, Bush and Aho (1986) investigated suggestions that there are fundamental differences between the communities of helminths in fish and bird hosts. To test the suggestions, they made use of some helminth species lists from the literature and concentrated on the helminths of the alimentary canal of adult hosts only. They confirmed that the distinction in helminth communities between fish and birds is truly justified and that the differences are valid and fundamental. They identified five factors as essential to the production of a diverse helminth community: (i) the complexity of the alimentary canal in terms of the number and variety of sites available for helminth occupation; (ii) host movement into different habitats containing a wide range of prey and hence potential intermediate hosts; (iii) a broad host diet; (iv) selective feeding on prey which serve as intermediate hosts for a wide variety of helminths, and (v) exposure of a host to direct life-cycle helminths which enter by penetration.

We suggest that intensive testing of the validity of these factors should continue both on freshwater and marine fish parasite communities and at three levels, namely:

(i) infracommunity level, being a study of all helminths of one host species in a particular locality;
(ii) component community level which involves understanding of ecology of the helminth fauna of one host species in different localities, and
(iii) compound community level, being an analysis of the relationships between the entire parasite community of every host species in one, and subsequently, if possible, other localities.

A number of studies have touched upon aspects of the first two named levels and, in particular, the second with regard to richness of parasite species and host or environmental factors which influence their distribution (Holmes, 1990). It was also emphasized by Holmes that relatively few studies had been carried out at the compound community level although several patterns had emerged from investigations by Noble (1973), Campbell (1983), Houston and Haedrick (1986) and Polyanski (1961b). Detailed work at infracommunity and component community levels were described by Kennedy (1990) for the helminths of *Anguilla anguilla* in the United Kingdom and Holmes (1990) for those of *Sebastes nebulosus* on the west coast of Vancouver Island, British Columbia. Before presenting summaries of their main findings we shall briefly mention other papers which either have already led to increasing interest in the community ecology of fish helminths or are likely to form interesting bases on which to plan further research.

Firstly we draw attention to the large number of papers on life-cycle patterns, site and host-specificity, zoogeography and phylogeny referred to elsewhere in this book. These topics are important aspects of parasite community studies. Secondly some investigations have concentrated either on more than one helminth species, multiple species studies in one host or on groups of helminths (multiple helminth group studies).

Multiple helminth species studies

Izyumova *et al.* (1985a,b) investigated the prevalence, abundance and distribution of up to four *Dactylogyrus* spp. on *Abramis brama*. Significantly-increased prevalences of infection were recorded for the four species when 1954–1956 results were compared with those for 1982. Intensities were also increased. Prevalence and intensity was similar in the Danube delta and Rybinsk reservoir but much lower in the Lake Issuk-Kul. In all water-bodies the right gills were more heavily parasitized than the left. Similar results have also been discussed by Izyumova, Zharikova, Karabekova and Asylbaeva (1985). Kuperman and Shul'man (1978) investigated *Dactylogyrus auriculatus, D. falcatus* and *D. wunderi* on *Abramis brama*. The parasites became sexually active and increased their numbers on transfer of host from natural cold temperatures to warmer conditions. Transfer from warmer natural summer temperatures to colder temperatures 7–10°C in May (when natural infection is increasing) lowered the numbers of *D. falcatus* and *D. wunderi* but not of *D. auriculatus*. However, in June and July (when natural infection is at its peak) reproduction was more active in experimental (colder) conditions than at natural (warmer) temperatures. These unexpected results may be due to increased resistance, not developed in May, but strong in June and July. In the laboratory the immune response was inhibited and led to increased parasite burdens. The different result for *D. auriculatus* is explained by its preference for colder waters which begins active life earlier than the other two and, therefore, found the experimental conditions optimal.

According to Zharikova (1984a, b), *Dactylogurus* spp. infections of *Abramis brama* are related to host sex and also to different behaviour of males and females during spawning and feeding. Males were significantly more heavily infected by each of five species. There were no seasonal differences in the relative abundance of the five species.

In Japan, *Dactylogyrus extensus* and *D. minutus* were found by Nakatsugawa and Muroga (1977) on carp of all ages in culture ponds throughout 1975–1976. Seasonal changes in abundance of both species were related to water temperature. They also showed peaks in incidence in spring and autumn, the increase being more marked for *D. extensus* in spring and for *D. minutus* in autumn. Incidence of both fell in midsummer. When water temperature was artifically increased from 11°C to 20°C, numbers of *D. extensus* increased but then fell sharply when the temperature reached 20–25°C.

Anderson (1981) investigated a situation in which young grey mullet, *Chelon labrosus* (up to two years old) were characterized by one host-specific monogenean, *Ergenstrema labrosi*, and older mullets (over four years) by another, *Ligophorus angustus*. Fishes of 2–4 years had both species. This was thought to be associated with difference in salinity, young mullet living in (tidal) brackish water and older mullet in higher salinities. Experimentally the two were shown to be adapted to different salinities. The behaviour of young and old mullet in aquaria differed markedly, the smaller fish spending most of their time swimming whereas older fish spent over half of their time hovering.

Shul'man (1977) found seasonal variations of *Gyrodactylus aphyae, G. macronychus, G. magnificus* and *D. phoxini* infections of *Phoxinus phoxinus*, the rate of infection being related to daylight and water temperatures, being very low in the perpetual Arctic night (November to mid-January), reaching a maximum in April and May (when there were, respectively, 20 and 24 hours of daylight, although snow and ice melted only in mid-May) and decreased in August and September, when water temperature reached its maximum.

Regarding gyrocotylideans Colin, Williams and Halvorsen (1986) and Williams, Colin and Halvorsen (1987) concluded that, until further and better evidence was available. '*Gyrocotyloides nybelini*' and *Gyrocotyle confusa* (both alleged to be rare parasites of *Chimaera monstrosa*) should be considered synonyms of *G. urna* (a very common parasite of the same host). Since Bandoni and Brooks (1987b) Bristow and Berland (1988) and Berland, Bristow and Grahl-Nielsen (1990) disagree with those views and suggest that two genera and three species of gyrocotylideans may occur in the gut of *Chimaera monstrosa*, this host and its gyrocotylideans and several other helminths will remain important and interesting for further investigations of population biology problems concerned with the rarity of some species, the niche segregation between species pairs, infrapopulation size and other aspects of community studies.

Three caryophyllidean species, *Glaridacris catostomus, G. laruei* and *Isoglaridacris bulbocirrus*, may occur in *Catostomus commersoni* in New Hampshire rivers, USA (Muzzall, 1980c). *G. catostomus* and *I. bulbocirrus* occupy the anterior intestine and *G. larvei* may occur throughout the intestine but most worms are in the posterior region. *G. catostomi* and *G. laruei* show intraspecific competition. The mean size of the gravid worms of these two species decreased as the population density of each increased and, as parasite density decreased, the mean size increased. The seasonal prevalence and mean intensities of infection of *G. laruei* and *G. catostomi* in *Catostomus commersoni* have been described by Williams (1979a). Further detailed work on the ecology of caryophyllideans is required for which the reviews of the group by Mackiewicz (1972, 1981, 1982a) and the work of Anderson (1974b) and Kennedy (1968, 1969b, 1971) cited elsewhere in this chapter will be invaluable.

The distribution of three tetraphyllidean, one diphyllidean and one trypanorhynch species of cestodes within the spiral intestine of *Raja naevus* was described by McVicar (1979). *Acanthobothrium quadripartitum, Phyllobothrium piriei* and *Echinobothrium* sp. occur in most parts of the spiral intestine with peak infections in tiers two and three. *Echinobothrium harfordi* was restricted to anterior tiers and *Grillotia erinaceus* showed little site-specificity. Details of the distribution patterns could be correlated with the dorso-ventral orientation of the gut in the body cavity, the internal anatomy of the spiral intestine and associated pathways of food movement, the surface topography of the mucosa, acidity, sugar gradients and intraspecific effects of high intensities of infection. The distribution of the larger ('mature') cestodes does not depend on the settling pattern of the larvae.

Reimchen (1982) reported *Cyathocephalus truncatus*, usually found in salmonids, and the plerocercoids of *Schistocephalus solidus* from *Gasterosteus aculeatus* in British Columbia, Canada. Adult *C. truncatus* were attached at the anterior end of the intestine adjacent to the pyloric sphincter, whereas in salmonids the pyloric caeca is the usual attachment site. Mean intensities of infection for *C. truncatus* and *S. solidus* were 2.7 (maximum 26) and 3.4 (maximum 87) respectively, with intensities increasing in larger fish. The highest prevalence of *C. truncatus* infection was from February to May (80 per cent) and of *S. solidus* from April to September (50 per cent). Infection rates for *C. truncatus* during different seasons and among different length-classes of fish were directly correlated with the relative abundance of amphipods (the intermediate hosts of *C. truncatus*) in the fish stomachs. Possible modification of host feeding behaviour is indicated by a relative increase in the consumption of amphipods by infected fish.

Evans (1978) investigated the occurrence of both larval and adult stages of the digenean *Asymphylodora kubanicum* in *Bithynia tentaculata* and *Rutilus rutilus* during a one-year period from 1974 to 1975. Whilst investigating *A. kubanicum* the presence of *Sphaerostoma bramae* in the fish intestine was noted and records kept of its occurrence (Evans, 1977). It became apparent that there were likely to be differences in the seasonal occurrence of the two species. The high winter and low summer level of infection for *S. bramae* was in direct contrast to the seasonal pattern exhibited by *A. kubanicum* in the same definitive host in the same locality. The high summer and low winter levels of infection with *A. kubanicum* were explained largely in terms of the feeding behaviour of the definitive host and seasonal fluctuations in environmental temperature; differences in the effects of these and possibly other variables may thus occur between the *R. rutilus*—*S. bramae* and *R. rutilus*—*A. kubanicum* systems. A positive association exists between *S. bramae* and *A. kubanicum*, with the two species occurring together more often than would be expected from chance alone. Since this is probably due to the feeding habits of the definitive host it is unlikely that the differences in the seasonal patterns of occurrence of the two parasites are accounted for by the host's feeding behaviour. The effect of temperature upon the *A. kubanicum*—*R. rutilus* system was somewhat different from that observed by others for several fish-helminth systems. The overdispersion of *A. kubanicum* and *S. bramae* is a feature which has been noted for several other fish-helminth systems. A likely causative factor is the restriction of infected intermediate hosts to localized regions of the habitat with the result that fewer fish will feed upon infective larval stages, but those that do may ingest relatively high densities of cysts. The two species showed distinct and different site preferences in the roach intestine, most *A. kubanicum* being found in the first limb and most *S. bramae* occurring in the second and third limbs. *A. kubanicum* excysts and establishes itself in the first limb and this is considered to be the basis of the observed distribution of the species in the host's intestine. The causative factors of the site preference of *S. bramae* are less well understood. The different site preferences for the two species was explained by inherent differences between them which dictate their responses to stimuli, bringing about localized encystment, establishment and ensure that any subsequent migrations are influenced by certain biochemical and physiochemical gradients along the host intestine. Site-selection was similar in single and concurrent infections of the two species. Competitive interaction between the two, however, was thought to be less striking than that found by MacKenzie and Gibson (1970) for two species of nematodes in the gut of *Platichthys flesus*. The non-pathogenic nature of *S. bramae* and *A. kubanicum* may, at least partly, explain the probable absence of any extensive competitive interaction between the species.

Previous work on the concept of the different species of parasites in a host being a community was discussed by Scott (1982) in a paper on four flatfishes, *Hippoglossoides platessoides, Limanda ferruginea, Glyptocephalus cynoglossus* and *Pseudopleuronectes americanus* as hosts for 13 digenean species. The parasites indicated overlap of feeding habits amongst the hosts but also considerable diversity in feeding behaviour. Differences in the species composition of parasites corresponded with differences in diet, and there was also considerable variation in prevalence in the different geographical areas investigated. There were also real differences in rates of parasite infection between summer and winter. Other references to papers by Scott on the digeneans of gadid and pleuronectid fish may be found on pages 246 and 247 of this chapter.

The incidence, intensity and frequency distributions of *Tylodelphys clavata* and *Diplostomum gasterostei* in *Perca fluviatilis* in a British lake was investigated by Kennedy and Burrough (1977). It was concluded that *T. clavata* has a life-span of one year or less and that the parasites die within the fish, mainly in summer. *D. gasterostei*, in contrast, has a life-span of at least one year and disappearance of the parasite from the population is probably due to death of some heavily infected fish, although it could not be determined if this was caused by the parasite.

A study of the distribution and abundance of *Bunodera sacculata, B. luciopercae* and *Crepidostomum cooperi* in yellow perch in Canada showed that they were more common in large than in small lakes (Cannon, 1972). The two *Bunodera* spp. increased with length of fish but *C. cooperi* did not. The incidence of the *Bunodera* spp. declined in summer but that of *C. cooperi* did not. Temperature was found to be an important factor accounting for seasonal changes of incidence. The microhabitat of juvenile *B. luciopercae* was the gall bladder, but for *B. sacculata* and *C. cooperi*, the intestine. Adults of *C. cooperi* were found predominantly in the intestinal caeca, of *B. sacculata* in the anterior intestine, and of *B. luciopercae* in the posterior intestine. Cannon discussed the coexistence of these three closely-related species in terms of the temporal and spatial differences of 'habitat' utilization.

We are not aware of intensive ecological studies of more than one species of nematode in one host species but elsewhere in this book we have cited references to nematode specialists whose works will form important bases for investigating this area. Papers by Moravec will be of particular importance in this context.

Kennedy and Moriarty (1987) investigated whether, and under what conditions, two species of acanthocephalans could coexist in the same host population, and specifically whether the species showed resource-partitioning in space and/or time. They used as material a population of *Anguilla anguilla* harbouring *Acanthocephalus lucii* and *A. anguillae* in Lough Derg, Ireland, over three years. The occurrence of two or more congeneric species of helminths in the same host species is fairly common amongst cestodes, digeneans and nematodes but exceptionally rare in acanthocephalans of fish according to Kennedy and Moriarty. *A. lucii* and *A. anguillae* use the same intermediate host, *Asellus aquaticus*, and eels appeared to be the only definitive hosts of *A. anguillae*. *A. lucii* was the dominant parasite, was overdispersed throughout the eel population and most frequently occurred as a single-species infection. *A. anguillae* was far less common, its dispersion was close to random at most times and usually occurred as a mixed-species infection. They concluded that congeneric acanthocephalans can coexist in fish without obvious interactions or resource partitioning. Another study of the rare occurrence of two species of acanthocephalans in one host species is that of Muzzall (1984a) on *Neoechinorhynchus limi* and *Fessisentis tichiganensis* in *Umbra limi* in a Michigan river. Both species were present in the spring and absent in the summer and autumn, a seasonal infection pattern which could be related to movements of the fish in and out of the sampling area, changes in temperature, the life-span of the acanthocephalans and possibly in the case of *N. limi*, to a change in the diet of the fish. *N. limi* occurred throughout the intestine, but *F. tichiganensis* was attached either in its anterior region or in the pyloric caeca.

Multiple helminth group studies

This is a little-known area of ecological research although a large number of published surveys (see Appendices) should be useful in planning further research on various groups of parasites in one host species and thus pave the way towards understanding one of the most challenging areas of community ecology. As examples of studies we are aware of, we shall refer to the papers by Adams, by Gupta *et al.* and by James and Srivastava. Adams (1985, 1986) discussed the interspecific interactions and community structure of six species of parasites on the gills of *Fundulus kansae*. Using species diversity as a measure of gill arch-preference, arches two and three had greater values than one and four, although calculations of niche breadth and overlap showed that the populations of parasites were capable of even distribution on the gill arches.

Gupta *et al.* (1984) studied several aspects of the population biology of six species of helminths in *Channa punctatus*. Infection was generally low from September to December in females and from October to January in males, and higher from January to August (with a peak in January) in females and February to September (with peaks in February and September) in males. *Pallisentis nagpurensis* had the highest relative density. All species were overdispersed and there was a negative interaction between *P. nagpurensis* and *Senga* sp., between *P. nagpurensis* and *Isoparorchis hypselobagri* and between *I. hypselobagri* or *Allocreadium* sp. and *Clinostomum giganticum*. It is suggested that this negative interaction may be due to non-reciprocal cross-immunity.

Because relatively little was then known on factors influencing the incidence and intensity of infection of fish helminths James and Srivastava (1967) examined 170 *Onos mustelus* for *Podocotyle atomon, Bothriocephalus scorpii, Contracaecum clavatum* and *Echinorhynchus gadi* over a period of one year. There was a general increase in the incidence and intensity of infection by *P. atomon* and *C. clavatum* with increase in the length (age) of the host. With *E. gadi* and *B. scorpii*, however, infection decreased in larger fish. Seasonal variations in the incidence and intensity of infection were influenced by tidal fluctuations, temperature and by the feeding habits and breeding cycle of *O. mustelus*. The frequency distribution of specimen numbers of the four helminth species showed that host specimens infected with only one parasite specimen are the most frequent. The extremely steep curves for *B. scorpii* and *E. gadi* underline the rarity of finding a host with more than one parasite. In contrast, the gradual slope and extended tail of the curve for *P. atomon* shows that host specimens with large numbers of this parasite are commonly found.

More detailed community ecology studies fall into several categories, e.g. low-diversity systems (Kennedy, 1978c; Kennedy *et al.*, 1986; Esch, *et al.*, 1988), food webs and transmission of parasites to fish (Curtis, 1982; Aho and Kennedy, 1987; Holmes *et al.*, 1977), host-specificity and exchange of parasites (Cone and Ryan, 1984; Leong and Holmes, 1981), shoaling fish (Nikolaeva 1963), and important infracommunity and component community investigations (Holmes, 1990, Kennedy, 1990). We shall, therefore, conclude this account of community ecology by discussing these various aspects.

Low diversity

Details of the parasites found and of their relative abundance in *Salvelinus alpinus* in three lakes in the Arctic were discussed by Kennedy (1978d). The differences in incidence and mean worm burden of species found in char in different lakes was probably associated with differences in their ecology, topography and bird-fish interactions. The parasite communities in lakes on different islands were often dominated by a single species, e.g. the parasite fauna of char on Bjornoya was dominated by *Diphyllobothrium ditremum*. It was concluded that the parasite fauna of char on offshore Arctic islands does not agree with the predictions of island biogeography theory. It appeared that the local conditions, especially the invertebrate species present that can act as intermediate hosts, are the major factors determining the composition of the parasite fauna on both the mainland lakes and on the islands, and not the size or position of the island.

Kennedy *et al.* (1986) emphasized that the number of parasite species that a fish species harbours varies widely from one host to another and from one locality to another. They discussed previous work in this area and, for comparison with Kennedy's previous work on Arctic islands (Kennedy, 1978d) analyzed the parasite community structures in the freshwater fish of Jersey. It was found that Jersey had a poor parasite fauna and consequently parasite communities exhibited low diversity and high dominance by a single species and the presence of many vacant niches. They found discrete eel, salmonid and cyprinid components of the parasite fauna. The most interesting feature was the dominance of the eel parasite community by the salmonid element of *Echinorhynchus truttae* and the absence of characteristic eel acanthocephalans such as *Acanthocephalus clavula*, *A. lucii* and *A. anguillae*. The results support the prediction of island biogeographical theory that island parasite communities will be less diverse than those on the mainland but do not support the predicted relationship between species richness and island size or distance from the mainland.

Esch *et al.* (1988) investigated patterns of helminth communities in six species of freshwater fish and found an erratic and unpredictable occurrence and distribution of many species. The observed patchy, spatial distribution of many helminth species could be explained in terms of host movement and ability to cross land- or sea-barriers and break down habitat isolation, and their period of residence in a locality, whether transient or permanent. In emphasizing that colonization plays an important part in structuring helminth communities Esch and colleagues also point out that it is important to distinguish between colonization potential and colonization ability. Because the three groups of fish investigated by Esch *et al.* have already received much attention (see Appendix) and their concepts and conclusions appear to be applicable on a world-wide basis, community ecologists should find their views invaluable in planning further work in this area of research.

Food webs and transmission of parasites

For parasitologists interested in host-parasite interactions at the population level, certain Arctic lakes are ideally suited for modelling. Curtis (1982) found that, although food webs in the lakes are relatively simple, alternative pathways are often available to reinforce parasite transmission to higher levels of the food web. Where fewer pathways exist the robustness of the links between predator and prey organisms assumes critical importance. Under such circumstances the dependence of a top predator, such as arctic char, on the limited number of species available as food, e.g. *Cyclops* and *Microcalanus*, is sufficient to ensure continued transmission of *Diphyllobothrium* spp. Alterations in the structure of basic food-webs of northern lakes, whether natural or artifical, have potentially substantial effects on parasite transmission. Thus experimental manipulation of food webs could be useful in the control of *Diphyllobothrium*.

Aho and Kennedy (1987) followed a study of the infrapopulation dynamics of *Cystidicoloides ephemeridarum* with one on its suprapopulation because the locale permitted quantitative measurements on the absolute population density of all life-history stages of the nematode and its hosts to be obtained relatively easily. They found that due to the greater importance of the mayfly, *Leptophlebia marginata*, in the diet of trout, between 73 and 98 per cent of the nematode infrapopulation circulated through *Salmo trutta*. Around 99 per cent of the parasite's eggs originated from trout, but of these, 90 per cent or more were not ingested by an insect. Larval *C. tenuissima* were found in 18 species of insects but could develop to the infective stage only in *L. marginata*. Two common but unsuitable species of insects harboured up to 80 per cent of the larval parasites. Thus less than 10 per cent actually circulated through *L. marginata*. Differences in circulation pattern between sites could be related more to differences in fish feeding preferences than to differences in fish or insect density, but monthly differences in transmission rate reflected both fish diet and insect abundance. The overall mean transmission rate of eggs to larvae in *L. marginata* varied from 0.25–0.87 per cent but transmission rates of these larvae to fish were from 10.8–39.8 per cent. Aho and Kennedy were, therefore, able to discuss in detail the relative importance of ecological factors, host community structure and parasite specificity in determining circulation routes and transmission efficiencies.

Holmes, Hobbs and Leong (1977) derived a deterministic mathematical model employing discrete generations to illustrate the flow rates of the acanthocephalan *Metechinorhynchus salmonis* in 10 species of salmonid and other fish in Cold Lake, Alberta, Canada. They concluded that a suprapopulation of a parasite (all the individuals of a single species in an ecosystem) can be regulated at the level of the host individual by a mechanism operating on infrapopulations (all the individuals of a single species in a single host individual) in only one of several host species.

Host-specificity and exchange of parasites

The metazoan parasite community of fishes (*Salvelinus fontinalis, Salmo salar* and *Gasterosteus aculeatus*) in a Canadian lake was studied by Cone and Ryan (1984). It consisted of three components: parasites of salmonids, parasites of sticklebacks and parasites shared by the two groups of fish. *S. fontinalis* carried almost all of the salmonid parasite populations. The loss of parasites due to the seaward migration of *S. salar* at 2.5 years of age was insignificant. Most of the population of *Cystidicoloides tenuissima, Metechinorhynchus lateralis* and *Ergasilius* sp. is carried by the more abundant small fishes while most of the population of *Discocotyle sagittata, Apophallus imperator, Tetracotyle* sp. 1 and *Salmincola edwardsii* is carried by less abundant large hosts. There is a common life-history pattern for each of the two groups of parasites and these two patterns suggest an explanation for the different distributions of parasite numbers in the two groups. Those in group one have annual life-cycles whereas those in group two have a tendency to be long-lived and to accumulate from year to year.

Communities of parasites found in 10 species of fish in a Canadian lake were described by Leong and Holmes (1981). Parasites of the numerically dominant salmonid fishes dominated the overall community, with over half of the individual parasites being *Metechinorhynchus salmonis*. Exchange of parasites between host species was greatest between related and/or abundant host species. Parasites of communities of *Coregonus* spp. were relatively rich in species and diverse compared with communities in these species in other lakes in North America, but communities in *Salvelinus* spp. and the non-salmonid fish were poor in species and low in diversity compared with communities in other lakes. The results supported the hypothesis of Wisniewski (1958) that parasite communities within an ecosystem are characterized by parasites of the numerically dominant hosts.

Shoaling fish

Nikolaeva (1963) investigated the parasite fauna of local stocks of anchovy, sprat and horse mackerel in the Black Sea in relation to the biology of the hosts, migrations of the fish, composition of their food, their age, season and the composition of the coexisting invertebrate fauna—as well as several other biotic and abiotic factors. She found that a feature of shoaling fish is the extreme uniformity of their parasite fauna. The species composition of the parasites of different local stocks of pelagic fish in the Black Sea is basically the same, the stocks differing primarily in the prevalence and intensity of infection. The application of this kind of community ecology strictly to the fishing industry is discussed.

Infra, component and compound communities

Holmes (1990) and Kennedy (1990) stress the lack of information at these three levels of study, in particular the first and last. Component communities are relatively better understood. Kennedy presented data indicating that helminth communities in eels, *Anguilla anguilla* are potentially rich but the richness was seldom approached and never achieved. Both infra and component communities were found to be generally species-poor with low diversity. A challenging conclusion for future researchers is that environmental and habitat conditions of a compound community also determine the helminth community structure and pattern in eels. Kennedy also suggests that most fish helminth communities are isolationist in nature, and rather more in the way of being stochastic assemblages than structural communities. The infracommunities are samples of the specialists in the host component community and generalists in the compound community. Interactions between species were thought to be uncommon or nonexistent.

Holmes (1990) suggested that marine fish have infracommunities of alimentary canal helminths which are highly variable in numbers of individuals and species, but frequently are more complex (contain more species for a given number of individuals) than communities in freshwater fish. The difference was explained by two factors, the greater vagility[1] of marine fish and the relatively broad specificity of marine teleost parasites. Several factors, however, suggest that host-specificity plays a minor role in determining some helminth communities of the alimentary canal of marine teleosts in cold waters. Host specificity is thought to play a much more important role in monogeneans and in communities of helminths of tropical marine teleosts and those of elasmobranchs. Factors acting at the host species level were, therefore, thought to be important determinants for some helminths and stochastic factors important for others.

Appendices

Introduction

At the beginning of this chapter we drew attention to difficulties in planning and arranging information on the ecology of fish helminths. Having presented our results under various headings we remained conscious of a number of papers, especially major taxonomic works containing ecological data, which could not easily be included in our discussions of these. Many other papers are checklists or surveys which, according to Petrushevskii and Petru-

---

[1] Vagility = relative movement of the host, i.e. either the movement of the host over large areas and diverse habitats or local movements of hosts over small areas providing these include different habitats. Kennedy, Bush and Aho (1986, pp. 211–212).

schevskaya (1960), Price and Clancy (1983), Mackiewicz (1972) and Kennedy, Bush and Aho (1986), can be valuable, if used cautiously, for further research in ecology. We have, therefore, compiled two lists as appendices to the chapter. These are intended to aid and encourage further work on the ecology of fish worms especially on such aspects as the coevolution of hosts and parasites, zoogeography and community ecology.

The **first appendix** is a selection of papers arranged according to families of fish which appear to have received most attention by parasitologists, using the classification by Nelson (1976). For convenience of use, families of the Osteichthyes are given in alphabetical order. An interesting feature of Appendix One is the relatively large amount of attention given to Cyprinidae and Salmonidae, for instance, and that so little has been done on living survivors of the Subclass Dipneusti, especially the lungfish genera, *Neoceratodus*, *Lepidosiren* and, *Protopterus*, the coelacanth *Latimeria chalumnae* and the paddlefish *Psephurus* and *Polyodon* (although a number of species have been recorded from *Polyodon*.)

Appendix I. Checklists, surveys and other references: Hosts.

PETROMYZONTIFORMES: Appy and Anderson (1982); Lethbridge, Potter, Bray and Hilliard (1983); Wilson and Ronald (1967).

HOLOCEPHALI: Dienske (1968); Williams, Colin and Halvorsen (1987).

CHRONDRICHTHYES: Bates (1990); Beveridge (1990a,b); Beveridge and Campbell (1988 a,b); Brooks and Mayes (1980); Orlowska (1979); Watson and Thorson (1976); Williams (1966, 1968c, 1969); Williams, McVicar and Ralph (1970).

FAMILIES OF THE CLASS OSTEICHTHYES

ACIPENSERIDAE: Appy and Dadswell (1978); Ibragimov (1985); Lyubarskaya and Lavrent'eva (1985).

ALEPOCEPHALIDAE: Zubchenko (1984).

ANARRHICHADIDAE: Bray (1987b); Zubchenko (1980)

ANGUILLIDAE: Buchmann (1988a,b,c,d,e); Buchmann, Mellergaard and Køie (1987); Carvalho-Varela *et al.* (1984); Conneely and McCarthy (1986); Crane and Eversole (1980, 1985); Cruz *et al.* (1986); Esch *et al.* (1988); Golovin and Shukhgalter (1979); Hanek and Threlfall (1970); Hellström *et al.* (1988); Hine (1980a,b); Hine and Francis (1980); Hristovski and Riggio (1977); Kennedy (1985a, 1990); Kennedy and Moriarty (1987); Kennedy and Fitch (1990); Køie (1988a,b); Lacey, Williams and Carpenter (1982); Molnár (1985); Peters and Hartmann (1986); Rid (1973); Seyda (1973); Taraschewski *et al.* (1987); Williams and Bolton (1985).

ATHERINIDAE: Nikitina (1983).

BAGRIDAE: (Bagrid catfish, Africa) Srivastava and Mukherjee (1976); Mbahinzireki (1980).

BLENNIIDAE: Lushchina (1985).

BRAMIDAE: Rokicki (1981).

CARANGIDAE: Cattan and Videla (1976); Gaevskaya and Kovaleva (1980a,b, 1982, 1985); Llewellyn (1962); Piasecki (1982).

CATOSTOMIDAE: Bell and Beverley-Burton (1980); Combs, Harley and Williams (1977); Hayunga (1980); Muzzall (1980a,b, 1982); Mergo and White (1984); Molnár, Chan and Fernando (1982).

CENTRARCHIDAE: Amin (1988); Cone and Anderson (1977); Esch, Johnson and Coggins (1975); Goude and Vaniceck (1985); Hanek (1974); Hanek and Fernando (1978a,b,c,d,e); Hazen and Esch (1978); Ingham and Dronen (1980); Joy (1984); McDaniel and Bailey (1974).

CERATODONTIDAE: Pichelin, Whittington and Pearson (1991).

CHANNIDAE: Gupta, Niyogi, Naik and Agarwal (1984); Mahajan *et al.* (1978).

CICHLIDAE: Nasir and Gomez (1977a); Paperna, van As and Basson (1983).
CLARIIDAE: Boomker (1982); Mashego and Saayman (1981); Niyogi *et al.* (1982, 1984); Shotter (1980).
CLUPEIDAE: Apprahamian (1985): Arthur and Arai (1980a,b); Bishop and Margolis (1955); Feijo *et al.* (1979); Gaevskaya and Shapiro (1981); Hubschman (1985); Ivanchenko and Grozdilova (1985a,b); Khalil (1969); Kulachkova (1977); MacKenzie (1985, 1987b); McGladdery (1985); McGladdery and Burt (1985); Parukhin (1976); Petrushevskii (1957).
COBITIDAE: Adamczyk (1980); Prost (1981); Robotham and Thomas (1982); Rumpus (1975)
COTTIDAE: Adamczyk (1979).
CRANOGLANIDAE: Chang and Ji (1981).
CYCLOPTERIDAE: Machida (1985).
CYPRINIDAE: Baturo (1978); Bibby (1972); Esch *et al.* (1988); Davydov *et al.* (1981); Evans (1977, 1978); Gattaponi and Corallini (1982); Gomonenko and Goncharenko (1981); Gurkina (1983); Hanzelova and Zitnan (1985); Heckmann *et al.* (1986); Izyumova *et al.* (1985b); Izyumova, Zharikova and Stepanova (1985); Kiskaroly (1987); Kiskaroly and Tafro (1983a,b,c, 1984a,b,c); Loseva (1983); Malhotra (1982); Mashtakov (1979); Moody and Gaten (1982); Moravec (1984b,c, 1985b, 1986); Nakajima and Egusa (1977a,b,c); Niewiadomska (1977); Oisboit and Yasyuk (1980); Pashkyavichyute (1981); Pereira-Bueno (1980); Pojmanska and Dzika (1987); Razmashkin *et al.* (1984); Riley (1978); Rautskis (1982, 1983); Reda (1987, 1988); Shulman (1977); Sous (1979); Timoshechkina (1978); Tokobaev and Chibichenko (1978); Uraev (1981); Vinikour (1977); Wierzbicka (1977, 1978); Willomitzer (1980a); Yunchis (1977); Zharikova (1981, 1984a,b); Zitnan and Hanzelova (1981, 1984b).
CYPRINODONTIDAE: Adams (1985, 1986); Barkman and James (1979); Janovy and Hardin (1987, 1988); Kozel and Whittaker (1985); Weisberg (1986).
EMBIOTOCIDAE: Arai (1967).
ENGRAULIDAE: Linnik (1980b); Rego *et al.* (1983).
ESOCIDAE: Malakhova (1982); Moravec (1979c); Thompson and Threlfall (1978); Watson and Dick (1980).
EXOCOETIDAE: Gichenok (1979); Kurochkin (1980).
GADIDAE: Appy & Burt (1982); Arthur, Margolis, Whitaker & McDonald (1982); Boje (1987); Buchmann (1986); Bussmann & Ehrick (1979); Grabda (1977b, 1978); Karasev (1983b, 1984a,b); Kelle (1977); Kulachkova & Timofeeva (1977); Kusz & Treder (1980); Lopukhina *et al.* (1979); MacKenzie (1979); Muzzall, Whelan & Peebles (1987); Scott (1981, 1985a); Srivastava (1966a,b,c,d); Shotter (1976); Timofeeva & Marsaeva (1984); Wirtz & Schreiber (1980).
GASTEROSTEIDAE: Chappell (1969a,b); Curtis (1981); Dartnall & Walkey (1979); Font (1983); Hotta *et al.* (1981); Pennycuick (1971a,b); Rau & Gordon (1978).
GOBIIDAE: Vaes (1978).
ICTALURIDAE: Bowser & Evans (1983).
LABRIDAE: Sekhar & Threlfall (1970a,b).
LATIMERIDAE: Dollfus & Campana-Rouget (1956); Kamegai (1971b).
LEPIDOSIRENIDAE: Berger & Thatcher (1983).
MACROURIDAE: Armstrong (1975); Zubchenko (1985).
MERLUCCIIDAE: Aleshkina (1982); Carvajal *et al.* (1979); Carvajal & Cattan (1985); Duran & Oliva (1980); Sankurathri, Kabata & Whitaker (1983); Scott (1987).
MUGILIDAE: Armas (1979); Lahav (1974); Moravec & Libosvarsky (1975); Paperna (1975); Paperna & Overstreet (1981); Rawson (1977); Rekharani & Madhavi (1985); Skinner (1975); Solonchenko & Tkachuk (1985).

MYCTOPHIDAE: Mordvinova & Parukhin (1987).

NOTOTHENIIDAE: Parukhin (1986); Parukhin & Lyadov (1981); Parukhin & Zaitsev (1984); Zdzitowiecki (1988).

OSMERIDAE: Fréchet, Dodson & Powles (1983); Kazakov (1967); Palsson & Beverley-Burton (1984).

PERCIDAE: Adamczyk (1978); Andersen (1978a); Andrews (1979); Bespyatova & Rumyantsev (1982); Izyumova & Mashtakov (1984); Kennedy & Burrough (1978); Kennedy (1981c, 1987); Poole & Dick (1985); Priemer (1979); Rautskis (1977); Starovoitov (1986); Tarmakhanov (1987); Tedla & Fernando (1969, 1970/71); Smith (1986); Tregubova (1972); Walter (1988); Wierzbicka (1970).

PLATYCEPHALIDAE: Hooper (1983).

PLEURONECTIDAE: Burn (1980b); MacKenzie & Gibson (1970); Hendrickson & Yindeepol (1987); Lauckner (1984); Scott (1975a,b, 1982, 1985b); Sulgostowska *et al.* (1987); Tsimabyuk (1978); Wickins & MacFarlane (1973); Zubchenko (1980).

POECILIIDAE: Aho, Gibbons & Esch (1976); Davis & Huffman (1978); Granath & Esch (1983a); Wright & Boyce (1986).

POLYODONTIDAE: Miyazaki, Rogers & Semmens (1988); Yu Yi & Wu Huisheng (1989).

POLYPTERIDAE: Shotter & Medaiyedu (1977).

PROTOPTERIDAE: Khalil & Thurston (1973).

SALMONIDAE: Alvarez-Pellitero (1976a,b, 1979a); Anikieva, Malakhova & Ieshko (1983); Arthur (1984); Bailey & Margolis (1987); Beverley-Burton (1978b); Bristol *et al.* (1984); Butorina (1975, 1978, 1980); Butorina & Kuperman (1981); Cone & Cusack (1988); Cone & Ryan (1984); Ching (1985); Dick & Belosevic (1981); Dick & Rosen (1981); Esch *et al.* (1988); Frandsen, Malmquist & Snorrason (1989); Frimeth (1987a); Greenwood & Baker (1987); Haldorson (1984); Halvorsen & Andersen (1984); Halvorsen & Hartvigsen (1989); Johnsen & Jensen (1986, 1988); Kakacheva-Avramova & Menkova (1979); Kashkovskii (1979); Kazic (1978); Kennedy (1977a, 1978d); Konovalov & Butorina (1985); Lockard, Parsons & Schaplow (1975); Malakhova (1976); Malakhova & Anikieva (1975); Malakhova, Titova & Potapova (1972); Mamyshev, Spiranti & Chernova (1981); Margolis (1982a,b); Margolis & Boyce (1990); Margolis & Evelyn (1987); Mitenev (1982); Moravec (1982a); Mpoane & Rinne (1984); Muzzall (1984b, 1986); Muzzall & Peebles (1986); Nagasawa (1985); Nilz (1984); Ogawa & Egusa (1985); Osipov (1984); Pennell, Becker & Scofield (1973); Permyakov & Rumyantsev (1982); Pickering & Christie (1980); Potapova *et al.* (1972); Poynton & Bennett (1985); Pronin & Tugarina (1971); Razmashkin *et al.* (1979, 1981); Rau, Gordon & Curtis (1979); Rimaila-Pärnänen & Wiklund (1987); Rintamäki & Valtonen (1988); Roytman (1968); Sattaur (1988); Sobenin (1975); Speed & Pauley (1984); Stables & Chappell (1986a,b,c); Thomas (1964); Vahida (1984); Valtonen & Valtonen (1978); Watson & Dick (1979); Whitaker (1985); Wootten & Smith (1980); Yakushev (1984).

SCIAENIDAE: Mergo & Crites (1986).

SCOMBRIDAE: Komasara & Lisinska (1986); Lester, Barnes & Habib (1985); MacKenzie (1990); Rego *et al.* (1985); Manooch & Hogarth (1983); Rohde (1987a,b).

SCORPAENIDAE: Bourgeois & Ni (1984); Muzzall & Sweet (1986); Scott (1988); Sekerak & Arai (1973, 1977).

SERRANIDAE: Ulrich (1983); Silan, Euzet & Maillard (1987).

SOLEIDAE: Olson (1978).

SPARIDAE: Roubal (1981).

SYNODONTIDAE: Jensen, Moser & Heckmann (1979); Sathyanarayana (1982).

XIPHIIDAE: Hogans *et al.* (1983).

ZEIDAE: Korotaeva (1982).

ZOARCIDAE: Graś-Wawrzyniak *et al.* (1979).

Appendix II. Check-lists and surveys: Countries.

In this appendix the importance of each country in world fisheries in ranking order is indicated by a number in brackets following its name. Thus Japan is the most important and Egypt the least. The word 'others' in brackets follows the name of 12 countries because they are of lesser importance than Egypt with regard to fish catches and are not given a ranking order in the FAO Year Book for 1988. The additional information which follows the ranking number indicates (a) the country's position in Wallace's classification of the zoogeographical regions of the world as given in Lincoln, Boxshall and Clark (1982), and (b) its location in the FAO world map of major inland and marine fishing areas. This information is intended to aid and encourage further research in biogeography on the assumption that most workers will wish to continue indicating the zoogeographical region and the country in which the work was carried out, the area investigated and, if possible, the precise location where the fish were caught. In publishing the results of survey work it may be more useful to indicate the FAO fishing areas than to refer to the traditional six biogeographical regions and their 24 subregions, but information on both would be preferable. Information on the different vertical regions under investigation is also important, ie. whether epipelagic, mesopelagic, bathypelagic, demersal or abyssal (see Collard, 1970; Noble & Collard, 1970; and Campbell, 1990).

An interesting feature of this list is the relatively large amount of work already carried out in some countries, eg., the United Kingdom, United States, USSR. Scandinavia and Canada, and that so little has been done in others which are fairly high in the FAO ranking order, eg. China, Chile, Thailand, South Korea, Peru, India, Indonesia and the Philippines.

Afghanistan (Others) — Siberian subregion (P2) and Mediterranean subregion (P3)

Moravec & Amin (1978) — Mediterranean subregion (P3)

Africa (Others) — East African subregion (E1)
Ethiopian region (E1-E4)
Malgasy subregion (E3)
South African subregion (E4)

Bray (1984, 1986a, 1987a); El-Naffar, Saoud & Hassan (1983); Ergens (1981, 1988b); Faisal (1988); Fischthal & Thomas (1968, 1972); Khalil (1969, 1974); Khalil & Thurston (1973); Mashego & Saayman (1989); Molnár & Mossalam (1985); Obiekezie, Möller & Anders (1988); Paperna (1979, 1980); Sarig (1976); van As & Basson (1984).

Argentina (30) & **Antarctica**: — Neotropical Chilean subregion (NT4)
America South-Inland waters (03)
Atlantic Southwest (41)
Atlantic Antarctica (48)

Gaevskaya, Kovaliova & Rodjuk (1985); Gibson (1976); Holloway & Spence (1980); Lyadov (1985); Parukhin (1973, 1986); Rodjuk (1985); Zdzitowiecki (1979, 1987a,b).

Australia and **New Zealand** (Others) — Australian subregion (A1)
& New Zealand subregion (A4)

Beumer *et al.* (1983); Beveridge (1990); Callinan (1986); Hine (1977b, 1978); Langdon & Humphrey (1986); Lester (1986); Munday (1986); Reddacliff (1986); Rodhe (1987a,b, 1988).

Austria (Others) — Palaeartic, European subregion (P3)

Kritscher (1988a,b).

Bangladesh (23)
Oriental, Indian subregion (01) and Indo-Chinese subregion (03)
Asian Inland waters (04)
Indian Ocean Eastern (57)

Brazil (24)
Neotropical Brazilian subregion (NT3)
America South Inland waters (03)
Atlantic Southwest (41)

Fernandes *et al* (1985); Kohn & Fernandes (1987); Rego (1975); Vicente, Rodrigues & Gomes (1985); Wallet & Kohn (1987).

Bulgaria (Others)
Palaearctic, European subregion (P2)

Kakacheva-Avramova (1977, 1983a,b); Nedeva (1988).

Canada (16)
Nearctic, Canadian subregion (NA1)
America North-Inland waters (02)
Arctic Sea (18)
Atlantic Northwest (21)
Pacific Northeast (67).

Appy & Burt (1982); Arai (1969, 1989); Arai & Mudry (1983); Arai, Kabata & Noakes (1988); Arthur (1984); Arthur & Arai (1980a,b); Beverley-Burton (1984); Bray (1979); Chinniah & Threlfall (1978); Houston & Haedrick (1986); Leong & Holmes (1981); Lester (1974); Mackie, Morton & Ferguson (1983); Margolis (1970b) Margolis & Arthur (1979); Mudry & Anderson (1977); Rosen & Dick (1983); Smith (1986); Stromberg & Crites (1975a,b); Szalai, Yang & Dick (1989).

Chile (5)
Neotropical region, Chilean subregion (NT4)
Pacific Southeast (87) and
Pacific Antarctic (88).

Torres (1990); Torres, Teuber & Miranda (1990), Torres *et al.* (1991).

China (3)
Indo-Chinese subregion (03)
Asia-Inland waters (04) and
Pacific, Northwest (61)

Chang (1983); Chang & Ji (1978, 1981); Jianying & Qizhi (1981); Lang (1981), Li *et al.* (1987), Liao & Liang (1987); Liu & Zhang (1987); Wang (1980a,b, 1981a, b, 1982a, b, 1984), Wang, Zhao & Chen (1978).

Czechoslovakia (Others)
Palaearctic, European subregion (P1)

Ergens (1981, 1988a, b); Ergens & Gelnar (1985); Gelnar (1987a, b, c.)
Lom (1986); See reference list for many papers by Moravec.

Denmark (14)
Palaearctic region (P1)
Europe-Inland waters (05)
Atlantic Northeast (27)

See many references to Køie.

Ecuador (26)
Neotropical Brazilian subregion (NT3)
America South Inland waters (03)
Pacific Eastern Central (77)
Pacific Southeast (87).

Egypt (44) — Palaearctic (P1-P4)
Mediterranean subregion (P3)
Africa, Inland waters
Mediterranean & Black Sea (37)

See under Africa.

Faroe Islands (39) — Palaearctic region (P1-P4)
European subregion (P1)
Atlantic Northeast (27)

France (22) — Palaearctic European subregion (P1)
Palaeartic Mediterranean subregion (P3)
Europe Inland waters (05)
Mediterranean and Black Sea (37).

See references to papers by Dollfus and by Euzet: also Robert, Boy & Gabrion (1990); Silan & Maillard (1986).

Germany (others) — Palaearctic, European subregion (P2)

Lux (1989); Smija (1982).

Ghana (38) — Ethiopian region (E2-E4)
West African subregion (E2)
Africa–Inland waters (01)
Atlantic, Eastern Central (34)

See under Africa.

Greece (Others) — Palaearctic, Mediterranean subregion (P3)
Papoutsoglou (1976)

Greenland (Others) — Palaearctic, European subregion (P1)

Iceland (15) — Palaearctic (P1);
Atlantic, Northeast (27)

India (7) — Indian subregion (01)
Indo-Chinese subregion (02) and
Ceylonese subregion (03)
Asia Inland waters (04)
Indian Ocean, Western (51)
Indian Ocean, Eastern (57) and
Pacific, Western Central.

Chauhan & Malhotra (1984); Gupta & Ahmad (1979); Maiti, Roy & Haldar (1987); Mahajan *et al.* (1978); Malhotra & Chauhan (1984); Niyogi, Gupta & Agarwal (1982); Radhakrishnan & Nair (1981); Soota (1983); Tkachuk (1985).

Indonesia (9) — Australian region (A2-A4)
Austro-Malayan subregion A2
Pacific, Western Central (71)

Iran (others) — Palaearctic, Mediterranean subregion (P3)
Mokhayer (1981)

Iraq (Others) — Palaearctic, Mediterranean subregion (P3)
Ali *et al.* (1987a,b,c, 1988); Khalifa *et al.* (1978); Khamees & Mhaisen (1988); Mhaisen *et al.* (1988).

Israel (Others) Palaearctic Mediterranean subregion (P3)

Fischthal (1980). See papers by Paperna in list of references.

Italy (31) Palaearctic Mediterranean subregion (P3)
Europe Inland waters (05)
Mediterranean and Black Sea (37).

Orecchia & Paggi (1978) and other references to these authors; Ulmer & James (1981).

Japan (1) Manchurian subregion (P4)
Pacific, Northwest (61)

Nagasawa, K (1986, 1987, 1988, 1989). These four newsletters cite 187 references to recent publications on fish parasites in Japan, many on parasitic worms. See also Kamegai & Ichihara (1972).

Malayasia (29) Oriental Indo-Malayan subregion (04)
Asia Inland waters (04)}
Indian Ocean Eastern (57)
Pacific Western Central (71)

Cheah & Rajamanickam (1987a,b) Fernando & Furtado (1963), Leong & Wong (1988); Seng *et al.* (1987); Zaman & Leong (1987, 1988).

Mexico (17) Neotropical, Mexican subregion (NT1)
Atlantic Western Central (31)
Pacific Eastern Central (77)

Bravo-Hollis (1984); Zhukov (1976).

Morocco (33) Palaearctic Mediterranean subregion (P3)
Africa Inland waters (01)
Atlantic Eastern Central (34)
Mediterranean and Black Sea (37)

Myanmar (Burma) (25) Oriental, Indian subregion (01) and
Indo-Chinese subregion (03)
Asian Inland waters (4)
Indian Ocean Eastern (57)

Namibia, South West Africa (32) Ethiopian South African subregion (E4)
Africa Inland waters (01)
Atlantic Southeast (47)

See under Africa.

North Korea (13) Palaearctic (P4)
Asia, Inland waters (04)
Pacific, Northwest (61).

Norway (12) Palaeartic region (P1)
Europe Inland waters (05)
Atlantic, Northeast (27)

See under Scandinavia

New Zealand (35) Australian New Zealand subregion (A4)
Pacific Southwest (81)
Pacific Antarctica (88)

See under Australia & New Zealand.

Pakistan (36)
Indian subregion (01)
Asia—Inland waters (04)
Indian Ocean, Western (51)

Peru (6)
Neotropical, Chilean subregion (NT4) and Pacific, Southeast (87)

Philippines (11)
Indo-Malayan subregion (04)
Pacific, Western Central (71)

Poland (27)
Palaearctic Siberian subregion (P2)
USSR Inland waters (07)

Dzika (1987); Grabda-Kazubska & Pilecka-Rapacz (1987); Prost (1984, 1988)

Portugal (37)
Mediterranean subregion (P3)
Europe—Inland waters (05)
Atlantic, Northwest (27)

Carvalho-Varela & Cunha-Ferreira (1987); Rodrigues, Noronha & Carvalho-Varela (1975); Rodrigues *et al.* (1972, 1973, 1975).

Romania (43)
Palaearctic region (P1-P4)
European and Mediterranean subregions (P1-P3)
Europe, Inland waters (05)
Mediterranean & Black Sea (37).

Scandinavia
Palaearctic European subregion P1;
Europe Inland waters (05)
Atlantic Northeast (27)

Andersen, (1978b); Bakke (1985); Berland (1987); Buchmann (1988a, c, d, e); Buchmann, Mellergaard and Køie (1987); Bylund (1987); Dolmen (1987); Fagerholm (1990); Halvorsen (1972); Halvorsen & Hartvigsen (1989); Kennedy (1978a, d); Køie (1987a,b); Körting (1987); Malmberg (1987a,b); Mo (1987a,b, 1989); Turovskij (1985); Valtonen (1980a, b, c, 1983a,b); Valtonen & Crompton (1990); Valtonen, Gibson & Kurttila (1984); Valtonen & Niinimaa (1983); Valtonen & Valtonen (1980); Vik (1954, 1958, 1959, 1963). See also list of references for other papers by these authors.

South Africa (20)
Ethiopian South African subregion E4
Africa Inland waters (01)
Atlantic, Southeast (47)
Atlantic, Antarctic (48)
Indian Ocean Western (51)
Indian Ocean Antarctic (58)

See under Africa

South Korea (8)
Palaearctic region P4
Asia, Inland waters (04) and
Pacific, Northwest (61)

Joo (1988)

Senegal (42)
Ethiopian region E1-E4
Africa, Inland waters (01)
Atlantic, Eastern Central (34)

See under Africa

Spain (18) — Palaearctic Mediterranean subregion P3
Europe Inland waters 05
Atlantic, Northeast (27)

Sanmartin Durán *et al.* (1989)

Tanzania (40) — Ethiopian region E1-E4
Africa Inland waters (01)
Indian Ocean, Western (51)

Thailand (10) — Indo-Chinese subregion (03)
Asia-Inland waters 04
Indian Ocean, Eastern (57)
Pacific, Western Central (71)

Chonchuenchob, Sumpawapol & Mearoh (1987)

The Netherlands (34) — Palaerctic P1
Europe Inland waters (05)
Atlantic Northeast (27)

Turkey (28) — Palaearctic Mediterranean subregion P3
Europe Inland waters (05)
Mediterranean & Black Sea (37)

United Kingdom (19) — Palaearctic European subregion P1
Europe—Inland waters (05)
Atlantic Northeast (27)

Brown, Chubb & Veltkamp (1986); Chappell & Owen (1969); Chubb (1970, 1977a, b, 1979, 1980, 1982); Coneely & McCarthy (1984); Crozier (1987); Kennedy (1974b, 1978b,c); Kennedy & Burrough (1978); Lee (1977); McCullough, Fairweather & Montgomery (1986); McGuigan & Sommerville (1985); Wootten & Waddell (1977). See also list of references for papers by Arme & Co-author, Bray, Gibson, Kearn, Kennedy, Llewellyn, MacKenzie, McVicar, Smith, Williams.

United States (4) — Nearctic region including Californian subregion NA2
Californian subregion NA2
Rocky Mountain subregion NA3 and Alleghany subregion NA4
America North-Inland waters (02)
Arctic Sea (18)
Atlantic Northwest (67)
Atlantic Western Central (31)
Pacific, Northeast (67) and
Pacific Eastern Central (77)

Adams (1985); Amin (1977, 1981, 1982, 1984, 1985a,b); Blouin *et al.* (1984); Bravo-Hollis & Deloya (1973); Campbell (1983); Campbell, Haedrick & Munroe (1980); Fischthal (1977); Gartner & Zwerner (1989); Goude & Vanicek (1985); Gruninger, Murphy & Britton (1977); Hendrix & Short (1965); Hendrix & Overstreet (1977); Joy (1984); Janovy & Hardin (1987); Joy, Tarter & Sheridan (1986); Love & Moser (1983); Lubieniecki (1973); Miller *et al.* (1973); Muzzall 1984a, b); Nahhas & Cable (1964); Nickol & Samuel (1983); Noble & Collard (1970); Overstreet (1969, 1978a); Sarabia (1982); Skinner (1978); Sogandares-Bernal (1959b); Underwood & Dronen (1984); Zhukov (1985).

USSR (2)
Palaearctic (P2)
Arctic Sea (18)
Pacific Northwest (61)

Andreyuk (1984); Bauer (1984); Bocharova & Nud'ga (1983); Bocharova, Golovo & Nikulina (1983); Chernova (1975, 1981); Dubinina (1980); Golovin & Shukhgalter (1979); Gurkina (1983); Ieshko, Malakhova & Golitzyna (1982); Iskov (1975, 1976); Izyumova *et al.* (1982a,b); Karaseva (1983); Linnik (1975, 1983); Mikailov & Ibragimov (1980); Mitenev (1979, 1984a,b); Osipov *et al.* (1981); Osmanov & Yusopov (1985); Parukhin (1975, 1976), Pugachev (1983, 1984a,b); Protasova (1977, 1982); Roitman (1981); Skryabina (1975, 1978); Rumyantsev (1982); Solonchenko (1982); Strelkov (1983); Strelkov *et al.* (1981); Strelkov & Shul'man (1971); Valovaya (1979) and very many other papers by Russian authors cited in this book.

West Indies
Neotropical region (NA1-NA4)
Antillian subregion (NT2)
Pacific Western Central

Sogandares-Bernal (1959)

Venezuela (41)
Neotropical region (NT2-NT4)
Brazilian subregion (NT3)
America, South Inland waters (03)
Atlantic, Western Central (31)

Alvarez & Conroy (1985); Bashirullah & Rado (1987); Nasir & Fuentes-Zambrano (1983); Nasir & Gomez (1977a,b).

Vietnam (21)
Indo-Chinese subregion (03)
Asia—Inland waters (04)
Indian Ocean Eastern (57)
Pacific Western Central (71)

Moravec & Sey (1989)

Yugoslavia (others)
Palaearctic, European subregion (P1)

Kiskaroly (1987); Kiskaroly & Tafro (1983a,b,c, 1984a,b,c).

4 Host–parasite relationships

Introduction

'Host-parasite relationships' has been used in the literature to encompass many aspects of the relationship between parasite and host, including ecology, life-cycles and host-parasite interactions (pathology and immunology). Another aspect is that of host-specificity. The present chapter concentrates on host-specificity and the factors which influence it.

Two broad categories of host-specificity have been recognized: phylogenetic host-specificity in which related parasites inhabit related hosts as a result of parallel evolution, and ecological specificity which is determined by the ecological requirements of the host and parasite and may result in related parasites infesting unrelated hosts which share the same habitat.

Phylogenetic host-specificity. Three rules have been applied to the parallel evolution of hosts and parasites. As expressed by Rohde (1982a) these are: (i) Fahrenholz's rule that the classification of some parasitic groups parallels that of their hosts, ie. ancestors of extant parasites must have parasitized ancestors of extant hosts and both groups must have evolved in parallel; (ii) Szidat's rule that primitive hosts have primitive parasites, and specialized hosts have specialized parasites, and (iii) Eichler's rule that large host groups have more genera of parasites than small groups. As Rohde (1982a) noted, these rules are not equally valid for all parasitic groups and are least relevant to groups with a low degree of phylogenetic host-specificity.

Ecological host-specificity. Ecological host-specificity is determined by the ecological requirements of the hosts and parasites rather than by parallel evolution. Hosts which share the same environment and have, for example, similar feeding requirements, are likely to harbour closely-related parasites although they are not closely related themselves. In order that a parasitic relationship shall develop, the distribution of the host and parasite must overlap spatially and temporally; the parasite must be able to cope with the physicochemical characteristics of the environment during the free-living stages of its life-cycle, and of the habitat within or upon the host during its parasitic stages. Infective stages must be capable of invading the host and, having entered it, of withstanding the host's natural or acquired resistance. The developing and reproductively mature parasite must withstand any adverse changes in its environment, including antigenic changes in the host, and must achieve transmission to the next host in the cycle (Rogers, 1962; Michajłow, 1985). Any of these factors may affect specificity. In parasitic associations of relatively recent origin, the ecology and behaviour of host and parasite are important in determining specificity, whereas in long-established associations physiological factors and natural or acquired host resistance may be more significant.

The factors which affect host-specificity have often been grouped for convenience into three categories: morphological, physiological and ecological. None of these operates in isolation from the others and ultimately specificity is determined by the genetic constitution of host and parasite. Morphological factors are most often linked to the relationship between a parasite and its host at the site of attachment: the attachment organs must be compatible in size and shape with the topography of the surface to which they are attached. This often results in both specificity to a host and also to the site occupied within the host. Physiological factors include the physicochemical characteristics of the habitat, the immune response of the host and the nutritional requirements of the parasite, which influence site-selection and specificity. Ecological factors incorporate the distribution and environment of the host, its diet and mode of feeding, and its behaviour.

A series of reviews by various authors on different ecological aspects of parasitology, edited by Kennedy (1976a), includes host-finding (MacInnes, 1976), host-selection (Holmes, 1976), invasion of the host and site-selection (Crompton, 1976), feeding (Arme, 1976), host responses (Wakelin, 1976) and reproduction and dispersal (Kennedy, 1976b). Among the habitats considered, the body surface of fish (Kearn, 1976) and the gills (Fernando and Hanek, 1976) are particularly relevant. Comprehensive reviews of the evolutionary biology of parasites, although drawing on few examples from the helminths, have elucidated the basic principles of host-parasite relationships (Price, 1977, 1980). The specificity of fish parasites has been reviewed by Shul'man (1954). Site selection has been reviewed by Holmes (1973) and niches and niche restriction by Rohde (1979a, 1981a, 1982a).

Phylogenetic relationships in helminths and host-specificity

The phylogenetic relationships amongst the parasitic helminths have long been the subject of interest and controversy. A number of theories have been advanced to explain the origin and evolution of parasitism in the platyhelminths, nematodes and acanthocephalans and, among other factors, host-specificity has been used as an indicator of host-parasite relationships and of the development of parasitic life-cycles involving hosts from invertebrate and vertebrate groups. A high degree of host-specificity has been taken to indicate the most ancient association between host and parasite. The degree of host-specificity varies widely between the different helminth groups and within groups at different stages of the life-cycle. The monogeneans, with direct cycles, show an extremely high degree of specificity to their vertebrate hosts, explicable only by parallel evolution. Some cestode groups exhibit high specificity to their vertebrate hosts in the adult stage (eg. trypanorhynchans, tetraphyllideans) but others are much less specific. Cestodes also differ in the degree of specificity to the first intermediate host. Digeneans, however, show a much higher degree of specificity to the molluscan first intermediate host than to their second intermediaries or their vertebrate definitive hosts. Nematodes exhibit more specificity towards their definitive than intermediate hosts. Acanthocephalans are considered moderately to strongly specific for their arthropod hosts but more weakly specific for their definitive hosts (Conway Morris and Crompton, 1982a).

Most schemes of classification which have been proposed for the various helminth groups, particularly the platyhelminths, are based on concepts of their phylogenetic relationships.

The following account of theories of the origin and evolution of parasitism in the Platyhelminthes, Nematoda and Acanthocephala is offered as a basis for an understanding of their host-parasite relationships and host-specificity.

Origin, evolution and phylogenetic relationships in the parasitic helminths

Phylum Platyhelminthes

The majority of authors favour the view that the ancestor of the parasitic platyhelminths resembled a rhabdocoel turbellarian. Malmberg (1974, 1986) has suggested that the ancestral form was acoelous and that cestodes never had a gut at any stage of their ontogeny, but Smith and Tyler (1985) have argued that the acoelous condition of modern acoels is secondary, and Brooks, O'Grady and Glenn (1985a) have argued that the absence of a gut in modern taxa is not evidence that it was never present in ancestral forms. Kulakovskaya and Demshin (1978) have, however, derived cestodes from acoels.

Opinions conflict as to whether all the parasitic platyhelminth groups evolved in a more or less linear sequence (eg. Llewellyn, 1965) or whether they arose independently at different times from similar ancestral stock (eg. Freeman, 1973). Many schemes proposed for the evolution and phylogenetic relationships of platyhelminths are based on the Cercomer Theory of Janicki (1920). While not all authorities support this theory, it has been fundamental to the discussion of relationships within the group.

Cercomer theory

Mackiewicz (1984) has reviewed Janicki's (1920) Cercomer Theory and subsequent amendments. According to this theory, the opisthaptor of monogenean oncomiracidia and adults, the cercomer of metacestodes and the tail of digenean cercariae are all homologous, and possession of this common feature allows these groups to be combined in the subclass Cercomeromorphae. The monogeneans were thought to be the most primitive group, which gave rise to the digeneans from which in turn the Cestoidea, including the Cestodaria, then evolved.

The first modification to the theory was the removal of the digeneans from the sequence following rejection of the homology of the cercarial tail with the cercomer of the other groups. Hyman (1951a) cited differences in life-cycles and morphology to show that digeneans had not evolved from monogeneans but that both had evolved independently from rhabdocoel ancestors. Bykhovskii (1957) considered that any resemblances between monogeneans and digeneans resulted from convergence rather than from phylogenetic relationships and separated the monogeneans as a class within a new superclass, again named the Cercomeromorphae, which also contained the Gyrocotyloidea, Cestodaria and Cestoidea. Bykhovskii's (1957) separation of the Monogenea from the Trematoda (Digenea plus Aspidogastrea) has been widely accepted although some workers (eg. Stunkard, 1963) prefer to group monogeneans and digeneans together. Stunkard's (1963) contentions that polyembryony in dactylogyrid monogeneans equates to the successive larval stages in digeneneans and that some monogeneans (*Polystoma*) utilize two hosts have not gained widespread acceptance.

As Mackiewicz (1984) pointed out, excluding the digeneans from the evolutionary sequence has strengthened the case for deriving cestodes from monogeneans, a major feature of the phylogenetic scheme proposed by Llewellyn (1965) (see below). Most subsequent phylogenies (considered in more detail below) have relied to greater or lesser degrees on the Cercomer Theory although the cercomer has been interpreted differently by various authors and the theory itself has not gone unchallenged. For example, Brooks, O'Grady and Glenn (1985a) interpreted the ventral sucker of digeneans, rather than the cercarial tail, and the adhesive disc of aspidogastreans, which they regard as derived from a posterior adhesive organ, as homologues of a cercomer. This interpretation of the ventral sucker has been strongly criticized by Lebedev (1988), who remarked that if the cercomer were to be interpreted in the traditional way, the phylogeny of Brooks, O'Grady and Glenn (1985a) would assume the basic features of those of Bykhovskii (1937, 1957) and Llewellyn (1965).

The Cercomer Theory has been challenged on a number of occasions. Freeman (1973) has pointed out that some cestodes lack cercomers and that *Diphyllobothrium* procercoids, on which Janicki based his observations, are unusual and unrepresentative of the larval stages of primitive cestodes and, therefore, should not be used to explain cestode evolution. Baer and Euzet (1961) and Stunkard (1975a) also rejected the Cercomer Theory. Mackiewicz (1984) suggested that rather than being a vestigial organ that once functioned in swimming or attachment, the cercomer merely represents the remains of the oncosphere and is a product of metamorphosis, the residue of a previous stage rather than an organ in itself. In the metamorphosis from the active burrowing oncosphere to the passive parenteral metacestode, hooks and tissue which are not resorbed are discarded as the cercomer. Any resemblance of the cercomer to a tail has no significance. If this interpretation is correct, the evolutionary significance of the cercomer and the Cercomer Theory should be reassessed. Previously, however, Jarecka, Michajłow and Burt (1981) and Krasnoshchekov (1980) confirmed the phylogenetic significance of the cercomer. Krasnoshchekov (1980) found that the cercomer is not a degenerate organ but is functional, with a role in nourishing the larva, protecting it against the immune response of the intermediate host and against damage in the anterior region of the digestive tract of the final or paratenic host.

Jarecka, Michajłow and Burt (1981) pointed to ultrastructural similarities of cestode larvae from different species and orders, describing the cercomer tegument as microvillar in contradistinction to that of the procercoid which is microtrichal. Subsequently, these similarities were reported for different larval types of various cyclophyllidean species and for a haplobothriidean species (Jarecka, Bance and Burt, 1984; Burt and Jarecka, 1984; MacKinnon, Jarecka and Burt, 1985). In each case, there is a microvillous homologue to the cercomer and to that part of the larval worm, covered with microtriches, which shows pre-adaptation to life in the definitive host.

The following account outlines the major schemes proposed for the origin, evolution and phylogeny of the platyhelminths but is not intended as a detailed coverage of the subject. Because most authorities accept that the digeneans and aspidogastreans are distinct from the other platyhelminth groups, theories on their development are considered after those on the cercomeromorphans (monogeneans and the cestodes).

Cercomeromorphans

Two alternative proposals on platyhelminth evolution, presented by Llewellyn (1965) and Freeman (1973), are outlined here as examples of different views. Llewellyn's (1965) scheme draws heavily on the work of Bykhovskii (1957). In this scheme, free-living rhabdocoel ancestors living on the substrate became ectoparasitic on the first fish (which were slow-moving), developed a posterior apparatus for attachment which became augmented with marginal hooks, then hamuli, or with suckers or sucker-clamps. A direct life-cycle with a vertebrate host was, therefore, established for the primitive monogeneans, which developed a high degree of specificity to their vertebrate hosts. From this ectoparasitic monogenean, the gyrocotylideans, caryophyllideans, cyathocephalideans, amphilinideans and the strobilate cestodes later evolved by the sequence of events outlined below (Figure 4.1).

The ectoparasitic monogeneans invaded the internal organs of the fish host; some modern taxa occupy internal sites, eg. *Calicotyle* (rectum, rectal gland), *Dictyocotyle* (coelom) and *Enterogyrus* (intestine). During evolution into the cestodes, they lost the gut, which was no longer necessary, and developed tanned eggs. This stage represents the protogyrocotylidean stock from which modern gyrocotylids descended and from which the monozoic caryophyllidean line developed by replacing the posterior attachment organ (cercomer) with an anterior one, loss of the gut, shedding of the cercomer so that the excretory system terminated posteriorly in a single terminal pore, and incorporation of an intermediate host into the life-cycle. This might have been accomplished if the eggs were eaten by microphages, hatched and penetrated through the gut wall into the coelom, the host eventually becoming an obligate intermediary. Strobilate cestodes evolved from a caryophyllidean ancestor by serial repetition of the genitalia to increase reproductive potential. Subsequent evolution in the strobilate cestodes proceeded along two lines, one of which lost the tanned eggshell so that the eggs hatched in response to the digestive enzymes of the intermediate host. In this line, which gave rise to the majority of strobilate cestodes, the eggs remained in the uterus to protect them from the enzymes of the definitive host, with the subsequent closure of the uterus and shedding of gravid proglottides by apolysis. The remaining strobilate cestodes, including the ancestors of the pseudophyllidean diphyllobothriids and ligulids, retained tanned eggs which hatched actively in the intermediary's gut, retained an open uterus and shed exhausted proglottides by pseudoapolysis. Second intermediaries were incorporated into the cycles of both lines through the food chains. Price

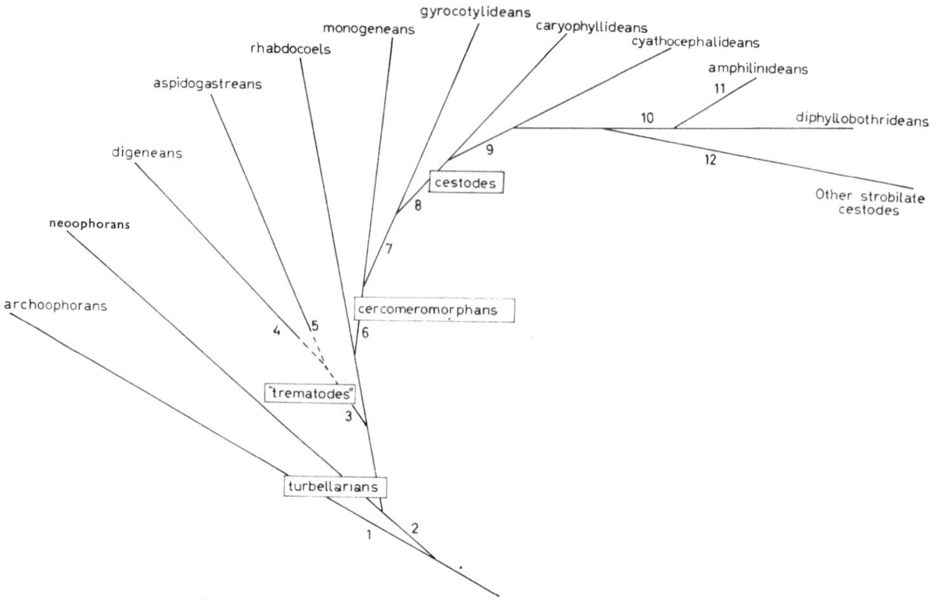

Figure. 4.1 Evolution of parasitic platyhelminths.
1. Spiral cleavage, undivided ovary, flame cells? 2. Irregular cleavage, germarium-vitellarium, flame cells present. 3. Endoparasitism in molluscs 4. Polyembryonic larval multiplication in mollusc; pre-adult actively leaves mollusc and parasitizes vertebrates. 5. No polyembryony; adults do not actively leave mollusc host. 6. Ectoparasitism on vertebrates. 7. Endoparasitism in intestine of vertebrates. 8. Two-host life-cycle; six-hooked larva. 9. Strobilization. 10. Incorporation of a second intermediate host into life-cycle. 11. Progenetic development in second intermediate host, loss of definitive host, four supplementary larval hooks. 12. Loss of tanning, closure of uterus, apolysis. From Llewellyn (1965).

(1967) has suggested a similar origin for cestodes from monogeneans through gyrocotylids but deriving cestodes from the amphilinids, which are parasitic in the coelom of vertebrates (but see Freeman (1973) below).

Objections to Llewellyn's (1965) scheme have been summarized by Mackiewicz (1984). Its opponents have pointed out that its derivation of cestodes from monogeneans does not take into account several basic differences between them. Stunkard (1975a) and Mackiewicz (1984) rejected the homology between the opisthaptor and the cercomer and considered that the oncomiracidium and oncosphere differed fundamentally in hook function and development and in the occurrence of metamorphosis and reversal of polarity in the oncosphere alone. Cestodes undergo metamorphosis from the oncosphere to the procercoid stage, during which polarity of the animal is said to be reversed. Because the oncosphere moves with its hooked end leading, that end has been held to be anterior. In later development, the scolex differentiates at the end opposite the hook-bearing cercomer; during movement, the scolex leads and is, therefore, considered anterior, so that the polarity of the organism has been reversed. Llewellyn (1986) has argued that cestode larvae have been seen to move with the hooked end trailing and that the evidence from embryological studies for reversal of polarity is not convincing.

Mackiewicz (1984) also drew attention to differences between monogeneans and cestodes in the morphology of the spermatozoon, considered a stable and conservative cell, which suggest that the two groups are phylogenetically distant from each other. This is covered in more detail later in this chapter.

A view differing from that of Llewellyn (1965, 1986) has been set out by Freeman (1973). The fundamental differences are that the monogeneans are not part of cestode evolution and that the original hosts of the cestodes are believed to be invertebrates rather than vertebrates. Freeman (1973, 1982a) derived the Platyhelminthes from a free-living rhabdocoel ancestor which gave rise in the Early Palaeozoic to three separate lines, one leading to the monogeneans and gyrocotylids, another to the pretrematodes and a third to the precestodes (Figure 4.2). The monogenean-gyrocotylid line developed in the Mid-Palaeozoic after the first vertebrates had evolved, whereas a precestode cycle could have become established in the Early Palaeozoic before the vertebrates appeared. Freeman (1982a) accepted the significance of reversal of polarity and rejected an evolutionary link between the monogeneans and cestodes.

The precestode cycle consisted of a tiny, free-living, possibly saprozoic adult with a single anterior sucker, a non-ciliated oncosphere which penetrated into and developed in the coelom of invertebrates, and a tailed, free-swimming plerocercoid-preadult stage. Invertebrates were the original hosts of cestodes and the adults never matured in the invertebrate gut lumen. The precestode probably lacked a gut but may have had a phagocytoblast. The question of whether the ancestors of cestodes ever had a gut has remained controversial. Mackiewicz (1981) and Freeman (1982a), for example, both considered that there was no evidence that cestodes had endoderm and that they had, therefore, not lost a gut by reduction, as they would have had to do to be descended from monogeneans. However, Brooks, O'Grady and Glenn (1985a) regarded the absence of a gut as a derived state rather than evidence of its absence in ancestral forms.

In the Early- to Mid-Palaeozoic, the precestode stock produced the amphilinids as an offshoot. The precestodes also gave rise to a protocestode stock from which other cestodes evolved. Following the appearance of the first vertebrates, precestode adults fell prey to them and some survived in the vertebrate gut. The protocestode adult thus became established in the vertebrate host, achieving a two-host life-cycle with a parenteral stage in the invertebrates, and protocestodes in the benthos eventually died out. Freeman (1973, 1982a) postulated a series of steps by which this might have occurred, beginning with a minute, free-living biradially symmetrical hermaphrodite precestode which lacked a gut but had a phagocytoblast, with a saprozoic mode of nutrition, which eventually gave rise to monozoic or polyzoic cestodes with a two-host life-cycle. He suggested that protocestodes originally lived in a freshwater environment and colonized the sea later. He also suggested that pre- and protocestodes may have existed as monozoic and polyzoic forms, ie. strobilization may have been foreshadowed in the free-living ancestors. By contrast, other authors, eg. Llewellyn (1965, 1986), Burt and Jarecka (1982) saw strobilization as following the development of a parasitic life-cycle, and Brooks, O'Grady and Glenn (1985a) and O'Grady (1985) regarded strobilization as an apomorphy acquired by terminal addition to the life-cycle. Ehlers (1985a,b) held a similar opinion. In Freeman's (1982a) view, the protocestode scolex was originally unifossate but di- and tetra-fossate forms appeared later. In the protocestode life-cycle, eggs developed into solid non-ciliated preoncospheres with hooks at the anterior end, which developed a sucker at the other end after reversal of polarity. They grew into monozoic or polyzoic sexually-reproducing adults. By the Pre-Cambrian, other invertebrates had evolved and predated upon protocestodes, initiating a one-host life-cycle. Some entered the parenteron, survived, and ultimately gave rise to monozoic forms adapted to freshwater annelids. Polyzoic forms became adapted to bottom-dwelling arthropods and, in both cases, the worms are assumed to have left the host to reproduce. The preoncosphere became primarily an invasive stage similar to modern hexacanths. Protocestodes never matured sexually in the invertebrate gut lumen although some monozoic forms may have approached it in the coelom of annelids. Vertebrates were colonized through the food chain as they appeared at the end of the Cambrian. Some protocestodes survived and matured in the gut, establishing a two-host life-cycle. The adult holdfast diversified and the lines which gave rise to the major cestode groups appeared. At a later stage, a second

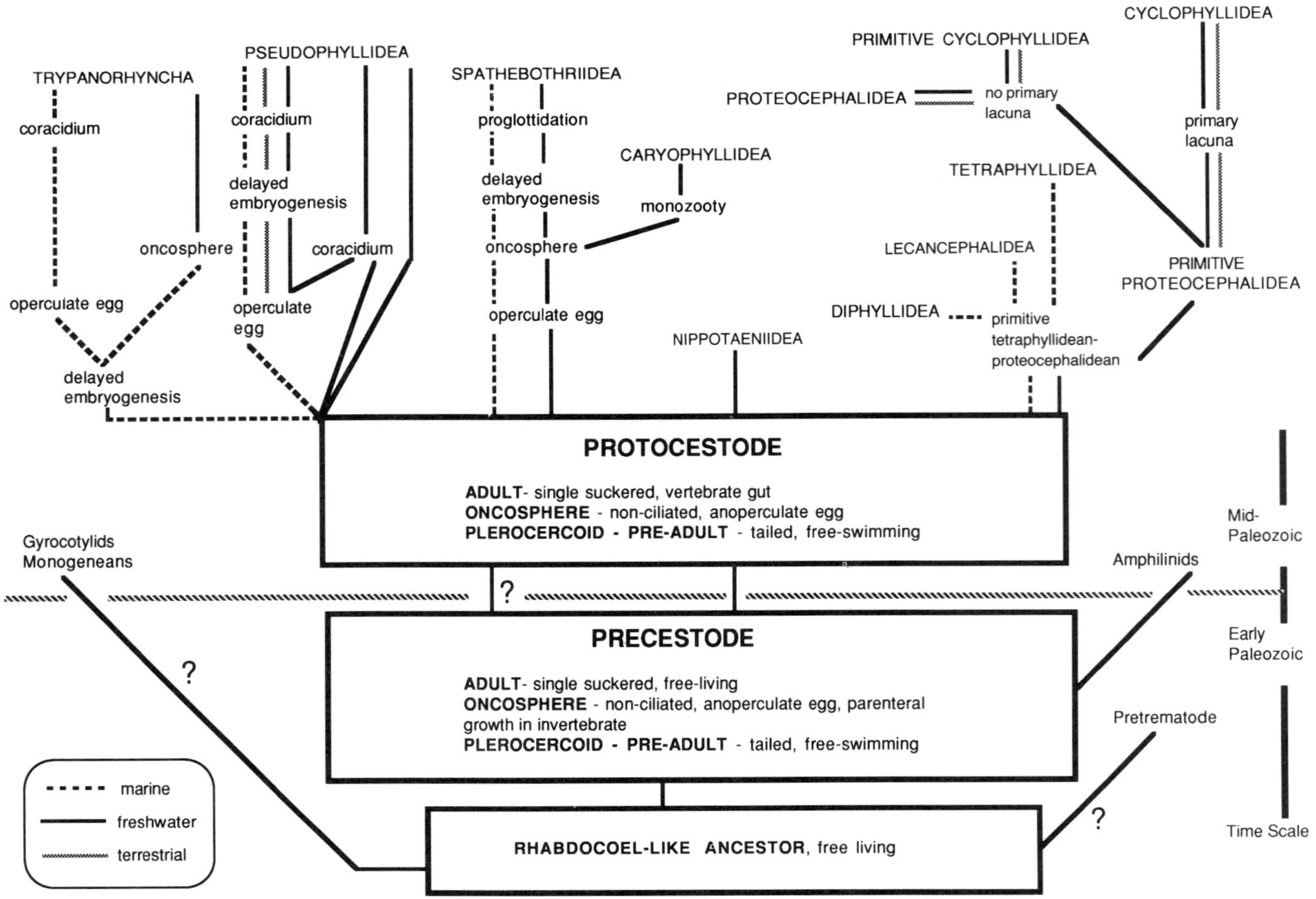

Figure. 4.2 Pattern of evolution from rhabdocoel-like ancestors to present-day orders of cestodes. From Freeman (1973).

intermediate host, a vertebrate, was added to the life-cycle as a derivative of a parenteric vertebrate host. Transfer between hosts was always passive and dependent on the food chains linking various host groups. The current life-cycles of cestodes involve two dissimilar habitats: the invertebrate parenteron, which is obligatory, and the vertebrate gut, which may not be. Invasion of the first site is the more primitive condition. There is no direct development of an oncosphere to a sexually mature adult in the gut of vertebrates.

Freeman's view that the original hosts of cestodes were invertebrates has received support from many authors, eg. Stunkard (1962b, 1967, 1983), Demshin (1971), Kulakovskaya and Demshin (1978), Dubinina (1980), Burt and Jarecka (1982), Malmberg (1974) and many others (see Mackiewicz, 1981) but others disagree (eg. Mamaev, 1975; Mackiewicz, 1981).

It is interesting that two of the most recent phylogenies of the Platyhelminthes, both the results of cladistic analyses, have produced very similar relationships between the parasitic groups, although the use of different terms for the same taxa, or the same term for groups of different composition, can be confusing. Both Ehlers (1984, 1985a, b, 1986) (Figure 4.3) and Brooks, O'Grady and Glenn (1985a) and O'Grady (1985) (Figure. 4.4) see the Trematoda (aspidogastreans and digeneans, sister groups in relation to each other) as the sister group of the remaining groups, the monogeneans and the cestodes *sensu lato* (gyrocotylids, amphilinids plus eucestodes), which all together constitute the Cercomeromorphae. Within the Cercomeromorphae, the monogeneans are the sister group of the cestodes *sensu lato*, termed the Cestoda by Ehlers (1986) and the Cestodaria by Brooks, O'Grady and Glenn (1985a). In both schemes, the gyrocotylids are included with the cestodes as the sister group of the amphilinids plus other cestodes. The amphilinids and the Cestoidea of Ehlers (1986) are sister groups and his Cestoidea comprises the caryophyllideans and the eucestodes as sister groups. Brooks *et al.* (1985a), however, include caryophyllideans as part of the eucestodes; the amphilinids are the sister group of the eucestodes and the amphilinids plus eucestodes together constitute the Cestoidea, not to be confused with the Cestoidea of Ehlers. Despite the differences in terminology, the similarities between the two schemes are very striking (Bray, 1986b: Gibson, 1987) and Brooks (1989) proposed a classification representing a compromise between the two systems.

Phylogenetic relationships as indicated by sperm morphology and protonephridial ultrastructure

Characters other than the traditional ones relating to morphological and life-cycle data have been brought into use in recent years as indicators of phylogenetic relationships, particularly in the Phylum Platyhelminthes (eg. Brooks, O'Grady and Glenn, 1985a; Ehlers, 1984, 1985a, b, 1986; Xylander, 1988b).

The morphology of the spermatozoon has been used to elucidate relationships between the free-living and parasitic platyhelminths and within the parasitic groups. A spermatozoon with two axonemes is considered the most primitive type because this type occurs in free-living platyhelminths (Euzet, Swiderski and Mokhtar-Maamouri, 1981; MacKinnon and Burt, 1985a; Williams, 1988). It occurs in kalyptorhynch rhabdocoels, in most polyopisthocotylean monogeneans except *Diplozoon*, in digeneans and aspidogastreans and in cestodes of the Pseudophyllidea, Trypanorhyncha, Tetraphyllidea (family Onchobothriidae) and Proteocephalidea. One axoneme is absent in the Caryophyllidea, Tetraphyllidea (Phyllobothriidae) and Cyclophyllidea. Spermatozoa have a column of peripheral microtubules which is incomplete in the aspidogastrids, digeneans and cestode groups with two axonemes. An important feature which distinguishes cestode sperm from those of all other groups is the absence of mitochondria; all other known sperm with incorporated axonemes and peripheral microtubules have rodlike or filiform mitochondrial derivatives (Williams, 1988).

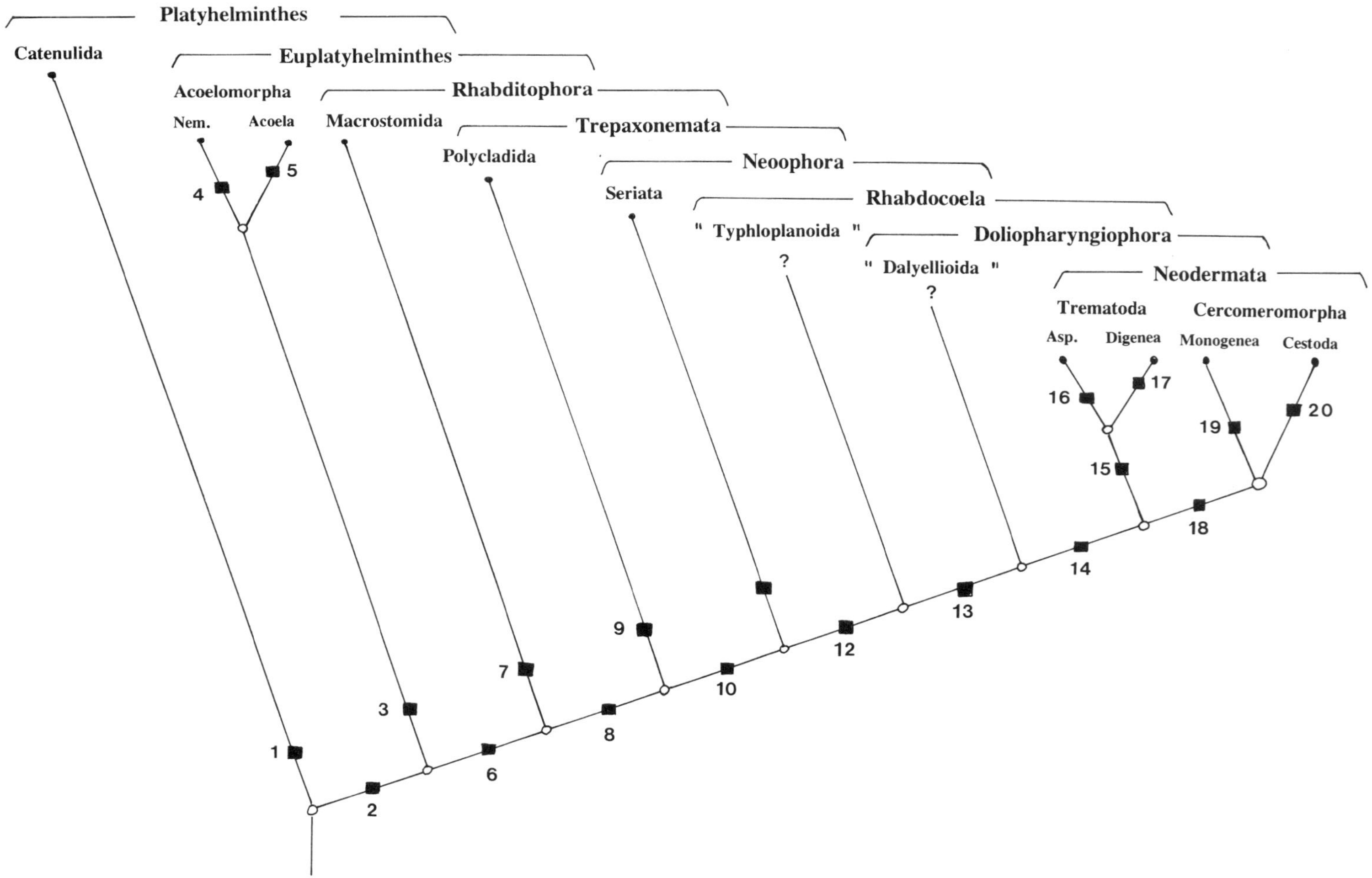

Figure. 4.3. Phylogenetic relationships among the main taxa of the Platyhelminthes (excluding the Lecithoepitheliata and Prolecithophora). Asp. Aspidobothrii; Nem. Nemertodermatida. Black squares are hypothesized autapomorphies, indicating the monophyly of the taxa. From Ehlers (1985b).

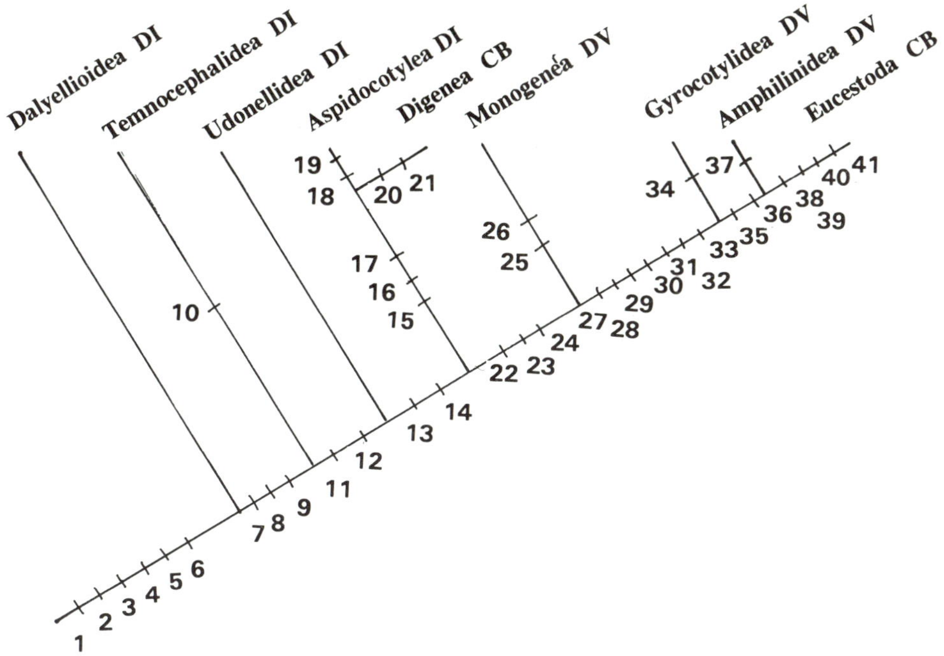

Figure. 4.4. Cladogram representing the hypothesized phylogenetic relationships of the parasitic platyhelminth taxa examined. Each slash mark postulates a synapomorphy or evolutionary novelty, common to all the taxa above that branch. In some cases subsequent evolution has modified the character, creating an homologous or transformation series. There are 41 postulated changes for 39 characters indicating two cases of parallel evolution: loss of copulatory stylet (No. 17 and No. 30) and acquisition of bifurcate gut (No. 20 and No. 26). The taxa are identified as having life-cycles that are directly in an invertebrate host (DI) or vertebrate host (DV) or complex with both an invertebrate and vertebrate host (CB).
From Brooks, O'Grady and Glenn (1985a).

The position of the nucleus in most free-living platyhelminths, in digeneans and monogeneans, has been interpreted as posterior by Justine, Lambert and Mattei (1985) and others because the nucleus migrates distally during spermiogenesis. Hendow and James (1988) examined the evidence, concluding that although in most parasitic platyhelminths the nucleus is distal to the region of differentiation, but proximal in other animals including some digeneans, it is probably correct to regard the head (anterior), middle and tail regions in all spermatozoa as homologous. In most spermatozoa, anterior is equivalent to proximal but in most parasitic platyhelminths it is distal. Polyopisthocotyleans were considered closer than monopisthocotyleans to the trematodes because their sperm have two axonemes plus microtubules. Although some cestodes and some monopisthocotyleans have a single axoneme, this should not be interpreted as a phylogenetic link because spermiogenesis is different in each case (Justine, Lambert and Mattei, 1985).

With regard to other helminth groups, the spermatozoa of acanthocephalans are more similar to those of rotifers, nematomorphs and nematodes than to those of cestodes (Whitfield, 1971).

Comparative spermatology has been applied particularly to relationships within the monogeneans and the cestodes.

Examination of the ultrastructure of the protonephridial system of the parasitic platyhelminths has shown that the monogeneans, aspidogastreans and digeneans all have the same type of flame bulb. In contrast, a different type is found in the gyrocotylideans, amphilinideans and other cestode groups, suggesting that these all have a monophyletic origin. On this basis, the gyrocotylideans are not more closely related to the monogeneans than to the various cestode groups as maintained by Llewellyn (1965, 1986). In addition, the protonephridia of *Udonella* resemble those of the cestodes, but differ from those of the monogeneans, in lacking a septate junction in the flame bulbs and capillaries. It also has characters different from those of all other parasitic platyhelminths, including cestodes. As a result, Rohde, Watson and Roubal (1989) and Rohde, Justine and Watson (1989) concluded that *Udonella* is not a monogenean and is unlikely to be the sister group of all the major parasitic platyhelminth groups, as suggested by Ehlers (1984, 1985a,b) and Brooks, O'Grady and Glenn (1985a) because such a group would be expected to have a protonephridial structure corresponding to that of the most original neodermatans (Figures 4.3, 4.4), the trematodes and monogeneans. The sperm structure of *Udonella*, which conforms to pattern 2 of the monogeneans described by Justine, Lambert and Mattei (1985) (see below) may have been achieved by convergent evolution.

Phylogenetic relationships within the monogeneans

Relationships within the monogeneans have been proposed on the basis of characteristics of the adult (Sproston, 1946) or larval features, the genital organs and host specificity (Bykhovskii, 1957), on larval features, ontogeny and the adult attachment apparatus (Llewellyn, 1970) and most recently on chaetotaxy and ciliated cells in the oncomiracidia (Lambert, 1980) and on sperm structure (Justine, Lambert and Mattei, 1985). It falls within the scope of this chapter to consider these relationships insofar as they shed light on the phylogenetic aspects of host specificity. Monogenean phylogeny has been most recently reviewed by Lebedev (1988).

Beverley-Burton (1984) has adopted the following names for three taxa of ordinal rank within the Monogenea: Polyopisthocotylida (= Polyopisthocotylea), Dactylogyrida and Gyrodactylida (together equivalent to the Monopisthocotylea). The names in this text are used as cited by the original authors.

Llewellyn (1970) has postulated that the main monogenean stock, having given off the protogyrocotylideans (regarded as monogeneans) diverged to form the four major monogenean groups, the dactylogyrideans, the viviparous gyrodactylids, the monocotylids and the polyopisthocotyleans. Apart from the gyrodactylids, the haptor of which has remained relatively unchanged, each evolutionary line is characterized by the development of the haptor. The haptor has been elaborated from a simple muscular disc firstly by the acquisition of marginal hooks, secondly by acquisition of hamuli and thirdly by further complication of the hamuli, or by subdivision of the haptor into loculi or by acquisition of a new type of attachment organ. From this basis, Llewellyn (1970) suggested the following evolutionary relationships between the monogeneans and their hosts. The holocephalans split off early from other vertebrates; they harbour the archaic gyrocotylids and chimaericolids and some later colonists (some hexabothriids, monocotylids and gyrocotylids). Elasmobranchs harbour microbothriids, acanthocotylids, amphibdellatids, monocotylids and hexabothriids. Modern chondrosteans host some capsalids and diclybothriids. In the teleosts, some monogenean lines persist only as relicts, eg. calceostomatids and tetraonchoidids while others have flourished, eg. dactylogyrids and microcotylids. Crossopterygians and their derivatives host polystomatids and some capsalids.

Sperm structure has been adopted as an indicator by some workers because they consider that, unlike the haptor, it would not alter in response to habitat. Justine, Lambert and Mattei (1985) identified four patterns of sperm structure within the monogeneans according to the presence of one or of two axonemes and the presence or absence of cortical microtubules.

Pattern 1: Two axonemes plus microtubules. Found only in polyopisthocotyleans, in all families. Also found in trematodes and turbellarians.

Three patterns occur in the monopisthocotyleans:

Pattern 2: Two axonemes, microtubules absent. In Capsalidae, Dionchidae, Gyrodactylidae, Udonellidae and *Euzetrema knoepffleri*. Derived from pattern 1 by loss of the microtubules.

Pattern 3: One axoneme plus one altered axoneme plus microtubules. In Monocotylidae (*Heterocotyle*) and Loimoisidae (*Loimosina*). One axoneme progressively altered during spermiogenesis; possibly derived from pattern 1 by alteration of an axoneme and reduced development of the microtubules. Intermediate between patterns 1 and 4.

Pattern 4: One axoneme, no microtubules. In Diplectanidae, Amphibdellatidae, Ancyrocephalidae and Calceostomatidae. Only one axoneme, even in the early stages of spermiogenesis.

Diplozoon is unique in the parasitic platyhelminths in possessing an aflagellate spermatozoon; this condition is related to the reproductive biology of the genus and has no phylogenetic significance (Justine, Le Brun and Mattei, 1985a, b). Table 4.1 summarizes current information on sperm structure in the monogeneans. Justine, Lambert and Mattei (1985) concluded that on the basis of comparative spermatology a clear separation existed between the polyopisthocotyleans and the monopisthocotyleans, the former being the more primitive (Figure 4.5). Because sperm structure in the polyopisthocotyleans is homogeneous, it is of little use in elucidating their interrelationships. Pattern 1 is believed to be the primitive pattern from which the others are derived by simplification but the authors make the point that phylogenetic relationships based on haptor structure can be interpreted differently depending on whether its evolution is regarded as progressive or regressive (Malmberg, 1982) and the same is true of comparative spermatology.

Table 4.1 Sperm structure in the monogeneans. (Data compiled from Justine, Lambert and Mattei, 1985; Justine and Mattei, 1985a, b, c, d; Le Brun, Lambert and Justine, 1986).

Polyopisthocotylea		
Pattern 1	Polystomatidae	*Polystoma, Polystomoides*
	Hexabothriidae	*Erpocotyle*
	Plectanocotylidae	*Plectanocotyle*
	Diclidophoridae	*Choricotyle, Diclidophora*
	Gotocotylidae	*Gotocotyla*
	Microcotylidae	*Microcotyle*
	Heteraxinidae	*Heteraxine, Heteraxinoides*
	Axinidae	*Axine*
	Cemocotylidae	*Cemocotyle*
	Mazocraeidae	*Kuhnia*
	Chauhaneidae	*Pseudomazocraes*
	Pyragraphoridae	*Pyragraphorus*
Monopisthocotylea		
Pattern 2	*Incertae sedis*	*Euzetrema*
	Gyrodactylidae	*Gyrodactylus*
	Udonellidae	*Udonella*
	Capsalidae	*Caballerocotyle, Megalocotyle, Trochopus*
	Dionchidae	*Dionchus*
Pattern 3	Monocotylidae	*Heterocotyle*
	Loimoidae	*Loimosina*
Pattern 4	Diplectanidae	*Diplectanum, Furnestinia, Lamellodiscus*
	Amphibdellatidae	*Amphibdella, Amphibdelloides*
	Ancyrocephalidae	*Cleitharticus*
	Calceostomatidae	*Calceostoma*
	Pseudodactylogyridae	*Pseudodactylogyrus*

The monopisthocotyleans are divided into three groups based on sperm patterns 2, 3 and 4. Pattern 2 incorporates two subgroups, the Capsalidae plus Dionchidae (this relationship agrees with the interpretation by Llewellyn, 1970) and the Udonellidae—Gyrodactylidae and *Euzetrema* (Figure 4.5). Sperm data suggest that the gyrodactylids belong with the monopisthocotyleans, which agrees with classical interpretations but not that of Lambert (1980) based on oncomiracidial characters (Figure 4.6). The sperm structure of the Udonellidae links them to the monopisthocotylean monogeneans, a view which contrasts with that of Brooks, O'Grady and Glenn (1985a) (Figure 4.4) who treated the Udonellidea as a superclass, a sister group to their new superclass Cercomeridea which included the trematodes and the cercomeromorphans. Beverley-Burton (1984) treats the udonellids as an order within the Turbellaria rather than as monogeneans. Justine, Lambert and Mattei (1985) link *Euzetrema* with the capsalids, as did Llewellyn (1970).

The monocotylids and loimoids are closely linked by possession of sperm pattern 3 (Figure 4.5); they have been regarded as a single family (Baer and Euzet, 1961). Sperm structure does not support linkage of the monocotylids to the capsalids.

The amphibdellatids, ancyrocephalids, calceostomatids and diplectanids are all monopisthocotyleans with uniflagellate sperm (pattern 4) and the last three are linked within the Dactylogyridea by their larval chaetotaxy (Lambert, 1980) and amphibdellatids are close to them in classical phylogenies (Baer and Euzet, 1961; Llewellyn, 1970). This agrees with the phylogeny of Lambert (1980) (Figure 4.6) who found a close relationship between the Dactylogyridea and Monocotylidae; all three have a tendency towards uniflagellate spermatozoa which is partial in the loimoids and monocotylids and total in the Dactylogyridea.

Phylogenetic relationships within the cestodes

The monozoic groups

In any discussion of the phylogeny and evolution of cestodes, the monozoic groups are regarded as particularly significant. Their exact nature is controversial; they are regarded by some as paedomorphic forms and by others as true, sexually-reproducing adults. The majority of authors regard them as primitive groups, closer than the strobilate cestodes to the pre- or protocestode stock.

Attention has been strongly focused on the gyrocotylideans, which some authors regard as endoparasitic monogeneans ancestral to the cestodes, and on the caryophyllideans which have several unique features that set them apart from other cestode groups. The gyrocotylideans and amphilinideans have frequently been grouped together as the Cestodaria, a grouping in which the caryophyllideans have at times also been included. Currently, all three groups are accorded a status separate from each other and from other cestodes. The gyrocotylideans and amphilinideans both have lycophore larvae although the hooks of the amphilinidean lycophore are not all identical. Caryophyllideans have a hexacanth larva, resembling other cestode groups in this feature. Mackiewicz (1981) considered that the gyrocotylideans and amphilinideans evolved along different lines from the caryophyllideans, with which they had nothing in common other than the monozoic body form.

Gyrocotylideans

The gyrocotylideans have been regarded as pivotal in the debate about the origin of the cestodes because, as outlined above, some authorities, eg. Llewellyn (1965, 1986) believe that they are endoparasitic monogeneans which lost the gut and subsequently gave rise to the cestodes. He argued that the larval hooks of monogeneans and gyrocotylideans are

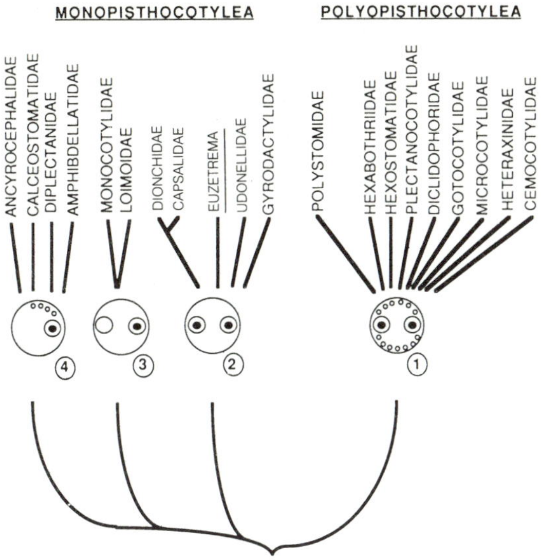

Figure. 4.5 Phylogeny of the Monogenea, drawn from the data of comparative spermatology. Numbers 1–4 refer to sperm patterns. From Justine, Lambert and Mattei (1985).

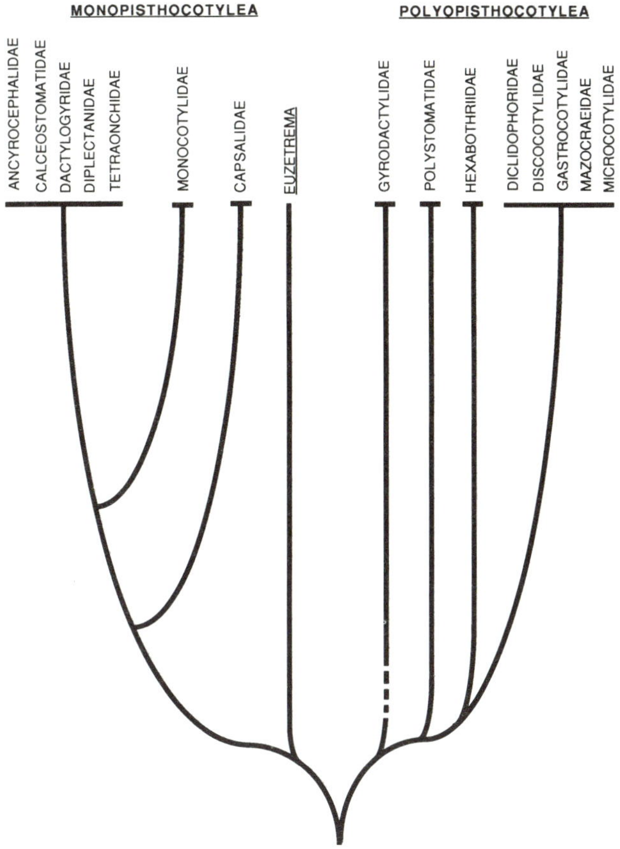

Figure. 4.6 Phylogeny of the Monogenea according to Lambert (1980) elaborated mainly from the study of the chaetotaxy and ciliated cells of the oncomiracidium. From Justine, Lambert and Mattei (1985).

homologous and that they further resemble each other in having paired anterior excretory organs and a posterior adhesive organ. Bandoni and Brooks (1987b), however, believed these to be symplesiomorphies (common possession of an ancestral character) and the resemblances to be superficial and therefore not suitable indicators of phylogenetic relationships.

Williams, Colin and Halvorsen (1987) concluded from a detailed review of the biology of gyrocotylideans and of previous speculations on their origins that there was little to support their derivation from monogeneans. On the contrary, they concluded that the group was probably further removed from both monogeneans and cestodes than hitherto believed, set apart by the absence of a gut, several unique morphological features of the adult and its strict specificity to a highly specialized and ancient host group, the holocephalans. They speculated that gyrocotylideans might have originated in Palaeozoic or pre-Cambrian times and that their relationship with their hosts had persisted almost unchanged for over 350 million years.

The life-cycle of gyrocotylideans has not yet been conclusively established although a direct cycle, possibly with a tissue-dwelling larval stage, has been postulated (Williams, Colin and Halvorsen, 1987). By contrast, Xylander (1988b), who grouped both gyrocotylideans and amphilinideans with the cestodes, has suggested that the life-cycle will be found to be indirect and that the gyrocotylideans probably have a crustacean intermediate host.

In recent works on classification, the gyrocotylideans have been given class status by Bazitov (1984) and are regarded by Brooks, O'Grady and Glenn (1985a) as an infrasubclass constituting a sister group to the infrasubclass Cestoidea which incorporates the amphilinids and the eucestodes. The last proposal is similar to that of Ehlers (1985a, b, 1986) although the terminology adopted is confusing because the Cestoidea of Brooks, O'Grady and Glenn (1985a) equates to the Nephroposticophora of Ehlers, not to his Cestoidea.

Amphilinideans

Opinions on the origin and host-parasite relationships of the amphilinideans differ very widely. Dubinina (1960) believed that they originated directly from monogeneans and has continued to maintain (1974) that they are closer to monogeneans than to the cestodes, and merit class status. She has also suggested (1982) that they originally became parasitic in crustaceans. Most authors believe that they originated from cestode stock. Freeman (1973) believed that their original hosts were invertebrates but derived them from precestode stock in the early Palaeozoic, diverging from the protocestode stock, eventually to mature in the coelom rather than the gut of vertebrates. They might have been parenteral parasites of invertebrates, preadapted for parenteral development in fish when they appeared. The group remained fairly small and stable because the vertebrate parenteron is a less easy habitat to exploit than the gut, in which the protocestode stock diversified. In Freeman's (1973) opinion, amphilinids probably evolved in fresh water and are currently associated with freshwater and euryhaline teleosts and turtles. Where the life-cycle is known, it involves freshwater crustacean intermediate hosts (Rohde and Georgi, 1983; Bandoni and Brooks, 1987a; Gibson, Bray and Powell, 1988).

The antiquity of the group has long been recognized. Bandoni and Brooks (1987a) considered that they had coevolved with their freshwater teleost hosts, originating before the break-up of Pangaea. Gibson, Bray and Powell (1988) regarded two morphologically very similar species, *Nesolecithus africanus* and *N. janickii*, from osteoglossomorph fishes of West Africa and of the Amazonian region, respectively, to be vicariant Gondwanaland relicts, isolated by continental drift 100 million years ago.

It has been suggested that amphilinidean worms producing eggs in the coelom of vertebrates are sexually mature plerocercoids, homologous with the plerocercoid larvae of other cestodes, which once had and have now lost, a strobilate adult in a vertebrate host which is now extinct, ie. are progenetic in the original second intermediate host (Janicki, 1930; Llewellyn, 1965). Alternatively, Dubinina (1982) suggested that amphilinideans were originally parasites of crustaceans. Vik (1981) did not believe that the amphilinid life-cycle was achieved by loss of a definitive host and Brooks, O'Grady and Glenn (1985a) and Bandoni and Brooks (1987a) rejected both the above hypotheses. They regarded the vertebrate host as the original (*plesiomorphic*) host of the cestoideans (= amphilinideans plus eucestodes) and the absence in amphilinideans of the characteristics of cestodes as plesiomorphic and not the result of secondary loss. The strobilate condition was regarded as a secondary acquisition. However, Gibson, Bray and Powell (1988) inclined to the view that adult amphilinids are derived from paedomorphic plerocercoids arguing that, as plerocercoids have a low host-specificity, this would explain the current diverse distribution of amphilinideans in chondrosteans, primitive and advanced teleosts and chelonians, adducing their location in the coelom as further evidence. Bandoni and Brooks (1987a) have explained the current patchy distribution of amphilinideans in part by the extinction of various teleost groups and, among extant host groups, by physiological or ecological barriers to infection, eg. amphilinids have a freshwater distribution, which excludes marine fish as hosts, and they are large, which imposes a physical constraint on host size.

As previously mentioned, the amphilinideans have been grouped with the gyrocotylideans in the Cestodaria: both taxa lack a gut, both have the monozoic body form and both have a lycophore larva, although in the amphilinids the ten hooks comprise six large hooks and four differently-shaped small hooks. Brooks, O'Grady and Glenn (1985a) have used the name Cestodaria (= subclass Cestodaria Monticelli, 1891) to refer to a different assemblage comprising the gyrocotylideans, amphilinideans and the eucestodes (polyzoic cestodes plus the caryophyllideans). In this scheme (Figure 4.4), the amphilinideans are seen as the sister group of the eucestodes. This closely resembles the scheme of Ehlers (1984, 1985a, b, 1986) although he separated the caryophyllideans from the strobilate cestodes.

Caryophyllideans

The caryophyllideans have many unique features which set them apart from other cestodes: they are monozoic with a single set of reproductive organs; they exhibit progenesis (which also occurs in some pseudophyllideans); their use of tubificid annelids as intermediaries and the capacity of a species of one genus (*Archigetes limnodrili*) to complete its life-cycle exclusively in this host, the only bothriomonid cestode known to do so in an invertebrate; the highly unusual phenomenon of conversion of the nucleus of vitelline cells from a regulatory organ to a glycogen store; possession of large chromosomes; and the occurrence of polyploidy and of parthenogenesis in some species (Mackiewicz, 1982a). They have been regarded as offshoots from the main line of cestode development by some workers and as the ancestors of the strobilate cestodes by others. They have remarkably little in common with other cestodes except the possession of a hexacanth embryo but they are morphologically so similar to the plerocercoids of pseudophyllideans that they have often been viewed as progenetic plerocercoids.

The origin of the caryophyllideans, their affinities with other groups and the significance of progenesis and of the monozoic body form remain controversial. Some authors derive caryophyllideans from a protocestode stock (eg. Freeman, 1973; Dubinina, 1980) which continued to develop independently or, alternatively, gave rise to the strobilate cestodes. Mackiewicz (1981, 1982a), however, has advanced the view that, as a group, caryophyllideans originated by progenesis from a line of polyzoic cestodes rather than, in contrast, having given rise to them.

Comprehensive reviews of the biology, evolution and systematics of the caryophyllideans by Mackiewicz (1972, 1981, 1982a) have emphasized the significance in evolution of the monozoic body form and of progenesis.

Possession of a monozoic body form has the following consequences: (i) a reduction in reproductive capacity compared with the strobilate body form; (ii) reduced morphological and developmental complexity so that only one developmental level is present at a time and the number of stages (metacestodes) present in the life-cycle is reduced; (iii) a tendency towards a small size; and possibly (iv) a short life-span because spent reproductive units are not replaced and the worm dies once egg production stops. Obligatory self-fertilization may result in populations with somewhat fixed genotypes, but evolution and speciation are more likely to occur in this type of population than one where random outbreeding is the norm because selection can act on all genes in the population and any genetic change can be rapidly expressed (Price, 1980). The monozoic condition is an evolutionary dead-end which has reached its full potential in the caryophyllideans, the only large monozoic group; they are a successful group despite the lack of high reproductive capacity in the adults or of any form of asexual reproduction in the metacestode stage to compensate for this (Mackiewicz, 1982a).

As mentioned previously, many authors have regarded the monozoic egg-producing stage in vertebrates, ie. fish, as a progenetic (or alternatively neotenic) form but this has been rejected by others (eg. Freeman, 1982a; Mackiewicz, 1982a) who considered it to be the terminal adult stage and do not think that a strobilate adult stage has been lost from each caryophyllidean life-cycle. Mackiewicz (1982a) does, however, regard as a progenetic procercoid the egg-producing stage of *Archigetes* species in tubificids because these forms retain the juvenile characters of non-functional gonopores and a cercomer. If, however, the procercoid is eaten by a fish before progenesis is achieved, the cercomer is shed and the gonopores become functional, ie. the juvenile characters are lost and the worm assumes the character of a characteristic caryophyllidean adult. The effect of progenesis in evolutionary terms is to shorten the life-cycle; this is most marked in *Archigetes* species where eggs are produced in the tubificid host by the procercoid, resulting in a life-cycle which is the shortest known for a cestode. *A. limnodrili* matures only in tubificids but *A. iowensis* can mature in tubificids or in fish. Mackiewicz (1982a) differs from some earlier workers in regarding the *Archigetes* cycle as the end of an evolutionary sequence rather than the beginning of a series of events which culminated in the development of the strobilate cestodes. In Mackiewicz's view, *Archigetes* is not more primitive than other caryophyllideans but has achieved its life-cycle by reduction, through progenesis, from a two-host cycle involving both vertebrates and tubificids.

Mackiewicz (1981, 1982a) has speculated that progenesis had a major role in the evolution of caryophyllideans as a group, *although not in the evolution of individual taxa*. He suggested that they might have evolved from parenteral plerocercoids in fish, in a life-cycle of the pseudophyllidean type, that became progenetic, possibly in response to host hormonal changes associated with regular migrations between marine and fresh water. Such changes had been noted by Szidat (1959). Mackiewicz (1982a) cited Szidat's (1959) suggestion that 'neotenic' (Szidat's term) development in caryophyllids might have been stimulated by hormonal fluctuations in the host, similar to the way in which some monogenean life-cycles are regulated by hormonal changes in their amphibian hosts. Hormonal changes, as shown by the presence of hormones in the intestine, accompany migrations in fish such as *Salmo* and *Gasterosteus*, which harbour pseudophyllideans with procercoid and plerocercoid stages which do exhibit some precocity. Migratory behaviour of the host would result in repeated exposure of developing cestode stages to high hormone levels. A precocious juvenile stage could have arisen many times until eventually it was selected for. Part of the life-cycle could have become adapted to a freshwater rather than marine environment and a 'new' form of the parasite could have evolved. The principal hosts of caryophyllideans are ostariophysan fish which are a primary freshwater group, probably dating from before the Tertiary age. Both the host group and the cestodes are considered primitive. Although most current hosts are freshwater fish, some North American hosts are migratory and the ancestral hosts of the caryophyllidean stock may have been migratory marine fish rather than freshwater fish, although Mackiewicz (1981) considered freshwater fish more probable.

Caryophyllideans are almost the only helminths with oligochaete intermediaries. As hosts, they offered a number of advantages for the evolution and expansion of a monozoic group in having a long, narrow coelom and in their longevity, thus providing both space and time for growth (Mackiewicz, 1982a). There was little competition from other parasites and the tubificids, as part of the benthos, offered access to benthophagous fish as potential hosts. The hosts of modern caryophyllideans are ostariophysan and siluriform fishes, all benthic feeders.

Mackiewicz (1981) reviewed recent proposals on caryophyllidean evolution and has stated his conviction that tubificids were not the original hosts of caryophyllideans and that, as outlined above, caryophyllideans as a group were secondarily derived from a progenetic plerocercoid and soon became primarily monozoic after losing the strobilate stage. In some modern caryophyllideans, primarily *Glaridacris*, the calcareous corpuscles are distributed in groups along the longitudinal axis of the body. This may reflect the presence of segments in an ancestral worm, ie. indicate that a polyzoic body plan has been lost. The strobilate ancestral stock may have resembled the spathebothriideans or pseudophyllideans.

Bazitov (1976) and Bazitov, Kulakovskaya and Shestakova (1979) rejected theories that caryophyllideans had developed by neoteny, considering that differences in spermatogenesis between caryophyllideans and the strobilate cestodes precluded a close phylogenetic link between them. Bazitov (1981) raised the group to class status and regarded them as either phylogenetically further from the strobilate cestodes than from other groups or as equidistant from the cercomeromorphan groups.

The polyzoic groups

Rees (1969) suggested an evolutionary sequence for the cestodes, based purely on morphological features of the adult, which is very similar to that of Llewellyn (1965). It incorporates the following sequence of events: a protocaryophyllidean with one set of genitalia, all genital pores on one surface, no special attachment organ (eg. caryophyllids) *to* repetition of the genitalia (eg. spathebothriids) *to* better development of the anterior musculature with genital pores on one surface (eg. cyathocephalids, diplocotylids) or uterine pore ventral, cirro-vaginal pore dorsal (eg. *Anantrum*, syn. *Acompsocephalum*) *to* the shedding of spent proglottides by pseudoapolysis *to* the development of anterior attachment organs (eg. Pseudophyllidea) *to* the beginning of segmentation (eg. Pseudophyllidea) *to* retention of eggs in an uterine sac (eg. Pseudophyllidea) *to* closure of the uterine pore and apolysis (eg. most other cestodes).

Regarding the further evolution of the cestodes, Freeman (1973) (Figure 4.2) has suggested an evolutionary sequence in which a unifossate monozoic stem gave rise to the caryophyllideans, and a polyzoic stem to several evolutionary lines. The first of these was a unifossate stock which gave rise to spathebothriideans and possibly nippotaeniideans. The second was a difossate pseudophyllidean stock, lacking a coracidium which gave rise to modern difossate cestodes with and without coracidia. The third line was a di- and tetrafossate stock with tentacles which developed into trypanorhynchs, the fourth a tetrafossate stock ancestral to the tetraphyllideans and the proteocephalideans, the latter giving rise to the cyclophyllideans.

Kulakovskaya and Demshin (1978) are among the minority who derive cestodes from acoelous ancestors but agree with Freeman (1973) and others that their original hosts were invertebrates. The acoels colonized aquatic invertebrates in the early Palaeozoic era, ancestral caryophyllids becoming specific to oligochaetes and colonizing vertebrates much later, after teleosts had appeared. Other cestode groups (pseudophyllideans, trypanorhynchs, tetraphyllideans) were derived later than the caryophyllideans from turbellarians which first became parasites of crustaceans and then transferred to vertebrates, retaining non-parasitic traits (free-swimming coracidium) lost by caryophyllids. This theory of transfer from crustaceans to vertebrates has not received widespread support although some authors (Stunkard, 1962b, 1967; Cameron, 1964) accept that protocestodes first parasitized arthropods, then invaded fish later. Freeman (1973) and others have pointed out that the coracidium is not a widely-occurring stage in the cestodes, present only in some but not all pseudophyllideans and trypanorhynchs and absent from the tetraphyllideans and proteocephalideans.

Dubinina (1980) (Figure 4.7) proposed a scheme based on morphology, ontogeny and host groups in which she derived three evolutionary lines from a tetrabothriate protocestode. The caryophyllideans form an independent line parasitic first in oligochaetes then in teleosts. Of the other two lines, one gives off the trypanorhynchs, diphyllideans and *Haplobothrium* near its base and culminates in the pseudophyllideans. The other gives off the tetraphyllideans and lecanicephalids as offshoots and progresses through the proteocephalids to the cyclophyllideans and aporideans.

Burt (1986) has suggested, on the basis of the early development of fertilized eggs, adult morphology and life-cycle data, that the cestode orders Pseudophyllidea, Haplobothrioidea and Trypanorhyncha, which are oviparous, are more primitive than the Tetraphyllidea, Proteocephalidea and Cyclophyllidea, which are viviparous. Oviparity was believed to be the more primitive ancestral condition and viviparity the more advanced or derived. The majority of monogeneans and digeneans are oviparous.

As with the monogeneans, sperm structure has been used to construct phylogenetic schemes within the cestodes. As previously outlined, possession of sperm with two axonemes is considered a primitive character (found in the free-living platyhelminths, Pseudophyllidea, Trypanorhyncha, onchobothriid Tetraphyllidea, Haplobothriidea and Proteocephalidea) and possession of sperm with one axoneme is more advanced (found in the Caryophyllidea, Diphyllidea, phyllobothriid Tetraphyllidea and Cyclophyllidea) (Table 4.2). The single-axoneme condition was found to be derived by early abortion of one axoneme in spermiogenesis in the phyllobothriids (Euzet, Swiderski and Mokhtar-Maamouri, 1981).

Several authors have proposed a diphyletic origin for the cestodes and many schemes show broad agreement (eg. Freeman, 1973 (Figure 4.2); Euzet, 1959, 1974 (Figures 4.8, 4.9); Dubinina, 1980 (Figure 4.7)). While these proposals were based on morphology, embryology or features of the life-cycle, Euzet, Swiderski and Mokhtar-Maamouri (1981) have suggested two hypotheses based on sperm structure (Figures 4.10, 4.11). In both, a common ancestor gives rise to two branches and, in both, one branch, A, has the same structure, linking the pseudophyllideans, trypanorhynchans and haplobothriideans. The sperm of all three groups are now known to have two axonemes, following the work of MacKinnon and Burt (1985a) on *Haplobothrium globuliforme*. The caryophyllideans, with a single axoneme, occupy an isolated position diverging early from this branch. Two hypotheses are offered for branch B, comprising either two evolutionary lines (Figure 4.10) or only one (Figure 4.11). In the first, the onchobothriid tetraphyllideans, parasites of elasmo-

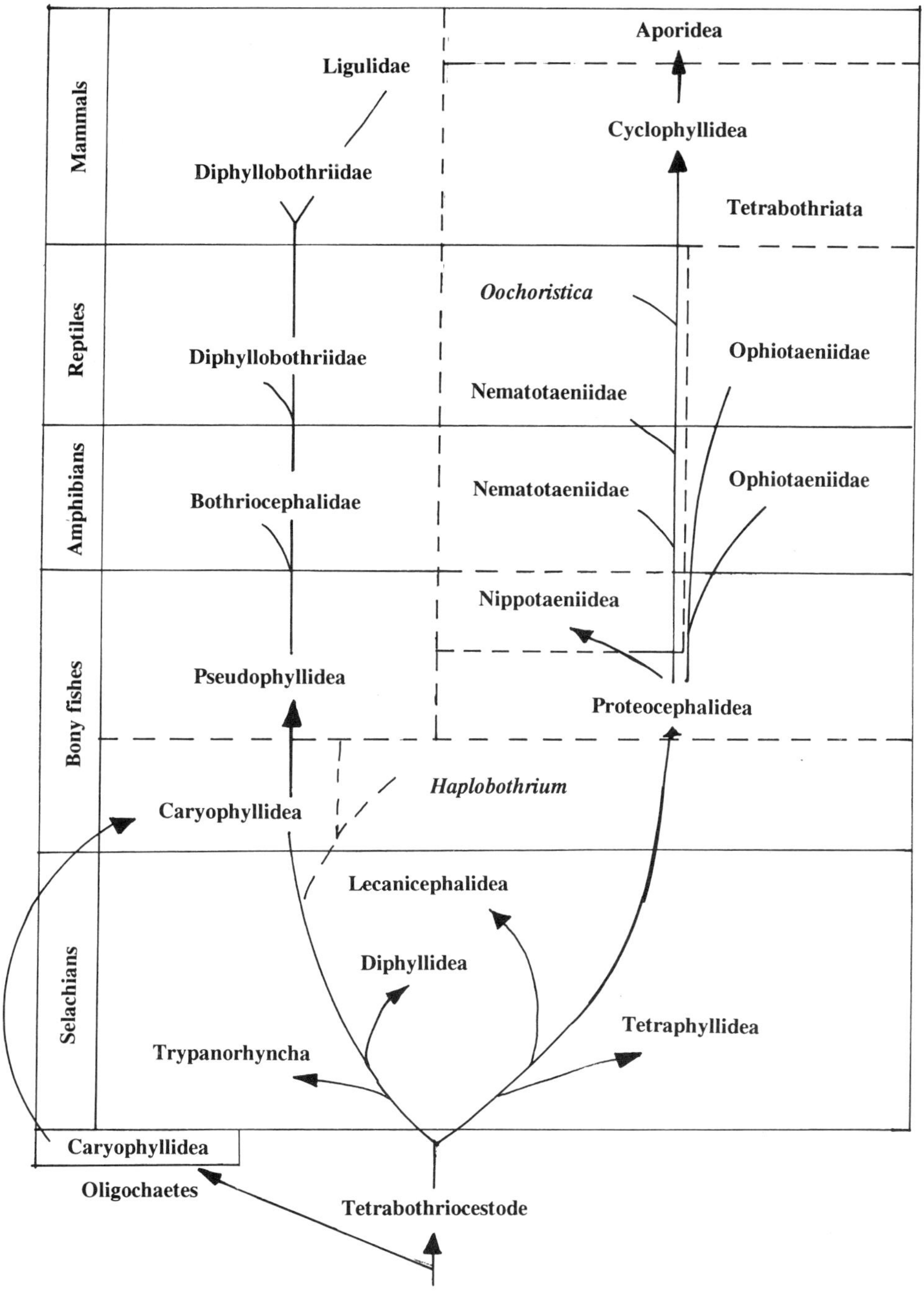

Figure. 4.7 Hypothetical evolutionary lines in the cestodes. From Dubinina (1980).

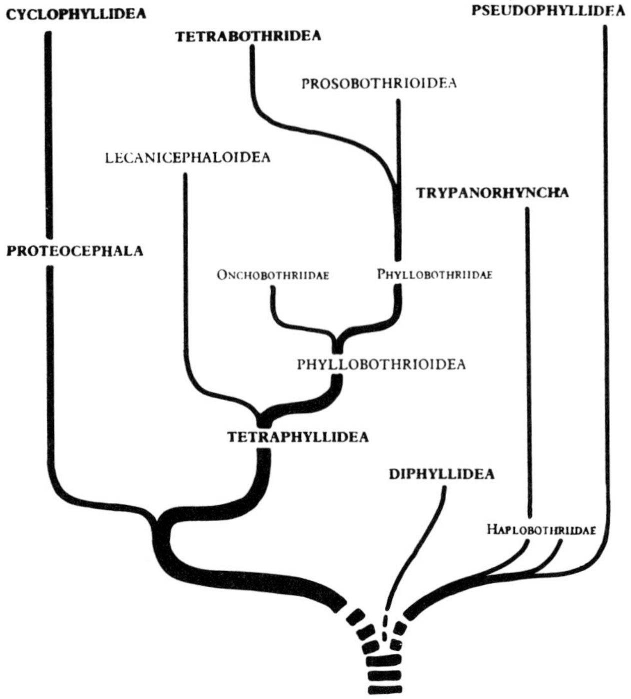

Figure 4.8 Phylogeny of the Cestoda. From Euzet (1959).

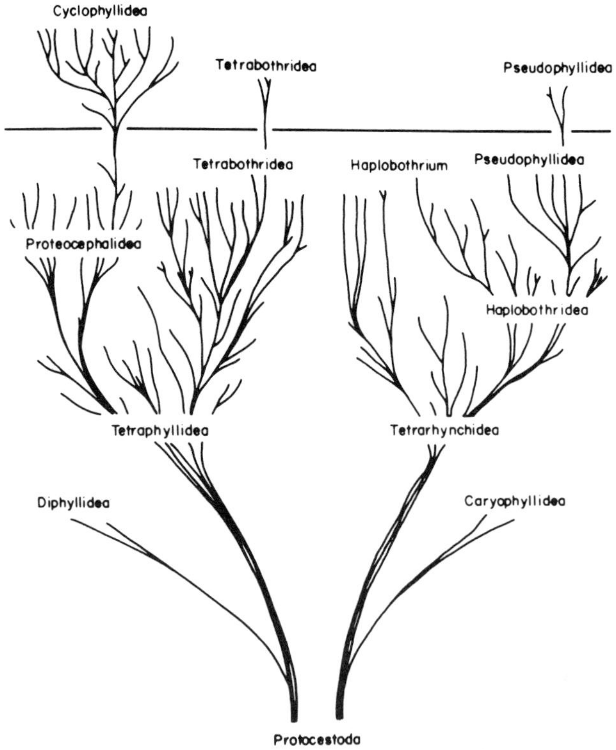

Figure 4.9 Phylogeny of the Cestoda. From Euzet (1974).

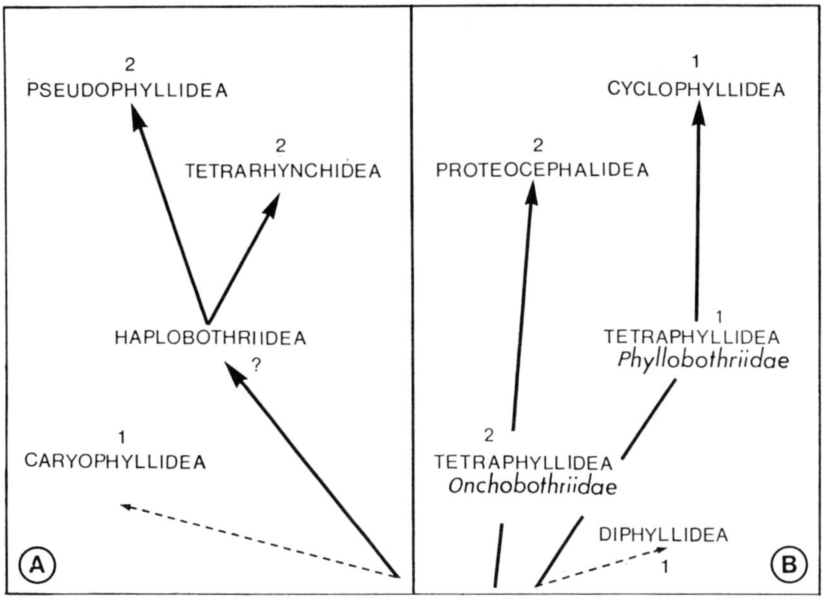

4.10

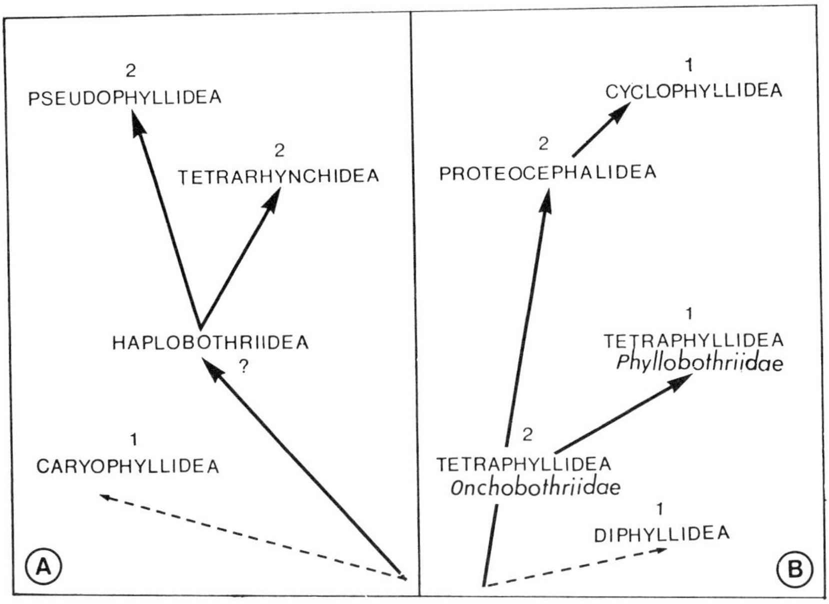

4.11

Figures 4.10, 4.11 Phylogenetic relationships between the different orders of cestodes, based on sperm ultrastructure. From Euzet, Swiderski and Mokhtar-Maamouri (1981).

Table 4.2 Sperm structure in cestodes of fishes (after Mackiewicz, 1981).

Species	No. of axonemes	References
Caryophyllidea		
Glaridacris catostomi	1	Swiderski and Mackiewicz (1976)
Haplobothridea		
Haplobothrium globuliforme	2	MacKinnon and Burt (1985a)
Pseudophyllidea		
Bothriocephalus clavibothrium	2	Swiderski and Mokhtar-Maamouri (1980)
Bothrimonus sturionis	2	MacKinnon and Burt (1984)
Diphyllobothrium latum	2	Bonsdorff and Telkka (1965)
Trypanorhyncha		
Lacistorhynchus tenuis	2	Swiderski (1976)
Diphyllidea		
Echinobothrium affine	1	Euzet, Swiderski and Mokhtar-
E. typus	1	Maamouri (1981)
Tetraphyllidea (Onchobothriidae)		
Acanthobothrium filicolle benedenii	2	Mokhtar-Maamouri and Swiderski (1975)
A.f. filicolle	2	Mokhtar-Maamouri (1976, 1982)
Tetraphyllidae (Phy-llobotriidae)		
Echeneibothrium beauchampi	1	Mokhtar-Maamouri and Swiderski (1976)
Phyllobothrium gracile	1	Mokhtar-Maamouri (1979)
Psuedanthobothrium hanseni	1	MacKinnon and Burt (1984)
Proteocephalidea		
Proteocephalus longicollis	2	Swiderski and Eklu-Natey (1978)

branchs, give rise to the proteocephalideans in teleosts and reptiles. The sperm of groups in this line have two axonemes. The other line, in which the sperm have a single axoneme, incorporates first the phyllobothriid tetraphyllideans (in elasmobranchs) and culminates in the cyclophyllideans (in terrestrial vertebrates). Alternatively (Figure 4.11), only a single line is postulated, containing the onchobothriids and proteocephalideans, with two axonemes, culminating in the cyclophyllideans with one axoneme. The phyllobothriids diverge from the onchobothriids. In both schemes, the diphyllideans separate at a very early stage. Although not figured, the Tetrabothriidea were grouped with the orders of branch B, a suggestion supported by the work of Hoberg (1987), who postulated that tetrabothriids had evolved from tetraphyllidean ancestors in elasmobranchs which had through predator-prey relationships colonized marine birds and mammals in which they became the dominant cestode group.

The use of sperm structure as an indicator of relationships in the cestodes should be approached cautiously. Euzet, Swiderski and Mokhtar-Maamouri (1981) emphasized that the sperm structure of many cestode orders and families remained to be discovered. Mackiewicz (1981) pointed out that reliance on this character would lead to some paradoxical conclusions. Possession of a sperm with a single axoneme would suggest a link between the caryophyllideans, generally held to be primitive, and the much more advanced cyclophyllideans with which they have little in common. It would not link them to the pseudophyllideans, with which they have often been grouped.

Phylogenetic relationships of the Trematoda

Aspidogastreans

The aspidogastreans and digeneans are regarded as sister groups by the majority of authors (eg. Rohde, 1972, 1973; Pearson, 1972; Cable, 1974, 1982; Schell, 1973; Tang and Tang, 1980; Gibson and Chinabut, 1984; Gibson, 1987; Brooks, O'Grady and Glenn, 1985a, b; Ehlers, 1984, 1985a, b). There is little support for suggestions that aspidogastreans were derived from digeneans by loss of the parthenogenetic intramolluscan generations. Aspidogastreans are regarded as archaic because of their direct development, use of molluscan hosts and various morphological features of the adults and cotylocidia (Rohde, 1972). Additionally, Rohde (1972) regards them as poorly adapted to parasitism, but Pearson (personal communication) disagrees.

Rohde (1972, 1973) has suggested that aspidogastreans are closely related to prodigeneans and that both groups had a simple one-host life-cycle in molluscs, without the alternation of generations, and were capable of survival in vertebrates if ingested. Both he and Gibson (1981, 1987) believed that ingestion of infected molluscs was the primary route by which vertebrates acquired digenean parasites, in contrast to, eg. Pearson (1972), Cable (1965, 1974) and others, who believe this is secondary. The cercaria is regarded as a dispersal stage by Rohde and Gibson but as the protodigenean adult by Pearson and Cable. Pearson (1972) suggested that the aspidogastreans diverged from protodigeneans before the cercaria evolved and before the protodigenean became a visceral parasite. Cable (1974) thought it more probable that the two groups either had a common monoxenous, monogenetic ancestor parasitic in archaic molluscs, or that the aspidogastreans evolved later from a digenetic and probably heteroxenous stock. Common descent from a turbellarian ancestor was considered less probable. He later (1982) suggested that the life-cycles of homoxenous aspidogastreans may have become simplified by precocious development in molluscs. Gibson (1987) suggested that prototrematodes diverged from the protocercomeromorphan line (monogeneans and cestodes) at the stage when their common ancestor had become ectocommensal on sedentary animals. Gibson and Chinabut (1984) have recognized two orders, Aspidogastrida and Stichocotylida, within the subclass Aspidogastrea and have emphasized that the two differ considerably in various aspects of their biology. This should be taken into account when speculating on phylogenetic relationships among the trematodes. The stichocotylids and aspidogastrids may have diverged after the acquisition of a mollusc host by the prototrematode, with the stichocotylids (currently with no known mollusc associations) transferring to crustaceans with which they are now associated, before acquiring a vertebrate host but after the protodigeneans had diverged from the prototrematode stock. Alternatively, the stichocotylids may have diverged earlier from the prototrematode ancestor before it acquired a mollusc host. The stichocotylids, which parasitize the more archaic elasmobranchs and holocephalans, are more primitive than the aspidogastrids which, like digeneans, parasitize the more recently evolved teleosts. Gibson (1987) proposed that protoaspidogastridans had probably diverged from the protodigeneans before either acquired vertebrate hosts and that each group colonized such hosts independently, perhaps more recently in the case of the aspidogastridans.

Digeneans

Contrasting views on the origin of parasitism in the digeneans, the nature and evolution of their larval stages and the development of their life-cycle patterns have been outlined in Chapter 2. Depending on their views of the nature of the original protodigenean adult, some workers emphasize the value of the cercaria as an indicator of phylogenetic relationships while others rely on the adult stage. Bray (1988), in a discussion of the use of the cercaria and its tail in phylogenetic speculation, concluded that tail features should be treated with great caution. He emphasized that, in deducing relationships from life-cycle data, it is important, when drawing comparisons between different organisms, to compare the same developmental stages of each. He also noted that each digenean life-cycle, as the term is usually understood, comprises three separate ontogenies (miracidium—mother sporocyst, daughter sporocyst or redia, and cercaria—metacercaria—adult) and that different authors have used the term 'life-cycle' to express different meanings (eg. Cable, 1982; O'Grady, 1985). The most satisfactory definition for application to the digenean life-cycle is that of Bonner (1974), who defined the life-cycle in sexually-reproducing organisms as a meiosis—to—meiosis (or fertilization—to—fertilization) cycling. Gibson (1987) has warned against over-reliance on life-cycle studies as a basis for phylogeny, considering that all digenean cycles are variations on the same basic theme and are subject to considerable plasticity.

One of the most detailed phylogenies recently proposed for the digeneans is that of Cable (1974, 1982) (Figure 4.12) which resembles that of La Rue (1957) in many respects. Cable holds that the furcocercaria is the most primitive larval type and the ancestral adult was furcocercous. He identified a number of radiating evolutionary lines rooted in a primitive two-host life-cycle in which furcocercariae escaped from snails and were swallowed by fish, but did not encyst or penetrate a host. Of existing families, the Bivesiculidae and Azygiidae retain such a cycle, but the azygiids are allied to the hemiuroid line which has three-host life-cycles in which cystophorous cercariae are swallowed by a crustacean second intermediary rather than a fish. The Transversotrematidae also have furcocercariae which lack penetration and cystogenous glands but they are not eaten; instead they establish beneath the scales of the fish. This is a phylogenetically isolated group. Two other lines developed from furcocercariae which acquired penetration glands. One led to the Bucephalidae, Fellodistomidae and Brachylaimidae and the other to the Clinostomidae and the strigeoids (Cyathocotylidae, Diplostomidae, Strigeidae), in which cercariae penetrate and encyst in fish intermediaries, and the schistosomatoids (Sanguinicolidae, Spirorchiidae, Schistosomatidae), in which cercariae penetrate the definitive host directly. For members of these families, fish serve as both second intermediaries and definitive hosts but, in the more evolved taxa, fish tend to be second intermediaries for species which parasitize higher vertebrate groups.

The furcocercous ancestors of the remaining major evolutionary line lost the caudal furcae and developed the ability to encyst in the open. Except for the Heronimidae, members of this line encyst rapidly on food or superficially in second intermediaries, or encyst without leaving the mollusc, to become metacercariae which are infective to the definitive host with little or no further development. The heronimids were viewed as exceptional in this line because the cercariae remained in the snail without encysting and were ingested within the snail by the turtle definitive host. The line culminated in the echinostomatoid families but, before that point, a line diverged from the main stem. This line was characterized by penetration and encystment in a second intermediary, most frequently a fish, with metacercariae which required more prolonged development to achieve infectivity. Where known, cercariae of the families in this line are biocellate with well-developed penetration glands. The line includes, in sequence, the acanthocolpids and campulids, the opisthorchioids, lepocreadioids and Monorchiidae, followed by a divergence (Figure 4.12) into the plagiorchioid and allocreadioid complexes, all of which have xiphidiocercariae.

Brooks, O'Grady and Glenn (1985b) applied cladistic analysis to digenean systematics and phylogeny. In their scheme, the heronimiforms are the most primitive order (cf. Cable, 1974, 1982) followed in ascending evolutionary order by the paramphistomiforms, echinostomatiforms, haploporiforms, hemiuriforms, strigeiforms, opisthorchiiforms, lepocreadiiforms and plagiorchiiforms. They noted that their phylogenetic tree and that of Cable (1974) would have yielded broadly similar results had both been rooted at the same point. Theirs was rooted at the heronimiforms and Cable's at the strigeiforms (Brooks, O'Grady and Glenn, 1985b). In both schemes, the aspidogastreans and heronimids were closely related. Gibson (1987) disputed both conclusions, regarding the heronimids as an aberrant rather than a primitive family. In his view, of the possible contenders for the most

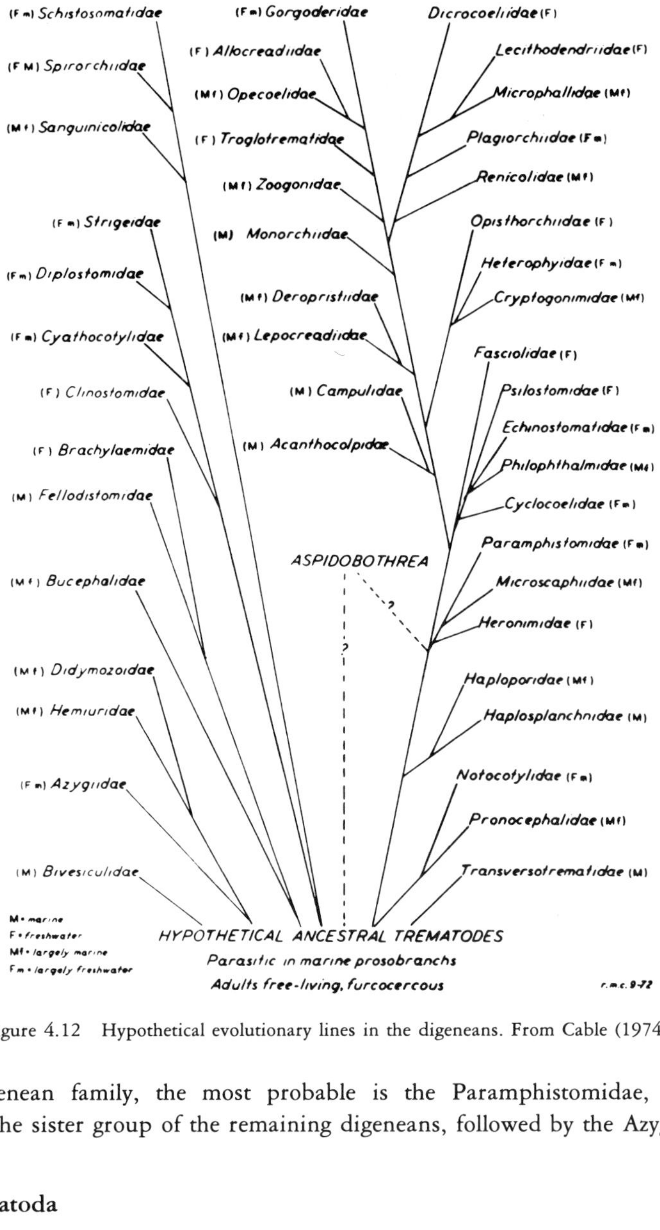

Figure 4.12 Hypothetical evolutionary lines in the digeneans. From Cable (1974).

primitive digenean family, the most probable is the Paramphistomidae, which would therefore, be the sister group of the remaining digeneans, followed by the Azygiidae.

Phylum Nematoda

Origin of parasitism

Perhaps because of their thick, decay-resistant cuticle, nematodes are surprisingly well-represented in the fossil record. The first known nematodes are from the Lower and Upper Carboniferous Periods about 345 million years ago (Conway Morris, 1981). The older fossils, *Scorpiophagus baculiformis* and *S. latus*, occurred between the cuticle layers of a large scorpion and were saprobiotic. Those from the Upper Carboniferous were free-living and aquatic. Sinusoidal markings from Eocene lacustrine sediments that are 54 million years old and from borings in Upper Cretaceous foraminifers have also been ascribed to free-living nematodes. Parasitic nematodes are known from the Eocene and Oligocene Periods of the Cainozoic Era: a single specimen projecting from the anus of a beetle from Eocene lignite, specimens of a related species closely associated with a dipteran preserved in Baltic amber of the Oligocene, and a larval nematode encysted in striated muscle of an Eocene beetle. Additionally, parasitic nematodes have been found in Pleistocene sousliks, woolly mammoths and a horse frozen in Siberian permafrost, and a nematode mould has been discovered in a kidney stone from a Pleistocene cave bear. Rausch (1985) cited reports of *Strongylus edentatus* in a horse dating from 33,000 years BP and of a *Cobboldina* sp. from a Siberian mammoth.

Some authorities believe that, in contrast to other helminth groups, the zooparasitic nematodes may have had a terrestrial rather than an aquatic origin. Anderson (1984) proposed that they were derived from soil nematodes and invaded vertebrates only after the tetrapods had appeared and colonized the land, estimated to have occurred in the Carboniferous Period, 350 million years ago. He also suggested that higher invertebrates were invaded by nematodes only after the ancestors of the former (annelids, molluscs and terrestrial arthropods) had colonized the land. The overwhelming majority (91.7 per cent of families, 97.7 per cent of genera) of nematode parasites of vertebrates are secernenteans which have evolved from ancestors resembling soil-dwelling rhabditids, which indicates that parasitism in the secernenteans originated in a terrestrial rather than an aquatic environment. The existence in the nematode life-cycle of the dauer larva (the third-stage larva enclosed in the second-stage cuticle, capable of prolonged survival in adverse conditions), which was probably already present when vertebrates first colonized the land, may have been significant in the development of parasitism. Dauer larvae had probably developed phoretic associations with some invertebrates and were transferred to the carnivorous early tetrapods. Secernenteans account for 79 per cent of all genera parasitic in invertebrates and 88 per cent of all families. Nematodes are common in terrestrial invertebrates but rare in aquatic invertebrates (polychaetes, molluscs and crustaceans) despite the ancient associations of these groups with marine adenophoreans. There is no clear evidence that free-living adenophoreans colonized the aquatic ancestors of vertebrates. Adenophorean parasites are derived from a minor line which evolved from the predominantly soil-dwelling dorylaimids and include the mermithoids, trichinelloids and dioctophymatoids.

During the course of evolution, nematodes which had invaded vertebrates and established a monoxenous life-cycle adopted intermediate hosts, presumably to protect free-living stages from the environment. The advantages of a heteroxenous cycle are that the intermediary protects the larvae from the environment, may assist transmission by channelling the parasite to the definitive host, and may extend the life of the parasite in time and space. Paratenic hosts were subsequently included to adapt heteroxenous life-cycles to the changing food habits of the hosts during the radiation of the vertebrates. Following the development of the heteroxenous life-cycle and of paratenesis in terrestrial vertebrates, nematodes transferred to the aquatic environment.

In the Spirurida, the paratenic host is interposed between the intermediate and final hosts but, in the ascaridoids, it occurs primitively at the beginning of the life-cycle. It is represented by an invertebrate which ingests an egg containing a second-stage larva (the first moult usually occurs in the egg) and transfers it to a vertebrate intermediate host in which development proceeds to the stage infective to a vertebrate final host. In aquatic ascaridoids, the invertebrate paratenic host may become an intermediate host but the vertebrate intermediate host is retained, eg. *Anisakis* in fishes, or the invertebrate paratenic host becomes an intermediate host and the vertebrate intermediary is dropped, eg. *Sulcascaris sulcata*, a parasite of marine turtles which can reach the infective stage in molluscs.

Anderson (1984) rejected the assumption that because fish are the most ancient vertebrate group, their nematode parasites must be the most primitive, having evolved in the water and colonized land along with the first terrestrial vertebrates. He regarded the nematode fauna of fish as limited; fish harbour no unique nematode orders or superfamilies and only five of the 17 families which occur in fish are unique to them. Only about 11 per cent of known nematode genera of vertebrates occur in fish and only about 12 per cent of species. Within the fish nematodes, heteroxenous groups (trichinelloids, spiruroids and ascaridoids) predominate. Only a few of the monoxenous Oxyuroidea occur in fish and none of the Rhabditida or Strongylida. These points suggest an acquired fauna in fish, transferred from terrestrial hosts mainly by heteroxeny and paratenesis. The nematodes of marine mammals are mainly heteroxenous and are closely related to those of terrestrial hosts, suggesting that nematodes made a second invasion of the aquatic environment with this host group.

The aquatic environment apparently did not, in Anderson's (1984) view, provide the conditions required to stimulate parasitism in the nematodes but a soil environment did so. Perhaps an aquatic environment would not have provoked the appearance of a dauer larva stage, which may have been significant in the development of parasitism. In addition, nematodes have no kinocilia and, unlike the platyhelminths, they lack an effective swimming stage, which may have limited their exploitation of the parasitic potential in aquatic habitats.

Sprent (1982, 1983) has suggested that the original hosts of the ascaridoid nematodes were crocodilians, although he did not rule out a concurrent line in marine teleosts (but not in elasmobranchs or freshwater fish). He pointed out that the ascaridoids of crocodilians are relatively small (a primitive character), are distributed throughout the geographical range of the group and have diversified into many genera and species. This indicates that even if the crocodiles were not the original hosts, extensive radiation took place in this group, probably in the deltas and estuaries of the great rivers of the Old World Tropics.

Evolution within the ascaridoids, one of the major nematode groups occurring in fish, has been discussed by Sprent (1983) and Gibson (1983). Sprent (1983) considered the Heterocheilinae to be the most primitive existing ascaridoid group; they occupy freshwater, fluviatile or coastal habitats whereas the Ascaridinae are terrestrial, the Raphidascaridinae are parasites of fish or fish-eating reptiles and the Anisakinae and Contracaecinae are parasites of fish or piscivorous birds and mammals. He suggested that heterocheiline-like forms in freshwater or estuarine hosts radiated in the Tertiary by host-range expansion and host-succession extension. In the first of these, hosts share common prey and therefore ingest the same larval stages, and in the second the adult nematodes live in both prey and predator. This has resulted in closely related ascaridoids occurring in hosts which have the same prey, and in predators at successive levels of the trophic pyramid. Host-parasite relationships are not phylogenetically determined. During the Tertiary, new forms appeared by host-range expansion in different ecological situations, eg. *Orneascaris* in anurans, *Paranisakis* in selachians and *Raphidascaris* in teleosts which gave rise by host-succession extension to Ascaridinae, Anisakinae and Contracaecinae, respectively.

Gibson (1983) differed in regarding the Acanthocheilidae, parasites of elasmobranchs, as the most primitive group. He suggested an evolutionary sequence (Figure 4.13) leading next to the Anisakidae, which parasitize elasmobranchs and other primitive fish. Their invasive larvae have a boring tooth, which Gibson (1983) related to the appearance in the life-cycle of a fish second intermediary, the gut wall of which had to be penetrated by the larvae. He suggested that the Goeziinae and Anisakinae developed from forms similar to *Paranisakis* and *Paranisakiopsis*, respectively. The Anisakinae extended into the reptiles and higher vertebrates. The Goeziinae, it is suggested, gave rise to the Ascarididae in the form of the Heterocheilinae, which may have developed in piscivorous reptiles or possibly teleosts. Of the terrestrial subfamilies, the angusticaecines may have developed from the heterocheilines but the origins of the ascaridines and toxocarines are unclear.

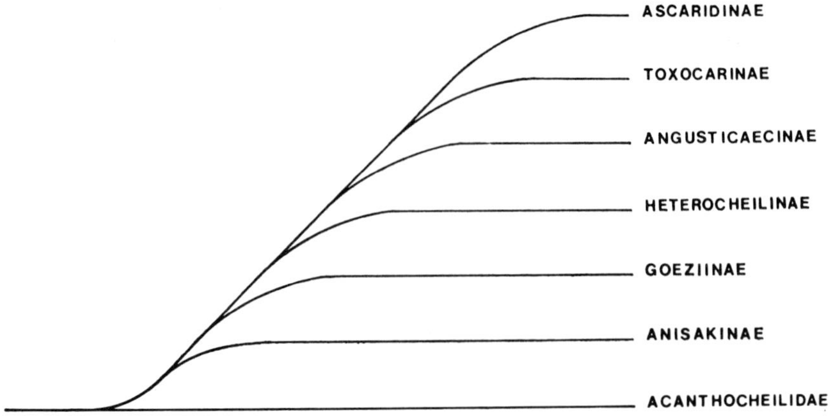

Figure 4.13 A suggested arrangement for the evolution of the ascaridoid groups. (From Gibson, 1983).

Phylum Acanthocephala

Origin of parasitism

The acanthocephalans are a small group, completely adapted to parasitism, whose relationships with other animal groups are problematical. Their closest relatives are among the aschelminths, notably the rotifers or priapulids (Conway Morris, 1981). Whitfield (1971) has shown that the spermatozoa of *Polymorphus minutus* are more similar to those of rotifers, nematomorphs and nematodes than to cestodes.

Two marine fossil priapulids, *Ottoia prolifica* and *Ancalagon minor*, are known from the Burgess Shale deposits of the mid-Cambrian period in Canada and are very similar to acanthocephalans. *A. minor*, in particular, is strikingly reminiscent of Golvan's (1958) hypothetical proto-acanthocephalan (Conway Morris, 1981; Conway Morris and Crompton, 1982a, b). They were burrowers with an armed proboscis, which suggested pre-adaptation to parasitism.

Conway Morris and Crompton (1982a) suggested that the evolution of a meiofaunal species transitional from priapulids to acanthocephalans might have occurred through progenesis, a strategy linked with a life-history showing the characteristics of r-selection. If progenesis occurred soon after acanthocephalans became parasitic rather than during their origins in the meiofauna, its advantage would lie in the development of a highly fecund reproductive system. Two major evolutionary steps in the development of the unique acanthocephalan reproductive system proposed by Conway Morris and Crompton (1982a) are the development of internal fertilization and associated modification of the reproductive system in meiofaunal forms, and continued specialization of the reproductive organs to maximize egg production in parasites.

The original hosts of acanthocephalans were probably arthropods, now the intermediaries, with the vertebrate and currently definitive hosts being interpolated later in the life-cycle. Parasitism might have originated by the consumption of eggs, larval stages or even adults of ancestral Acanthocephala by bottom-feeding arthropods in the mid-Cambrian, by which time a diverse arthropod fauna had already evolved. Vertebrates could have been added to the life-cycle by predation on infected arthropods. The earliest vertebrates (fish) appeared in the Upper Cambrian but may not have been invaded until later. Conway Morris and Crompton (1982b) believed that a vertebrate host would have been adopted by the mid-Mesozoic, possibly much earlier. After their incorporation into the life-cycle, vertebrates eventually became the definitive hosts. The transmission of acanthocephalans depends entirely on food chains and many current definitive hosts are at or near the top of trophic pyramids. Paratenic hosts are important links between different levels of the pyramid.

Evolution

On anatomical grounds, the Archiacanthocephala could be considered the most primitive acanthocephalan order but the host-parasite relationships of some Palaeacanthocephala suggest that this order is the most primitive (Conway Morris and Crompton, 1982b). Palaeacanthocephalans have aquatic crustacean intermediaries and, usually, teleost definitive hosts. In addition, some occur in the more primitive and ancient fishes. *Acanthocephaloides* has been reported, unusually, from an elasmobranch (*Raja*) but this probably represents an accidental infection. *Acanthocephaloides* and *Echinorhynchus* have been found in lampreys (*Petromyzon*) and several palaeacanthocephalans have been recorded from sturgeon (*Acipenser*) and bowfin (*Amia*) (Conway Morris and Crompton, 1982a).

Additional factors affecting host specificity

Morphological factors

Morphology has a role in the maintenance of host-specificity because the close contact between a parasite and host at the site of attachment results in adaptations which may be highly specialized (Llewellyn, 1956; Williams, 1960d, 1961a, 1970). These adaptations often contribute to limiting a parasite to a particular host species and sometimes to a particular site in the host, and are linked to the genetic constitution of the parasite. Among fish parasites, the best documented examples are to be found in the monogeneans and in the tetraphyllidean cestodes of elasmobranchs. Holmes (1973) has reviewed factors influencing site-selection by helminths.

Monogeneans

Morphological compatibility between the attachment organs of a monogenean and the site of attachment in the host has been proposed as one factor maintaining the strict host-specificity found in this group. Several authors have shown that monogenean oncomiracidia may invade hosts other than those which they normally parasitize but they usually persist for only a short time and fail to reach maturity. This has been demonstrated for several species of *Dactylogyrus* on different species of cyprinid fish (Izyumova, 1956, 1969, 1970; Musselius and Ptasuk, 1970) and it has been suggested that the relationship between the haptoral elements of *Dactylogyrus* species and the gill structure of the host is crucial for their establishment (Gusev, 1976; Gerasev, 1977 and others). In a series of hybridization experiments between common carp (*Cyprinus carpio*), silver carp (*Hypophthalmichthys molitrix*), grasscarp (*Ctenopharyngodon idella*) and bighead carp (*H.* (= *Aristichthys*) *nobilis*), Molnár, Bakos and Krasznai (1984) showed that in most hybrids only the *Dactylogyrus* species characteristic of the female parental line became established. This is because hybrid carp resemble the female parent species more than the male in their morphological and physiological characters, including gill structure.

Adaptation to different sites with consequent modifications of the haptor was regarded as a major factor in the evolution of the Monocotylidae (Timofeeva, 1985; Lambert, 1982).

Four groups are believed to have evolved from an ancestral monocotylid in the branchial cavity of Mesozoic Batoidea (Timofeeva, 1985). The haptor has a central loculus from which radiate septa demarcating eight loculi; there are 14 marginal hooks and a pair of hamuli. The Monocotylinae, the most primitive subfamily, are gill cavity parasites of Myliobatiformes and Rajiformes, in which body size is secondarily reduced and the basic haptoral plan is modified by the loss or addition of radial septa. The Dendromonocotylinae, parasites of the body surface of Myliobatiformes, developed from a *Monocotyle*-like ancestor. The Calicotylinae tend towards endoparasitism, living on the skin or in the cloaca, gut or reproductive system of Rajiformes. They have a strongly muscular anterior body and a muscular haptor. The Merizocotylinae, parasites of the nasal cavity of Myliobatiformes and Rajiformes, have a haptor with an additional circle of marginal alveoli, adapted to the surface of the nasal cavity. Lambert (1982) has illustrated the haptoral modifications encountered in this family (Chapter One, Figure 1.14). In addition to those already mentioned, they include the loss of the hamuli in the Dendromonocotylinae and some Merizocotylinae and the appearance of sclerites on the septa and on the dorsal surface of the haptor.

Llewellyn (1956, 1959, 1964) has provided many examples of the close relationship between the attachment organs of gill monogeneans and their sites of attachment, and has also elucidated the effect of the gill ventilating current on the orientation and morphology of the parasites. Monogeneans are attached with the haptor upstream with respect to the ventilating current and, in some species, the body is facultatively or permanently asymmetrical in a manner determined by the incidence of the current on the body. *Discocotyle sagittata*, attached to the secondary gill lamellae of trout, is subjected to a continuous flow of water which passes into the space between two hemibranchs of the same gill arch (Figures 4.14–4.16) (Llewellyn and Owen, 1960). To reduce resistance to this unilaterally incident current, the worm aligns with its anterior end inclined towards the inter-hemibranchial space. The haptor becomes asymmetrical, with all four clamps of one side attached anteriorly to those of the other. The more posterior set of clamps is always on the same side of the body to which its longitudinal axis is inclined, and nearer to the inter-hemibranchial space; thus the clamps are arranged in the best position to resist the flow of the current. The worm could be asymmetrical to right or left depending on whether it was attached to an inner or outer hemibranch or to the dorsal or ventral surface of a primary lamella. The clamps are too large to allow both members of a clamp pair to attach to the same secondary lamella. Asymmetry in *D. sagittata* is facultative; parasites which became detached could re-attach with the opposite alignment and moribund or dead specimens were always bilaterally symmetrical (Llewellyn and Owen, 1960). In species of *Anthocotyle, Gastrocotyle, Pseudaxine, Axine* and others, however, obligate unilateral asymmetry is the case (Llewellyn, 1956).

Llewellyn (1959, 1964) has shown that the oncomiracidium of *Gastrocotyle trachuri* is bilaterally symmetrical but, once established on the secondary gill lamellae of *Trachurus trachurus*, clamps develop only on one side of the body. *Pseudaxine trachuri* shows a similar development pattern but it is not clear whether the side of the asymmetry in these species is determined by the side of incidence of the gill ventilating current or whether it is genetically determined (Llewellyn, 1959).

An asymmetrical habit has been described for *Atriospinum* (= *Microcotyle*) *labracis* on the gills of *Dicentrarchus labrax* by Winch, (1983).

Relationships between morphology and site have been demonstrated for parasites of *Seriolella brama* by Rohde, Roubal and Hewitt (1980). A monogenean, *Eurysorchis australis*, attaches to the flat surface of the smooth part of the gill arch by an open clamp with a secondarily developed sucker and another monogenean, *Neogrubea seriolellae*, has clasp-like

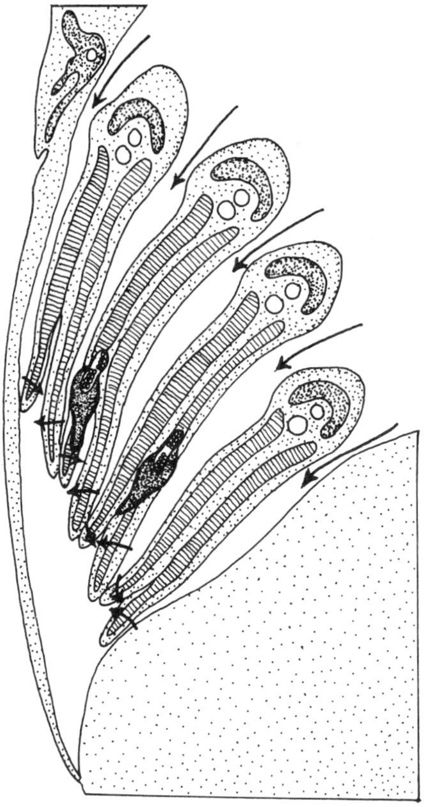

Figure 4.14 Horizontal section through one side of branchial region of *Salmo trutta* to show typical sites of attachment of *Discocotyle sagittata* and paths of gill ventilating currents.

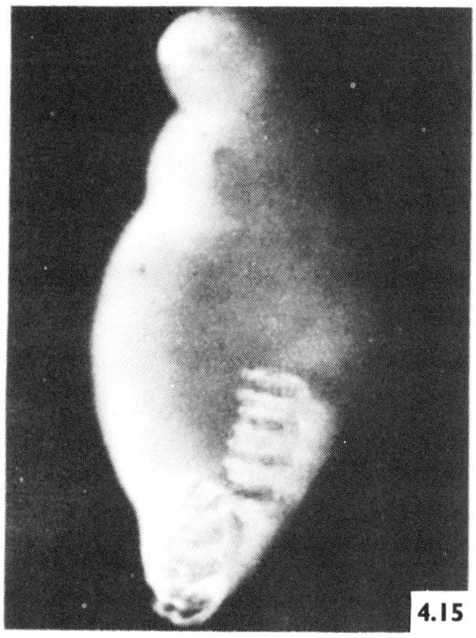

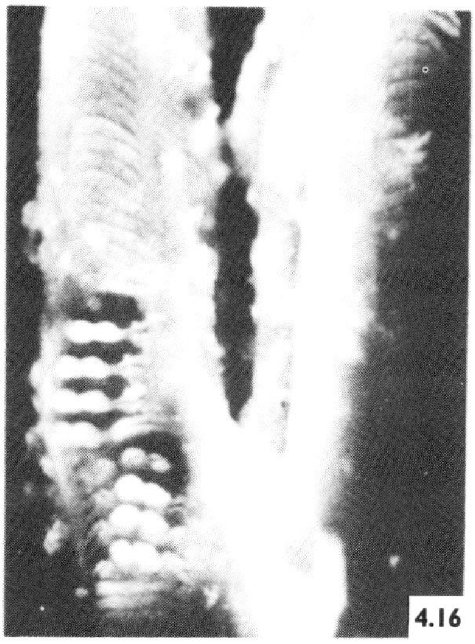

Figure 4.15 *Discocotyle sagittata* in ventral view, showing asymmetrical disposition of posterior adhesive organs; specimen fixed in its characteristic adhesive attitude.

Figure 4.16 Interbranchial septum region of gill of *Salmo trutta* with 'footprints' of *D. sagittata*. From Llewellyn and Owen (1960).

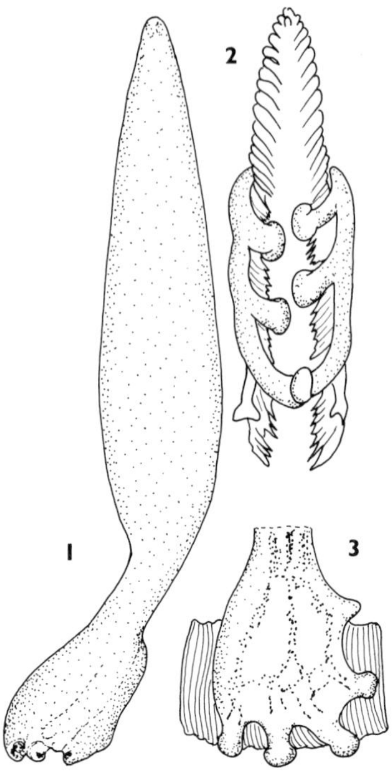

Figure 4.17 (1–3) *Macruricotyle clavipes*. 1. Entire specimen. 2. Haptor attached to primary gill filament showing press-stud arrangement. 3. Attached haptor, lateral view. From Mamaev and Avdeev (1984).

clamps with which it grasps the gill filaments (see the digenean section later for comments on an unusual digenean which lives on the spiny parts of the gills of this host).

Llewellyn and Simmons (1984) described the curious attachment mechanism of *Callorhynchicola multitesticulatus* to the gills of its holocephalan host, *Callorhynchus milii*. Juveniles attach to the secondary gill filaments by their clamps but the haptor of adult worms invades the host tissue. The elongated stalk connecting the haptor to the body has annulations with anteriorly-directed flanges that sink into a sheath of host tissue, produced by an inflammatory response around the stalk, serving as the effective means of attachment. The stalk is thought to be a preadaptation exploited for partial endoparasitism. A related species, *Chimaericola leptogaster*, has a similar stalk but remains ectoparasitic.

A monogenean ectoparasitic on sharks has an interesting mode of attachment in which a cup-shaped opisthaptor fits over a placoid scale and secretes a cement over it. *Dermophthirius carcharhini* on *Carcharhinus galapagensis* erodes the epidermis and ruptures adjacent goblet cells; scales can be lost and ulcers formed at the site. The ventral surface of the opisthaptor bears furrows that complement the patterns on the surface of the scale (Rand, Wiles and Odense, 1986).

Mamaev and Avdeev (1984) have described the interesting mode of attachment of *Macruricotyle clavipes* to the gills of *Coryphaenoides* sp. The opisthaptor wraps itself around the primary gill filament (Figure 4.17.1–3) with the two anterior pairs adhering by suction to the surface of the spine of the filament. The posterior pair of clamps attach to secondary gill filaments on either side but the third pair attach to each other by means of a connective tissue callus on one which exactly fits the cavity of the other. This press-stud arrangement remains intact even if all other clamps lose their grip.

Cestodes

Some of the most convincing evidence for the role of morphological factors in contributing to host-specificity comes from the tetraphyllidean cestodes, which are exclusively parasites of elasmobranchs and are highly host-specific. Euzet (1957) found that 36 of 61 species in Mediterranean selachians occurred in only one host species. The tetraphyllidean scolex is characterized by four sessile or pedunculate bothridia which are highly mobile and often complicated in structure, frequently complemented by accessory attachment organs such as suckers. A myzorhynchus is present in many representatives of the group. In a series of papers published during the 1960s, Williams has shown that the shape and structure of the bothridia is closely correlated with the topography of the intestinal mucosa of the host.

The first indication of the role of morphological factors in determining host and site-specificity in tetraphyllideans was given by Williams (1960b, 1961a) in describing the relationship between *Echeneibothrium maculatum* and *Raja montagui*. The surface of the mucosa in skates and rays may form a close network of ridges and depressions, or may be a very open network of ridges separated by depressions with a very finely reticulated surface, or possibly only fine reticulations may be present. The mucosa may be divided into loculi, the shape of which varies according to the species (Williams, 1959a, 1960c). In sharks, the intestinal mucosa can form very characteristic complicated shapes and in *Hexanchus griseus* is raised into well-developed folds.

The genus *Echeneibothrium* is specific to *Raja* and its members are highly species-specific. It has a myzorhynchus and four pedunculated bothridia. Williams (1966) identified species groups within the genus according to whether the myzorhynchus is hemispherical when contracted or is always cylindrical, and according to the shape of the bothridia, the number of loculi and whether the septa between the loculi are superficial or not. The structure of these organs is correlated with the structure of the mucosa, the exact site of attachment and the degree of damage, if any, inflicted on the host gut, the lack of which is itself an indicator of host-specificity. These points can best be illustrated by an account of the genera *Echeneibothrium, Pseudanthobothrium, Phyllobothrium* and *Acanthobothrium*.

E. maculatum occurs in the spiral intestine of *R. montagui*, the mucosal surface of which is thrown into folds. It locates in the second and third tiers of the intestine where the primary mucosal folds are deepest and where it presumably derives the greatest protection and possibly the greatest nourishment (Williams, 1961a, 1966; Figures 4.18–4.20). The bothridia are well-adapted for attachment to the folded mucosa of *R. montagui* but not to the highly villous mucosal surface of *R. naevus*.

A folded mucosa like that of *R. montagui* appears to provide a more suitable site for attachment of species with a large myzorhynchus, eg. *Echeneibothrium* sp. of Williams (1966) in *R. montagui*, than that of *R. clavata* which is relatively flat overall but is thrown into deep tubular crypts. *Discobothrium fallax*, a species with a large myzorhynchus, occurs in *R. clavata* but is fairly rare, in contrast to the abundance of *Echeneibothrium* sp. in *R. montagui*. Williams (1966) believed this was due to differences in mucosal structure between the hosts. *D. fallax* provoked a much greater tissue reaction in the host than did *Echeneibothrium* sp. in *R. montagui*. An *Echeneibothrium* sp. found in *R. clavata* provoked no reaction but in this species the myzorhynchus was non-functional as an attachment organ, being retracted into the scolex when the bothridia were attached. Williams (1966) concluded that species with a large myzorhynchus cause the most damage and are therefore less successful as parasites and consequently more rare than those with a small myzorhynchus which do little harm and are much more common. Species which occur in intermediate numbers provoke some reaction but do not cause serious damage, eg. *E. variabile* in *R. clavata* (Figure 4.21). *E. variabile* has a moderately developed myzorhynchus, the

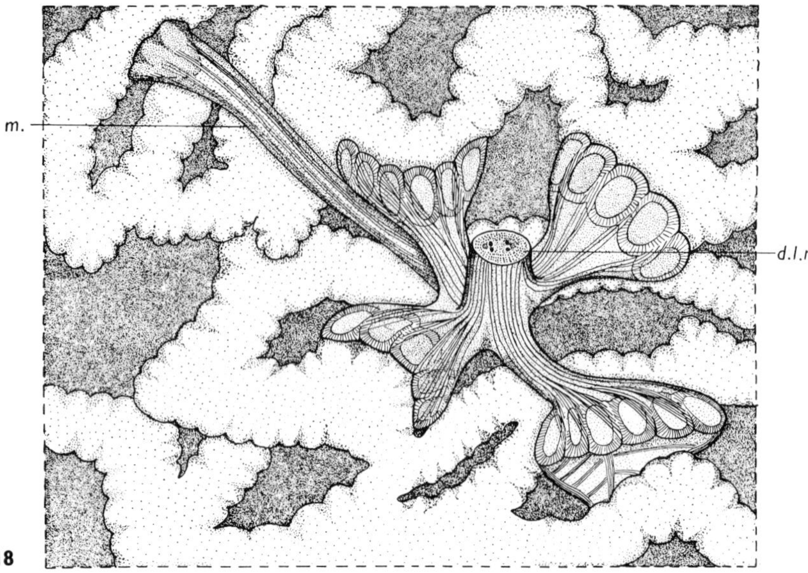

4.18

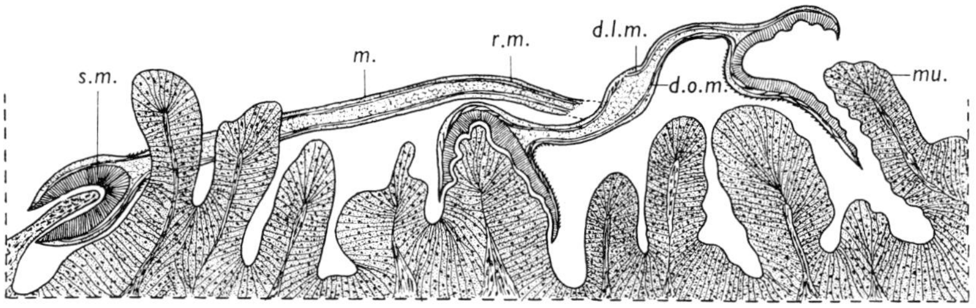

4.19

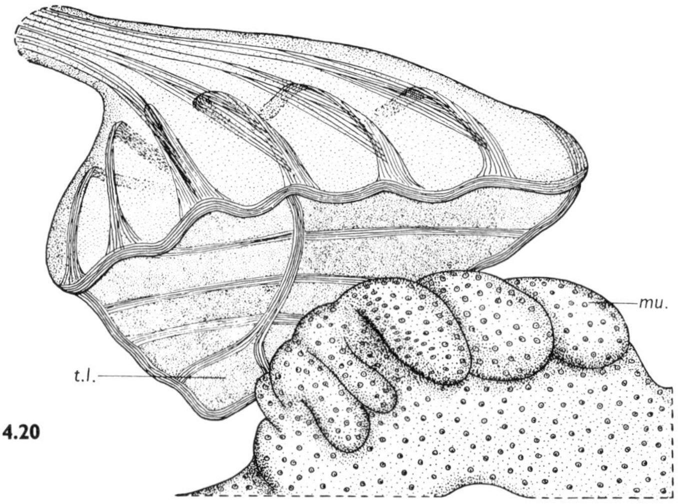

4.20

Figures 4.18–4.20 *Echeneibothrium maculatum* in *Raja montagui.*

Figure 4.18 Showing orientation of bothridium and myzorhynchus while attached to the mucosa of the intestine. 4.19 Section through scolex attached to intestine. 4.20 Showing natural position and shape of bothridium when attached to a portion of the mucosa which, in this case, shows a distinct imprint of some of the posterior loculi. Abbreviations: d.l.m. deep longitudinal muscles; d.o.m., deep oblique muscles; m., myzorhynchus; mu., mucosa; r.m., retractor muscles; s.m., apical sucker of myzorhynchus; t.l., anterior terminal loculus of bothridium. From Williams (1961a).

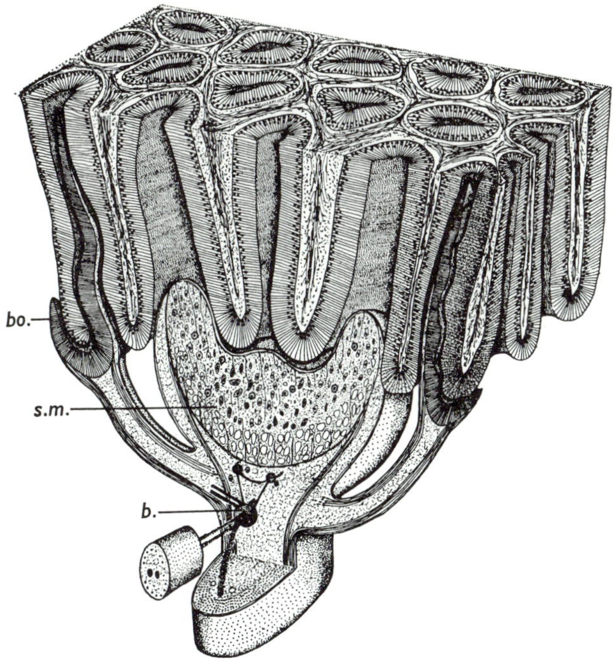

Figure 4.21 *Echeneibothrium variabile*. Diagram showing an early stage in the attachment of the scolex to mucosa of *Raja clavata*; there is no host-tissue reaction. From Williams (1966).

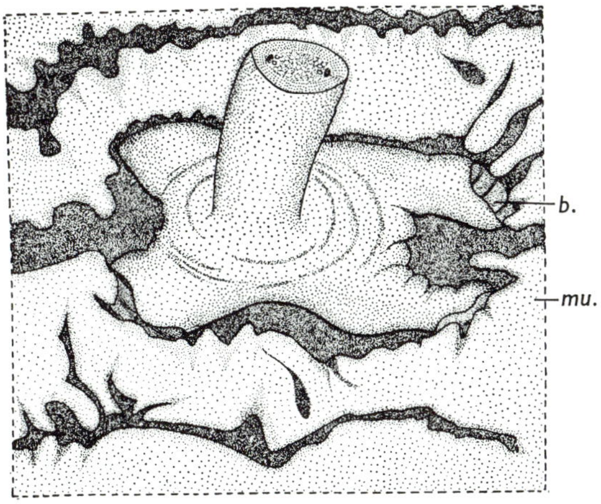

Figure 4.22 *Echeneibothrium* sp. attached between folds of mucosa of *Raja montagui*.

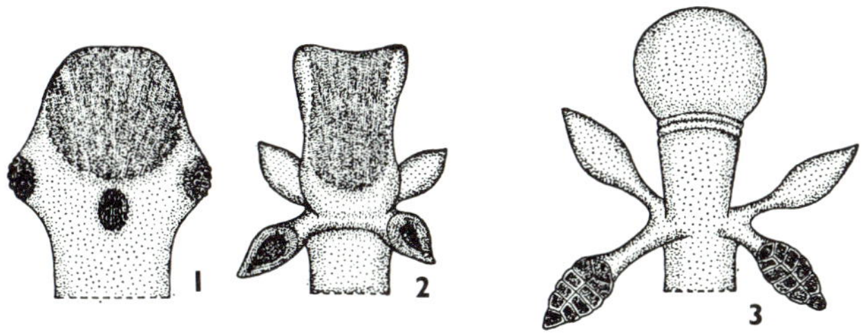

Figure 4.23.1–3 Scolex of *Echeneibothrium* sp. from *R. montagui*. From Williams (1966).

exact function of which is unknown, but the powerfully developed myzorhynchus of *D. fallax* and *Echeneibothrium* sp. from *R. montagui* (Figures 4.22 and 4.23) may function as an eroding sucker and provokes a strong response. In contrast, *E. maculatum* and *E. minutum* have small myzorhynchi which cause little damage. Williams also observed an inverse relationship between the degree of development of the myzorhynchus and the number of loculi per bothridium, eg. there are none in the bothridia of *D. fallax* but 20 in those of *E. minutum*. Species with many loculi per bothridium, with the exception of *E. dubium*, are adapted for attachment to a relatively large area of mucosa. In species with a cylindrical myzorhynchus, the correlation is between the length of the myzorhynchus and the number of loculi per bothridium. The shorter the myzorhynchus, the greater the number of loculi. *Pseudanthobothrium hanseni* has the longest myzorhynchus and its bothridia are cup-shaped, lacking loculi altogether. This species occurs in *R. radiata*, attaching to a region of the intestine covered with closely-packed long villi. Each bothridium fits over the apex of a single villus and the myzorhynchus probes between the villi, penetrating the wall of a villus with its terminal sucker (Figure 4.24). This species would be unable to attach effectively to a surface covered with shorter villi. In species which attach to shorter villi, the number of loculi per bothridium tends to increase, eg. *Echeneibothrium* sp. from *R. naevus* has five loculi per bothridium (Williams, 1966). Shallow open bothridia with many clearly-defined loculi are well-adapted for probing between the mucosal folds, eg, *Echeneibothrium* sp. from *R. clavata* (Figures 4.25, 4.26) in which the bothridia penetrate into deep tubular crypts (Williams, 1966).

Williams (1966, 1968a,b,c, 1970) has repeatedly emphasized that the effects of morphological factors are among the most difficult to elucidate and cannot be assessed independently of ecological and physiological factors. An illustration of this within *Echeneibothrium* has been provided by McVicar's (1977a) discovery of two morphological types of the genus in *Raja naevus*, one with a large saucer-like myzorhynchus and the other with a protrusible myzorhynchus. The former occupied tiers one and two of the spiral intestine, and the latter tiers two to four. It was confirmed that both morphological types developed from a common larval form, suggesting that the myzorhynchus is capable of varying its shape in response to differing conditions in different sites of attachment.

Members of the genus *Phyllobothrium* have very highly folded, frilly bothridia, each with an accessory sucker, adapted for attachment to an intestinal mucosa covered in villi. The scolex of living *P. sinuosiceps* attached to the mucosa of the six-gilled shark, *Hexanchus griseus*, covers an area up to 12 mm in diameter, penetrating between and interdigitating with the net-like folds of the host mucosa (Williams, 1959a). The very highly folded nature of the scolex, which is not clearly divisible into bothridia, corresponds very closely with the finely reticulate mucosal surface and the folds in the scolex are the same depth as those of the mucosa (Williams, 1968a). The scolex of *P. piriei* in the intestine of *Raja radiata* (Figure 4.27) covers a large area of the mucosa, each highly-branched and horizontally-flattened bothridium being in contact with about 12 villi (Williams, 1968b; Williams, McVicar and Ralph, 1970).

Phyllobothrium and some closely-related genera provide further good illustrations of the morphological basis for host- and site-specificity (Williams, 1968c). Of two cestodes found in *Mustelus mustelus, P. lactuca* is attached at the anterior end of the spiral valve where the mucosal surface forms long, straight primary folds which branch into secondary and tertiary folds. *P. lactuca* has highly folded bifid bothridia up to 3 mm long; *Orygmatobothrium musteli*, a species with much smaller bothridia, occupies the posterior region of the intestine where the folds are shallower. The presumption is that *P. lactuca*, with larger bothridia, could not attach itself in the posterior region because the folds would not be sufficiently deep. *P. britannicum* has bothridia which spread out over a circular area and attach to a

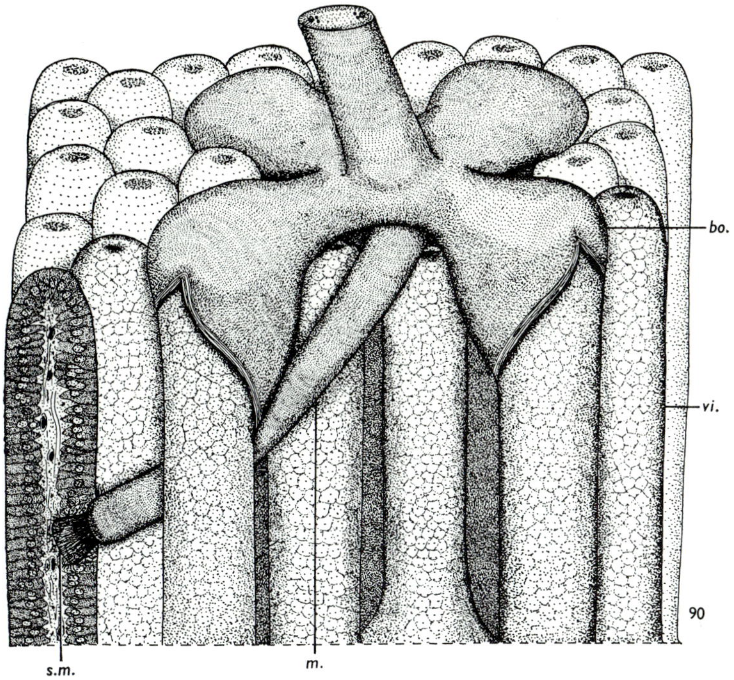

Figure 4.24 ***Pseudanthobothrium hanseni*** attached to the gut of ***Raja radiata***. Scolex ***in situ*** on mucosa. bo, bothrium; m, myzorhynchus; sm, apical sucker of myzorhynchus; vi, villus. From Williams (1966).

4.25

4.26

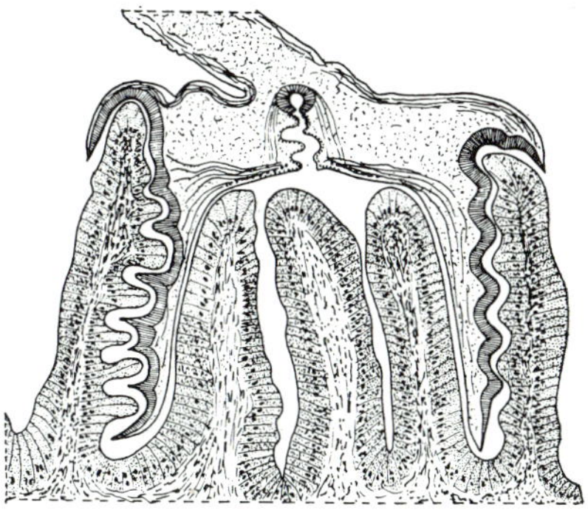

Figures 4.25, 4.26 Diagrammatic representation of mode of attachment of *Echeneibothrium* sp. to mucosa of *Raja clavata*. 4.25 Scolex drawn while attached to the mucosa with anterior regions of bothridia deeply embedded in mucosal crypts. 4.26 Section of specimen showing retracted myzorhynchus which is non-functional for attachment purposes. From Williams (1966).

surface (in *R. montagui*) composed of more randomly-orientated folds than in *M. mustelus* in which the intestinal folds coalesce to form a network (Williams, 1968c).

A *Crossobothrium (?Phyllobothrium)* sp. from *Squalus acanthias* has large flat leaf-like bothridia suitable for attachment to the finely-folded, comparatively flat but uneven mucosal surface encountered in this host (Williams, 1968c). In *Scyliorhinus stellaris* and *S. canicula*, the width of clefts between the wavy parallel folds of the mucosa may correspond with the depth of the scolex of *Monorygma (?Phyllobothrium)* sp.. *P. thridax*, a large species, has a tiny scolex which penetrates spaces in the very fine mesh-like mucosa of *Squatina squatina*. The surface of the strobila also appears to be attached along its length to the surface of the mucosa.

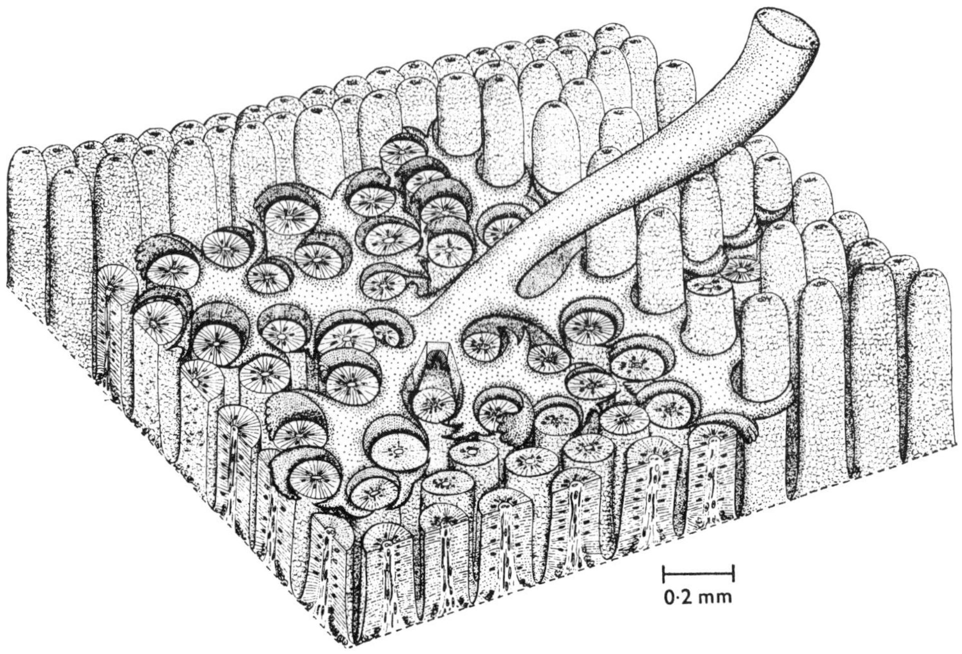

Figure 4.27 *Phyllobothrium piriei*. Specimen drawn attached to gut mucosa of *Raja radiata*. From Williams (1968b).

Williams, McVicar and Ralph (1970) have highlighted an interesting example of site-selection and specificity in *Pseudanthobothrium hanseni* and *Phyllobothrium piriei* in *Raja radiata*. The two species occupy different regions of the spiral intestine, the former being restricted to tiers 1–3 and the latter to tiers 5–7. In the first three tiers, the villi are very long, providing the depth to which the long myzorhynchus and elongated cup-like bothridia are adapted (Figure 4.24) whereas in tiers 5–7 the villi are comparatively short, providing a suitable surface for the much more extensive bothridia of *Phyllobothrium piriei*, which has a much shallower attachment apparatus (Figure 4.27).

Species of *Acanthobothrium* from *Raja* and *Dasyatis* are highly host-specific (Williams, 1969), for which the immune response of the host may be largely responsible (McVicar and Fletcher, 1970—see Chapter Five, (p. 365). Nevetheless, Williams (1969) has shown that *Acanthobothrium* species also have a degree of morphological adaptation to the mucosa. *Acanthobothrium* sp. from *R. circularis*, *A. edwardsi* from *R. fullonica* and *A. quadripartitum* from *R. naevus* (Figures 4.28, 4.29) all have bothridial hooks with markedly curved lateral prongs, probably increasing efficiency of attachment to the villous mucosal surface in these hosts. *A. coronatum* is attached in narrow clefts between the closely arranged parallel mucosal ridges in the spiral intestine of *Scyliorhinus stellaris*. The scolex is embedded vertically or horizontally within a cleft, which it dilates, with the hooks penetrating the epithelium. The hooks are bifid with strongly curved lateral prongs and inflict considerable damage on the mucosa (Rees and Williams, 1965).

Williams (1968c) has drawn parallels between the mode of attachment in *Phyllobothrium* and in the cestodarian *Gyrocotyle*, the posterior attachment organ or rosette of which strongly resembles the frilly scolex found in some *Phyllobothrium* species. *Gyrocotyle* are located mainly in the anterior tier of the spiral valve of *Chimaera monstrosa* where the mucosa is villous. The highly folded surface of the rosette, and the lateral folds on the body, are capable of grasping the villi. As the rosette increases in size and complexity as

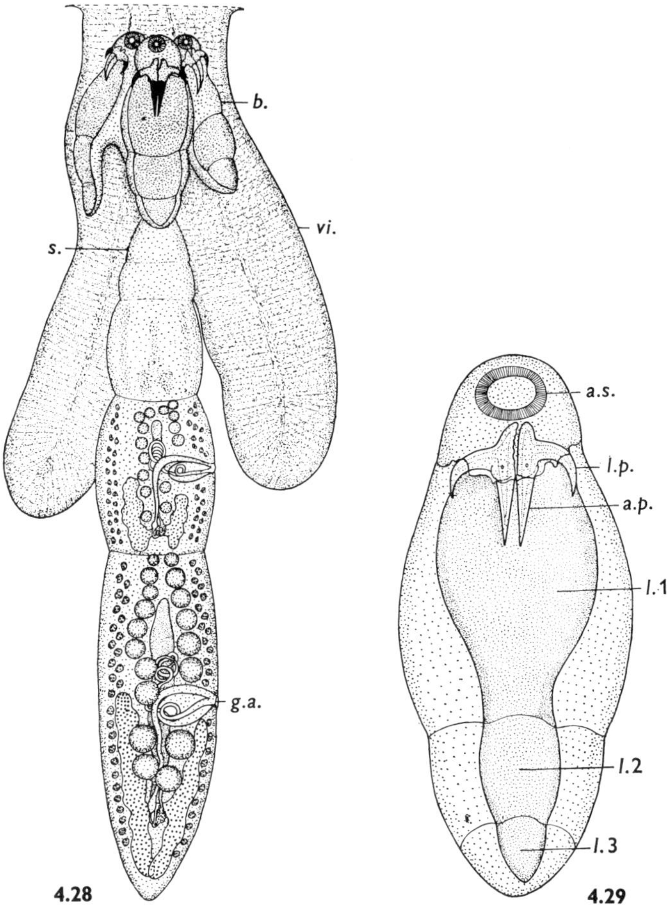

Figures 4.28, 4.29 *Acanthobothrium quadripartitum* 4.28 Scolex and strobila, relative to depth of villi of mucosa. 4.29 Hooks and bothridium. From Williams (1968a)

the worm develops, it attaches to a greater number of villi. More posteriorly, the villous area is reduced and the mucosa becomes ridged. Additionally, the volume of the anterior tier has been calculated as at least equal to the combined volume of the other three tiers and is probably the best site for food and space for growth (Halvorsen and Williams, 1968). Physiological factors which influence site-selection have been investigated by Laurie (1971) and Simmons and Laurie (1972) (see p. 299).

The diphyllideans *Echinobothrium brachysoma* and *E. affine* have armed scolices with two groups of apical hooks and a pair of flattened oval bothridia. The scolex is very well suited to attach deep within the tubular crypts characteristic of the mucosa of the host, *Raja clavata*, where it maintains position by elevating the apical hooks so that they penetrate the epithelium (Rees, 1961a).

Digeneans

Examples among fish digeneans of morphological adaptations to site to the extent of influencing host-specificity are comparatively rare. However, some species are adapted to occupying particular sites in fish, eg. *Otodistomum cestoides* which is almost circular in transverse section when alive and occupies the spaces between the folds and ridges in the cardiac stomach of *Raja radiata* (Williams, McVicar and Ralph, 1970).

Among the digeneans, it is rare to find examples of adults which occupy sites other than within the internal organs or tissues of the host, but some instances are known. Rohde, Roubal and Hewitt (1980) described *Copiatestes filiferus* (= *Syncoelium filiferum*) which lives on the spiny parts of the gill arches of its hosts, *Seriolella brama* and other Pacific marine fish, attaching itself to one or two spines by means of a strong pedunculated ventral sucker.

The Transversotrematidae are unusual in occupying an external surface of the fish host. *Transversotrema patialense* will infect any tropical fish species which offers a subscale micro-habitat of the appropriate size. Under experimental conditions, it failed to infect only three of 24 species; these three had a totally unsuitable surface morphology (Whitfield, Anderson and Bundy, 1986).

Bucephaloides gracilescens adults are highly specific to *Lophius piscatorius* and the encysted metacercariae occur only in gadoid fish. Their specificity to gadoids may be related to their location in the well-developed acoustico-lateralis system in these fish, to which they are almost entirely restricted. In *Merluccius merluccius*, they occur only in this system which they probably reach by entering the pores of the lateral line canals of the head and passing from there along the nerves (Matthews, 1974; Crofton and Frazer, 1954).

Adults of the homalometrine *Crassicutis archosargi* occur only in the sheepshead, *Archosargus probatocephalus*. Small specimens attached to the intestinal epithelium in the usual way by means of the suckers but, in larger worms, the tegument became modified and served as a very effective attachment organ, the acetabulum being withdrawn (Overstreet, 1976). The modified tegument of the ventral surface was five times thicker than normal tegument; it filled the space between the body of the worm and the mucosal surface and also cemented many of the intestinal crypts together.

Bartoli (1987) has described morphological adaptations which enable the digeneans of *Sarpa salpa* to inhabit the unusual digestive tract of this sparid, which feeds exclusively on vegetation and has a very smooth intestinal wall with very few villi. The digeneans, of the families Mesometridae and Robdollfusidae, have a high degree of specificity to this host. In some of the mesometrids, a monostomatous family, the body is elongate and narrow posteriorly but with a wider anterior end which is cup-shaped and provided with concentric circular muscle fibres that enable it to function as a sucker, eg. *Centroderma spinosissima, Elstia stossichianum, Wardula capitellata*. Spines are present on the dorsal and ventral surfaces. In *W. capitellata*, the lateral margins are flexed ventrally so as to enhance the sucking action of the anterior region by maintaining close contact with the gut surface. In *Mesometra orbicularis* and *M. brachycoelia*, the body is round, flattened with thin margins, and ventrally concave with concentric muscle fibres which enable the entire body to act as a sucker. Other modifications related to the diet of the host include the presence in the oral sucker of the mesometrids and *Robphildollfusium fractum* of ridges bearing denticles which filter vegetable debris in the intestinal contents of the host on which the parasites feed.

The mesometrids have a powerful oesophageal bulb (or pharynx, according to opinion) which functions as a pump, drawing the intestinal contents of the host past the filtering mechanism of the oral sucker. Additionally, these species have developed a reticulate excretory system which ramifies throughout the parenchyma and may have evolved to cope with the products of the fermentation processes in the intestine of the host.

Nematodes

There are few studies of any correlation between morphology and host- or site-specificity in fish nematodes. Williams and Richards (1968) have described the attachment of *Pseudanisa-*

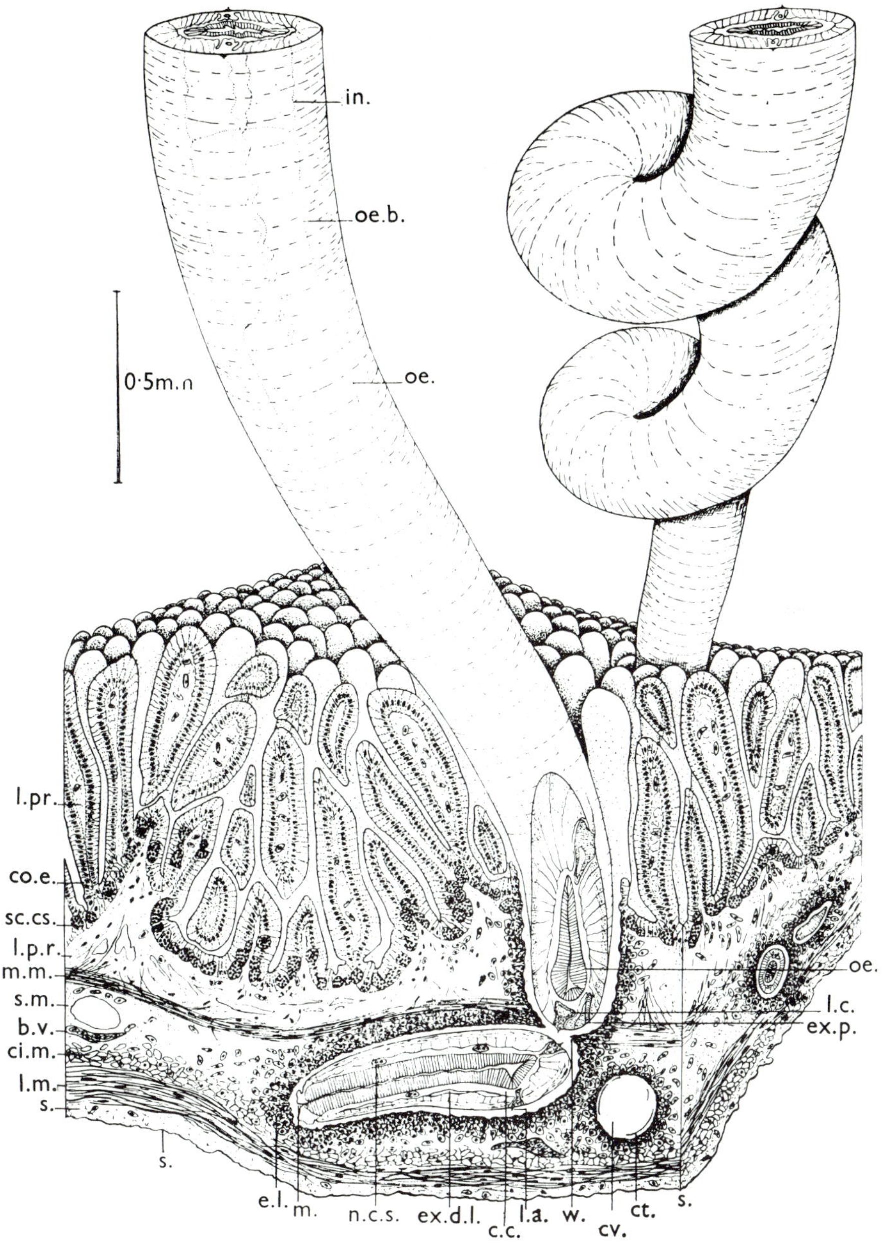

Figure 4.30 *Pseudanisakis rotundata. In situ* drawing showing anterior end of nematode embedded in the intestinal wall of *Raja radiata*. From Williams and Richards (1968).

kis rotundata in the intestine of *Raja radiata* (Figure 4.30) where they occur in pairs, usually near the opening of the pancreatic duct. Most worms have a characteristic constriction in the region of the nerve ring which corresponds to the position of the mucosa muscularis layer around the anterior end of the worm while it is attached. This layer produces a strong tissue reaction around this region of the worm, which may help to anchor it in position.

Physiological factors

Compared with other vertebrate groups, relatively little information is available on the relationship between the physiology of fish and their parasites. Reviews are available on aspects of the digestive tract of vertebrates as a habitat for helminths (eg. Read, 1950; Crompton, 1973; Pery, 1982) but data on fish have been sparse. Accounts of the morphology and physiology of the digestive tract of teleosts and elasmobranchs have been given by Bishop and Odense (1966), Sullivan (1908), Williams (1960c, 1968c), Williams, McVicar and Ralph (1970) and Williams and Halvorsen (1971). Williams, McVicar and Ralph (1970) identified pH, enzyme activity, bile, urea concentration and antibodies in intestinal mucus as factors influencing the parasite population.

Chemical hatching stimulants and attractants

Chemical factors have a strong influence in maintaining host-specificity in some monogeneans in that components of fish mucus have been found to stimulate hatching in the eggs of species specific to that host and to attract oncomiracidia to the host (see also Chapter Two). Urea in the skin mucus of sole and rays stimulates the eggs of their parasites, *Entobdella soleae* and *Acanthocotyle lobianchi* respectively, to hatch (Kearn and Macdonald, 1976; Macdonald, 1974). Mucus from other fish species also stimulates hatching of *E. soleae* eggs but the oncomiracidia always choose sole scales in preference to those of other fish, regardless of the source of the mucus (Kearn, 1973). Sole mucus will not stimulate eggs of *A. lobianchi* to hatch but they will hatch in response to mucus of *Raja* species other than the original host (Kearn and Macdonald, 1976). Hatching usually occurs within minutes but is particularly rapid in *A. lobianchi* eggs which hatch in 2–4 seconds; the non-ciliated and therefore non-swimming oncomiracidium is presumed to hatch quickly when a ray settles on sediment containing eggs, in order to maximize the chance of infesting the host. Eggs of *Squalonchocotyle marmorata* from *Torpedo marmorata*, of *Microcotyle salpae* from *Sarpa salpa* and of *A. greeni* from *Raja* species all hatch in response to host mucus but eggs of other monogenean species do not, eg. *E. hippoglossi* from *Hippoglossus hippoglossus*, *Diclidophora merlangi* from *Merlangius merlangus* and *D. luscae* from *Trisopterus luscus* (Euzet and Raibaut, 1960; Ktari, 1969; Macdonald and Llewellyn, 1980; Kearn, 1974a; Macdonald, 1975). Cone and Burt (1982a) found that oncomiracidia of *Urocleidus adspectus* would always become established on *Perca flavescens* but not on other species. They were attracted to isolated scales of *P. flavescens* in preference to those of *Lepomis gibbosus*. Chemical attraction was postulated but the active substance was not determined. In contrast, however, Paling (1969) has shown that *Discocotyle sagittata* oncomiracidia are not chemically attracted to trout.

Cercariae of several digenean species that use fish as second intermediate hosts have been found to be chemically attracted to their hosts. Cercariae of *Acanthostomum brauni* will encyst in fish, frogs or snails. They become attached to *Poecilia reticulata* (guppy) in response to long-chain glycoproteins in the skin, an unusually complex stimulus. The stimulus may be required to be complex in this instance because the cercariae are shed in a muddy habitat and the chemicals in the environment might interfere with a response to a more simple stimulus. Other known attachment stimuli are small molecules, eg. CO_2 and H_2CO_3 for *Diplostomum spathaceum* (Haas, 1974, 1975; Haas and Ostrowski de Nunez, 1988).

Immune response

The immune response of the host has been held responsible for the failure of some helminths to establish in an abnormal host. The immune response is covered in Chapter

Five and only a few examples are presented here as an indication of its possible role in maintaining host-specificity.

A component of host skin mucus is probably responsible for preventing the establishment of some monogeneans on non-susceptible hosts. *Benedenia melleni* survived longer *in vitro* in mucus from susceptible hosts or in sea water than in mucus from immunized susceptible hosts or from non-susceptible elasmobranchs (Nigrelli, 1937).

Serum antibodies secreted in intestinal mucus are probably responsible for maintaining the host-specificity of the tetraphyllidean cestode *Acanthobothrium quadripartitum*, which occurs in *Raja naevus* in Scottish waters but not in *R. radiata*, although both species occupy the same habitat and eat the same food. The cestode survived much longer in serum from the susceptible host than from the non-susceptible host; skin and intestinal mucus from *R. radiata* was not toxic to *A. quadripartitum in vitro*, however, probably because the serum factors were too dilute. Williams (1969) has suggested that each British species of *Raja* may have its own species of *Acanthobothrium*.

Oliver (1977b) linked differences in the incidence of the monogeneans *Microcotyle labracis* and *Diplectanum aequans* on two species of *Dicentrarchus* to biochemical differences between the host species, expressed through the blood. *M. labracis* is common on *D. punctatus* but rare on *D. labrax*, while the opposite is true of *Diplectanum aequans*. *M. labracis* normally feeds on blood while *D. aequans* only feeds on it occasionally. Similar findings have been made for other diplectanids on species of *Diplodus* and *Serranus*, and are presumed to be due to biochemical or immunological factors (Oliver, 1982).

Miscellaneous physiological factors

The composition of bile has a role in the host specificity of some cyclophyllidean cestodes (eg. *Echinococcus granulosus*) in mammals (Smyth, 1962; Smyth and Haslewood, 1963). Mackiewicz (1982a) has suggested that differences in bile acids may account for the absence of caryophyllideans from catfishes in North America and South America, although they are the predominant hosts of these cestodes in the Ethiopian and Oriental zoogeographical areas.

Site-selection in species of *Gyrocotyle* may be related to variations in tegumental permeability. *G. fimbriata* usually inhabits the anterior third of the spiral intestine of *Hydrolagus colliei* and *G. parvispinosus* the middle third. The permeability of the two species is different, *G. fimbriata* absorbing some carbohydrates at a lower rate. As there is no shortage of space in the intestine, one or more components of the intestinal contents may determine the site occupied by each species (Laurie, 1971; Simmons and Laurie, 1972).

Physiological factors may influence the development of species of the pseudophyllidean *Bothriocephalus* in its teleost hosts. *B. acheilognathi*, which does not have strict host-specificity, will infect almost all carp species but its successful development depends on the physiological characteristics of the intestine of the host. It will establish in hybrids of its most common hosts, common and grass carp (common carp × silver carp, bighead × grass carp, grass carp × bighead) (Molnár, Bakos and Krasznai, 1984).

The application of chemotaxonomic techniques has shown that some species believed to infest a wide range of hosts are actually species complexes, the members of which show strict host-specificity. *Bothriocephalus scorpii*, reported from over 50 teleost species, is such an example. Distinct species, *B. gregarius* and *B. barbatus* have been shown to infest *Psetta maxima* and *Scophthalmus rhombus*, respectively. Similarly, the littoral fishes *Ciliata mustela* and *Taurulus* (= *Myoxocephalus*) *scorpius* harbour two distinct species, *B. funiculus* and *B. scorpii*, respectively. Physiological barriers may be responsible for maintaining host-specificity (Renaud, Gabrion and Pasteur, 1983, 1986; Renaud and Gabrion, 1984; Renaud, Gabrion and Romestand, 1984; Euzet, Renaud and Gabrion, 1984).

Acanthocephalans do not occur in the gut of elasmobranch fish. These hosts have a very high concentration of urea in the body tissues and fluids and urea also occurs in the gut, at concentrations toxic to parasites of other vertebrate groups. Williams, McVicar and Ralph (1970) suggested that failure to adapt to high urea concentrations may account for the absence of acanthocephalans from these fish.

Ecological factors

Host diet and feeding behaviour

In order for parasites to be transferred from host to host through predator-prey relationships, a prerequisite is that the prey, the infected intermediate host, must occur in the same environment as the predator and must be a component of its diet. Paratenic hosts are utilized by members of various parasitic groups to bridge gaps in the food chain. Carnivorous fish at the apex of their trophic pyramid, particularly those which feed on many prey species, have a more varied parasite fauna in terms of numbers of species and individuals than predators that feed on fewer prey species or are planktonivorous (various authors in Rohde, 1982a).

The feeding habits of fish often determine host-specificity of their parasites. The fish hosts (*Salarias meleagris*) of the digenean *Paucivitellosus fragilis* graze on the surface of the substratum, to which *P. fragilis* cercariae become attached. Other fish in the same environment with different feeding habits do not become infected (Pearson, 1968). Herbivorous fish ingest cercariae of the families Haploporidae, Haplosplanchnidae, Megaperidae and Mesometridae which encyst on vegetation (see Table 2.3 for references). The majority of the hosts of caryophyllidean cestodes (ostariophysan fish, principally Cyprinidae and Catostomidae) are benthophagous, a prerequisite for ingesting the benthic tubificid intermediaries. An unusual benthophagous salmonid, *Salmo ischchan*, with an inferior mouth unique in the salmonids, occasionally becomes infected with *Khawia armeniaca* in Lake Sevan, Armenia (Mackiewicz, 1982a). Nevertheless, Mackiewicz (1982a) did not attribute caryophyllidean host-specificity solely to the feeding habits of the host.

The distribution of anisakid nematodes in marine fish off Queensland, Australia, has been found to vary according to host distribution and diet. *Contracaecum* and *Thynnascaris*, which are from shallow and intermediate water, respectively, occur in benthophagous host species whereas *Terranova* and *Anisakis*, intermediate and open water types, respectively, occur in nekton feeders (Cannon, 1977).

Molluscivorous hosts harbour parasite species of all groups which use molluscs as intermediaries, eg. the dioecotaeniid, tetraphyllidean, lecanicephalid and diphyllidean cestodes of elasmobranchs (Cake, 1976, 1977; Cake and Menzel, 1980) and the gnathostomatid nematodes of elasmobranchs and other fish. Non-molluscivorous hosts fail to acquire these parasites, eg. on the Great Barrier Reef, *Trachinotus blochi* becomes infected with the aspidogastrean *Lobatostoma manteri* but other fish in the same environment do not. *Cerithium moniliferum*, the mollusc host of *L. manteri*, is a major component of the diet of *T. blochi* alone (Rohde, 1973). Fish which become molluscivorous at a particular stage in their lives will acquire new parasites. *Rutilus rutilus* in a canal in Britain became infected with *Asymphylodora kubanicum* in and after their third year, when molluscs became their principal source of food (Evans, 1978). Among the digenean families, fish acquire some gorgoderids, monorchiids, allocreadiids, lepocreadiids and opecoelids by eating molluscs.

Different ecological groups of a single host species with different feeding habits may exhibit differences in infectivity for a parasite. In three ecological groups of *Perca fluviatilis* in a reservoir in the USSR, the cestode *Triaenophorus nodulosus* was most prevalent in the planktonivorous group, less so in benthophagous fish and least prevalent in predatory fish (Kuperman, 1981a). *Triaenophorus* did not show any specificity for the calanoid or cyclopoid first intermediary under experimental conditions although, in nature, the range of potential hosts was restricted by the mode of feeding of the copepod and whether it occurred in shallow water where the eggs were deposited and hatched. The degree of host-specificity increased at each successive stage in the life-cycle, being greatest at the adult stage with all species except *T. amurensis* being restricted to species of *Esox*. The second intermediaries are planktonivorous fish, and the range of hosts is comparatively wide, varying with the species. Three are confined to a single family: *T. amurensis* in Cyprinidae, *T. meridionalis* in Gobiidae and *T. orientalis* in Eleotridae. *T. crassus* and *T. stizostedionis* have wider ranges and *T. nodulosus*, in 17 families, has the greatest range.

The habitat of the host and its diet are the main factors determining specificity in acanthocephalans of fish. *Pomphorhynchus laevis* occurred in a number of species in the River Avon, Britain, but only two were preferred hosts. Changes in the abundance of *P. laevis* along the river were related to the distribution of these hosts, *Leuciscus cephalus* and *Barbus barbus*, and the importance in their diet of the intermediary, *Gammarus pulex* (Hine and Kennedy, 1974b).

Asellus hilgendorfi, the intermediate host of *Acanthocephalus opsariichthydis*, a parasite of salmonids in Japan, inhabits the littoral zone of lakes where it lives in vegetation and feeds on dead fish. *Salvelinus fontinalis* also occupies the littoral zone and *Asellus* forms a major part of its diet. The rate of infection of *S. fontinalis* with *Acanthocephalus opsariichthydis* was higher than in two species of *Oncorhynchus* which had wider distributions in the lake and fed on other fish or planktonic cladocerans (Nagasawa, Egusa, Hara and Yagisawa, 1983).

Host-specificity can vary widely among related taxa. The cystidicolid nematodes *Spinitectus carolini* and *S. gracilis* lacked specificity to their invertebrate intermediaries, resulting in an expansion of the host range at this stage of the life-cycle (Jilek and Crites, 1981). In contrast, another cystidicolid, *Capillospirura pseudoargumentosa*, was highly specific to its intermediaries, *Gammarus tigrinus* and *G. fasciatus*, and to shortnose sturgeon, *Acipenser brevirostrum*, as definitive host; *G. tigrinus* is a major component of the diet of this fish (Appy and Dadswell, 1983).

Behavioural factors
(i) Hatching and host-finding behaviour

The monogeneans provide some of the best examples of the correlation between the behaviour of hosts and parasites. The hatching patterns of oncomiracidia of several species have been shown to follow an endogenous circadian rhythm related to the activity patterns of the host fish in such a way that the chances of an oncomiracidium locating and infecting the right host are maximized (see also Chapter Two). *Entobdella soleae* eggs hatch in the first few hours after dawn, a time when soles cease to be active and descend to the sediment where they spend the daylight hours (Kearn, 1973). Both *E. soleae* and *E. hippoglossi* (from halibut) have a hatching rhythm with a periodicity of 24 hours but, in the case of *E. hippoglossi*, the eggs hatch in the first two hours of successive periods of darkness, suggesting that halibut, the behaviour of which is largely unknown, may be inactive during the night (Kearn, 1974a). Similarly, eggs of *Diclidophora merlangi* hatch around dawn when the host, *Merlangius merlangus*, sinks to the sea bottom where it spends the daylight hours. In contrast, *Pollachius virens* rests on the bottom at night and *D. denticulata* eggs hatch around dusk (Macdonald, 1975). Eggs of *Diplozoon homoion gracile* hatch around and after dusk; the host, *Barbus meridionalis*, is inactive at night (Macdonald and Jones, 1978).

The responses of newly-hatched oncomiracidia (not chemically stimulated to hatch) of

Entobdella soleae stimulate movements which amount to a search strategy for the host. They are initially photopositive and swim upwards, but become increasingly photonegative as they age, swimming downwards. These vertical movements, with horizontal movements caused by water currents, may assist in locating a host. Larvae which hatch in light conditions in response to urea are initially photonegative, which would increase their chances of contacting a fish resting on the sea bottom. In darkness, however, larvae stimulated to hatch with urea are geonegative and capable of emerging from sediment. The result would be an upward migration which would bring them into contact with the lower surface of a resting sole (Kearn, 1980).

Eggs of *Entobdella diadema*, like those of *Acanthocotyle lobianchi*, hatched extremely rapidly about three seconds after illumination was switched off under experimental conditions (Kearn, 1982) and the oncomiracidia were geonegative in darkness. The host, the stingray (*Dasyatis pastinaca*), inhabits shallow water and prefers sandy bottoms, where turbidity is probably low. Little is known about the behaviour patterns and distribution of stingrays but if the fish tend to swim intermittently with short rest periods on the bottom, the sudden hatching of *E. diadema* eggs in response to darkness and the geonegativity of the oncomiracidia would enhance the chances of contact with the host.

In contrast to the monogeneans, little is known of the behaviour of free-swimming digenean larvae of species which infest fish as intermediate or definitive hosts. Cable's (1972) review of digenean behaviour makes some references to the generally photopositive behaviour of cercariae that invade fish. Some opisthorchioid cercariae tend to swim when shadowed as if by a fish swimming above them. *Diplostomum spathaceum* cercariae, however, cease swimming when shadowed. *Cryptocotyle lingua* cercariae are initially positively phototaxic and swim upwards in the water; they then sink slowly with intermittent swimming periods, penetrating the skin of any fish with which they establish random contact. They are not actively attracted towards fish tissue (Grainger, 1977).

Cercariae of *Proterometra macrostoma* and *P. edneyi* develop in the same species of snail, *Goniobasis semicarinata* in Kentucky, USA; both species are ingested by their fish hosts as prey but their hosts belong to different groups. Those of *P. macrostoma* are sunfish (Centrarchidae) and those of *P. edneyi* are darters (Percidae). Lewis, Welsford and Ugelm (1989) have shown that *P. macrostoma* cercariae emerge from the snail during the night but those of *P. edneyi* emerge early in the day. Since most darters feed diurnally, *P. edneyi* cercariae may increase their chances of being eaten by emerging at this time. In contrast, the feeding habits of sunfish vary widely according to age, habitat, season and presence of competitors and predators and *P. macrostoma* cercariae, by emerging at night, may enhance their chances of reaching an appropriate host by reducing their susceptibility to predatory diurnal-feeding non-host fish.

(ii) Behavioural changes in fish hosts

Recent studies have shown that helminths, regardless of the site they occupy, alter the behaviour of their hosts in various ways, often such as to increase their susceptibility to predation and so enhance the chances of completion of the life-cycle of the parasite.

Sluggishness, lethargy and poor swimming performance are among the most frequently reported effects on fish hosts, eg. *Gyrodactylus bullatarudis* on *Poecilia reticulata* (Rehulka, 1978), *G. salmonis* on *Salvelinus fontinalis* (Cusack and Cone, 1986), *Capingentoides moghei* on *Heteropneustes fossilis* (Jain, Pandey and Pandey, 1976), *Eubothrium salvelini* on *Oncorhynchus nerka* (Smith, 1973), *Schistocephalus solidus* in *Gasterosteus aculeatus* (Arme and Owen, 1967), and *Triaenophorus crassus* in *Salmo gairdneri* (Rosen and Dick, 1984b). *Nanophyetus salmincola* metacercariae reduced the swimming ability of juvenile salmonid fish (Baldwin, Millemann and Knapp, 1967; Butler and Millemann, 1971) and caused erratic swimming behaviour, effects attributed to the damage done by migrating cercariae. There are, however, several instances where infections have had no adverse effects. Sogandares-Bernal, Hietała and Gunst (1979) found that *Ornithodiplostomum ptychocheilus* metacercariae in the brain and viscera of *Richardsonius balteatus* did not impair speed or stamina. Brassard, Rau and Curtis (1982b) cited several other examples in which encysted metacercariae did not affect swimming performance. *Salmo gairdneri* infected experimentally with 5, 10, 20, 40 or 80 *Truttaedacnitis truttae* had swimming abilities similar to those of uninfected trout and the maximum swimming speed and the time taken to become fatigued at cruising speeds were correlated with fish size rather than with the worm burden (Russell, 1980).

In *Gasterosteus aculeatus* infested with *Schistocephalus solidus*, the reduction in swimming ability was attributed to the distention of the abdomen by the plerocercoids with a consequent loss of flexibility of the body so that the fish could propel itself only by means of the pectoral fins and movement of the post-anal region (Arme and Owen, 1967). The fish swam very slowly and often sank to and stayed on the bottom but could move rapidly for short distances when disturbed. Spawning behaviour was also disrupted. Grossly distended males failed to construct nests and most grossly distended females were unresponsive to males which attempted to mate. One which did respond was too large to enter the nest, which its great bulk destroyed.

Loss of equilibrium occurred in *Channa punctata* harbouring *Isoparorchis hypselobagri* in the swim-bladder (Mahajan, Agrawal, John and Katta, 1979) and in *Salmo gairdneri* with *Triaenophorus crassus* plerocercoids in the skeletal muscles (Rosen and Dick, 1984b).

An increase in the risk of predation is often a consequence of altered host behaviour. Any reduction in the swimming ability of a fish would tend to make it vulnerable to predators (Brassard, Rau and Curtis, 1982b) and might account for the disappearance of heavily-infested fish from a population (Arme and Owen, 1967; Lester, 1984). Juvenile *Oncorhynchus nerka* infested with *Eubothrium salvelini* swam very poorly and exhibited aberrant behaviour which increased their susceptibility to predation (Smith, 1973).

In addition to reducing the swimming efficiency of its host, *Schistocephalus solidus* also curtailed its fright reaction (Giles, 1983). Infested *Gasterosteus aculeatus* made a faster recovery from a frightening overhead stimulus than uninfested fish and often resumed feeding shortly afterwards, which uninfested fish did not. Curtailment of the fright reaction was related to worm burden and was suggested to be adaptively advantageous for the parasite. Milinski (1985) confirmed the loss of the fright reaction in *G. aculeatus* infested with *S. solidus*. When offered *Tubifex* at different distances from a live fish predator, infested sticklebacks fed at all distances equally often and at the same rate as when the predator was absent. Uninfested fish fed at a lower rate in the presence of a predator and always chose the *Tubifex* positioned farthest away from it. The fearless behaviour of infested fish gave them an advantage in competition for food but also increased the risk of predation, both beneficial to the parasite.

The above authors suggested that the increased metabolic demand for oxygen in sticklebacks infected with *S. solidus* was responsible for an increased sensitivity to aquatic hypoxia, stimulating changes in behaviour which resulted in more time spent at the water surface where the level of dissolved oxygen was higher. Smith and Kramer (1987) tested this with *Pungitius pungitius* infected with *Schistocephalus* sp. and found that under hypoxic conditions infected fish spent more time than uninfected fish at the surface performing aquatic surface respiration. Their fright response to a simulated aerial predator was curtailed. They also had a higher lethal oxygen level than uninfected fish, suggesting that an increased oxygen demand is the factor responsible for increased surface respiration.

Godin and Sproul (1988) found a curtailed fright response in *Gasterosteus aculeatus* infected with *S. solidus*. Infected fish had a higher energy requirement and exhibited greater willingness to risk predation in order to feed. Risk-taking increased with increasing parasite burden, especially when the potential reward was high, i.e. high food density. On average, infected fish consumed more food as food density increased, but did so by spending more time foraging while at greater risk of predation rather than by feeding faster. The feeding rate of individual parasitized fish was not positively correlated with parasite load and food density, suggesting that the presence of the parasite constrains the foraging effort of the host.

Diplostome metacercariae in the eyes have a pronounced effect in that fish with partially or completely impaired vision have difficulty in feeding. *Leuciscus leuciscus* infested with *Diplostomum spathaceum* spent more time feeding than uninfested fish but, nevertheless, their rate of predation on *Gammarus pulex* actually decreased (Crowden and Broom, 1980). Because they spent more time than uninfested fish in surface waters, the risk of predation by birds increased. Brassard, Rau and Curtis (1982a) found that *Salvelinus fontinalis* predating on *Lebistes reticulatus* consistently took more guppies which had been exposed to infection with 100 cercariae of *D. spathaceum* than uninfested fish. Exposure to as few as 25 cercariae, resulting in an average of one metacercaria per fish, was sufficient to increase the rate of predation significantly. Infested fish were always less active than uninfested controls and tended to swim in circles or erratically and to spend time immobile near the top or bottom of an aquarium. It was noted, however, that while this made fish more vulnerable to piscivorous birds in which the parasite could complete its life-cycle, it also made them susceptible to predators which were not suitable hosts for further development.

(iii) Behavioural changes in invertebrate hosts

The modification of the behaviour of intermediate hosts, including invertebrates, by helminths has been reviewed by Holmes and Bethel (1972). The effect of infection in invertebrate intermediaries, as with vertebrates including fish, is often to render the host more conspicuous and to alter its behaviour so that it becomes more exposed to predation and less capable of avoiding predators.

Acanthocephalan infections of arthropods have provided many examples. *Acanthocephalus jacksoni* in the isopods *Asellus intermedius, Lirceus garmani* and *L. lineatus* interfered with the metabolic pathway for ommochrome pigmentation; infected isopods were conspicuously lighter, which may increase their vulnerability to predation by fish (Oetinger and Nickol, 1982). *Asellus aquaticus* infected with *Acanthocephalus lucii* had intensely darkened respiratory opercula and the number of infected isopods eaten by perch (*Perca fluviatilis*) under experimental conditions was significantly greater than uninfected isopods (Brattey, 1983).

Zoogeographical factors

The current geographical distribution of some helminths and their fish hosts indicates the origin of fish and parasite stocks, particularly when applied to parasites with a high degree of phylogenetic host-specificity. The formation of the monogenean fauna of freshwater fish in different biogeographical regions has been reviewed by Gusev (1976).

Cystidicola stigmatura is a parasite of *Salvelinus namaycush* (lake trout) and *S. alpinus* (Arctic char) in Canada. Its current distribution is confined to the Great Lakes and St. Lawrence River drainage basins, in lakes along an arc from Northwest Ontario to Great Bear Lake and into the Arctic Archipelago. The nematode appears to have had a preglacial distribution in the upper Mississippi River region and to have dispersed postglacially through glacial water connections between this and other regions. It first colonized Arctic char during its northward migration. The restricted preglacial distribution and recent history of *Mysis relicta*, the intermediary, suggests that *C. stigmatura* evolved as a species during the last glaciation (Black, 1983b). Similarly, in the north-west and north-east Pacific regions, currently isolated but connected during the late Pliocene-early Quaternary, ten species of game fish from each region shared 22 of their 25 trematode species (Korotaeva and Kirillova, 1979). The parasite fauna of fish in four rivers of the north-eastern Asian region of the USSR includes four North American and one Sino-Indian species, as well as typically Siberian species. Its formation has been influenced by geological events including glaciations, the disappearance of the Bering Straits and advances by the sea (Khokhlov and Pugachev, 1979).

Harris (1985b) has suggested that several strictly host-specific species of *Gyrodactylus* common to freshwater fish in Britain and continental Eurasia had colonized Britain with their hosts across the Doggerland bridge after the Devensian glaciation.

In the Gulf of Mexico, 66 species of fish in coastal waters around Cuba, currently an isolated area, harboured a highly host-specific digenean fauna, indicating an ancient association between host and parasite. Some were common to other geological areas, mainly the Pacific Ocean off Central America, but also in the Red Sea, off Japan and Hawaii and off West Africa (Zhukov, 1976).

There are many examples of parallel speciation in the monogeneans. The presence in Australian waters of four species of *Acanthopagrus* harbouring four species of *Polylabroides* suggests parallel speciation, possibly during temporary geographical isolation of the hosts. Each *Polylabroides* species occurs on two of *Acanthopagrus* but shows a distinct preference for one host species (Byrnes, 1985).

5 Host–parasite interactions

Introduction

The pathological effects of helminths on their hosts and the immune response of the fish to infestation are two aspects of the host-parasite interaction. As yet, relatively little is known of the immune response of fish to helminth infestation but information on pathology is much more extensive. The effects of parasites on individual fish have been documented in some detail and many helminths have been found to have little significant effect except in cases of massive infestation. Recent reviews (Rohde, 1982a, 1984c; Sindermann, 1986) have pointed out the difficulties of assessing the effect which helminth infestation has on a population rather than on individuals. It is particularly difficult to determine the role of helminths in the population dynamics of their hosts in the natural environment, especially those of marine fish in which even mass mortality could go undetected. In populations of farmed fish, more accurate assessments can be achieved. Several attempts have been made to quantify parasite-related mortality, eg. Adjei, Barnes and Lester (1986) on *Callitetra-*

rhynchus gracilis in *Saurida tumbil* and *S. undosquamis*, and Kennedy (1984) on several species of diplostomes in freshwater fish. Methods for estimating mortality due to parasites in wild fish populations have been reviewed by Lester (1984) who concluded that while they do not provide definitive answers, they do indicate the existence of parasite-related mortality and can sometimes provide an estimate of its contribution to the total mortality rate.

In restricted environments, some parasitic groups are potentially much more pathogenic than in the wild because the opportunities to infect fish are enhanced and massive infestations may build up. Examples are the monogeneans with their direct life-cycles, and some digenean infestations when fish, enclosed in a restricted area with snails shedding cercariae, may acquire massive numbers of metacercariae. In general, helminths have been regarded as less pathogenic to fish than protozoans, bacteria, fungi and viruses. With the exception of the monogeneans, many of the most pathogenic helminths are not those which mature in fish but the larval stages of species which, as adults, parasitize other fish or piscivorous birds or mammals. The pathology of helminth infestations of fish has been reviewed by Bauer (1958a), Bauer, Musselius and Strelkov (1973, 1981), Petrushevskii and Shul'man (1958), Williams (1967), Margolis (1970a), Hoffman (1975a), Paperna (1974), Paperna, Diamant and Overstreet (1984), Roberts (1978b), Needham and Wootten (1978).

Pathology of organ systems

The skin and gills

The monogeneans are the most important and most numerous helminth group parasitizing the external surfaces of fish. Fernando and Hanek (1976), reviewing the skin of fish as a habitat for parasites, reported 243 species from the gills of freshwater fish in North America alone. Particularly in farmed fish, monogeneans have been responsible for major epizootics with serious consequences, although Molnár (1986) concluded that some groups, eg. dactylogyrids on carp, had lost much of their economic importance due to the availability of adequate techniques for treatment and control.

The text and Table 5.1 provide examples selected from the very extensive literature to illustrate the main features of monogenean infestations, which have been reviewed by Paperna (1986), Paperna, Diamant and Overstreet (1984), and Molnár (1986).

Some mechanical injury is usually inflicted by the haptor of monogeneans at the point of attachment to the skin or gills. The degree of damage varies from minimal to severe. Four species of *Gyrodactylus* studied by Cone and Odense (1984) produced a small depression in the skin of the host (see Table 5.1) at the point of contact but only minimal tissue damage resulted. However, the marginal hooks of *G. salmonis* penetrated deeply into the epidermis of *Salvelinus fontinalis* fry, causing damage to the fins and discoloration of the skin. This species caused 15–44 per cent mortality in *S. fontinalis* fry and, in heavy infections, the epidermis was very thin with fewer goblet cells than normal (Cusack and Cone, 1986). This species also caused kidney lesions with tubular degeneration and necrosis. Moribund fish were cachexic, lethargic and dark in colour. Neither Cone and Odense (1984) nor Cusack and Cone (1986) found evidence of secondary bacterial invasion of the wounds. Cusack and Cone (1986) attributed death of the fry to disruption of the osmotic permeability of the epidermis resulting from damage caused by the attachment and feeding activity of the worms. Although *G. bullatarudis* (now *G. turnbulli*) caused little tissue damage to *Poecilia reticulata* according to Cone and Odense (1984), Řehulka (1978) found that infected *P. reticulata* were restless, swam sluggishly and had an elevated oxygen requirement.

Of the other skin-dwelling monogeneans, some members of the Capsalidae are highly pathogenic. The effect of *Benedenia monticellii* on *Liza carinata* and other species of mullet has been described by Paperna, Diamant and Overstreet (1984). The parasites attached to the skin of the head around the mouth and eyes and to the dorsal fin but extended to the whole body as the intensity of infection increased. The epithelium of the snout and lips was eroded, often to the bone; haemorrhagic ulcerative lesions developed in the mouth; the cornea and adipose eyelid thickened and became white and the epithelial layers sloughed off. The epithelium around the pharyngeal lesions, on and at the base of the dorsal fin and on the body surface, became hyperplastic and necrotic with some haemorrhage and sloughing. In heavy infestations, the dermal and collagenous layers were exposed. The inflammatory response was severe, involving the dermal and subdermal layers, with congestion, oedema, and massive infiltration by eosinophils, macrophages and lymphocytes. Secondary bacterial invasion of wounds was common. Other species of mullet were also susceptible to *B. monticelli* but the pathology varied according to host species or even stock. Some species (*L. subviridus*, *L. ramada*) could be killed by only a few parasites concentrated in the mouth, by septicaemia developing in bacteria-infected wounds. The pathological effects of *B. monticelli* showed similarities to those of *Neobenedenia melleni* on aquarium fish and *B. seriolae* on *Seriola quinqueradiata*.

A well-known example of a capsalid causing mass mortality is that of *Nitzschia sturionis* which, when transferred with its host, *Acipenser stellatus*, from the Caspian to the Aral Sea, almost obliterated the local sturgeon, *A. nudiventris*, which had not previously been exposed to it. Reviewing extensive studies by Russian authors, Rohde (1984a) summarized its pathological effects as reduction in fat content, emaciation, jaundice and discoloration of organs adjacent to the gall-bladder. The opisthaptor caused mechanical irritation of gill tissue followed by epithelial hyperplasia, hypertrophy of connective tissue and atrophy of the gill capillaries and secondary gill layers. Death was attributed to reduction or total cessation of gas exchange.

The pathology of dactylogyrid, diplectanid, tetraonchid and other monogenean groups involves mechanical damage due to attachment and/or feeding activity. The host response includes increased mucus production, epithelial hyperplasia, loss of lamellar structure, clubbing or fusion of the gill filaments, haemorrhage, aneurysms and secondary invasion by fungi or bacteria (Table 5.1). Mechanical damage from hooks of 80–120 *Ancylodiscoides vistulensis* on the gills of *Silurus glanis* killed the fish in 2–4 days from loss of respiratory function (Molnár, 1980). *Pseudogyrodactylus anguillae* and *P. bini* produced a similar effect on cultured *Anguilla anguilla* (Chan and Wu, 1984). *Cleidodiscus robustus* on *Lepomis macrochirus* provoked the formation of granulation tissue which obliterated the structure of the gill lamellae and became necrotic when not well-vascularized (Thune and Rogers, 1981).

Water pollution may exacerbate the effects of monogeneans on their hosts. Skinner (1982) found that the pathology of *Neodiplectanum wenningeri* on *Gerres cinereus*, of *Ancyrocephalus* sp. on *Lutjanus griseus* and of *A. parvus* on *Strongylurus timucu* was worse in polluted than in clean waters, the pollutants acting as irritants which stressed the fish and reduced their resistance to infection.

The feeding activity of monogeneans also generates a certain amount of mechanical damage. Kearn (1963c, 1976) considered that, in the case of *Entobdella soleae* on *Solea solea*, this was minimal since only the epidermis was eroded at the feeding site and the dermis remained intact. In addition, the rapid regeneration of the epidermis limited the exposure of the subepidermal tissues to secondary infections. The feeding activity of *Benedenia monticelli* exposed the dermis of *Liza* spp. to bacterial invasion with resulting ulcerative lesions, inflammation and necrosis. Paperna, Diamant and Overstreet (1984) suggested that the worms abandoned necrotic areas to seek new feeding sites and that some were lost during this process, leading to a gradual decline in infestation.

Table 5.1 Pathological effects of some monogenean parasites of fish (Figure 5.1)

Parasite	Host	Pathology	Reference
Gyrodactylidae			
Gyrodactylus adspersi	*Tautogolabrus adspersus* (skin)	Haptor compresses epidermis into a small depression at point of contact. Tissue damage minimal	Cone and Odense (1984)
G. avalonia	*Gasterosteus aculeatus* (skin)		
G. bullatarudis	*Poecilia reticulata* (skin)		
Gyrodactylus sp.	*Carassius auratus* (skin)		
G. bullatarudis	*Poecilia reticula* (fins)	Restlessness, sluggish swimming, increased oxygen requirement	Řehulka (1978)
G. salaris	*Salmo salar* parr	Mortality	Johnsen (1978)
G. salmonis	*Salvelinus fontinalis* fry	15–44% mortality. Thinning of epidermis, fewer goblet cells, and kidney lesions in heavy infections. Moribund fish cachexic, lethargic, dark in colour. No secondary bacterial or fungal infection. Death may be due to disrupted osmotic permeability of epidermis	Cusack and Cone (1986)
Dactylogyridae			
Ancylodiscoides vistulensis	*Silurus glanis* fry (skin and gills)	Mechanical damage; gill lesions; proliferation of epithelial cells, adhesion of filaments; mortality	Molnár (1980)
Ancyrocephalus sp. *A. parvus*	*Lutjanus griseus* (gills) *Strongylurus timucu* (gills)	Excessive mucus; epithelial hyper-plasia, fusion of lamellae, clubbing and fusion of filaments; aneurysms	Skinner (1982)
Cleidodiscus robustus	*Lepomis macrochirus* (gills)	Gill lesions, epithelial hyper-plasia, loss of lamellar structure, replaced by granulation tissue	Thune and Rogers (1981)
Dactylogyrus anchoratus	*Cyprinus carpio* (gills)	Excessive mucus production; regressive changes, atrophy, necrosis, haemorrhage; pathogenicity related to depth of penetration of anchors.	Prost (1963)
D. extensus	*C. carpio* (gills)	Excessive mucus production; pathology may involve either regressive, degenerative changes causing damage and atrophy of gill bars, or hypertrophy and thickening of mucosa with haemorrhagic foci.	Prost (1963)
D. vastator	*C. carpio* (gills)	Extensive epithelial hyperplasia with loss of respiratory function counteracted in rapidly growing fish by regeneration. Overcrowded fish show retarded growth and little or no regeneration, resulting in mass mortalities.	Paperna (1963b)
Pseudodactylogyrus anguillae; P. bini	*Anguilla japonica* (gills)	Hypertrophy and haemorrhage, loss of respiratory function	Chan and Wu (1984)

Table 5.1 continued

Parasite	Host	Pathology	Reference
Diplectanidae			
Diplectanum aequans	*Dicentrarchus labrax* (gills)	Hyperplasia and haemorrhage at point of attachment	Oliver (1977a)
Neodiplectanum wenningeri	*Gerres cinereus* (gills)	Excessive mucus, epithelial hyperplasia, fusion of lamellae, clubbing and fusion of filaments, aneurysms.	Skinner (1982)
Tetraonchidae			
Tetraonchus rauschi	*Thymallus arcticus* (gills)	Epithelial hyperplasia, infiltrating lymphocytes, haemorrhage, fungal invasion	Wobeser *et al.* (1976)
T. alaskensis	*Coregonus* spp. (gills)	Gill necrosis; secondary infections; mortality at 1,500–2,000 parasites per fish	Razmashkin and Kashkovskii (1977)
Diplozoidae			
Diplozoon nipponicum	*Carassius carassius* (gills)	Hypochromic microcytic anaemia	Kawatsu (1978)
Capsalidae			
Benedenia monticelli	*Liza carinata* (skin, mouth, eyelids, fins)	Erosion and degradation of epithelium; severe haemorrhagic ulcers, intense inflammatory reaction; bacterial invasion	Paperna (1975); Paperna *et al.* (1984)
Nitzschia sturionis	*Acipenser nudiventris* (gills, lips, mouth)	Reduced fat content; emaciation; jaundice; irritation; inflammation; hypertrophy of connective tissue; atrophy of capillaries and secondary gill layers; reduced gas exchange leading to mortality	Various authors cited by Rhode (1984)
Microcotylidae			
Allobivagina sp.	*Siganus luridus* (gills)	Emaciation, anaemia	Paperna *et al.* (1984)

Anaemia is a consequence of some monogenean infestations, eg. in *Siganus luridus* infested with the microcotylid *Allobivagina* sp. (Paperna, Diamant and Overstreet, 1984). In wild populations, the worm burden was up to 10 per fish but neither these nor the few fish with 40–200 parasites showed any ill effects. Fish in a culture pond, however, had from 50–1,133 worms and all those with more than 150 worms were emaciated and anaemic with pale gills and a haematocrit level less than 10 per cent. The erythrocytes in blood smears were predominantly immature. Mortality associated with massive infestations occurred naturally and could be induced experimentally, but surviving fish gradually became free of parasites and those with fewer than 50 worms made a recovery with their haematocrit levels returning to the normal value of over 30 per cent. Interestingly, *S. rivulatus* and *S. argenteus* were resistant to mass infestation and never accumulated more than 10 worms, nor did *S. luridus* longer than 200 mm in total length. Based on their own findings and those of Eto, Sakamoto, Fujii and Yone (1976) on *Axine heterocerca* on *Seriola quinqueradiata*, Paperna, Diamant and Overstreet (1984) suggested that anaemia was characteristic of microcotylid gill infestations. They could not, however, attribute all the blood loss to ingestion of erythrocytes by the worms, which was observed, or to excessive haemorrhage, which was not. The diplozoid *Diplozoon nipponicum* provoked hypochromic, microcytic anaemia in crucian carp (*Carassius carassius*), with haemoglobin levels declining in inverse relation to increasing number of worms (Kawatsu, 1978). All haematological

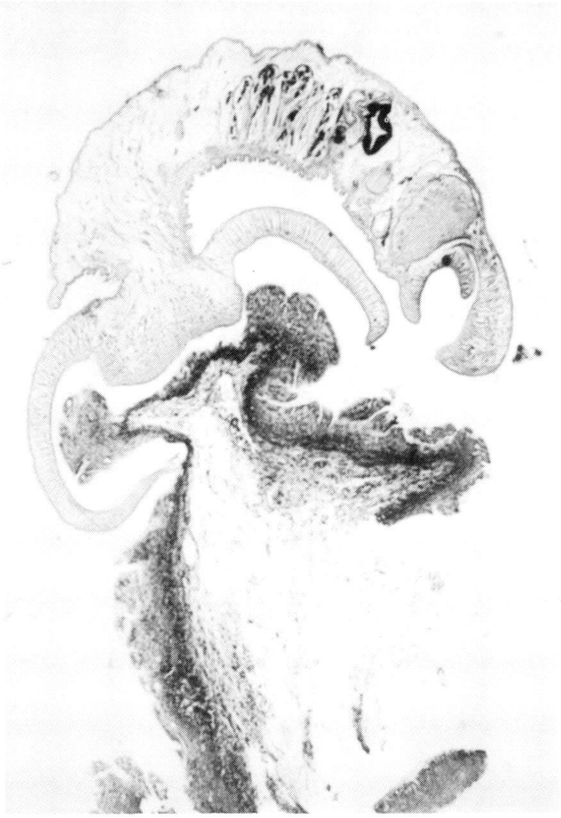

Figure 5.1 Longitudinal section of a *Calicotyle* sp. (Monogenea) attached to the wall of the cloaca of *Raja radiata* and showing mechanical injury to its mucosal lining.
From Williams Ralph and McVicar (1970)
Similar effects have been described for gill monogeneans, eg. by Cosgrove (1975), Prost (1963) and Buchmann (1988b).

indices were below normal values and, as with *Allobivagina* sp. infestations of *Siganus luridus*, there was a very high proportion of immature erythrocytes in blood smears.

Digenean metacercariae and philometrid nematodes are the other major pathogens of the skin and gills.

The digenean *Posthodiplostomum cuticola* in the skin of *Hypophthalmichthys molitrix* caused weight loss, blood changes, granular dystrophy in the kidney and liver, deposition of haemosiderin in the spleen and of neutral lipids in various organs, and increased acid mucopolysaccharide levels in the intestine (Denisov, 1979). The blood picture returned to normal one month after the end of clinical disease but the other changes did not resolve. Hunter and Hunter (1938) attributed significant weight loss in *Micropterus dolomieui* to infestations of *Uvulifer ambloplitis* metacercariae in the fins, tail base and muscle. Unidentified metacercariae caused proliferation of cartilage resulting in loss of respiratory epithelium and occasional fusion of the primary lamellae in naturally infected *Xiphophorus maculatus*, *X. helleri*, *Hyphessobrycon* and *Brachydanio rerio* (Blazer and Gratzek, 1985). Haemorrhage was marked only in heavily-infested fish.

The heterophyid *Cryptocotyle lingua* causes black spot disease in the skin of many species of marine fish and can kill small fish (Sindermann and Rosenfield, 1954). According to Mawdesley-Thomas (1975), this species provokes no host reaction when it occurs in skeletal muscle.

Table 5.2 Pathological effects of some cestode parasites of fish (Figures 5.2–5.14)

Parasite	Host	Pathology	References
Caryophyllidea	(Figures 5.2.1–6, 5.3)		
Capingentidae			
Capingentoides moghei	*Heteropneustes fossilis* (stomach)	Inappetance, sluggishness, weight loss; local damage at point of attachment	Jain, Pandey and Pandey (1976)
Caryophyllaeidae			
Caryophyllaeus fimbriceps	carp (intestine)	Occlusion of gut, inflammation, mortality	Ivasik (1952); Bauer (1968)
Glaridacris catostomi and *Hunterella nodulosa*	*C. commersoni* (intestine)	Necrotic pancreas, submucosal granulomata	Hayunga (1980)
G. catostomi and *G. laruei*	,,	Pathology minimal	Hayunga (1979)
H. nodulosa	,,	Loss of epithelium and lamina propria; chronic inflammatory response characterized by lymphocyte infiltration and extensive hyperplasia of submucosa	,,
,,	,,	Nodules formed at point of attachment	Mackiewicz and McCrae (1962)
Wenyonia virilis	*Synodontis schall*	Degenerative changes in gut wall, liver and pancreas	Banhawy *et al.* (1975)
Lytocestidae			
Djombangia indica	*Clarias batrachus* ('duodenum')	Diverticulum formed at point of attachment, thickening of muscle layers	Satpute and Agarwal (1974)
D. penetrans	*C. batrachus* (anterior intestine)	Nodule formation, occasional penetration of gut wall	Ahmed and Sanaullah (1979)
Khawia sinensis	*Cyprinus carpio* (intestine)	No significant effect at levels below 10 per fish	Kapustina (1978)
,,	,,	Irritation, inflammation and perforation of of mucosa. Fatalities at 35–45 worms per fish	Musselius, Ivanova, Laptev and Apazidi (1963)
,,	,,	Blood changes; decreases in haemoglobin, erythrocytes, monocytes, polymorphs and neutrophils; decrease in total serum protein, albumin and alpha and beta globulins; increase in gamma globulin	Sapozhnikov (1969)
,,	,,	Changes in activity of alanine and aspartate aminotransferases	Lozinska-Gabska (1981)
Lytocestoides fossilis	*Heteropneustes fossilis* (intestine)	Hyperplasia and hypertrophy of villi, degeneration of mucosal epithelium and vacuolation of submucosal cells	Kanth and Srivastava (1984)
Lytocestus indicus	*Clarias batrachus* (intestine)	Reduced haemoglobin; eosinophilia; anisocytosis, macrocytosis and poikilocytosis of erythrocytes, suggestive of macrocytic pernicious anaemia	Sircar and Sinha (1974)
,,	,,	Hyperplasia and hypertrophy of villi, proliferation of connective tissue cells, oedema, vacuolation of submucosal cells, infiltration of mononuclear cells, localized giant cell reaction in submucosa	Sircar and Sinha (1980)

Table 5.2 continued

Parasite	Host	Pathology	References
Lytocestus indicus	*Clarias batrachus* (intestine)	Ulceration; scolex rarely penetrated to serosa; mechanical obstruction	Ahmed and Sanaullah (1979)
L. parvulus	,,	Denudation of mucosa and submucosa	Ahmed and Sanaullah (1979)
Caryophyllids (unnamed)	*Clarias batrachus*	Changes in serum amino acids; blood changes; depletion of blood sugar and of liver and muscle glycogen	Kadav and Agarwal (1982; 1983a,b)
Spathebothriidea			
Cyathocephalus truncatus	trout (pyloric caeca)	Distention and perforation of caeca; mortality	Vik (1958)
Tetraphyllidea			
Oncobothriidae	(Figures 5.4.1–9)		
Acanthobothrium coronatum	*Scyliorhinus stellaris* (spiral valve)	Enlargement of intervillar crypt, erosion of epithelium, exposure of stratum compactum; haemorrhage	Rees and Williams (1965)
Scolex pleuronectis	*Trichiurus lepturus*	Blood changes suggestive of anaemia, depletion of basic energy reserves	Radhakrishnan *et al.* (1984)
Diphyllidea			
Echinobothriidae			
Echinobothrium brachysoma (Figures 5.7.1–2)	*Raja clavata* (spiral valve)	Local flattening and tearing of epithelium	Rees (1961a)
Proteocephalidea			
Proteocephalidae	(Figure 5.10)		
Proteocephalus ambloplitis plerocercoids	*Micropterus dolomieui* (ovaries)	Reproductive capacity reduced by loss of oogenic tissue caused by scarring of ovary, fibrosis and direct oocyte destruction	Esch and Huffines (1973); McCormick and Stokes (1982)
Pseudophyllidea			
Amphicotylidae			
Abothrium gadi (Figure 5.11.1)	*Gadus callarias* (pyloric caeca and intestine)	Discoloration of intestinal wall; fibrosis of mucosa, submucosa and underlying layer	Williams (1960b)
Eubothrium rugosum	*Lota lota* (pyloric caeca)	Desquamation of caecal epithelium in heavy infection	Davydov (1977)
E. salvelini	*Oncorhynchus nerka*	Retarded growth, poor swimming performance; aberrant behaviour increasing susceptibility to predation	Smith (1973)
,,	,,	Reduced ability to adapt to sea water; elevated plasma sodium levels	Boyce and Clarke (1983)
,,	,,	Increased susceptibility to zinc	Boyce and Yamada (1977)
,,	*O. nerka* fry	Reduction in growth rate and stamina	Boyce (1979)
,,	*O. nerka* smolts	Reduction in growth, stamina and resistance to stress	Smith and Margolis (1970)

Table 5.2 continued

Parasite	Host	Pathology	References
E. salvelini	*Salvelinus alpinus* (pyloric caeca)	Reduced condition factor; decrease in packed-cell volume, red blood cells and haemoglobin; increase in white blood cell count; haemosiderosis in spleen	Hoffman, Kennedy and Meder (1986)
Parabothrium gadipollachii (Figure 5.11.2)	*Gadus pollachius* (pyloric caeca and intestine)	Discoloration of caeca, breakdown of caecal wall so that worm protrudes into body cavity; effect on intestinal wall similar to that of *Abothrium gadi* (see above)	Williams (1960b)
Bothriocephalidae	(Figures 5.12.1–5)		
Bothriocephalus acheilognathi (syn. *B. gowkongensis*, *B. opsariichthydis*)	*Gambusia affinis* (intestine)	Increased susceptibility to high temperature	Granath and Esch (1983b)
,,	*Ctenopharyngodon idella* (intestine)	Slightly elevated oxygen consumption in fish with low infections (1.2 worms)	Kititsina (1985)
,,	*C. idella, Notemigonus crysoleucas, Pimephales promelas* (intestine)	Focal pressure necrosis; mucus production; lymphocyte infiltration, haemorrhage; occlusion	Scott and Grizzle (1979)
,,	*Cyprinus carpio* fry	Lower activities of intestinal trypsin and chymotrypsin. No effect on α-amylase activity	Matskási (1978, 1984)
,,	*C. carpio* (intestine)	Burdens over 4 worms per fish affected carbohydrate and protein metabolism and reduced nutritional status. Heavily infected fish emaciated. Lysis of large areas of mucosa, with necrotic changes, in heavy infections	Davydov (1977, 1978a)
,,	,,	Reduction in kidney, liver and spleen weight. Greatest reduction in kidney	Balakhnin (1979)
,,	,,	Reduction in growth, weight and production	Yashchuk, Sventsitskii and Rudoi (1978)
,,	,,	Histopathological changes in intestine; degenerative changes in liver, kidneys and intestine attributed to parasite toxins. Mortality	Lozanov and Kolarova (1979)
,,	,,	Elevated leucocyte count; damage to gut in infections of more than 15 worms	Pár (1978)
,,	,,	No significant effect on various physiological indices; elevated leucocyte count attributed to inflammation of gut	Svobodova (1978)
,,	,,	Burdens over 5 per fish reduced total serum proteins, elevated leucocyte and phagocyte counts and reduced haemoglobin and total blood volume	Kudryashov (1970)
,,	,,	Blood changes; appearance of giant lymphocytes	Kirichenko and Kosareva (1972)
,,	,,	Fry infected with both *B. acheilognathi* and *Lernaea cyprinacea* showed decreased body weight and fat content.	Kurovskaya (1984)

Table 5.2 continued

Parasite	Host	Pathology	References
		Alkaline phosphatase activity depressed but acid phosphatase, amylase and protease activity increased	
Bothriocephalidae *Bothriocephalus acheilognathi* (syn. *B. gowkongensis*, *B. opsariichthydis*)	*C. carpio* (intestine)	Changes in activities of aspartate and alanine amino-transferases in liver, related to season. Deaths in winter attributed to impaired protein synthesis combined with the effect of poor environmental conditions.	Lozinska-Gabska (1981)
,,	,,	Thinning of intestinal wall, attributed to increased acid phosphatase activity	Sekretaryuk (1983)
B. scorpii	*Scophthalmus maximus*, *S. rhombus*, *Ciliata mustela*, *Pseudophycis bacchus* (intestine)	Local damage at site of attachment; collagenous connective tissue collar formed around scolex in *P. bacchus*	Jones (1972, 1975); Rees (1958); Davey and Peachey (1968); MacKinnon and Featherstone (1982)
Diphyllobothriidae			
Digramma interrupta plerocercoids	*Carassius* sp.	Reduction of total weight, flesh weight and nutrition coefficient. Muscle fat content and calorific content reduced in heavy infections.	Lyubina (1980)
,,	*C. auratus*	Almost 90% of fish infected as fry died. Losses due to mortality and weight loss estimated as over 40% of annual catch	Razmashkin and Shirshov (1983)
Diphyllobothrium cordiceps plerocercoids	*Salmo clarki* (serosa, liver, spleen, testis)	Plerocercoids encapsulated with connective tissue; compression and displacement of pancreatic and testicular tissue; liver necrosis, with oedema; necrosis of myofibrils near parasites with oedema and fatty infiltration	Otto and Heckman (1984)
D. latum plerocercoids	*Lota lota*, *Acerina cernua*, *Esox lucius* (liver, muscle)	Encapsulation; non-specific reactions (necrosis of muscle, hypertrophy of connective tissue, inclusions in fatty tissue, and resorption of eggs), more severe in *L. lota* than *E. lucius*.	Davydov (1978b)
,,	*Lota lota*	Changes in lipid composition	Gur'yanova (1980)
,,	*Esox lucius* (muscle)	Haemorrhage and oedema	Davydov (1981)
,,	*Lota lota* (muscle and liver)	Reduction of total lipids and phospholipids in muscle and liver; triglycerides reduced only in liver	Gur'yanova and Sidorov (1985)
,,	*Acerina* sp.	Metabolites inhibit chemotaxis of host leucocytes	Berezantsev and Operin (1976)
D. vogeli plerocercoids	*Gasterosteus aculeatus* (liver)	Reduction in total lipids, phospho-lipids and triglycerides	Gur'yanova and Sidorov (1985)
Diphyllobothriid plerocercoids (unidentified)	*Salvelinus fontinalis* (heart and viscera)	Mortality of small fish from haemorrhage caused by migrating plerocercoids	Hoffman and Dunbar (1961)
Ligula intestinalis plerocercoids	*Abramis brama*	Retarded growth; scale damage	Safonov (1976)
,,	,,	Reduced weight of parietal body wall relative to total weight; reduction in number of muscle	Richards and Arme (1981a)

Table 5.2 continued

Parasite	Host	Pathology	References
		fibres in heavily parasitized fish; no increase in girth	
Ligula intestinalis plerocercoids	*Rutilus rutilus*	Growth rate not markedly reduced but fish failed to reach sexual maturity	Sweeting (1976)
,,	,,	Reduced levels of free amino acids, resembling effects of starvation	Soutter, Walkey and Arme (1980)
,,	,,	No evidence for production of sex steroids by the plerocercoids	Arme, Griffiths and Sumpter (1982)
,,	*Notropis hudsonius*	Sterility, poor condition, stunting	Mahon (1976)
Ligula sp.	*Ctenopharyngodon idella*	Respiration rate considerably reduced even in single-worm infections	Kititsina (1985)
Schistocephalus solidus plerocercoids (Figures 5.13.1–4).	*Gasterosteus aculeatus* (body cavity)	Distention of abdomen; poor swimming performance; displacement of heart and liver; reduction in liver weight and in packed-cell volume of erythrocytes; delay in oocyte maturation; inhibition of spawning in heavy infections	Arme and Owen (1967)
,,	,,	Progressive curtailment of fright reaction	Giles (1983); Milinski (1985)
,,	,,	Increased susceptibility to stress due to infection, reduced diet and exposure to cadmium; reduction in weight of gonads and liver	Pascoe and Cram (1977); Pascoe and Mattey (1977); Pascoe and Woodworth (1980)
,,	,,	Infection rate less than 5% in gravid females but over 40% in sexually mature but non-gravid females; no difference in prevalence between breeding and non-breeding males	McPhail and Peacock (1983)
,,	,,	Metabolites inhibit chemotaxis of host leucocytes	Berezantsev and Operin (1976)
Triaenophoridae			
Triaenophorus sp. plerocercoids	*Salmo gairdneri* (liver and mesentery)	Weight reduced by about 6%	Mirle *et al.* (1985)
T. amurensis plerocercoids	*Leuciscus amurensis* (liver and peritoneum)	Plerocercoid encapsulated; slight parenchymal hyperaemia and perivascular leucocytic infiltration are the only liver changes; strong pathological reaction to parasites in atypical sites (eg. bile-duct)	Pronina (1979)
T. crassus plerocercoids (Figures 5.14.1–3)	*S. gairdneri* (muscles, body cavity)	Muscle haemorrhage and lesions caused by migrating worms; granuloma formation; decreased swimming activity and loss of equilibrium	Rosen and Dick (1984b)
,,	*Coregonus* (muscle)	Muscle tissue reduced to homogenous mass by migrating worms; host tissue reaction weak	Davydov (1981)
T. crassus adults	*Esox lucius* (intestine)	Inflammation and disruption of mucosa; rupture of blood vessels; infiltration of macrophages, plasmacytes and fibroblasts, fibrous tissue surrounds scolex	Davydov (1977)

Table 5.2 continued

Parasite	Host	Pathology	References
T. nodulosus plerocercoids	*E. lucius* fry (liver)	Destruction of liver tissue, reduction in capacity of hepatocytes to synthesize RNA and protein; increased susceptibility to oxygen starvation	Pronin *et al.* (1976)
,,	*Perca fluviatilis* (liver)	Encapsulation of plerocercoids accompanied by inflammation, destruction of hepatocytes and thickening of capillary walls	Pronina (1977)
,,	*Salvelinus alpinus* (liver)	Liver damage indicated by elevated plasma concentrations of aspartate and alanine amino-transferases and lactate dehydrogenase	Scheinert (1984)
,,	*Lota lota* (liver)	Reduction in total lipids, phospholipids and triglycerides	Gur'yanova and Sidorov (1985)
,,	,,	Lysis of liver tissue	Davydov (1981)
,,	*Morone chrysops* (mesentery, liver)	Haemorrhage; acute inflammatory response; liver necrosis, squamous metaplasia, fibrosis, displacement of liver tissue	Stromberg and Crites (1974)
T. nodulosus adults	*E. lucius* (stomach and intestine)	Histopathological changes at point of attachment and throughout gut; no mortality even at 200 worms per fish	Pronina and Pronin (1982a,b)
,,	,,	Inflammation and disruption of mucosa, rupture of blood vessels, infiltration by macrophages, plasmacytes and fibroblasts	Davydov (1977)
Ptychobothriidae			
Polyonchobothrium clarias	*Clarias mossambicus* (gall-bladder, gut)	Nodules formed in gall-bladder	Wabuke-Bunoti (1980)

Subgravid female *Philometroides huronensis* migrating into the fins of *Catostomus commersoni* prior to release of the larvae disrupted and compacted the subepidermal tissue (Uhazy, 1978). Gravid female nematodes became enclosed in a connective tissue capsule which was disrupted when the female burst and discharged the larvae through a lesion in the skin. This provoked an acute inflammatory response and spent females were eventually resorbed. *P. lusiana* infestations of *Cyprinus carpio* stimulated an increase in levels of aspartate and alanine aminotransferases in the serum, muscle, skin, hepatopancreas and brain. Alanine, but not aspartate, aminotransferase activity increased in the kidney and aldolase activity was elevated in the serum, muscle, kidney, hepatopancreas and brain (Sekretaryuk, 1980).

The skeletal muscles

Pseudophyllidean plerocercoids are the cestodes most frequently found in the musculature. *Diphyllobothrium latum* occurs in other sites beside the muscles but *Triaenophorus crassus* is characteristically found in this site. In *Lota lota*, *Acerina cernua* and *Esox lucius*, *D. latum* plerocercoids became encapsulated; they stimulated non-specific reactions which included muscle necrosis, proliferation of connective tissue, inclusions in fatty tissue and partial resorption of eggs (Davydov, 1978). The intensity of the response was host-related and was more severe in *Lota lota* than in *Esox lucius*. Migrating plerocercoids cause haemorrhage and oedema (Davydov, 1981). In *L. lota*, total lipids and phospholipids were reduced in muscle

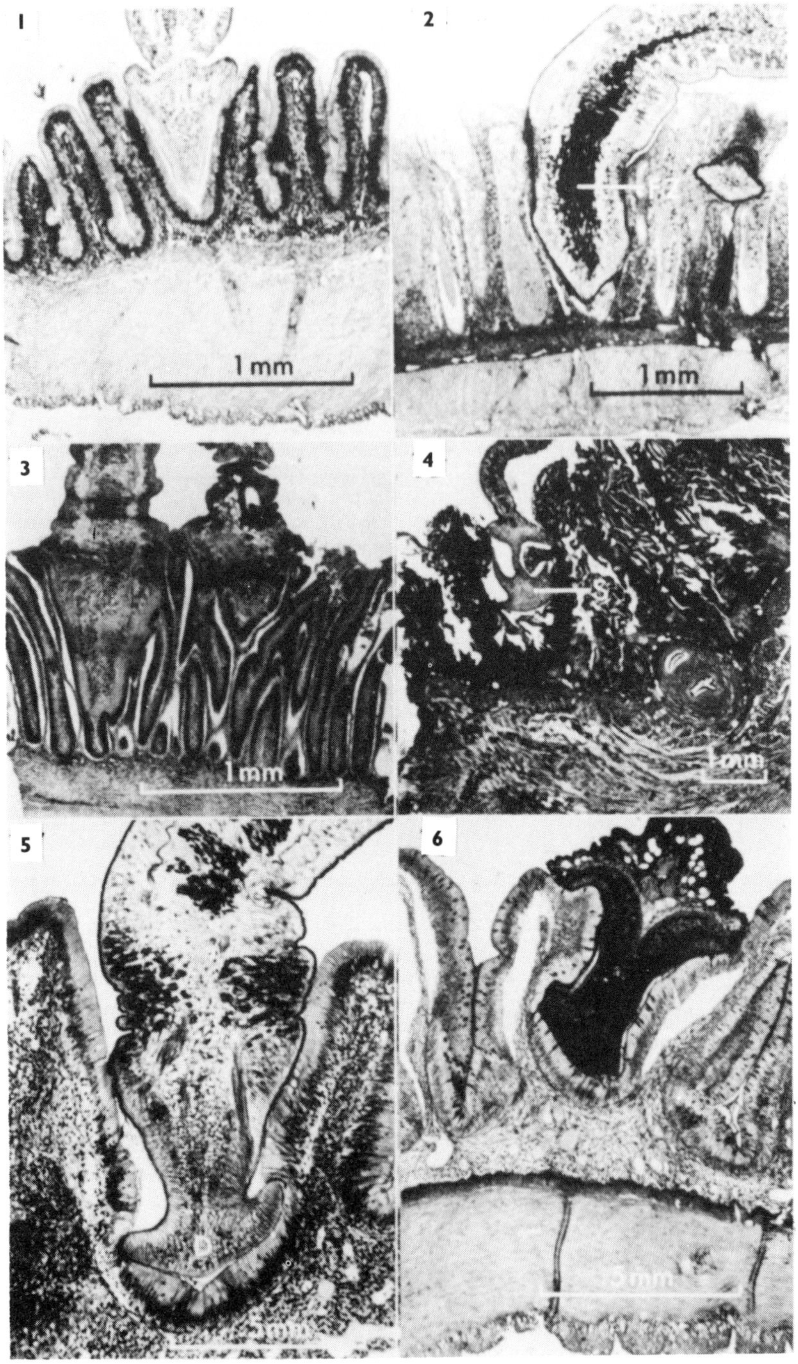

Figures 5.2.1–5.2.6 Relationship of pathology to scolex morphology among caryophyllid cestodes.
5.2.1 *Spartoides wardi*; illustrating the mechanical displacement of the mucosa by the scolex. 5.2.2 *A. huronensis*; note the large concentration of 'Faserzellenstrange' (FZ) in the scolex and neck (PAS and diastase). 5.2.3 *Caryophyllaeus laticeps*; illustrating the adhesion of scolex folds to mucosa. 5.2.4 *Capingens singularis*; low-power micrograph illustrating the grasping function of the bothria in the deep folds of the 'stomach' mucosa. 5.2.5 *Glaridacris confusus*; note terminal disc (D) and mechanical displacement of mucosa. 5.2.6 *Biacetabulum infrequens*; acetabular suckers firmly attached to epithelium (PAS).
From Mackiewicz, Cosgrove and Gude (1972).

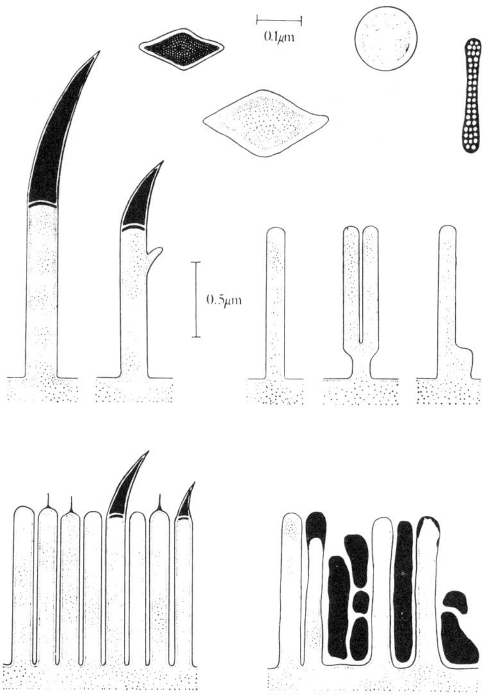

Figure 5.3 Section through nodule in the fish gut showing *Hunterella nodulosa in situ*. Note the intimate contact between parasite and host tissue and how the posterior part of the worm protrudes into the lumen of the gut. From Hayunga and Mackiewicz (1975).
See Halvorsen and Macdonald (1972), and Hermanns and Korting (1987) for the effects of *Cyathocephalus truncatus* (Spathebothriidea) on rainbow trout.

and liver tissue (Gur'yanova, 1980; Gur'yanova and Sidorov, 1985). In *Coregonus*, migrating *T. crassus* plerocercoids reduced the muscle tissue to a homogenous mass, provoking a weak host reaction (Davydov, 1981). The damage caused by *Triaenophorus* to muscle of *Coregonus clupeaformis* are shown in Figure 5.14.3. Haemorrhage, lesions and granuloma formation were observed in *Salmo gairdneri* infected with *T. crassus* plerocercoids (Rosen and Dick, 1984b). Swimming behaviour was affected and several trout died.

Metacercariae from several digenean families invade the muscle and skin of fish. One of the best known is the diplostome, *Uvulifer ambloplitis*, one of the causes of black spot disease. The metacercariae became encapsulated by host tissue and melanophores invaded the outer layers, producing the characteristic black spots (Hunter and Hamilton, 1941). Infected fish lost weight and condition as a result of metabolic dysfunction, their lipid content decreased and their oxygen requirement was elevated until the parasites had become encapsulated, after which it declined (Hunter and Hunter, 1938; Lemly and Esch, 1984b). Heavily infected *Lepomis macrochirus* began winter in a state of lipid depletion which became critical for survival when total lipids fell to 5 per cent of the dry weight or less. From 10–20 per cent of the young of the year died each winter (Lemly and Esch, 1984b).

Heterophyid metacercariae also cause black spot, eg. *Cryptocotyle lingua* (see previous section) which locates predominantly in the skin. *C. concavum* metacercariae in the skin and muscles of *Gobius melanostomus* reduced total lipids, particularly triglycerides, in muscle and liver and contributed to winter mortality (Shchepkina, 1981). Migrating cercariae of *Haplorchis pumilio* caused focal haemorrhage in the skeletal muscles of fry of *Sarotherodon spilurus*, *S. mossambicus* and *S. galilaea* (Figures 5.23.1–4); invasion by many cercariae killed

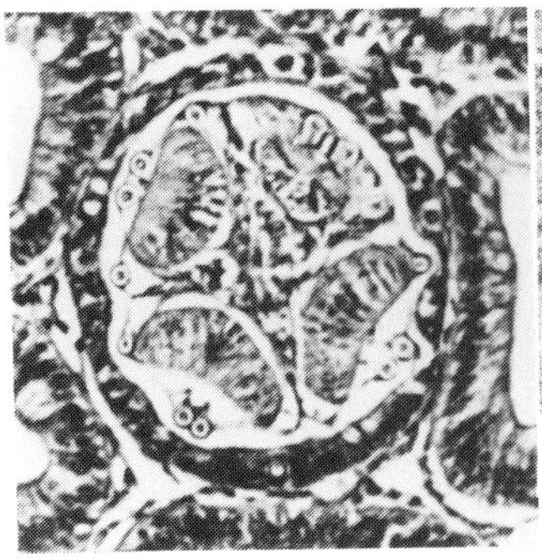

1

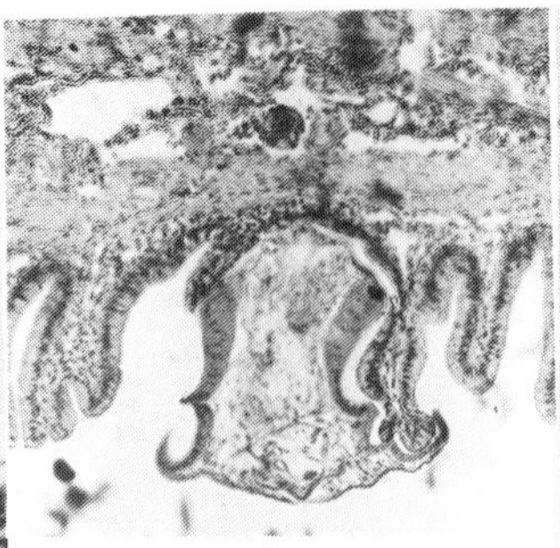

2

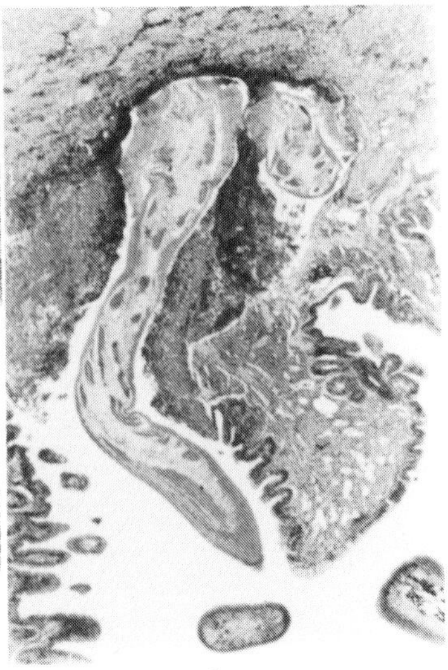

3

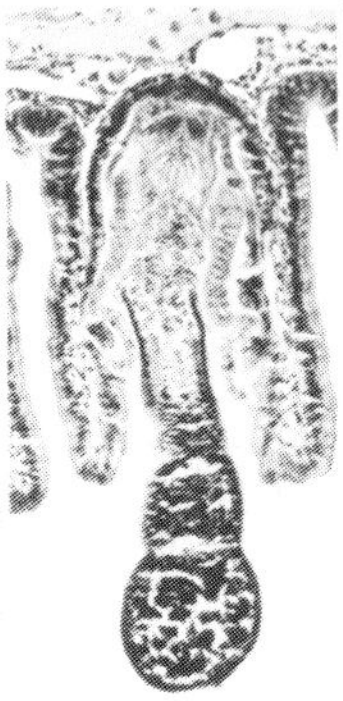

4

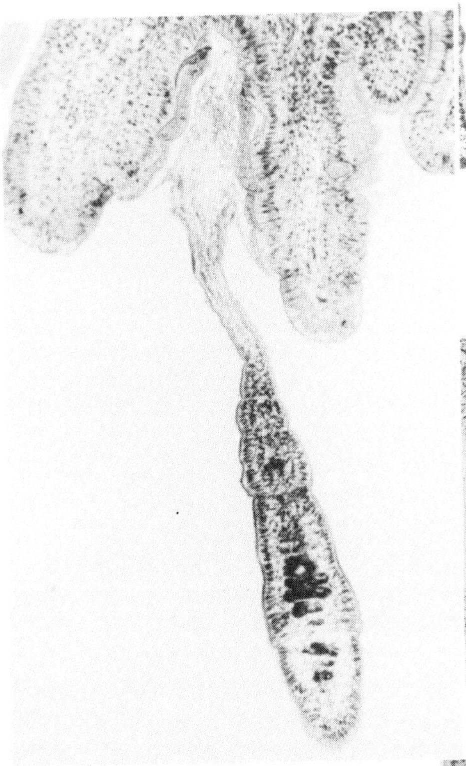

5

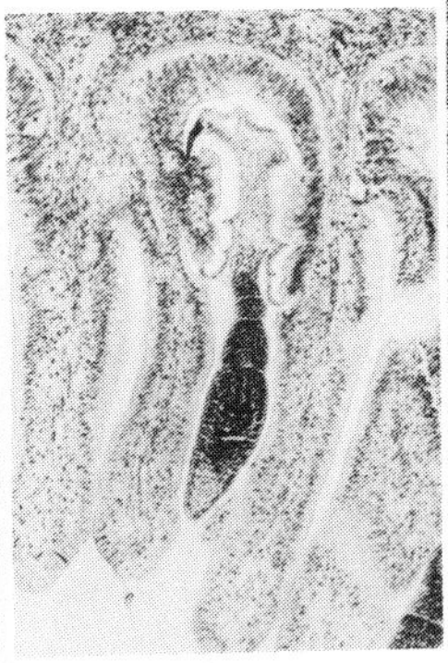

6

7

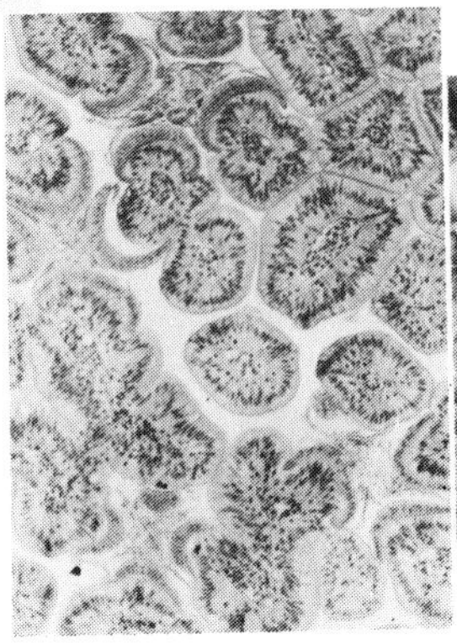

8

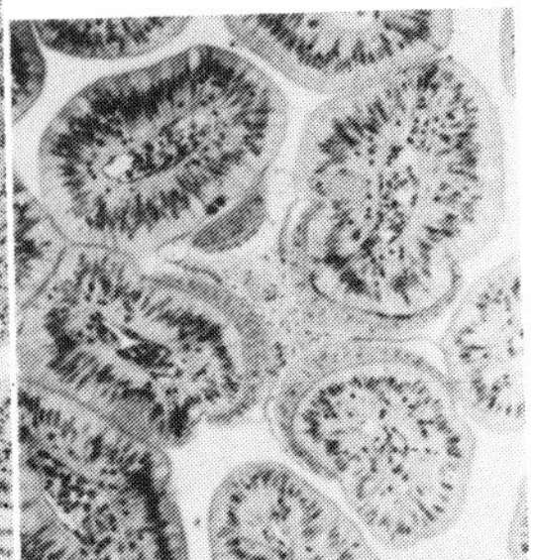

9

the fish within hours. Growth studies showed no significant differences between infected and uninfected fish in specific growth, food conversion ratios and condition (Sommerville, 1982b). The host response to encysted *Apophallus brevis* metacercariae involved deposition of a capsule histochemically almost identical with host bone. Melanocytes and connective tissue containing lipid droplets surrounded the capsule (Pike and Burt, 1983).

Penetrating and migrating cercariae of the bucephalids *Bucephalus polymorphus* and *Rhipidocotyle illense* in a variety of freshwater fish cause damage mechanically and by the action of proteolytic secretions (Baturo, 1980). Hyperaemia, haemorrhage, oedema and extensive extravasation occurred in various organs and tissues, with necrotic changes. *Nanophyetus salmincola* (Nanophyetidae), which occurs in all tissues of salmonid fish but especially the kidneys, muscles, fins and gills, causes exophthalmia, damage to the retina, cornea, fins, tail and gills, intestinal prolapse and local haemorrhages (Figures 5.21.1–6) (Millemann and Knapp, 1970). Heavily infected fish died.

In *Rutilus rutilus* infested with metacercariae of the cyathocotylid *Paracoenogonimus ovatus*, total amino acid levels were elevated in muscle and reduced in the blood of Group III fish heavily infested with 180–396 metacercariae. The levels of various amino acids in the muscle of Group I and Group III fish also changed (Linnik and Litvyak, 1980).

Stephanochasmus baccatus metacercariae encysted in the fins and muscles of *Pleuronectes platessa*, *Scophthalmus maximus*, *Solea solea* and *Limanda limanda* provoked a chronic granulomatous inflammatory reaction, the rate and intensity of which depended on the host species (Figures 5.22.1–5). In turbot alone, all the encysted cercariae died. Large 0-group fish were minimally affected but, in small 0-group plaice, the relatively large cysts disrupted organ function (Sommerville, 1981a).

Clinostome metacercariae, one of the causative agents of yellow-grub disease, inhabit the skin, viscera and muscles of their hosts. Metacercariae migrating from the body wall can cause haemorrhage and tissue damage, eg. in *Clinostomum complanatum* infestation of *Misgurnus anguillicaudatus* as described by Lo, Chen and Wang (1985). In this instance, only massive infestations had serious consequences. In *Channa punctata*, *Euclinostomum heterostomum* metacercariae reduced the levels of total free amino acids, caused a slight increase in total proteins and raised aspartate aminotransferase levels in the liver. Alanine aminotransferase levels decreased (Gupta and Agarwal, 1984). Changes in the free amino acid composition of infected liver tissue were also noted by Bhagavathiammai and Ramalingam (1983).

Larvae of the nematode *Eustrongylides* in the muscles of various species of cultured freshwater fish homogenized the adjacent tissue which became infiltrated with fibroblasts (Paperna, 1974).

Nervous system and sense organs

The majority of the pathogens reported from the brain and sense organs are strigeoid metacercariae. Many have little ill-effect, eg. *Ornithodiplostomum ptychocheilus* metacercariae in the brain and viscera of *Richardsonius balteatus* did not impair swimming performance and

---

Figures 5.4.1–9 The pathological effects of Tetraphyllidea (Oncobothriidae) on the gut mucosa of elasmobranchs 5.4.1. A scolex of an attached specimen of *Acanthobothrium tripartitum* in transverse section showing effect on the epithelial lining of the mucosal tubular crypt. 5.4.2 *A. dujardini* in longitudinal section with posterior loculi of bothridium grasping the ridge between tubular crypts. 5.4.3 Penetration of *A. rajaebatis* to muscular layers of intestinal wall. 5.4.4 *A. tripartitum* with most of strobila in tubular crypt of mucosa. Figs. 5.4.1–5.4.4 from Williams (1969). 5.4.5 *Acanthobothrium* sp. in *Raja montagui*: longitudinal section of entire specimen with its scolex attached to mucosa. 5.4.6 Longitudinal section of immature *A. quadripartitum* between villi of mucosa. (Williams, 1968a, p.111). 5.4.7 Longitudinal section through bothridium attached to villus, showing one of a pair of hooks penetrating the epithelium. (Williams, 1968a). 5.4.8 Transverse sections through four scolices attached to villi. (Williams, 1968a). 5.4.9 Transverse section of scolex showing attachment to four villi. (Williams, 1968a).

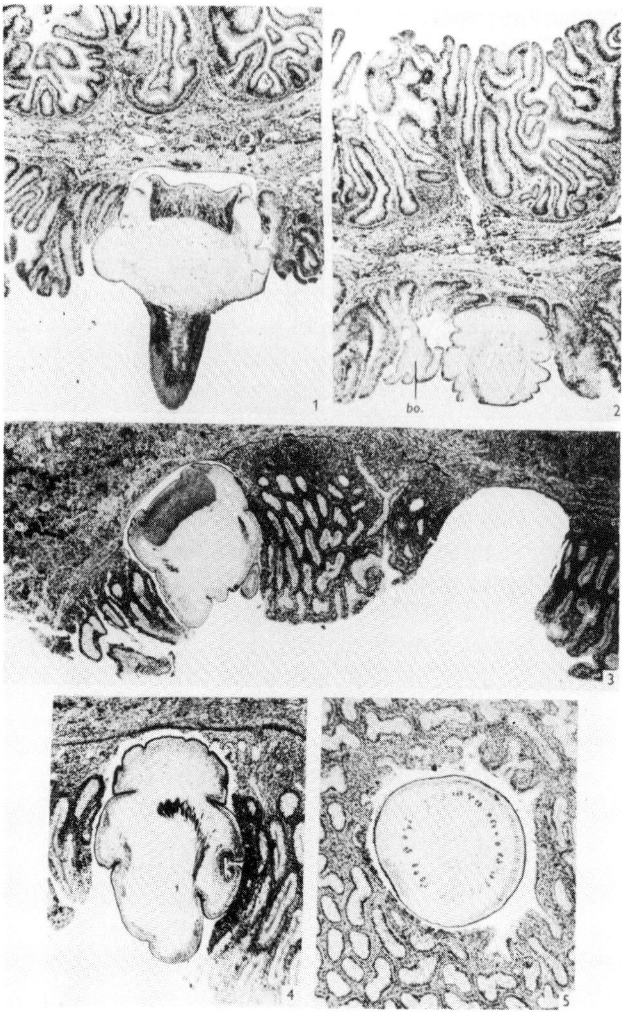

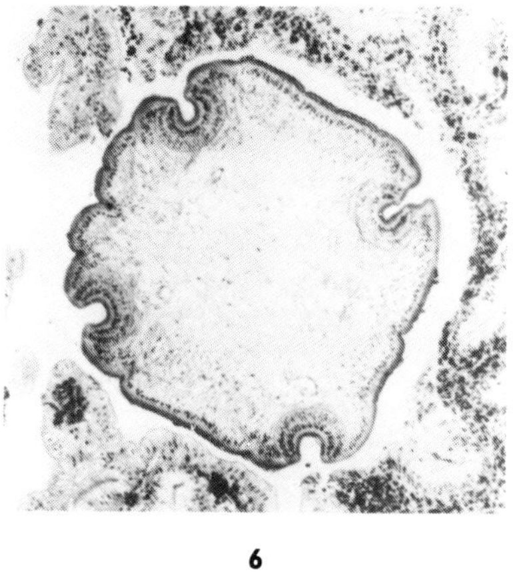

6

stamina (Sogandares-Bernal, Hietała and Gunst, 1979). However, *Diplostomum baeri eucaliae* in the brain of *Eucalia inconstans* provoked a tumour-like growth (Hoffman and Hoyme, 1958).

Fatal outbreaks of whirling disease in *Engraulis japonica* and *Seriola quinqueradiata* have been caused by metacercariae of the heterophyid *Galactosomum* (Kimura and Endo, 1979). Mortalities of *Sparus auratus* fry in hatcheries were attributed to the acanthostomid *Acanthostomum inbutiforme* and many fish showed torsion of the body with an encysted metacercaria at the point of torsion. Because of their small size, fry were more severely affected than adults which apparently tolerated infection without ill-effect (Maillard, Lambert and Raibaut, 1980).

Diplostome metacercariae are the most important pathogens of the eyes. Cercariae migrating to the eyes damage the tissues through which they pass and cercarial invasion can kill very young fish. Rumyantsev (1976b) found that invasion by more than three cercariae per fish was lethal to three to 10-day-old *Cyprinus carpio* fry. Most cercariae penetrated the gills and pharynx, reaching the lens within an hour. The number of metacercariae which impaired sight varied. Eyes of *Salmo gairdneri* were slightly clouded by 30–40 *Diplostomum* sp. metacercariae per eye, but severely or totally clouded by 120 per eye, while *S. salar* with *D. pusillum* in the vitreous humour had normal vision at 200 per eye but were blinded by 400 per eye (Sato, Hoshina and Horiuchi, 1976; Ieshko and Shustov, 1982). Infection of the lens was the most pathogenic: *Phoxinus phoxinus* were blinded by 100 *D. commutatum* per lens although 25 per lens had no effect (Ieshko and Shustov, 1982). Opacity of the lens in *Rutilus rutilus* infected with *D. spathaceum* was caused by destruction of the fibres of the cortical layer, and spaces between the fibres became filled with liquid containing excretory products of the parasites (Grevtseva, 1977). Subcapsular cataracts have been reported in acute infections of *Diplostomum* spp. in *S. gairdneri*, while chronic cases showed lens deterioration, retinal detachment, capsular rupture with phacogenic uveitis, and formation of Elschnig's pearls and Wedl cells (Shariff, Richards and Sommerville, 1980). Corneal haemorrhage also occurred (Sato, Hoshina and Horiuchi, 1976).

Chronically-infected fish are often blind, exophthalmic and emaciated because of their inability to catch enough food. Also, the behaviour of heavily-infected fish may be modified in ways which render them more susceptible to predation by piscivorous birds. *Leuciscus leuciscus* infected with *D. spathaceum* was less effective at catching *Gammarus pulex* despite spending more time feeding, and remained in surface water longer than uninfected fish, which increased the risks of predation (Crowden and Broom, 1980).

The alimentary canal

Relatively few of the very large number of helminth species which occupy sites in the alimentary canal of fish are pathogenic. Pathological effects are often the result of mechan-

---

Figures 5.5.1–5.5.6 The effects of *Echeneibothrium* and *Discobothrium* (Tetraphyllidea) on the host intestinal mucosa.

5.5.1 and 5.5.2 *Echeneibothrium* sp. from *Raja montagui*.

5.5.1 Median longitudinal section through scolex attached to gut mucosa of the host to show myzorhynchus. 5.5.2 Lateral longitudinal section of scolex of specimen shown in Figure 5.5.1 showing mode of function of the loculate bothridium. 5.5.3–5.5.5 *Discobothrium fallax* from *Raja clavata*.

5.5.3 Longitudinal section through scolex attached to the mucosa showing 'scar' where this specimen had been previously attached. 5.5.4 Longitudinal section of scolex to show 'sucker'. 5.5.5 Transverse section of scolex behind region of the 'suckers' showing retractor muscles of the myzorhynchus. 5.5.6 *Discobothrium fallax* from *Raja clavata*. Transverse section through scolex showing acetabular nature of four 'suckers' (bothridia?)

From Williams (1966).

Abbreviations bo. bothridium

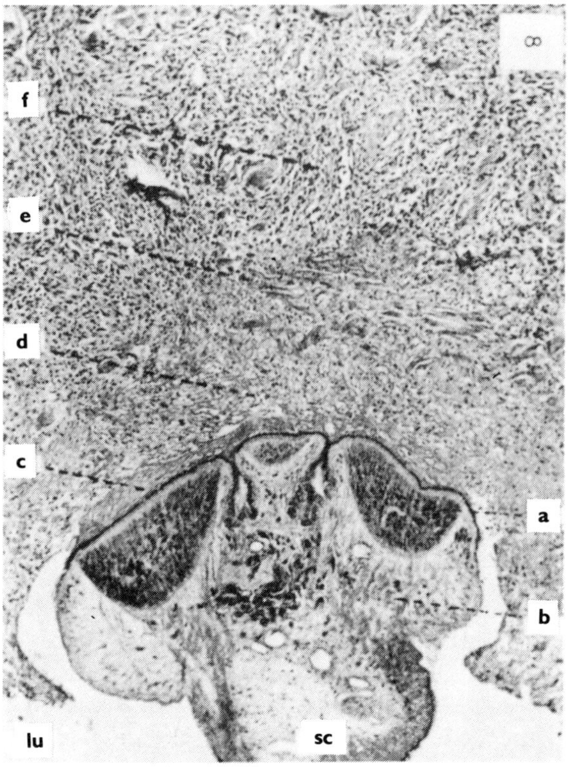

1

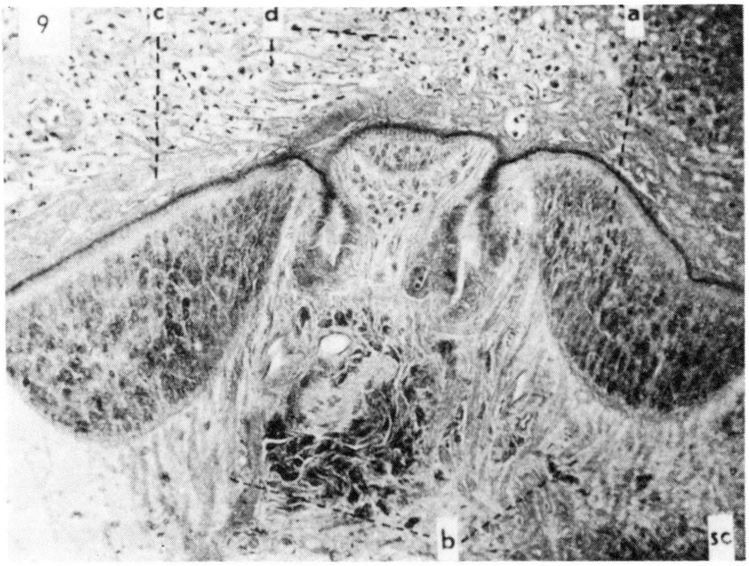

2

Figures 5.6.1–5.6.2 The effects of ***Prosobothrium armigerum*** (Cestoda: Tetraphyllidea) on the intestine of *Carcharhinus glaucus*.
5.6.1 Transverse section of the scolex attached to the intestinal mucosa. 5.6.2 Detail of the glandular suckers and of the intestinal mucosa. Abbreviations: a, glandular sucker; b, muscles; c and e, lysis at point of contact with scolex; d, inflamed zone; f, 'zone oedemateuse'; lu, intestinal lumen; sc, scolex.
From Baer and Euzet (1956).

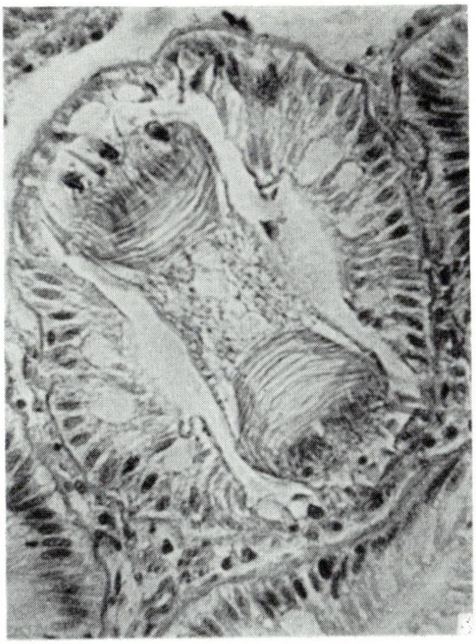

1

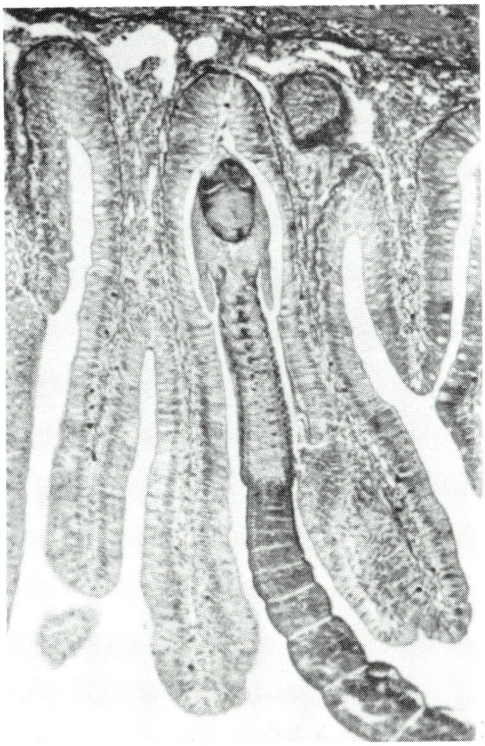

2

Figures 5.7.1–5.7.2 *Echinobothrium brachysoma* in the tubular crypts of the mucosa of *Raja clavata*.
5.7.1 Transverse section through crypt and contained scolex. 5.7.2 Vertical section through the scolex attached to the mucosal wall of the spiral valve showing local flattening and tearing of epithelium.
From Rees (1961a).

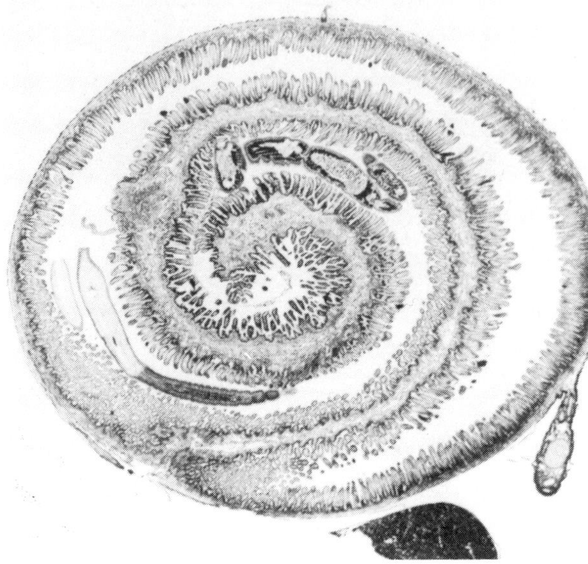

1

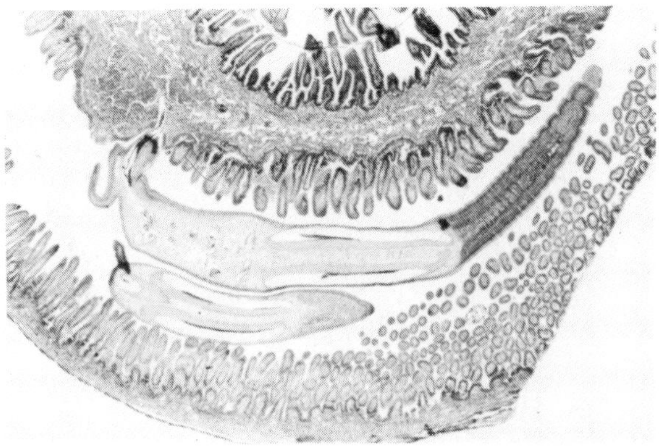

2

Figures 5.8.1–5.8.2 The effects of the scolex of an adult trypanorhynch tapeworm on the mucosal lining of the intestine of a *Raja radiata*.
5.8.1 Transverse section of the anterior part of the spiral intestine of *R. radiata* showing the point of attachment of *Grillotia* sp. and the orientation of the strobila around the spiral at right angles to the long axis of the intestine. 5.8.2 Detail of a scolex tentacle with hooks penetrating the intestinal wall.
From Williams, McVicar and Ralph (1970).

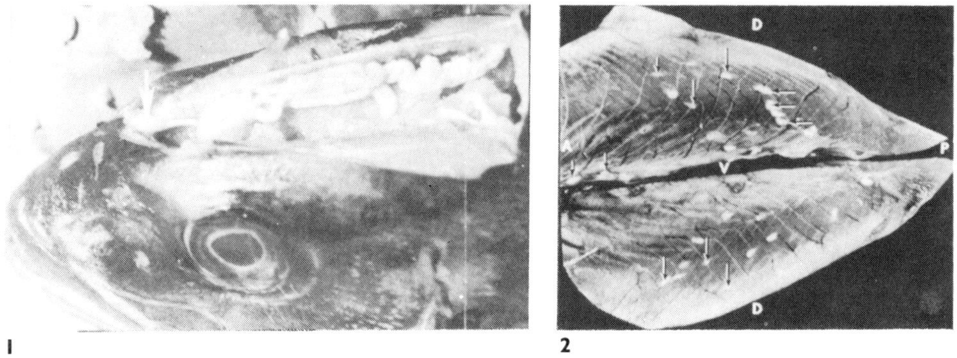

Figures 5.9.1 and 5.9.2 Larval trypanorhynch tapeworms in the tissues of teleost fish.
5.9.1 The head of *Seriola dumerili* with the skin and some muscle removed to show a coiled trypanorhynch blastocyst in the tissue. The saclike end (arrow) contains the infective plerocercoid, possibly a member of the genus *Pseudogrillotia*. 5.9.2 Abdominal or belly musculature of *Katsuwonus pelamis* showing about 30 encysted trypanorhynch plerocerci (arrows) of *Tentacularia coryphaenae*. A similar infection to that shown in Figure 5.9.2 is described by Rae (1958) for the plerocercoid larvae of *Grillotia erinaceus* in the muscle of *Hippoglossus hippoglossus*.
From Deardorff, Raybourne and Mattis (1984).

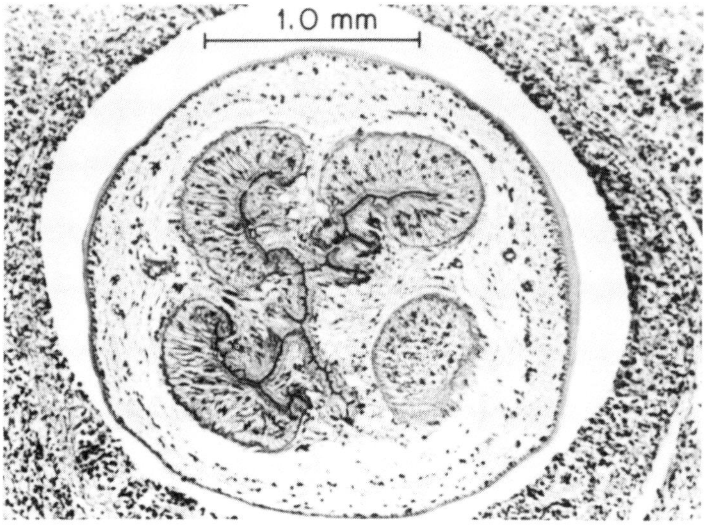

Figure 5.10 A transverse section of the scolex of a larval tapeworm, *Proteocephalus* sp. (Proteocephalidae) in the liver of small-mouth bass.
From Cosgrove (1975).

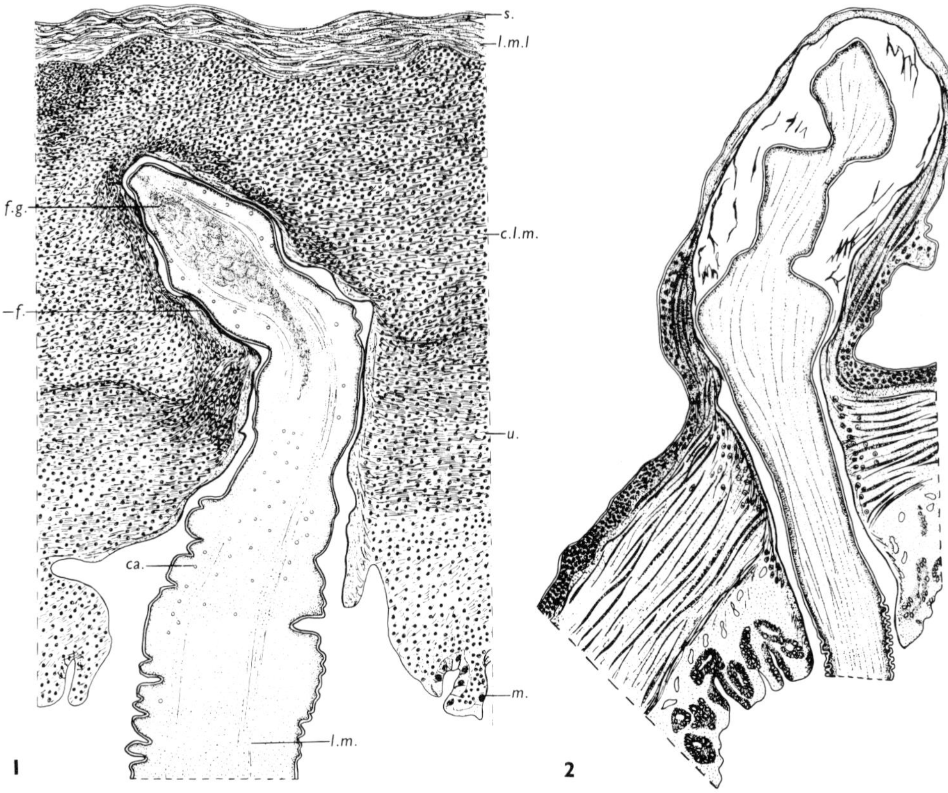

Figures 5.11.1–5.11.2 Pseudophyllidean cestodes in which the scolex has degenerated in the host tissues. 5.11.1 *A. gadi* from *Gadus aeglifinus*: sagittal section through the anterior end embedded in the wall of the intestine. 5.11.2 *Parabothrium gadi-pollachii*: section through the scolex deformatus embedded in the wall of the intestine. Abbreviations: ca. calcareous corpuscle; cae. caecum; f. fibrous layer; f.g. 'frontal gland'; i. intestine; l.m. longitudinal muscles; l.m.l. longitudinal muscle layer; m. mucosa; s. serosa; s.d. scolex deformatus; u. unidentified layer of tissue.
From Williams (1960b).

ical damage by the attachment organs, with an inflammatory response at the site of attachment and sometimes in the adjacent tissue. The intensity of the response is directly related to the depth to which the parasite penetrates the gut wall, those which penetrate deeply provoking the greater response. The gut wall may be completely perforated. Some parasites obstruct the gut lumen, particularly in young fish. In addition to mechanical damage, gut parasites cause functional disturbances, blood and metabolic changes, retarded growth and weight loss, alterations in behaviour and increased susceptibility to a variety of stresses.

Although adult digeneans are usually non-pathogenic, intestinal lesions and chronic inflammation have been caused by *Crepidostomum farionis* in salmonids (Davis, 1937; Mitchum, 1965; Wales, 1958b), by *Azygia lucii* in *Hucho hucho* (Brglez, Rakovec and Snoj, 1966) and by *Asymphylodora tincae* in *Tinca tinca* (Wierzbicka, 1970). Abrasion may be attributable to feeding activity, eg. *A. tincae* (above) and the nematode *Truttaedacnitis truttae* in *Salmo gairdneri* (Dunn, Russell and Adams, 1983). Body spines of *Spinitectus* sp. caused lesions and inflammation in *Lepomis macrochirus* (Hoffman, 1975a).

Cestodes and acanthocephalans may cause mechanical obstruction. It has been reported for *Bothriocephalus acheilognathi* in three fish species by Scott and Grizzle (1979), *Caryophyllaeus fimbriceps* in carp by Ivasik (1952) and Bauer (1968), and *Lytocestus indicus* in

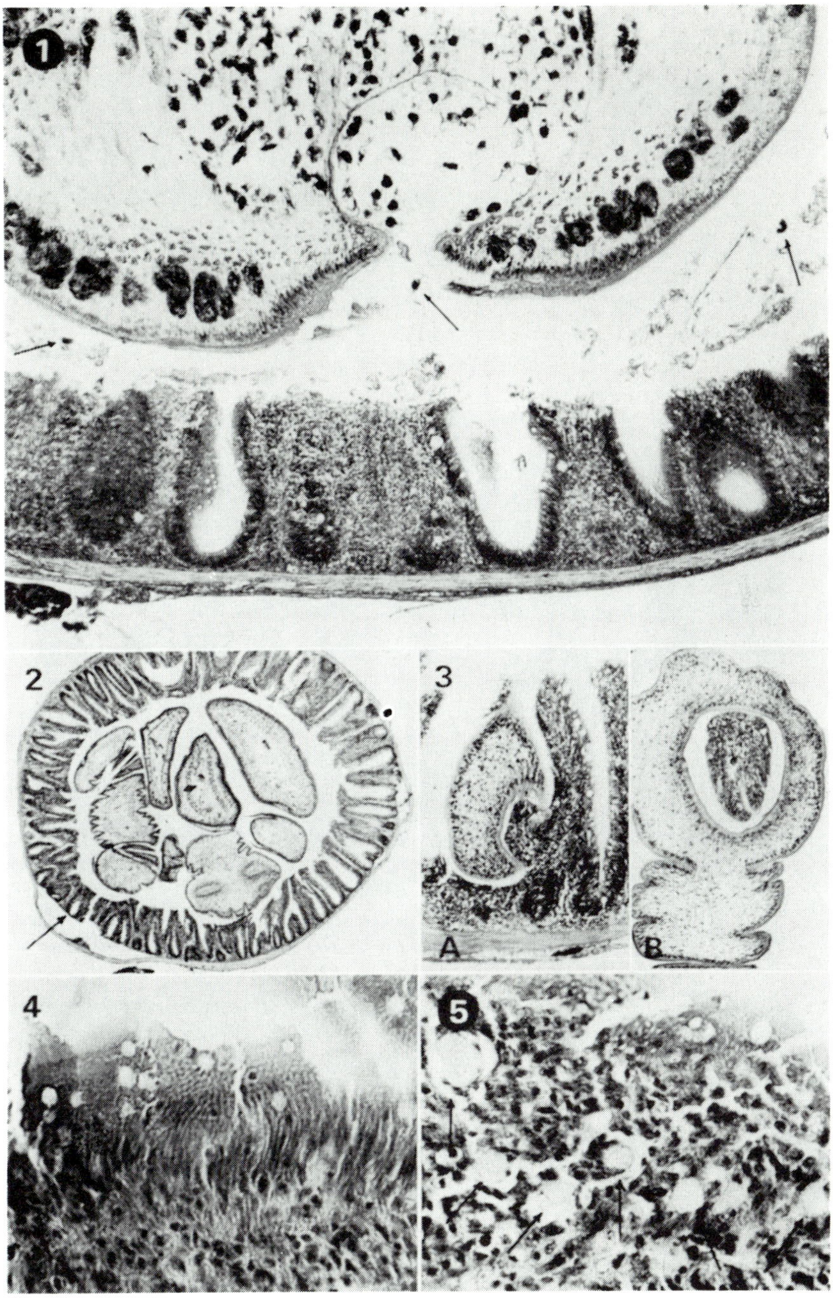

Figure 5.12.1–5 The effects of *Bothriocephalus* (Psuedophyllidea) on the gut of cultured carp.
5.12.1. Transverse section of the small intestine of a 4-month-old carp. The intestinal microvilli are missing. The mucosa is visible but damaged. Eggs are being released into the intestinal lumen via the worm's uterine pore. 5.12.2. A transverse section of the intestine with a worm infection but the mucosal lining is not so damaged as that shown in Figure 5.12.1. 5.12.3. The scolex is firmly attached to the gut wall. 5.12.4. The intestinal microvilli are seriously destroyed and hardened. 5.12.5. Coagulated materials including tissue debris on the surface of the hardened microvilli.
From Nakajima and Egusa (1974b).

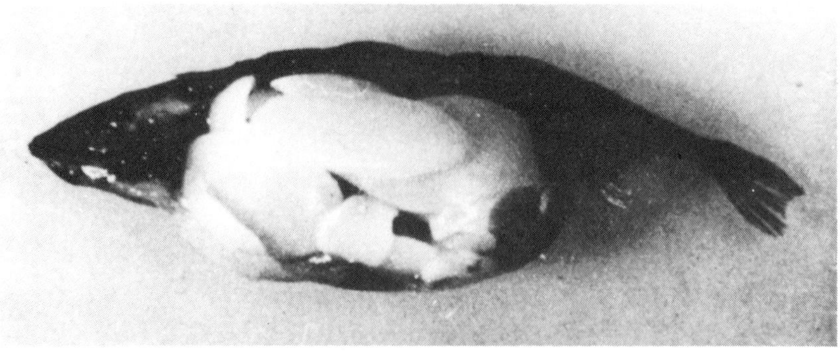

5.13.1

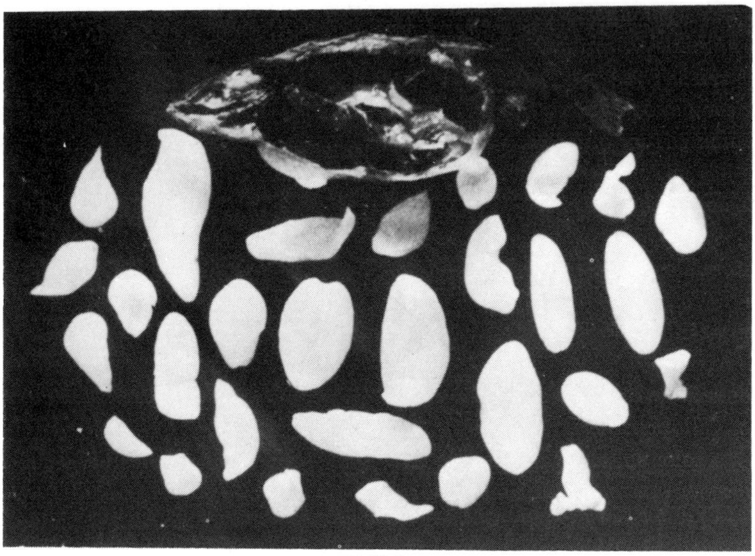

5.13.2

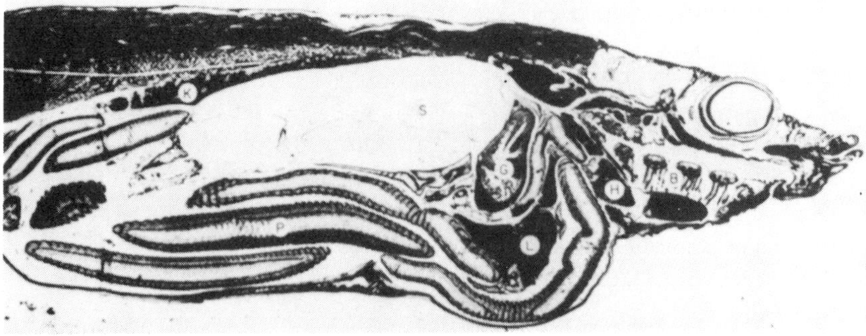

5.13.3

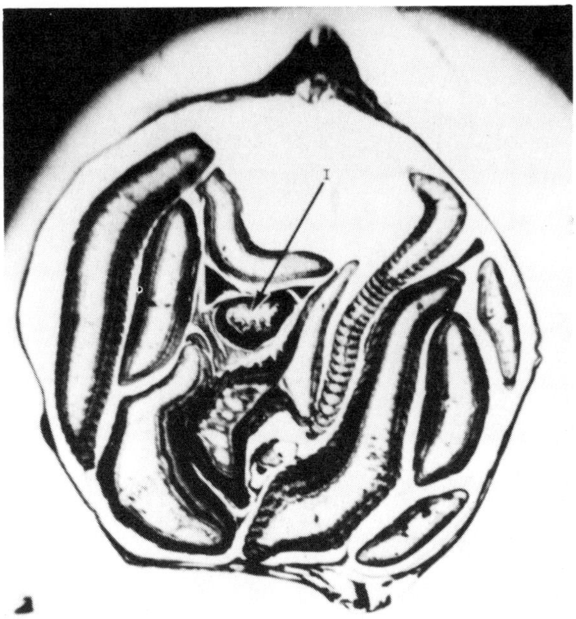

5.13.4

Figures 5.13.1–5.13.4 *Schistocephalus solidus* (Pseudophyllidea) in the body cavity of a teleost fish.
5.13.1 Body cavity opened to show the densely-packed plerocercoids. 5.13.2 Plerocercoids removed from the fish shown in 5.13.1. 5.13.3 and 5.13.4 Longitudinal and transverse sections of infected fish.
From Reichenbach-Klinke and Elkan (1965).

Clarias batrachus by Ahmed and Sanaullah (1979). *Neoechinorhynchus prolixus* partly blocked the gut of *Carpiodes carpio* when 80 or more worms were present (Self and Timmons, 1955) and occlusion has also resulted from infections of *Echinorhynchus veli* in *Synaptura orientalis* (George and Nadakal, 1982). Reduction of the gut lumen as a result of the host reaction to migrating larvae of *Philometroides lusiana* in the mucosa has been described by Sekretaryuk (1983a).

There are many descriptions of the changes which take place in the mucosa and underlying layers in response to parasites. These include displacement and flattening of the villi, mechanical injury by the hooks of armed parasites, focal pressure necrosis, loss of the epithelium at or near the point of attachment, hyperplasia and metaplasia of the mucosa, haemorrhage, and the destruction of the mucosa and, to varying degrees, the underlying layers, usually accompanied by a host cellular reaction. The resulting histopathological effects are the formation of lesions, ulcers, nodules or diverticula, calcareous, collagenous or fibrous collars or capsules around parasites penetrating or within the gut wall, fibrosis and the appearance of granulomata.

The host cellular reaction, described for teleosts by Roberts (1978), is a non-specific inflammatory response, the rate of which is temperature-dependent. It provides, at the site of injury, the cells and tissue fluids best able to preserve homeostasis, and is characterized by infiltration of the injured tissue by polymorphonuclear leucocytes, mononuclear macrophages, lymphocytes and thrombocytes. The latter activate the formation of fibrin which results in the delimitation of lesions by fibrous tissue.

With few exceptions, a host reaction of this nature has been described for most gut parasites which injure the gut wall. Bullock (1961), however, found no cellular reaction in infections of the acanthocephalans *Atactorhynchus verecundus* and *Neoechinorhynchus cristatus*, although they destroyed the epithelium and lamina propria at the site of attachment.

Fibrosis is a common reaction and may cause the production of a fibrous capsule

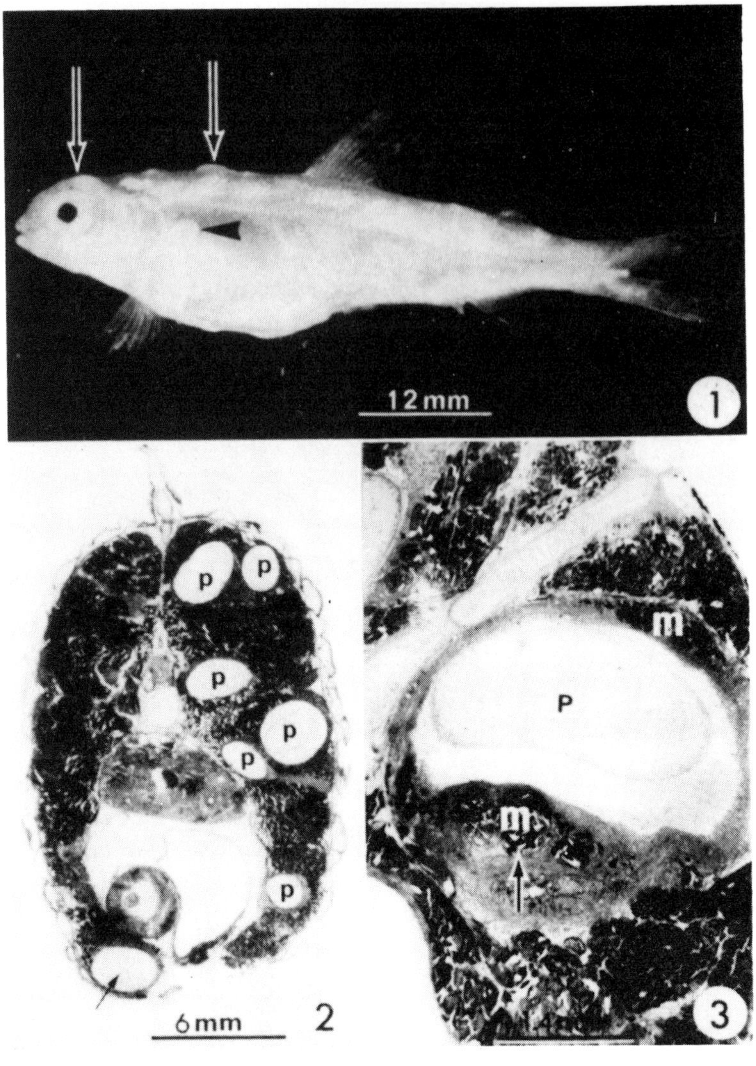

Figures 5.14.1–5.14.3 Infections of whitefish with *Triaenophorus crassus* (Pseudophyllidea).
5.14.1 Whitefish infected with *Triaenophorus crassus* at day 63 postinfection. This fish had three plerocercoids. Black and white arrows indicate positions of developing plerocercoids; black arrow indicates a worm *in situ*. 5.14.2 Cross-section through fish at posterior edge of the dorsal fin. Arrow indicates portions of plerocercoid (p) in hypaxial muscles. Note enlargement of right side of fish. 5.14.3 Section through epaxial muscles. m, muscle. Note that cyst wall has incorporated host muscle tissue (arrow).
From Dick and Rosen (1982).

around the proboscis or scolex of a parasite. This has been reported for species of *Pomphorhynchus* by McDonough and Gleason (1981) (Figures 5.33.1–4), Chubb (1965) and Wanstall, Robotham and Thomas (1986). Prakash and Adams (1960) described the formation of a collagenous capsule around *Echinorhynchus lageniformis* in the intestine of *Platichthys stellatus*, while 'collars' are formed around the neck region of *Microsentis wardae* and the scolex of *Bothriocephalus scorpii* (Martin and Multani, 1966; MacKinnon and Featherstone, 1982). Martin and Multani (1966) speculated that the collar might eventually make it impossible for *M. wardae* to retain its hold.

The hooks of armed parasites, ie. acanthocephalans and some cestodes, cause traumatic

Table 5.3 Pathological effects of some digenean parasites of fish (Figures 5.15–5.24)

Parasite	Host	Pathology	Reference
Diplostomidae (metacercariae)			
Diplostomum spp.	*Salmo gairdneri* (eyes)	Acute: subcapsular cataracts. Chronic: lens dislocation, capsular rupture or duplication, formation of Elschnig's pearls and Wedl cells; phacogenic uveitis in fish with capsular rupture; retinal detachment. Chronic cases were blind, exophthalmic and emaciated.	Shariff, Richards and Sommerville (1980)
Diplostomum sp.	*Cyprinus carpio* fry	Infection rate of over 3 cercariae/fish was lethal	Rumyantsev (1976)
Diplostomum sp.	*S. gairdneri* (eyes)	Eyes containing 30–40 metacercariae slightly clouded, those with up to 120 severely or entirely clouded. No deaths, and fish continued to feed but were thin and dark in colour; 15% had corneal haemorrhage.	Sato, Hoshina and Horiuchi (1976)
D. baeri eucaliae	*Eucalia inconstans* (brain)	Hyperplasia of choroid plexus, 'tumour-like' growth, macrophage accumulation, brain tissue resorption followed by hyperplasia of ependyma.	Hoffman and Hoyme (1958)
D. commutatum	*Phoxinus phoxinus* (lens)	Vision unimpaired by up to 25 larvae but deteriorated with increasing numbers and was lost at 100 per lens.	Ieshko and Shustov (1982)
D. pusillum	*Salmo salar* (vitreous body)	Sight normal at 200 per eye but lost at over 400	Ieshko and Shustov (1982)
D. spathaceum	*Leuciscus leuciscus* (eyes)	Heavily-infected fish spent more time feeding (but caught fewer *Gammarus*) and more time in surface water, increasing their vulnerability to predators.	Crowden and Broom (1980)
,,	*Rutilus rutilus* (lens)	Fibres of cortical layer of lens destroyed by 25–35 days p.i.; lens becomes opaque.	Grevtseva (1977)
Ornithodiplostomum ptychocheilus	*Richardsonius balteatus* (brain, viscera)	Swimming performance and stamina not affected.	Sogandares-Bernal, Hietała and Gunst (1979)
Posthodiplostomum cuticola	*Hypophthalmichthys molitrix* (skin)	Retardation of growth; weight loss; reduction of haemoglobin and erythrocyte numbers; increase in erythrocyte sedimentation rate and leucocyte numbers; granular dystrophy in liver and kidneys, haemosiderin deposits in spleen; disturbed lipid metabolism and increase in acid mucopoly-saccharide levels.	Denisov (1979)
P. minimum	*Lepomis macrochirus*	Significant reduction in growth of fingerlings with over 353 metacercariae; increased mortality.	Smitherman (1968)
,,	,, (heart)	Additional serum proteins (β and τ-globulins); decreased albumin levels.	Meade and Harvey (1969)
,,	*Pimephales promelas* (liver and other organs)	Hyperaemia at fin bases; no nervous symptoms; mortality in heavy infections.	Hoffman (1958)

Table 5.3 continued

Parasite	Host	Pathology	Reference
P. minimum	*Roccus saxatilis* (orbit, musculature: unusual locations in centrarchids)	Body deformed; exophthalmia; swelling of muscle tissue; mortality	Hoffman and Hutcheson (1970)
P. minimum and *Proteocephalus* plerocercoids	*L. macrochirus* (liver and viscera)	Severe melanotic liver and visceral fibroses	Mitchell, Ginal and Bailey (1983)
Uvulifer ambloplitis	*Micropterus dolomieui* (fins, tail base, muscles)	Significant weight loss attributed to metabolic disturbance.	Hunter and Hunter (1938)
,,	*M. dolomieui, Eupomotis gibbosus, Ambloplitis rupestris* (muscles)	Metacercariae encapsulated by a reaction similar to inflammatory or immune responses in higher vertebrates; melanophores invade outer layers of cyst.	Hunter and Hamilton (1941)
,,	*L. macrochirus* (muscles)	Increased susceptibility to decreasing temperature; lipid reduction; loss of condition; increased O_2 consumption until metacercariae encapsulated; death of 10–20% of 0-group fish each winter.	Lemly and Esch (1984b)
Allocreadiidae			
Crepidostomum farionis	*Salmo gairdneri, Salvelinus fontinalis* (intestine)	Intestinal lesions; inflammation; mass mortality.	Davis (1937) Mitchum (1965) Wales (1958b)
Azygiidae			
Azygia lucii	*Hucho hucho* (intestine)	Haemorrhagic lesions	Brglez, Rakovec and Snoj (1966)
Monorchiidae			
Asymphylodora kabanicum	*Rutilus rutilus* (kidneys, atypical site)	Kidney damage, provoking abdominal dropsy	Bykhovskaya-Pavlovskaya and Bykhovskii (1940)
A. tincae	*Tinca tinca* (intestine)	Abrasion, causing chronic inflammation	Wierzbicka (1970)
Opisthorchiidae			
Opisthorchis pedicellata	*Rita rita* (gall-bladder)	Reduced levels of protein, cholesterol, calcium, glucose and alkaline phosphatase in blood; reduced levels of protein, glycogen, cholesterol, glucose, alkaline phosphatase in liver; HSI reduced; elevated lactic acid levels in blood and liver.	Joshi (1979)
Cryptogonimidae			
Acetodextra amiura	Catfish (ovary)	Destruction of eggs which young worms invade; older worms consume yolk.	Perkins (1950)
Isoparorchiidae			
Isoparorchis hypselobagri	*Channa punctatus* (swim-bladder)	Necrosis of fins; scale loss; depigmentation; damage to viscera and abdominal muscles; anaemia; decreases in blood glucose level and liver glycogen; behavioural changes	Mahajan *et al.* (1979)
Gorgoderidae			
Phyllodistomum folium	*Blicca bjoerkna* (urinary system)	Acute or chronic inflammation with lesions of bladder, ureters and, rarely, kidney canaliculi; worms feed on host tissue causing damage mechanically	Wierzbicka and Einszporn-Orecka (1985)

and by lysis.

Table 5.3 continued

Parasite	Host	Pathology	Reference
P. lysteri	*Minytrema melanops* (archinephric ducts, proximal region)	Occlusion of ducts; mechanical damage due to feeding and attachment	Gleason, Christensen and Chung (1983)
P. superbum	*Hypentelium nigricans* (archinephric ducts, distal region)	,,	,,
,,	*H. nigricans* (Wolffian ducts)	Flattening of columnar epithelium, erosion and loss of epithelium; increase in number of goblet cells; leucocyte infiltration; lesions caused by oral sucker.	Chung (1981)
Sanguinicolidae			
Sanguinicola inermis	*Cyprinus carpio* fry	Lethal infection level in 8-day-old fish was 10 cercariae but the lethal dose was 2,000–2,500 in 3-month-old fish.	Sapozhnikov (1976)
,,	*C. carpio* fry (gills, heart, kidney)	Gill inflammation, necrosis and septicaemia; 80–90% mortality.	Hłond, Kozlowski and Szaryk (1977)
S. davisi	*Salmo gairdneri kamloops* (gill capillaries)	Fish weak, growth reduced, respiration impeded; gill filaments pale, flaccid, full of eggs; haemorrhage when miracidia escape from gill tissue; heavy mortalities.	Wales (1958b)
S. klamathensis	*S. clarki henshawi*		
Nanophyetidae (metacercariae)			
Nanophyetus salmincola (Figures 5.21.1–6)	Salmonid fish (all tissues but most frequently kidneys, muscles, fins and gills	Pathology includes exophthalmia, intestinal prolapse; damage to fins, tail, gills, retinas and corneas; local haemorrhage; growth and swimming performance impaired, mortality in heavy infections.	Millemann and Knapp (1970)
Bucephalidae (metacercariae)			
Bucephalus polymorphus; Rhipidocotyle illense	Fry of *Rutilus rutilus, Scardinius erythrophthalmus, Alburnus alburnus, Blicca bjoerkna, Tinca tinca, Gobio gobio, Ctenopharyngodon idella* (all organs and tissues)	Mechanical damage by penetrating cercariae and proteolytic secretions; hyperaemia; haemorrhage; oedema; extravasation; necrosis; simultaneous regenerative changes.	Baturo (1980)
Acanthostomidae (metacercariae)			
Acanthostomum imbutiforme	*Sparus auratus* (fry) (adjacent to spinal column)	Lateral or dorso-ventral torsion of body due to encysted metacercaria at point of torsion.	Maillard, Lambert and Raibaut (1980)
Cyathocotylidae (metacercariae)			
Paracoenogonimus ovatus	*Rutilus rutilus* (muscle)	Total amino acid levels elevated in muscle and reduced in blood.	Linnik and Litvyak (1980)
Acanthocolpidae (metacercariae)			
Stephanochasmus baccatus (Figures 5.22.1–5)	*Scophthalmus maximus, Solea solea, Limanda limanda, Pleuronectes platessa*	Chronic granulomatous inflammatory encapsulation	Sommerville (1981)

(skin, muscles)

Table 5.3 continued

Parasite	Host	Pathology	Reference
Unidentified metacercariae (family unknown)			
	Xiphophorus maculatus, X. helleri, Hyphessobrycon, Brachydanio rerio, Poecilia reticulata, Carassius auratus (gills)	Extensive proliferation of cartilage, loss of respiratory epithelium, fusion of primary lamellae, haemorrhage.	Blazer and Gratzek (1985)
Heterophyidae (metacercariae)			
Apophallus brevis	*Perca flavescens* (myotomes)	Intense tissue response producing capsule similar to host bone; layers of melanocytes and connective tissue surround capsule.	Pike and Burt (1983)
Cryptocotyle concavum	*Gobius melanostomus* (skin, muscles)	Reduction of total lipids in muscle and liver may contribute to mortality in winter.	Shchepkina (1981)
C. lingua	Many marine fish (skin)	Black spot disease; mortality in young fish	Sindermann and Rosenfield (1954)
Galactosomum sp.	*Engraulis japonica, Seriola quinqueradiata* (brain)	Fatal whirling disease	Kimura and Endo (1979)
Haplorchis pumilio (Figures 5.23.1–4)	*Sarotherodon* spp. fry (connective tissue associated with skeletal structures)	Focal haemorrhage in skeletal muscle caused by migrating cercariae; mortality in heavy infections.	Sommerville (1982)
Clinostomidae (metacercariae)			
Clinostomum complanatum	*Misgurnus anguillicaudatus*	Metacercariae migrating from body wall cause congestion, haemorrhage and tissue damage.	Lo, Chen and Wang (1985)
Euclinostomum heterostomum	*Channa punctata*	Decrease in total free amino acids; slight increase in total protein; lower alanine aminotransferase activity in liver.	Gupta and Agarwal (1984)
,,	,,	Changes in free amino acid composition in liver tissue.	Bhagavathiammai and Ramalingam (1983)

injury to host tissues. Bullock (1961) found that acanthocephalans with a small proboscis caused little damage (see above) but that species with large proboscides, eg. *Pomphorhynchus bulbocolli* or *Acanthocephalus* sp., penetrated to or beyond the submucosa, provoking a marked connective tissue reaction.

Most reports of the pathology caused by armed adult cestodes relate to *Triaenophorus*, a genus significant in the context of fish farming. *T. nodulosus* and *T. crassus* both caused destruction of the intestinal mucosa and blood vessels of pike and stimulated a marked inflammatory response, but *T. nodulosus*, which has small hooks and an annual life-cycle, was less pathogenic than *T. crassus*, which has large hooks and remains in the intestine for longer periods (Davydov, 1977). The tetraphyllidean *Acanthobothrium coronatum*, which eroded the epithelium of the spiral valve of *Scyliorhinus stellaris* to expose the stratum compactum, was rather more pathogenic than the diphyllidean *Echinobothrium brachysoma*, which produced local flattening and tearing of the epithelium (Rees, 1961a; Rees and Williams, 1965).

In infections with unarmed cestodes, damage is usually restricted to the point of attachment but may be more extensive, particularly in heavy infections. *Eubothrium rugosum* caused desquamation of the caecal epithelium of *Lota lota* when present in large numbers (Davydov, 1977). The bothria of *Bothriocephalus acheilognathi* caused focal pressure necrosis,

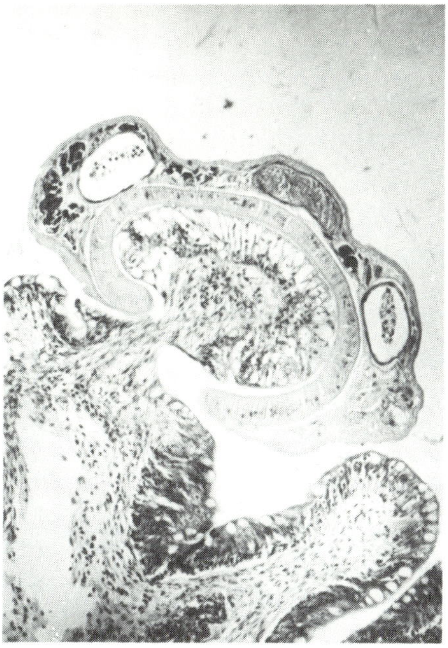

Figure 5.15. Transverse section of an adult *Podocotyle atomon* attached by its ventral sucker to the intestinal mucosa of *Platichthys flesus*
(Courtesy of David Gibson).

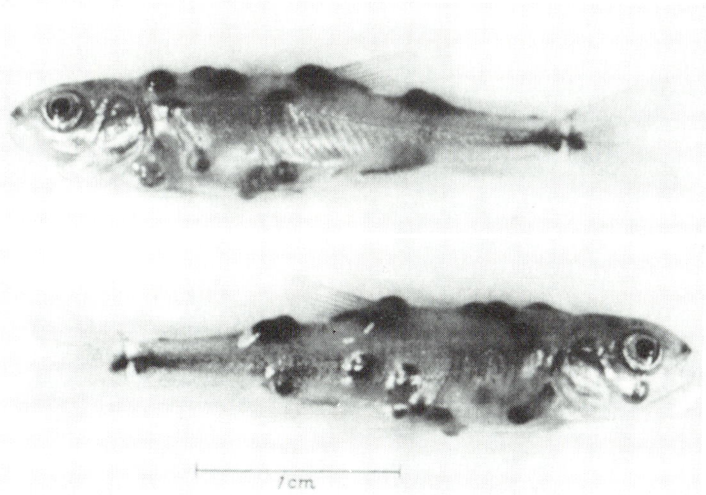

Figure 5.16. The skin and underlying tissues of *Phoxinus laevis* infected with the larvae of *Posthodiplostomum cuticola*.
From Donges (1964).

increased mucus production, lymphocyte infiltration and haemorrhage (Scott and Grizzle, 1979) and, in heavy infections in *Cyprinus carpio*, provoked lysis of large areas of the mucosa with necrotic changes (Davydov, 1977). Inflammation of the gut and thinning of the intestinal wall attributed to increased acid phosphatase activity, have also been described for this species (Pár, 1978; Svobodova, 1978; Sekretaryuk, 1983a). Rees (1958) and other authors (see Table 5.2) found that the scolex of *B. scorpii* caused only local erosion of the mucosa and proliferation of fibrous tissue at the site of attachment in

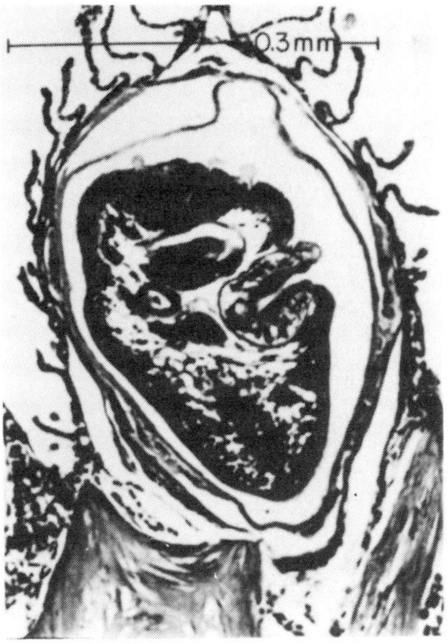

5.17

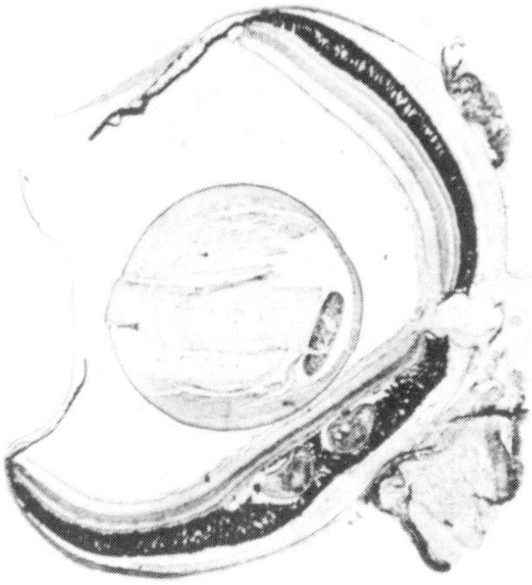

5.18

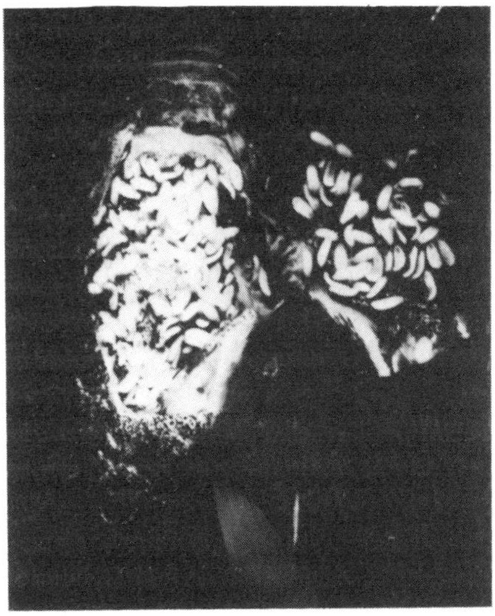

5.19

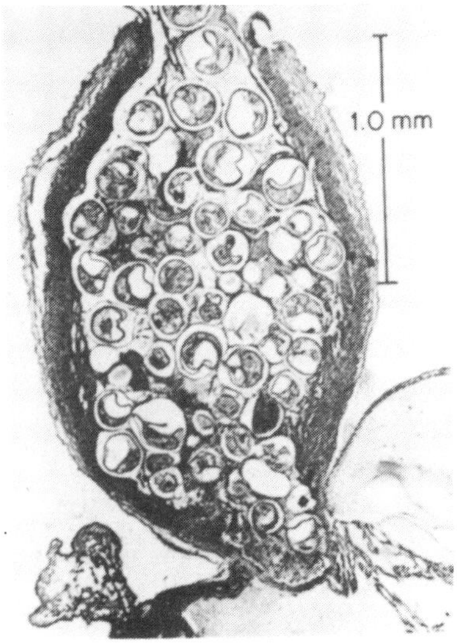

5.20

Figure 5.17 Larval digenetic trematode in a cyst in the gill of a flyingfish. From Cosgrove (1975).

Figure 5.18 Section through an eye of an infected stickleback showing two macercariae in the pigment layer and one in the lens. From Berrie (1960).

Figure 5.19 Fish, *Orestias* sp., with deformed head due to *Diplostomulum* sp. Roof of cranium, opened to show parasites. From Hoffman (1975a).

Figure 5.20 Mass of encysted larval trematodes (*Ascocotyle* sp.) in the bulbus arteriosis of the heart of a green molly. From Cosgrove (1975).

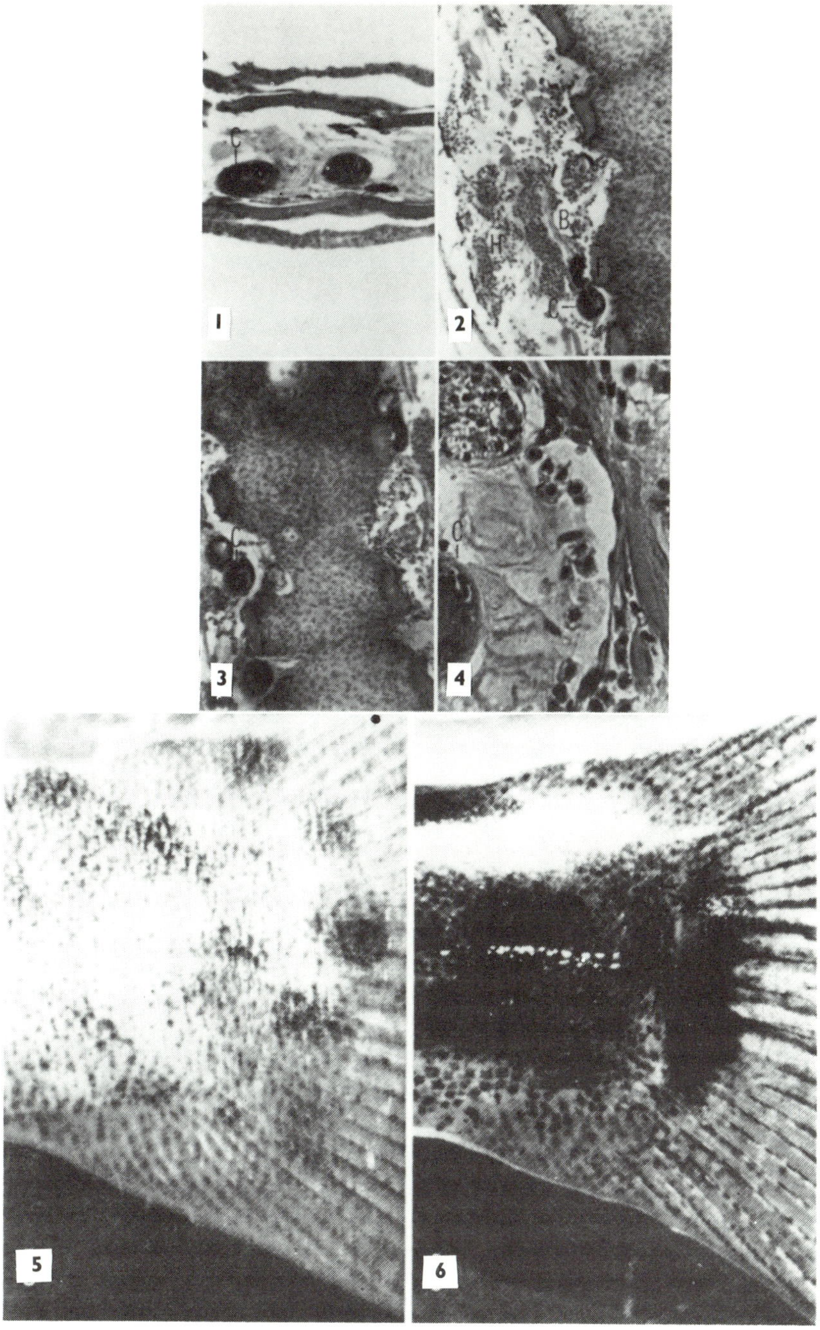

Figures 5.21.1 to 5.21.6. The effects of the cercariae of *Nanophyetus Salmincola* on cutthroat trout. 1. Cross section through the caudal fin of a Lahontan cutthroat trout, showing two cercariae (C) inside a fin ray. × 58.7. 2. Cross section through the base of the caudal fin of a Lahontan cutthroat trout 52 mm long exposed to 500 cercariae of *N. salmincola*, showing hemorrhagic area (H) in loose connective tissue and a cercaria (C) near a fin ray (F) and a blood vessel (B). × 58.7. 3. Cross section through the base of the caudal fin of a Lahontan cutthroat trout, showing one or two cercariae (C) of *N. salmincola* inside a blood vessel and two other cercariae in the subcutaneous tissue. × 58.7. 4. Higher magnification of Figure 3, showing cercariae (C) in a blood vessel. × 253. 5. Caudal region of an uninfected Lahontan cutthroat trout 45 mm long. × 3.5. 6. Photograph of caudal region of a Lahontan cutthroat trout exposed to 700 cercariae of *Nanophyetus salmincola* 24 hr before the picture was taken, showing large hemorrhagic area. The entire dark area shown was red in color. The chromatophore pigment was dispersed, but most of the darkness was due to hemorrhage in the dermis. × 3.5. From Baldwin, Millemann and Knapp (1967).

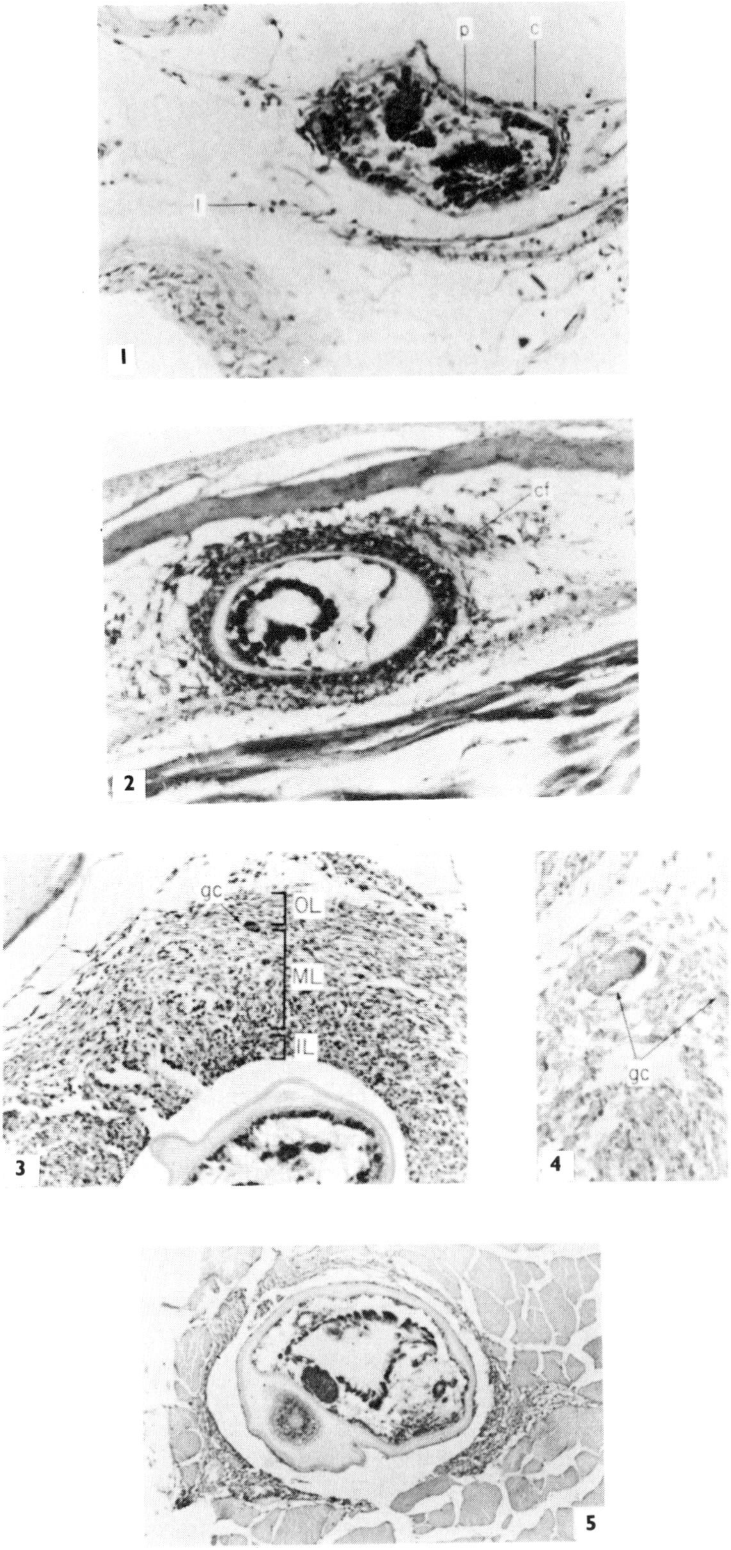

5.22

Scophthalmus maximus but MacKinnon and Featherstone (1982) reported the formation of a collagenous collar around the scolex in *Pseudophycis bacchus*. Williams (1960b) described discoloration of the intestinal wall of *Gadus callarias* infected with *Abothrium gadi* with fibrosis in the mucosa and deeper layers. *Parabothrium gadipollachii* had a similar effect in *G. pollachius* and, if attached in the pyloric caeca, sometimes broke through the caecal wall to protrude into the body cavity.

The caryophyllideans include some highly pathogenic species which have attracted attention because they can present a serious problem in farmed fish. The degree of pathology is closely related to the structure of the scolex (Figures 5.2.1–6). The most pathogenic species are those which lack specialized attachment organs, ie. have only terminal introverts, feeble loculi or have no such organs (eg., *Atractolytocestus huronensis, Hunterella nodulosa, Khawia iowensis* and *Caryophyllaeus laticeps*), while those which have loculi, bothria or acetabula have less effect (Mackiewicz, Cosgrove and Gude, 1972). Examples of species with loculi include *Spartoides wardi, Isoglaridacris folius, Glaridacris catostomi* and *Biacetabulum biloculoides*. *Monobothrium hunteri* and *M. ulmeri* have a terminal introvert, *M. ingens* has an introvert and shallow bothria, *G. confusus* and *Capingens singularis* have well-developed bothria and accessory loculi, and *Biacetabulum infrequens* and *B. carpiodi* have acetabula. Of the species lacking specialized attachment organs, the scolices of the first two are deeply buried in the mucosa. *A. huronensis, K. iowensis* and *Caryophyllaeus laticeps*, while causing mechanical displacement and loss or compression of adjacent tissue, do not provoke ulceration or inflammation. *H. nodulosa* (Figure 5.3), however, produces very large nodules with chronic inflammation; the epithelium and lamina propria are lost and there is extensive lymphocyte invasion and hyperplasia of the submucosa (Mackiewicz and McCrae, 1962; Mackiewicz, Cosgrove and Gude, 1972; Hayunga, 1979). *B. biloculoides, M. hunteri, M. ulmeri* and *M. ingens* provoke nodule formation with a pronounced inflammatory reaction and necrotic debris; *M. hunteri* penetrates to the lamina propria and *M. ulmeri* to the stratum compactum. Of the remaining species, only *Glaridacris catostomi*, in heavy infections, produces lesions. The effect of the other species is limited to local epithelial erosion, without lesions or an inflammatory response. Mackiewicz, Cosgrove and Gude (1972) suggested that a lytic secretion used by the worm to penetrate into the gut wall may provoke the marked host response elicited by some species.

The formation of nodules and diverticula and denudation of the mucosa have also been reported for *Djombangia* spp. and *Lytocestus indicus* in *Clarias batrachus* (Satpute and Agarwal, 1974; Ahmed and Sanaullah, 1979). *Lytocestoides fossilis* in *Heteropneustes fossilis* and *Lytocestus indicus* in *Clarias batrachus* provoked hyperplasia and hypertrophy of the villi, mucosal degeneration, inflammation and fibrosis (Kanth and Srivastava, 1984; Sircar and Sinha, 1980) and *Khawia sinensis* caused inflammation in the gut of carp (Musselius *et al.*, 1963).

The spathebothriid *Cyathocephalus truncatus* distended the pyloric caeca of trout, which became perforated at the tips, allowing worms to escape into the body cavity, often with fatal consequences (Vik, 1958).

Figures 5.22.1–5.22.5 The response of plaice and dab to *Stephanochasmus baccatus*. From Sommerville (1981a). 5.22.1 Plaice. Cyst at 48 h showing early capsule development; c, capsule; l, leucocytes; p, parasites. (H and E, × 320). 5.22.2 Plaice cyst at 11 days showing a well-defined, compact capsule and the appearance of collagen fibres in the outer capsule layers; cf, collagen fibres (MSB, × 200). 5.22.3 Plaice. Section through a capsule at 28 days postinfection showing three layers; the inner layer (IL) consisting of a matrix of necrotic ground substance in which are embedded karyorrhectic nuclei; a middle layer (ML) composed of degenerating epithelioid tissue on the periphery of which occur giant cells; the outermost layer (OL) is comprised largely of fibroblasts and collagen; gc, giant cells (H and E × 175). 5.22.4 Plaice. Section showing giant cells at the periphery of the middle layer of a plaice capsule at 28 days postinfection (H and E, × 286). 5.22.5 Dab. Section through a muscle cyst 55 days postinfection showing the cellular components of the capsule continuous with the fibrocyte replacement tissue in the surrounding muscle (H and E, × 65).

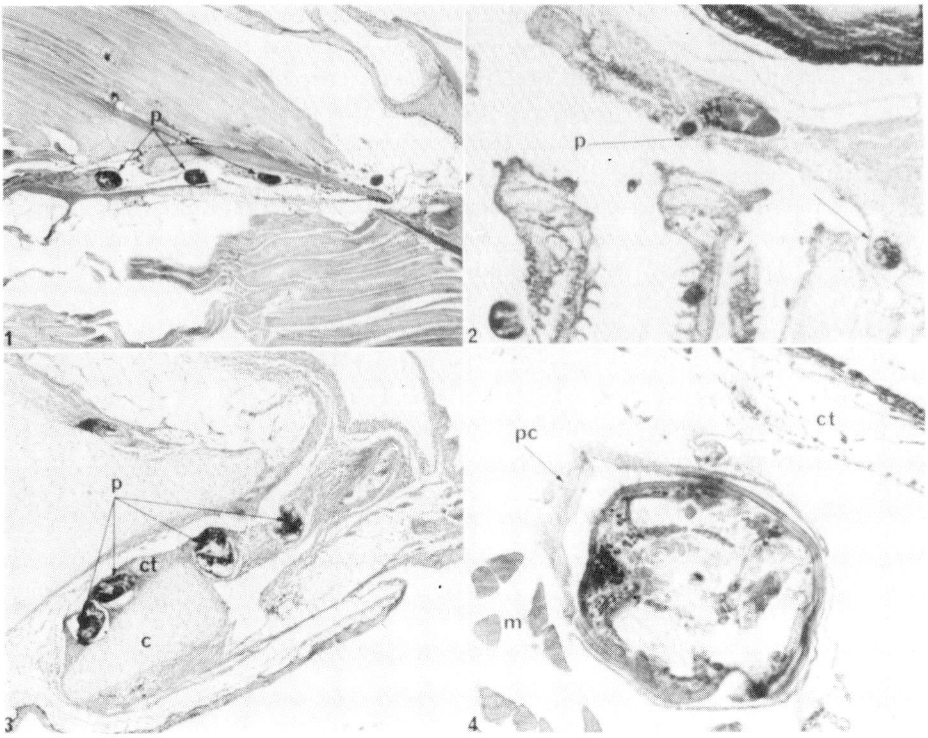

Figures 5.23.1–5.23.4 The pathology of *Haplorchis pumilio* infections in cultured tilapias.
5.23.1 Cercariae migrating through connective tissue of *Sarotherodon galilaea*; p, parasite (H and E, × 5). 5.23.2 Cercariae migrating through the wall of the branchial cavity of *Sarotherodon spilusus*; p, parasite (H and E, × 5). 5.23.3 Mature cysts in the connective tissue binding cartilage in the head of *Sarotherodon spilurus* with consequent thickening of the membrane; p, parasite; c, cartilage; ct, connective tissue (H and E, × 5). 5.23.4 A mature cyst of 3 months from *Sarotherodon spilurus* illustrating the lack of inflammatory response; pc, parasite cyst wall; m, muscle; ct, connective tissue (H and E, × 40).
From Sommerville (1982b).

Nematodes have similar effects. *Spinitectus carolini* larvae in the intestinal lumen of *Lepomis macrochirus* caused aseptic traumatic enteritis but those which penetrated the wall elicited intense tissue reactions (Jilek and Crites, 1982c). Larval *Cucullanus minutus* in the gut wall of plaice and flounder produced similar results (Janiszewska, 1938) but the adults in the gut lumen had relatively little effect. *Pseudanisakis rotundata* and *Proleptus obtusus* caused inflammatory reactions at the point of attachment in elasmobranchs (Williams and Richards, 1968; Schuurmans-Stekhoven and Botman, 1932) and larvae of *Anisakis* and *Contracaecum rigidum* caused tumours in the stomach of *Lophius piscatorius* (Baylis and Jones, 1933; Arai, 1969). *Raphidascaris acus* larvae provoked chronic granulomata and became enclosed in collagenous capsules in the gut wall of various freshwater fish and may also disrupt the function of the alimentary tract (Poole and Dick, 1984; Eiras and Reichenbach-Klinke, 1982). *Acanthocheilus nidifex* in the stomach of *Galeocerda cuvieri* penetrated sufficiently deeply to provoke ulcers and a strong inflammatory response (Linton, 1900).

In addition to the histopathological consequences of infection, parasites of the alimentary canal provoke changes in the blood and metabolism. Clinical signs of infection include reduction in growth rate, inappetance and emaciation, sometimes culminating in mortalities. Infected fish may also exhibit behavioural changes, which may tend to make them more vulnerable to predation, and their resistance to stress may be reduced.

Much of the available information on blood and metabolic changes relates to infections of farmed fish species with caryophyllideans and with pseudophyllidean metacestodes. *Khawia sinensis* in *Cyprinus carpio* and *Lytocestus indicus* in *Clarias batrachus* both reduced haemoglobin levels. *K. sinensis* reduced the number of erythrocytes, monocytes, polymorphs and neutrophils (Sapozhnikov, 1969). The blood changes caused by *L. indicus* included eosinophilia and were suggestive of macrocytic pernicious anaemia (Sircar and Sinha, 1974). A reduction in haemoglobin and packed-cell volume, and eosinophilia were described in *C. batrachus* infected with unidentified caryophyllideans by Kadav and Agarwal (1983a). Reduced haemoglobin and total blood volume, elevated leucocyte and phagocyte counts, and appearance of giant lymphocytes have been reported in *Cyprinus carpio* infected with *Bothriocephalus acheilognathi* (Pár, 1978; Svobodova, 1978; Kudryashova, 1970; Kirichenko and Kosareva, 1972). Opinions vary as to the intensity of infection necessary to induce pathogenic effects: Kudryashova (1970) found the above effects in fish with more than five worms but Svobodova (1978) found no significant effect on various physiological indices in fish harbouring from 1–21 worms and attributed the elevated leucocyte count to inflammation of the gut. According to Pár (1978), fish with 1–29 worms had an elevated leucocyte count and those with more than 15 showed marked damage to the gut. *Eubothrium salvelini* caused blood changes in *Salvelinus alpinus* similar to those described for *K. sinensis* in *C. carpio* and there was a significant correlation between the degree of haemosiderosis in the spleen and the intensity of infection (Hoffman, Kennedy and Meder, 1986). Blood changes suggestive of anaemia occurred in *Trichiurus lepturus* infested with the tetraphyllidean metacestode *Scolex pleuronectis* (Radhakrishnan *et al.*, 1984).

A reduction in total serum proteins has been found in *Cyprinus carpio* infested with *Bothriocephalus acheilognathi* or with *Khawia sinensis* (Kudryashova, 1970; Sapozhnikov, 1969; Kurovskaya, 1984), and of serum amino acids in *Clarias batrachus* infested with unnamed caryophyllideans (Kadav and Agarwal, 1982). Both *B. acheilognathi* and *K. sinensis* disrupt intestinal and liver enzyme activity in *Cyprinus carpio.* Morbidity and mortality in winter were attributed to changes in alanine and aspartate aminotransferase activity which interfered with protein synthesis (Lozinska-Gabska, 1981). Intestinal trypsin and chymotrypsin activities were lower in carp fry infested with *B. acheilognathi* than in uninfected fish and their growth and development were depressed (Matskási, 1978, 1984). Sekretaryuk (1983) attributed thinning of the intestinal wall to elevated acid phosphatase activity in infested *C. carpio.* Davydov (1978a) found that more than four *B. acheilognathi* in *C. carpio* disrupted carbohydrate and protein metabolism and affected the nutritional status of the host. Kurovskaya (1984) found reduced levels of intestinal alkaline phosphatase but increases in acid phosphatase, amylase and protease, confirming the effect of infestation on the nutritional status of the fish. In *Ctenopharyngodon idella*, even light infestations of *B. acheilognathi* elevated oxygen consumption (Kititsina, 1985).

Reduction in growth rate, weight loss, inappetance and emaciation culminating in some cases in heavy mortalities have been reported for fish infected with various cestodes, nematodes and acanthocephalans. Of the caryophyllideans, *Capingentoides moghei* caused inappetance, sluggishness and weight loss in *Heteropneustes fossilis* (Jain, Pandey and Pandey, 1976). *Khawia sinensis* in *Cyprinus carpio* did not affect growth at levels below 10 per fish (Kapustina, 1978). Infections of 35–45 per fish were fatal (Musselius *et al.*, 1963). Of the pseudophyllideans, *Eubothrium salvelini* depressed the growth rate of *Oncorhynchus nerka* fry (Smith, 1973; Boyce and Clarke, 1983; Boyce, 1979; Smith and Margolis, 1970) and *B. acheilognathi* that of *Cyprinus carpio* (Yashchuk, Sventsitskii and Rudoi, 1978). *B. acheilognathi* also caused a fall in body weight and fat content (Kurovskaya, 1984) and a decrease in kidney, liver and spleen weight (Balakhnin, 1979). Emaciation and subsequent mortality have been reported for infections of the intestinal capillarids *Capillaria brevispicula* in *Puntius tetrazona* and of *C. pterophylli* in *Cichlasoma octofasciatum*, an aquarium fish (Moravec, Ergens and Repová, 1984;

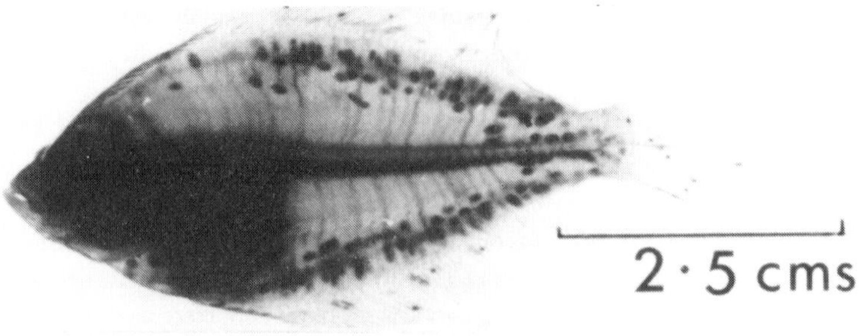

1

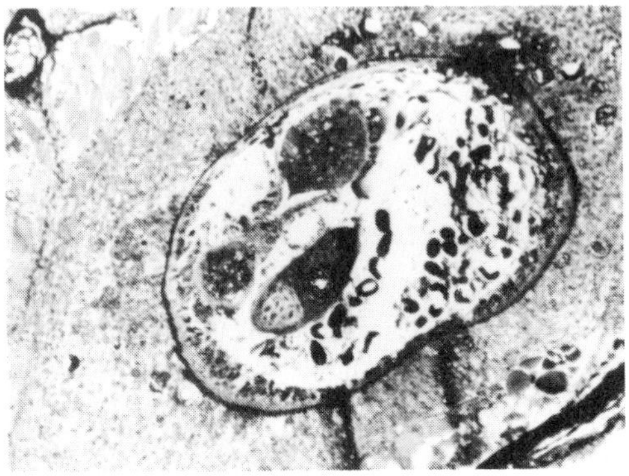

2

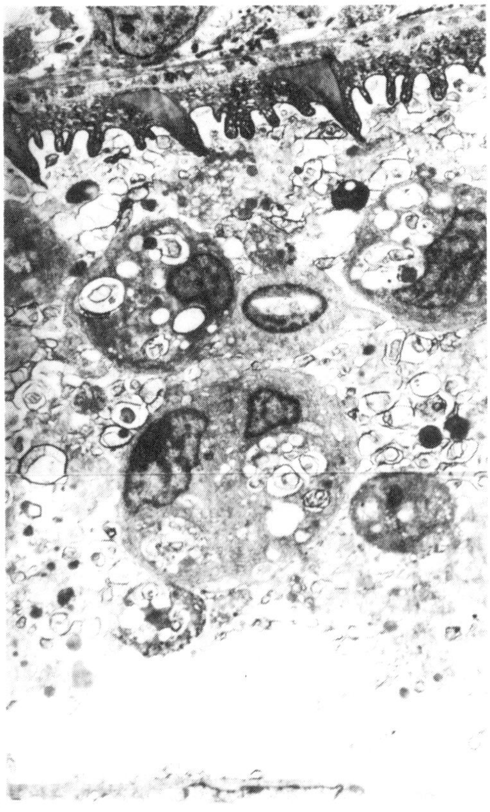

3

5.24

Moravec, 1983b). The life cycle of *Capillaria pterophylli* is direct and it therefore poses a potential problem in aquarium fish. The salmonid hosts of the acanthocephalans *Neoechinorhynchus rutili* and *Echinorhynchus truttae* were up to 17% underweight (Mann, 1971). *E. salmonis* retarded the growth of *Coregonus lavaretus ludoga* (Bauer and Nikolskaya, 1957), caused emaciation in *Salmo gairdneri*, producing up to 80 per cent mortality in the latter species, maximal in young fish (Bertocchi and Francálanci, 1963). *Pomphorhynchus laevis* has been reported to cause emaciation in *Leuciscus cephalus* by Chubb (1965) and weight loss and mortality in *Barbus* sp. by Wurmbach (1937). In contrast, Kennedy (1967) and Hine and Kennedy (1974a) concluded that it did not affect the growth rate or cause mortality in *L. cephalus* or *B. barbus* and was not a serious pathogen in British waters.

The gall-bladder

There are very few reports of pathology caused by helminths in this site. Wabuke-Bunoti (1980) reported nodular outgrowths in the gall-bladder of *Clarias mossambicus* infested with *Polyonchobothrium clarias*. The worms were most common in the gall-bladder but also occurred throughout the gut although their numbers decreased towards the rectum.

Opisthorchis pedicellata in the gall-bladder of *Rita rita* reduced levels of protein, cholesterol, calcium, glucose and alkaline phosphatase in the blood. All of these, plus the level of glycogen, were reduced in the liver and the hepatosomatic index fell. The changes were attributed to the reduction in the availability of bile. Lactic acid values in the blood and liver increased (Joshi, 1979).

Viscera, skeletal musculature and the peritoneal cavity

The majority of species which occupy these sites in fish are larval digeneans, metacestodes and larval nematodes, parasitic as adults in piscivorous birds or mammals, or tissue-dwelling philometrid nematodes. Some species may occupy two or more habitats, eg. plerocercoids of *Diphyllobothrium latum* localize in the liver and musculature, those of *Triaenophorus amurensis* in the liver, muscle or serosa, and metacercariae of *Nanophyetus salmincola* inhabit all tissues but especially the kidneys, muscles, fins and gills. For this reason, each habitat cannot always be considered in isolation.

Liver

Pseudophyllidean metacestodes of the genera *Diphyllobothrium* and *Triaenophorus* are among the most important parasites of the liver, in which they become encapsulated by host connective tissue. Pathology and host reaction may vary in different species. Liver necrosis with oedema occurred in *Salmo clarki* infected with *D. cordiceps* plerocercoids (Otto and Heckman, 1984) but Pronina (1979) found only slight parenchymal hyperaemia and perivascular leucocytic invasion in *Leuciscus amurensis* infected with *T. amurensis*. *T. nodulosus* plerocercoids in several fish species (see Table 5.2) became encapsulated with concurrent destruction of liver tissue (Pronin *et al.*, 1976; Davydov, 1981; Stromberg and Crites,

---

Figures 5.24.1–5.24.3 The cellular response of the plaice to *Rhipidocotyle johnstonei* at ultrastructural level.
5.24.1 'O' group plaice with infection of *Rhipidocotyle johnstonei* metacercariae in the interspinal zone at the base of the ventral dorsal and caudal fins. 5.24.2 Section through a capsule containing a metacercaria in the fin musculature of an 'O' group plaice. 5.24.3 Macrophages closely associated with the metacercarial surface.
From Pulsford and Mathews (1984).

Table 5.4 Pathological effects of some nematode parasites of fish (Figures 5.25–5.30)

Parasite	Host	Pathology	Reference
Habronematoidea			
Cystidicolidae			
Cystidicola farionis	*Salmo gairdneri* (swim-bladder)	Aerocystitis	Otto and Körting (1973)
,,	,,	Raised ulcers sometimes surrounded by hyperaemic mucosa and sometimes with hard ochre-coloured material covering a central crater.	Lankester and Smith (1980)
C. stigmatura (= *C. cristivomeri*)	*Salvelinus* (swim-bladder)	Raised ulcers as for *C. farionis* above.	Black (1984b); Lankester and Smith (1980)
Spinitectus sp.	*Lepomis macrochirus* (intestine)	Body spines abrade mucosa causing lesions and inflammation; some nematodes burrowed into mucosa; morbidity.	Hoffman (1975a)
S. carolini	,,	Larvae in intestinal lumen cause aseptic traumatic enteritis; those in submucosa cause simple infectious enteritis with inflammatory infiltrations. The most intense tissue reactions were provoked by larvae which completely penetrated the gut wall.	Jilek and Crites (1982c)
Seuratoidea			
Cucullanidae			
Truttaedacnitis truttae	*Salmo gairdneri* (caeca)	No significant effect on growth rate, food consumption and food conversion efficiency; no effect on swimming ability.	Russell (1980)
,,	,,	Loss of epithelium at site of attachment, mucosal hyperplasia and haemorrhage, fibrosis in lamina propria, all attributed to feeding activity.	Dunn, Russell and Adams (1983)
Cucullanus minutus	*Pleuronectes platessa*, *Platichthys flesus* (intestine)	Larvae in gut wall provoke swelling in sub-mucosa with localized necrosis, haemorrhage, inflammation and hyaline degeneration in the serosa.	Janiszewska (1938)
Dracunculoidea			
Philometridae			
Philometra sp.	*Chrysophrys auratus* (gonads)	Swelling; lesions composed of fibrous inflammatory tissue around worms; later granulomata composed of epithelioid and fibrous layers with macrophages surrounded by loose tissue containing necrotic foci.	Hine and Anderson (1982)
Philometra sp. (Figures 5.26. 1–3)	*Canthigaster jactator* (body cavity)	Protrusion of abdomen	Deardorff and Stanton (1983)
P. obturans (Figures 5.27. 1–2)	*Esox lucius* (blood system)	Females in bulbus arteriosus, ventral aorta and gill arteries obstruct the circulation, feed on blood and perforate gill arteries to release larvae. Parietal thrombi formed in bulbus, endothelial hypertrophy in ventral aorta. Highly pathogenic.	Moravec and Dyková (1978)

Table 5.4 continued

Parasite	Host	Pathology	Reference
P. rubra larvae	*Morone saxatilis* (body-cavity)	Peritonitis, with serosal adhesions	Paperna and Zwerner (1976)
P. saltatrix	*Pomatomus saltatrix* (heart, pericardial cavity)	Inflammation, blood stasis and necrosis	Cheung, Nigrelli and Ruggieri (1984)
Philometroides huronensis	*Catostomus commersoni* (skin)	Mechanical disruption and compaction of subepidermal tissue by migrating females; enclosure of gravid females in fibrous capsule; disruption of skin and capsule during release of larvae, with acute local inflammatory response.	Sekretaryuk (1980)
P. lusiana	*Cyprinus carpio* (skin and viscera)	Elevated activities of aspartate and alanine aminotransferases in serum, muscle, skin, hepatopancreas and brain, of alanine aminotransferase in kidney and of aldolase in serum, muscle, kidney, hepatopancreas and brain.	Sekretaryuk (1980)
P. lusiana larvae	,,	Reduction of intestinal lumen by thickened areas formed in mucosa around larvae; some damage to various organs by migrating larvae but no severe functional disturbance; adult males in swim bladder had only slight effects.	Sekretaryuk (1983)
,,	,,	Larvae disrupt functioning of swim-bladder; fish lose equilibrium and may starve.	Bauer, Musselius and Strelkov (1973)
Philonema agubernaculum	*Salmo salar, Salvelinus fontinalis* (body-cavity and viscera)	Multiple mesenteric or serosal adhesions preventing spawning or stripping.	Meyer (1960)
Dioctophymatoidea			
Dioctophymatidae			
Eustrongylides sp. larvae (with *Hedruris spinigera*: Habronematoidea)	Trout	Adverse effect on growth rate, especially in spawning season.	Stockell (1936)
Eustrongylides	*Clarias mossambicus, Haplochromis* spp., *Bagrus docmac* (mesentery, spleen, muscles, gonads)	Fatty changes in spleen; muscles around unencysted worms digested and infiltrated with fibroblasts; gonads deformed and degenerate.	Paperna (1974)
Trichinelloidea			
Trichuridae			
Capillaria (= *Pseudocapillaria*) *brevispicula*	*Puntius tetrazona* (intestine)	Mortality	Moravec, Ergens and Řepová (1984)
Capillaria pterophylli	*Cichlasoma octofasciatum* (intestine)	Inappetance, emaciation, mortality.	Moravec (1983)
Physalopteroidea			
Physalopteridae			
Proleptus obtusus	*Scyliorhinus caniculus* (intestine)	Inflammation, lymphocyte infiltration, haemorrhage at point of attachment.	Schuurmans-Stekhoven and Botman (1932)

Table 5.4 continued

Parasite	Host	Pathology	Reference
Ascaridoidea			
Acanthocheilidae			
Acanthocheilus nidifex	*Galeocerda cuvieri* (stomach)	Worms penetrated submucosa and formed crypts surrounded by inflamed, haemorrhagic tissue, with hyperplasia of connective tissue.	Linton (1900)
Pseudanisakis rotundata	*Raja radiata* (intestine)	Head penetrates mucosa provoking inflammatory response in sub-mucosa.	Williams and Richards (1968)
Raphidascaris acus larvae	*Perca flavescens* (liver)	Distortion or destruction of blood vessels by migrating larvae; formation of collagenous capsules around worms.	Poole and Dick (1984)
R. acus larvae	*Salmo trutta fario* (intestine)	Chronic granulomata on or in intestinal wall around larvae; 19–25 nodules per fish.	Eiras and Reichenbach-Klinke (1982)
Anisakidae			
Anisakis sp. larvae (Figure 5.29)	*Clupea harengus pallasi* (viscera, serosa)	Mechanical compression of pancreas and liver; granulomatous inflammation and necrosis of liver; trauma to muscularis externa of pyloric caeca; exudates in lesions contained histiocytes and lymphocytes.	Hauck and May (1977)
Anisakis sp. larvae	*Ophiodon elongatus* (stomach)	Larvae penetrate to serosal layer, provoke ulcers; host emaciated.	Arai (1969)
Anisakis sp. and *Contracaecum* sp. larvae	*Merlangius merlangus* (liver)	*Contracaecum* larvae found inside liver parenchyma, *Anisakis* larvae under liver capsule. Both provoke cellular response characterized by neutrophils, macrophages and proliferating fibroblasts. Larvae encapsulated, melanin deposited around capsules.	Elarifi (1982)
C. aduncum larvae	*Engraulis encrasicholus ponticus*	Decrease in total lipids, particularly triglycerides, and non-esterified fatty acids.	Shchepkina (1978, 1980)
C. rigidum	*Lophius piscatorius* (stomach)	Worms found in stomach tumours	Baylis and Jones (1933)
Contracaecum sp. larvae	*Clupea harengus* larvae (peritoneal cavity)	Distention of abdomen, external compression and eventual permanent distortion of gut, cessation of peristalsis and feeding, with death from damage caused by activity of larva.	Rosenthal (1967)
Lappetascaris lutjani	*Hilsa ilisha* (intestine)	Worms embedded in intestinal wall, causing erosion of mucosa and haemorrhage	Rasheed (1965)
Thynnascaris sp. larvae	Lizard fish (liver)	Pathology related to numbers present. In heavy infections (>5 larvae) liver decreased in size, became darker and necrosis occurred in liver cells, blood vessels and bile-ducts.	Awadalla *et al.* (1982)

5.25 *Parophrys vetulus* (English sole) with 'blisters' containing *Philometra americana* in the fins. From Margolis (1970a).

1974c). An acute inflammatory response was described in *Perca fluviatilis* and *Morone chrysops* (Pronina, 1977; Stromberg and Crites, 1974c). In addition, Pronina (1977) reported destruction of hepatocytes and thickening of capillary walls, while squamous metaplasia, fibrosis and displacement of liver tissue were observed by Stromberg and Crites (1974c). In *Salvelinus alpinus*, liver damage was associated with elevated plasma levels of aspartate and alanine aminotransferases (Scheinert, 1984). Reductions in total lipids, phospholipids and triglycerides occurred in the liver of *Lota lota* infected with *D. latum* and *T. nodulosus*, and in that of *Gasterosteus aculeatus* infected with *D. vogeli* (Gur'yanova and Sidorov, 1985). Plerocercoids also compressed or displaced the visceral organs, eg. in *D. cordiceps* infection in *Salmo clarki* (Otto and Heckman, 1984).

Weight loss and mortality, particularly in young fish, are serious consequences of infection with pseudophyllidean plerocercoids in several farmed fish species, eg. *Digramma interrupta* infection in *Carassius* spp. In their third year, 90 per cent of *C. auratus* infected as fry had died and total loss from deficient weight and mortality were equivalent to over 40 per cent of the annual catch in fisheries in the Tyumen area of the USSR (Lyubina, 1968; Razmashkin and Shirshov, 1983). *Triaenophorus* sp. plerocercoids reduced the weight of *Salmo gairdneri* by about 6 per cent, about half the fish examined over three years were infected, each with from 1–40 plerocercoids (Mirle *et al.*, 1985).

Of the digenean larvae which occur in the liver and other visceral organs, metacercariae of *Posthodiplostomum minimum* in *Lepomis macrochirus* contributed to severe melanosis of the liver and visceral fibrosis (Mitchell, Ginal and Bailey, 1983). It also reduced growth and contributed to mortality of fingerlings (Smitherman, 1968); elevated β and τ-globulin serum levels, but decreased albumin levels were reported by Meade and Harvey (1969). *P. minimum* caused the death of heavily-infected *Pimephales promelas* (Hoffman, 1958). Clinostome metacercariae also occur in the liver, among other locations (see below).

Larval ascaridoid nematodes of the genera *Anisakis, Terranova, Thynnascaris* and *Contracaecum* are the most serious pathogens of the liver of marine fish and have received the most attention. Margolis (1970a) has reviewed this group. They usually become encapsulated in connective tissue, the thickness of the capsule depending in part on the age of the infection and the relative abundance of connective tissue in the site of infection. Histopathologically, worms which penetrate into the liver parenchyma cause more damage

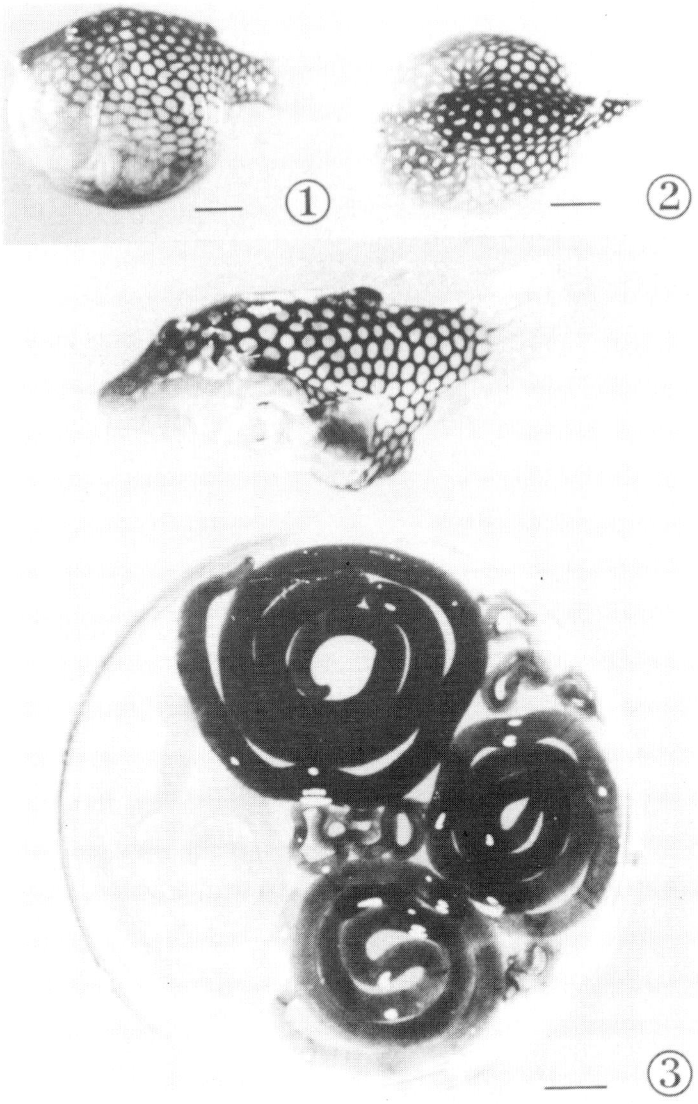

Figures 5.26.1–5.26.3 *Canthigaster jactator* infected with *Philometra* sp. in its body cavity. Bars = 1 cm. 5.26.1 Lateral view of host showing distended abdomen. 5.26.2 Dorsal view of host. Note asymmetry of the expanded abdomen. 5.26.3 Nematodes removed from the same host.
From Deardorff and Stanton (1983).

than those which remain at its surface (Petrushevskii and Shul'man, 1955). The majority of *Anisakis* larvae remain under the connective tissue capsule but those which penetrate more deeply destroy the parenchyma in their vicinity. In *Gadus morhua, Melanogrammus aeglefinus, Myoxocephalus scorpius* and *Sebastes marinus, Anisakis* larvae destroyed liver cells, blood vessels and bile-ducts, becoming encapsulated within the parenchyma (Kahl, 1938; Mikhailova, Prazdnikov and Prusevich, 1964). Elarifi (1982) reported that both *Anisakis* larvae under the liver capsule and *Contracaecum* larvae in the liver parenchyma of *Merlangius merlangus* produced a similar cellular response, characterized by invasion by neutrophils, macrophages and fibroblasts, and melanin deposition around the connective tissue capsules. *Terranova* larvae, although often found encapsulated, are more active than *Anisakis* larvae

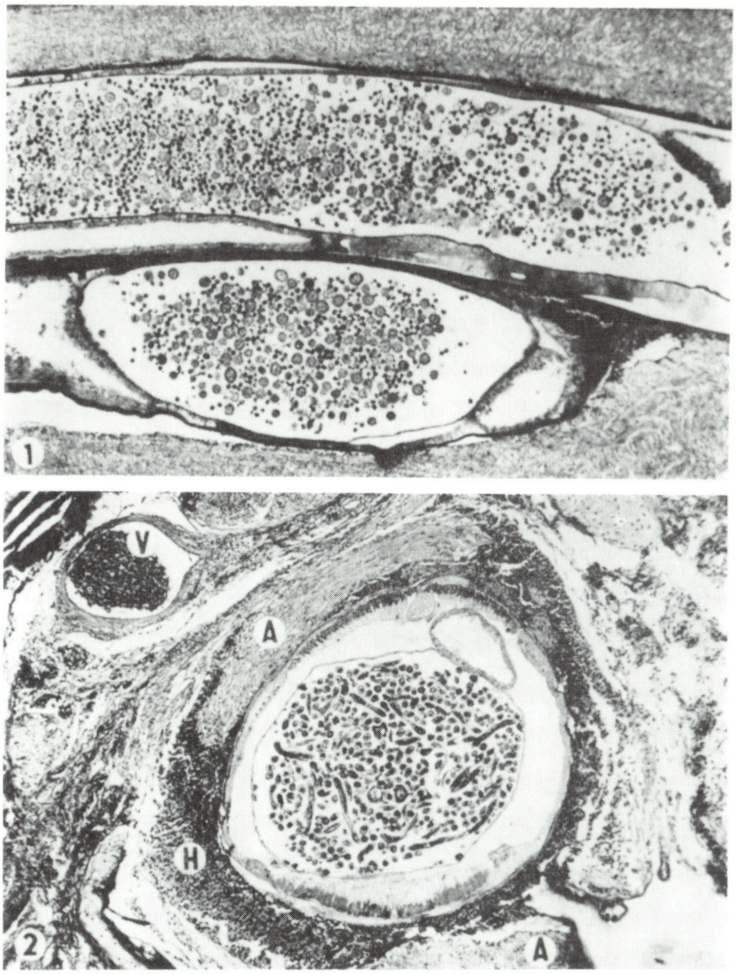

Figures 5.27.1–5.27.2 *Philometra obturans* in *Esox lucius.*
5.27.1 Longitudinal section of the bulbus arteriosus of the heart containing a female worm. 5.27.2 Rupture of the gill artery caused by the female worm. Remnants of the gill artery (A), haemorrhage (H) around the gill artery, with a cross section through the worm; gill vein (V).
From Moravec and Dykova (1978).

and migrate through liver tissue, destroying liver cells and blood vessels, eg. *T. decipiens* in *Myoxocephalus scorpius* (Mikhailova, Prazdnikov and Prusevich, 1964).

Russian workers (reviewed by Margolis, 1970a) have concluded that in gadoids infected with *Contracaecum aduncum* larvae the absolute size and weight of the liver, its weight relative to that of the body, liver fat content, total body weight and condition factor all decrease with increasing intensity of liver infestation. However, Margolis (1970a) has pointed out that the degree of pathology in this species is not directly or constantly related to the intensity of infestation but that other factors should be considered. Large worms, for example, have a greater effect on the host than small worms. Many studies do not provide sufficient information on factors which may have a bearing on pathogenicity. Awadalla, Mansour, Khalil and Guirgis (1982) considered that in lizard fish infested with *Thynnascaris* sp., pathology was related to the number of larvae present. In heavy infestations (more than five larvae per fish), the size of the liver decreased, with necrosis of the liver cells, blood-

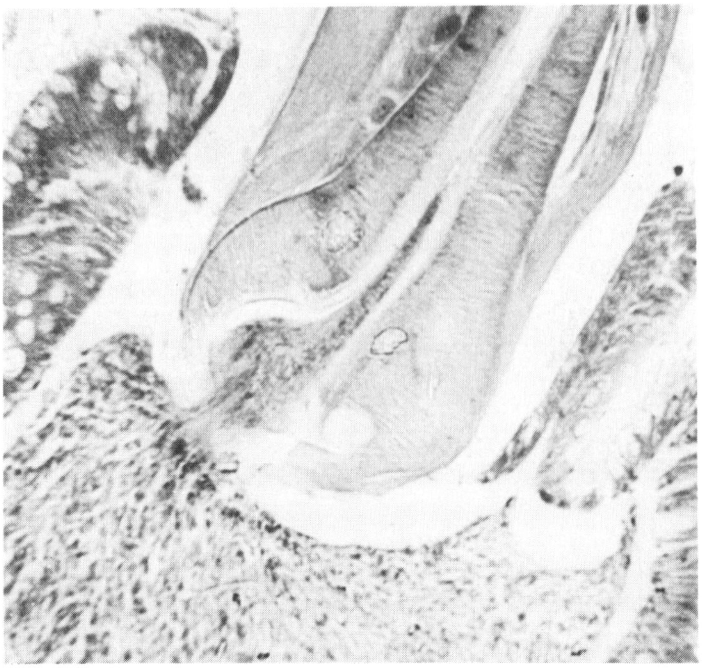

Figure 5.28 Section of *Cucullanus heterochrous* attached to and feeding upon the rectal wall of the flounder. From Mackenzie and Gibson (1970).
(See also Chapter 4, Figure 4.30, for *Pseudanisakis tricupola* which is more deeply embedded in the gut wall of *Raja radiata*.)

vessels and bile-ducts and a darkening in colour of the liver. Infested livers often show a change in colour from pink or pale yellow to reddish-brown with signs of hyperaemia, eg., *C. aduncum* infestations in gadids and *Anisakis* sp. in various hosts (in Margolis, 1970a).

The fat content of gadoid livers infested with *C. aduncum* may drop from about 57 per cent in uninfested fish to 14.5 per cent in severely-affected fish, with a consequent loss of oil yield (Margolis, 1970a). Shchepkina (1978, 1980) found a decrease in total lipids, particularly triglycerides and non-esterified fatty acids, in the liver of *Engraulis encrasicholus ponticus* infested with *C. aduncum* larvae.

Peritoneal cavity

Two of the best-known parasites of the peritoneal cavity of freshwater fish are the plerocercoids of *Ligula intestinalis* and *Schistocephalus solidus*, adults of which parasitize piscivorous birds. Both species induce distention of the body and reductions in liver weight and in packed cell volume of erythrocytes (Arme and Owen, 1967, 1968). *L. intestinalis* stunted growth in *Abramis brama* resulting in scale damage (Safonov, 1976) and in *Notropis hudsonius*, infected specimens of which were sterile (Mahon, 1976). Sweeting (1976) found no reduction in the growth rate of *Rutilus rutilus* but most infected fish failed to reach sexual maturity. They disappeared from the population as a result of predation by *Esox lucius*. In *A. brama* the weight of the parietal body wall relative to total body weight fell and the decrease was attributed to a reduction in muscle weight and of the number of muscle fibres in infested fish (Richards and Arme, 1981a). No reduction occurred in the tail muscles and no increase in girth was noted in this case. Soutter, Walkey and Arme (1980) found that amino acid levels in infested *Rutilus rutilus* were consistent with those characteristic of starvation.

Figure 5.29 Portion of liver from a gadoid fish heavily infected with encapsulated *Anisakis* larvae. From Smith and Wootten (1978a).

In 2-year-old *Ctenopharyngodon idella*, even a single specimen of *Ligula* sp. significantly reduced the respiration rate, suggesting that this host-parasite system has undergone relatively little adaptation (Kititsina, 1985).

Disruption of the reproductive process is one of the major consequences of infestation with plerocercoids of these species (Arme and Owen, 1967, 1968; Owen and Arme, 1965). *Ligula* infestation in most host species results in severe regression of the gonads associated with cytological changes in the meso-adenohypophysis of the pituitary gland which could be induced in previously uninfested fish surgically implanted with plerocercoids. In *Gasterosteus aculeatus* infested with *Schistocephalus solidus*, delayed oocyte maturation was not associated with changes in the pituitary, which appeared to function normally (Arme and Owen, 1967). Heavily infested fish did not spawn and it was postulated that the delay in maturation of oocytes and the reduction in packed-cell volume and liver weight were symptomatic of the metabolic drain imposed on the resources of the host by the parasite. Arme and Owen (1967) noted that spermatogenesis appeared to proceed normally in infected male *G. aculeatus* and McPhail and Peacock (1983) found no difference in prevalence between breeding and non-breeding males. Over 40 per cent of mature but non-gravid females were infested but less than 5 per cent of gravid females, and the major impact of infestation was thought to be directed at postreproductive fish, a strategy regarded as an evolutionary adaptation.

The mechanism by which ligulid plerocercoids inhibit gonad development in their hosts is not clear. It appears to result from suppression of presumed gonadotrophin-producing cells in the pituitary gland and it has been postulated that the plerocercoids produce a sex steroid. However, Arme, Griffiths and Sumpter (1982) failed to detect sex steroids in *Ligula intestinalis* plerocercoids and suggested several alternative explanations. The metacestode might produce a fish-specific non-steroidal antigonadotrophin or a substance which acts directly on the gonad rather than the pituitary, or metabolizes or neutralizes fish hormones. Alternatively, the stress reaction to infestation might generate high levels of plasma corticosteroids which disturb the hypothalamus-pituitary-gonadal axis of the host. Finally, the metacestode might produce serotonin. In mammals, this reacts with the endocrine system and regulates the secretion of pituitary hormones, including gonadotrophin. A similar mechanism might operate in fish.

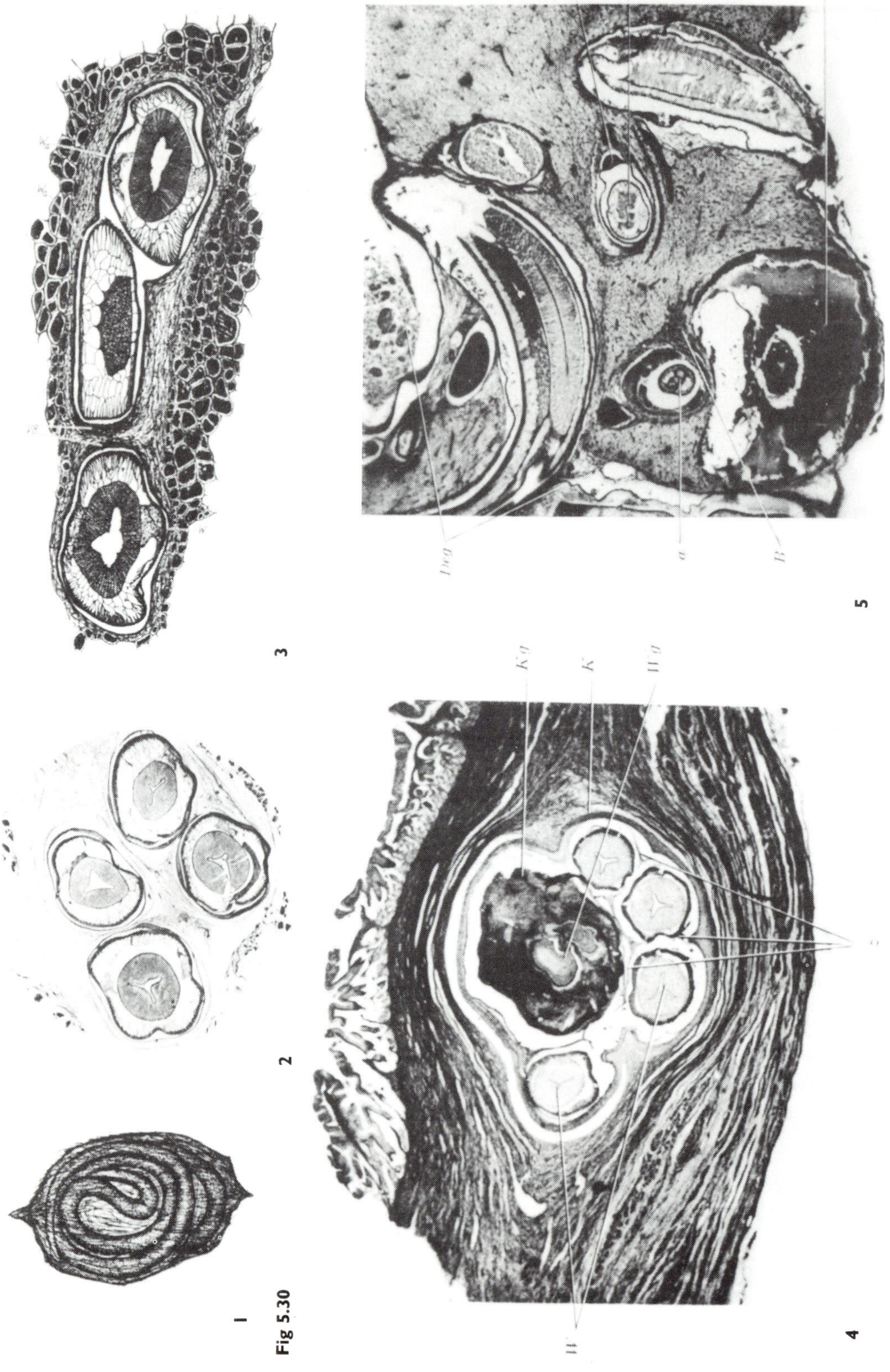

Fig 5.30

1 2 3 4 5

Another pseudophyllidean metacestode which, although localized in the peritoneal cavity, affected the reproductive processes of the host, is that of *Diphyllobothrium latum*. In *Lota lota, Acerina cernua* and *Esox lucius*, it provoked non-specific reactions which included partial resorption of the eggs (Davydov, 1978b).

Contracaecum sp. larvae in the peritoneal cavity of larval *Clupea harengus* distended the abdomen and coiled around the gut, causing irreversible distortions and compression which prevented feeding and defecation. In addition, the larvae might have fed on the tissues and fluids of the host, which died after about 11 days (Rosenthal, 1967).

Swim-bladder

The most common parasites of the swim-bladder are cystidicolid and philometrid nematodes, of which the cystidicolids are the more pathogenic. Otto and Körting (1973) recorded aerocystitis in *Salmo gairdneri* infected with *Cystidicola farionis*. Both *C. farionis* in *S. gairdneri* and *C. stigmatura* in *Salvelinus namaycush* provoked raised ulcers, sometimes surrounded by hyperaemic mucosa and with a hard plug over a central crater. These effects occurred only when unusually large numbers of worms were present and were attributable to accumulation of worms in order to breed (Lankester and Smith, 1980).

Infections of *Channa punctatus* with the isoparorchid trematode *Isoparorchis hypselobagri* produced a variety of pathological effects including fin necrosis, scale loss, blood changes resulting in hypochromic-macrocytic and normochromic-normocytic anaemia and elevated leucocyte numbers, and decreases in blood glucose level and liver glycogen. Behavioural changes also occurred (see below) (Mahajan *et al.*, 1979).

Reproductive organs

Some pseudophyllidean and proteocephalidean metacestodes are responsible for damaging the gonads during migration through the host tissues or by localization in the gonads. *Proteocephalus ambloplitis* plerocercoids in the ovaries of *Micropterus dolomieui* considerably reduced the reproductive capacity of the fish (Esch and Huffines, 1973; McCormick and Stokes, 1982). The anterior ends of the plerocercoids entered the advanced vitellogenic oocytes, perhaps assisted by secretions from the apical gland, and possibly used the nutrients in the yolk for their own growth. The reproductive potential of the fish was reduced by direct oocyte destruction, by accumulation of scar tissue and by fibrosis (McCormick and Stokes, 1982). *Diphyllobothrium cordiceps* plerocercoids, which localized in the serosa, liver, spleen and testes of *Salmo clarki* compressed the testes, but the histological structure appeared to remain normal and the parasite was not considered to have a serious deleterious effect on the host.

Ligula intestinalis and *Schistocephalus solidus* which, although not localized in the gonads have a profound effect on reproductive capacity, have been considered above as parasites of the peritoneal cavity.

Acetodextra amiuri occurs in the ovaries of catfish, invading and destroying the ova. The parasites are expelled from the host during spawning and discharge their own eggs by rupturing on contact with water (Perkins, 1950, 1951, 1956).

Philometrid nematodes in the gonads, other visceral organs or serosa have an effect on

---

Figures 5.30.1–5.30.5 The effects of nematode larvae on fish tissues. From Kahl (1938).
5.30.1 A *Porrocaecum decipiens* larva encapsulated in host connective tissue. 5.30.2 Section through a *P. decipiens* larva in its capsule. 5.30.3 Section through three convolutions of an encapsulated larva of *P. decipiens*. 5.30.4 Section through stomach wall of *Gadus morrhua* in which larvae of 'Anacanthocheilus' are encapsulated. 5.30.5 Section through the liver of *G. morhua* heavily infected with *Anacanthocheilus* larvae.

Table 5.5 Pathological effects of some acanthocephalan parasites of fish (Figures 5.31.1–5.32.4)

Parasite	Host	Pathology	Reference
Eoacanthocephala			
Neoechinorhynchidae			
Atactorhynchus verecundus, *Neoechinorhynchus cristatus*		Destruction of epithelium and lamina propria at site of attachment; no cellular reaction	Bullock (1961)
N. cylindratus and *Leptorhynchoides thecatus* (Rhadinorhynchidae)	*Huro salmoides* (intestine)	Mechanical damage to mucosa and submucosa; lymphocyte infiltration; necrosis; some secondary bacterial invasion.	Venard and Warfel (1947)
N. hutchinsoni	*Noemacheilus kashmirensis* (intestine)	Destruction of mucosa and underlying layers; marked cellular reaction; perforation of gut wall.	Raina and Koul (1984)
N. prolixus	*Carpiodes carpio* (intestine)	Partial occlusion of gut in infections of 80 or more worms but no other significant effect.	Self and Timmons (1955)
N. rutili and *Echinorhynchus truttae* (Echinorhynchidae)	*Salmo trutta* (intestine)	Gill lesions; shortening of abdominal fins; curvature of spine; fish 6–10% under-weight.	Mann (1971)
,,	*Salmo gairdneri* (intestine)	No lesions but fish 17% underweight	,,
Microsentis wardae	*Gillichthys mirabilis* (intestine)	Nodules formed in gut wall; wall thinned over proboscis; calcareous collar formed around neck of worm	Martin and Multani (1966)
Quadrigyridae			
Pallisentis nagpurensis juveniles	*Ophiocephalus striatus* (liver)	Tissue damage during migration provoking encapsulation; worm enveloped by lymphocytes, macrophages and other cells; focal degeneration of adjacent liver cells.	George and Nadakal (1983)
Palaeacanthocephala			
Echinorhynchidae			
Echinorhynchus lageniformis	*Platichthys stellatus* (intestine)	Intestinal lesions; intense cellular reaction; formation of granulation tissue; collagenous capsule formed formed immediately adjacent to worm.	Prakash and Adams (1960)
E. lateralis	*Salvelinus fontinalis* (intestine)	Virtually non-pathogenic	Fantham and Porter (1948)
E. salmonis	*Coregonus lavaretus ludoga* (intestine)	Acute inflammation of gut wall; growth retarded	Bauer and Nikolskaya (1957)
,,	*Salmo gairdneri* (intestine)	Gut lesions, emaciation; 80% mortality, maximal in young fish.	Bertocchi and Francálanci (1963)
,,	*S. irideus*	Curvature of spine	Reichenbach-Klinke and Elkan (1965)
E. veli	*Synaptura orientalis* (intestine)	Destruction of tissue in region of attachment; nodule formation; marked cellular reaction; gut occlusion.	George and Nadakal (1982)
Rhadinorhynchidae			
Leptorhynchoides thecatus	*Micropterus dolomieui* (pyloric caeca)	Erosion of mucosa; fibrosis; chronic inflammation	Esch and Huffines (1973)

Table 5.5 continued

Parasite	Host	Pathology	Reference
Serrasentis nadakali	*Rachycentron canadus* (intestine)	Destruction of tissue at site of attachment; degeneration and necrosis of mucosa; invasion by lymphocytes and macrophages.	George and Nadakal (1981)
Pomphorhynchidae			
Pomphorhynchus bulbocolli (Figure 5.32.1–4)	*Etheostoma caeruleum* (intestine)	Flattening and compression of mucosa; erosion of epithelium; marked cellular reaction; penetration of intestinal wall provoking formation of fibrous capsule around proboscis	McDonough and Gleason (1981)
P. laevis (Figure 5.31.1–2)	*Salmo gairdneri* (intestine)	Compression and abrasion of mucosa; penetration of gut wall with formation of fibrous capsule; marked cellular reaction.	Wanstall, Robotham and Thomas (1986)
,,	*Leuciscus cephalus* (intestine)	Emaciation, formation of fibrous capsule around proboscis and neck	Chubb (1965)
,,	*Barbus* (intestine)	Weight loss; mortality.	Wurmbach (1937)
,,	*Leuciscus cephalus, Barbus barbus* (intestine)	Local intestinal pathology but no effect on growth rate; no mortality.	Hine and Kennedy (1974)

the reproductive process. *Philometra* sp. in the gonads of *Chrysophrys auratus* stimulated an intense inflammatory reaction around the worms, culminating in granuloma formation (Hine and Anderson, 1982). An unidentified species of *Philometra* virtually destroyed the ovaries of an *Otolithus argenteus* (Annigeri, 1962, in Margolis, 1970a). *P. saltatrix* invaded the gonads of sexually-mature *Pomatomus saltatrix*, reducing reproductive potential (Sindermann, 1986). Both *Philonema oncorhynchi* in *Oncorhynchus nerka* and *P. agubernaculum* in *Salmo salar* and *Salvelinus fontinalis* cause peritoneal adhesions which may reduce reproductive capacity by enveloping the gonads in connective tissue, so preventing release of the eggs. In *O. nerka*, pathogenicity is associated with the migration of the fish. While the fish are at sea and before they return to spawn, larvae migrate from the tissues into the peritoneal cavity, setting up a host reaction which results in multiple adhesions (Margolis, 1970a). In this instance, the effect must be transitory since maturing salmon have few adhesions and mature salmon have none, but in landlocked or freshwater salmonids parasitized by *P. agubernaculum* the adhesions may be severe enough to prevent spawning or manual stripping (Meyer, 1960; Margolis, 1970a).

Eustrongylides sp. larvae (Dioctophymatidae) caused deformation and degeneration in the gonads of *Clarias mossambicus, Haplochromis* spp. and *Bagrus docmac* (Paperna, 1974).

Excretory system

The gorgoderid genus *Phyllodistomum* (*P. folium, P. lysteri, P. superbum*) provokes inflammation and lesions of the urinary bladder and other parts of the urinary tract (Table 5.3). The worms feed on host tissue, which is mechanically damaged by the action of the oral sucker and by lysis of host cells. Other effects include obstruction of the archinephric ducts and erosion of the epithelial lining.

Asymphylodora kubanicum infection in the kidneys of *Rutilus rutilus* caused damage resulting in abdominal dropsy (Bykhovskaya-Pavlovskaya and Bykhovskii, 1940).

1

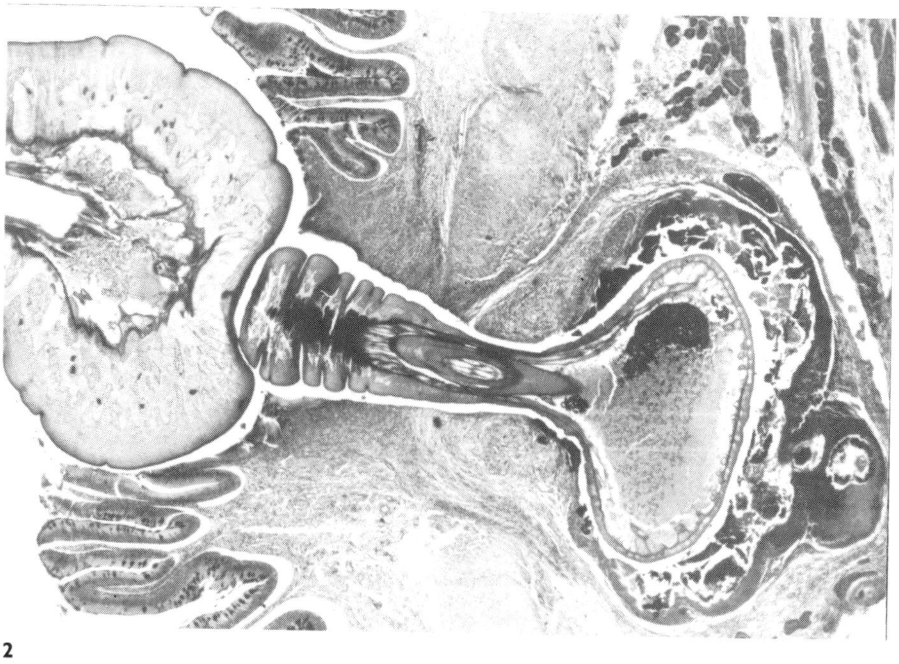

2

Cardiovascular system

Sanguinicolid and heterophyid digeneans and philometrid nematodes are the most pathogenic of the few helminths which occur in the heart and vascular system.

The effects of sanguinicolids on their hosts have been reviewed by Smith (1972) who suggested that they reduce the capacity of the blood to carry and exchange gases by mechanical obstruction, by altering the number and type of circulating blood cells and by causing haemorrhage. Unless otherwise stated, the authorities for the following information are given by Smith (1972).

Smith's (1972) summary of the literature showed that *Sanguinicola inermis* caused acute disease in *Cyprinus carpio* when eggs blocked the gill capillaries, causing thrombosis and consequent necrosis of gill tissue. Lack of oxygen was probably a more immediate cause of death than gill necrosis, which developed later in the infection. Eggs carried to the kidney caused renal dysfunction, resulting in chronic disease and tumours, or capsules containing eggs appeared in the gills, kidneys or heart. Haemorrhage followed by septic necrosis occurred after miracidia which hatched in the gills escaped through the host tissue. Hłond, Kozlowski and Szaryk (1977) described losses of 80–90 per cent of infected *C. carpio* fry with severe gill inflammation. *S. klamathensis* and *S. davisi* were responsible for large-scale mortality of *Salmo gairdneri* fingerlings at a Californian hatchery, due to haemorrhage caused by escaping miracidia. The blood changes caused by sanguinicolids have received little attention. *Sanguinicola inermis* reduced the haemoglobin content and erythrocyte count but elevated the leucocyte count in *C. carpio; S. ornata* in *Tinca tinca* doubled the polymorphonuclear agranulocyte count and raised the monocyte and neutrocyte counts slightly.

In sufficiently large numbers, invading sanguinicolid cercariae can kill fish fingerlings, eg. *Salmo clarki* and *S. gairdneri* fingerlings died within hours of massive infection by *Sanguinicola alseae* cercariae (Meade and Pratt, 1965). Ten cercariae of *S. inermis* were sufficient to kill 8-day-old *C. carpio* fry but 3-month-old fish succumbed only to invasion by 2,000–2,500 cercariae (Sapozhnikov, 1976).

The heterophyid *Ascocotyle tenuicollis* in the bulbus arteriosus of *Gambusia affinis* produced morbidity and could occlude the lumen of the bulbus. *A. leighi* encysted in *Cyprinodon variegatus* and *Mollienesia latipinna* provoked fibrosis which often caused partial occlusion of the bulbus (Hoffman, 1975a). (See Figure 5.20.)

Two philometrid nematodes, *Philonema saltatrix* in *Pomatomus saltatrix* and *Philometra obturans* in *Esox lucius*, occupy the cardiovascular system of their hosts. *P. saltatrix*, in the heart and pericardial cavity of 0-group fish, causes inflammation, blood stasis and necrosis (Cheung, Nigrelli and Ruggieri, 1984); prevalence approached 80 per cent and survival of the year class may have been endangered. *P. obturans* is highly pathogenic: females located in the bulbus arteriosus, ventral aorta and gill arteries obstruct blood flow and feed on the blood cells (Figure 5.27. 1–2). They also perforate the gill arteries when releasing larvae and provoke thrombi in the bulbus arteriosus and hypertrophy in the ventral aorta (Moravec and Dyková, 1978).

Figures 5.31.1–5.31.2 *Pomphorhynchus laevis* (Acanthocephala) attached to the intestine of chub. Original photographs kindly provided by Dr. J.C. Chubb. See Chubb (1967).

5.31.1 *Pomphorhynchus* firmly attached to the intestinal walls of the fish. 5.31.2 Section through proboscis embedded in the gut wall.

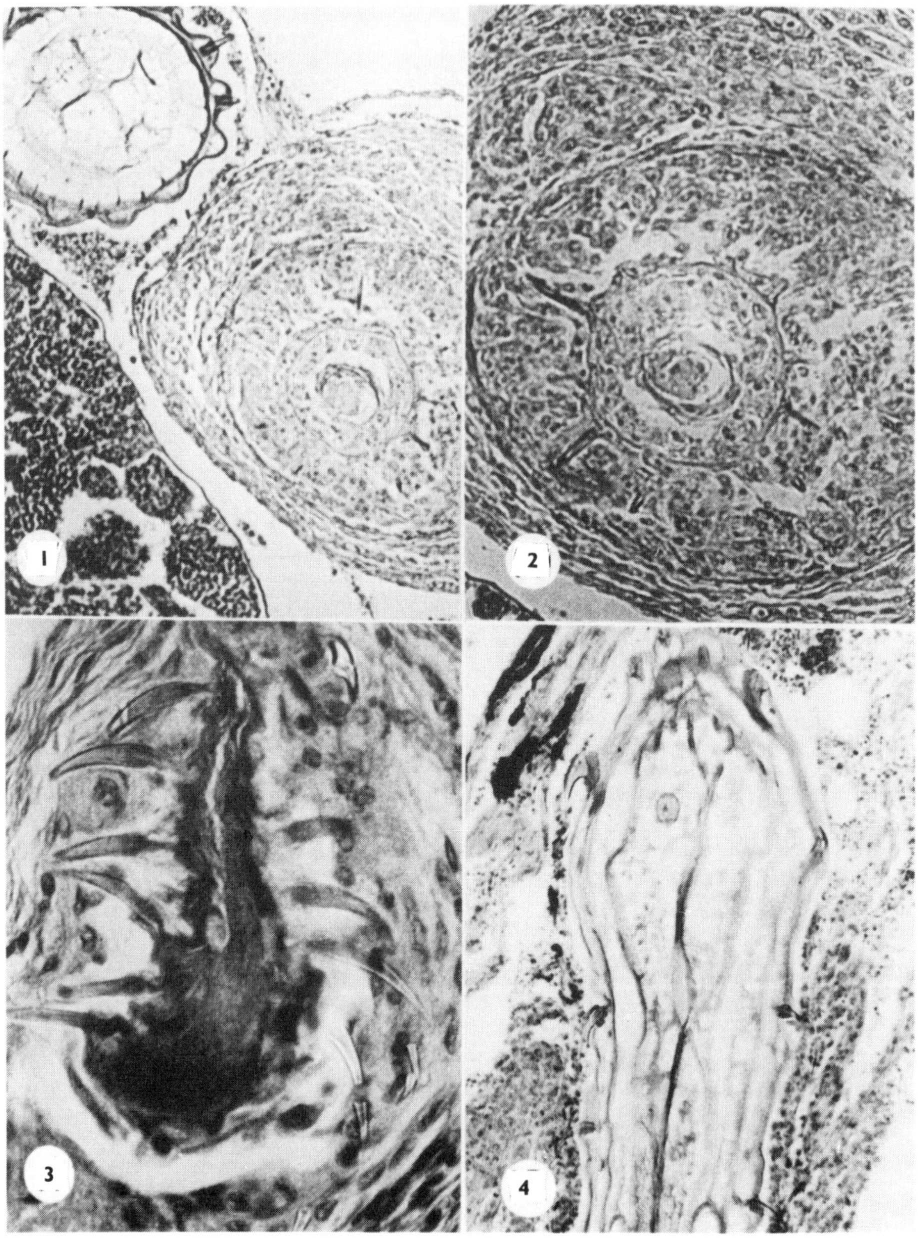

Figures 5.32.1–5.32.4 Histopathology of *Etheostoma caeruleum* infected with *Pomphorhynchus bulbocolli*. 5.32.1 Cross-section through two proboscides of *P. bulbocolli*. The lower proboscis is that of a dead worm and is heavily encapsulated. The upper proboscis is that of a living worm and has less encapsulation. × 185. 5.32.2 Same proboscis as in Figure 5.31.1 showing the infiltration of host cells into the proboscis. × 235. 5.32.3 Encapsulated proboscis of dead worm showing the loss of integrity of the proboscis tissue, host fibrous connective tissue, and mild infiltration of leucocytes. × 450. 5.32.4 Proboscis of worm that was completely within the coelomic cavity showing host response of fibroblasts and leucocytes which was more pronounced in this region than in the trunk region. × 140.
From McDonough and Gleason (1981).
(See also Wanstall, Robotham and Thomas (1986) and Wanstall, Thomas and Robotham (1988) for the effects of *P. laevis* on the gut of *Salmo gairdneri* and *Noemacheilus barbatulus*.)

Susceptibility to stress

Parasitic infestations have been recognized as one of a number of stressful factors which may lower the resistance of the host and its ability to adapt to changing environmental conditions, e.g. of temperature, salinity or pollution with trace elements. Most of the helminths so far investigated in this context are cestodes.

Eubothrium salvelini both retards the growth and reduces the stamina of *Oncorhynchus nerka* fry (Smith and Margolis, 1970; Boyce, 1979) but also affects the ability of migrant fish to adapt to salt water. When tested in laboratory conditions, infested fish caught at the outlet of Babine Lake, Canada, in 1979 when the prevalence of infestation was 60 per cent, died in significantly greater numbers than uninfested fish. They also had higher plasma sodium levels than uninfested fish although the difference was not statistically significant (Boyce and Clarke, 1983). In 1980, when the prevalence was 30 per cent, both infested and non-infested fish adapted better to sea water than fish caught the previous year; there was no significant difference in mortality but infested fish had a significantly higher plasma sodium level. Infestation would considerably reduce survival at sea.

Infestation with *E. salvelini* also considerably increased the susceptibility of *O. nerka* smolts to zinc pollution (Boyce and Yamada, 1977). *Schistocephalus solidus* infestation reduced the period of survival of *Gasterosteus aculeatus* exposed to cadmium when compared with uninfested controls and infested fish also died sooner when their diet was restricted (Pascoe and Cram, 1977; Pascoe and Mattey, 1977). In a study of various combinations of these stresses, Pascoe and Woodworth (1980) found that fish exposed to all three stresses (ie. *S. solidus* infestation, cadmium, and dietary restriction) died soonest (12.8 days) followed by infested fish exposed to dietary stress (23.5 days), those exposed only to dietary stress (36 days) and, finally, those exposed both to cadmium and dietary restriction (44 days).

Triaenophorus nodulosus plerocercoids reduced the ability of *Esox lucius* fry to withstand oxygen starvation (Pronin, Pronina and Shigaev, 1976). At an oxygen level below 1 mg/litre, all infested fry died, although uninfested fish could survive for long periods. Susceptibility increased with increasing worm burden since fry with one or two plerocercoids could survive at 1.5 mg/litre but those with four could not.

Compared with uninfested controls, *Bothriocephalus acheilognathi* reduced the survival of *Gambusia affinis* in changing conditions of temperature. Survival fell by 15 per cent at 20°C, 27 per cent at 25°C and 55 per cent at 30°C. At 25°C, large fish with lower parasite densities survived longer than smaller fish with higher parasite densities (Granath and Esch, 1983b).

Metacercariae of *Uvulifer ambloplitis* also affected the survival of the host in relation to temperature. *Lepomis macrochirus* heavily infested with more than 50 metacercariae per fish died when the temperature decreased but not when it remained elevated (20–25°C) (Lemly and Esch, 1984b).

Immunology

Comparatively little is known of the immune mechanisms of fish compared with higher vertebrates but the growth of the fish-farming industry in recent years has stimulated interest in the subject. Most work has been done on the teleosts, principally the most commonly farmed species (Cyprinidae and Salmonidae) but some has been done on the Agnatha, Chondrichthyes, Holocephali and Dipnoa, all of which differ in their immune mechanisms. The most recent account of the immune mechanisms of teleosts is that by Ellis (1982) and the following introduction to the topic relies heavily on his review.

Fish have all the characteristics of both humoral and cell-mediated adaptive immunity. Ellis (1982) reviewed the origin of stem cells and the structure and function of the lymphoid organs. Lymphocyte subpopulations with functional dichotomy appear to have evolved at the level of fishes but it is uncertain whether they originate in organs equivalent to the mammalian thymus and bone-marrow. Some teleost species have T- and B-like lymphocytes in the thymus. This organ lacks antigen-producing cells but antigen-binding cells, possibly with a T-helper function, proliferate in the thymus in response to low-dose antigen stimulations, and Ellis suggests that antigens may gain access to the thymus. All lymphocytes in adult fish, regardless of whether they originate in the thymus, kidney, spleen or blood, have membrane-bound immunoglobulin (mIg); fish are the only vertebrate group in which this is the case. It further suggests that the antigen receptor of T-cells is Ig.

Immunoglobulins are found in most fish tissue fluids (plasma, lymph, skin and gut mucus and bile) and form 40–50 per cent of the total serum proteins in fish blood. The blood vascular system is permeable to serum Ig. Fish have only one class of immunoglobulin, which resembles mammalian IgM. Teleost IgM is tetrameric but some also have a 7S monomeric form. Chondrichthyes have only IgM but in the monomeric and pentameric forms. Only dipnoans, the most phylogenetically-advanced group, have a second immunoglobulin class. With one known exception, there is no shift from high- to low-molecular weight Ig in fish during the immune response. Fish Ig is structurally heterogeneous with subclasses which may be functionally specialized, fulfilling the roles of different Ig classes in higher vertebrates.

Fish antibody will lyse cells in the presence of normal serum as a complement source; therefore, fish Ig can activate complement. Agglutinating activity occurs in fish but antibody precipitation does not. Possibly, different functional subclasses of IgM have different functions of complement fixation, agglutinating activity and precipitating activity.

Work on immunoglobulin in fish secretions has shown that in plaice, mucus IgM is similar to serum IgM. In sheepshead, IgM in mucus and bile is not derived by exudation from serum IgM and, although antigenically identical, is structurally dissimilar. IgM may mediate in the release of histamine in fish, important in inflammation and allergic responses, a function in mammals of IgE which is apparently absent in fish. Mechanisms for hypersensitivity responses in fish are not understood but antigen-specific sensitization with lethal consequences has been reported.

Several factors affecting the immune response in fish have been investigated. Temperature is one of the major factors, influencing both humoral and cell-mediated immune responses. Within the physiologically tolerable temperature range of the fish, higher temperatures accelerate the onset of the immune response and enhance its magnitude while, at lower temperatures, the response is slower with reduced antibody titres, or absent. The temperature critical for the development of the immune response is determined by the natural environmental temperature of the fish.

Only some phases of the response are temperature-dependent. Carp immunized with BSA gave a normal secondary response at low temperature provided that primary stimulation had occurred at high temperatures. Carp produced antibodies to sheep red blood cells at 8–28°C but did not give a typical secondary response if maintained below 18°C. However, there is evidence of acclimatization to low temperatures. The nature of the antigen may also be important: it has been suggested that T-dependent antigens may be more affected by temperature than T-independent antigens. The mechanism of the effect of temperature is believed to involve a temperature-dependent phase of interaction between T- and B-like cells, possibly with a block of T-helper function or an increase in T-suppressor activity, followed by a temperature-independent period of multiplication and differentiation of activated antibody-producing cells (B-like cells) with a final temperature-dependent phase involving the level of antibody production by plasma cells.

The effect of temperature acclimatization on the responsiveness of lymphocytes is important for developing vaccination programmes. A good antibody response with long-term memory formation could be obtained by raising the temperature at some times of the year when vaccinating fish. Exposure to antigen at low winter temperatures may cause suppression of immunity and a common consequence on fish farms is disease in the spring.

Several other factors affect the immune response. The magnitude of the primary and secondary response is affected by the dose of antigen and the route by which it is administered. The nature of the antigen is also a factor; some antigens, particularly soluble protein antigens, are not effective immunogens.

Endogenous rhythms cause seasonal variation in the antibody response, even when fish are maintained under constant conditions of light and temperature. This variability has significant implications for vaccination schemes.

In crowded conditions, some fish produce a crowding pheromone with immune-suppressive effects, possibly linked to stress.

The complement system occurs in fish. Trout IgM will fix and activate trout complement. The rainbow trout's complement system has much in common with the mammalian system, including an alternative pathway of complement activation which requires only Mg^{2+} ions. Minor differences exist in the optimum temperature for activity, in the proportion of the protein component, and in its heat-labile properties. The optimum temperature for inactivation of complement in fish sera differs between fish species. More information is available on the shark complement system. The antibody-mediated lytic activity of *Ginglymostoma cirratum* serum is similar to that of mammals but shark complement differs in the number of its components, their compatibility with the components of mammalian complement, the absence of labile intermediate complexes and the absence of an alternative pathway for activation.

Ellis (1982) included reviews of phagocytic cells and inflammation and of hypersensitivity. Macrophages are present in many tissue sites in fish, and are highly phagocytic, but little is known of their relationship to the specific immune response or complement system. Neutrophils are often found in inflammatory lesions but they do not appear to be phagocytic and their role is not clear. They may have an extracellular rather than intracellular bactericidal role.

In mammals, phagocytozing neutrophils produce free radicals extracellularly, which cause tissue damage. If fish neutrophils perform similarly tissue damage would be expected and does, in fact, occur in inflammatory reactions. In teleosts, cells containing dark pigments (mainly melanin but occasionally ceroid or haemosiderin) occur in lymphoid tissues and inflammatory lesions. Melanin is able to quench free-radical reactions and thus protect fish tissue from the damage that occurs during inflammation. Visceral melanin may protect from free-radical damage, similar to its protective function against radiation-induced free radicals in the skin. In lymphoid tissue, it may protect against free radicals generated by phagocytozing neutrophils. Fish tissue lipids are readily oxidised to ceroids by free-radical reactions. Melanosomes in macrophages which phagocytoze tissue rich in free radicals may protect them from damage. Melanin may protect against lytic enzyme systems. It may protect against tissue invasion by some parasites and against potentially harmful mechanisms which are stimulated when the host's defence system is activated.

Hypersensitivity in fish is a controversial subject. Fish apparently lack mast cells, high tissue histamine levels and IgE, but antigen-specific lethal anaphylactic responses and immediate hypersensitivity skin reactions are known to occur in some species. The role of histamine is controversial. The responsive cell in hypersensitivity reactions is not known but analogues of the mammalian mast cell have been proposed for some groups. An understanding of the mechanisms of the immune hypersensitivity responses in fish is essential for formulating vaccination procedures.

Reviews of the immune response to metazoan parasites have remarked that only recently has immunity to helminths other than monogeneans been demonstrated (Sinderman, 1966). It has been established for some time that fish produce specific antibodies in response to parenteral injections of antigens. Temperature-dependency is characteristic of the immune response in fish and the rate and intensity of antibody production are lower at low than at high temperatures and are generally considered to correlate with the normal environmental temperature range that each species experiences. Carp showed no immune response in 30 days at or below 10–12°C, but the response at 18°C was good (Snieszko, 1969). Parasites which live in the gut have been regarded as less immunogenic than tissue parasites (Weinmann, 1966) and it has been suggested that the response provoked by gut parasites depends on their ability to damage or penetrate the gut wall (Chaicharn and Bullock, 1967; Kennedy and Walker, 1969; Harris, 1972).

Monogeneans

The best-known example of immunity to a monogenean is that against the capsalid *Benedenia* (= *Epibdella*) *melleni* which is responsible for heavy and often fatal infections of marine fish of the order Acanthopteri (Jahn and Kuhn, 1932). Nigrelli and Breder (1934) found that some fish acquired total immunity, some partial immunity and some naturally-immune species became slightly infected in epidemic conditions. Some species were always susceptible to infection. *Trachinotus carolinus* and *T. falcatus*, both highly-susceptible species, became permanently or partially resistant after several exposures to infection but immunity could not be induced by injection of a suspension of dried or fresh worms, or serum from immunized fish (Nigrelli, 1937). *B. melleni* died more rapidly when exposed *in vitro* to mucus from non-susceptible elasmobranchs or immunized susceptible fish than they did when exposed to mucus from susceptible fish or to sea water.

Vladimirov (1971) has shown that hybrid and crucian carp developed immunity lasting over two months to *Dactylogyrus vastator*, with specific antibody formation. Immunization with precipitated antigen from *D. vastator* considerably decreased the susceptibility of fish to infection. Saprykin and Kashkovskii (1979) reported a correlation between the polymorphous serum transferrin system in carp and the prevalence of *D. extensus* infection. High and low prevalence appeared to be genetically controlled. Scott and Robinson (1984) found that various parameters in challenge infections of *Gyrodactylus bullatarudis* in *Poecilia reticulata* were significantly lower than those in the initial infections or in non-treated unchallenged controls. Establishment success, population size, parasite burden, time to peak burden and duration of infection were all reduced. Madhavi and Anderson (1985) found that four inbred strains of *P. reticulata* could be categorized as innately resistant or susceptible to *G. bullatarudis* infection. The biological basis of resistance or susceptibility was not determined but the *G. bullatarudis*-guppy system was considered a useful model for the study of genetic factors in host-parasite interactions.

Khalil (1964) suggested that *Polypterus senegalensis* developed an immunity or resistance to reinfection with *Macrogyrodactylus polypteri*. Fish which survived very heavy infestations did not initially acquire new infestations, possibly because of excessive mucus production which prevented attachment. This was, however, short-lived and new infestations soon became established.

Cestodes

Several species of intestinal cestodes provoke an immune response which has, in some instances, been attributed to non-specific immune mechanisms, eg. *Caryophyllaeus laticeps* in species of *Leuciscus*, and *Acanthobothrium quadripartitum* in *Raja naevus*. Kennedy and Walker (1969) showed that in dace (*L. leuciscus*) experimentally infected with *C. laticeps*, an initial period of establishment was followed by rejection. The parasites did not grow or mature and only a decreased number established after reinfection, suggesting an acquired immune reaction by the host. Attempts to demonstrate circulating antibodies failed but the fact that a rise in temperature led to a decrease in the period of establishment and an increase in the speed of rejection was consistent with antibody formation. *C. laticeps* was more successful in establishing itself and survived for longer periods at low than at high temperatures, and it was suggested that temperature-dependent changes in the immune response controlled the cycle of incidence in the field (Kennedy and Walker, 1969; Kennedy, 1969b, 1970). Although *C. laticeps* larvae were present in the intermediate host throughout the year except in August, dace became infected only between December and March and the infection was lost by July. This cycle was independent of the availability of larvae and of host feeding behaviour but was correlated with water temperature.

Survival of *C. laticeps* in experimentally-infected farm-reared orfe (*L. idus*) depended on temperature (Kennedy, 1971). At and below 12°C, those which survived an initial rejection remained for up to a month, but at 18°C they were rejected and killed after three days. At 18°C, transplanting the cestodes to fresh orfe at two-day intervals prolonged survival only slightly and suggested that the factors responsible for rejection must have become operative within two days after infection. The inhibition of growth and development in dace and orfe, and the time lag before rejection, were typical of an immune response to intestinal helminths, and the increased speed of elimination was consistent with antibody formation in fish. However, the response by orfe was too rapid to be attributed to specific antibody formation, nor could specific antibodies to *C. laticeps* be detected in dace serum. The involvement of non-specific antibodies was suggested by Kennedy (1971), a proposal already advanced by McVicar and Fletcher (1970) to explain the rejection of a tetraphyllidean cestode by a ray.

Acanthobothrium quadripartitum is specific to *Raja naevus* and has not been recorded from *R. radiata*, although both species occur in the same localities and have the same feeding habits. It survived for over 24 hours in fresh serum from *R. naevus* but 80 per cent died within two hours when exposed to fresh serum from *R. radiata* (McVicar and Fletcher, 1970). The toxicity of fresh *R. radiata* serum could be destroyed by heating for 30 minutes at 40°C, dilution with elasmobranch saline, storage at low temperatures and freeze-drying, but could be restored by the addition of fresh *R. naevus* serum. Since these procedures are known to inactivate complement, it was suggested that both rays had a similar complement system in fresh serum but only *R. radiata* had a natural serum antibody toxic to *A. quadripartitum*. Intestinal and dermal mucus of *R. radiata* was not toxic *in vitro*, probably because the serum factors were too dilute. *R. radiata* was thought to secrete both antibody and complement into the intestine via the mucus. As immunogenicity has been correlated with tissue damage, it is interesting that *Acanthobothrium* plerocercoids, which infect rays, have no hooks and therefore cannot damage the mucosa. Nevertheless, the earliest stage found in *R. naevus*, which had partly-developed bifid hooks, showed the same response to *R. radiata* serum *in vitro* as the adults. *Acanthobothrium* plerocercoids have never been found developing in *R. radiata*. The incidence of another intestinal cestode, *Triaenophorus nodulosus*, in *Esox lucius* was thought to be determined by an immune response. Although plerocercoids were available in prey fish throughout the year, pike acquired infection only in

May-June, which was attributed by Scheuring (1929) to the presence of non-specific antibody (cf. *C. laticeps* above).

The immune response to *Bothriocephalus acheilognathi* (= *B. gowkongensis*), an important pathogen of farmed cyprinids, has received increasing attention from Russian workers. Kozinenko (1981) found that a delayed hypersensitivity reaction occurred in carp sensitized with *Bothriocephalus* antigen and then challenged, although specific antibodies were not detected in the serum. Unsensitized infected carp also responded positively, demonstrating the potential of the test for detecting infection. A group of carp with a low prevalence of light infections exhibited a delayed hypersensitivity reaction when tested after abdominal injection of extracts of young and adult worms (Kozinenko and Balakhnin, 1981) and, in this instance, antibodies were detected in 72 per cent by the indirect haemagglutination test. The reaction was also found in 35–42 per cent of carp infested with less than five *B. acheilognathi* and some previously infected but free when tested (Balakhnin and Kozinenko, 1981). Uninfected fish and those with a burden of more than five worms failed to respond, suggesting an immunological paralysis in heavily-infected fish. Antibodies were detected in 42.6 per cent of fish two weeks after immunization with *B. acheilognathi* antigen (Balakhnin and Kozinenko, 1981). The immune response could be altered by anthelmintic treatment. Phenasal reduced the number of fish which showed a delayed hypersensitivity reaction after immunization with *Bothriocephalus* extract (Balakhnin, Kozinenko and Kurovskaya, 1982). Antibody titres were significantly lower in fish given phenasal compared with untreated fish, although the number in each group which produced antibodies did not differ.

Carp serum inhibits the movement of *B. acheilognathi in vitro* whereas that of *Silurus glanis*, in which this cestode does not occur, stimulated its activity (Balakhnin and Davydov, 1975).

Extracts and secretions of a related marine species, *Bothriocephalus scorpii* in *Scophthalmus maximus*, have antigenic determinants reactive with teleost C-reactive protein (CRP) and with antiserum to phosphoryl choline (Fletcher, White and Baldo, 1980). CRP is present in fish at naturally high levels. Its production is provoked by an inflammatory stimulus but *B. scorpii* does not cause a chronic inflammatory reaction. There was no evidence for a humoral immune response to *B. scorpii*. Turbot serum was not toxic to *B. scorpii in vitro*; CRP precipitated with *B. scorpii* excretions but was not obviously protective against the adult worm. Intradermal injection of parasite C-substance induced a skin erythema which was not due to tissue damage or direct mast cell degranulation. It was believed to be related to the amount of circulating CRP rather than to a previous infection with the cestode. CRP is an acute-phase protein in man but not necessarily in fish. Its functional significance in fish is not yet known.

Plerocercoids in the peritoneal cavity and skeletal musculature provoke a strong response. Two of the earliest reports of an immune response to cestodes by fish are those of Scheuring (1929) and Vogt (1938) on *Triaenophorus nodulosus* plerocercoids in the peritoneal cavity of trout. Overstreet (1977) suggested that the sciaenid *Cynoscion nebulosus* showed an immune response to *Poecilancistrium caryophyllum* metacestodes because, although the prevalence of infection increased with increasing length (age) of the host, the intensity of infection did not.

Plerocercoids of the pseudophyllidean *Schistocephalus solidus* died after penetrating the abdominal wall of sticklebacks (Vik, 1954) and the presence of encapsulated and dead metacestodes in the peritoneal cavity suggested that the fish had responded against them and might have acquired a temporary immunity to reinfection. Bråten (1966) and Orr, Hopkins and Charles (1969) found that plerocercoids from one species did not survive when transplanted to another, indicating a degree of host specificity. In primary infections, plerocercoids grew more slowly in *Pungitius pungitius* than in the natural host, *Gasterosteus aculeatus*, and did not survive beyond 14 days at 10°C. Challenge infections established in *G. aculeatus* but were rejected by *P. pungitius* within 3–5 days; the rejection period is consistent with the operation of an immune mechanism, involving antibody formation.

Specific antibodies have been detected in sera of *Abramis brama* in response to *Ligula intestinalis* plerocercoids (Berczi and Molnár, 1965; Molnár and Berczi, 1965). Sweeting (1977) did not find precipitating antibodies to this species in the serum of infected roach (*Rutilus rutilus*) but an increase in τ-globulin was associated with infection. The cellular response of fish to this species has been described in detail by Hoole and Arme (1982; 1983a,b; 1986). They speculated (1983b) that the metacestode adsorbs host proteins onto its surface to avoid the host response, as suggested for *Bothriocephalus scorpii* by Fletcher, White and Baldo (1980). This appeared to happen in naturally-infected gudgeon (*Gobio gobio*) which, in contrast to other cyprinids, did not show an intense cellular response to the parasite. Metacestodes from gudgeon would, if exposed to roach plasma *in vitro*, provoke a response if transplanted back into gudgeon. Gudgeon responded to parasites transplanted directly from roach but not to those exposed *in vitro* to gudgeon proteins. The adherence of roach leucocytes (macrophages, neutrophils and lymphocytes) to the surface of live worms occurred only in the presence of normal roach serum and of heat-inactivated normal roach serum, suggesting the involvement of a non-specific factor, probably a complement protein, and of an antibody ligand (Hoole and Arme, 1986). Normal roach serum is unlikely to contain an antibody to *L. intestinalis*, so leucocyte adherence to the worms probably derives from activation of complement via the alternative pathway. Complement-mediated leucocyte adherence may be more important than antibody production at low temperature when antibody production is reduced.

Ligula affected the lymphoid organs of roach by provoking an increase in the number of macrophages and in the presence of 'Type B' granulocytes in the spleen; the pronephros of infected and uninfected fish contained mainly macrophages and neutrophils. 'Type B' granulocytes were found only in infected fish. In gudgeon, there were no differences between the spleen and pronephros of infected and uninfected fish (Taylor and Hoole, 1987).

Digeneans

There are very few reports of immune responses to adult digeneans in the gastrointestinal tract. McCoy (1930) found that a heavy infection with *Hamacreadium gulella* or *H. mutabile* in *Lutjanus griseus* did not confer resistance to reinfection but, in a challenge infection of *H. gulella*, fewer worms became established in previously-infected fish than in controls and they were smaller in size. Fish experimentally infected with several hundred *H. gulella* lost nearly all of them within four weeks while light infestations (10–15 cysts) persisted. McCoy (1930) attributed the expulsion to a host reaction, stimulated by the presence of many worms, rather than to a crowding effect.

Precipitating antibodies have been detected by agar gel diffusion and passive haemagglutination in serum and gut mucus of *Anguilla australis schmidtii* and *A. dieffenbachi*, respectively, in response to the intestinal digenean *Telogaster opisthorchis* (McArthur, 1978). The antibody was a macroglobulin analogous to IgM and eel serum was also found to contain naturally-occurring agglutinins and lysins against sheep erythrocytes. McArthur and Sengupta (1982) detected antibodies to *T. opisthorchis* in eel sera by ELISA and found the titres were highest in larger, ie. older eels.

Precipitating antibodies have also been demonstrated in response to tissue parasites. Metacercariae of *Cryptocotyle lingua and Rhipidocotyle johnstonei* in plaice (*Pleuronectes platessa*) provoked the production of specific antibodies; these were immunoglobulins resembling mammalian IgM and the rate and magnitude of antibody production were temperature-dependent (Cottrell, 1977). Fish showed no response at 5°C but produced antibodies at 15°C and above. Increasing temperature reduced the induction period between infection and the primary antibody response. Cottrell (1977) noted that *R. johnstonei*, the metacercariae of which were not encysted, provoked only a mild antibody response although their direct contact with host tissue might be expected to generate a strong response. He suggested that selection pressure during evolution of the host-parasite relationship might have favoured the reduction of antigenic stimulation by the parasite. The cellular response of plaice to *R. johnstonei* metacercariae is an intense chronic inflammatory response culminating in granuloma formation (Figures 5.24.1–3) (Pulsford and Matthews, 1984). Mullet (*Chelon labrosus*) exposed to 20,000 *Cryptocotyle lingua* cercariae produced humoral antibody, sensitized pronephric leucocytes and cytotoxic serum factors. Antibody titres measured by passive haemagglutination were not reliable as a rapid test. Pronephric leucocytes, sensitive to cercarial antigen, were found during weeks 1–6 with a peak in the second week. An increase in polarization *in vitro* was found when cells were incubated with antigen, but did not differ significantly between infected and uninfected fish. Cercaricidal activity was demonstrated in tests *in vitro* with antibody at 1:200 collected in the fourth week, and was associated with structural damage to the tegument. Pronephric leucocytes from immune fish did not adhere to cercariae or metacercariae *in vitro* (Wood and Matthews, 1987).

Immune responses to diplostomes have been reported in cyprinids and salmonids. When *Cyprinus carpio* were exposed to *Diplostomum spathaceum* or *D. paracaudum* cercariae and challenged with the same or the other species, establishment of the challenge infection was lower in all cases, except when *D. paracaudum* was challenged by *D. spathaceum*, when the difference was statistically insignificant (Razmashkin, 1985).

Attempts have been made to immunize salmonid fish against diplostomiasis but these have been hampered because the metacercariae locate in the lens, an immunologically-privileged site. Nevertheless, some promising results have been achieved. Bortz *et al.* (1984) detected circulating antibodies against *D. spathaceum* in *Salmo gairdneri* by ELISA three weeks after immunizing fish with sonicated metacercarial antigen; the titres declined by six weeks after immunization but an enhanced secondary response was obtained following a booster at nine weeks. Serum samples from naturally-infected wild fish were positive by ELISA. It is interesting that an anti-metacercarial antibody was produced, although it might have been expected that metacercariae, in an immunologically-privileged site, might be protected from the host immune system (Whyte, Allan, Secombes and Chappell, 1987). When *S. gairdneri* of three different size categories were immunized with an antigen containing the equivalent of 10, 50 or 100 metacercariae per ml, immunized fish in all size categories survived four months longer after challenge with cercariae than non-immunized fish (Speed and Pauley, 1985). Survival was greater in large than in small fish and in those that received the two higher concentrations of antigens. All those given the equivalent of 100 metacercariae per ml were alive at the end of the 12-month period of the experiment.

Stables and Chappell (1986) found a significant decrease in *D. spathaceum* infection in *S. gairdneri* immunized with suspensions of dead cercariae but failed to detect humoral antibody despite using a variety of techniques. Further investigation of fish immunized with 250 sonicated cercariae or diplostomula in Freund's complete adjuvant, boosted after two weeks with the same dose in Freund's incomplete adjuvant, showed that circulating antibodies to cercariae and diplostomula could be detected by ELISA at 6–7 weeks after immunization (Whyte, Allan, Secombes and Chappell, 1987). Anti-cercarial sera cross-reacted with diplostomulum antigen, although less strongly than the cross-reaction between anti-diplostomulum serum and cercarial antigen. Positive results were obtained with immunofluorescence tests using anti-cercarial and anti-diplostomulum sera. The tail region of cercariae fluoresced more strongly than the body, indicating that antibodies were produced mainly against the tail. Its absence in diplostomula might account for the reduced cross-reactivity between anti-cercarial sera and diplostomula. The development of a vaccine would be an important step in the control of diplostomiasis in the fish-farming environment. Whyte, Chappell and Secombes (1988) described a technique for obtaining post-penetration larvae of *D. spathaceum* by *in vitro* penetration of rainbow trout skin, and succeeded in maintaining the diplostomula *in vitro* for a short time. About 80 per cent survived for 24–48 hours in L-15 medium supplemented with 5 per cent foetal calf serum. This will permit studies of the migratory stage of the parasite.

Metacercarial cysts of *Posthodiplostomum minimum* in bluegill (*Lepomis macrochirus*) had components which reacted both with rabbit anti-fish serum and rabbit anti-metacercarial serum, indicating the dual origin of the cyst wall (Crider and Meade, 1975). Some antigenic properties were common to the metacercaria and the cyst wall but affinities between the cyst wall and the fish, although present, were less distinct. Undiluted serum from infected or uninfected *Lepomis* sp. lysed *P. minimum* cercariae (Latimer and Meade, 1979). At dilutions of and greater than 1:4, lysis failed to occur but a pericercarial envelope was formed. Because this developed only when cercariae were exposed to serum from infected fish, Latimer and Meade (1979) attributed the cercaricidal effect to an antigen-antibody interaction.

Antibodies have been detected in fish mucus in response to helminths of several groups (Nigrelli and Breder, 1934; Harris, 1972) and although Cottrell (1977) failed to detect any in plaice, this might be attributable to dilution of antibody during transudation from the serum to the epidermal mucus. Extracts of the epidermal mucus of *Carassius carassius* and *Cyprinus carpio nudus* are toxic to cercariae, encysted metacercariae and adults of *Clonorchis sinensis* (Rhee, Baek, Ahn and Park, 1980a; Rhee, Kim, Baek, Lee and Ahn, 1982).

Nematodes

An immune response to fish nematodes has been demonstrated for plaice (*Pleuronectes platessa*) by Harris and Cottrell (1976). Plaice serum contained precipitins to the spiruroid nematode *Proleptus obtusus*, a common parasite of dogfish which has never been recorded from plaice. Harris and Cottrell (1976) originally supposed that C-reactive protein (CRP) might be involved, as Baldo and Fletcher (1973) had demonstrated its presence in plaice sera and Fletcher and Baldo (1974) had shown that it reacted with antigenic extracts of a wide range of organisms including *Ascaris lumbricoides*. Further investigation indicated that the precipitin was an immunoglobulin because the precipitin bands did not dissolve in sodium citrate, EDTA or phosphate-buffered saline, showed β-mobility on electrophoresis and had characteristics usually associated with teleost immunoglobulin-mercaptoethanol sensitivity, ie., β-globulin electrophoretic mobility, a high sedimentation coefficient and comparable gel filtration behaviour. Conversely, CRP precipitins from plaice serum dissolved in EDTA and sodium citrate and produced an arc in the α^2 position on electrophoresis (Baldo and Fletcher, 1973). The immunoglobulin is analogous to human IgM. To explain why it precipitated with antigens from a parasite unknown in plaice, Harris and Cottrell (1976) pointed out that serological cross-reactions do occur between different nematode species and that two species of *Cucullanus*, a spiruroid genus closely related to *Proleptus*, are common in plaice. The fish may therefore be responding to an antigen of a species it commonly harbours and the antibody is precipitating with the serologically-related *P. obtusus* antigen. Both species attach themselves to the intestinal mucosa of their respective hosts and secrete enzymes which damage the gut wall and may be the source of the antigen.

Acanthocephalans

Very little information is available on immune responses in fish to worms of this phylum. In an early observation suggestive of an immune response in fish, Cross (1934) reported that, in ciscoes infected with both the cestode *Proteocephalus exiguus* and an acanthocephalan ('*Neoicanthorinchus*'), fish with a heavy infection of one species harboured very few of the other. As they attached in different regions of the gut, a crowding effect was not considered responsible and Cross suggested that non-specific immunity limited either species when the other was present in large numbers.

Serum and mucus precipitins produced in response to antigens of *Pomphorhynchus laevis* have been demonstrated in naturally and experimentally-infected chub (*Leuciscus leuciscus*) by Harris (1970, 1972), but there was no evidence that they conferred resistance to current infections or to reinfections. Although *P. laevis* occurred naturally in five other species of fish, it matured only in chub and only chub responded to it by humoral antibody formation. The antigen was thought to be an excretory or secretory product, probably produced only by mature adults. The antibody had characteristics similar to those of IgM and was the same in serum and mucus. The presence of antibody in mucus suggested the presence of a secretory antibody system. As previously mentioned, attempts to demonstrate antibody to *Caryophyllaeus laticeps* in dace had failed (Kennedy and Walker, 1969) and the contrasting situation with *P. laevis* might reflect differences in the attachment of the two species (Harris, 1973). *C. laticeps* is much less pathogenic than *P. laevis* which causes extensive lesions and provokes formation of a fibrous tissue capsule around its proboscis. *P. laevis* has a more pronounced effect on the gut of chub than on dace (*L. leuciscus*) or grayling (*Thymallus thymallus*) which do not produce antibodies to it. Prakash and Adams (1960) and Chaicharn and Bullock (1967) have shown that the histological reaction of the host is determined by the degree of penetration of the gut wall and immunogenicity may be similarly determined (Harris, 1972). It was not possible to demonstrate skin-sensitizing or anaphylactogenic antibodies in dace and chub immunized with various protein antigens or infected with *C. laticeps* or *P. laevis* (Harris, 1973).

6 Fish worms and man

Introduction

Fish worms are of particular concern to man when: (i) they cause disease and/or death of their hosts in nature, (ii) they affect captive fish, (iii) they are detrimental to human and animal health and (iv) their abundance and prevalence are influenced by man. For these four reasons much attention has been given to the control of fish worms through chemical and a variety of other means. In contrast to man's concern about the pathogenic effects and accidental spread of fish worms, however, is the particular interest shown in these parasites as: (a) biological indicators in fisheries research, (b) indicators of changes in the environment—natural or man-made, (c) biological control agents and (d) 'models' for use in teaching and research. Our aim in this final chapter is to review recent publications on the four areas of concern, the control of fish worms and the four areas where they are of interest or value in extending biological knowledge of either the worms themselves or their hosts.

Fish worm diseases in nature

Since 1886 there has been increased interest in academic parasitology as well as in medical and economic aspects of diseases caused by parasites (Rogers, 1986). We agree with Rogers, although research on helminths as important fish pathogens has lagged considerably behind that on birds and mammals as hosts. This was mentioned by Williams (1961b, 1967), especially with regard to the nematodes of fish. He pointed out that until about 1950 most researchers on fish helminths had dismissed them as being of little or no importance as agents of disease. From this date, greater prominence has been given to fisheries helminthology and fish culture with an accompanying degree of caution in interpreting the effects of helminths on their fish hosts. It should, however, be mentioned that there were notable exceptions amongst fish helminthologists during the first half of this century who also believed these helminths to be potentially dangerous. Some of these researchers were

*Important footnote
The treatments mentioned under control in this chapter are based on a summary of the literature and therefore intended only as a guide to the chemicals which have been tried. We emphasize that all dosages cited should be confirmed from original sources before they are used.

referred to in Williams (1967). Others, including Hofer (1906), drew attention to the importance of the monogeneans *Dactylogyrus, Gyrodactylus, Tetraonchus, Diplozoon* and '*Octobothrium*' (= *Mazocraes alosae*?); the tapeworms *Bothriocephalus, Caryophyllaeus, Ligula* and *Schistocephalus*; many digenean and nematode species now known under different names and several species of the acanthocephalan '*Echinorhynchus*'. Ward (1911) referred to T.H. Bean's plea to the American Fisheries Society in 1910 for a 'systematic study of fish diseases' and stated that the breaking-out of parasite epizootics among wild and cultured fish was already well-known. Ward (1933) suggested that the vastness of oceans and the rapidity with which dead or dying fish are either consumed by scavengers or dispersed by ocean currents had, perhaps, concealed evidence of epizootics due to parasites amongst oceanic fish.

Hubbs (1927) seems to have been the first to point out that fish worms can modify the specific characters of a host fish to such an extent that infected individuals appear to represent a different species. Infections of a cyprinid fish, the minnow *Platygobio gracilis*, with the cestode *Proteocephalus* caused retardation in early growth, retention of late embryonic characters, poor development of melanophores with resultant loss of colour, incomplete differentiation of muscle and fin-rays resulting in soft consistency of body and fins, a pot-bellied condition, eye enlargement, a retarded development of the mouth, snout and nostrils, rudimentary development of the lateral line and the absence of scales in the smaller abnormal young. The presence of digeneans may have enhanced these effects.

It was not until the 1950s and early 1960s that a full response to the views of Hofer, Ward and Hubbs was witnessed when helminthologists in the USSR, various European countries, Malaysia, Canada and the USA began stressing the need for full parasitological investigations not only of all fish but also of all animals closely associated with them. This is evident from a number of sources, including Davis (1953), Prost (1953), Schäperclaus (1954), Duijn (1956), Bauer (1958a), Amlacher (1976), Dogiel, Petrushevski and Polyanski (1961), Furtado (1962), Lyaiman (1963), Kocylowski (1963), Freeman (1964b), Reichenbach-Klinke and Elkan (1965), Roberts (1975 1978a,b), Ribelin and Migaki (1975) and Kabata (1985). Direct and circumstantial evidence from these and other publications and their cited references suggest that we should now accept that fish helminths are potentially dangerous pathogens even among fish in nature, but only with due regard to a number of influential biotic and abiotic factors which may trigger their pathogenicity and to whether the consequent effects are lethal or sublethal. The following selected sources from about 1950 to 1986 support this view.

An early and dramatic account of the effects of worms on fish is given by Miller (1952) from which one must conclude that natural mortality, directly or through predation, is inevitable. Miller refers to the pseudophyllidean tapeworm *Triaenophorus crassus* and its great economic importance with regard to Canadian fishes. One of its larval stages, the plerocercoid, occurs encysted in the muscles of coregonids (whitefish) which are the second intermediate hosts and in which their presence is aesthetically objectionable. So numerous are these cysts that they have interfered with the whitefish industry. Many lakes, potential producers of large quantities of whitefish, were not fished at all or, at best, only lightly so, because the fish were too heavily parasitized to be acceptable on the American market.

The problems of understanding such outbreaks as *Triaenophorus* infestations are many and were emphasized by Grabda and Kozicka (1961) in linking fish diseases in nature with ecological and other biological data, not only with the hosts and parasites directly involved, but also with those of amphibians, reptiles, birds and mammals which are closely connected with the same aquatic habitats. Abiotic factors such as water quality in different lakes, reservoirs, rivers, estuaries and seas, must also be considered. Biotic and abiotic factors in relation to ecology are discussed in Chapter Three of this book.

The predictions of Hoffman and Sindermann (1962) that there were many harmful parasites of fish including *Discocotyle, Clinostomum, Cryptocotyle, Diphyllobothrium, Ligula, Proteocephalus* and *Eustrongylides* was soon followed by good supportive evidence. Thus Chubb (1969) referred to the possibility of serious losses or damage to fish in nature, eg. of whitefish infested with *Triaenophorus*, roach with *Ligula intestinalis*, *Salmo trutta* with *Diphyllobothrium* and of various freshwater fish with *Diplostomum spathaceum*. Chubb (1967) suggested that *Pomphorhynchus laevis* on a cyclical basis was a 'killer parasite' of chub in the Hampshire Avon. The lethal effects of a larval *Contracaecum* sp.[1] in artificially-reared herring, as described by Rosenthal (1967), led Margolis (1970a) to suggest that herring larvae in nature must be subject to the same infection, presumably with the same consequences. A more recent account of parasites as pathogens of herring is given by MacKenzie (1987a).

The precise effects of larval digeneans, especially the diplostomatids, on fish in nature remains controversial largely because of the taxonomic problem as to the exact number of species involved, the host species each is capable of invading and, indeed, the site or sites a particular species may prefer. Haen and Ryan (1967) stated that *Diplostomum spathaceum* is known to cause many deaths among fish. The species has been known since 1832 when Nordmann found 370 worms in the eye of a rudd, *Scardinius erythophthalmus*.[2] Dogiel *et al.* (1961) regarded cyprinid fish as the most common hosts with up to 900 metacercariae being recorded in one eye. It is inconceivable that such infections do not eventually lead to death of the host. But with another strigeoid genus and species, *Posthodiplostomum minimum*, the effects may not be so drastic. Thus Spall and Summerfelt (1970) state that the metacercariae of the strigeid fluke *P. minimum* are generally so numerous in the liver, kidney, heart and other viscera of fish that many researchers have suggested that they must be histopathogenic. They did, however, challenge the claims of Hughes in 1928 and Hunter in 1937 of fish dying in large numbers in nature when heavily infected with the fluke. This is largely because in wild fish, several hundreds of encysted metacercariae in the liver, sinus venosus, heart and kidneys are often observed to have no debilitating effects. Bamber *et al.* (1983) and Bamber and Henderson (1985) found that the population of sand smelt, *Atherina presbyter*, breeding in two areas at Fawley, Southampton and in the Fleet, an elongate lagoon inside Chesil Bank, Dorset, U.K., shows a high infection with a *Neodiplostomum* sp. All age classes are infected. None of the other sand smelt populations, where reasonable numbers were examined, showed levels of infection approaching those of Fawley and Fleet. There was, however, a differential intensity of diplostomiasis between the Fleet and Fawley populations and this is most likely attributable to differential host density at two levels in the life-cycle, ie. mollusc and bird hosts. Kennedy (1984a) followed up suggestions by Gordon and Rau (1982) and Anderson and Gordon (1982) based on two examples from natural fish populations that, subject to important qualifications, a peaked host age-parasite abundance curve, concomitant with a decline in the degree of overdispersion of parasites in the older age classes of hosts, can provide evidence of parasite-induced host mortality. Kennedy used the metacercariae of *Diplostomum spathaceum*, *D. gasterostei*, *Tylodelphys clavata* and *T. podocipina*, the first-named of which occurs in the lenses of the eyes of roach, *Rutilus rutilus*, and the other three in the humour of perch, *Perca fluviatilis*. He could find no clear and unambiguous evidence of host mortality induced by heavy infections of any of the four species. Some digeneans, however, are, according to Millemann and Knapp (1970) pathogenic in salmonid fish in nature. For example, *Nanophyetus salmincola* causes exophthalmia, prolapse of the intestine and may damage the

---

[1] Now a controversial generic name and may be what is now known as *Hysterothylacium*

[2] Bouillon (1987) found 2852 diplostomula in one *Salvelinus alpinus*.

fins, tails, gills, retinas, corneas, heart ventricles, muscle fibres, kidney tubules, pancreas and the gall bladder. Histopathological evidence suggests that practically every organ of the fish is weakened physiologically.

The lack of supporting evidence on the effects of roundworms on fish in nature is well described by Margolis (1970a) who said that few parasitologists have given their attention to the problems caused by marine fish nematodes, the neglect or lack of interest being evident from the lack of samples cited in reviews on tissue reactions to nematodes of vertebrates. Yet the available information shows that nematodes may harm their fish hosts in a variety of ways, from causing a little inconvenience to outright mortality. Direct mortality probably occurs only in certain infections of larval fish, eg. *Clupea harengus*, or in the case of massive infections. Little is known of the sublethal effects which may subsequently lead to reduction in the fish's reproductive potential, greater susceptibility to predation or to other pathogens and to stresses, loss of weight and inhibition of growth. Margolis advocated more field and laboratory observations on the effects of nematodes on blood and other tissues, growth, stamina, performance, ability and other physiological parameters, but it was not until 1976 that a comprehensive fish health programme was announced for the Pacific region of Canada (British Columbia and the Yukon Territory). This was initiated at the Pacific Biological Station, Nanaimo, with priority given to salmonids as the most economically important group of fish. Bell and Margolis (1976) reported many well-known threatening disease agents that were widely distributed in both cultured and wild stocks of the Pacific Region, but surprisingly only *Gyrodactylus* had been responsible for epizootics and then only among fish in totally artificial conditions.

If we are to understand the role of helminths on fish in nature it is to be hoped that comprehensive fish health programmes along the lines mentioned by Bell and Margolis (1976) will be established rapidly in a number of other countries. This is a prerequisite for understanding disease in captive fish. It is also necessary because present knowledge is so scanty and patchy in most countries. Thus Kinne (1980) was able to say that much of what is known of the diseases of marine animals stems from species which: (i) are of commercial interest; (ii) live near the coast or are otherwise easily accessible; (iii) are often used in cultivation and laboratory experiments. Such selection tends to affect, if not distort, the overall picture. It is encouraging to note that more countries are responding to this situation, eg. in Finland, in the first of a proposed series of papers, Bylund, Fagerholm, Calenius, Wikgren and Wikstrom (1979) underline the fact that attitudes towards fish macroparasites had undergone a rapid change. They were of the view that an increasing number of parasites were significant pathogens of fish populations in nature. The results of such a team approach, when considered along with those of isolated publications in various countries, seem to support the need for greater and more coordinated efforts in studying the natural effects of worms on fish. Amongst recent publications which support this view are: (i) Benz and Pohley (1980), who reported exophthalmia in *Lepomis macrochirus* due to a *Philometra* sp. in the oculo-orbit, some orbits having two worms and when this occurred the remaining orbit contained at least one worm. Such infections cannot possibly be conducive to the future health and survival of the host; (ii) Cheung, Nigrelli and Ruggieri (1984), who suggested that *Philometra saltatrix* affected the survival rate of O-class blue fish, *Pomatomus saltator*, in nature; (iii) Paperna, Diamant and Overstreet (1984), who reported mass mortalities in wild populations of grey mullets, *Liza carinata*, in the Gulf of Suez, due to skin and mouth infections of *Benedenia monticelli* (fish with up to 40 individuals of this relatively large worm had eroded snouts and lesions in the mouth and some were emaciated); (iv) Yamamoto, Takagi and Matsuaka (1984) referred to mass mortality in wild anchovies (*Engraulis japonica*) caused by *Pseudanthocotyloides* sp.; (v) Hoffman, Kennedy and Meder (1986), who clearly showed that *Salvelinus alpinus* in South Bavaria is affected by *Eubothrium salvelini*, the condition factor and composition of the blood being altered as intensity of infection per fish increased. Hopefully, such isolated publications will lead to more bibliographies on fish diseases such as that published by Mann (1978) for students, librarians, researchers and practitioners. Mann listed books and symposia on the topic published before 1978 and a list of primary journals and abstracting services. One other reason for intensifying and collating research on marine fish helminths was emphasized by Lester (1978) who drew attention to their effects on the marketability of commercial fish.

Developments in the 1980s towards a greater and better understanding of ichthyoparasitology have been encouraging, as was seen at the First International Conference of the European Association of Fish Pathologists held at Plymouth, U.K. (1983) and the First and Second International Symposia of Icthyoparasitology held in Czechoslovakia in 1983 and in Hungary in 1987. Two new journals have appeared, namely the 'Journal of Fish Diseases' and 'Diseases of Aquatic Organisms'. The latter will cover disease phenomena in marine, limnic and brackish waters for pathologists, parasitologists, veterinarians, physicians, ecologists, aquaculturists, zoologists, botanists and microbiologists. Two recent reviews by Sindermann (1986) and by Dobson and May (1986) indicate our present state of knowledge and the difficulties in interpreting the effects of worms on fish in nature.

Sindermann (1986) was of the view that a generally-held perception of fish worms is that most seem to have little effect on the host, except in rare instances of massive infection. With exceptions, he felt that this perception is to a large extent correct. He advocated approaching the topic by distinguishing lethal and sublethal parasites and also effects on wild populations of fish from those held in farmed or captive conditions. Sindermann refers to the methods suggested by Lester (1984) in estimating mortality caused by parasites in wild fish populations, namely (i) using a statistical approach based on observed prevalences over time in natural populations; (ii) detailed autopsies; (iii) determining the frequency of infections known to be eventually lethal; (iv) observing a decrease in prevalence of a long-lived parasite with host age; (v) observing a decrease in variance/mean ratio for the parasites with host age; (vi) comparing the observed frequency of a combination of two independent events with the calculated probability of their occurrence; and (vii) the long-established method in fisheries biology described by Munro *et al.* (1983) comparing recovery rates of tagged, healthy and diseased fish. It seems that not one of these seven methods, when used alone, provides a solution. With regard to sublethal effects, Sindermann referred to *Philometra saltatrix* causing cardiac dysfunction and reproductive impairment in blue-fish, *Pomatomus saltatrix*, as described by Cheung *et al.* (1984); and to *Anisakis* sp. causing impaired liver function in blue-whiting, *Micromesistius poutassou*, as described by Wootten and Smith (1976), Smith and Wootten (1978b) and Karasev *et al.* (1981).

Dobson and May (1986) stated that parasitic worms may be important in reducing fish numbers by mortality, eg. the Aral sea sturgeon by the monogenean *Nitzschia sturionis*, or by reducing the weight of individual fish, eg. various marine teleosts by the larval tapeworm *Grillotia erinaceus* or the roundworm *Porrocaecum decipiens*, now *Pseudoterranova decipiens*. It also seems likely, according to Dobson and May, that freshwater fish will be more susceptible to invasions by parasites if the hosts have no previous experience of infestation. They discussed the classic example of the collapse in the mid-1930s of a highly lucrative '400 ton per year harvest' of the sturgeon *Acipenser nudiventris* in the Aral Sea after the introduction from the Caspian Sea of another species, *A. stellatus*, infected with *Nitzschia sturionis* which was transmitted to and multiplied on *A. nudiventris*. On the basis of this information they predict that the introduction of new species into Lake Malawi could lead to similar introductions of novel pathogens which might destroy the large multispecies fishery based in this lake and also disturb the structure of a unique and highly-evolved ecosystem. The adult stages of parasitic helminths are usually only mildly pathogenic to fish. However, where parasites use fish as the paratenic host in a complex life-cycle, they are common sources of both mortality and economic loss to fisheries. Dobson and May cite *Pseudoterranova decipiens* and *Anisakis* sp. as good examples. They also suggested that parasites may be important in determining the structure of fish communities where multispecies fisheries are involved, as had been shown by Kennedy and his co-workers on the effects of introducing the pseudophyllidean cestode *Ligula intestinalis* into a lake containing roach and rudd. The chance introduction of the parasite reduced the numbers of roach and apparently increased the numbers of rudd. There was also a decrease in the levels of stunting observed in roach.

Worm disease in captive fish

Captive freshwater and marine fish have, for a long time, been successfully maintained in small-to-large aquaria and also in ornamental and commercial ponds of varying size. In many cases species have been bred and reared in these situations. Consequently, a vast literature has accumulated on fish aquaculture and on diseases of captive fish. This is particularly so for fish-farming in fresh water. Much less is known of farming in a marine environment. It is thought that the annual production of farmed fish will increase from a few million 'tonnes' at present to about 40 to 50 million 'tonnes' by 2000 AD. Bauer, Egusa and Hoffman (1981), in reviewing the literature on parasites of cultivated fish for the period 1974–78, estimated that, in 1975, 8–9 per cent of the world's total catches of fish and shellfish was based on aquaculture. In 1972 the Indo-Pacific region accounted for 88 per cent of the world's 3,657,373 tonnes of fish produced through aquaculture (Kabata, 1985).

A world-wide increase in fish-farming has been accompanied by an intensification of effort to control parasites, especially in the USSR, Europe, North America, the Near and Middle East, the Indo-Pacific region, Africa and in Central and South America. Not surprisingly, therefore, information on aquaculture and the diseases of captive fish is already widely disseminated in various academic books and journals and also in more popular sources aimed at fish-farmers. This knowledge on captive fish can be categorized under the following headings: (i) general sources covering fish-farming and/or all fish diseases; (ii) studies on fish hosts in various countries; (iii) intensive studies on particular helminths; (iv) mariculture; (v) aquarium and ornamental fish; (vi) spread of disease within countries and transcontinentally; (vii) control measures; and (viii) host-parasite interactions (pathology, pathogenesis and immunity). The last area of investigation has already been discussed in Chapter 5, whilst the 'spread of disease' and 'control measures' are treated separately in this chapter. Our aim here is, therefore, to refer to selected publications on topics (i) to (v) of those listed above.

General sources of information

Amongst a number of books on aquaculture are those by Brown and Gratzek (1980), Huet (1986), Lewis (1979), Milne (1972), Sedgwick (1978), Kabata (1985) and Spannhof, Reimer and Jürss (1989). Journals which cover the subject include *Aquaculture*, *Biologist*, *FAO Aquaculture Bulletin*, *Fish Farming International*, *Fish Pathology*, *Journal of Fish Biology*, *The American Fish Farmer*, *The Commercial Fish Farmer* and *Aquaculture*, *the Progressive Fish Culturist* and the *Journal of Fish Diseases*. Particular aspects of the subject are treated in papers by Ayles (1977), Elson (1971), Finlay (1978), Jones (1972), Needham and Wootten (1978), Roberts and Sommerville (1982), Shepherd (1978) and Varley (1977). Many publications covering the scientific and technical disciplines applicable to the field of aquaculture are listed by Mann *et al.* (1982) and Huet (1986).

It is immediately apparent from many of these sources on aquaculture that worm infections which have had the most impact in fish-farming are those caused by monogeneans, larval cestodes and digeneans and, to a lesser extent, adult cestodes and acanthocephalans. Monogeneans, having a direct life-cycle, rapidly produce infections of gigantic proportions on fish in confined areas. Tapeworms such as species of *Bothriocephalus* (the only intermediate hosts of which, in freshwater, are Copepoda) are also serious pathogens and very difficult to eliminate from ponds and hatcheries. A repeatedly recurring theme is that gut parasites, being less pathogenic than others, have received comparatively little attention. Many of the accounts of fish diseases in the context of fish-farming, however, refer only briefly to helminths or, at best, give very short accounts of their life-cycles with, perhaps, some indication of prophylaxis and treatment. We draw attention, therefore, to Table 6.1 for more specific information on various helminthiases amongst captive fish:

Table 6.1 Helminthiases amongst captive fish

Topic	References
General accounts of fish diseases and icthyopathology	Amlacher (1976); Bauer *et al.* (1977); Dogiel, Petrushevski and Polyanski (1961); Duijn (1973), Grabda (1977a), Lucky (1977), Lyaiman (1963), Möller and Anders (1986), Needham and Wootten (1978), Osetrov (1978), Reichenbach-Klinke and Elkan (1965), Roberts (1978a), Schäperclaus (1954), Sindermann (1970a, b) and Sniezsko, Hoffman and McAllister (1979);
Cestode and nematode infections	(Freeman, 1964b);
Diseases of carp, trout and white Amur and silver carp	Verbitskaya *et al.* (1972)
Parasites of fish reared in reservoirs in various countries throughout the world	Hoffman and Bauer (1971)
Aquaculture and fish diseases of zoonotic significance	Ghittino (1972)
General review of fish diseases and treatment of gastrointestinal helminths	Pearson (1972)
Diseases of farmed fish generally	Hill (1977), Mikityuk *et al* (1984), Musselius and Golovin (1984), Odening (1989) and Osetrov (1978)

These various sources show that much of the work done on parasites of farmed fish has been carried out in the USSR. Some useful reviews by Russian authors include accounts of fish parasites in hydroelectric reservoirs by Bauer and Stolyarov (1958); of fish acclimatized to new environments by Petrushevski (1958b); and species most commonly farmed in the USSR by Bauer (1958a,b,c, 1959b, 1968); fish in natural water bodies of the USSR by Petrushevskii and Shul'man (1958); and an historical account by Bykhovskii and Gusev (1964) which discusses the biology of some of the most pathogenic worms.

Some host fish studies in various countries

Important studies on parasites of particular fish groups and species are indicated in Table 6.2.

In addition to the above, many surveys of fish parasites have been made in particular geographical areas or countries, or of economically-important fish species, largely with fish-farming in mind. Only a few examples can be given here to indicate the type of work carried out on diseases of fish in Portugal by Carvalho-Varela et al (1981); in Finland by Christensen, Jensen and Rasmussen (1963); in Italy by Ghittino (1963); in Denmark by

Table 6.2 Some host fish studies in various countries

Carp and trout in Europe	Körting (1975, 1984a,b)
Carp in Japan	Nakajima and Egusa (1974a,b,c., 1977a,b,c,d)
Carp in Malaysia	Furtado (1961)
Carp in the Volgograd and Vinnitsa areas of the USSR and in the Ukraine	Kirichencho (1974), Yashchuk (1974) Shcherban (1965)
Channel catfish *Ictalurus punctatus* in the USA	Hoffman (1979)
Cichlid fish in Southern Africa	Paperna, van As and Basson (1983)
Coregonids in the USSR	Bauer (1970)
Eels in Europe	Buchmann Mellergaard and Køie (1987), Kennedy and Fitch (1990), Koops and Hartman (1989). Taraschewski *et al.* (1987)
Fish in Africa	Sarig (1976)
Fish in the Alta territory, USSR	Nikulyna and Kirillova (1975)
Fish in East Germany	Kulow (1982)
Fish in Japan	Egusa and Nakajima (1973)
Fish in the USA	Ghittino (1972)
Salmo, Oncorhynchus, Thymallus and *Salvelinus* in various countries	Roberts and Shepherd (1974)
Fish in warm-water ponds in North America	Meyer (1968)
Flounder in Britain	MacKenzie and Gibson (1970)
Grass carp, *Ctenopharyngodon idella* in the USA	Riley (1978)
Plaice in Britain	MacKenzie (1968), MacKenzie and Gibson (1970) MacKenzie, McVicar and Waddell (1976)
Rutilus rutilus and *Stizostedion lucioperca* in the Azovo-Kubanskii region of the USSR	Terekhov (1976)
Salmo salar in the Leningrad area	Bauer (1957)
Salmonidae and freshwater fish in Czechoslovakia	Pacak (1957), Dyk (1956)
Salmonids in Britain	Richards (1983), Stables and Chappell (1986a,b,c)
Salmonids in Spain	Cordero del Campillo (1974)
Salmonids in Norway	Halvorsen and Hartvigsen (1989), Johnsen and Jensen (1988), Mo (1989)
Three species of *Mugil* (mullet) in Israel	Paperna (1974)
Trout and other fish in the USA	Davis (1953)
Whitefish (*Coregonus*) in the USSR	Bogdanova (1982).

Kocylowski (1963); in Poland by Ojala (1963); in the Vologda region of the USSR by Kudryavtseva (1961); and on parasites of fish in the Surkhandariya river basin in the Uzbek SSR and those of *Salmo gairdneri* on fish farms on the rivers Pliva and Buna in Yugoslavia by Zitnan and Cankovic (1970/1971). Many other examples will be found throughout this book, especially in the chapters on ecology and life-histories, because it is now accepted that more information on these two disciplines in particular is essential to the control of parasites of captive fish.

Intensive studies on particular helminths

With the exception of aspidogastreans, members of all major groups of helminths referred to in Chapter One are implicated in causing diseases of fish under captive conditions. Examples from amongst these will, therefore, be given below and in the same order as they were introduced in Chapter One.

Turbellaria

Ichthyophaga subcutanea, according to Menitskii (1963), attacks the anal and branchial regions of *Cephalopholus pachycentron, Bero elegans* and *Hexagrammos decagrammus* in marine aquaria. Cannon and Lester (1988) list all records of turbellarians on fish. From a recent

(1988) personal communication with Professor D.A. Conroy, University Central de Venezuela, we have been made aware of a turbellarian as the possible cause of mortalities of 'pompanos' in post-larvae and young fish.

Monogenea

Paperna (1960a) stressed the importance of monogeneans as parasites of bred fish in that they were a frequent cause of mortality especially among fry and fingerlings raised in artificial conditions of hatcheries or fish ponds. He emphasized the importance of understanding a range of ecological factors and the degree of specificity of monogeneans before initiating a fisheries operation. Of particular importance are members of *Dactylogyrus* and *Gyrodactylus* and, as research progresses, new pathogenic species of these two are being discovered as well as other potentially harmful genera. Prost (1963) carried out intensive investigations on the development and pathogenicity of *D. anchoratus* and *D. extensus* on carp, *Cyprinus carpio*, raised on fish farms in Poland. She described the eggs of the species, their resistance to drying and freezing, the effect of temperature on the number of eggs laid, the fate of the laid eggs, the development of the larvae within an egg, the viability of the free-living larvae, the route of infection of carp, the age of carp and their susceptibility to infection, the development of larvae and the survival of the parasite on carp gills. It was on the basis of such an intensive study that practical recommendations were possible on the control of dactylogyrosis in culture ponds; and these are largely based on the fact that eggs cannot overwinter in a viable state at the bottom of drained ponds. Paperna (1964a) reported on interesting and significant differences in the ecological adaptations of *D. extensus* in Israel compared with those it adopts in the USSR and elsewhere. Unlike its behaviour in the USSR, the species in Israel thrives at normal summer temperatures of 24–28°C (optimum in the USSR is 17°C) and the percentage of eggs that hatch is undiminished at these relatively high temperatures (100 per cent hatch at 28°C). However, unlike the northern *D. extensus*, the Israeli form is also resistant to prolonged oxygen deficiency even in water with a dissolved oxygen content of 0.4 mg/litre, a concentration which is harmful to the host carp fingerlings. In Israel, as in eastern Europe, *D. extensus* can tolerate relatively high salinity.

Carp, *Cyprinus carpio*, the host of *D. extensus*, was first introduced from eastern Europe into Israel in about 1920. It seems that the present Israeli population of *D. extensus* has been able to survive and multiply in ecological conditions which are considerably different from their former habitat. Natural selection seems to have established, in a relatively short time, a population of *D. extensus* morphologically indistinguishable from its recent ancestors, but which appears to be quite different in its response to environmental factors.

Bauer, Musselius and Strelkov (1973) listed *D. vastator, D. extensus, D. anchoratus, D. achmerovi* and *D. minutus* as causes of dactylogyrosis on carp, *Cyprinus carpio*. Only *Dactylogyrus vastator* and *D. extensus*, however, were thought to be of epizootiological importance. *D. vastator* thrives in water of 22–28°C, ie. at a temperature in which carp may die due to oxygen deficiency, and at a pH of 6.5–7.1. Thus it will thrive in small, shallow, warm-water bodies. It is especially dangerous to young carp fry of about 2–5 cm long. The susceptibility of carp to *D. vastator* and the pathology of the disease is described in Chapter Five. Intensive research on these aspects continues in the USSR as may be seen, for instance, from Golovin (1987) and Golovina (1987). In contrast to *D. vastator, D. extensus* occurs in large numbers on young and adult fish. It is more pathogenic than *D. vastator* but is more sensitive to a deficiency of oxygen and high temperature. Thus, *D. extensus* is less dangerous than *D. vastator*.

Species of *Dactylogyrus* infest only one, rarely two, host species of the genus *Cyprinus*. The

Table 6.3 Other recent researches on *Dactylogyrus* on captive fish

Species and host	Author
D. vastator on *Cyrpinus carpio*	Bohl (1975)
Dactyloqyrus sp. and other helminths in fish farms	Ghittino (1979),
D. nobilis and *D. lamellatus* on *Hypophthalmichthys molitrix, H. nobilis* (big head carp) and *Ctenopharyngodon idella*	Lucky (1984)
D. lamellatus on *C. idella*	Razmashkin (1984)
Dactylogyrus in fish farms	Shoshkov and Georgiev (1983)
D. vastator and *D. extensus* on *Cyprinus carpio*	Samman (1987)
D. extensus on carp	Skomorokhova and Kashkovskii (1979),
D. vastator on carp	Uspenskaya (1961)

pond-reared grass carp, *Ctenopharyngodon idella*, can host both *D. lamellatus* and *D. ctenopharyngodontis* and these two species have spread extensively to other countries with the transport of *C. idella* from China and the Amur, USSR. Similarly, two other species of pond-reared plant-feeding carp, the silver carp, *Hypophthalmichthys molitrix*, and the bighead, *H. nobilis*, may be hosts to more than one species of *Dactylogyrus*, namely *D. hypophthalmichthys, D. suchengtaii, D. nobilis* and/or *D. aristichthys*. Other work on *Dactylogyrus* is referred to in Table 6.3.

Gyrodactylus

We have estimated from Yamaguti (1963a) and Malmberg (1970) that over 300 species of *Gyrodactylus* were known at that time. In 1988, 400 were listed in the catalogue of the Parasitic Worms Section, British Museum (Natural History), London (R. Bray, personal communication). They are common ectoparasites on the gills and body surface of marine and freshwater fish and frogs. Some species are potentially serious pests in fish hatcheries as may be seen, for example, from the reports of Ghittino (1979), Krasilshchikov and Lyubina (1984) and Shoshkov and Georgiev (1983). Whilst there are no difficulties in distinguishing *Gyrodactylus* from *Dactylogyrus*, differentiation between species of *Gyrodactylus* is not easy. As will be seen later, this taxonomic problem is an important aspect of gyrodactyliasis in aquaculture.

Four species, *G. elegans, G. medius, G. cyprini* and *G. sprostonae* attack carp, *Cyprinus carpio*, in ponds and another species, *G. ctenopharyngdontis* infests *Ctenopharyngodon idella* (Bauer, Musselius and Strelkov, 1973). *Gyrodactylus katharineri* may be found in numbers of up to one million on 1-year-old carp reared in cages in water receiving heated effluent. Parasite numbers increased dramatically when temperature ranged from over 5°C to about 20°C but at about 25°C they die (Bauer, Egusa and Hoffman, 1981). The detailed studies of Gelnar (1987a) and Ergens and Gelnar (1985) on *G. katharineri* are discussed in Chapter Three, p. 263. Rawson and Rogers (1973) found that populations of *Gyrodactylus macrochiri* were at high levels on bluegill, *Lepomis macrochirus*, and largemouth bass, *Micropterus salmoides*, during the host's spawning in spring and at temperatures similar to the spawning period in autumn and winter. During mid-summer, at temperatures above 28°C, the populations on both host species were at their lowest levels. The authors emphasize that *Gyrodactylus* populations, even in nature, expand rapidly under optimal conditions and are, therefore, a constant threat in fish hatcheries. They suggest that the periods of abundance of *G. macrochiri* are the times when maximum benefit can be obtained from prophylactic treatments to reduce the stress on adult stock and to prevent infection of young fish. Ogawa and Hioki (1986) described two new species of *Gyrodactylus, G. egusa* and *G. joi*, on greenhouse pond cultures of *Anguilla japonica* in the Yoshida area, the centre of eel culture in Japan where the annual production was over 3,000 tonnes in 1984. They studied changes in the occurrence of these two species and of *G. nipponensis* on cultured eels between 1980 and 1985. Gyrodactylids became less frequent throughout the period and were also less frequent in summer than in other seasons. The quality of the pond water had deteriorated markedly throughout the period because the high cost of oil used for boilers to heat water prohibited frequent changes.

Species of *Gyrodactylus* which attack Salmonidae have become a matter of concern in recent years but, as suggested earlier, many difficulties occur in determining the species involved and whether a particular highly-pathogenic species has been introduced into a country or has never been detected there before due to a lack of interest in fish helminths. Malmberg (1987a,b) refers to a *G. salaris* group of species and to a *G. truttae* group. He suggested that *G. truttae* occurs on trout in Britain and other countries whilst a closely-related species, *G. derjavini*, infests *Salmo gairdneri* in Sweden, Norway, Denmark and Italy. *Salmo salar* in Scotland harbours a species similar to but not identical with *G. derjavini*. Halvorsen and Hartvigsen (1989) provide an invaluable review of the *Gyrodactylus* problem in Norway.

According to Ogawa (1986) a hitherto unidentified *Gyrodactylus* is a common parasite of cultured fish of the salmonid species *Oncorhynchus masu, O. rhodurus* and *Salmo gairdneri* in Japan. He named the species *G. masu*. It is suggested that its occurrence on rainbow trout, first introduced to Japan from America in 1877, is an example of transfaunation to trout after its arrival in the form of live eggs in Japan. The devastating effects of a *Gyrodactylus* sp. on salmon may be seen by taking *G. salaris* in Norway as an example. From 1980 to 1985, Johnsen and Jensen (1986) investigated 212 Norwegian rivers for the occurrence of *G. salaris*. It was found in 26 rivers and also in six salmon hatcheries. There was a clear connection between the distribution of *G. salaris* in Norwegian rivers and known deliveries of stock fish from infected hatcheries. Infestation with *G. salaris* caused a great reduction and near-extermination of populations of salmon parr. Infestations were characterized by sudden outbreaks, often with thousands of parasites on a single fish combined with fungus attacks resulting in death of salmon parr. A primary goal of Norwegian fishery authorities is the prevention of further spread of *G. salaris* and its extermination in as many rivers and hatcheries as possible. It is claimed that extermination was accomplished in the River Vikja in Sogn by rotenone treatment in 1981 and 1982. *Gyrodactylus salaris* was not found in the river after treatment and the population of salmon parr is growing. An infested hatchery has also been successfully disinfected by drying. According to Mo (1987a), *G. salaris* was first observed in a Norwegian river in 1975 and it was estimated that the loss in that year to the Norwegian salmon fisheries was about 300 tonnes. Mo thought the species was probably imported into Norway with live salmon parr and smolts to meet the demands of a rapidly-increasing fish-farming industry. Within 10 years it is thought to have spread to 28 rivers and 11 hatcheries along the west and north-west coast of Norway from Tromsø county to Sogn og Fjordane. Although not reported in Britain, it has become a notifiable parasite in the country (personal communication, Dr. R. Bray).

Margolis and Evelyn (1987) confirmed that species of *Gyrodactylus* cause mortalities in freshwater salmonid hatcheries. These authors cited *G. nerkae* and *G. salmonis* in British Columbia as examples, but pointed out that they were not the focus of research because they are readily controlled by chemotherapy.

Other monogeneans known to have caused problems under aquaculture conditions are listed in Table 6.4. Because all those listed are marine species, they are referred to again below under mariculture.

Table 6.4 Other problematical monogeneans in aquaculture

Parasite	Host and locality	Reference
Benedenia sp.	Species of mullet in Israel	Reiss and Paperna, 1975
	Hybrids of *Oreochromis mosambicus* and *O. aureus*	Khalil, Robinson and Hall, 1988
Benedenia sp. and *Bicotylophora trachinoti*	'Pompano'	Sindermann, 1974
Bivagina tai	cultured red sea bream, *Pagrus major*, in Japan	Ogawa, 1988
Diplectanum aequans	*Dicentrarchus labrax* at Sete and Brest, France	Paperna and Laurencin, 1979
	Dicentrarchus labrax	Silan and Maillard, 1986
Diclidophora tetrodontis	'Puffers'	Bardach, Ryther and McLarney, 1972
Entobdella soleae	Dover Sole, *Solea solea*, in Scotland	Anderson and Conroy, 1968
Heteraxine heterocerca	Yellowtail in Japan	Hoshina, 1968
Microcotyle sp.	Fingerling *Siganus* in Israel	Reiss and Paperna, 1975
Polylabris tubicirrus	Fingerlings of *Sparus aurata* in the Mediterranean	Silan, Cabral and Maillard, 1985

Eucestoda

Adults of the genera *Caryophyllaeus, Khawia, Cyathocephalus, Bothriocephalus* and *Proteocephalus* and larvae belonging to *Diphyllobothrium, Digramma, Ligula, Triaenophorus* and *Schistocephalus* may be problematical in captive fish. According to Bauer, Musselius and Strelkov (1973), *Caryophyllaeus fimbriceps, K. sinensis* and *B. acheilognathi* are found in pond cultured Cyprinidae. The first two species occur in *Carassius carassius* (Crucian carp) and its hybrids and the third in *Cyprinus carpio, Carassius carassius, Ctenopharyngodon idella* and other Cyprinidae. Adults of *Cyathocephalus truncatus* infest trout and other Salmonidae. The larvae of *Triaenophorus nodulosus* inhabit the liver of trout and cause mortality on trout farms; the adult occurs naturally in the intestine of pike, *Esox lucius*. The plerocercoids of *Ligula intestinalis* and of *Digramma interrupta* occur in the bodycavity of many freshwater fish, especially members of the Cyprinidae. The adults inhabit the intestine of fish-eating birds. Some of the genera and species named above are the subject of more detailed discussion below.

Khawia sinensis

Carp in fish farms in central Russia were once thought to be infested to a harmful degree with *Caryophyllaeus laticeps*, but Musselius, Ivanova, Laptev and Apazidi (1963) identified the parasite as *Khawia sinensis*. The rate of infestation in 2-year old carp increased up to June and fell sharply by late August. Thirty-five to 45 specimens of *K. sinensis* per fish were often fatal. Körting (1975) reviewed the distribution of *K. sinensis* in Europe and thought its importance in carp was increasing, and it was also thought to be epizootiologically important by Skomorokhova and Kashkovskii (1979). *Khawia sinensis* infection of carp in breeding ponds in four regions of Poland continued to rise between 1977 and 1981, the average infection rate varying from 21.2–32.4 per cent according to Niemczuk (1984). The prevalence of *K. sinensis* on carp farms in the GDR was reviewed by Weirowski (1979) who thought it caused little loss among the fish although it delayed growth. There were usually 2–3 cestodes per fish and 10–20 per cent of carp were infested.

Bothriocephalus acheilognathi (= *B. gowkongensis, B. opsariichthydis*)

As with diplostome digeneans the literature on *Bothriocephalus acheilognathi* is extensive but no detailed accessible reviews of the species have been published. Our discussion of the species is highly selective. The world-wide spread of *B. acheilognathi* amongst Chinese carp and other fish species in warm water was mentioned by Bauer, Egusa and Hoffman (1981). The species is indigenous to the Amur River and southern China where it infects the Chinese silver carp *Hypophthalmichthys molitrix*, grass carp *Ctenopharyngodon idella*, and other cyprinids (Scott and Grizzle, 1979). *Bothriocephalus acheilognathi* has spread into eastern and central Europe, sometimes causing mortalities as high as 100 per cent of carp, *Cyprinus carpio*, in German pond hatcheries (Körting, 1975). The tapeworm was found in the 1970s in grass carp, golden shiner *Notemigonus crysoleucas*, and fathead minnow *Pimephales promelas*, in hatcheries in the south-eastern United States. Scott and Grizzle (1979) described the pathology caused by the tapeworm in these fish.

Chubb (1980a) recorded its occurrence in the British Isles while Shoshkov and Georgiev (1983) referred to it as the cause of economic losses to the fishing industries in Bulgaria. Over a five-year period the occurrence of *B. acheilognathi* was studied in 220 carp farms in the GDR and reported by Weirowski (1984). The prevalence of infestation ranged from 8.4–29.1 per cent (average 14.5 per cent) of the fish stocks and the intensity of infection was 1.6–3.4 (average 2.6) worms per fish. On highly intensive farms the prevalence of infection reached 45–80 per cent and for short periods an intensity of 10–25 worms per fish. It was estimated, however, that carp production suffered losses of only 1–1.5 per cent due to *B. acheilognathi* infestation. Krotenkov (1985) is among a number of Russian helminthologists who have described the epizootiology of bothriocephaliasis and its occurrence throughout the year, in farmed carp in the USSR. Priemer (1987), however, found that prevalence and intensity of *B. acheilognathi* in carp in the GDR was seasonal and depended additionally on water temperature. The length of the tapeworms also showed the same dependence. Consequently there was a high prevalence and intensity from May to October/November with peaks in May and November. In artificial warm-water systems the tapeworms also produced eggs in early Spring.

Proteocephalus spp.

The stocking of small water bodies with larvae and fingerlings of valuable fish species to culture marketable specimens has led to the realization that well-known adult tapeworms, previously thought to be harmless, can be highly pathogenic under artificial conditions. A good example is *Proteocephalus exiguus* in *Coregonus peled*. The tapeworm is widely distributed in North America, Europe and Asia. Concentrated stocking of the fish is known to have led to increased burdens of *P. exiguus* (up to 1,800 per fish) and resulted in exhaustion, anaemia, decrease in fat content and, in some cases, mortality of the host (Bauer, Egusa and Hoffman, 1981). Of greater significance has been the spread of adult *Proteocephalus* spp. from indigenous hosts such as *Coregonus* to cultured salmonids or of their plerocercoid stages which, in some species, are notable for their lack of host-specificity to salmonids. An example of the former situation was noted by Priemer (1987) and Zitnan and Hanzelova (1987) in the GDR and Czechoslovakia respectively, and of the latter by Becker and Brunson (1968). Priemer and Goltz (1986) described species-determination problems in *Proteocephalus*, especially those from Salmonidae and Coregonidae and reported, for the first time, *P. exiguus* as a parasite of *Salmo gairdneri*. Previously this species was thought to be a typical parasite of coregonids and restricted to this family of fish. Its ability to establish in trout and complete the life-cycle was confirmed by Priemer (1987). Zitnan and Hanzelova (1987) stated that, in 1980, they found

P. neglectus for the first time in rainbow trout in Czechoslovakia and that, by 1986, it seemed to have spread to 17 localities of various water types in the country. They concluded that proteocephalosis was an epizootiologically and economically important disease of rainbow trout. Its epizootiological importance had, in fact, been reported by Becker and Brunson (1968) following their researches on *P. ambloplitis*, the bass tapeworm, and the problems it causes in North America where it is a common parasite of the Centrarchidae, the bass family. Concentration of larvae in the reproductive organs of bass often resulted in parasitic castration. The tapeworm requires largemouth bass, *Micropterus salmoides*, as a definitive host. The plerocercoids are notably lacking in host-specificity.

Centrarchids, including black basses (three species of *Micropterus*) are not endemic to the American Pacific Northwest but were introduced in the nineteenth century (Becker and Brunson, 1968). Undoubtedly *P. ambloplitis* was introduced into Washington waters along with these fish. There was also an early introduction of these exotic fish into California, Oregon and Washington by the US Fish Commission, not realizing that some of the bass harboured a cestode infective to native fish. The establishment of the parasite was aided by: (i) the successful adaptation of black bass, (ii) the absence of terrestrial animals in the life-cycle, (iii) the low specificity of the procercoids and plerocercoids for crustacean and teleost intermediaries respectively, and (iv) a sufficiently warm summer temperature for the tapeworm to mature in its final host. Consequently, Becker and Brunson (1968) concluded that most trout species may become infected by the bass tapeworm. It does not, however, follow that *Proteocephalus* spp. are always harmful. Ingham and Arme (1973) found rainbow trout, *Salmo gairdneri*, from a fish farm, to be infected with two cestodes, *Eubothrium crassum* and *Proteocephalus* sp. They suggested that the parasites had no effect on fish growth or on the intestinal adsorption of L-leucine and D-glucose. This is fortunate because the elimination of these parasites from fish farms fed by a river water source would be costly and extremely difficult to achieve.

Larval tapeworms

The larvae of three genera of tapeworms in particular, *Diphyllobothrium, Ligula* and *Triaenophorus*, are thought to be important in causing diseases of fish (Williams, 1967). For this and other reasons many references to these genera are found throughout this book, including this chapter. There are few records, however, of these tapeworms being problematical in captive fish. Wootten and Smith (1979), Pogosyan and Grigoryan (1983), Georgescu *et al.* (1981) and Bauer and Solomatova (1984) are examples. The last of these authors discussed *T. crassus* as a serious pathogen of pond-cultured trout with a prevalence of up to 75 per cent during July in young fish cultured in cages in power-station effluent at a water reservoir in the Upper Volga. Much is known of the biology and life-cycles of these three tapeworm genera and, therefore, control is possible by preventing their natural final hosts coming into close contact with cultured fish. The taxonomy even of the hitherto difficult genus *Diphyllobothrium* is now much better understood as can be seen from Anderson and Gibson (1988).

Digenea

Although we shall refer to only two families of digeneans here, the Sanguinicolidae and Diplostomidae as well-known pathogens of captive fish, it is probable that these will be added to in future as research intensifies on the parasitic worms of farmed fish. Some likely candidates are *Acetodextra, Crepidostomum* and various didymozoids as indicated by Williams (1967). Other digeneans are mentioned later in this chapter under mariculture.

Sanguinicola

Adults of four species of this genus may be found in the blood system of cultured fish in the USSR but only one, *S. inermis* in *Cyprinus carpio*, was of epizootiological importance (Bauer, Musselius and Strelkov, 1973). According to Lee, Lewis and Sweeting (1987), this fluke was recorded for the first time in the UK in 1977 when it caused mass mortality in carp. They thought it had been introduced into the country in the early 1960s and then spread through the restocking of fisheries and angling sites during the expansion of the carp-farming industry and had become a serious problem. It is not host-specific to carp as initially thought but is now recorded with increased frequency in tench. There are many references to this important genus throughout this book for those who wish to seek further information on other aspects of its biology.

Larval digeneans

Larval Digenea are well-known pathogens of fish as may be seen from Williams (1967) and Chapter Five of this book. According to the species or groups of species, they will attack the skin, muscle, nervous system, eyes or the alimentary canal and its associated organs. Often they will seek specific sites within these tissues or organs.

Intensive studies have been carried out for a long time on *Diplostomum* and related genera with particular attention being given to solving taxonomic and other biological problems with the aim of better methods of control. The literature on diplostomes is extensive and widely scattered in a range of journals. For this reason, our treatment of the subject could only be superficial. Hopefully, detailed accessible reviews on diplostomes will be written by experts on these digeneans in the near future. Bauer, Musselius and Strelkov (1973) thought that the larvae of at least four species of *Diplostomum* occurred in pond fish and caused disease. Adults inhabit the intestine of birds, mainly gulls and terns. Three of the four species attack the lens of the eye of fish and the fourth penetrates the musculature. These authors cautioned against confusing *Diplostomum* with *Posthodiplostomum*, the cause of a disease known as 'black spot'. Attention was also drawn to six species of *Tetracotyle* attacking the serous membranes, intestine, heart and peritoneum of cultured pond fish, and although three were deemed to be important, control was impossible because so little was known of their life-cycles. Eight years later, Bauer, Egusa and Hoffman (1981) were able to refer to 13 species of *Diplostomum* (eye-fluke) in USSR waters with the comment that in Europe and North America they had often been erroneously identified as *D. spathaceum*. Bauer, Vladimirov and Mindel (1964) reported that species of *Coregonus* are heavily attacked by the cercariae of *D. spathaceum*. Experimentally, young *Coregonus* (weight 1.5–2 g) died within an hour of receiving an infestation of 120–150 cercariae to every millilitre of water. Their results were of significance to Czechoslovakia and the GDR because there large-scale production of *Coregonus* had been started. *Posthodiplostomum cuticola* was also found to be important with regard to four species of the Cyprinidae. Shigin (1986) collates data on *Diplostomum* metacercariae found in the USSR and neighbouring countries.

In the light of the foregoing information and knowledge that massive infestations of diplostomid cercariae can be lethal to fish (Paperna and Lengy, 1963) it is not surprising that research on this group has intensified. It is becoming increasingly apparent from some helminth surveys of freshwater fish that diplostomid metacercariae dominate the fauna, as was found, for instance, by McGuigan and Sommerville (1985). Other researches on the group are listed in Table 6.5

Table 6.5 Other work on larval digeneans

Reference	Topic
Deufel (1975)	Several hosts in the life-cycle of *Diplostomum (Proalaria) spathaceum* and on *Posthodiplostomum cuticola* in Germany
Ghittino (1979)	*D. spathaceum* in Italy
Hoffman, Scheinert and Bibelriether (1991)	*Apatemon cobitidis* in *Cottus gobio*, two sites of infestation
Heckmann (1983)	*D. spathaceum* in various fish.
Hendrickson (1978a,b, 1979, 1986)	*D. spathaceum, D. scheuringi* and *Ornithodiplostomum* experimentally and in natural hosts in the USA
Pike (1987)	Experimental infections of perch, rainbow trout and sticklebacks with *D. gasterostei* and *D. spathaceum* in Scotland
Zhatkanbayeva (1987)	Diplostomiasis in carp in the USSR
Razmashkin (1984)	Dominance of *Diplostomum* spp. including *D. paracaudum, D. spathaceum* and *D. commutatum* in grass carp, *Ctenopharyngodon idella* in the lake fisheries of the Tyumen region, USSR
Stables and Chappell (1986a,b,c.)	Distribution and abundance of *D. spathaceum* in farmed *Salmo gairdneri* in Scotland
Shigin (1980)	Intensive 7-year study of fish as hosts for *Diplostomum* spp. and role of *Larus ridibundus* and four mollusc species in the life-cycle
Vismanis *et al.* (1984)	Epizootiological importance of *Diplostomum* in *Salmo gairdneri, Oncorhynchus kisutch* and their hybrids bred in marine fisheries of the Baltic coast of Estonia and Latvia, USSR.

Nematoda

The study of pathogenic effects of nematodes is one of the most neglected fields in fisheries helminthology (Williams, 1967). At that time it was suggested that this was partly due to our ignorance of the occurrence, distribution and taxonomy of fish nematodes. Moravec (1987a,b), in a review of capillariids, has confirmed this by taking capillariid nematodes as an example. They are widely distributed in freshwater and marine fish and attack not only the digestive tract but also other organs and tissues, including muscles and skin. Some species are highly pathogenic and cause serious damage to cultured fish. Moravec, a leading authority on nematodes, emphasized also our lack of knowledge of other aspects of the biology of capillariids, including their life-cycles and the possible need for an obligatory intermediate host. This applies to the four species *Pseudocapillaria tormentosa, P. salvelini, Piscicapillaria tuberculata* and *Schulmanella petruschowski*, all of which parasitize European freshwater fish. Three of these nematodes are now frequently recorded in the pond breeding of carp, tench, salmonids and other species including introduced fish, eg. grass carp. Ghittino (1979) included *Capillaria* among other helminths as a cause of a major source of loss in Italian fish farms.

Ascaridoids are also becoming a focus of increased attention in hatcheries and fish farms. Gaines and Rogers (1971) reported a *Goezia* sp. for the first time from freshwater fish in North America. Extensive mortalities amongst striped bass (*Morone saxatilis*) resulted from the damage caused by these ascarid nematodes. It was thought that the worm was introduced into young striped bass at Richloam State Hatchery, Florida, when they were fed frozen marine herring. They cite a precedent described by Dollfus (1935) when *Goezia ascaroides* established itself in a 3-year-old rainbow trout fed raw fish meal made from marine fish. They were of the view that control measures could not be instituted until more was known of the biology of the worm. Berland and Egidius (1980) drew attention to the dangers of feeding rainbow trout on fish farms with fresh sprats each containing 4–5 larvae of *Thynnascaris adunca* (now *Hysterothylacium aduncum*) and capable of developing into sexually-mature nematodes in the pylorus of trout. Kakacheva-Avramova (1983a) confirmed that heavy infections of *Contracaecum aduncum* (now *Hysterothylacium aduncum*) (= *T. adunca*) normally a parasite of marine fish, caused heavy inflammation of the digestive tract in trout and suggested that the worm may be able to complete its life-cycle in fresh water; *H. aduncum* was also considered by Vismanis *et al.* (1984) to be amongst the most important causes of disease amongst *Salmo gairdneri, Oncorhynchus kisutch* and their hybrids bred in marine fisheries of the Baltic coast of Estonia and Latvia, USSR.

Bauer, Egusa and Hoffman (1981) referred to two new species of the family Skrjabillanidae (which use the crustacean *Argulus* as an intermediate host and have no free-living phase in water) having extended to other countries in the three species of carp traditionally cultured in China. *Cyprinus carpio* and its hybrids also harbour in their swim-bladder, *Philometra lusiana*, which is reported to cause sluggishness, emaciation, dullness and retarded growth in its hosts (Bauer, Musselius and Strelkov, 1973). *Philometra lusiana*, according to Krasilshchikov and Lyubina (1984), is a troublesome nematode of *Ctenopharyngodon idella* on the fish farms of the Omsk region, USSR. According to Nakajima (1976), *P. lusiana* is a synonym of *P. cyprini*.

Finally, in recent years members of the genus *Anguillicola* have become a cause for greater concern in Italy and other European countries, including the UK. Massimo (1987) states that the first case of anguillicoliasis in eel-rearing in Italy occurred in Spring 1985 but the first record of the genus in that country is by Paggi, Orecchia, Minervini and Mattiucci (1982). By 1987, all eels on Italian farms had a 10–50 per cent prevalence. Being haematophagous, *Anguillicola* spp. cause a general weakening and a decreased resistance to bacterial and parasitic diseases; at high temperatures they might cause losses on farms. A revision of the genus, including descriptions of two new species, was given by Moravec and Taraschewski (1988) and references to its introduction into Europe may be seen in Table 6.2 under eels in Europe.

Acanthocephala

Williams (1967) and Chapter Five of this book refer to acanthocephalans as fish pathogens but few have been reported as problematical on fish farms. An exception is the report by Bertocchi and Francálanci (1963), who stated that a trout-breeding farm in Italy had lost 80 per cent of its fish in two months due to infection with *Echinorhynchus truttae. Gammarus pulex* was the only intermediate host present. Mortality was at a maximum among the youngest fish and especially among those in an area where *G. pulex* was plentiful. Trout longer than 3–4 cm survived with 5–10 parasites in the intestine, loss of weight being the only symptom. Ghittino (1979) regarded *Metechinorhynchus truttae* (= *Echinorhynchus truttae*) as being amongst the major causes of loss due to parasites on Italian fish farms.

Mariculture

Most of the authors referred to in this section have emphasized that mariculture, ie. the artificial culture of marine organisms including fish, as compared with freshwater culture, is in its infancy. The importance of helminths in mariculture, however, is already evident in many publications, eg. Hargis (1985), Rosenfield, Drucker and Sindermann (1985) and Sindermann and Lightner (1988). Steele (1966) suggested that there were three problems to be considered with marine fish-farming in temperate regions, namely predation, disease and, most importantly, food supply. He predicted from preliminary experimental work in Scotland at the time, that disease was likely to be troublesome. Sindermann (1974) was of the view, however, that worm diseases, with the possible exception of those produced by monogeneans, have not yet appeared to be significant in mariculture. This was probably due largely to their complex life-cycles and the difficulties of completing such cycles in culture systems. He made the general point that worm diseases which are normally benign or for which pathology is unknown in natural populations may emerge in culture situations, a statement which was soon to be confirmed by Ghittino (1974). He mentioned *Benedenia seriolae, Microcotyle tai, Philometroides seriolae, Callotetrarhynchus nipponicum* and *Longicollum pagrosomi* as causes of disease in cultured marine fish. According to Williams and Phelps (1974) the total number of parasite species decreased on cultured *Trachinotus cardinus, Morone saxatilis, Mugil cephalus* and *Leiostomus xanthurus* but the numbers of some species increased, eg. monogeneans which lightly infected preculture *T. cardinus* were present on 90 per cent of post-culture fish. Among the first comprehensive reviews of diseases and parasite problems associated with the culture of marine fish was that of McVicar and MacKenzie (1977). They discussed the effects on marine fish parasites of different monoculture systems, based on their results on the parasites of flatfish, principally plaice, *Pleuronectes platessa*, and turbot, *Scophthalmus maximus* (= *Psetta maxima*), reared in the White Fish Authority's Scottish marine cultivation units. They cited references to stress factors in relation to diseases of fish and noted that the immune responses of only a few species of marine teleosts have been studied, and only fragmentary evidence is available on species of economic importance. Monogeneans were considered to be amongst the most serious pathogens of cultivated marine fish, especially in situations of high fish population density. Those involved are listed in Table 6.4. Diet, artificial and natural, was also thought to be an important factor in transmitting infections, *Hysterothylacium* (= *Contracaecum*) *aduncum* and *Hemiurus communis* in plaice being cited as examples. The attraction to cultivation units of wild disease-carrying fish was also thought to be important by McVicar and MacKenzie, eg. pipe fish, *Syngnathus* sp. carry a *Bothriocephalus* sp., possibly *B. scorpii*, and may transmit the infection to caged turbot. Other examples of transmission to caged fish are yellowtail in Japan acquiring the larval tetrarhynch *Callitetrarhynchus* sp. through being given anchovy, a natural host, as food (Nakajima and Syuzo, 1969), and herring reared in aquaria being fed with wild plankton and acquiring the nematode *Hysterothylacium* sp. with 10 per cent mortality resulting (Rosenthal, 1967). Whilst rearing White Sea herring in aquaria, Ivanchenko and Grozdilova (1985a,b) observed that the intestinal digeneans *Brachiphallus crenatus* and *Lecithaster gibbosus* killed larval fish.

The importance of siting of enclosures was demonstrated by MacKenzie, McVicar and Waddel (1976) with caged plaice in two locations, an exposed beach with a sandy bottom and a sheltered bay with a muddy bottom. Heavy infestations of *Cryptocotyle lingua* metacercariae and *Hysterothylacium* infestations were acquired at the muddy bay location and no live plaice were recovered. It was concluded that parasitism at least contributed to the mortality. In contrast, the metacercariae of two other digeneans, *Stephanostomum baccatum* and *Rhipidocotyle minimum*, were common in wild plaice but absent in cultivated stock (MacKenzie, McVicar and Waddell, 1976). Such variations were explained in terms of behavioural differences of the free-living cercariae of the different species involved and siting of the cages in a floating situation rather than on the bottom.

Musselius (1984) confirmed some of the helminths referred to by McVicar and MacKenzie as agents of disease of farmed marine fish, whilst Euzet and Raibaut (1985) drew attention to potentially serious pathological problems of parasitic worm origin on marine fish farms. They cite examples to show the complexity as well as the variety of worms which are involved during all rearing stages of fish. Consequently they recommend accurate epidemiological studies not only on marine fish farms but also in the surrounding natural environment. Their studies in these directions had already allowed them to determine potential pathogens and possible ways of prevention. Three species of monogeneans, *Diplectanum aequans, D. laubieri*, both gill parasites of *Dicentrarchus labrax*, and *Benedenia monticelli*, which attacks members of the Mugilidae, are potentially dangerous. The cercariae of the digeneans *Timoniella imbutiforme*, the adults of which are found naturally in the rectum of *Dicentrarchus labrax*, may attack and kill young *Sparus aurata* if *Hydrobia acuta*, which carry the cercariae, are introduced into their enclosure. Euzet and Raibaut also briefly discuss the following examples from the literature: *Heteraxine heterocerca* infesting *Seriola quinqueradiata; Gyrodactylus unicopula* on plaice, *Pleuronectes platessa; Cryptocotyle lingua* in turbot, *Scophthalmus maximus*; and the larvae of *Callitetrarhynchus gracilis*, found as adults in the sharks *Scoliodon walbeehmi* and *Triakis scyllium*, attacking anchovies, *Engraulis japonica*, under natural conditions but *Seriola quinqueradiata* in marine cages. The danger of transferring fish, a topic discussed earlier in this chapter, was stressed by Euzet and Raibaut.

Aquarium and ornamental fish

Richards (1977a, b) referred to *Gyrodactylus*, *Dactylogyrus*, *Caryophyllaeus laticeps*, *Schistocephalus*, *Ligula intestinalis*, *Camallanus*, *Capillaria*, *Diplostomulum* and *Sanguinicola*, metacercariae and acanthocephalans as disease agents in aquaria. Others who have written on this topic are Reichenbach-Klinke (1978), Phalempin (1979), Korzyukov (1979), Michel (1981), Moravec and Gut (1982), Moravec and Řehulka (1987), Reddacliff (1985) and Sommerville (1981b).

As has been seen for all the topics discussed in this account, the worm diseases of captive fish, aquarium and ornamental fish seem to be receiving most attention and more species, nematodes in particular, are being identified as potential pathogens. Rychlinski and Deardorff (1982), for instance, reported that the nematode *Spirocamallanus* has a worldwide distribution in tropical and subtropical fish. They suggested that marine aquarists, in particular, should be aware of these potentially harmful parasites. The nematodes are found either free in the lumen or attached to the mucosa of the intestine and pyloric caeca of marine, brackish and freshwater fish. They found 10 *Spirocamallanus* in one black-and-white butterfly fish, *Heniochus acuminatus*, and stated that from 5–8 nematodes are common in other planktivorous chaetodontids. Since a closely-related genus, *Camallanus*, has been implicated in aquarium fish disease and death, they believed the pathology induced by the two genera to be similar.

Pseudocapillaria brevispicula, a common nematode of cyprinids in Europe, was found for the first time in aquarium fish in Czechoslovakia by Moravec, Ergens and Repová (1984). Adults, including gravid females, were recovered from the intestine of *Puntius tetrazona* and probably were the cause of three deaths. Larvae were found in the liver of *Hyphessobrycon innesi*. The parasites were introduced into the aquarium with live oligochaetes during feeding. Moravec, Gelnar and Řehulka (1987) reported a new nematode species, *Capillostrongyloides ancistri*, from the intestine of an aquarium-reared catfish, *Ancistrus dolichopterus*, in Czechoslovakia. It had probably been introduced from South America. This worm appears to be highly pathogenic and lethal to aquarium-reared *A. dolichopterus*. In autumn 1985 heavy infections of *Raillietnema synodontisi* (150–200 worms per fish) were found in the upside-down catfish, *Synodontis eupterus*, bred in an aquarium in Czechoslovakia (Moravec and Řehulka, 1987). It was thought that this intestinal nematode caused stomach and intestinal atrophy and emaciation of the host but was not considered to be as pathogenic as *Capillaria*, *Pseudocapillaria*, *Capillostrongyloides* and *Camallanus*, all of which have been recorded from aquarium-reared exotic fish in Europe. The mass development of certain monogeneans on marine aquarium fish has been known since MacCallum (1927) demonstrated the importance of *Neobenedenia melleni* as a significant fish pathogen. Möller and Anders (1986) cite *Leptocotyle minor* on the skin of *Scyliorhinus canicula* as a good example. Nigrelli (1943) reported that it had remained a major hazard of New York Aquarium fish, having been responsible for 18 per cent of the deaths in 1939 and 20 per cent in 1940. Cheung, Nigrelli, Ruggieri and Cilia (1982) reported that endemic ectoparasitic monogenean infestations of a *Dermophthirius* sp. named *D. nigrelli* by Cheung and Ruggieri (1983), are common on lemon sharks, *Negaprion brevirostris*, kept in aquaria. The disease usually develops within a month at 22°C in recirculating sea water systems at a pH of 7.8–8.2 and salinity of 30–32 per cent. Early manifestations of the infestation are erratic swimming, flashing and rubbing on the bottom of the tank. Later, greyish patches and open wounds appear on the skin with the placoid scales detached at the site of monogenean attachment. The ulcerated lesions become secondarily infected with bacteria of the *Vibrio*-complex. It was thought by Grimes, Gruber and May (1985) that *D. nigrelli* is a vector for this bacterium. An outbreak of monogenean infection in the 'Shark Tank' of the New York Aquarium caused the death of a dusky shark, *Carcharhinus obscurus* and several lemon sharks, *Negaprion brevirostris*, according to Cheung and Ruggieri (1985). *Dermophthirius carcharhini* and *D. nigrelli* were identified as the causative agents.

Human infections with fish worms

General account

It is established that helminths which are natural parasites of man are responsible for loss of appetite, inhibition of digestion and absorption in the gut and may compete directly with the host's intestinal mucosa for nutrients (Ash, Crompton and Keymer, 1984). Consequently the accidental ingestion by humans of helminth species which are naturally parasites of other animals and yet can survive in such an abnormal host are likely to be equally, if not more, pathogenic. Witenberg (1932) was amongst the first to highlight the fact that 14 species of digeneans, five of cestodes and one of nematodes could be acquired by humans through eating uncooked fish. The potential detrimental effects of such unusual helminth infections on human health have, therefore, remained a matter of interest for fish biologists, helminthologists and physicians.

The advent of marine fish-farming in the 1960s led Sindermann (1966) to write that an inescapable aspect of any discussion of fish parasites is their possible relation to human disease. The role of freshwater fish in transmitting parasites to humans had been known for a long time, although few examples could be found for a similar role for marine fish. Janssen (1970), in reviewing the marine origin of bacterial infections in man, stressed the need for much more research on fish as possible vectors of human disease because of the increasing use of fish as food, increasing contamination of the aquatic environment with human wastes and increasing contact between man and the aquatic environment.

Sprent (1969) thought the term zoonosis should be retained in the broad sense, ie. a disease of animals, but he suggested the term anthropozoonosis for human infections in which other vertebrates are definitely involved in the development, transmission or epidemiology of disease irrespective of the biological role played by man or animal. Sprent subdivided anthropozoonosis into categories depending on (i) whether man is the usual or essential host, (ii) man with other animals are epidemiologically important alternative hosts, (iii) man is an intermediate, paratenic or accidental host, and (iv) man is an incidental definitive host of parasites normally maturing in other vertebrates. The 39 species of helminths listed by Sprent known to infect man through ingestion of fish, usually raw or partially-cooked, cover these four categories. Sprent's review of zoonosis has remained an important impetus for many investigations of anthropozoonosis. Here we are concerned with the role of fish in anthropozoonosis.

It was largely in response to the views of Sprent, Sindermann, Janssen and others that Williams and Jones (1976, 1977) discussed a number of marine fish helminths which were known either to have established themselves in man or were potentially capable of doing so. These included 10 species of cestode belonging to the genera *Diphyllobothrium* or *Diplogonoporus*, 12 digeneans (mostly heterophyids), and cercarial dermatitis caused by members of the Schistosomatidae and nematodes belonging to the genera *Anisakis*, *Hysterothylacium*, *Pseudoterranova* and *Trichinella*. The life-cycles of these worms are illustrated in this chapter (Figures 6.1–6.4.)

Since 1976 others have also encouraged further research into anthropozoonosis in which animals, including fish, are involved in disease transmission to man. Giordano (1976) briefly discussed important digeneans, cestodes and nematodes which are transmitted to man by the consumption of fish, molluscs and crustaceans, and included information on aetiology, source of human infection, life-cycle, means of transmission, disease in man, control and geographical distribution. A list of the parasites of man reported in Canada from 1830–1978 was given by Croll and Gyorkos (1978) and included some species that had been transmitted with food, including fish. Healey and Juranek (1979) included the following fish helminths in their account of human parasite infections transmitted in food: *Gnathostoma spinigerum*, 'anisakine' larvae, *Diphyllobothrium latum*, *Clonorchis sinensis*, *Opisthorchis* sp., *Metagonimus yokagawi* and *Heterophyes heterophyes*. The role of livestock, fish and shellfish in transmitting helminths which are hazardous to human health was discussed by Ruitenberg, van Knapen and Weiss (1979). The helminths of fish in Poland which may infest man are listed with notes on their hosts by Prost (1979). Largely from their experience of the food habits of man in Japan, especially when uncooked marine fish and squid fillets are consumed, Oshima (1984) and Oshima and Kliks (1986) reviewed anisakiasis, gnathostomiasis and diphyllobothriasis. Research publications by Japanese authors on parasites of fish which can infect man were also briefly reviewed by Ichihara (1983). The foregoing comments show that certain groups of cestodes, digeneans and nematodes of fish are well known as potentially dangerous pathogens of man. The remainder of this account will, therefore, concentrate on these groups.

Cestoda—diphyllobothriids (Figure 6.1)

Records of human infection with cestodes associated with fish are largely restricted to countries where fish is sometimes eaten raw, marinaded or lightly cooked, eg. in Scandinavia, the USSR, Alaska, Canada, USA, Chile, Peru and Japan. *Diphyllobothrium latum* infection in humans in Finland was linked with the consumption of raw or semi-cooked fish as long ago as 1747 (von Bonsdorff, 1977). This infection in humans has attracted particular attention since the 1880s when it was found that it caused tapeworm pernicious anaemia (TPA). Some patients are particularly predisposed to TPA and it is thought that complicated factors which promote genuine PA also promote the development of vitamin B_{12} deficiency and tapeworm PA in carriers of *D. latum* (von Bonsdorff, 1977). He suggested that *D. latum*, in contrast to other intestinal helminths, requires large amounts of vitamin B_{12} because the parasite surpasses other comparable worm species in growth rate and egg-production. Von Bonsdorff devotes about 90 pages to the clinical manifestations of *D. latum* infection and the pathogenesis of tapeworm pernicious anaemia and his account is of the greatest interest to all those concerned with the medical problems of diphyllobothriasis. A 48-page review of TPA is also given by Totterman (1976) and it includes a discussion of von Bonsdorff's hypothesis and recent research on vitamin B_{12}

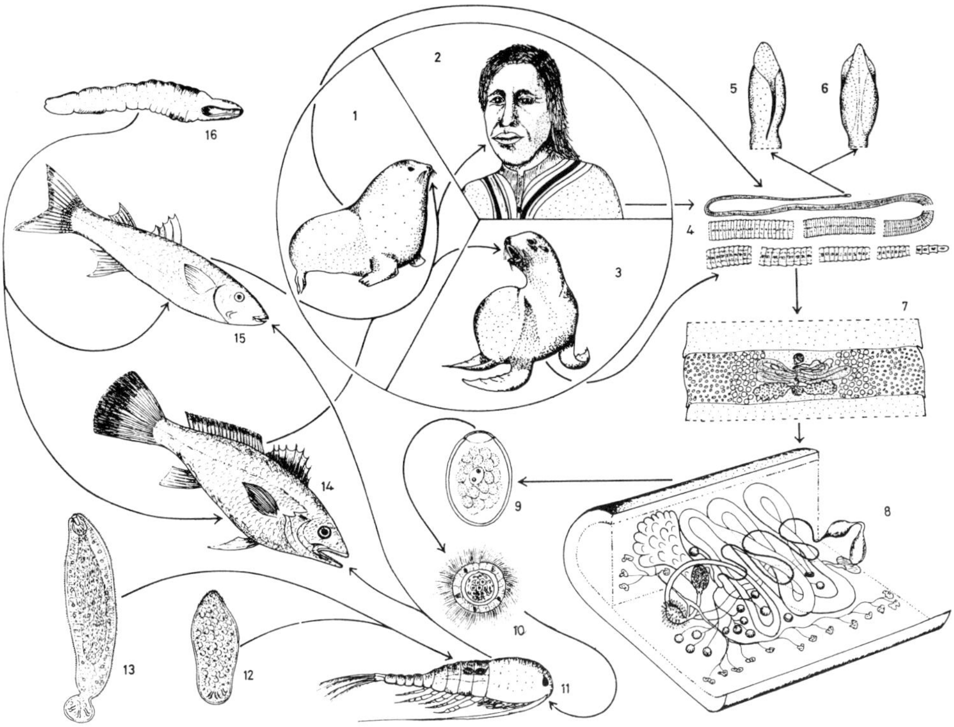

Figure 6.1 Life-cycle of *Diphyllobothrium*.
Figs 1–3. Fur seal, man, and sea-lion, some of the definitive hosts of *D. pacificum*. Figs 4–8. Anatomical details of the adult cestode in the intestine of these hosts. Fig. 4. Adult cestode. Figs. 5,6. Scolex, Fig. 7. One proglottis, Fig. 8. Reproductive organs. Fig. 9. Egg. Fig. 10. Ciliated free-living coracidium larva emerges from the egg and is eaten by a marine copepod (Fig. 11), the first intermediate host. Figs. 12 and 13. Early and late stages of development of the procercoid larva in the copepod. Figs. 14 and 15. *Sciaena* and *Mugil*, two of the second intermediate hosts eat infected copepods containing procercoids which develop into pleroceroids (Fig. 16), infective to the definitive hosts, in the body cavity of the fish.
From William and Jones (1976).

deficiency. Although the plerocercoids have been recorded from very many fish species representing a number of families, it is now generally accepted that the four species of fish which are most important in transmitting *D. latum* to man in Eurasia are pike, *Esox lucius*, burbot, *Lota lota*, perch, *Perca fluviatilis*, and ruffe, *Acerina cernua* (= *Gymnocephalus cernuus*).

In North America the following species are generally considered to be important carriers of *D. latum*: great northern pike, *Esox lucius*; wall-eyed pike, *Stizostedion vitreum*; sand pike, *S. canadense griseum*; burbot, *Lota maculosa*, and yellow perch, *Perca flavescens*. Various species of the salmonid genus *Oncorhynchus* are also regarded as important transmitters along the Pacific coast region of North America.

In Japan plerocercoids of *D. latum* are found especially in *Oncorhynchus masu* and *O. gorbuscha*, but also in *O. keta* and *O. nerka*.

Three other species of *Diphyllobothrium*, *D. ditremum*, *D. dendriticum* and *D. vogeli* occur in northern Europe and western Siberia but only one, *D. dendriticum*, is known to be infective to man. According to Andersen and Gibson (1988), *D. vogeli* is a synonym of *D. ditremum*. That other species remain to be discovered and described is evident from Andersen (1977), who found a type of *Diphyllobothrium* plerocercoid in a marine fish, blue whiting, *Micromesistius poutassou*.

In parts of the world, especially the USSR, North America, South America and Japan, the following 15 species which also use fish as second intermediate hosts may be found in man (Williams and Jones, 1976; von Bonsdorff, 1977): *D. skrjabini, D. giliacicum*, *D. luxi*, *D. theileri*, *D. trinitatis*, *D. cordatum*, *D. colymbi*, *D. alascense*, *D. dalliae*, *D. ursi*, *D. lanceolatum*, *D. pacificum*, (see Figure 6.1), *Diplogonoporus grandis*, *D. balaenopterae*, *D. fukuokaensis*. Some of these species were not accepted by Andersen, Ching and Vik, (1987). To begin to understand the complicated epidemiology of each of these 15 diphyllobothriid species, a round-table discussion on *Diphyllobothrium* research was arranged in August 1986 at the Sixth International Congress of Parasitology held in Brisbane, Queensland (personal communications from Goran Bylund and Hans-Peter Fagerholm, Abo, Finland). At this discussion it was agreed that progress on *Diphyllobothrium* taxonomy, especially of marine species, has been slow. A matter of special interest and concern is to determine which plerocercoid larvae from fish can be linked to adult worms from birds and mammals and, in the context of the present discussion, which of those may infect humans. Comparison of research results, however, now in progress between the USA, Canada, Norway, Finland and Japan, may provide results of fundamental importance. Important recent contributions towards solving this once formidable taxonomic problem were made by Andersen, Ching and Vik (1987) on *D. dendriticum* and *D. ditremum*, and by Dick and Poole (1985) on *D. dendriticum* and *D. latum* in Canadian freshwater fish.

A reliable figure concerning the global status of human *Diphyllobothrium* infections seemed unattainable in 1986. The prevalence in Finland at 1.2 per cent was constantly decreasing but, in contrast, in Japan the prevalence was increasing with more than 100 new cases a year being recorded. The origin of human infections in Japan seems to be mainly marine or anadromous fish. In Canada approximately 25 human cases per year were recorded, at least some from marine fish.

It is with regard to this kind of increasing interest in the population dynamics of *Diphyllobothrium*, and also the changing patterns in prevalence, that the following selected publications are briefly mentioned to update comments already made by Williams and Jones (1976), von Bonsdorff (1977), Bylund (1982) and von Bonsdorff and Bylund (1982).

Human *Diphyllobothrium latum* or *Diphyllobothrium* sp. infections in the USA are mentioned by Craun (1977) and Barclay (1981) and are discussed in greater detail by Ruttenber *et al.* (1984). Of the 52 cases of diphyllobothriasis in the west coast states of the USA in 1980, salmon was implicated as the carrier in 82 per cent. Changes in fish marketing practices and the popularity of uncooked or slightly-cooked fresh salmon were responsible for the epidemic. Ching (1984) thought that, although the risk of diphyllobothriasis is generally low in Canada, it seemed to have become more frequent in recent years. Three of the five species of *Diphyllobothrium* reported in Canadian fishes are known to infest humans. They are *D. latum, D. dendriticum* and *D. ursi*. *Diphyllobothrium latum* in Canada is found in the body cavity of such freshwater fish as 'northern pike', 'wall-eye', 'sanger' or 'yellow perch'; *D. dendriticum* occur in the liver of 'rainbow trout', 'short-jaw cisco', 'lake trout', 'Dolly Varden' and 'Kokance'; *D. ursi* (= *D. dendriticum*) occur in the flesh of salmon.[3] Following the discovery of *D. pacificum* in humans in Peru by Baer, Miranda, Fernandez and Medina (1967) the species has received the attention of Escalante (1983) in Peru and of Sagua, Fuentes, Soto and Delano (1979), Torres (1983), and Ferreira, Araújo, Confalonieri and Nuñez (1984) in Chile. The last group of researchers discovered the ova of *D. pacificum* in human coprolites from archaeological material dated at 4110–1950 BC at Iquique, Northern Chile. *Diphyllobothrium pacificum* is naturally a tapeworm of two species of otariid seals, *Otaria byroni* and *Arctocephalus australis*, off the coast of Peru. It also occurs in the northern fur seal, *Callorhinus ursinus*, and, therefore, the

[3] See also Ching (1988)

tapeworm is a potential human parasite in the northern hemisphere. Sagua *et al.* (1979) reported the first record of human infection with *D. pacificum* in Chile while Torres (1983) reviewed the prevalence and epidemiology of *D. latum*, *D. pacificum* and *D. dendriticum* in Chile. Diphyllobothriids were found by Escalante (1983) in four species of fish used for human consumption, namely *Sciaena callaensis, Cynoscion analis, Merluccius gayi peruanus* and *Genypterus maculatus*.

In the USSR Klebanovskii, Smirnov, Klebanovskaya and Obgol'ts (1977) found that 12.4 per cent of the indigenous population of the Khatangskii region had diphyllobothriasis. Of 33 people treated, 22 passed *D. latum*, 13 *D. dendriticum* and two both species. *Coregonus* spp. were thought to be the fish hosts of *D. dendriticum* and both *Lota lota* and *Esox lucius* of *D. latum*. From self-infestation experiments Klebanovskii (1983) reported that *D. ditremum* could not mature in man but some *D. dendriticum* did to such an extent that 3.1 per cent of the ova passed produced coracidia. The complicated epidemiology of diphyllobothriasis in the USSR, as was emphasized at the 6th International Congress of Parasitology workshop on the disease for other countries involved, is evident from an 8-year study by Serdyukov (1979) on the *Diphyllobothrium* spp. of man and terrestrial animals in West Siberia. He referred to 34 species of the genus of which *D. latum, D. dendriticum* and *D. ditremum* are valid and *D. vogeli, D. ursi, D. alascense* and *D. microcordiceps* are provisionally valid. Serdyukov described his experiments on the development of several plerocercoid types in golden hamsters and man (self experiment). *D. latum* and *D. dendriticum* are the most important infections of man, an observation confirmed by Obgol'ts (1983) for the Taimyr Peninsula, USSR, and no natural *D. ditremum* infections were seen in man. Muratov (1983) found a relatively low rate of infection with *D. latum* in three districts of the Khabarovsk territory, Far-Eastern USSR, but Romanenko *et al.* (1986) found it in 6.3 per cent of over 1,000 inhabitants examined near the Krasnoyansk reservoir, USSR.

The eleventh volume of Principles of Cestodology by Delyamure, Skryabin and Serdyukov (1985) is devoted entirely to the diphyllobothriids of man, mammals and birds.

Digenea

Laird (1961) stated that about 60 species of digeneans have been identified from man, 37 of them as accidental parasites of the intestine, infecting people of the eastern borders of the Oriental and Palaearctic regions who commonly eat raw or inadequately-cooked fish. Witenberg (1932), Sprent (1969) and Williams and Jones (1976) suggested that members of about 25 genera of digeneans were known to be acquired by man as a result of eating raw or inadequately-cooked fish. Most of these were heterophyids, which have been the subject of extensive taxonomic studies by Pearson (1964, 1973) and Pearson and Ow-Yang (1982). Aspects of the life-cycle of a heterophyid are shown in Figure 6.2.

Heterophyidae

Members of this family comprise small digeneans most of which are found, naturally, in the intestines of birds and mammals. The metacercariae are usually found in fish (Figure 6.2). Humans accidentally infected with digeneans belonging to the family may contract a disease known as heterophyidiasis. Cheng (1973) listed those species which are transmissible to man by eating raw or poorly-cooked fish, and Ito (1964) concentrated on human heterophyid infections in Japan. The latter described the morphology, taxonomy, life-cycle and epidemiology of heterophyid species belonging to *Metagonimus, Centrocestus, Stellantchasmus, Haplorchis, Procerovum* and of *Pygidiopsis summa*. Heterophyidiasis may involve more than infection of the alimentary canal in man. Heterophyid eggs may pass into the

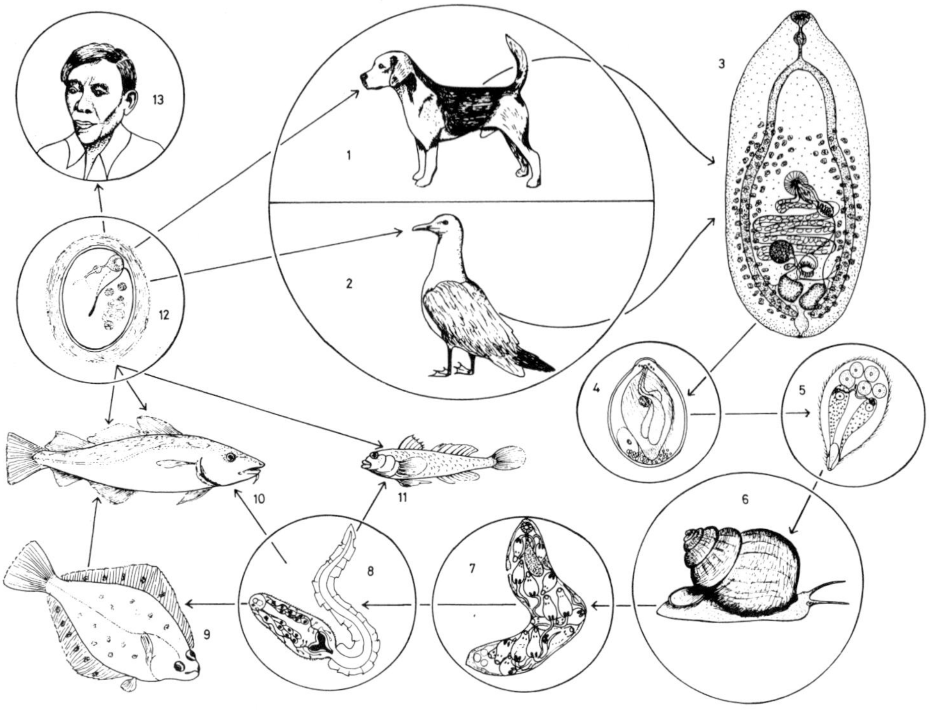

Figure 6.2 Life-cycle of a heterophyid.
Fig. 1. Dog, a suitable experimental definitive host. Fig. 2. Herring-gull, a natural definitive host. Fig. 3. Adult from the intestine of the definitive host. Fig. 4. Egg containing the ciliated larva or miracidium. Fig. 5. Free-swimming miracidium. Fig. 6 *Littorina littorea*, first intermediate host of *Cryptocotyle lingua*. Fig. 7. Redia containing cercariae at various stages of development. Fig.8. Free-swimming cercaria. Figs. 9, 10 and 11. Plaice cod and common goby, three common second intermediate hosts of *C. lingua*. Fig 12. Metacercaria encysted in skin of fish. Fig. 13. Man, a common accidental host of many heterophyids.
From Williams and Jones (1976).

intestinal wall and be carried in the blood stream to the heart where they become encapsulated in cardiac muscle. Such a situation has been suspected as the cause of death among many people of Philippine descent. Kean and Breslau (1964) reported that 14.6 per cent of cardiac failures in the Philippines are due to heterophyid myocarditis. Heterophyidiasis in animals and humans was discussed by Velasquez (1982) with regard to *Heterophyes* spp., *Haplorchis* spp., *Procerovum* spp., *Metagonimus yokogawai, Stellantchasmus falcatus, Centrocestus caninus* and *Carneophallus brevicaeca*.

Apophallus muehlingi, normally found in *Larus* spp. and also in cats and dogs, may encyst in the flesh and fins of *Perca fluviatilis* and also in cyprinids, especially *Abramis brama, Leuciscus rutilus, Blicca björkna* and *Scardinius erythrophthalmus*. Species of the genus have a world-wide distribution. It has been suggested by Laird (1961) that one species, *A. venustus*, may infect man in the Tokelau Islands in the Pacific and in Canada, but it was thought to be non-pathogenic by Cheng (1986).

Ascocotyle coleostoma was reported from a woman with chronic colitis by Roussett, Baufine-Ducrocq, Rabia and Benoit (1983) and was believed to be the first recorded case of human infection with this heterophyid. It was suggested that the infection had been acquired while she was attending a conference in Egypt when she ate large quantities of raw marine fish.

Centrocestus armatus (synonym *Stamnosoma armatum*) occurs in humans in Japan according to Witenberg (1964), where cyprinids are thought to be the fish hosts (Yamaguti, 1975); *C.*

cuspidatus has been recorded from *Gambusia affinis* by Martin (1959) and *C. formosanus* from *G. affinis* in Egypt and from *Xiphophorus kelleri* and *Mugil cephalus* by Martin (1958b) in Hawaii. *Cryptocotyle lingua* metacercariae occur in the skin of several species of fish whilst adults occur in the intestine of seagulls throughout Europe and North America. Human infections have been recorded in Greenland (Witenberg, 1964).

Galactosomum lacteum normally occur as adults in *Phalacrocorax* spp. and *Ardea cinerea* whilst the metacercariae encyst in the brain of *Myoxocephalus scorpius* and *Cottus bubalis*. *Haplorchis* spp. metacercariae occur in *Cyprinus carpio, Carassius auratus*, mullet and other fish representing the Cyprinidae, Siluridae and Cobitidae according to Yamaguti (1975). In the Philippines and Indonesia, four species, *H. taichui, H. calderoni, H. yokogawai* and *H. vanassima* have been found in humans (Witenberg, 1964). *H. taichui* metacercariae were found by Martin (1958b) in *Gambusia affinis, Mugil cephalus* and *Mollienesia formosus* in Hawaii and he also recorded *H. yokogawai*, experimentally, from *Clarias fuscus*. Cats were experi-mentally infested with *Euhaplorchis californicus* by Bamberger and Martin (1951) to clarify heterophyidiasis. In her account of the disease, Velasquez (1982) refers to *Haplorchis* spp.

Members of the genus *Heterophyes* are the most common heterophyids of man and many species of fish act as carrier hosts with *Mugil* spp. the most abundantly infected (Witenberg, 1964). *Heterophyes heterophyes*, *H. dispar* and *H. equalis* occur in man in the Mediterranean, Far East and Egypt. The three species are endemic in Israel in cats and dogs but, according to Witenberg (1964), only *H. heterophyes* occurred in man. Every mullet examined was found to have metacercariae. *Mugil cephalus* is the vector host according to Yamaguti (1975) and Martin (1959) who experimentally infected this fish in Egypt with *H. heterophyes* and *H. aequalis*. *Heterophyes heterophyes* is said to have been introduced into Hawaii. Rousset and Pasticier (1972) also found *Mugil cephalus* to be the vector of *Heterophyes heterophyes* in Egypt and Rifaat *et al.* (1980) extended the host range to include *M. auratus, M. capita, Tilapia nilotica, T. zilli, Sciaena equalla* and *Solea vulgaris*. *Heterophyes expectans* was recorded by Vasquez-Colet and Africa (1938) from *Hemiramphus georgei* and *Ambassis buruensis* in the Philippines. Chai, Seo and Lee (1984) refer to *H. heterophyes nocens* in man in Korea. *Heterophyes aequalis* and *H. dispar* were reported by Taraschewski and Nicolaidou (1987) in Greece with the claim that it is the only country in Europe where data on the prevalence of *Heterophyes* spp. have been collected. It follows, therefore, that nothing is known of its prevalence in other European countries. Velasquez (1982) discussed *Heterophyes* spp. generally.

Heterophyopsis continua adults occur in dog and man in Japan according to Yamaguti (1975) and the metacercariae are encysted in the muscles of *Mugil cephalus*. In Korea, Seo, Lee, Chai and Hong (1984b) found the metacercariae in 63 per cent of *Lateolabrax japonicus* and 29.4 per cent of *Acanthogobius flavimanus*. Two Korean men were found to be infected while two rats and one dog were experimentally infected.

Metagonimus yokogawai is the second most common heterophyid of man and is prevalent in the Far East, Central Europe and Spain (Witenberg, 1964). Its distribution in other warm-blooded animals is more widespread. Ito and Mochizuki (1968) found metacercariae in *Mugil cephalus* in Japan, whilst Yamaguti (1975) stated that *Plecoglossus, Carassius, Cyprinus, Leuciscus, Acheilognathus* and many other genera can act as intermediate hosts. *Rutilus rutilus caspicus* and *Abramis brama* were quoted as hosts by Mikailov and Kazieva (1981) and the spread of this digenean was predicted with a consequent need for surveillance of populations at risk. The species was discussed by Velasquez (1982).

Procerovum varium metacercariae were found by Velasquez (1973) in *Platycephalus indicus, Mollienesia latipinna, Scatophagus argus, Chanos chanos, Oxyurichthys microlepis, Hemiramphus georgei* and *Anabas testudinaceus*, whilst *P. calderoni* occurred in *Butis amboinensis, C. chanos, M.*

latipinna, Platycephalus indicus, Mugil dussumieri, Pelates quadrilineatus and *H. georgei*. Vasquez-Colet and Africa (1938) discussed the life-cycle of *P. calderoni*.

Pygidiopsis summa adults occur naturally in dogs in China and Japan and its metacercariae occur in the wall of the oesophagus or intestine of *Mugil cephalus* and *Glossogobius brunneus*. *Pygidiopsis genata* and *P. marvilli* were found in *Mugil* sp. by Vasquez-Colet and Africa (1938) in the Philippines. *Stellantchasmus falcatus* was recorded as *Diorchitrema pseudocirrata* from *Mugil* sp. in the Philippines by Vasquez-Colet and Africa (1938). *Mugil cephalus* has been recorded as a host in Japan (Yamaguti, 1975) and also in Hawaii by Martin (1958). The species was described in detail by Seo, Lee, Chai and Hong (1984a), who reported adult worms from two Korean men. They had eaten raw brackish-water fish. *Stellantchasmus falcatus* is of particular interest in that its metacercariae were found by Fukuda and Yamamoto (1981) to harbour a neorickettsia-like organism (SF strain). A dog inoculated with the strain developed a mild fever but the pathogenicity of the SF strain appeared to be much lower than that of *Neoricketssia helminthoeca* or the Elokomin fluke fever agent. *Stellantchasmus* is also discussed by Velasquez (1982).

Stictodora querreroi and *S. manilensis* metacercariae were found in *Hepsetia balabacensis, Ambassis buruensis* and *Mugil* sp. by Vasquez-Colet and Africa (1938) in the Philippines. They also recorded *S. querreroi* in *Hemiramphus georgi*. *Stictodora tridactyla* metacercariae occur in *Aphanius fasciatus* and *Gambusia affinis* in Egypt (Martin and Kuntz, 1955).

Opisthorchiidae

Members of this family are delicate digeneans with relatively small suckers and are natural parasites of the biliary system of reptiles, birds and mammals. Two species, *Clonorchis sinensis* and *Opisthorchis tenuicollis*, are of considerable medical importance as parasites of the bile-ducts of fish-eating mammals including cats, dogs and humans.

Clonorchis sinensis, the Chinese liver fluke, is widely distributed in Korea, Japan, China, Taiwan and Vietnam and, in 1947, was estimated to infect 19 million people. Due to the increased consumption of fish it is thought that this figure may not have decreased. The fluke is able to survive in man for 25 years. Kobayashi (1933) listed about 40 species of fish as carriers of the disease, 24 of which occurred in China. *Ctenopharyngodon idella* and *Mylopharyngodon aetiops* were the most heavily infested. In Japan the most heavily infested fish was *Pseudorasbora parva*. The disease is also prevalent in Korea (Witenberg, 1964). It is now thought that about 80 species of fish are involved as carriers of *Clonorchis* (Cheng, 1986). Clonorchiasis was reviewed by Komiya (1966).

Opisthorchis tenuicollis, also known as *O. felineus*, causes a disease known as northern opisthorchiasis and is second in importance to *C. sinensis* as a parasite of the human liver. In 1947 it was estimated that 1.7 million people were infected. It is most common in southern central and eastern Europe, Turkey, the southern Soviet Union, Vietnam, India and Japan but occurs also in Puerto Rico and possibly other Caribbean Islands. Metacercariae have been found in *Cyprinus carpio, Tinca tinca, Barbus barbus, Abramis brama, Blicca björkna, Leuciscus rutilus, Idus idus* and *Scardinius erythrophthalmus* (Yamaguti, 1975). *Opisthorchis viverrini* is thought to be the cause of opisthorchiasis in the Far East. Rim (1982) discussed the two species of *Opisthorchis* with regard to natural and alternative hosts, distribution, the disease it causes in man and animals, epidemiology, prevention and control. Other opisthorchiids thought to infect man only rarely are *Amphimerus norverca, Metorchis conjunctus* and *Pseudamphistomum truncatum* (Witenberg, 1964). *Metorchis conjunctus* was thought to be a focus of metorchiasis at Lake Edward, Canada, around 1940, and has remained one according to Watson (1981). *Catostomus commersoni, C. catostomus, Semotilus corporalis* and *Salvelinus fontinalis* are given as the fish carrier hosts.

Nanophyetidae

Cribb and Pearson (1988) recognized the Nanophyetidae. Two well-known species are potentially infective to humans, *Nanophyetes salmincola* and *N. schikhobalowi. Nanophyetus* infections generally were reviewed by Philip (1955) and Millemann and Knapp (1970). No human case of *N. salmincola* infection had been recorded according to Witenberg (1964) but, in the Soviet Far East, *N. schikhobalowi* may infect the entire population in some areas. Thus Filimonova (1963) found it in dogs, cats and most of the population of a village on the Khor river in the Khabarovsk region of the USSR. Of locally caught fish, 97 per cent of *Thymallus arcticus*, 96 per cent of *Brachymystax lenok* and small numbers of *Hucho taiman, Oncorhynchus keta, Phoxinus phoxinus* and *Mesocottus haitey* were infected with *N. schikhobalowi* metacercariae. Until 1966 it was thought that only salmonids carried the metacercariae of *N. salmincola* but Gebhardt, Millemann, Knapp and Nyberg (1966) extended the natural and experimental fish host list to include *Cottus perplexus, Lampetra richardsoni, L. tridentata, Richardsonius balteatus, Carassius auratus, Catostomus macrocheilus, Lepomis macrochirus, Gasterosteus aculeatus aculeatus, G. a. microcephalus* and *Gambusia affinis*. Human infestations of *N. salmincola* were reported by Eastburn, Fritsche and Terhune (1986).

Clinostomatidae

A few species may rarely infect man: eg. *Clinostomum complanatum* and *C. marginatum*, both of which use a very large number of fish as intermediate hosts (Witenberg, 1964; Yamaguti, 1975). Britz, van As and Saayman (1984, 1985) reported that *Clinostomum tilapiae* and *Euclinostomum heterostomum* are a potential hazard to humans if *Oreochromis mossambicus* (an important component of the fish-harvesting programme in Venda and Lebowa, South Africa) is not properly cooked or only partially dried in the sun. The rejection, for human consumption, of *O. mossambicus* infected with *Clinostomum* sp. had already been reported by Kabunda and Sommerville (1984).

Echinostomatidae

Several echinostomes infest man via freshwater fish according to Rim (1982). *Echinostoma ilocanum* occur as adults in humans, rats and dogs (Cheng, 1986) and *Echinochasmus perfoliatus* uses about 12 species of fish belonging to several families as carrier hosts (Yamaguti, 1975; Witenberg, 1964).

Cyathocotylidae

Prohemistomum vivax has been recorded once in man in Egypt (Witenberg, 1964). It occurs naturally in fish-eating birds and experimentally in cats and dogs, and its metacercariae occur in *Gambusia affinis* and *Tilapia nilotica* (Yamaguti, 1975). *Prohemistomum chandleri* adults also occur in piscivorous birds, with metacercariae in *Huro salmoides*.

Isoparorchidae

According to Witenberg (1964), *Isoparorchis hypselobagri*, a parasite of the swim-bladder of freshwater fish, is fairly common in man in the Manipur Valley, India. This is a surprising record which requires confirmation.

Didymozoidae

Kamegai (1971a) recovered a species of *Gonapodasmis* in the flying fish, *Cypselurus heterus doderleim*, from the Pacific coast of Japan. Eggs of the parasite have been found in human faeces in Japan and its incidence in humans coincides with the migration of the fish along the Pacific coast (Williams and Jones, 1976).

Renicolidae

On the basis of the fish-eating habits of the bird final hosts of *Renicola* and the lack of its host-specificity generally, Williams and Jones (1976) suggested that such unusual worms might be capable of establishing themselves in humans.

Microphallidae

Velasquez (1982) discussed the life-cycle of *Carneophallus brevicaeca* and its crustacean intermediate host, stating that continuous consumption of the raw, naturally-infected, shrimps was likely to lead to infection of the heart, spinal cord and other vital organs in humans.

Nematoda

It is well established that a few species belonging to three groups of nematodes (ascaridoids, gnathostomatids and *Trichinella*) which use fish as intermediate hosts and birds and mammals as final hosts will survive in humans with pathogenic consequences. The only

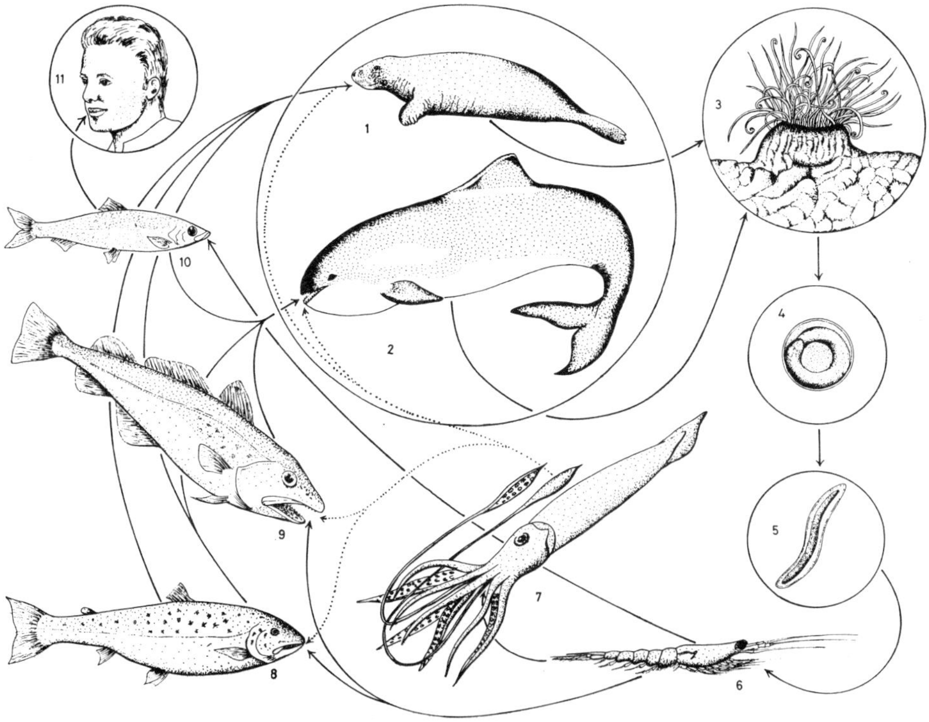

Figure. 6.3 Life-cycle of *Anisakis*.
Figs. 1 and 2. Grey seal and porpoise, natural definitive hosts. Fig. 3. Nodular parasitic granuloma caused by adult *Anisakis* attached to the lining of the stomach of a porpoise. Fig 4. Egg which is passed in the faeces of marine mammals and embryonates in sea water. Fig. 5. Ensheathed free-living second-stage larva which emerges from the egg and is swallowed by *Thysanoessa inermis* and *T. longicaudata* (Fig. 6), first intermediate hosts in British waters. The larva moults, becoming the third-stage larva in the haemocoel. Figs. 7, 8, 9 and 10. Squid, salmon, cod and herring are second intermediate hosts for third-stage larvae. Fig. 11. Man becomes infected in Europe by eating herring but the larvae do not mature in man.
From Williams and Jones (1976).

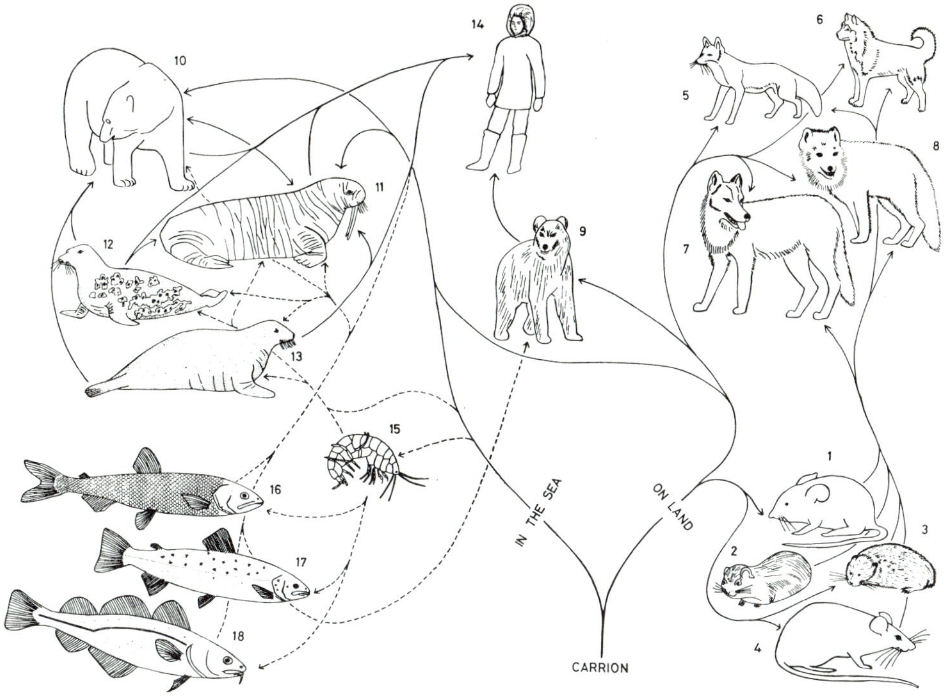

Figure 6.4 Possible transmission routes of *Trichinella spiralis* in the Arctic. Carrion containing infective larvae is eaten on land by rodents (Figs. 1, 2, 3 and 4. Mouse, lemming, vole and rat) which in turn fall prey to carnivores (Fig. 5, 6, 7 and 8. Fox, husky, wolf and Arctic fox) which also scavenge directly. Carrion is also eaten on land by black or brown bear (Fig. 9), polar bear (Fig. 10,) walrus (Fig. 11), ringed seal (Fig.12) and bearded seal (Fig. 13), all of which are consumed by man (Fig. 14) and all of which, except black of brown bears, also eat carrion in the sea. Amphipods (Fig. 15. *Gammarus*) may transmit larvae from carrion to marine mammals and fish. (Figs 16, 17 and 18. Smelt, salmon and cod).

known mode of infection is through eating raw or inadequately-cooked fish. The nematodes will be discussed below under four headings: *Anisakis, Pseudoterranova, Gnathostoma* and *Trichinella*. There is insufficient information on other fish nematodes which have either been recorded from humans or are thought to be potential human pathogens, eg. a *Philometra* sp. from *Caranx melampygus* in Hawaii invading a wound in a fisherman's hand (Deardorff *et al*., 1986) and species of *Contracaecum, Hysterothylacium, Raphidascaris* and *Porrocaecum* which are occasionally mentioned in publications dealing with anisakiasis. Overstreet and Meyer (1981) found experimentally that *Hysterothylacium* type MB, which normally matures in a fish, penetrated the stomach wall, causing haemorrhage and attracting eosinophils, in a rhesus monkey, *Macaca mulatta*. Cheng (1982, 1986) stated that those species of *Porrocaecum* and of *Contracaecum* occurring as adults in marine mammals and as third-stage larvae in fish hosts can cause human anisakiasis. Until more taxonomic work of the kind referred to by Gibson (1983) and Soleim (1984) is done on some of these genera, an open mind will be required in discussing the identity of fish nematodes that are infective to humans. In view of the many unresolved problems we have confined our account to four genera.

Anisakis, Dujardin 1845

It will become apparent in the following account that this genus and the species *Anisakis simplex* has attracted considerable research attention since 1955. Ruitenberg (1970) discussed (i) the history of *A. simplex* and anisakiasis and the disease it causes in humans from the time the first case was detected in Holland in July, 1955; (ii) research progress from 1955, the classification of *A. marina* (as it was then called), together with aspects of its life-cycle which involved different marine mammals and fish in Europe and Japan; (iii) pathogenesis of anisakiasis as described in the literature and also from experimental work on the rabbit; (iv) serodiagnosis in man, and also mentioned (v) *in vitro* work on the behaviour of *Anisakis* and methods of preventing anisakiasis, which is discussed further on page 427 of this book. As a copy of Ruitenberg's thesis may now be difficult to obtain, it may be useful to point out that an extended summary of his main results may be found in Williams and Jones (1976). Although ascaridoid nematodes had become the focus of intensive investigations from about 1965 in the Netherlands, Japan and other countries, Van Thiel (1976) drew attention to unresolved problems in anisakiasis research with regard to (a) its epidemiology, (b) global distribution, including differences in frequency of anisakiasis in the Netherlands and Japan (2.4 and 74.5 per cent respectively), (c) marine mammal hosts of adult worms, (d) first and second intermediate hosts (see Figure 6.3), (e) taxonomic problems in identifying *A. simplex* and also in accepting it as the correct name, (f) other anisakine worms which might cause the disease, and (g) procedures for control by killing larvae which occur in fish hosts.

A consequence of the discovery of anisakiasis was the accumulation of an enormous literature on the general biology of *Anisakis* and on the accidental infection of man and other animals with *A. simplex* (see Smith and Wootten, 1978a). These authors cited almost 300 references in a classical review of the subject, and it is highly significant that all except about 20 were a world-wide response to Thiel, Kuipers and Roskam (1960), who described how a nematode normally parasitic in herring could establish itself in the alimentary canal of humans and cause acute abdominal syndromes. It is important to appreciate that Smith and Wootten concerned themselves with *Anisakis* (*sensu stricto*) and not with the other marine ascaridoid nematodes, including *Pseudoterranova* (sealworm = *Terranova* = *Phocanema* = *Porrocaecum*), *Contracaecum, Hysterothylacium* and *Phocascaris*, all of which are potentially infective to humans. Smith and Wootten were also selective in their choice of references on *Anisakis* and explained that the literature explosion had occurred in two stages. The first was a response from about 1945 to 1960 to the anxiety expressed, for example by Myers (1960), about the occurrence of large numbers of larval nematodes in the flesh of cod, and the second was a response, as mentioned earlier, to the paper by Van Thiel and his colleagues in 1960. Since Smith and Wootten cited 14 review-type papers on *Anisakis* or anisakiasis, including Williams and Jones (1976), our summary of the topic in this book is restricted to their main conclusions and to selectively updating the literature from about 1975.

Smith and Wootten thought the term 'herring-worm' disease was misleading and that it should be avoided in future in at least the European North Atlantic, because it had been confirmed that while *Anisakis* was common in herring it is equally, if not more, abundant in other fish species including blue whiting, *Micromesistius poutassou*. Many taxonomic, *in vitro* culture and life-history problems remained to be resolved in 1978 and our knowledge of the behaviour, physiology and biochemistry of different stages in the life-history of *Anisakis* was sadly lacking. More physiological work of the kind already done for *Phocanema* was advocated for *Anisakis*. Volatile ketones and alcohols are produced by larval *Phocanema* and it had been thought that these acted as a local anaesthetic as the worm burrows through fish muscle. *Anisakis* was suggested as a useful experimental model for studying physiological functions, especially the excretory system. Man's possible allergic predisposition to severe gastro-intestinal reaction was in need of investigation and it was important to improve differential diagnosis of human infections with *Anisakis* and other ascaridoid nematodes.

The knowledge explosion based on *Anisakis* and anisakiasis research has continued unabated since 1978 to the present day. Some of the results will be mentioned here whilst other publications will be referred to later, under fish worms as biological indicators and under fish worms as models in teaching and research. An attempt will be made wherever possible, as recommended by Smith and Wootten, to deal with *Anisakis, Pseudoterranova* and other ascaridoid infections separately.

Summarizing and presenting such a volume of recent research is problematical but, since Thiel (1976) predicted that anisakiasis might well occur in countries other than Japan and the Netherlands, it was decided to do so by referring to continents or countries, in which it has been reported, in alphabetical order.

(i) Europe

In Poland, Grabda (1976b) reported investigations on the Baltic cod, *Gadus morhua callarias*, as a host for the third stage larvae of *A. simplex*, the source being western herring entering the Baltic to spawn. Man is susceptible through eating not only cod flesh but also the liver and gonads. The highest prevalence and intensity of infection occurred in cod from the Western Baltic. No *Anisakis* larvae were found in cod from the Gdansk Bay. Grabda (1977b) found 11 species of helminths in *Theragra chalcogramma* imported into Poland from the USSR; 38.8 per cent of the fish harboured *Anisakis simplex*. Grabda (1978) found the European blue whiting, *Micromesistius poutassou*, heavily infected with *Anisakis* larvae but the southern blue whiting, *M. australis*, was not; it harboured instead, heavy infections of the protozoan *Kudoa* sp. All fish were infected with up to 148 *Anisakis* larvae per host. She disagreed with Reimer *et al.* (1971), who had said that the flesh of European blue whiting could be commercially processed. The discovery, around 1980, of *A. simplex* in smoked herring on sale in a wholesale market in France prompted more reviews, by Lagoin (1980), covering the life-cycle, epidemiology, pathogenicity, diagnosis, treatment and prophylaxis; and by Carletti (1981), who discussed anisakiasis generally in North European countries and described the life-cycle, transmission and pathogenicity of *A. simplex*. Methods which had been used to destroy infective larvae in the flesh of fish were listed. Godeau, Danis, Bouchareine and Nozais (1985) drew attention to an increase in the number of human *Anisakis* infections reported in France and the high prevalence (80–90 per cent) of *Anisakis*-infected fish in Paris markets being a cause for public concern. Huang (1988) also surveyed anisakid larvae in commercial fish from the Paris district.

Meanwhile, expressions of surprise had occurred in the Netherlands when Van Thiel and Bakker (1981) reported the discovery of a worm granuloma and an adenocarcinoma in the stomach of a patient. Since 1968 in the Netherlands, government regulations had required all raw herring to be frozen to at least $-20°C$ and kept at that temperature for at least 24 hours before being released for human consumption. It seems that from about 1971, no further cases of anisakiasis had been known to the Public Health Inspectorate in Holland. Consequently, van Thiel and Bakker (1981) discussed the possibility of *Anisakis simplex* occurring again in the Netherlands in the light of a growing number of travellers to Japan. They reaffirmed that well-cooked fish prevented human infection with *Pseudoterranova* and *Hysterothylacium* with consequential gastric granulomata.

A general interest in *Anisakis* has been retained in Norway. For example Berland (1981) reported briefly on the behaviour of *A. simplex* in the stomach of cod which had massive infections of the species in clusters embedded in the stomach wall. He suggested that this was because previous infections raised the immunity to such a level that new infections are blocked from migrating further by a tissue reaction in the submucosa. It is interesting that *A. simplex* should attack the stomach of cod whereas, in humans, it seems to prefer the ileal intestine. On the other hand, the sealworm (*Pseudoterranova*) prefers the human stomach.

The findings of researchers from the USSR, UK and other European countries, are also of interest to Norway and other North European countries. For instance, Lorenz (1981) and Karasev (1983a) found every specimen of *Micromesistius poutassou* over 4-years-old from the Norwegian Sea to be infected with *A. simplex*. UK research findings have been well-summarized by Smith and Wootten (1978b, 1984b). In the light of the Netherlands government's regulations for treating freshly-caught herring, it is significant that Panebianco and Schiavo (1985) found 10 herring containing live *A. simplex* larvae in their mesenteries, intestines and gonads in a sample of 40 salted and 30 smoked fish from a consignment imported into Italy from the Netherlands. Before this discovery, interest in anisakiasis had already extended to South-West Europe, eg. Carvalho-Varela and Cuna-Ferreira (1984) discussed records of ascaridoid larvae in Portuguese fish in relation to public health and reported a 51.4 per cent prevalence of *Anisakis, Hysterothylacium* and *Thynnascaris* larvae in *Trachurus trachurus* and a 28.4 per cent prevalence of *Hysterothylacium* or *Thynnascaris* larvae in *Sardina pilchardus*.

An interest in *Anisakis* in Germany extends from about 1970, but apparently the first case of anisakiasis was not reported in the country until 1982 (Knöfler and Lorenz, 1982).[4] Five cases were reported and in one the nematode larvae had bored through the intestinal wall into the abdominal cavity. Consequently, Reimer (1983) published a leaflet on anisakiasis. Priebe (1986) refers to the Alaska pollack, *Theragra chalcogramma*, as the most frequently caught fish species in the world and its fillets are popular in the Federal Republic of Germany. He recorded and described *Anisakis* sp., *Pseudoterranova decipiens* and blastocycts of the trypanorhynch *Nybelinia surmenicola* from these fillets. Wikerhauser (1986) suggested that, so far, no case of the disease had occurred in Yugoslavia.

(ii) Hawaii

Although no cases of human anisakiasis had been identified in Hawaii, Deardorff *et al.* (1984b) surveyed its marine fish for larval nematodes with a view to predicting which species were potentially hazardous. This was done largely because consumption of large quantities of raw seafood dishes such as sashimi, poki, sushi and lomi lomi salmon on the island greatly increased the local consumer's exposure to infection with these nematodes. Larval nematodes of the genera *Anisakis* (two types), *Hysterothylacium* (three types), *Raphidascaris* (one type) and *Terranova* (two types) were identified, described, illustrated and keyed. Because nearly all anisakine larvae found were in non-edible tissue, they thought the risk of human infection was probably slight. They did, however, succeed in establishing *Anisakis* type II, '*Terranova* type and Hawaii (HA)' *Hysterothylacium* type HB in rats with pathogenic consequences (Deardorff et al., 1982). Deardorff, Kliks and Desowitz (1983) established '*Terranova* HA' in the stomachs of rats and they suggested, therefore, that it was a potential pathogen of humans although there were still no known cases of human anisakiasis from Hawaii.

Deardorff, Raybourne and Desowitz (1984a) identified a *Terranova* type HA larva from the viscera of the blue-green snapper, *Aprion virescens*, as being closely related to but different from *Pseudoterranova* (= *Phocanema/Terranova decipiens*), and also different from six other larval types belonging to the genus *Terranova* which had already been adequately described. Deardorff Raybourne and Desowitz (1984b) had become more cautious in that they were now suggesting that human health risk does exist for people who eat raw or inadequately-cooked Hawaiian snappers or imported rockfishes infected with *Terranova* type

4 See also Möller and Schroder (1987).

HA or *Anisakis* type I larvae. A method of preventing the infection is suggested and this is further discussed under control. Their fears were justified because, according to Oshima (1987), Kliks in 1986 recorded the first case of duodenal anisakiasis due to an adult *A. simplex* in Hawaii. It is unusual for the species to settle in the duodenum, so the observation calls for further research.

(iii) Indonesia

Although human anisakiasis had not yet been reported from Indonesia, Hadidjaja *et al.* (1978) were prompted to examine three species of fish for anisakids. *Anisakis* type I and *Terranova* type B larvae were found in 719 of 1,459 *Rastrelliger kanagurta*, 445 of 884 *Decapterus russelli* and 217 of 531 *Sardinella sirm*. They concluded that the *Anisakis* larvae found were a potential source of infection for humans. These researches were extended by Hutomo, Burhanuddin and Hadidjaja (1978). The monthly prevalence of anisakid infections in *R. kanagurta* ranged from 4–87.7 per cent and the average worm burden from 1.5–13. In general, prevalence and intensity of infection increased with fish length.

(iv) Japan

A number of cases of anisakiasis have been reported in Japan from about 1970 and these have prompted further surveys of marine animals, especially fish and mammals, experimental infections of laboratory animals with anisakids and refinement of diagnostic techniques. For example, a review of 1531 cases of human ileitis by Hayasaka, Ishikura and Takayama (1971) suggested that 77 were due to intestinal anisakiasis while 196 cases of gastric anisakiasis were detected. The experimental infection of rabbits suggested that intestinal anisakiasis may be an Arthus-type allergic reaction caused by reinfection with *Anisakis* larvae. Shiraki (1974) found five ascaridoid genera (*Anisakis, Terranova* (= *Phocanema*), *Contracaecum, Raphidascaris* and *Hysterothylacium*) in marine fish in the northern coastal waters of Japan, including *Anisakis* type I larvae in a brackish water fish (*Tribolodon hakonensis*) and a freshwater species (*Hemibarbus barbus*). *Contracaecum osculatum* was also found in *T. hakonensis* and was successfully used in experimentally infecting rabbits. This led to the suggestion that *C. osculatum* may be pathogenic to man, thus drawing attention once again to the complicated taxonomic problems which can arise from studying anisakiasis. Suzuki, Ishida, Asaishi and Nishino (1976) continued with their series of investigations on the immunodiagnosis of anisakiasis; Torisu *et al.* (1983) reported on its pathogenesis and Kagei *et al.* (1978) traced a human infection to perhaps eating a raw fish (*Katsuwonus pelamis*) and raw squid (*Todarodes pacificus*). It was the first report of infection due to *Anisakis* type II and again, this demonstrated the complex nature of the disease. Four years later, Fukuda *et al.* (1982) examined 20 species of marine fish from the Seto Inland Sea and two, *Scomber trapeinocephalus* and *Trachurus japonicus* from the open sea. These two species showed a high prevalence of *Anisakis* infection. Non-migratory fish caught in the centre of the Inland Sea were thought to be fit for human consumption if eaten raw but some of the migratory species were not. An *Anisakis*-like larva attacking a cancerous stomach growth, described by Tsutsumi and Fujimoto (1983), and the possible relationships of such an attack to local changes in pH is interesting in relation to the account by Berland (1981) of *Anisakis* in the stomach of cod and to our lack of knowledge of the tolerance of helminths to wide ranges in pH as discussed by Williams, McVicar and Ralph (1970). Fujino, Ooiwa and Ishii (1984) showed that the greatest number of 150 cases of gastric anisakiasis studied from December 1968 to June 1983 had occurred between February and May with the least from June to August. High prevalence was correlated with periods of high consumption of common and horse mackerel (winter to spring). The foregoing summary of progress in anisakiasis research in Japan is well discussed by Oshima and Kliks (1986) and Oshima (1987). The first-named reference reviews the role of marine mammals as carriers of nematodes infective to man with emphasis on *Anisakis simplex*. In the second reference Oshima emphasized that *A. simplex* normally migrate to the human ileum and cause intestinal anisakiasis. In south west Japan, the main source of human infection with *Anisakis simplex* was the common mackerel, *Scomber japonicus*, a fish which remains the most important source of anisakiasis in Japan because it is '*cheap*' and often used *at home* to prepare vinegar-pickled sashimi or sushi. Other 'cheap' fish used in this way are halibut, cod, pollack, greenling, salmon and herring. Oshima concludes, however, that the world-famous Japanese dish sushi made by commercial sushi shops uses high-class and expensive fish such as raw fillet of blue-fin tuna, big-eye tuna, striped marlin, broadbill swordfish, porgy or sea bream and horse mackerel; most have no sealworm or *Anisakis*. Oshima points out that the clinical appearance of anisakiasis is much milder in the States compared with Japan and Holland and, despite suggestions that this might be due to different strains of worms involved, the question remains enigmatic. His warning that the pattern of the disease has dramatically changed in recent years warrants further research.

(iv) Korea

Anisakiasis was being reported from Korea in the early 1980s (Appelby *et al.*, 1982; Seo *et al.*, 1984).

(v) South America

From 1975–76 a total of 86 species of marine food fish caught near Montevideo, Uruguay, was examined by Botto *et al.* (1976) for *Anisakis* sp. Torres *et al.* (1981) recorded the following ascaridoid larvae for the first time in fish caught off the southern coast of Chile: *Anisakis* type I in *Cilus montti; Terranova* in *Hippoglossina macrops* and in *Schoroederichthys chilensis*; and *Contracaecum* in *Notothenia cornucala* and *N. angustata*. Cattan and Carvajal (1984) observed the migration of *A. simplex* in *Merluccius gayi* whilst Carvajal and Rego (1985), in concluding that the most important causes were *Pseudoterranova decipiens* and *Anisakis simplex*, also accepted that *Hysterothylacium* and *Pseudoterranova* are potentially pathogenic. Rego, Carvajal and Schaeffer (1985) recorded *Hysterothylacium, Raphidascaris* and *Terranova* larvae from the liver of red porgies, *Pagrus pagrus*, caught off the coast of Brazil.

(vi) USA

Myers (1979) discussed the high prevalence of *Anisakis* sp. larvae in Pacific coastal waters and attributed it to the large population of whales in the area. A low prevalence of *Phocanema* (= *Pseudoterranova*) sp. larvae was linked to the small population of seals. Anisakiasis in the USA was reported by Valdiserri (1981), Kliks (1983), Rushovich *et al.* (1983), Heckmann and Otto (1985), McKerrow and Deardorff (1988), Sakanari *et al.* (1988) and Sakanari and McKerrow (1989). Hauck (1977) and Hauck and May (1977) refer to the occurrence of *Anisakis* in Pacific herring and its effects on the host.

(vii) USSR

From published information, Khadshiiski (1980) discussed the morphology, biology and resistance of anisakid larvae to heat and cold and drew attention to their potential danger to humans, whilst Todorov (1982) reported *Anisakis* larvae from *Merluccius merluccius, M. bilinearis, Trachurus trachurus, Dentex macrophthalmus, Scomber japonicus* and *Genypterus capensis* caught in various zones of the Atlantic. The prevalence was highest in the *Merluccius* spp. but all larvae were found to be dead as the fish had been kept at −18 to −20°C for up to two months.

Pseudoterranova Mosgovoi, 1950

Adult roundworms known as *Pseudoterranova decipiens* (= *'Terranova decipiens'* = *'Porrocaecum decipiens'* = *'Phocanema decipiens'*) are cosmopolitan in the stomachs of seals, sea lions and walruses. Hence they are commonly known as sealworms. Eggs from the worms are passed in the faeces and hatch in the sea as second-stage larvae which are probably eaten by small crustaceans. Infected crustaceans are, in turn, eaten by fish, and the larvae migrate from their intestines into the flesh where they rest as third-stage larvae. Probably a very large number of fish species are infected but, in the northern hemisphere, they occur most frequently in cod. Hence the worm is, perhaps, better known as 'cod worm' rather than 'seal worm'. Since *Pseudoterranova*, like *Anisakis*, belongs to the family Anisakidae of the order Ascaridida the disease it causes has been called anisakiasis, 'phocanemal (or terranoval) anisakiasis' or 'codworm anisakiasis'. Although Oshima (1987) suggested adopting the term 'cod-worm anisakiasis', probably to avoid further confusion, we feel that eventually the more precise approach suggested by Margolis (1977) and Smith and Wootten (1978a) should be universally adopted because the vernacular names 'codworm' and 'herringworm' disease will become increasingly misleading. This is obvious from our earlier account of *Anisakis* and anisakiasis, where it proved difficult to segregate information on *A. simplex* (*sensu stricto*) from that on other ascaridoid infections, including *Psuedoterranova*.

Williams and Jones (1976) reviewed information on human infections with *Pseudoterranova* but referred to it then as *Terranova*. Margolis (1977) also reviewed 'codworm' infections, preferring to use the name *Phocanema decipiens*. We have already recommended adopting Margolis' approach to nomenclature of human disease caused by *Anisakis* and *Pseudoterranova*. Margolis gave a tabulated summary of confirmed or presumed human 'codworm' infection in chronological order beginning with the first in 1921 and the fourty-sixth case in 1976. He also drew attention to an undetermined number of cases in Japan. According to Oshima (1987), the apparent sudden increase in 'codworm anisakiasis' in northern Japan after 1972 was not an epedemiological outbreak but was due to better and increased diagnosis by endoscopy. The most important fish carriers in Japan have been Pacific cod, *Gadus morhua macrocephala*, halibut, *Hippoglossus stenolepis*, and Japanese greenling, *Hexagrammos otakii*. The human disease seems to be seasonal, at Hakodate from March to July through eating raw cod fillet, and at the Aomori Prefecture from January to April through eating greenling and cod. Until 1984, a total of 3,141 cases of stomach anisakiasis had been reported in Japan. Of particular importance are Margolis' views on future research needs into the biology, prevention, diagnosis, treatment and epidemiology of *P. decipiens*. He suggested that the absence of any records of clinical cases in western nations may be due to unfamiliarity among western physicians and surgeons of Japanese work based on a few thousand cases of the syndromes of stomach and intestinal anisakiasis.

Very many other factors probably account for the differences and these will be slowly revealed through further experimental infections. For example, Gibson (1970) showed that the stomach of the rat on a fish diet provided a more suitable environment for larval development of 'herringworm' through to the preadult stage than either the guinea-pig or the rat stomach without the fish diet.

Gnathostoma, Owen, 1836

Species of *Gnathostoma* have been found in all continents except Europe and Australia (Cheng, 1986). Adults of this nematode usually occur in cats and dogs. Eggs passed in the faeces give rise to free-swimming larvae in about seven days at an optimum temperature, and these are infective to *Mesocyclops leuckarti, Eucyclops serrulatus, Cyclops strenuus* or *C. vicinus.* These species of cyclops, if ingested, transmit the worm to fish, reptile or amphibian second intermediate hosts. Fish may carry from 1–150 infective larvae. For a monograph on the genus see Daengsvang (1980).

Although there are about 12 species of *Gnathostoma, G. spinigerum* is the cause of virtually all cases of gnathostomiasis reported in humans (Daengsvang, 1982). Uncooked or partially-cooked fish is a common source of infection. From the human stomach, the larva will migrate through its wall to the liver and from there it will proceed into muscles and connective tissue and come to rest subcutaneously. Larva migrans may, however, occur when the larva continues to move under the skin. Recently, Bhaibulaya and Charoenlarp (1983) described a tortuous eruption of *G. spinigerum* on the left thigh above the knee of a Thai woman. She had eaten fermenting salted freshwater fish, known locally as Sam Fak, 20 days earlier. It is thought that recent cases in Japan referred to by Akahane, Iwata and Miyazaki (1982) were due to eating raw loaches, *Misgurnus anguillicaudatus*, imported from China into Japan and containing infective stages of another species, *G. hispidum.*

Trichinella Railliet, 1895 (= *Trichina* Owen, 1835) (Figure 6.4)

According to Williams and Jones (1976) evidence had gradually accumulated to indicate that *Trichinella spiralis*, one of the best known roundworms of man and animals, is capable of being transmitted to man in a marine environment. Although prevalence of trichinelliasis is considerably less today than the 28 million human cases estimated in 1947, the disease remains a threat to human health because of its wide distribution in domestic and wild animals which occasionally may include fish. Oshima and Kliks (1986), however, doubted the suggestion of Myers (1970) and Williams and Jones (1976) that trichinelliasis is a zoonotic disease cycling between eskimos and other animals in the Arctic region. The epidemiology of the disease remains an enigma, as emphasized throughout our previous review and again by Oshima and Kliks (1986). This is also obvious from information on possible transmission routes of *Trichinella spiralis* (Fig. 6.4). The possibility of transmission of *Trichinella* by cold-blooded animals was investigated experimentally by Tomasovicova (1981). The only likely agents are fish and by using *Cyprinus, Perca, Gymnocephalus* and *Alburnus*, this suggestion was confirmed. It was, however, concluded that although fish can act as reservoir hosts they are not of great importance as transmission agents in nature.

Acanthocephala

There are about 1,000 species of acanthocephalans, very many of which occur as adults in fish (Crompton and Nickol, 1985). Normally, fish and other vertebrate hosts become infected through eating the intermediate host which is usually an arthropod. Rarely, fish can act as paratenic hosts for encysted cystacanths, the infective stage. Generally, human infections with acanthocephalans are rare and these are not considered to be acquired through eating fish although *Acanthocephalus rauschi*, normally found in fish, has been reported from the peritoneum of an Alaskan eskimo. *Corynosoma strumosum*, a parasite of seals, has also been reported from humans, and since this species may be found encysted in about 90 genera of fish it is reasonable to assume that fish were the source of infection. This view is supported by the recent discovery by Tada *et al.* (1983) of a *Bolbosoma* sp. in Japanese fishermen who had become infected by eating raw fish (sashimi), perhaps bluefin tuna. *Bolbosoma*, like *Corynosoma*, is known from a number of genera of economically important food fish including *Thynnus*.

Fish worms and animal health

It is difficult to exclude any phylum of the animal kingdom for which it can be said that none of its members has any association with one or more species of fish worm. Consequently, the topic of fish worms and animal health is all-embracing and is touched upon in this and other chapters of this book. Members of the Platyhelminthes, Nematoda and Acanthocephala which, as adults, parasitize fish, may use species representing many other groups of animals to complete their life-cycles (see Chapter Two). Annelids, molluscs, arthropods and fish themselves are particularly susceptible to infestation with larval stages of fish worms. Occasionally mammals are also involved as carrier hosts. Additionally, a very large number of helminth species which mature in amphibians, reptiles, birds and mammals use fish as hosts of infective stages in the life-cycle. This applies in particular to helminths of fish-eating birds and mammals. The complex nature of the subject is well-shown in the accompanying diagrams (Figures 6.5, 6.6) from Nikolsky (1963). An example of such a complex situation was given by Krotas *et al.* (1984) who discussed the role of aquatic vertebrates in the formation of foci of helminthiases and the additional roles of leeches, molluscs and benthic invertebrates in spreading infections. Papers published between 1974 and 1977 on the parasites of marine mammals, aquatic birds, fishes and aquatic invertebrates were reviewed by Chubb (1981). It is, perhaps, initially simpler to refer to the few groups of animals which are unlikely to become infested with a fish helminth, eg. the Protozoa and Porifera at one end of a scale and, at the other, highly-selective feeders from amongst domestic and wild animals, eg. horses, sheep, giant pandas and koala bears. As already emphasized, most groups of other animals are susceptible to attack by one species or another. There are suggestions in the literature, mostly unconfirmed, of fish worms having important associations with viruses, bacteria and protozoans, and occasionally it has been implied that these are not only of nutritional importance to the worm but are also transmitted by fish worms as disease agents of other animals. On the other hand, it has been implied that such associations may be detrimental to the worm and are of possible use, therefore, as biological control agents. Rarely, fish helminths have been known to attack other helminths, be it members of other species or of their own, eg. the larvae of *Gyrocotyle* invade other individuals of their kind, possibly the parent worm (Williams, Colin and Halvorsen, 1987).

In view of the foregoing remarks which highlight our ignorance of a complex subject and especially our lack of knowledge of the effects of fish worms on other animals, this brief account is highly selective and concentrates only on giving a few key or recent references on major groups of animals which are susceptible to attack.

Invertebrates generally

Dollfus (1923, 1960, 1963, 1964a,b, 1967, 1974) discussed the helminths of marine invertebrates and cited many references to those of coelenterates, annelid worms, gastropods and lamellibranchs. A more modern treatise on the diseases of marine invertebrates from the Protozoa through to the Mollusca is that of Kinne (1980). Invertebrates, especially molluscs and arthropods, are constantly being recorded as hosts of helminths, as may

Figure 6.5 The main forms of biotic links in inland water-masses. 1, Sea-eagle capturing a fish; 2, fish hawk, and 3, bear with captured fishes; 4, beavers constructing a dam, which has altered the conditions in the water-body; 5 otter with a captured fish; 6, spawn laid on a submerged plant; 7, sea-gull capturing a fish; the sea-gull's droppings fall into the water and form the source of infection of *Cyclops* with helminths; 8, fishes eating the infected *Cyclops*; 9, pike-perch hunting for small fishes; 10, merganser capturing a fish; 11, grass-snake capturing a fish; 12, heron and 13, frog with captured fishes; 14, bladderwort with fish larvae in its bladders; 15, stickleback at its nest constructed of plants; 16, ruffe and bream eating the same food; 17 female bitterling spawning into a mollusc; 18, dragonfly nymph; and 19, dytiscid larva with captured fish larvae.

Figure 6.6 The main forms of biotic links in the sea. 1, Blue whale and herrings, both feeding on the same food, 2, sea-gulls eating fishes; 3, dolphins hunting for fishes, 4, storm-petrel diving for fishes; 5, fishes dwelling under the umbra of a medusa; 6, medusa with captured fishes; 7, squid seizing a fish; 8, humpback whale hunting for herrings; 9, *Careproctus* laying its spawn underneath the claw of a crab; 10, haddock and cod consuming benthic and epibenthic invertebrates; 11, *Fierasfer* inhabiting the cloaca of a holothurian.

be seen at a glance through the Annual Indices of *Helminthological Abstracts* Series A, Animal and Human Helminthology. The following references are, therefore, intended only to draw attention to key works or recent papers of potential significance from a zoonotic aspect.

Annelida

Margolis (1971) reviewed the role of polychaetes as intermediate hosts of helminths and Demshin (1984) stated that 129 species of helminths (digeneans, cestodes and nematodes) used oligochaetes as intermediate hosts, and emphasized the special link between caryophyllaeid tapeworms and oligochaetes. Mackiewicz (1972, 1981) reviewed the life-cycles of caryophyllaeids.

Mollusca

The diseases of molluscs were discussed by Lauckner (1983) whilst Cheng (1967) listed and reviewed the helminths of commercially-important marine molluscs and discussed the significance of understanding their biology from three major viewpoints: (i) that of the fisheries biologists and shell fishermen, (ii) that of the public health officer and (iii) the more fundamental one of the biologist. With regard to the first category, he stressed that very many species of worms found as larval stages in molluscs develop into adults in fish including the economically-important species. The helminths of the squid, *Ommastrephes bartrami* were discussed by Gaevskaya (1976) and the digeneans of cephalopods by Overstreet and Hochberg (1975).

A recently discovered example which explains the importance of molluscs is given in the report of Beveridge (1987), who found *Echinocephalus overstreeti* in 90 per cent of 59 Port Jackson shark, *Heterodontus portusjacksoni*, examined. Larval stages of this nematode occurred in the adductor muscles of 62 per cent of the queen scallops, *Chlamys bifrons*, and 80 per cent of king scallops, *Pecten albus*, examined.

Arthropoda

Hall (1929) remains an important reference work for all researches into arthropods as intermediate hosts of helminths. A good example of the role of economically-important arthropods in the transmission of fish worms is given by Overstreet (1978a) in his discussion of the valuable penaeid shrimp fishery in the northern Gulf of Mexico. Penaeids may harbour the larval digenean *Opecoeloides fimbriatus* in the coelom and these mature in the gut of a member of the fish family Sciaenidae. The shrimps also contain larval tapeworms in two principal sites, the digestive gland and the intestine. Up to 1,000 of the intestinal forms may be present and, although not yet identified with an adult tapeworm, it is almost certainly a tetraphyllidean species which matures in the gut of an elasmobranch fish. The digestive gland form is a trypanorhynch tapeworm, *Prochristianella hispida*, which matures in the gut of stingrays, *Dasyatis sabina* and *D. sayi*. Three larval nematodes, *Thynnascaris* type MA, *Thynnascaris* type MB and *Spirocamallanus cricotus*, are also found in three shrimp species and all mature in fish hosts.

Culturing crustaceans is quickly gaining economic significance on a world-wide scale (Overstreet, 1986). Overstreet (1973, 1983) reviewed a number of helminths which can infect crustaceans. The nervous system of these animals is an especially favourable medium for helminth growth and is, therefore, a vulnerable site for heavy infections, eg. with the tapeworm *Polypocephalus*, which matures in an elasmobranch host.

Fish

The pathogenic effects of helminths on fish are discussed in Chapter Five. Many of the species mentioned complete their life-cycles in fish, birds or mammals. Of particular importance are the strigeid and heterophyid digeneans; tapeworms belonging to the Trypanorhyncha and Pseudophyllidea, eg. *Poecilancistrum caryophyllum* (see Overstreet, 1978a), *Grillotia erinaceus* (see Rae, 1958), *Ligula* sp. (see Dubinina, 1980) and *Triaenophorus nodulosus* (see Arthur *et al.*, 1976); and anisakid and gnathostomatid nematodes. These and other species are referred to frequently throughout this chapter and elsewhere in the book.

Birds

Fish play an important role in transmitting helminths to birds (Ryzhikov *et al.* 1985; Sonin, 1985). The monograph by Barus *et al.* (1978) on the Nematoda was the first of a series of three volumes on the helminths of fish-eating birds. It included a host/parasite list and references to 700 publications. Two other volumes were promised covering the digeneans (see Sonin, 1985), tapeworms and acanthocephalans of fish-eating birds (Ryzhikov *et al*, 1985). Little is known of the pathogenic effects on birds but the gulls *Larus ridibundus* and *L. argentatus* die from hyperinfections of two species of the strigeid digenean *Cotylurus, C. variegatus* and *C. platycephalus*. The metacercariae of *C. variegatus* occur in *Gymnocephalus cernua*, *Stizostedion lucioperca* and *Perca fluviatilis*, mainly on the wall of the swimbladder, and the metacercariae of *C. platycephalus* occur mainly in *G. cernua*, *S. lucioperca, Abramis brama, Leuciciscus cephalus*, but also in *Osmerus eperlanus* and *Rutilus rutilus*, especially in the pericardium and mesentery (Swennen, Heesen and Hocker, 1979).

Birds and mammals

The heterophyid fluke *Phagicola longus* matures in the intestine of mammals and also in birds such as herons and pelicans. Brown pelicans, for example, in Louisiana have been known to harbour up to 12 000 adult flukes per bird (Overstreet, 1978a). Further research is required on the pathogenicity of such infestations in birds and mammals. *P. longus* infests the flesh and viscera of mullet along the Gulf of Mexico and southeastern Atlantic states. Natural infections occur in the largemouth bass, *Micropterus salmoides*, redear sunfish, *Lepomis microlophus*, bluegill, *L. macrochirus*, and orange spotted sunfish, *L. humilis*.

Geraci and St. Aubin (1986) discussed the effects of various helminths, originating from fish, on marine mammals and suggested that *Diphyllobothrium* spp. in extreme circumstances can cause debilitation and death. They also refer to the ubiquitous distribution of larval tetraphyllideans of the genera *Phyllobothrium* and *Monorygma* in the tissues of cetaceans and pinnipeds; and the role of sharks, which harbour the adult cestodes, in this relationship.

The pathogenic role of the following digeneans, transmitted by fish to marine mammals, is little known: *Braunia cordiformis*, *Pholeter gastrophilus*, *Campanula*, *Microphallus pirum*, *Zalophotrema*, *Oschmarinella*, *Orthosplanchnus*, *Nasitrema* and *Hunterotrema*. Dailey (1985), Lauckner (1985a,b) and Margolis and Arai (1989) have made important contributions in this area of research.

The most cosmopolitan of marine mammal parasites are the ascaridoid nematodes, belonging to the genus *Phocascaris* and *Pseudoterranova decipiens*, and several species of *Contracaecum* and *Anisakis*. Other nematode genera include *Stenurus*, *Pseudalius*, *Torynurus*, *Halocercus*, *Parafilaroides*, *Otostrongylus*, *Dipetalonema*, *Uncinaria*, *Placentonema* and *Crassicauda*. The importance of ascaridoid nematodes in zoonoses has focused attention on marine mammals as carriers of nematodes.

Young (1972) investigated the grey seal, *Halichoerus grypus*, and the common seal, *Phoca vitulina*, as transmitters of anisakine larvae to cod. He also examined *Balaenoptera acutorostrata*, *B. physalis*, *Physeter catodon*, *Globicephala melaena* and *Phocaena phocaena* for their nematode parasites. Only seals were significantly parasitized by adult *Terranova* (= *Pseudoterranova*) *decipiens*. *Anisakis* sp. adults were absent from common seals but were found in all species of Cetacea and also in grey seals, but the latter are not hosts of significance to species of *Anisakis*. Howard, Britt and Matsumoto (1983) discussed the more common and significant lesions caused by worms in marine mammals.

Valtonen and Helle (1986) examined 84 ringed seals, *Phoca hispida bothnica*, and eight grey seals, *Halichoerus grypus*, from the northernmost part of the Baltic Sea. *Corynosoma semerme* and *C. strumosum* were found in *P. hispida bothnica*, the former being 10 times more numerous than the latter and with maximum numbers of 2,692 and 324 per seal. Young Saimar ringed seals, *P. hispida saimensis*, from Lake Saimar in south east Finland, however, harbored up to 1,657 *C. strumosum* only. Of the three cestodes found in these seals, only *Schistocephalus solidus* was found to be mature. *Diphyllobothrium vogeli* and *D. dendriticum* did not seem to mature in the ringed seals. Mature *Contracaecum osculatum* were found in abundance in the grey seals but they did not mature in ringed seals. More detailed accounts of *Corynosoma* spp and *Contracaecum osculatum* in seals is given by Valtonen and Helle (1986, 1988) and Valtonen, Fagerholm and Helle (1988) respectively.

Dawbin (1986) examined 100 humpback whales, *Megaptera noviaeangliae* and found *Anisakis* sp. in the stomach, *Crassicauda* sp. in the kidney and many thousands of individuals of a *Bolbosoma* sp. along the length of the gut. Tapeworms were relatively infrequent. The digenean *Ogmogaster* sp. occurred in 20 per cent of the whales. Adams (1986) examined 275 ringed seals, *Phoca hispida*, and found three species of *Corynosoma*, *Diphyllobothrium* and *Diplogonoporus*, and that *Contracaecum* was more prevalent than *Phocanema* (= *Pseudoterranova*). Taraschewski (1986) found that *Heterophyes heterophyes*, *H. aequalis* and *H. dispar* matured in Canidae and Felidae and can be transmitted by euryhaline fish. The comprehensive works of Delyamure (1969) and Yablokov, Bel'kovich and Borisov (1972) remain standard references for work on marine mammals and helminth zoonoses. (See also Margolis and Arai, 1989).

Man's influence on the abundance and prevalence of fish worms

Factors responsible for the present-day global distribution and numbers of fish worms are the same as those which have so often been discussed, eg. by Cockburn (1977) for such important human helminths as *Schistosoma*, *Paragonimus*, *Trichuris*, *Ascaris* and *Ancylostoma*. They include the continental drift, the origin, evolution and migrations of man, the ending of the Ice Age, the development of agriculture and aquaculture, the domestication of animals, an increase in human populations and urbanization. The spread of *Diphyllobothrium latum* plerocercoids in fish is a good example. Its distribution in subarctic and temperate zones of the world relates to the availability of (i) a suitable copepod first intermediate host, (ii) suitable second intermediate hosts, in particular pike, *Esox lucius*, and burbot, *Lota lota*, (iii) final hosts such as fish-eating mammals or a human population with a habit of eating raw or insufficiently-cooked fish, and (iv) fresh or brackish water which can be contaminated with faeces from the final hosts. Building operations and industrialization, especially those connected with the construction of dams, reservoirs and hydroelectric power stations, often contribute to extending the foci of infection with *D. latum*. In 1973, it was estimated that of the nine million human carriers of *D. latum*, five million were in Europe, four million in Asia and less than one hundred thousand in America (Bylund, 1982). A detailed account of the spread of the species in the Soviet Union, Europe,

America and Greenland, the Far East and Australia, is given by von Bonsdorff (1977). Man's influence on the abundance and prevalence of fish worms other than *Diphyllobothrium* and possibly such well-known forms as *Opisthorchis*, *Clonorchis* and *Gnathostoma* has, however, only recently been recognized from three areas of research: (i) the consequences of the transfer of diseased fish from one country or continent to another, (ii) man-made changes in lakes and rivers, especially those connected with hydroelectricity and the construction of artificial water bodies for fish culture, and (iii) the pollution of aquatic environments with organic and inorganic chemicals and, unwittingly, with introduced pathogenic species belonging to all major groups of parasites including the viruses, bacteria, helminths and arthropods. Although the second and third topics often overlap, eg. man-made changes in lakes and rivers may lead to thermal pollution and eutrophication, the three areas will be discussed below under the headings, Transfer of Diseased Fish, Man-made Changes in Water Bodies, and Pollution.

Transfer of diseased fish

Most of the transfers of fish worms within countries or continents and intercontinentally can be attributed to man's activities. According to Hoffman (1970) this has been done for three main reasons: (a) the need to establish highly desirable species in new locations, (b) to rear edible species such as trout, catfish, carp and *Tilapia* in other countries, and (c) for the aquarists' trade in attractive fish. In achieving these aims, however, too little thought was given to the transfer of fish harbouring parasitic worms. Consequently, by 1970 it was known that 31 species of monogeneans, five digeneans, three nematodes and one acanthocephalan had been widely spread for this reason. Details are given by Hoffman on these transfers, eg. from the USA to South Africa, South America to the USA, Africa to the USA, the USA to the USSR, Europe to Israel and USA, and South America to Europe. The transcontinental dissemination of parasites has been equally disturbing. The list of seven monogeneans, one digenean, two cestodes and one acanthocephalan given by Hoffman is probably only a very small percentage of the total number which has been transferred. Also disturbing has been the fact that when fish are introduced into an area not previously occupied by that species, they may acquire new parasites. Hoffman lists *Diplostomum spathaceum*, *Bothriocephalus gowkongensis* (now *B. acheilognathi*), *Capillaria eupotomis* and *Raphidascaris* sp. as good examples from amongst freshwater fish helminths. He speculated that because of the similarities of the fauna of North America, Europe and northern Asia, it is most likely that more fish worms would be transfaunated. The ease with which helminths are established in new areas depends on the complexity of the life-cycle (Bauer and Hoffman, 1976). Monogeneans which do not require intermediate hosts have been the most numerous to be transferred. It is suggested that *Dactylogyrus extensus*, although not discovered and described in the USA until 1932 had, in fact, been transported from Germany to the USA on carp, *Cyprinus carpio*, around 1870. *Dactylogyrus minutus* and *Gyrodactylus cyprini* may have arrived at the same time from Europe but they were not discovered in the USA until the late 1960s. Four species of *Dactylogyrus* of goldfish, *Carassius carassius*, are thought to have been transported from Japan to the USA. Species of *Urocleidus* and *Cleidodiscus* are thought to have been transferred with North American fishes to Europe but this was not realized until the 1960s. Again in the 1960s, many monogeneans were transferred from East Asian rivers to the fish farms of the European and Central Asian parts of the USSR, Rumania, Hungary and to other continents including the USA. The classic example of the transfer of *Nitszchia sturionis* from one species of sturgeon to another, from the Caspian to the Aral Sea, with disastrous results is well known (see page 376). The origin and spread of *Bothriocephalus gowkongensis* from South China to many countries,

Khawia sinensis from the Amur extensively into Europe, *Philometroides lusiana* from the Amur to Latvian rivers, *Philometra sanguinea* from Japan to the USA, and *Amurotrema dombrovskajae* from East Asia to European USSR and Central Asia seem to be generally accepted although little is known of the epidemiological factors responsible for their successful establishment in other countries. Less is known of the extension of *Gyrodactylus*, *Sanguinicola*, *Proteocephalus* and *Anguillicola*. Some of these genera have already been discussed in this chapter under 'Worm diseases of Captive fish' and others are further referred to below.

Bothriocephalus acheilognathi is a particularly good example of a dangerous worm pathogen carried into new areas by imported fish. It was recorded in fish farms in the Belorussian SSR for the first time in 1965 and was thought by Emelyanov (1971) to have been introduced with imported fish. It was first detected in carp fry and fingerlings in Yugoslavia in 1972, having probably been imported with grass carp, silver carp and bighead broadfish. It then spread to several other fish farms (Kezic, Fijan and Kajanga, 1975). *Bothriocephalus acheilognathi* was recorded for the first time in Poland in carp by Panczyk and Zelazny (1974) and in Britain on *Cyprinus carpio* at three fish farms in Essex, Lincolnshire and Yorkshire by Andrews *et al.* (1981). Actually, *B. gowkongensis* [*sic*] species from grass carp had been registered in Britain in 1970 (R. Bray, personal communication) Its appearance and spread in the USA has already been mentioned. Grigoryan and Pogosyan (1983) referred to *B. acheilognathi*, introduced from the Amur basin to fish farms in Armenia, having spread to natural waters and acquired a new host, *Barbus capito*.

Malmberg, at the Scandinavian Society of Parasitology meeting in 1972, discussed the complicated factors involved in understanding the spread of *Gyrodactylus* spp., namely (i) the difficulties in identifying species which appeared identical but were, in fact, different in Europe and North America, and (ii) the differing behaviour patterns of species when introduced to different areas in fish hatcheries: they may disappear, infest one or more species, although normally highly host-specific, or even change into morphologically-different strains on the same host in different hatcheries or hosts. Malmberg (1973,1989) discussed four closely-related species on five species of *Salmo* and *Salvelinus*: *G. salaris* on *Salmo salar*; *Gyrodactylus* sp. 2 on *Salvelinus alpinus*, *S. fontinalis*, *Salmo trutta* and *S. gairdneri* in hatcheries and naturally on *S. trutta* and *Salvelinus fontinalis* in small streams in Sweden and Scotland—it may be species specific to *Salmo trutta* in nature; *Gyrodactylus* sp. 3 which may fatally infest *Salvelinus fontinalis* and *Salmo gairdneri*, and *Gyrodactylus* sp. 4 on *S. gairdneri* in hatcheries in Sweden and Denmark. Heggberget and Johnsen (1982) attribute heavy mortalities in *Salmo salar* (caused by a *Gyrodactylus salaris*-like worm) to environmental factors consequent upon man-made changes in river flow and pollution due to cattle farming or to infection from recently-established hatcheries. Halvorsen and Hartvigsen (1989) thoroughly reviewed the biogeography and epidemiology of *G. salaris*.

Three species of the nematode *Anguillicola* have been known for a long time as parasites of East Asian and Australian eels (Peters and Hartmann, 1986). They are *A. globiceps* from *Anguilla japonica* (since 1927), *Anguillicola australiensis* from *Anguilla reinhardtii* in Australia (since 1940), and *Anguillicola crassa* found in 1974 in *Anguilla anguilla* and *A. japonica* in Japanese eel farms. These nematodes were reported as parasites of *A. anguilla* in Europe for the first time in 1982 and have since spread throughout Europe and were thought to have reached Britain in 1987 (C. Kennedy, personal communication). They are especially common in young eels in which the prevalence can be as high as 70 per cent. Because the supply of eels for culture must be obtained from wild populations, this worm is a serious threat to the eel industry. For updated information on *Anguillicola* readers are referred to Moravec and Taraschewski (1988) and to Bauer (1991) on the spread of *Anguillicola* and other fish helminths.

Man-made changes in water bodies

It has already been seen that the construction of artificial ponds and lakes for fish-farming purposes have important implications for the spread of disease (p. 377) and that this danger is often enhanced by introducing diseased fish from other areas into man-made water bodies (p. 413). Among some of the early investigations of such problems were those of Ivasik (1960) and Ivasik *et al.*, (1969) of phytophagous fish imported from China to the Ukraine; Bogdanova (1972) in the USSR; Diarova (1971, 1975) in Kazakhstan, and Skripchenko (1973) in West Siberia.

Holloway and Hagstrom (1981) concluded that four interrelated factors affect the development of fish helminth faunas in man-made lakes and reservoirs in North Dakota, USA. They are: (i) degree of environmental change at the impoundment, (ii) water mix and flow between the impoundment and river systems on which they are based, (iii) degree of contact with potential sources of new parasites, and (iv) ability of individual parasite species to continue reproducing in a new environment. Changes in the helminth fauna of fish may also arise from the regulation of river flows, diversion of rivers and lakes, construction of canals, dams and reservoirs, and increased water temperatures caused by outflows from hydroelectric power stations. Osmanov and Yusopov (1985) described an interesting effect of a man-made change on the helminths of fish in the Aral Sea. The inflow of freshwater into the Aral dropped by 40 per cent in the 1969–1979 period due to the demands of agriculture. Consequently the water level dropped, salinity increased from 9.6–16.4 per cent, the invertebrate fauna became poorer and catches of commercial fish fell drastically. The parasite fauna declined from 105–29 species.

The regulation of water flow through a tunnel in a Colorado river and the consequent changes in water temperature had the effect of flushing away available intermediate hosts at high water, and increased feeding by fish at higher temperatures during low water. The prevalence of *Crepidostomum farionis* and *Rhabdochona* sp. in *Salmo trutta* was greater during low water and higher temperatures (Voth, Anderson and Kleinschuster, 1974). Amongst the consequences of regulating the river flow in connection with Kakhovka reservoir, USSR, were a slowing of the current, increased depth, redistribution of shallow zones and pollution caused by blooming and silting (Iskov, 1976). These factors changed the species composition of the free-living animals involved in parasite life-cycles to such an extent that the number of digenean species was reduced from 53 to 32, and only nine of these occurred near the dam. In contrast, the percentage of fish infested with monogeneans had doubled to almost 100 per cent. Izyumova (1977, 1984) reported that pathogenic species such as *Ligula, Diplostomum*, philometrids and diphyllobothriids seem to be more problematical following the regulation of water flow to form reservoirs. An intensive study on *Dactylogyrus falcatus, D. wunderi, D. zandti* and *D. auriculatus* on *Abramis brama* was carried out by Izyumova and Zharikova (1984) in relation to different areas and water temperatures, some of which were connected to the warm outflow of a hydroelectric plant: *D. wunderi* and *D. zandti* predominated and seemed to be very numerous at 26.4°C but much less so at 29.8°C.

Arthur, Margolis and Arai (1976) speculated on changes which would, perhaps, occur in the fish helminth fauna if plans were to proceed in diverting Stevens Lake at the headwaters of the Yukon River system (draining into the Bering Sea) into Aishihik Lake at the headwaters of the Alsek River system (draining into the Pacific Ocean). Among other biological problems this raised the possibility of the transfaunation of detrimental fish parasites into Aishihik Lake and throughout the lower reaches of the Alsek River system. Forty-four species of parasites were found in Aishihik Lake and 23 in fishes from Stevens Lake. Although their study appears to be the first to examine the potential parasitological consequences of connecting two water bodies, the authors say that Lindsey, in 1957, had speculated on the possible effects of a number of proposed water diversions in British Columbia and warned of the dangers of spreading the pathogenic cestode *Triaenophorus crassus*. Twenty-three species of fish parasites in Aishihik Lake did not occur in Stevens Lake and only two species of the Stevens Lake parasites were not found in Aishihik Lake. Three known pathogenic species were found: *Discocotyle sagittata* and *Neoechinorhynchus rutili* in Aishihik Lake and *Triaenophorus nodulosus* in Stevens Lake. The scheme discussed by Arthur and Margolis was eventually abandoned for reasons unrelated to possible parasite transfer. Bogdanova and Kotova (1984) also attempted to forecast changes consequent on the partial southwards diversion of the Sukhona river, USSR. At the time, about 90 helminth species had been recorded from Lake Kubenskoe and the Sukhona river and although some of the more common species were considered with regard to ecological factors affecting prevalence and intensity, no clear-cut view of the future situation emerged.

Khramov, Pavlyukov and Shelikhanova (1984) warned of the dangers of constructing a canal along a route where from 1–100 per cent of the fish were infested (depending on the host species, host age and locality) with *Opisthorchis felineus* metacercariae and *Diphyllobothrium* plerocercoids. *Leuciscus idus* and *Esox lucius* were the most frequently infected with *O. felineus* and *Diphyllobothrium* respectively.

Following the damming of White River, Arkansas, USA to form the Beaver Reservoir, Becker *et al.* (1978) found that 'incidence' of infection with monogeneans increased while that of digeneans, cestodes, acanthocephalans and nematodes decreased. In about six years, however, the parasite community in the reservoir had increased to become much larger and more complex than it was in White River. Dontsov and Markov (1981) compared the parasites of *Stizostedion lucioperca* in the Tsimlyansk reservoir in the River Don catchment area with previously published information on this fish in the River Don and found that the reservoir fish harbored 34 species and those in the river only 12. Amongst the important introductions of helminth species were *Rhipidocotyle illense, Ichthyocotylurus platycephalus, Bucephalus polymorphus* and *I. pileatus*. It was noted that there were closer relationships between the parasite faunas of *S. luciopercae* in the Tsimlyansk and Volgograd reservoirs than between the fish in the rivers Don and Volga. A similar study by Dontsov (1979) on the Volga and its seven reservoirs found that, in general, the number of helminth species increased from its upper reaches to the delta.

The effects on fish helminths of artificially increasing water temperatures (often referred to as 'thermal pollution') have been discussed by a number of authors including Aho, Gibbons and Esch (1976), Boxrucker (1979), Bauer and Solomatova (1978), Pojmanska (1980), Mitenev and Shulman (1980), Granath and Esch (1983a, b) and Albetova and Michurin (1984). In the cooling tank of a hydroelectric station in the Ukranian SSR, *Diplostomum spathaceum, D. clavatum* and *Tetracotyle percae-fluviatilis* were widespread in 17 fish species due to the presence of large numbers of mollusc intermediaries and avian final hosts. In contrast, those digeneans which used fish as final hosts were represented by only five species and infection rates were low. There were only six species of cestodes, all light infections, except for *Ligula*. Monogeneans, except for *Dactylogyrus difformis* and *D. crucifer*, were recorded infrequently (Kulakovskaya, 1975). According to Aho, Gibbons and Esch (1976) the density of *Diplostomum scheuringi* encysted in the body cavity of *Gambusia affinis* decreased as temperature increased above ambient temperature. The density of *Ornithodiplostomum ptychocheilus*, encysted in the brain and eyes of *G. affinis*, was higher in fish from areas receiving thermal effluent than in fish from ambient temperature areas. Three hypotheses are proposed to account for the relationship between thermal loading and the observed abundance and distribution of these two species: (i) a direct differential influence of temperature on the physiological tolerance of the various life-cycle stages of the species, (ii) the same factor influencing the mollusc intermediate hosts, and (iii) temperature influencing the abundance and distribution of the definitive hosts within the study area.

Boxrucker (1979) investigated two helminths of the black bullhead, *Ictalurus melas*, in a thermal outfall area and an unaltered area of Lake Monona, Wisconsin, USA. There was a slightly greater prevalence of the monogenean *Cleidodiscus* from the period April to September in the unaltered area and it was significantly more abundant in August in the same area. Similarly the prevalence of the acanthocephalan *Pomphorhynchus bulbocolli* was fairly constant but the average number per fish rose to a peak in August and declined in autumn. A seasonal cycle of abundance was not apparent in the thermal outfall area. The general effects of thermal pollution were discussed by Bauer and Solomatova (1978) in a short communication during the Fourth International Congress of Parasitology, and is of interest in that it cites about 20 publications (authors' names only) in this field of research in the 1970s.

Pronin *et al.* (1985) studied the effects of the outflow of warm water from a hydroelectric station on the parasites of *Perca fluviatilis* and *Esox lucius*, and found reduced numbers of *Ancyrocephalus percae, Triaenophorus nodulosus, Proteocephalus percae and Bunodera luciopercae* on perch and of *Raphidascaris acus* and *Rhipidocotyle illense* on pike. Significantly increased prevalences were observed for *Tetraonchus monenteron* on *E. lucius* and of *Apatemon anuligerum* on perch.

Male salmon ascending the River Tuloma, USSR from the Barents Sea and trapped near a hydroelectric reservoir because they had failed to negotiate a salmon run, were infected with 12 helminth species, mostly marine forms, but they also harboured *Triaenophorus crassus* which had been acquired the previous year. Of the 19 parasite species collected from males returning to the sea, 14 were freshwater forms acquired during active feeding near a dam negotiated on the return run. A focus of *T. crassus* infection at the reservoir was temperature-related (Mitenev and Shul'man, 1980). Granath and Esch (1983a, b) investigated the seasonal dynamics of *Bothriocephalus acheilognathi* in *Gambusia affinis* from a thermally-altered location and an ambient temperature location in Belwes Lake, a North Carolina cooling reservoir. At both stations, lowest prevalences and densities were observed during summer months. Densities rose sharply in autumn and peaked by early winter, after which they declined. Parasite recruitment was also seasonal, beginning in late spring and continuing into October. Recruitment at the thermally-altered site, however, began about two weeks sooner, lasted about two weeks longer and was interrupted for several weeks in late summer when water temperature exceeded 35°C. Laboratory studies showed that growth and development were stimulated at water temperatures above 25°C. At 35°C coracidium activity was depressed. It was thought that the lower hatching success and briefer period of coracidium motility at 35°C might partly explain differences in the population biology of the cestode at the two sites during part of the summer months. Albetova and Michurin (1984) found that *Bothriocephalus opsariichthydis, Gyrodactylus sprostonae* and *G. cyprini* were potentially dangerous helminths of carp found in artificially-heated warm waters of a hydroelectric station.

Pollution[5]

Khan and Kiceniuk (1983) and Möller (1986) have drawn attention to the lack of information on the interaction of pollutants and parasites in fish. These authors refer to cases where increased disease prevalences may occur if fish living under natural conditions are

[5] See also Khan and Thulin (1991).

subjected to pollutants but also note that there are few examples to substantiate these claims. Most of the evidence available seems to support Möller (1986) who stated that polluted water may act directly on larval free-living stages of parasites or on adult ectoparasites on the skin or gills. It may also act directly on intermediate and definitive hosts. Generally, the use of pesticides, man-made alterations in water temperature, salinity and chlorination have a greater effect on parasites than on their hosts. Experimentally, it has been found that water polluted with heavy metals seems to have a greater effect on parasitized fish, leading to mortality, than on non-parasitized fish. The effects of industrial waste discharge, including that from paper mills, and of sewage effluents, seem to vary according to the species of helminths involved. In some cases of pollution with trace metals, pesticides, high concentrations of ammonia and crude oil, it is not clear whether the effects on parasite populations are primary or secondary. Published papers on these aspects of pollutants and parasitism are further discussed below.

Overstreet and Howse (1977) investigated and briefly discussed the helminths of the Atlantic croaker, *Micropogonias undulatus*, living in waters suspected or known to be polluted. Two samples of fish were taken, one from the Pascagoula River, which was considerably polluted, and another near Ocean Springs, Mississippi. Although the results are inconclusive for four helminths, the monogenean *Macrovalvitrematoides micropogoni*, the acanthocephalan *Dollfusentis chandleri*, the larval tetraphyllidean '*Scolex polymorphus unilocularis* (*minor*)' and the digenean *Diplomonorchis leiostomi*, they suggest that prevalence and intensity increased as pollution decreased. Generally, however, some other non-helminth diseases they had investigated showed an increase in prevalence with increased pollution. The results of Couch (1985) are inconclusive in that he found no grossly-evident epizootics that could be related to pollution in estuarine fish of the Gulf of Mexico. It is perhaps appropriate to mention that McVicar, Bruno and Fraser (1988) found that there were lower disease levels (non-helminths) on sewage dumping sites.

Boyce and Yamada (1977) studied experimentally the effects of zinc on sockeye salmon harbouring *Eubothrium salvelini* and in those without the parasite. The mortality rate was significantly greater in infected salmon when compared with the uninfected fish. Similarly, Pascoe and Cram (1977) and Pascoe and Woodworth (1980) found that cadmium chloride had the same effect on *Gasterosteus aculeatus* infected with *Schistocephalus solidus*.

Kostarev (1980) found that industrial waste discharge into two reservoirs had the effect of reducing the number of species of monogeneans and the parasite burden with *Proteocephalus, Triaenophorus, Ligula, Digramma, Acanthocephalus, Neoechinorhynchus, Philometra* and *Rhabdochona* because of the reduction in intermediate host numbers. There was, however, a rise in the numbers of *Diplostomum, Cotylurus* and *Posthodiplostomum cuticola* due to increased bird populations attracted by dying fish. High numbers of *Caryophyllaeus laticeps* were due to increased oligochaete populations. Where household sewage was discharged into the reservoirs there were increased prevalences of *Diphyllobothrium latum* and *Opisthorchis felineus*. Anikieva (1982b) compared the helminth fauna of 11 species of fish from three zones of a lake differing in the degree of pollution caused by a paper and pulp plant. Twenty, 21 and 41 species of helminths were collected in the most heavily-polluted zone, the moderately-polluted zone and the clean zone, respectively. The monogeneans showed the most striking differences with 10 species in the clean zone and only *Diplozoon paradoxum* in the most polluted. *Diplozoon homoion* and *Dactylogyrus amphibothrium* occurred in the moderately-polluted zones. Generally, *Caryophyllaeus laticeps, Diphyllobothrium latum, Cotylurus* metacercariae, *Bunodera luciopercae* and *Raphidascaris acus* were the most tolerant of polluted waters.

Skinner (1982) suggested that the effects of high concentrations of ammonia, trace metals and pesticides on three species of fish in a Florida bay had irritated the gills and stressed the fish with resulting physical and physiological changes which made them more susceptible to infections with *Neodiplectanum wenningeri* (on *Gerres cinereus*), *Ancyrocephalus* sp. (on *Lutjanus griseus*) and *A. parvus* (on *Strongylura timucu*).

Hodgins *et al.* (1977) discuss the effects of oil in relation to fish diseases whilst Khan and Kiceniuk (1983) tried to ascertain experimentally the effect of crude oils on selected parasites occurring naturally in the lumen of the gastro-intestinal tract of the flounder, *Pseudopleuronectes americanus*, naturally infested with the digenean *Steringophorus furciger*, and of the cod, *Gadus morhua*, harbouring the acanthocephalan *Echinorhynchus gadi*. In all cases prevalence and intensity of infections were lower in oil-treated fish. This might be due to the direct toxic effect induced by drinking polluted water and/or modification of the gut environment brought about by changes in host physiology. In this context Khan and Kiceniuk (1984) stated that in oil-exposed Atlantic cod there was an increased number of mucus-producing epithelial cells, capillary dilatation, lamellar hyperplasia and fusion of adjacent filaments in gills, microvesicular formation in hepatocytes, delayed spermatogenesis and an increase in melanomacrophage centres in the spleen and kidney.

Valtonen and Koskivaara (1987) and Valtonen *et al.* (1987) studied the effects of effluents from pulp mills and the paper industry and found that they reduced the number of intermediate hosts with a consequent significant reduction (in *Rutilus rutilus* and *Perca fluviatilis*) of *Tylodelphys clavata, Diplostomum* spp., *Ichthyocotylurus variegatus*, and *Sphaerostoma globiporum*. *Bunodera luciopercae* and *Phyllodistomum folium* did not seem to be affected but there was a massive infection with the metacercariae of *Rhipidicotyle illense* in the fins and gill arches of roach. The reduction from 12 to four parasite species found in *Platichthys flesus* in the Gulf of Gdansk, Poland, is attributed by Sulgostowska and Styczynska-Jurewicz (1987) to municipal and industrial sewage causing catastrophic eutrophication in the area. Pojmanska and Dzika (1987) found that over a period of about 10 years five species of helminths had disappeared from *Abramis brama* in Lake Goslawskie, Poland, but they recorded three species new to the locality. The frequency of 4 species had decreased while that of six had increased. These changes were thought to be due to long-term thermal pollution.

Control of fish worms

An inevitable consequence of helminths causing disease in fish, of some fish worms being potentially harmful to human and animal health, and of man having spread many dangerous fish pathogens, has been the need to implement control measures. These comprise prevention, protection, treatment, eradication and legislation. A comprehensive account of these five basic principles for the proposed control of bacterial, viral, fungal and protozoan diseases of freshwater and anadromous fish in Canada has been prepared by Bell and Margolis (1972).

Attempts to apply the same principles to control fish worms have been numerous and varied, as can be seen from a number of general publications. They include Bauer (1968), Braun (1975), Demidov and Potemkina (1980), Duijn (1973), Dupree (1981), Engashev (1969), Furtado (1961), Hoffman (1979), Hoffman and Meyer (1974), Hoffman and Mitchell (1978), Kalyuga and Kalashnik (1982), Kiskaroly and Tafro (1984c), Kolovarova (1983), Linnik (1980b), Meyer (1979), Poddubnaya and Ivanova (1984), Poupard (1978), Sniezsko (1973), Vasilkov and Engashev (1981) and Verbitskaya *et al.* (1972). There are, in addition, a very large number of publications which concentrate on one or more treatments, usually chemical, in controlling the developing egg, free-living larvae or adult of a particular monogenean, cestode, digenean, nematode or acanthocephalan. Prevention has often focused on the treatment of fish for human or animal consumption (including the consumption of fish meal on fish farms) by cooking, freezing, use of chemicals including salting, gamma-irradiation, refining methods such as candling in detecting worms in fish flesh, and identifying foci of infection within fishing grounds with the view to avoiding or rejecting infected fish. Protection has involved biological control and other methods of eradicating infective stages as well as invertebrate and vertebrate carrier hosts, and legislation. The immune responses of fish, as discussed in Chapter Five, and the use of resistant species, if only as biological control agents, are also important in the context of prevention.

To illustrate the principles of prevention, protection, treatment and eradication, comments are made below on recent publications on the most problematical of fish worms. They are the monogeneans *Gyrodactylus* and *Dactylogyrus*, the tapeworms *Bothriocephalus, Proteocephalus, Eubothrium, Khawia, Caryophyllaeus* and *Diphyllobothrium*, the digeneans *Diplostomum* and *Sanguinicola*, the nematodes *Philometroides* and *Anisakis* or related genera, and the acanthocephalan *Neoechinorhynchus*. In view of the increasing importance of biological control and the enormous potential of this topic for further research it will be discussed separately. A brief comment on legislation concludes this section on the control of fish worms.

Monogenea

Gyrodactyliasis (fin disease) and dactylogyriasis (gill disease) caused by *Gyrodactylus* and *Dactylogyrus* spp. respectively remain difficult to control. Guberlet, Hansen and Kavanagh (1927) recommended dipping fish with *Gyrodactylus* in a 4.5–5 per cent common salt solution for 1.5–2.5 minutes, with transfer to rapidly running water. They also tested very dilute ammonium hydroxide and acetic acid and advocated great caution in their use. Although the number of chemicals tried since 1927 has increased considerably to include formalin, potassium permanganate, pyridylmercuric acetate and various anthelmintics (all greatly diluted) prevention remains more effective than trying to effect cures. The effects of temperature, salinity and chlorine, however, remain important factors in understanding control procedures. McVicar and MacKenzie (1977) discussed these factors with reference to *Gyrodactylus unicopula*, a marine form found on plaice, and suggested that it was sensitive to chlorine. *Gyrodactylus unicopula* succumbed quickly to treatment with fresh water which left the plaice apparently unharmed. The use of fresh water to control *Benedenia seriolae* and of water of higher salinity to control *Heteraxine heterocerca*, both on cultivated yellowtail in Japan, was also mentioned by McVicar and MacKenzie. Table 6.6 summarizes some recent work on controlling *Gyrodactylus, Dactylogyrus* and other monogeneans.

Cestoda

Bothriocephalus acheilognathi is the species against which most work appears to have been done. Attempts to eliminate this cestode in fingerlings of *Ctenopharyngodon idella* (grass carp) in fish farms in the Turkmen SSR involved drainage of the ponds in spring to eliminate invertebrates, disinfection with chlorinated lime and treatment of the fish with 100 g of kamala given by intestinal probe (Babaev and Shcherbakova, 1963). Mass excretion of worms followed and the intensity of infection was dramatically reduced. A similar régime has been recommended against *B. acheilognathi, Caryophyllaeus fimbriceps* and *Khawia sinensis* in cyprinids in the Ukraine (Shcherban, 1965). Di-*n*-butyltin oxide has replaced kamala as an anthelmintic (Herman, 1970). Compounds tested against *Bothriocephalus* in grass carp include ground conifer needles (successful at 250 g per day per 400 fish in ponds) and onion and horse-radish leaves (Klenov, 1969a, b). Phenasal, an analogue of yomesan, which was found to be more effective than kamala, dichlorophen or bithionol against fish

Table 6.6 Control of Monogenea

Parasite and host species	Treatment	Reference
Gyrodactylus on wild and cultivated carp	Malachite green (1g/m^3) and 10% limewater in rearing tanks, limewater treatment repeated regularly	Antsyshkina, Vazderkina and Kalyuga (1981)
Gyrodactylus on *Fundulus parvapinnis*	Neguvon (0,0-dimethyl 2,2,2-trichloro-1-hydroxethylphosphonate) at 7.5 mg/litre on day one, 3.25/litre on day 5 and 0.75 mg/litre on day 9 (each for one hour) effective.	Puffer and Beal (1981)
Gyrodactylus elegans and *Dactylogyrus vastator* on *Carassius auratus*	Combination of mebendazole at 0.4 mg/litre and trichlorphon at 1.8 1.8 mg/litre completely effective	Goven and Amend (1982)
Gyrodactylus stancovici, G. kathartneri, G. sprostonae and *Gyrodactylus* sp. on *Cyprinus carpio*	5% salt solutions, 0.1% ammonium solutions and calcium hypochlorite effective	Kashkovskii and Skomorokhova (1982)
Pseudogyrodactylus anguillae and *P. bini* on *Anguilla**	Mebendazole 100 mg/litre for 10 min	Székely and Molnár (1987)
Gyrodactylus elegans	Acquires resistance to mebendazole	Goven, Gilbert and Gratzek (1980)
Dactylogyrus	One dose only of 0.3 mg/litre of copper ammoniate effective.	Avdos'ev *et al.* (1978)
Pseudodactylogyrus microrchis on *Anguilla anguilla**	Two 24-hour baths (at 3-day intervals) trichlorfon at 0.5 ppm effective and harmless.	Imada and Muroga (1979)
Dactylogyrus on *Ctenopharyngodon idella*	Dipping fry fingerlings for 45 sec in a potassium permanganate bath at 1:1000 was 25% effective	Willomitzer (1980b).
Dactylogyrus spp. on yearling *Cyprinus carpio*	0.04 ammonium baths, alternated by clear water baths effective	Zharikova (1979).
Dactylogyrus spp. on *Carassius auratus* and *Cyprinus carpio*	Fish kept in chlorophos solutions (10 or 50 mg/litre) at 15 or 19°C for 5 days became less resistant to infections than untreated fish.	Zharikova and Flerov (1981)
Dactylogyrus vastator, D. extensus and *Diplozoon paradoxum*	Praziquantel causes tegumental damage in the three species and suggested therapy for *Dactylogyrus* on carp fingerlings is 10 µg/ml for 3 h at 22°C.	Schmahl and Mehlhorn (1985)
Dactylogyrus spp. on *Cyprinus carpio*	Chlorophos at high concentrations and temperatures increased infections	Flerov, Mikryakov and Kuperman (1982)
Dactylogyrus extensus on *Cyprinus carpio*	Copper ammoniate at 0.5 mg/litre effective	Kashkovskii and Skomorokhova (1982).
Dactylogyrus anchoratus on *C. carpio*	Numbers decreased when fish cooled for 20 days at 5°C (from 17 to 18°C)	
Monogeneans on *C. carpio*:	Neguvon at 2 g/1000 1 damage liver and kidneys of fish	Dzubic *et al.* (1981).
Dermophthirius on *Negaprion brevirostris*	Dylox (trichlorphan) at 0.5 ppm (3 times in 10 days) eliminated parasites	Cheung *et al.* (1982).

* See also Buchmann (1987), Buchmann and Mellergaard (1988) and Chan and Wu (1984)

cestodes, was found to be completely effective, without side effects, at 0.015 g per fish daily given at 1 per cent medicated food on 10 consecutive days (Muzikovski and Vasilkov, 1969). It was toxic at 3 g/kg body-weight and above. In another trial, phenasal was found to be more effective in pelleted feed than in mash and was 93.1 per cent effective at the non-toxic dose of 1 g/kg of feed; kamala was less effective and phenothiazine ineffective (Muzikovskii and Vasilkov, 1969). A spectrophotometric method for determining the amount of phenasal in medicated food given to carp has been described (Koscheleva and Muzikovskii, 1969). A different compound, niclosamid, has been tested against *B. acheilognathi* in carp in Hungary (Molnár, 1970a, b). As stated earlier, this tapeworm, originally imported from China to the USSR, was first recorded in Hungary in 1970; it was apparently imported to Eastern Europe from Asia in phytophagous fish (Molnár, 1970a, b; Shcherban, 1965). Recommended treatment was niclosamide in feed at 100–200 mg/kg body-weight. The compound was not toxic to cestode eggs so reinfection frequently followed successful treatment. It was, however, toxic to fish in aqueous solutions at concentrations as low as 10^{-6} and should, therefore, only be used in large ponds or those with a good through-flow of water. Of several compounds tested against *B. acheilognathi* in carp, two doses each of yomesan at 0.1 g/kg and of scolaban at 0.2 g/kg were effective.

The control of *B. acheilognathi* received increased attention in the 1970s and the complexity of the programmes involved is well seen in Yashchuk (1979) with regard to infection of *Cyprinus carpio* on fish farms in the Ukrainian SSR. It is based on a knowledge of the life-cycle of the tapeworm in fish and intermediate hosts of the genus *Cyclops*. Ponds are treated with unslaked lime and allowed to dry in freezing temperatures and fish food is medicated with phenasal when infected intermediaries are most numerous. In spring, mustard mixed with food is thought to speed elimination of tapeworms through increased peristalsis, improved digestion and a boosting of the host's defence reaction. Table 6.7 summarizes the range of chemical and other treatments used against *B. acheilognathi* and other fish tapeworms since the 1970s.

Digenea

In view of their complicated life-cycles involving one or more intermediate hosts, including molluscs and fish, a variety of control methods has been adopted as can be seen from Table 6.8 and the account of biological control (Table 6.10). Species of *Diplostomum, Sanguinicola inermis* and larval heterophyids in fish flesh have been the main focus of attention.

Nematoda

As with digeneans, the control of nematodes is in its infancy with isolated attempts on *Philometroides*, ascaridoids and a few other species (Table 6.9).

Acanthocephala

Few attempts have been made to treat acanthocephalans. Tetrafinol (carbon tetrachloride, mesocaine and liquid paraffin) at 1 mg/kg body-weight has been found to be effective against *Neoechinorhynchus rutili* in carp in Czechoslovakian fish farms. When applied to a 1.74 hectare pond containing 280 carp, at 1 ml/carp in spring, tetrafinol prevented reinfection with *N. rutili* in the autumn. The mean worm burden was one, compared with up to 153 in untreated fish and the weight-gain was over 1,400 g compared with 390 g (Tesarvík 1971, 1972).

Biological control

Biological control is an ecological and a human economic problem where it concerns outlay and benefit—the benefit often much more difficult to assess than the outlay (Lie *et al.*, 1968). These authors referred to the time-honoured and often successful method of introducing, or perhaps repeatedly introducing, a selected pathogen or predator until it becomes adapted to local conditions and 'takes', multiplying its own and reducing the target population, confining its distribution or even, on rare occasions, eradicating it. They also mentioned a new type of biological control involving two species of digenean flukes capable

Table 6.7 Control of Cestoda

Parasite and host species	Treatment	Reference
B. acheilognathi in carp	Tobacco leaf powder	Avdos'ev (1973)
B. gowkongensis	p,p' DDT; p,p' DDD and M.P. DDD at 0.1 mg/litre toxic experimentally	Davydov *et al.* (1976).
Bothriocephalus in *Cyprinus carpio*	10 g phenasal to 1 kg feed, 88.2% of effective phenasal + trichlorophen not eaten by fish	Muzykovskii (1977).
Bothriocephalus in *C. carpio*	Phenasal at 1.5% in food 86.1% effective	Karanikolov (1977)
B. acheilognathi in *C. carpio*	5% tobacco powder (with 1.12% nicotine) for 10–15 days reduced prevalence and intensity	Kolovarova (1978)
B. acheilognathi	Two doses of dibutyl tin oxide (interval of 3 days) at 0.12 g/kg body-weight at 18–21°C was 83% effective in group 0 fish and 100% effective in 0 + 1 fish	Muzykovskii, Vasil'kov and Borisova (1977)
B. acheilognathi in *C. carpio*	Mansonil (the piperazine salt of niclosamide) and yomesan, both at 1.0 g/kg body-weight were fully effective. Taenifugin granulate (a medicinal form of monsomil) also effective	Pár, Párová and Prouza (1977)
Bothriocephalus in *Ctenopharyngodon idella*	Phenasal-efficacy related to level of fish serum proteins	Davydov (1978c)
B. acheilognathi eggs	Slaked lime, calcium hypochlorite, potassium permanganate and copper sulphate at certain concentrations and temperatures effective	Kakacheva-Avramova and Menkov (1978)
Bothriocephalus in *Ctenopharyngodon idella* and *Aristichthys nobilis*	Various—grills for water channels, exclusion of wild fish, regular disinfection of ponds, anthelmintic treatment (phenasal orkamala in feed) but breeders treated separately	Laptev (1980)
Bothriocephalus in O-group carp	200 mg/kg phenasal in food *ad libitum*; 60–64.5% cure rate	Davydov, Muzykovskii and Vasil'kov (1980)
B. acheilognathi in fish	A new formulation—taenifugin carp (0.7% piperazine salt of niclosamide, 10% ground limestone, 4.8% coating and extending compounds made up to 100% with wheat flour) effective	Král *et al.* (1980)
Bothriocephalus sp. in *Cyprinus carpio* and *Barbus kimberleyensis*	Lintex (2, 5-dichloro-4-nitrosalicylanilide) in fish pellets at 50 Tmg active ingredient/kg per fish daily for 7 days, successful	Brandt *et al.* (1981)
B. acheilognathi in vitro	Niclosamide effects tegumental enzymes	Kurovskaya (1981).
B. acheilognathi in O-group *Cyprinus carpio*	Sulphene or halosphene in one dose of 0.5 g/kg body weight in food granules 100% effective. 'Okside' only 22.2% effective	Skachkov (1978).
B. acheilognathi eggs and coracidia	Pollena Jod K the most effective chemical killing all embryos at 0.75 parts/thousand	Zelazny (1979)
B. acheilognathi in *C. carpio*	Phenothiazine, zestocarp, savermin and sodium sulphate fully effective at certain doses in laboratory and field tests	Zelazny (1980)
B. acheilognathi in *C. carpio*	Repeated treatment with taenifugin carp at the correct dose gave full control in a short time	Zitnan *et al.* (1981)
B. acheilognathi in *C. carpio* fry and yearlings	Halosphene at 0.5 g/kg body-weight readily eaten removed all tapeworms and far more effective than bithionol, BMC and lopatol	Skachkov (1981)
B. acheilognathi in *C. carpio* fry	Phenasal not detrimental from haematological examination	Skachkov, Kozachenko and Skvortsova (1981)
Bothriocephalus in *C. carpio* fingerlings	Cyprinocestin (10% phenasol medicated food) 100% effective and two doses on consecutive days of halosphene (each of 0.5 g/kg body weight) 85.7% effective	Skachkov *et al.* (1982)

Table 6.7 continued

Parasite and host species	Treatment	Reference
B. acheilognathi in *Ctenopharyngodon idella*	Praziquantel via stomach tube at 35–100 mg/kg completely effective. Praziquantel in pelleted food at 105 and 125 mg/kg body-weight—no worms found after treatment. Niclosamide and Mebendazole ineffective	Pool, Ryder and Andrews (1984)
B. acheilognathi	Exposure to 10^5 µg/litre for 24h not lethal	Pool (1985)
Caryophyllaeus laticeps in *Cyprinus carpio*	99 to 100% efficacy with 3 doses, each of 0.1 g Kamala, given as oral suspension: same result with 0.2 g in fish food	Nowack and Pietrzak (1971)
Khawia sinensis in fish	Introduction of 500 mg kamala into intestine effective	Musselius *et al.* (1963)
K. sinensis in fish	Phenosal and niclosamid effective	Sapozhnikov (1979), Schrenkenbach (1975)
Khawia in *C. carpio*	50 ml Coriban/kg food at 6% total fish weight daily for 10 days reduced worm burdens	Didenko *et al.* (1979)
Proteocephalus neglectus in *Salmo gairdneri*	Granulated feed with 0.5% concentration piperazine salt of niclosamide effective	Král (1977)
P. ambloplitis larvae in *Micropterus salmoides*	Intraperitoneal and and capsule implantation of 200 mg/kg of mebendazole reduced infection by 95%	Boonyaratpalin and Rogers (1984)
P. ambloplitis in *M. salmoides* and *M. dolomieui*	Rearing of fry or fingerlings in bass-free waters; use of piscicides (rotenone, toxaphene and antimycin) to eliminate infected fish and copepods; elimination of surviving tapeworm eggs	Becker and Brunson (1968)
Bothriocephalus scorpii in eels	Prophylactic treatment of elvers with 100 mg/litre formalin for 1 hr before stocking ponds. Unsuccessful but reduced *Trichodina* infections	Field and Eversole (1982)
Triaenophorus nodulosus in *Esox lucius*	DDT and its metabolites accumulate in tapeworm and is perhaps toxic	Davydov *et al.* (1976)
T. nodulosus in perch	High concentrations of chlorophos increased susceptibility to infection	Flerov, Mikryakov and Kuperman (1982)
T. nodulosus in trout	91 mg praziquantel per kg bodyweight (or two doses at 58 mg/kg) effective	Mirle *et al.* (1985)
T. nodulosus coracidia	Chlorophos at 12.5 mg/litre and 72h exposure inhibited hatching. Polychlorpin toxic	Flerov, Mikryakov and Kuperman (1982)
Dilepis in *Cyprinus carpio*	50 ml coriban/kg food at 6% total fish weight daily for 10 days reduced worm burdens	Didenko *et al.* (1979)
D. unilateralis in *C. carpio*	Reducing number of nesting herons recommended	Sapozhnikov (1975b)
D. unilateralis in *Cyprinus carpio*	Chlorophos (used against monogenean infection) incidentally decreased the tapeworm by 10–30% due to destruction of intermediate hosts	Sapozhnikov (1975b)
Valipora campylancristrota in *Cyprinus carpio*	Phenacetin at 200 mg/kg body-weight killed all cysticerci. Acemidophen (diamfenetide), praziquantel and panacur (fenbendazole) less effective	Skvortsova, Kozachenko and Nikulina (1985)
Diphyllobothrium latum in pike, burbot and perch	General review of control. Heat treatment of fish flesh at 55°C for 10–30 min and freezing to −18°C for 24h kill plerocercoids. Large unpalatable quantities of salt (10–12%) effective. Reduce number of infected fish by treating sewage	Bylund (1976)
D. latum in humans	Construction of sewage plants	Yakhad, Mukhina and Gracheva (1979)

Table 6.8 Control of Digenea

Parasite and host species	Treatment	Reference
Clinostomum complanatum in *Plecoglossus altivelis*	Dipterex causes excystation and released metacercariae killed fish	Lo, Huber and Kou (1981)
Diplostomum spathaceum in *Salmo gairdneri*	Prevent re-entry of more snails through drains	Fritzsche (1975)
Diplostomum in *Salmo gairdneri*	Flectron (which contains 16.5% N-trity (morpholine) at 0.8 mg active substancel more effective than trichlorphon and chlorkalk against *Lymnaea balthica ovata* and *Planorbarius corneus*	Fritzsche (1975).
D. spathaceum in *S. gairdneri*	Remove water fowl Molluscicide (copper sulphate) treatment	Fritzsche (1975), Stables and Chappell (1986b)
Diplostomum in trout fry above 3 gm in weight	Malachite green in 0.3 g/m^3 water	Lavrovskii (1977)
Diplostomum in trout fry	Closed water system in breeding tanks from artesian wells	Lavrovskii (1977)
Diplostomum in *Coregonus* spp	Prophylaxis best form of control	Rumyantsev (1978)
Diplostomum in *Cyprinus carpio*	50 ml coriban/kg food at 6% of total fish weight daily for 10 days reduced worm burden	Didenko *et al.* (1979)
D. spathaceum in age 2 + *Salmo gairdneri*	Droncit (praziquantel) inhibited cercarial penetration and/or migration for 48h	Bylund and Sumari (1981)
Diplostomiasis in *Coregonus peled*	Molluscicide 5,4^1-dichlorosalicylanilide at 1:5000000 killed all *Lymnaea stagnalis*, *L. ovata*, *L. auricularia* and *L. palustris* within 2 days. Treated molluscs non-toxic to fish-eating birds	Sapozhnikov (1981).
Diplostomum on fish farms	200 g of ozone/h in water entering fish ponds considered effective against cercariae	Linnik and Yanchenko (1982)
Digenean metacercariae in aquarium fish	Piperazine and yomesan recommended	Dulin (1977)
Sanguinicola in *Cyprinus carpio*	Fuadin at 10 ml/kg feed given 6 times cured 45% and 70% at 30ml/kg. Zanil, miracil D less effective. Ambilhar ineffective and toxic to fish	Sapozhnikov (1975a)
Sanguinicola in *C. carpio*	Of 9 anthelmintics tested only miracil D (at 0.2 g/kg body-weight) and antimosan (15 ml/kg) were effective	Sapozhnikov (1977)
Sanguinicola in *C. carpio*	Acemidophen (diamphenethide), 0.8 g/kg/day for 6 days or 0.48 g/kg/day for 10 days reduced prevalence by about 65% and worm burden by 81.8–73.9%	Sapozhnikov (1979)
S. inermis in *Cyprinus carpio*	50ml coriban/kg food at 6% total fish weight daily for 10 days reduced worm burden by 82.6 and 78.26%	Didenko *et al.* (1979)
Sanguinicola sp.	Various invertebrates, fish and frogs eat cercariae and miracidia. Introducing resistant fish and drying of ponds also suggested	Sapozhnikov (1978)
Pygidiopsis genata metacercariae in *Tilapia*	Control—gamma radiation	Youssef *et al.* (1981)
Heterophyes metacercariae in *Mugil cephalus*, *M. capita*, *M. auratus*, *Tilapia nilotica*, *Sciaena asuilla* and *Solea vulgaris*	Gamma radiation—7.5 × 10^{5}r killed larvae in all fish	Hamed and Elias (1975)

Table 6.9 Control of Nematoda

Parasite and host species	Treatment	Reference
Anguillicola globiceps	Dipterex (90%) at 0.2–0.4 ppm kill intermediate hosts	Huang (1981)
Philometroides carassii first-stage larvae	Killed by drying, compression, heating, ultraviolet light, acrinol, benzethonium chloride, sodium dichlorisocyanurate, bithionol and dithiazanine	Nakajima and Egusa (1977d)
Nanophyetes salmincola in Oncorhynchus tshawytscha	Praziquantel at 10, 20 or 100 mg/kg body weight evaluated	Foreyt and Gorham (1988)
Philometroides lusiana	Krasverm (a Soviet analogue of Nilverm insoluble in water) 300 mg/kg body weight daily in food	Borisova (1980)
P. lusiana in carp	Tests using mebendazole, levamisole and 'preparation No. 1' cured from 40% to 85% of fish. Methyl-N-2-benzimidazolyl carbamate, benacil, peperazine, napthamon and moranetl tartrate at high doses had little effect	Borisova, Kozachenko and Vasil'kov (1981a)
Philometroides in carp	'Preparation No. 1' in food, 500 mg/kg daily for 5 days, 88.8 to 85% efficacy respectively against larval and immature stages	Borisova, Kozachenko and Vasil'kov (1981b)
P. lusiana in *Cyprinus carpio*	Control—chemical	Vasil'kov (1980, 1981)
Philometroides free-living larvae	LC50 and LC100 determined for tetramisole, dithiazanine, tiazon (dazomet), carbofos (malathion), dertil (niclofolan) and esulan	Pirus (1982)
Philometroides in *C. carpio*	Dazomet (25 mg/kg) considerably reduced intensity of infection	Pirus (1982)
Anisakine larvae in teleost muscle	Freeze at −40°C for 24 h store at −10°C	Dailey (1975)
Anisakis sp. I larvae from *Theragra chalcogramma*	Salt processing to required State standards kill larvae. Two weeks salting recommended for home-salted fish	Leont'eva (1976)
Anisakis and *Thynnascaris* in fish food on freshwater fish farms	Levamisole (decaris vet.) at 250 mg/500 kg feed for 10 tonne fish for 3 consecutive days	Christensen and From (1978)
Anisakis, *Terranova* and *Contracaecum* in fish	Survey work with the view to control	Leong (1980)
Terranova sp. and *Anisakis simplex* in fish flesh	−20°C for 1–5 days for Hawaiian snappers and *Sebastes* spp.	Deardorff, Raybourne and Desowitz (1984a)
Phocanema (= *Pseudoterranova*) *decipiens* in cod fillets	Candling	Valdimarsson, Einarsson and King (1985)
Procamallanus sp. in *Heteropneustus fossilis*	Potassium permanganate more toxic to the worm than copper sulphate, sodium sulphate and hydrochloric acid solutions	Lal and Mithal (1979)
Sterliadochona pedispicula in *Salmo gairdneri*	Significant reductions in worm burdens with dichlorvos and its analogue SD 1836, diethyl 2-chlorovinyl phosphate. SD 1836 100% effective at 5–100 µl/litre	Maggenti (1973a, b)

of infecting the same mollusc species, but one was antagonistic and actually preyed on the other. This led to the theoretical possibility of being able to decrease snail numbers and the cercariae of pathogenic flukes. Lie *et al.* also mentioned how snails of the genus *Bulinus* were often eradicated by catfish. The interesting potential and considerable challenge for further research in the biological control of parasites as discussed by Lie *et al* (1968) for schistosomiasis is already apparent in fisheries helminthology. Topics for further research include cleaning symbiosis as discussed by Losey (1971); the role of aquatic invertebrates in eating cercariae, miracidia and tapeworm eggs and larval stages; fish, especially annual fishes, as predators of miracidia, cercariae, molluscs and other carrier hosts; fish which

Table 6.10 Biological control of fish helminths

Monogeneans on roach	Influence of vegetation	Yunchis, Nesterenko, Kononov and Khokhlova (1983)
Pseudoleptobothrium aptychotremae on *Aptychotrema banksii*	Cleaned by *Platycephalus* spp. and/or *Paramonacanthus oblongus*	Kearn (1978a)
Diplostomum spathaceum cercariae	The cladoceran *Moina macrocopa* plays important role in the biological control of *D. spathaceum*	Shigin and Gorovaya (1974)
Sanguinicola miracidia and cercariae	Reduced by cladocera, copepoda, *Chaetogaster*, insect larvae, frogs and resistant fish, e.g. *Hypophthalmichthys moditrix, Aristichthys nobilis, Carassius auratus* and *Ctenopharyngodon idella*	Sapozhnikov and Petrov (1980)
Diplostomum spathaceum cercariae	Destroyed by the phyllopods *Apus cancriformes* and *Leptestheria* sp.	Gorovaya (1975)
D. spathaceum and other cercariae	Eaten by carp fry	Sudarikov and Shigin (1975)
D. spathaceum	By hyperparasitism	Palmieri and Heckmann (1976)
Ova and coracidia of tapeworms	Eaten by *Artemia*	Solonchenko and Radchenko (1987)
Aquatic invertebrates	Eaten by the annual fishes *Cynolebias bellottii, C. elongatus, C. ladigesi* and *Nothobranchias guentheri*	Hildemann and Walford (1963)
Diplostomiasis in *Hypophthalmichthys molitrix, Ctenopharyngodon idella, Aristichthys nobilis* and *Cyprinus carpio*	50 *Mylopharyngodon piceus* (black carp)/ha considerably reduced mollusc intermediate hosts	Zobel (1975); Denisov (1982)
Physa and *Lymnaea*	*Lepomis microlophus* eat these molluscs	Carothers and Allison (1968)
Diagramma interrupta	*Coregonus peled* and *Esox lucius* eat infected *Carassius auratus*	Popov, Razhmashkin and Shirshov (1983)
Ligula and *Digramma* in indigenous fish	Introduction of *Coregonus albula ladogensis, C. peled* and *C. albula* reduce infections	Sapozhnikov and Antonov (1979)
Miracidia (of *Schistosoma mansoni*)	Eaten by guppies *Lebistes reticulatus*; therefore a possibility for control of miracidia of fish worms	Bunnag, Freitas and Scott (1977)
Six diplostomatid species in 20 species of fish	Introduction of three coregonid fish species to eat mollusc intermediate hosts	Antonov and Sapozhnikov (1981)
Clonorchis sinensis	Wormicidal substance in epidermal mucus of *Cyprinus carpio* and *C. carpio nudus*	Rhee and co-authors (1980–1984)
Diplostomum sp.	*Nosema* hyperparasite of	Bauer, Egusa and Hoffman (1981)
Bucephalus longicornutus	Problems in the use of its hyperparasite, *Urosporidium constantae*, as a biological control agent	Howell (1966)
Deropristis inflata in eels	*Hexamita* in reproductive system of fluke	Hunninen and Wichterman (1938)

might have helminthicidal substances in their skin; the hyperparasites, especially Microsporida of helminths; nematodes which might attack arthropods as discussed by Nickle (1980) and Platzer (1980) for mosquitoes; antagonism of the kind discussed by Lie *et al.* (1968) and aquatic 'medicinal' plants which might contain antiparasitic substances (Lozoya, 1977). Preliminary work in some of these areas is indicated in Table 6.10. A detailed review of the 'elimination' of helminths in a freshwater environment is given by Shigin (1981). Of particular importance, however, for future progress in the biological control of helminths is the need to research viruses, bacteria, fungi, protozoa, helminths and arthropods which attack them; many are cited in Dollfus (1946) and Lauckner (1980).

Legislation

Indigenous species of fish must be protected from introduced diseased individuals of the same or other species, from other vectors of diseases (invertebrates, birds and mammals), and from changes in water quality. Very many countries in the world have their own Fisheries Acts governing protection and conservation and these include a requirement for thorough disinfection of eggs before transportation. These Acts usually extend to cover catching, loading, landing, handling, transporting, possession, disposal and the export and import of fish. Bell and Margolis (1972) reviewed control regulations for fish disease in Canada and other countries and referred to an important publication by F.B. Zenny, '*Comparative Study of Laws and Regulations Governing International Traffic in Live Fish and Fish Eggs*' (European Inland Fisheries Advisory Commission, EIFAC, Technical Paper No. 10, 1969, 52p). At that time, 42 governments (12 European) had laws and regulations which had some control of traffic in live fish and fish eggs. Margolis and Evelyn (1987) described Canada's national Fish Health Protection Regulations which are designed to prevent the spread of salmonid diseases into Canada and between its provinces.

Helminths as indicators of fish population biology

General account

Parasites have been used as biological tags to separate populations of terrestrial hosts, especially birds, for over a hundred years, but the first attempt to use them as tags for fish was that of Dogiel and Bychovskii (1939) who distinguished between two sturgeon stocks in the Caspian Sea using several tags, including the monogenean *Nitszchia sturionis* and the cestode *Eubothrium acipenserinum*. *Sebastes* spp. were also among the first fish species to be examined by using biological tags, partly because of their tendency to die on being trawled from deep water (Herrington *et al.*, 1939). Mechanical tags are not appropriate for such delicate species.

The topic was reviewed by MacKenzie (1983, 1986)[6] and the following account is largely based on these reviews. There has been a remarkable recent growth of interest in research on the use of parasites as biological or natural 'tags', indicators or markers of fish populations. In the 1950s, nine papers were published, over 30 in the 1960s, over 50 in the 1970s and over 60 from 1980 to 1985. Biological tags have certain advantages over artificial markers in that they can be used on delicate and deepwater species, including crustaceans, and are cheaper. The capture and handling of fish for artificial tagging may cause abnormal behaviour and so give misleading results. Ideally, however, in fisheries research, data from biological tagging should be combined with those from artificial markers and the two sets of data should be used to complement information from meristic, morphometric and biochemical approaches to studying fish biology. Guidelines for selecting biological tags have been discussed by Sindermann (1961, 1983), Kabata (1963), Konovalov (1975), Biocca and Khalil (1981) and MacKenzie (1983, 1986). Although MacKenzie (1986) pointed out that many useful tags have a complex life-cycle, the suggested guidelines include: (i) the need to choose preferably, but not necessarily, a direct life-cycle parasite, (ii) parasite species which are easily seen and identified, (iii) species with a high site-specificity and which can be found with a minimum of dissection (iv) those which have no marked pathological effects on the host (otherwise there may be selective mortalities and/or behavioural changes), (v) species with a long lifespan, and (vi) species which have a geographic variation in levels of intensity of infection.

Biological tags have been used for the stock separation of freshwater, anadromous and

[6] Updated by Williams, Mackenzie and McCarthy (1992).

marine fish, for recruitment studies on anadromous and marine fish and to study seasonal migrations. Different stocks of a fish species may have different spawning, nursery, or feeding grounds and other forms of behaviour. A knowledge of the mixing and separation of such stocks is important in fisheries management. It is not usually necessary to select parasites with long life-spans to study stock separation of these kinds. The stock separation of freshwater and anadromous fish was reviewed by Shul'man and Shul'man (1983) who used *Diphyllobothrium* to study *Oncorhynchus* and *Salvelinus* spp., and adult hemiurid digeneans, juvenile and adult acanthocephalans and ascaridoid nematode larvae to investigate *Salvelinus* and *Coregonus* spp.

Trypanorhynch metacestodes, five families of digeneans and larval ascaridoids in particular, have been used for the stock separation of marine fish, and a biological tag approach to the study of elasmobranchs and deep-water fish is a recent development (Caira, 1990; Campbell, 1990).

Recruitment studies of fish populations involve following adolescent hosts from their nursery grounds until they arrive at the feeding and spawning grounds of adults. There are two additional requirements in choosing a biological tag for recruitment studies: (i) the parasite species must infect young fish on the nursery ground, and (ii) it must have a long enough life-span to remain identifiable in fully mature adult hosts. Recruitment studies of particular interest are those of Margolis (1963) on *Oncorhynchus nerka* (sockeye salmon) using pseudophyllidean plerocercoids; of Konovalov (1975) on sockeye using pseudophyllidean plerocercoids and diplostomatid metacercariae; of Jennings and Hendrickson (1982) on *O. tschawytcha* and *O. kisutch* using diplostomatid metacercariae; and Hare and Burt (1976) using *Discocotyle sagittata* and adult acanthocephalans on *Salmo salar*. For recruitment studies on marine fish, MacKenzie (1985) used renicolid metacercariae in Scottish inshore waters and a trypanorhynch plerocercus in the eastern North Sea for *Clupea harengus*. Zubchenko (1985) used an adult cestode to study recruitment in *Coryphaenoides rupestris*.

Seasonal migration studies using biological tags have been carried out on *Clupea harengus* moving from high salinity areas (where they characteristically have marine anisakid larvae and hemiurid digeneans) to spawn in low salinity areas (Grabda, 1974; Gaevskaya and Shapiro, 1981). Grabda (1981) used a trypanorhynch plerocercus and ascaridoid larvae to study the migration of *Belone belone* from the Baltic Sea (where they spawn) to feeding grounds in the North Sea, while MacKenzie and Mehl (1984) studied *Scomber scombrus* migrations in the North-East Atlantic using the trypanorhynch *Grillotia angeli* as a tag.

MacKenzie (1986) listed the total number of parasites which had so far been used in biological tag studies. This included metacercariae of 20 digeneans, 37 adult digeneans, 15 metacestodes (ie. larval tapeworms), 10 adult tapeworms, 2 juvenile acanthocephalans, 15 larval nematodes and 8 adult nematodes. For the Salmonoidei, for instance, one study had involved a monogenean species, 11 had used 12 species of digeneans, 14 were based on 11 cestode species, 8 papers were based on 4 acanthocephalan species and 8 were on seven nematode tags. MacKenzie gave similar information on the following groups of fish: Atherinoidei, Clupeoidei, Exocetoidei, Gadoidei, Macrouroidei, Percoidei, Pleuronectoidei, Scombroidei and Scorpaenoidei. He cited 67 references in the use of parasites as biological tags and suggested further work on the use of adult helminths in fish guts for seasonal migration studies, trypanorhynch metacestodes in recruitment studies of marine teleosts, and parasites to study elasmobranch and deep-water marine fish populations. A small number of studies have concentrated on the use of parasites in separating sympatric species of fish, eg. Davis and Huffman (1977) on the ecological differences between *Gambusia affinis* and *G. geigeri*; Cloutman (1976) on *Campostoma anomallum pullum* and *C.*

Table 6.11 Separation of fish stocks

Host	Parasite	Area	Reference
Theragra chalcogramma	13 spp. parasite	NE Pacific	Arthur (1984)
Clupea harengus	*Thynnascaris adunca* Bucephalidae spp. *Anisakis simplex* larvae	British Columbia	Arthur and Arai (1980a,b)
Podonema longipes	*Nybelinia surmenicola*, *Diphyllobothrium* sp.	Kurils, Japan	Avdeev (1985)
Salmo salar	*Anisakis simplex* larvae	West Greenland	Beverley-Burton (1978) Beverley-Burton and Pippy (1978)
Oncorhynchus nerka	*Anisakis* sp. larvae	Kamchatka River Basin	Bourgeois and Ni (1984)
Oncorhynchus nerka	*Diphyllobothrium* sp. plerocercoids	Kamchatka River Basin	Bugaev (1982)
Liopsetta putnami	*Cryptocotyle lingua*	New Hampshire	Burn (1980b)
Salvelinus spp.	Various helminth species	Kamchatka Bay, USSR	Butorina (1980)
Oncorhynchus spp. and other teleosts	Cestodes	Kamchatka Bay	Butorina and Kuperman (1981)
Osmerus mordax	*Echinorhynchus salmonis* *Diphyllobothrium sebago* *Glugea hertwigi*	Quebec	Fréchet, Dodson and Powles (1983)
Mallotus villosus	*Eubothrium parvum*	Barents Sea	Kennedy (1979)
Limanda limanda	Digeneans	Danish and adjacent waters	Køie (1983)
Perca fluviatilis	*Triaenophorus nodulosus*	Rybinsk reservoir, USSR	Kuperman (1981a)
Melanogrammus aeglifinus	*Grillotia erinaceus*	North Sea	Lubieniecki (1977)
Oncorhynchus spp.	Various helminth species	North America	Margolis (1982c),
Clupea harengus	Various parasite species	Various areas	MacKenzie (1987a)
Salmo salar and *Salmo trutta*	Various parasite species	Barents Sea and White Sea	Mitenev (1984a,b)
Notothenia spp. and *Dissosthychus eleginoides*	Various helminth species	Antarctic sector of Indian Ocean	Parukhin and Lyadov (1982)
Argentina silus	Digeneans unsuitable as tags	Nova Scotia and Newfoundland	Scott (1969)
Melanogrammus aeglefinus	19 species including *Lepidapedon rachion*	Scotian Shelf	Scott (1981)
4 spp. flatfish	13 spp. digeneans	Scotian Shelf	Scott (1982)
Pollachius virens	*Derogenes varicus* *Hemiurus levinseni* *Echinorhynchus gadi* *Anisakis* sp. larvae	Scotian Shelf	Scott (1985a)
4 channichthyid spp.	*Eubrachiella antarctica* *Contracaecum* sp.	Antarctic Shelf	Siegel (1980)
Macrourus rupestris	*Dolichoenterum* sp. and other helminths	Labrador Sea	Szuks (1980)
Atherina presbyter	*Neodiplostomum sp.*	The Fleet, Dorset, Southampton Water	Turnpenny *et al.* (1981); Bamber *et al.* (1983); Bamber and Henderson (1985)
Coryphaenoides rupestris	25 species of parasites	North Atlantic	Zubchenko (1985)

oligolepis; and Lambert and Romand (1984) on two groups of cyprinids which hybridize freely in nature.

Tables 6.11 and 6.12 refer to some of the hosts, helminth species, geographical areas and authors who have used fish worms in studies of different fish stocks and their changing patterns, recruitment migrations from nursery grounds to adult feeding grounds and seasonal or other kinds of fish migrations.

Fish stocks

Helminths with direct or complicated life-cycles may be used in stock separation studies and in such studies, age determination of the fish is recommended. Care is necessary in interpreting the significance of different prevalences of a particular helminth species amongst different fish stocks to verify that it is a useful tag. It follows that a species with a direct lifecycle is simpler to use and is likely to give early results but we are aware of only one publication, the work of Smith (1972) on the use of *Diclidophora esmarkii* as a tag for Norway pout, *Trisopterus esmarkii*.

Helminths have been used as tags not only to distinguish between fish stocks but also to examine long-term changes in such stocks which may be related to season, recruitment, age-dependent migrations and extension of stock boundaries (Table 6.12).

Recruitment migrations

Recruitment migrations involve the journeys of young fish, often over long distances, from nursery to adult feeding and spawning grounds. A species which infects fish on only some nursery grounds can provide information about the origin of fish and the connection between different juvenile and adult fish populations. The tag is of value only if the fish is susceptible to infection only on the nursery grounds, and subsequently loses its susceptibility to further infection, preferably becoming resistant when it migrates away from a nursery ground. Otherwise an infection picked up by adult fish could mistakenly identify them as coming from a particular nursery ground. The tag must also remain either alive or in an identifiable form within the fish for the rest of the host's life—a far longer life span than is necessary in stock separation or seasonal migration studies. These criteria severely limit the types and numbers of suitable tags, as can be seen from Table 6.13.

Seasonal migrations

Many fish species migrate seasonally to feeding or spawning grounds or in pursuit of prey. Helminth species which infest such fish in one area and are thus carried to another, where the fish is unlikely to be infected with the same species, are likely to be successful tags. Helminths with a life-span of less than a year may also be used. Species which have been used to study seasonal migrations are shown in Table 6.14.

Age-dependent migration

Older fish tend to move to different areas for feeding, overwintering and possibly spawning. Such migrations can be investigated by using tags similar to those for recruitment migrations, except that infestation must occur at the young adult rather than the

Table 6.12 Helminths used to study changing patterns in stock migration

Host	Parasite	Area	Reference
Oncorhynchus nerka	*Diphyllobothrium* sp. *Diplostomum* sp. *Tetracotyle* sp. *Philonema* sp. *Salmincola californiensis*	North Pacific	Groot, Margolis and Bailey (1984)
Clupea harengus	*Cercaria pythionike* *C. doricha* *Lacistorhynchus* sp.	North Sea	MacKenzie (1985, 1987a)
C. harengus	*Anisakis* larvae	Northwest Atlantic	McGladdery and Burt (1985)
C. harengus	*Anisakis* larvae	North Sea	Van Banning and Becker (1978)
C. harengus	Various parasites	Various areas	MacKenzie (1987a)

Table 6.13 Tags for studying fish recruitment migrations

Host	Parasite	Area	Reference
Coregonus lavaretus	*Thynnascaris adunca*	Northern Baltic	Fagerholm and Valtonen (1980)
Belone belone	*Lacistorhynchus tenuis* *Anisakis simplex* larvae	Baltic Sea	Grabda (1981)
Scomber scombrus	*Grillotia smaris-gora* = *G.angeli*	Eastern North Atlantic	MacKenzie (1980, 1983) MacKenzie and Mehl (1984); MacKenzie, Smith and Williams (1984)
Clupea harengus	*Lacistorhynchus* sp. *Cercaria pythionike*	Eastern North Atlantic	MacKenzie (1985, 1987a)
Salmo salar	*Pomphorhynchus laevis*	British Isles	Pippy (1969)
Salmo salar	43 parasite species	Open Atlantic and Greenland coastal waters	Pippy (1980)
Parophrys vetulus	*Echinorhynchus lageniformis* *Philonema americana* *Zoogonus dextrocirrus* *Otodistomum veliporum*	Oregon, USA	Olson and Pratt (1973)

Table 6.14 Helminths used to study seasonal migrations

Host	Parasite	Area	Reference
Salvelinus alpinus	*B. crenatus*	Mosie River system	Black (1981) Black, Montgomery and Whoriskey (1983)
Oncorhynchus nerka	*E. salvelini*	Babine Lake, British Columbia	Boyce and Clarke (1983)
Salvelinus alpinus	*Diphyllobothrium* spp. *Eubothrium salvelini* *Proteocephalus longicollis* *Brachyphallus crenatus* *Bothrimonus sturionis* *Prosorhynchus squamosus*	Baffin Island	Dick and Belosevic (1981)
Salvelinus alpinus	*Cystidicola cristivomeri*	Ontario and Northwest Territories, Canada	Eddy and Lankester (1978) Black and Lankester (1981)
Clupea harengus	*Scolex pleuronectis* *Anisakis simplex* *Hysterothylacium aduncum* *Cryptocotyle lingua*	Northwest Atlantic	McGladdery and Burt (1985)
Vimba vimba	Entire parasite fauna	Baltic	Rautskis (1983)
Pseudopleuronectes americanus	Digenean spp.	Passamaquoddy Bay, Canada	Scott (1985b)

juvenile stage. Thus, Hislop and MacKenzie (1976) used the larva of a trypanorhynch to distinguish coastal and offshore stocks of whiting *Merlangus merlangius* in the North Sea and Avdeev (1985) used trypanorhynch and diphyllobothriid larvae to identify separate stocks of *Podonema longipes* in the northwest Pacific. Others who have referred to ecological groups of fish according to spawning and/or feeding migrations are Gibson (1972a) for *Platichthys flesus* in the North Sea; Karasev (1984b) for *Micromesistius poutassou* in the Norwegian Sea; Rumyantsev (1976a) for *Coregonus* spp. in various lakes of different regions of the USSR; and Zubchenko (1984) for *Alepocephalus bairdi* and *A. agassizi* in the North Atlantic and Frimeth (1987b) for *Salvelinus fontinalis* in the Tabusintac river and eastern seaboard of Canada.

Fish worms as indicators of pollution

Earlier in this chapter the effects of pollution on fish worms were discussed. It was seen that polluted waters: (i) reduced the abundance and prevalence of some species (Overstreet and Howse, 1977; Kostarev, 1980; Anikieva, 1982a; Khan and Kiceniuk, 1983; Valtonen and Koskivaara, 1987; Valtonen *et al.* 1987; Sulgostowska and Styczynska-Jurewicz, 1987), (ii) may have a more lethal effect on some parasitized fish than those without infestation (Boyce and Yamada, 1977, Pascoe and Cram, 1977), (iii) did not seem to affect some fish worm species (Anikieva, 1982b), (iv) may increase the susceptibility of some fish species to worm infestations (Skinner, 1982), and (v) may actually increase the abundance and prevalence of a few species (Kostarev, 1980).

From these and other researches on man's influence on the distribution and spread of fish worms, a list of genera can now be compiled for further research into their potential use as indicators of man-made changes in an aquatic environment. This possibility was predicted by Williams in 1978 (see Cole 1979) and the following are now thought to be of potential value: *Neodiplectanum, Ancyrocephalus, Cleidodiscus, Rhipidocotyle, Tylodelphys, Icthycotylurus, Bucephalus, Steringophorus, Posthodiplostomum, Opisthorchis, Bunodera, Sphaerostoma, Zoogonus, Opecoeloides, Diplomonorchis, Crepidostomum, Steringophorus, Proteocephalus, Caryophyllaeus, Ligula, Digramma, Diphyllobothrium, Camallanus, Philometra, Rhabdochona, Ascarophis, Hysterothylacium, Echinorhynchus, Neoechinorhynchus, Dollfusentis, Acanthocephalus* and *Pomphorhynchus*. The best examples from amongst these will be those for which most is known of their life-cycles. This research area, however, is in its infancy as suggested by Möller and Anders (1986). They, and Anikieva (1982a), postulated that ectoparasites in particular might be useful in detecting man-made oxygen deficiencies, the effects of increasing salinities in freshwater ecosystems and the presence of certain heavy metal salts in water.

Fish worms as biological control agents

The use of parasites as biological control agents is a field of great current interest (Anderson and Canning, 1982); these authors present an over-all review of attempts to use pathogens from all major groups of parasites including helminths, in biological control programmes.

Whilst many digeneans and nematodes have been tested, work on fish helminths seems to have been confined to a few species which might be useful in controlling mollusc and arthropod vectors of other diseases. Combes (1982, 1983) discussed two ways in which digeneans can be used in the control of other digeneans that transmit diseases of medical or veterinary importance. A species may interfere with the reproductive capacity of its snail host or be antagonistic to another target digenean species within the same snail host. Consequently, there are six essential criteria in choosing a suitable digenean for biological control: (i) complete and permanent sterility of the snail host; (ii) clear dominance over the target digenean species; (iii) strong infectivity to molluscs of all ages; (iv) high egg productivity in the definitive host; (v) lack of pathogenicity in man and domestic animals, and (vi) ease and low cost of maintenance of the life-cycle. Of the digeneans mentioned by Combes which have already been tested with the view to biological control, only a few use fish as second intermediate hosts. They are *Clinostomum golvani* in *Poecilia reticulata*; *Petasiger caribbensis* in *Tilapia mossambica, Poecilia reticulata, Xiphophorus helleri* and *Gambusia affinis*, and *Ribeiroia marini* in *Carassius auratus*, *Tilapia mossambica* and *Poecilia reticulata*. Two other fish digeneans which use molluscs as intermediate hosts are mentioned: *Zoogonus rubellus* and *Lepocreadium* against *Nassarius obsoletus*.

The only other example we are aware of is the control of mosquito fish *Gambusia affinis* using the tapeworm *Bothriocephalus acheilognathi* (Granath and Esch, 1983c).

The use of fish worms as biological control agents is, therefore, an area for intensified research. A better knowledge of associations between fish helminths and other parasites, viruses, bacteria and protozoa, might be invaluable, especially if it can be demonstrated that the worms are dominant.

Fish worms as models in teaching and research

Intensive field-work over a number of years backed up by laboratory-based research on a few species of fish worms has contributed much, not only to our understanding of the biology of fish worms themselves, but also to parasitology and biology generally. It has become fashionable to refer to such species as useful field or laboratory models (Smyth, 1969; Whitfield, 1983). Various publications by Kearn since 1967 on many aspects of the biology of *Entobdella soleae* reveal the enormous potential for advancing knowledge by selecting one species as a model for intensive and continuous study. Some species have been used to understand the broad topic of parasite reproduction, for instance, and the many questions it poses for molecular biologists, developmental biologists or behaviorists interested in the tactics of hosts and parasites within their lifecycles (Whitfield, 1983). Others have been the subject of ultrastructural investigations of the body surface to understand aspects of biochemistry, nutrition (including active transport phenomena), growth, respiration, excretion and sensitivity of animal parasites generally. Species for which the entire life-cycles are well-known have been extensively used in understanding the many facets of ecology. It is now accepted that an interpretation of the evolution and coevolution of parasitic worms can only be achieved through knowledge of major groups of fish helminths which are not found in any other group of vertebrates. Such knowledge has enabled more detailed researches into factors responsible for host-specificity, because not only are some helminth groups restricted to fish but very many species within these groups are absolutely restricted to one host species.

In sharp contrast, many species can infect a wide range of host species often with pathogenic and immunological consequences and these have been used for research into host-parasite interactions and into the control of fish worms.

The foregoing observations indicate, in general terms, how fish worms have been used to advance knowledge in every important teaching discipline in biology and also areas where their use is essential in teaching parasitology. For this reason, this chapter will end with a list of various biology disciplines and some well-known species of fish worms which have been used as models for investigating parasitological problems in these disciplines (Table 6.15). The list is not based on a comprehensive search of the literature for all 'models' that have been used and is, therefore, intended only as initial guidance for those in search of suitable material for teaching or research purposes. Consequently, we are aware of major gaps in our coverage of well-known parasites, for instance *Clonorchis, Opisthorchis, Metagonimus* and other human heterophyids, *Diphyllobothrium, Schistocephalus, Ligula* and *Gnathostoma*. These are adequately covered by Morishita, Komiya and Matsubayashi (1964), Rim (1982), von Bonsdorff (1977), Daengsvang (1980, 1982) and in some of the references cited in our introduction. In the introduction we stated that because of the need to limit the length of this book we could not include a chapter on physiology and biochemistry. Towards filling this gap in our coverage we have included an appendix to this chapter as a guide to literature on the physiology, biochemistry and immunology, *in vitro* culture, chemotaxonomy and ultrastructure of fish worms.

Table 6.15 Some fish helminths used in teaching and research arranged by discipline

Reproduction	
Monogenea	Halton, Stranock and Hardcastle (1976) for *Diclidophora merlangi*, *Diplozoon paradoxum* and *Calicotyle kroyeri*; Halton and Hardcastle (1976) for *D. merlangi*; Harris (1983) for *Oogyrodactylus farlowellae*; Harris (1985a) for *Gyrodactylus gasterostei*; Kearn (1970b, 1973, 1974a, b, 1975a, 1985) for *Entobdella soleae*; Kearn (1986), Lambert (1982), Llewellyn (1981) for various monogeneans; and Lester and Adams (1974a, b) for *Gyrodactylus alexanderi*; Macdonald (1974, 1975) and Macdonald and Caley (1975) for *Diclidophora merlangi*; Macdonald and Llewellyn (1980) for *Acanthocotyle greeni*; Scott (1982), Scott and Anderson (1984) and Scott (1985) for *Gyrodactylus turnbulli*; Silan, Euzet and Maillard (1983) for *Diplectanum aequans*; Whittington (1987) for *Hexabothrium appendiculatum* and *Leptocotyle minor*.
Gyrocotylidea:	Williams, Colin and Halvorsen (1987) for *Gyrocotyle*
Caryophyllidea:	Mackiewicz (1981) for several species
Cestoda:	Burt (1986) for egg and larval development in platyhelminths, mainly cestodes; MacKinnon and Burt (1984) and MacKinnon, Jarecka and Burt (1985) for *Haplobothrium globuliforme*; Whitfield (1983) for various papers which cite references to work on cestode reproduction.
Digenea:	Bundy (1981a, b, c) for *Transversotrema patialense*; Whitfield and Evans (1983) for *Azygia longa*, *T. patialense*, *Cryptocotyle lingua* and *Diplostomum flexicaudum*; Lasee, Font and Sutherland (1988) for *Culaeatrema inconstans*
Nematoda:	Lewis, Jones and Adams (1974) for *Philometra oncorhynchi*; Bashirullah and Adams (1983) for *P. oncorhynchi*
Acanthocephala:	Stranack (1972) for *Pomphorhynchus laevis*; Crompton (1985) for a review of reproduction in acanthocephalans
Ecology (See also Chapter Three)	
Monogenea:	Frankland (1955) for *Diclidophora denticulata*; Paling (1965) for *Discocotyle sagittata*; Scott (1985) for *Gyrodactylus turnbulli*
Gyrocotylidea:	Williams, Colin and Halvorsen (1987) for *Gyrocotyle urna* and other species
Caryophyllidea:	Anderson (1974b) for *Caryophyllaeus laticeps*
Cestoda:	Andersen (1972), Halvorsen and Andersen (1974) for *Diphyllobothrium dendriticum*; Black and Fraser (1984) for *Ligula intestinalis*; McCullough, Fairweather and Montgomery (1986) for *Trilocularia acanthiaevulagaris*
Digenea:	Anderson and Whitfield (1975), Anderson, Whitfield and Mills (1977) for *Transversotrema patialense*; Kennedy (1987) for *Tylodelphys podicipina*; Stables and Chappell (1986a, b, c) for *Diplostomum spathaceum*
Nematoda:	Muzzall (1986), Aho and Kennedy (1984, 1987) for *Cystidicoloides tenuissima*
Acanthocephala:	Kennedy and Moriarty (1987) for *Acanthocephalus lucii* and *A. anguillae*; Kennedy (1975) and Kennedy and Rumpus (1977) for *Pomphorhynchus laevis*; Amin (1987a, b, c) for *P. bulbocolli*; Kennedy (1985d) for many references on population dynamics
Behaviour (see also Chapter Four)	
Monogenea:	Kearn (1984a, 1986, 1988a, b), Lambert (1982) Llewellyn (1972, 1981), Lyons (1972) for various species; Whittington (1987) for *Hexabothrium appendiculatum* and *Leptocotyle minor*; Whittington and Kearn (1986, 1989) for *Rajonchocotyle emarginata* and *Plectanocotyle gurnardi*
Cestoda:	Brown and Thompson (1986) for *Schistocephalus* and *Ligula*; Giles (1983) for *S. solidus*; Orr (1966) for *Ligula*
Digenea:	Brown and Thompson (1986) for *Diplostomum*; Cable and Hunninen (1942) for *Cryptocotyle lingua*; Dönges (1963, 1964) for *Posthodiplostomum cuticola*; Grainger (1977) for *C. lingua*; Haas (1969) for *D. spathaceum*; Ieshko and Kaufman (1984) for *Bunodera luciopercae*; Køie (1985b) for many references to various species; Lewis *et al.* (1989), Millemann and Knapp (1970) for *Nanophyetus salmonicola*; Pearre (1979) for *Derogenes varicus*; Rau, Gordon and Curtis (1979) for *Diplostomum*; Whitfield, Anderson and Bundy (1977) for *Transversotrema patialense*
Nematoda:	Adams (1969) for *Philonema oncorhynchi*; Deardorff, Raybourne and Desowitz (1984a) for *Terranova* (= *Pseudoterranova*) sp. and *Anisakis simplex*; Gavrilyuk (1978) and Kakacheva-Avramova (1983a) for *Hysterothylacium aduncum*; Leont'eva (1976) for *Anisakis* sp.; Lyadov (1976) for *Pseudoterranova* and *Anisakis* larvae; Nakajima and Egusa (1977d) for *Philometroides carassii*; Nikolaeva, Gavrilyuk and Shchenkina (1976) for *Anisakis*; Ronald (1960) for *Pseudoterranova decipiens*; Smith (1984) for *Raphidascaris acus*; Smith (1984b) for *A. simplex*; Smith and Wootten (1975) for *Anisakis* sp.; Uhazy (1976) for *Philometroides huronensis*
Acanthocephala:	Brown and Thompson (1986) for *Acanthocephalus* sp. and *Pomphorhynchus laevis*; Pilecka-Rapacz (1986) for *Acanthocephalus anguillae* and *A. lucii*; Schmidt G.D. (1985) and Dobson and Keymer (1985) on development, life-cycles and life-cycle models of acanthocephalans

Table 6.14 continued

Biochemistry and physiology

Monogenea:	Arme and Walkey (1970); Barrett (1981) for *Diclidophora merlangi*; Smyth and Halton (1983) for *Amphibdella, Calicotyle, Dactylogyrus, Diclidophora, Diplectanum, Discocotyle* and *Gyrodactylus*
Gyrocotylidea:	Barrett (1981) for *Gyrocotyle* spp.
Cestoda:	Andersen (1978b) for *Diphyllobothrium* sp.; Arme (1966) for *Ligula intestinalis*; Barrett (1981) for *Bothriocephalus gowkongensis, Khawia sinensis, Ligula intestinalis, Lacistorhynchus tenuis, Pterobothrium lintoni, Schistocephalus solidus, Triaenophorus crassus* and *T. nodulosus*; Beis and Barrett (1980) for *S. solidus*; Buteau *et al.* (1969, 1971) for tetraphyllideans; Carvajal, Barros and Santander (1982) for *Rhodobothrium mesodesmatum*; Hopkins (1950, 1951), McCaig and Hopkins (1965), Orr and Hopkins (1969), Orr, Hopkins and Charles (1969) and Sinha and Hopkins (1967) for *S. solidus*; Read and Simmons (1963) for many papers by Read and by Simmons on biochemistry of elasmobranch tapeworms; Smyth (1990) for *Schistocephalus solidus* and *Ligula intestinalis*; Ward, Allan and McKerr (1986a, b) for *Grillotia erinaceus*
Digenea:	Barrett (1981) for *Cryptocotyle lingua, Clonorchis sinensis* and *Diplostomum phoxini*; Smyth and Halton (1983) for *Allopodocotyle, Diplostomum, Posthodiplostomum, Transversotrema, Zoogonoides* and *Zoogonus*
Nematoda:	Smith and Wootten (1978a) for various references to anisakids; Carvajal *et al.* (1981) for anisakids
Acanthocephala:	Starling (1985) on *Echinorhynchus salmonis, Pomphorhynchus bulbocolli* and *Neoechinorhynchus*

Pathology and epizootiology (see also Chapters Five and Six)

Monogenea:	Eller (1975) for *Dactylogyrus vastator* and *D. solidus*; Scott (1985) for *Gyrodactylus* turnbulli
Cestoda:	Deardorff, Raybourne and Mattis (1984) for Trypanorhynch plerocerci; Dick and Rosen (1982) for *Triaenophorus crassus*; Gonzalez *et al.* (1978) for *Diphyllobothrium* spp; Hermanns and Körting (1987) for *Cyathocephalus truncatus*; Hoffman (1975b) for *Proteocephalus, Diphyllobothrium* spp. and *Triaenophorus* spp.; Shostak and Dick (1986) for *T. crassus*; Taylor and Hoole (1987) for *Ligula intestinalis.*
Digenea:	Eller (1975) for *Sanguinicola davisi* and *S. klamathensis;* Lee, Lewis and Sweeting (1987) for *S. inermis*; Dukes (1975) for *Diplostomum flexicaudatum*; Higgins (1980) for *Bucephalus haimeanus*; Hoffman (1975b) for *S. inermis, Diplostomulum baeri eucaliae, D. flexicaudum* (= *D. spathaceum*?), *Posthodiplostomulum minimum, Neascus* sp., *Neogogatea kentuckiensis, Bucephalus polymorphus* and *Ascocotyle tenuicollis*; Kennedy (1984a) for *Diplostomum spathaceum, D. gasterostei, Tylodelphys clavata* and *T. podicipina*; Pike and Burt (1983) for *Apophallus brevis* and Rodgers and Burke (1988) for *Prototransversotrema steeri* and a possible link with 'redspot' disease (vibriosis)
Nematoda:	Croll *et al.* (1980) for *Phocanema* (= *Pseudoterranova) decipiens*; Gibson (1970) for *Anisakis* sp.; Deardorff, Kliks and Desowitz (1983) and Deardorff *et al.* (1982) for *Terranova* (= *Pseudoterranova*) sp.; Hoffman (1975a) for *Philonema agubernaculum, Philometra* spp.; *Hepaticola petrushewskii* and *Spinitectus* spp.; Margolis and Beverley-Burton (1977) for *A. simplex*; Oshima (1972) for anisakiasis; Overstreet and Meyer (1981) for piscine ascaridoid nematodes; Raybourne *et al.* (1983) for *A. simplex* and *Terranova* (= *Pseudoterranova*) sp.; Smith and Wootten (1978a) for anisakiasis; Torres *et al.* (1988) for *Anisakis* sp.
Acanthocephala:	Amin (1987a,b,c) for *Pomphorhynchus bulbocolli*; Hoffman (1975) for *Echinorhynchus salmonis* and *Acanthocephalus jacksoni*; Nickol (1985) for epizootiology of acanthocephalans generally

Evolution, coevolution, host-specificity, experimental taxonomy (see also Chapter Four)

Platyhelminths (evolution and coevolution)	Andrews *et al.* (1988); Bray (1986b); Brooks (1982, 1986); Brooks *et al.* (1985a, b, 1989); Cable (1982); Gibson (1987); Llewellyn (1986); Mackiewicz (1982b, 1984); Malmberg (1982, 1986); Williams, Halvorsen and Colin (1987); Robert, Renaud, Mathieu and Gabrion (1988); Maillard and Aussel (1988); Euzet, Agnèse and Lambert (1989)
Nematoda: (evolution and coevolution)	Anderson (1984); Black (1983a, b, c) for *Cystidicola farionis* and *C. stigmatura*; Gibson (1983); Sprent (1983); Adamson (1986)
Acanthocephala: (evolution and coevolution)	Crompton (1975); Conway-Morris and Crompton (1982a, b); Buron and Golvan (1986); Golvan and Buron (1988)
Host-specificity:	Maillard and Aussel (1988) and many references in Chapter Four
Experimental taxonomy	Berland, Bristow and Grahl-Nielson (1990); Bristow and Berland (1986, 1988) for *Gyrocotyle*; Bray and Rollinson (1985) for *Fellodistomum, Steringotrema* and *Steringophorus*; Seo *et al.* (1980) for *Heterophyes metacercariae*; Nascetti *et al.* (1986) for *Anisakis simplex* and Orechia *et al.* (1986) for *A. simplex* and *A. physeteris*

Table 6.15 continued

Ultrastructure for investigating various disciplines	
Monogenea:	Lyons (1977) for many monogeneans; Halton (1982a, b) for *Diclidophora merlangi*; Rohde (1982b, c) for Monogenea
Gyrocotylidea:	Allison (1980) for *Gyrocotyle rugosa*; Xylander (1986a, b, 1987a, b, c, d, 1988a, 1989)
Cestoda:	Andersen (1977, 1987) for *Diphyllobothrium*; Andersen (1979b) for *Eubothrium, Bothriocephalus* and *Triaenophorus*; Andersen (1979a) for *Proteocephalus*; Charles and Orr (1968) for *Ligula intestinalis* and *Schistocephalus solidus*; Granath, Lewis and Esch (1983) for *Bothriocephalus acheilognathi*; Gustafsson and Wikgren (1981) for *Diphyllobothrium*; Hamilton-Attwell, Tiedt and van As (1980) for *B. acheilognathi*; McVicar (1972) for tetraphyllideans; Ward, Allen and McKerr (1986a) for *Grillotia erinaceus*; Swiderski and Eklu-Natey (1978) Swiderski (1985, 1986) for *Proteocephalus longicollis*
Digenea:	Halton (1982c); Køie (1985b) for references to various digeneans studied; Simon-Martin and Rojo-Vasquez (1984) for *Sanguinicola inermis*
Nematoda:	Uhazy (1976) for *Philometroides huronensis*; Williams and Richards (1968, 1978) for *Pseudanisakis rotundata* and *Proleptus mackenziei*
Acanthocephala:	Miller and Dunagan (1985) for functional morphology of acanthocephalans generally

Appendix: Publications on physiology, biochemistry and immunology, *in vitro* culture, chemotaxonomy and ultrastructure.

BOOKS AND/OR REVIEWS COVERING SOME OR ALL OF THESE TOPICS

Arme (1983)
Arme and Walkey (1970)
Barrett (1981)
Brand, von (1979)
Bryant (1982a,b.)
Chappell (1980a, b., 1982)
Crompton (1970a, 1975, 1977)
Erasmus (1972)
Kearn (1971)
Krotov (1971)
Lee (1965, 1977, 1982)
Lethbridge (1980)
Lyons (1977, 1978)
Mettrick and Podesta (1974)
Pappas and Cain (1981)
Platzer (1977)
Podesta (1982)
Read (1971)
Read and Simmons (1963)
Rechcigl (1977)
Smyth (1966, 1969, 1990)
Smyth and Halton (1983)
Smyth and McManus (1989)
Taylor and Baker (1987)

PHYSIOLOGY: REPRODUCTION

Abele and Gilchrist (1977)
Anya (1976)
Bazitov (1978, 1979)
Bazitov, Kulakovskaya and Shestakova (1979)
Bazitov and Lyapkalo (1979, 1980, 1981)
Beveridge and Smith (1985)
Brun, Lambert and Justine (1986)
Bundy (1981b)
Calow (1983)
Chu, Kang, Chu and Sung (1977)
Clark (1978)
Crompton (1983a,b)
Dailey and Overstreet (1973)
Davis and Roberts (1983a, b)
Dhar and Sharma (1984)
Erwin and Halton (1983)
Euzet, Swiderski and Mokhtar-Maamouri (1981)
Finlayson (1982)
Foor (1983a, b)
Gerasev and Khotenovskii (1985)
Ginetskinskaya, Palm, Besedina and Timofeeva (1971)
Grey and Mackiewicz (1980)
Halton and Hardcastle (1977)
Halton, Stranock and Hardcastle (1976)
Harris (1985a, 1989)
Jones and Mackiewicz (1969)
Justine, LeBrun, Mattei (1985a, b)
Justine and Mattei (1982, 1984a, b, 1985a, b, c, d)
Kamegai (1970)
Kearn (1986)
Kennedy (1976b)
Kuperman and Shul'man (1978)
Lebedev (1979b)
Llewellyn (1981, 1983)
Llewellyn and Anderson (1984)
Lyukshina (1977)
MacDonald and Jones (1978)
MacDonald and Llewellyn (1980)
MacKinnon and Burt (1984, 1985a)
Marchand (1981)
Marchand and Mattei (1976a, b, 1978, 1980a, b)
Mills (1980a)
Mills, Anderson and Whitfield (1979)
Mohandas (1983)
Mokhtar-Maamouri (1976, 1979, 1982)
Mokhtar-Maamouri and Swiderski (1975, 1976)
Nellaiappan and Ramalingham (1980a, b)
Nollen (1983)
Oshmarin and Prokhorova (1978)
Parshad and Crompton (1981)
Ramalingam (1972)
Scott (1982)
Serov (1984)
Swiderski and Mokhtar-Maamouri (1974)
Tinsley (1983)
Wharton (1983)
Whitfield and Evans (1983)
Whitfield, Anderson and Bundy (1986)
Williams and McVicar (1968a, b)

PHYSIOLOGY: NUTRITION

PHYSIOLOGY: GROWTH

PHYSIOLOGY: RESPIRATION

Nizami and Siddiqi (1976, 1978, 1979)
Vernberg (1968)

PHYSIOLOGY: MOVEMENT

Bundy (1981c)
Cone and Burt (1982b)
Davydov (1973)
Taranenko, Davydov and Sinits'kii (1976)
Whitfield, Anderson and Bundy (1977)
Zharikova (1980)

PHYSIOLOGY: SENSITIVITY

Arvy (1975)
Bakke and Lien (1978)
Desser (1984)
Desser and Weller (1982)
Gabrion and Euzet-Sicard (1979)
Goh and Davey (1976a, b)
Granath, Lewis and Esch (1983)
Gustafsson (1984, 1985)
Gustafsson, Jukanen and Wikgren (1983)
Kearn (1978b, 1980, 1984b)
Keenan and Koopowitz (1982)
Keenan, Koopowitz and Solon (1984)
Klochkova (1974)
Kotikova (1977, 1983)
Kotikova and Kuperman (1978a, b)
Krishna and Simha (1980)
Lambert (1979)
Llewellyn (1981)
Lyons (1973)
Malakhov and Valovaya (1984)
Niewiadomska and Moczon (1982, 1984)
Ramasamy and Hanna (1985)
Ramulu and Rama Krishna (1982)
Richards and Arme (1982a, b)
Rohde (1982c)
Rohde and Garlick (1985a, b)
Rohde, Watson and Garlick (1986)
Shaw (1981)
Shaw (1982)
Simon-Martin and Rojo-Vasquez (1984)
Thulin (1980b)
Uglem and Prior (1983)
Venkatanarsaiah (1981)
Venkatanarsaiah and Kulkarni (1980)
Xylander (1984)

PHYSIOLOGY: EXCRETION AND OSMOREGULATION

Arthur and Sanborn (1969)
Dailey (1975)
Davydov (1975)
Gavrilyuk (1978)
Gibson (1973b)
Khlebovich and Mikhailova (1976)
Leonteva (1976)
Lewis, Jones and Adams (1974)
Lyadov (1976)
Malmberg (1969)
Möller (1978)
Rohde (1982b)
Shishkova-Kasatochkina, Koloskova and Sokhina (1976)
Vinogradov, Davydov and Kuperman (1980, 1982a, b)

PHYSIOLOGY: CATABOLIC PATHWAYS

Dryuchenko (1982)
Dubovskaya (1979, 1982)
Dubovskaya and Shishova-Kasatochkina (1984)
McManus (1975)
Rasheed (1981)
Rasheed and Simha (1982a)
Siddiqui and Kulkarni (1985)
Syamsunder and Aruna (1984)
Walkey and Körting (1985)

BIOCHEMISTRY AND IMMUNOLOGY

Ackman (1976)
Arme (1975, 1977)
Arme and Whyte (1975)
Bailey and Rock (1975)
Balakhnin and Kozinenko (1978)
Baron and Appleton (1977)
Bogoyavlenskii and Nikitina (1979)
Buteau *et al.* (1969, 1971)
Chandra, Hanumantha-lao and Shyamasundari (1983)
Dabrowski (1980)
Davey (1976)
Davydov (1975, 1979)
Fisher and Starling (1970a)
Gupta and Gargs (1977)
Gupta and Gupta (1977)
Gur'yanova (1977a, b)
Gur'yanova and Freze (1981)
Gur'yanova, Sidorov and Freze (1976)
Halton and Hendrix (1978)
Haque and Siddiqi (1982)

Jakutowicz and Korpaczewska (1976)
Kayton (1983)
Kennedy, G.Y. (1979)
Ko and Ling (1980)
Kritsky, Leiby and Kayton (1978)
Lee and Smith (1965)
Marra and Esch (1970)
Matskási (1980)
McCullough and Fairweather (1987)
Mohsin and Farooqi (1979)
Niyogi and Agarwal (1983)
Pronina, Davydov and Kuperman (1985)
Pronina, Logachov and Pronin (1981)
Pronina and Pronin (1979)
Ramalingham (1973)
Rasheed (1980)
Ravindranathan and Nadakal (1971)
Sekretaryuk (1983b)
Shchepkina (1980)
Sidorov and Gur'yanova (1981)
Sidorov and Smirnov (1980)
Sidorov, Vysotskaya and Bykhovskaya-Pavlovskaya (1972)
Simmons (1969)
Smirnov (1983)
Smirnov and Sidorov (1978, 1979, 1981, 1984)
Soutter, Walkey and Arme (1980)
Strazhnik (1980)
Strazhnik and Davydov (1971)
Susuki and Ishida (1979)
Tandon and Misra (1980, 1985)
Terenina (1983)
Uglem and Beck (1972)
Vykhrestyuk, Yarygina and Nikitenko (1977)
Vysotskaya and Sidorov (1973)
Vysotskaya, Sidorov and Bykhovskaya-Pavlovskaya (1972)
Walker and Barrett (1983)
Whyte *et al.* (1987)
Wood and Matthews (1987)
Yarygina, Vykhrestyuk and Klochkova (1982)

IN VITRO CULTURE
Awakura (1980)
Banning (1971)
Davey (1979)
Fried (1978)
Johnston and Halton (1981)
Matthews (1982)
Mills (1979b)
Nizami and Siddiqi (1976)
Schroeder, Johnson and Mohammad (1981)
Smyth (1990)
Sommerville and Davey (1976)
Whyte, Chappell and Secombes (1988)

CHEMOTAXONOMY
Agatsuma (1982)
Bray and Rollinson (1985)
Buron, Renaud and Euzet (1986)
Cain and Raj (1980)
Euzet, Renaud and Gabrion (1984)
Freze, Dalin and Sergeeva (1985a,b)
Niyogi, Gaur and Agarwal (1985)
Renaud, Gabrion and Pasteur (1983, 1986)
Renaud, Gabrion and Romestand (1984)
Romanenko and Shigin (1977)
Simmons, Buteau, MacInnis and Kilejian (1972)
Susuki and Ishida (1979)

ULTRASTRUCTURE
Arme and Threadgold (1976)
Bazitov, Kulakovskaya, Yukhimenko, Lyapkalo and Shestakova (1978)
Bazitov and Lyapkalo (1984)
Biserova (1984)
Biserova and Smetanin (1982)
Bogolepova (1979)
Bogoyavlenskii, Nikitina and Khatkevich (1975)
Cho (1981)
Coggins (1980a, b)
Cone and Beverley-Burton (1981)
Crompton (1970b)
Davydov and Biserova (1985)
Davydov and Kuperman (1979)
El-Naggar and Kearn (1983a, b)
Euzet, Swiderski and Mokhtar-Maamouri (1981)
Fischer and Freeman (1969)
Fredericksen and Specian (1981)
Granath, Lewis and Esch (1983)
Halton (1979, 1982a)
Hayunga (1979a)

Higgins, Wright and Matthews (1977)
Jilek and Crites (1982c)
Kritsky (1978)
Kuperman (1976, 1980, 1981a)
Kuperman and Davydov (1980, 1982)
Le Brun, Lambert and Justine (1986)
Lynn, Suriano and Beverley-Burton (1981)
Lyons (1973)
Mackiewicz and Ehrenpris (1980)
MacKinnon and Burt (1985a, b, c)
MacKinnon, Jarecka and Burt (1985)
Mamaev and Avdeev (1984)
Mankes and Mackiewicz (1972)
McCullough and Fairweather (1983, 1989)
McVicar (1977b)
Poddubnaya, Davydov and Kuperman (1984)
Swiderski and Eklu-Natey (1978)
Swiderski, Eklu-Natey, Subilia and Huggel (1978)
Tedesco and Coggins (1980)
Richards and Arme (1981b, 1982a, b)
Rohde (1979b, 1981b)
Rohde and Garlick (1985a, b)
Sekretaryuk (1984)
Shaw (1979a, 1980)
Thompson, Hayton and JueSue (1980)
Thulin (1980b)
Timoshechkina (1984)
Whitfield (1984)
Whittaker, Apkarian, Curless and Carvajal (1985)
Yazaki, Fukumoto, Maeda, Maejima and Kamo (1985)

References

Abele, L. G. and Gilchrist, S, 1977, Homosexual rape and sexual selection in acanthocephalan worms. *Science*, 197, 81–83.

Ackman, R. G, 1976, Volatile ketones and alcohols of codworms (*Terranova decipiens*) from axenic culture and fresh fish muscle. *J. Fish. Res. Bd Can.*, 33, 2819–2821.

Adamcyzk, L. H, 1978, [Helminth fauna of fish in an open and closed lake in the Laczna-Wlodawa Lake district (Lublin district).] *Annls Univ. Mariae Curie-Sklodowska, C (Biol.)*, 33, 499–515. [In Polish.]

Adamczyk, L. H, 1979, [Contribution to the knowledge of helminth parasites of *Cottus gobio*.] *Annls Univ. Mariae Curie-Sklodowska, C (Biol.)*, 34, 279–288. [In Polish.]

Adamczyk, L. H, 1980, [Helminth fauna of *Nemachilus barbatulus* from the Wieprz river basin.] *Annls Univ. Mariae Curie-Sklodowska, C (Biol.)*, 35, 365–380. [In Polish.]

Adams, A. M, 1985, Parasites on the gills of the plains killifish, *Fundulus kansae*, in the South Platte River, Nebraska. *Trans. Am. Microsc. Soc.*, 104, 278–284.

Adams, A. M, 1986, Distribution of helminths in ringed seals *Phoca hispida* Schreber in northern Alaska. In: *Parasitology-Quo Vadit? Proceedings of the Sixth International Congress of Parasitology* (Ed. Howell, M. J.), Canberra: Australian Academy of Science, p. 78.

Adams, J. R, 1969, Migration route of invasive juvenile *Philonema onchorhynchi* (Nematoda: Philometridae) in young salmon. *J. Fish. Res. Bd Can.*, 26, 941–946.

Adams, J. R, 1974, Development of *Philonema agubernaculum* in the fish host. In: *Third International Congress of Parasitology. Munich, August 25–31, 1974. Proceedings, Vol. 1. Vienna: FACTA Publication*, pp. 482–483.

Adamson, M. L, 1986, Modes of transmission and evolution of life histories in zooparasitic nematodes. *Can. J. Zool.*, 64, 1375–1384.

Adjei, E. L., Barnes, A. and Lester, R. J. G, 1986, A method for estimating possible parasite-related host mortality, illustrated using data from *Callitetrarhynchus gracilis* (Cestoda: Trypanorhyncha) in lizardfish (*Saurida* spp.). *Parasitology*, 92, 227–243.

Agatsuma, T, 1982, Electrophoretic studies on glucosephosphate isomerase and phosphoglucomutase in two types of *Anisakis* larvae. *Int. J. Parasit.*, 12, 35–39.

Ahmed, A. T. A. and Sanaullah, M, 1979, Pathological observations of the intestinal lesions induced by caryophyllid cestodes in *Clarias batrachus* (Linnaeus) (Siluriformes: Clariidae). *Fish Path.*, 14, 107.

Aho, J. M. and Kennedy, C. R, 1984, Seasonal population dynamics of the nematode *Cystidicoloides tenuissima* (Zeder) from the River Swincombe, England. *J. Fish Biol.*, 25, 473–489.

Aho, J. M. and Kennedy, C. R, 1987, Circulation pattern and transmission dynamics of the suprapopulation of the nematode *Cystidicoloides tenuissima* (Zeder) in the river Swincombe, England. *J. Fish Biol.*, 31, 123–141.

Aho, J. M., Camp, J. W. and Esch, G. W, 1982, Long-term studies on the population biology of *Diplostomulum scheuringi* in a thermally altered reservoir. *J. Parasit.*, 68, 695–708.

Aho, J. M., Gibbons, J. W. and Esch, G. W, 1976, Relationship between thermal loading and parasitism in the mosquitofish. In: *Thermal ecology* (Eds Esch, G. W. and McFarlane, R. W.) *Georgia: ERDA Symposium Series 40*, pp. 213–218.

Aisa, E, 1976, Plerocercoid larval cysts of *Triaenophorus lucii* (Müller, 1776) in the ovaries of *Tinca tinca* (L). and *Esox lucius* (L). *Atti. Soc. Ital. Sci. Vet.*, 29, 613–620.

Akahane, H., Iwata, K. and Miyazaki, I, 1982, [Studies on *Gnathostoma hispidum* Fedchenko, 1872 parasitic in loaches imported from China.] *Jap. J. Parasit.*, 31, 507–516. [In Japanese.]

Akasaka, Y., Kizu, M., Aoike, A. and Kawai, K, 1979, Endoscopic management of acute gastric anisakiasis. *Endoscopy*, **11**, 158–162.

Albetova, L. M. and Michurin, S. M, 1984, [Parasites and diseases of carp in the warm waters of the Surgut hydroelectric station.] *Sb. Nauch. Trud. Gosud. Nauchno-Issled. Inst. Ozern. Rech. Ryb. Khozyaist. (Bolez. parazit. ryb vod. Zapad. Sib.)*, No. 226, 3–15. [In Russian.]

Aleshkina, L. D, 1982, [The parasite fauna of *Merluccius capensis* and the dependence of its species composition upon the age of the host.] *Gidrobiol. Zhurnal.*, **18**, 65–68. [In Russian.]

Ali, N. M., Abul-Eis, E. S. and Abdul-Ameer, K. N, 1988, Study on the parasites of common carp *Cyprinus carpio* and other freshwater fishes in Habbaniyah Lake, Iraq. *J. Biol. Sci. Res.*, **19**, 395–407.

Ali, N. M., Al-Jafery, A. R. and Abdul-Ameer, K. N, 1987a, Parasitic fauna of freshwater fishes in Diyala River, Iraq. *J. Biol. Sci. Res.*, **18**, 163–181.

Ali, N. M., Salih, N. E. and Abdul-Ameer, K. N, 1987b, Parasitic fauna of some freshwater fishes from Tigris river, Baghdad, Iraq. III. Cestoda. *J. Biol. Sci. Res.*, **18**, 25–33.

Ali, N. M., Salih, N. E. and Abdul-Ameer, K. N, 1987c, Parasitic fauna of some freshwater fishes from Tigris river, Baghdad, Iraq. IV. Nematoda. *J. Biol. Sci. Res.*, **18**, 35–45.

Allaniyazova, T, 1975, [Intermediate hosts of *Bothriocephalus gowkongensis* (Cestoda: Pseudophyllidea) in the lower reaches of the Amudar'ya.] *Vest. Karakalpaksk. Fil. Akad. Nauk Uzbek. SSR, No. 1 (59)*, 27–32. [In Russian.]

Allee, W. C., Emerson, O., Park, I. and Schmidt, K. P, 1949, *Principles of Animal Ecology*. Philadelphia: Saunders, xii + 837pp.

Allison, F. R, 1980, Sensory receptors of the rosette organ of *Gyrocotyle rugosa*. *Int. J. Parasit.*, **10**, 341–353.

Alvarez, J. D. and Conroy, D. A, 1985, Lista preliminar de las principales enfermedades y parasitos reportados para pampanos (Teleostei, Carangidae, *Trachinotus*) en aguas americanas. *Rev. Fac. Ciens. Vets. U.C.V.*, **32**, 113–124.

Alvarez-Pellitero, M. P, 1976a, [Seasonal variations in infestation of trout by *Crepidostomum farionis* and *C. metoecus* in Leon rivers.] *An. Fac. Vet. Leon*, **22**, 95–153. [In Spanish.]

Alvarez-Pellitero, M. P, 1976b, [Seasonal variations in *Cystidicoloides tenuissima* infestation of trout in Leon rivers.] *An. Fac. Vet. Leon*, 22, 155–180. [In Spanish.]

Alvarez-Pellitero, M. P, 1979a, [The helminth coenosis of the digestive tract of trout in the rivers of Leon.] *Institucion 'Fray Bernardino de Sahagun' de la Excma. Diputacion Provincial (C.S.I.C.)*, 269 pp. [In Spanish.]

Alvarez-Pellitero, M. P, 1979b, Observaciones sobre el ciclo vital de *Raphidascaris acus* en los ambientes naturales de los vios de Leon. *An. Fac. Vet. Leon*, **25**, 129–154.

Amin, O. M, 1975a, *Acanthocephalus parksidei* sp.n. (Acanthocephala: Echinorhynchidae) from Wisconsin fishes. *J. Parasit.*, **61**, 301–306.

Amin, O. M, 1975b, Host and seasonal associations of *Acanthocephalus parksidei* Amin, 1974 (Acanthocephala: Echinorhynchidae) in Wisconsin fishes. *J. Parasit.* **61**, 318–329.

Amin, O. M, 1977, Helminth parasites of some southwestern Lake Michigan fishes. *Proc. Helminth. Soc. Wash.*, **44**, 210–217.

Amin, O. M, 1978a, On the crustacean hosts of larval acanthocephalan and cestode parasites in southwestern Lake Michigan. *J. Parasit.*, **64**, 842–845.

Amin, O. M, 1978b, Effect of spawning on *Echinorhynchus salmonis* (Muller, 1784) (Acanthocephala: Echinorhynchidae) maturation and localization. *J. Fish Dis.*, **1**, 195–197.

Amin, O. M, 1981, The seasonal distribution of *Echinorhynchus salmonis* (Acanthocephala: Echinorhynchidae) among rainbow smelt, *Osmerus mordax* Mitchell, in Lake Michigan. *J. Fish Biol.*, **19**, 467–474.

Amin, O. M, 1982, *Acanthocephala*. In: *Synopsis and classification of living organisms* (Ed. Parker, S. P.). New York: McGraw-Hill Book Company, pp. 934–940.

Amin, O. M, 1984, Camallanid and other nematode parasites of lake fishes in southeastern Wisconsin. *Proc. Helminth. Soc. Wash.*, **51**, 78–84.

Amin, O. M, 1985a, Classification. In: *Biology of the Acanthocephala* (Ed. Crompton, D. W. T. and Nickol, B. B.). Cambridge: Cambridge University Press, pp. 27–72.

Amin, O. M, 1985b, The relationship between the size of some salmonid fishes and the intensity of their acanthocephalan infections. *Can. J. Zool.*, **63**, 924–927.

Amin, O. M, 1985c, Hosts and geographic distribution of *Acanthocephalus* (Acanthocephala: Echinorhynchidae) from North American freshwater fishes, with a discussion of species relationships. *Proc. Helminth. Soc. Wash.*, **52**, 210–220.

Amin, O. M, 1985d, Acanthocephala from lake fishes in Wisconsin: *Neoechinorhynchus robertbaueri* n.sp. from *Erimyzon sucetta* (Lacépède), with a key to species of the genus *Neoechinorhynchus* Hamann, 1892 from North American freshwater fishes. *J. Parasit.*, 71, 312–318.

Amin, O. M, 1987a, Key to the families and subfamilies of Acanthocephala, with the creation of a new class (Polyacanthocephala) and a new order (Polyacanthorhynchida). *J. Parasit.*, 73, 1216–1219.

Amin, O. M, 1987b, Acanthocephala from lake fishes in Wisconsin: ecology and host relationships of *Pomphorhynchus bulbocolli* (Pomphorhynchidae). *J. Parasit.*, 73, 278–279.

Amin, O. M, 1987c, Acanthocephala from lake fishes in Wisconsin: morphometric growth of *Pomphorhynchus bulbocolli* (Pomphorhynchidae). *J. Parasit.*, 73, 278–279.

Amin, O. M, 1988, Acanthocephala from lake fishes in Wisconsin: on the ecology of *Leptorhynchoides thecatus* (Rhadinorhynchidae). *Proc. Helminth. Soc. Wash.*, 55, 252–255.

Amin, O. M., Burns, L. A. and Redlin, M. J, 1980, The ecology of *Acanthocephalus parksidei* Amin, 1975 (Acanthocephala: Echinorhynchidae) in its isopod intermediate host. *Proc. Helminth. Soc. Wash.*, 47, 37–46.

Amlacher, E, 1976, *{Pocket book of fish diseases for veterinarians and biologists.} Third edition*. Jena, Germany: VEB Gustav Fischer Verlag, 394 pp. [In German.]

Anantaraman, S, 1964, A juvenile *Echinocephalus uncinatus* (Molin, 1858) in the marine gastropod, *Hemifusus pugilinus* (Born) with notes on the genus *Echinocephalus* (Nematoda: Gnathostomidae). *Ann. Mag. Nat. Hist.*, 7, 101–105.

Andakulova, N. A, 1976, [The spread of the cestode *Gryporhynchus cheilancristrotus* (Wedl 1855).] In: *Kazakhstan Proceedings of the Fifteenth Science Conference*, pp. 227–229. [In Russian.]

Andakulova, N. A, 1978, [The study of the cestode *Gryporhynchus cheilancristrotus* in Kazakhstan.] *Trudy Vses. Nauchno-Issled. Inst. Prudov. Ryb. Khozyaist. (Paraz. bolez. ryb profilakt.)*. 27, 26–30. [In Russian.]

Andersen, K. I, 1972, Studies on the helminth fauna of Norway. XXIV. The morphology of *Diphyllobothrium ditremum* (Creplin 1825) from golden hamster (*Mesocricetus auratus* Waterhouse 1839) and a comparison with *D. dendtiricum* (Nitzsch, 1824) and *D. latum* (L. 1758) from the same final *Norw. J. Zool.*, 20, 255–264.

Andersen, K. I, 1977, A marine *Diphyllobothrium* plerocercoid (Cestoda, Pseudophyllidea) from blue whiting (*Micromesistius poutassou*). *Z. ParasitKde*, 52, 289–296.

Andersen, K. I, 1978a, The helminths in the gut of perch (*Perca fluviatilis* L.) in a small oligotrophic lake in southern Norway. *Z. ParasitKde*, 56, 17–27.

Andersen, K. I, 1978b, The development of the tapeworm *Diphyllobothrium latum* (L. 1756) (Cestoda: Pseudophyllidea) in its definitive hosts, with special references to the growth patterns of *D. dendriticum* (Nitzsch, 1824) and *D. ditremum* (Creplin, 1827). *Parasitology*, 77, 111–120.

Andersen, K. I, 1979a, Variation in scolex morphology within and between some species of the genus *Proteocephalus* Weinland (Cestoda: Proteocephala) with references to strobilar morphology. *Z. Scripta*, 8, 241–248.

Andersen, K. I, 1979b, Studies on the scolex morphology of *Eubothrium* spp. with emphasis on characters usable in species discrimination and with brief references on the scolices of *Bothriocephalus* sp. and *Triaenophorus* spp. (Cestoda: Pseudophyllidea). *Z. ParasitKde.*, 60, 147–156.

Andersen, K. I, 1979c, Abnormal growth pattern of *Diphyllobothrium dendriticum* in rainbow trout. *J. Helminth.*, 53, 39–40.

Andersen, K. I, 1987, A redescription of *Diphyllobothrium stemmacephalum* Cobbold, 1858 with comments on other marine species of *Diphyllobothrium* Cobbold, 1858. *J. Nat. Hist.*, 21, 411–427.

Andersen, K. I. and Gibson, D. I, 1988, A key to three species of larval *Diphyllobothrium* Cobbold, 1858 (Cestoda: Pseudophyllidea) occurring in European and North American freshwater fishes. *Syst. Parasit.*, 13, 3–9.

Andersen, K. I., Ching, H. L. and Vik, R, 1987, A review of freshwater species of *Diphyllobothrium* with redescriptions and the distribution of *D. dendriticum* (Nitzsch, 1824) and *D. ditremum* (Creplin, 1825) from North America. *Can. J. Zool.*, 65, 2216–2228.

Anderson, D. J. and Kikkawa, J, 1986, Developments of concepts. In: *Community Ecology: Patterns and Process* (Eds Kikkawa, J. and Anderson, R. M.). Oxford: Blackwell Scientific Publications, pp. 3–16.

Anderson, J. I. W. and Conroy, D. A, 1968, Extensive infestation by *Entobdella soleae* in brood stock of Dover sole. *Bull. Off. Int. Epizoot.*, 69, 1129–1137.

Anderson, M, 1981, The change with host age of the composition of the ancyrocephaline (monogenean) populations of parasites on thick-lipped grey mullets at Plymouth. *J. Mar. Biol. Ass. U.K.*, 61, 833–842.

Anderson, M. G. and Anderson, F. M, 1963, Life history of *Proterometra dickermani* Anderson, 1962. *J. Parasit.*, 49, 275–280.

Anderson, R. C, 1984, The origins of zooparasitic nematodes. *Can. J. Zool.*, 62, 317–328.

Anderson, R. C, 1988, Nematode transmission patterns. *J. Parasit.*, 74, 30–45.

Anderson, R. C. and Bain, O, 1982, Keys to genera of the superfamilies Rhabditoidea, Dioctophymatoidea, Trichinelloidea and Muspiceoidea. CIH Keys to the nematode parasites of vertebrates. *Farnham Royal: England: Commonwealth Agricultural Bureaux.* No. 9, 26pp.

Anderson, R. C., Chabaud, A. G. and Willmott, S, 1974, CIH Keys to the nematode parasites of vertebrates. *Farnham Royal, Bucks, England: Commonwealth Agricultural Bureaux, No. 1*, iv + 18pp.

Anderson, R. M, 1974a, *Mathematical models of host-helminth parasite interactions.* In: *Ecological stability* (Eds Usher, M. B. and Williamson, M. H.). London, UK: Chapman and Hall Ltd., pp. 43–69.

Anderson, R. M, 1974b, Population dynamics of the cestode, *Caryophyllaeus laticeps* (Pallas, 1781) in the bream (*Abramis brama* L.). *J. Anim. Ecol.*, 43, 305–321.

Anderson, R. M, 1974c, An analysis of the influence of host morphometric features on the population dynamics of *Diplozoon paradoxum* (Nordmann, 1832). *J. Anim. Ecol.*, 43, 873–887.

Anderson, R. M, 1976a, Dynamic aspects of parasite population ecology. In: *Ecological aspects of parasitology* (Ed. Kennedy, C. R.). Amsterdam, The Netherlands: North-Holland Publishing Company, pp. 431–462.

Anderson, R. M, 1976b, Seasonal variation in the population dynamics of *Caryophyllaeus laticeps. Parasitology*, 72, 281–305.

Anderson, R. M, 1978, The regulation of host population growth by parasitic species. *Parasitology*, 76, 119–157.

Anderson, R. M, 1979, The influence of parasitic infection on the dynamics of host population growth. In: *Population dynamics (Symposium of the British Ecological Society)* (Eds Anderson, R. M., Turner, B. D. and Taylor, L. R.) Oxford, UK, Blackwell Scientific Publications, pp. 245–281.

Anderson, R. M, 1982a, *Population dynamics of infectious diseases. Theory and applications. Series: Population and Community Biology series* (Eds Usher, M. B. and Rosenzweig, M. L.). London, UK: Chapman and Hall, xix + 400pp.

Anderson, R. M, 1982b, Host-parasite population biology. In: *Parasites—their world and ours. Proceedings of the Fifth International Congress of Parasitology, Toronto, Canada* (Eds Mettrick D. F. and Desser, S. S.). Amsterdam, Netherlands: Elsevier Biomedical Press, pp. 303–312.

Anderson, R. M, 1982c, Transmission dynamics and control of infectious disease agents. In: *Population biology of infectious diseases. (Report of the Dahlem Workshop on Population Biology of Infectious Disease Agents; Berlin)* (Eds Anderson R. M. and May, R. M.) Berlin: Springer-Verlag, pp. 149–176.

Anderson, R. M, 1982d, Parasite dispersion patterns: generative mechanisms and dynamic consequences. In: *Aspects of Parasitology* (Ed. Meerovitch, E.). Montreal: The Institute of Parasitology, McGill University, pp. 1–41.

Anderson, R. M, 1986, The role of mathematical models in helminth population biology. In: *Parasitology-Quo Vadit? Proceedings of the Sixth International Congress of Parasitology* (Ed. Howell, M. J.). Canberra: Australian Academy of Science, pp. 519–529.

Anderson, R. M. and Canning, E. U. (Eds) 1982, Parasites as Biological Control Agents. *Symp. Soci Parasit.*, 19, 1–296.

Anderson, R. M. and Gordon, D. M, 1982, Processes influencing the distribution of parasite numbers within host populations with special emphasis on parasite-induced host mortalities. *Parasitology*, **85**, 373–398.

Anderson, R. M. and Kikkawa, J, 1986, Development of concepts. In: *Community Ecology: Patterns and Process* (Eds Kikkawa, J. and Anderson, R. M.) London: Blackwell Scientific Publications, pp. 3–16.

Anderson, R. M. and May, R. M, 1978, Regulation and stability of host–parasite population interactions. I. Regulatory processes. *J. Anim. Ecol.*, 47, 219–247.

Anderson, R. M. and May, R. M, 1979, Population biology of infectious diseases: part 1. *Nature* **280**, 361–367.

Anderson, R. M. and May, R. M, 1981, The population dynamics of micro-parasites and their invertebrate hosts. *Phil. Trans. R. Soc., B*, **291**, 451–524.

Anderson, R. M. and May, R. M, 1982a, Coevolution of hosts and parasites. *Parasitology*, **85**, 411–426.

Anderson, R. M. and May, R. (Eds) 1982b, *Population biology of infectious diseases Report of the Dahlem workshop on population biology of infectious disease agents, Berlin 1982*), Berlin: Springer-Verlag, viii + 314 pp.

Anderson R. M. and Whitfield, P. J, 1975, Survival characteristics of the free-living cercarial population of the ectoparasitic digenean *Transversotrema patialensis* (Soparker, 1924). *Parasitology*, 70, 295–310.

Anderson, R. M., Whitfield, P. J. and Dobson, A. P, 1978, Experimental studies of infection dynamics: infection of the definitive host by the cercariae of *Transversotrema patialense. Parasitology*, 77, 189–200.

Anderson, R. M., Whitfield, P. J. and Mills, C. A, 1977, An experimental study of the population dynamics of an ectoparasitic digenean *Transversotrema patialense*: the cercarial and adult stages. *J. Anim. Ecol.*, 46, 555–580.

Andreeva, G. N. and Tarakanov, V. I, 1973, [Growth factor in nematodes.] *Trudy Vses. Inst. Gel'mi. K. I. Skryabina*, 20, 13–16. [In Russian.]

Andrews, C, 1979, Host specificity of the parasite fauna of perch *Perca fluviatilis* L. from the British Isles with special reference to a study at Llyn Tegid (Wales). *J. Fish Biol.*., 15, 195–209.

Andrews, C. and Chubb, J. C, 1980, Observations on the development of *Bunodera luciopercae* (Müller, 1776) (Trematoda: Allocreadiidae) under field and laboratory conditions. *J. Fish. Dis.*, 3. 481–493.

Andrews, C., Chubb, J. C., Coles, T. and Dearsley, A, 1981, The occurrence of *Bothriocephalus acheilognathi* Yamaguti, 1934 (*B. gowkongensis*) (Cestoda: Pseudophyllidea) in the British Isles. *J. Fish Dis.*, 4, 89–93.

Andrews, R. H., Beveridge, I., Adams, M. and Baverstock, P. R, 1988, Identification of life cycle stages of the nematode *Echinocephalus overstreeti* by allozyme electrophoresis. *J. Helminth.*, 62, 153–157.

Andreyuk, G. I, 1984 [The epizootiology of diplostomiasis and prophylaxis of clinical infection in fish in lake fisheries of the Tyumen region.] *Sb. Nauch. Trud. Gosud. Nauchno-Issled. Inst. Ozern. Rech. Ryb. Khozyaist. (Bolez. parazit. ryb vod. Zapad. Sib.), No. 226*, 16–24. [In Russian.]

Andryuk, L. V, 1979a, [The developmental cycle of *Acanthocephalus anguillae*.] *Zool. Zh.*, 58, 168–174. [In Russian.]

Andryuk, L. V, 1979b, [On the life-cycle of acanthocephalans from the genus *Acanthocephalus* Koelreuther.] *Dokl. Timir. Sel'skokhoz. Akad.*, 255, 101–104. [In Russian.]

Andryuk, L. V, 1979c, [The life-cycle of *Acanthocephalus lucii* (Echinorhynchidae).] *Parazitologiya*, 13, 530–539. [In Russian.]

Angel, L. M, 1960, *Cercaria haswelli* Dollfus, 1927; a re-examination of Haswell's material, with discussion of the genus *Tergestia. Libro Homenaje al Dr. Eduardo Caballero y Caballero, Jubileo 1930–1960*, 75–86.

Angel, L. M, 1969, *Prototransversotrema steeri* gen. nov., sp. nov. (Digenea: Transversotrematidae) from a south Australian fish. *Parasitology*, 59, 719–724.

Angel, L. M, 1971, *Burnellus* gen. nov. (Digenea: Fellodistomidae): the life history of the type-species *B. trichofurcatus* (Johnston and Angel, 1940), and a note on a related species, *Tandanicola bancrofti* Johnston, 1927, both from the Australian freshwater catfish, *Tandanus tandanus. Parasitology*, 62, 375–384.

Anikieva, L. V, 1982a, [The use of helminthological data in the assessment of the status of a water-body.] In: *Ekologiya paraziticheskikh organizmov v. biogeotsenozakh severa. Petrozavodsk, USSR: Karel'skii Filial Akademi Nauk SSSR, Institut Biologii*, pp. 72–83. [In Russian.]

Anikieva, L. V, 1982b, [The influence of waste waters of Segezhsk paper-and-pulp plant on the helminth fauna of fish in lake Vygozero.] In: *Ekologiya paraziticheskikh organizmov v biogeotsenozakh severa. Petrozavodsk, USSR: Karel'skii Filial Akademi Nauk SSSR, Institut Biologii*, pp. 83–94. [In Russian.]

Anikieva, L. V, 1982c, [The development of *Proteocephalus exiguus* in intermediate hosts.] In: *Ekologiya paraziticheskikh organizmov v biogeotsenozakh severa. Petrozavodsk, USSR: Karelskii Filial Akademi Nauk SSSR, Institut Biologii*, pp. 114–128. [In Russian.]

Anikieva, L. V. and Malakhova, R. P, 1975, [Aspects of the development of the cestode *Proteocephalus exiguus* related to its environmental conditions.] In: *Osnovy bioproduktivnosti vnutrennikh vodoemov Pribaltiki. Vilnius USSR, Ref. Zh. Biol.*, 3K, 134. [In Russian.]

Anikieva, L. V. and Malakhova, R. P, 1982, [Distribution of *Proteocephalus exiguus* according to host age and sex.] In: *Gel'minty v presnovodnykh biotsenozakh. Moscow, USSR: 'Nauka'*, pp. 68–73. [In Russian.]

Anikieva, L. V., Malakhova, R. P. and Ieshko, E. P, 1983, [Ecological analysis of the parasites of whitefish.] *Leningrad, USSR: Nauka, Leningradskoe Otdelenie*, 168 pp. [In Russian.]

Annigeri, G. G, 1962, A viviparous nematode, *Philometra* sp. in the ovaries of *Otolithus argenteus* (Cuvier). *J. Mar. Biol. Assoc. India*, 3, 263–265.

Antonov, P. P. and Sapozhnikov, T. I, 1981, [Control of diplostomatid and ligulid infection by the introduction of coregonid fish.] In: *Bor'ba s invazionnymi boleznyami. Ufa, USSR: Ul'yanovskii Sel'skokhozyaistvennyi Institut*, pp. 52–54. [In Russian.]

Antsyshkina, L. M., Vazderkina, S. A. and Kalyuga, N. V, 1981, [Parasites and parasitic diseases of wild and cultivated carp hybrids and their control in the Pridneprovsk warm water farm.] *Materialy II Respublikanskot Nauchnoi Konferentsii, Kiev, 1980*, 419–423. [In Russian.]

Anya, A. O, 1976, Physiological aspects of reproduction in nematodes. *Adv. Parasit.*, 14, 267–351.

Appleby, D., Kapoor, W., Karpf, M. and Williams, S, 1982, Anisakiasis—nematode infestation producing small-bowel obstruction. *Arch. Surg.*, 117, 836.

Apprahamian, M. W, 1985, The effect of the migration of *Alosa fallax fallax* (Lacépède) into fresh water, on branchial and gut parasites. *J. Fish Biol.*, 27, 521–532.

Appy, R. G. and Anderson, R. C, 1982, The parasites of lampreys. In: *The biology of lampreys* (Ed Hardisty, M. W. and Potter, I. C.). London, UK: Academic Press, 3, pp. 1–42.

Appy, R. G. and Burt, M. D. B, 1982, Metazoan parasites of cod, *Gadus morhua* L., in Canadian Atlantic waters. *Can. J. Zool.*, 60, 1573–1579.

Appy, R. G. and Dadswell, M. J, 1978, Parasites of *Acipenser brevirostrum* LeSueur and *Acipenser oxyrhynchus* Mitchill (Osteichthyes: Acipenseridae) in the Saint John River Estuary, N. B. with a description of *Caballeronema pseudoargumentosus* sp.n. (Nematoda: Spirurida). *Can. J. Zool.*, 56, 1382–1391.

Appy, R. G. and Dadswell, M. J, 1983, Transmission and development of *Capillospirura pseudoargumentosa* (Appy and Dadswell, 1978) (Nematoda: Cystidicolidae). *Can. J. Zool.*, 61, 848–859.

Arai, H. P, 1962, Trematodos digeneos de peces marinos de Baja California, México. *An. Inst. Biol. Univ. Méx.*, 33, 113–130.

Arai, H. P, 1967, Ecological specificity of parasites of some embiotocid fishes. *J. Fish. Res. Bd Canada*, 24, 2161–2168.

Arai, H. P, 1969, Preliminary report on the parasites of certain marine fishes of British Columbia. *J. Fish. Res. Bd Can.*, 26, 2319–2337.

Arai, H. P, 1989, Acanthocephala. In: L. Margolis and Z. Kabata eds. Guide to the parasites of fishes of Canada. Part III. *Can. Spec. Publ. Fish. Aquat. Sci.* 107: pp. 1–90.

Arai, H. P. and Mudry, D. R, 1983, Protozoan and metazoan parasites of fishes from the headwaters of the Parsnip and McGregor rivers, British Columbia: a study of possible parasite transfaunations. *Can. J. Fish. Aquat. Sci.*, 40, 1676–1684.

Arai, H. P., Kabata, Z. and Noakes, D, 1988, Studies on seasonal changes and latitudinal differences in the metazoan fauna of the shiner perch, *Cymatogaster aggregata*, along the west coast of North America. *Can. J. Zool.*, 66, 1514–1517.

Argumedo, R. L, 1963, Estudio de algunos monogeneos y digeneos. Parasitos de peces del Pacifico Mexicano. Thesis, Universidad Nacional Autonoma de Mexico, 111pp.

Armas, G, 1979, Observations on diseases and parasites of mullet alevins *Mugil cephalus* L. from the Rio Moche coastal lagoon of Peru. *J. Fish Dis.*, 2, 543–547.

Arme, C, 1966, Histochemical and biochemical studies on some enzymes of *Ligula intestinalis* (Cestoda: Pseudophyllidea). *J. Parasit.*, 52, 63–68.

Arme, C, 1968, Effects of the plerocercoid larva of a pseudophyllidean Cestode, *Ligula intestinalis*, on the pituitary gland and gonads of its host. *Biol. Bull.*, 134, 15–25.

Arme, C, 1975, Chemical composition of *Diclidophora merlangi* (Monogenea). *Z. ParasitKde*, 47, 211–215.

Arme, C, 1976, Feeding. In: *Ecological aspects of parasitology* (Ed. Kennedy, C. R.). Amsterdam, Netherlands: North-Holland Publishing Co., pp. 75–97.

Arme, C, 1977, Amino acids in eight species of Monogenea. *Z. ParasitenKde*, 51, 261–263.

Arme, C, 1982, Nutrition. In: *Modern parasitology. A textbook of parasitology* (Ed. Cox, F. E. G.). Oxford, UK: Blackwell Scientific Publications, pp. 148–172.

Arme, C, 1983, Aspects of tapeworm physiology. *J. Biol. Educ.*, 17, 352–357.

Arme, C. and Fox, M. G, 1974, Oxygen uptake by *Diclidophora merlangi* (Monogenea). *Parasitology*, 69, 201–205.

Arme, C. and Halton, D, 1972, Observations on the occurrence of *Diclidophora merlangi* (Trematoda: Monogenea) on the gills of whiting, *Gadus merlangus*. *J. Fish Biol.*, 4, 27–32.

Arme, C. and Owen, R, 1967, Infection of the three-spined stickleback *Gasterosteus aculeatus* L., with the plerocercoid larvae of *Schistocephalus solidus* (Müller, 1776) with special reference to pathological effects. *Parasitology*, 57, 301–314.

Arme, C. and Owen, R, 1968, Occurrence and pathology of *Ligula intestinalis* infections in British fishes. *J. Parasit.*, 54, 272–280.

Arme, C. and Owen, R, 1970, Observations on a tissue response within the body cavity of fish infected with the plerocercoid larvae of *Ligula intestinalis* (L.) (Cestoda: Pseudophyllidea). *J. Fish Biol.*, 2, 35–37.

Arme, C. and Pappas, P. W. (Eds) 1983, *Biology of the Eucestoda*. London, Academic Press, 1, 1–126, 2, 297–628.

Arme, C. and Threadgold, L. T, 1976, A unique tegumentary cell type and unicellular glands associated with the scolex of *Eubothrium crassum* (Cestoda: Pseudophyllidea). *Rice Univ. Stud.*, 62, 21–34.

Arme, C. and Walkey, M, 1970, The physiology of fish parasites. In: *Aspects of fish parasitology* (Eds Taylor, A. E. R. and Muller, R.). *Symposium of the British Society for Parasitology (8th)*. Oxford: Blackwell Scientific Publications, pp. 79–101.

Arme, C. and Whyte, A, 1975, Amino acids of *Diclidophora merlangi* (Monogenea). *Parasitology*, 70, 39–46.

Arme, C., Griffiths, D. V. and Sumpter, J. P, 1982, Evidence against the hypothesis that the plerocercoid larva of *Ligula intestinalis* (Cestoda: Pseudophyllidea) produces a sex steroid that interferes with host reproduction. *J. Parasit.*, **68**, 169–171.

Armstrong, H. W, 1975, A study of the helminth parasites of the family Macrouridae from the Gulf of Mexico and Caribbean Sea: their systematics, ecology and zoogeographical implications. *Diss. Abstr. Inter.*, **36B**, 123–124.

Arthur, E. J. and Sanborn, R. C, 1969, Osmotic and ionic regulation in nematodes. In: *Chemical Zoology Vol. III Echinodermata, Nematoda and Acanthocephala* (Eds Florkin, M. and Scheer, B. T.). New York and London: Academic Press Inc., pp. 429–464.

Arthur, J. R, 1984, A survey of the parasites of walleye pollock (*Theragra chalcogramma*) from the northeastern Pacific Ocean off Canada and a zoogeographical analysis of the parasite fauna of this fish throughout its range. *Can. J. Zool.*, 62, 675–684.

Arthur, J. R. and Arai, H. P, 1980a, Studies on the parasites of Pacific herring (*Clupea harengus pallasi* Valenciennes): survey results. *Can. J. Zool.*, **58**, 64–70.

Arthur, J. R. and Arai, H. P, 1980b, Studies on the parasites of Pacific herring (*Clupea harengus pallasi* Valenciennes): a preliminary evaluation of parasites as indicators of geographical origin for spawning herring. *Can. J. Zool.*, **58**, 521–527.

Arthur, J. R. and Margolis, L, 1975, Revision of the genus *Haplonema* Ward and Magath, 1917 (Nematoda: Seuratoidea). *Can. J. Zool.*, **53**, 736–747.

Arthur, J. R., Margolis, L. and Arai, H. P, 1976, Parasites of fishes of Aishihik and Stevens Lakes, Yukon Territory, and potential consequences of their interlake transfer through a proposed water diversion for hydroelectrical purposes. *J. Fish. Res. Bd Can.*, 33, 2489–2499.

Arthur, J. R., Margolis. L., Whitaker, D. J. and McDonald, T. E, 1982, A quantitative study of economically important parasites of walleye pollock (*Theragra chalcogramma*) from British Columbian waters and effects of postmortem handling on their abundance in the musculature. *Can. J. Fish. Aquat. Sci.*, 39, 710–726.

Arvy, L, 1975 [Les yeux des trématodes digénétiques] *Vie et Milieu, C (Biol. Terr.)*, 25, 203–235.

Arya, S. N, 1979, On a new nematode, *Comephoronema johnsoni* sp.nov. (Family: Rhabdochonidae Skrjabin, 1946) from a marine fish *Cybium guttatum*. *Ind. J. Helm.*, **30**, 17–20.

Arya, S. N, 1980, On the larval forms of a nematode *Dujardinascaris cybii* Arya and Johnson, 1978. *Rev. Bras. Biol.*, 40, 751–753.

As, J. G. van and Basson, L, 1984, Checklist of freshwater fish parasites from southern Africa. *S. Afr. J. Wildl. Res.*, 14, 49–61.

Ash, C., Crompton, D. and Keymer, A, 1984, Nature's unfair food tax. *New Scientist*, **1400**, 17–20.

Avdeev, V. V, 1985, Plerocercoids of some Cestoda as bioindicators of the population structure of *Podonema longipipes*. In: *(Ed. Hargis, W. J. J.) Parasitology and pathology of marine organisms of the world ocean. NOAA tech. Rep. NFMS, No. 25.* pp. 79–82.

Avdos'ev, B. S, 1973, [The use of tobacco dust in the control of *Bothriocephalus gowkongensis* infection in carp.] *Rybnoe Khozyaistvo, Kiev, Issue 17*, 109–118. [In Russian.]

Avdos'ev, B. S., Ben'ko, K. I., Syara, Ya. I. and Shemchuk, V. K, 1978, [Accumulation of copper in water, silt, plants, plankton and fish when copper ammoniate is introduced into ponds.] *Trudy Vses. Nauchn-Issled. Inst. Prud. Ryb. Khozyaist. (Parazit., bolez. ryb profilak.)*, 27, 21–26. [In Russian.]

Awachie, J. B. E, 1963, On the development and life-history of *Metechinorhynchus truttae* (Schrank, 1788) Petrochenko, 1956 (Acanthocephala). [Abstract]. *Parasitology*, 53, 3P.

Awachie, J. B. E, 1965, The ecology of *Echinorhynchus truttae* Schrank 1788 (Acanthocephala) in a trout stream in North Wales. *Parasitology*, 55, 747–762.

Awachie, J. B. E, 1966a, Observations on *Cyathocephalus truncatus* Pallas 1781 (Cestoda: Spathebothriidea) in its intermediate and definitive hosts in a trout stream, North Wales. *J. Helminth.*, 40, 1–10.

Awachie, J. B. E, 1966b, The development and life history of *Echinorhynchus truttae* Schrank, 1788 (Acanthocephala). *J. Helminth.*, 40, 11–32.

Awachie, J. B. E, 1968, On the bionomics of *Crepidostomum metoecus* (Braun, 1900) and *Crepidostomum farionis* (Muller, 1784) (Trematoda: Allocreadiidae). *Parasitology*, 58, 307–324.

Awachie, J. B. E, 1973, Ecological observations on *Metabronema truttae* Baylis, 1935, and *Cystidicola farionis* Fischer V. Waldhei, 1798 (Nematoda, Spiruroidea) in their intermediate and definitive hosts, in Afon Terrig. *Acta Parasit. Pol.*, 21, 661–670.

Awadalla, H. N., Mansour, M. A., Khalil, A. I. and Cuirgis, R, 1982, Pathological changes in the liver tissue of the lizard fish caused by larvae *Thynnascaris*. *J. Egypt. Soc. Parasit.*, 12, 389–394.

Awakura, T, 1980, [Cultivation of *Digramma alterans* in vitro.] *Sci. Rep. Hokkaido Fish Hatch.*, No. 35, 63–71. [In Japanese.]

Ax, P, 1963, Relationships and phylogeny in the Turbellaria. In: *The lower metazoa* (Eds E. C. Dougherty, Z. N. Brown, E. D. Hanson and W. D. Hartman). Berkeley and Los Angeles, CA.: University of California Press, pp. 191–224.

Ayles, G. B, 1977, Trout farming. In: *Miscellaneous farm enterprises. Principles and practices of commercial farming. 5th Edition. Winnipeg: Faculty of Agriculture, University of Manitoba. F. W. I.*, 492, 448–451.

Babaev, B. and Shcherbakova, A, 1963, [The control of *Bothriocephalus gowkongensis* from *Ctenopharyngodon idella* (Vallenciennes).] *Izvestiya Akademii Nauk Turkmenskoi SSR. Seriya Biologicheskikh Nauk, No. 4*, 86–87. [In Russian.]

Baer, J. G, 1952, *Ecology of animal parasites*. Urbana, Illinois: The University of Illinois Press, 224pp.

Baer, J. G, (Ed.) 1957, First symposium on host specificity among parasites of vertebrates. Neuchâtel: Paul Attinger, 324pp.

Baer, J. G. and Euzet, L, 1956, *Prosobothrium armigerum* Cohn 1902 (Cestoda). Historique, synonymie, déscription et position systèmatique. *Recl. des Trav. Lab. Bot. Géol. Zool. Univ. Montpellier,* Sér. Zool., pp. 44–55.

Baer, J. G. and Euzet, L, 1961, Classe des monogènes. In: *Traité de Zoologie* (Ed. P. Grassé). Paris: Masson et Cie, 4, pp. 243–325.

Baer, J. G. and Joyeux, C, 1961, Plathelminthes, mésozoaires, acanthocéphales, nemertiens. In: *Traité de Zoologie* (Ed. Grassé, P.), Paris: Masson et Cie, 4, pp. 561–692.

Baer, J. G., Miranda, H. C., Fernandez, R. and Medina, J. T, 1967, Human diphyllobothriasis in Peru. *Z. ParasitKde*, 28, 277–289.

Bagrov, A. A, 1982, [Morphological variability of nematode larvae of the genus *Anisakis* (Nematoda, Anisakidae)]. *Parazitologiya*, 16, 469–475. [In Russian.]

Baidel'dinova, T. U, 1983, [The parasite fauna of commercial fish in the Bukhtarminsk reservoir]. In: *Biologicheskie osnovy rybnogo khozyaistva Zapadnoi Sibiri (Materialy XVIII ob 'edinennogo plenuma Zapadno-Sibirskogo otdeleniya Ikhtiologicheskoi Komissii Ministerstva Rybnogo Khozyaistva SSSR i Biologicheskogo Instituta SO AN SSR)* [Eds Ioganzen, B. G. and Krivoshchekov, ?.], pp. 118–120. [In Russian.]

Bailey, H. H. and Rock, C. O, 1975, The lipid composition of *Aspidogaster conchicola* Von Baer, 1826. *Proc. Okla. Acad. Sci.*, 55, 97–100.

Bailey, R. E. and Margolis, L, 1987, Comparison of parasite fauna of juvenile sockeye salmon (*Oncorhynchus nerka*) from southern British Columbia and Washington State Lakes. *Can. J. Zool.*, 65, 420–431.

Bakke, T. A, 1980, A scanning electron microscope study of the microtopography of *Aphalloides timmi* Reimer, 1970 (Digenea: Cryptogonimidae). *Fauna Norvegica, A* 1, 38–44.

Bakke, T. A, 1984, A redescription of adult *Phyllodistomum umblae* (Fabricius) (Digenea, Gorgoderidae) from *Salvelinus alpinus* (L.) in Norway. *Zool. Scripta*, 13, 89–99.

Bakke, T. A, 1985, *Phyllodistomum conostomum* (Olsson, 1876) (Digenea, Gorgoderidae): a junior subjective synonym for *P. umblae* (Fabricius, 1780). *Zool. Scripta*, 14, 161–168.

Bakke, T. A. and Bailey, R. E, 1987, *Phyllodistomum umblae* (Fabricis) (Digenea, Gorgoderidae) from British Columbia salmonids: a description based on light and scanning electron microscopy. *Can. J. Zool.*, 65, 1703–1712.

Bakke, T. A. and Lien, L, 1978, The tegumental surface of *Phyllodistomum conostomum* (Olsson, 1876) (Digenea), revealed by scanning electron microscopy. *Int. J. Parasit.*, 8, 155–161.

Bakker, K. E. and Diegenbach, P. C, 1973, The ultrastructure of spermatozoa of *Aspidogaster conchicola* Baer, 1826 (Aspidogastridae, Trematoda). *Netherl. J. Zool.*, 23, 345–346.

Balakhnin, I. A, 1979, [Changes in the correlations between the internal organs of carp with bothriocephaliasis.] *Gidrobiol. Zh.*, 15, 45–49. [In Russian.]

Balakhnin, I. A. and Davydov, O. N, 1975 [The effect of the blood serum of some fish species on the motile activity of *Bothriocephalus gowkongensis*.] In: *Problemy parazitologii. Materially VIII nauchnoi konferentsii parazitologov UkSSr. Chast' 1.* Kiev, USSR: Izdatel'stvo 'Naukova Dumka', pp. 53–54. [In Russian.]

Balakhnin, I. A. and Kozinenko, I. I, 1978, [Aspects of the host–parasite relationships in fish helminths.] In: *1st All-Union Meeting of Parasitocoenologists (Poltava, September 1978), Abstracts. Part I. Akad. Nauk Ukrain, SSR.* Kiev: 'Naukova Dumka', pp. 26–27. [In Russian.]

Balakhnin, I. A. and Kozinenko, I. I, 1981, [Immunological reaction in carp with bothriocephaliasis and after antigenic stimulation.] *Rybnoe Khozyaistvo, Kiev, No. 33, Zool. Inst., Acad. of Sci. of the Ukrainian SSR, Kiev, USSR.* 52–56. [In Russian.]

Balakhnin, I. A., Kozinenko, I. I. and Kurovskaya, I. Ya, 1982 [The influence of phenasal on the immune response in carp.] *Veterinariya, Mosk., No. 8*, 28–29. [In Russian.]

Baldo, B. A. and Fletcher, T. C, 1973, C-reactive protein-like precipitins in plaice. A C-reactive protein-like serum component was found in *Pleuronectes platessa* in agar gel diffusion tests against aqueous extracts of bacteria, fungi, dermatophytes and *Ascaris lumbricoides*. *Nature, Lond.*, 246, 145–146.

Baldwin, N. L., Millemann, R. E. and Knapp, S. E, 1967, 'Salmon poisoning' disease. III. Effect of experimental *Nanophyetus salmincola* infection on the fish host. *J. Parasit.*, 53, 556–564.

Ball, I. R. and Khan, R. A, 1976, On *Micropharynx parasitica* Jägerskiöld, a marine planarian ectoparasitic on thorny skate *Raja radiata* Donovan, from the North Atlantic Ocean. *J. Fish Biol.*, 8, 419–426.

Bamber, R. N. and Henderson, P. A, 1985, Diplostomiasis in sand smelt, *Atherina presbyter* Cuvier, from the Fleet, Dorset and its use as a population indicator. *J. Fish Biol.*, 26, 223–229.

Bamber, R. N., Glover, R., Henderson, P. A. and Turnpenny, A. W. H, 1983, Diplostomiasis in the sand smelt, *Atherina presbyter* (Cuvier), population at Fawley Power Station. *J. Fish Biol.*, 23, 201–210.

Bamberger, J. W. and Martin, W. E, 1951, Effect of size of infective dose, partial ileectomy, and time on intensity of experimental infections of *Euhaplorchis californiensis* Martin, 1950 (Trematoda) in the cat. *J. Parasit.*, 37, 387–391.

Bandoni, S. M. and Brooks, D. R, 1987a, Revision and phylogenetic analysis of the Amphilinidea Poche, 1922 (Platyhelminthes: Cercomeria: Cercomeromorpha). *Can. J. Zool.*, 65, 1110–1128.

Bandoni, S. M. and Brooks, D. R, 1987b, Revision and phylogenetic analysis of the Gyrocotylidea Poche, 1926 (Platyhelminthes: Cercomeria: Cercomeromorpha). *Can. J. Zool.*, 65, 2369–2389.

Banhawy, M. A., Saoud, M. F. A., Anwar, I. M. and El-Naffar, M. K, 1975, The histopathological effects of the parasitic tapeworm *Wenyonia virilis* on the ileum and liver of the silurid fish *Synodontis schall*. *Ann. Zool.*, 11, 83–101.

Banning, P. van, 1971, Some notes on a successful rearing of the herring-worm *Anisakis marina* L. (Nematoda: Heterocheilidae). *J. Cons. Perm. Int. Explor. Mer*, 34, 84–88.

Baozan, C. and Baowa, W, 1984, [Studies on the pathogenicity, biology and treatment of *Pseudoactylogyrus* for the eels in fish farms]. *Acta Zool. Sin.*, 30, 173–180. [In Chinese.]

Barclay, W. R, 1981, Diphyllobothriasis [Editorial]. *J. Am. Med. Ass.*, 246, 2483.

Bardach, J. E., Ryther, J. H. and McLarney, W. O, 1972, *Aquaculture. The farming and husbandry of freshwater and marine organisms.* New York: Wiley-Interscience, xii + 868 pp.

Barkman, L. L. and James, H. A, 1979, A population study of monogenetic trematodes from the killifish, *Fundulus heteroclitus* (Linnaeus) in Connecticut. *Iowa St. J. Res.*, 54, 77–81.

Baron, P. J. and Appleton, T, 1977, Calcification in an ageing *Ligula intestinalis* (L.) pleroceroid from a bream (*Abramis brama* L.). *Z. ParasitKde*, 53, 239–246.

Barrett, J, 1977, Energy metabolism and infection in helminths. In: *Parasite invasion* (Eds Taylor, A. E. R. and Muller, R.) Oxford, UK: Blackwell Scientific Publications, pp. 121–144.

Barrett, J, 1981, *Biochemistry of parasitic helminths*. London, UK; Macmillan Publishers Ltd., 250pp.

Barrett, J, 1984, The anaerobic end-products of helminths. *Parasitology*, 88, 179–198.

Barrett, J. and Körting, W, 1977, Lipid catabolism in the plerocercoids of *Schistocephalus solidus* (Cestoda: Pseudophyllidea). *Int. J. Parasit.*, 7, 419–422

Barrett, J. and Lloyd, G. M, 1981, A novel phosphagen phosphotransferase in the plerocercoids of *Schistocephalus solidus* (Cestoda: Pseudophyllidea). *Parasitology*, 82, 11–16.

Bartoli, P, 1966, Sur le cycle vital de *Lepocreadium pegorchis* (M. Stossich, 1900) M. Stossich, 1903 (Trematoda, Digenea, Lepocreadiidae). *C. R. Hebd. Séanc. Acad. Sci., Serie D*, 263, 1398–1400.

Bartoli, P, 1967, Étude du cycle évolutif d'un trématode peu connu: *Lepocreadium pegorchis* (M. Stossich (1900) (Trematoda, Digenea). *Annls Parasit. Hum. Comp.*, 42, 605–619.

Bartoli, P, 1983, Populations ou espèces? Récherches sur la signification de la transmission de Trématodes Lepocreadiinae (T. Odhner, 1905) dans deux écosystèmes marins. *Annls Parasit. Hum. Comp.*, 58, 117–139.

Bartoli, P, 1987, Caractères adaptifs originaux des digènes intestinaux de *Sarpa salpa* (Teleostei, Spaxidae) et leur interprétation en termes d'évolution. *Annls Parasit. Hum. Comp.*, 62, 542–576.

Bartoli, P. and Prévot, G, 1978, Le cycle biologique de *Holorchis pycnoporus* M. Stossich, 1901 (Trematoda, Lepocreadiidae). *Z. ParasitKde*, 58, 73–90.

Barus, V. and Libosvársky, J, 1984, On phenetic and cladistic relations among genera of family Capillariidae. *Folia Parasit.*, 31, 227–240.

Barus, V., Sergeeva, T. P., Sonin, M. D. and Ryzhikov, K. M, 1978, *Helminths of fish-eating birds of the Palaearctic Region. I. Nematoda* (Eds R. Rysavy and K. M. Ryzhikov) The Hague, The Netherlands: Dr. W. Junk, b.v., Publishers, 318 pp.

Bashirullah, A. K. M, 1973, Infeccion causada por *Philonema* sp. entre trucha y salmón. [XXIII Conv. An. AsoVAC. Abstract], *Acta Cient. Venez.*, (*Suppl. 1*), 24, 57–58.

Bashirullah, A. K. M. and Adams, J. R, 1983, *Philonema oncorhynchi*: effect of hormones on maturation in anadromous sockeye, *Oncorhynchus nerka*. *Int. J. Parasit.*, 13, 261–265.

Bashirullah, A. K. M. and Ahmed, B, 1976a, Development of *Camallanus adamsi* Bashirullah (Nematoda: Camallanidae) in cyclopoid copepods. *Can. J. Zool.*, 54, 2055–2060.

Bashirullah, A. K. M. and Ahmed, B, 1976b, Larval development of *Spirocamallanus intestinecolas* (Bashirullah, 1973) Bashirullah, 1974 in copepods. *Riv. Parassit.*, 37, 303–311.

Bashirullah, A. K. M. and Rado, N. E, 1987, Co-occurrence of three species of *Choricotyle* (Monogenea: Diclidophoridae) in the grunt *Orthopristis ruber* and their host-specificity. *J. Fish Biol.*, 30, 419–422.

Bates, R. M, 1987, Aspects of the biology of trypanorhynch tapeworms and investigations on their use as biological tags. *Ph.D. thesis, The Open University* 371 pp.

Bates, R. M, 1990, A checklist of the Trypanorhyncha (Platyhelminthes: Cestoda) of the World (1935–1985). *Cardiff, National Museum of Wales Zoological Series No. 1*, 1–218.

Baturo, B, 1977, *Bucephalus polymorphus* Baer, 1827 and *Rhipidocotyle illense* (Zeigler, 1883) (Trematoda, Bucephalidae): morphology and biology of developmental stages. *Acta Parasit. Pol.*, 24, 203–220.

Baturo, B, 1978, Larval bucephalosis in artificially heated lakes of the Konin region, Poland. *Acta Parasit. Pol.*, 25, 307–321.

Baturo, B, 1980, Pathological changes in cyprinid fry infected by *Bucephalus polymorphus* Baer, 1827 and *Rhipidocotyle illensis* (Ziegler, 1883) metacercariae (Trematoda, Bucephalidae). *Acta Parasit. Pol.*, 27, 241–246.

Bauer, O. N, 1957 [Parasite fauna of young *Salmo salar* during their development.] *Tr. Leningr. Obshch. Estest., Otdel. Zool.*, 73, 159–163. [In Russian.]

Bauer, O. N, 1958a, Relationship between host fishes and their parasites. In: *Parasitology of fishes*. (Eds Dogiel, V. A., Petrushevski, G. K. and Polyanski, Y. I. English translation, Z. Kabata.) Edinburgh and London: Oliver and Boyd, pp. 84–103.

Bauer, O. N, 1958b, Parasitic diseases of cultured fishes and methods of their prevention and treatment. In: *Parasitology of fishes* (Eds Dogiel, V. A., Petrushevski, G. K. and Polyanski, Y. I., English translation, Z. Kabata.) Edinburgh and London: Oliver and Boyd, pp. 265–298.

Bauer, O. N, 1958c, Fishes as carriers of human helminthoses. In: *Parasitology of fishes* (Eds Dogiel, V. A., Petrushevski, G. K. and Polyanski, Y. I., English translation, Z. Kabata.) Edinburgh and London: Oliver and Boyd, pp. 320–335.

Bauer, O. N, 1959a [The influence of environmental factors on reproduction of fish parasites.] *Vopr. Ekol. (Izdat. Kiev. Univ.)*, 3, 132–141. [In Russian.]

Bauer, O. N, 1959b, The ecology of parasites of freshwater fish. *Bull. State Sci. Res. Inst. Lake Riv. Fish., Nat. Sci. Found., Wash.*, 49, 3–215.

Bauer, O. N, 1968, Control of carp diseases in the USSR. *FAO Fish. Rep., No. 44*, 344–352.

Bauer, O. N, 1970, Parasites and diseases of USSR coregonids. In: *Biology of coregonid fishes*. (Eds Lindsey, C. C. and Woods, C. S.). Canada: University of Manitoba Press, Winnipeg, pp. 267–278.

Bauer, O. N, 1976, [The study of the parasites and diseases of fish in the water bodies of the Polar Sea Province]. In: *Bolezni i parazity ryb Ledovitomorskoi provintsii (v predelakh SSSR)*. Sverdlovsk, USSR; Sredne-Ural'skoe Knizhnoe Izdatel'stvo, pp. 4–12. [In Russian.]

Bauer, O. N, 1977, [The epizootiological significance of monogeneans]. In: *Issledovaniya monogenei v*

SSSR. (Materialy vsesoyuznogo Simposiuma po monogeneyam, 16–18 Noyabrya, 1976, Leningrad). Leningrad, USSR; Akademiya Nauk SSSR, Zoologicheskii Institut, pp. 131–136. [In Russian.]

Bauer, O. N, 1979, [Parasitic diseases of cultivated fish]. *Parazitologiya*, **13**, 377–385. [In Russian.]

Bauer, O. N, 1982, [Regulation of parasite numbers in freshwater ecosystems]. In: *Gel'minty v presnovodnykh biotsenozakh.* Moscow, USSR: 'Nauka', pp. 4–16. [In Russian.]

Bauer, O. N, 1984a, Soviet investigations on the population biology of fish parasites. *J. Fish Biol.*, 25, 545–550.

Bauer, O. N, 1984b, *{A guide to the freshwater fish parasites of the U.S.S.R.} Volume 2. Metazoan parasites (Part 1).* Leningrad: Soviet Academy of Sciences, 425 pp. [In Russian.]

Bauer, O. N, 1986, [Population ecology of fish parasites. Some results and prospects.] *Trudy Zool. Inst. (Morf. sist. faun. parazit. zhivot.), No. 155*, 4–12. [In Russian.]

Bauer, O. N. [Ed.] 1987, *{A guide to the freshwater fish parasites of the U.S.S.R.} Volume 3. Metazoan parasites (Part 2).* Leningrad: Soviet Academy of Sciences, 583 pp. [In Russian.]

Bauer, O. N, 1991, Spread of parasites and diseases of aquatic organisms by acclimatization: a short review. *J. Fish Biol.*, **89**, 679–686.

Bauer, O. N. and Hoffman, G. L, 1976, Helminth range extension by translocation of fish. In: (Page, L. A. [Editor]) Wildlife diseases. (Proc. 3rd Int. Wildl. Dis. Conf., Munich, 1975). New York: Plenum Press, pp. 163–172.

Bauer, O. N. and Mirzoeva, L. M, 1984, [Parasitic diseases of Salmonidae, their prophylaxis and treatment]. *Rybokhozyaist. Ispol'zov. Vnutr. Vodoemov. Obzorb. Informatsiya, No. 2, 64pp.* [In Russian.]

Bauer, O. N. and Nikolskaya, N. P, 1957, The dynamics of the parasite fauna of *Coregonus lavaretus ladoga* and its epizootiological significance. *Bull. Inst. Freshw. Fish., Leningrad, No. 42*, 224–238

Bauer, O. N. and Solomatova, V. P, 1984, The cestode *Triaenophorus crassus* (Pallas) (Pseudophyllidea: Triaenophoridae) as a pathogen of cage-reared salmonids. *J. Fish. Dis.*, 7, 501–504.

Bauer, O. N. and Solomatova, V. P, 1978, Influence of thermal pollution on fish parasites. In: *4th International Congress of Parasitology 19–26 August, 1978, Warsaw. Short communications, Section H.* Warsaw, Poland: Organizing Committee, 22.

Bauer, O. N. and Stolyarov, V. P, (1958), Formation of the parasite fauna and parasitic diseases of fishes in hydroelectric reservoirs. In: *Parasitology of fishes* (Eds Dogiel, V. A., Petrushevski, G. K. and Polyanski, Y. I., English translation Z. Kabata.) Edinburgh and London: Oliver and Boyd, pp. 247–255.

Bauer, O. N., Egusa, S. and Hoffman, G. L, 1981, Parasitic infections of economic importance in fishes. In: *Review of advances in parasitology. (Proc. 4th Internat. Congr. Parasit. (ICOPA IV), Warsaw, 19–26 Aug. 1978).* (Editor Slusarski, W. J.) Warsaw, Poland: PWN-Polish Scientific Publishers, pp. 425–441.

Bauer, O. N. Musselius, V. A. and Strelkov, Yu. A. 1973, *Diseases of pond fishes.* Eng. Transl.; Jerusalem: Israel Progr. Scient. Transl. Moscow: Izdatel'stvo 'Kolos', 355pp, 220pp.

Bauer, O. N., Musselius, V. A. and Strelkov, Yu. A. 1981, *{Diseases of freshwater fish}.* Moscow, USSR: Legkaya i Pischchevaya Promyshlennost. 319 pp. [In Russian.]

Bauer, O. N., Musselius, V. A., Nikolaeva, V. M. and Strelkov, Yu. A. 1977, *Ichthyopathology.* Moscow, USSR; Pishchevaya Promyshlennost, 431 pp. [In Russian.]

Bauer, O. N. and Stolyarov, V. P, 1958, Formation of the parasite fauna and parasitic diseases of fishes in hydroelectric reservoirs. In: *Parasitology of fishes* (Eds Dogiel, V. A., Petrushevski, G. K. and Polyanski, Y. I. English translation, Z Kabata.) Edinburgh and London: Oliver and Boyd, pp 247–255.

Bauer, O. N., Vladimirov, V. L. and Mindel, 1964, N. V. New knowledge about the biology of Strigeata causing mass diseases of fishes. In: *Parasitic worms and aquatic conditions.* Ergens, R. and Rysavy, B. (Eds) Proc. Symp. Prague, Oct. 29–Nov. 2, 1962. Prague: Czechoslovak Academy of Sciences, pp. 77–82.

Baylis, H. A. and Jones, E. I, 1933, Some records of parasitic worms from marine fishes at Plymouth. *J. Mar. Biol. Ass., U.K.*, **18**, 627–634.

Bazikalova, A, 1932 [Data on the parasitology of Merman fishes]. *Sb. Nauch.—Pronyslov. Rabot na Murmane.* Moscow and Leningrad: Snabtechnizdat, 136–153. [In Russian.]

Bazitov, A. A, 1976 [The status of Caryophyllidea in the Platyhelminthes]. *Zool. Zh.*, 55, 1779–1787. [In Russian.]

Bazitov, A. A, 1978, [Spermatogenesis in the cestode *Bothriocephalus scorpii* (Müller, 1775) (Pseudophyllidea, Cestoda)]. *Biologiya Morya, No. 2*, 87–91. [In Russian.]

Bazitov, A. A, 1979 [Spermatogenesis in *Caryophyllaeus fimbriceps* and *Khawia sinensis* (Caryophyllidea).] *Mater. Nauch. Konf. Vses. Obschch. Gel'm. (Tsestody i tsestodozy)*, 31, 3–9. [In Russian.]

Bazitov, A. A, 1981 [Caryophyllids, their origins and position in the phylum Platyhelminthes]. *Zh. Obshch. Biol.*, 42, 920–927. [In Russian.]

Bazitov, A. A, 1984 [Taxonomic relationships of parasitic Platyhelminthes]. *Zool. Zh.*, 63, 818–827. [In Russian.]

Bazitov, A. A. and Lyapkalo, E. V, 1979 [Spermatogenesis in *Amphilina foliacea* (Amphilinidea).] *Mater. Nauch. Konf. Vses. Obshch. Gel'm. (Tsestody i tsestodozy)*, 31, 9–15. [In Russian.]

Bazitov, A. A. and Lyapkalo, E. V, 1979 [Embryonic development of *Amphilina japonica* (Amphilinidea). 1. Structure of the zygote and cleavage.] *Zool. Zh.*, 59, 645–655. [In Russian.]

Bazitov, A. A, and Lyapkalo, E. V, 1981 [The embryonic development of *Amphilina japonica* (Amphilinoidea. 2. Formation of the decacanth larva.] *Zool. Zh.*, 60, 805–816. [In Russian.]

Bazitov, A. A, and Lyapkalo, E. V, 1984. [Structure of the body wall and parenchyma of Amphilinida.] *Zool. Zh.*, 63, 338–347. [In Russian.]

Bazitov, A. A., Kulakovskaya, O. P. and Shestakova, K. A, 1979, [The spermatogenesis of *Biacetabulum appendiculatum* (Szidat, 1937) Janiszewska, 1950 (Cestoda, Caryophyllidea).] *Vest. Zool., No. 2*, 20–24. [In Russian.]

Bazitov, A. A., Shestakova, K. A., Lyapkalo, E. V, and Moreva, L. S, 1978, [The embryonic development of *Bothriocephalus scorpii* (Cestoda, Pseudophyllidea).] *Zool. Zh.*, 57, 653–657. [In Russian.]

Bazitov, A. A., Kulakovskaya, O. P., Yukhimenko, S. S., Lyapkalo, E. V. and Shestakova, K. A, 1978, [The tegument of cestodes and Amphilinoidea and questions of host–parasite relationships.] In: *I Vsesoyuznyi s''ezd parazitotsenologov (Poltava, Sentyabr' 1978), Tezisy dokladov. Chast' 1.* Kiev, USSR: 'Naukova Dumka', pp. 24–26. [In Russian.]

Beamish, F. W. H, 1978, Swimming capacity. In: *Fish Physiology Vol. VII* [Ed. Hoar, W. S. and Randall, D. J] London: Academic Press pp. 101–187.

Becker, C. D, and Brunson, W. D, 1968, The bass tapeworm: a problem in northwest trout management. *Progve Fish Cult.*, 30, 76–83.

Becker, D. A., Carr, W. D., Clouteman, D, G., Evans, W. A., Heard, R. G. and Holmes, P, 1978. Pre- and post-impoundment ichthyoparasite succession in a new Arkansas reservoir. *Publication, Arkansas Water Resources Research Center, University of Arkansas, No. 54,* iv + 85.

Befus, A. D. and Freeman, R. S, 1973a, *Corallobothrium parafimbriatum* sp. n. and *Corallotaenia minutia* (Fritts, 1959) comb. n. (Cestoda: Proteocephaloidea) from Algonquin Park, Ontario, *Can. J. Zool.*, 51, 243–248.

Befus, A. and Freeman, R. S, 1973b, Life cycles of two corallobothriin cestodes (Proteocephaloidea) from Algonquin Park, Ontario, *Can. J. Zool.*, 51, 249–257.

Begon, M., Harper, J. L. and Townsend, C. R, 1986, *Ecology: Individuals, populations and communities.* Oxford: Blackwell Scientific Publications, xii + 876 pp.

Begoyan, Zh. T, 1977, [Some data on the morphogenesis of *Khawia armeniaca* (Cholodkowsky, 1915) during its development in the final host]. *Biol. Zh. Armenii*, 30, 79–84. [In Russian.]

Beilfuss, E. R, 1954. The life histories of *Phyllodistomum lohrenzi* Loewen, 1935, and *P. caudatum* Steelman, 1938 (Trematoda: Gorgoderinae). *J. Parasit.*, 40(5) (Sect. 2), 44.

Beis, I. and Barrett, J, 1980, Oxidative enzymes in the plerocercoids of *Schistocephalus solidus* (Cestoda: Pseudophyllidea). *Int. J. Parasit.*, 10, 151–153.

Beis, I. and Theophilidis, G, 1982, Phosphofructokinase in the plerocercoids of *Schistocephalus solidus* (Cestoda: Pseudophyllidea). *Int. J. Parasit.*, 12, 389–393.

Bell, D. A. and Beverley-Burton, M, 1980, Prevalence and intensity of *Capillaria catostomi* (Nematoda, Trichuroidea) in white sucker (*Catostomus commersoni*) in southern Lake Huron, Canada. *Env. Biol. Fish.*, 5, 267–271.

Bell, D. A. and Beverley-Burton, M, 1981, The taxonomy of *Capillaria* spp. (Nematoda: Trichuroidea) in North American Freshwater fishes. *Syst. Parasit.*, 2, 157–159.

Bell, E. J. and Hopkins, C. A, 1956, The development of *Diplostomum phoxini* (Strigeida, Trematoda). *Ann. Trop. Med. Parasit.*, 50, 275–282.

Bell, G. R. and Margolis, L. (Co-Chairmen) 1972, Considerations and recommendations for the control of fish diseases in Canada. *Fish. Res. Bd Can. Miscellaneous Special Publication No. 16*, i–iv + 52 pp.

Bell, G. R. and Margolis, L, 1976, The fish health program and the occurrence of fish diseases in the Pacific region of Canada. *Fish Path.*, 10, 115–122.

Belova, S. V, 1977, [Variability of the chitinous structures of *Dactylogyrus hypophthalmichthys* (Monogenoidea) related to host size and to season.] In: *Issledovaniya monogenei v SSSR. (Materially vsesoyuznogo Simposiuma po monogeneyam, 16–18 Noyabrya, 1976, Leningrad).* Leningrad, USSR: Akademiya Nauk SSSR, Zoologicheskii Institut, pp. 41–47. [In Russian.]

Benz, G. W. and Pohley, W. J, 1980, A conspicuous *Philometra* sp. (Nematoda: Philometridae) from the occulo-orbits of centrarchid fishes. *Proc. Helm. Soc. Wash.*, 47, 264–266.

Berczi, I. and Molnár, K, 1965, Parazitaspecifikus ellenanyagok kimutatása halak vérébol agargel-precipitációs próbával. *Mag. Allatorv. Lap.*, 20, 540–542.

Berezantsev, Yu. A. and Operin, E. N, 1976, [The 'telergons' of cestode larvae: inhibitors of chemotaxis of host leucocytes]. *Dokl. Akad. Nauk SSSR, Biol. Nauk.*, 226, 1236–1239. [In Russian.]

Berger, W. A. and Thatcher, U. E, 1983, *Kalipharynx piramboae* gen. et sp. n. (Trematoda: Fellodistomidae) Parasita do peixe pulmonado Amazonico *Lepidosiren paradoxa*. *Acta Amazonica* 13(1), 171–175.

Berland, B, 1961, Nematodes from some Norwegian marine fishes. *Sarsia*, 2, 50pp.

Berland, B, 1970, On the morphology of the head in four species of the Cucullanidae (Nematoda). *Sarsia*, 43, 15–64.

Berland, B, 1977, Fiskeparasitter med naeringsmiddel—hygienisk betydning. In: *Kontroll med fisk of fiskeprodukter*. (Edited by: Aalvik, B.) (Veterinaerhygienisk forening's arlige etterutdanningskurs, Bergen, 15–18 Mars 1977). Norway: Norske Veterinaerforening, pp.122–134.

Berland, B, 1981, Mass occurrence of *Anisakis simplex* larvae in stomach of cod (*Gadus morhua* L.). In: IV *Wissenschaftliche Konferenz zu Fragen der Physiologie, Biologie und Parasitologie von Nutzfischen. 3 September 1980*, pp.125–128.

Berland, B, 1983, Redescription of *Cucullanus elongatus* Smedley, 1933 (Nematoda: Seuratoidea) from the lingcod *Ophiodon elongatus* Girard, 1854 from the Pacific coast of Canada. *Can. J. Zool.*, 61, 385–395,

Berland, B, 1987, Helminth problems in sea-water aquaculture. In: *Parasites and diseases in natural waters and aquaculture in Nordic countries. Proceedings of a Symposium, 2–4 December 1986, Stockholm, Sweden* [Eds Stenmark, A. and Malmberg, G.]. Stockholm, Sweden: Zoo-tax, Naturhistoriska riksmuseet, pp.56–62.

Berland, B. and Egidius, E, 1980. Rundmark i oppdrettsfisk—større problem enn antatt? *Norsk Fiskeoppdrett*, 5, 16.

Berland, B., Bristow, G. A and Grahl-Nielsen, O, 1990, Chemotaxonomy of *Gyrocotyle* (Platyhelminthes: Cercomeria) species, parasites of chimaerid fish (Holocephali), by chemometry of their fatty acids. *Mar. Biol.*, 105, 185–189.

Berrie, A. D, 1960a, The influence of various definitive hosts on the development of *Diplostomum phoxini* (Strigeida, Trematoda). *J. Helminth.*, 34, 205–210.

Berrie, A. D, 1960b, Two diplostomulum larvae (Strigeida, Trematoda) in the eyes of sticklebacks (*Gasterosteus aculeatus* L.). *J. Helminth.*, 34, 211–216.

Bertocchi, D. and Francálanci, G, 1963, Grave infestazione da *Echinorhynchus truttae* Schrank in trote iridee di allevamento (*Salmo gair{d}nerii*). *Veterinaria Ital.*, 14, 475–481.

Bespyatova, L. A. and Rumyantsev, E. A, 1982. [The parasite fauna of *Lucioperca lucioperca* in lake Vedlozero.] In: *Ekologiya paraziticheskikh organizmov v biogeotsenozakh severa*. Petrozavodsk, USSR: Karel'skii Filial Akademii Nauk SSSR, Institut Biologii, pp. 94–98. [In Russian.]

Beumer, J. P., Ashburner, L. D., Burbury, M. E., Jette, E. and Latham, D. J, 1983, *A checklist of the parasites of fishes from Australia and its adjacent Antarctic territories*. Tech. Comm., Commonw. Inst. Parasit., No. 48, Farnham Royal, UK: Commonwealth Agricultural Bureaux, viii + 99pp.

Beveridge, I, 1987, *Echinocephalus overstreeti* Deardorff and Ko, 1983 (Nematoda: Gnathostomatoidea) from elasmobranchs and molluscs in South Australia. *Trans. R. Soc. S. Aust.*, 111, 79–92.

Beveridge, I, 1990a, Revision of the family Gilquinidae Dollfus (Cestoda: Trypanorhyncha) from elasmobranch fishes. *Aust. J. Zool.*, 37, 481–520.

Beveridge, I, 1990b, Taxonomic revision of Australian Eutetrarhynchidae Guiart (Cestoda: Trypanorhyncha). *Invert. Tax.*, 4, 785–845.

Beveridge, I. and Campbell, R. A, 1987, *Trimacracanthus* gen. nov. (Cestoda: Trypanorhyncha: Eutetrarhynchidae), with redescriptions of *T. aetobatidis* (Robinson, 1959) comb. nov. and *T. binuncus* (Linton, 1909) comb. nov. *Trans. R. Soc. S. Aust.*, 111, 163–171.

Beveridge, I. and Campbell, R. A, 1988a, A review of the Tetrarhynchobothriidae Dollfus, 1969 (Cestoda: Trypanorhyncha) with description of two new genera, *Didymorhynchus* and *Zygorhynchus*. *Syst. Parasit.*, 12, 3–29.

Beveridge, I. and Campbell, R. A, 1988b, *Cetorhinicola* n.g., *Shirleyrhynchus* n.g. and *Stragulorhynchus* n.g., three new genera of trypanorhynch cestodes from elasmobranchs in Australian waters. *Syst. Parasit.*, 12, 47–60.

Beveridge, I. and Campbell, R. A, 1989, *Chimaerarhynchus* n.g. and *Patellobothrium* n.g., two new genera of trypanorhynch cestodes with unique poeciloacanthus armatures, and a reorganisation of the poeciloacanthus trypanorhynch families. *Syst. Parasit.*, 14, 209–225.

Beveridge, I. and Sakanari, J. A, 1987, *Lacistorhynchus dollfusi* sp. nov. (Cestoda: Trypanorhyncha) in elasmobranch fishes from Australian and North American coastal waters. *Trans. R. Soc. S. Aust.*, 111, 147–154.

Beveridge, I. and Smith, K, 1985, An ultrastructural study of the cirrus and vagina of *Phyllobothrium vagans* (Cestoda: Tetraphyllidea). *Z. ParasitKde*, 71, 609–616.

Beverley-Burton, M, 1978a. Population genetics of *Anisakis simplex* (Nematoda: Ascaridoidea) in Atlantic salmon (*Salmo salar*) and their use as biological indicators of host stocks. *Environ. Biol. Fish.*, 3, 369–377.

Beverley-Burton, M, 1978b, Metazoan parasites of arctic char (*Salvelinus alpinus* L.) in a high arctic, landlocked lake in Canada. *Can. J. Zool.*, 56, 365–368.

Beverley-Burton, M, 1984, Monogenea and Turbellaria. In: L. Margolis and Z. Kabata (Eds). *Guide to the parasites of fishes of Canada. Part I. Canadian Special Publication of Fisheries and Aquatic Sciences, No. 74*, pp. 5–209.

Beverley-Burton, M. and Early, G, 1982, *Deretrema philippinensis* n.sp. (Digenea: Zoogonidae) from *Anomalops katoptron* (Beryciformes: Anomalopidae) from the Philippines. *Can. J. Zool.*, 60, 2403–2408.

Beverley-Burton, M. and Margolis, L, 1982, *Ophioxenos lampetrae* sp. nov. (Digenea: Paramphistomidae) from ammocoetes of the western brook lamprey (*Lampetra richardsoni*) (Vladykov and Follett) in British Columbia, with comments on lamprey host–parasite relationships. *Can. J. Zool.*, 60, 2514–2520.

Beverley-Burton, M. and Pippy, J. H. C, 1978, Distribution, prevalence and mean numbers of larval *Anisakis simplex* (Nematoda: Ascaridoidea) in Atlantic salmon, *Salmo salar* L. and their use as biological indicators of host stocks. *Environ. Biol. Fish.*, 3, 211–222.

Bhagavathiammai, A. and Ramalingam, K, 1983, Free amino acid pattern in infected and uninfected liver of *Channa punctata* vis-à-vis the metacercaria of *Euclinostomum heterostomum. Riv. Parassit.*, 44, 317–324.

Bhaibulaya, M. and Charoenlarp, P, 1983, Creeping eruption caused by *Gnathostoma spinigerum. SE. Asian J. Trop. Med. Publ. Hlth*, 14, 266–268.

Bibby, M. C, 1972, Population biology of the helminth parasites of *Phoxinus phoxinus* L. the minnow, in a Cardiganshire lake. *J. Fish Biol.*, 4, 359–360.

Bier, J. W., Payne, W. L. and Jackson, G. J, 1974, Seasonal periodicity of a *Philometra* sp. (Nematoda) infection in the rockfish, *Morone saxatilis*. In: *Proc. Third International Congress of Parasitology, Munich 25–31 August, 1974. Vol. 3*, pp. 1631–1632.

Biocca, E. and Khalil, L. F, 1988, Parasites as biological tags. In: *Review of advances in parasitology. (Proceedings of the Fourth International Congress of Parasitology, Warsaw 19–26 Aug. 1978).* [Ed. Slusarski, W. J.). Warsaw, Poland: PWN—Polish Scientific Publishers, pp. 873–879.

Birgi, E. and Euzet, L, 1983, Monogènes parasites des poissons des eaux douces du Caméroun. Présence des genres *Cichlidogyrus* et *Dactylogyrus* chez *Aphyosemion* (Cyprinodontidae). *Bull. Soc. Zool. Fr.*, 108, 101–106.

Biserova, N. M, 1984, [Morphological and functional differentiation of the microtriches of the cestode *Acanthobothrium dujardini* (Tetraphyllidea).] *Biol. Vnutr. Vod. Inf. Byull. No. 62*, 62, 51–55. [In Russian.]

Biserova, N. M. and Smetanin, M. M, 1982, [On the accuracy of the estimation of increases in the body surface of *Acanthobothrium dujardini* (Cestoda: Tetraphyllidea).] *Trudy Inst. Biol. Vnutr. Vod (Ots. progresh metod. gidrobiol. ikhtiol. issled.) No. 49*, 156–161. [In Russian.]

Bishop, C. and Odense, P. H, 1966, Morphology of the digestive tract of cod, *Gadus morrhua. J. Fish. Res. Bd Can.*, 23, 1607–1615.

Bishop, Y. M. N. and Margolis, L, 1955, A statistical examination of *Anisakis larvae* (Nematoda) in herring *Clupea pallasi* off British Columbia coast. *J. Fish. Res. Bd Can.*, 12, 571–592.

Bjørge, A. J, 1979, An isopod as intermediate host of cod-worm. *Fiskeridirektoratets Skrifter Serie Havundersokelser*, 16, 561–565.

Black, G. A, 1981, Metazoan parasites as indicators of movements of anadromous brook charr (*Salvelinus fontinalis*) to sea. *Can. J. Zool.*, 59, 1892–1896.

Black, G. A, 1983a, Taxonomy of a swimbladder nematode, *Cystidicola stigmatura* (Leidy) and evidence of its decline in the Great Lakes. *Can. J. Fish. Aquat. Sci.*, 40, 643–647.

Black, G. A, 1983b, Origin, distribution, and postglacial dispersal of a swimbladder nematode, *Cystidicola stigmatura. Can. J. Fish. Aquat. Sci.*, 40, 1244–1253.

Black, G. A, 1983c, *Cystidicola farionis* (Nematoda) as an indicator of lake trout (*Salvelinus namaycush*) of Bering ancestry. *Can. J. Fish. Aquat. Sci.*, 40, 2034–2040.

Black, G. A, 1984a, Morphometrics of *Cystidicola stigmatura* (Nematoda) in relation to its glacial history. *J. Parasit.*, 70, 967–974.

Black, G. A, 1984b, Swimbladder lesions in lake trout (*Salvelinus namaycush*) associated with mature *Cystidicola stigmatura*. (Nematoda). *J Parasit.*, 70, 441–443.

Black, G. A, 1985, Reproductive output and population biology of *Cystidicola stigmatura* (Leidy) (Nematoda) in Arctic char, *Salvelinus alpinus* (L.) (Salmonidae). *Can. J. Zool.*, 63, 617–622.

Black, G. A. and Fraser, J. M, 1984, Dynamics of prevalence of *Ligula intestinalis* (L.) in *Catostomus commersoni* (Lacépède). *J Fish Biol.*, 25, 139–146.

Black, G. A. and Lankester, M. W, 1980, Migration and development of swim-bladder nematodes, *Cystidicola* spp. (Habronematoidea), in their definitive hosts. *Can. J. Zool.*, 58, 1997–2005.

Black, G. A. and Lankester, M. W, 1981, The transmission, life span and population biology of *Cystidicola cristivomeri* White, 1941 (Nematoda: Habronematoidea) in char, *Salvelinus* spp. *Can. J. Zool.*, 59, 498–509.

Black, G. A. and Lankester, M. W, 1984, Distribution and biology of swimbladder nematodes, *Cystidicola* spp. (Habronematoidea) in charr, *Salvelinus* spp. In: *Biology of the Arctic charr: Proceedings of the International Symposium on Arctic charr*. (Eds Johnson, L. and Burns, B.). Manitoba, Canada: University of Manitoba Press, pp. 413–429.

Black, G. A., Montgomery, W. L. and Whoriskey, F. G, 1983, Abundance and distribution of *Salmincola edwardsii*, (Copepoda) on anadromous brook trout, *Salvelinus fontinalis* (Mitchill) in the Moisie River System, Quebec. *J. Fish Biol.*, 22, 567–575.

Blazer, V. S. and Gratzek, J. B, 1985, Cartilage proliferation in response to metacercarial infections of fish gills. *J. Comp. Path.*, 95, 273–280.

Blouin, E. F., Johnson, A. D., Dunlap, D. G. and Spiegel, D. K, 1984, Prevalence of black spot (*Neascus pyriformis: Trematoda: Diplostomatidae) of fishes in Brule Creek, South Dakota. Proc. Helminth. Soc. Wash.*, 51, 357–359.

Bocharova, T. A. and Nud'ga, I, 1983, [Fish parasites in catchment lakes of the Tomi basin.] In: *Biologicheskie osnovy rybnogo khozyaistva Zapadnoi Sibiri (Materialy XVIII ob 'edinennogo plenuma Zapadno-Sibirskogo otdeleniya Ikhtiologicheskoi Komissii Ministerstva Rybnogo Khozyaistva SSR i Biologicheskogo Instituta SO AN SSSR* [Eds Ioganzen, B. G. and Krivoschchekov, G. M.], pp. 65–68 [In Russian.]

Bocharova, T. A., Golovko, G. I. and Nikulina, V. N, 1983, [Changes in the parasite fauna of fish in the River Tom' from 1946 to 1978.] *Problemy Ekologii*. Tomsk, USSR: Tomskii Universitet, 5, 118–124. [In Russian.]

Boczon, K. and Bier, J. W, 1986, *Anisakis simplex*; uncoupling of oxidative phosphorylation in muscle mitochondria of infected fish. *Expl. Parasit.*, 61, 270–279.

Bogdanova, E. A, 1972, [Changes in the parasitological situation in inland waters in the European part of the USSR.] In: *Problemy parazitologii. Trudy VII Nauchnoi Konferentsii Parazitologov USSR. Part 1. Kiev, USSR*, pp. 87–90. [In Russian.]

Bogdanova, E. A, 1975, [Changes in morphological and biological indices in young whitefish caused by diplostomatid infections and their economic assessment.] *Rybokhozyaistvennoe Izuch. Vnytr. Vod., No. 16*, 52–53. [In Russian.]

Bogdanova, E. A, 1982, [Review of parasitological investigations of farmed whitefish fry.] *Sb. Nauch. Trud. Gosud. Nauchno-Issled. Inst. Ozer. Rech. Ryb. Khozyaist., No. 181*, 92–99. [In Russian.]

Bogdanova, E. A. and Kotova, L. I, 1984, [Features of the helminth fauna of fish in the waters of the Severnaya Dvina basin in relation to changes in the hydrological regime.] *Vodnye Resursy, No. 4*, 151–156. [In Russian.]

Bogolepova, I. I, 1979, [Morphological-functional characteristics of the intestine of some nematodes parasitic in fish.] [Abstract]. In: *VII Vsesoyuznoe Soveshchanie po parazitam i boleznyam ryb, Leningrad, Sentyabr', 1979 g. (Tezisy dokladov)*. Leningrad, USSR; 'Nauka', pp. 14–15. [In Russian.]

Bogoyavlenskii, Yu. K. and Nikitina, R. V, 1979, [Comparative histochemical study of the amounts and distribution of glycogen, RNA and lipids in the tegument of some acanthocephalans.] *Trudy Gel'm. Lab.*, 29, 14–16. [In Russian.]

Bogoyavlenskii, Yu. K., Nikitina, R. V. and Khatkevich, L. M, 1975, [The fine structure of the integument of the acanthocephalan *Metechinorhynchus salmonis* (Muller, 1780).] *Trudy (1) Moskovskogo Meditsinskogo Instituta im. I. M. Sechenova (Aktual'nye voprosy sovremennoi parazitologii*), 84, 31–34. [In Russian.]

Bohl, M, 1975, Die Kiemenwurmkrankheit (Dactylogyrose), die gefürchtetste Parasitose bei der Karpfenbrut. Die Bedeutung der Parasiten für die Produktion vol Süsswasserfischen. *Fisch und Umwelt, No. 1*, 67–80.

Boje, J, 1987, Parasites as natural tags on cod (*Gadus morhua* L.) in Greenland waters. In: *Parasites and diseases in natural waters and aquaculture in Nordic countries. Proceedings of a Symposium, 2–4*

December 1986, Stockholm, Sweden (Editors: Stenmark, A. and Malmberg, G.). Stockholm, Sweden: Zoo-tax, Naturhistoriska riksmuseet, pp. 94–101,

Bone, L. W, 1982a, Pheromone communication in parasitic helminths. In: *Parasites—their world and ours. Proc. 5th Int. Congr. of Parasit., Toronto, Canada, 7–14 August, 1982, under the auspices of the World Federation of Parasitologists.* [Eds Mettrick, D. F. and Desser, S. S.] Amsterdam, Netherlands; Elsevier Biomedical Press, pp. 138–140.

Bone, L. W, 1982b, Reproductive chemical communication of helminths. I. Platyhelminthes. *Int. J. Invert. Repro.*, 5, 261–268.

Bonner, J. T, 1965, *Size and cycle, an essay on the structure of biology.* Princeton, New Jersey: Princeton University Press, 219 pp.

Bonner, J. T, 1974. *On Development.* Cambridge, USA: Harvard University Press, 282 pp.

Bonsdorff, B. von, 1977, *Diphyllobothriasis in man.* London, UK and New York, USA: Academic Press Inc., xii–189 pp.

Bonsdorff, B. von and Bylund, G, 1982, The ecology of *Diphyllobothrium latum. Ecol. Dis.*, 1, 21–26.

Bonsdorff, B. von, Nyberg, W. and Grasbeck, R, 1960, Vitamin B12 deficiency in carriers of the fish tapeworm *Diphyllobothrium latum.* (*Proc. 7th Congr. Europ. Soc. Haemat., London 1959.) Acta Haemat.*, 14, 15–19.

Bonsdorff, C. H. and Telkka, A, 1965, The spermatozoon flagella in *Diphyllobothrium latum* (fish tapeworm). *Z. Zellforsch. Mikrosk. Anat.*, 66, 643–648.

Boomker, J, 1982, parasites of South African freshwater fish. I. Some nematodes of the catfish (*Clarias gariepinus* (Burchell, 1822) from the Hartbeespoort Dam. *Onderstepoort J. Vet. Res.*, 49, 41–51.

Boomker, J., Huchzermeyer, F. W. and Naude, T. W, 1980, Bothriocephalosis in the common carp in the Eastern Transvaal. *J. S. Afr. Vet. Ass.*, 51, 263–264.

Boonyaratpalin, S. and Rogers, W. A, 1984, Control of the bass tapeworm, *Proteocephalus ambloplitis* (Leidy), with mebendazole. *J. Fish Dis.*, 7, 449–456.

Borgström, R, 1970, Studies of the helminth fauna of Norway XVI. *Triaenophorus nodulosus* (Pallas, 1760) (Cestoda) in Bogstad Lake. III. Occurrence in pike, *Esox lucius* L. *Nytt Mag. Zool.*, 18, 209–216.

Borgström, R. and Halvorsen, O, 1968, Studies of the helminth fauna of Norway. XI. *Caryophyllaeides fennica* (Schneider) (Cestoda: Caryophyllidea) in Lake Bogstad. *Nytt Mag. Zool.*, 16, 20–23.

Borisova, M. N, 1980, [Anthelmintic treatment of *Philometroides* infection in carp.] *Byull. Vses. Inst. Edsper. Vet., No. 38*, 72–73. [In Russian.]

Borisova, M. N, 1981, [Experimental *Philometroides* infections and superinfections in carp.] *Byull. Vses. Inst. Gelm. K. I. Skryabina, No.* 30, 10–12. [In Russian.]

Borisova, M. N., Kozachenko. N. G. and Vasil'kov, G. V, 1981a, [A new anthelmintic against *Philometroides* infection in carp.] *Byull. Vses. Inst. Eksper. Vet., No. 41*, 21–24. [In Russian.]

Borisova, M. N., Kozachenko, N. G. and Vasil'kov, G. V, 1981b, [Nematicidal preparations tested against *Philometroides* infections in carp.] *Byull. Vses. Inst. Gel'm. K. I. Skryabina, No.* 29, 7–11. [In Russian.]

Bortz, B. M., Kenny, G. E., Pauley, G. B. and Bunt-Milam, A. H, 1988, Prevalence of two site-specific populations of *Diplostomum* spp. in eye infections of rainbow trout, *Salmo gairdneri* Richardson, from lakes in Washington State, USA. *J. Fish Biol.*, 33, 31–43.

Bortz, B. M., Kenny, G. E., Pauley, G. B., Garcia-Ortigoza, E. and Anderson, D. P, 1984, The immune response in immunized and naturally infected rainbow trout (*Salmo gairdneri*) to *Diplostomum spathaceum* as detected by enzyme-linked immunosorbent assay (ELISA). *Dev. Comp. Immunol.*, 8, 813–822.

Bose, K. C. and Sinha, A. K, 1977, Haematological indices in *Clarias batrachus* (L) and *Heteropneustes fossilis* (BL.) carrying helminthic infections. In: *Abstracts of the 1st National Congress of Parasitology, Baroda, 24–26 February, 1977.* India: Indian Society for Parasitology, pp. 1–2.

Bose, K. C. and Sinha, A. K, 1980, Studies on cestode migration and pH changes in *Clarias batrachus* (Linn) infected by *Lytocestus indicus* (Moghe). *Curr. Sci.*, 49, 683.

Bose, K. C. and Sinha, A. K, 1981, Studies on high secretion of neutral polysaccharide in the stomach of *Chana gachua* (Ham.) infected by *Genarchopsis goppo* (Ozaki). *Curr. Sci.*, 50, 873–874.

Botto, C., Osimani, J. and Mane Garzon, F, 1976, Sobre la presencia de larvas de *Anisakis* sp. en peces de la costa Atlántica Uruguaya y su patogenicidad experimental para el perro y el gato. *Revta. Urug. Patol. Clin. Microbiol.*, 14, 49–62.

Bouillon, D. R, 1987, Diplostomiasis (Trematoda: Strigeidae) in Arctic charr (*Salvelinus alpinus*) from Charr Lake, Northern Labrador. *J. Wildlife Dis.*, 23, 502–505.

Bourgeois, C. E. and Ni, I. H, 1984, Metazoan parasites of Northwest Atlantic redfishes (*Sebastes* spp.). *Can. J. Zool.*, 62, 1879–1885.

Bovet, J, 1967, Contribution à la morphologie et à la biologie de *Diplozoon paradoxum* v. Nordmann, 1832. *Bull. Soc. Neuchâtel. Sci. Nat.*, **90**, 63–159.

Bowen, W. D, [Ed.] 1990, Population biology of sealworm (*Pseudoterranova decipiens*) in relation to its intermediate and seal hosts. *Can. Bull. Fish. Aquat. Sci., No. 222*, 306 pp.

Bowser, P. R. and Evans, B. A, 1983, A bibliography of the diseases and parasites of the channel catfish (*Ictalurus punctatus*, Rafinesque) 1947–1980. *Bull. Miss. Agric. For. Expl Stn, No. 924*, 6 pp.

Boxrucker, J. C, 1979, Effects of a thermal effluent on the incidence and abundance of the gill and intestinal metazoan parasites of the black bullhead. *Parasitology*, **78**, 195–206.

Boyce, N. P, 1974, Biology of *Eubothrium salvelini* (Cestoda: Pseudophyllidea) a parasite of juvenile sockeye salmon (*Oncorhynchus nerka*) of Babine Lake, British Columbia. *J. Fish. Res. Bd Can.*, **31**, 1735–1742.

Boyce, N. P, 1979, Effects of *Eubothrium salvelini* (Cestoda: Pseudophyllidea) on the growth and vitality of sockeye salmon, *Oncorhynchus nerka. Can. J. Zool.*, **57**, 597–602.

Boyce, N. P. and Clarke, C. W, 1983, *Eubothrium salvelini* (Cestoda: Pseudophyllidea) impairs seawater adaptation of migrant sockeye salmon yearlings (*Oncorhynchus nerka*) from Babine Lake, British Columbia. *Can. J. Fish. Aquat. Sci.*, **40**, 821–824.

Boyce, N. P. and Yamada, S. B, 1977, Effects of a parasite, *Eubothrium salvelini* (Cestoda: Pseudophyllidea), on the resistance of juvenile sockeye salmon, *Oncorhynchus nerka*, to zinc. *J. Fish. Res. Bd Can.*, **34**, 706–709.

Bradshaw, T. and Mortimer, M, 1986, *Evolution in communities*. In: *Community Ecology: Patterns and Process*. (Eds Kikkawa, J. and Anderson, D. J.) Oxford: Blackwell Scientific Publications, pp. 309–342.

Brand, T. von, 1979, *Biochemistry and physiology of endoparasites*. Amsterdam, The Netherlands: Elsevier/North-Holland Biomedical Press, xv + 477 pp.

Brandt, F. de W., Van As, J. G., Schoonbee, H. J. and Hamilton-Attwell, V. L, 1981, The occurrence and treatment of bothriocephalosis in the common carp, *Cyprinus carpio* in fish ponds with notes on its presence in the largemouth yellowfish *Barbus kimberleyensis* from the Vaal Dam, Transvaal. *Water SA*, 7, 34–42.

Brassard, P., Curtis, M. A. and Rau, M. E, 1982, Seasonality of *Diplostomum spathaceum* (Trematoda: Strigeidae) transmission to brook trout (*Salvelinus fontinalis*) in northern Quebec, Canada. *Can. J. Zool*, **60**, 2258–2263.

Brassard, P., Rau, M. E. and Curtis, M. A, 1982a, Infection dynamics of *Diplostomum spathaceum* cercariae and parasite-induced mortality of fish hosts. *Parasitology*, **85**, 489–493.

Brassard, P., Rau, M. E. and Curtis, M. A, 1982b, Parasite-induced susceptibility to predation in diplostomiasis. *Parasitology*, **85**, 495–501.

Bråten, T, 1966, Host specificity in *Schistocephalus solidus. Parasitology*, **56**, 657–664.

Brattey, J, 1983, The effects of larval *Acanthocephalus lucii* on the pigmentation, reproduction and susceptibility to predation of the isopod *Asellus aquaticus. J. Parasit.*, **69**, 1172–1173.

Braun, M, 1879–93, Platyhelminthes. I. Trematodes. In: H. G. Bronn, *Klassen und Ordnungen des Thier-Reichs*, Leipzig: Akademische Verlagsgesellschaft MBH, 4, pp. 303–925.

Braun, M, 1894–1990, Platyhelminthes. II. Cestodes. In: H. G. Bronn, *Klassen und Ordnungen des Thier-Reichs*, Leipzig: Akademische Verlagsgesellschaft MBH, 4, pp. 926–1731.

Braun, F, 1975, Die Bedeutung der Zwischenwirte und Überträger für Fischparasiten. [Die Bedeutung der Parasiten für die Produktion von Süsswasserfischen.] *Fisch und Umwelt*, 1, 147–150.

Bravo-Hollis, M, 1984, Monogenea (Van Beneden, 1858) Carus, 1863 de peces de litoral Mexicano de Golfo de México y del mar caribe. X. Nueva localidades de colecta de seis especies conocidas. *An. Inst. Biol. Univ. Nac. Autón México. Zoologia*, **55**, 61–71.

Bravo-Hollis, M. and Deloya, J. C, 1973, Catalogo de la Coleccion helmintologica del Instituto de Biologia Universidad, Nacional Autonoma de Mexico. *Publ. Esp. 2*, 138pp.

Bravo-Hollis, M. and Manter, H. W, 1957, Trematodes of marine fishes of Mexican waters. X. Thirteen Digenea, including nine new species and two new genera from the Pacific coast. *Proc. Helminth. Soc. Wash.*, 24, 35–48.

Bray, R. A, 1979, Digenea in marine fishes from the eastern seaboard of Canada. *J. Nat. Hist.*, 13, 399–431.

Bray, R. A, 1982, Two new species of *Bacciger* Nicoll, 1914 (Digenea: Fellodistomidae) from mullet in Australia. *J. Nat. Hist.*, 16, 23–29.

Bray, R. A, 1983, On the fellodistomid genus *Proctoeces* Odhner, 1911 (Digenea), with brief comments on two other fellodistomid genera. *J. Nat. Hist.*, 17, 321–339.

Bray, R. A, 1984, Some helminth parasites of marine fishes and cephalopods of South Africa: Aspidogastrea and the digenean families Bucephalidae, Haplosplanchnidae, Mesometridae and Fellodistomidae. *J. Nat. Hist.*, 18, 271–292.

Bray, R. A, 1985a, *Macvicaria taksengi* n.sp. (Digenea: Opecoelidae) in marine teleosts fron Pinang, Malaysia. *Syst. Parasit.*, 7, 75–80.

Bray, R. A, 1985b, Some helminth parasites of marine fishes of South Africa: families Gorgoderidae, Zoogonidae, Cephaloporidae, Acanthocolpidae and Lepocreadiidae (Digenea). *J. Nat. Hist.*, 19, 377–405.

Bray, R. A, 1986a, Some helminth parasites of marine fishes of South Africa: families Enenteridae, Opistholebetidae and Pleorchiidae (Digenea). *J. Nat. Hist.*, 20, 471–488.

Bray, R. A, 1986b, Patterns in the evolution of marine helminths. In: (Howell, M. J., Ed.), *Parasitology—Quo Vadit? Proc. ICOPA VI, Brisbane, Australia 24–29 Aug., 1986*. Canberra, Australia: Australian Academy of Science, pp. 337–344.

Bray, R. A, 1987a, Some helminth parasites of marine fishes of South Africa: family Opecoelidae (Digenea). *J. Nat. Hist.*, 21, 1049–1075.

Bray, R. A, 1987b, A study of the helminth parasites of *Anarhichas lupus* (Perciformes: Anarhichadidae) in the North Atlantic. *J. Fish Biol.*, 31, 237–264.

Bray, R. A, 1988, A discussion of the status of the subfamily Baccegerinae Yamaguti, 1958 (Digenea) and the constitution of the family Fellodistomidae Nicoll, 1909. *Syst. Parasit.*, 11, 97–112.

Bray, R. A. and Gibson, D. I, 1977, The Accacoeliidae (Digenea) of fishes from the north-east Atlantic. *Bull. Br. Mus. Nat. Hist.*, 31, 53–99.

Bray, R. A. and Gibson, D. I, 1980, The Fellodistomidae (Digenea) of fishes from the north-east Atlantic. *Bull. Br. Mus. Nat. Hist.*, 37, 199–293.

Bray, R. A. and Gibson, D. I, 1986, The Zoogonidae (Digenea) of fishes from the north-east Atlantic. *Bull. Br. Mus. Nat. Hist.*, 51, 127–206.

Bray, R. A. and Rollinson, D, 1985, Enzyme electrophoresis as an aid to distinguishing species of *Fellodistomum, Steringotreme* and *Steringophorus* (Digenea: Fellodistomidae). *Int. J. Parasit.*, 15, 255–263.

Brett, J. R, 1979, Environmental factors and growth. In: *Fish Physiology* (Eds Hoar, W. S., Randall, D. J. and Brett, J. R.) Volume VIII Bioenergetics and Growth. New York, London, San Francisco: Academic Press, pp. 559–667.

Brglez, J., Rakovec, R. and Snoj, N, 1966, The digenetic trematode *Azygia lucii* (Müller, 1776) Lühe, 1909 in the intestine of salmon (*Hucho hucho* L.) from the Savinja and Ljubljanica rivers. *Biol. Vest.*, 14, 105–107.

Brill, R. W., Bourne, R., Brock, J. A. and Dailey, M. D, 1987, Prevalence and effects of infection of the dorsal aorta in yellowfin tuna, *Thunnus albacares*, by the larval cestode, *Dasyrhynchus talismani*. *Fishery Bulletin*, 85, 767–776.

Brinkman, A. Jr, 1952, Fish trematodes from Norwegian waters. 1. The history of fish trematode investigations in Norway and the Norwegian species of the order Monogenea. *Universitetet i Bergen Arbok 1952. Naturvitenskapelig rekke, No. 1*, 134 pp.

Bristol, J. R., Mayberry, L. F., Huber, D. and Ehrlich, I, 1984, Endoparasite fauna of trout in the Plitvice Lakes National Park. *Vet. Arch.*, 54, 5–11.

Bristow, G. A. and Berland, B, 1988, A preliminary electrophoretic investigation of the gyrocotylid parasites of *Chimaera monstrosa* L. *Sarsia*, 73, 75–77.

Britz, J., As, J. G. van and Saayman, J. E, 1984, Notes on the morphology of the metacercaria and adult of *Clinostomum tilapiae* Ukoli, 1966 (Trematoda: Clinostomatidae). *S. Afr. J. Wildl. Res.*, 14, 69–72.

Britz, J., As, J. G. van and Saayman, J. E, 1985, Occurrence and distribution of *Clinostomum tilapiae* Ukoli, 1966 and *Euclinostomum heterostomum* (Rudolphi 1809) metacercarial infections of freshwater fish in Venda and Lebowa, southern Africa. *J. Fish Biol.*, 26, 21–28.

Britz, J., Saayman, J. E. and As, J. G. van 1984, Anatomy of the metacercaria and adult of *Euclinostomum heterostomum* (Rudolphi, 1809) (Trematoda: Clinostomatidae). *S. Afr. J. Zool.*, 19, 91–96.

Broek, E. van den and Jong, N, de 1979, Studies on the life cycle of *Asymphylodora tincae* (Modeer, 1790) (Trematoda: Monorchiidae) in a small lake near Amsterdam. Part I. The morphology of various stages. *J. Helminth.*, 43, 79–89.

Bromage, N. R. and Whitehead, C, 1980, Control of reproduction in rainbow trout. *NERC Newsjournal*, 2, 10–11.

Brooks, D. R, 1977, A new genus and two new species of trematodes from characid fishes in Colombia. *Trans. Am. Microsc. Soc.*, 96, 267–270.

Brooks, D. R, 1978, Evolutionary history of the cestode order Proteocephalidea. *Syst. Zool.*, 27, 312–323.

Brooks, D. R, 1979a, Testing hypotheses of evolutionary relationships among parasites: The digeneans of crocodilians. *Amer. Zool*, 19, 1225–1238.

Brooks, D. R, 1979b, Testing the context and extent of host–parasite coevolution. *Syst. Zool.*, **28**, 299–307.

Brooks, D. R, 1980, Allopatric speciation and non-interactive parasite community structure. *Syst. Zool.*, 29, 192–203.

Brooks, D. R, 1981, Hennig's parasitological method: a proposed solution. *Syst. Zool.*, **30**, 229–249.

Brooks, D. R, 1982, Higher level classification of parasitic platyhelminthes and fundamentals of cestode classification. In: *Parasites—their world and ours. Proc. 5th Int. Congr. Parasit., Toronto, Canada, 7–14 August 1982, under the auspices of the World Federation of Parasitologists.* (Eds Mettrick, D. F., and Desser, S. S.). Amsterdam, Netherlands: Elsevier Biomedical Press, pp. 189–193.

Brooks, D. R, 1985, Phylogenetics and the future of helminth systematics. *J. Parasit.*, 71, 719–727.

Brooks, D. R, 1986, Analysis of host-parasite co-evolution. In: *Parasitology Quo Vadit? {Editor: Howell, M. J.} Proceedings of the VI International Congress of Parasitology.* Canberra, Australia: Australian Academy of Sciences, Canberra, pp. 291–301.

Brooks, D. R, 1989, A summary of the database pertaining to the phylogeny of the major groups of parasitic platyhelminths, with a revised classification. *Can. J. Zool.*, 67, 714–720.

Brooks, D. R. and Deardorff, T. L, 1980, Three proteocephalid cestodes from Colombian siluriform fishes, including *Nomimoscolex alovaris* sp.n. (Monticelliidae: Zygobothriinae). *Proc. Helm. Soc. Wash.*, 47, 15–21.

Brooks, D. R. and Deardorff, T. L, 1988, *Rhinebothrium devaneyi* n. sp. (Eucestoda: Tetraphyllidea) and *Echinocephalus overstreeti* (Deardorff and Ko, 1983, (Nematoda: Gnathostomatidae) in a thorny back ray, *Urogymnus asperrinus*, from Enewetok Atoll, with phylogenetic analysis of both species groups. *J. Parasitol.*, 74, 459–465.

Brooks, D. R. and Glen, D. R, 1982, Pinworms and primates: a case study in coevolution. *Proc. Helm. Soc. Wash.*, 49, 76–85.

Brooks, D. R. and Mattis, T. E, 1978, Redescription of *Nagmia floridensis* Markell, 1953 with discussion of the composition of the Anaporrhutinae Looss, 1901 (Digenea: Gorgoderidae). *Proc. Helm. Soc. Wash.*, 45, 169–171.

Brooks, D. R. and Mayes, M. A, 1975, *Phyllodistomum scrippsi* sp.n. (Digenea: Gorgoderidae) and *Neobenedenia girellae* (Hargis 1955) Yamaguti, 1963 (Monogenea: Capsalidae) from the California sheephead, *Pimelometopon pulchrum* (Ayres) (Pisces: Labridae). *J. Parasit.*, 61, 407–408.

Brooks, D. R. and Mayes, M. A, 1980, Cestodes in four species of euryhaline stingrays from Colombia. *Proc. Helm. Soc. Wash.*, 47, 22–29.

Brooks, D. R., O'Grady, R. T. and Glen, D. R, 1985a, The phylogeny of the Cercomeria Brooks, 1982 (Platyhelminthes). *Proc. Helm. Soc. Wash.*, 52, 1–20.

Brooks, D. R., O'Grady, R. T. and Glen, D. R, 1985b, Phylogenetic analysis of the Digenea (Platyhelminthes: Cercomeria) with comments on their adaptive radiation. *Can. J. Zool.*, 63, 411–443.

Brooks, D. R., Thorson, T. B. and Mayes, M. A, 1981, Freshwater stingrays (Pomatotrygonidae) and their helminth parasites: testing hypotheses of evolution and coevolution. In: *Advances in Cladistics, Vol. 1.* (Eds Funk, V. A. and Brooks, D. R.) New York: New York Botanical Gardens, pp. 147–75.

Brooks, D. R., Bandoni, S. M., Macdonald, C. A. and O'Grady, R. T, 1989, Aspects of the phylogeny of the Trematoda Rudolphi, 1808 (Platyhelminthes: Cercomeria). *Can. J. Zool.*, 67, 2609–2634.

Brooks, G. L, 1934, Some new ectoparasitic trematodes (Onchocotylinae) from the gills of American sharks. *Parasitology*, 26, 259–267.

Brown, A. F, 1986, Evidence for density-dependent establishment and survival of *Pomphorhynchus laevis* (Müller, 1776) (Acanthocephala) in laboratory-infected *Salmo gairdneri* Richardson and its bearing on wild populations in *Leuciscus cephalus* (L.) *J. Fish Biol.*, 28, 659–669.

Brown, A. F. and Thompson, D. B. A, 1986, Parasite manipulation of host behaviour: acanthocephalans and shrimps in the laboratory. *J. Biol. Education*, 20, 121–127.

Brown, A. F., Chubb, J. C. and Veltkamp, C. J, 1986, A key to the species of Acanthocephala parasitic in British freshwater fishes. *J. Fish Biol.*, **28**, 327–334.

Brown, E. E, and Gratzek, J. B, 1980, *Fish farming handbook. Food, bait, tropicals and goldfish.* Westport, Connecticut, USA: AVI Publishing Company, Inc., xvi + 391 pp.

Brown, M. E, (Editor) 1957, *The Physiology of Fishes.* New York: Academic Press, Vol. I, i–xiii + 447 pp; Vol II. i–xi + 525 pp.

Brun, N. Le., Lambert, A. and Justine, J. L, 1986, Oncomiracidium, morphogenèse du hapteur et ultrastructure du spermatozoide de *Pseudodactylogyrus anguillae* (Yin et Sproston, 1948) Gussev, 1965 (Monogenea, Monopisthocotylea. Pseudodactylogyridae n. fam.). *Annls Parasit. Hum. Comp.*, 61, 273–284.

Bryant, C, 1981, Biochemical adaptation in helminth and protozoan parasites. *J. Biol. Educ.*, 15, 254–258.

Bryant, C, 1982a, Biochemistry. In: *Modern parasitology. A textbook of parasitology.* (Ed. Cox, F. E. G.) Oxford, UK: Blackwell Scientific Publications, pp. 84–115.

Bryant, C, 1982b, The biochemical origins of helmith parasitism. In: *Biology and control of endoparasites. McMaster Laboratory 50th Anniversary Symposium in Parasitology, University of Sydney, 506 November 1981.* (Eds Symons, L. E. A., Donald, A. D. and Dineen, J. K.). North Ryde, Sydney, NSW, Australia: Academic Press, Australia, pp. 29–52.

Buchmann, K, 1986, Prevalence and intensity of infection of *Cryptocotyle lingua* (Creplin) and *Diplostomum spathaceum* (Rudolphi)—parasitic metacercariae of Baltic cod (*Gadus morhua* L.). *Nord. Vet.*, **38**, 303–307.

Buchmann, K, 1987, The effects of praziquantel on the monogenean gill parasite *Pseudodactylogyrus bini*. *Acta Vet. Scand.*, **28**, 447–450.

Buchmann, K, 1988a, Spatial distribution of *Pseudodactylogyrus anguillae and P. bini* (Monogenea) on the gills of European eel, *Anguilla anguilla*. *J. Fish Biol.*, 32, 801–802.

Buchmann, K, 1988b, Feeding of *Pseudodactylogyrus bini* (Monogenea) from *Anguilla anguilla*. *Bull. Eur. Ass. Fish Path.*, **8**, 79–81.

Buchmann, K, 1988c, Epidemiology of pseudodactylogyrosis in an intensive eel-culture system. *Dis. Aquat. Org.*, 5, 81–85.

Buchmann, K, 1988d, Temperature-dependent reproduction and survival of *Pseudodactylogyrus bini* (Monogenea) on the European eel, *Anguilla anguilla*. *Parasit. Res.*, 75, 162–164.

Buchmann, K, 1988e, Interactions between the gill-parasitic monogeneans *Pseudodactylogyrus anguillae* and *P. bini* and the fish host *Anguilla anguilla*. *Bull. Eur. Ass. Fish. Pathol.*, **8**, 98–100.

Buchmann, K. and Mellergaard, S, 1988, Histochemical demonstration of the inhibitory effect of Nuvan and Neguvon on cholinesterase activity in *Pseudodactylogyrus anguillae* (Monogenea). *Acta Vet. Scand.*, **29**, 51–55.

Buchmann, K., Køie, M. and Prento, P, 1987, The nutrition of the gill parasite monogenean *Pseudodactylogyrus anguillae*. *Parasit. Res.*, 73, 532–537.

Buchmann, K., Mellergaard, D. and Køie, M, 1987, *Pseudodactylogyrus* infections in eel: a review. *Dis. Aquat. Org.*, **3**, 51–57.

Buckner, R. L., Overstreet, R. M. and Heard, R. W, 1978, Intermediate hosts for *Tegorhynchus furcatus* and *Dollfusentis chandleri* (Acanthocephala). *Proc. Helm. Soc. Wash.*, **45**, 195–201.

Bugaev, V. F, 1982, [Prevalence of *Diphyllobothrium* sp. plerocercoids in *Oncorhynchus nerka* in the Kamchatka river basin.] *Vopr. Ikhtiol.*, 22, 489–497. [In Russian.]

Bullock, W. L, 1961, A preliminary study of the histopathology of Acanthocephala in the vertebrate intestine. *J. Parasit.*, 47(4)(**Sect.** 2), 31.

Bullock, W. L, 1963, Intestinal histology of some salmonid fishes with particular reference to the histopathology of acanthocephalan infections. *J. Morph.*, **112**, 23–44.

Bullock, W. L, 1970, The zoogeography and host relations of the eoacanthocephalan parasites of fishes. In: *A symposium on diseases of fishes and shell-fishes.* (Ed. Sniesko, S. F.), Spec. Publ. Amer. Fish. Soc., No. 5, pp. 161–173.

Bundy, D. A. P, 1981a Survival characteristics of the free-living miracidial population of *Transversotrema patialense* (Soparkar) (Digenea: Transversotrematidae). *J. Parasit.*, 67, 531–534.

Bundy, D. A. P, 1981b The surface ultrastructure of the egg capsule of *Transversotrema patialense* (Transversotrematidae: Digenea). *Int. J. Parasit.*, 11, 19–22.

Bundy, D. A. P, 1981c Swimming behaviour of the cercaria of *Transversotrema patialense*. *Parasitology*, **82**, 319–334.

Bunnag, T., Freitas, J. R. de and Scott, H. G, 1977, The predatory activity of *Lebistes reticulatus* (Peters, 1859) on *Schistosoma mansoni* miracidia in laboratory experiments. *Trop. Geogr. Med.*, 29, 411–414.

Burenina, E. A, 1977, [The carbohydrate metabolism of *Bothriocephalus scorpii*.] *Trudy Biol.Poch. Inst. (Parazit. svobodnozh. chervi fauny Dal'nego (Vostoka), Novaya Seriya*, 47(**150**), 106–108. [In Russian.]

Burn, P. R, 1980a, Density dependent regulation of a fish trematode population. *J. Parasit.*, **66**, 173–174.

Burn, P. R, 1980b, The parasites of smooth flounder, *Liopsetta putnami* (Gill), from the Great Bay Estuary, New Hampshire. *J. Parasit.*, **66**, 532–541.

Buron, I. and Golvan, Y. J, 1986, Les Hôtes des Acanthocéphales I—Les Hôtes intermédiaires. *Ann. Parasitol. Hum. Comp.*, 61, 581–592.

Buron, I. de, Renaud, F. and Euzet, L, 1986, Speciation and specificity of acanthocephalans. Genetic and morphological studies of *Acanthocephaloides geneticus* sp. nov. parasitizing *Arnoglossus laterna* (Bothidae) from the Mediterranean littoral (Sète–France). *Parasitology*, 92, 165–171.

Burreson, E. M. and Olson, R. E, 1974, Seasonal variations in the populations of two hemiurid trematodes from the pacific staghorn sculpin, *Leptocottus armatus* Girard in an Oregon estuary. *J. Parasit.*, 60, 764–767.

Burrough, R. J, 1978, The population biology of two species of eyefluke, *Diplostomum spathaceum* and *Tylodelphys clavata* in roach and rudd. *J. Fish Biol.*, 13, 19–32.

Burt, M. D. B, 1981, Biology of Platyhelminthes and Nemathelminthes. II. Cestoda. In: *Review of advances in parasitology. Proceedings of the Fourth International Congress of Parasitology, Warsaw, 19–26 Aug. 1978.* (Ed. Slusarski, W.) Warsaw, Poland: PWN Polish Scientific Publishers, pp. 37–41.

Burt, M. D. B, 1986, Early morphogenesis in the Platyhelminthes with special reference to egg development and development of cestode larvae. In: Parasitology–Quo Vadit? Proceedings of the Sixth International Congress of Parasitology 24–29 Aug. 1986. [Editor: M. J. Howell], Canberra, Australia: Australian Academy of Science, pp. 241–253.

Burt, M. D. B. and Jarecka, L, 1982, Phylogenetic host specificity of cestodes. [Deux. Symp. spec. parasit. parasites des vertébrés, 13–17 Avril, 1981. Colloque internat. CNRS.] *Mém. Mus. Nat. Hist. Nat., Sér. A, Zool.*, 123, 47–51.

Burt, M. D. B. and Jarecka, L, 1984, Studies on the life-cycle of *Hymenolepis ductilis* Linton, 1927 and on the ultrastructure of its cysticercoid tegument. *Acta Parasit. Pol.* 29, 35–42.

Burt, M. D. B. and Sandeman, I. M, 1969, Biology of *Bothrimonus* (= *Diplocotyle*) (Pseudophyllidea: Cestoda). Part I. History, description, synonymy, and systematics. *J. Fish. Res. Bd Can.*, 26, 975–996.

Burt, M. D. B. and Sandeman, I. M, 1974, The biology of *Bothrimonus* (= *Diplocotyle*) (Pseudophyllidea: Cestoda): detailed morphology and fine structure. *J. Fish. Res. Bd Can.*, 31, 147–153.

Bussmann, B. and Ehrich, S, 1979, Investigations on infestation of blue whiting (*Micromesistius poutassou*) with larval *Anisakis* sp. (Nematoda: Ascaridida). *Arch. Fischereiwiss.*, 29, 155–165.

Buteau, G. H., Simmons, J. E. and Fairbairn, D, 1969, Lipid metabolism in helminth parasites. IX. Fatty acid composition of shark tapeworms and of their hosts. *Exp. Parasit.*, 26, 209–213.

Buteau, G. H., Simmons, J. E., Beach, D. H., Holz, G. G. and Sherman, I. W, 1971, The lipids of cestodes from Pacific and Atlantic coast triakid sharks. *J. Parasit.*, 57, 1272–1278.

Butler, J. A. and Milleman, R. E, 1971, Effect of the 'salmon poisoning' trematode *Nanophyetus salmincola* on swimming ability of juvenile salmonid fishes. *J. Parasit.*, 57, 860–865.

Butorina, T. E, 1975, [Dynamics of the parasite fauna of different forms of *Salvelinus alpinus* in lake Azabach'e basin (USSR).] *Parazitologya*, 9, 237–246. [In Russian.]

Butorina, T. E, 1978, [Parasitological examination of loach (*Salvelinus*) in Lake Azabach' on Kamchatka.] *Nauch. Soobsh. Inst. Biol. Morya (Biolog. Issled. Dal'nevost. Mor.), No. 3*, 12–16. [In Russian.]

Butorina, T. E, 1980, [Ecological analysis of the parasite fauna of char (*Salvelinus*) in the river Kamchatka.] *Sb. Rabot, Inst. Biol. Morya, Dal'nevost. Tsentr Akad. Nauk SSR (Populyat. biol. sistemat. los., No. 18*, 65–81. [In Russian.]

Butorina, T. E. and Kuperman, B. I, 1981, [Ecological analysis of the prevalence of cestode infections in freshwater fish of the Kamchatka.] In: *Biologiya i sistematika gel'mintov zhivotnykh Dal'nego Vostoke.* Vladivostok, USSR: AN SSSR Dal'nevostochnyi Nauchnyi Tsentr, pp. 86–100. [In Russian.]

Buttner, A, 1951a, La progénèse chez les trématodes digénètiques (suite). Technique et recherches personnelles. *Ann. Parasitol. Hum. Comp.*, 26, 19–66.

Buttner, A, 1951b, La progénèse chez les trématodes digénétiques (suite). Recherches personnelles sur deux espèces progénétiques déjà connues: *Ratzia joyeuxi* (E. Brumpt, 1922) et *Pleurogenes medians* (Olsson, 1876). *Ann. Parasitol. Hum. Comp.*, 26, 138–189.

Buttner, A, 1951c, La progénèse chez les trématodes digénétiques (fin). Étude de quelques métacercaires à évolution inconnue et de certaines formes de développement voisines de la progénèse. Conclusions générales. *Ann. Parasitol. Hum. Comp.*, 26, 279–322.

Bykhovskaya-Pavlovskaya, I. E, 1964, To the methods and problems of parasitological investigations of animals bound to aquatic environment. In: *Parasitic worms and aquatic conditions. Proceedings of symposium, Prague, Oct. 29–Nov. 2, 1962.* (Eds Ergens, R. and Rysavy, B.). Prague: Czechoslovak Academy of Sciences, pp. 29–36.

Bykhovskaya-Pavlovskaya, I. E. and Bykhovskii, B. E, 1940, The parasite fauna of the fishes of the Akhtatinskie limans (delta of the Kuban, Sea of Azov). *Mag. Parasit.*, **8**, 131–161.

Bykhovskii, B. E, 1933, Die Bedeutung der monogenetischen Trematoden für die Erforschung der systematischen Beziehung der Karpenfische. *Zool. Anz.*, **102**, 243–251.

Bykhovskii, B. E, 1937, [Ontogenesis and phylogenetic inter-relationships of parasitic flatworms.] *Bull. Acad. Sci. URSS, Sér. Biol.*, 1353–1383. [In Russian.]

Bykhovskii, B. E, 1957, *Monogenetic trematodes, their classification and phylogeny*. Academy of Sciences, USSR, Moscow and Leningrad, [English translation by W. J. Hargis and P. C. Oustinoff (1961). American Institute of Biological Sciences, Washington, xx + 627 pp.], 509 pp. [In Russian.]

Bykhovskii, B. E. and Gusev, A. V, 1964, The study of the fauna of parasites of fishes in the Soviet Union and its theoretical and practical results. *Academy of Sciences of the USSR. Report presented at the 1st International Congress of Parasitology*, 11 pp.

Bykhovskii, B. E. and Mikhailov, T. K, 1975, Zoogeographical analysis of the parasite fauna of fishes in Azerbaijan water bodies. *Folia Parasit.*, **22**, 97–103.

Bylund, G, 1969, Experimental undersokning av *Diphyllobothrium dendriticum* (= *D. norvegicum*) fran norra Finland. *Tiedoksianto - Information*, **10**, 3–17.

Bylund, G, 1972, Pathogenic effects of a diphyllobothriid plerocercoid on its host fishes. *Comment. Biol. Soc. Sci. Fenn.*, **53**, 1–11.

Bylund, G, 1973, Observations on the taxonomic status and the biology of *Diphyllobothrium ditremum* (Creplin, 1825) (= *D. osmeri*) (von Linstow, 1878)). *Acta Acad. Abo., Ser. B.*, **33**, 18 pp.

Bylund, G, 1975a, The taxonomic significance of embryonic hooks in four European *Diphyllobothrium* species. (Cestoda, Diphyllobothriidae). *Acta. Zool. Fenn.*, **142**, 22 pp.

Bylund, G, 1975b, Delimitation and characterization of European *Diphyllobothrium* species. *Academic Dissertation, Abo Akademi*, 23 pp.

Bylund, G, 1975c, Studies on the taxonomic status and biology of *Diphyllobothrium vogeli* Kuhlow, 1953. *Comment. Biol. Soc. Sci. Fenn.*, **79**, 22 pp.

Bylund, G, 1976, Lapamatotartunnan leviaminen ja ennaltaehkaisy. *Duodecim*, **92**, 646–648.

Bylund, G, 1982, Diphyllobothriasis. In: *CRC handbook series in zoonoses. Section C: parasitic zoonoses. Volume I.* (Eds Jacobs, L., Arambulo, P., III) Boca Raton, Florida, USA: CRC Press, Inc., pp. 217–225.

Bylund, G, 1987, Aspects of fish disease problems in Scandinavian countries. In: *Parasites and diseases in natural waters and aquaculture in Nordic countries. Proceedings of a Symposium, 2–4 December 1986, Stockholm, Sweden* (Eds A. Stenmark and G. J. Malmberg). Stockholm, Sweden: Zoo-tax Naturhistoriska riksmuseet, pp. 102–111.

Bylund, G. and Sumari, O, 1981, Laboratory tests with Droncit against diplostomiasis in rainbow trout, *Salmo gairdneri* Richardson. *J. Fish Dis.*, **4**, 259–264.

Bylund, G. and Wikgren, B. J, 1984, The retention and inactivation of tapeworm eggs in modern purification plants. *Abstracts of the Fourth European Multicolloquium of Parasitology, Izmir*, p. 243.

Bylund, G., Fagerholm, H. P., Calenius, G., Wikgren, B. H. and Wistrom, M, 1979, Parasites of fish in Finland. I. Introduction. *Acta Acad. Abo., Ser. B.*, **40**, 14 pp.

Bylund, G., Fagerholm, H. P., Calenius, G., Wikgren, B. J. and Wikstrom, M, 1980, Parasites of fish in Finland. II. Methods for studying parasite fauna in fish. *Acta Acad. Abo., Ser. B.*, **40**, 1–23.

Byrnes, T, 1985, Four species of *Polylabroides* (Monogenea: Polyopisthocotylea: Microcotylidae) on Australian bream, *Acanthopagrus* spp. *Aus. J. Zool.*, **33**, 729–742.

Caballero, C. E, 1952, Revision de los generos y especies que integran la familia Acanthocolpidae Lühe, 1909. *Revta Med. Vet. Parasit.*, **11**, 231 pp.

Caballero, C. E, 1977, [*Parasitological papers published in memory of Doctor Eduardo Caballero y Caballero.*] Mexico: Universadid Nacional Autónoma de México, 553 pp.

Caballero, C. E. and Bravo-Hollis, M, 1952, *Ichthyotrema vogelsangi* n.g. n.sp. (Trematoda: Digenea) en peces marinos de aguas Mexicanas. *An. Inst. Biol.*, **23**, 155–165.

Caballero, C. E. and Bravo-Hollis, M, 1965, Trematoda Rudolphi, 1808 de peces marinos del litoral mexicano del Golfo de Mexico y del Mar Caribe. I. *Revta Biol. Trop.*, **13**, 297–301.

Cable, R. A, 1954a, Studies on marine digenetic trematodes of Puerto Rico. The life-cycle in the family Haplosplanchnidae. *J. Parasit.*, **40**, 71–76.

Cable, R. M, 1954b, Studies on marine digenetic trematodes of Puerto Rico. The life cycle in the family Megaperidae. *J. Parasit.*, **40**, 202–208.

Cable, R. M, 1954c, A new marine cercaria from the Woods Hole region and its bearing on the interpretation of larval types in the Fellodistomatidae (Trematoda: Digenea). *Biol. Bull.*, **106**, 15–20.

Cable, R. M, 1962, A cercaria of the trematode family Haploporidae. *J. Parasit.*, **48**, 419–422.

Cable, R. M, 1965, 'Thereby hangs a tail.' *J. Parasit.*, 51, 3–12.

Cable, R. M, 1972, Behaviour of digenetic trematodes. In: *Behavioural aspects of parasite transmission.* (Eds E. U. Canning and C. A. Wright). *Suppl. No. 1, Zool. J. Linn. Soc. Lond.*, 51, 1–18.

Cable, R. M, 1974, Phylogeny and taxonomy of trematodes with reference to marine species. In: *Symbiosis in the sea.* (Ed. Vernberg, W. B.) *{The Belle V. Baruch library in marine science, no. 2}*, Columbia, USA: University of South Carolina Press, pp. 173–193.

Cable, R. M, 1982, Phylogeny and taxonomy of the malacobothrean flukes. In: *Parasites—their world and ours. Proceedings of the Fifth International Congress of Parasitology, Toronto, Canada, 7–14 Aug. 1982, under the auspices of the World Fed. of Parasitologists.* (Eds D. F. Mettrick and S. S. Desser), Amsterdam, Netherlands: Elsevier Biomedical Press, pp. 194–197.

Cable, R. M. and Dill, W. T, 1967, The morphology and life history of *Paulisentis fractus* Van Cleave and Bangham, 1949 (Acanthocephala: Neoechinorhynchidae). *J. Parasit.*, 53, 810–817.

Cable, R. M. and Hunninen, A. V, 1942, Studies on *Deropristis inflata* (Molin), its life history and affinities to trematodes of the family Acanthocolpidae. *Biol. Bull*, 82, 292–312.

Cable, R. M. and Isseroff, H, 1969, A protandrous haploporid cercaria, probably the larva of *Saccocoelioides sogandaresi* Lumsden, 1963. *Proc. Helminth. Soc. Wash.*, 36, 131–135.

Cable, R. M. and Nahhas, F. M, 1962a, *Lepas* sp., second intermediate host of a didymozoid trematode. *J. Parasit.*, 48, 34.

Cable, R. M. and Nahhas, F. M, 1962b, *Bivesicula caribbensis* sp. n. (Trematoda: Digenea) and its life history. *J. Parasit.*, 48, 536–538.

Cable, R. M. and Nahhas, F. M, 1963, The cercaria of *Dichadena acuta* Linton, 1910 (Trematoda: Hemiuridae). *Proc. Helminth. Soc. Wash.*, 30, 206–210.

Cain, G. D. and Raj, R. K, 1980, *Anisakis, Phocanema, Contracaecum*, and *Sulcascaris* spp.: electrophoresis and thermostability of alcohol and malate dehydrogenases from larvae. *Expl Parasit.*, 49, 56–67.

Caira, J. N, 1989, A revision of the North American papillose Allocreadiidae (Digenea) with independent cladistic analyses of larval and adult forms. *Bull. Univ. Nebraska State Mus.*, 11, 1–58.

Caira, J. N, 1990, Metazoan parasites as indicators of elasmobranch biology. In: *Elasmobranchs as living resources. Advances in the biology, ecology, systematics and the status of fisheries.* (Eds Pratt, H. L., Gruber, S. H. and Taniuchi, T.). *NOAA Technical Report*, 90, 71–96.

Cake, E. W, 1976, A key to larval cestodes of shallow-water benthic molluscs of the Northern Gulf of Mexico. *Proc. Helminth. Soc. Wash.*, 43, 160–171.

Cake, E. W, 1977, Larval cestode parasites of edible molluscs of the northeastern Gulf of Mexico. *Gulf Res. Rep.*, 6, 108.

Cake, E. W, 1979, *Polypocephalus* sp. (Cestoda: Lecanicephalidae): a description of tentaculo-plerocercoids from bay scallops of the northeastern Gulf of Mexico. *Proc. Helminth. Soc. Wash.*, 46, 165–170.

Cake, E. W. and Menzel, R. W, 1980, Infections of *Tylocephalum* metacestodes in commercial oysters and three predaceous gastropods of the eastern Gulf of Mexico. *Proc. Nat. Shellfish. Ass.*, 70, 94–104.

Calentine, R. L, 1964, The life-cycle of *Archigetes iowensis* (Cestoda: Caryophyllaeidae). *J. Parasit.*, 50, 454–458.

Calentine, R. L, 1965, The biology and taxonomy of *Biacetabulum* (Cestoda: Caryophyllaeidae). *J. Parasit.*, 51, 243–248.

Calentine, R. L. and Williams, D. D, 1967, Larval development of *Glaridacris confusa* (Cestoda: Caryophyllaeidae). *J. Parasit.*, 53, 692–693.

Callinan, R. B, 1986, Diseases of native Australian fishes. In: *Diseases of Australian fish and shellfish. Proceedings of the Australian Workshop on diseases of fish and shellfish, Benalla, Victoria, Australia, 27–30 May 1985* (Eds Humphrey, J. D. and Langdon, J. S.) Dep. Agric. New South Wales Reg. Vet. Lab., Wollongbar, NSW, Australia, pp. 102–117.

Calow, P, 1983, Pattern and paradox in parasite reproduction. In: *The reproductive biology of parasites*, Symposia of the Brit. Soc. for Parasit., Vol. 20 (Ed. Whitfield, P. J.) *Parasitology*, 86, 197–207.

Calow, P. and Jennings, J. B, 1974, Calorific values in the phylum Platyhelminthes: the relationship between potential energy, mode of life, and the evolution of entoparasitism. *Biol. Bull.*, 147, 81–94.

Cameron, T. W. M, 1964, Host specificity and the evolution of human parasites. *Adv. Parasit.*, 2, 1–34.

Camp, J. W. and Huizinga, H. W, 1980, Seasonal population interactions of *Acanthocephalus dirus* (Van Cleave 1931) in the creek chub, *Semotilus atromaculatus*, and isopod, *Asellus intermedius. J. Parasit.*, 66, 299–304.

Camp, J. W., Aho, J. M. and Esch, G. W, 1982, A long-term study on various aspects of the population biology of *Ornithodiplostomum ptychocheilus* in a South Carolina cooling reservoir. *J. Parasit.*, 68, 709–718.

Campana-Rouget, Y, 1961a, Rémarques sur le cycle évolutif de *Camallanus lacustris* (Zoega, 1776) et la phylogenie des Camallanidae. *Annls Parasit. Hum. Comp.*, 36, 425–434.

Campana-Rouget, Y, 1961b, Nématodes de poissons. Résultats scientifiques. Exploration hydrobiologique des Lacs Kivu, Edouard et Albert (1952–1954). *Inst. R. Sci. Nat. Belg.*, 3, 1–61.

Campana-Rouget, Y., Petter, A. J., Kremer, M., Molet, B. and Miltgen, F, 1976, Présence du nématode *Camallanus fotedari* dans le tube digestif de poissons d'aquarium de diverses provenances. *Bull. Acad. Vet. Fr.*, 49, 205–210.

Campbell, R. A, 1975a, *Hudsonia agassizi* gen. et sp. n. (Zoogonidae: Hudsoniinae subf. n.) from a deep-sea fish in the Western North Atlantic. *J. Parasit.*, 61, 409–412.

Campbell, R. A, 1975b, *Steringophorus pritchardae* gen. et sp. n. (Digenea: Fellodistomidae) from the deep-sea fish, *Alepocephalus agassizi* Goode and Bean 1883. *J. Parasit.*, 61, 661–664.

Campbell, R. A, 1977a, *Degeneria halosauri* (Bell 1887) gen. et comb. n. (Digenea: Gorgoderidae) from the deep-sea teleost *Halosauropsis macrochir*. *J. Parasit.*, 63, 76–79.

Campbell, R. A, 1977b, A new family of pseudophyllidean cestodes from the deep-sea teleost *Acanthochaenus lutkenii* Gill 1884. *J. Parasit.*, 63, 301–305.

Campbell, R. A, 1979, Two new genera of pseudophyllidean cestodes from deep-sea fishes. *Proc. Helminth. Soc. Wash.*, 46, 74–78.

Campbell, R. A, 1983, Parasitism in the deep sea. In: *Deep-sea Biology, The Sea, vol. 8* (Ed. Rowe, G. T.). New York: Wiley, pp. 473–552.

Campbell, R. A, 1990, Deep water parasites. *Annls Parasit. hum. comp.*, 65, 65–68.

Campbell, R. A. Haedrich, R. L. and Munroe, T. A, 1980, Parasitism and ecological relationships among deep-sea benthic fishes. *Mar. Biol.*, 57, 301–313.

Cannon, L. R. G, 1971, The life cycles of *Bunodera sacculata* and *Bunodera luciopercae* (Trematoda: Allocreadiidae) in Algonquin Park, Ontario. *Can. J. Zool.*, 49, 1417–1429.

Cannon, L. R. G, 1977, Some ecological relationships of larval ascaridoids from south-eastern Queensland marine fishes. *Int. J. Parasit.*, 7, 227–232.

Cannon, L. R. G. and Lester, R. J. G, 1988, Two turbellarians parasitic in fish. *Dis. Aquatic Organisms*, 5, 15–22.

Cannon, L. R. G, 1972, Studies on the ecology of the papillose allocreadiid trematodes of the yellow perch in Algonquin Park, Ontario. *Can. J. Zool.*, 50, 1231–1239.

Carletti, G, 1981, L'anisakiasi o malattia del verme del'aringa. Una zoonosi larvale elmintica poco nota di origine alimentare. *Nuovo Prog. Vet.*, 36, 395–399.

Carothers, J. L. and Allison, R, 1968, Control of snails by the redear (shellcracker) sunfish. [Proc. of the FAO World Symp. on warm-water pond fish culture, Rome, Italy, 18–25 May 1966, Volume 5.]. *FAO Fish. Rep. (1966, publ. 1968)*, 44, 399–406.

Carrara, O. and Grimaldi, E, 1960, Su di una enzoozia parassitaria a decorsa mortale in un allevamento di trote iridee (*Salmo gairdneri*). *Atti Soc. Ital. Sci. Vet.*, 14, 423–426.

Carvajal, J. and Cattan, P. E, 1978, Occurrence of the plerocercus of *Grillotia dollfusi* Carvajal 1971 (Cestoda: Trypanorhyncha) in the Chilean hake *Merluccius gayi* (Guichenot, 1848). *J. Parasit.*, 64, 695.

Carvajal, J. and Cattan, P. E, 1985, A study of the anisakid infection in the Chilean hake, *Merluccius gayi* (Guichenot, 1848). *Fish. Res.*, 3, 245–250.

Carvajal, J. and Rego, A. A, 1985, Anisaquiase: uma enfermidade de origem marinha pouco conhecida. *Cienc. Cult.*, 37, 1847–1849.

Carvajal, J., Barros, C. and Santander, G, 1982, *In vitro* culture of *Rhodobothrium mesodesmatum* (Cestoda: Tetraphyllidea), parasite of a Chilean clam. *Proc. Helminth. Soc. Wash.*, 49, 226–230.

Carvajal, J., Barros, C., Santander, G. and Alcalde, 1981, C. *In vitro* culture of larval anisakid parasites of the Chilean hake *Merluccius gayi*. *J. Parasit.*, 67, 958–959.

Carvajal, J., Cattan, P. E., Castillo, C. and Schatte, P, 1979, Larval anisakids and other helminths in the hake, *Merluccius gayi* (Guichenot) from Chile. *J. Fish Biol.*, 15, 671–677.

Carvalho-Varela, M. and Cunha-Ferreira, V, 1984, Larva migrans visceral por *Anisakis* e outros ascarídeos: helmintozoonoses potenciais por consumo de peixes marinhos em Portugal. *Revta Port. Cienc. Vet.*, 79, 299–309.

Carvalho-Varela, M. and Cunha-Ferreira, V, 1987, Helminth parasites of the common sole, *Solea solea*, and the Senegalese sole, *Solea senegalensis*, on the Portuguese continental coast. *Aquaculture*, 67, 135–138.

Carvalho-Varela, M., Cunha-Ferreira, V., Cruz, E., Silva, M. P. and Grazina-Freitas, M. S, 1984,

Sobre a parasitofauna de enguia-europeia (*Anguilla anguilla* (L.)) em Portugal. *Repos. Trab. Lab. Nac. Invest. Vet.*, 16, 143–150.

Carvalho-Varela, M., Cunha-Ferreira, V., Cruz E., Silva, M. P., Monteiro, M. T. and Grazina-Freitas, 1981, Parasites and parasitosis in fish culture in Portugal. *J. Wld Mariculture Soc.*, 12, 9–14.

Catalini, N., Orecchia, P., Paggi, L. and Todaro, P, 1978, Parassitofauna di *Salmo trutta* L. del fiume Tirino. Nota III. Osservazioni istologiche sui parassiti reperiti a livello intestinale. (IX Congr. Naz., Soc. Ital. Parassit., Ravenna, 30 Mar–1 Apr, 1978.) *Parassitologia*, 20, 169–173.

Cattan, P. E. and Carvajal, J, 1984, A study of the migration of larval *Anisakis simplex* (Nematoda: Ascaridida) in the Chilean hake, *Merluccius gayi* (Guichenot). *J. Fish Biol.*, 24, 649–654.

Cattan, P. E. and Videla, N. N, 1976, [The presence of *Anisakis* sp. larvae in the Pacific jack mackerel, *Trachurus murphyi*. (Some comments on their relation with human eosinophilic granuloma).] *Boletin Chileno de Parasitología*, 31, 71–74. [In Spanish.]

Chabaud, A. G, 1965, Systematique des Nématodes Sous-classe des Secernentea. Ordre des Ascaridida. Superfamilies des Acaridoidea, Heterakoidea et Subuluroidea. In: (Editor: Grassé, P. P.) *Traité de Zoologie, I: Nemanthelminthes (nematodes, gordiaces) rotifères, gastrotriches, kinorhynches.* Paris: Masson et Cie, IV, 988–1025.

Chabaud, A. G, 1975a, Keys to genera of the Order Spirurida. Part 1. Camallanoidea, Dracunculoidea, Gnathostomatoidea, Physalopteroidea, Rictularioidea and Thelazioidea. *CIH Keys to the nematode parasites of vertebrates.* Farnham Royal, England: Commonwealth Agricultural Bureaux, No. 3, 1–27.

Chabaud, A. G, 1975b, Keys to genera of the Order Spirurida. Part 2. Spiruroidea, Habronematoidea and Acuarioidea. *CIH Keys to the nematode parasites of vertebrates.* Farnham Royal, England: Commonwealth Agricultural Bureaux, No. 3, 29–58.

Chabaud, A. G, 1978, *Keys to genera of the superfamilies Cosmocercoidea, Seuratoidea, Heterakoidea and Subuluroidea. CIH Keys to the nematode parasites of vertebrates.* Farnham Royal, England: Commonwealth Agricultural Bureaux, No. 6, 1–71.

Chabaud, A. G, 1982a, Evolution et taxonomie des nématodes—Révue. In: *Parasites—Their World and Ours. Proceedings of the Fifth International Congress of Parasitology, Toronto, Canada, 7–14 August 1982. Under the auspices of the World Federation of Parasitologists.* (Eds D. F. Mettrick, and S. S. Desser), pp. 216–222.

Chabaud, A. G, 1982b, Spèctre d'hôtes et évolution des nématodes parasites de vertébrés. [Deux. Symp. spec. parasit. parasites des vertébrés, 13–17 Avril, 1981. Colloque internat. CNRS.]. *Méms Mus. Natn. Hist. Nat., Sér. A, Zool.*, 123, 73–76.

Chabaud, A. G. and Biguet, J, 1959, étude d'un trématode hémiuroide à métacercaire progénètique. I. Développement chez le mollusque. II. Infestation du copépode. III. Développement chez le copépode. *Annls Parasit. Hum. Comp.*, 29, 527–545.

Chabaud, A. G. and Buttner, A, 1959, Note complémentaire sur le *Bunocotyle* (Trématode Hémiuroïde) de l'étang du Canet. *Vie Milieu*, 10, 204–206.

Chabaud, A. G. and Campana-Rouget, Y, 1959, Notes sur le trématode hémiuroïde *Sterrhurus fusiformis* Lühe 1901 et sur sa cercaire (? *Cercaria vaullegeardi* Pelseneer 1906). *Vie Milieu*, 10, 168–175.

Chai, J.-Y., Seo, B.-S. and Lee, S.-H, 1984, Studies on intestinal trematodes in Korea. XI. Two cases of human infection by *Heterophyes heterophyes nocens. Korean J. Parasit.*, 22, 37–42.

Chaicharn, A. and Bullock, W. L, 1967, The histopathology of acanthocephalan infections in suckers with observations on the intestinal histology of two species of catastomid fishes. *Acta Zool., Stockholm*, 48, 19–42.

Chan, B. Z. and Wu, B. W, 1984, [Studies on the pathogenicity, biology and treatment of *Pseudodactylogyrus* for eels in fishfarms.] *Acta Zool. Sin.*, 30, 173–180. [In Chinese.]

Chandra, K. J., Hanumantha Rao, K. and Shyamasundari, K, 1983, Histochemical nature of the cyst wall of a larval cestode, *Dasyrhynchus* sp. *Zool. Anz.*, 211, 129–136.

Chang, C. Y, 1983, [Families and genera of monogenetic trematodes from China.] *Annual Bulletin of the Society of Parasitology, Guangdong Province* 415 (December), 65–66. [In Chinese.]

Chang, C. Y. and Ji, G. L, 1978, [A preliminary report on monogenetic trematodes of commercial fishes from the Lake Hong-hu, with description of two new species.] *Acta Hydrobiological Sin.*, 6, 353–363. [In Chinese.]

Chang, C. Y. and Ji, G. L, 1981, [Five new species of monogenetic trematodes from freshwater catfishes of China.] *Acta Zootax. Sin.*, 6, 4–12 [In Chinese.]

Chappell, L. H, 1969a, The parasites of the three-spined stickleback *Gasterosteus aculeatus* L. from a Yorkshire pond. I. Seasonal variation of the parasite fauna. *J. Fish Biol.*, 1, 137–152.

Chappell, L. H, 1969b, Competitive exclusion between two intestinal parasites of the three-spined stickleback *Gasterosteus aculeatus* L. *J. Parasit.*, 55, 775–778.

Chappell, L. H, 1980a, *Physiology of parasites. {Series: Tertiary level biology.}*. Glasgow: Blackie, UK. 230 pp.

Chappell, L. H, 1980b, The biology of the external surfaces of helminth parasites. [Proc. Symp. 'Surface ecosystems and the interactions within them which overcome skin defence mechanisms'. Ayr, Scotland, Sept. 1979.] *Proc. R. Soc. Edinb., B (1979, publ. 1980)*, 79, 145–172.

Chappell, L. H, 1982, Physiology. In: *Modern parasitology. A textbook of parasitology.* (Ed. F. E. G. Cox). Oxford, UK: Blackwell Scientific Publications, pp. 116–147.

Chappell, L. H. and Owen, R. W, 1969, A reference list of parasite species recorded in freshwater fish from Great Britain and Ireland. *J. Nat. Hist.*, 3, 197–216.

Charles, G. H. and Orr, T. S. C, 1967, The fine structure of the developing pseudophyllid tegument with reference to *Ligula intestinalis* and *Schistocephalus solidus*. Proc. British Soc. Parasit. [Abstract.] *Parasitology*, 57, 8p.

Charles, G. H. and Orr, T. S. C, 1968, Comparative fine structure of outer tegument of *Ligula intestinalis* and *Schistocephalus solidus*. *Expl Parasit.*, 22, 137–149.

Chauhan, R. S. and Malhotra, S. K, 1984. Association of temperature with establishment and survival of pseudophyllidean cestodes in hill-stream fish. *Curr. Sci.*, 53, 208–209.

Cheah, T. S. and Rajamanickam, C, 1987a, Parasite fauna nilotica in Malaysia. In: *Actual Problems in Fish Parasitology*. The Second International Symposium of Ichthyoparasitology, Tihany, Hungary. p. 11.

Cheah, T. S. and Rajamanickam, C, 1987b, The first record of *Pallisentis ophiocephali* (Thapar, 1930) in a fresh-water fish (*Ophiocephalus striatus*) in Malaysia Kajian *Kajian—Veterinar*. 19, 81–83.

Cheng, T. C, 1964, *The biology of animal parasites*. Philadelphia and London: W. B. Saunders Company, 727 pp.

Cheng, T. C, 1967, Marine molluses as hosts for symbioses, with a review of known parasites of commercially important species. In: *Advances in Marine Biology*, London: Academic Press. Vol. 5, pp. xiii + 424.

Cheng, T. C, 1973, Human parasites transmissible by seafood and related problems. In: *Microbial Safety of Fishery Procs* (Eds Chichester, C. O. and Graham, H.) New York: Academic Press. pp. 163–169.

Cheng, T. C, 1976a, The natural history of anisakiasis in animals. *J. Milk Food Tech.*, 39, 32–46.

Cheng, T. C, 1976b, Liver and other digestive organs. In: (Ed. Kennedy, C. R.), *Ecological aspects of parasitology*. Amsterdam, The Netherlands: North Holland Publishing Company, pp. 287–302.

Cheng, T. C, 1982, Anisakiasis. In: *CRC Handbook Series in Zoonoses*. (Ed. Steele, J. H.) *Section C*. Boca Raton, Florida: CRC Press, II, pp. 37–61.

Cheng, T. C, 1986, *General parasitology*, Harcourt Brace Jovanovitch, Orlando, pp. 1–827.

Cherepanov, A. A, 1982, [Fish-rearing—biological reservoirs for the eradication of helminth eggs from pig liquid manure.] *Byull. Vses. Inst. Gel'm., K. I. Skryabina, No. 32*, 72–73. [In Russian.]

Chernogorenko, M. I. and Ivantsiv, V. V, 1980, [Life-cycle of the trematode *Rhipidocotyle illense* (Ziegler, 1883).] In: *Vopr. parazit. vod. bespoz. zhivot., (Temat. Sb.)* Vilnius, USSR: Inst. Zool. Parazit. An. Lit. SSR, pp. 105–106. [In Russian.]

Chernogorenko, M. I., Komarova, T. I. and Kurandina, D. P, 1978, [The life-cycle of the trematode *Plagioporus skrjabini* Kowal, 1951 (Allocreadiata, Opecoelidae).] *Parazitologiya*, 12, 479–486. [In Russian.]

Chernogorenko-Bidulina, M. I. and Bliznyuk, I. D, 1960, [The life-cycle of *Sphaerostoma bramae* Müller, 1776.] *Dokl. Akad. Nauk SSSR*, 134, 237–240. [In Russian.]

Chernova, T. N, 1975, [The effect of salinity on the parasite fauna of fish in lake Paleostomi.] *Trudy Vses. Nauchno-Issled. Inst. Morsk. Ryb. Khozyaist. Okean. (Biol. osn. produkt. vod. Gruzinskoi SSR)*, 105, 121–127. [In Russian.]

Chernova, T. N, 1981, [The parasite fauna of benthos-feeding fish in lakes Paleostomi and Dzhapana.] In: *Rybokhozyaistvennye issledovaniya vnutrennikh vodoemov Gruzii. (Sbornik Nauchnykh Trudov)*. Moscow, USSR, pp. 27–47. [In Russian.]

Cheung, P. J. and Nigrelli, R. F, 1983, *Dermophthirioides pristidis* n. gen., n. sp. (Microbothriidae) from the skin and *Neoheterocotyle ruggierii* n. sp. (Monocotylidae) from the gills of the smalltooth sawfish *Pristis pectinata. Trans. Am. Microsc. Soc.*, 102, 366–370.

Cheung, P. J. and Ruggieri, G. D, 1983, *Dermophthirius nigrelli* n. sp. (Monogenea: Microbothriidae), an ectoparasite from the skin of the lemon shark, *Negaprion brevirostris. Trans. Am. Microsc. Soc.*, 102, 129–134.

Cheung, P. J., Nigrelli, R. F. and Ruggieri, G. D, 1984, *Philometra saltatrix* infecting the heart of the O-class bluefish, *Pomatomus saltatrix* (L.) from the New York coast. In: *S. F. Snieszko Commemoration Fish Disease Workshop. Joint Workshop of Fish Health Section, AFS and Midwest Fish Disease Group, Little Rock, Ark.*, p. 27.

Cheung, P. J., Nigrelli, R. F., Ruggieri, G. D. and Cilia, A, 1982, Treatment of skin lesions in captive lemon sharks, *Negaprion brevirostris* (Poey), caused by monogeneans (*Dermophthirius* sp.). *J. Fish Dis.*, 5, 167–170.

Ching, H. L, 1979, The life cycle of *Podocotyle enophrysi* Park, 1937 (Trematoda: Opecoelidae). *Can. J. Zool.*, 57, 1341–1344.

Ching, H. L, 1982, Report of the metacestode of *Paradilepis simoni* Rausch, 1949 (Cyclophyllidea: Dilepidae) from fish in British Columbia. *Can. J. Zool.*, 60, 184–186.

Ching, H. L, 1984, Fish tapeworm infections (diphyllobothriasis) in Canada, particularly British Columbia. *Can. Med. Ass.*, 130, 1125–1127.

Ching, H. L, 1985, Occurrence of the eyefluke, *Diplostomum (Diplostomum) baeri bucculentum* Dubois et Rausch, 1948, in salmonid fishes of northern British Columbia. *Can. J. Zool.*, 63, 396–399.

Ching, H. L, 1988, The distribution of plerocercoids of *Diphyllobothrium dendriticum* (Nitzch) in sockeye salmon (*Oncorhynchus nerka*) smolts from Great Central Lake, British Columbia. *Can. J. Zool.*, 66, 850–852.

Chinniah, V. C. and Threlfall, W, 1978, Metazoan parasites of fish from the Smallwood Reservoir, Labrador, Canada. *J. Fish Biol.*, 13, 203–213.

Cho, B. C, 1981, [Electron microscopical and histochemical studies on the epicuticle of *Echinorhynchus gadi* (Acanthocephala).] *Korean J. Parasit.*, 19, 45–54. [In Korean.]

Choi, D. W, 1978, Prevalence of *Clonorchis sinensis* in vicinity of Seongju, Kyungpook Province, Korea. *Korean J. Parasit.*, 16, 140–147.

Chomicz, L, 1980, [The effect of host sex on the course of parasitic infections.] *Wiad. Parazyt.*, 26, 13–21. [In Polish.]

Chonchuenchob, P., Sumpawapol, S. and Mearoh, A, 1987, Diseases of cage-cultured sea bass (*Lates calcarifer*) in South-western Thailand. *ACIAR Proceedings Series, Australian Centre for Int. Agr. Res.*, 20, 194–197.

Choquette, L. P. E, 1954, A note on the intermediate hosts of the trematode, *Crepidostomum cooperi* Hopkins, 1931, parasitic in speckled trout (*Salvelinus fontinalis* (Mitchell)) in some lakes and rivers of the Quebec Laurentide Park. *Can. J. Zool.*, 32, 375–377.

Choquette, L. P. E, 1955, The life history of the nematode *Metabronema salvelini* (Fujita, 1920) parasitic in the speckled trout, *Salvelinus fontinalis* (Mitchill), in Quebec. *Can. J. Zool.*, 33, 1–4.

Christensen, N. Ø, 1978, A method for the *in vivo* radiolabelling of *Diplostomum spathaceum, Hypoderaeum conoideum*, Plagiorchiidae sp. and *Notocotylus attenuatus* cercariae with radioselenium. *Z. ParasitKde*, 57, 155–162.

Christensen, N. Ø., Jensen, M. and Rasmussen, C. I, 1963, Fish diseases in Denmark. Report from the Danish delegation. *Bull. Off. Int. Epizoot.*, 59, 21–29.

Christensen, S. and From, J, 1978, [Nematodes and their control in fish farming.] *Ferskvansfiskeribladet*, 76, 94–100. [In Danish.]

Chu, J. K., Kang, S. Y., Chu, J. P. and Sung, D. W, 1977, [Histochemical studies on *Echinorhynchus gadi* (Acanthocephala).] *Korean J. Parasit.*, 15, 36–42. [In Korean.]

Chubb, J. C, 1963a, Seasonal occurrence of *Triaenophorus nodulosus* (Pallas 1781) in pike *Esox lucius* of Llyn Tegid., *Parasitology*, 53, 419–433.

Chubb, J. C, 1963b, Preliminary observations on the occurrence of *Echinorhynchus clavula* (Acanthocephala) in the fish of Llyn Tegid. *Parasitology*, 53, 2–3P.

Chubb, J. C, 1963c, Preliminary observations on the occurrence of *Echinorhynchus clavula* (Acanthocephala) in the fish of Llyn Tegid. *Parasitology*, 53, 2–3P.

Chubb, J. C, 1963d, Some parasite problems in fishery management. *Proceedings of the First British Coarse Fish Conference, University of Liverpool, 17–18th April 1963*, pp. 30–34.

Chubb, J. C, 1964, Occurrence of *Echinorhynchus clavula* (Dujardin, 1845) *nec* Hamann, 1892 (Acanthocephala) in the fish of Llyn Tegid (Bala Lake), Merionethshire. *J. Parasit.*, 50, 52–59.

Chubb, J. C, 1965, Mass occurrence of *Pomphorhynchus laevis* (Müller, 1776), Monticelli 1905 (Acanthocephala) in the chub *Squalius cephalus* L. from the River Avon, Hampshire, *Parasitology*, 55, 5P.

Chubb, J. C, 1967, *Pomphorhynchus laevis. Proceedings of 3rd British Coarse Fish Conference*, Liverpool, 21–23 March 1967, pp. 20–23.

Chubb, J. C, 1968, Host specificity of some Acanthocephala of freshwater fishes. *Helminthologia*, 9, 63–70.

Chubb, J. C, 1970, The parasite fauna of British freshwater fish. In: *Aspects of Fish Parasitology* (Ed. Taylor, A. E. R.). Oxford: Blackwell Scientific Publications, pp. 119–144.

Chubb, J. C, 1977a, *A review of the parasite fauna of the fishes of the River Dee system. A report for the Central Water Planning Unit, July 1977*. Reading, UK: Central Water Planning Unit, 197 pp.

Chubb, J. C, 1977b, Seasonal occurrence of helminths in freshwater fishes. Part I. Monogenea. *Adv. Parasit.*, 15, 133–199.

Chubb, J. C, 1979, Seasonal occurrence of helminths in freshwater fishes. Part II. Trematoda. *Adv. Parasit.*, 17, 141–313.

Chubb, J. C, 1980a, *Bothriocephalus acheilognathi* Yamaguti, 1934 (Cestoda: Pseudophyllidea) in the British Isles. Part 1. Identification and taxonomy. [Proc. Spring Meet., Brit. Soc. Parasit., Univ. Leeds 24–26 March 1980. Abstract of Session 10C]. *Parasitology*, **81**, xlvi.

Chubb, J. C, 1980b, Seasonal occurrence of helminths in freshwater fishes. Part III. Larval Cestoda and Nematoda. *Adv. Parasit.*, 18, 1–120.

Chubb, J. C, 1980c, Observations on the development of *Bunodera luciopercae* (Muller, 1776) (Trematoda: Allocreadiidae) under field and laboratory conditions. *J. Fish Dis.*, 3, 481–493.

Chubb, J. C, 1981, Parasites of animals in aquatic environments. 1. Literature survey 1974–1977. In: *Review of advances in parasitology. (Proc. 4th Internat. Congr. of Parasit. (ICOPA IV), Warsaw, 19–26 Aug. 1978)* (Editor: Slusarski, W. J.). Warsaw, Poland: PWN—Polish Scientific Publishers, pp. 909–923.

Chubb, J. C, 1982, Seasonal occurrence of helminths in freshwater fishes. Part IV. Adult Cestoda, Nematoda and Acanthocephala. *Adv. Parasit.*, 20, 1–292.

Chubrik, G. K, 1952, [The life-cycle of *Prosorhynchus squamatus* Odhner, 1905.] *Dokl. Akad. Nauk SSSR*, **83**, 327–329. [In Russian.]

Chubrik, G. K, 1966, [Fauna and ecology of larval trematodes in molluscs of the Barents and White Seas.] *Tr. Murmansk. Morsk. Biol. Inst.*, 10, 78–158. [In Russian.]

Chung, Y. T, 1981, A study on the histopathology in the Wolffian ducts of *Hypentelium nigricans* (Osteichthyes: Catostomidae) caused by *Phyllodistomum superbum* (Trematoda: Gorgoderidae). *Quart. J. Taiwan Mus.*, 34, 237–240.

Clark, R. B, 1979, Radiation of the Metazoa. In: *The origin of major invertebrate groups* (Ed. House, M. R.). Systematics Association special volume No. 12. London: Academic Press, pp. 55–101.

Clark, W. C, 1974, Interpretation of life history pattern in the Digenea. *Int. J. Parasit.*, 4, 115–123.

Clark, W. C, 1978, Hermaphroditism as a reproductive strategy for metazoans; some correlated benefits. *N. Z. J. Zool.*, 5, 769–780.

Cleave, H. J. van and Mueller, J. F, 1932, Parasites of the Oneida Lake fishes. Part I. Descriptions of new genera and new species. *Roosevelt Wild Life Annals*, **3**, 72 pp.

Cloutman, D, 1976, Parasitism in relation to taxonomy of the sympatric sibling species of stonerollers, *Campostoma anomalum pullum* (Agassiz) and *C. oligolepsis* Hubbs and Green, in the White River, Arkansas. *Southwest. Nat.*, 21, 67–70.

Cloutman, D. C, 1978, Abundance of *Cleidodiscus pricei* Müller (Monogenea: Dactylogyridae) on the flat bullhead, *Ictalurus platycephalus* (Girard), in Lake Norman, North Carolina. *J. Parasit.*, 64, 170–172.

Cockburn, A, 1977, Where did our infectious diseases come from? The evolution of infectious disease. In: *Health and disease in tribal societies. (Ciba Foundation Symposium 49 (new series), London, 28–30 Sept. 1976.)* Amsterdam, The Netherlands: Elsevier/Excerpta Medical North-Holland, 103–112.

Coggins, J. R, 1980a, Apical end organ structure and histochemistry in plerocercoids of *Proteocephalus ambloplitis*. *Int. J. Parasit.*, 10, 97–101.

Coggins, J. R, 1980b, Tegument and apical end organ fine structure in the metacestode and adult *Proteocephalus ambloplitis*. *Int. J. Parasit.*, 10, 409–418.

Coil, W. H. and Kuntz, R. E, 1963, Observations on the histochemistry of *Syncoelium spathulatum* n. sp. *Proc. Helminth. Soc. Wash.*, 30, 60–65.

Coil, W. H., Reid, W. A. and Kuntz, R. E, 1965, *Paucivitellosus fragilis* gen. et sp. nov. (Bivesiculidae: Digenea), a parasite of *Chelon troscheli* from Formosa. *Trans. Am. Microsc. Soc.*, 84, 365–368.

Cole, H. A, 1979, The assessment of sub-lethal effects of pollutants in the sea. *Trans. Roy. Soc. Lond B.*, 286, 399–424.

Colin, J. A., Williams, H. H. and Halvorsen, O, 1986, One or three gyrocotylideans (Platyhelminthes) in *Chimaera monstrosa* (Holocephali)? *J. Parasit.*, 72, 10–21.

Collard, S. B, 1970, Some aspects of host-parasite relationships in mesopelagic fishes. In: *A Symposium of the American Fisheries Society on diseases of fishes and shellfishes.* (Ed. Sniesko, S.). American Fisheries Society, pp. 41–56.

Collins, M. R., Marshall, M. J. and Lanciani, C. A, 1984, The distribution of *Poecilancistrium caryophyllum* (Trypanorhyncha) plerocercoids in spot, *Leiostomus xanthurus* Lacèpéde, and spotted seatrout, *Cynoscion nebulosus* (Cuvier). *J. Fish Biol.*, 25, 63–68.

Combes, C, 1982, Trematodes: antagonism between species and sterilizing effects on snails in biological control. *Parasitology*, **84**, 151–175.

Combes, C. I., 1983. Usage des trematodes sterilisants—dominants pour le controle de certaines maladies parasitaires. *Symbioses*, **15**, 93–125. [In French with English summary.]

Combs, D. L., Harley, J. P. and Williams, J. C, 1977, Helminth parasites of the spotted sucker and golden redhorse from the Kentucky river. *Trans. Kentucky Acad. Sci.*, **38**, 128–131.

Cone, D. K, 1979, Hatching of *Urocleidus adspectus* Müeller, 1936 (Monogenea: Ancyrocephalinae). *Can. J. Zool.*, **57**, 833–837.

Cone, D. K. and Anderson, R. C, 1977, Parasites of pumpkinseed (*Lepomis gibbosus* L.) from Ryan Lake, Algonquin Park, Ontario. *Can. J. Zool.*, **55**, 1410–1423.

Cone, D. K. and Beverley-Burton, M, 1981, The surface topography of *Benedenia* sp. (Monogenea: Capsalidae). *Can. J. Zool.*, **59**, 1941–1946.

Cone, D. K. and Burt, M. D. B, 1981, The invasion route of the gill parasite *Urocleidus adspectus* Müeller, 1936 (Monogenea: Ancyrocephalinae). *Can. J. Zool.*, **59**, 2166–2171.

Cone, D. K. and Burt, M. D. B, 1982a, The host specificity of *Urocleidus adspectus* Müeller, 1936 (Monogenea: Ancyrocephalinae). *J. Parasit.*, **68**, 1168–1170.

Cone, D. K. and Burt, M. D. B, 1982b, The behaviour of *Urocleidus adspectus* Müeller, 1936 (Monogenea) on the gills of *Perca flavescens*. *Can. J. Zool.*, **60**, 3237–3240.

Cone, D. K. and Burt, M. D. B, 1985, Population biology of *Urocleidus adspectus* Müeller, 1936 (Monogenea) on *Perca flavescens* in New Brunswick. *Can. J. Zool.*, **63**, 272–277.

Cone, D. K. and Cusack, R, 1988, A study of *Gyrodactylus colemanensis* Mizelle and Kritsky, 1967 and *Gyrodactylus salmonensis* (Yin and Sproston, 1948) (Monogenea) parasitizing captive salmonids in Nova Scotia. *Can. J. Zool.*, **66**, 409–415.

Cone, D. K. and Odense, P. H, 1984, Pathology of five species of *Gyrodactylus* Nordmann, 1832 (Monogenea). *Can. J. Zool.*, **62**, 1084–1088.

Cone, D. K. and Ryan, P. M, 1984, Population sizes of metazoan parasites of brook trout (*Salvelinus fontinalis*) and Atlantic salmon (*Salmo salar*) in a small Newfoundland lake. *Can. J. Zool.*, **62**, 130–133.

Conneely, J. J. and McCarthy, T. K, 1984, The metazoan parasites of freshwater fishes in the Corrib catchment area, Ireland. *J. Fish Biol.*, **24**, 363–375.

Conneely, J. J. and McCarthy, T. K, 1986, Ecological factors influencing the composition of the parasite fauna of the European eel, *Anguilla anguilla* (L.), in Ireland. *J. Fish Biol.*, **28**, 207–219.

Conroy, D. A, 1981, The importance of fish diseases in relation to the development of salmonid culture in South America. *Riv. Ital. Pisc. Ittiopatol.*, **16**, 61–62, 64–68.

Conroy, G. A, De, 1986, *Crocodilicola pseudostoma* (Willemoes-Suhm, 1870) Poche, 1925 (Trematoda: Proterodiplostomatidae), endoparásito del bagre pimelódido *Rhamdia hilarii* Val., 1840 del Estado de Sao Paulo, Brasil. *Revta. Ibér. Parasit.*, **46**, 35–38.

Conway Morris, S, 1981, Parasites and the fossil record. *Parasitology*, **82**, 489–509.

Conway Morris, S. and Crompton, D. W. T, 1982a, Origin and evolution of acanthocephalan worms. [Deux. symp. spec. parasit. parasites des vertébrés, 13–17 Avril, 1987. Colloque. internat. CNRS.] *Méms Mus. Natn. Hist. nat.*, **123**, 61–66.

Conway Morris, S. and Crompton, D. W. T, 1982b, The origins and evolution of the Acanthocephala. *Biol. Rev. Camb. Phil. Soc.*, **57**, 85–115.

Cooper, C. L., Crites, J. L. and Sprinkle-Fastkie, D. J, 1978, Population biology and behaviour of larval *Eustrongylides tubifex* (Nematoda: Dioctophymatida) in poikilothermous hosts. *J. Parasit.*, **64**, 102–107.

Corbel, M. J, 1975, The immune response in fish: a review. *J. Fish Biol.*, **7**, 539–563.

Cordero del Campillo, M, 1981, Alimentación, nutrición y enfermedades parasitarias. *Aliment. Mejora Anim. (Madrid)*, **22**, 3–10.

Cordero del Campillo, M. and Alvarez-Pellitero, M. P, 1974, Fish disease problems in Spain and measures for control. *Fish Farm. Int.*, **2**, 124–131.

Corkum, K. C. and Beckerdite, F. W, 1975, Observations on the life history of *Alloglossidium macrobdellensis* (Trematoda: Macroderoididae) from *Macrobdella ditetra* (Hirudinea: Hirudinidae). *American Midland Naturalist*. Baton Rouge: Louisiana State Univ., USA, **93**, 484–491.

Cort, W. W. and Ameel, D. J, 1944, Further studies on the development of the sporocyst stages of plagiorchiid trematodes. *J. Parasit.*, **30**, 37–56.

Cosgrove, G. E, 1975, Parasites in tissue sections: recognition and reaction. In: *The pathology of fishes* (Eds Ribelin, W. E. and Migaki, G.) Wisconsin: The University of Wisconsin Press, pp. 205–245.

Cottrell, B, 1977, The immune response of plaice (*Pleuronectes platessa* L.) to the metacercariae of *Cryptocotyle lingua* and *Rhipidocotyle johnstonei*. *Parasitology*, 74, 93–107.

Couch, J. A, 1985, Prospective study of infectious and noninfectious diseases in oysters and fishes in three Gulf of Mexico estuaries. *Dis. Aquat. Org.*, 1, 59–82.

Cox, F. E. G, (Ed.) 1982, *Modern Parasitology. A textbook of parasitology*. London: Blackwell Scientific Publications, 346 pp.

Crane, J. S. and Eversole, A. G, 1980, Ectoparasitic fauna of glass eel and elver stages of American eel (*Anguilla rostrata*). *Proc. Wld Maricult. Soc.*, 11, 275–280.

Crane, J. S. and Eversole, A. G, 1985, Metazoan ectoparasitic fauna of American eels from brackish water. In: *Proc. 39th Ann. Conf., Southeastern Ass. Fish and Wildl. Agencies, October 27–30, 1985, Lexington, Kentucky* (Eds Sweeney, J. M. and Sweeney J. R.. USA. Southeastern Ass. Fish and Wildl. Agencies pp. 248–254.

Crane, J. W, 1972, Systematics and new species of marine Monogenea from California. *Wasmann J. Biol.*, 30, 109–166.

Craun, G. F, 1977, Waterborne outbreaks of human disease. *J. Water Pollut. Cont.*, 49, 1268–1279.

Cribb, T. H, 1988, Life-cycle and biology of *Prototransversotrema steeri* Angel, 1969 (Digenea: Transversotrematidae). *Aust. J. Zool.*, 36, 111–129.

Cribb, T. H. and Pearson, J. C., 1988, *Baiohelmins elegans* n.g., n.sp. (Digenea: Nanophyetidae) from the Australian water rat, *Hydromys chrysogaster*, *Syst. Parasitol.*, 12, 117–121.

Crider, C. R. and Meade, T. G, 1975, Immunological studies on the origin of the cyst wall of *Posthodiplostomum minimum* (Trematoda: Diplostomidae). *Proc. Helminth. Soc. Wash.*, 42, 21–24.

Crofton, H. D, 1961, *Nematodes*. Hutchinson University Library, London, pp. 1–160.

Crofton, H. D, 1971a, A quantitative approach to parasitism. *Parasitology*, 62, 179–193.

Crofton, H. D, 1971b, A model of host parasite relationships. *Parasitology*, 63, 343–364.

Crofton, H. D. and Fraser, P. G, 1954, The mode of infection of the hake, *Merluccius merluccius* (L.) by the trematode *Bucephalopsis gracilescens* (Rud.). *Proc. Zool. Soc. Lond.*, 124, 105–109.

Croll, N. A. and Gyorkos, T. W, 1978, The parasites of man in Canada. *Canada Dis. Week. Rep.*, 4, 69–72.

Croll, N. A. and Soyza, K. de, 1980, Comparative calorie values of nematodes. *J. Nematol.*, 12, 132–135.

Croll, N. A., Ma, K., Smith, J. M., Sukhdeo, M. V. K. and Wild, G, 1980, *Phocanema decipiens*: intestinal penetration in the laboratory rat. *Expl Parasit.*, 50, 145–154.

Crompton, D. W. T, 1969, On the environment of *Polymorphus minutus* (Acanthocephala) in ducks. *Parasitology*, 59, 19–28.

Crompton, D. W. T, 1970a, *An ecological approach to acanthocephalan physiology. Cambridge Monogr. Exper. Biol., 17*, Cambridge: Cambridge University Press, 125 pp.

Crompton, D. W. T, 1970b, Functional aspects of acanthocephalan surface layers. (Proceedings of the Second International Congress of Parasitology, 6–12 September 1970, Washington, DC.) *J. Parasit.*, 56, 64.

Crompton, D. W. T, 1973, The sites occupied by some parasitic helminths in the alimentary tract of vertebrates. *Biol. Rev. Camb. Phil. Soc.*, 48, 27–83.

Crompton, D. W. T, 1975, Relationships between acanthocephala and their hosts. In: *Symp. Soc. Expl Biol., No. 29, Symbiosis*. (Eds Jennings, D. H. and Lee, D. L.) Cambridge UK: Cambridge University Press, pp. 467–504.

Crompton, D. W. T, 1976, Entry into the host and site selection. In: *Ecological aspects of parasitology*. [Ed: Kennedy, C. R.] Amsterdam, The Netherlands: North Holland Publishing Company, pp. 41–73.

Crompton, D. W. T, 1977, Qualitative requirements and utilization of nutrients: Acanthocephala. In: *CRC handbook series on nutrition and food. Section D: nutritional requirements, Vol. 1, comparative and qualitative requirements*. (Ed. Rechcigl, M.). Cleveland, Ohio: CRC Press, Inc., pp. 289–294.

Crompton, D. W. T, 1983a, Acanthocephala. In: *Reproductive biology of invertebrates, Vol. 1. Oogenesis, oviposition and oosorption*. (Eds Adiyodi, K. G. and Adiyodi, R. G.) Chichester, UK: John Wiley and Sons Ltd., pp. 257–268.

Crompton, D. W. T, 1983b, Acanthocephala. In: *Reproductive biology of invertebrates. Vol. 2. Spermatogenesis and sperm function*. (Eds Adiyodi, K. G. and Adiyodi, R. G.) Chichester, UK: John Wiley and Sons, Ltd., pp. 257–267.

Crompton, D. W. T, 1984, Influence of parasitic infection on food intake. *Fed. Proc.*, 43, 239–245.

Crompton, D. W. T, 1985, Reproduction. In: *Biology of the Acanthocephala*. (Eds Crompton, D. W. T. and Nickol, B. B.). Cambridge: Cambridge University Press, pp. 213–263.

Crompton, D. W. T. and Nesheim, M. C, 1976, Host–parasite relationships in the alimentary tract of domestic birds. *Adv. Parasit.*, **14**, 95–194.

Crompton, D. W. T. and Nickol, B. B. (Eds) 1985, *Biology of the Acanthocephala*. Cambridge: Cambridge University Press, 519 pp.

Cross, S. X, 1934, A probable case of non-specific immunity between two parasites of ciscoes of the Trout Lake region of northern Wisconsin. *J. Parasit.*, **20**, 244–245.

Crowden, A. E. and Broom, D. M, 1980, Effects of the eyefluke, *Diplostomum spathaceum*, on the behaviour of dace (*Leuciscus leuciscus*). *Animal Behaviour*, **28**, 287–294.

Crozier, W. W, 1987, Occurrence of *Contracaecum clavatum* Rudolphi in angler-fish (*Lophius piscatorius* L.) from the north Irish Sea. *Fish. Res.*, **5**, 83–90.

Crusz, H, 1957, A new anaporrhutine trematode, *Staphylorchis parisi* sp. nov. from the shark, *Scoliodon walbeehmi*. *Ceylon J. Sci.*, **25**, 193–195.

Crusz, H., Ratnayake, W. E. and Sathananthan 1964, Observations on the structure and life-cycle of the digenetic fish-trematode *Transversotrema patialense* (Soparkar). *Ceylon J. Sci.*, **5**, 8–17.

Cruz, E., Silva, M. P., Ventura, M. T. and Grazina Freitas, M. S, 1986, Binómio patologia/maneio numa anguilicultura portuguesa. *Reposit. Trab. Lab. Nac. Invest.*, **18**, 41–48.

Curtis, M. A, 1981, Observations on the occurrence of *Diplostomum spathaceum* and *Schistocephalus* sp. in ninespine sticklebacks (*Pungitius pungitius*) from the Belcher Islands, Northwest Territories, Canada. *J. Wildl. Dis.*, **17**, 241–246.

Curtis, M. A, 1982, Host–parasite interaction in arctic and subarctic lakes. In: *Aspects of Parasitology* (Ed. Meerovitch, E.), The Institute of Parasitology, McGill University, Montreal. pp. 41–48.

Curtis, M. A. and Rau, M. E, 1980, The geographical distribution of diplostomiasis (Trematoda: Strigeidae) in fishes from northern Quebec, Canada, in relation to the calcium ion concentrations of lakes. *Can. J. Zool.*, **58**, 1390–1394.

Cusack, R. and Cone, D. K, 1985, A report of bacterial microcolonies on the surface of *Gyrodactylus* (Monogenea). *J. Fish Dis.*, **8**, 125–127.

Cusack, R. and Cone, D. K, 1986, *Gyrodactylus salmonis* (Yin and Sproston, 1948) parasitizing fry of *Salvelinus fontinalis* (Mitchill). *J. Wildl. Dis.*, **22**, 209–213.

Dabrowski, K. R, 1980, Amino-acid composition of *Ligula intestinalis* (L.) (Cestoda) plerocercoids and of the host parasitized by these cestodes. *Acta Parasit. Pol.*, **27**, 45–48.

Daengsvang, S, 1980, *A monograph on the genus Gnathostoma and gnathostomiasis in Thailand.* Tokyo, Japan: Southeast Asian Medical Information Center; International Medical Foundation of Japan, ix + 85pp.

Daengsvang, S, 1982, Gnathostomiasis. In: *CRC handbook series in zoonoses. Section C: parasitic zoonoses,* (Ed. Schultz, M. G.) Boca Raton, Florida, USA: CRC Press, Inc. **II**, 147–180.

Dailey, M. D, 1966, *Biology and morphology of Philometroides nodulosa (Thomas, 1929) n. comb. (Philometridae: Nematoda) in the western white sucker (Catostomus commersoni).* Ph.D. Thesis, Colorado State University, Fort Collins.

Dailey, M. D, 1969, *Litobothrium alopias* and *L. coniformis*. Two new cestodes representing a new order from elasmobranch fishes. *Proc. Helminth. Soc. Wash.*, **36**, 218–224.

Dailey, M. D, 1975, Investigations on the viability of larval helminths after freezing. *Aquatic Mammals*, **3**, 22–25.

Dailey, M. D, 1985, Mammalia: Cetacea. In: *Diseases of Marine Animals, IV (2)* (Ed. Kinne, O.) Hamburg: Biologische Anstalt Helgoland, pp. 805–847.

Dailey, M. D. and Overstreet, R. M, 1973, *Cathetocephalus thatcheri* gen. et. sp. n. (Tetraphyllidea: Cathetocephalidae fam. n.) from the bull shark: a species demonstrating multistrobilization. *J. Parasit.*, **59**, 469–473.

Dartnall, H. J. G. and Walkey, M, 1979, Parasites of marine sticklebacks. *J. Fish Biol.*, **14**, 471–474.

Davey, J. T, 1971, A revision of the genus *Anisakis* Dujardin, 1845 (Nematoda: Ascaridata). *J. Helminth.*, **45**, 51–72.

Davey, J. T. and Peachey, J. E, 1968, *Bothriocephalus scorpii* (Cestoda: Pseudophyllidea) in turbot and brill from British coastal waters. *J. Mar. Biol. Ass. UK*, **48**, 335–340.

Davey, K. G, 1976. Hormones and hormonal effects in parasitic nematodes. In: *Biochemistry of parasites and host–parasite relationships*. (Ed. Van den Bossche, H.) Amsterdam, The Netherlands: North-Holland Publishing Company, pp. 359–372.

Davey, K. G, 1979, Molting in a parasitic nematode, *Phocanema decipiens*: role of water uptake. *Int. J. Parasit.*, **9**, 121–125.

Davies, E. H, 1967, Parasite fauna of the fish of the river Lugg, a tributary of the river Wye, Herefordshire. Ph.D. thesis, University of Liverpool.

Davies, W. and Jackson, H, 1977, Chemosterilant action of niridazole on *Tilapia mossambica* (*Sarotherodon mossambicus*). *General Pharmacology*, **8**, 31–35.

Davis, H. S, 1937, Care and diseases of trout. *US. Dept. Commerce, But. Fish. Inv. Rep.*, 1–35.

Davis, H. S, 1953, *Culture and diseases of game fishes*. University of California Press, USA, 322 pp.

Davis, J. R. and Huffman, D. G, 1977. A comparison of the helminth parasites of *Gambusia affinis* and *Gambusia geiseri* (Osteichthyes: Poeciliidae) from the upper San Marcos River. *SWest Nat.*, 22, 359–366.

Davis, J. R. and Huffman, D. G, 1978, Some factors associated with the distribution of helminths among individual mosquitofish, *Gambusia affinis*. *Texas J. Sci.*, **30**, 43–53.

Davis, R. E. and Roberts, L. S, 1983a, Platyhelminthes—Eucestoda. In: *Reproductive biology of invertebrates*. Vol. 1. *Oogenesis, oviposition,* and *oosorption.* (Eds Adiyodi, K. G. and Adiyodi, R. G.). Chichester, UK: John Wiley and Sons Ltd. pp. 109–133.

Davis, R. E. and Roberts, L. S, 1983b, Platyhelminthes—Eucestoda. In: *Reproductive biology of invertebrates*. Vol. 2. *Spermatogenesis and sperm function.* (Eds Adiyodi, K. G. and Adiyodi, R. G.). Chichester, UK: John Wiley and Sons Ltd. pp. 131–149.

Davydov, O. N, 1972, [The nutrition of *Ligula intestinalis* plerocercoids.] In: *Problemy parazitologii. Trudy VII Nauchnoi Conferentsii Parazitologov USSR. Part I.*, Kiev, USSR: Izdatel'stvo 'Naukova-Dumka', pp. 245–246. [In Russian.]

Davydov, O. N, 1973, [The effect of ions from the environment on the locomotor activity of the cestode *Bothriocephalus gowkongensis*.] *Parazitologiya*, 7, 357–363. [In Russian.]

Davydov, O. N, 1975, [Osmotic regulation in some fish cestodes.] *Gidrobiol. Zh.*, 11, 75–79. [In Russian.]

Davydov, O. N, 1978a, [The effect of *Bothriocephalus gowkongensis* on the morpho-physiological indices of carp.] *Gidrobiol. Zh.*, 14, 59–64. [In Russian.]

Davydov, O. N, 1978b, [Growth, development and fecundity of *Bothriocephalus gowkongensis* (Yeh, 1955), a parasite of cyprinid fish.] *Gidrobiol. Zh.*, 14, 70–77. [In Russian.]

Davydov, O. N, 1978c, The influence of the physiological condition of fish on the efficacy of phenasal treatment. In: *I Vses. s"ezd parasitotseno. Poltava, Chast' 3. Kiev, USSR*. pp. 26–28. [In Russian.]

Davydov, O. N. and Kosenko, L. Ya, 1972, [Membrane digestion of *Ligula intestinalis* plerocercoids.] *Parazitologiya*, 6, 269–273. [In Russian.]

Davydov, O. N. and Kurovskaya, L. Ya, 1978a, [Some aspects of the trophic links in the *Bothriocephalus*-carp system.] In: *I Vses. s"ezd. parazitoseno. Poltava. Chast' 3. Kiev, USSR*. pp. 104–106. [In Russian.]

Davydov, O. N. and Kurovskaya, L. Ya, 1978b, The influence of the diet of the carp on the protein level in the cestode body. In: *I Vses. s"ezd. parazitoseno. Poltava. Chast' 3. Kiev, USSR*. pp. 28–29. [In Russian.]

Davydov, O. N., Muzykovskii, A. M. and Vasil'kov, G. V, 1980, [Granulated feed with activated phenasal for *Bothriocephalus* infection in carp.] *Byull. Vses. Inst. Gel'm. K. I. Skryabina*, **25**, 17–18. [In Russian.]

Davydov, O. N., Perovozchenko, I. I., Braginskii, P. and Bal'on, Ya. G, 1976, [The accumulation, desorption and effect of chlorinated pesticides in fish cestodes.] *Parazitologiya*, **10**, 238–244. [In Russian.]

Davydov, O. N., Sepegina, L. Ya., Komarova, T. I., Strizhnik, L. V. and Kurovskaya, L. Ya, 1981, [Species composition of parasites of fish in tanks of the cooling reservoir of the Kiev Heat and Power Plant 5.] *Osvoenie teplykh vod energetiches-kikh ob"ektov dlya intensivnogo rybovodstva. Materially II Respublikanskoi Nauchnoi Konferentsii, Kiev, 1980*, Kiev, USSR: Naukova Dumka, pp. 405–409. [In Russian.]

Davydov, V. G, 1975, [Stored nutrients at all phases of the life-cycle of *Triaenophorus nodulosus*] *Inst. Biol. Vnytr. Vod. Inf. Byull. No. 27*, 17–20. [In Russian.]

Davydov, V. G, 1977, [Host tissue reaction to different types of cestode attachment.] *Biol. Vnut. Vod Inf. Byull., No. 33*, 45–48. [In Russian.]

Davydov, V. G, 1978, [Differences in the tissue reactions of fish in response to infection with *Diphyllobothrium latum* (L., 1758) plerocercoids.] *Biol. Vnut. Vod. Inf. Byull., No. 39*, 68–71. [In Russian.]

Davydov, V. G, 1979, [Histochemical study of pseudophyllid cestodes.] *Trudy Inst. Biol. Vnut. Vod. (Fiziol. parazit. presnovod. zhivot.)*, **38**, 189–200. [In Russian.]

Davydov, V. G, 1979, [Trophic relationships between parasite and host and host reaction to penetration by *Triaenophorus nodulosu* (Pallas, 1781) plerocercoids.] *Biol. Vnut. Vod. Inf. Byull., No. 44*, 54–58. [In Russian.]

Davydov, V. G, 1981, [The penetration of the plerocercoids of some cestodes into the host tissues.] *Biol. Bnut. Vod. Inf. Byull., No. 52*, 57–62. [In Russian.]

Davydov, V. G. and Biserova, N. M, 1985, [Morphology of frontal glands of *Grillotia erinaceus* (Cestoda, Trypanorhyncha.] *Parazitologiya*, 19, 32–38. [In Russian.]

Davydov, V. G. and Kuperman, B. I, 1979, [The structure of the frontal glands of representatives of three orders of cestodes.] *Trudy Inst. Biol. Vnyt. Vod. (Fiziol. parazit. presnovod. zhivot.), No. 38 (41)*, 177–188. [In Russian.]

Dawbin, W. H, 1986, The relative abundance of helminth parasites in specified categories of humpback whales, *Megaptera nova-angliae*. In: *Parasitology Quo Vadit?* (Ed. Howell, M. J.) Proceedings of the VI I.C.O.P.A. Canberra Australia: Australian Academy of Science, Canberra. Abstract 174, p. 120.

Dawes, B, 1946, *The Trematoda. With special reference to British and other European forms*. Cambridge: Cambridge University Press, 644 pp.

Dawes, B, 1947, *The Trematoda of British Fishes*. Royal Society of London. London: Royal Society, 364 pp.

Dawes, Ben (Ed.), 1977. *Advances in parasitology*. London, UK: Academic Press 15, 409 pp.

De, N. C. and Moravec, E, 1979, Some new data on the morphology and development of the nematode *Cystidicoloides tenuissima* (Zeder, 1800), *Folia Parasit.*, 26, 231–237.

Deardorff, T. L, 1982, What's bugging the billfishes? *Pacif. Gamefish Res. News*, 4.

Deardorff, T. L. and Ko, R. C, 1983, *Echinocephalus overstreeti* sp. n. (Nematoda: Gnathostomatidae) in the stingray, *Taeniura melanopilos* Bleeker, from the Marquesas Islands, with comments on *E. sinensis* Ko, 1975. *Proc. Helminth. Soc. Wash.*, 50, 285–293.

Deardorff, T. L. and Overstreet, R. M, 1978, *Thynnascaris rhacodes* sp. n. (Nematoda: Ascaridoidea) in fishes from the Israeli Mediterranean coast. *Annls Parasit. Hum. Comp.*, 53, 519–525.

Deardorff, T. L. and Overstreet, R. M, 1980a, Taxonomy and biology of North American species of *Goezia* (Nematoda: Anisakidae) from fishes, including three new species. *Proc. Helminth. Soc. Wash.*, 47, 192–217.

Deardorff, T. L. and Overstreet, R. M, 1980b, *Contracaecum multipapillatum* (= *C. robustum*) from fishes and birds in the Northern Gulf of Mexico. *J. Parasit.*, 66, 853–856.

Deardorff, T. L. and Overstreet, R. M, 1981, Review of *Hysterothylacium* and *Iheringascaris* (both previously = *Thynnascaris*) (Nematoda: Anisakidae) from the Northern Gulf of Mexico. *Proc. Biol. Soc. Wash.*, 93, 1035–1079.

Deardorff, T. L. and Stanton, F. G, 1983, Nematode-induced abdominal distention of the Hawaiian puffer fish, *Canthigaster jactator* (Jenkins). *Pacif. Sci.*, 37, 45–47.

Deardorff, T. L., Brooks, D. R. and Thorson, T. B, 1981, A new species of *Echinocephalus* (Nematoda: Gnathostomidae) from neotropical stingrays with comments on *E. diazi*. *J. Parasit.*, 67, 433–439.

Deardorff, T. L., Kliks, M. M. and Desowitz, R. S, 1983, Histopathology induced by larval *Terranova* (Type HA) (Nematoda: Anisakinae) in experimentally infected rats. *J. Parasit.*, 69, 191–195.

Deardorff, T. L., Raybourne, R. B. and Desowitz, R. S, 1984a, Behavior and Viability of Third-Stage Larvae of *Terranova* sp. (Type HA) and *Anisakis simplex* (Type 1) under coolant conditions. *J. Food Prot.*, 47, 49–52.

Deardorff, T. L., Raybourne, R. B. and Desowitz, R. S, 1984b, Description of a third-stage larva, *Terranova* Type Hawaii A (Nematoda: Anisakinae), from Hawaiian fishe. *J. Parasit.*, 70, 829–831.

Deardorff, T. L., Raybourne, R. B. and Mattis, T. E, 1984, Infections with trypanorhynch plerocerci (Cestoda) in Hawaiian fishes of commercial importance. *Sea Grant Quarterly*, 6, 1–6.

Deardoff, T. L., Overstreet, R. M., Okihiro, M. and Tam, R, 1986, Piscine adult nematode invading an open lesion in a human hand. *Am. J. Trop. Med. Hyg.*, 35, pp. 827–830.

Deardoff, T. L., Kliks, M. M., Rosenfield, M. E., Rychlinski, R. A. and Desowitz, R, 1982, Larval ascaridoid nematodes from fishes near the Hawaiian Islands with comments on pathogenicity experiments. *Pacif. Sci.*, 36, 187–201.

Dechtiar, A. O, 1969, Two new species of monogenetic trematodes (Trematoda: Monogenea) from nasal cavities of catostomid fishes. *J. Fish. Res. Bd Can.*, 26, 865–869.

De Giusti, D. L, 1949, The life cycle of *Leptorhynchoides thecatus* (Linton), an acanthocephalan of fish. *J. Parasit.*, 35, 437–460.

De Guisti, D. L, 1962, Ecological and life history notes on the trematode *Allocreadium lobatum* (Wallin, 1909) and its occurrence as a progenetic form in amphipods. *J. Parasit.*, 48 (2, Sect. 2), 22.

Dekker, W. and Willigen, J. van, 1989, Short note on the distribution and abundance of *Anguillicola* in the Netherlands. *J. Appl. Ichthyol.*, 5, 46–47.

Delyamure, S. L, 1968, (Study of the helminth fauna of marine mammals of the world oceans carried out by the Department of Zoology of the Crimean Pedagogical Institute from 1955 to 1965. In: *Biology of seas. No. 14, Parasites of marine animals* (Ed. Bodyanitski, V. A.) Kiev: 'Naukova Dumka', pp. 14–24.

Delyamure, S. L, 1969, Helminthfauna of marine mammals (ecology and phylogeny). *Jerusalem: Israel Program for Scientific Translations.* London: H. A. Humphrey Ltd., ix + 522 pp.

Delyamure, S. L., Skryabin, A. S. and Serdyukov, A. M, 1985, [*Principles of cestodology. Vol. II. Diphyllobothriids—tapeworms of man, mammals and birds.*] Moscow, USSR: 'Nauka', 200 pp. [In Russian.]

Demidov, N. V. and Potemkina, V. A, 1980, [*Handbook on the therapy and prophylaxis of animal helminthiases.*] Moscow, USSR: 'Kolos', 240 pp. [In Russian.]

DeMont, D. J. and Corkum, K. C, 1982, The life-cycle of *Octospiniferoides chandleri* Bullock, 1957 (Acanthocephala: Neochinorhynchidae) with some observations on parasite-induced photophilic behavior in ostracods. *J. Parasit.*, 68, 125–130.

Demshin, N. I, 1971, [Phylogenesis of cestode life-cycles.] In: *Biol. Med. Issled. Dal'n. Vostoke, Vladivostok*: Akad. Nauk SSSR, Dal'nev. Nauch. Tsentr., pp. 280–296. [In Russian.]

Demshin, N. I, 1978, [The biology of *Khawia japonensis* (Caryophyllidae, Cestoda), a parasite of the Amur carp.] *Parazitologiya*, 12, 493–496. [In Russian.]

Demshin, N. I, 1984, [Historical development of host–parasite relationships between oligochaetes and helminth larvae.] *Ekol. Gel'm., No. 4,* 16–26. [In Russian.]

Demshin, N. I. and Dvoryadkin, V. A, 1980, [The biology of *Khawia sinensis* Hsü, 1935 (Caryophyllidea, Cestoda), a parasite of the Amur carp.] *Gidrobiol. Zh.*, 16, 77–82. [In Russian.]

Demshin, N. I. and Dvoryadkin, V. A, 1981, [The development of *Markevitschia sagittata* (Cestoda: Caryophyllidae) a parasite of *Cyprinus carpio haematopterus* in the external environment and in the intermediate host.] *Parazitologiya*, 15, 113–117. [In Russian.]

Denisov, A. I, 1977, [The epizootiology of *Posthodiplostomum* infection in fish in the northern Caucasus.] *Mater. Nauch. Konf. Vses. Obshch. Gel'm. (Trem.trematod.), No. 29,* 42–52. [In Russian.]

Denisov, A. I, 1979, [The pathogenic effect of *Posthodiplostomum cuticola* infection on silver carp.] *Sb. Nauch. Trud. Vses. Nauchno-Issled. Inst. Prud. Ryb. Khozyaist. (Bolez. ryb bor'ba s nimi), No. 23,* 117–123. [In Russian.]

Denisov, A. I, 1980, [Development of *Diplostomum* larvae in the fish eye.] In: *IX Konferentsiya Ukrainskogo Parazitologicheskogo Obshchestva. Tesizy dokladov. Chast' 2.* Kiev, USSR: 'Naukova Dumka', pp. 15–16. [In Russian.]

Denisov, A. I, 1982, [The epizootiology and prophylaxis of diplostomiasis.] *Veterinariya, Mosk., No. 8,* 40–42. [In Russian.]

Despommier, D, 1976, Musculature. In: *Ecological aspects of parasitology* (Ed. Kennedy, C. R.) Amsterdam: North-Holland Publishing Company, pp. 269–285.

Desser, H. S, 1984, Transmission electron microscopy of the tegumentary sense organs of *Cotylogaster occidentalis* (Trematoda: Aspidogastrea). *J. Parasit.*, 70, 563–575.

Desser, H. S. and Weller, I, 1982, *Cotylogaster occidentalis* (Trematoda: Aspidogastrea): scanning electron microscopic observations of sense organs and associated surface structures. *Trans. Am. micro. Soc.*, 101, 253–261.

Deufel, J, 1975, Der Wurmstar (*Diplostomum*-Krankheit) und die Schwarzfleckenkrankheit der Fische. [Die Bedeutung der Parasiten für die Produktion von Susswasserfischen-Vortr. gehalt. Müchener fisch.-biol. Semin. 5–7 Dec. 1973]. *Fisch und Umwelt, No. 1,* 97–104.

Dhar, V. N. and Sharma, G. P, 1984, Behaviour of chromosomes during gametogenesis and fertilization in *Paradistomoides orientalis* (Digenea: Trematoda). *Caryologia*, 37, 207–218.

Diamond, J. M. and Case, T. J. (Eds), 1986, *Community Ecology.* New York Harper and Row, xxii + 665 pp.

Diarova, G. S, 1971, [The parasite fauna of fry of *Ctenopharyngodon idella* and common carp in the ponds of southern Kazakhstan under conditions of factory farming.] *Trudy Vses. Nauchno-Issled. Inst. Prud. Ryb. Khozyaist. (Vopr. prud. ryb.)*, 18, 63–65. [In Russian.]

Diarova, G. S, 1975, [The role of coarse fish in the dissemination of some diseases on fish farms in Kazakhstan, USSR.] In: *Ekologiya parazitov. vodnykh zhivotnykh.* Alma-Ata, USSR: Izdatel'stvo 'Nauka' Kazakhskoi SSR, pp. 52–62. [In Russian.]

Dick, T. A. and Belosevic, M, 1981, Parasites of Arctic charr *Salvelinus alpinus* (Linnaeus) and their use in separating sea-run and non-migrating charr. *J. Fish Biol.*, 18, 339–347.

Dick, T. A. and Poole, B. C, 1985, Identification *Diphyllobothrium dendriticum* and *Diphyllobothrium latum* from some freshwater fishes of central Canada. *Can. J. Zool.*, 63, 196–201.

Dick, T. A. and Rosen, R, 1981, Identification of *Diplostomum* spp. from the eyes of lake whitefish,

Coregonus clupeaformis (Mitchill), based on experimental infection of herring gull chicks, *Larus argentatus* Pontoppidan. *Can. J. Zool.*, 59, 1176–1179.

Dick, T. A. and Rosen, R, 1982, Experimental infections of whitefish, *Coregonus clupeaformis* (Mitchill), with *Triaenophorus crassus* Forel. *J. Fish Dis.*, 5, 83–86.

Dickerman, E. E, 1946, Studies on the trematode family Azygiidae. III. The morphology and life cycle of *Proterometra sagittaria* n. sp. *Trans. Am. Microsc. Soc.*, 65, 37–44.

Dickerman, E. E, 1954, *Paurorhynchus hiodontis*, a new genus and species of trematode (Bucephalidae: Paurorhynchinae n. subfam.) from the mooneye fish, *Hiodon tergisus. J. Parasit.*, 40, 311–315.

Di Conza, J. J. and Cooper, R. C, 1969, Occurrence of the nematode, *Philometra americana*, in the English sole, *Parophrys vetulus*, in San Francisco Bay. *Calif. Fish Game*, 55, 327–329.

Didenko, P. P., Sapozhnikov, G. I., Babii, Yu, A., Verbovyi, G. N. and Parozonyak, P. I, 1979, [Coriban in *Sanguinicola* infection.] *Veterinariya, Mosk., No. 5*, 49–50. [In Russian.]

Dienske, H, 1968, A survey of the metazoan parasites of the rabbit-fish, *Chimaera monstrosa* L. (Holocephali). *Netherl. J. Sea Res.*, 4, 32–58.

Dillon, W. A and Hargis, W. J., Jr, 1965a, Monogenetic trematodes from the Southern Pacific Ocean. 1. Monopisthocotyleids from New Zealand fishes. Biology of the Antarctic Seas II. *Antarctic Research Series, No. 5,* Washington, DC, USA: American Geophysical Union, 229–249.

Dillon, W. A. and Hargis, W. J., Jr, 1965b, Monogenetic trematodes from the Southern Pacific Ocean. 2. Polyopisthocotyleids from New Zealand fishes: The families Discocotylidae, Microcotylidae, Axinidae and Gastrocotylidae(1). Biology of the Antarctic Seas II. *Antarctic Research Series No. 5,* Washington, DC, USA: American Geophysical Union, 251–280.

Dillon, W. A. and Hargis, W. J., Jr, 1968, Helminth parasites of antarctic vertebrates. Part III. Monogenetic trematodes from Antarctic Fishes: The superfamily Tetraonchoidea Yamaguti, 1963. Antarctic Research Series. No. 5 Biology of the Antarctic Seas III. Washington, DC., USA: American Geophysical Union, 101–112.

Dobrovolny, C. G, 1939, Life history of *Plagioporus sinitsini* Mueller and embryology of new cotylocercous cercariae (Trematoda). *Trans. Am. Microsc. Soc.*, **58**, 121–155.

Dobson, A. P, 1985, The population dynamics of competition between species. *Parasitology*, **91**, 317–347.

Dobson, A. P. and Keymer, A. E, 1985, Life history models. In: *Biology of the Acanthocephala* (Eds Crompton, D. W. T. and Nickol, B. B.) Cambridge: Cambridge University Press, pp. 347–384.

Dobson, A. P. and May, R. M, 1986, The effects of parasites on fish populations—theoretical aspects. In: *Parasitology-Quo Vadit?* Proceedings of the Sixth International Congress of Parasitology. (Ed. Howell, M. J.). Canberra, Australia: Australian Academy of Science, pp. 363–371.

Dobzhansky, T, 1937, Genetics and the origin of species. *New York*, Columbia University Press, xvi + 364 pp.

Dogiel, V. A, 1961, Ecology of the parasites of freshwater fishes. In: *Parasitology of fishes.* (Eds Dogiel, V. A., Petrushevski, G. K. and Polyanski, Yu. I.). Edinburgh and London: Oliver and Boyd, pp. 1–48.

Dogiel, V. A, 1962, *{General parasitology.}* Leningrad: Izdatel'stvo Leningradskogo Universiteta, 3rd edition, 464 pp. [In Russian.]

Dogiel, V. A, 1964, *General parasitology*. Edinburgh and London: Oliver and Boyd, [translated from Russian by Z. Kabata.] 1x + 516 pp.

Dogiel, V. A. and Bychovsky, B. E, 1939, [Parasites of the fishes of the Caspian sea]. *Trudy Kompleksnoi izucheniyu Kaspiikogo morya*, 7, 1–150. [In Russian.]

Dogiel, V. A. and Petrushevski, G. K, 1935, [An attempt at an ecological study of parasite fauna of salmon in the White Sea.] *Probl. Ecol. Biocen. Leningrad*, 1, 137–169. [In Russian.]

Dogiel, V. A, Petrushevski, G. K, and Polyanski, Y. I. (Eds), 1961, *Parasitology of fishes*. Edinburgh and London: Oliver and Boyd, (Translation from Russian of: Basic problems of the parasitology of fishes 1958.) Leningrad: Izdatel'stvo Leningradskogo Universiteta, x +384 pp.

Dolgikh, A. V, 1968, [Some characteristics of the biology of cercariae of *Bacciger bacciger* (Rudolphi, 1819).] In: Bodyanitski, V. A. (Ed.), *Biology of Seas No. 14. Parasites of marine animals*. Kiev, USSR: 'Naukova Dumka', pp. 127–132. [In Russian.]

Dollfus, R. P, 1923, Énumération des cestodes du plancton et des invertébrés marins. *Annls Parasit. Hum. Comp.*, 1, 276–300

Dollfus, R. P, 1935, Nématode du genre *Goezia* chez une truite arc-en-ciel (*Salmo irideus*) d'élévage. *Bull. Soc. Zool. Fr.*, **60**, 244–265.

Dollfus, R. P, 1946, Parasites (animaux et vegetaux) des helminthes. Hyperparasites, ennermis et predateurs des helminthes parasites et des helminthes libres. *Encyclopedie Biologique,* 27, 1–482.

Dollfus, R. P, 1949, Sur une cercaire ophthalmoxiphidiocerque, *Cercaria isopori* A. Looss 1894, et sur la délimitation des Allocreadioidea. *Annls Parasit. Hum. Comp.*, 24, 424–435.

Dollfus, R. P, 1953, Parasites animaux de la morue Atlanto-Arctique (*Gadus callarias* L. (= *morhua* L.). *Encycl. Biol.*, *1–423.*

Dollfus, R. P, 1958, Cours d'helminthologie. 1. Trématodes sous-classe Aspidogastrea. *Annls Parasit. Hum. Comp.*, 33, 20–395

Dollfus, R. P, 1960, Distomes des Chaetognathes. *Bulletin I. P. M. M*, No. 4, 1–27.

Dollfus, R. P, 1963, Liste des Coelentères marins, Palearctiques et Indiens ou ont été trouvés des trématodes digenetiques. *Bulletin I. P. M. M*, No. 9–10, 33–57.

Dollfus, R. P, 1964a, Sur le cycle évolutif d'un cestode diphyllide. Identification de la larve chez *Carcinus maenas* (L. 1758), hôte intermédiaire. *Annls Parasit. Hum. Comp.*, 39, 235–241.

Dollfus, R. P, 1964b, Énumération des cestodes du plancton et des invertébrés marins, *Annls Parasit, Hum. Comp.* 39 329–379.

Dollfus, R. P, 1967, Énumération des Cestodes du plancton et des invertébrés marins. *Annls Parasit. Hum. Comp.* 42, 155–178.

Dollfus, R. P, 1971, De quelques trématodes digenetiques de selaciens du Senegal et considerations anatomiques sur les Anaporrhutinae A. Looss, 1901. *Bulletin IFAN*, 33, 347–370.

Dollfus, R. P, 1974, Énumération des Cestodes du plancton et des invertébrés marins. *Annls Parasit. Hum. Comp.*, 49, 381–410.

Dollfus, R. P. and Euzet, L, 1973, Nouvelles données sur *Paracyclocotyla cherbonnieri* Dollfus, 1970 (Monogenea). *Bull. Mus. Natn. Hist. Nat., Paris, 3(e) ser., No. 137, Zoologie*, 101, 815–819.

Dollfus, R. P. and Campana-Rouget, Y, 1956, Helminthes found in the digestive system of the coelacanth. *Mem. Inst. Sci., Madagascar*, 11, 33–41.

Dollfus, R. P., Chabaud, A. G. and Golvan, Y. J, 1957, Helminthes de la region de Banyuls. V. Nouveau distome *Aphalloides coelomicola* n. gen. n. sp. de la cavité générale d'un gobius d'eau saumatre. *Annls Parasit. Hum. Comp.*, 32, 28–40.

Dolmen, D, 1987, *Gyrodactylus salaris* (Monogenea) in Norway: infestations and management. In: *Parasites and diseases in natural waters and aquaculture in Nordic countries. Proc. Symp., 2–4 December 1986, Stockholm, Sweden* (Eds Stenmark, A. and Malmberg, G.). Stockholm, Sweden: Zoo-tax, Naturhistoriska Riksmuseet, pp. 63–69.

Dombroski, E, 1955, Cestode and nematode infection of sockeye smolts from Babine Lake, British Columbia. *J. Fish. Res. Bd Can.*, 12, 93–96.

Dönges, J, 1963, Reizphysiologische Untersuchungen an der Cercariae von *Posthodiplostomum cuticola* (v. Nordmann 1832) Dubois 1936, dem Erreger des Diplostomatiden—Melanoms der Fische. *Verh. Dt. Zool. Ges., Year 1963*, 216–223.

Dönges, J, 1964, Der Lebenszyklus von *Posthodiplostomum cuticola* (v. Nordmann 1831) Dubois 1936 (Trematoda, Diplostomatidae). *Z. ParasitKde*, 24, 160–248.

Dönges, J, 1967, Parasites support continental drift. *New Scientist*, 28 Sept, 1967, p. 687.

Dontsov, Yu. S, 1979, [Effect of the regulated flow of the Volga river on the helminth fauna of fish in the Volga cascade revervoirs.] In: *Fauna, sistematika, biologiya i ekologiya gel'mintov i ikh promezhutochnykh khozyaev, 1979 (Respublikanskii sbornik).* Gor'kii, USSR: Gor'kovskii Gosudarstvennyi Pedagogicheskii Institut, pp. 13–40. [In Russian.]

Dontsov, Yu. S. and Markov, S, 1981, [The effect of the regulation of the flow of the Don on the parasite fauna of *Stizostedion lucioperca*.] In: *Fauna, sistematika, biologiya i ekologiya gel'mintov i ikh promezhutochnykh khozyaev*. Gor'kii, USSR: Izdatel'stvo Gor'kovskogo Gosudarstvennogo Pedagogicheskogo Instituta, pp. 17–24. [In Russian.]

Dossou, C. and Euzet, L, 1984a, Parasites de poissons d'eau douce du Benin II. Espèces nouvelles du genre *Bouixella* (Monogenea), parasites de Mormyridae. *Bull. Mus. Natn. Hist. Nat., Paris, 4(e) ser., 6, Section A*, 1, 41–47.

Dossou, C. and Euzet, L, 1984b, Les monogènes du genre *Eutrianchoratus paperna*, 1969 parasites branchiaux *d'Ophiocephalus obscuris* au Benin (Afrique). *Bull. Soc. Zool. Fr.*, 239–247.

Dryuchenko, E. A, 1982, [Formation of cadaverine in some fish helminths and in the liver and intestine of their hosts.] In: *Gel'minty v presnovodnykh biotsenozakh*. Moscow, USSR: 'Nauka', pp. 81–85. [In Russian.]

Dubey, N. K, Dubey, U. and Pandey, P. K, 1981, Worm cataract of *Heteropneustes fossilis* (Bloc). *Curr. Sci.*, 50, 651.

Dubinina, M. N, 1960, [The morphology of Amphilinidae (Cestoda) in relation to their position in the classification of Platyhelminthes. *Dokl. Akad. Nauk SSSR*, 135, 501–504. [In Russian.]

Dubinina, M. N, 1966, [*Ligulidae (Cestoda) of the USSR*.] Moscow and Leningrad: Izdat. 'Nauk'

(English translation 1980, New Delhi, India: Amerind Publishing Co. Pvt. Ltd., vii + 320 pp), 261 pp. [In Russian.]

Dubinina, M. N, 1974, [The development of *Amphilina foliacea* (Rud.) at all stages of its life-cycle and the position of Amphilinidea in the systematics of Platyhelminthes.] *Parazit. Sb., No. 26*, 9–38 [In Russian.]

Dubinina, M. N, 1980, [The importance of organs of attachment in the phylogeny of tapeworms.] *Parazit. Sb., No. 29*, 29, 65–83. [In Russian.]

Dubinina, M. N, 1982, [Parasitic worms of the Class Amphilinida (Platyhelminthes).] *Tr. Zool. Inst., Leningrad*, 100, 1–143. [In Russian.]

Dubovskaya, A. Ya, 1979, [A study of the arginase activity in cestode plerocercoids and adults.] *Trudy Gel'm. Lab. (Gel'm. zhivot. rast.)*, 29, 46–49. [In Russian.]

Dubovskaya, A. Ya, 1982, [Study of the properties of arginase in adult and larval cestodes, parasitic in freshwater fish.] In: *Gel'minty v presnovodnykh biotsenozakh*. Moscow, USSR: 'Nauka', pp. 85–90. [In Russian.]

Dubovskaya A. Ya, and Shishova-Kasatochkina, O. A, 1984, [Adaptation and pre-adaptation of helminths to the thermal homeostasis of their hosts.] *Trudy Gel'mint. Lab. (Biokhim. fiziol. gel'm. immun. gel'mint.)*, 32, 24–38. [In Russian.]

Duijn, van C, 1956, *Diseases of fishes*. Dorset House, London: Dorset House, 174 pp.

Duijn, van C, 1973, *Diseases of fishes. Third edition*. London: Iliffe Books, 392pp.

Dukes, T. W, 1975, Ophthalmic pathology of fishes. In: *The pathology of fishes* (Eds Ribelin, W. E. and Migaki, G.) Wisconsin: University of Wisconsin Press, pp. 383–395.

Dulin, M. P, 1977, Aquatic animal medicine: a new speciality for veterinarians. *Vet. Med. S. Anim. Clin.*, 72, 1060–1065.

Duniec, H, 1980, Parasitic fauna of the grey gurnard *Trigla gurnardus* (L.) from Shetland Islands fishing grounds. *Acta Ichth. Pisc.*, 10, 65–77.

Dunn, I. J., Russell, L. R. and Adams, J. R, 1983, Cecal histopathology caused by *Truttaedacnitis truttae* (Nematoda: Cucullanidae) in rainbow trout, *Salmo gairdneri*. *Int. J. Parasit.*, 13, 441–445.

Dupree, H. K, 1981, An overview of the various techniques to control infectious diseases in water supplies and water reuse aquacultural systems. In: *Proc. Bio-Eng. Symp. fish cult., Traverse City, Michigan, 16–18 October 1979*. (Eds Allen, L. J, and Kinney, E. C.) Bethesda, Maryland, USA: American Fisheries Society, Fish Culture Section, pp. 83–89.

Duran, L. E, and Oliva, M, 1980, Estudio parasitológico en *Merluccius gayi peruanus* Gingsburg 1954. *Bol. Chil. Parasit.*, 35, 18–21.

Durette-Desset, M. C, 1983, Keys to genera of the superfamily Trichostrongyloidea. CIH *Keys to the nematode parasites of vertebrates*. Farnham Royal, England: Commonwealth Agricultural Bureau, No. 10, 83 pp.

Dyk, V, 1956, [Parasitofauna ryb tatranskych ples, Ceskoslovenska.] *Cslká Parasit.*, 3, 33–42. [In Czech.]

Dyk, V, 1961, Ekonomicznie wazne parazytozy ryb w Czechoslowacji. *Wiad. Parazyt.*, 7, 802–807.

Dyk, V. and Lucky, Z, 1956, Parasitofauna ryb reky Moravice. *Prirodovedecky Sbornik Ostravskeho Kraje, Opava*, 17, 571–580.

Dzhalilov, U. D, 1976, [A new monogenean from the nasal cavity of *Schizopygopsis stoliczkai* Steindachner, 1866).] *Dokl. Akad. Nauk Tadzhik. SSR*, 19, 64–67. [In Russian.]

Dzhalilov, U. D, 1976b, [Parasites of some commercial fish in Kairakkumskoe reservoir (Tadzhik, SSR) (from 1972 data).] *Proceedings of the Fifth Science Congress*, pp. 279–293.

Dzika, E, 1987, Annual occurrence dynamics of common monogeneans on the gills of bream from the lake Goslawskie (Poland). *Acta Parasit. Pol.*, 32, 121–137.

Dzubic, A., Kiskaroly, M., Ibrovic, M. and Sabic, S, 1981. [Experimental investigation into the use of some antiparasitic drugs in carp ponds, with particular reference to their action on liver tissue.] *Veterinaria, Yugoslavia*, 30, 257–262.

Eastburn, R. L., Fritsche, T. R. and Terhune, C. A., 1986, Human infections with the salmon fluke *Nanophyetus salmincola* in North America. In: *Parasitology-quo Vadit? Handbook of the Sixth International Congress of Parasitology, Brisbane, Australia*. (Ed. Howell, M. J.). Canberra: Australian Academy of Science. p. 127

Eddy, S. B. and Lankester, M. W, 1978, Feeding and migratory habits of arctic char, *Salvelinus alpinus*, indicated by the presence of the swimbladder nematode *Cystidicola cristivomeri* White. *J. Fish. Res. Bd Can.*, 35, 1488–1491.

Egidius, E, 1984, Diseases of salmonids in aquaculture. *Diseases of marine organisms. Internat. Helgoland Symp. 11–16 Sept. 1983, Helgoland, Hamburg, GFR*. (Eds Kinne, O. and Bulnheim, H. P.). Helgoländer Meeresuntersuchungen, 37, 547–569.

Egusa, S. and Nakajima, K, 1973, [A bibliography of diseases of fish in Japan. I. Parasitic diseases and parasites.] *Fish Path.*, 7, 137–229. [In Japanese.]

Ehlers, U, 1976, Phylogenetic relationships within the Platyhelminthes. II. *Zoologisches Institut und Museum der Universitat, Gottingen, Federal Republic of Germany*, pp. 143–158.

Ehlers, U, 1984, Phylogenetisches System der Plathelminthes. *Verh. Naturwiss. Ver. Hamburg*, 27, 291–294.

Ehlers, U, 1985a, *Das Phylogenetische System der Plathelminthes*. Stuttgart: G. Fischer Verlag, 317 pp.

Ehlers, U, 1985b, Phylogenetic relationships within the Platyhelminthes. In: (Eds Conway Morris, S., George, D. G., Gibson, R. and Platt, H. M.) *The origin and relationships of lower invertebrates*. Oxford: Oxford University Press, pp. 144–158.

Ehlers, U, 1986, Comments on a phylogenetic system of the Platyhelminthes. *Hydrobiologia*, 132, 1–12.

Eiras, J. C. and Reichenbach-Klinke, H. H, 1982, Nematoden als Ursache von Darmknoten bei Süsswasserfischen. [Gesundheitsprobleme des Menschen im Zusammenhangmit Fischen. Beiträge zur Fischereibiologie]. Stuttgart, GFR: Gustav Fischer Verlag. *Fisch. Umwelt, No. 11* 47–55.

El-Naffar, M. K., Saoud, M. F. and Hassan, I. M, 1983, A general survey of the helminth parasites of some fishes from Lake Nasser at Asswan, A. R. Egypt. *Assiut Vet. Med. J.*, 11, 141–148.

El-Naggar, M. M. and Kearn, G. C, 1980, Ultrastructural observations on the anterior adhesive apparatus in the monogeneans *Dactylogyrus amphibothrium* and *D. hemiamphibothrium*. *Z. ParasitKde*, 61, 223–241.

El-Naggar, M. M. and Kearn, G. C, 1983a, Glands associated with the anterior adhesive organs and body margins in the skin-parasitic monogenean, *Entobdella soleae*. *Int. J. Parasit.*, 13, 67–81.

El-Naggar, M. M. and Kearn, G. C, 1983b, The tegument of the monogenean gill parasites *Dactylogyrus amphibothrium* and *D. hemiamphibothrium*. *Int. J. Parasit.*, 13, 579–592.

Elarifi, A. E, 1982, The histopathology of larval anisakid nematode infections in the liver of whiting, *Merlangius merlangus* (L.), with some observations on blood leucocytes of the fish. *J. Fish Dis.*, 5, 411–419.

Eller, L. L, 1975, Gill lesions in freshwater teleosts. In: *The pathology of fishes*. (Eds W. E. Ribelin and G. Migaki.) Madison, USA: University of Wisconsin Press. pp. 305–330.

Ellis, A. E, 1982, Differences between the immune mechanisms of fish and higher vertebrates. In: *Microbial diseases of fish*. (Ed. Roberts, R. J.). London, UK: Academic Press, pp. 1–29.

Elson, K. G. R, 1971, Fish farming of rainbow trout (*Salmo gairdneri*) in Denmark. *Scottish Fish. Bull. No. 35*, 15–20.

Elton, C., Ford, E. B., Baker, J. R. and Gardner, A. D, 1931, The health and parasites of a wild mouse population. *Proc. Zool. Soc., Lond.*, Part 3, 657–721.

Emel'yanov, V. S, 1971, [The spread of *Bothriocephalus gowkongensis* Yeh in the fish farms of the Belorussian SSR. (Preliminary report).] *Trudy Vses. Nauchno-Issled. Inst. Prud. Ryb. Khozyaist. (Vopr. prud. ryb.)*, **18**, 66–68. [In Russian.]

Engashev, V. G, 1969 [Methods of antiparasitic treatment of fish in ponds.] In: *Fish Farming and Diseases of Fish*, [Ed. Martyshev, F. G.) Moscow: Izdatel'stvo 'Kolos' pp. 243–247. [In Russian.]

Engelbrecht, H, 1957, Uber die Haufigkeit des *Echinorhynchus gadi* (Zoega) Muller 1776 in Dorschen (*Gadus morrhua* L.) der Ostsee sowie seine Langenkorrelation zu den Wirtsfischen, *Wiss. Z. Ernst Moritz Arndt-Univ. Greifswald, Math.-Naturwiss. Reihe*, 6, 385–389.

Erasmus, D. A, 1958, Studies on the morphology, biology and development of a strigeid cercaria (*Cercaria X* Baylis 1930). *Parasitology*, **48**, 312–335.

Erasmus, D. A, 1959, The migration of *Cercaria X* Baylis (Strigeida) within the fish intermediate host. *Parasitology*, **49**, 173–190.

Erasmus, D. A, 1962, Studies on the adult and metacercaria of *Holostephanus lukei* Szidat, 1936. *Parasitology*, **52**, 353–374.

Erasmus, D. A, 1972, *The biology of trematodes*. London: Edward Arnold, 312 pp.

Ergens, R, 1981, Nine species of the genus *Cichlidogyrus* Paperna, 1960 (Monogenea: Ancyrocephalinae) from Egyptian fishes. *Folia Parasit.*, **28**, 205–214.

Ergens, R, 1988a, *Paraquadriacanthus nasalis* gen. et sp.n. (Monogenea: Ancyrocephalidae) from *Clarias lazera* Cuvier et Valenciennes. *Folia Parasit.*, **35**, 189–191.

Ergens, R, 1988b, Four species of the genus *Annulotrema* Paperna et Thurston, 1969 (Monogenea: Ancyrocephalinae) from Egyptian freshwater fish. *Folia Parasit.*, **35**, 209–215.

Ergens, R. and Gelnar, M, 1985, Experimental verification of the effect of temperature on the size of hard parts of the opisthaptor of *Gyrodactylus katharineri* Malmberg, 1964 (Monogenea). *Folia Parasit.*, 32, 377–380.

Ergens, R. and Repova, R, 1984, First record of the nematode *Pseudocapillaria brevispicula* (Linstow 1873) from aquarium fishes. *Folia Parasit.*, 31, 241–245.

Erwin, B. E. and Halton, D. W, 1983, Fine structural observations on spermatogenesis in a progenetic trematode, *Bucephaloides gracilescens*. *Int. J. Parasit.*, 13, 413–426.

Escalante, A. H, 1983, Larvas plerocercoides de Diphyliobothridae Lühe, 1910: hallazgo en peces marinos de consumo humano en la costa peruana. *Bol. Chil. Parasit.*, **38**, 50–52.

Esch, G. W, (Ed.), 1977, Regulation of parasite populations. Proc. Symp. sponsored by Amer. Microscop. Soc. and Amer. Soc. Parasit., New Orleans, 10–14 Nov 1975. London, UK: Academic Press, Inc., xi + 253 pp.

Esch, G. W., Gibbons, J. W. and Bourque J. E, 1975, An analysis of the relationship between stress and parasitism. *Amer. Midl. Natur.* **93**, 339–353.

Esch, G. W. and Huffines, W. J, 1973, Histopathology associated with endoparasitic helminths in bass. *J. Parasit.*, **59**, 306–313.

Esch, G. W, 1982, Abiotic factors: an overview. In: *Parasites—their world and ours. Proceedings of the Fifth International Congress of Parasitology, Toronto, Canada, 7–14 August, 1982.* (Eds Mettrick, D. F. and Desser, S. S.) Amsterdam: Elsevier Biomedical Press, pp. 279–288.

Esch, G. W., Bush, A. O. and Aho, J. M. (Eds) 1990, *Parasite Communities: Patterns and Processes.* London: Chapman and Hall, i–xi + 335 pp.

Esch, G. W., Hazen, T. C. and Aho, J. M, 1977, Parasitism and r- and K-selection. In: (Ed. Esch, G. W.) *Regulation of parasite populations.* (Proc. Symp. sponsored by Am. Microsc. Soc. and Am. Soc. Parasit., New Orleans, 10–14 Nov. 1975). London: Academic Press Inc., pp. 9–62.

Esch, G. W., Johnson, W. C. and Coggins, J. R, 1975, Studies on the population of *Proteocephalus ambloplitis* (Cestoda) in the smallmouth bass. *Proc. Okla. Acad. Sci.*, 55, 122–127.

Esch, G. W., Kennedy, C. R., Bush, A. O. and Aho, J. M, 1988, Patterns in helminth communities in freshwater fish in Great Britain: alternative strategies for colonization. *Parasitology*, **96**, 519–532.

Essex, H. E, 1927, Early development of *Diphyllobothrium latum* in northern Minnesota. *J. Parasit.*, 14, 106–109.

Essex, H. E, 1928, On the life-history of *Bothriocephalus cuspidatus* Cooper, 1917, a tapeworm of the wall-eyed pike. *Trans. Am Microsc. Soc.*, 47, 348–355.

Essex, H. E, 1929a, *Crepidobothrium fragile* n. sp., a tapeworm of the channel catfish, *Parasitology*, 21, 164–167.

Essex, H. E, 1929b, A report from the Mississippi River and other waters with respect to infestation by *Diphyllobothrium latum*. *Minn. Med.*, 12, 149–150.

Essex, H. E, 1929c, The life-cycle of *Haplobothrium globuliforme* Cooper, 1914. *Science*, **69**, 677–678.

Eto, A., Sakamoto, S., Fujii, M. and Yone, Y, 1976, Studies on an anaemia of yellowtail parasitized by a trematode, *Axine (Heteraxine) heterocerca*. *Rep. Fish. Res. Lab., Kyushu Univ.*, 3, 45–51.

Eure, H, 1976, Seasonal abundance of *Proteocephalus ambloplitis* (Cestoidea: Proteocephalidea) from largemouth bass living in a heated reservoir. *Parasitology*, 73, 205–212.

Euzet, L, 1957, Cestodes des selaciens. *Symposium on host specificity among parasites of vertebrates (1st), University of Neuchâtel, April 15–18, 1957. University of Neuchâtel*, pp. 259–269.

Euzet, L, 1959, Récherches sur les cestodes tétraphyllides des sélaciens Côtes de France. *Thesis, University of Montpellier*, 263 pp.

Euzet, L, 1974, Essai sur la phylogénèse des cestodes à la lumière de faits nouveaux. In: *Proceedings of the Third International Congress of Parasitology, Munich, Aug. 25–31, 1974, Vol. I. Vienna, Austria: FACTA Publication*, pp. 378–379.

Euzet, L, 1982, Problèmes posés par la spécificité parasitaire des cestodes Protocephalidea et Pseudophyllidea parasites de poissons. (*Deux Symp. spec. parasit. parasites des vertébrés, 13–17 Avril, 1981. Colloque internat. CNRS.) Mems Mus. Natn. Hist. Nat., Paris, Ser. A, Zool.*, 123, 279–287.

Euzet, L, 1984, Diplectanidae (Monogenea) parasites de poissons des Iles Kerkennah (Tunisie). *Archs Inst. Pasteur Tunis*, 61, 463–474.

Euzet, L. and Birgi, E, 1975, *Heterobothrium fluviatilis* n. sp. (Monogenea, Diclidophoridae) parasite bran-chial de *Tetraodon fahaka* Bennett, 1834 (Teleostei au Tchad). *Bull. Soc. Zool. Fr.*, **100**, 411–420.

Euzet, L. and Cauwet, A, 1967, *Polylabris diplodi* n.g., n. sp. (Monogenea, Microcotylidae) parasite de téléostéens du genre *Diplodus* (Sparidae). *Bull. Mus. Natn. Hist. Nat., Paris, 2(e) Serie No. 39*, 213–220.

Euzet, L. and Combes, C, 1980, Les problèmes de l'éspèce chez les animaux parasites. In: *Les problèmes de l'éspèce dans le règne animal. Tome III. Paris, France: Société Zoologique de France, No. 40. (Méms Soc. Zool. Fr.*, pp. 239–283.

Euzet, L. and Dossou, C, 1975, Parasites de poissons d'eau douce du Dahomey. I. Espèces nouvelles du genre *Heteronchocleidus* (Monogenea), parasites d'Anabantidae. *Bull. Mus. Natn. Hist. Nat., Paris, 3(e) ser. no. 282, Zool.*, 23–34.

Euzet, L. and Dossou, C, 1976, Découverte de monogènes (*Bouixella* n.g.) les Mormyridae (Teleostei) au Benin. *C. R. Acad. Sci., Ser. D.*, **283**, 1413–1416.

Euzet, L. and Durette-Desset, M. C, 1973, *Diplectanum cayennensis* n. sp. (Monogenea) parasite branchial de *Plagioscion auratus* (Castelnau, 1855) (Teleosteen, Sciaenidae) sur les côtes de Guyane. *Bull. Mus. Natn. Hist. Nat., Paris, 3 ser, no. 137, Zool.*, **101**, 789–794.

Euzet, L. and Ktari, M. H, 1970a, *Pyragraphorus hollisae* sp. nov. (Monogenea) parasite de *Lichia glauca* (L., 1758) (Carangidae) en Mediterranée. *An. Inst. Biol. Univ. Nac. Auton. Mex., Ser. Zool.*, 41, 61–72.

Euzet, L. and Ktari, M. H, 1970b, *Heteraxinoides hannibali* n. sp. (Monogenea, Polyopisthocotylea), parasite branchial de *Pomadasys incisus* (Bowdich, 1825) (Teleostei) dans le Golfe de Tunis. *Bull. Mus. Natn. Hist. Nat., Paris, 2(e) Serie*, 41, 269–279.

Euzet, L. and Ktari, M. H, 1971, *Aspinatrium gallieni* n. sp. (Monogenea, Polyopisthocotylea) parasite de *Strongylura acus* Lacépède, 1803 en Mediterranée. *Bull. Soc. Zool. Fr.*, **96**, 509–517.

Euzet, L. and Maillard, C, 1967, Parasites de poissons de mer ouest-Africains, recoltés par J. Cadenat. *Bull. Inst. Fond. Afr. Noire*, **29**, 1435–1493.

Euzet, L. and Maillard, C, 1973, Sur deux Microcotylidae (Monogenea) parasites branchiaux de téléostéens du genre *Diplodus* (Sparidae). *Bull. Mus. Natn. Hist. Nat., Paris, 3(e) Sér. No. 137, Zool.*, 795–805.

Euzet, L. and Maillard, C, 1974, Les monogènes Hexabothriidae Price, 1942. Historique, systèmatique, phylogenese. *Bull. Mus. Natn. Hist. Nat., 3(e) Serie No. 206, Zool.* **136**, 113–141.

Euzet, L. and Noisy, D, 1978/9, *Microcotyle chrysophrii* van Beneden et Hesse, 1863 (Monogenea, Microcotylidae), parasite du téléostéen *Sparus aurata*: précisions morpho-anatomiques sur l'adulte et l'oncomiracidium. *Vie Milieu*, **28/29**, 569–578.

Euzet, L. and Oliver, G, 1965a, *Lamellodiscus serranelli* n. sp. (Monogenea) parasite de téléostéens du genre *Serranus*. *Annls Parasit. Hum. Comp.*, **40**, 261–264.

Euzet, L. and Oliver, G, 1965b, Diplectanidae (Monogenea) de téléostéens de la Mediterranée Occidentale. II. Parasites d'*Epinephelus gigas* (Brunnich, 1768) (1). *Annls Parasit. Hum. Comp.*, **40**, 517–523.

Euzet, L. and Oliver, G, 1966, Diplectanidae (Monogenea) des téléostéens de la Mediterranée Occidentale. III. Quelques *Lamellodiscus* Johnston and Tiegs 1922, parasites de poissons du genre *Diplodus (Sparidae)*. *Annls Parasit. Hum. Comp.*, 41, 573–598.

Euzet, L. and Raibaut, A, 1960, Le développement post-larvaire de *Squalonchocotyle torpedinis* (Price 1942) (Monogenea, Hexabothriidae). *Bull. Soc. Neuchâtel. Sci. Nat.*, **83**, 101–108.

Euzet, L. and Raibaut, A, 1985, Les maladies parasitaires en pisculture marine. *Symbioses*, **17**, 51–68.

Euzet, L. and Sanfilippo, D, 1983, *Ligophorus parvicirrus* n. sp. (Monogenea, Ancyrocephalidae) parasite de *Liza ramada* (Risso, 1826) (Teleostei, Mugilidae). *Annls Parasit. Hum. Comp.*, **58**, 325–335.

Euzet, L. and Vala, J. C, 1975, Monogènes de poissons marins de côtes de Maroc. Description de *Calceostoma herculanea* n. sp. parasite d'*Umbrina canariensis* Valenciennes, 1845. *Vie Milieu*, **25**, 277–288.

Euzet, L. and Vala, J. C, 1977, Monogènes parasites de Mullidae (Teleostei) des côtes de la Guadeloupe. *Inst. Biol. Publs Espec.*, 4, 35–44.

Euzet, L. and Wahl, E, 1970a, Biologie de *Rhinecotyle crepitacula* Euzet et Trilles, 1960 (Monogenea) parasite de *Sphyraena piscatorum* Cadenat, 1964 (Teleostei) dans la lagune Ebrie (Côte d'Ivoire). *Rev. Suisse Zool.*, 77, 687–703.

Euzet, L. and Wahl, E, 1970b, Parasites de poissons de mer ouest-Africains, récoltes par J. Cadenat. VII. Sur un monogène de *Hynnis goreensis* Cuv. et Val. (Téléostéens, Carangidae). *Bull. IFAN*, **32**, 73–82.

Euzet, L. and Wahl, E, 1977, *Bicotylophora baeri* n. sp. (Monogenea) parasite branchial du télóstéen *Trachinotus falcatus* (L., 1758) en Côte-d'Ivoire. *Rev. Suisse Zool.*, **84**, 71–79.

Euzet, L. and Williams, H. H, 1960, A re-description of the trematode *Calicotyle stossichii* Braun, 1899, with an account of *Calicotyle palombi* sp. nov. *Parasitology*, **50**, 21–30.

Euzet, L., Agnèse, J. F. and Lambert, A, 1989, Value of parasites as criteria for identifying the host species. Convergent demonstration by a parasitological study of branchial monogeneans and by genetic analysis of their hosts. *C. R. Acad. Sci., III (Sci. Vie)*, **308**, 385–388.

Euzet, L. Renaud, F. and Gabrion, C, 1984, Le complexe *Bothriocephalus scorpii* (Mueller, 1776) (Cestoda): différenciation à l'aide des méthodes biochimiques de deux espèces parasites du turbot (*Psetta maxima*) et de la barbue (*Scophthalmus rhombus*). *Bull. Soc. Zool. Fr.*, **109**, 343-346.

Euzet, L., Swiderski, Z. and Mokhtar-Maamouri, F, 1981, Ultrastructure comparée du spermatozoide des cestodes. Rélations avec la phylogénèse. *Annls Parasit. Hum. Comp.*, 56, 247-259.

Evans, N. A, 1977, The site preference of two digeneans, *Asymphylodora kubanicum* and *Sphaerostoma bramae* in the intestine of the roach. *J. Helminth.*, 51, 197-204.

Evans, N. A, 1978, The occurrence and life-history of *Asymphylodora kubanicum* (Platyhelminthes: Digenea: Monorchidae) in the Worcester-Birmingham canal, with special reference to the feeding habits of the definitive host, *Rutilus rutilus*. *J. Zool., Lond.*, **184**, 143-153.

Fagerholm, H. P, 1982, Parasites of fish in Finland. VI. Nematodes. *Acta Acad. Abo., Ser. B, Math. Phys. Mat. Naturvet. Tek.*, **40**, 128 pp.

Fagerholm, H. P, 1990, Systematic position and delimitation of ascaridoid nematode parasites of the genus *Contracaecum* with a note on the superfamily Ascaridoidea. *Thesis: Abo Akad. Kopieringscentral, (ISBN 051-649-730-6).*

Fagerholm, H. P. and Valtonen, E. T, 1980, Metazoan parasites of migratory whitefish (*Coregonus lavaretus* L.) from two areas in the northern Baltic region separated by a salinity gradient. *Bothnian Bay Rep.*, 2, 67-33.

Fagerholm, H. P., Kuusela, K. and Valtonen, E. T, 1982, On the occurrence of *Cystidicoloides ephemeridarum* (Nematoda: Spiruroidea) in grayling (*Thymallus thymallus*) in the Oulanka and Kitkajoki rivers. *Memoranda Soc. Fauna Flora Fennica*, **58**, 67-70.

Faisal, M, 1988, Prevalence of *Macrogyrodactylus clarii* among the African catfish *Clarias lazera* (C and V) and possibilities of its treatment. *Vet. Med. Rev.*, 59, 82-87

Fallis, A. M. (Ed.) 1971, Ecology and physiology of parasites. *Symposium, University of Toronto, 19-20 February 1970.* Toronto, Canada: University of Toronto Press, x +258 pp.

Fallon, M. E. and Wallace, D. C, 1977, The occurrence of *Phyllodistomum undulans* in the urinary bladder of the mottled sculpin, *Cottus bairdi*. *Trans. Am. Fish. Soc.*, **106**, 189-191.

Fantham, H. B. and Porter, A, 1948, The parasitic fauna of vertebrates in certain Canadian fresh waters, with some remarks on their ecology, structure and importance. *Proc. Zool. Soc. Lond.*, 117, 609-649.

Fedorov, K. P, 1981, [Mathematical methods in the study of parasite populations.] *Itogi Nauki i Techniki, Zooparazitologiya, Volume 7, Popyulyatsionnye i biotsenoticheskie aspekty izucheniya gel'mintov*, 134-184. [In Russian.]

Feijo, L. M. F., Oliveira Rodrigues, H. de and Sodre Rodrigues, S, 1979, Contribuiscao ao estudo da fauna helmintológica de sardinhas (*Sardinella* sp.) do litoral do estado do Rio de Janeiro. *Atas Soc. Biol. Rio de J.*, 20, 23-28.

Ferguson, M. S, 1943, Migration and localisation of an animal parasite within the host. *J. Exp. Zool.*, 93, 375-400.

Fernandes, B. M. M., Kohn, A. and Magalhaes Pinto, R, 1985, Aspidogastrid and digenetic trematodes parasites of marine fishes of the coast of Rio de Janeiro State, Brazil. *Revta Bras. Biol.*, 45, 109-116.

Fernando, C. H. and Furtado, J. I, 1963, Helminth parasites of some Malayan fresh-water fishes. *Bull. Nat. Mus., Singapore, No. 32*, 45-71.

Fernando, C. H. and Hanek, C, 1976, Gills. In: *Ecological aspects of parasitology*. (Ed. Kennedy, C. R.) Amsterdam, The Netherlands: North-Holland Publishing Company, pp. 209-226.

Ferreira, L. F., Araújo, A. J. G. de, Confalonieri, U. E. C. and Nuñez, L, 1984, The finding of eggs of *Diphyllobothrium* in human coprolites (4,100-1,950 B.C.) from Northern Chile. *Mem. Inst. Oswaldo Cruz*, 79, 175-180.

Field, D. W. and Eversole, A. G, 1982, Parasite levels of stocked glass eel and elver stages of American eel. *(13th Ann. Meeting, World Mariculture Soc. 1982.) J. Wld Maricult. Soc.*, 13, 268-273.

Filimonova, L. V, 1963, [Life-cycle of the trematode *Nanophyetus schikhobalowi*.] *Trudy Gel'm. Lab.*, 13, 347-357. [In Russian.]

Finlay, J, 1978, Disinfectants in fish farming. *Fish. Manag.*, **9**, 18-21.

Finlayson, J. E, 1982, The alleged alternation of sexual phases in *Kuhnia scombri*, a monogenean of *Scomber scombrus*. *Parasitology*, **84**, 303-311.

Fischer, H, 1968, The life-cycle of *Proteocephalus fluviatilis* Bangham (Cestoda) from smallmouth bass, *Micropterus dolomieui* Lacépède. *Can. J. Zool.*, 46, 569-579.

Fischer, H. and Freeman, R. S, 1969, Penetration of parenteral plerocercoids of *Proteocephalus ambloplitis* (Leidy) into the gut of smallmouth bass. *J. Parasit.*, 55, 766-774.

Fischer, H. and Freeman, R. S, 1973, The role of plerocercoids in the biology of *Proteocephalus ambloplitis* (Cestoda) maturing in smallmouth bass. *Can. J. Zool.*, 51, 133-141.

Fischthal, J. H, 1951, Rhopalocercariae in the trematode sub-family Gorgoderinae. *Amer. Midl. Nat.*, 46, 395–443.

Fischthal, J. H, 1972, Zoogeography of digenetic trematodes from West African marine fishes. *Proc. Helminth. Soc. Wash.*, 39, 192–203.

Fischthal, J. H, 1977, Some digenetic trematodes of marine fishes from the barrier reef and reef lagoon of Belize. *Z. Scripta*, 6, 81–88.

Fischthal, J. H, 1978a, Allometric growth in three species of digenetic trematodes of marine fishes from Belize. *J. Helminth.*, 52, 29–39.

Fischthal, J. H, 1978b, Allometric growth in four species of digenetic trematodes of marine fishes from Belize. *Z. Scripta*, 7, 13–18.

Fischthal, J. H, 1980, Some digenetic trematodes of marine fishes from Israel's Mediterranean coast and their zoogeography, especially those from Red Sea immigrant fishes. *Z. Scripta*, 9, 11–23.

Fischthal, J. H. and Thomas, J. D, 1968, Digenetic trematodes of marine fishes from Ghana: Families Acanthocolpidae, Bucephalidae, Didymozoidae. *Proc. Helminth. Soc. Wash.*, 35, 237–247.

Fischthal, J. H. and Thomas, J. D, 1972, Digenetic trematodes of fish from the Volta River drainage system in Ghana prior to the construction of the Volta Dam at Akosombo in May 1964. *J. Helminth.*, 46, 91–106.

Fischthal, J. H., Carson, D. O. and Vaught, R. S, 1982a, Comparative allometry of size of the digenetic trematode *Bucephalus gorgon* (Linton, 1905) Eckmann, 1932 (Bucephalidae) in two sites of infection in the marine fish *Seriola dumerili* (Risso). *J. Parasit.*, 68, 173–174.

Fischthal, J. H., Carson, D. O. and Vaught, R. S, 1982b, Comparative size allometry of the digenetic trematode *Lissorchis attenuatus* (Monochiidae) at four intensities of infection in the white sucker. *J. Parasit.*, 68, 314–318.

Fischthal, J. H., Carson, D. O. and Vaught, R. S, 1982c, Size allometry of the caryophyllidean cestode *Glaridacris laruei* from the white sucker. *J. Parasit.*, 68, 1175–1177.

Fischthal, J. H., Fish, B. L. and Vaught, R. S, 1980, Comparative allometric growth of the digenetic trematode *Metadena globosa* (Linton 1910) Manter 1947 (Cryptogonimidae) in three species of Caribbean fishes. *J. Parasit.*, 66, 642–644.

Fisher, F. M. Jr. and Read, C. P, 1971, Transport of sugars in the tapeworm *Calliobothrium verticillatum*. *Biol. Bull.*, 140, 46–62.

Fisher, F. M. Jr. and Starling, J. A, 1970, The metabolism of L-valine by *Calliobothrium verticillatum* (Cestoda: Tetraphyllidea): identification of α-ketoisovaleric acid. *J. Parasit.*, 56, 103–107.

Flerov, B. A., Mikryakov, V. R. and Kuperman, B, 1982, [Parasitic and infective processes in fish under toxic action.] In: *Gel'minty v presnovodnykh biotsenozakh*. Moscow, USSR: 'Nauka', pp. 58–68. [In Russian.]

Fletcher, T. C. and Baldo, B. A, 1974, Immediate hypersensitivity responses in flatfish. *Science*, 185, 360

Fletcher, T. C., White, A. and Baldo, B. A, 1980, Isolation of a phosphorylcholine-containing component from the turbot tapeworm, *Bothriocephalus scorpii* (Müller), and its reaction with C-reactive protein. *Parasite Immunol.*, 2, 237–248.

Font, W. F, 1983, Seasonal population dynamics of five species of intestinal helminths of the brook stickleback, *Culaea inconstans*. *Can. J. Zool.*, 61, 2129–2137.

Font, W. F., Heard, R. W. and Overstreet, R. M, 1984, Life cycle of *Ascocotyle gemina* n. sp., a sibling species of *A. sexidigita* (Digenea: Heterophyidae). *Trans. Am. Microsc. Soc.*, 103, 392–407.

Font, W. F., Overstreet, R. M. and Heard, R. W, 1984, Taxonomy and biology of *Phagicola nana* (Digenea: Heterophyidae). *Trans. Am. Microsc. Soc.*, 103, 408–422.

Foor, W. E, 1983a, Nematoda. In: *Reproductive biology of invertebrates. Oogenesis, oviposition and oosorption.* (Eds Adiyodi, K. G. and Adiyodi, R. G.). Chichester, UK: John Wiley and Sons, Ltd. pp. 223–256.

Foor, W. E, 1983b, Nematoda. In: *Reproductive biology of invertebrates. Spermatogenesis and sperm function.* (Eds Adiyodi, K. G, and Adiyodi, R. G.). Chichester, UK: John Wiley and Sons, Ltd. pp. 221–256.

Foreyt, W. J. and Gorham, J. R, 1988, Preliminary evaluation of praziquantel against metacercariae of *Nanophyetus salmincola* in chinook salmon (*Oncorhynchus tshawytscha*). *J. Wildl. Dis.*, 24, 531–554.

Forster, R. P. and Goldstein, L, 1969, Formation of excretory products. In: *Fish Physiology* (Eds Hoar, W. S. and Randall, W. S.), *Vol. 1*. London: Academic Press, pp. 313–345.

Frandsen, F., Malmquist, H. J. and Snorrason, S. S, 1989, Ecological parasitology of polymorphic Arctic charr, *Salvelinus alpinus* (L.), in Thingvallavatn, Iceland. *J. Fish Biol.*, 34, 281–297.

Frankland, H. M. T, 1955, The life history and bionomics of *Diclidophora denticulata* (Trematoda: Monogenea). *Parasitology*, 45, 313–351.

Frayha, G. J, 1983, Lipid metabolism in parasitic helminths. *Adv. Parasit.*, 22, 309–387.

Fréchet, A., Dodson, J. J. and Powles, H, 1983, Les parasites de l'éperlan d'amérique (*Osmerus mordax*) anadrome du Québec et leur utilité comme étiquettes biologiques. *Can. J. Zool.*, 61, 621–626.

Fredericksen, D. W, 1978, The fine structure and phylogenetic position of the cotylocidium larva of *Cotylogaster occidentalis* Nickerson 1902 (Trematoda: Aspidogastridae). *J. Parasit.*, 64, 961–978.

Fredericksen, D. W. and Specian, R. D, 1981, The value of cuticular fine structure in identification of juvenile anisakine nematodes. *J. Parasit.*, 67, 647–655.

Freeman, R. S, 1964a, On the biology of *Proetocephalus parallacticus* MacLulich (Cestoda) in Algonquin Park, Canada. *Can. J. Zool.*, 42, 387–408.

Freeman, R. S, 1964b, Flatworm problems in fish. Can. Fish. Cult., No. 32, 11–18.

Freeman, R. S, 1973, Ontogeny of cestodes and its bearing on their phylogeny and systematics. *Adv. Parasit.*, 11, 481–557.

Freeman, R. S, 1982a, How did tapeworms get that way? *Bull. Can. Soc. Zool.*, 13, 5–8.

Freeman, R. S, 1982b, Do any *Anonchotaenia, Cyathocephalus, Echeneibothrium* or *Tetragonocephalum* (= *Tylocephalum*) (Eucestoda) have hookless oncospheres or coracidia? *J. Parasit.*, 68, 737–743.

Freze, V. I., Dalin, M. V. and Sergeeva, E. G, 1985a, Coracidia agglutination reaction of pseudophyllid cestodes and its application in cestode studies. *Angew. Parasit.*, 26, 85–90.

Freze, V. I., Dalin, M. V., Sergeeva, E. G, 1985b, [Agglutination reaction in pseudophyllidean coracidia and its use in cestodological studies.] *Parazitologiya,* 19, 257–263. [In Russian.]

Fried, B, 1978, Trematoda. In: *Methods of cultivating parasites in vitro.* (Eds Taylor, A. E. R. and Baker, J. R.), London, UK: Academic Press pp. 151–192.

Frimeth, J. P, 1987a, A survey of the parasites of nonanadromous and anadromous brook charr (*Salvelinus fontinalis*) in the Tabusintac River, New Brunswick, Canada. *Can. J. Zool.*, 65, 1354–1362.

Frimeth, J. P, 1987b, Potential use of certain parasites of brook charr (*Salvelinus fontinalis*) as biological indicators in the Tabusintac River, New Brunswick, Canada. 65, 1989–1995.

Fritzche, S, 1975, Untersuchungen zur Einschränkung des Diplostomum-Befalls der Regenbogenforellensetzlinge (*Salmo gairdneri*) in der Teichwirtschaft Dobbin. *Z. Binnenfisch. DDR,* 22, 243–346.

Fuhrmann, O, 1928, Trematoda. In: *Handbuch der Zoologie*, Berlin (Eds Kükenthal and Krumbach) 2, pp. 1–140.

Fuhrmann, O, 1931, Cestoidea. In: *Handbuch der Zoologie*, Berlin (Eds Kükenthal and Krumbach) 2, pp. 141–416.

Fujino, T., Ooiwa, T. and Ishii, Y, 1984, [Clinical, epidemiological and morphological studies on 150 cases of acute gastric anisakiasis in Fukuoka prefecture.] *Jap. J. Parasit.*, 33, 73–92. [In Japanese.]

Fukuda, T., Tongu, Y., Aji, T., Shung Lai, J., Shing, H. L., Shimono, K. and Inatomi, S, 1982, [Anisakidae larvae from some fishes in the Seto Inland sea.] *Jap. J. Parasit.*, 31, 171–176. [In Japanese.]

Fukuda, T. and Yamamoto, S, 1981, Neorickettsia-like organism isolated from metacercaria of a fluke, *Stellantchasmus falcatus*. *Jap. J. Med. Sci. Biol.*, 34, 103–107.

Furtado, J, 1961, A note on intestinal parasites in some Chinese carps. *Malay. Agric. J.*, 43, 239–241.

Furtado, J, 1962, Fish worms. *Malay. Nat. J.*, 16, 49–53.

Fusco, A. C, 1980, Larval development of *Spirocamallanus cricotus*. (Nematoda: Camallanidae). *Proc. Helminth. Soc. Wash.*, 47, 63–71.

Gabrion, C. and Euzet-Sicard, S, 1979, Étude du tégument et des récepteurs sensoriels du scolex d'un plérocercoide de cestode Tetraphyllidea à l'aide de la microscopie électronique. *Annls Parasit. Hum. Comp.*, 54, 573–583.

Gaevskaya, A. V, 1976, [Helminth fauna of Atlantic squids, *Ommastrephes bartrami* le Sueur.] *Trudy Inst. Rybr. Khozyaist. Okean. (Atlant. NIRO) (Biol. Rybokhozyaist. issled. Atlantich. Ok.), Issue* 69, 89–96. [In Russian.]

Gaines, J. L. and Rogers, W. A, 1971, Fish mortalities associated with *Goezia* sp. (Nematoda: Ascaroidea) in Central Florida. *Proc. S. E. Ass. Game and Fish Commissioners*, 496–497.

Gaevskaya A. V. and Kovaleva, A. A, 1980a, [The reasons for the similarities and differences between the parasite faunas of two subspecies of *Trachurus trachurus* in the Atlantic Ocean.] *Nauch. Doke. Vyssh. Shkol. Biol. Nauk., No.* 6, 52–56. [In Russian.]

Gaevskaya, A. V. and Kovaleva, A. A, 1980b, [Ecological-geographical characteristics of the parasite fauna of *Trachurus trachurus trachurus* in the Atlantic Ocean.] In: *Issled. biol. resurs. Atlantich. Ok.* Kaliningrad: USSR, pp. 18–24. [In Russian.]

Gaevskaya, A. V. and Kovaleva, A. A, 1982, [The trematode fauna of Atlantic *Trachurus* and its special features.] *Gidrobiol. Zh.*, **18**, 60–65. [In Russian.]

Gaevskaya, A. V. and Kovaleva, A. A, 1985, [The parasite fauna of the oceanic mackerel, *Trachurus picturatus picturatus* and the ecological and geographical aspect of its formation.] *Ekol. Mor., Kiev, No. 20*, 80–94. [In Russian.]

Gaevskaya, A. V. and Shapiro, L. S, 1981, [The question of the local nature of the Baltic herring (*Clupea harengas membras* L.) in the Vistula lagoon of the Baltic Sea.] In: *Stock State and Principles of the Rational Fishery in the Atlantic.* Kaliningrad, U.S.S.R.: Atlant NIRO, pp. 11–79. [In Russian.]

Gaevskaya, A. V., Kovaliova, A. A. and Rodjuk, G. N, 1985, Parasitofauna of the Fishes of the Falkland-Patagonian region. In: *Parasitology and pathology of marine organisms of the world ocean.* (Ed. Hargis, W. J.) NOAA Technical Report NMFS 25, pp. 25–28.

Gallegos, O. M, 1982, Cestoda. In: (Eds S. H. Hurlbert and A. Villalobos-Figueroa, A.) *Aquatic biota of Mexico, Central America and the West Indies.* San Diego, California: San Diego State University, pp 90–106.

Garrity, R. D, 1975, Studies on the fine structure, and selected uptake and digestive processes of the gut of *Cotylaspis insignis* Leidy, 1857 (Trematoda: Aspidobothria). *Diss. Abstr. Int.*, **36B**, 2574.

Gartner, J. V. Jr. and Zwerner, D. E, 1989, The parasite faunas of meso- and bathypelagic fishes of Norfolk Submarine Canyon, western North Atlantic. *J. Fish Biol.*, **34**, 79–95.

Gattaponi, P. and Corallini, S. C, 1982, Nuove acquisizioni sul parassitismo e sul regime dietetico di *Scardinius erythrophthalmus* L. del Lago Trasimeno. (XXXVI Convegno, Società Italiana delle Scienze Vet., Sanremo, 22–25 Settembre 1982.) *Atti Soc. Ital. Sci. Vet.*, **36**, 672–675.

Gaur, A. S. and Agarwal, S. M, 1981, Non-specific phospho monoesterases in three species of caryophyllids from *Clarias batrachus* (Linn). *Proc. Ind. Acad. Parasit.*, **2**, 71–75.

Gavrilyuk, L. P, 1978, [The effect of NaCl on the survival of *Contracaecum aduncum* (Red., 1802) larvae in *Engraulis encrasicholus ponticus*.] *Biol. Morya, Kiev (Parazit. zhivot. Yuzh. mor.), Issue 45*, 15–20. [In Russian.]

Gear, J. H. S. (Ed.) 1977, Medicine in a tropical environment. *Proceedings of the International Symposium, 19–23 July, 1976, Pretoria, South Africa.* Rotterdam, Netherlands: A. A. Balkema, xiii + 817 pp.

Gebhardt, G. A., Millemann, R. E., Knapp, S. E. and Nyberg, P. A, 1966, 'Salmon poisoning' disease. II. Second intermediate host susceptibility studies. *J. Parasit.*, **52**, 54–59.

Geller, E. R, 1957, [Epizootiology of *Contracaecum* infection of sterlet, *Acipenser ruthenus*.] *Zool. Zh.*, **36**, 1441–1447. [In Russian.]

Gelnar, M, 1987a, Experimental verification of the effect of water temperature on micropopulation growth of *Gyrodactylus katharineri* Malmberg, 1964 (Monogenea) parasitizing carp fry (*Cyprinus carpio* L.) *Folia Parasit.*, **34**, 19–23.

Gelnar, M, 1987b, Experimental verification of the effect of physical conditions of *Gobio gobio* (L.) on the growth rate of micropopulations of *Gyrodactylus gobiensis* Glaser 1974 (Monogenea). *Folia Parasit.*, **34**, 211–217.

Gelnar, M, 1987c, Experimental verification of the effect of the form (scaly or scaleless) of carp host (*C. carpio*L.) on the growth rate of micropopulations of *Gyrodactylus katharineri Malmberg*, 1964 (Monogonea). *Folia Parasit.* **34**, 305–309.

George, P. V. and Nadakal, A. M, 1973, Studies on the life-cycle of *Pallisentis nagpurensis* Bhalerao, 1931, (Pallisentidae, Acanthocephala) parasitic in the fish *Ophiocephalus striatus* (Bloch). *Hydrobiologia*, **42**, 31–43.

George, P. V. and Nadakal, A. M, 1981, Observations on the intestinal pathology of the marine fish, *Rachycentron canadus* (Gunther) infected with the acanthocephalid worm, *Serrasentis nadakali* (George and Nadakal, 1978). *Hydrobiologia*, **78**, 59–62.

George, P. V, and Nadakal, A. M, 1982, Histopathological changes in the intestine of the fish, *Synaptura orientalis* (Bl. and Sch.) parasitised by an acanthocephalid worm, *Echinorhynchus veli* (George and Nadakal, 1978). *Jap. J. Parasit.*, **31**, 99–103.

George, P. V, and Nadakal, A. M, 1983, Studies on encapsulation of immature juvenile of the acanthocephalid worm, *Pallisentis nagpurensis* Bhalerao, 1931 in the liver of definitive host, *Ophiocephalus striatus* (Bloch.). *Jap. J. Parasit.*, **32**, 387–391.

Georgescu, R., Mihai, S., Petica, M., Angelescu, N. and Dascalescu, P, 1981, [*Hepaticola petruschewskii* and *Triaenophorus nodulosus* in fish from nursery ponds and natural waters.] *Bul. Cerc. Pisc.*, **3**(34), 123–130. [In Romanian.]

Geraci, J. R, and St. Aubin, D. J, 1986, Effects of marine parasites on marine mammals. In: (Ed. Howell, M. J.) *Parasitology Quo Vadit? Proceedings of the VI ICOPA Canberra, Australia*: Canberra: Australian Academy of Science, pp. 407–415.

Gerasev, P. I, 1977, [The mechanism of adhesion to the gills of its host in *Dactylogyrus extensus* and *D. achmerowi* (Monogenoidea).] *Parazitologiya*, **11**, 513–519. [In Russian.]

Gerasev, P. I, and Khotenovskii, I. A, 1985, [Some characteristics of the initial stages of morphogenesis of the reproductive system of *Paradiplozoon rutili* (Monogenea, Diplozoidae).] *Parazitologiya*, **19**, 386–393. [In Russian.]

Ghittino, P, 1963, Les maladies des poissons en Italie. *Bull. Off. Int. Epizoot.*, **59**, 59–87.

Ghittino, P, 1972, Aquaculture and associated diseases of fish of public health importance. *J. Am. Vet. Med. Ass.*, **161**, 1476–1485.

Ghittino, P, 1974, Present knowledge of the principal diseases of cultured marine fish. *Riv. Ital. Pisc. Ittiopatol.*, **9**, 51–56.

Ghittino, P, 1979, Principali problemi di parassitologia nelle piscicolture Italiane. (Atti IX Congr. Naz., Soc. Italiana di Parassit., Ravenna, 30 Marz.–1 Apr. 1978). *Parassitologia*, **21**, 27–33.

Gibson, D. I, 1970, Aspects of the development of 'herringworm' (*Anisakis* sp. larva) in experimentally infected rats. *Nytt Mag. Zoo.*, **18** (2), 175–187.

Gibson, D. I, 1972a, Flounder parasites as biological tags. *J. Fish Biol.*, **4**, 1–9.

Gibson, D. I, 1972b, Contributions to the life histories and development of *Cucullanus minutus* Rudolphi, 1819 and *C. heterochrous* Rudolphi, 1802 (Nematoda: Ascaridida). *Bull. Br. Mus. Nat. Hist., Zool.*, **22**, 153–170.

Gibson, D. I, 1973a, The genus *Pseudanisakis* Layman & Borovkova, 1926 (Nematoda: Ascaridida). *J. Nat. Hist.*, **7**, 319–340.

Gibson, D. I, 1973b, Some ultrastructural studies on the excretory bladder of *Podocotyle staffordi* Miller, 1941, (Digenea). *Bull. Br. Mus. Nat. Hist., Zool.*, **24**, 461–465.

Gibson, D. I, 1976, Monogenea and Digenea from fishes. *Discovery Reports*, **36**, 179–266.

Gibson, D. I, 1981, Evolution of digeneans. In: *Evolution of helminths (Workshop 13, EMoP 3)*. (Ed. Willmott, S.) *Parasitology*, **83**, 161–163.

Gibson, D. I, 1983, The systematics of ascaridoid nematodes—a current assessment. In: (Eds Stone, A. R., Platt, H. M. and Khalil, L. F.) Systematics Association Special Vol. 22, *Concepts in Nematode Systematics*. London: Academic Press, pp. 321–338.

Gibson, D. I, 1987, Questions in digenean systematics and evolution. *Parasitology*, **95**, 429–460.

Gibson, D. I. and Bray, R. A, 1977, The Azygiidae, Hirudinellidae, Ptychogonimidae, Sclerodistomidae and Syncoeliidae of fishes from the north-east Atlantic. *Bull. Br. Mus. Nat. Hist., Zool.*, **32**, 167–245.

Gibson, D. I. and Bray, R. A, 1979, The Hemiuroidea: terminology, systematics and evolution. *Bull. Br. Mus. Nat. Hist., Zool.*, **36**, 35–146.

Gibson, D. I. and Bray, R. A, 1982, A study and reorganization of *Plagioporus* Stafford, 1904 (Digenea: Opecoelidae) and related genera, with special reference to forms from European Atlantic waters. *J. Nat. Hist.*, **16**, 529–559.

Gibson, D. I. and Bray, R. A, 1986, The Hemiuridae (Digenea) of fishes from the northeast Atlantic. *Bull. Br. Mus. Nat. Hist. Zool.* **51**, 1–125.

Gibson, D. I. and Chinabut, S, 1984, *Rohdella siamensis* gen. et. sp. nov. (Aspidogastridae: Rohdellinae subfam. Nov.) from freshwater fishes in Thailand, with a reorganization of the classification of the subclass Aspidogastrea. *Parasitology,* **88**, 383–393.

Gibson, D. I., Bray, R. A. and Powell, C. B, 1987, Aspects of the life-history and origins of *Nesolecithus africanus* (Cestoda: Amphilinidea). *J. Nat. Hist.*, **21**, 785–794.

Gibson, D. I., Mackenzie, K. and Cottle, J, 1981, *Halvorsenius exilis* gen. et sp. nov., a new didymozoid trematode from the mackerel *Scomber scombrus* L. *J. Nat. Hist.*, **15**, 917–929.

Gibson, D. I., Rollinson, D. and Matthews, B. E, 1985, The potential use of enzyme studies in elucidating aspects of the systematics and life-history of marine digeneans. (12th Scandinavian Symposium of Parasitology). *Information (Jbo Akademi)*, **18**, 8–9.

Gichenok, L. A, 1979, [Monogeneans from flying fish of the genus *Exocoetus* (Beloniformes) from the Indian and Pacific oceans.] *Zool. Zh.*, 58, 958–968. [In Russian.]

Giles, N, 1983, Behavioural effects of the parasite *Schistocephalus solidus* (Cestoda) on an intermediate host, the three-spined stickleback, *Gasterosteus aculeatus* L. *Animal Behaviour*, **31**, 1192–1194.

Ginetsinskaya, T. A, 1968, [*Trematodes, their life-cycles, biology and evolution.*] Leningrad: Izdatestvo 'Nauka', 411 pp. [In Russian.]

Ginetskinskaya, T. A., Palm, V., Besedina, V. V., 1971, and Timofeeva, T. A, [Accumulation of reserve substances in the yolk glands of platyhelminths.] *Parazitologiya*, **5**, 147–154. [In Russian.]

Giordano, G, 1976, Igiene del pesce, dei crostacei e dei molluschi. *Annali Sanità Pubbl.*, **37**, 3–31.

Gladunko, I. I, 1981, [Biology of *Sanguinicola* in the pre-Carpathian area (Ukrainian SSR).] *Visnik L'vivs'kogo Universitetu (Ekosistemy Karpats'kogo Visokogir'ya, ikh optimizatsiya i okhrana), No. 12*, 100–105. [In Ukrainian.]

Gleason, L. N., Christensen, B. M, and Chung, Y. T, 1983, Archinephric duct lesions caused by *Phyllodistomum superbum* and *P. lysteri* (Digenea: Gorgoderidae) in catostomid fishes. *J. Wildl. Dis.*, 19, 277–279.

Godeau, R., Danis, M., Bouchareine, A. and Nozais, J. P, 1985, Une cause inhabituelle d'oedèmes segmentaires: l'anisakiase. *Press Médicale*, 14, 1246–1247.

Godin, J. G. Y. and Sproul, C. D, 1988, Risk taking in parasitized sticklebacks under threat of predation: effects of energetic need and food availability. *Can. J. Zool.*, 66, 2360–2367.

Goh, S. L. and Davey, K. G, 1976a, Localization and distribution of catecholaminergic structures in the nervous system of *Phocanema decipiens* (Nematoda) *Int. J. Parasit.*, 6, 403–411.

Goh, S. L. and Davey, K. G, 1976b, Selective uptake of noradrenaline, dopa, and 5-hydroxytryptamine by the nervous system of *Phocanema decipiens* (Nematoda): a light autoradiographic and ultrastructural study. *Tissue Cell*, 8, 421–435.

Goil, M. M. and Harpur, R. P, 1978, Studies on sorbitol dehydrogenase from the parasitic nematode larvae of *Phocanema decipiens*. *Z. ParasitKde*, 57, 117–120.

Goil, M. M. and Harpur, R. P, 1979, A comparison of the non-specific acid phosphomonoesterase activity in the larva of *Phocanema decipiens* (Nematoda) with that of the muscle of its host the codfish (*Gadus morhua*). *Z. ParasitKde*, 60, 177–183.

Golovin, P. P, 1987, Young carp susceptibility to *Dactylogyrus vastator* infection. In: *Actual problems in fish parasitology. 2nd International Symposium of Ichthyoparasitology 17th Sept–3rd Oct., Tihany, Hungary*, p. 21.

Golovin, P. P. and Shukhgalter, O. A, 1979, [The biology of the monogenean parasites of eels from the genus *Pseudogyrodactylus*.] *Sb. Nauch. Trud., Vses. Nauch-Issled. Inst. Prud. Ryb. Khoz., No. 23*, 107–116. [In Russian.]

Golovina, N. A, 1987, [Effect of *D. vastator* on differential blood count of carp.] *Sb. Nauch. Trud., Vses. Nauch.-Issled. Inst. Prud. Ryb. Khoz., No. 23*, 22. [In Russian.]

Golvan, Y. J, 1958, Le phylum des Acanthocephala. Première note. Sa place dans l'échelle zoologique. *Annls Parasit. Hum. Comp.*, 33, 538–602.

Golvan, Y. J, 1959, Le Phylum des Acanthocephala. Deuxième note. La classe de Eoacanthocephala (Van Cleave, 1936) *Annis Parasit. Hum. Comp.*, 34, 5–52.

Golvan, Y. J, (1960–1961), Le Phylum des Acanthocephala. Troisième note. La Classe des Palaecanthocephala (Meyer, 1931). *Annis Parasit. Hum. Comp.*, 35, 76–91, 138–65, 350–86, 573–93, 713–23; 36, 76–91, 612–47, 717–36.

Golvan, Y. J, 1962, Le Phylum Acanthocephala (Quatrième note). La Classe des Archiacanthocephala (A. Meyer, 1931). *Annis Parasit. Hum. Comp.*, 37, 1–72.

Golvan, Y. J. and Buron, I, 1988, Les hôtes des acanthocéphales. II. Les hôtes definitifs. I. Poissons. *Annls Parasit. Hum. Comp.*, 63, 349–375.

Gomonenko, N. F. and Goncharenko, L. A, 1981, Parasites and parasitic diseases of carp reared in the basin of an experimental-industrial warm water fish farm at the Kiev Heat and Power Plant 5. *Osvoenie teplykh vod energeticheskikh ob "ektov dlya intensivnogo rybovodstva. Materialy II Respublikanskoi Nauchnoi Konferentsii, Kiev, 1980*. Kiev, USSR: 'Naukova Dumka', 409–418. [In Russian.]

Gonzalez, H., Torres, P., Figueroa, L., Contreras, B. and Franjola, R, 1978, Researches on Pseudophyllidea (Carus, 1813) in the south of Chile. II. Hepatic and splenic pathology by plerocercoids infections [sic] of *Diphyllobothrium* spp. in *Salmo gairdnerii* Richardson, 1836 of Calafquen lake. *Indian J. Parasit.*, 2, 127–129.

Gonzalez-Lanza, C. and Alvarez-Pellitero, P, 1982, Description and population dynamics of *Dactylogyrus legionensis* n. sp. from *Barbus barbus bocagei* Steind. *J. Helminth.*, 56, 263–273.

Gordon, D. M. and Rau, M. E, 1982, Possible evidence for mortality induced by the parasite *Apatemon gracilis* in a population of brook sticklebacks (*Culaea inconstans*). *Parasitology*, 84, 41–47.

Gorovaya, T. V, 1975, [Experimental study of the role of Phyllopoda in the elimination of *Diplostomum* (Strigeidida, Diplostomatidae) cercariae.] *Trudy Gel'm. Lab.*, 25, 17–26. [In Russian.]

Goto, S. and Ishii, N, 1936, On a new cestode species, *Amphilina japonica*. *Jap. J. Exp. Med.*, 14, 81–83.

Goto, S. and Nobutaro, I, 1936, On a new cestode species, *Amphilina japonica*. *Jap. J. Exp. Med.*, 14, 81–83.

Goto, S. and Ozaki, Y, 1929, Brief notes on new trematodes. II. *Jap. J. Zool.*, 2, 369–383.

Goude, C. C. and Vanicek, C. D, 1985, Parasites of the Sacramento perch, *Archoplites interruptus*. *Calif. Fish Game*, 71, 246–250.

Gould, S. J, 1977, *Ontogeny and Phylogeny*. Cambridge, Mass.: Harvard University Press, i–ix + 501 pp.

Goven, B. A., Gilbert, J. P. and Gratzek, J. B, 1980, Apparent drug resistance to the organophosphate dimethyl(2,2,2-Trichloro-l hydroxyethyl)phosphonate by monogenetic trematodes. *J. Wildlife Dis.* 16, 343–346.

Goven, B. A. and Amend, D. F, 1982, Mebendazole-trichlorfon combination: a new anthelmintic for removing monogenetic trematodes from fish. *J. Fish Biol.*, 20, 373–378.

Grabda, J, 1974, The dynamics of the nematode larvae, *Anisakis simplex* (Rud.) invasion in the South-Western Baltic herring (*Clupea harengus* L.). *Acta Ichthyol. Pisc.*, 4, 3–21.

Grabda, J, 1976a, Studies on the life cycle and morphogenesis of *Anisakis simplex* (Rudolphi, 1809) (Nematoda: Anisakidae) cultured *in vitro*. *Acta Ichthyol. Pisc.*, 6, 119–141.

Grabda, J, 1976b, The occurrence of anisakid nematode larvae in Baltic cod (*Gadus morhua callarias* L.) and the dynamics of their invasion. *Acta Ichthyol. Pisc.*, 6, 3–22.

Grabda, J, 1977a, {*An outline of parasitology of marine fishes.*} Zarys morskiej parazytologii rybackiej. Szczecin, Poland: Akademia Rolnicza, 190 pp. [In Polish.]

Grabda, J, 1977b, Studies on parasitisation and consumability of Alaska pollack, *Theragra chalcogramma* (Pall.). *Acta Ichthyol. Pisc.*, 7, 15–34.

Grabda, J, 1978, Studies on parasitic infestation of blue whiting (*Micromesistius* sp.) with respect to the fish utilization for consumption. *Acta Ichthyol. Pisc.*, 8, 29–41.

Grabda, J, 1981, Parasitic fauna of garfish *Belone belone* (L.) from the Pomeranian Bay (southern Baltic) and its origin. *Acta Ichthyol. Pisc.*, 11, 75–85.

Grabda, J, 1991, Marine fish parasitology. An outline. Warsaw. PWN-Polish Scientific Publishers, I–VII + 306pp.

Grabda, E. and Kozicka, J, 1961, Parasitological problems of a Polish fishery. *Wiad. parazyt.*, 7, 795–801.

Grabda-Kazubska, B. and Pilecka-Rapacz, M, 1987, Parasites of *Leuciscus idus* (L.), *Aspius aspius* (L.) and *Barbus barbus* (L.) from the river Vistula near Warszawa. *Acta Parasit. Pol.*, 31, 219–230.

Grabda-Kazubska, B., Baturo-Warszawaska, B. and Pojmanska, T, 1987, Dynamics of parasite infestation of fish in lakes Dgal Wielki and Warniak in connection with introduction of phytophagous species. *Acta Parasit. Pol.*, 32, 1–28.

Grabiec, S., Guttowa, A., Malzahn, E. and Michajlow, W, 1969, Investigations on the oxidation-reduction activity in embryos and coracidia of *Triaenophorus nodulosus* (Pall.) by the chemiluminescence method. *Bull. Acad. Pol. Sci., Cl. II. Série Sci. Biol.*, 17, 609–612.

Grainger, R. C, 1977, Occurrence of the metacercariae of the digenetic trematode *Cryptocotyle lingua* (Creplin, 1825) on the dorsal surfaces of inshore fishes in relation to the behaviour of the cercariae. (Proc. BSP Spring Meet., Univ. Dundee, 5–7 April, 1977). *Parasitology*, 75, viii.

Granath, W. O. Jr. and Esch, G. W, 1983a, Seasonal dynamics of *Bothriocephalus acheilognathi* in ambient and thermally altered areas of a North Carolina cooling reservoir. *Proc. Helminth. Soc. Wash.*, 50, 205–218.

Granath, W. O. Jr. and Esch, G. W, 1983b, Temperature and other factors that regulate the composition and infrapopulation densities of *Bothriocephalus acheilognathi* (Cestoda) in *Gambusia affinis* (Pisces). *J. Parasit.*, 69, 1116–1124.

Granath, W. O. Jr. and Esch, G. W, 1983c, Survivorship and parasite-induced host mortality among mosquitofish in a predator-free, North-Carolina cooling reservoir. *Amer. Midl. Nat.*, 110, 315–323.

Granath, W. O. Jr., Lewis, J. C. and Esch, G. W, 1983, An ultrastructural examination of the scolex and tegument of *Bothriocephalus acheilognathi* (Cestoda: Pseudophyllidea). *Trans. Am. Microsc. Soc.*, 102, 240–250.

Gras-Wawrzyniak, B., Grawinski, E. and Wawrzyniak, W, 1979, Parazytofauna wegorzycy *Zoarces viviparus* (L.) z. Zatoki Puckiej. *Med. Wet.*, 35, 557–561.

Greenwood, S. J. and Baker, M. R, 1987, *Cystidicoloides ephemeridarum* (Linstow, 1872) (Nematoda) in speckled trout *Salvelinus fontinalis*, from southern Ontario. *Can. J. Zool.*, 65, 2589–2593.

Greer, G. J. and Corkum, K. C, 1979, Life cycle studies of three digenetic trematodes, including descriptions of two new species (Digenea: Cryptogonimidae). *Proc. Helminth. Soc. Wash.*, 46, 188–200.

Greer, G. J. and Corkum, K. C, 1980, Notes on the biology of three trematodes (Digenea: Cryptogonimidae). *Proc. Helminth. Soc. Wash.*, 47, 47–51.

Grevtseva, M. A, 1977, [Changes in the crystalline lens of fish experimentally infected with *Diplostomum*.] *Parazitologiya*, 11, 260–263. [In Russian.]

Grey, A. J. and Hayunga, E. G, 1980, Evidence for alternative site selection by *Glaridacris laruei* (Cestoda: Caryophyllidea) as a result of interspecific competition. *J. Parasit.*, 66, 371–372.

Grey, A. J. and Mackiewicz, J. S, 1980, Chromosomes of caryophyllidean cestodes; diploidy, triploidy, and parthenogenesis in *Glaridacris catostomi. Int. J. Parasit.*, 10, 397–407.

Grigoryan, D. A. and Pogosyan, S. B, 1983, [Comparative faunistic analysis of the parasites of fish in natural waters and pond fisheries of the Ararat plain.] *Biologicheskii Zhurnal Armenii*, 36, 884–889. [In Russian.]

Grimes, D. J., Gruber, S. H. and May, E. B, 1985, Experimental infection of lemon shark *Negaprion brevirostris* (Poey) with *Vibrio* species. *J. Fish Dis.*, 8, 173–180.

Grimes, L. R. and Miller, G. C, 1976, Seasonal periodicity of three species of caryophyllaeid cestodes in the creek chubsucker, *Erimyzon oblongus* (Mitchill), in North Carolina. *J. Parasit.*, 62, 434–441.

Groot, C., Margolis, L. and Bailey, R, 1984, Does the route of seaward migration of Fraser River sockeye salmon (*Oncorhynchus nerka*) smolts determine the route of return migration of the adults? In: *Mechanisms of migration in fishes*. (Eds McCleave, J. D., Arnold, G. P., Dodson, J. J. and Neill, W. H.). New York, USA: Plenum Publishing Corporation, pp. 283–292.

Gruninger, T. L., Murphy, C. E. and Britton, J. C, 1977, Macroparasites of fish from Eagle Mountain Lake, Texas. *Southwest. Nat.*, 22, 525–535.

Guberlet, J. E., Hansen, H. A. and Kavanagh, J. A, 1927, Studies on the control of *Gyrodactylus. Publications in Fisheries, University of Washington. College of Fisheries*, 2, 17–29.

Guegan, J. F., Lambert, A. and Euzet, L, 1988, Étude des monogènes des Cyprinidae du genre *Labeo* en Afrique de l'Ouest. I. Genre *Dactylogyrus* Diesing, 1850. *Rev. Hydrobiol. Trop.*, 21, 135–151.

Gupta, A. K. and Agarwal, S. M, 1983, Host parasite relationships in *Channa punctatus* and *Euclinostomum heterostomum*. *Curr. Sci.*, 52, 474–476.

Gupta, A. K. and Agarwal, S. M, 1984, Host parasite relationships in *Channa punctatus* and *Euclinostomum heterostomum*. III. Transaminases and total proteins and free amino acids. *Curr. Sci.*, 53, 710–711.

Gupta, A. K., Niyogi, A., Naik, M. L. and Agarwal, S. M, 1984, Population dynamics of endohelminths of *Channa punctatus* at Raipur, India. *Jap. J. Parasit.*, 33, 105–118.

Gupta, A. N, 1970, Histochemical localization of alkaline phosphatase and its functional significance in the gut of digenetic trematodes. *Acta Biol. Acad. Sci. Hung.*, 21, 91–97.

Gupta, N. K. and Gargs, V. K, 1977, Free amino acids in *Paranisakis* sp. *Ind. J. Parasit.*, 1, 103.

Gupta, S. P. and Gupta, R. C, 1977, Free amino acid composition of *Opisthorchis pedicellata* (Trematoda) from the gall-bladder of a freshwater fish, *Rita rita* (Ham.) *Z. ParasitKde*, 54, 83–87.

Gupta, S. P. and Srivastava, A. B, 1975a, Studies on three new species of the genus *Camallanus* from Pentkota, Puri, Orissa. *Indian Science Congress Association. Proc. of the 62nd Indian Congress, Delhi (1975). Part III. Abstracts*, pp. 216.

Gupta, S. P. and Srivastava, A. B, 1975b, Studies on two new species of the genus *Indocucullanus*, from Pentkota, Puri, Orissa. *Indian Science Congress Association. Proc. of the 62nd Indian Congress, Delhi (1975). Abstracts*, 217.

Gupta, V. and Ahmad, J, 1979, Digenetic trematodes of marine fishes. VIII. On four new species of the genus *Haplosplanchnus Looss*, 1902 from marine fishes of Puri, Orissa. *Helminthologia*, 16, 185–193.

Gurkina, R. A, 1983, [Seasonal variations in the parasite fauna of roach in Lake Vrevo.] *Sb. Nauch. Trud. Gosud. Nauchno-Issled. Inst. Ozern. Rech. Khozyaist. (Probl. ekol. parazit. ryb)*, 85–89. [In Russian.]

Gur'yanova, S. D, 1977a, [The amino-acid composition of the total proteins of plerocercoids of the genus *Diphyllobothrium* and of their hosts.] *Sravnitel'naya biokhimiya ryb i ikh gel'mintov. Lipidy, fermenty, belki. (Sbornik statei. Otv.red.: V. S. Sidorov). Petrozavodsk*, USSR, 159 pp. [In Russian.]

Gur'yanova, S. D, 1977b, [Quarternary ammonium bases in fish and their helminths.] *Sravnitel'naya biokhimiya ryb i ikh gel'mintov. Lipidy, fermenty, belki. (Sbornik statei. Otv.red.: V. S. Sidorov). Petrozavodsk*, USSR, 109–116. [In Russian.]

Gur'yanova, S. D, 1980, [The lipid composition of some tissues of *Lota lota* infected with helminths.] In: *Biokhimiya presnovodnykh ryb Karelii*. Petrozavodsk, USSR, pp. 36–40. [In Russian.]

Gur'yanova, S. D. and Freze, V. I, 1978, [Lipid content in different eco- and host-forms of *Diphyllobothrium* plerocercoids.] In: *Ekologicheskaya biokhimiya zhivotnykh*. [Edi Sidorov, V. S.]. Petrozavodsk, USSR: Institut Biologii, Karel'skii Filial AN SSSR, pp. 24–33. [In Russian.]

Gur'yanova, S. D. and Freze, V. I, 1981, [Lipid composition of larval and adult forms of some cestodes.] In: *Spravnitel'nye aspekty biokhimii ryb i hekotorykh drugikh zhivotnykh*. (Ed. Sidorov, V. S.). Petrozavodsk, USSR: Institut Biologii, Karel'skii Filial AN SSSR, pp. 116–121. [In Russian.]

Gur'yanova, S. D. and Sidorov, V. S, 1985, [The effect of some cestodes on the lipid composition of the tissues of the turbot and the stickleback.] *Parazitologiya*, 19, 152–155. [In Russian.]

Gur'yanova, S. D., Sidorov, V. S. and Freze, V. I, 1976, [The amino acid composition of the plerocercoids of *Diphyllobothrium* related to ecological factors.] In: *Parazitologicheskie issledovaniya Karel'skoi* ASSR *i Murmanskoi oblasti*. Petrozavodsk, USSR: Karel'skoi filial Akademii Nauk SSSR, Institut Biologii, pp. 222–230. [In Russian.]

Gusev, A. V, 1976, [The origin of freshwater fish monogeneans.] *Izv. gosud. nauch.-issled. Inst. ozer. rech. Ryb. Khoz. (Probl. izuch. parazit. bolez. ryb)*, 105, 69–75. [In Russian.]

Gusev, A. V., Roitman, V. A, and Shul'man, S. S, 1984, [Results and prospects in the studies of freshwater fish parasites.]. In: *Biologicheskie osnovy rybovodstva: parazity i bolezni ryb. (Biologicheskie resursy gidrosfery i ikh ispol'zovanie.)* (Eds Bauer, O. N., Musselius, V. A. and Skryabina, E. S.) Moscow, USSR: 'Nauka', pp. 63–78. [In Russian.]

Gustafson, P. V, 1942, Life cycle studies on *Spinitectus gracilis* and *Rhabdochona* sp. (Nematoda: Thelaziidae). *J. Parasit.*, 25, 12–13.

Gustafsson, M. K. S, 1984, Synapses in *Diphyllobothrium dendriticum* (Cestoda). An electron microscopical study. *Ann. Zool. Fenn.*, 21, 167–175.

Gustafsson, M. K. S, 1985, Cestode neurotransmitters. *Parasitology Today*, 1, 72–75.

Gustafsson M. K. S. and Wikgren, M. C, 1981, Release of neurosecretory material by protrusions of bounding membranes extending through the axolemma, in *Diphyllobothrium dendriticum* (Cestoda). *Cell and Tissue Res.* 220, 473–479.

Gustafsson, M. K. S., Jukanen, A. E, and Wikgren, M. C, 1983, Activation of the peptidergic neurosecretory system in *Diphyllobothrium dendriticum* (Cestoda) at suboptimal temperatures. *Z. ParasitKde*, 69, 279–282.

Haas, W, 1969, Reizphysiologische untersuchungen an Cercarien von *Diplostomum spathaceum*. *Z. Vergleich. Physiol.*, 64, 254–287.

Haas, W, 1974, Analyse der Invasionsmechanismen der Cercarie von *Diplostomum spathaceum*: II. Chemische Invasionsstimuli. *Int. J. Parasit.*, 4, 321–330.

Haas, W, 1975, Einfluss von CO_2 und pH auf das Fixationsverhalten der Cercarie von *Diplostomum spathaceum* (Trematoda). *Z. ParasitKde*, 46, 53–60.

Haas, W. and Ostrowski de Nunez, M, 1988, Chemical signals of fish skin for the attachment response of *Acanthostomum brauni* cercariae. *Parasit. Res.*, 74, 552–557.

Hadidjaja, P., Ilahude, H. D., Mahfudin, H. and Burhanuddin, Hutomo, M, 1978, Larvae of Anisakidae in marine fish of coastal waters near Jakarta, Indonesia. *Am. J. Trop. Med. Hyg.*, 27, 51–54.

Haen, P. J. and Ryan, B, 1967, *Diplostomulum spathaceum* (Rudolphi) a helminth new to Ireland. *Fish Natural. J.*, 15, 270–272.

Haensly, W. E. J. M., Neff, J. M., Sharp, J. R., Morris, A. C., Bedgood, M. F. and Boem, P. D, 1982, Histopathology of *Pleuronectes platessa* L. from Aber Wrach and Aber Benoit, Brittany, France: long term effects of the Amoco Cadiz crude oil spill. *J. Fish. Dis.*, 5, 365–391.

Haldorson, L, 1984, A seasonal survey of metazoan parasites of Arctic cisco (*Coregonus autumnalis*) from Alaskan Arctic coastal waters. *Proc. helminth. Soc. Wash.*, 51, 245–247.

Hall, M. C, 1929, Arthropods as intermediate hosts of helminths. *Smithsonian Misc. Coll.*, 81, 1–77.

Halton, D. W, 1974, Haemoglobin absorption in the gut of a monogenetic trematode, *Diclidophora merlangi*. *J. Parasit.*, 60, 59–66.

Halton, D. W, 1975, Intracellular digestion and cellular defecation in a monogenean, *Diclidophora merlangi*. *J. Parasit.*, 70, 331–340.

Halton, D. W, 1976, *Diclidophora merlangi*: sloughing and renewal of hematin cells. *Expl Parasit.*, 40, 41–47.

Halton, D. W, 1978, Trans-tegumental absorption of L-alanine and L-leucine by a monogenean, *Diclidophora merlangi*. *Parasitology*, 76, 29–37.

Halton, D. W, 1979, The surface topography of a monogenean, *Diclidophora merlangi* revealed by scanning electron microscopy. *Z. ParasitKde*, 61, 1–12.

Halton, D. W, 1982a, Morphology and ultrastructure of parasitic helminths. In: *Parasites—their world and ours. Proceedings of the Fifth International Compress of Parasitology, Toronto, Canada, 7–14 Aug, 1982, under the auspices of the World Fed. Parasit.* (Eds Mettrick, D. F. and Desser, S. S.) Amsterdam, Netherlands: Elsevier Biomedical Press, pp. 60–69.

Halton, D. W, 1982b, X-ray microanalysis of pigment granules in the gut of *Diclidophora merlangi* (Monogenea). *Z. ParasitKde*, 68, 113–115.

Halton, D. W, 1982c, An unusual structural organization to the gut of a digenetic trematode, *Fellodistomum fellis*. *Parasitology*, 85, 53–60.

Halton, D. W. and Hardcastle, A, 1976, Spermatogenesis in a monogenean, *Diclidophora merlangi*. *Int. J. Parasit.*, 6, 43–53.

Halton, D. W. and Hardcastle, A, 1977, Ultrastructure of the male accessory ducts and prostate gland of *Diclidophora merlangi* (Monogenoidea). *Int. J. Parasit.*, 7, 393–401.

Halton, D. W. and Hendrix, S. S, 1978, Chemical composition and histochemistry of *Lobatostoma ringens* (Trematoda: Aspidogastrea). *Z. ParasitKde*, 57, 237–241.

Halton, D. W. and Jennings, J. B, 1965, Observations on the nutrition of monogenetic trematodes. *Biol. Bull.*, 129, 257–272.

Halton, D. W. and McCrae, J. M, 1985, Development of the tegument and alimentary tract in a digenetic trematode, *Fellodistomum fellis*. *Parasitology*, 90, 193–204.

Halton, D. W. and Stranock, S. D, 1976, The fine structure and histochemistry of the caecal epithelium of *Calicotyle kroyeri* (Monogenea: Monopisthocotylea). *Int. J. Parasit.*, 6, 253–263.

Halton, D. W., Dermott, E. and Morris, G. P, 1968, Electron microscope studies of *Diclidophora merlangi* (Monogenea: Polyopisthocotylea). I. Ultrastructure of the cecal epithelium. *J. Parasit.*, 54, 909–916.

Halton, D. W., Morris, G. P. and Hardcastle, A, 1974, Gland cells associated with the alimentary tract of a monogenean, *Diclidophora merlangi*. *Int. J. Parasit.*, 4, 589–599.

Halton, D. W., Stranock, S. D. and Hardcastle, A, 1976, Fine structure observations on oocyte development in monogeneans. *Parasitology*, 73, 13–23.

Halvorsen, O, 1968, Studies of the helminth fauna of Norway. XII. *Azygia lucii* (Müller, 1776) (Digenea, Azygiidae) in pike (*Esox lucius* L.) from Bogstad lake, and a note on its occurrence in lake and river habitats. *Nytt Mag. Zool.*, 16, 29–38.

Halvorsen, O, 1969, Studies of the helminth fauna of Norway. XIII. *Diplozoon paradoxum* Nordmann 1832, from roach, *Rutilus rutilus* (l.) bream, *Abramis brama* (L.) and hybrid of roach and bream. Its morphological adaptability and host specificity. *Nytt Mag. Zool.*, 17, 93–103.

Halvorsen, O, 1970, Studies of the helminth fauna of Norway. XV. On the taxonomy and biology of plerocercoids of *Diphyllobothrium* Cobbold, 1858 (Cestoda, Pseudophyllidea) from North-western Europe. *Nytt Mag. Zool.*, 18 (2), 113–174.

Halvorsen, O, 1971, Studies of the helminth fauna of Norway XVIII: On the composition of the parasite fauna of coarse fish in the river Glomma, southeastern Norway. *Norw. J. Zool.*, 19, 181–192

Halvorsen, O, 1972, Studies of the helminth fauna of Norway. XX. Seasonal cycles of fish parasites in the River Glomma. *Norw. J. Zool.*, 20, 9–18.

Halvorsen, O, 1976, Negative interaction amongst parasites. In: (Ed. Kennedy, C. R.) *Ecological aspects of parasitology*. Amsterdam, The Netherlands: North-Holland Publishing Company 99–114.

Halvorsen, O. and Andersen, K, 1974, Some effects of population density in infections of *Diphyllobothrium dendriticum* (Nitzsch) in golden hamster (*Mesocricetus auratus* Waterhouse) and common gull (*Larus canus* L.). *Parasitology*, 69, 149–160.

Halvorsen, O. and Andersen, K, 1984, The ecological interaction between arctic charr, *Salvelinus alpinus* (L.), and the plerocercoid stage of *Diphyllobothrium ditremum*. *J. Fish. Biol.*, 25, 305–316.

Halvorsen, O. and Hartvigsen, R, 1989, A review of the biogeography and epidemiology of *Gyrodactylus salaris*. NINA *Utredning*, 2, 1–41.

Halvorsen, O. and Macdonald, S, 1972, Studies on the helminth fauna of Norway. XXVI. The distribution of *Cyathocephalus truncatus* (Pallas) in the intestine of the brown trout (*Salmo trutta*) (L.). *Norw. J. Zool.*, 20, 265–272.

Halvorsen, O. and Williams, H. H, 1968, Studies of the helminth fauna of Norway. IX. *Gyrocotyle* (Platyhelminthes) in *Chimaera monstrosa* from Oslo Fjord, with emphasis on its mode of attachment and a regulation in the degree of infection. *Nytt Mag. Zool.*, 15, 130–142.

Hamed, M. G. E. and Elias, A. N, 1975, Feasibility of using gamma radiation for the control of the trematode *Heterophyes* sp. in the flesh of fish caught from brackish waters. *Egyptian J. Food Sci. (1974, publ. 1975)*, 2, 135–140.

Hamilton-Attwell, V. L., Tiedt, L. R, and van As, J. G, 1980, A sem and tem study of the integument of *Bothriocephalus gowkongensis* Yeh 1955. *Electron Microscopy Society of Southern Africa Proceedings*, 10, 105–106.

Hanek, G, 1974, Micro-ecology and spatial distribution patterns of the gill parasites infesting *Lepomis gibbosus* (L.) and *Ambloplites rupestris* (Raf.) in the Bay of Quinte area, Ontario. *Diss. Abstr. Int.*, 34B, 4834.

Hanek, G. and Fernando, C. H, 1978a, Spatial distribution of gill parasites of *Lepomis gibbosus* (L.) and *Ambloplites rupestris* (Raf.). Can. J. Zool., 56, 1235–1240.

Hanek, G. and Fernando, C. H, 1978b, Seasonal dynamics and spatial distribution of *Urocleidus ferox* Mueller 1934, a gill parasite of *Lepomis gibbosus* (L.). *Can. J. Zool.*, 56, 1241–1243.

Hanek, G. and Fernando, C. H, 1978c, Seasonal dynamics and spatial distribution of *Cleidodiscus stentor* Mueller 1937 and *Ergasilus centrarchidarum* Wright 1882, gill parasites of *Ambloplites rupestris* (Raf.). *Can. J. Zool.*, 56, 1244–1246.

Hanek, G. and Fernando, C. H, 1978d, The role of season, habitat, host age, and sex on gill parasites of *Lepomis gibbosus* (L.). *Can. J. Zool.*, 56, 1247–1250.

Hanek, G. and Fernando, C. H, 1978e, The role of season, habitat, host age, and sex on gill parasites of *Ambloplites rupestris* (Raf.). *Can. J. Zool.*, 56, 1251–1253.

Hanek, G. and Threlfall, W, 1970, Metazoan parasites of the American eel (*Anguilla rostrata* (Le Sueur)) in Newfoundland and Labrador. *Can. J. Zool.*, 48, 597–600.

Hanzelova, V. and Zitnan, R, 1982, The seasonal dynamics of the invasion cycle of *Gyrodactylus katharineri* Malmberg, 1964 (Monogenea). *Helminthologia,* 19, 257–265.

Hanzelova, V. and Zitnan, R, 1983, The seasonal dynamics of the invasion of *Dactylogyrus vastator* Nybelin, 1924 (Monogenea) in the carp fry. *Helminthologia,* 20, 137–150.

Hanzelova, V. and Zitnan, R, 1985, Epizootiologic importance of the concurrent monogenean invasions in the carp. *Helminthologia,* 22, 277–283.

Haque, M. and Siddiqi, A. H, 1982, The biochemical composition of digenetic trematodes. *Ind. J. Parasit.*, 6, 37–41.

Hare, G. M. and Burt, M. D. B, 1976, Parasites as potential biological tags of Atlantic salmon (*Salmo salar*) in the Miramichi River system, New Brunswick. *J. Fish. Res. Bd Can.*, 33, 1139–1143.

Hargis, W. J, 1959, The host specificity of monogenetic trematodes. *Expl. Parasit.* 6, 610–625.

Hargis, W. J, (Ed.). 1985, *Parasitology and pathology of marine organisms of the world ocean*. NOAA Technical Report NMFS 25, iv + 135 pp.

Hargis. W. J., Lawler, A. R., Morales-Alamo, R and Zwerner, D. E, 1969, Bibliography of the monogenetic trematode literature of the world 1758 to 1969. Virginia, U.S.A. Virginia Institute of Marine Science Special Scientific Report No. 55, i–v + 195pp.

Hargis, W. J., Lawler, A. R., Morales-Alamo, R. and Zwerner, D. E, 1971, Bibliography of the monogenetic trematode literature of the world 1758–1969. Supplement 2 (with errata). Virginia, U.S.A. Virginia Institute of Marine Science Special Scientific Report, No. 55, + 25pp.

Hargis, W. J. Jr. and Thoney, D. A, 1983, *Bibliography of the Monogenea. Literature of the world 1758–1982.* Virginia, USA: College of William and Mary in Virginia viii + 384 pp.

Harms, C. E, 1965, The life cycle and larval development of *Octospinifer macilentis* (Acanthocephala: Neoechinorhynchidae). *J. Parasit.*, 51, 286–293.

Harris, C. C, 1978, *Fish farming*. Pelham Books Ltd., 111 pp.

Harris, J. E, 1970, Precipitin production by chub, *Leuciscus cephalus*, to an intestinal helminth. *J. Parasit.*, 56, 1035.

Harris, J. E, 1972, The immune response of a cyprinid fish to infections of the acanthocephalan *Pomphorhynchus laevis*. *Int. J. Parasit.*, 2, 459–469.

Harris, J. E, 1973, Some aspects of the immune response of cyprinid fish, with particular reference to their helminth parasites. *Ph.D. Thesis, University of Exeter*.

Harris, J. E. and Cottrell, B. J, 1976, Precipitating activity in the sera of plaice, *Pleuronectes platessa* L. to a helminth antigen. *J. Fish Biol.*, 9, 405–410.

Harris, P. D, 1983, The morphology and life-cycle of the oviparous *Oogyrodactylus farlowellae* gen. et. sp. nov. (Monogenea, Gyrodactylidea). *Parasitology*, 87, 405–420.

Harris, P. D, 1985a, Observations on the development of the male reproductive system in *Gyrodactylus gasterostei* Gläser, 1974 (Monogenea, Gyrodactylidae). *Parasitology*, 91, 519–529.

Harris, P. D, 1985b, Species of *Gyrodactylus* von Nordmann, 1832 (Monogenea: Gyrodactylidae) from freshwater fishes in southern England, with a description of *Gyrodactylus rogatensis* sp. nov. from the bullhead *Cottus gobio* L. *J. Nat. Hist.*, 19, 791–809.

Harris, P. D, 1989, Interactions between population growth and sexual reproduction in the viviparous monogenean *Gyrodactylus turnbulli* Harris, 1986 from the guppy, *Poecilia reticulata* Peters. *Parasitology*, 98, 245–251.

Hartwich, G, 1974, Keys to genera of the Ascaridoidea. *CIH Keys to the nematode parasites of vertebrates.* Farnham Royal, England: Commonwealth Agricultural Bureaux, No. 2, 15 pp.

Hauck, A. K, 1977, Occurrence and survival of the larval nematode *Anisakis* sp. in the flesh of frozen, brined and smoked Pacific herring *Clupea harengus pallasi*. *J. Parasit.*, 63, 515–519.

Hauck, A. K. and May, E. B, 1977, Histopathologic alterations associated with *Anisakis* larvae in Pacific herring from Oregon. *J. Wildl. Dis.*, 13, 290–293.

Hayasaka, H., Ishikura, H. and Takayama, T, 1971, Acute regional ileitis due to *Anisakis* larvae. *Int. Surg.*, 55, 8–14.

Hayunga, E. G, 1979a, The structure and function of the scolex glands of three caryophyllid tapeworms. *Proc. Helminth. Soc. Wash.*, 46, 171–179.

Hayunga, E. G, 1979b, Observations on the intestinal pathology caused by three caryophyllid tapeworms of the white sucker *Catostomus commersoni* Lacépède. *J. Fish Dis.*, 2, 239–248.

Hayunga, E. G, 1980, Two atypical lesions from white suckers *Catostomus commersoni* Lacépède infected by caryophyllid tapeworms. *J. Fish Dis.*, 3, 167–172.

Hayunga, E. G, 1991, Morphological adaptations of intestinal helminths. *J. Parasit.*, 77, 865–873.

Hayunga, E. G. and Mackiewicz, J. S, 1975, An electron microscope study of the tegument of *Hunterella nodulosa* Mackiewicz and McCrae, 1962 (Cestoidea: Caryophyllidea). *Int. J. Parasit.*, 5, 309–319.

Hazen, T. C. and Esch, G. W, 1978, Observations on the ecology of *Clinostomum marginatum* in largemouth bass (*Micropterus salmoides*). *J. Fish Biol.*, 12, 411–420.

Healy, G. R. and Juranek, D, 1979, Parasitic infections. In: *Food-borne infections and intoxications. (Food Science and Technology, monographs)* (Eds Riemann, H. and Bryan, F. L.). 2nd Edn. New York, USA: London, UK: Academic Press, pp. 343–385.

Heckmann, R, 1983, Eye fluke (*Diplostomum spathaceum*) of fishes from the Upper Salmon River near Obsidian, Idaho. *Great Basin Nat.*, 43, 675–683.

Heckmann, R. and Otto, T, 1985, Occurrence of anisakid larvae (Nematoda, Ascaridida) in fishes from Alaska and Idaho. *Great Basin Nat.*, 45, 427–431.

Heckmann, R. A., Deacon, J. E., and Greger, P. D, 1986, Parasites of the woundfin minnow, *Plagopterus argentissimus*, and other endemic fishes from the Virgin River, Utah. *Great Basin Nat.*, 46, 662–676.

Heggberget, T. G. and Johnsen, B. O, 1982, Infestations by *Gyrodactylus* sp. of Atlantic salmon, *Salmo salar* L., in Norwegian rivers. *J. Fish Biol.*, 21, 15–26.

Hellström, A., Ljungberg, O. and Bornstein, S, 1988, *Anguillicola*, en ny alparasit i Sverige. *Svensk Veterinartidning*, 40, 211–213.

Hendow, H. T. and James, B. L, 1988, Ultrastructure of spermatozoon and spermatogenesis in *Maritrema linguilla* (Digenea: Microphallidae). *Int. J. Parasit.*, 18, 52–63.

Hendrickson, G. L, 1978a, Observations on strigeoid trematodes from the eyes of southeastern Wyoming fish. I. *Diplostomulum spathaceum* (Rudolphi, 1819). *Proc. Helminth. Soc. Wash.*, 45, 60–64.

Hendrickson, G. L, 1978b, Observations on strigeoid trematodes from the eyes of southeastern Wyoming fish. II. *Diplostomulum scheuringi* Hughes 1929; *Neascus ptychocheilus* Faust 1917; and other types. *Proc. Helminth. Soc. Wash.*, 45, 64–68.

Hendrickson, G. L, 1979, *Ornithodiplostomulum ptychocheilus*: migration to the brain of the fish intermediate host, *Pimephales promelas*. *Expl Parasit.*, 48, 245–258.

Hendrickson, G. L, 1986, Observations on the life cycle of *Ornithodiplostomum ptychocheilus* (Trematoda: Diplostomatidae). *Proc. Helminth. Soc. Wash.*, 53, 166–172.

Hendrickson, G. L. and Yindeepol, W, 1987, Parasites of dover sole, *Microstomus pacificus* (Lockington), from northern California. *Proc. Helminth. Soc. Wash.*, 54, 111–114.

Hendrix, S. S, 1978, The life history and biology of *Plagioporus hypentelii* Hendrix 1973 (Trematoda: Opecoelidae). *J. Parasit.*, 64, 606–612.

Hendrix, S. S. and Overstreet, R. M, 1977, Marine aspidogastrids (Trematoda) from fishes in the northern Gulf of Mexico. *J. Parasit.*, 63, 810–817.

Hendrix, S. S. and Short, R. B, 1965, Aspidogastrids from Northeastern Gulf of Mexico river drainages. *J. Parasit.*, 51, 561–569.

Henricson, J, 1977, The abundance and distribution of *Diphyllobothrium dendriticum* (Nitzsch) and *D. ditremum* (Creplin) in the char *Salvelinus alpinus* (L.) in Sweden. *J. Fish Biol.*, 11, 231–248.

Henricson, J, 1978, The dynamics of infection of *Diphyllobothrium dendriticum* (Nitzsch) and *D. ditremum* (Creplin) in the char *Salvelinus alpinus* (L.) in Sweden. *J. Fish Biol.*, 13, 51–71.

Herman, R. L, 1970, Chemotherapy of fish diseases: a review. *J. Wildl. Dis.*, 6, 31–34.

Hermanns, von W. and Körting, W, 1987, Histopathologie des *Cyathocephalus truncatus*—Befalls bei der Regenbogenforelle. *Dtsch. tierarztl. Wschr.*, 94, 49–136.

Herrington, W. C., Bearse, H. M. and Firth, F. E, 1939, Observations on the life history, occurrence and distribution of the redfish parasite *Sphyrion lumpi*. *United States Bureau of Fisheries Special Report, No. 5,* 1–18.

Herzog, P. H, 1969, Untersuchungen über die parasiten der süsswasserfische des Iraq. *Arch. Fish.*, 20, 132–147.

Hewitt, G. C. and Hine, P. M, 1971, Checklist of parasites of New Zealand fishes and their hosts. *N.Z.J. Mar. Freshw. Res.,* 6, 69–114.

Hickman, C. P. and Trump, B. F, 1969, The kidney. In: *Fish Physiology, Volume I.* (Eds Hoar, W. S. and Randall, D. J.) London: Academic Press, pp. 91–227.

Hickman, J. L, 1960, Cestodes of some Tasmanian Anura. *Ann. Mag. Nat. Hist. Ser. 13,* 3, 1–23.

Hickman, V. V, 1934, On *Coitocaecum anaspidis* sp. nov., a trematode exhibiting progenesis in the fresh-water crustacean *Anaspides tasmaniae* Thomson. *Parasitology,* 26, 121–128.

Higgins, J. C, 1979, The role of the tegument of the metacercarial stage of *Bucephalus haimeanus* (Lacaze-Duthiers, 1854) in the absorption of particulate material and small molecules in solution. *Parasitology,* 78, 99–106.

Higgins, J. C, 1980, Formation of the cyst wall and related changes in the structure of the tegument of *Bucephalus haimeanus* (Lacaze-Duthiers, 1854) during its metamorphosis from the cercarial to the metacercarial stage. *Parasitology,* 81, 47–59.

Higgins, J. C., Wright, D. E. and Matthews, R. A, 1977, The ultrastructure and histochemistry of the cyst wall of *Bucephalus haimeanus* (Lacaze-Duthiers, 1854). *Parasitology,* 75, 207–214.

Hildemann, W. H. and Walford, R. L, 1963, Annual fishes—promising species as biological control agents. *J. Trop. Med. Hyg.,* 66, 163–166.

Hill, B. J, 1977, The diseases of farmed freshwater fish and the methods used to control them. In: *Fish Farming Feasibility and Finance, 3-day-Study Course, Highcliffe Hotel, Bournemouth, Jan. 17–19, 1977.* Sutton, Surrey, UK: IPC Business and Industrial, pp. 46–50.

Hine, P. M, 1977a, New species of *Nippotaenia* and *Amurotaenia* (Cestoda: Nippotaeniidae) from New Zealand freshwater fishes. *J. R. Soc. N.Z.,* 7, 143–155.

Hine, P. M, 1977b, Two new digenean trematodes from New Zealand freshwater fishes. *J. R. Soc. N.Z.,* 7, 163–170.

Hine, P. M, 1978, [Distribution of some parasites of freshwater eels in New Zealand. *N.Z. Jl. Marine and Freshwater Research* 12, 179–187.

Hine, P. M, 1980a, Distribution of helminths in the digestive tracts of New Zealand freshwater eels. 1. Distribution of digeneans. *N. Z. J. Mar. Freshw. Res.,* 14, 329–338.

Hine, P. M, 1980b, Distribution of helminths in the digestive tracts of New Zealand freshwater eels. 2. Distribution of nematodes. *N. Z. J. Mar. Freshw. Sci.,* 14, 339–347.

Hine, P. M. and Anderson, C. D, 1982, Diseases of the gonads and kidneys of New Zealand snapper, *Chrysophrys auratus* Forster (F. Sparidae). In: *Wildlife diseases of the Pacific Basin and other countries. (Proc. 4th Internat. Conf. of the Wildlife Dis. Assoc., Sydney, Australia, 25–28 Aug. 1981).* (Ed. Fowler, M. E.) USA: Wildlife Disease Association, pp. 166–170.

Hine, P. M. and Francis, R. I. C. C, 1980, Distribution of helminths in the digestive tracts of New Zealand freshwater eels. 3. Interspecific associations and conclusions. *N. Z. J. Mar. Freshw. Res.,* 14, 349–356.

Hine, P. M. and Kennedy, C. R, 1974a, Observations on the distribution, specificity and pathogenicity of the acanthocephalan *Pomphorhynchus laevis* (Müller). *J. Fish Biol.,* 6, 521–535.

Hine, P. M. and Kennedy, C. R, 1974b, The population biology of the acanthocephalan *Pomphorhynchus laevis* (Müller) in the River Avon. *J. Fish Biol.,* 6, 665–679.

Hirschfield, M. F., Morin, R. P. and Hepner, D. J, 1983, Increased prevalence of larval *Eustrongyloides* (Nematoda) in the mummichog, *Fundulus heteroclitus* (L.) from the discharge canal of a power plant in the Chesapeake Bay. *J. Fish Biol.,* 23, 135–142.

Hislop, J. R. G. and Mackenzie, K, 1976, Population studies of the whiting *Merlangius merlangus* (L.) of the northern North Sea. *J. Cons. Perm. Int. Explor. Mer.,* 37, 98–110.

Hłond, S., Kozlowski, F. and Szaryk, A, 1977, Zgorzel skrzel narybku karpia na tle zarazenia przywra *Sanguinicola inermis* Plehn. *Roczn. Nauk Roln. H,* 98, 65–76.

Hnath, J. G, 1969, Transfer of an adult acanthocephalan from one fish host to another. *Trans. Am. Fish. Soc.,* 98, 332.

Hoar, W. S, 1969, Reproduction and growth, bioluminescence, pigments, and poisons. In: *Fish Physiology Vol. III.* (Eds Hoar, W. S. and Randall, D. J.) London: Academic Press, pp. 1–485.

Hoar, W. S. and Randall, D. J. (Eds) 1971, Sensory systems and electric organs. *Fish Physiology,* Vol. V. London: Academic Press, pp. –600.

Hoar, W. S., Randall, D. J. and Brett, J. R. (Eds) 1979, Bioenergetics ansd Growth. *Fish Physiology, Vol. VIII.* London: Academic Press, 786pp.

Hoberg, E. P, 1987, Recognition of larvae of the Tetrabothriidae (Eucestoda): implications for the origin of tapeworms in marine homeotherms. *Can. J. Zool.,* 65, 997–1000.

Hodgins, H. O., McCain, B. B. and Hawkes, J. W, 1977, Marine fish and invertebrate diseases, host disease resistance, and pathological effects of petroleum. In: *Effects of petroleum on Arctic and*

Subarctic marine environments and organisms Vol. II. Biological effects. Malins, D. C.}. New York, USA: Academic Press, Inc., pp. 95–173.

Hoeden, J. van der (Ed.) 1964, Helminthozoonoses (Wittenberg). *From: Zoonoses. Amsterdam: Elsevier Publishing Company,* VI, 190 pp.

Hofer, B, 1906, *Handbuch Der Fischkrankheiten.* Stuttgart: Schweizerbartshche Verlagsbuchhandlung, 359 pp.

Hoffman, G. L., 1958, Experimental studies on the cercaria and metacercaria of a strigeoid trematode, *Posthodiplostomum minimum. Epl. Parasit.*, 7, 23–50.

Hoffman, G. L, 1967, *Parasites of North American freshwater fishes.* Los Angeles: University of California Press, viii + 486 pp.

Hoffman, G. L, 1970, Intercontinental and transcontinental dissemination and transformation of fish parasites with emphasis on whirling disease (*Myxosoma cerebralis*). A *symposium on diseases of fishes and shell fishes.* (Ed. Sniesko, S. F.) Washington, D.C. Am. Fish. Soc., Spec. Publ. 5, pp. 69–81.

Hoffman, G. L, 1975a, Lesions due to internal helminths of freshwater fishes. In: (Eds Ribelin, W. E. and Migaki, G.) *The Pathology of Fishes.* Madison, Wisconsin, USA: University of Wisconsin Press, pp. 151–186.

Hoffman, G. L, 1975b, What's bugging that fish? *An angler's guide to fish diseases and parasites. Nebrask Game and Parks Commission Project F-4-R*, pp. 93–95.

Hoffman, G. L, 1976a, The Asian tapeworm, *Bothriocephalus gowkongensis*, in the United States, and research needs in fish parasitology. In: *Proc. 1976 Fish Farm. Conf. Ann. Conv. Catfish Farmers Texas, Texas A and M Univ. Coll. Stn, USA*, pp. 84–90.

Hoffman, G. L, 1976b, Fish diseases and parasites in relation to the environment. *Fish Path.*, 10, 123–128.

Hoffman, G. L, 1979, Helminthic parasites. In: (Ed. Plumb, J. A.) *Principal diseases of farm raised catfish.* Southern Co-operative Series No. 225, pp. 40–58.

Hoffman, G. L. and Bauer, O. N, 1971, Fish parasitology in water reservoirs: a review. *Reservoir Fisheries and Limnology, Special Publication of the American Fisheries Society*, **8**, 495–511.

Hoffman, G. L. and Dunbar, C. E, 1961, Mortality of eastern brook trout caused by plerocercoids (Cestoda: Pseudophyllidea: Diphyllobothriidae) in the heart and viscera. *J. Parasit.*, 47, 399–400.

Hoffman, G. L. and Hoyme, J. B, 1958, The experimental histopathology of the tumor on the brain of the stickleback caused by *Diplostomum baeri eucaliae. J. Parasit.*, 44, 374–378.

Hoffman, G. L. and Hutcheson, J. A, 1970, Unusual pathogenicity of a common metacercaria of fish. *J. Wildl. Dis.*, 6, 109.

Hoffman, G. L. and Meyer, F. P, 1974, *Parasites of freshwater fishes. A review of their control and treatment.* Neptune City, N.J., USA: T. F. H. Publications, Inc., 224 pp.

Hoffman, G. L. and Mitchell, A. J, 1978, Fish disease diagnosis and control. When to ask for help and how to go about it. *Commercial Fish Farmer*, 4, 20–23.

Hoffman, G. L. and Schubert, G, 1984, Some parasites of exotic fishes. In: *Distribution, biology and management of exotic fishes.* (Eds Courtenay, W. R. Jr. and Stauffer, J. R.) Baltimore, USA and London, UK: John Hopkins University Press, pp. 233–261.

Hoffman, G. L. and Sindermann, C. J, 1962, Common parasites of fishes. *United States Department of the Interior Fish and Wildlife Service Circular 144*, 17pp.

Hoffmann, R., Kennedy, C. R. and Meder, J, 1986, Effects of *Eubothrium salvelini* Schrank, 1790 on arctic charr, *Salvelinus alpinus* (L.) in an alpine lake. *J. Fish Dis.,* 9, 153–157.

Hoffmann, R. W., Scheinert, P. and Bibelriether, J, 1991, Histological studies on the effects of *Apatemon cobitidis* in its second intermediate host, the bullhead (*Cottus gobio*). *Angew. Parasit.,* 32, 27–32.

Hogans, W. E., Brattey, J., Uhazy, L. S. and Hurley, P. C. F, 1983, Helminth parasites of swordfish (*Xiphias gladius* L.) from the northwest Atlantic Ocean. *J. Parasit.,* 69, 1178–1179.

Holliman, R. B, 1963, *Gyrodactylus shorti*, a new species of monogenetic trematode from the brood pouch of the southern pipefish, *Syngnathus scovelli* (Evermann and Kendall). *Tulane Stud. Zool.,* 10, 83–86.

Holloway, H. L. Jr. and Hagstrom, N. T, 1981, Comparison of four north Dakota impoundments and factors affecting the development of impoundment parasitofauna. *Prairie Nat.,* 13, 85–93.

Holloway, H. L. Jr. and Spence, J. A, 1980, Ecology of animal parasites in McMurdo Sound, Antarctica. *Comp. Physicol. Ecol.,* 5, 262–284.

Holmes, J. C, 1961, Effects of concurrent infections on *Hymenolepis diminuta* (Cestoda) and *Moniliformis dubius* (Acanthocephala). I. General effects and comparison with crowding. *J. Parasitol.,* 47, 209–216.

Holmes, J. C, 1962, Effects of concurrent infections on *Hymenolepis diminuta* (Cestoda) and *Moniliformis dubius* (Acanthocephala) II. Effects on growth. *J. Parasitol.,* 48, 87–96.

Holmes, J. C, 1971, Habitat segregation in sanguinicolid blood flukes (Digenea) of scorpaenid rockfishes (Perciformes) on the Pacific coast of North America. *J. Fish. Res. Bd Can.*, **28**, 903–909.

Holmes, J. C, 1973, Site selection by parasitic helminths: interspecific interactions, site segregation and their importance to the development of helminth communities. *Can. J. Zool.*, **51**, 333–347.

Holmes, J. C, 1976, Host selection and its consequences. In: *Ecological aspects of parasitology*. (Ed. Kennedy, C. R.) Amsterdam, Netherlands: North-Holland Publishing Co. pp. 21–39.

Holmes, J. C, 1979, Parasite populations and host community structure. In: *Host–parasite interfaces. (Proc. Symp., Univ. of Nebraska-Lincoln, October 5–7.* (Ed. Nickol, B. B.). London: Academic Press Inc. pp. 27–46.

Holmes, J. C, 1982, Impact of infectious disease agents on the population growth and geographical distribution of animals. In: *Population biology of infectious diseases. (Report of the Dahlem Workshop on Population Biology of Infectious Disease Agents, Berlin, March 14–19). (Life Science Research Report, 25)* (Eds Anderson, R. M. and May, R. M.). Berlin: Springer Verlag, pp. 37–51.

Holmes, J. C, 1983, Evolutionary relationships between parasitic helminths and their hosts. In: *Coevolution* (Eds Futuyma, D. J. and Slatkin, M.), 9.3, 9.3.2. Sunderland, Mass.: Sinauer, pp. 161–185.

Holmes, J. C, 1986, The structure of helminth communities. In: (Ed. Howell, M. J.). Parasitology-Quo Vadit? Proceedings of the Sixth International Conpress of Parasitology, Canberra, Australia. Canberra: Australian Academy of Sciences, pp. 203–209.

Holmes, J. C, 1990, Helminth communities in marine fishes In: *Parasite communities: patterns and processes.* (Eds Esch, G. W., Bush, A. O. and Aho, J. M.). London: Chapman and Hall, pp. 101–130.

Holmes, J. C. and Bethel, W. M, 1972, Modification of intermediate host behaviour by parasites. In: *Behavioural aspects of parasite transmission.* (Eds Canning, E. U. and Wright, C. A.). London, UK: Academic Press, pp. 123–149.

Holmes, J. C. and Price, P. W, 1980, Parasite communities: the roles of phylogeny and ecology. *Syst. Zool.*, 29, 203–213.

Holmes, J. C. and Price, P. W, 1986, Communities of parasites. In: *Community Ecology: Pattern and Process*, (Eds Kikkawa, J. and Anderson, D. J.). Oxford: Blackwell Scientific Publications, pp. 187–213.

Holmes, J. C., Hobbs, R. P. and Leong, T. S, 1977, Populations in perspective: community organization and regulation of parasite populations. In: *Regulation of parasite populations. (Proc. Symp. sponsored by Amer. Microscop. Soc. and Amer. Soc. Parasit., New Orleans, 10–14 Nov. 1975).* (Ed. Esch, G.W.). London, UK: Academic Press, Inc., pp. 209–245.

Holton, A. L, 1983, Observations on the life history of *Deretrema minutum* Manter, 1954 (Digenea: Zoogonidae) in freshwater crustacean and fish hosts from Canterbury, New Zealand. *N. Z. J. Mar. Freshw. Res.*, 17, 373–376.

Hoole, D. and Arme, C, 1982, Ultrastructural studies on the cellular response of roach, *Rutilus rutilus* L., to the plerocercoid larva of the pseudophyllidean cestode, *Ligula intestinalis. J. Fish Dis.*, 5, 131–144.

Hoole, D. and Arme, C, 1983a, Ultrastructural studies on the cellular response of fish hosts following experimental infection with the plerocercoid of *Ligula intestinalis* (Cestoda: Pseudophyllidea). *Parasitology*, 87, 139–149.

Hoole, D. and Arme, C, 1983b, *Ligula intestinalis* (Cestoda: Pseudophyllidea): an ultrastructural study on the cellular response of roach fry, *Rutilus rutilus. Int. J. Parasit.*, 13, 359–363.

Hoole, D. and Arme, C, 1986, The role of serum in leucocyte adherence to the plerocercoid of *Ligula intestinalis* (Cestoda: Pseudophyllidae). *Parasitology*, 92, 413–424.

Hooper, J. N. A, 1983, Parasites of estuarine and oceanic flathead fishes (family Platycephalidae) from northern New South Wales. *Aust. J. Zool, Suppl. Ser. No. 90*, 1–69.

Hopkins, C. A, 1950, Studies on cestode metabolism. I. Glycogen metabolism in *Schistocephalus solidus in vivo. J. Parasit.*, 36, 384–390.

Hopkins, C. A, 1951, Studies on cestode metabolism. II. The utilization of glycogen by *Schistocephalus solidus in vitro. Expl Parasit.*, 1, 196–213.

Hopkins, C. A, 1959, Seasonal variations in the incidence and development of the cestode *Proteocephalus filicollis* (Rud. 1810) in *Gasterosteus aculeatus* (L. 1766). *Parasitology*, 49, 529–542.

Hopkins, C. A. and McCaig, M. L. O, 1963, Studies on *Schistocephalus solidus*. I. The correlation of development in the plerocercoid with infectivity to the definitive host. *Expl Parasit.*, 13, 235–243.

Hopkins, C. A., Law, L. M. and Threadgold, L. T, 1978, *Schistocephalus solidus*: pinocytosis by the plerocercoid tegument. *Expl Parasit.*, 44, 161–172.

Hopkins, S. H, 1933, The morphology, life histories and relationships of the papillose Allocreadiidae (Trematoda). Preliminary report. *Zool. Anz.*, 103, 65–74.

Hopkins, S. H, 1937, A new type of allocreadiid cercaria: the cercariae of *Anallocreadium* and *Microcreadium*. *J. Parasit.*, 23, 94–97.

Hoshina, T, 1968, On the monogenetic trematode *Benedenia seriolae* parasitic on yellow tail *Seriola quinqueradiata*. *Bull. Off. Int. Epizoot.*, 69, 1179–1191.

Hotta, T., Hasegawa, H., Sekikawa, H., Hashimoto, T. and Otsuru, M, 1981, Studies on the diphyllobothriid cestodes in northern Japan. (5) Further survey on *Diphyllobothrium dendriticum* in Hokkaido. *Acta Med. Biol.*, 28, 133–141.

Houlihan, D. F. and Macdonald, S, 1979, *Diclidophora merlangi* and *Entobdella soleae*: egg production and oxygen consumption at different oxygen partial pressures. *Expl Parasit.*, 48, 109–117.

Houston, K. A. and Haedrich, R. L, 1986, Food habits and intestinal parasites of deep demersal fishes from the upper continental slope east of Newfoundland, northwest Atlantic Ocean. *Mar. Biol.*, 92, 563–574.

Howard, E. B., Britt, J. O. Jr. and Matsumoto, G, 1983, Parasitic diseases. In: *Pathobiology of marine mammal diseases* (Ed. Howard, E. B.). *Boca Raton, Florida*, USA: kCRC *Press, Inc.*, 1, 119–232.

Howell, M, 1966, A contribution to the life history of *Bucephalus longicarnatus* (Manter, 1954). *Zoo. Pubs V. U. W.*, 40.

Howell, M. J, 1976, The peritoneal cavity of vertebrates. In: *Ecological Aspects of Parasitology*. (Ed. Kennedy, C. R.). Amsterdam: North Holland Publishing Co., pp. 243–260.

Hristovski, N. D. and Riggio, S, 1977, A contribution to the study of the parasitic helminthofauna associated with populations of the common eel *Anguilla anguilla* L. from fresh-water biota of Macedonia and northern Sicily. *Acta Parasit. Iugol.*, 8, 111–113.

Hsu, H. F. and Hoeppli, R, 1933, Die Oesophagusdrusen einer *Proleptus* sp. und von *Thelazia callipaeda* (Nematoda). *Z. ParasitKde.*, 6, 273–276.

Huang, L. F, 1981, [Studies on the life-cycle of *Anguillicola globiceps*.] *Zool. Mag. (Dongwuxue Zazhi)*, No. 1, 24–25. [In Chinese.]

Huang, W, 1988, Anisakides et anisakidoses humaines. Deuxième partie: Enquête sur les anisakides de poissons commerciaux du marché parisien. *Annls Parasit. Hum. Comp.*, 63, 197–208.

Hubbs, C. L, 1927, The related effects of a parasite on a fish. *J. Parasit.*, 14, 75–84.

Hubschman, J. H, 1985, *Tanaorhamphus longirostris* (Acanthocephala) in gizzard shad from Caesar Creek Lake, Ohio. *Ohio J. Sci.*, 85, 7–11.

Huet, M, 1986, *Textbook of fish culture—breeding and cultivation of fish*. Second edition. London: Fishing News (Books) Ltd. 1–428 pp.

Hughes, R. C, 1928, Studies on the trematode family Strigeidae (Holostomidae) No. IX *Neascus* van *cleavei* (Agersborg). *Trans. Am. microsc. Soc.* 47, 320–341.

Hunninen, A. V. and Cable, R. M., 1943, The life-history of *Lecithaster confusus* Odhner (Trematoda: Hemiuridae). *J. Parasit.*, 29, 71–79.

Hunninen, A. V. and Wichterman, R, 1938, Hyperparasitism: a species of *Hexamita* (Protozoa, Mastigophora) found in the reproductive system of *Deropristis inflata* (Trematoda) from marine eels. *J. Parasit.*, 24, 95–101.

Hunter, G. C. and Kille, R. A., 1950, Some observations on *Dictyocotyle coeliaca* Nybelin, 1941 (Monogenea). *J. Helminth.*, 24, 15–22

Hunter, G. W, 1937, Parasitism of fishes in the lower Hudson area. In: *A biological survey of the lower Hudson watershed*. Biological Survey No. XI (1936); Suppl. to the Twenty-sixth Ann. Rep. N. Y. State Cons. Dept., pp. 264–273.

Hunter, G. W. and Hamilton, J. M., 1941, Studies on host–parasite reactions to larval parasites. IV. The cyst of *Uculifer ambloplitis* (Hughes). *Trans. Am. Microsc. Soc.*, 60, 498–507.

Hunter, G. W. and Hunter, W. S, 1929, Further studies on the bass tapeworn, *Proteocephalus ambloplitis* (Leidy). In: *A Biological Survey of the Erie-Niagara system*. Ann. Rep. of N. Y. State Conserv. Dept. (1928), Suppl., pp. 198–207.

Hunter, G. W. and Hunter, W. S, 1938, Studies on host reactions to larval parasites. I. The effect on weight. *J. Parasit.*, 24, 475–481.

Hutchinson, G. E., 1958, Concluding remarks. *Cold Spring Harbor Symp. Quant. Biol.*, 22, 415–427.

Hutomo, M., Burhanuddin and Hadidjaja, P., 1978, Observations on the incidence and intensity of infection of nematode larvae (Fam. Anisakidae) in certain marine fishes of waters around Panggang Island, Seribu Islands. *Mar. Res. Indonesia*, No. 21, 49–60.

Hutton, R. F, 1957, Preliminary notes on trematoda (Heterophyidae and Strigeoidea) encysted in the heart and flesh of Florida mullet, *Mugil cephalus* L. and *M. curema* Cuvier and Valenciennes. *Bulletin, Dade County Med. Ass., Contribution No. 4.*

Hutton, R. F. and Sogandares-Bernal, F, 1958, Further notes on trematoda encysted in Florida mullets. *Quart. J. Fla. Acad. Sci.*, 21, 329–334.

Hyman, L. H, 1951a, *The Invertebrates: Platyhelminthes and Rhynchocoela, the acoelomate Bilateria. Vol.* II. New York: McGraw-Hill Book Co., Inc., 550 pp.

Hyman, L. H, 1951b, *The Invertebrates: Acanthocephala, Aschelminthes, and Entoprocta. Vol.* II. New York: McGraw-Hill Book Co., Inc., 572 pp.

Ibragimov, Sh. R, 1985, [Parasite fauna of acipenserid fish in the Caspian Sea.] *Izv. Akad. Nauk. Azerb SSR, No. 2*, 47–51. [In Russian.]

Ibragimov, Sh. R, 1988, [The parasite fauna of fish in the Turkmenskit bay of the Caspian Sea.] *Izv. Akad. Nauk Turkmen SSR {Ser. biol. Nauk}, No. 2*, 51–55. [In Russian.]

Ichihara, A, 1983, [Parasites of aquatic species.] *New Food Industry*, 25, 56–57. [In Japanese.]

leshko, E. P. and Anikaeva, L. V, 1980, The polymorphism of *Proteocephalus exiguus* (Cestoidea: Proteocephalidae, a mass parasite of coregonid fish. *Parazitologiya*, 14, 422–426.

Ieshko, E. P. and Golitsina, N. B, 1984, [Analysis of the spatial structure of a population of *Bunodera luciopercae* from *Perca fluviatilis*.] *Parazitologiya*, **18**, 374–382. [In Russian.]

Ieshko, E. P. and Kaufman, B. Z, 1984, Emergence of the mature trematode *Bunodera luciopercae* from infected fish under different light conditions. *J. Zool., Lond.*, **203**, 537–539.

Ieshko, E. P. and Shustov, Yu. A, 1982, [Effect of diplostome infection (Trematoda, Diplostomidae) on sight acuity in fish.] *Parazitologiya*, **16**, 81–83. [In Russian.]

Ieshko, E. P., Malakhova, R. P. and Golitsyna, N. B, 1982, [Ecological aspects of the formation of the parasite fauna of fish in the lakes of the river Kammenaya system.] *In: Ekologiya paraziticheskikh organizmov v biogeotsenozakh severa*. Petrozavodsk, USSR: Karelsk. Fil. Akad. Nauk SSR, Inst. Biol., pp. 5–25. [In Russian.]

Ilyushina, T. L, 1982, [Trematode cercariae—food objects of caddis flies.] In: *Gel'minty v presnovodnykh biotsenozakh*. Moscow, USSR: 'Nauka', pp. 90–98. [In Russian.]

Imada, R. and Muroga, K, 1978, [*Pseudodactylogyrus microrchis* (Monogenea) on the gills of cultured eels. II. Oviposition, hatching and development on the host.] *Bull. Jap. Soc. Sci. Fish.*, 44, 571–576. [In Japanese.]

Imada, R. and Muroga, K, 1979, [*Pseudodactylogrus microrchis* (Monogenea) on the gills of cultured eels. III. Experimental control by trichlorfon. *Bull. Jap. Soc. Sci. Fish.*, 45, 25–29. [In Japanese.]

Ingham, L. and Arme, C, 1973, Intestinal helminths in rainbow trout, *Salmo gairdneri* (Richardson): absence of effect of nutrient absorption and fish growth. *J. Fish Biol.*, 5, 309–313.

Ingham, R. E. and Dronen, N. O., Jr, 1980, Endohelminth parasites from largemouth bass, *Micropterus salmoides*, in Belton and Livingstone Reservoirs, central Texas. *Proc. Helminth. Soc. Wash.*, 47, 140–142.

Inglis, W. G, 1983, An outline classification of the Phylum Nematoda. *Aust. J. Zool.*, 31, 245–255.

Inglis, W. G, 1985, Evolutionary waves: patterns in the origins of animal phyla. *Aust. J. Zool.*, 33, 153–178.

Ip, H. S. and Desser, S. S, 1984, Transmission electron microscopy of the tegumentary sense organs of *Cotylogaster occidentalis* (Trematoda: Aspidogastrea). *J. Parasit.*, 70, 563–575.

Ip, H. S., Desser, S. S. and Weller, I, 1982, *Cotylogaster occidentalis* (Trematoda: Aspidogastrea): scanning electron microscopic observations of sense organs and associated surface structures. *Trans. Am. Microsc. Soc.*, 101, 253–261.

Ishii, S, 1933, On a Filaria parasitic in the caudal fin of *Carassius auratus* L. from Japan. *Fifth Pacific Science Congress*, 4141–4143.

Iskov, M. P, 1975, [Main factors determining the formation of the parasite fauna of fish in the Kakhov reservoir.] [Abstract.] In: *VIII Nauchnaya Konferentsiya Parazitologov Ukrainy. (Tezisy dokladov). Donetsk, Sentyabr' 1975*. Kiev, USSR: Ukr. Nauchno-Issled. Inst., Nauchno-Tekh. Inf., pp. 61–64. [In Russian.]

Iskov, M. P, 1976, Factors responsible for the changes in development of fish parasites in the Kakhovka Reservoir. *Hydrobiological J.* 11, 54–58.

Iskov, M. P, 1978, [The parasite fauna and epizootiological situation at the Sulinskoe fish breeding farm on the Kremenchug reservoir and methods of improving the existing situation.] In: *Problemy gidro-parazitologii. Kiev*, USSR: 'Naukova Dumka', pp. 59–71. [In Russian.]

Ito, J, 1964, *Metagonimus* and other human heterophyid trematodes. In: *Progress of Medical Parasitology in Japan*. (Eds Morishita, K., Komiya, Y. and Matsubayashi, H.). Volume I. Meguro Parasitological Museum, Tokyo, pp. 317–393.

Ito, J. and Mochizuki, H, 1968, [An epidemiologic study of human helminths in Shizuoka Prefecture. VI. The metacercarial fauna in fresh and brackish water fish.] *Jap. of Parasit.*, 17 (2), 69–74.

Ivanchenko, O. F. and Grozdilova, T. A, 1985a, Infestation rate of the young of White Sea herring, reared under experimental conditions and caught in the sea, by trematodes, and their pathogenic effect. In: *NOAA Technical Report NMFS25* (Ed. W. J. Hargis), p. 65

Ivanchenko, O. F. and Grozdilova, T. A, 1985b, [Comparative parasitological study of *Clupea pallasi marisalbi* fry raised in the laboratory and caught in the sea.] *Parazitologiya*, 19, 264–267. [In Russian.]

Ivantsiv, V. V. and Chernogorenko, M. I, 1984, [The life-cycle of *Rhipidocotyle illense* (Trematoda, Bucephalidae).] *Vest. Zool., No. 2*, 66–69. [In Russian.]

Ivashkin, V. M. and Khromova, L. A, 1964, [Biological characters of nematodes of the suborder Camallanata Chitwood, 1936.] *Tr. Gel'm. Lab.*, 14, 98–104. [In Russian.]

Ivasik, V. M, 1952, [Some observations on pathogenicity of *Caryophyllaeus fimbriceps* to carp.] *Trudy nauchno-issled. Inst. Prud. Ozerno-rech. Ryb. Khozyaist, Kiev*, 8, 127–130. [In Russian.]

Ivasik, V. M, 1960, [The importance of other fish in the spread of parasites in carp ponds.] *Zool. Zh.*, 39, 299–302. [In Russian.]

Ivasik, V. M., Kulakovskaya, O. P. and Vorona, N. I, 1969, Exchange of parasites between phytophagous fishes and carp in ponds of the west Ukrainian provinces. *Hydrobiological J.*, 5, 68–71.

Iversen, E. S. and Hoven, E. E, 1958, Some trematodes of fishes from the Central Equatorial Pacific. *Pacific Science, Honolulu*, 12, 131–134.

Iwasik, W. and Swirepo, B, 1967, Zagadnienie pochodzenia karpia (*Cyprinus carpio* L.) w swietle wynikow badan parazytologicznych i biochemicznych rodzaju *Cyprinus*. *Wiad. parazyt.*, 13, 271–273.

Izyumov, Yu. G. and Kas'yanov, A. N, 1981, [Stable morphogenesis and resistance of bream to ligulid infections.] *Parazitologiya*, 15, 174–178. [In Russian.]

Izyumova, N. A, 1956, [On the specificity of *D. vastator* and *D. solidus* to their hosts.] *Parazit. Sb.*, 16, 217–228. [In Russian.]

Izyumova, N. A, 1969, [On the biology and specificity of *Dactylogyrus chranilowi* Bychowsky, 1931.] *Parazit. Sb.*, 24, 128–133. [In Russian.]

Izyumova, N. A, 1970, [On the specificity of some species of *Dactylogyrus* Diesing, 1850.] *Parazitologiya*, 4, 466–472. [In Russian.]

Izyumova, N. A, 1977, [Parasitological studies on the Gor'kov water-reservoir.] Ekologiya *Gel'mintov, No. 1.* Yaroslavl', USSR, 34–45. [In Russian.]

Izyumova, N. A, 1984, [Parasites of fish in conditions of regulated waters.] In: *Biologicheskie resursy vodokhranishch*. Moscow, USSR, pp. 243–251. [In Russian.]

Izyumova, N. A. and Mashtakov, A. V, 1984, [Morphological variability of the trematode *Phyllodistomum angulatum* in relation to the living conditions of the host.] *Ekol. Gel'mi. (Vop. ekol. gel'm.)*, No. 4, 42–49. [In Russian.]

Izyumova, N. A. and Zelentsov, N. I, 1969, [Observations on the development of *Dactylogyrus extensus* Muller and Van Cleave, 1932.] *Parazitologiya*, 3, 528–531. [In Russian.]

Izyumova, N. A. and Zharikova, T. I, 1982, [On some features of the distribution *Dactylogyrus anchoratus* and *D. chranilowi* (Monogenoidea (Beneden) Bychowski, 1937; Dactylogyridea Bychowsky, 1937) on the gills of *Carassius auratus, Cyprinus carpio* and *Abramis ballerus*.] *Trudy Inst. Biol. Vnutr. Vod (Gidrobiol. Kharakter. vodokhran. Volzhsk.bass., No. 46*, 89–100. [In Russian.]

Izyumova, N. A. and Zharikova, T. I, 1984 [Numbers and species composition of dactylogyrids on bream in different zones of the Gor'kovskoe reservoir.] *Mater. Nauch. Konf. Vses. Obshch. Gel'mi. (Biol. takson. gel'mi. zhivot. chelov.), No. 34*, 24–28. [In Russian.]

Izyumova, N. A., Mashtakov, A. V. and Stepanova, M. A, 1982a, [Comparative characteristics of the helminth faunas of bream, pike-perch and sterlet.] *Biolog. Vnutr. Vod, Inf. Byull.', No. 54*, 52–58. [In Russian.]

Izyumova, N. A., Mashtakov, A. V. and Stepanova, M. A, 1982b, [The helminth faunas of pike, bream and pike-perch in the zone of the outflow of warm water of the Kostroma hydroelectric station on the Gorkovskii reservoir.] *Trudy Inst. Biol. Vnutr. Vod (Gidrobiol. Kharakter. vodokhran. Volzhsk. bass., No. 46*, 101–108. [In Russian.]

Izyumova, N. A., Zharikova, T. I. and Stepanova, M. A, 1985a, [Species composition, abundance and distribution of *Dactylogyrus* on the gills of *Abramis brama* in the Dunai river.] *Biol. Vnutr. Vod. Inf. Byull., No. 67*, 19–22. [In Russian.]

Izyumova, N. A., Zharikova, T. I., and Karabekova, D. U. and Asylbaeva, Sh. M, 1985b, [Comparative data on the abundance of *Dactylogyrus* on *Abramis brama* in the Rybinsk reservoir and Lake Issyk-Kul'.] *Biol. Vnutr. Vod. Inf. Byull.*, 66, 34–38 [In Russian.]

Jackson, G. J., Bier, J. W., Payne, W. L. and McClure, F. D, 1981, Recovery of parasitic nematodes from fish by digestion or elution. *Appl. Environ Microbiol.* 41, 912–914.

Jackson, H. C, 1987, The role of blood in helminth nutrition. *Helminth. Abstr.*, 56, 427–434.

Jahn, T. L. and Kuhn, L. R, 1932, The life-history of *Epibdella melleni* MacCallum 1927, a monogenetic trematode parasitic on marine fishes. *Biol. Bull.*, 62, 89–111.

Jain, S. P., Pandey, K. C. and Pandey, A. K, 1976, Some histopathological observations on the stomach wall of *Heteropneustes fossilis* (Bloch.) infected with a cestode. *Agra. Univ. J. Res.*, 25, 1–3.

Jakutowicz, K. and Korpaczewska, W, 1976, Some trace elements in plerocercoid and adult forms of *Ligula intestinalis* (L., 1758) (Cestoda: Diphyllobothriidae). *Bull. Acad. Pol. Sci., Sci. Biol.*, 24, 525–527.

James, B. L. and Bowers, E. A, 1967, Reproduction in the daughter sporocyst of *Cercaria bucephalopsis haimeana* (Lacaze-Duthiers, 1854) (Bucephalidae) and *Cercaria dichotoma* Lebour, 1911 (non Müller) (Gymnophallidae). *Parasitology*, 57, 607–625.

James, B. L. and Srivastava, L. P, 1967, The occurrence of *Podocotyle atomon* (Red., 1802) (Digenea), *Bothriocephalus scorpii* (Müller, 1776) (Cestoda), *Contracaecum clavatum* (Red., 1809) (Nematoda), and *Echinorhynchus gadi* Zoegra, in Müller, 1776 (Acanthocephala) in the five-bearded rockling, *Onos J. Nat. Hist.*, 1, 363–372.

Jamieson, B. G. M, 1966a, *Parahemiurus bennettae* n. sp. (Digenea), a hemiurid trematode progenetic in *Salinator fragilis* Lamarck (Gastropoda, Amphibolidae). *Proc. R. Soc.* 77, 73–80.

Jamieson, B. G. M, 1966b, Larval stages of the progenetic trematode *Parahemiurus bennettae* Jamieson, 1966 (Digenea: Hemiuridae) and the evolutionary origin of cercariae. *Proc. R. Soc. Qld.*, 77, 81–92.

Janicki, C, 1920, Grundlinien einer 'Cercomer Theorie' zur Morphologie der Trematoden und Cestoden. *Festschrift für Zschokke, Basel, No. 30*, 22 pp.

Janicki, C, 1928, Die Lebensgeschichte von *Amphilina foliacea* G. Wagen., Parasiten des Wolga-Sterlets, nach Beobachtungen und Experimenten. *Arb. Biol. Wolga-Statopm.* 10, 97–134.

Janicki, C, 1930, Ueber die jüngsten Zustände von *Amphilina foliacea* in der Fischleibeshöhle, sowie Generalles zur Auffassung des Genus *Amphilina* Wagener. *Zool. Anz.*, 90, 190–205.

Janicki, C. and Rasin, K, 1930, Bemerkung uber *Cystoopsis acipenseri* des Wolga-Sterlets, sowie uber die Entwicklung dieses Nematoden im Zwischenwirt. *Z. Wiss. Zool.*, 136, 1037.

Janiszewska, J, 1938, Studien über die Entwicklung und die Lebenweise der parasitischen Wurmer in der Flunder (*Pleuronectes flesus* L.). *Mem. Acad. Polonaise Sci. Lettres*, 14, 1–68.

Janovy, J. and Hardin, E. L, 1987, Population dynamics of the parasites in *Fundulus zebrinus* in the Platte river of Nebraska. *J. Parasit.*, 73, 689–696.

Janovy, J. and Hardin, E. L, 1988, Diversity of the parasite assemblage of *Fundulus zebrinus* in the Platte river of Nebraska. *J. Parasit.*, 74, 207–213.

Janssen, W. A, 1970, Fish as potential vectors of human bacterial diseases. In: *A symposium on diseases of fishes and shell fishes* (Ed. Sniesko, S. F.) Special Publication, *Am. Fish. Soc., No. 5*, pp. 284–290.

Jara, Z, and Szerow, D, 1981a, [Histopathological changes and localization of the cestode *Khawia sinensis* in the intestine of carp (*Cyprinus carpio*).] *Wiad. parazyt.*, 27, 695–703. [In Polish.]

Jara, Z, and Szerow, D, 1981b, [Electrophoretic examination of the serum of carp (*Cyprinus carpio*) infected with the cestode *Caryophyllaeus* sp.).] *Wiad. Parazyt.*, 27, 713–716. [In Polish.]

Jarecka, L, 1964, Cycle évolutif à un seul hôte intermédiaire chez *Bothriocephalus claviceps* (Goeze, 1782), cestode de *Anguilla anguilla* L. *Annls Parasit. Hum. Comp.*, 39, 149–156.

Jarecka, L, 1975, Ontogeny and evolution of cestodes. *Acta Parasit. Pol.* 23, 93–114.

Jarecka, L., Bance, G. N. and Burt, M. D. B, 1984, On the life cycle of *Anomotaenia micracantha dominicana* (Railliet et Henry, 1912), with ultrastructural evidence supporting the definition of cercoscolex for dilepidid larvae (Cestoda, Dilepididae). *Acta Parasit. Pol*, 29, 27–34.

Jarecka, L., Michajlow, W. and Burt, M. D. B, 1976, Comparative ultrastructure of cestode larvae and Janicki's cercomer theory. [Abstract]. *In: Ann. Meet. Can. Soc. Zool., Univ. Regina, Saskatchewan, 8–11 June, 1976. Abstr. Symp., contrib. pap., Canada,* p. 25.

Jarecka, L., Michajlow, W. and Burt, M. D. B, 1981, Comparative ultrastructure of cestode larvae and Janicki's cercomer theory. *Acta Parasit. Pol.* 28, 65–72.

Jennings, J. B, 1974, Symbioses in the Turbellaria and their implications in studies on the evolution of parasitism. In: *Symbiosis in the Sea.* (Ed. Vernberg, W. B.). Columbia, South Carolina, USA: University of South Carolina Press, pp. 127–160.

Jennings, M. R. and Hendrickson, G. L, 1982, Parasites of chinook salmon (*Oncorhynchus tschawytscha*) and coho salmon (*O. kisutch*) from the Mad River and vicinity, Humboldt County, California. *Proc. Helminth. Soc. Wash.*, 49, 279–284.

Jensen, L. A., Moser, M. and Heckmann, R. A, 1979, The parasites of the California lizardfish, *Synodus lucioceps. Proc. Helminth. Soc. Wash.*, 46, 281–284.

Jianying, Z. and Qizhi, G, 1984, [Three new species of *Dactylogyrus* from gills of *Xenocypris argentel*]. *Acta Zool Sinica,* 30, 247–253. [In Chinese.]

Jilek, R. and Crites, J. L, 1981, Observations on the lack of specificity of *Spinitectus carolini* and *Spinitectus gracilis* (Spirurida: Nematoda) for their intermediate hosts. *Can. J. Zool.*, 59, 476–477.

Jilek, R. and Crites, J. L, 1982a, The life cycle and development of *Spinitectus gracilis* (Nematoda: Spirurida). *Trans. Am. Microsc. Soc.*, 101, 75–83.

Jilek, R. and Crites, J. L, 1982b, The life cycle and development of *Spinitectus carolini* Holl, 1928 (Nematoda: Spirurida). *Amer Midl. Nat.*, 107, 100–106.

Jilek, R. and Crites, J. L, 1982c, Intestinal histopathology of the common bluegill, *Lepomis macrochirus* Rafinesque, infected with *Spinitectus carolini* Holl, 1928 (Spirurida: Nematoda). *J. Fish Dis.*, 5, 75–77.

Job, S. V, 1961a, *Didymozoon tetragynae*: a digenetic trematode of the family Didymozoidae. *J. Madras Univ.*, 31, 311–314.

Job, S. V, 1961b, New record of a digenetic trematode of the family Didymozoidae. *The Presidency College Zoology Magazine*, 8, 12–14.

Job, S. V, 1962, A new record of a digenetic trematode of the genus *Platocystis* (Family: Didymozoidae). *J. Zool. Soc. India*, 13, 143–147.

Job, S. V, 1964, Description of a new species of digenetic trematode (Family: Didymozoidae) and some histochemical observations on the same. *Proc. Indian Acad. Sci.*, 60, 128–134.

Johansen, K, 1970, Air breathing in fishes. In: *Fish Physiology*. (Eds Hoar, W. S. and Randall, D. J.). Volume 4. London: Academic Press, pp. 361–408.

Johnsen, B. O, 1978, The effect of an attack by the parasite *Gyrodactylus salaris* on the population of salmon parr in the river Lakselva, Misvaer in northern Norway. *J. Arctic Biol.*, 11, 7–9.

Johnsen, B. O. and Jensen, A. J, 1986, Infestations of Atlantic salmon *Salmo salar* by *Gyrodactylus salaris* in Norwegian rivers. *J. Fish Biol.*, 29, 233–241.

Johnsen, B. O. and Jensen, A. J, 1988, Introduction and establishment of *Gyrodactylus salaris* Malmberg, 1957, on Atlantic salmon *Salmo salar* L., fry and parr in the River Vefsna, northern Norway. *J. Fish Dis.*, 11, 35–45.

Johnston, B. R. and Halton, D. W, 1981, Excystation *in vitro* of *Bucephaloides gracilescens* metacercaria (Trematoda: Bucephalidae). *Z. ParasitKde*, 85, 71–78.

Jones, A, 1972, The biology of helminth parasites of littoral fishes, including morphology, histochemistry and host–parasite relationships. *Ph.D. thesis, University of Wales* 239 pp.

Jones, A, 1975, The morphology of *Bothriocephalus scorpii* (Müller) (Pseudophyllidea, Bothriocephalidae) from littoral fishes in Britain. *J. Helminth.*, 49, 251–261.

Jones, A, 1980, *Proteocephalus pentastoma* (Klaptocz, 1906) and *Polyonchobothrium polypteri* (Leydig, 1853) from species of *Polypterus* Geoffroy, 1802 in the Sudan. *J. Helminth.*, 54, 25–38.

Jones, A, and Leong, T. S, 1986, Amphistomes from Malaysian fishes, including *Osteochilotrema malayae* gen. nov. sp. nov. (Paramphistomidae: Osteochilotrematinae subfam. nov.). *J. Nat. Hist.*, 20, 117–129.

Jones, A. W. and Mackiewicz, J. S, 1969, Naturally occurring triploidy and parthenogenesis in *Atractolytocestus huronensis* Anthony (Cestoidea: Caryophyllidea) from *Cyprinus carpio* L. in North America. *J. Parasit.*, 55, 1105–1118.

Joo, C. Y, 1988, Changing patterns of infection with digenetic larval trematodes from fresh-water fish in River Taewha, Kyongnam Province. *Korean J. Parasit.*, 26, 263–274.

Joshi, B. D, 1979, Biochemical changes in the liver and blood of a freshwater fish, *Rita rita*, infested with a trematode *Opisthorchis pedicellata*. *Folia Parasit.*, 26, 143–144.

Joshi, B. N. and Sathyanesan, A. G, 1981, A report on the presence of trematode parasites in the brain ventricles of the teleost *Channa punctatus* (Bloch). *Mikroskopie*, 38, 253–255.

Joy, J. E, 1984, Prevalence of the Ancyrocephalinae (Monogenea) on largemouth and spotted basses in Beech Fork Lake, West Virginia. *Proc. Helminth. Soc. Wash.*, 51, 168–170.

Joy, J. E., Tarter, D. C. and Sheridan, M, 1986, *Pomphorhynchus rocci* (Acanthocephala: Echinorhynchidea) from the freshwater drum, *Aplodinotus grunniens* in West Virginia. *Proc. Helminth. Soc. Wash.*, 53, 140–141.

Juhasz, S, 1979, Studies on the nature of the protease inhibitor of *Ligula intestinalis*. *Helminthologia*, 16, 293–298.

Justine, J. L. and Mattei, X, 1982, Présence de spermatozoides a un seul axonème dans trois familles de monogènes Monopisthocotylea: Ancyrocephalidae, Diplectanidae et Monocotylidae. *Annls Parasit. Hum. Comp.*, 57, 419–420.

Justine, J. L. and Mattei, X, 1983, A spermatozoon with two 9 + O axonemes in a parasitic flatworm, *Didymozoon* (Digenea: Didymozoidae). *J. Submicrosc. Cytol.*, 15, 1101–1105.

Justine, J. L. and Mattei, X, 1984a, Comparative ultrastructural study of spermiogenesis in monogeneans (flatworms). 4. *Diplectanum* (Monopisthocotylea Diplectanidae). *J. Ultrastr. Res.*, 88, 77–91.

Justine, J. L. and Mattei, X, 1984b, Atypical spermiogenesis in a parasitic flatworm, *Didymozoon* (Trematoda: Digenea: Didymozoidae). *J. Ultrastr. Res.*, 87, 106–111.

Justine, J. L. and Mattei, X, 1985a, A spermatozoon with undulating membrane in a parasitic flatworm, *Gotocotyla* (Monogenea, Polyopisthocotylea, Gotocotylidae). *J. Ultrastr. Res.*, 90, 163–171.

Justine, J. L. and Mattei, X, 1985b, Ultrastructure de la spermiogenèse et du spermatozoïde de *Loimosina wilsoni* et affinités phylètiques des Loimoidae (Platyhelminthes, Monogenea, Monopisthocotylea). *Zool. Scripta*, 14, 169–175.

Justine, J. L. and Mattei, X, 1985c, Ultrastructure du spermatozoïde de trois monogènes Polyopisthocotylea: *Cemocotyle, Heteraxine* et *Heteraxinoides*. *Annls Parasit. Hum. Comp.*, 60, 659–663.

Justine, J. L. and Mattei, X, 1985d, Particularités ultrastructurales des spermatozoïdes de quelques monogènes Polyopisthocotylea. *Annls Sci. Nat., Zool., Biol. Anim.*, 7, 143–152.

Justine, J. L. and Mattei, X, 1987, Phylogenetic relationships between the families of Capsalidae and Dionchidae (Platyhelminthes, Monogenea, Monopisthocotylea) indicated by the comparative ultrastructural study of spermiogenesis. *Zool. Scripta.*, 16, 111–116.

Justine, J. L., Lambert, A. and Mattei, X, 1985, Spermatozoon ultrastructure and phylogenetic relationships in the monogeneans (Platyhelminthes). *Int. J. Parasist.*, 15, 601–608.

Justine, J. L., Le Brun, N. and Mattei, X, 1985a, Première observation d'un spermatozoide aflagellé chez un plathelminthe parasite, trouvé chez le monogène *Diplozoon gracile* (Polyopisthocotylea, Diplozoidae). *Annls Parasit. Hum. Comp.*, 60, 761–762.

Justine, J. L., Le Brun, N. and Mattei, X, 1985b, The aflagellate spermatozoon of *Diplozoon* (Platyhelminthes: Monogenea: Polyopisthocotylea): a demonstrative case of relationship between sperm ultrastructure and biology of reproduction. *J. Ultrastr. Res.*, 92, 47–54.

Kabata, Z, 1963, Parasites as biological tags. *Special Publications of the International Commission for the Northwest Atlantic Fisheries*, 4, 31–37.

Kabata, Z, 1985, *Parasites and diseases of fish cultured in the tropics*. London, UK: Taylor and Francis Ltd, 318 pp.

Kabata, Z. and Ho, J. S, 1981, The origin and dispersal of hake genus (genus *Merluccius*: Pisces: Teleostei) as indicated by its copepod parasites. *Oceanogr. mar. Biol. Ann. Rev.*, 19, 381–404.

Kabunda, M. Y. and Sommerville, C, 1984, Parasitic worms causing the rejection of tilapia (*Oreochromis* species) in Zaire. *Brit. Vet. J.*, 140, 263–268.

Kadav, M. and Agarwal, S. M, 1982, Amino acid picture (qualitative and quantitative) of host serum of uninfected and infected *Clarias batrachus* parasitized with caryophyllids. *Indian J. Helminth.*, 33, 79–86.

Kadav, M. and Agarwal, S. M, 1983a, Parasitic effects on haematology of *Clarias batrachus* infected with caryophyllids. *Indian J. Helminth.*, 33, 137–143.

Kadav, M. and Agarwal, S. M, 1983b, Parasitic effects on carbohydrate metabolism of *Clarias batrachus* parasitized by caryophyllids. *Indian J. Helminth.*, 33, 153–155.

Kaeding, L. R, 1981, Observations on *Eustrongylides* sp. infection of brown and rainbow trout in the Firehole River, Yellowstone National Park. *Proc. Helminth. Soc. Wash.*, 48, 98–101.

Kagei, N., Sano, M., Takahashi, Y. and Tamura, Moto M, 1978, A case of acute abdominal syndrome caused by *Anisakis* type-II larva. *Jap. J. Parasit.*, 27, 427–431.

Kahl, W, 1938, Nematoden in Seefischen. II. Erhebungen über den Befall von Seefischen mit Larven von *Anacanthocheilus rotundatus* (Rudolf) und die durch diese Larven bervorgerufenen Reaktionen des Wirtsgewebes. *Z. ParasitKde*, 10, 513–534.

Kakacheva-Avramova, D, 1976, The influence of pollution of the Bulgarian section of the river Danube on the occurrence of helminthiases in fish. (Proc. Int. Symp., Environmental parasitology in the programme MAB, Warsaw, 6–8 Mar. 1975). *Wiad. Parazyt.*, 22, 429–431.

Kakacheva-Avramova, D, 1977, [The effect of pollution on the fish helminths in the Bulgarian section of the river Danube.] *Ribno Stopanstvo*, 24, 11–12. [In Bulgarian.]

Kakacheva-Avramova, D, 1983a, [New helminth infections in freshwater fish in Bulgaria.] *Vet. Sbir., Sof.*, 81, 27–29. [In Bulgarian.]

Kakacheva-Avramova, D, 1983b, *Helminths of freshwater fishes in Bulgaria*. Sofia: Bulgarian Academy of Sciences, 1–261 pp.

Kakacheva-Avramova, D. and Menkov, V, 1978, The effects of some chemical compounds on the viability of *Bothriocephalus gowkongensis* Yeh, 1955. *Khelmintologiya, Sofia*, 5, 28–38.

Kakacheva-Avramova, D. and Menkova, I, 1979, [Helminths of Salmonidae.] *Vet. Sbir., Sof.*, 77, 15–17. [In Bulgarian.]

Kalyuga, N. V. and Kalashnik, V. I, 1982, [Therapy and prophylaxis of parasitic diseases of fish in the Pridneprov warm-water fish farm.] *Rybnoe Khozyaistvo, Kiev*, USSR, No. 35, 56–60. [In Russian.]

Kamegai, S, 1970, On *Diplozoon nipponicum* Goto, 1891. Part III. The seasonal development of the reproductive organs of *Diplozoon nipponicum* parasitic on *Cyprinus carpio. Res. Bull. Meguro Parasit. Mus.*, 3, 21–25.

Kamegai, S, 1971a, [The determination of a generic name of a parasite of flying fishes' muscle, a didymozoid whose ova have occasionally been found in human faeces in Japan.] *Jap. J. Parasit.*, 20, 170–176. [In Japanese.]

Kamegai, S, 1971b, On some parasites of a coelacanth (*Latimeria chalumnae*) a new monogenean, *Dactylogyrus latimeris* n. g. n. sp. (Dactylodiscidae n. fam.) and two larval helminths. *Res. Bull. Meguro Parasit. Mus.* 5, 1–5.

Kamegai, S. and Ichiara, A, 1972, A checklist of the helminths from Japan and adjacent areas. Part I. Fish parasites reported by S. Yamaguti from Japanese waters and adjacent areas. *Res. Bull. Meguro Parasit. Mus.* 6, 1–143.

Kamegai, S, and Shimazu, T, 1982, *Ovarionematobothrium saba* n. sp. (Didymozoidae), a new trematode parasite of the ovary of a marine fish *Pneumatophorus japonicus*, from Japan. *Res. Bull. Meguro Parasit. Mus.*, 8, 35–38.

Kanth, L. K. and Srivastava, L. P, 1984, Host–parasite relations in monozoic tapeworm, *Lytocestoides fossilis* infection of freshwater fish, *Heteropneustes fossilis* (Bloch). *Curr. Sci.*, 53, 607–608.

Kapoor, B. G., Smith, H. and Verighina, I. A, 1975, The alimentary canal and digestion in teleosts. *Adv. Mar. Biol.*, 13, 109–239.

Kapustina, N. I, 1978, [Host–parasite relationships in the system *Khawia sinensis*—carp in low intensity infection.] *Trudy Vses. Inst. Prud. Ryb. Khozyaist. (Parazit., bolez. ryb ikh parazity)*, 27, 75–87. [In Russian.]

Karanikolov, Y, 1977, [Industrial trials for the control of bothriocephaliasis in carp.] *Izv. Inst. Ribna Promishlenost, Filial Sladkovodno Ribarstvo, Plovdiv*, 13, 77–82. [In Bulgarian.]

Karasev, A. B, 1983a, [*Anisakis* infection in *Gadus poutassou* in the Norwegian Sea.] In: *Biologiya i promysel pelagichskikh ryb Severnogo basseina*. Murmansk, USSR, pp. 81–92. [In Russian.]

Karasev, A. B, 1983b, [Infection with the nematode *Contracaecum aduncum* Rud., 1802 and some biological indices in cod fry.] In: *Issledovaniya biologii, morfologii i fiziologii gidrobiontov*. (Ed. Matishov, G. G.). Apatity, USSR: Kol'ski Filial AN SSSR, pp. 60–64. [In Russian.]

Karasev, A. B, 1984a, [Parasitological investigations on O-group cod in the Barents Sea in the autumn and winter of 1980/81.] *Annls Biol.*, 38, 91.

Karasev, A. B, 1984b, [Ecological characteristics of *Micromesistius poutassou*.] In: *Ekologo-parazitologicheskie issledovaniya severnykh morei*. Apatity, USSR: Kol'skii Filial Akademii Nauk SSSR, pp. 82–88. [In Russian.]

Karasev, A. B., Mitenev, V. K., Zubchenko, A. V. and Bezgachina, T. V, 1981, Some data on infestation of blue whiting by nematode *Anisakis* sp. larvae in the northeast Atlantic. *ICES CM 1981/H:20*. 6 pp.

Karaseva, T. A, 1983, [The fish parasite fauna in experimental nursery lakes in the Murmansk region.] In: *Issledovaniya biologii, morfologii i fiziologii gidrobiontov*. (Ed. Matishov, G. G.). Apatity, USSR: Kol'skii Filial AN SSSR, pp. 50–53. [In Russian.]

Karling, T. G, 1974, On the anatomy and affinities of the turbellarian orders. In: *Biology of the Turbellaria* (Eds N. W. Riser and M. P. Morse). New York, McGraw-Hill, pp. 1–16.

Karyakarte, P. P. and Kulkarni, H. S, 1980, Phosphatase activity in *Lecithochirium acutum* Chauhan, 1945 (Trematoda: Digenea). *Proc. Indian Acad. Parasit.*, 1, 7–11.

Kashkovskii, V. V, 1979, [The parasites of coregonids in the southern and mid-Urals and in the Zaural'e.] *Sb. Nauch. Trud., Gosud. Nauchno-Issled. Inst. Ozern. Rech. Ryb. Khozaist., Ural'skoe otdel., No. 10*, 145–151. [In Russian.]

Kashkovskii, V. V, 1982, [Seasonal variations of the age composition of a population of *Dactylogyrus amphibothrium* (Monogenea, Dactylogyridae).] *Parazitologiya*, 16, 35–40. [In Russian.]

Kashkovskii, V. V. and Skomorokhova, N. K, 1982, [Control of carp diseases in winter.] *Veterinariya, Mosk., No. 12*, 60. [In Russian.]

Kassim, M. H., Rahemo, Z. I. and Warsi, A. A, 1977, The influence of season and the sex of the host on the intensity of some parasites infecting *Cyprinion macrostomus* and *Acanthobrama marmid* fishes from the river Tigris Mosul (Iraq). *Mesopotamia J. Agric.*, 12, 133–144.

Kawatsu, H, 1978, Studies on the anaemia of fish. IX. Hypochronic microcytic anemia of crucian carp caused by infestation with a trematode, *Diplozoon nipponicum*. *Bull. Jap. Soc. Sci. Fish.*, 44, 1315–1319.

Kayton, R. J, 1983, Histochemical and X-ray elemental analysis of the sclerites of *Gyrodactylus* spp. (Platyhelminthes: Monogenoidea) from the Utah chub, *Gila atraria* (Girard). *J. Parasit.*, 69, 862–865.

Kazakov, B. E, 1967, [New data on helminths of *Osmerus eperlanus* and a short ecological analysis of its helminth fauna. (Collected papers on the helminth fauna of fish and birds).] *Izdatelstvo Akademii Nauk SSSR*, 18–31. [In Russian.]

Kazic, D, 1978, Endohelminths of Salmonida from the artificial lake Piva, Montenegro, Yugoslavia, (Congress 1977, Denmark). *Verh. Int. Verein. Theor. Angew. Limnol.*, **20**, 2154–2158.

Kean, B. H. and Breslau, R. C, 1964, *Parasites of the human heart.* New York: Grune and Stratton, pp. 95–103.

Kearn, G. C, 1962, Breathing movements in *Entobdella soleae* (Trematoda, Monogenea) from the skin of the common sole. *J. Mar. Biol. Ass.* UK, 42, 93–104.

Kearn, G. C, 1963a, The life cycle of the monogenean *Entobdella soleae*, a skin parasite of the common sole. *Parasitology*, 53, 253–263.

Kearn, G. C, 1963b, The egg, oncomiracidium and larval development of *Entobdella soleae*, a monogenean skin parasite of the common sole. *Parasitology*, 53, 435–447.

Kearn, G. C, 1963c, Feeding in some monogenean skin parasites: *Entobdella solea* on *Solea solea* and *Acanthocotyle* sp. on *Raja clavata. J. Mar. Biol. Ass. UK*, 43, 747–766.

Kearn, G. C, 1967a, The life-cycles and larval development of some acanthocotylids (Monogenea) from Plymouth rays. *Parasitology*, 57, 157–167.

Kearn, G. C, 1967b, Experiments on host-finding and host-specificity in the monogenean skin parasite *Entobdella soleae. Parasitology*, 57, 585–605.

Kearn, G. C, 1967c, Observations on monogenean parasites from the nasal fossae of European rays: *Empruthotrema raiae* (MacCallum 1916) Johnston and Tiegs, 1922 and *E. torpedinis* sp. nov. from *Torpedo marmorata. Transl. of Russian paper from 'Studies on the monogeneans'. In memory of Academician B. E. Bychowsky. Proc. Inst. Biol. and Ped., Far-East Sci. Centre, Vladivostok, USSR*, 34, 18 pp.

Kearn, G. C, 1968a, The larval development of *Merizocotyle* sp., a monocotylid monogenean from the nasal fossae of *Raia undulata. Parasitology*, **58**, 921–928.

Kearn, G. C, 1968b, The development of the adhesive organs of some diplectanid, tetraonchid and dactylogyrid gill parasites (Monogenea). *Parasitology*, **58**, 149–163.

Kearn, G. C, 1970a, The oncomiracidia of the monocotylid monogeneans *Dictyocotyle coeliaca* and *Calicotyle kröyeri. Parasitology*, **61**, 153–160.

Kearn, G. C, 1970b, The production, transfer and assimilation of spermatophores by *Entobdella soleae*, a monogenean skin parasite of the common sole. *Parasitology*, **60**, 301–311.

Kearn, G. C, 1971, The physiology and behaviour of monogenean skin parasite *Entobdella soleae* in relation to its host (*Solea solea*). In: *Ecology and Physiology of Parasites*, (Ed. A. M. Fallis). Toronto, Canada: University of Toronto Press, pp. 161–186.

Kearn, G. C, 1973, An endogenous circadian hatching rhythm in the monogenean skin parasite *Entobdella soleae*, and its relationship to the activity rhythm of the host (*Solea solea*). *Parasitology*, 66, 101–122.

Kearn, G. C, 1974a, Nocturnal hatching in the monogenean skin parasite *Entobdella hippoglossi* from the halibut, *Hippoglossus hippoglossus. Parasitology*, **68**, 161–172.

Kearn, G. C, 1974b, The effects of fish skin mucus on hatching in the monogenean parasite *Entobdella soleae* from the skin of the common sole (*Solea solea*). *Parasitology*, **68**, 173–188.

Kearn, G. C, 1974c, A comparative study of the glandular and excretory systems of the oncomiracidia of the monogenean skin parasites *Entobdella hippoglossi*, *E. diadema* and *E. soleae. Parasitology*, 69, 257–269.

Kearn, G. C, 1975a, The mode of hatching of the monogenean *Entobdella soleae*, a skin parasite of the common sole (*Solea solea*). *Parasitology*, 71, 419–431.

Kearn, G. C, 1975b, Hatching in the monogenean parasite *Dictyocotyle coeliaca* from the body cavity of *Raja naevus. Parasitology*, 70, 87–93.

Kearn, G. C, 1976, Body surface of fishes. In: *Ecological aspects of parasitology.* (Ed. Kennedy, C. R.). Amsterdam, The Netherlands: North-Holland Publishing Company, pp. 185–208.

Kearn, G. C, 1978a, Predation on a skin-parasitic monogenean by a fish *J. Parasit.*, **64**, 1129–1130.

Kearn, G. C, 1978b, Eyes with, and without, pigment shields in the oncomiracidium of the monogean parasite *Diplozoon paradoxum. Z. ParasitKde*, 57, 35–47.

Kearn, G. C, 1979, Studies on gut pigment in skin-parasitic monogeneans, with special reference to the monocotylid *Dendromonocotyle kuhlii. Int. J. Parasit.*, 9, 545–552.

Kearn, G. C, 1980, Light and gravity responses of the oncomiracidium of *Entobdella soleae* and their role in host location. *Parasitology*, **81**, 71–89.

Kearn, G. C, 1982, Rapid hatching induced by light intensity reduction in the monogenean *Entobdella diadema, J. Parasit.*, **68**, 171–172.

Kearn, G. C, 1984a, The migration of the monogenean *Entobdella soleae* on the surface of its host, *Solea solea*. *Int. J. Parasit.*, 14, 63–69.

Kearn, G. C, 1984b, A possible ciliary photoreceptor in a juvenile polyopisthocotylean monogenean *Sphyranura* sp. *Int. J. Parasit.*, 14, 357–361.

Kearn, G. C, 1985, Observations on egg production in the monogenean *Entobdella soleae*. *Int. J. Parasit.*, 15, 187–194.

Kearn, G. C, 1986, The eggs of monogeneans. *Adv. Parasit.*, 25, 175–273.

Kearn, G. C, 1987, The site of development of the monogenean *Calicotyle kroeyeri*, a parasite of rays. *J. Mar. Biol. Ass. UK*, 67, 77–87.

Kearn, G. C, 1988a, The monogenean skin parasite *Entobdella soleae*: movement of adults and juveniles from host to host (*Solea solea*). *Int. J. Parasit.*, 18, 313–319.

Kearn, G. C, 1988b, Orientation and locomotion in the monogenean parasite *Entobdella soleae* on the skin of its host (*Solea solea*). *Int. J. Parasit.*, 18, 753–759.

Kearn, G. C. and Green, J. E, 1983, *Squalotrema llewellyni* gen. nov., sp. nov., a monocotylid monogenean from the nasal fossae of the spur-dog, *Squalus acanthias*, at Plymouth. *J. Mar. Biol. Ass., UK*, 63, 17–25.

Kearn, G. C. and Macdonald, S, 1976, The chemical nature of host hatching factors in the monogenean skin parasites *Entobdella soleae* and *Acanthocotyle lobianchi*. Int. *J. Parasit.*, 6, 457–466.

Keenan, L. and Koopowitz, H, 1982, Physiology and *in situ* identification of putative aminergic neurotransmitters in the nervous system of *Gyrocotyle fimbriata*, a parasitic flatworm. *J. Neurobiol.*, 13, 9–21.

Keenan, L., Koopowitz, H. and Solon, M. H, 1984, Primitive nervous systems: electrical activity in the nerve cords of the parasitic flatworm, *Gyrocotyle fimbriata*, *J. Parasit.*, 70, 131–138.

Kelle, W, 1977, Unterschiedlich starker Parasitenbefall der Wittlinge *Merlangius merlangus des Neuwerker Fahrwassers in Sommer 1974*. *Arch. Fischereiwissenschaft*, 28, 65–68.

Kennedy, C. R, 1965a, The life-history of *Archigetes limnodrili* (Yamaguti) (Cestoda: Caryophyllaeidae) and its development in the invertebrate host. *Parasitology*, 55, 427–437.

Kennedy, C. R, 1965b, Taxonomic studies on *Archigetes* (Leuckart, 1878) (Cestoda: Caryophyllaeidae). *Parasitology*, 55, 439–451.

Kennedy, C. R, 1967, *Pomphorhynchus laevis* in dace of the River Avon. *Proc. 3rd Brit. Coarse Fish Conf.*, 24–26.

Kennedy, C. R, 1968, Population biology of the cestode *Caryophyllaeus laticeps* (Pallas, 1781) in dace, *Leuciscus leuciscus* L. of the River Avon. *J. Parasit.*, 54, 538–543.

Kennedy, C. R, 1969a, The occurrence of *Eubothrium crassum* (Cestoda: Pseudophyllidea) in salmon *Salmo salar* and trout *S. trutta* of the River Exe. *J. Zool., Lond.*, 157, 1–9.

Kennedy, C. R, 1969b, Seasonal incidence and development of the cestode *Caryophyllaeus laticeps* (Pallas) in the River Avon. *Parasitology*, 59, 783–794.

Kennedy, C. R, 1970, The population biology of helminths of British freshwater fish. In: *Aspects of fish parasitology*, (Eds Taylor, A. E. R. and Muller, R.), Symposium of the British Society for Parasitology (8th) London, November 7, 1969. Oxford: Blackwell Scientific Publications, pp. 145–159.

Kennedy, C. R, 1971, The effect of temperature upon the establishment and survival of the cestode *Caryophyllaeus laticeps* in orfe, *Leuciscus idus*. *Parasitology*, 63, 59–66.

Kennedy, C. R, 1972a, Parasite communities in freshwater ecosystems. In: *Essays in Hydrobiology* (Eds Clarke, R. B. and Wootton, E. J.). University of Exeter Press, pp. 53–68.

Kennedy, C. R, 1972b, The effects of temperature and other factors upon the establishment and survival of *Pomphorhynchus laevis* (Acanthocephala) in goldfish, *Carassius auratus*. *Parasitology*, 65, 283–294.

Kennedy, C. R, 1974a, The importance of parasite mortality in regulating the population size of the acanthocephalan *Pomphorhynchus laevis* in goldfish. *Parasitology*, 68, 93–101.

Kennedy, C. R, 1974b, A checklist of British and Irish freshwater fish parasites with notes on their distribution. *J. Fish Biol.*, 6, 613–644.

Kennedy, C. R, 1975, *Ecological Animal Parasitology*. Oxford: Blackwell Scientific Publications, 163 pp.

Kennedy, C. R, 1976a, *Ecological aspects of parasitology*. Amsterdam, Netherlands: North-Holland Publishing Co., x +474 pp

Kennedy, C. R, 1976b, Reproduction and dispersal. In: *Ecological aspects of parasitology*. (Ed. Kennedy, C. R.). Amsterdam, The Netherlands: North-Holland Publishing Company, pp. 143-160.

Kennedy, C. R, 1977a, Distribution and zoogeographical characteristics of the parasite fauna of char *Salvelinus alpinus* in Arctic Norway, including Spitsbergen and Jan Mayen Islands. *Astarte*, 10, 49–55.

Kennedy, C. R, 1977b, The regulation of fish parasite populations. In: *Regulation of parasite populations. (Proc. Symp. sponsored by Amer. Microscop. Soc. and Amer. Soc. Parasit., New Orleans, 10-14 Nov. 1975).* (Ed. Esch, G. W.), London, UK: Academic Press Inc., pp. 61–109.

Kennedy, C. R, 1978a, Studies on the biology of *Eubothrium salvelini* and *E. crassum* in resident and migratory *Salvelinus alpinus* and *Salmo trutta* and in *S. salar* in North Norway and the islands of Spitsbergen and Jan Mayen. *J. Fish Biol.*, 12, 147–162.

Kennedy, C. R, 1978b, The biology, specificity and habitat of the species of *Eubothrium* (Cestoda: Pseudophyllidea), with reference to their use as biological tags: a review. *J. Fish Biol.*, 12, 393–410.

Kennedy, C. R, 1978c, An analysis of the metazoan parasitocoenoses of brown trout *Salmo trutta* from British Lakes. *J. Fish Biol.*, 13, 255–263.

Kennedy, C. R, 1978d, The parasite fauna of resident char *Salvelinus alpinus* from Arctic islands, with special reference to Bear Island. *J. Fish Biol.*, 13, 457–466.

Kennedy, C. R, 1978e, The status of brown and rainbow trout, *Salmo trutta* and *S. gairdneri* as hosts of the acanthocephalan, *Pomphorhynchus laevis*. *J. Fish Biol.*, 13, 265–275.

Kennedy, C. R, 1979, The distribution and biology of the cestode *Eubothrium parvum* in capelin, *Mallotus villosus*, (Pallas) in the Barents Sea, and its use as a biological tag. *J. Fish Biol.*, 15, 223–236.

Kennedy, C. R, 1981a, The occurrence of *Eubothrium fragile* (Cestoda: Pseudophyllidea) in twaite shad, *Alosa fallax* (Lacépède) in the River Severn. *J. Fish Biol.*, 19, 171–177.

Kennedy, C. R, 1981b, [Dynamics of the parasitic coenoses in British freshwater ecosystems.] *Trudy. Zool. Inst., Leningr.*, 108, 9–22 [In Russian].

Kennedy, C. R, 1981c, Long term studies on the population biology of two species of eyefluke, *Diplostomum gasterostei* and *Tylodelphys clavata* (Digenea: Diplostomatidae), concurrently infecting the eyes of perch, *Perca fluviatilis*. *J. Fish Biol.*, 19, 221–236.

Kennedy, C. R, 1981d, The establishment and population biology of the eyefluke *Tylodelphys podicipina* (Digenea: Diplostomatidae) in perch. *Parasitology*, 82, 245–255.

Kennedy, C. R, 1982, Biotic factors. In: *Parasites—their world and ours. Proceedings of the Fifth International Congress of Parasitology., Toronto, Canada, 7–14 August 1982, under the auspices of the World Federation of Parasitologists* (Eds Mettrick, D. F. and Desser, S. S.). Amsterdam, The Netherlands: Elsevier Biomedical Press, pp. 293–302.

Kennedy, C. R, 1984a, The use of frequency distributions in an attempt to detect host mortality induced by infections of diplostomatid metacercariae. *Parasitology*, 89, 209–220.

Kennedy, C. R, 1984b, The dynamics of a declining population of the acanthocephalan *Acanthocephalus clavula* in eels *Anguilla anguilla* in a small river. *J. Fish Biol.*, 25, 665–667.

Kennedy, C. R, 1985a, Site segregation by species of Acanthocephala in fish, with special reference to eels, *Anguilla anguilla*. *Parasitology* 90, 375–390.

Kennedy, C. R, 1985b, Interactions of fish and parasite populations: to perpetuate or pioneer? In: *Ecology and Genetics of Host-Parasite Interactions* (Eds Rollinson, D. and Anderson, R. M.). London: Academic Press, pp. 1–21.

Kennedy, C. R, 1985c, [Population biology of parasites: present state and perspectives.] *Parazitologiya*, 19, 347–356. [In Russian.]

Kennedy, C. R, 1985d, Regulation and dynamics of acanthocephalan population. In: *Biology of the Acanthocephala* (Eds Crompton, D. W. T. and Nickol, B. B.). Cambridge: Cambridge University Press. pp. 385–416.

Kennedy, C. R, 1987, Long-term stability in the population levels of the eyefluke *Tylodelphys podicipina* (Digenea: Diplostomatidae) in perch. *J. Fish Biol.*, 31, 571–581.

Kennedy, C. R, 1990, Helminth communities in freshwater fish: structured communities or stochastic assemblages. In: *Parasite communities: Patterns and Processes.* (Eds Esch, G., Bush, A. and Aho,). London: Chapman and Hall, pp. 1–327.

Kennedy, C. R. and Burrough, R, 1977, The population biology of two species of eyefluke, *Diplostomum gasterostei* and *Tylodelphys clavata*, in perch. *J. Fish Biol.*, 11, 619–633.

Kennedy, C. R. and Burrough, R. J, 1978, Parasites of trout and perch in Malham Tarn. *Field Studies*, 4, 617–629.

Kennedy, C. R. and Burrough, R. J, 1981, The establishment and subsequent history of a population of *Ligula intestinalis* in roach *Rutilus rutilus* (L.). *J. Fish Biol.*, 19, 105–126.

Kennedy, C. R. and Fitch, D. J, 1990, Colonization, larval survival and epidemiology of the nematode *Anguillicola crassus* parasitic in the eel, *Anguilla anguilla*, in Britain. *J. Fish Biol.*, 36, 117–131.

Kennedy, C. R. and Hine, P. M, 1969, Population biology of the cestode *Proteocephalus torulosus* (Batsch) in dace *Leuciscus leuciscus* L. of the River Avon. *J. Fish Biol.*, 1, 209–219.

Kennedy, C. R. and Lord, D, 1982, Habitat specificity of the acanthocephalan *Acanthocephalus clavula* (Dujardin, 1845) in eels *Anguilla anguilla* L. *J. Helminth.*, 56, 121–129.

Kennedy, C. R. and Moriarty, C, 1987, Co-existence of cogeneric species of Acanthocephala: *Acanthocephalus lucii* and *A. anguillae* in eels *Anguilla anguilla* in Ireland. *Parasitology*, 95, 301–310.

Kennedy, C. R. and Rumpus, A, 1977, Long-term changes in the size of the *Pomphorhynchus laevis* (Acanthocephala) population in the River Avon. *J. Fish Biol.*, 10, 35–42.

Kennedy, C. R. and Walker, P. J, 1969, Evidence for an immune response by dace, *Leuciscus leuciscus* to infections by the cestode *Caryophyllaeus laticeps*. *J. Parasit.*, 55, 579–582.

Kennedy, C. R., Broughton, P. F. and Hine, P. M, 1976, The sites occupied by *Pomphorhynchus laevis* in the alimentary canal of fish. *Parasitology*, 72, 195–206.

Kennedy, C. R., Broughton, P. F. and Hine, P. M, 1978, The status of brown and rainbow trout, *Salmo trutta* and *S. gairdneri*, as hosts of the acanthocephalan *Pomphorhynchus laevis*. *J. Fish Biol.*, 13, 265–275.

Kennedy, C. R., Bush, A. O. and Aho, J. M, 1986, Patterns in helminth communities: why are birds and fish different? *Parasitology*, 93, 205–215.

Kennedy, C. R., Laffoley, D. d'A., Bishop, G., Jones, P. and Taylor, M, 1986, Communities of parasites of freshwater fish of Jersey, Channel Islands. *J. Fish Biol.*, 29, 215–226.

Kennedy, G. Y, 1979, Pigments of marine invertebrates. In: *Advances in marine biology, Volume 16*, (Eds Russell, F. S. and Yonge, M.). London: Academic Press Inc., pp. 309–381.

Kennedy, M. J, 1979, The responses of miracidia and cercariae of *Bunodera mediovitellata* (Trematoda: Allocreadiidae) to light and to gravity. *Can. J. Zool.*, 57, 603–609.

Kent, M. L, 1981, Disease prevention and control. The life-cycle and treatment of a turbellarian disease of marine fishes. *Freshwater and Marine Aquarium*, 4, 11–13.

Kent, M. L. and Olson, A. C, 1986, Interrelationships of a parasitic turbellarian (*Paravortex* sp.) (Graffillidae, Rhabdocoela) and its marine fish hosts. *Fish Path.* 21, 65–72.

Keppner, E. J, 1975, Life cycle of *Spinitectus micracanthus* Christian. 1972 (Nematoda: Rhabdochonidae) from the bluegill, *Lepomis macrochirus* Rafinesque, in Missouri with a note on *Spinitectus gracilis* Ward and Magath, 1917. *Amer. Midl. Nat.*, 93, 411–423.

Keymer, A., Crompton, D. W. T. and Walters, D. E, 1983, Parasite population biology and host nutrition: dietary fructose and *Moniliformis* (Acanthocephala). *Parasitology*, 87, 265–278.

Kezic, N., Fijan, N. and Kajanga, L, 1975, Bothriocephaliasis of carp in S. R. Croatia. *Veterinarski Arhiv.*, 45, 289–291.

Khadzhiiski, Z, 1979, [Liguliasis and some aspects of its pathology.] *Ribno Stopanstvo, No. 4*, 11–13. [In Bulgarian.]

Khadzhiiski, Z, 1980, [The biology and medical significance of anisakid nematodes, parasites of marine fish.] *Ribno Stopanstvo*, 27, 14–16. [In Bulgarian.]

Khalifa, K. A., Hassan, F. K., Attiah, H. H. and Latif, B. M. A, 1978, Parasitic infestation of fishes in Iraqi waters. *Iraq J. Biol. Sci.*, 6, 58–63.

Khalil, L. F, 1963, On a redescription of *Brevicaecum niloticum* McClelland, 1957 (Trematoda, Paramphistomidae), and the erection of a new subfamily. *J. Helminth.*, 37, 215–220.

Khalil, L. F, 1964, On the biology of *Macrogyrodactylus polypteri* Malmberg, 1956, a monogenetic trematode on *Polypterus senegalus* in the Sudan. *J. Helminth.*, 38, 219–222.

Khalil, L. F, 1965, On a new philometrid nematode, *Thwaitia bagri* sp. nov. from a freshwater fish in the Sudan. *J. Helminth*, 34, 309–312.

Khalil, L. F, 1969, Larval nematodes in the herring (*Clupea harengus*) from British coastal waters and adjacent territories. *J. Mar. Biol. Ass. UK*, 49, 641–659.

Khalil, L. F, 1970, Further studies on *Macrogyrodactylus polypteri*, a monogenean on the African freshwater fish *Polypterus senegalus*. *J. Helminth.*, 44, 329–348.

Khalil, L. F, 1971, *Ichthybothrium ichthybori* gen. et. sp. nov. (Cestoda: Pseudophyllidea) from the African freshwater fish *Ichthyborus besse* (Joannis, 1835). *J. Helminth.*, 44, 371–379.

Khalil, L. F, 1974, Some nematodes from the freshwater fishes of Rhodesia with the description of a new species *Cithariniella petterae* n. sp. *Annls Parasit. Hum. Comp.*, 48, 811–818.

Khalil, L. F, 1977, *Gyrometra kunduchi* n. sp., a cestodarian from *Plectorhinchus pictus* (Thunberg, 1792) from the Indian Ocean. *J. Fish Biol.*, 11, 15–19.

Khalil, L. F, 1981, *Australotrema brisbanensis* n.g., n.sp. (Paramphistomidae: Dadaytrematinae) from the Australian freshwater mullet *Trachystoma petardi* (Castlenau). *Syst. Parasit.*, 3, 65–70.

Khalil, L. F. and Abdul-Salam, J, 1989, *Macrobothridium rhynchobati* n.g., n.sp. from the elasmobranch *Rhynchobatus granulatus*, representing a new family of diphyllidean cestodes, the Macrobothrididae. *Syst. Parasit.*, 13, 103–109.

Khalil, L. F. and Thurston, J. P, 1973, Studies on the helminth parasites of freshwater fishes of Uganda including the descriptions of two new species of digeneans. *Rev. Zool. Bot. Afr.*, 87, 209–248.

Khalil, L. F., Robinson, R. D. and Hall, R. N, 1988, Monogenea causing mortality of hybrid cichlids cultured in coastal waters of Southern Jamaica. In: *Program and abstracts of the Vth European Multicolloquium of Parasitology. September 4–9, Budapest.* Hungary, European Federation of Parasitologists, p. 191.

Khamees, N. R. and Mhaisen, F. T, 1988, Ecology of parasites of the cyprinid fish *Carasobarbus luteus* from Mehaijeran Creek, Basrah. *J. Biol. Sci. Res.*, 19, 409–419.

Khan, R. A. and Kiceniuk, J, 1983, Effects of crude oils on the gastrointestinal parasites of two species of marine fish. *J. Wildl. Dis.*, 19, 253–258.

Khan, R. A. and Kiceniuk, J, 1984, Histopathological effects of crude oil on Atlantic cod following chronic exposure. *Can. J. Zool.*, 62, 2038–2043.

Khan, R. A. and Thulin, J, 1991, Influence of pollution on parasites of aquatic animals. *Adv. Parasit.*, 30, 201–239.

Khlebovich, V. V. and Mikhailova, O. Yu, 1976, [Osmotic tolerance and adaptation of *Echinorhynchus gadi* (Acanthocephala, Echinorhynchidae) to hypotonic conditions.] *Parazitologiya,* 10, 444–448. [In Russian.]

Khokhlov, P. P. and Pugachev, O. N, 1979, [The formation of the parasite fauna of fish in north-eastern Asia.] In: *Bolezni i parazity ryb Ledovitomorskoi provintsii (v predelakh SSSR).* (Eds Gundrizer, A. N. and Bauer, O. N.). Tomsk, USSR: Izdatel'stvo Tomskogo Universiteta, 43–55. [In Russian.]

Khotenovskii, I. A, 1977, The life-cycle of some species of monogeneans from the genus *Diplozoon. Parazit. Sb.,* 27, 35–43. [In Russian.]

Khotenovskii, I. A, 1985, [Suborder Octomacrinea Khotenovskly. (Fauna of the USSR. Monogenea. New Series No. 132).] Leningrad, USSR: 'Nauka', 262 pp. [In Russian.]

Khramov, A. A., Pavlyukov, I. A. and Shelikhanova, R. M, 1984, [Infection of fish with the agents of opisthorchiasis and diphyllobothriasis in the areas of the Tyumen region that will be affected by the projected partial re-routing of Siberian rivers.] *Sb. Nauch. Trud. Gosud. Nauchno-Issled. Inst. Ozern. Rech. Rybnogo Khozyaist. (Bolez. parazit. ryb vod. Zapad. Sib.), No. 226,* 66–69. [In Russian.]

Khromova, L. A, 1975, [Development of *Dacnitis sphaerocephalus caspicus* (Nematoda, Cucullanidae).] *Zool. Zh.,* 54, 449–452. [In Russian.]

Kikkawa, J. and Anderson, D. J. (Eds) 1986, *Community Ecology. Patterns and Process.* London: Blackwell Scientific Publications, i–xi + 432 pp.

Kimura, M. and Endo, M, 1979, [Whirling disease caused by trematode metacercariae.] *Fish Path.* 13, 211–213. [In Japanese.]

Kinne, O, 1980, Diseases of marine animals: general aspects. In: *Diseases of marine animals, Vol. I. General Aspects, Protozoa to Gastropoda* (Ed. Kinne, O.). Chichester, UK: John Wiley and Sons, pp. 13–73.

Kinne, O. (Ed.) 1984, *Diseases of marine animals Vol. IV, Part 1 Introduction, Pisces.* Hamburg: Biologische Anstalt Helgoland i–xiii + 501 pp.

Kirby, J. M, 1981, Seasonal occurrence of the ectoparasite *Gyrodactylus atratuli* on spotfin shiners. *Trans. Am. Fish. Soc.,* 110, 462–464.

Kirichenko, L. M, 1974, [Age dynamics of the parasite fauna of carp in the fish farms of the Volgograd region (USSR).] In: *Voprosy parazitologii zhivotnykh Yugo-Vostoka SSSR.* Volgogradskii Pedagogicheskii Institut im. A. S. Serafimovicha, Volgograd, USSR, pp. 18–25. [In Russian.]

Kirichenko, L. M. and Kosareva, N. A, 1972, [Effect of *Bothriocephalus* on yearling carp.] In: *Vopr. Morf. Ekol. Parazit. zhiv.* Volgograd, USSR: Pedagogicheskii Institut, pp. 123–127. [In Russian.]

Kisielewska, K. (Chairman) 1974, Session 2. Parasite ecology and the problem 'Man and his environment'. Round Table talks, Symp. on 'Modern ecological problems in parasitology', Parasit. Inst. Polish Acad. Sci and Ecol. Inst., 4–5 Apr. 1973, Dziekanów Lésny, Warsaw. Wiad. Parazyt., 20, 775–786.

Kiskaroly, M, 1987, [Investigations on the parasite fauna of fishes in freshwater ponds in Bosnia and Hercegovina I. Cyprinid ponds A. Monogenetic trematodes 5.] *Veterinaria, Yugoslavia*, 36, 391–397. [In Croat.]

Kiskaroly, M. and Tafro, A, 1983a, [Prevalence of diplostomiasis—a frequent parasitic disease in cyprinid fish ponds in Bosnia–Hercegovina (Yugoslavia). I.] *Veterinaria, Yugoslavia*, 32, 97–103. [In Croat.]

Kiskaroly, M. and Tafro, A, 1983b, [Prevalence of diplostomiasis—a frequent parasitic disease of cyprinid fish ponds in Bosnia–Hercegovina (Yugoslavia). II. (Distribution of diplostomiasis in the Sanicani fish ponds).] *Veterinaria, Yugoslavia*, 32, 195–202. [In Croat.]

Kiskaroly, M. and Tafro, A, 1983c, [Prevalence of diplostomiasis, the most common parasitosis in cyprinid ponds of Bosnia and Hercegovina. III. (Prevalence of diplostomiasis on the Prnjavor fishfarm).] *Veterinaria, Yugoslavia*, 32, 511–520. [In Croat.]

Kiskaroly, M. and Tafro, A, 1984a, [Treatment and prophylaxis of the most frequent parasitoses of fish in freshwater fish farms.] *Veterinarski Glasnik*, 38, 317–324. [In Croat.]

Kiskaroly, M. and Tafro, A, 1984b, [Prevalence of diplostomiasis, the most frequent parasitic infection in cyprinid fish ponds in Bosnia–Hercegovina. V. Diplostomiasis prevalence among so-called 'wild fish'.] *Veterinaria*, 33, 351–358. [In Croat.]

Kiskaroly, M. and Tafro, A, 1984c, [Prevalence of diplostomiasis, the most frequent parasitic infection in cyrpinid fish ponds in Bosnia-Hercegovina. IV. Diplostomiasis prevalence in the 'Bardaca' fish farm.] *Veterinaria*, 33, 55–67. [In Croat.]

Kiskaroly, M., Ibrovic, M., Dzubic, A. and Obratil, S, 1981, [The parasite fauna of marsh birds associated with cyprinid fish farms in Bosna and Hercegovina and their role in the epizootiology of parasitic fish diseases.] *Veterinaria, Yugoslavia*, 30, 291–303. [In Croat.]

Kititsina, L. A, 1985, [Respiratory intensity of 2-year-old grass carp infected with helminths. [Abstract].] In: *VI Vsesoyuznaya konferentsiya po ekologicheskoi fiziologii i biokhimii ryb, Sentyabr' 1985 g. Tezisy dokladov.* (*VI All-Union conference on ecological physiology and biochemistry of fish, Sept. 1985*. Chief Editor: Virbitskas, Yu. B.). Vilnius, USSR; Institu Zoologii i Parazitologii, pp. 194–195. [In Russian.]

Kititsyna, L. A, 1984, [The utilization of oxygen by the cestode *Bothriocephalus acheilognathi* from the intestine of carp fry.] *Vest. Zool.*, 5, 44–46. [In Russian.]

Klebanovskii, V. A, 1983, [Epidemiological significance of diphyllobothriid plerocercoids from whitefish.] *Vop. Prirod. Ochagov. Bolez., No. 13*, 141–147. [In Russian.]

Klebanovskii, V. A., Smirnov, P. L., Klebanovskaya, I. A. and Obgol'ts, A. A, 1977, [Human helmin-thiases in eastern Taimyr (Katangskii region).] In: *Probl. epidem. profilak. prirod. bolez..Zapolyar'e. (Sb. nauch. rabot)*. Omsk, USSR: Omskii Meditsinskii Institut., pp. 144–164. [In Russian.]

Klenov, A. P, 1969a, [Coniferous needles tested against *Bothriocephalus* in grass carp.] *Veterinariya*, 46, 56–57. .[In Russian.]

Klenov, A. P, 1969b, [Testing 'phytoncides' (onion and leaves of horse-radish) against *Bothriocephalus* infection in grass carp.] *Veterinariya*, 46, 59. [In Russian.]

Kliks, M. M, 1983, Anisakiasis in the western United States: four new case reports from California. *Am. J. Trop. Med. Hyg.*, 32, 526–532.

Klochkova, E. A, 1974, [The sensory apparatus of *Hysteromorpha triloba* (Diplostomatidae) cercariae.] *Trudy Gel'mi. Lab. (Ekol. geogr. gel'm.)*, 24, 56–61. [In Russian.]

Knöfler, H. and Lorenz, G, 1982, Akutes Abdomen durch Nematodenlarvenbefall (Anisakiasis). *Deutsche Gesundheitswesen*, 37, 189–192.

Ko, R. C, 1986, A preliminary review of the genus *Ascarophis* van Beneden, 1871 (Nematoda: Cystidicolidae) of the gastrointestinal tract of fishes, Hong Kong. *Department of Zoology, University of Hong Kong*, 54 pp.

Ko, R. C. and Adams, J. R, 1969, The development of *Philonema oncorhynchi* (Nematoda: Philometridae) in *Cyclops bicuspidatus* in relation to temperature. *Can. J. Zool.*, 47, 307–312.

Ko, R. C. and Ling, J, 1980, A histochemical study of the cephalic-cervical system of a gnathostomatid nematode, *Echinocephalus sinensis*, parasite of oysters and rays. *Z. ParasitKde*, 63, 59–63.

Ko, R. C., Margolis, L. and Machida, M, 1985, *Pseudascarophis kyphosi* n. gen., n. sp. (Nematoda: Cystidicolidae) from the stomach of the fish *Kyphosus cinerascens* (Forskal) from Japan. *Can. J. Zool.*, 63, 2684–2688.

Ko, R. C., Morton, B. and Wong, P. S, 1975, Prevalence and histopathology of *Echinocephalus sinensis* (Nematoda: Gnathostomatidae) in natural and experimental hosts. *Can. J. Zool.*, 53, 550–559.

Kobayashi, I, 1933, Recent researches on Japanese fish which serve as the intermediate hosts of helminths. *The Proceedings of the Fifth Science Congress of the Pacific*, 5, 4157–4163

Kocyłowski, B, 1963, État actuel des maladies des poissons. Organisation de l'inspection des poissons et de leurs produits de consommation en Pologne. *Bull. Off. Int. Épizoot.*, 59, 89–109.

Kohn, A. and Fernandes, B. M. M, 1987, Estudo comparativo dos helmintos parasitos de peixes do Rio Mogi Guassu, coletados nas excursoes realizadas entre 1927 e 1985. *Mems. Inst. Oswaldo Cruz*, 82, 483–500.

Køie, M, 1971a, On the histochemistry and ultrastructure of the redia of *Neophasis lageniformis* (Lebour, 1910) (Trematoda, Acanthocolpidae). *Ophelia* 9, 113–143.

Køie, M, 1971b, On the histochemistry and ultrastructure of the daughter sporocyst of *Cercaria buccini* (Lebour, 1911). *Ophelia*, 9, 145–163.

Køie, M, 1971c, On the histochemistry and ultrastructure of the tegument and associated structures of the cercaria of *Zoogonoides viviparus* in the first intermediate host. *Ophelia*, 9, 165–206.

Køie, M, 1973, The host–parasite interface and associated structures of the cercaria and adult *Neophasis lageniformis* (Lebour, 1910). *Ophelia*, 12, 205–219.

Køie, M, 1975, On the morphology and life-history of *Opechona bacillaris* (Molin, 1859) Looss, 1907 (Trematoda, Lepocreadiidae). *Ophelia*, 13, 63–68.

Køie, M, 1976, On the morphology and life-history of *Zoogonoides viviparus* (Olsson, 1868) Odhner, 1902 (Trematoda, Zoogonidae). *Ophelia*, 15, 1–14.

Køie, M, 1977, Stereoscan studies of cercariae, metacercariae, and adults of *Cryptocotyle lingua* (Creplin 1825) Fischoeder 1903 (Trematoda, Heterophyidae). *J. Parasit.*, 63, 835–839.

Køie, M, 1978, On the morphology and life-history of *Stephanostomum caducum* (Looss 1901) Manter 1934 (Trematoda, Acanthocolpidae). *Ophelia*, 17, 121–133.

Køie, M, 1979a, On the morphology and life-history of *Monascus* [= *Haplocladus*] *filiformis* (Rudolphi, 1819) Looss, 1907 and *Steringophorus furciger* (Olsson, 1868) Odhner, 1905 (Trematoda, Fellodistomidae). *Ophelia*, 18, 113–132.

Køie, M, 1979b, On the morphology and life-history of *Derogenes varicus* (Müller, 1784) Looss, 1901 (Trematoda, Hemiuridae). *Z. ParasitKde*, 59, 67–78.

Køie, M, 1980, On the morphology and life-history of *Steringotrema pagelli* (Van Beneden, 1871) Odhner, 1911 and *Fellodistomum fellis* (Olsson, 1868) Nicoll, 1909 [syn. *S. ovacutum* (Lebour, 1908) Yamaguti, 1953] (Trematoda, Fellodistomidae). *Ophelia*, 19, 215–236.

Køie, M, 1981, On the morphology and life-history of *Podocotyle reflexa* (Creplin, 1825) Odhner, 1905, and a comparison of its developmental stages with those of *P. atomon* (Rulolphi, 1802) Odhner, 1905 (Trematoda, Opecoelidae). *Ophelia*, 20, 17–43.

Køie, M, 1982, The redia, cercaria and early stages of *Aporocotyle simplex* Odhner, 1900 (Sanguinicolidae)—a digenetic trematode which has a polychaete annelid as the only intermediate host. *Ophelia*, 21, 115–145.

Køie, M, 1983, Digenetic trematodes from *Limanda limanda* (L.) (Osteichthyes, Pleuronectidae) from Danish and adjacent waters, with special reference to their life-histories. *Ophelia*, 22, 201–228.

Køie, M, 1984, Digenetic trematodes from *Gadus morhua* L. (Osteichthyes, Gadidae) from Danish and adjacent waters, with special reference to their life-histories. *Ophelia*, 23, 195–222.

Køie, M, 1985a, On the morphology and life-history of *Lepidapedon elongatum* (Lebour, 1908) Nicoll, 1910 (Trematoda, Lepocreadiidae). *Ophelia*, 24, 135–153.

Køie, M, 1985b, The surface topography and life-cycles of digenetic trematodes in *Limanda limanda* (L.) and *Gadus morhua* L. *D.Sc. thesis, Copenhagen University*. Copenhagen: Olson and Olson, 20 pp. + folding table.

Køie, M, 1987a, Eel parasites in freshwater and marine habitats in Denmark. In: *Actual problems in fish parasitology. Proceedings of the Second International Symposium of Ichthyoparasitology, Tihany, Hungary*, p. 46.

Køie, M, 1987b, Eel parasites in natural waters and aquaculture in Denmark. In: *Parasites and diseases in natural waters and aquaculture in Nordic countries. Proceedings of a Symposium, 2–4 December 1986, Stockholm, Sweden* (Eds Stenmark, A. and Malmberg, G.). Stockholm, Sweden; Zoo-tax, Naturhistoriska riksmuseet, pp. 38–45.

Køie, M, 1988a, Parasites in eels, *Anguilla anguilla* (L.), from eutrophic Lake Esrum (Denmark). *Acta Parasit. Pol.*, 33, 89–100.

Køie, M, 1988b, Parasites in European eel *Anguilla anguilla* (L.) from Danish freshwater, brackish and marine localities. *Ophelia*, 29, 92–118.

Køie, M, 1989, On the morphology and life-history of *Lecithaster gibbosus* (Rudophi, 1802) Luhe, 1901 (Digenea, Hemiuroidea). *Parasit. Res.*, 75, 361–367.

Køie, M. and Lester, R. J. G, 1985, Larval didymozoids (Trematoda) in fishes from Moreton Bay, *Proc. Helminth. Soc. Wash.*, 52, 196–203.

Kolovarova, V, 1978, [The use of tobacco dust in the control of *Bothriocephalus* infection in carp.] *Ribno Stopanstvo*, 25, 10–11. [In Bulgarian.]

Kolovarova, V, 1983, [Achievements in the control of bothriocephaliasis in carp.] *Ribno Stopanstvo*, 29, 7–9. [In Bulgarian.]

Komarova, T. I, 1976, [The changes in the parasite fauna of carp larvae and fry in the Kremenchug reservoir.] *Gibdrobiol. Zh.*, 12, 93–96. [In Russian.]

Komarova, T. I, 1978, [The influence of Diplostomatidae metacercariae on the survival of young fish in experimental conditions.] In: *Probl. gidro-parazit. Kiev.* USSR, 'Naukova Dumka', pp. 87–93. [In Russian.]

Komasara, P. and Lisinska, K, 1986, Pasozyty zewnetrzne ryb: koliasa (*Scomber colias* Gmelin, 1788) i makreli (*Scomber scombrus* L. 1758). *Wiad. Parazyt.*, 32, 497–499.

Komiya, V, 1966, Clonorchis and clonorchiasis. *Adv. Parasit.*, 4, 53–106.

Konovalov, S. M, 1975, Differentiation of local populations of sockeye salmon *Oncorhynchus nerka* (Walbaum). *University of Washington Publications in Fisheries, New Series 6. ('Nauka' Publishing House, Leningrad, 1971*),

Konovalov, S. M. and Butorina, T. E, 1985, Parasites as indicators of specific features of fish ecology. In: *Parasitology and pathology of marine organisms of the World Ocean* (Ed. Hargis, W. J. Jr.). Seattle, USA, NOAA Technical Report NMFS, 25, 35–38.

Koops, H. and Hartmann, E, 1989, *Anguillicola*—infestations in Germany and in German eel imports. *J. Applied Ichthyol.*, 5, 41–45.

Korotaeva, V. D, 1975, [The incidence of helminths in the muscle of food fish of the Antarctic and sub-Antarctic.] In: *Probl. Parazit., Mater. VIII nauch. Konf. parazit. UKSSR. Chast 1.* Kiev, USSR: Naukova Dumka, 255–257. [In Russian.]

Korotaeva, V. D, 1982, [Trematode fauna of fish of the order Zeiformes.] *Parazitologiya*, 16, 464–468. [In Russian.]

Körting, W, 1975, Aspekte zum Bandwurmbefall der Fische. Die Bedeutung der Parasiten für die Produktion von Süsswasserfischen Vortr. gehalt. Münchener Fisch.-biol. Semin 5–7 Dec. 1973. *Fisch und Umwelt, No. 1*, 81–87.

Körting, W, 1976a, Metabolism in parasitic helminths of freshwater fish. In: *Biochemistry of parasites and host–parasite relationships* (Ed. Van den Bossche, H.). Amsterdam, The Netherlands: North-Holland Publishing Company, pp. 95–100.

Körting, W, 1976b, Zum Stoffwechsel von *Bothriocepahlus gowkongensis* (Cestoda: Pseudophyllidea) aus Süsswasserfischen. *Z. ParasitKde*, 50, 186–187.

Körting, W, 1977, The host reaction against some fish parasites. Die Reaktion des Wirts gegenüber einigen Fischparasiten. (Beiträge zur Histopathologie der Fische—Vortr. gehalt. Münchener Fisch.-biol. Semin. 26–28 Oct. 1976). *Fisch und Umwelt*, 4, 37–48.

Körting, W, 1984a, Economically important parasitic diseases in aquaculture of fishes. *Bull. European Ass. Fish Pathol.*, 4, 70–71.

Körting, W, 1984b, Larval cyclophyllidean cestodes in carp and tench. *Bull. European Ass. Fish Pathol.*, 4, 40–41.

Körting, W, 1987, Fish diseases in Central Europe. Their importance and control. In: *Parasites and diseases in natural waters and aquaculture in Nordic countries. Proceedings of a Symposium, 2–4 December 1986, Stockholm, Sweden* (Eds Stenmark, A. and Malmberg, G.). Stockholm, Sweden: Zootax, Naturhistoriska riksmuseet, pp. 21–37.

Körting, W. and Barrett, J, 1977, Carbohydrate catabolism in the plerocercoids of *Schistocephalus solidus* (Cestoda: Pseudophyllidea). *Int. J. Parasit.*, 7, 411–417.

Körting, W. and Barrett, J, 1978, Studies on beta-oxidation in the plerocercoids of *Ligula intestinalis* (Cestoda: Pseudophyllidea). *Z. ParasitKde*, 57, 243–246.

Korzyukov, Yu. A, 1979, [*Diseases of aquarium fish.*] Moscow, USSR: 'Kolos'. 175 pp. [In Russian.]

Koscheleva, L. I. and Muzikovskii, A. M, 1969, Quantitative evaluation of phenasal in medicated granulated food for fish. *Byulleten Vsesoyuznogo Instituta Gelmintologii im. K. I. Skryabina*, 3, 62–64.

Kostarev, G. F, 1980, [The effects of pollution on the fish parasite fauna in the Kama reservoirs.] In: *Biologicheskie resursy vodoemov zapadnogo Urala. Mezhyuzovskii Sbornik Nauchnykh Trudov.* Perm, USSR: Permskii Gosudarstvennyi Universitet, pp. 147–153. [In Russian.]

Kotikova, E. A, 1977, [The evolution of the nervous system of cestodes and the laws governing changes in the number of nerve stems.] In: *Znachenie protsessov polimerizatsii i oligomerizatsii v evolyutsii Sbornik Nauchnykh Rabot.* Leningrad, USSR: Akademiya Nauk SSSR, Zoologicheskii Institut, pp. 39–41. [In Russian.]

Kotikova, E. A, 1983, [Aspects of the structure of the nervous system of Diplozoonidae (Monogenea).] *Trudy Zool. Insti., Leningrad (Issled. morf. faun. parazit. cherv.*), 121, 12–17. [In Russian.]

Kotikova, E. A. and Kuperman, B. I, 1978a, [The anatomy of the nervous system of cestodes from the families Amphicotylidae and Diphyllobothriidae (Pseudophyllidea).] *Parazitologiya*, 12, 210–217. [In Russian.]

Kotikova, E. A. and Kuperman, B. I, 1978b, [New data on the structure of the nervous system in Pseudophyllidea.] *Biologiya Morya, Vladivostok, No. 6*, 41–46. [In Russian.]

Kozel, T. R. and Whittaker, F. H, 1985, Ectoparasites of the blackstripe topminnow, *Fundulus notatus*, from Harrods Creek, Oldham County, Kentucky. *Proc. Helminth. Soc. Wash.*, 52, 314–315.

Kozicka, J, 1958, Diseases of fishes of Druzno Lake (parasitofauna of the biocoenosis of Druzno Lake—part VII.) *Acta Parasit. Pol.*, 6, 393–432.

Kozinenko, I. I, 1981, [The use of immunoserological reactions in *intra vitam* diagnosis of *Bothriocephalus* in carp.] In: *Ekol. morf. osob. zhiv. sred. obit.* Kiev, USSR: 'Naukov Dumka', pp. 127–129. [In Russian.]

Kozinenko, I. I. and Balakhnin, I. A, 1981, [Reactivity of carp to *Bothriocephalus antigens.*] *Mater. Nauch. Konf. Vses. Obstich. Gel'm.*, 33, 22–27. [In Russian.]

Král, J, 1977, [Use of piperazine salt of niclosamide for treatment of *Proteocephalus* infection of rainbow trout.] *Veterinaria SPOFA*, 19, 75–83. [in Czech.]

Král, J., Sevcik, B., Prouza, A. and Vondrka, K, 1980, [Taenifugin carp., medicated granulate for the treatment of *Bothriocephalus gowkongensis* infections in fish.] *Biologizace e Chemizace Zivocisné Vyroby, Veterinaria*, 16, 183–192. [In Czech.]

Krasil'shchikov, M. S. and Lyubina, T. V, 1984, [Parasites and diseases of fish in pond and lake fish farms in the Omsk region.] *Sb. Nauch. Trud. Gosud. Nauchno-Issled. Inst. Ozern. Rech. Ryb. Khozyaist., No. 226*, 25–31. [In Russian.]

Krasnoshchekov, G. P, 1980, [The cercomer, the larval organ of cestodes.] *Zh. Obshch. Biol.*, 41, 615–627. [In Russian.]

Krebs, C. J, 1972, *Ecology. The experimental analysis of distribution and abundance.* New York, Evanston, San Francisco, London: Harper and Row, 694 pp.

Krishna, G. V. R. and Simha, S. S, 1980, Histochemical localization of esterases in *Gangesia* sp. *Proc. Indian Acad. Parasit.*, 1, 4–6.

Kritscher, E, 1988a, Die Fische des Neusiedlersees und ihre Parasiten. VI. Cestoidea. *Annalen des Naturhistorischen Museums in Wien. Serie B für Botanik und Zoologie*, 90, 183–192.

Kritscher, E, 1988b, Die Fische des Neusiedlersees und ihre Parasiten. VII. Trematoda: Monogenea under Zusammenfassung. *Annalen des Naturhistorischen Museums in Wien. Seri B für Botanik und Zoologie*, 90, 407–421.

Kritsky, D. C, 1978, The cephalic glands and associated structures in *Gyrodactylus eucaliae* Ikezaki and Hoffman, 1957 (Monogenea: Gyrodactylidae). *Proc. Helminth. Soc. Wash.*, 45, 37–49.

Kritsky, D. C., Boeger, W. A. and Thatcher, V. E, 1988, Neotropical Monogenea. 11. *Rhinoxenus*, new genus (Dactylogyridae: Ancyrocephalinae) with descriptions of three new species from the nasal cavities of Amazonian Characoides. *Proc. Biol. Soc. Wash.*, 101, 87–94.

Kritsky, D. C., Leiby, P. D. and Kayton, R. J, 1978, A rapid stain technique for the haptoral bars of *Gyrodactylus* species (Monogenea). *J. Parasit.*, 64, 172–174.

Kritsky, D. C., Thatcher, V. E. and Boeger, W. A, 1986, Neotropical Monogenea. 8. Revision of *Urocleidoides* (Dactylogyridae, Ancyrocephalinae). *Proc. Helminth. Soc. Wash.*, 53, 1–37.

Krotas, R., Kiselene, V. (Kiseliene, V.) and Volskis, G, 1984, [Parasitological situation in the cooling tank of the Lithuanian hydroelectric station.] In: *Teploenergetika i okruzhayushchaya sreda: Funktionirovanie populyatsii i soobshchesty vodnykh zhivotnykh v okhladitele Litovskoi GRES. Tom 4. Vilnius, USSR; Mokslas, Institut Zoologii i Parazitologii AN Litovskoi SSR*, pp. 45–72. [In Russian.]

Krotenkov, V. P, 1985, [Sources and dissemination of *Bothriocephalus* infection in carp.] *Veterinariya, Mosk., USSR, No. 12*, 50–52. [In Russian.]

Krotov, A. I, 1971, [Comparative physiology of helminths.] *Trudy Vses. Inst. Gel'mi. K. I. Skryabina*, 17, 141–147. [In Russian.]

Ktari, M. H, 1969, Récherches sur l'anatomie et la biologie de *Microtyle salpae* Parona et Perugia, 1890 parasite de *Box salpa* L. (Téléostéen). *Annls Parasit. Hum. Comp.*, 44, 425–440.

Ktari, M. H, 1977, Le parasitisme d'*Echeneis naucrates* L. (poisson, teleostéen) par deux monogènes (Monopisthocotylea) du genre *Dionchus*: *D. agassizi* Goto, 1899 et *D. remorae* MacCallum, 1916. In: *Excerta parasitologica en memoria del Doctor Eduardo Caballero y Caballero.* Mexico: Universidad Nacional Autonoma de Mexico, pp. 61–67.

Kudentsova, R. A, 1979, [Ecological analysis of the parasite fauna of trash fish and farmed fish in pond farms of different types.] *Sb. Nauch. Trud. Gosud. Nauchno-Issled. Inst. Ozern. Rech. Ryb. Khozyaist. (Ekol. parasit. ryb), No. 140*, 48–107. [In Russian.]

Kudinova, M. A, 1979, [The correlative links of morphological systems and organs of *Phyllodistomum maritae.*] *Trudy Gel'mintologicheskoi Laboratorii (Gel'minty zhivotnykh i rastenii)*, 29, 80–88. [In Russian.]

Kudryashova, Yu. V, 1970, [The effect of *Bothriocephalus gowkongensis* on the haematological indicators of 2-year-old carp.] *Dok. Mosk.Sel'skokhoz. Akad. K. A. Timiryazeva, No. 164*, 345–349. [In Russian.]

Kudryavtseva, E. S, 1961, [The more important parasitic diseases of fish in the Sukhona river and Kubenskoye lake.] *Sb. Rab. Vologod. Nauchno-issled. Vet. Opyt. Sta.*, 5, 109–117. [In Russian.]

Kulachkova, V. G, 1977, Gyrodactylidae as indicator of local herring (*Clupea harengus pallasi marisalbi* Berg) shoals in the White Sea. *Parazit. Sb., No. 27*, 27–34. [In Russian.]

Kulachkova, V. G, and Timofeeva, T. A, 1977, [The acanthocephalan *Echinorhynchus gadi* (Zoega) from relict cod in Lake Mogil'noe.] *Parazitologiya*, 11, 316–320. [In Russian.]

Kulakovskaya, O. P, 1962, [The development of Caryophyllaeidae (Cestoda) in the intermediate host.] *Zool. Zh.*, 41, 986–992. [In Russian.]

Kulakovskaya, O. P, 1975, [The structure of the parasitic coenosis of fish in the cooling tank of the Burshtyn hydro-electric station.] [Abstract]. In VIII Nauchnaya Konferentsiya Parazitology Ukrainy. (Tezisy dokladov). Donetsk, Sentyabr 1975. Kiev, USSR; Ukrainskiĭ Nauchno—Issledovatel'skiĭ] Institut, Nauchno-Tekhnicheskoĭ Informatsii. 85–88. [In Russian.]

Kulakovskaya, O. P. and Demshin, N. I, 1978, [Origins and phylogeny of caryophyllids (Cestoda: Caryophyllidea).] *In: Probl. gidro-parazit.* Kiev, USSR: 'Naukova Dumka', pp. 95–104. [In Russian.]

Kulemina, I. V, 1977, [The role of temperature in the formation of the opisthaptor of *Gyrodactylus* sp. from *Carassius*.] *Vestnik Leningradskogo Universiteta No. 9*, 12–18. [In Russian.]

Kulow, H, 1982, Die wichtigsten Krankheiten der Fische in der DDR und die sich daraus ergebenden Aufgaben für die Fischkrankheitsforschung. *Monats.Vet. Med.*, 37, 470–472.

Kuntz, S. M. and Font, W. F, 1984, Seasonal dynamics of *Allopodocotyle boleosomi* (Pearse, 1924) n.comb. (Digenea: Opecoelidae) in Wisconsin darters (Etheostomatinae). *Can. J. Zool.*, 62, 2666–2672.

Kuperman, B. I, 1973, [Tapeworms of the genus *Triaenophorus* parasites of fishes.] *Academy of Sciences of the USSR, Institute of Biology of Inland Waters, Leningrad*, viii + 222 pp. [In Russian.]

Kuperman, B. I, 1976, Electron-microscope study of the tegument of the cestodes *Bothriocephalus scorpii* and *Eubothrium crassum* [Abstract]. In: *Kratkie tezisy dokladov II Vsesoyuznogo simpoziuma po parazitam i boleznyam morskikh zhivotnykh. Kaliningrad, USSR; Ministerstvo Rybnogo Khozaistva SSSR, AtlantNIRO*, pp. 37–39. [In Russian.]

Kuperman, B. I, 1978a, [The biology and development cycle of *Eubothrium rugosum* (Cestoda: Pseudophyllidea).] In: *Probl. Gidro-biol.* Kiev, USSR: 'Naukova Dumka', pp. 105–112. [In Russian.]

Kuperman, B. I, 1978b, [Aspects of the life-cycle and biology of cestodes from Salmonidae in the Kamchatka.] *Biol. Mor., No. 4*, 53–60. [In Russian.]

Kuperman, B. I, 1980, [Ultrastructure of the cestode integument and its importance in systematics.] *Parazito. Sb., No. 29*, 84–95. [In Russian.]

Kuperman, B. I, 1981a. [Ultrastructure of the tegument and glandular apparatus of cestodes in ontogenesis.] *Biol. Vnutr. Vod. Inf. Byull., No. 51*, 29–36. [In Russian.]

Kuperman, B. I, 1981b. [Demonstration of ecological groups of perch by means of parasites.] In: *Ekologiya gel'mintov*. Yaroslavl', USSR: Yaroslavskii Gosudarstvennyi Universitet, pp. 29–35. [In Russian.]

Kuperman, B. I, 1981c. *Tapeworms of the genus Triaenophorus parasites of fishes*. [Translated from Russian by B. R. Sharma]. New Delhi, India: Amerind Publishing Co. Pvt. Ltd., for the US Department of the Interior and the National Science Foundation, viii + 222 pp.

Kuperman, B. I. and Davydov, V. G, 1980, Ultrastructure and evolution of the gland system in pseudophyllidean cestodes in the early stages of their development. In: *Voprosy parazitologii vodnykh bespozvonochnykh zhivotnykh. (Tematicheskii Sbornik.). Vilnyus, USSR: Akademiya Nauk Litovskoi SSR, Institut Zoologii i Parazitologii*, pp. 61–63. [In Russian.]

Kuperman, B. I. and Davydov, V. G, 1982, The fine structure of frontal glands in adult cestodes. *Int. J. Parasit.*, 12, 285–293.

Kuperman, B. I. and Shul'man, R. E, 1978, [Experimental study of factors influencing the multiplication and numbers of *Dactylogyrus* parasites of *Abramis brama*.] *Parazitologiya*, 12, 101–107. [In Russian.]

Kupryanova, R. A, 1954, [The biology of the fish nematodes *Camallanus lacustris* (Zoega, 1776) and *Camallanus truncatus* (Rudolphi, 1814) (Nematoda: Spirurida).] *Dokl. AN SSSR*, 97, 373–376. [In Russian.]

Kurochkin, Yu. V, 1980, [The parasite fauna of flying fish (family Exocoetidae) in the Pacific Ocean.] *Trudy Inst Okeanogr. im. P. P. Shirshova (Sarganoobraznye ryby Mirovogo Okeana)*, 97, 276–296. [In Russian.]

Kurochkin, Yu. V, 1981, [*Nybelinia* infection and the use of *Theragra chalcogramma* as food.] In: *Ekologicheskie zapasy i promysel mintaya*. Vladivostok, USSR, pp. 116–124. [In Russian.]

Kurochkin, Yu. V, 1984, [Applied and scientific aspects of marine parasitology.] In: *Biologicheskie osnovy rybovodstva: parazity i bolezni ryb. (Biologicheskie resursy gidrosfery i ikh ispol'zovanie)* (Eds Musselius, V. A. and Skryabina, E. S.). Moscow, USSR: 'Nauka', pp. 180–188. [In Russian.]

Kurochkin, Yu. V, 1985, Applied and scientific aspects of marine parasitology. In: *Parasitology and pathology of marine organisms of the world ocean* [Ed: Hargis, W. J.] NOAA Technical Report, NMFS 25, pp. 15–18.

Kurovskaya, L. Ya, 1981, [Effect of the anthelmintic phenasal on the phosphatases of *Bothriocephalus gowkongensis* at different developmental stages.] In: *Ekologo-morfologicheskie osobennosti zhivotnykh i sreda ikh obitaniya*. Kiev, USSR: 'Naukova Dumka', pp. 129–132. [In Russian.]

Kurovskaya, L. Ya, 1984, [The influence of parasite infection of carp fry on the activity of their intestinal enzymes.] *Sb. Nauch. Trud. Vses. Nauchno-Issled. Inst. Prud. Ryb. Khozyaist., No. 40*, 68–73. [In Russian.]

Kusz, W. and Treder, A, 1980, Parasitic fauna of European blue whiting, *Micromesistius poutassou* (Risso, 1810). *Acta Ichthyol. Pisc.*, **10**, 45–58.

Kuz'mina, V. V. and Kuperman, B. I, 1983, [Comparative characteristics of membrane digestion in cestodes and their fish hosts.] *Parazitologiya*, 17, 436–441. [In Russian.

Lacey, S. M., Williams, I. C. and Carpenter, A. C, 1982, A note on the occurrence of the digenetic trematode *Sphaerostoma bramae* (Müller) in the intestine of the European eel, *Anguilla anguilla* (L.). *J. Fish Biol.*, **20**, 593–596.

Lagoin, Y, 1980, Données actuelles sur une nématodose larvaire de l'homme, l'anisakiase ou 'maladie du ver du hareng'. *Bull. Acad. Vét. France*, **53**, 139–146.

Lahav, M, 1974, The occurrence and control of parasites infesting Mugilidae in fish ponds in Israel. *Bamidgeh*, **26**, 99–103.

Laird, M, 1961, Distomiasis in Tokelau Islanders. *Can. J. Zool.*, **39**, 149–152.

Lal, S. S. and Mithal, R. P, 1979, On the control of some nematode parasites of fishes. *Bioresearch*, **3**, 15–17.

Lambert, A, 1975, Cycle biologique de *Caryophyllaeus brachycollis* Janiszewska, 1953 (Cestoda: Caryophyllaeidae) parasite de *Barbus meridionalis* Risso, 1826 dans le Sud de France. *Acta Trop.*, 32, 296–303.

Lambert, A, 1979, Evolution post-larvaire de l'appareil sensoriel chez les Dactylogyridea (Monogenea). *Z. ParasitKde*, **58**, 259–263.

Lambert, A, 1980, Oncomiracidiums et phylogenèse des Monogenea (Plathelminthes). Deuxième partie: structures argyrophiles des oncomiracidiums et phylogenèse des Monogenea. *Annls Parasit. Hum. Comp.*, 55, 281–326.

Lambert, A, 1982, Specificité parasitaire et évolution des Monogenea Monocotylidae, plathelminthes parasites des sélaciens. [*Deux. Symp. spec. parasit. parasites des vertébrés, 13–17 avril 1981. Colloque internat. CNRS.*]. *Méms Mus. Nat. Hist. Nat., Ser. A, Zool.*, **123**, 303–311.

Lambert, A. and Romand, R, 1984, Les Monogènes Dactylogyridae marqueurs biologiques des Cyprinidae? *Cybium*, **8**, 9–14.

Lambert, M, 1976, Cycle biologique de *Parasymphylodora markewitschi* (Kulakovskaya, 1947) (Trematoda, Digenea, Monorchiidae). *Bull. Mus. Nat. Hist. Nat., 3e Sér., No. 407, Zool.*, 1107–1114.

Lang Suo, 1981, [Monogenetic fauna of freshwater fishes of Hainan Island.] *Acta Zoologica Sinica*, 27, 75–86. [In Chinese.]

Langdon, J. S. and Humphrey, J. D, 1986, Appendix II: A checklist of the disease status of Australian fish and shellfish. In: *Diseases of Australian fish and shellfish. Proceedings of the First Australian Workshop on Diseases of Fish and Shellfish held at Benella, Victoria, May 27–30, 1985* (Eds Humphrey, J. D. and Langdon, J. S.). Benella, Victoria, Australia: Australian Fish Health Reference Laboratory, pp. 233–246.

Lankester, M. W. and Smith, J. D, 1980, Host specificity and distribution of swim bladder nematodes, *Cystidicola farionis* (Fischer, 1978) and *Cystidicola cristivomeri* (White, 1941) (Habronematoidea) in salmonid fishes of Ontario. *Can. J. Zool.*, **58**, 1298–1305.

Laptev, V. I, 1980, The control of *Bothriocephalus* infection. *Veterinariya, Mosk., No. 7*, 40–41. [In Russian.]

Larsh, J. E. Jr, 1941, *Corallobothrium parvum* n. sp., a cestode from the common bullhead, *Ameiurus nebulosus* Le Sueur. *J. Parasit.*, 27, 221–227.

La Rue, G. R, 1957, The classification of digenetic Trematoda: a review and a new system. *Expl. Parasit.*, 6, 306–349.

La Rue, G. R., Butler, E. P. and Berkhout, P. G, 1926, Studies on the trematode family Strigeidae (Holostomidae). No. IV. The eye of fishes, an important habitat for larval Strigeidae. *Trans. Am. microsc. Soc.*, 45, 282–288.

Lasee, B. A., Font, W. F. and Sutherland, D. R, 1988, *Culaeatrema inconstans* gen. n., sp.n. (Digenea: Allocreadiidae) from the brook stickleback (*Culaea inconstans*) in Wisconsin and observations on parthenogenetic populations. *Can. J. Zool.*, 66, 1328–1333.

Latimer, D. C. and Meade, T. G, 1979, Pericercarial envelope formation in *Posthodiplostomum minimum* (Trematoda: Diplostomidae). *Texas J. Sci.*, 31, 53–58.

Lauckner, G, 1980, Diseases of helminths. In: *Diseases of marine animals Volume 1. General Aspects Protozoa to Gastropoda* (Ed. Kinne, O.). New York: John Wiley and Sons, pp. 279–302.

Lauckner, G, 1983, Diseases of Mollusca. In: *Diseases of Marine Animals*, (Ed. Kinne, O.) Vol. 2. pp. 477–961, 963–977, 979–983.

Lauckner, G, 1984, Impact of trematode parasitism on the fauna of a North Sea tidal flat. *Diseases of marine organisms, Internat. Helgoland Symp., 11–16 Sept. 1983, Helgoland, Hamburg, GFR.* (Eds Kinne, O. and Bulnheim, H. P.). Helgoländer Meeresuntersuch., 37, 185–188.

Lauckner, G, 1985a, Diseases of Mammalia: Carnivora. In: *Diseases of Marine Animals. IV (2)* (Ed. Kinne, O.). Hamburg: Biologische Anstalt Helgoland. pp. 645–682.

Lauckner, G, 1985b, Diseases of Mammalia: Pinnipedia. In: *Diseases of Marine Animals, IV (2)* (Ed. Kinne, O.). Hamburg: Biologische Anstalt Helgoland. pp. 683–793.

Laurie, J. S, 1971, Carbohydrate absorption by *Gyrocotyle fimbriata* and *Gyrocotyle parvispinosa* (Platyhelminthes). *Expl Parasit.*, 29, 375–385.

Lavrovskii, V. V, 1977, [Trout breeding techniques and disease prophylaxis.] *Veterinariya, Mosk., No. 5*, 67–69. [In Russian.]

Lawler, G. H, 1969, Aspects of the biology of *Triaenophorus nodulosus* in yellow perch, *Perca flavescens*, in Heming Lake, Manitoba. *J. Fish. Res. Bd Can.*, 28, 821–831.

Lawler, G. H, 1970, Parasites of coregonid fishes. In: *Biology of coregonid fishes.* (Eds Lindsey, C. C. and Woods, C. S.) Winnipeg: University of Manitoba Press, pp. 279–309.

Lebedev, B. I, 1969, [Basic regularities in the distribution of monogeneans and trematodes of marine fishes in the world ocean]. *Zool. J.*, 48, 41–50. [In Russian.]

Lebedev, B. I, 1978, [Some aspects of monogenean existence.] *Folia Parasit.*, 25, 131–136. [In Russian]

Lebedev, B. I, 1979a, Faunistic aspects of studies into the higher Monogenoidea of marine fishes. *Zool. Anz.*, 202, 99–104.

Lebedev, B. I, 1979b, [Some aspects of monogenean ecology and evolution.] *Zh. Obshch. Biol.*, 40, 271–281. [In Russian.]

Lebedev, B. I, 1981, [The concept of niche in parasitology.] In: *Biologiya i sistematika gel'mintov zhivotnykh Dal'nego Vostoka. Vladivostok, USSR: AN SSSR, Dal'nevostochnyi Nauchnyi Tsentr*, pp. 107–110. [In Russian.]

Lebedev, B. I, 1983, Taxonomic diversity, evolution and structure of taxa (exemplified by higher monogeneans). *Zh. Obshch. Biol.*, 44, 202–213. [In Russian.]

Lebedev, B. I, 1984a, [New taxa of gastrocotylid-like monogeneans.] In: *Parazity zhivotnykh i rastenii.* Vladivostok, USSR: Akademiya Nauk SSSR, pp. 17–24. [In Russian.]

Lebedev, B. I, 1984b, [Classification of monogeneans of the suborder Gastrocotylinea.] In: *Parazity zhivotnykh i rastenii.* Vladivostok, USSR: Akademiya Nauk SSR, Dal'nevostochnyi, pp. 3–16. [In Russian.]

Lebedev, B. I, 1986, [*Monogeneans of the sub-group Gastocotylinea.*] Leningrad: Soviet Academy of Sciences, pp. 1–200. [In Russian.]

Lebedev, B. I, 1988, Monogenea in the light of new evidence and their position among platyhelminths. *Angew. Parasit.*, 29, 149–167.

Lebedev, B. I. and Mamaev, U. L. (Eds) 1976, Studies on the monogeneans. *Proceedings of the Institute of Biology and Pedology, Far-East Science Centre, Academy of Sciences of the U.S.S.R. New Series.* Volume 34 (137), Vladivostok, pp. 1–222.

Le Brun, N., Lambert, A. and Justine, J. L, 1986, Oncomiracidium, morphogenèse du hapteur et ultrastructure du spermatozöide de *Pseudodactylogyrus anguillae* (Yin et Sproston, 1948) Gussev, 1965 (Monogenea, Monopisthocotylea, Pseudodactylogyridae n. fam.). *Annls Parasit. Hum. Comp.*, 61, 273–284.

Le Brun, N., Renaud, F. and Lambert, A, 1988, The genus *Diplozoon* (Monogenea, Polyopisthocotylea) in southern France: speciation and specificity. *Int. J. Parasit.*, 18, 395–400.

Le Brun, N., Renaud, F., Lambert, A. and Euzet, S, 1985, Le problème de l'espèce chez *Diplozoon* (Plathelminthe, Monogenea, Polyopisthocotylea). *Bull. Soc. Fr. Parasit., No. 1*, 105–109.

Lee, D. L, 1965, *The Physiology of Nematodes.* Edinburgh: Oliver and Boyd, 154 pp.

Lee, D. L, 1977, The nematode epidermis and collagenous cuticle, its formation and ecdysis. *Proc. Symp. Zool. Soc. London, 30–31 Oct. 1975.* (Ed. Spearman, R. I. C.) *Comparative biology of skin. (Symposia of the Zoological Society of London No. 39).* London, UK: Academic Press Inc. pp. 145–170.

Lee, D. L, 1982, Ecological physiology of parasites. In: *Parasites—their world and ours. Proc. 5th Int. Congr. Parasit., Toronto, Canada, 7–14 August 1982, under the auspices of the World Federation of*

Parasitologists (Eds Mettrick, D. F. and Desser, S. S.). Amsterdam, Netherlands: Elsevier Biomedical Press, pp. 125–134.

Lee, D. L. and Smith, M. H, 1965, Haemoglobins of parasitic animals. *Expl Parasit.*, 16, 392–424.

Lee, R. L. G, 1977, The Serpentine fish and their parasites. *London Nat.*, **56**, 57–70.

Lee, R. L. G, 1981, Ecology of *Acanthocephalus lucii* (Müller, 1776) in perch, *Perca fluviatilis*, in the Serpentine, London, UK. *J. Helminth.*, 55, 149–154.

Lee, R. S., Lewis, J. W. and Sweeting, R. A, 1987, The epidemiology of *Sanguinicola inermis* (Plehn, 1905) in *Cyprinus carpio* in the UK. 2nd International Symposium of Ichthyoparasitology. Actual problems in fish parasitology, Tihany, Hongary Sept 27–Oct 3, p. 51.

Lees, E. and Bass, L, 1960, Sex hormones as a possible factor influencing the level of parasitisation in frogs. *Nature, Lond.*, **188**, 1207–1208.

Lemly, A. D. and Esch, G. W, 1983, Differential survival of metacercariae of *Uvulifer ambloplitis* (Hughes, 1927) in juvenile centrarchids. *J. Parasit.*, **69**, 746–749.

Lemly, A. D. and Esch, G. W, 1984a, Population biology of the trematode *Uvulifer ambloplitis* (Hughes, 1927) in juvenile blue-gill sunfish *Lepomis macrochirus*, and largemouth bass, *Micropterus salmoides*. *J. Parasit.*, 70, 466–474.

Lemly, A. D. and Esch, G. W, 1984b, Effects of the trematode *Uvulifer ambloplitis* on juvenile blue-gill sunfish, *Lepomis macrochirus*: ecological implications. *J. Parasit.*, **70**, 475–492.

Lemly, A. D. and Esch, G. W, 1985, Black-spot caused by *Uvulifer ambloplitis* (Trematoda) among juvenile centrarchids in the Piedmont area of North Carolina. *Proc. Helminth. Soc. Wash.*, **52**, 30–35.

Leong, T. S, 1980, A survey of anisakid larvae in marine fish in Penang, Malaysia. *South East Asian J. Trop. Med. Publ. Hlth*, 11, 493–495.

Leong, T. S, 1986, Seasonal occurrence of metazoan parasites of *Puntius binotatus* in an irrigation canal, Pulau Pinang, Malaysia. *J. Fish Biol.*, **28**, 9–16.

Leong, T. S. and Holmes, J. C, 1981, Communities of metazoan parasites in open water fishes of Cold Lake, Alberta. *J. Fish Biol.*, **18**, 693–713.

Leong, T. S. and Wong, S. Y, 1988, A comparative study of the parasite fauna of wild and cultured grouper (*Epinephelus malabaricus* Bloch et Schneider) in Malaysia. *Aquaculture*, **68**, 203–207.

Leont'eva, V. G, 1976, The effect of salt solutions on the viability of *Anisakis* sp. larvae. *Izvest. Tikhook. Nauchno-Issled. Inst. Ryb. Khozyaist. Okeanogr. (TINRO)*, **99**, 173–175. [In Russian.]

Lester, R. J. G, 1974, Parasites of *Gasterosteus aculeatus* near Vancouver, British Columbia. *Syesis*, 7, 195–200.

Lester, R. J. G, 1977, An estimate of the mortality in a population of *Perca flavescens* owing to the trematode *Diplostomum adamsi*. *Can. J. Zool.*, 55, 288–292.

Lester, R. J. G, 1978, Marine parasites costly for fisherman. *Australian Fisheries*, 37, 32–33.

Lester, R. J. G, 1979, Description of two new didymozoids from Australian fishes. *J. Parasit.*, 65, 904–908.

Lester, R. J. G, 1980, Host–parasite relations in some didymozoid trematodes. *J. Parasit.*, **66**, 527–531.

Lester, R. J. G, 1984, A review of the methods for estimating mortality due to parasites in wild fish populations. *Helgoländer Meeresuntersuch.*, 37, 53–64.

Lester, R. J. G, 1986, Parasites and parasitic diseases of aquatic animals. In: *Diseases of Australian fish and shellfish. Proc. 1st Austr. workshop on diseases of fish and shellfish, Benalla, Victoria, Australia, 27–30 May 1985* (Eds Humphrey, J. D. and Langdon, J. S.). Australia: Australian Fish Health Reference Laboratory. pp. 87–101.

Lester, R. J. G. and Adams, J. R, 1974a, *Gyrodactylus alexanderi*; reproduction, mortality, and effect on its host *Gasterosteus aculeatus*. *Can. J. Zool.*, 52, 827–833.

Lester, R. J. G. and Adams, J. R, 1974b, A simple model of a *Gyrodactylus* population. *Int. J. Parasit.*, 4, 497–506.

Lester, R. J. G., Barnes, A. and Habib, G, 1985, Parasites of skipjack tuna *Katsuwonus pelamis*: fishery implications. *Fishery Bull.*, **83**, 343–356.

Lethbridge, R. C, 1980, The biology of the oncosphere of cyclophyllidean cestodes. *Helm. Abstr. Series A*, 49, 59–72.

Lethbridge, R. C., Potter, I. C., Bray, R. A. and Hilliard, R. W, 1983, The presence of helminths in a southern hemisphere lamprey (*Geotria australis* Gray), with a discussion of the significance of feeding mechanisms in lampreys in relation to the acquisition of parasites. *Acta Zool.*, **64**, 79–83.

Levin, S. A, 1970, Community equilibria and stability, and an extension of the competitive exclusion principle. *The American Naturalist*, 104, 413–423.

Lewis, D. H, 1979, Bacterial diseases. In: *Principal diseases of farm raised catfish*. (Ed. Plumb, J. A.). Southern Cooperative Series, 225, 15–24.

Lewis, J. W., Jones, D. R. and Adams, J. R, 1974, Functional bursting by the dracunculoid nematode *Philonema oncorhynchi*. *Parasitology*, 69, 417–427.

Lewis, M. C., Welsford, I. G. and Uglem, G. L, 1989, Cercarial emergence of *Proterometra macrostoma* and *P. edneyi* (Digenea: Azygiidae): contrasting responses to light: dark cycling. *Parasitology,* 99, 215–223.

Li Minmin, 1984, [Parasites of the mullets *Mugil cephalus* (Linnaeus) and *Liza haematocheila* (Temminck et Echlegel) in the areas of Bohai Gulf. I. Hangu area.] *Acta Zool. Sin.*, 30, 153–158. [In Chinese.]

Li Minmin, 1984, [Parasites of the mullets *Mugil cephalus* (Linnaeus) and *Liza haematocheila* (Temminck et Schlegel) in the areas of Bohai Gulf. II. Penglai area.] *Acta Zool. Sin.*, 30, 231–242. [In Chinese.]

Li, Q. K. *et al.* 1987, Digenetic trematode fauna of marine fishes in China. *Sichuan J. Zool.*, 6, 4–7 [In Chinese.]

Liao, X. H. and Liang, Z. X, 1987, Distribution of ligulid tapeworms in China. *J. Parasit.*, 73, 36–48.

Lie, K. I., Basch, P. F., Heyneman, D., Beck, A. J. and Audy, J. R, 1968, Implications for trematode control of interspecific larval antagonism within snail hosts. *Trans. R. Soc. Trop. Med. Hyg.*, 62, 299–319.

Lien, L, 1970, Studies of the helminth fauna of Norway XIV: *Triaenophorus nodulosus* (Pallas, 1760) (Cestoda) in Bogstad Lake. II. Development and life span of the plerocercoids in perch *Perca fluviatilis*) L. (1758). *Nytt Mag. Zool.*, 18, 85–86.

Lien, L. and Borgström, R, 1973, Studies of the helminth fauna of Norway. XXXI. Distribution and seasonal occurrence of *Proteocephalus* sp. Weinland, 1858 (Cestoda: Proteocephala) in brown trout, *Salmo trutta* L., from Southern Norway. *Norwegian J. Zool.,* 21, 293–297.

Lincoln, R. J., Boxhall, G. A. and Clark, P. F, 1982, *A dictionary of ecology, evolution and systematics*. Cambridge: Cambridge University Press. 298 pp.

Lindroos, P, 1984, Observations on the extracellular spaces and intercellular junctions in *Diphyllobothrium dendriticum* (Cestoda). *Acta Zool.*, 65, 153–158.

Lindsey, C. C, 1957, Possible effects of water diversion on fish distribution in British Columbia. *J. Fish. Res. Bd Can.*, 14, 651–668.

Linnik, V. Ya, 1975, [Diseases and parasites of fish in the Lyubanskoe and Oispovichskoe Reservoirs.] *Trudy (Nauchnye Trudy) Nauchno-Issled. Vet. Inst. Belorusskoi SSSR (Veterinarnaya Nauka—Proizvodstvu),* 13, 137–142. [In Russian.]

Linnik, V. Ya, 1980a, [Helminth zoonoses and other parasitic infections in fish in lake Osveisk (USSR).] *Dostizheniya Veterinarnoi Nauki i Peredovogo Opyta—Zhivotnovstvu, Minsk, USSR*, No. 5, 88–90. [In Russian.]

Linnik, V. Ya, 1980b, [The prophylaxis of fish helminth zoonoses.] *Veterinariya, Mosk., No. 9*, 41–42. [In Russian.]

Linnik, V. Ya, 1983, [Helminth zoonoses and other fish parasites in Lake Bobrovichskoe.] *Veterinarnaya Nauka—Proizvodstvu,* 21, 94–96. [In Russian.]

Linnik, V. Ya. and Litvyak, V. S, 1980, [Amino acid composition of the muscle and blood of fish infected with trematode metacercariae.] *Trudy Beloruss. Nauchno-Issled. Inst. Eksper. Vet. (Veterinarnaya Nauka—Proizvodstvu),* 18, 120–123. [In Russian.]

Linnik, V. Ya. and Yanchenko, V. F, 1982, [Ozone in the prophylaxis of diplostomiasis in fish.] *Veterinariya, Mosk., No. 7*, 46 [In Russian.]

Linton, E, 1914, On the seasonal distribution of fish parasites. *Trans. Am. Fish. Soc.*, 44, 48–56.

Liu, M. C. and Zhang, J. Y, 1987, [Monogenetic trematodes of freshwater fishes from Tianjin.] *Ann. Bull. Soc. Parasit. Guandong Province,* 8–9, 140–142. [In Chinese.]

Llewellyn, J, 1956, The host specificity, micro-ecology, adhesive attitudes, and comparative morphology of some trematode gill parasites. *J. Mar. Biol. Ass. UK,* 35, 113–127.

Llewellyn, J, 1957a, The larvae of some monogenetic trematode parasites of Plymouth fishes. *J. Mar. Biol. Ass. UK,* 36, 243–259.

Llewellyn, J, 1957b, Host-specificity in monogenetic trematodes. *Symposium on host specificity among parasites of vertebrates (1st), University of Neuchâtel, April 15–18, 1957*. Neuchâtel, Switzerland: University of Neuchâtel, pp. 211–212.

Llewellyn, J, 1959, The larval development of two species of gastrocotylid trematode parasites from the gills of *Trachurus trachurus*. *J. Mar. Biol. Ass. UK,* 38, 431–467.

Llewellyn, J, 1960, Amphibdellid (monogenean) parasites of electric rays (Torpedinidae). *J. Mar. Biol. Ass. UK,* 39, 561–589.

Llewellyn, J, 1962, The life histories and population dynamics of monogenean gill parasites of *Trachurus trachurus* (L.). *J. Mar. Biol. Ass. UK,* 42, 587–600.

Llewellyn, J, 1963, Larvae and larval development of monogeneans. *Adv. Parasit.,* 1, 287–326.

Llewellyn, J, 1964, The effects of the host and its habits on the morphology and life-cycle of a monogenean parasite. In: (Eds Ergens, R. and Rysavy, B.) *Parasitic worms and aquatic conditions.* Proc. Symp., Prague, Oct. 29–Nov. 2, 1962. Prague, Czechoslovakia: Czechoslovak Academy of Sciences, pp. 147–152.

Llewellyn, J, 1965, The evolution of parasitic platyhelminths. In: *Evolution of parasites. Third Symposium of the British Society for Parasitology.* (Ed. A. E. R. Taylor). Oxford: Blackwell Scientific Publications, pp. 47–78.

Llewellyn J, 1968, Larvae and larval development of monogeneans. *Adv. Parasit.,* 6, 373–383.

Llewellyn, J, 1970, Monogenea. *J. Parasit.,* 56, (Sect. II) (3), 493–504.

Llewellyn, J, 1972, Behaviour of monogeneans. In: *Behavioural Aspects of Parasite Transmission* (Eds Canning, E. U. and Wright, C. A.) *Zool. J. Linn. Soc.,* 51, (Suppl.1), pp. 19–30.

Llewellyn, J, 1981, Biology of monogeneans. (Workshop Proceedings, EMOP 3). *Parasitology,* 82, 57–68.

Llewellyn, J, 1982, Host specificity and corresponding evolution in monogenean flatworms and vertebrates. [*Deux. Symp. spéc. parasit. parasites des vertébrés, 13–17 avril, 1981. Colloque internat. CNRS.*] *Mems Mus. Nat. Hist. Nat., Ser. A, Zool.,* 123, 289–293.

Llewellyn, J, 1983, Sperm transfer in the monogenean gill parasite *Gastrocotyle trachuri. Proc. R. Soc. Lond., B (Biol. Sci.),* 219, 439–446.

Llewellyn, J, 1986, Phylogenetic inference from platyhelminth life-cycle stages. In: *Parasitology—Quo Vadit?* (Ed. Howell, M. J.). Proceedings of the VI International Congress of Parasitology, Canberra, Australia. Canberra: Australian Academy of Sciences, pp. 281–291.

Llewellyn, J. and Anderson, M, 1984, The functional morphology of the copulatory apparatus of *Ergenstrema labrosi* and *Ligophorus angustus,* monogenean gill parasites of *Chelon labrosus. Parasitology,* 88, 1–7.

Llewellyn, J. and Owen, I. L, 1960, The attachment of the monogenean *Discocotyle sagittata* Leuckart to the gills of *Salmo trutta* L. *Parasitology,* 50, 51–59.

Llewellyn, J. and Simmons, J. E, 1984, The attachment of the monogenean parasite *Callorhynchicola multitesticulatus* to the gills of its holocephalan host *Callorhynchus milii. Int. J. Parasit.,* 14, 191–196.

Llewellyn, J. and Tully, C. M, 1969, A comparison of speciation in diclidophorinean monogenean gill parasites and in their fish hosts. *J. Fish. Res. Bd Can.,* 26, 1063–1074.

Llewellyn, J. MacDonald, S. and Green, J. E, 1980, Host-specificity and speciation in diclidophoran (monogenean) gill parasites of trisopteran (gadoid) fishes at Plymouth. *J. Mar. Biol. Ass. UK,* 60, 73–79.

Lloyd, J. H, 1928, On the life-history of the common nematode of the dogfish *Scyllium canicula. Ann. Mag. Nat. Hist. (10),* 1, 712–715.

Lo, C. F., Chen, S. C. and Wang, C. H, 1985, The study of *Clinostomum complanatum* (Rud., 1814) V. The influences of metacercaria of *Clinostomum complanatum* on fish. (Proc. Internat. Seminar on fish pathology, Japanese Society of Fish Pathology.) *Fish Path.*, 20, 305–312.

Lo, C. F., Huber, F. and Kou, G. H, 1981, Studies on *Clinostomum complanatum* (Red., 1819). (Proc. Internat. Seminar on Fish Diseases., Tokyo, Japan, Nov. 8–9, 1980). *Fish Path.*, 15, 219–227.

Lockard, L. L., Parsons, R. R. and Schaplow, B. M, 1975, Some relationships between internal parasites and brown trout from Montana streams. *Great Basin Nat.*, 35, 442–448.

Lom, J, 1986, Ichthyoparasitology in Czechslovakia. *Bull. Europ. Ass. Fish Path.*, 6, 61–63.

Lomakin, V. V. and Chernova, T. N, 1980, [A new nematode genus *Garakavillanus* n. g. (Camallanata: Skrjabillanidae).] *Gel'm. Vodnykh Nazemnykh Biot. Semozov., No. 30,* 45–51. [In Russian.]

Lopukhina, A. M., Yunchis, O. J., Chernysheva, N. B. and Voronin, V. N, 1979, [Ecological analysis of the parasite fauna of sexually mature burbot in Lake Verkhnee Vrevo.] *Sb. Nauch. Trud. Gosud. Nauchno-Issled. Inst. Ozern. Rech. Ryb. Khozyalist. (Ekol. parazit. ryb), No. 140,* 26–47. [In Russian.]

Lorenz, H, 1981, Einige Aspekte der Parasitierung von *Micromesistius poutassou. Angew. Parasit.*, 22, 221–224.

Lorenzen, S, 1985, Phylogenetic aspects of pseudocoelomate evolution. In: *The origins and relationships of Invertebrates* (Eds Conway-Morris, S. *et al.*). Oxford: Clarendon Press. *The Systematics Association Special Volume. No. 28,* pp. 210–223.

Loseva, T. G, 1983, [Seasonal dynamics of the parasite fauna of *Blicca bjoerkna* in Lake Vrevo.] *Sb. Nauch. Trud. Gosud. Nauchno-Issled. Inst. Ozern. Rech. Khozyaist. (Probl. ekol. parazit. ryb).*, *No. 197*, 74–84. [In Russian.]

Losey, G. S, 1971, Communication between fishes in cleaning symbiosis. In: *Aspects of the biology of symbiosis* (Ed. Cheng, T. C.). Baltimore, London: University Park Press. pp. 45–76.

Love, M. S. and Moser, M, 1983, *A checklist of parasites of California, Oregon, and Washington marine and estuarine fishes*. Seattle, Washington, USA: US Department of Commerce, NOAA Technical Report NMFS SSRF-777. iii + 577 pp.

Lozanov, L. and Kolarova, V, 1979, [The pathology and pathogenesis of *Bothriocephalus gowkongensis* infection in carp.] *Obshch. Sravnitel. Patol.*, 7, 127–134. [In Bulgarian.]

Lozinska-Gabska, M, 1981, [Aspartate and alanine aminotransferase activity in the intestine of carp (*Cyprinus carpio*) infected with the cestodes *Bothriocephalus gowkongensis* or *Khawia sinensis*.] *Wiad. Parazyt.*, 27, 717–743. [In Polish.]

Lozoya, X, 1977, Balance between man and nature. *World Health*, November, 8–11.

Lubieniecki, B, 1973, Note on the occurrence of larval *Anisakis* in adult herring and mackerel from Long Island to Chesapeake Bay. *Res. Bull. ICNAF, No. 10*, 79–81.

Lubieniecki, B, 1977, The plerocercus of *Grillotia erinaceus* as a biological tag for haddock *Melanogrammus aeglefinus* in the north sea and north-east Atlantic. *J. Fish Biol.*, 11, 555–565.

Lucky, E, 1977, *Methods for the diagnosis of fish diseases*. (Ed. G. L. Hoffman.). New Delhi, India: Amerind Publishing Co. Pvt. Ltd. *For*: Fish and Wildlife Serv., US Dept. Interior and Natnl Sci. Foundn, Washington, DC, 140 pp.

Lucky, Z, 1984, Investigation on invasive diseases of the herbivorous fish fry and their treatment. In: *Fish, pathogens and environment in European polyculture. (Proc. Internat. Seminar, 23–27 June 1981, Szarvas, Hungary. Simposia Biologica Hungarica* Vol. 24. (Ed. Olah, J.). Budapest, Hungary: Akademiai Kiado, pp. 173–180.

Lumsden, R. D. and Murphy, W. A, 1980, Morphological and functional aspects of the cestode surface. In: *Cellular interactions in symbiosis and parasitism*. (Eds Cook, C. P., Pappas, P. W. and Rudolph, E. D.). Columbus, USA: Ohio State University Press, pp. 95–130.

Lushchina, V. G, 1985, [The helminth fauna of blenniid fish in the Black Sea.] *Ekol. Morya, Kiev, No. 20*. 43–48. [In Russian.]

Lux, E, 1989, Zum Artenbestand und zur Dynamik ektoparasitischer Helminthen (Pectobothrii) bei Karpfen und Forelle, Teil I. *Z. Binnenfisch. DDR*, 36, 128–134.

Lyadov, V. N, 1976, [Effect of different water temperatures and salinities on the survival of larvae of anisakid nematodes.] [Abstract]. In: *Kratkie tezisy dokladov II Vsesoyuznogo simpoziuma po parazitam i boleznyam morskikh zhivotnykh*. Kaliningrad, USSR: Ministerstvo Rybnogo Khozyaistva SSSR, AtlantNIRO, pp. 41–42. [In Russian.]

Lyadov, N. V, 1985, Zoogeographical characteristics of the helminths of fishes from the Antarctic zone of the world ocean. In: *Parasitology and pathology of marine organisms of the World Ocean* (Ed. Hargis, W. J., Jr.). Seattle, USA: Department of Commerce, National Oceanic and Atmospheric Administration, pp. 41–43.

Lyaiman, E. M, 1963, *[Diseases of fish.]* Moscow: Selkhozizdat, 295 pp. [In Russian.]

Lynn, D. H., Suriano, D. M. and Beverley-Burton, M, 1981, Chatton-Lwoff silver impregnation: an improved technique for the study of oncomiracidia (Platyhelminthes: Monogenea) chaetotaxy. *Syst. Parasit.*, 3, 21–23.

Lyons, K. M, 1972, Sense organs of monogeneans. In: *Behavioural aspects of parasite transmission*. (Eds Canning, E. U. and Wright, C. A.). Zool J. Linn. Soc. 51, (Suppl. 1), pp. 181–199.

Lyons, K. M, 1973, The epidermis and sense organs of the Monogenea and some related groups. *Adv. Parasit.*, 11, 193–232.

Lyons, K. M, 1977, Epidermal adaptations of parasitic platyhelminths. *Proc. Symp. Zool. Soc. London, 30–31 Oct. 1975*. (Ed. Spearman, R. I. C.). *Comparative biology of skin. Symp. Zool. Soc. Lond. No. 39*. London, UK: Academic Press Inc., pp. 97–144.

Lyons, K. M, 1978, The biology of helminth parasites. *The Institute of Biology's Studies in Biology No. 102*. London, UK: Edward Arnold (Publishers) Ltd., iv + 60 pp.

Lyubarskaya, O. D. and Lavrent'eva, Yu. I, 1985, [The parasite fauna of *Acipenser ruthenus* in the mid-Volga and the Kuibyshev reservoir.] *Parazitologiya*, 19, 320–323. [In Russian.]

Lyubina, T. V, 1968, [Data on the epizootiology and pathogenesis of liguliasis of fish in water-reservoirs in the Omsk region.] *All-Union Meeting (Vth) on diseases and parasites of fish and aquatic invertebrates*. Leningrad: Izdat 'Nauka', pp. 75–76. [In Russian.]

Lyubina, T. V, 1980, [The effect of ligulid infection on the fat and calorie content of fish muscle.] *Sb. Rabot Sibirsk. Nauchno-Issled. Vet. Inst., No. 38*, 162–164. [In Russian.]

Lyukshina, L. M, 1977, [Seasonal and age dynamics of the formation of egg-shell material in representatives of *Diplozoon* (Monogenoidea).] In: *Issled. monog. SSSR. (Materialy vses Simp. monog., 16–18 Noyabrya, 1976, Leningrad)*. Leningrad, USSR: Akademiya Nauk SSSR, Zoologicheskii Institut., pp. 79–83. [In Russian.]

Macdonald, G, 1965, The dynamics of helminth infections, with special reference to Schistosomes. *Trans. Roy. Soc. Trop. Med. and Hyg.* 59, 489–506.

MacCallum, G. A, 1921, Studies in helminthology. *Zoopathologica, New York,* 1, 3–38.

MacCallum, G. A, 1927, A new ectoparasitic trematode, *Epibdella melleni*, sp. nov. *Zoopathologica, New York,* 1, 291–300.

MacDonald, G, 1957, The epidemiology and control of malaria. London: Oxford University Press.

Macdonald, S, 1974, Host skin mucus as a hatching stimulant in *Acanthocotyle lobianchi*, a monogenean from the skin of *Raja* spp. *Parasitology*, **68**, 331–338.

Macdonald, S, 1975, Hatching rhythms in three species of *Diclidophora* (Monogenea) with observations on host behaviour. *Parasitology*, 71, 211–228.

Macdonald, S. and Caley, J, 1975, Sexual reproduction in the monogenean *Diclidophora merlangi*: tissue penetration by sperms. *Z. ParasitKde,* 45, 323–334.

Macdonald, S. and Jones, A, 1978, Egg-laying and hatching rhythms in the monogenean *Diplozoon homoion gracile* from the southern barbel (*Barbus meridionalis*). *J. Helminth.*, 52, 23–28.

Macdonald, S. and Llewellyn, J, 1980, Reproduction in *Acanthocotyle greeni* n. sp. (Monogenea) from the skin of *Raia* spp. at Plymouth. *J. Mar. Biol. Ass., UK,* **60**, 81–88.

Macfarlane, W. V, 1939, Life cycle of *Coitocaecum anaspidis* Hickman, a New Zealand digenetic trematode. *Parasitology*, 31, 172–184.

Macfarlane, W. V, 1951, The life cycle of *Stegodexamene anguillae* n.g., n. sp., an allocreadiid trematode from New Zealand. *Parasitology,* 41, 1–10.

Machida, M, 1984a, Two new trematodes from gallbladder of tropical marine fishes, *Myripristis* and *Abudefduf. Bull. Natn. Sci. Mus., Japan, A. (Zool),* 10, 1–5.

Machida, M, 1984b, Two new trematodes from tropical marine fishes of South western Japan. *Bull. Nat. Sci. Mus., Japan, A. (Zool.),* 10, 51–55.

Machida, M, 1985, Helminth parasites of cyclopterid fish *Aptocyclus ventricosus*, caught off northern Japan. *Bull. Nat. Sci. Mus., Japan, A (Zool.)*, 11, 123–128.

Machida, M., Takahashi, K. and Masuuchi, S, 1978, *Thynnascaris haze* n. sp. (Nematoda, Anisakidae) from goby in the Bay of Tokyo. *Bull. Nat. Sci. Mus. Japan, A. (Zool.)*, 4, 241–244.

Machkevskii, V. K, 1982, Development and biology of the parthenitae of *Proctoeces maculatus* (Trematoda) in Black Sea mussels. *Zool. Zh.*, 61, 1635–1642. [In Russian.]

Machkevskii, V. K, and Parukhin, A. M, 1981, The biology of trematodes from the family Fellodistomatidae, parasitic in Black Sea mussels. *Parazitologiya*, 15, 181–185. [In Russian.]

Maclnnis, A. J, 1976, How parasites find hosts: some thoughts on the conception of host–parasite integration. In: (Ed. Kennedy, C. R.). *Ecological aspects of parasitology*. Amsterdam, The Netherlands: North-Holland Publishing Co., pp. 3–20.

MacKenzie, K, 1965, The plerocercoid of *Gilquinia squali* Fabricius, 1794. *Parasitology*, 55, 607–615.

MacKenzie, K, 1968, Some parasites of O-group plaice, *Pleuronectes platessa*, L., under different environmental conditions. *Mar. Res.*, 3, 23pp.

MacKenzie, K, 1974, The use of parasites in tracing herring recruitment migrations. *J. Parasit., International Council for the Exploration of the Sea, Paper CM1974/H:31*, 4 pp.

MacKenzie, K, 1975a, *Renicola* metacercariae (Digenea: Renicolidae) in clupeoid fish: new host records. *J. Fish Biol.*, 7, 359–360.

MacKenzie, K, 1975b, Some aspects of the biology of the plerocercoid of *Gilquinia squali* Fabricius 1794 (Cestoda: Trypanorhyncha). *J. Fish Biol.*, 7, 321–327.

MacKenzie, K, 1979, Some parasites and diseases of blue whiting, *Micromesistius poutassou* (Risso), to the north and west of Scotland and at the Faroe Islands. *Scottish Fish. Res. Rep., DAS, No. 17*, 14 pp.

MacKenzie, K, 1980, First report of the plerocercus of *Grillotia angeli* Dollfus 1969 (Cestoda: Trypanorhyncha). *J. Parasit.*, **66**, 175–176.

MacKenzie, K, 1983, Parasites as biological tags in fish population studies. *Adv. Appl. Biol.*, 7, 251–331.

MacKenzie, K, 1985, The use of parasites as biological tags in population studies of herring (*Clupea harengus* L.) in the North Sea and to the north and west of Scotland. *J. Cons. Int. Explor. Mer, Paper CM 1984/H:42*, 33–64.

MacKenzie, K, 1986, Parasites as indicators of host populations. In: *Parasitology—Quo Vadit?*

Proceedings of the Sixth International Congress of Parasitology (Ed. Howell, M. J.). Australian Academy of Science, Canberra, pp. 345–352.

MacKenzie, K, 1987a, Relationships between the herring, *Clupea harengus* L., and its parasites. *Adv. Mar. Biol.*, 24, 264–319.

MacKenzie, K, 1987b, Long term changes in the prevalence of two helminth parasites (Cestoda: Trypanorhyncha) infecting marine fish. *J. Fish Biol.*, 31, 83–87.

MacKenzie, K, 1990, Cestode parasites as biological tags for mackerel (*Scomber scombrus* L.) in the Northeast Atlantic. *J. Cons. perm. int. Explor. Mer*, 46, 155–166.

MacKenzie, K. and Gibson, D. I, 1970, Ecological studies of some parasites of plaice, *Pleuronectes platessa* L. and flounder *Platichthys flesus* (L.). In: *Aspects of Fish Parasitology*, (Eds Taylor, A. E. R. and Müller, R.), *Symposia of the British Society for parasitology*. Oxford: Blackwell Scientific Publications, pp. 1–42.

MacKenzie, K. and Liversidge, J. M, 1975, Some aspects of the biology of the cercaria and metacercaria of *Stephanostomum baccatum* (Nicoll, 1907) Manter, 1934 (Digenea: Acanthocolpidae). *J. Fish Biol.*, 7, 247–256.

MacKenzie, K. and Mehl, S, 1984, The cestode parasite *Grillotia angeli* as a biological tag for mackerel in the eastern North Atlantic. *J. Cons. Int. Explor. Mer, Paper CM 1984/H:52*, 14 pp.

MacKenzie, K., McVicar, A. H, and Waddell L. F, 1976, Some parasites of plaice *Pleuronectes platessa* L. in three different farm environments. *Scottish Fish. Res. Rep., No. 4*, 14 pp.

MacKenzie, K., Smith, R. M. and Williams, H. H, 1984, Aspects of the biology of *Grillotia angeli* Dollfus, 1969 (Cestoda: Trypanorhyncha) and its use as a biological tag for mackerel and other fish. *J. Cons. Int. Explor. Mer, Paper CM 1984/H:53*, 15 pp.

Mackie, G. L., Morton, W. B. and Ferguson, M. S, 1983, Fish parasitism in a new impoundment and differences upstream and downstream. *Hydrobiologia*, **99**, 197–205.

Mackiewicz, J. S, 1965, Redescription and distribution of *Glaridacris catostomi Cooper*, 1920 (Cestoidea: Caryophyllaeidae). *J. Parasit.*, 51, 554–560.

Mackiewicz, J. S, 1972, Caryophyllidea (Cestoidea): A review. *Expl Parasit.*, 31, 417–512.

Mackiewicz, J. S, 1981, Caryophyllidea (Cestoidea): evolution and classification. *Adv. Parasit.*, 19, 139–206.

Mackiewicz, J. S, 1982a, Caryophyllidea (Cestoidea): perspectives. *Parasitology*, **84**, 397–417.

Mackiewicz, J. S, 1982b, Parasitic platyhelminth evolution and systematics: perspectives and advances since ICOPA IV, 1978.. In: *Parasites—their world and ours. Proc. ICOPA V, Toronto, Canada, 7–14 Aug., 1982*, (Eds Mettrick, D. F. and Desser, S. S.). Amsterdam, The Netherlands: Elsevier Biomedical Press, pp. 179–188.

Mackiewicz, J. S, 1984, Cercomer theory: significance of sperm morphology, oncosphere metamorphosis, polarity reversal, and the cercomer to evolutionary relationships of Monogenea to Cestoidea. *Acta. Parasit. Pol.*, 29, 11–21.

Mackiewicz, J. S, 1988, Cestode transmission patterns. *J. Parasit.*, 74, 60–71.

Mackiewicz, J. S. and Ehrenpris, M. B, 1980, Calcareous corpuscle distribution in caryophyllid cestodes: possible evidence of cryptic segmentation. *Proc. Helminth. Soc. Wash.*, 47, 1–9.

Mackiewicz, J. S. and McCrae, R. C, 1962, *Hunterella nodulosa* gen. n., sp. n. (Cestoidea: Caryophyllaeidae) from *Catostomus commersoni* (Lacépède) (Pisces: Catostomidae). *J. Parasit.*, **48**, 798–806.

Mackiewicz, J. S., Cosgrove, G. E. and Gude, W. D, 1972, Relationship of pathology to scolex morphology among caryophyllid cestodes. *Z. ParasitKde*, 39, 233–246.

MacKinnon, A. D. and Featherstone, D. W, 1982, Location and means of attachment of *Bothriocephalus scorpii* (Müller) (Cestoda: Pseudophyllidea) in the red cod, *Pseudophycis bacchus* (Forster in Bloch and Schneider), from New Zealand Waters. *Aust. J. Mar. Freshw. Res.*, 33, 595–598.

MacKinnon, B. M. and Burt, M. D. B, 1984, The comparative ultrastructure of spermatozoa from *Bothrimonus sturionis* Duv. 1842 (Pseudophyllidea), *Pseudanthobothrium hanseni* Baer, 1956 (Tetraphyllidea), and *Monoecocestus americanus* Stiles, 1895 (Cyclophyllidea). *Can. J. Zool.*, 62, 1059–1066.

MacKinnon, B. M. and Burt, M. D. B, 1985a, Ultrastructure of spermatogenesis and the mature spermatozoon of *Haplobothrium globuliforme* Cooper, 1914 (Cestoda: Haplobothrioidea). *Can. J. Zool.*, 63, 1478–1487.

MacKinnon, B. M. and Burt, M. D. B, 1985b, The comparative ultrastructure of the plerocercoid and adult primary scolex of *Haplobothrium globuliforme* Cooper, 1914 (Cestoda: Haplobothrioidea). *Can. J. Zool.*, 63, 1488–1496.

MacKinnon, B. M. and Burt, M. D. B, 1985c, Histological and ultrastructural observations on the secondary scolex and strobila of *Haplobothrium globuliforme* (Cestoda: Haplobothrioidea). *Can. J. Zool.*, 63, 1995–2000.

MacKinnon, B. M., Jarecka, L. and Burt, M. D. B, 1985, Ultrastructure of the tegument and penetration glands of developing procercoids of *Haplobothrium globuliforme* Cooper, 1914 (Cestoda: Haplobothrioidea). *Can. J. Zool.*, 63, 1470–1477.

Madhavi, R, 1968, A didymozoid metacercaria from the copepod, *Paracalanus aculeatus* Giesbrecht, from Bay of Bengal. *J. Parasit.*, 54, 629.

Madhavi, R, 1976, Miracidium of *Allocreadium fasciatusi* Kakaji 1969 (Trematoda: Allocreadiidae). *J. Parasit.*, 62, 410–412.

Madhavi, R, 1978a, Life history of *Allocreadium fasciatusi* Kakaji, 1969, (Trematoda: Allocreadiidae) from the freshwater fish *Aplocheilus melastigma* McClelland. *J. Helminth.*, 52, 51–59.

Madhavi, R, 1978b, Life history of *Genarchopsis goppo* Ozaki, 1925 (Trematoda: Hemiuridae) from the freshwater fish *Channa punctata*. *J. Helminth.*, 52, 251–259.

Madhavi, R, 1980, Life history of *Allocreadium handiai* Pande, 1937 (Trematoda: Allocreadiidae) from the freshwater fish *Channa punctata* Bloch. *Z. ParasitKde*, 63, 89–97.

Madhavi, R, 1982, Didymozoid trematodes (including new genera and species) from marine fishes of the Waltair coast, Bay of Bengal. *Syst. Parasit.*, 4, 99–124.

Madhavi, R. and Anderson, R. M, 1985, Variability in the susceptibility of the fish host, *Poecilia reticulata*, to infection with *Gyrodactylus bullatarudis* (Monogenea). *Parasitology*, 91, 531–544.

Magath, T. B, 1918, The morphology and life-history of a new trematode parasite, *Lissorchis fairporti* nov. gen. et nov. spec. from the buffalo fish, *Ictiobus*. *J. Parasit.*, 4, 58–69.

Maggenti, A. R, 1973a, Anthelmintics for nematode parasites of fish: I. Control of *Sterliadochona pedispicula* in *Salmo gairdnerii* by diethyl 2-chlorovinyl phosphate and 2,2 dichlorovinyl dimethyl phosphate pellets. *Proc. Helminth. Soc. Wash.*, 40, 94–97.

Maggenti, A. R, 1973b, Anthelmintics for nematode parasites of fish: II. Aqueous anthelmintic bath treatments using diethyl 2-chlorovinyl phosphate and 2,2- dichlorovinyl dimethyl phosphate for control of *Sterliadochona pedispicula* in *Salmo Gairdneri*. *Proc. Helminth. Soc. Wash.*, 40, 97–101.

Mahajan, C. L., Agrawal, N. K., John, M. J. and Katta, V. P, 1978, Parasitization of *Isoparochis hypselobagri* Billet in *Channa punctatus* Bloch. *Curr. Sci.*, 47, 835–836.

Mahajan, C. L., Agrawal, N. K., John, M. J. and Katta, V. P, 1979, Effect of a digenean *Isoparorchis hypselobagri* (Billet, 1898) on an airbreathing fish *Channa punctatus* (Bloch) with particular reference to biochemical and haematological changes. *J. Fish Dis.*, 2, 519–528.

Mahon, R, 1976, Effect of the cestode *Ligula intestinalis* on spottail shiners, *Notropis hudsonius*. *Can. J. Zool.*, 54, 2227–2229.

Maillard, C, 1976, Distomatoses de poissons en milieu lagunaire. *Thesis, Univ. Sci. Tech. Languedoc, Montpellier, France*, [14] + iv + 383 pp.

Maillard, C, 1982, Spécificité des trématodes de poissons. [*Deux. Symp. spéc. parasit. parasites des vertébrés, 13–17 avril 1981. Colloque internat.* CNRS.] *Mems Mus. Nat. Hist. Nat., Ser. A Zoologie*, 123, 313–319.

Maillard, C. and Ausel, J, 1988, Host specificity of fish trematodes investigated by experimental ichthyophagy. *Int. J. Parasit.*, 18, 493–498.

Maillard, C., Lambert, A. and Raibaut, A, 1980, Nouvelle forme de distomatose larvaire. Etude d'un trématode pathogène pour les alevins de daurade (*Sparus aurata* L., 1758) en écloserie. *C. R. Acad. Sci., Paris, D*, 290, 535–538.

Maiti, S. S., Roy, R. N. and Haldar, D. P, 1987, Infection by a ciliate and a fluke on the fry of Indian major carps. *Curr. Sci.*, 56, 1029–1030.

Maitland, P. S, 1972, A key to the freshwater fish of the British Isles with notes on their distribution and ecology. *Sci. Publ. Freshwater Biol. Ass.*, 27, 1–139.

Malakhov, V. V. and Valovaya, M. A, 1984, Comparative morphological analysis of the cephalic structures of nematodes in the suborder Cucullanina. *Parazitologyia*, 18, 445–450. [In Russian.]

Malakhova, R. P, 1976, [Parasites of fish in the salmon river, Pista (Lake Kuito basin).] In: *Lososevye (Salmonidae) Karelii. (Sbornik). Petrozavodsk, USSR*, pp. 122–130. [In Russian.]

Malakhova, R. P, 1982, [Features of the parasite faunas of local populations of *Esox lucius* in lake Syamozero.] In: *Ekol. parazit, organizmov biogeotsen. sev.* Petrozavodsk, USSR: Karel'skii Filial Akademii Nauk SSSR, Institut Biologii, pp. 26–39. [In Russian.]

Malakhova, R. P. and Anikieva, L. V, 1975, [The biology of *Proteocephalus exiguus* in Coregoninae.] In: *Parazit. issled. Karel'sk. ASSR Murmansk. obl.* Petrozavodsk: Karel'skii filial Akademii Nauk SSSR, Institut Biologii, pp. 168–175. [In Russian.]

Malakhova, R. P., Titova, V. F. and Potapova, O. I, 1972, [Data on the correlation between the intensity of *Proteocephalus exiguus* infection in *Coregonus albula* and the biological indicators of the host. (Communication 1.).] In: *Lososevye (Salmonidae) Karelii. Vypusk 1. Eologiya, parazitofauna, biokhimiya*. Petrozavodsk, USSR: Karel'skii filial Akademii Nauk SSSR, pp. 84–89. [In Russian.]

Malhotra, S. K, 1982, Bionomics of hill-stream cyprinids. III. Food, parasites and length–weight relationship of Garhwal mahaseer, *Tortor* (Ham.). *Proc. Indian Acad. Sci., Anim. Sci.*, 91, 493–499.

Malhotra, S. K. and Chauhan, R. S, 1984, Helminth infection in hillstream fishes. *Indian J. Parasit.*, **8**, 303–305.

Malmberg, G, 1956, On a new genus of viviparous monogenetic trematodes. *Arkiv. Zool.*, **10**, 317–329.

Malmberg, G, 1969, The protonephridia of the genus *Gyrodactylus* (Monogenoidea). *Parazit. Sb.*, 24, 166–168.

Malmberg, G, 1970, The excretory systems and the marginal hooks as a basis for the systematics of *Gyrodactylus* (Trematoda, Monogenea). *Arkiv. Zool.*, 23, 235 pp.

Malmberg, G, 1973, *Gyrodactylus* infestations on species of *Salmo* in Danish and Swedish hatcheries. *{Proc. Scand. Soc. Parasitol., Copenhagen 10–13 Dec. 1972. Abstract.}. Norwegian Journal of Zoology*, 21 (4), 325–326.

Malmberg, G, 1982, On evolutionary processes of Monogenea, though basically from a less traditional viewpoint. In: *Parasites—their world and ours. Proc. ICOPA V, Toronto, Canada, 7–14 Aug., 1982* (Eds Mettrick, D. F. and Desser, S. S.). Amsterdam, The Netherlands: Elsevier Biomedical Press, pp. 198–202.

Malmberg, G, 1986, The major parasitic platyhelminth classes—progressive and regressive evolution? *Hydrobiologia*, 132, 23–29.

Malmberg, G, 1987a, *Gyrodactylus salaris* Malmberg, 1957 and *G. truttae* Gläser, 1974—two problematic species. *Proc. XIII Symp. Scandinavian Society for Parasitology (Helsinki, Finland 1987), Åbo Akademi Inf.*, 19, p. 34.

Malmberg, G, 1987b, *Gyrodactylus*—a monogenean of economic interest to fish farmers. *Vattenbruk*, 5, 15–20.

Malmberg, G, 1989, Salmonid transports, culturing and *Gyrodactylus* infections in Scandinavia. In: *Parasites of Freshwater fishes of North-west Europe. Materials of the Int. Symp. within the program of the Soviet-Finnish Cooperation 10–14 January 1988. Institute of Biology USSR, Academy of Sciences, Karelian Branch and Zoological Institute USSR Academy of Sciences*, pp. 88–104.

Malmberg, G, 1990, On the ontogeny of the haptor and the evolution of the Monogenea. *Syst. Parasit.*, 17, 1–65.

Malmberg, G. and Fernholm, B, 1989, *Myxinidocotyle* gen. n. and *Lophocotyle* Braun (Platyhelminthes, Monogenea, Acanthocotylidae) with descriptions of three new species from hagfishes (Chordata, Myxinidae). *Zool. Scr.*, 18, 187–204.

Malmberg, G. and Malmberg, M, 1987, *Gyrodactylus* in salmon and rainbow trout farms. In: *Parasites and diseases in natural waters and aquaculture in Nordic countries. Proc. Symp., 2–4 Dec. 1986, Stockholm, Sweden* (Eds Stenmark, A. and Malmberg, G.). Stockholm, Sweden: Zoo-tax, Naturhistoriska riksmuseet, pp. 199–204.

Mamaev, Yu. L, 1975, [Hypotheses on the origin of cestodes from *Archigetes*-like ancestors, parasites of oligochaetes.] *Zool. Zh.*, 54, 1277–1283. [In Russian.]

Mamaev, Yu. L, and Avdeev, G. V, 1984, [Peculiar attachment of the monogenean *Macruricotyle clavipes* Mamaev Lyadov, 1975, to the gills of its host.] *Mater. Nauch. Konf. Vses. Obshch. Gel'm. (Biol. takson. gel'm. zhivot. chelov.), No. 34*, 35–40. [In Russian.]

Mamyshev, I. G., Spiranti, N. O. and Chernova, T. N, 1981, [The parasite fauna of salmonids in some rivers of western Georgia.] In: *Rybokhozyaist. issled. vnutr. vodoem. Gruzii. (Sb. Nauch. Trud.)*. Moscow, USSR: Vses Nauchno-Issled. Inst. Morsk Ryb. Khozyaist Okeanogr., pp. 48–56. [In Russian.]

Mankes, R. F, and Mackiewicz, J. S, 1972, Calcareous corpuscles of *Glaridacris laruei* (Lamont) (Cestoidea: Caryophyllidea). *Proc. Helminth. Soc. Wash.*, 39, 177–181.

Mann, H, 1971, 4. Binnenfischerei. Schadwirkungen bei forellen durch Befall mit Kratzern (Acanthocephalen). *Inf. Fischwirtschaft*, 18, 60–61.

Mann, J. A, 1978, Diseases and parasites of fishes: an annotated bibliography of books and symposia, 1904–1977. *Fish Disease Leaflet, Fish and Wildlife Service, United States Department of the Interior, No. 53*, 28 pp.

Mann, J. A., Catrow, V. J. and Engle, F. V, 1982, Fish culture. *An annotated bibliography of publications of the National Fisheries Centre, Leetown, 1972–1980*. 124 pp.

Manooch, C. S, III and Hogarth, W. T, 1983, Stomach contents and giant trematodes from wahoo, *Acanthocybium solanderi*, collected along the South Atlantic and Gulf Coasts of the United States. *Bull. Mar. Sci.*, 33, 227–228.

Manter, H. W, 1933, A new family of trematodes from marine fishes. *Trans. Am. Microsc. Soc.*, **52**, 233–242.

Manter, H. W, 1945, *Dermadena lactophrysi* n. gen. n. sp. (Trematoda: Lepocreadiidae) and consideration of the related genus *Pseudocreadium*. *J. Parasit.*, 31, 411–417.

Manter, H. W, 1955, The zoogeography of the trematodes of marine fishes. *Exp. Parasitol.*, 4, 62–86.

Manter, H. W, 1957, Host specificity and other host relationships among the digenetic trematodes of marine fishes. *Symposium on host specificity among parasites of vertebrates (1st), University of Neuchâtel, April 15–18, 1957.* University of Neuchâtel, Neuchâtel, pp. 185–196.

Manter, H. W, 1961, Studies on digenetic trematodes of fishes of Fiji. I. Families Haplosplanchnidae, Bivesiculidae, and Hemiuridae. *Proc. Helminth. Soc. Wash.*, **28**, 67–74.

Manter, H. W, 1963, Studies on the digenetic trematodes of fishes of Fiji. II. Families Lepocreadiidae, Opistholebetidae and Opecoelidae. *J. Parasit.*, 49, 99–113.

Manter, H. W, 1967, Some aspects of the geographical distribution of parasites. *J. Parasitol.*, 53, 1–9.

Manter, H. W, 1970, A new species of *Transversotrema* (Trematoda: Digenea) from marine fishes of Australia. *J. Parasit.*, 56, 486–489.

Manter, H. W. and Pritchard, M. H, 1961, Studies on digenetic trematodes of Hawaiian fishes: families Monorchiidae and Haploporidae. *J. Parasit.*, 47, 483–492.

Manter, H. W. and Pritchard, M. H, 1969, Some digenetic trematodes of Central Africa, chiefly from fishes. *Rev. Zool. Bot. Afr.*, **80**, 51–61.

Marchand, B, 1981, Ultrastructure du spermatozoïde et de la fécondation chez les acanthocéphales. *Afrique Médicale*, 20, 417–419.

Marchand, B. and Mattei, X, 1976a, Présence de flagelles spermatiques dans les sphères ovariennes des Eocanthocèphales. *J. Ultrastruct. Res.*, 56, 331–338.

Marchand, B. and Mattei, X, 1976b, La spermatogenèse des acanthocéphales. 1. L'appareil centriolaire et flagellaire au cours de la spermiogenèse d'*Illiosentis furcatus* var *africana* Golvan, 1956 (Palaeacanthocephala, Rhadinorhynchidae). *J. Ultrastruct. Res.*, 54, 347–358.

Marchand, B. and Mattei, X, 1978, La spermatogenèse des acanthocéphales. V. Flagellogenèse chez un Eoacanthocephala: mise en place et désorganisation de l'axonème spermatique. *J. Ultrastruct.* Res., 63, 41–50.

Marchand, B. and Mattei, X, 1980a, Présence d'une gouttière nucléocytoplasmique dans les spermatides d'un Acanthocéphale: *Pallisentis golvani. C. R. Séanc. Soc. Biol.*, 174, 933–936.

Marchand, B. and Mattei, X, 1980b, Fertilization in Acanthocephala. II. Spermatozoon penetration of oocyte, transformation of gametes and elaboration of the 'fertilization membrane'. *J. Submicrosc. Cytol.*, 12, 95–105.

Marcogliese, D. J. and Esch, G. W, 1989, Experimental and natural infection of planktonic and benthic copepods by the Asian tapeworm, *Bothriocephalus acheilognathi. Proc. Helminth. Soc. Wash.*, 56, 151–155.

Maren, M. J. van 1979, The amphipod *Gammarus fossarum* Koch (Crustacea) as intermediate host for some helminth parasites, with notes on their occurrence in the final host. *Bijdragen Tot de Dierkunde*, 48, 97–110.

Margolis, L, 1962, *Lampritrema nipponicum* Yamaguti (Trematoda) from new hosts in the North Pacific Ocean, the relationship of *Distomum miescheri* Zschokke, and the status of the family Lampritrematidae. *Can. J. Zool.*, 40, 941–950.

Margolis, L, 1963, Parasites as indicators of the geographical origin of sockeye salmon, *Onchorhynchus nerka* (Walbaum) occurring in the North Pacific Ocean and adjacent areas. *Fish. Res. Bd Can. INPFC Doc. No. 466, Bull. 11*, 101–156.

Margolis, L, 1964, *Paurorhynchus hiodontis* Dickerman, 1954 (Trematoda: Bucephalidae): a second record involving a new host and locality in Canada. *Can. J. Zool.*, 42, 716.

Margolis, L, 1965, Parasites as an auxiliary source of information about the biology of Pacific salmon (genus *Oncorhynchus*). *J. Fish. Res. Bd Can.*, 22, 1387–1395.

Margolis, L, 1967a, Blood feeding in *Salvelinema walkeri* (Nematoda: Cystidicolinae), a parasite of coho salmon (*Oncorhynchus kisutch*). *Can. J. Zool.*, 45, 1295–1296.

Margolis, L, 1967b, The swimbladder nematodes (Cystidicolinae) of Pacific salmon (Genus *Oncorhynchus*). *Can. J. Zool.*, 45, 1183–1199.

Margolis, L, 1968, Review of the Japanese species of *Cystidicola, Metabronema* and *Rhabdochona* (Nematoda) from salmonoid fishes. *Bull. Meguro Parasit. Mus.*, 2, 23–44.

Margolis, L, 1969, Special Edition for Journal of the Fisheries Research Board of Canada, dedicated to Professor T. W. M. Cameron. *J. Fish. Res. Bd Can.*, 26, 1111 pp.

Margolis, L, 1970a, Nematode diseases of marine fish. In: *A symposium on diseases of fishes and shellfishes.* Ed. S. F. Snieszko), Am. Fish. Soc. Spec. Publ. No. 5, pp. 190–208.

Margolis, L, 1970b, A bibliography of parasites and diseases of fishes of Canada: 1879–1969. *Fish. Res. Bd Can., Tech. Rep. 185*, 38 pp.

Margolis, L, 1971, Polychaetes as intermediate hosts of helminth parasites of vertebrates: a review. *J. Fish. Res. Bd Can.*, **28**, 1385–1392.

Margolis, L, 1977, Public health aspects of 'codworm' infection: a review. *J. Fish. Res. Bd Can.*, 34, 887–898.

Margolis, L, 1982a, Pacific salmon and their parasites: a century of study. Eighth Invitational Wardle Lecture, Ann. Meeting Canadian Soc. Zool., Univ. of British Columbia, Vancouver, May 17, 1982. *Bull. Can. Soc. Zool.*, 13, 7–11.

Margolis, L, 1982b, Parasitology of Pacific salmon—an overview. In: *Aspects of parasitology. A Festschrift dedicated to the 50th anniversity of the Institute of Parasitology of McGill University*, 1932–1982. (Ed. Meerovitch, E.) Montreal, Canada: Institute of Parasitology, McGill University, pp. 135–226.

Margolis, L, 1982c. Stock identification. In: *Annual Report* (1980) *of the salmonid enhancement programme. Gov. Can. Fish. Oce. Vancouver and province of British Columbia Ministry of the Environment, Victoria*, pp. 156.

Margolis, L. and Arai, H, 1989, Synopsis of the parasites of vertebrates of Canada. Parasites of marine mammals. *Alberta Agriculture Animal Health Division*, 1–26.

Margolis, L. and Arthur, J. R, 1979, Synopsis of the parasites of fishes of Canada. *Bull. Fish. Res. Bd Can.*, 199, 1–269

Margolis, L. and Beverley-Burton, M, 1977, Response of mink (*Mustela vison*) to larval *Anisakis simplex (Nematoda: Ascaridida). Int. J. Parasitol.*, 7, 269–273.

Margolis, L. and Boyce, N. P, 1969, Life span, maturation, and growth of two hemiurid trematodes, *Tubulovesicula lindbergi* and *Lecithaster gibbosus*, in Pacific salmon (genus *Oncorhynchus*). *J. Fish. Res. Bd Can.*, 26, 893–907.

Margolis, L. and Boyce, N. P, 1990, Helminth parasites from north Pacific anadromous chinook salmon, *Oncorhynchus tshawytscha*, established in New Zealand. *J. Parasit.*, 76, 133–135.

Margolis, L. and Evelyn, T. P. T, 1987, Aspects of disease and parasite problems in cultured salmonids in Canada, with emphasis on the Pacific region, and regulatory measures for their control. In: *Parasites and diseases in natural waters and aquaculture in Nordic countries. Proc. of a Symposium, 2–4 Dec. 1986, Stockholm, Sweden* (Eds Stenmark, A. and Malmberg, G.). Stockholm, Sweden: Zootax, Naturhistoriska riksmuseet, pp. 4–19.

Margolis, L. and Kabata, Z, 1984, General introduction. In: (Eds L. Margolis and Z. Kabata) *Guide to the parasites of fishes of Canada. Part I. Can. Publ. Fish. Aquat. Sci.*, 74, 1–4.

Margolis, L. and Moravec, F, 1982, *Ramellogammarus vancouverensis* Bousfield (Amphipoda) as an intermediate host for salmonid parasites in British Columbia. *Can. J. Zool.*, **60**, 1100–1104.

Margolis, L., Esch, G. W., Holmes, J. C., Kuris, A. M. and Schad, G. A, 1982, The use of ecological terms in parasitology (report of an *ad hoc* committee of the American Society of Parasitologists) *J. Parasit.*, **68**, 131–133.

Markevich, A. P, 1951, *The parasite fauna of fresh-water fish in the Ukrainian SSR*. Kiev: Izdat. Akad. Nauk Ukr. SSR, 376 pp. [In Russian.]

Markowski, S, 1937, Über die Entwicklungsgeschichte und Biologie des Nematoden *Contracaecum aduncum* (Rudolphi, 1802). *Bull. Int. Acad. Pol. Sci. Lett., Cl. Sci. Math. Nat.*, 2, 227–247.

Marra, M. and Esch, G. W, 1970, Distribution of carbohydrates in adults and larvae of *Proteocephalus ambloplitis* (Leidy, 1887). *J. Parasit.*, 56, 398–400.

Martin, S. and Vasquez, R, 1984, Biology and behaviour of the cercariae of a *Sanguinicola* sp. in the River Cilloruelo (Salamanca, Spain). *Annls Parasit. Hum. Comp.*, 59, 231–236.

Martin, W. E, 1940, Studies on the trematodes of Woods Hole. III. The life cycle of *Monorcheides cumingiae* (Martin) with special reference to its effect on the invertebrate host. *Biol. Bull.*, 79, 131–144.

Martin, W. E, 1945, Two new species of marine cercariae. *Trans. Am. Microsc. Soc.*, 64, 203–212.

Martin, W. E, 1958a, Hawaiian helminths 1. *Trigonocryptus conus* n. gen. n. sp. (Trematoda: Fellodistomidae). *Pacific Sci.*, 12, 251–254.

Martin, W. E, 1958b, The life histories of some Hawaiian heterophyid trematodes. *J. Parasit.*, 44, 305–323.

Martin, W. E, 1959, Egyptian heterophyid trematodes. *Trans. Am. microsc. Soc.*, **78**, 172–181.

Martin, W. E. and Kuntz, R. E, 1955, Some Egyptian heterophyid trematodes. *J. Parasit.*, 41, 374–382.

Martin, W. E. and Multani, S, 1966, *Microsentis wardae* n. g., n. sp. (Acanthocephala) in the marine fish *Gillichthys mirabilis* Cooper. *Trans. Am. Microsc. Soc.*, **85**, 536–540.

Martin, W. E. and Zam, S. G, 1967, *Vasorhabdochona cablei* gen. et. sp. n. (Nematoda) from blood vessels of the marine fish, *Gillichthys mirabilis* Cooper. *J. Parasit.*, **53**, 389–391.

Mashego, S. N. and Saayman, J. E, 1981, Observations on the prevalence of nematode parasites of the catfish, *Clarias gariepinus* (Burchell 1822), in Lebowa, South Africa. *S. Afr. J. Wildl. Res.*, 11, 46–48.

Mashego, S. N. and Saayman, J. E, 1989, Digenetic trematodes and cestodes of *Clarias gariepinus* (Burchell, 1822) in Lebowa, South Africa, with taxonomic notes. *S. Afr. J. Wildl. Res.*, 19, 17–20.

Mashtakov, A. V, 1979, [Ecological analysis of the parasite fauna of *Abramis brama* in the Gorkii reservoir basin (USSR).] *Trudy Inst. Biol. Vnutr. Vod (Fiziol. parazitol. presnovod. zhivotr.)*, **38**, 168–176. [In Russian.]

Massimo, S, 1987, Attempts to treat *Anguillicola australiensis* in reared eels. In: *Actual problems in fish parasitology. 2nd International Symposium of Ichthyoparasitology. Sept. 27–Oct. 3. Tihany, Hungary.*

Matskási, I, 1978, The effect of *Bothriocephalus acheilognathi* Yamaguti, 1934 infection of the protease activity in the gut of carp fry. *Parasit. Hung.*, 11, 51–56.

Matskási, I, 1980, *Ligula intestinalis* (Cestoda: Pseudophyllidea): studies on the α-amylase activity of plerocercoid larvae. *Parasit. Hung.*, **13**, 27–34.

Matskási, I, 1984, The effect of *Bothriocephalus acheilognathi* infection on the protease and α-amylase activity in the gut of carp fry. In: *Fish, pathogens and environment in European polyculture. (Proc. Internat. Seminar, 23–27 June 1981, Szarvas, Hungary. Symposia Biologica Hungarica Vol. 23).* (Ed. Olah, J.). Budapest, Hungary: Akademiai Kiado, pp. 119–125.

Matskási, I. and Juhasz, S, 1977, *Ligula intestinalis* (L.), 1758): investigation of plerocercoids and adults for protease and protease inhibitor activities. *Parasit. Hung.*, **10**, 51–60.

Matskási, I. and Nemeth, I, 1979, *Ligula intestinalis* (Cestoda: Pseudophyllidea): studies on the properties of proteolytic and protease inhibitor activities of plerocercoid larvae. *Int. J. Parasit.*, 9, 221–227.

Matthews, B. F, 1980, *Cercaria vaullegeardi* Pelseneer, 1906 (Digenea: Hemiuridae); the daughter sporocyst and emergence of the cercaria. *Parasitology*, **81**, 61–69.

Matthews, B. F, 1981a, *Cercaria vaullegeardi* Pelseneer, 1906 (Digenea: Hemiuridae); development and ultrastructure. *Parasitology*, **83**, 575–586.

Matthews, B. F, 1981b, *Cercaria vaullegeardi* Pelseneer, 1906 (Digenea: Hemiuridae); the infection mechanism. *Parasitology*, **83**, 587–593.

Matthews, B. F, 1982, Behaviour and enzyme release by *Anisakis* sp. larvae (Nematoda: Ascaridida). *J. Helminth.*, **56**, 177–183.

Matthews, B. F. and Matthews, R. A, 1988, The ecsoma in Hemiuridae (Digenea: Hemiuroidea): tegumental structure and function in the mesocercaria and the metacercaria of *Lecithochirium furcolabiatum* (Jones, 1933) Dawes, 1947. *J. Helminth.*, **62**, 317–330.

Matthews, R. A, 1973, The life-cycle of *Prosorhynchus crucibulum* (Rudolphi, 1819) Odhner, 1905, and a comparison of its cercaria with that of *Prosorhynchus squamatus* Odhner, 1905. *Parasitology*, 66, 133–164.

Matthews, R. A, 1974, The life-cycle of *Bucephaloides gracilescens* (Rudolphi, 1819) Hopkins, 1954 (Digenea: Gasterostomata). *Parasitology*, **68**, 1–12.

Mawdesley-Thomas, L. E, 1975, Some diseases of muscle. In: *The Pathology of Fishes* (Eds Ribelin, W. E. and Migaki, G.). Madison, Wisconsin, USA: University of Wisconsin Press, pp. 343–361.

May, R. M, 1988, How many species are there on earth? *Science*, **241**, 1441–1449.

May, R. M. and Anderson, R. M, 1978, Regulation and stability of host–parasite population interactions. II Destabilizing processes. *J. Anim. Ecol.*, 47, 249–268.

May, R. M. and Anderson, R. M, 1979, Population biology of infectious diseases: Part II. *Nature, Lond.*, **280**, 455–461.

Mbahinzireki, G. B, 1980, Observations on some common parasites of *Bagrus docmac forskahl* (Pisces: Siluroidea) of Lake Victoria. *Hydrobiologia*, **75**, 273–280.

McArthur, C. P, 1978, Humoral antibody production by New Zealand eels, against the intestinal trematode *Telogaster opisthorchis* Macfarlane, 1945. *J. Fish Dis.*, 1, 377–387.

McArthur, C. P. and Sengupta, S, 1982, A rapid micromethod for screening eel sera for antibodies against the digenean *Telogaster opisthorchis* Macfarlane, 1945. *J. Fish Dis.*, 5, 67–70.

McCaig, M. L. O. and Hopkins, C. A, 1965, Studies on *Schistocephalus solidus* III. The *in vitro* cultivation of the plerocercoid. *Parasitology*, 55, 257–268.

McClelland, W. F. J, 1957, Two new genera of amphistomes from Sudanese freshwater fishes. *J. Helminth.*, **31**, 247–256.

McCormick, J. H. and Stokes, G. N, 1982, Intraovarian invasion of smallmouth bass oocytes by *Proteocephalus ambloplitis* (Cestoda). *J. Parasit.*, **68**, 973–975.

McCoy, O. R, 1930, Experimental studies on two fish trematodes of the genus *Hamacreadium* (Family Allocreadiidae). *J. Parasit.*, 17, 1–13.

McCullough, J. S. and Fairweather, I, 1983, A SEM study of the cestodes *Trilocularia acanthiaevulgaris, Phyllobothrium squali* and *Gilquinia squali* from the spiny dogfish. *Z. ParasitKde*, 69, 655–665.

McCullough, J. S. and Fairweather, I, 1987, The structure, composition, formation and possible functions of calcareous corpuscles in *Trilocularia acanthiaevulgaris. Parasitol. Res.*, 74, 175–182.

McCullough, J. S. and Fairweather, I, 1989, The fine structure and possible functions of scolex gland cells in *Trilocularia acanthiaevulgaris* (Cestoda, Tetraphyllidea). *Parasitol. Res.*, 75, 575–582.

McCullough, J. S., Fairweather, I. and Montgomery, W. I, 1986, The seasonal occurrence of *Trilocularia acanthiaevulgaris* (Cestoda: Tetraphyllidea) from spiny dogfish in the Irish Sea. *Parasitology*, 93, 153–162.

McDaniel, J. B. and Bailey, H. H, 1974, Seasonal population dynamics of some helminth parasites of centrarchid fishes. *Southwestern Nat*, 18, 403–416.

McDonough, J. M. and Gleason, L. N, 1981, Histopathology in the rainbow darter, *Etheostoma caeruleum*, resulting from infections with the acanthocephalans, *Pomphorhynchus bulbocolli* and *Acanthocephalus dirus. J. Parasit.*, 67, 403–409.

McGladdery, S. E, 1985, Studies of the parasite fauna of Atlantic herring (*Clupea harengus* L.) from the northwestern Atlantic Ocean. Diss. Abstr. Int. B. (Sci. Eng.), 45, 3142.

McGladdery, S. E. and Burt, M. D. B, 1985, Potential of parasites for use as biological indicators of migration, feeding, and spawning behaviour of northwestern Atlantic herring (*Clupea harengus*). *Can. J. Fish. Aquat. Sci.*, 42, 1957–1968.

McGregor, E. A, 1963, Publications on fish parasites and diseases, 330 B.C.–A.D. 1923. United States Department of the Interior, Special Scientific Report—Fisheries No. 474. 84 pp.

McGuigan, J. B. and Somerville, C, 1985, Studies on the effects of cage culture of fish on the parasite fauna in a lowland freshwater loch in the west of Scotland. *Z. ParasitKde*, 71, 673–682.

McKerrow, J. H. and Deardorff, T. L, 1988, Anisakiasis: revenge of the sushi parasite. *New England J. Med.*, 319, 1228–1229.

McKinnon, A. D. and Featherston, D. W, 1982, Location and means of attachment of *Bothriocephalus scorpii* (Müller) (Cestoda: Pseudophyllidea) in red cod, *Pseudophycis bacchus* (Forster in Bloch and Schneider), from New Zealand waters. *Aust. J. Mar. Freshw. Res.*, 33, 595–598.

McManus, D. P, 1975, Tricarboxylic acid cycle enzymes in the plerocercoid of *Ligula intestinalis* (Cestoda: Pseudophyllidea). *Z. ParasitKde*, 45, 319–322.

McManus, D. P. and Sterry, P. R, 1982, *Ligula intestinalis*: intermediary carbohydrate metabolism in plerocercoids and adults. *Z. ParasitKde*, 67, 73–85.

McMullen, D. B, 1935a, Life history of *Macroderoides typicus. J. Parasit.*, 20, 135.

McMullen, D. B, 1935b, The life histories and classification of two allocreadid-like plagiorchids from fish *Macroderoides typicus* (Winfield) and *Alloglossidium corti* (Lamont). *J. Parasit.*, 21, 369–380.

McPhail, J. D. and Peacock, S. D, 1983, Some effects of the cestode (*Schistocephalus solidus*) on reproduction in the threespine stickleback (*Gasterosteus aculeatus*): evolutionary aspects of a host-parasite interaction. *Can. J. Zool.*, 61, 901–908.

McVicar, A. H, 1972, The ultrastructure of parasite-host interface of three tetraphyllidean tapeworms of the elasmobranch *Raja naevus. Parasitology*, 65, 77–88.

McVicar, A. H, 1976, *Echinobothrium harfordi* sp. nov. (Cestoda: Diphyllidea) from *Raja naevus* in the North Sea and English Channel. *J. Helminth.*, 50, 31–38.

McVicar, A. H, 1977a, Intestinal helminth parasites of the ray, *Raja naevus* in British waters. *J. Helminth.*, 51, 11–21.

McVicar, A. H, 1977b, The bothridial hooks of *Acanthobothrium quadripartitum*, Williams, 1968 (Cestoda: Tetraphyllidea); their growth and use in taxonomy. *Int. J. Parasit.*, 7, 439–442.

McVicar, A. H, 1979, The distribution of cestodes within the spiral intestine of *Raja naevus* Müller and Henle. *Int. J. Parasit.*, 9, 165–176.

McVicar, A. H. and Fletcher, T. C, 1970, Serum factors in *Raja radiata* toxic to *Acanthobothrium quadripartitum* (Cestoda: Tetraphyllidea), a parasite specific to *R. naevus. Parasitology*, 61, 55–63.

McVicar, A. H. and Gibson, D. I, 1975, *Pancreatonema torriensis* gen. nov. sp. nov. (Nematoda: Rhabdochonidae) from the pancreatic duct of *Raja naevus. Int. J. Parasit.*, 5, 529–535.

McVicar, A. H. and MacKenzie, K, 1977, Effects of different systems of monoculture on marine fish parasites. In: *Origins of pest, parasite, disease and weed problems (18th Symp. Brit. Ecol. Soc., Bangor, 12–14 Apr. 1976)* (Eds Cherrett, J. M. and Sagar, G. R.) Oxford, UK: Blackwell Scientific Publications, pp. 163–182.

McVicar, A. H., Bruno, D. W. and Fraser, C. O, 1988, Fish diseases in the North Sea in relation to sewage dumping. *Marine Pollution Bulletin*, 19, 169–173.

Meade, T. G. and Harvey, J. S., Jr, 1969, Effect of helminth parasitism of *Posthodiplostomum minimum* on serum proteins of *Lepomis macrochirus* and observations on piscine serological taxonomy. *Copeia, No. 3*, 638–641.

Meade, T. G. and Pratt, I, 1965, Description and life-history of *Cardicola alseae* sp. n. (Trematoda: Sanguinicolidae). *J. Parasit.*, 51, 575–578.

Mehra, H. R, 1960, On a new trematode *Staphylorchis scoliodonii* n. sp. (Family Gorgoderidae Looss, 1901; subfamily Anaporrhutinae Looss, 1901) from south Indian shark *Scoliodon sorrakowah* with a phylogenetic discussion and classification of the family. *Proc. Nat. Acad. Sci., India, Sect. B*, 30, 143–165.

Meinkoth, N. A, 1947, Notes on the life cycle and taxonomic position of *Haplobothrium globuliforme* Cooper, a tapeworm of *Amia calva* L. 66, 256–261.

Menitskii, Y. L, 1963, Structure and systematic position of the turbellarian *Ichthyophaga subcutanea* Syromjatnikova, 1949 parasitizing fish. *Parazit. Sbornik.*, 21, 245–258. [In Russian].

Mercer, J. G., Munn, A. E., Smith, J. W. and Rees, H. H, 1986, Cuticle production and ecdysis in larval marine ascaridoid nematodes *in vitro*. *Parasitology*, 92, 711–720.

Mergo, J. C., Jr. and Crites, J. L, 1986, Prevalence, mean intensity, and relative density of *Lintaxine cokeri* Linton 1940 (Monogenea: Heteraxinidae) on freshwater drum (*Aplodinotus grunniens*) in Lake Erie (1984). *Ohio J. Sci.*, 86, 101–105.

Mergo, J. C., Jr. and White, A. M, 1984, A survey of monogeneans on the gills of catostomid fishes from Ohio (1983). *Ohio J. Sci.*, 84, 33–35.

Merritt, S. V. and Pratt, I, 1964, The life history of *Neoechinorhynchus rutili* and its development in the intermediate host (Acanthocephala: Neoechinorhynchidae). *J. Parasit.*, 50, 394–400.

Mettrick, D. F, 1970, The microcosm of intestinal helminths—Discussion. *Proc. Int. Symp. Ecology and Physiology of Parasites* (Ed. Fallis, A. M.). Toronto: University of Toronto Press, pp. 197–200.

Mettrick, D. F, 1971a. *Hymenolepis diminuta*: pH changes in rat intestinal contents and worm migration. *Expl Parasit.* 29, 386–401.

Mettrick, D. F, 1971b. *Hymenolepis diminuta*, the microbiota, nutritional and physico-chemical gradients in the small intestine of uninfected and parasitized rats. *Can J. Physiol. Pharmacol.*, 49, 972–984.

Mettrick, D. F. and Podesta, R. B, 1974, Ecological and physiological aspects of helminth–host interactions in the mammalian gastrointestinal canal. *Adv. Parasit.*, 12, 183–278.

Meyer, A, 1932, Acanthocephala. In: H. G. Bronn's *Klassen und Ordnungen des Thier-Reichs*. Leipzig: Akademische Verlagsgesellschaft MBH. 4, pp. 1–332.

Meyer, A, 1933, Acanthocephala. In H. G. Bronn's *Klassen und Ordunungen des Thier-Reichs*. Leipzig: Akademishe Verlagsgesellschaft MBH. 4, pp. 333–582.

Meyer, F. P, 1968, A review of the parasitic diseases of fish in warm water ponds in North America. *FAO Fish Report, No. 44*, 290–319.

Meyer, F. P, 1979, The role of stress in fish diseases. In: *Principal diseases of farm raised catfish* (Ed. Plumb, J. A.). *Southern Cooperative Series No. 225*, pp. 7–9.

Meyer, M. C, 1960, Notes on *Philonema agubernaculum* and other related dracunculoids infecting salmonids. In: *Libro Homenaje al Dr. Eduardo Caballero cf. Caballero, Jubileo 1930–1960*, pp. 487–492.

Mhaisen, F. T., Al-Salim, N. K. and Khamees, N. R, 1988, Occurrence of parasites of the freshwater mugilid fish *Liza abu* (Heckel) from Basrah, southern Iraq. *J. Fish Biol.*, 32, 525–532.

Michajlow, W, 1951, 'Stadialnosc' rozwoju niectôrych tasiemcow (Cestoda). *Ann. Univ. Mar. Curie-Sklodowska. Lublin, Polonica, VI (Sect. C)*, 6, 77–147.

Michajlow, W, 1962, Species of the genus *Triaenophorus* (Cestoda) and their hosts in various geographical regions. *Acta parasit. pol.*, 15, 1–38.

Michajlow, W, 1985, [Morphological problems of coevolution in the animal world.] In: *Morfologicheskie issledovaniya zhivotnykh*. Moscow, USSR: 'Nauka' pp. 112–128. [In Russian.]

Michel, W, 1981, Parasitologische Untersuchungen an importierten tropischen Zierfischen. *Inaugural-Dissertation, Tierärztliche Hochschule, Hannover, GFR*, 113 pp.

Mikailov, T. K. and Ibragimov, Sh. R, 1980, [The ecology and zoogeography of the parasites of fish in the water-bodies of the Lenkoran natural region.] Baku, USSR: Elm, 114 pp. [In Russian.]

Mikailov, T. K. and Kazieva, N. Sh, 1981, [Prevalence of *Metagonimus yokogawai* Katsurada, 1912, in commercial fish of the Varvarinsk reservoir.] *Aktual'nye voprozy meditzinskoi parazitologii i tropicheskoi meditsiny. Chast' 2. (Mater. Nauch. Konf., 24–25 dekabrya 1981 b.)*. Baku, USSR: NII Meditsinskoi Parazitologii i Tropicheskoi Meditsiny im. S. M. Kirova, 38–40. [In Russian.]

Mikhailova, I. G., Prazdnikov, E. V. and Prusevich, T. O, 1964, [Morphological changes in fish tissue around the larvae of some parasitic worms.] *Trudy Murmansk. Morsk. Biol. Inst. Akad. Nauk SSSR*, 5, 251–264. [In Russian.]

Mikityuk, P. V., Osadchaya, E F., Pogorel'Tseva, T. P., Ryagin, S. T. and Trokhanchuk, V. A, 1984, *{Pocket book on diseases of fish farmed in ponds.}* (Ed. Mikityuk, P. V.). Kiev, Ukrainskaya SSR, USSR: Urozhai, 248 pp. [In Russian.]

Mikryakov, V. P. and Stepanova, M. A, 1983, [Infection of *Abramis ballerus* with *Dactylogyrus chranilowi* (Monogenoidea, Dactylogyridea Bykhovskii, 1937) correlated to the antimicrobial activity of host serum.] *Biol. Vnut. Vod, Inf. Byull., No. 57,* 34–36. [In Russian.]

Milbrink, G, 1975, Population biology of the cestode *Caryophyllaeus laticeps* (Pallas) in bream, *Abramis brama* (L.) and the feeding of fish on oligochaetes. *Rep. Inst. Freshw. Res., Drottningholm, No. 54,* 36–51.

Milinski, M, 1985, Risks of predation of parasitized sticklebacks (*Gasterosteus aculeatus* L.) under competition for food. *Behaviour,* 93, 203–216.

Millemann, R. E. and Knapp, S. E, 1970, Pathogenicity of the 'salmon poisoning' trematode, *Nanophyetus salmonicola,* to fish. In: *Diseases of fishes and shellfishes* (Ed. Snieszko, S. F.). Special Publication No. 5, American Fisheries Association, Washington DC, pp. 209–217.

Millemann, R. E, 1963, Studies on the taxonomy and life-history of echinocephalid worms (Nematoda: Spiruroidea) with a complete description of *Echinocephalus pseudouncinatus* Millemann, 1951. *J. Parasit.,* 49, 754–764.

Miller, D. M. and Dunagan, T. T, 1985, Functional morphology. In: *Biology of the Acanthocephala* (Eds Crompton, D. W. T. and Nickol, B. B.). Cambridge: Cambridge University Press, pp. 73–111.

Miller, R. B, 1952, A review of the *Triaenophorus* problem in Canadian lakes. *Fish. Res. Bd Can., Bull. No. 95,* 42 pp.

Miller, R. L., Olson, A. C. and Miller, L. W, 1973, Fish parasites occurring in thirteen southern California reservoirs. *Calif. Fish Game,* 59, 196–206.

Mills, C. A, 1979a, Attachment and feeding of the adult ectoparasitic digenean *Transversotrema patialense* (Soparker, 1924) on the zebra fish *Brachydanio rerio* (Hamilton-Buchanan). *J. Fish Dis.,* 2, 443–447.

Mills, C. A, 1979b, The influence of differing ionic environments on the cercarial, post-cercarial and adult stages of the ectoparasitic digenean *Transversotrema patialense. Int. J. Parasit.,* 9, 603–608.

Mills, C. A, 1980a, Temperature-dependent survival and reproduction within populations of the ectoparasitic digenean *Transversotrema patialense* on the fish host. *Parasitology,* 81, 91–102.

Mills, C. A, 1980b, Age- and density-dependent growth within populations of the ectoparasitic digenean *Transversotrema patialense* on the fish host. *Int. J. Parasit.,* 10, 287–291.

Mills, C. A., Anderson, R. M. and Whitfield, P. J, 1979, Density-dependent survival and reproduction within populations of the ectoparasitic digenean *Transversotrema patialense* on the fish host. *J. Anim. Ecol.,* 48, 383–399.

Milne, P. H, 1972, *Fish and shellfish farming in coastal waters.* London: Fishing News (Books) Ltd., 169 pp.

Mirle, Ch., Nickel, S., Schultka, H. and Hiltner, R, 1985, Untersuchungen zu Vorkommen, Apizootiologie, Schadwirkung und Bekämpfung der Larven-Triaenophorose der Regenbogenforelle. *Monats. Vetmed.,* 40, 138–142.

Mitchell, A. J. and Hoffman, G. L, 1980, Important tapeworms of North American freshwater fishes. *Fish Dis. Leafl. Fish Widl. Serv., US Dept. Int., No. 59,* 18 pp.

Mitchell, L. G., Ginal, J. and Bailey, W. C, 1983, Melanotic visceral fibrosis associated with larval infections of *Posthodiplostomum minimum* and *Proteocephalus* sp. in bluegill, *Lepomis macrochirus* Rafinesque, in central Iowa, USA. *J. Fish Dis.,* 6, 135–144.

Mitenev, V. K, 1979, [Ecological and geographical features of the parasite fauna of fish in the Kola peninsula (USSR).] In: *Bolezni i parazity ryb Ledovitomorskoi provintsii (v predelakh SSSR* (Eds Gundrizer, A. N. and Bauer, O. N.). Tomsk, USSR: Izdatel'stvo Tomskogo Universiteta, pp. 119–132. [In Russian.]

Mitenev, V. K, 1982, [Ecological peculiarities of the parasite fauna of *Salvelinus alpinus* in the European North.] In: *Plankton pribrezhnykh vod Vostochnogo Murmana.* Apatity, USSR, pp. 105–119. [In Russian.]

Mitenev, V. K, 1984a, [Problems of ecology of fish parasites on the Kola Peninsula.] In: *Ekologiya biologicheskikh resursov Severnogo basseina i ikh promyslovoe ispol'zovanie.* Murmansk, USSR, pp. 49–57. [In Russian.]

Mitenev, V. K, 1984b, [The parasite faunas of migratory *Salmo salar* and *S. trutta* in the Kol'skit Peninsula.] In: *Ekologoparazitologicheskie issledovaniya severnykh morei.* Apatity, USSR: Kol'skii Filial Akademii Nauk SSSR, pp. 88–97. [In Russian.]

Mitenev, V. K. and Shul'man, B. S, 1980, [The effect of hydroconstructions and reservoirs on the parasite fauna of the Atlantic salmon (*Salmo salar*).] *Parazitologiya,* 14, 97–102. [In: Russian.]

Mitter, C. and Brooks, D. R, 1983, Phylogenetic aspects of coevolution. In: *Coevolution* (Eds Futuyma, D. J. and Slatkin, M.) Sinauer, Sunderland, Mass. pp. 65–98.

Miyazaki, T., Rogers, W. A. and Semmens, K. J, 1988, Gastro-intestinal histopathology of paddlefish, *Polyodon spathula* (Walbaum), infected with larval *Hysterothylacium dollfusi* Schmidt, Leiby and Kritsky, 1974. *J. Fish Dis.*, 11, 245–250.

Mo, T. A, 1987, Seasonal variations in prevalence, intensity and the hard parts of the opisthaptor of *Gyrodactylus truttae* (Monogenea) on brown trout, *Salmo trutta* and salmon, *S. salar* in the river Sandvikselva, Norway. In: *Parasites and diseases in natural waters and aquaculture in Nordic countries. Proceedings of a Symposium, 2–4 December 1986. Stockholm, Sweden* (Eds Stenmark, Ak. and Malmberg, G.). Stockholm, Sweden: Zoo-tax, Naturhistoriska riksmuseet, pp. 70–74.

Mo, T. A, 1989, Parasites of the genus *Gyrodactylus* cause problems in fish culture and fish management. *Norsk Veterinaertidsskr.*, 101, 523–527.

Moczon, T, 1980a, Oxidoreductase histochemistry in the mature stage of *Triaenophorus nodulosus* (Pallas, 1781), and some remarks on the regulation of the respiratory metabolism in the ontogeny of pseudophyllidean cestodes. *Acta Parasit. Pol.*, 27, 359–366.

Moczon, T, 1980b, Oxidoreductase histochemistry in larval stages of pseudophyllidean cestodes. III. The plerocercoid. *Acta Parasit. Pol.*, 27, 367–371.

Mohandas, A, 1983, Platyhelminthes Trematoda. In: *Reproductive biology of invertebrates. Vol. 2. Spermatogenesis and sperm function* (Eds Adiyodi, K. G. and Adiyodi, R. G.). Chichester, UK: John Wiley and Sons Ltd., pp. 105–129.

Mohsin, S. and Farooqi, H. U, 1979, Histochemical observations on *Pallisentis ophiocephali* (Thapar, 1930). *Geobis, India*, 6, 60–63.

Mokhayer, B, 1981, [Parasites of fish in the Sefid Rud Basin.] *J. Vet. Faculty, University of Tehran*, 36 (4), 61–75.

Mokhtar-Maamouri, F, 1976, Étude ultrastructurale de la gamétogénèse et des premiers stades du développement de deux cestodes Tetraphyllidea. *Thesis, Univ. Sci. Tech. Languedoc, Montpellier, France*, xvi + v + 224 pp.

Mokhtar-Maamouri, F, 1979, Étude en microscopie électronique de la spermiogénèse et du spermatozoïde de *Phyllobothrium gracile* Wedl, 1855 (Cestoda, Tetraphyllidea, Phyllobothriidae). *Z. ParasitKde*, 59, 245–258.

Mokhtar-Maamouri, F, 1982, Étude en microscopie électronique de la spermiogénèse de *Acanthobothrium filicolle* var. *filicolle* Zschokke, 1888. *Annls Parasit. Hum. Comp.*, 57, 429–442.

Mokhtar-Maamouri, F. and Swiderski, Z, 1975, étude en microscopie électronique de la spermiogénèse de deux cestodes *Acanthobrium filicolle benedenii* Loennberg, 1889 et *Onchobothrium uncinatum* (Rud., 1819) (Tetraphyllidea, Oncobothriidae). *Z. ParasitKde*, 47, 269–281.

Mokhtar-Maamouri, F. and Swiderski, Z, 1976, Ultrastructure du spermatozoïde d'un cestode Tetraphyllidea Phyllobothriidae: *Echereibothrium beauchampi*, Euzet, 1959). *Annls Parasit. Hum. Comp.*, 51, 673–674.

Möller, H, 1978, The effects of salinity and temperature on the development and survival of fish parasites. *J. Fish Biol.*, 12, 311–323.

Möller, H, 1986, Pollution and parasitism on the aquatic environment. In: *Parasitology—Quo Vadit? Brisbane, Australia, Aug.* 24–29, 1986 (Ed. Howell, M. J.). Canberra, Australia: Australian Academy of Science, pp. 353–361.

Möller, H. and Anders, K, 1986, *Diseases and parasites of marine fishes*. Kiel, FRG: Verlag Moller, 365 pp.

Möller, H, and Schröder, S, 1987, New aspects of anisakiasis in Germany. *Arch. Lebensmittelhyg.*, 38, 123–128.

Molnár, K, 1966a, Life-history of *Philometra ovata* (Zeder, 1803) and *Ph. rischta* Skrjabin, 1917. *Acta Vet. Acad. Sci. Hung.*, 16, 227–242.

Molnár, K, 1966b, On some little-known and new species of the genera *Philometra* and *Skrjabillanus* from fishes in Hungary. *Acta Vet. Acad. Sci. Hung.*, 16, 143–158.

Molnár, K, 1967, Morphology and development of *Philometra abdominalis* Nybelin, 1928. *Acta Vet. Acad. Sci. Hung.*, 17, 293–300.

Molnár, K, 1970a, Pontyfelek bothriocephalosisa Magyaroszägon. *Magyar Allatorvosok Kapja*, 25, 606–608.

Molnár, K, 1970b, An attempt to treat fish bothriocephalosis with Devermin. Toxicity for the host and anti-parasitic effect. *Acta Veterinaria Academiae Scientiarum Hungaricae*, 20, 325–331.

Molnár, K, 1976, Data on the developmental cycle of *Philometra obturans* (Prenant, 1886) (Nematoda: Philometridae). *Acta Vet. Acad. Sci. Hung.*, 26, 183–188.

Molnár, K, 1980, A histological study on ancylodiscoidosis in the sheatfish (*Silurus glanis*). *Helminthologia*, 17, 117–126.

Molnár, K, 1985, Az angolna parazitás betegségei. *Halászat*, 31, 180–181.

Molnár, K, 1986, Solving parasite-related problems in cultured freshwater fish. In: *Parasitology—Quo Vadit? Proceedings of the sixth International Congress of Parasitology* (Ed. Howell, M. J.). Australia, Canberra: Australian Academy of Science, pp. 319–326.

Molnár, K. and Berczi, I, 1965, Flachweis von parasiten spezifischen Antikörpern in Fischblut mittels der Agar-Gel-Präzipitationsprobe. *Z. Imm. -Allerg.*, 129, 263–267.

Molnár, K. and Fernando, C. H, 1975, Morphology and development of *Philometra cylindracea* (Ward and Magath, 1916) (Nematoda: Philometridae). *J. Helminth.*, 49, 19–24.

Molnár, K. and Mossalam, I, 1985, Monogenean parasites from fishes of the Nile in Egypt. *Parasit. Hung.*, **18**, 5–9.

Molnár, K., Bakos, J. and Krasznai, Z, 1984, Parasites of hybrid fishes. *Parasit. Hung.*, 17, 29–34.

Molnár, K., Chan, G. L. and Fernando, C. H, 1982, Some remarks on the occurrence and development of philometrid nematodes infecting the white sucker, *Catostomus commersoni* Lacépède (Pisces: Catostomidae), in Ontario. *Can. J. Zool.*, **60**, 443–451.

Moody, J. and Gaten, E, 1982, The population dynamics of eyeflukes *Diplostomum spathaceum* and *Tylodelphys clavata* (Digenea: Diplostomatidae) in rainbow and brown trout in Rutland Water: 1974–1978. *Hydrobiologia*, **88**, 207–209.

Moorthy, V. N, 1938, Observations on the life history of *Camallanus sweeti*. *J. Parasit.*, 24, 323–342.

Moravec, F, 1969a, Observations on the development of *Camallanus lacustris* (Zoega, 1776) (Nematoda: Camallanidae). *Vest. Cesk. Spol. Zool.*, **33**, 15–33.

Moravec, F, 1969b, On the early development of *Bunodera luciopercae* (Müller, 1776) (Trematoda: Bunoderidae). *Vest. Cesk. Spol. Zool.*, **33**, 229–237.

Moravec, F, 1970a, Studies on the development of *Raphidascaris acus* (Bloch, 1779) (Nematoda: Heterocheilidae). *Vest. Cesk. Spol. Zool.*, 34, 33–49.

Moravec, F, 1970b, On the life-history of the nematode *Raphidascaris acus* (Bloch, 1779) in the natural environment of the River Bystrice, Czechoslovakia. *J. Fish Biol.*, 2, 313–322.

Moravec, F, 1971a, On the problem of host specificity, reservoir parasitism and secondary invasions of *Camallanus lacustris* (Nematoda: Camallanidae). *Helminthologia*, **10**, 107–114.

Moravec, F, 1971b, Some notes on the larval stages of *Camallanus truncatus* (Rudolphi, 1814) and *Camallanus lacustris* (Zoega, 1776) (Nematoda: Camallanidae). *Helminthologia*, **10**, 129–135.

Moravec, F, 1971c, Studies on the development of the nematode *Cystidicoloides tenuissima* (Zeder, 1800). *Vest. Cesk. Spol. Zool.*, 35, 43–55.

Moravec, F, 1971d, On the life history of the nematode *Cystidicoloides tenuissima* (Zeder, 1800) in the River Bystrice, Czechoslovakia. *Folia Parasit.*, **18**, 107–112.

Moravec, F, 1971e, Nematodes of fishes in Czechoslovakia. *Acta Sci. Nat. Brno*, 5, 1–49.

Moravec, F, 1972, Studies on the development of the nematode *Rhabdochona (Filochona) ergensi* Moravec, 1968. *Folia Parasit.*, 19, 321–333.

Moravec, F, 1974a, The development of *Paracamallanus cyathopharynx* (Baylis, 1923) (Nematoda: Camallanidae). *Folia Parasit.*, 21, 333–343.

Moravec, F, 1974b, Some remarks on the development of *Paraquimperia tenerrima* Linstow, 1878 (Nematoda: Quimperiidae). *Scr. Fac. Sci. Nat. Univ. Purk. Brun., Biol. 5*, 4, 135–142.

Moravec, F, 1975, The development of *Procamallanus laeviconchus* (Wedl, 1862) (Nematoda: Camallanidae). *Vest. Cesk. Spol. Zool.*, 39, 23–38.

Moravec, F, 1976a, Occurrence of the encysted larvae of *Cucullanus truttae* (Fabricius, 1794) in the brook lamprey, *Lampetra planeri* (B1.). *Scr. Fac. Sci. Nat. Univ. Purk. Brun., Biol. 1*, 6, 17–20.

Moravec, F, 1976b, Observations on the development of *Rhabdochona phoxini* Moravec, 1968 (Nematoda: Rhabdochonidae). *Folia Parasit.*, 23, 309–320.

Moravec, F, 1977a, Life history of the nematode *Rhabdochona phoxini* Moravec, 1968 in the Rokytka Brook, Czechoslovakia. *Folia Parasit.*, 24, 97–105.

Moravec, F, 1977b, The life history of the nematode *Philometra abdominalis* in the Rokytka Brook, Czechoslovakia. *Vest. Cesk. Spol. Zool.*, 41, 114–120.

Moravec, F, 1977c, The development of the nematode *Philometra abdominalis* Rybelin, 1928 in the intermediate host. *Folia Parasit.*, 24, 237–245.

Moravec, F, 1978a, Redescription of the nematode *Philometra obturans* (Prenant, 1886) with a key to the philometrid nematodes parasitic in European freshwater fishes. *Folia Parasit.*, 25, 115–124.

Moravec, F, 1978b, The development of the nematode *Philometra obturans* (Prenant, 1886) in the intermediate host. *Folia Parasit.*, 25, 303–315.

Moravec, F, 1979a, Observations on the development of *Cucullanus (Truttaedacnitis) truttae* (Fabricius, 1794) (Nematoda: Cucullanidae). *Folia Parasit.*, 26, 295–307.

Moravec, F, 1979b, Redescription of the nematode *Spinitectus inermis* parasitic in eels, *Anguilla anguilla*, of Europe. *Vest. Cesk. Spol. Zool.*, 43, 35–42.

Moravec, F, 1979c, Occurrence of the endoparasitic helminths in pike (*Esox lucius* L.) from the Mácha Lake fishpond system. *Vest. Cesk. Spol. Zool.*, 43, 174–193.

Moravec, F, 1980a, Development of the nematode *Philometra ovata* (Zeder, 1803) in the copepod intermediate host. *Folia Parasit.*, 27, 29–37.

Moravec, F, 1980b, Biology of *Cucullanus truttae* (Nematoda) in a trout stream. *Folia Parasit.*, 27, 217–226.

Moravec, F, 1980c, Revision of nematodes of the genus *Capillaria* from European freshwater fishes. *Folia Parasit.*, 27, 309–324.

Moravec, F, 1980d, The lamprey *Lampetra planeri* as a natural intermediate host for the nematode *Raphidascaris acus. Folia Parasit.*, 27, 347–348.

Moravec, F, 1981, The systematic status of *Filaria ephemeridarum* Linstow, 1872, *Folia Parasit.*, 28, 377–379.

Moravec, F, 1982a, A contribution to the bionomics of *Crepidostomum metoecus* (Trematoda: Allocreadiidae). *Vest. Cesk. Spol. Zool.*, 46, 15–24.

Moravec, F, 1982b, Proposal of a new systematic arrangement of nematodes of the family Capillariidae. *Folia Parasit. (Praha)*, 29, 119–132.

Moravec, F, 1983a, Observations on the bionomy of the nematode *Pseudocapillaria brevispicula* (Linstow, 1873). *Folia Parasit.*, 30, 229–241.

Moravec, F, 1983b, Some remarks on the biology of *Capillaria pterophylli* Heinze, 1933. *Folia Parasit.*, 30, 129–130.

Moravec, F, 1984a, General aspects of the bionomics of the parasitic nematodes of freshwater fishes. *Prague, Czechoslovakia: Academica Studie VSAV*, 4, 114 pp [In Czech.].

Moravec, F, 1984b, Seasonal occurrence and maturation of *Neoechinorhynchus rutili* (Müller, 1780) (Acanthocephala) in carp (*Cyprinus carpio* L.) of the Macha Lake fishpond system, Czechoslovakia. *Helminthologia*, 21, 55–65.

Moravec, F, 1984c, Occurrence of endoparasitic helminths in carp (*Cyprinus carpio*) from the Mácha Lake fishpond system. *Vest. Cesk. Spol. Zool.*, 48, 261–278.

Moravec, F, 1985a, Occurrence of endoparasitic helminths in eels (*Anquilla anguilla* (L.)) from the Mácha Lake fishpond system, Czechoslovakia. *Folia Parasit.*, 32, 113–125.

Moravec, F, 1985b, Occurrence of the endoparasitic helminths in tench (*Tinca tinca*) from the Mácha Lake fishpond system. *Vest. Cesk. Spol. Zool.*, 49, 32–50.

Moravec, F, 1986, Occurrence of endohelminths in three species of cyprinids (*Abramis brama, Rutilus rutilus* and *Scardinius erythrophthalmus*) of the Mácha Lake fishpond system, Czechoslovakia. *Vest. Cesk. Spol. Zool.*, 50, 49–69.

Moravec, F, 1987, Revision of capillariid nematodes (sub family Capillariinae) parasitic in fishes. *Studie CSAV 3. Praha, Czechoslovakia: Ceskoslovenská Akademie Ved*, 7–141.

Moravec, F. and Amin, A, 1978, Some helminth parasites, excluding Monogenea, from fishes of Afghanistan. *Acta Scientiarum Naturalium Academiae Scientiarum Bohemoslovacae, Brno*, 12, 1–45.

Moravec, F. and Arai, H. P, 1971, The North and Central American species of *Rhabdochona* Railliet, 1916 (Nematoda: Rhabdochonidae) of fishes, including *Rhabdochona canadiensis* sp. nov. *J. Fish. Res. Bd Can.*, 28, 1645–1662.

Moravec, F. and De, N. C, 1982, Some new data on the bionomics of *Cystidicoloides tenuissima* (Nematoda: Cystidicolidae). *Vest. Cesk. Spol. Zool.*, 56, 100–108.

Moravec, F. and Dyková, I, 1978, On the biology of the nematode *Philometra obturans* (Prenant, 1886) in the fishpond system of Mácha Lake, Czechoslovakia. *Folia Parasit.*, 25, 231–240.

Moravec, F. and Gut, J, 1982, Morphology of the nematode *Capillaria pterophylli* Heinze, 1933, a pathogenic parasite of some aquarium fishes. *Folia Parasit.*, 29, 227–231.

Moravec, F. and Libosvarsky, J, 1975, Intestinal helminths from grey mullets, *Mugil capito* Cuv. and *M. cephalus* L., of Lake Borullus, A.R.E. *Folia Parasit.*, 22, 279–281.

Moravec, F. and Malmqvist, B, 1977, Records of *Cucullanus truttae* (Fabricus 1794) (Nematoda: Cucullanidae) from Swedish brook lamprey, *Lampetra planeri* (Bloch). *Folia Parasit.*, 24, 323–329.

Moravec, F. and McDonald, T. E, 1981, *Capillaria margolisi* sp. nov. (Nematoda: Capillariidae) from a marine fish, *Scorpaenichthys marmoratus* (Ayres), from the west coast of Canada, *Can. J. Zool.*, 59, 88–91.

Moravec, F. and Řehulka, J, 1987, First record of the cosmocercoid nematode *Raillietnema synodontisi* Vassiliades, 1973 from the aquarium-reared upside-down catfish *Synodontis eupterus* Boulenger. *Folia Parasit.*, 34, 163–164.

Moravec, F. and Sey, O, 1989, Acanthocephalans of freshwater fishes from North Vietnam, *Vest. Cesk. Spol. Zool.*, 53, 89–106.

Moravec, F. and Taraschewski, H, 1988, Revision of the genus *Anguillicola* Yamaguti, 1935 (Nematoda: Anguillicolidae) of the swimbladder of eels, including descriptions of two new species, *A. novaezelandiae* sp. n. and *A. papernai* sp. n. *Folia Parasit.*, 35, 125–146.

Moravec, F., Ergens, R. and Repová, R, 1984, First record of the nematode *Pseudocapillaria brevispicula* (Linstow, 1873) from aquarium fishes. *Folia Parasit.*, 31, 241–244.

Moravec, F., Gelnar, M. and Řehulka, J, 1987, *Capillostrogyloides ancistri* sp. n. (Nematoda: Capillariidae) a new pathogenic parasite of aquarium fishes in Europe. *Folia Parasite*, 34, 157–161.

Moravec, F., Margolis, L. and Boyce, N. P, 1981, *Some nematodes of the genus Rhabdochona* (Spirurida) from fishes of Japan. *Vest. Cesk. Spol. Zool.*, 45, 277–290.

Moravec, F., Margolis, L. and McDonald, T. E, 1981, Two new species of nematodes of the genus *Capillaria (C. freemani* sp. nov. and *C. parophrysi* sp. nov.) from marine fishes of the Pacific coast of Canada. *Can. J. Zool.*, 59, 81–87.

Mordvinova, T. N. and Parukhin, A. M, 1987, [The helminth fauna of ocean fish from the family Myctophidae in different zones.] *Gidrobiol. Zh.*, 23, 57–60. [In Russian.]

Morishita, K., Komiya, Y. and Matsubayashi, H. (Eds), 1964, *Progress of Medical Parasitology in Japan.- Vol. I.* Meguro Parasitological Museum: Tokyo. 736pp.

Mozgovoi, A. A., Shakhmatova, V. I. and Semenova, M. K, 1971, [Life-cycle of *Goezia ascaroides* (Ascaridata: Goeziidae), nematodes of freshwater fish.] *Sb. Rabot. Gel'm. pos. 90-let. dnya rozh. Akad. K. I. Shryabina.* Moscow: Izdat. KOLOS, 259–265. [In Russian.]

Mpoane, M. and Rinne, J. N, 1984, Helminths of apache (*Salmo apache*), gila (*S. gilae*) and brown (*S. trutta*) trouts. *Southwestern Nat.*, 29, 505–506.

Mudry, D. R. and Anderson, R. S, 1977, Helminth and arthropod parasites of fishes in the mountain national parks of Canada. *J. Fish Biol.*, 11, 21–33.

Mudry, D. R. and Arai, H. P, 1973, The life cycle of *Hunterella nodulosa* Mackiewicz and McCrae, 1962 (Cestoidea: Caryophyllaeidae). *Can. J. Zool.*, 51, 781–786.

Mudry, D. R. and Dailey, M. D, 1969, *Phlyctainophora squali* sp. nov. (Nematoda, Philometridae) from the spiny dogfish, *Squalis acanthias. Proc. Helminth. Soc. Wash.*, 36, 280–284.

Mudry, D. R. and Dailey, M. D, 1971, Postembryonic development of certain tetraphyllidean and trypanorhynchan cestodes with a possible alternative life cycle for the order Trypanorhyncha. *Can. J. Zool.*, 49, 1249–1253.

Mueller, J. F. and Van Cleave, H. J, 1932, Parasites of Oneida Lake fishes. Part II. Descriptions of new species and some general taxonomic considerations, especially concerning the trematode family Heterophyidae. *Roosevelt Wild Life Annals*, 3, 154 pp.

Müller, M, 1975, Ein Seltener Fund: *Bucephalus*-Cercarien. *Mikrokosmos*, 64, 179–183.

Munday, B. L, 1986, Diseases of salmonids. I. In: *Diseases of Australian fish and shellfish. Proceedings of the first Australian workshop on diseases of fish and shellfish, Benalla, Victoria, Australia, 27–30 May 1985* (Eds Humphrey, J. D. and Langdon, J. S.), Benella, Victoria, Australia: Australian Fish Health Reference Laboratory 1986, pp. 127–141.

Munro, A. L. S., McVicar, A. H. and Jones, R, 1983, The epidemiology of infectious disease in commercially important wild marine fish. *Rapport of Proces-Verbaux des Réunions Conseil Permanent International pour l'Exploration de la Mer.*, 182, 21–32.

Munro, M. A., Whitfield, P. J. and Diffley, R, 1989, *Pomphorhynchus laevis* (Müller) in the flounder, *Platichthys flesus* L., in the tidal River Thames: population structure, microhabitat utilization and reproductive status in the field and under conditions of controlled salinity. *J. Fish Biol.*, 35, 719–735.

Munroe, T. A., Campbell, R. A. and Zwerner, D. E, 1981, *Diclidophora nezumiae* sp. n. (Monogenea: Diclidophoridae) and its ecological relationships with the macrourid fish *Nezumia bairdii* (Goode and Bean, 1877). *Biol. Bull.*, 161, 281–290.

Muratov, I. V, 1983, [Epidemiology of diphyllobothriasis in the eastern part of the Baikal-Amur Mainline Railway.] In: *Prirodnoochagovye infektsii v rayonakh narodnokhozyaistvenogo osvoeniya Sibiri i Dal'nego Vostoke. (Respublikanskii Sbornik Nauchnykh Trudov).* (Ed. Subbotina, L. S.). Omsk, USSR: Omskii Foausearcwnnyi Mwsirainakii Inarirur im. M. I. Kalinina, pp. 158–166. [In Russian.]

Musselius, V. A, 1984, [Diseases in mariculture and their prophylaxis.] In: *Biologicheskie osnovy rybovodstva: parazity i bolesni ryb. (Biologicheskiej resursy gidrosfery i ikh ispol'zovanie)* (Eds Bauer, O. N., Musselius, V. A. and Skryabina, E. S.). Moscow, USSR: 'Nauka', pp. 170–179. [In Russian.]

Musselius, V. A. and Golovin, P. P, 1984, [Contemporary methods in commercial fisheries and problems in the control of fish diseases.] In: *Biologicheskie osnovy rybovodstva: parazity i bolezni ryb.*

(Biologicheskie resursy gidrosfery i ikh ispol'zovanie.) (Eds Bauer, O. N., Musselius, V. A. and Skryabina, E. S.). Moscow, USSR: 'Nauka', pp. 20–28. [In Russian.]

Musselius, V. A. and Ptasuk, S. V, 1970, [On the development and specificity of *Dactylogyrus lamellatus* (Monogenoidea, Dactylogyridae).] *Parazitologiya*, 4, 125–132. [In Russian.]

Musselius, V., Ivanova, N., Laptev, V. and Apazidi, L, 1963, [*Khawia sinensis* Hsü, 1935 of carp.] *Ribovodstvo i Ribolovstvo, No. 3*, 25–27. [In Russian.]

Muzykovskii, A. M, 1972, [Cestode larvae in the muscles of sword-fish.] *Rybnoe Khozyaistvo, Moscow, No. 12*, p. 33. [In Russian.]

Muzykovskii, A. M, 1977, [Trials of anthelmintic premixes against *Bothriocephalus* infection in the carp.] *Byull. Vses Inst. Gel'm. K. I. Skryabina, No. 20*, 37–38. [In Russian.]

Muzikovskii, A. M. and Vasil'kov, G. V, 1969, Worming of carp against *Bothriocephalus. Veterinariya, Moscow*, 46, 55–56.

Muzykovskii, A. M., Vasil'kov, G. V. and Borisova, M. N, 1977, Tests with dibutyl tin oxide against *Bothriocephalus gowkongensis* in carp. *Byull. Vses. Inst. Gel'm. K. I. Skryabina, No. 21*, 39–41. [In Russian.]

Muzzall, P. M, 1980a, Ecology and seasonal abundance of three acanthocephalan species infecting white suckers in SE New Hampshire. *J. Parasit.*, 66, 127–133.

Muzzall, P. M, 1980b, Population biology and host–parasite relationships of *Triganodistomum attenuatum* (Trematoda: Lissorchiidae) infecting the white sucker, *Catostomus commersoni* (Lacépède). *J. Parasit.*, 66, 293–298.

Muzzall, P. M, 1980c, Seasonal distribution and ecology of three caryophyllaeid cestode species infecting white suckers in SE New Hampshire. *J. Parasit.*, 66, 542–550.

Muzzall, P. M, 1982, Comparison of the parasite communities of the white sucker (*Catostomus commersoni*) from two rivers in New Hampshire. *J. Parasit.*, 68, 300–305.

Muzzall, P. M, 1984a, Observations on two acanthocephalan species infecting the central mudminnow, *Umbra limi* in a Michigan river. *Proc. Helminth. Soc. Wash.*, 51, 92–97.

Muzzall, P. M, 1984b, Parasites of trout from four lotic localities in Michigan. *Proc. Helminth. Soc. Wash.*, 51, 261-266.

Muzzall, P. M, 1986, Parasites of trout from the Sable River, Michigan, with emphasis on the population biology of *Cystidicoloides tenuissima*. *Can. J. Zool.*, 64, 1549–1554.

Muzzall, P. M. and Bullock, W. L, 1978, Seasonal occurrence and host–parasite relationships of *Neoechinorhynchus saginatus* Van Cleave and Bangham, 1949 in the fallfish *Semotilus corporalis* (Mitchill). *J. Parasit.*, 64, 860–865.

Muzzall, P. M. and Peebles, C. R, 1986, Helminths of pink salmon, *Oncorhynchus gorbuscha* from five tributaries of Lake Superior and Lake Huron. *Can. J. Zool.*, 64, 508–511.

Muzzall, P. M. and Rabalais, F. C, 1975, Studies on *Acanthocephalus jacksoni* Bullock, 1962 (Acanthocephala: Echinorhynchidae). I. Seasonal periodicity and new host records. *Proc. Helminth. Soc. Wash.*, 42, 31–34.

Muzzall, P. M. and Sweet, R. D, 1986, Parasites of mottled sculpins, *Cottus bairdi*, from the Au Sable River, Crawford County, Michigan. *Proc. Helminth. Soc. Wash.*, 53, 142–143.

Muzzall, P. M., Whelan, G. E. and Peebles, C. R, 1987, Parasites of burbot, *Lota lota* (Family Gadidae), from the Ford River in the Upper Peninsula of Michigan. *Can. J. Zool.*, 65, 2825–2827.

Myers, B. J, 1960, On the morphology and life history of *Phocanema decipiens* (Krabbe, 1878) Myers 1959 (Nematoda: Anisakidae). *Can. J. Zool.*, 38, 331–344.

Myers, B. J, 1970, Nematodes transmitted to man by fish and aquatic mammals. *J. Wildl. Dis.*, 6, 266–271.

Myers, B. J, 1979, Anisakine nematodes in fresh commercial fish from waters along the Washington, Oregon and California coasts. *J. Food Prot.*, 42, 380–384.

Nagasawa, K, 1985, Prevalence of visceral adhesions in sockeye salmon, *Oncorhynchus nerka*, in the Central North Pacific Ocean. *Fish Path.*, 20, 313–321.

Nagasawa, K, 1986, Nippotaenia. *Newsletter of the Ichthyoparasitological group of Japan*, 1, 1–12.

Nagasawa, K, 1987, Nippotaenia. *Newsletter of the Ichthyoparasitological group of Japan*, 2, 1–9.

Nagasawa, K, 1988, Nippotaenia. *Newsletter of the Ichthyoparasitological group of Japan*, 3, 1–17.

Nagasawa, K, 1989, Nippotaenia. *Newsletter of the Ichthyoparasitological group of Japan*, 4, 1–19.

Nagasawa, K., Egusa, S., Hara, T. and Yagisawa, I, 1983, [Ecological factors influencing the infection levels of salmonids by *Acanthocephalus opsariichthydis* (Acanthocephala: Echinorhynchidae) in Lake Yunoko, Japan.] *Fish Path.*, 18, 53–60. [In Japanese.]

Nagaty, H. F, 1937, Trematodes of fishes from the Red Sea. Part I. Studies on the family Bucephalidae Poche, 1907. *The Egyptian University, Faculty of Medicine, Publ. No. 12*, 172 pp.

Nahhas, F. M. and Cable, R. M, 1964, Digenetic and aspidogastrid trematodes from marine fishes of Curacao and Jamaica. *Tulane Stud. Zool.*, 11, 169–228.

Nakajima, K, 1976, [Synonym of *Philometroides cyprini* (Ishii).]. *Fish Path.* 11, 97–99. [In Japanese.]

Nakajima, K. and Egusa, S, 1972a, [Studies on a new trypanorhynchan larva, *Callotetrarhynchus* sp., parasitic on cultured yellowtail. X. Movement of bladder worms and procercus in the gut of *Triakis scyllia*.] *Bull. Jap. Soc. Sci. Fish.*, 38, 939–944. [In Japanese.]

Nakajima, K. and Egusa, S, 1972b, [Studies on a new trypanorhynchan larva, *Callotetrarhynchus* sp., parasitic on cultured yellowtail. XI. Growth of the adult in the valvular intestine of *Triakis scyllia*.] *Bull. Jap. Soc. Sci. Fish.*, 38, 945–954. [In Japanese.]

Nakajima, K. and Egusa, S, 1973, [Studies on a new trypanorhynchan larva, *Callotetrarhynchus* sp., parasitic on cultured yellowtail. XIII. Morphology of the adult and its taxonomy.] *Bull. Jap. Soc. Sci. Fish.*, 39, 149–158. [In Japanese.]

Nakajima, K. and Egusa, S, 1974a, [*Bothriocephalus opsariichthydis* Yamaguti (Cestoda: Pseudophyllidea) found in the gut of cultured carp, *Cyprinus carpio* (Linné)—I. Morphology and taxonomy.] *Fish Path.*, 9, 31–39. [In Japanese.]

Nakajima, K. and Egusa, S, 1974b, [*Bothriocephalus opsariichthydis* Yamaguti (Cestoda: Pseudophyllidea) found in the gut of cultured carp, *Cyprinus carpio* (Linné)—II. Incidence and histopathology.] *Fish Path.*, 9, 40–45. [In Japanese.]

Nakajima, K. and Egusa, S, 1974c, [*Bothriocephalus opsariichthydis* Yamaguti (Cestoda: Pseudophyllidea) found in the gut of cultured carp, *Cyprinus carpio* (Linné)—III. Anthelmintic effects of some chemicals.] *Fish Path.*, 9, 46–49. [In Japanese.]

Nakajima, K. and Egusa, S, 1977a, [Studies on the philometrosis of crucian carp. IV. Invasion and growth of larvae in cyclops.] *Fish Path.*, 12, 191–197. [In Japanese.]

Nakajima, K. and Egusa, S, 1977b, [*Bothriocephalus opsariichthydis* Yamaguti (Cestoda: Pseudophyllidea) found in the gut of cultured carp, *Cyprinus carpio* (Linné). VI. Some ecological changes of the adult worm wintered within the host.] *Fish Path.*, 12, 1–2. [In Japanese.]

Nakajima, K. and Egusa, S, 1977c, [Studies on the philometrosis of crucian carp, 2: some ecological observations of gravid female.] *Fish Path.*, 12, 115–120. [In Japanese.]

Nakajima, K. and Egusa, S, 1977d, [Studies on philometrosis of crucian carp. III. Activity and resistance of the first-stage larvae.] *Fish Path.*, 12, 185–189. [In Japanese.]

Nakajima, K. and Egusa, S, 1979, [*Philometra* sp. found on the gonad of cultured red sea bream.] *Fish Path.*, 13, 197–200. [In Japanese.]

Nakajima, K. and Syuzo, E, 1969, Studies on a new trypanorhynchan larva, *Callotetrarhynchus* sp. parasitic on cultured yellow tail. II. On the source and route of infection. *Bull. Jap. Soc. Sci. Fish.*, 35, 351–357.

Nakatsugawa, T. and Muroga, K, 1977, [Seasonal changes in abundance of *Dactylogyrus extensus* and *D. minutus* (Monogenea) on cultured carp.] *Fish Path.*, 12, 99–104. [In Japanese.]

Nascetti, G., Paggi, L., Orecchia, P., Smith, J. W., Mattiucci, S. and Bullini, L, 1986, Electrophoretic studies on the *Anisakis simplex* complex (Ascaridida: Anisakidae) from the Mediterranean and North-East Atlantic. *Int. J. Parasit.*, 16, 633–640.

Nasir, P. and Fuentes-Zambrano, J. L, 1983, Algunos trematodos monogeneticos Venezolanos. *Riv. Parassit.*, 44, 335–380.

Nasir, P. and Gomez, Y, 1977a, *Carassotrema tilapiae* n. sp. (Haploporidae Nicoll, 1914) from the freshwater fish, *Tilapia mossambica* (Peters), in Venezuela. *Riv. Parassit.*, 37, 207–228.

Nasir, P. and Gomez, Y, 1977b, Digenetic trematodes from Venezuelan marine fishes. *Riv. Parassit.*, 38, 53–73.

Nedeva, I, 1988, On the biology of *Bothriocephalus acheilognathi* Yamaguti, 1934 (Bothriocephalidae). *Khelmintologiya*, 26, 32–38. [In Bulgarian.]

Needham, T. and Wootten, R, 1978, The parasitology of teleosts. In: *Fish Pathology* [Ed. Roberts, R. J.). London, UK: Baillière Tindall, pp. 144–182.

Nellaiappan, K. and Ramalingham, K, 1980a, Prophenoloxidase and its activation in *Paraplerurus sauridae* (Digenea: Hemiuridae). *Mol. Biochem. Parasit.*, 2, 109–112.

Nellaiappan, K. and Ramalingham, K, 1980b, Stabilization of egg-shell in *Paraplerurus sauridae* (Digenea: Hemiuridae). *Parasitology*, 80, 1–7.

Nelson, J. S, 1976, *Fishes of the World.* New York: John Wiley and Sons, 416 pp.

Nicholas, W. L, 1967, The biology of the Acanthocephala. *Adv. Parasit.*, 5, 205–246.

Nicholas, W. L, 1973, The biology of the Acanthocephala. *Adv. Parasit.*, 11, 671–706.

Nichols, K. C, 1975, Observations on lesser-known flatworms: *Udonella. Int. J. Parasit.*, 5, 475–482.

Nickle, W. R, 1980, Possible commercial formulations of insect-parasitic nematodes. *Biotechnol. Bioeng.*, 22, 1407–1414.

Nickol, B. B, 1985, Epizootiology. In: *Biology of the Acanthocephala* (Eds Crompton, D. W. T. and Nickol, B. B.). Cambridge: Cambridge University Press, pp. 307–346.

Nickol, B. B. and Samuel, N, 1983, Geographical distribution of Acanthocephala. *Trans. Nebraska Acad. Sci.*, 11, 31–52.

Niemczuk, W, 1984, Wystepowanie tasiemca *Khawia sinensis* u karpi na terenie dzialainosci Wroclawskiej Pracowni Chorób Ryb. *Gospodarka Rybna*, 36, 14–15.

Niewiadomska, K, 1977, Pasozyty wylegu i narybku niektórych gatunków ryb z jezior Koninskich. *Roczn. Nauk Roln. H.*, 97, 45–59.

Niewiadomska, K, 1984, Present status of *Diplostomum spathaceum* (Rudolphi, 1819) and differentiation of *Diplostomum pseudospathaceum* nom. nov. (Trematoda: Diplostomatidae). *Syst. Parasit.*, 6, 81–86.

Niewiadomska, K. and Moczon, T, 1982, The nervous system of *Diplostomum pseudopathaceum* Niewiadomska, (Digenea: Diplostomatidae). I. Nervous system and chaetotaxy in the cercaria. *Z. ParasitKde*, 68, 295–304.

Niewiadomska, K. and Moczon, T, 1984, The nervous system of *Diplostomum pseudospathaceum* Niewiadomska 1984 (Trematoda, Diplostomatidae). II. Structure and development of the nervous system in metacercaria. *Z. ParasitKde*, 70, 537–548.

Nigrelli, R. F, 1937, Further studies on the susceptibility and acquired immunity of marine fishes to *Epibdella melleni*, a monogenetic trematode. *Zoologica, N.Y.*, 22, 185–192.

Nigrelli, R. F, 1943, Causes of diseases and death of fishes in captivity. *Zoologica*, N.Y., 28, 203–216.

Nigrelli, R. F. and Breder, C. M, 1934, The suceptibility and immunity of certain marine fishes to *Epibdella melleni*, a monogenetic trematode. *J. Parasit.*, 20, 259–269.

Nikitina, E. N, 1983, [Helminth fauna of fish in the Bay of Krasnovodsk. I. The helminths of *Atherina mochon pontica* natio *caspia*.] In: *Biologicheskie resursy Kaspiiskogo Morya* (Eds Gilyarova, M. S. and Zevina, G. B.). Moscow, USSR: Izdat Mosk. Univ., pp. 147–162. [In Russian.]

Nikolaeva, V. M, 1963, [The parasite fauna of local shoals of certain pelagic fish of the Black Sea.] *Trudy sevastopol' biol. Sta*, 16, 387–438. [In Russian.]

Nikolaeva, V. M, 1985, Trematodes—Didymozoidae fauna, distribution and biology. In: *Parasitology and pathology of marine organisms of the World Ocean* (Ed. Hargis, W. J. Jr.). Seattle, USA: US Department of Commerce, National Oceanic and Atmospheric Administration, pp. 67–72.

Nikolaeva, V. M. and Dubina, V. R, 1978, New species of didymozoids from Indian Ocean fish. *Biologiya Morya, Kiev (Parazitofauna zhivotnykh Yuzhnykh morei), Issue 45*, 71–90. [In Russian.]

Nikolaeva, V. M. and Dubina, V. R, 1985, [On the Didymozoidae of fish in the western Indian Ocean.] *Ekologiya Morya, Kiev, No. 20*, 13–26. [In Russian.]

Nikolaeva, V. M. and Tkachuk, L. P, 1979, A new trematode genus in the family Didymozoidea from mackerel in the Indian Ocean. *Parazitologiya*, 13, 552–555. [In Russian.]

Nikolaeva, V. M. and Tkachuk, L. P, 1982, [Didymozoidae in mackerels in the Indian Ocean.] *Ekologiya Morya, No. 10*, 44–49. [In Russian.]

Nikolaeva, V. M., Gavrilyuk, L. P. and Shchenkina, A. M. 1976. [Survival of Barents Sea nematodes in fish during different processing methods.] *Biologiya Morya Kiev (Voprosy èkologii ryb i Kal'marov), No. 38*, 85–90. [In Russian.]

Nikolsky, G. V, 1963, *The ecology of fishes.* Translated from the Russian by L. Birkett. London: Academic Press, xv + 352 pp.

Nikulyna, V. N. and Kirillova, L. P, 1975, Fish cestodes in the natural waters of the Alta territory. *Nauchyne Omskogo Veterinarnogo Instituta (Gel'mintozy domashnikh zhivotnykh zapadnoi Sibiri i mery bor'by s nimi)* 31 81–84.

Nilz, J, 1984, Parasitologische Untersuchungen an Salmoniden in einer Forellenteichwirtschaft in Südniedersachsen. *Inaug. Diss., Tier. Hochschule, Hannover, GFR*, 118 pp.

Niyogi, A. and Agarwal, S. M, 1983, Free and protein amino acids in *Lytocestus indicus, Introvertus raipurensis* and *Lucknowia indica* parasitizing *Clarias batrachus* (Linn.). (Cestoda: Caryophyllidea). *Jap. J. Parasit.*, 32, 341–345.

Niyogi, A., Gaur, A. S. and Agarwal, S. M, 1985, Protein profiles as an aid to taxonomy among caryophyllidean cestodes. *Curr. Sci.*, 54, 277–278.

Niyogi, A., Gupta, A. K. and Agarwal, S. M, 1982, Population dynamics of caryophyllids parasitizing *Clarias batrachus* at Raipur. *Geobios New Reports* 1, 81–93.

Niyogi, A., Gupta, A. K., Naik, M. L. and Agarwal, S. M, 1984, Frequency distribution of caryophyllaeids parasitizing *Clarias batrachus* at Raipur *Proc. Ind. Acad. Parasit.*, 5, 5–9.

Nizami, W. A. and Siddiqi, A. H, 1976, Qualitative analysis of substances excreted by *Isoparorchis hypselobagri* (Trematoda) during aerobic *in vitro* culture. *Z. ParasitKde*, 50, 53–56.

Nizami, W. A. and Siddiqi, A. H, 1978, Effect of temperature and pH on oxygen consumption of digenetic trematodes. *Indian J. Parasit.*, 2, 171–174.

Nizami, W. A. and Siddiqi, A. H, 1979, Influence of TCA cycle intermediates on oxygen consumption of *Isoparorchis hypselobagri* (Trematoda). *Indian J. Parasit.*, 3, 189–191.

Noble, E. R, 1973, Parasites and fishes in a deep-sea environment. *Adv. Mar. Biol.*, 11, 121–195.

Noble, E. R. and Collard S. B, 1970, The parasites of midwater fishes. In: *Symposium on diseases of fishes and shellfishes* (Ed. Sniesko, F.). Washington: American Fisheries Society, pp. 57–68.

Noble, E. R. and King, R. E, 1960, The ecology of the fish *Gillichthys mirabilis* and one of its nematode parasites. *J. Parasitol.*, 46, 679–685.

Noble, E. R., Noble, G. A., Schad, G. A. and MacInnes, A. J, 1989, *Parasitology: the biology of animal parasites.* 6th edn, Philadelphia, PA, USA: Lea and Febiger, x + 574 pp.

Noisi, D. and Euzet, L, 1979, Microhabitat branchial de deux Microcotylidae (Monogenea) parasites de *Diplodus sargus* (Teleostei, Sparidae). *Rev. Ibér. Parasit.*, 39, 81–93.

Noisi, D. and Maillard, C, 1980, Microhabitat branchial préférentiel de *Microcotyle chrysophrii* van Beneden et Hesse, 1863 (Monogenea, Microcotylidae), parasite de la daurade (*Sparus aurata* L., 1758). *Annls Parasit. Hum. Comp.*, 55, 33–40.

Nollen, P. M, 1983, Patterns of sexual reproduction among parasitic platyhelminths. In: *The reproductive biology of parasites. Symposia of the Brit. Soc. for Parasit. Vol. 20.* (Ed. Whitfield, P. J.). *Parasitology*, 86, 99–120.

Norris, D. E. and Overstreet, R. M, 1976, The public health implications of larval *Thynnascaris* nematodes from shellfish. *J. Milk Fd Technol.*, 39, 47–54.

Novotny, A. J. and Uzmann, J. R, 1960, A statistical analysis of the distribution of a larval nematode (*Anisakis* sp.) in the musculature of chum salmon (*Oncorhynchus keta* Walbaum). *Expl Parasit.*, 10, 245–262.

Nowak, T. and Pietrzak, B., 1971, The effect of kamala in the treatment of carp infected with *Caryophyllaeus laticeps* Pall. *Wiad. Parazyt.*, 17, 411–416.

Nybelin, O, 1922, Anatomisch-systematische Studien veber Pseudophyllidien. *Göteborgs Kgl. Vetenskap-Akad. Handl.*, 26, 1–228.

O'Connor, G. R, 1976, Parasites of the eye and brain. In: *Ecological aspects of parasitology.* (Ed. Kennedy, C. R.). Amsterdam, The Netherlands: North Holland Publishing Company, pp. 327–347.

O'Grady, R. T, 1985, Ontogenetic sequences and the phylogenetics of parasitic flatworm life-cycles. Cladistics, 1, 159–170.

Obgol'ts, A. A, 1983, [On the types of diphyllobothriasis foci in the Taimyr.] In: *Prirod. infekts. rayon. narodnok. osvoen. Sib. Dal' Vostok. (Respublik Sb. Nauch. Trud.)* (Ed. Subbotina, L. S.). Omsk, USSR: OmskiI GosudarstvennyI MeditsinskiI Institut im. M. I. Kalinina, pp. 167–170. [In Russian.]

Obiekezie, A. I., Möller, H. and Anders, K, 1988, Diseases of the African estuarine catfish *Chrysichthys nigrodigitatus* (Lacépède) from the Cross River estuary, Nigeria. *J. Fish Bio.*, 32, 207–221.

Odening, K, 1974, Verwandtschaft, System und zyklo-ontogenetische Besonderheiten der Trematoden. *Zool. Jahrb. Abt. Syst. Ökol. Geogr. Tiere*, 101, 345–396.

Odening, K, 1976, Conception and terminology of hosts in parasitology. *Adv. Parasit.*, 14, 1–93.

Odening, K, 1978, The nature and development of the parthenitae in the trematode genus *Azygia.* Proc. EMOP II, 1–6 September 1975, Trogir, Yugoslavia, Belgrade: Association of Yugoslav Parasitologists. pp. 111–118.

Odening, K, 1989, New trends in parasitic infections of cultured freshwater fish. *Vet. Parasit.*, 32, 73–100.

Odening, K. and Bockhardt, I, 1976, Zum jahreszeitlichen Auftreten von *Azygia lucii* (Trematoda) bei *Esox lucius* (Pisces). *Zool. Anz.*, 196, 182–188.

Oetinger, D. F. and Nickol, B. B, 1974, A possible function of the fibrillar coat in *Acanthocephalus jacksoni* eggs. *J. Parasit.*, 60, 1055–1056.

Oetinger, D. F. and Nickol, B. B, 1982, Developmental relationships between acanthocephalans and altered pigmentation in freshwater isopods. *J. Parasit.*, 68, 463–469.

Ogawa, K, 1984a, [*Pseudodactylogyrus haze* sp. nov., a gill monogenean from the Japanese goby, *Acanthogobius flavimanus.*] *Jap. J. Parasit.*, 33, 403–405. [In Japanese.]

Ogawa, K, 1984b, [Development of *Benedenia hoshinai* (Monogenea) with some notes on its occurrence on the host.] *Bull. Jap. Soc. Sci. Fish.*, 50, 2005–2011. [In Japanese.]

Ogawa, K, 1986, A monogenean parasite *Gyrodactylus masu* sp. n. (Monogenea: Gyrodactylidae) of salmonid fish in Japan. *Bull. Jap. Soc. Sci. Fish.*, 52, 947–950.

Ogawa, K, 1988, Infection of cultured Red Sea bream, *Pagrus major*, with a monogenean, *Bivagina tai* (Yamaguti, 1938). *Bull. Jap. Soc. Sci. Fish.*, 54, 65–70

Ogawa, K. and Egusa, S, 1977, Redescription of *Heteraxine heterocerca* (Monogenea: Heteraxinidae). *Jap. J. Parasit.*, 26, 388–396.

Ogawa, K. and Egusa, S, 1978, Two new species of the genus *Tetraonchus* (Monogenea: Tetraonchidae) from cultured *Oncorhynchus masou*. *Bull. Jap. Soc. Sci. Fish.*, 44, 305–312.

Ogawa, K. and Egusa, S, 1979, Redescription of *Dactylogyrus extensus* (Monogenea: Dactylogyridae) with a special reference to its male terminalia. *Jap. J. Parasit.*, 28, 121–124.

Ogawa, K. and Egusa, S, 1980, Two species of microcotylid monogeneans collected from Black Sea bream, *Acanthopagrus schlegeli* (Bleeker) (Teleostei: Sparidae). *Jap. J. Parasit.*, 29, 455–462.

Ogawa, K. and Egusa, S, 1981, The systematic position of the genus *Anoplodiscus* (Monogenea: Anoplodiscidae). *Syst. Parasit.*, 2, 253–260.

Ogawa, K. and Egusa, S, 1985, *Tetraonchus* infection of masu salmon, *Oncorhynchus masou. Fish Path.*, 19, 215–223.

Ogawa, K. and Hioki, M, 1986, Two new species of *Gyrodactylus* (Monogenea: Gyrodactylidae) of eel, *Anguilla japonica*, with some data on the occurrence of gyrodactylids in greenhouse culture at Yoshida, Shizuoka Prefecture, Japan. *Fish Path.*, 21, 89–94.

Oisboit, M. I. and Yasyuk, V. P, 1980, [Nidi of ligulid infections in fish.] *Veterinariya, Mosk.*, 6, 49–50. [In Russian.]

Ojala, O, 1963, Fish diseases in Finland. *Bull. Off. Int. Epizoot.*, 59, 31–42.

Oliver, G, 1977a, Effet pathogène de la fixation de *Diplectanum aequans* (Wagener, 1857) Diesing, 1858 (Monogenea, Monopisthocotylea, Diplectanidae) sur les branchies de *Dicentrarchus labrax* (Linnaeus, 1758), (Pisces, Serranidae). *Z. ParasitKde*, 53, 7–11.

Oliver, G, 1977b, Biologie et écologie de *Microcotyle labracis* Van Beneden et Hesse, 1863 (Monogenea, Polyopisthocotylea) chez *Dicentrarchus labrax* (Linné, 1758) et *Dicentrarchus punctatus* (Bloch, 1792) (Pisces, Serranidae) des Côtes de France. In: *Excerta parasitologica en memoria del Doctor Eduardo Caballero y Caballero*. Mexico: Universidad Nacional Autonoma de Mexico, pp. 91–98.

Oliver, G, 1978, Déscription de deux cas d'ovoviviparité chez les Diplectanidae Bychowsky, 1957 (Monogenea, Monopisthocotylea). *Z. ParasitKde*, 57, 247–250

Oliver, G, 1982, Quelques aspects de la spécificité parasitaire chez les Diplectanidae Bychowsky, 1957 (Monogenea: Monopisthocotylea). [Deux. Symp. Spéc. Parasit. parasites vertebrés, 13–17 avril, 1981. Coll. Int. CNRS.] *Mems Mus. Natn. Hist. Nat., Nouv. Ser., Ser. A, Zool.*, 123, 295–301.

Oliver, G, 1984, *Microcotyle chrysophrii* Van Beneden et Hesse, 1863 (Monogenea, Polyopisthocotylea, Microcotylidae) parasite de *Sparus aurata* Linnaeus, 1758 (Teleostei, Sparidae) dans les etangs littoraux du Languedoc-Roussillon (France). *Bull. Soc. Zool. France*, 109, 113–118.

Olsen, O. W, 1974, *Animal parasites, their life-cycles and ecology*. Baltimore: University Park Press, 562 pp.

Olson, R. E, 1978, Parasites of silver (Coho) salmon and king (Chinook) salmon from the Pacific Ocean off Oregon. *Calif. Fish Game*, 64, 117–120.

Olson, R. E. and Pratt, I, 1971, The life-cycle and larval development of *Echinorhynchus lageniformis* Ekbaum, 1938 (Acanthocephala: Echinorhynchidae). *J. Parasit.*, 57, 143–149.

Olson, R. E. and Pratt, I, 1973, Parasites as indicators of English sole (*Parophrys vetulus*) nursery grounds. *Trans. Am. Fish. Soc.*, 102, 405–411.

Orecchia, P. and Paggi, L, 1978, [Systematics and ecology of helminth parasites of marine fish studied in the Institute of Parasitology of the University of Rome.] [*IX Congr. Naz. Soc. Ital. Parassit., Ravenna, 30 Mar.–1 Apr. 1978.*] *Parassitologia*, 20, 73–89. [In Italian.]

Orecchia, P., Paggi, L., Castagnolo, L., Della Seta, G. and Minervini, R, 1975, Ricerche sperimentali sul ciclo biologico di *Phyllodistomum elongatum* Nybelin, 1926 (Digenea: Gorgoderidae Looss, 1901). *Parassitologia*, 17, 95–101.

Orecchia, P., Paggi, L., Mattiucci, S., Smith, J. W., Nascetti, G. and Bulini, L, 1986, Electrophoretic identification of larvae and adults of *Anisakis* (Ascaridida: Anisakidae). *J. Helminth.*, 60, 331–339.

Orlowska, K, 1979, Parasites of North Sea spiny dogfish *Squalus acanthias* L. (Selachiiformes, Squalidae). *Acta Ichthyol. Pisc.*, 9, 33–44.

Orr, T. S. C, 1966, Spawning behaviour of rudd, *Scardinius erythrophthalmus* infested with plerocercoids of *Ligula intestinalis*. *Nature, Lond.*, 212, 736.

Orr, T. S. C, 1967, Distribution of the plerocercoid of *Ligula intestinalis*. *J. Zool., Lond.*, 153, 91–97.

Orr, T. S. C, 1968a, Distribution and specificity of the plerocercoid of *Ligula intestinalis* (L.) in the Northamptonshire area. *J. Helminth.*, 13, 117–124.

Orr, T. S. C, 1968b, Anomalous positions of the plerocercoid of *Ligula intestinalis* (Linnaeus, 1758). *J. Helminth.*, 42, 363–366.

Orr, T. S. C. and Hopkins, C. A, 1969, Maintenance of *Schistocephalus solidus* in the laboratory with observations on rate of growth of and proglottid formation in the plerocercoid. *J. Fish. Res. Bd Can.*, 26, 741–752.

Orr, T. S. C., Hopkins, C. A, and Charles, G, H, 1969, Host specificity and rejection of *Schistocephalus solidus*. *Parasitology*, 59, 683–690.

Osetrov, V. S, (Editor) 1978, [*Handbook on the diseases of fish.*] Moscow, USSR: 'KOLOS', 352 pp. [In Russian.]

Oshima, T, 1972, *Anisakis* and anisakiasis in Japan and adjacent area. In: *Progress of medical parasitology in Japan. Vol. 4.* (Eds Morishita, K., Komiya, Y. and Matsubayashi, H.). *Tokyo*, Japan: Meguro Parasitological Museum, pp. 301–393.

Oshima, T, 1984, Anisakiasis, diphyllobothriasis and creeping disease—changing pattern of parasitic diseases in Japan. In: *Current perspectives in parasitic disease. Proceedings of the Southeastern Asian Symposium on parasitology and modern medicine held in Hong Kong, 9–12 December 1983.* (Ed. Ko, R. C.). Hong Kong: Department of Zoology, University of Hong Kong, 93–102.

Oshima, T, 1987, Anisakiasis—is the sushi bar guilty? *Parasitology Today*, 3, 44–48.

Oshima, T. and Kliks, M, 1986, Effects of marine mammal parasites on human health. In: *Parasitology—Quo Vadit? I, Brisbane, Australia, 24–29 August, 1986.* (Ed. Howell, M. J.). Canberra, Australia: Australian Academy of Science, pp. 415–421.

Oshmarin, P. G. and Prokhorova, I. M, 1978, [The biological significance of some types of reproduction in cestodes.] *Mater. Nauch. Konf. Vses. Obshch. Gel'm. (Biol. osnov. bor'by gel'm. cheloveka zhivot., No. 30,* 117–125. [In Russian.]

Osipov, A. S, 1984, [The parasite fauna of Siberian cisco in different areas of the northern Tyumen region.] *Sb. Nauch. Trud. Gosud. Nauchno-Issled. Inst. Ozern. Rech. Ryb. Khozyaist. (Bolez. paraz. ryb vod. Zapad. Sib.), No. 226,* 32–35. [In Russian.]

Osipov, A. S., Al'Betova, L. M. and Shirshov, V. Ya, 1981, [Epizootic status of lakes in the taiga—swamp zone of the Tyumen region.] *Sbornik Nauchnykh Trudov Gosudarstvennogo N. I. I. Ozernogo i Rechnogo Rybnogo Khozyaistva (Rybnoe khozyaistvo na vodoemakh zapadnoi Sibiri), No. 171,* 84–89. [In Russian.]

Osmanov, S. O. and Urazbaev, A. N, 1980, [Control of parasites and fish diseases in Karakalpakyan ponds (USSR).] In: *Parazity ryb i vodnykh bespozvonochnykh nizov'ev Amudar'i.* Tashkent, USSR: 'Fan' Uzbekskoi SSR., pp. 80–150. [In Russian.]

Osmanov, S. O. and Yusupov, O, 1985, [Influence of increasing salinity of the Aral Sea on the fish parasite fauna.] *Parazit. Sb.*, 33, 14–43. [In Russian.]

Osmanov, S. O., Urazbaev, A. N. and Arystanov, E. A, 1980, [Prognosis of the species composition of fish parasites and prophylaxis of diseases in the Tuyamuyun reservoir complex.] In: *Parazity tyb i vodnykh bespozvonochnykh nizov'ev* Amudar'i Tashkent, USSR: 'Fan' Uzbekskoi SSR, pp. 13–61. [In Russian.]

Otto, F. and Körting, W, 1973, Report on post mortem findings in an outbreak of endoparasitism in rainbow trout. *Vet. Med. Rev., No. 2,* 99–106.

Otto, T. N. and Heckmann, R. A, 1984, Host tissue response for trout infected with *Diphyllobothrium cordiceps* larvae. *Great Basin Nat.*, 44, 125–131.

Overstreet, R. M, 1968, Parasites of the inshore lizardfish *Synodus foetens*, from South Florida including a description of a new genus of Cestoda. *Bull. Mar. Sci.,* 18, 444–470.

Overstreet, R. M, 1969, Digenetic trematodes of marine teleost fishes from Biscayne Bay, Florida. *Tulane Studies in Zoology and Botany,* 15, 119–176.

Overstreet, R. M, 1973, Parasites of some penaeid shrimps with emphasis on reared hosts. *Aquaculture,* 2, 105–140.

Overstreet, R. M, 1977, *Poecilancistrium caryophyllum* and other trypanorhynch cestode plerocercoids from the musculature of *Cynoscion nebulosus* and other sciaenid fishes in the Gulf of Mexico. *J. Parasit.,* 63, 780–789.

Overstreet, R. M, 1978a, Marine maladies? Worms, germs, and other symbionts from the Northern Gulf of Mexico. *Mississippi-Alabama Sea Grant Consortium-78–021, 140 pp.*

Overstreet, R. M, 1978b, Trypanorhynch infections in the flesh of sciaenid fishes. *Mar. Fish. Rev.,* 40, 37–38.

Overstreet, R. M, 1982, Abiotic factors affecting marine parasitism. In: *Parasites—their world and ours. Volume II. Plenary and discipline lectures and abstracts, 5th Int. Congr. of Parasitology, Toronto, Canada, 7–14 August, 1982, under the auspices of the World Fed. of Parasitologists.* (Eds Mettrick, D. F. and Desser, S. S.). Toronto, Canada, pp. 36–39.

Overstreet, R. M, 1983, Metazoan symbionts of crustaceans. In: *The biology of Crustacea. (Chief Ed: Bliss, D. E.). Volume 6. pathobiology.* (Ed. Provenzano, A. J., Jr.) New York, USA: Academic Press, Inc., pp. 155–250.

Overstreet, R. M, 1986, Solving parasite-related problems in cultured crustacea. In: *Parasitology—Quo*

Vadit? Proceedings of the sixth International Congress of Parasitology (Ed. Howell, M. J.). Canberra: Australian Academy of Science. pp. 309–318.

Overstreet, R. M. and Hochberg, F. G. Jr, 1975, Digenetic trematodes in cephalopods. *J. Mar. Biol. Ass U.K.* 55, 893–910.

Overstreet, R. M. and Howse, H. D, 1977, Some parasites and diseases of estuarine fishes in polluted habitats of Mississippi. *Conference on aquatic pollutants and biologic effects with emphasis on neoplasia, N. Y. Acad. Sci., Sept. 27–29, 1976.* (Eds Kraybill, H. F., Dawe, C. J., Harshbarger, J. C. and Tardiff, R. G.). *Ann. Acad. Sci.,* 298, 427–462.

Overstreet, R. M. and Meyer, G. W, 1981, Hemorrhagic lesions in stomach of rhesus monkey caused by a piscine ascaridoid nematode. *J. Parasit.,* 67, 226–235.

Owen, I. L, 1959, Studies on the helminth parasites of some British freshwater fishes and Amphibia. *Ph.D. Thesis, University of Wales, UK,* 192 pp.

Owen, R. W. and Arme, C, 1965, Some observations on the distribution of *Ligula* plerocercoids in British freshwater fishes. *Parasitology,* 55, 6P.

Pacak, S, 1957, Prispevok k stúdiu parazitofauny salmonidov v potaku Demänová. *Cesk. Parasit.*, 4, 239–247.

Paggi, L., Orecchia, P., Minervini, R. and Mattiucci, S, 1982, Sulla comparsa di *Anguillicola australiensis* Johnston e Mawson, 1940 (Dracunculoidea: Anguillicolidae) in *Anguilla anguilla* del Lago di Bracciano. *Parassitologia,* 24, 139–144.

Paling, J. E, 1965, The population dynamics of the monogenean gill parasite *Discocotyle sagittata* Leuckart on Windermere trout, *Salmo trutta,* L. *Parasitology,* 55, 667–694.

Paling, J. E, 1969, The manner of infection of trout gills by the monogenean parasite *Discocotyle sagittata. J. Zool., Lond.,* 159, 293–309.

Palmieri, J. and Heckmann, R, 1976, Potential biological control of diplostomatosis (*Diplostomum spathaceum*) in fishes by hyperparasitism. *Proc. Utah Acad. Sci. Arts Lett.*, 53, 17–19.

Palombi, A, 1930, Il ciclo biologico di *Diphtherostomum brusinae* Stossich. *Pubbl. Staz. Zool. Napoli,* 9, 237–292.

Palombi, A, 1933, *Cercaria pectinata* Huet e *Bacciger bacciger* (Rud.). Rapporti genetica e biologia. *Boll. Zool. Torino,* 4, 1–11.

Palombi, A, 1934, Gli stadi larvali dei trematodi del Golfo di Napoli. I. Contributo allo studio della morfologia, biologia e sistematica delle cercarie marine. *Pubbl. Staz. Zool. Napoli,* 14, 51–94.

Palombi, A, 1937, La cercaria di *Mesometra orbicularis* (Rud.) e la sua transformazione in metacercaria. Appunti sul ciclo evolutivo. *Riv. Parassit.*, 1, 13–17.

Palombi, A, 1940, Gli stadi larvali dei trematodi del Golfo di Napoli. III. Contributo allo studio della morfologia, biologia e sistematica della cercarie marine. *Riv. Parassit.*, 4, 7–30.

Palombi, A, 1941, *Cercaria dentalii* Pelseneer, forma larvale di *Ptychogonimus megastoma* (Red.) Nota previa. *Riv. Parassit.*, 5, 127–128.

Palombi, A, 1942a, Notizie ed osservazioni sui normali ed accidentali ospitatori definitive di *Ptychogonimus megastoma* (Rud.). *Ann. Mus. Zool. Univ. Napoli,* 7, 3 pp.

Palombi, A, 1942b, Il ciclo biologico di *Ptychogonimus megastoma* (Rud.). Osservazioni sulla morfologia e fisiologia delle forme larvali e considerazioni filogenetiche. *Riv. Parassit.*, 6, 117–172.

Palombi, A, 1955, Addattamenti biologici dei trematodi digenetici ai fini della conservazioni della especie. *Bol. Lab. Clin. 'Luis Razetti',* 16, 719–730.

Pálsson, J. and Beverley-Burton, M, 1984, Helminth parasites of capelin, *Mallotus villosus,* (Pisces: Osmeridae) of the North Atlantic. *Proc. Helminth. Soc. Wash.*, 51, 248–254.

Panczyk, J. and Zelazny, J, 1974, *Khawia sinensis* and *Bothriocephalus gowkongensis* infections in carp—new parasitic diseases found in Poland. *Gospodarka Rybna,* 26, 10–13.

Panebianco, A. and Schiavo, A. Lo, 1985, Indagine sulla presenza di larve anisakidi in aringhe salate e affumicate del commercio. Considerazioni d'ordine ispettivo. *Clin. Vet.*, **108**, 180–184.

Paperna, I, 1959, Studies on monogenetic trematodes in Israel. 1. Three species of monogenetic trematodes of reared carp. *Bamidgeh,* 11, 51–67.

Paperna, I, 1960a. The influence of monogenetic trematodes on fish breeding economy. *Bamidgeh,* 12, 54–55.

Paperna, I, 1960b, Studies on monogenetic trematodes in Israel. 2. Monogenetic trematodes of cichlids *Bamidgeh,* 12, 2–15.

Paperna, I, 1963a, *Enterogyrus cichlidarum* n. gen. n. sp., a monogenetic trematode parasitic in the intestine of a fish. *Bull. Res. Council Israel,* **11B**, 183–187.

Paperna, I, 1963b, Some observations on the biology and ecology of *Dactylogyrus vastator* in Israel. *Bamidgeh,* 15, 8–28.

Paperna, I, 1963c, Dynamics of *Dactylogyrus vastator* Nybelin (Monogea) populations on the gills of carp fry in fish ponds. *Bamidgeh*, 15, 31–50.

Paperna, I, 1964a, Adaptation of *Dactylogyrus extensus* (Müeller and Van Cleave, 1932) to ecological conditions of artificial ponds in Israel. *J. Parasit.*, 50, 90–93.

Paperna, I, 1964b, Parasitic helminths of inland-water fishes in Israel. *Israel J. Zool.*, 13, 1–20.

Paperna, I, 1964c, The metazoan parasite fauna of Israel inland water fishes. *Bamidgeh*, 16, 3–66.

Paperna, I, 1974, Hosts, distribution and pathology of infections with larvae of *Eustrongylides* (Dioctophymidae, Nematoda) in fishes from East African lakes. *J. Fish Biol.*, 6, 67–76.

Paperna, I, 1975, Parasites and diseases of the grey mullet (Mugilidae) with special reference to the seas of the Near East. *Aquaculture*, 5, 65–80.

Paperna, I, 1979, Monogenea of inland water fish in Africa. *Annales Musée Royal de l'Afrique Centrale—Sciences Zoologiques, Serie In-8, No. 226,* viii + 131.

Paperna, I, 1980, Parasites, infections and diseases of freshwater fishes in Africa. *Rome, Italy: Department of Fisheries, CIFA, FAO, No. 7,* ix + 216 pp.

Paperna, I, 1986, Solving parasite-related problems in cultured marine fish. *Parasitology—Quo Vadit?* Proceedings of the Sixth International Congress of Parasitology (Ed. Howell, M. J.). Canberra, Australia: Australian Academy of Science, pp. 327–336.

Paperna, I. and Kohn, A, 1964, Studies on the host-parasite relations between carps and populations of protozoa and monogenetic trematodes in mixed infestations. *Rev. Brasil. Biol.*, 24, 269–276.

Paperna, I. and Laurencin, F. B, 1979, Parasitic infections of sea bass, *Dicentrarchus labrax*, and gilt head sea bream, *Sparus aurata*, in mariculture facilities in France. *Aquaculture*, 16, 173–175.

Paperna, I. and Lengy, J, 1963, Notes on a new subspecies of *Bolbophorus confusus* (Krause, 1914) Dubois 1935 (Trematoda, Diplostomatidae), a fish-transmitted bird parasite. *Israel J. Zool.*, 12, 171–182.

Paperna, I. and Overstreet, R. M, 1981, Parasites and diseases of mullets (Mugilidae). In: *Aquaculture of grey mullets*. (Ed. Oren, O. H.) Cambridge, UK: Cambridge University Press, pp. 411–493.

Paperna, I. and Zwerner, D. E, 1976, Parasites and diseases of striped bass, *Morone saxatilis* (Wallbaum), from the lower Chesapeake Bay. *J. Fish Biol.*, 9, 267–281.

Paperna, I., As, J. G. van and Basson, L., 1983, Review of diseases affecting cultured cichlids. *International Symposium on Tilapia in Aquaculture*, 174–184.

Paperna, I., Diamant, A. and Overstreet, R. M, 1984, Monogenean infestations and mortality in wild and cultured Red Sea fishes. *{Diseases of marine organisms. Internat. Helgoland, Hamburg, GFR.* (Eds Kinne, O. and Bulnheim, H. P.). *Helgoländer Meeresuntersuchungen*, 37, 445–462.

Papoutsoglou, S. E, 1976, Metazoan parasites of fishes from Saronicos Gulf, Athens, Greece. *Thalassographica* 1, 69–102.

Pappas, P. W, 1988, The relative roles of the intestines and external surfaces in the nutrition of monogeneans, digeneans and nematodes. *Parasitology*, 96, S105–S121.

Pappas, P. and Cain, G. D, 1981, Physiology and biochemistry of parasitic helminths. In: *Review of advances in parasitology. (Proc. 4th Internat. Congr. of Parasit. (ICOPA IV), Warsaw, 19–26 Aug. 1978).* (Ed. Slusarski, W.) Warsaw, Poland: PWN—Polish Scientific Publishers, pp. 747–763.

Pappas, P. W. and Read, C. P, 1972, Sodium and glucose fluxes across the brush border of a flatworm (*Calliobothrium verticillatum*, Cestoda). *J. Comp. Physiol.*, 81, 215–228.

Pár, O, 1978, Vliv nízké intenzity invaze tasemnice *Bothriocephalus gowkongensis* na kondicní a fyziologické ukazatele zdravotního stavu kapru. *Buletin VURH (Vyzkumny Ustav Rybársky a Hydrobiologicky) Vodnany, CSSR*, 14, 26–33.

Pár, O., Párová, J. and Prouza, A, 1977, Mansonil-Ucinne anthelmintikum pri lécení botriocefalózy kapra. *Buletin VURH (Vyzkumny Ustav Rybársky a Hydrobiologicky) Vodnany, CSSR, No. 1*, 17–25.

Parshad, V. R. and Crompton, D. W. T, 1981, Aspects of acanthocephalan reproduction. *Adv. Parasit.*, 19, 73–138.

Parukhin, A. M, 1973, Fish parasites and their importance. [Nematodes from Southern Sea fish.] *Trans. Am. Fish. Soc., Biologiya Morya, Kiev (Issledovaniya biologii i parazitofauny ryb i golovonogikh molloskov) No. 31*, 162–177. [In Russian.]

Parukhin, A. M, 1975, [The spread of the nematodes of fish from southern seas to the Pacific Ocean.] *Vest. Zool., No. 1*, 33–38. [In Russian.]

Parukhin, A. M, 1976, [*Parasitic worms of food fishes in the southern seas.*] Kiev, USSR: 'Naukova Dumka', 184 pp. [In Russian.]

Parukhin, A. M, 1986, [The helminth fauna of commercial nototheniid fish in the subantarctic section of the Indian Ocean.] *Vest. Zool., No. 3*, 6–9. [In Russian.]

Parukhin, A. M. and Lyadov, V. N, 1981, [The parasite fauna of Notothenioidei in the waters of the Atlantic and Pacific Oceans.] *Vest. Zool., No. 3*, 90–94. [In Russian.]

Parukhin, A. M. and Lyadov, V. N, 1982, [The helminth fauna of commercial fish from the family Nototheniidae in the Kergelen sub-zone.] *Ekol. Morya, No. 10,* 49–56. [In Russian.]

Parukhin, A. M. and Zaitsev, A. K, 1984, [Helminth infection in *Notothenia squamifrons* of various ages in the subantarctic sector of the Indian Ocean.] *Nauch. Dokl. Vyssh. Shkoly, Biol. Nauki, No. 10,* 34–37. [In Russian.]

Pascoe, D. and Cram, P, 1977, The effect of parasitism on the toxicity of cadmium to the three-spined stickleback, *Gasterosteus aculeatus* L. *J. Fish Biol.*, 10, 467–472.

Pascoe, D. and Mattey, D, 1977, Dietary stress in parasitized and non-parasitized *Gasterosteus aculeatus* L. *Z. ParasitKde,* 51, 179–186.

Pascoe, D. and Woodworth, J, 1980, The effects of joint stress on sticklebacks. *Z. ParasitKde,* 62, 159–163.

Pashkyavichyute, A. S, 1981, [The structure of the parasitocoenosis of bream in the Kursiu Marios Gulf in 1976–78.] *Lietuvos TSR Mokslu Akademijos Darbai (Trudy Akademii Nauk Litovoskoi SSR), C,* 2, 127–134. [In Russian.]

Pearre, S., Jr, 1979, Niche modification in Chaetognatha infected with larval trematodes (Digena). *Int. Rev. der Gesamt. Hydrobiol.* 64, 193–206.

Pearse, J. S. and Timm, R. W, 1971, Juvenile nematodes (*Echinocephalus pseudouncinatus*) in the gonads of sea urchins (*Centrostephanus coronatus*) and their effect on host gametogenesis. *Biol. Bull.*, 140, 95–103.

Pearson, J. C, 1964, A revision of the sub-family Haplorchinae Looss, 1899 (Trematoda: Heterophyidae). I. The *Haplorchis* group. *Parasitology*, 54, 601–676.

Pearson, J. C, 1968, Observations on the morphology and life-cycle of *Paucivitellosus fragilis* Coil, Reid and Kuntz, 1965 (Trematoda: Bivesiculidae). *Parasitology*, 58, 769–788.

Pearson, J. C, 1972, A phylogeny of life cycle patterns of Digenea. *Adv. Parasit.*, 10, 153–189.

Pearson, J. C, 1973, A revision of the sub-family Haplorchinae Looss, 1899 (Trematoda: Heterophyidae). II. Genus *Galactosomum. Phil. Trans. Roy. Soc. Lond. B (Biol Sci.*), 266, 341–447.

Pearson, J. C. and Ow-Yang, C. K, 1982, New species of *Haplorchis* from Southeast Asia, together with keys to the *Haplorchis*-group of heterophyid trematodes of the region. *Southeast Asian J. Trop. Med. Public Health*, 13, 35–60.

Pennell, D. A., Becker, C. D. and Scofield, N. R, 1973, Helminths of sockeye salmon (*Oncorhynchus nerka*) from the Kvichak River system, Bristol Bay, Alaska. 4 *Fish. Bull. U.S. Nat. Oceanic Atmos. Admin.*, 71, 267–277.

Pennycuick, L, 1971a, Seasonal variations in the parasite infections in a population of three-spined sticklebacks, *Gasterosteus aculeatus* L. *Parasitology*, 63, 373–388.

Pennycuick, L, 1971b, Differences in the parasite infections in three-spined sticklebacks (*Gasterosteus aculeatus* L.) of different sex, age and size. *Parasitology*, 63, 407–418.

Pereira, C., Vianna Dias, M. and Azevedo, P, 1936, Biologia de nematoide *Procamallanus cearensis* n. sp. *Arch, Inst. Biol.*, 7, 209–226.

Pereira-Bueno, J. M, 1980, Helmintocenosis del tracto digestivo de los Ciprínidos de los ríos de León. *Thesis, Fac. Vet. Univ. Oviedo, León, Spain*, viii + 404 pp.

Perkins, K. W, 1950, Studies on the biology of *Acetodextra amiuri* (Stafford, 1900) (Trematoda: Heterophyidae). *J. Parasit.*, 36, Suppl. p. 27.

Perkins, K. W, 1951, Observations of *Acetodextra amiuri*, a digenetic trematode from the ovary of catfish. [Abstract]. *Proc. Indiana Acad. Sci, Year 1950*, 60, 312–313.

Perkins, K. W, 1956, Studies on the morphology and biology of *Acetodextra amiuri* (Stafford) (Trematoda: Heterophyidae). *Amer. Midl. Nat.*, 55, 139–161.

Permyakov, E. V. and Rumyantsev, E. A, 1982, [Ecological and faunistic analysis of the parasites of *Coregonus lavaretus.*] In: *Ekologiya parazitícheskikh organizmov v biogeotsenozakh severa.* Petrozavodsk, USSR: Karel'skii Fil. Akad. Nauk SSSR, Inst. Biol., pp. 51–59. [In Russian.]

Pery, P, 1982, Effect of the host immune response on the physiology of gastro-intestinal parasites and on their environment. In: *Parasites—their world and ours. Proceedings of the Fifth International Congress of Parasitology Toronto, Canada, 7–14 Aug. 1982* (Eds Mettrick, D. F. and Desser, S. S.). Amsterdam, Netherlands: Elsevier Biomedical Press, pp. 141–143.

Peters, G. and Hartmann, F, 1986, *Anguillicola*, a parasitic nematode of the swim bladder spreading among eel populations in Europe. *Dis. Aquat. Org.*, 1, 229–230.

Peters, L. E, 1960, The systematic position of the genus *Dihemistephanus* Looss, 1901 (Trematoda: Digenea) with the redescription of *D. lydiae* (Stossich, 1896) from the South Pacific. *Proc. Helminth. Soc. Wash.*, 27, 134–138.

Petrushevskii, G. K, 1954, [Changes in the parasite fauna of fish in relation to their acclimatization.] *Trudy Probl. Temat. Sovesh. Akad. Nauk SSSR*, 4, 29–38. [In Russian.]

Petrushevskii, G. K, 1957, [Parasite fauna of the Black Sea herring.] *Izvest. Vses Nauchno-Issled Inst. Ozern. Rech. Ryb. Khozyaist.*, 42, 304–314. [In Russian.]

Petrushevskii, G. K, 1958a, Changes in the parasite fauna of fish on their acclimatization. In: *Basic problems of the parasitology of fishes* (Eds Dogiel, V. A., Petrushevski, G. K. and Polyanski, Y. I.). Izdatelstvo Leningradskogo Universiteta, pp. 256–266. [In Russian.]

Petrushevskii, G. K, 1958b, [Parasitic diseases of salmonoid fishes in pisciculture.] *Vop. Ikhtiol., No. 10*, 162–171. [In Russian.

Petrushevskii, G. K. and Petrushevskaya, M. G, 1960, [The accuracy of quantitative indices relating to the study of parasite faunas of fishes.] *Parazit. Sb.*, 19, 333–343. [In Russian.]

Petrushevskii, G. K. and Shul'man, S. S, 1955, [Infection of the liver of Baltic cod with roundworms.] *Liet. TSR Moks. Akad. Darb. (Trudy Akad. Nauk Lit. SSR), Ser. B*, 2, 119–125. [In Russian.]

Petrushevskii, G. K. and Shul'man, S. S, 1958, [Parasitic diseases of fish in water reservoirs in the USSR.] In: *Basic problems of the parasitology of fishes* (Eds Dogel, V. A. Petrushevskii, G. K. and Polyanski, Y. I.). Leningrad: Izdatelstvo Leningradskogo Universiteta, pp. 301–320. [In Russian.]

Petter, A. J, 1969a, Enquête sur les nématodes des sardines pechées dans la région Nantaise. Rapport possible avec les granulomes éosinophiles observés chez l'homme dans la région. *Annls Parasit. Hum. Comp.*, 44, 25–36.

Petter, A. J, 1969b, Enquête sur les nématodes des poissons de la région Nantaise. *Annls Parasit. Hum. Comp.*, 44, 559–580.

Petter, A. J, 1978, Quelques nématodes Camallanidae parasites de poissons en Malaisie. *Bull. Mus. Natn. Hist. Nat., 3rd série, No. 515, Zoologie 354*, 319–330.

Petter, A. J. and Baudin-Laurencin, F, 1986, Deux espèces du genre *Philometra* (Nematoda, Dracunculoidea) parasites de thons. *Bull. Mus. Natn. Hist. Nat.*, 8, 769–775.

Petter, A. J. and Quentin, J. C, 1976, Keys to the genera of the Oxyuroidea. *CIH Keys to the nematode parasites of vertebrates*. Farnham Royal, England: Commonwealth Agricultural Bureaux, No. 4, 30 pp.

Phalempin, P, 1979, Les parasitoses des poissons d'aquarium. *Thesis, école Natn. Vét. Alfort, France*, v + 99 pp.

Philip, C. B, 1955, There's always something new under the 'parasitological' sun. (The unique story of helminth-borne salmon poisoning disease). *J. Parasit.*, 41, 125–148.

Pianka, E. R, 1981, Competition and niche theory. In: *Theoretical Ecology* (Ed. May, R. M.). Oxford: Blackwell Scientific Publications, pp. 167–196.

Piasecki, W, 1982, Parasitofauna of Cape horse mackerel, *Trachurus trachurus capensis* Castelnau, 1861. *Acta Ichthyol. Pisc.*, 12, 43–56.

Pichelin, S., Whittington, I. and Pearson, J, 1991, *Concinnocotyla* (Monogenea: Polystomatidae). a new genus for the polystome from the Australian lungfish, *Neoceratodus forsteri*. *Syst. Parasit.*, 18, 81–93.

Pickering, A. D. and Christie, P, 1980, Sexual differences in the incidence and severity of ectoparasitic infestation of the brown trout, *Salmo trutta* L. *J. Fish Biol.*, 16, 669–683

Pike, A. W, 1987, The infectivity and development of *Diplostomum gasterostei* and *D. spathaceum* in perch, rainbow trout and sticklebacks. *2nd International Symposium of Icthyoparasitology. Actual problems in fish parasitology, Tihany, Hungary, Sept. 27–Oct. 3*, p. 74.

Pike, A. W. and Burt, M. D. B, 1983, The tissue response of yellow perch, *Perca flavescens* Mitchill to infections with the metacercarial cyst of *Apophallus brevis* Ransom, 1920. *Parasitology*, 87, 393–404.

Pilecka-Rapacz, M, 1986, On the development of acanthocephalans of the genus *Acanthocephalus* Koelreuther, 1771, with special attention to their influence on intermediate host, *Asellus aquaticus* L. *Acta parasit. pol.*, 30, 233–250.

Pippy, J. H. C, 1969, *Pomphorhynchus laevis* (Zoega) Müller, 1776 (Acanthocephala) in Atlantic salmon (*Salmo salar*) and its use as a biological tag. *J. Fish. Res. Bd Can.*, 26, 909–919.

Pippy, J. H. C, 1980, The value of parasites as biological tags in Atlantic salmon at West Greenland. *Rapp. Proc. Verb. Réunions—Cons. Int. Expl. Mer.*, 176, 76–81.

Pirus, R. I, 1982, [Testing nematicides against *Philometroides* infection in carp.] *Ryb. Khozyaist. Kiev, UkrSSR, No. 34*, 70–73. [In Russian.]

Platzer, E. G, 1977, Culture media for nematodes. In: *CRC handbook series in nutrition and food. Section G: diets, culture media, food supplements. Vol. 2. Food habits of, and diets for, invertebrates and vertebrates—zoo diets*. Cleveland, Ohio, USA: CRC Press, Inc., 2, 29–59.

Platzer, E. G, 1980, Nematodes as biological control agents. *Calif. Agric.*, 34, 27.

Platzer, E. G. and Adams, J. R, 1967, The life history of a dracunculoid, *Philonema oncorhynchi* in *Oncorhynchus nerka*. *Can. J. Zool.*, 45, 31–43.

Plumb, J. A. (Editor) 1979, Principal diseases of farm-raised catfish. *Bull. South. Coop. Ser., Alabama Agric. Exper. Stn, Auburn Univ., USA, No. 225*, 92 pp.

Poddubnaya, A. V. and Ivanova, N. S, 1984, [Chemoprophylaxis and chemotherapy in commercial fisheries.] In: *Biologicheskie osnovy rybovodstva: parazity i bolezni ryb. (Biologicheskie resursy gidrosfery i ikh ispol'zovanie)*. (Eds Bauer, O. N., Musselius, V. A. and Skryabina, E. S.). Moscow, USSR: 'Nauka', pp. 108–125. [In Russian.]

Poddubnaya, L. G., Davydov, V. G. and Kuperman, B. I, 1984, [Morpho-functional study of *Archigetes sieboldi* Leuckart, 1878 (Cestoda: Caryophyllidea), related to peculiarities of its life cycle.] *Dokl. Akad. Nauk SSSR*, 276, 1010–1013. [In Russian.]

Podesta, R. B, 1982, Membrane physiology of helminths. In: *Membrane physiology of invertebrates* (Eds Podesta, R. B., Dean, L. L. McDiarmid, S. S., Timmers, S. F. and Young, B. W.). New York, USA: Marcel Dekker, pp. 121–177.

Pogosyan, S. B. and Grigoryan, D. A, 1983, [The cestode *Ligula intestinalis* in the pond fisheries of the Ararat plain.] *Biol. Zh. Armenii*, 36, 1086–1087. [In Russian.]

Poinar, G. O. and Thomas, G. M, 1976, Occurrence of *Ascarophis* (Nematoda: Spiruroidea) in *Callianassa californiensis* Dana and other decapod crustaceans. *Proc. Helminth. Soc. Wash.*, 43, 28–33.

Pojmanska, T, 1980, Pasozyty niektórych gatunków ryb w jeziorach okolic Konina. *Gospodarka Rybna*, 32, 12–14.

Pojmanska, T, 1984, An analysis of seasonality of incidence and maturation of some fish parasites, with regard to thermal factor. III. *Bunodera luciopercae* (Müller, 1776). *Acta Parasit. Pol.*, 29, 313–321.

Pojmanska, T, 1985a, An analysis of seasonality of incidence and maturation of some fish parasites, with regard to thermal factor. IV. *Bucephalus polymorphus* Baer, 1827. *Acta Parasit. Pol.*, 30, 25–34.

Pojmanska, T, 1985b, An analysis of seasonality of incidence and maturation of some fish parasites, with regard to thermal factor. V. Digeneans of the genus *Sphaerostoma* Rudolphi, 1809. General conclusions. *Acta Parasit. Pol.*, 30, 35–46.

Pojmanska, T. and Dzika, E, 1987, Parasites of bream (*Abramis brama* L.) from the lake Goslawskie (Poland) affected by long-term pollution. *Acta Parasit. Pol.*, 32, 139–161.

Pojmanska, T., Grabda-Kazubska, B., Kazubski, S. L., Machalska, J. and Niewiadomska, K, 1980, Parasite fauna of five fish species from the Konin lakes complex, artificially heated with thermal effluents, and from Goplo lake. *Acta Parasit. Pol.*, 27, 319–357.

Polyanski, Y. I, 1957, [Some questions on the parasitology of fishes of the Barents Sea.] *Trudi Murmanskoi Biologicheskoi Stansii*, 3, 175–183. [In Russian.]

Polyanski, Yu. I, 1958, Ecology of parasites of marine fishes. In: *Parasitology of Fishes* (Eds Dogiel, V. A., Petrushevski,, G. K. and Polyanski, Y. I.). Leningrad University Press, pp. 1–384.

Polyanski, Yu. I, 1961a, Zoogeography of parasites of the USSSR marine fishes. In: *Parasitology of Fishes. (English translation)*. (Eds Dogiel, V. A., Petrushevskii, G. K. and Polyanski, Yu. I.). Edinburgh and London: Oliver and Boyd, pp. 230–246.

Polyanski, Yu. I, 1961b, Ecology of parasites of marine fishes. In: *Parasitology of fishes. (English translation)*. (Eds Dogiel, V. A., Petrushevski, G. K. and Polyanski, Yu. I.). Edinburgh and London: Oliver and Boyd, pp. 48–84.

Pool, D. W, 1985, The effect of praziquantel on the pseudophyllidean cestode *Bothriocephalus acheilognathi in vitro*. *Z. ParasitKde*, 71, 603–608.

Pool, D., Ryder, K. and Andrews, C, 1984, The control of *Bothriocephalus acheilognathi* in grass carp, *Ctenopharyngodon idella*, using praziquantel. *Fish. Management*, 15, 31–33.

Poole, B. C. and Dick, T. A, 1983, Metacercarial distribution of *Apophallus brevis* (Heterophyidae) in yellow perch (*Perca flavescens*) from the Heming Lake study area. *Can. J. Zool.*, 61, 2104–2109.

Poole, B. C. and Dick, T. A, 1984, Liver pathology of yellow perch, *Perca flavescens* (Mitchill), infected with larvae of the nematode *Raphidascaris acus* (Bloch, 1779). *J. Wildl. Dis.*, 20, 303–307.

Poole, B. C. and Dick, T. A, 1985, Parasite recruitment by stocked walleye, *Stizostedion vitreum vitreum* (Mitchill), fry in a small boreal lake in central Canada. *J. Wildl. Dis.*, 21, 371–376.

Popiel, I. and James, B. L, 1978, The ultrastructure of the tegument of the daughter sporocyst of *Microphallus similis*. (Jag., 1900) (Digenea: Microphallidae). *Parasitology*, 76, 359–367.

Popov, N. Ya., Razmashkin, D. A. and Shirshov, V. Ya, 1983, [Changes in the epizootiology of *Digramma* infection in Lake Fomintsevo since it has been used as a nursery of coregonid fish.]

Sb. Nauch. Trud. Gosud. Nauchno-Issled. Inst. Ozern. Rech. Khozyaist. (Prob. ekol. parazit. ryb, No. 197, 107–112. [In Russian.]

Popova, T. I. and Gichenok, L. A, 1978, [On the possible occurrence of an intermediate host in the life-cycle of monogeneans.] In: *Nauchnye i priklad nye problemy gel'mintologii.* Moscow: 'Nauka', pp. 79–84. [In Russian.]

Popova, T. I., Mozgovoi, A. A. and Dmitrenko, M. A, 1964, [Biology of Ascaridata of animals from the White Sea.] *Trudy Gel'mint. Lab.*, 14, 163–169. [In Russian.]

Potapova, O. I., Malakhova, R. P., Silivanova, N. P, and Sterligova, O. P, 1972 [Data on the biology and parasite fauna of *Coregonus albula* L. in Lake Nasonovskoe.] In: *Lososevye (Salmonidae) Karelii. Vypusk l. Ekologiya, parazitofauna, biokhimiya.* Petrozavodsk, USSR: Karel'skii filial Akademii, pp. 74–83. [In Russian.]

Poupard, C. J, 1978, Therapy of fish diseases. In: *Fish pathology* (Ed. Roberts, R. J.). London, UK: Baillière Tindall, pp. 268–275.

Poynton, S. L. and Bennett, C. E, 1985, Parasitic infections and their interactions in wild and cultured brown trout and cultured rainbow trout from the River Itchen, Hampshire. In: *Fish and Shellfish Pathology* (Ed. Ellis, A. E.). London, UK: Academic Press Inc., pp. 353–357.

Prakash, A. and Adams, J. R, 1960, A histopathological study of the intestinal lesions induced by *Echinorhynchus lageniformis* (Acanthocephala: Echinorhynchidae) in the starry flounder. *Can. J. Zool.*, 38, 895–897.

Pratt, I. and Herrmann, R, 1962, *Nitzschia quadritestes* sp. n. (Monogenea: Capsalidae) from the Columbia River sturgeon. *J. Parasit.*, 48, 291–292.

Prévot, G, 1966, Sur deux trématodes larvaires d'*Antedon mediterranea* Lmk. (Echinoderme): *Metacercaria* sp. (Monorchiidae Odhner, 1911) et métacercaire de *Diphtherostomum brusinae* Stoss, 1904 (Zoogonidae Odhner, 1911). *Annls Parasit. Hum. Comp.*, 41, 233–242.

Price, C. E, 1967, The phylum Platyhelminthes: a revised classification. *Riv. Parassit.*, 28, 249–260.

Price, P. W, 1977, General concepts on the evolutionary biology of parasites. *Evolution*, 31, 405–420.

Price, P. W, 1980, Evolutionary biology of parasites. *Monographs in population biology 15.* (Ed. May, R. M.). Princeton, New Jersey, USA: Princeton University Press, xi–239 pp.

Price, P. W, 1986, Evolution in parasite communities. In: *Parasitology—Quo Vadit?* (Ed. Howell, M. J.) *Proceedings of the VI International Congress of Parasitology*, Canberra, Australia: Australian Academy of Sciences, Canberra, pp. 209–215.

Price, P. W. and Clancy, K. M, 1983, Patterns in number of helminth parasites in freshwater fishes. *J. Parasit.*, 69, 449–454.

Priebe, K. von, 1986, Muskelparasiten des Alaska-pollocks *Theragra chalcogramma. Arch. Lebensmittelhyg.*, 37, 129–156.

Priemer, J, 1979, Darmhelminthen von *Perca fluviatilis* L. und *Acerina cernua* (L.) (Pisces) und Gewässern des Berliner Randgebietes. *Zool. Anz.*, 203, 241–253.

Priemer, J, 1980, Zum Lebenszyklus von *Proteocephalus neglectus* (Cestoda) aus Regenbogenforellen *Salmo gairdneri. Angew. Parasit.*, 21, 125–133.

Priemer, J, 1987, On the life cycle of *Proteocephalus exiguus* (Cestoda) from *Salmo gairdneri* (Pisces). *Helminthologia*, 24, 75–85.

Priemer, J. and Goltz, A, 1986, *Proteocephalus exiguus* (Cestoda) also Parasit von *Salmo gairdneri (Pisces). Angew. Parasit.*, 27, 157–168.

Pronin, N. M. and Tugarina, P. Ya, 1971, [The comparative analysis of the parasite fauna of Baikal graylings.] In: *Issledovanie gidrobiologicheskogo rezhima vodoemov Vostochnoi Sibiri.* Irkutsk, USSR, pp. 76–81. [In Russian.]

Pronin, N. M., Pronina, S. V. and Shigaev, S. Sh, 1976, [Larval *Triaenophorus* infection in young-of-the-year *Esox lucius.*] *Mater. Nauch. Konf. Vses. Obshch. Gel'm. (Teoret. priklad. probl. gel'm.), No. 28,* 118–126. [In Russian.]

Pronin, N. M., Tarmakhanov, G. D. and Rusinek, O. T, 1985, [The influence of the warm waters from the Lake Gusinoe hydroelectric station on the parasite faunas of *Perca fluviatilis* and *Esox lucius.*] In: *Gidrobiologiya i gidroparazitologiya Pribaikal'ya i Zabaikal'ya* (Eds Alimov, A. F. and Pronin, N. M.). Novosibirsk, USSR: Nauka, Sibirskoe Otdelenie, pp. 30–44. [In Russian.]

Pronina, S. V, 1977, [The encapsulation of *Triaenophorus nodulosus* plerocercoids in the liver of *Perca fluviatilis* (histomorphology and pathogenesis).] *Trudy. Buryat. Inst. Est. Nauk. (Faunisticheskie, ekologicheskie issledovaniya v Zabaikal'e) No. 15, Zool.*, 46–51. [In Russian.]

Pronina, S. V, 1979, [Interrelationship between *Triaenophorus amurensis* plerocercoids and the liver tissues of *Leuciscus amurensis.*] In: *Zooparazitologiya basseina ozera Baikal.* Ulan-Ude, USSR: Akademiya Nauk SSSR, Sibirskoe Otdelenie, Buryatskii Filial, pp. 77–82. [In Russian.]

Pronina, S. V. and Pronin, N. M, 1979, [Polysaccharides of plerocercoids of *Triaenophorus nodulosus*

(Pseudophyllidea, Cestoda), their composition and distribution in capsules and in the liver of different hosts.] In: *Parazity zhivotnykh i vrediteli rastenii Pribaikal'ya i Zabaikal'ya*. Ulan-Ude, USSR: Akademiya Nauk SSSR, Sibirskoe Otdelenie, Buryatskii Filial, pp. 103–112. [In Russian.]

Pronina, S. V. and Pronin, N. M, 1982a, The effect of cestode (*Triaenophorus nodulosus*) infestation on the digestive tract of pike (*Esox lucius*). *J. Ichthyol.*, 22, 105–113.

Pronina, S. V. and Pronin, N. M, 1982b, [Structure of the digestive tract of *Esox lucius* infected and uninfected with *Triaenophorus nodulosus* (Pallas) (Pseudophyllidea, Triaenophoridae).] *Voprosy Ikhtiologii*, 22, 641–648. [In Russian.]

Pronina, S. V., Davydov, V. G. and Kuperman, B. I, 1985, [Histochemical studies of some caryophyllaeid, pseudophyllid and proteocephalid cestodes.] In: *Gidrobiologiya i gidroparazitologiya Pribaikal'ya i Zabaikal'ya* (Eds Alimov, A. F. and Pronin, N. M.). Novosibirsk, USSR: Nauka, Sibirskoe Otdelenie, pp. 153–167. [In Russian.]

Pronina, S. V., Logachov, E. D. and Pronin, N. M, 1981, [Carbohydrate-containing biopolymers in cestode plerocercoids of the order Pseudophyllidea and the capsules surrounding them.] *Izvestiya Sibirskogo Otdeleniya Akademii Nauk SSR, Biologicheskikh Nauk, No. 10*, 121–127. [In Russian.]

Prost, M, 1953, Nowe poglady na metodyke badan i problematyke ichtioparazytologiczna. *Medycyna Wet.*, 9, 156–158.

Prost, M, 1963, Investigations on the development and pathogenicity of *Dactylogyrus anchoratus* (Duj., 1845) and *D. extensus* Müeller et v. Cleave, 1932 for breeding carps. *Acta Parasit. Pol.*, 11, 19–47.

Prost, M, 1979, [Fish as a source of human infection.] *Vet. Sbirka*, 77, 22–25. [In Bulgarian.]

Prost, M, 1981, Fish Monogenea of Poland. VI. Parasites of *Nemachilus barbatulus* (L.) and *Misgurnus fossilis* (L.) *Acta Parasit. Pol.*, 28, 1–10.

Prost, M, 1984, Fish Monogenea of Poland. VII. Parasites of *Gobio gobio* (L.) and *Leucaspius delineatus* (Heck.). *Acta Parasit. Pol.*, 29, 291–297.

Prost, M, 1988, Fish Monogenea of Poland. VIII. Parasites of *Barbus meridionalis petenyi* (Heck.). *Acta Parasit. Pol.*, 33, 1–6.

Protasova, E. N, 1982, Study of the cestode fauna of fish of some lakes in the Lithuanian SSR. *Trudy Gel'm. Lab. (Gel'm. vod. zhivot.)*, 31, 87–98. [In Russian.

Protasova, E. N, 1977, [Principles of cestodology, edited by K. M. Ryzhikov. Volume VIII. Bothriocephalata—cestodes of fish.] Moscow, USSR: Izdatel'stvo 'Nauka', 298 pp. [In Russian.]

Prudhoe, S. and Bray, R. A, 1982, *Platyhelminth parasites of the Amphibia*. British Museum (Natural History), Oxford: Oxford University Press, 217 pp.

Puffer, H. W. and Beal, M. L, 1981, Control of parasitic infestations in killifish (*Fundulus parvipinnis*). *Lab. Anim. Sci.*, 31, 200–201.

Pugachev, O. N, 1983, [Monogenea of freshwater fish in north-east Asia.] *Trudy Zool. Inst. Leningr. (Issled. morf. faun. parazit. cherv.)*, 121, 22–34. [In Russian.]

Pugachev, O. N, 1984a, [Nematodes of freshwater fishes in north-eastern Asia.] *Trudy Zool. Inst. Akad. Nauk SSSR (Ekol-geogr. issled. nematod)*, 126, 10–19. [In Russian.]

Pugachev, O. N, 1984b, [Parasites of freshwater fishes of north-east Asia.] *Parazity presnovodnykh ryb severo-vostoka Azii. Leningrad, USSR: Zoologicheskii Institut, Akademiya Nauk SSSR*, 156 pp. [In Russian.]

Pulsford, A. and Matthews, R. A, 1984, An ultrastructural study of the cellular response of the plaice, *Pleuronectes platessa* L., to *Rhipidocotyle johnstonei* nom. nov. (pro *Gasterostomum* sp. Johnstone, 1905) Matthews, 1968 (Digenea: Bucephalidae). *J. Fish Dis.*, 7, 3–14.

Pybus, M. J., Uhazy, L. S. and Anderson, R. C, 1978, Life cycle of *Truttaedacnitis stelmioides* (Vessichelli, 1910) (Nematoda: Cucullanidae) in American brook lamprey (*Lampetra lamottenii*). *Can. J. Zool.*, 56, 1420–1429.

Radhakrishnan, S. and Nair, N. B, 1981, *Tetrochetus coryphaenae* (Digenea: Accacoeliidae) infection of *Diodon hystrix* (Pisces: Diodontidae). *Proc. Indian Nat. Sci. Acad. B*, 47, 47–52.

Radhakrishnan, S., Nair, N. B. and Balasubramanian, N. K, 1984, Nature of infection of *Trichiurus lepturus* Linnaeus (Pisces: Trichiuridae) by *Scolex pleuronectis* Müeller (Cestoda: Tetraphyllidea). *Arch. Hydrobiol.*, 99, 254–267.

Rae, B. B, 1958, The occurrence of plerocercoid larvae of *Grillotia erinaceus* (van Beneden) in halibut. *Mar. Res., No. 4.*, 31 pp.

Rahkonen, R. and Valtonen, E. T, 1987, Occurrence of *Phyllodistomum umblae* (Fabricius, 1780) in the ureters of coregonids of lake Yli-Kitka in northeastern Finland. *Folia Parasit.*, 34, 145–155.

Rahkonen, R., Valtonen, E. T. and Gibson, D. I, 1984, Trematodes in northern Finland II: The occurrence of *Bunodera luciopercae* in three different water-bodies. *Bothnian Bay Reports*, 3, 55–66.

Rai, S. L, 1964a, Morphology and life-history of *Aspidogaster indica* Dayal, 1943 (Trematoda: Aspidogastridae). *Indian J. Helminth*, 16, 100–141.

Rai, S. L, 1964b, Observations on the life-history of *Phyllodistomum srivastavai* sp. nov. (Trematoda: Gorgoderidae). *Parasitology*, 54, 43–51.

Raina, M. K. and Koul, P. L, 1984, The histopathology of *Neoechinorhynchus hutchinsoni* Datta, 1936 (Neoacanthocephala: Neoechinorhynchidae) infection in *Nemacheilus kashmirensis* Hora. *J. Helminth.*, 58, 165–168.

Rakova, V. M, 1954, [Infestations of *Leuciscus idus* and their influence on that fish.] *Report Mosk. tekhn. in-t ryb. prom. khoz.*, [In Russian.]

Ramadevi, P, 1976, The life cycle of *Senga visakhapatnamensis* Ramadevi and Hanumantha Rao, 1973 (Cestoda: Pseudophyllidea). *Riv. Parassit.*, 37, 79–90.

Ramalingam, K, 1972, Studies on vitelline cells of Monogenea III. Nature of phenolic material and a possible alternative mechanism involved in hardening of egg-shell in helminths. *Acta Histochem.*, 44, 71–76.

Ramalingam, K, 1973, Chemical nature of monogenean sclerites. I. Stabilization of clamp-protein by formation of dityrosine. *Parasitology*, 66, 1–7.

Ramasamy, P. and Hanna, R. E. B, 1985, The surface topography of *Pseudothoracocotyla indica* (Unnithan, 1956) (Monogenea) from the gills of *Scomberomorus commerson. Z. ParasitKde*, 71, 575–581.

Ramulu, G. R. and Rama Krishna, G. V, 1982, The neuroanatomy of *Lytocestus indicus* (Cestoda). *Proc. Indian Acad. Parasit.*, 3, 50–53.

Rand, T. G. and Burt, M. D. B, 1985, Seasonal occurrence, recruitment and maturation of *Allocreadium lobatum* Wallin, 1909 (Digenea: Allocreadiidae) in the fallfish, *Semotilus corporalis* Mitchell, in a New Brunswick, Canada, lake system. *Can. J. Zool.*, 63, 612–616.

Rand, T. G., Wiles, M. and Odense, P, 1986, Attachment of *Dermophthirius carcharhini* (Monogenea: Microbothriidae) to the Galapagos shark *Carcharhinus galapagensis. Trans. Am. Microsc. Soc.*, 105, 158–169.

Randall, D. J, 1970, Gas exchange in fish. In: *Fish Physiology Vol. IV* (Eds Hoar, W. S. and Randall, D. J.) London: Academic Press, pp. 253–292.

Rao, K. H, 1954, A new bothriocephalid parasite (Cestoda) from the gut of the fish *Saurida tumbil* (Bloch). *Curr. Sci.*, 23, 333–334.

Rao, P. S. and Simha, S. S, 1983, Phosphatase activity in *Isoparorchis hypselobagri* of the fresh water cat fish, *Wallagonia att. Proc. Indian Acad. Parasit.*, 4, 33–35.

Rasheed, S, 1963, A revision of the genus *Philometra* Costa, 1845. *J. Helminth.*, 37, 89–130.

Rasheed, S, 1965, On a remarkable new nematode, *Lappetascaris lutjani* gen. et sp. nov. (Anisakidae: Ascaridoidea) from marine fishes of Karachi and an account of *Thynnascaris inquies* (Linton, 1901) n. comb. and *Goezia intermedia* n. sp. *J. Helminth.*, 39, 313–342.

Rasheed, U, 1980, Histochemical demonstration of glycogen, proteins and lipids in *Lytocestus indicus*, Moghe, 1931. *Proc. Indian Acad. Parasit.*, 1, 159–162.

Rasheed, U, 1981, Transaminase activity in *Lytocestus indicus* and its host. *Proc. Indian Acad. Parasit.*, 2, 115–116.

Rasheed, U. and Simha, S. S, 1982a, Non-specific phosphomonoesterase in *Lytocestus indicus* and its host, *Clarius batrachus. Indian J. Parasit.*, 6, 135–136.

Rasheed, U. and Simha, S. S, 1982b, Cholesterol content in *Lytocestus indicus* and its host. *Proc. Indian Acad. Parasit.*, 3, 46–47.

Rasin, K, 1931, Beiträge zur postembryonalen Entwicklung der *Amphilina foliacea* (Rud.), rebat einer Bermerkung über die Laboratoriumskultur von *Gammarus pulex* (L.) *Z. Wiss. Zool.*, 138, 555–579.

Ratcliffe, L. H., Taylor, H. M., Whitlock, J. H. and Lynn, W. R, 1969, Systems analysis of a host parasite interaction. *Parasitology*, 59, 649–661.

Rau, M. E. and Gordon, D. M, 1978, The frequency distribution of tetracotyles of *Apatemon gracilis pellucidus* (Yamaguti, 1933) in stickleback *Culaea inconstans* (Kirtland) populations of homogeneous age and size structure. *J. Fish Dis.*, 1, 259–263.

Rau, M. E., Gordon, D. M. and Curtis, M. A, 1979, Bilateral asymmetry of *Diplostomum* infections in the eyes of whitefish *Coregonus clupeaformis* (Mitchill) and a computer simulation of the observed metacercarial distribution. *J. Fish Dis.*, 2, 291–297.

Rausch, R. L, 1985, Parasitology: retrospect and prospect. *J. Parasit.*, 71, 139–151.

Rautskis, E. Yu, 1977, [Seasonal dynamics of parasitic infections in *Perca fluviatilis* in the different thermal regimes of lakes Obyaliya, Shlavantas and Galtas.] *Liet. TSR Mokslu Akad. Darbai (Trudy Akad. Nauk Lit. SSR) C*, 4, 63–73. [In Russian.]

Rautskis, E. Yu, 1982, [Dynamics of the parasite fauna of bream in the river Nyavezhis in 1975–1979.] *Liet. TSR Mokslu Akad. Darbai (Trudy Akad. Nauk Lit. SSR), C*, 3, 89–96. [In Russian.]

Rautskis, E. Yu, 1983, Prevalence of parasites in the Baltic-Nemunas population of *Vimba vimba* related to spawning migration. *Acta Parasit. Lit.* 20, 33–40. [In Russian.]

Ravindranathan, R. and Nadakal, A. M, 1971, Carotenoids in an acanthocephalid worm *Pallisentis nagpurensis*. *Jap. J. Parasit.*, 20, 1–5.

Rawson, M. V. Jr, 1977, Population biology of parasites of striped mullet, *Mugil cephalus* L. Crustacea. *J. Fish Biol.*, 10, 441–451.

Rawson, M. V. and Rogers, W. A, 1973, Seasonal abundance of *Gyrodactylus macrochiri* Hoffman and Putz 1964 on bluegill and largemouth bass. *J. Wildl. Dis.*, 9, 174–177.

Raybourne, R., Desowitz, R. S., Kliks, M. M. and Deardorff, T. L, 1983, *Anisakis simplex* and *Terranova* sp.: inhibition by larval excretory-secretory products of mitogen-induced rodent lymphoblast proliferation. *Expl Parasit.*, 55, 289–298.

Razarihelisoa, M, 1959, Sur quelques trematodes digenes de poissons de Nossibe (Madagascar). *Bull. Soc. Zool. Fr.*, 84, 421–434.

Razmashkin, D. A, 1984, [The parasite fauna and diseases of *Ctenopharyngodon idella* in the lake fisheries of the Tyumen region.] *Sb. Nauch. Trud. Gosud. Nauchno-IOssled. Inst. Ozern. Rech. Ryb. Khozyaist. (Bolez. parazit. ryb vod. Zapad. Sib.), No. 226*, 50–55. [In Russian.]

Razmashkin, D. A, 1985, [Adaptability of *Diplostomum* cercariae to carp and the effect on it of previous infection.] *Parazitologiya*, 19, 44–48. [In Russian.]

Razmashkin, D. A. and Kashkovskii, V. V, 1977, [The epizootiological significance of *Tetraonchus alaskensis* Price, 1937.] *Parazitologiya*, 11, 247–251. [In Russian.]

Razmashkin, D. A. and Shirshov, V. Ya, 1983, [Losses caused by *Digramma* infection in *Carassius* in lake fisheries of the Tyumen region.] *Sb. Nauch. Trud. Gosud. Nauchno-Issled. Inst. Ozern Rech. Khozyaist. (Probl. ekol. parasit. ryb), No. 197*, 10–106. [In Russian.]

Razmashkin, D. A, Osipov, A. S., Shirshov, V. Ya. and Al'Betova, L. M, 1979, [The parasite fauna and diseases of *Coregonus peled* in the water-bodies of the Tyumen region (USSR).] In: *Bolezni i parazity ryb Ledovitomorskoi provintsii (v predelakh SSSR)*. (Eds Gundrizer, A. N. and Bauer, O. N.). Tomsk, USSR: Izdatel'stvo Tomskogo Universiteta, pp. 94–100. [In Russian.]

Razmashkin, D. A., Osipov, A. S., Shirshov, V. Ya. and Al'Betova, L. M, 1984, [The parasite fauna and invasive diseases of whitefish in the lakes of West Siberia.] In: *Biologicheskie osnovy rybovodstva: parazity i bolezni ryb (Biologicheskie resursy gidrosfery i ikh ispol'zovanie)* (Eds Bauer, O. N., Musselius, V. A. and Skryabina, E. S.). Moscow, USSR: 'Nauka', pp. 89–108. [In Russian.]

Razmashkin, D. A., Kashkovskii, V. V., Osipov, A. S., Shirshov, V. Ya. and Kolesova, V. E, 1981, [The parasite fauna of whitefish in the lower Ob' and its Ural tributaries.] *Sb. Nauch. Trud. Gosud N. I. I. Ozer. Rech. Ryb. Khozyaist. (Ryb. khozyaist vod. zapad. Sib.), No. 171*, 72–83. [In Russian.]

Read, C. P, 1950, The vertebrate small intestine as an environment for parasitic helminths. *Rice Inst. Pamph.*, 37, 1–94.

Read, C. P, 1968, Some aspects of nutrition in parasites. *Am. Zool.*, 8, 139–149.

Read, C. P, 1971, The microcosm of intestinal helminths. In: *Ecology and Physiology of Parasites*. (Ed. Fallis, A. M.). London: Adam Hilger, pp. 188–200.

Read, C. P. and Simmons, J. E., Jr, 1963, Biochemistry and physiology of tapeworms. *Physiol. Rev.*, 43, 263–305.

Read, C. P., Douglas, L. T. and Simmons, J. E., Jr, 1959, Urea and osmotic properties of tapeworms from elasmobranchs. *Expl Parasit.*, 8, 58–75.

Read, C. P., Rothman, A. H. and Simmons, J. E, 1963, Studies on membrane transport, with special reference to parasite–host integration. *Ann. N.Y. Acad. Sci.*, 13, 154–205.

Rebecq, J. and Leray, C, 1960, Observations sur trois trématodes hébergés par quelques poissons des calanques marseillaises. *Bull. Inst. Oceanogr., Monaco, No. 1190*, 15 pp.

Rechcigl, M. Jr. (Ed.) 1977, *CRC handbook series in nutrition and food. Section D: nutritional requirements, Volume 1—comparative and qualitative requirements*. Cleveland, Ohio, USA: CRC Press, Inc., xiii + 551 pp.

Reda, E. S. A, 1987, An analysis of parasite fauna of bream, *Abramis brama* (L.), in Vistula near Warszawa in relation to the character of fish habitat. I. Review of parasite species. *Acta Parasit. Pol.*, 32, 309–326.

Reda, E. S. A, 1988, An analysis of parasite fauna of bream, *Abramis brama* (L.), in Vistula near Warszawa in relation to the character of fish habitat. II. Seasonal dynamics of infestation. *Acta Parasit. Pol.*, 33, 35–58.

Reddacliff, G. L, 1985, Diseases of aquarium fishes. A practical guide for the Australian veterinarian. *Vet. Rev.*, 25, ix + 116 pp.

Reddacliff, G. L, 1986, Diseases of ornamental fish, including carp. In: *Diseases of Australian fish and shellfish. Proceedings of the first Australian workshop on diseases of fish and shellfish, Benalla, Victoria, Australia, 27–30 May 1985* (Eds Humphrey, J. D. and Langdon, J. S.), pp. 118–126.

Rees, G, 1953a, Some parasitic worms from fishes off the coast of Iceland. I. Cestoda. *Parasitology*, 43, 4–14.

Rees, G, 1953b, Some parasitic worms from fishes off the coast of Iceland. II. Trematoda (Digenea). *Parasitology*, 43, 15–26.

Rees, G, 1953c, Some parasitic worms from fishes off the coast of Iceland. III. Monogenea, Nematoda, Acanthocephala. *Parasitology*, 43, 193–198.

Rees, G, 1958, A comparison of the structure of the scolex of *Bothriocephalus scorpii* (Müller 1776) and *Clestobothrium crassiceps* (Rud. 1819) and the mode of attachment of the scolex to the intestines of the host. *Parasitology*, 48, 468–492.

Rees, G, 1959, *Ditrachybothridium macrocephalum* gen. nov., sp. nov., a cestode from some elasmobranch fishes. *Parasitology*, 49, 191–209.

Rees, G, 1961a, Studies on the functional morphology of the scolex and of the genitalia in *Echinobothrium brachysoma* Pintner and *E. affine* Diesing from *Raja clavata* L. *Parasitology*, 51, 193–226.

Rees, G, 1961b, *Echinobothrium acanthinophyllum* n. sp. from the spinal valve of *Raja montagui* Fowler. *Parasitology*, 51, 407–414.

Rees, G, 1968, *Macrolecithus papilliger* sp. nov. (Digenea: Allocreadiidae, Stossich, 1904) from *Phoxinus phoxinus* (L.). Morphology, histochemistry and egg capsule formation. *Parasitology*, **58**, 855–878.

Rees, G, 1969, Cestodes from Bermuda fishes and an account of *Acompsocephalum tortum* (Linton, 1905) gen. nov. from the lizard fish *Synodus intermedius* (Agassiz). *Parasitology*, 59, 519–548.

Rees, G, 1970, Some helminth parasites of fishes of Bermuda and an account of the attachment organ of *Alicornis carangis* MacCallum, 1917 (Digenea: Bucephalidae). *Parasitology*, 60, 195–221.

Rees, G. and Williams, H. H, 1965, The functional morphology of the scolex and the genitalia of *Acanthobothrium coronatum* (Red.) (Cestoda: Tetraphyllidea). *Parasitology*, 55, 617–651.

Rees, J. A. and Kearn, G. C, 1984, The anterior adhesive apparatus and an associated compound sense organ in the skin-parasitic monogenean *Acanthocotyle lobianchi*. *Z. ParasitKde*, 70, 609–625.

Rego, A. A, 1975, Estudos de cestoides de peixes do Brasil. 2.a Nota: Revisao do genero *Monticellia* La Rue, 1911 (Cestoda, Proteocephalidae). *Revta Bras. Biol.*, 35, 567–586.

Rego, A. A., Carvajal, J. and Schaeffer, G, 1985, Patogenia del higado de peces (*Pagrus pagrus* L.) provocada por larvas de nematodos Anisakidae. *Parasit. al Dia*, 9, 75–79.

Rego, A. A., Carvalho-Varela, M., Mendonca, M. M. and Afonso-Roque, M. M, 1985, Helmintofauna da sarda (*Scomber scombrus* L.) peixe da costa continental portuguesa. *Mems Inst. Oswaldo Cruz*, **80**, 97–100.

Rego, A. A., Vincente, J. J., Santos, C. P. and Wekid, R. M, 1983, [Parasites of the anchovy *Pomatomus saltatrix* from Rio de Janeiro.] *Ciência e Cultura* 35, 1329–1336.

Rehana, R. and Bilqees, F. M, 1979, Three nematode species of genus *Procamallanus* (Baylis, 1923) including two new species from the fishes of Kalri lake, Sind, Pakistan. *Pakist. J. Zool.*, 11, 281–293.

Rehulka, J, 1978, Gyrodaktyloza u *Poecilia reticulata* Peters, 1859 v akvarijnim chovu. *Buletin VURH (Vyzkumny Ustav Rybarsky a Hydrobiologicky), Vodnany, CSSR*, 14, 34–36.

Reichenbach-Klinke, H. H, 1978, *Krankheiten der Aquarienfische. Mit besondere Berüksichtigung tropischer Arten*. Stuttgart, GFR: Alfred Kernen Verlag, 3rd amendment. 143 pp.

Reichenbach-Klinke, H. H. and Elkan, H, 1965, *The principal diseases of lower vertebrates*. London: Academic Press, 600 pp.

Reimchen, T. E, 1982, Incidence and intensity of *Cyathocephalus truncatus* and *Schistocephalus solidus* infection in *Gasterosteus aculeatus*. *Can. J. Zool.*, 60, 1091–1095.

Reimer, L. V, 1977, [Cestode larvae in plankton invertebrates from the Atlantic Ocean near the coast of North-West Africa.] *Parazitologiya*, 11, 309–315. [In Russian.]

Reimer, L. W, 1980, Description of the female of a new didymozoid: *Gonapodasmius microovatus* n. sp. *Angew. Parasit.*, 21, 26–31.

Reimer, L. W, 1981, Zur Darstellung der Verwandtschaftsverhältnisse der Fischgattung *Merluccius* auf Grund des Parasitenbefalls. *Angew. Parasit.*, 22, 25–32.

Reimer, L. W, 1983, Die Heringswürmer (*Anisakis simplex*) und verwandte Arten. *Angew. Parasit.*, 24, 1–16.

Reimer, L. W., Berger, C., Heuer, B., Lainka, C. and Rosental, I, 1971, O rasprostranemi licinok gel'mintov v planktonnyh zivotnych severnogo Norja. *Parazitologiya*, 6, 542–550.

Reiss, Z. and Paperna, I, 1975, Studies on diseases of marine fish. *Fourth Rep. H. Steinitz Mar. Biol. Lab., Elat, Israel*, 55–69.

Rekharani, Z. and Madhavi, R, 1985, Digenetic trematodes from mullets of Visakhapatnam (India). *J. Nat. Hist.*, 19, 929–951.

Renaud, F. and Gabrion, C, 1984, Polymorphisme enzymatique de populations du groupe *Bothriocephalus scorpii* (Mueller, 1776) (Cestoda: Pseudophyllidea). Etude des parasites de divers téléostéens des côtes du Finistère., *Bull. Soc. Fr. Parasit.*, 2, 95–99.

Renaud, F., Gabrion, C. and Pasteur, N, 1983, Le complexe *Bothriocephalus scorpii* (Mueller, 1776): différenciation par électrophorèse enzymatique des espèces parasites du turbot (*Psetta maxima*) et de la barbue (*Scophthalmus rhombus*). *C. R. Séanc. Acad. Sci. Sér. III*, 296, 127–129.

Renaud, F., Gabrion, C. and Romestand, B, 1984, Le complèxe *Bothriocephalus scorpii* (Mueller, 1776). Différenciation des éspèces parasites du turbot (*Psetta maxima*) et de la barbue (*Scophthalmus rhombus*). étude des fractions protéiques et des complèxes antigéniques. *Annls Parasit. Hum. Comp.*, 59, 143–149.

Renaud, F., Gabrion, C. and Pasteur, N, 1986, Geographical divergence in *Bothriocephalus* (Cestoda) of fishes demonstrated by enzyme electrophoresis. *Int. J. Parasit.*, 16, 553–558.

Rhee, J. K. and Lee, S. B, 1984, [The wormicidal substance of fresh water fishes on *Clonorchis sinensis.* V. Purification and chemical characterization of clonorchicidal substance from epidermal mucus of *Cyprinus carpio.*] *Korean J. Parasit.*, 22, 127–134. [In Korean.]

Rhee, J. K., Baek, B. K., Ahn, B. Z. and Park, Y. J, 1980a, [The anthelmintic action of substances of freshwater fishes on *Clonorchis sinensis* II. Preliminary research on the anthelmintic substances from mucous substances of various freshwater fishes.] *Korean J. Parasit.*, 18, 98–104. [In Korean.]

Rhee, J. K., Baek, B. K., Ahn, B. Z. and Park, Y. J, 1980b, [Anthelmintic action of substances from freshwater fishes on *Clonorchis sinensis.* III. Seasonal variation of the anthelmintic activity of mucous substances of *Carassius carassius.*] *Korean J. Parasit.*, 18, 179–184. [In Korean.]

Rhee, J. K., Kim, P. G., Baek, B. K. and Lee, S. B, 1983, [Experimental infection of *Clonorchis sinensis* in *Cyprinus carpio nudus.*] *Korean J. Parasit.*, 21, 11–19. [In Korean.]

Rhee, J. K., Kim, P. G., Baek, B. K., Lee, S. B. and Ahn, B. Z, 1982, Clavate cells of epidermis in *Cyprinus carpio nudus* with reference to its defence activity to *Clonorchis sinensis. Korean J. Parasit.*, 20, 201–203. [In Korean.]

Rhee, J. K., Kim, P. G., Baek, B. K. and Lee, S. B, 1983, [The wormicidal substance of freshwater fishes on *Clonorchis sinensis.* IV. Preliminary research on the wormicidal substance from mucus of *Cyprinus carpionudus.*] *Korean J. Parasit.*, 21, 21–26. [In Korean.]

Rhee, J. K., Lee, S. B. and Baek, B. K, 1984, Seasonal effects on clonorchicidal substances from epidermal mucus of *Cyprinus carpio, Ophicephalus argus* and *Parasilurus asotus. Korean J. Parasit.*, 22, 135–137.

Ribelin, W. E. and Migaki, G, 1975, *The pathology of fishes.* Madison, Wisconsin: University of Wisconsin Press, x + 1004 pp.

Richards, K. S. and Arme, C, 1981a, The effects of the plerocercoid larva of the pseudophyllidean cestode *Ligula intestinalis* on the musculature of bream (*Abramis brama*). *Z. ParasitKde*, 65, 207–215.

Richards, K. S. and Arme, C, 1981b, Observations on the microtriches and stages in their development and emergence in *Caryophyllaeus laticeps* (Caryophyllidea: Cestoda). *Int. J. Parasit.*, 11, 369–375.

Richards, K. S. and Arme, C, 1982a, Sensory receptors in the scolex-neck region of *Caryophyllaeus laticeps* (Caryophyllidea: Cestoda). *J. Parasit.*, 68, 416–423.

Richards, K. S. and Arme, C, 1982b, The microarchitecture of the structured bodies in the tegument of *Caryophyllaeus laticeps* (Caryophyllidea: Cestoda). *J. Parasit.*, 68, 425–432.

Richards, R, 1977a, Diseases of aquarium fish—2: Skin diseases. *Vet. Rec.*, 101, 132–135.

Richards, R, 1977b, Diseases of aquarium fish—3: Disease of the internal organs. *Vet. Rec.*, 101, 149–150.

Richards, R. H, 1983, Diseases of farmed fish: salmonids. *Vet. Rec.*, 112, 124–126.

Rid, L. E, 1973, Helminth parasites of the long-finned eel, *Anguilla dieffenbachii*, and the short-finned eel, *A. australis. Mauri Ora*, 1, 99–106.

Rifaat, M. A., Salem, S. A., El Kholy, S. I., Hegazi, M. M. and Yousef, M. El-M, 1980, Studies on the incidence of *Heterophyes heterophyes* in Dakahlia governorate. *J. Egypt. Soc. Parasit.*, 10, 369–373.

Riggs, R. and Esch, G. W, 1987, The suprapopulation dynamics of *Bothriocephalus acheilognathi* in a North Carolina reservoir: abundance, dispersion, and prevalence. *J. Parasit.*, 73, 877–892.

Riggs, R., Lemly, A. D. and Esch, G. W, 1989, The growth, biomass, and fecundity of *Bothriocephalus acheilognathi* in a North Carolina cooling reservoir. *J. Parasit.*, 73, 893–900.

Riley, D. M, 1978, Parasites of grass carp and native fishes in Florida. *Trans. Am. Fish. Soc.*, 107, 207–212.

Rim, H. J, 1982, Opisthorchiasis. In: *CRC handbook series in zoonoses. Section C: parasitic zoonoses. Volume III.* (Eds Hillyer, G. V. and Hopla, C. E.) Boca Raton, Florida, USA: CRC Press, Inc., pp. 109–121.

Rimaila-Pärnänen, E. and Wiklund, T, 1987, *Gyrodactylus salaris*—loismadon levin-neisyydestä makeanveden kalanviljelylaitoksissamme. *Suomen Eläinlääkärilehti*, 93, 506–507.

Rintamäki, P. and Valtonen, E. T, 1988, Seasonal and size-bound infection of *Proteocephalus exiguus* in four coregonid species in northern Finland. *Folia Parasit.*, 35, 317–328.

Riser, N. W, 1956, Early larval stages of two cestodes from elasmobranch fishes. *Proc. Helminth. Soc. Wash.*, 23, 120–124.

Riser, N. W. and Morse, M. P, 1974, *Biology of the Turbellaria.* New York: McGraw-Hill Book Co., Inc., 530 pp.

Robert, F., Boy, V. and Gabrion, C, 1990, Biology of parasite populations: population dynamics of bothriocephalids (Cestoda—Pseudophyllidea) in teleostean fish. *J. Fish Biol.*, 37, 327–342.

Robert, F., Renaud, F., Mathieu, E. and Gabrion, C, 1988, The biology and evolution of parasite populations: the importance of the paratenic host in the complex *Bothriocephalus scorpii* (Cestoda: Pseudophyllidea). *Int. J. Parasitol*, 18, 611–621.

Roberts, R. J, 1975, The effect of temperature on disease and its histopathological manifestation in fish. In: *The pathology of fishes* (Eds Ribelin, W. E. and Migaki, G.) Madison: University of Wisconsin Press, pp. 477–497.

Roberts, R. J, (Ed.) 1978a, *Fish pathology*. London, UK: Baillière Tindall, x + 318 pp.

Roberts, R. J, 1978b, The pathophysiology and systematic pathology of teleosts. In: *Fish Pathology* (Ed. Roberts, R. J.). London, UK: Baillière Tindall, pp. 55–91.

Roberts, R. J. and Shepherd, C. J, 1974, *Handbook of trout and salmon diseases*. Farnham, UK: Fishing News, 172 pp.

Roberts, R. J. and Sommerville, C, 1982, Diseases of tilapias. In: *The biology and culture of tilapias.* (Eds R. S. V. Pullin and R. H. Lowe-McConnell). ICLARM Conference Proceedings, International Center for Living Aquatic Resources Management, Manila, Philippines, 7, 247–263.

Robotham, P. W. J. and Thomas, J. S, 1982, Infection of the spined loach, *Cobitis taenia* (L.) by the digenean, *Allocreadium transversale* (Rud.). *J. Fish Biol.*, 21, 699–703.

Rodgers, L. J. and Burke, J. B, 1988, Aetiology of 'red spot' disease (vibriosis) with special reference to the ectoparasitic digenean *Prototransversotrema steeri* (Angel) and the sea mullet, *Mugil cephalus* (Linnaeus). *J. Fish Biol.*, 32, 655–663.

Rodjuk, G. N, 1985, Parasitic fauna of the fishes of the Atlantic part of the Antarctic (South Georgia Island and South Shetland Isles). In: *Parasitology and pathology of marine organisms of the World Ocean* (Ed. Hargis, W. J. Jr.). Seattle, USA: Department of Commerce, National Oceanic and Atmospheric Administration, pp. 31–32.

Rodrigues, H. O., Noronha, D. and Carvalho-Varela, M, 1975, Some acanthocephalans of Atlantic Ocean fishes—Portuguese continental coast and North African continental coast. *Mems. Inst. Oswaldo Crusz*, 73, 209–214.

Rodrigues, H. O., Rodrigues, S. S., Cristofaro, R. and Carvalho-Varela, M, 1972, [Some digenetic trematodes of Atlantic Ocean fishes—Portuguese continental coast and African continental coast.] *Atas. Soc. Biol. Rio de J.*, 15, 87–93. [In Portuguese.]

Rodrigues, H. O., Rodrigues, S. S., Cristofaro, R. and Carvalho-Varela, M, 1973, Some nematodes of Atlantic Ocean fishes—Portuguese continental coast and North African continental coast. *Mems. Inst. Oswaldo Cruz*, 71, 247–259.

Rodrigues, H. O., Rodrigues, S. S., Cristofaro, R. and Carvalho-Varela, M, 1975, New contribution for the study of nematodes of Atlantic Ocean fishes—Portuguese continental coast and North African continental coast. *Mems. Inst. Oswaldo Cruz*, 73, 127–134.

Rogers, W. P, 1962, *The nature of parasitism*. New York and London: Academic Press, ix + 287 pp.

Rogers, W. P, 1986, Advances in parasitology: 1886–1986. In: *Parasitology—Quo Vadit? Proceedings of the Sixth International Congress of Parasitology* (Ed. Howell, M. J.) Canberra: Australian Academy of Science, pp. 1–13.

Rogers, W. P. and Sommerville, R. E, 1969, Chemical aspects of growth and development. In: *Chemical Zoology, Vol. III. Echinodermata, Nematoda and Acanthocephala*, (Eds Florkin, M. and Scheer, B. T.) London: Academic Press Inc., pp. 465–499.

Rohde, K, 1971, Phylogenetic origin of trematodes. *Parasit. Schrift.*, 21, 17–27.

Rohde, K, 1972, The Aspidogastrea, especially *Multicotyle purvisi* Dawes, 1941. *Adv. Parasit.*, 10, 77–151.

Rohde, K, 1973, Structure and development of *Lobatostoma manteri* sp. nov. (Trematoda: Aspidogastrea) from the Great Barrier Reef, Australia. *Parasitology*, 66, 63–83.

Rohde, K, 1975a, Fine structure of the Monogenea, especially *Polystomoides* Ward. *Adv. Parasit.*, 13, 1–33.

Rohde, K, 1975b, Early development and pathogenesis of *Lobatostoma manteri* Rohde (Trematoda: Aspidogastrea). *Int. J. Parasit.*, 5, 597–607.

Rohde, K, 1978a, Latitudinal differences in host-specificity of marine Monogenea and Digenea. *Mar. Biol.*, 47, 125–134.

Rohde, K, 1978b, Latitudinal gradients in species diversity and their causes. II. Marine parasitological evidence for a time hypothesis. *Biol. Zentralbl.*, 97, 405–418.

Rohde, K, 1979a, A critical evaluation of intrinsic and extrinsic factors responsible for niche restriction in parasites. *Am. Nat.*, 114, 648–671.

Rohde, K, 1979b, The buccal organ of some Monogenea Polyopisthocotylea. *Zool. Scr.*, 8, 161–170.

Rohde, K, 1980a, Diversity gradients of marine Monogenea in the Atlantic and Pacific Oceans. *Experientia*, 36, 1368–1369.

Rohde, K, 1980b, Host specificity indices of parasites and their application. *Experientia*, 36, 1370–1371.

Rohde, K, 1981a, Niche width of parasites in species-rich and species-poor communities. *Experientia*, 37, 359–361.

Rohde, K, 1981b, Ultrastructure of the buccal organs and associated structures of *Zeuxapta seriolae* (Meserve, 1938) Price, 1962 and *Paramicrocotyloides reticularis* Rohde, 1978 (Monogenea, Polyopisthocotylea). *Zool. Anz.*, 206, 279–291.

Rohde, K, 1982a, *Ecology of marine parasites*. St. Lucia, Queensland: University of Queensland Press, xvi + 245 pp.

Rohde, K, 1982b, The flame cells of a monogenean and an aspidogastrean, not composed of two interdigitating cells. *Zool. Anz.*, 209, 311–314.

Rohde, K, 1982c, The nervous system of parasitic helminths. In: *Parasites—their world and ours. Proc. 5th Int. Congr. Parasit., Toronto, Canada, 7–14 Aug. 1982, under the auspices of the World Fed. of Parasitologists.* (Eds Mettrick, D. F. and Desser, S. S.) Amsterdam, Netherlands: Elsevier Biomedical Press, pp. 70–72.

Rohde, K, 1984a, Ecology of marine parasites. In: *Diseases of marine organisms. Internat. Helgoland Symp., 11–16 Sept. 1983, Helgoland, Hamburg, GFR.* (Eds Kinne, O. and Bulnheim, H. P.). Helgoländer Meeresuntersuchungen, 37, 5–33.

Rohde, K, 1984b, Zoogeography of marine parasites. In: *Diseases of marine organisms. Internat. Helgoland Symp., 11–16 Sept. 1983, Helgoland, Hamburg, GFR.* (Eds Kinne, O. and Bulnheim, H. P.). Helgoländer Meeresuntersuchunge, 37, 35–52.

Rohde, K, 1984c, Diseases caused by Metazoans: Helminths. In: *Diseases of marine animals, Vol. IV* (Ed. Kinne, O.) Hamburg: Biol. Anst. Helgoland, pp. 193–319; 435–501.

Rohde, K, 1986a, Marine parasitology in Australia. *Parasitology Today*, 2, 520.

Rohde, K, 1986b, Differences in species diversity of Monogenea between the Pacific and Atlantic Oceans. *Hydrobiologia*, 137, 21–28.

Rohde, K, 1987a, Different populations of *Scomber australasicus* in New Zealand and south-eastern Australia, demonstrated by a simple method using monogenean sclerites. *J. Fish Biol.*, 30, 651–657.

Rohde, K, 1987b, *Grubea australis* n. sp. (Monogenea, Polyopisthocotylea) from *Scomber australasicus* in southeastern Australia, and *Grubea cochlear* Diesing, 1858 from *S. scombrus* and *S. japonicus* in the Mediterranean and western Atlantic. *Systematic Parasit.* 9, 29–38.

Rohde, K, 1988, Gill Monogenea of deepwater and surface fish in southeastern Australia. *Hydrobiologia*, 16, 271–283.

Rohde, K. and Garlick, P. R, 1985a, Ultrastructure of the posterior sense receptor of larval *Austramphilina elongata* (Amphilinidae). *Int. J. Parasit.*, 15, 399–402.

Rohde, K. and Garlick, P. R, 1985b, A multiciliate 'starcell' in the parenchyma of the larva of *Austramphilina elongata* (Amphilinidea). *Int. J. Parasit.*, 15, 403–407.

Rohde, K. and Georgi, M, 1983, Structure and development of *Austramphilina elongata Johnston, 1931 (Cestodaria: Amphilinidea). Int. J. Parasit.*, 13, 273–287.

Rohde, K., Justine, J. L. and Watson, N, 1989, Ultrastructure of the flame bulbs of the monopisthocotylean Monogenea *Loimosina wilsoni* (Loimoidae) and *Calceostoma herculanea* (Calceostomatidae). *Annls Parasit. Hum. Comp.*, 64, 433–442.

Rohde, K., Roubal, F. and Hewitt, G. C, 1980, Ectoparasitic Monogenea, Digenea and Copepoda from the gills of some marine fishes of New Caledonia and New Zealand. *N. Z. J. Mar. Fresher. Res.*, 14, 1–13.

Rohde, K., Watson, N. and Garlick, P. R, 1986, Ulstrastructure of three types of sense receptors of larval *Austramphilina elongata* (Amphilinidea). *Int. J. Parasit.*, 16, 245–251.

Rohde, K., Watson, N. and Roubal, F, 1989, Ultrastructure of the protonephridial system of *Dactylogyrus* sp. and an unidentified ancyrocephaline (Monogenea: Dactylogyridae). *Int. J. Parasit.*, 19, 859–864.

Roitman, V. A, 1968, [On certain species of nematodes from the swim bladder of salmonid fishes of the genera *Oncorhynchus* and *Salvelinus*] *Parazity Zhivotnykh i Rastenii*, No. 4, 144–150. [In Russian.]

Roitman, V. A, 1981, [Population biology of helminths of freshwater biocoenoses.] *Itogi Naui i Techniki, Zooparazitologiya, Volume 7, Popyulyatsionnye i biotsenoticheskie aspekty izucheniya gel'mintov*, 43–88. [In Russian.]

Rokicki, J, 1981, The ectoparasites (Crustacea and Monogenea) of *Brama raii* from the shelf of north-west Africa. *Acta Parasit. Pol.*, 28, 85–90.

Rollinson, D. and Anderson, R. M, (Editors) 1985, Ecology and genetics of host-parasite Interactions. *Linnean Society Symposium Series No. 11*, London: Academic Press, 266 pp.

Romanenko, L. N. and Shigin, A. A, 1977, [The chromosome complex of trematodes of the genera *Diplostomum* and *Tylodelphys* (Strigeidida, Diplostomatidae) and its taxonomic significance.] *Parazitologiya*, 11, 530–536. [In Russian.]

Romanenko, N. A, *et al.* 1986, [Emergence and current status of a diphyllobothriasis focus in the Krasnoyarsk water reservoir.] *Medskaya Parazit., No. 1*, 69–73. [In Russian.]

Ronald, K, 1960, The effects of physical stimuli on the larval stage of *Terranova decipiens* (Krabbe, 1878) (Nematoda: Anisakidae). I. Temperature. *Can. J. Zool.*, 38, 623–642.

Rosa-Molinar, E. and Williams, C. S, 1983, Larval nematodes (Philometridae) in granulomas in ovaries of black-tip sharks, *Carcharinus limbatus* (Valenciennes). *J. Wildl. Dis.*, 19, 275–277.

Rosen, R. and Dick, T. A, 1983, Development and infectivity of the procercoid of *Triaenophorus crassus* Forel and mortality of the first intermediate host. *Can. J. Zool.*, 61, 2120–2128.

Rosen, R. and Dick, T. A, 1984a, Growth and migration of plerocercoids of *Triaenophorus crassus* Forel and pathology in experimentally infected whitefish, *Coregonus clupeaformis* (Mitchill). *Can. J. Zool.*, 62, 203–211.

Rosen, R. and Dick, T. A, 1984b, Experimental infections of rainbow trout *Salmo gairdneri* Richardson, with plerocercoids of *Triaenophorus crassus* Forel. *J. Wildl. Dis.*, 20, 34–38.

Rosenfield, A., Drucker, B. and Sindermann, C. J, (Conveners) 1985, Symposium: The role of diseases in marine fisheries management. *Reprinted from the Transactions of the fiftieth North American Wildlife and Natural Resources Conference, March 15–20, 1985. NOAA Technical Memorandum NMFS FINWR-16*, pp. 603–674.

Rosenthal, H, 1967, Parasites in larvae of the herring (*Clupea harengus* L.) fed with wild plankton. *Mar. Biol.*, 1, 10–15.

Roubal, F. R, 1981, The taxonomy and site specificity of the metazoan ectoparasites on the black bream, *Acanthopagrus australis* (Günther), in northern New South Wales. *Austr. J. Zool., Supplementary Series No. 84*. 100 pp.

Rousset, J. J. and Pasticier, A, 1972, A propos de deux cas de distomatose à hetérophyidés. Aspects épidémiologiques et cliniques. Importance de la notion d'opsopathologie. *Annls Parasit. Hum. Comp.*, 47, 465–474.

Rousset, J. J., Baufine-Ducrocq, H., Rabia, M. and Benoit, A, 1983, Distomatose colique à *Ascocotyle coleostoma*. Premier cas mondial ou les suites d'un congrès de pathologie tropicale en Egypte. *Presse Médicale*, 12, 2331–2332.

Ruitenberg, E. J, 1970, *Anisakiasis. Pathogenesis, serodiagnosis and prevention.* Thesis: Rijksuniversiteit, Utrecht, 138 pp.

Ruitenberg, E. J., Van Knapen, F. and Weiss, J. W, 1979, Food-borne parasitic infections—a review. *Vet. Parasit.*, 5, 1–10.

Rumpus, A. E, 1975, The helminth parasites of the bullhead *Cottus gobio* (L.) and the stone loach *Noemacheilus barbatulus* (L.) from the River Avon, Hampshire. *J. Fish Biol.*, 7, 469–483.

Rumyantsev, E. A, 1975, Ecological comparative analysis of parasite fauna of *Rutilus rutilus* L., and *Coregonus albula* L. from waters of the European part of the Arctic Ocean Province. *Folio Parasit.* 22, 337–340.

Rumyantsev, E. A, 1976a, [Parasites as ecological indicators of some species of Coregoninae.] In: *Lososevye (Salmonidae) Karelii. (Sbornik)*. Petrozavodsk, USSR, pp. 118–121. [In Russian.]

Rumyantsev, E. A, 1976b, [A study of *Diplostomum* (Trematoda Rudolphi, 1808: Strigeida La Rue, 1926) infection in fish.] In: *Parazitologicheskie issledovaniya Karel'skoi ASSR i Murmanskoi oblasti.* Petrozavodsk: Karel'skii filial Akademii Nauk SSR, Institut Biologii, pp. 186–190. [In Russian.]

Rumyantsev, E. A, 1978, [*Diplostomum* infection in fish from lake fisheries and its control.] *Parazitologiya*, 12, 487–492. [In Russian.]

Rumyantsev, E. A, 1982, [The parasites of fish in the Veshkelitskie lakes.] In: *Ekologiya paraziticheskikh organizmov v biogeotsenozakh severa*. Petrozavodsk, USSR: Karel'skii Filial Akademii Nauk SSSR, Institut Biologii, pp. 39–50. [In Russian.]

Rushovich, A. M., Randall, E. L., Captrini, J. A. and Westenfelder, G. O, 1983, Omental anisakiasis: a rare mimic of acute appendicitis. *Am. J. Clin. Path.*, **80**, 517–520.

Russell, L. R, 1980, Effects of *Truttaedacnitis truttae* (Nematoda: Cucullanidae) on growth and swimming of rainbow trout, *Salmo gairdneri*. *Can. J. Zool.*, **58**, 1220–1226.

Ruttenber, A. J., Weniger, B. G., Sorvillo, F., Murray, R. A. and Ford, S. L, 1984, Diphyllobothriasis associated with salmon consumption in Pacific Coast states. *Am. J. Trop. Med. Hyg.*, **33**, 455–459.

Ruyck, R. De and Chabaud, A. G, 1960, Un cas de parasitisme attribuable à des larves de *Phlyctainophora lannae* Steiner chez un sélacier, et cycle évolutif pobable de ce nématode, *Vie et Milieu*, Paris, 11(3), 386–389.

Rybak, V. F, 1982, [The formation of the parasite fauna of fish in the Vygozero reservoir.] In: *Ekologiya paraziticheskikh organizmov v biogeotsenozakh severa*. Petrozavodsk, USSR: Karel'skii Filial Akademii Nauk SSSR, Institut Biologii, pp. 59–72. [In Russian.]

Rychlinski, R. A. and Deardorff, T. L, 1982, *Spirocamallanus*: a potential fish health problem. *Freshw. Mar. Aquarium.*, **5**, 22–23, 79, 82.

Ryzhikov, K. M., Rysavy, B., Khokilova, I. G., Tolkatcheva, L. M. and Kornyushin, V. V, 1985, *Helminths of fish-eating birds of the Palaearctic region. 2. Cestoda and Acanthocephales* (Eds Ryzhikov, K. M. and Rysavy, B.) Prague: Academia Publishing House, 412 pp.

Safonov, N. N, 1976, [Destructive changes in the scales of bream infected with *Ligula intestinalis*.] *Byull. Vses. Inst. Gel'm. K. I. Skryabina*, 18, 60–62. [In Russian.]

Safonov, N. N. and Vasil'kov, G. V, 1971, Incidence of cysticerciasis in Cyprinidae. *Byull. Vses. Inst. Gel'm K. I. Skryabina*, 5, 95–96. [In Russian.]

Sagua, H., Fuentes, A., Soto, J. and Delano, B, 1979, Difilobotriasis humana por *Diphyllobothrium pacificum* en Chile. Experiencia con 11 casos. *Revta Med. Chile*, **107**, 16–19.

Sakaguchi, S., Kuniyuki, K. and Ueda, K, 1980, [Ecological observations and morphological characteristics of the parasitic nematode *Thynnascaris* found in juvenile Red Sea bream, *Chrysophrys major*.] *Bull. Nat. Res. Inst. Aqua Japan*, 1, 95–106. [In Japanese.]

Sakanari, J, 1989, *Grillotia heroniensis* sp. nov. and *G. overstreeti*, sp. nov., (Cestoda: Trypanorhyncha) from Great Barrier Reef fishes. *Aust. J. Zool.*, **37**, 81–87.

Sakanari, J. A., Loinaz, H. M., Deardorff, T. L., Raybourne, R. B., McKerrow, J. H. and Frierson, J. G, 1988, Intestinal anisakiasis. A case diagnosed by morphologic and immunologic methods. *Am. J. Clin. Path.*, **90**, 107–113.

Sakanari, J. A. and McKerrow, J. H, 1989, Anisakiasis. *Clin. Microbiol. Rev.*, **2**, 278–284.

Sakanari, J. and Moser, M, 1985a, Salinity and temperature effects on the eggs, coracidia and procercoids of *Lacistorhynchus tenuis* (Cestoda: Trypanorhyncha) and induced mortality in a first intermediate host. *J. Parasit.*, **71**, 583–587.

Sakanari, J. and Moser, M, 1985b, Infectivity of, and laboratory infection with, an elasmobranch cestode, *Lacistorhynchus tenuis* (Van Beneden, 1858). *J. Parasit.*, **71**, 788–791.

Sakanari, J. and Moser, M, 1989, Complete life-cycle of the elasmobranch cestode, *Lacistorhynchus dollfusi* Beveridge and Sakanari, 1987 (Trypanorhyncha). *J. Parasit.*, **75**, 806–808.

Samman, A, 1987, Incidence of gill parasites in polycultural fisheries. In: *Acute problems in fish parasitology. 2nd International Symposium of Ichthyoparasitology, Sept 27th–Oct 3rd., Tihany, Hungary*, p. 83

Samuel, G. and Bullock, W. L, 1981, Life cycle of *Paratenuisentis ambiguus* (Van Cleave, 1921) Bullock and Samuel, 1975 (Acanthocephala: Tenuisentidae). *J. Parasit.*, **67**, 214–217.

Sandeman, I. M. and Burt, M. D. B, 1972, Biology of *Bothrimonus* (= *Diplocotyle*) (Pseudophyllidea: Cestoda): ecology, life cycle, and evolution; a review and synthesis. *J. Fish. Res. Bd Can.*, **29**, 1381–1395.

Sankurathri, C. S., Kabata, Z. and Whitaker, D. J, 1983, Parasites of the Pacific hake, *Merluccius productus* (Ayres, 1855) in the Strait of Georgia, in 1974–1975. *Syesis*, **16**, 5–22.

Sanmartin Durán, M. L., Quinteiro Alonso, P., Rodriguez, A. and Fernández, J. A, 1989, Some Spanish cestode fish parasites. *J. Fish Biol.*, **34**, 977–978.

Sapozhnikov, G. I, 1969, Testing of phenasal against *Khawia* in carp fry. In: Probl. Parazit. (Ed. Markevich, A. P.). Trudy nauch. Konf. Parazit. UkSSR (6th) II, pp. 399–401. [In Russian.]

Sapozhnikov, G. I, 1975a, [Testing of preparations for *Sanguinicola* infection in carp.] In: *Probl. Parazit. Mater. VIII nauch konf. parazit. UkSSR Chast' 2*. Kiev, USSR: 'Naukova Dumka', pp. 154–156. [In Russian.]

Sapozhnikov, G. I, 1975b, [*Dilepis unilateralis* in carp.] In: *VIII Nauchnaya Konferentsiya Parazitologov*

Ukrainy (Tezisy dokladov). Donetsk, Sentyabr' Kiev, USSR: Ukrainskii Nauchno-Issledovatel'skii Institut, Nauchno-Teknicheskoi Informatsii, pp. 132–135. [In Russian.]

Sapozhnikov, G. I, 1976, [*Sanguinicola inermis* infection in fish fry.] *Veterinariya, Mosk., No. 10*, 53–54. [In Russian.]

Sapozhnikov, G. I, 1977, [Trials of some preparations against *Sanguinicola* infection in carp.] *Mater. Nauch. Konf. Vses. Obshch. Gel'm. (Trematody i trematodozy), No. 29*, 133–135. [In Russian.]

Sapozhnikov, G. I, 1978, [Ecological bases for controlling *Sanguinicola*,] *1st All-Union Meeting of parasitocoenologists (Poltavo, September 1978. Abstracts, Part 3.) Kiev*: Naukova Dumka, 142–144. [In Russian.]

Sapozhnikov, G. I, 1979, [Trials of acemidophen for *Sanguinicola* infection in carp.] *Byull'. Vses. Inst. Eksper. Vet., No. 37*, 57–58. [In Russian.]

Sapozhnikov, G. I, 1981, [The prophylaxis of diplostomiasis in *Coregonus peled.*] *Byull' Vses. Inst. Eksper. Vet., No. 41*, 26–30. [In Russian.]

Sapozhnikov, G. I. and Antonov, P. P, 1979, [Control of *Ligula* and *Digramma* infections by propagating coregonid fish.] *Izv. Akad. Nauk. Tadzhik. SSR (Ahboroti Akademijai Fanhoi RSS Tocikiston), Biol. Nauk, No. 4 (77)*, 103–104. [In Russian.]

Sapozhnikov, G. I. and Petrov, Yu. F, 1980, [The role of the living components of pond biocoenoses in the eradication of *Sanguinicola*.] *Veterinariya, Mosk., No. 9*, 45–47. [In Russian.]

Saprykin, V. G. and Kashkovskii, V. V, 1979, [Correlation of serum transferrins with the prevalence of *Dactylogyrus oxtensus* and *Ichthyophthirius multifiliis* infection.] *Sb. Nauch. Trud. Gosud. Nauch.-Issled. Inst. Ozer. Rech. Ryb. Khoz. Perm. Lab., No. 2*, 85–88. [In Russian.]

Sarabia, D. O, 1982, Contribución al estudio parasitologico de las especies de peces nativas e introducidas en la presa Adolfo Lopez Mateos 'El Infiernillo'. *Thesis, Universidad Nacional Autonoma de México, Mexico, D. F.*, x + 194 pp.

Sarig, S, 1976, The status of information on fish diseases in Africa and possible means of their control. *Suppl. 1, Rep. Symp. 'Aquaculture in Africa', Accra, Ghana, 30 Sept. - 2 Oct. 1975. Reviews and Experience Papers*. Rome, Italy: FAO, *CIFA Technical Paper No. 4, Suppl. 1*, 715–721.

Sathyanarayana, M. C, 1982, Incidence of trematode parasite, *Paraplerurus sauridae*, in relation to season, sex and length of the marine fish, *Saurida undosquamis*. *Indian J. Mar. Sci.*, 11, 188–189.

Sato, T., Hoshina, T. and Horiuchi, M, 1976, On worm cataract of rainbow trout in Japan. *Bull. Jap. Soc. Sci. Fish.*, 42, 249.

Satpute, L. R. and Agarwal, S. M, 1974, 'Diverticulosis' of the fish duodenum infested with cestodes. *Indian J. Exp. Biol.*, 12, 373–375.

Sattaur, O, 1988, Parasites prey on wild salmon in Norway. *New Scientist, No. 120*, 21.

Schäperclaus, W, 1954, *Fischkrankheiten. Third Edition*. Berlin: Academie-Verlag, 708 pp.

Scheer, D, 1934, Die Jugendform des Acanthocephalen *Echinorhynchus truttae* scht. undihr Vorkommen in *Gammarus pulex*. *Z. ParasitKde*, 7, 440–442.

Scheinert, P, 1984, Klinisch-chemische Untersuchungen an durch *Triaenophorus nodulosus* befallenen Seesaiblingen (*Salvelinus alpinus* L.) des Königssees. *Inaugural Dissertation, Ludwig-Maximilians-Universität, GFR*, 67 pp.

Schell, S. C, 1967, The life-history of *Phyllodistomum staffordi* Pearse, 1924 (Trematoda: Gorgoderidae Looss, 1901). *J. Parasit.*, 53, 569–576.

Schell, S. C, 1970, *How to know the trematodes*. Dubuque, Iowa: W. C. Brown Co., Publishers, vol. 1, vii + 355 pp.

Schell, S. C, 1973, *Rugogaster hydrolagi* gen. et sp. n. (Trematoda: Aspidobothrea: Rugogastridae fam. n. from the ratfish, *Hydrolagus colliei* (Lay and Bennett, 1839). *J. Parasit.*, 59, 803–805.

Schell, S. C, 1974, The life history of *Sanguinicola idahoensis* sp. n. (Trematoda: Sanguinicolidae), a blood parasite of steelhead trout, *Salmo gairdneri* Richardson. *J. Parasit.*, 60, 561–566.

Schell, S. C, 1975, The miracidium of *Lecithaster salmonis* Yamaguti, 1934 (Trematoda: Hemiuroidea). *J. Parasit.*, 61, 562–563.

Schell, S. C, 1985, *Trematodes of North America north of Mexico*. Moscow, USA: University of Idaho Press, iii + 263 pp.

Scheuring, L, 1929, Beobachtungen zur Biologie des Genus *Triaenophorus* und Betrachtungen ueber die jaherzeitliche Auftreten von Bandwürmern. *Z. ParasitKde*, 2, 157–177.

Schmahl, G. and Mehlhorn, H, 1985, Treatment of fish parasites. 1. Praziquantel effective against Monogenea (*Dactylogyrus vastator, Dactylogyrus extensus, Diplozoon paradoxum*). *Z. ParasitKde*, 71, 727–737.

Schmidt, G. D, 1969, *Dioecotaenia cancellata* (Linton, 1890) gen. et comb. n., a dioecious cestode (Tetraphyllidea) from the cow-nosed ray, *Rhinoptera bonasus* (Mitchell), in Chesapeake Bay, with the proposal of a new family, Dioecotaeniidae. *J. Parasit.*, 55, 271–275.

Schmidt, G. D, 1970, *How to know the tapeworms*. Dubuque, Iowa: W. C. Brown Co., Publishers, xii + 266 pp.

Schmidt, G. D, 1973, *Acanthobothrium urolophi* sp. n., a tetraphyllidean cestode (Oncobothriidae) from an Australian stingaree. *Proc. Helminth. Soc. Wash.*, **40**, 91–93.

Schmidt, G. D, 1985, Development of life cycles. In: *Biology of the Acanthocephala* (Eds Crompton, D. W. T. and Nickol, B. B.). Cambridge, UK: Cambridge University Press, pp. 273–286.

Schmidt, G. D, 1986, *CRC Handbook of tapeworm identification*. Boca Raton, Florida: CRC Press Inc., 675 pp.

Schmidt, G. D. and Beveridge, I, 1990, *Cathetocephalus australis* n. sp. (Cestoidea: Cathetocephalidae) from Australia, with a proposal for Cathetocephalidea n. ord. *J. Parasit.*, **76**, 337–339.

Schmidt, G. D. and Roberts, L. S, 1981, *Foundations of parasitology*. Second edition. London: C. V. Mosby Company, 795 pp.

Scholz, T, 1986, Observations on the ecology of five species of intestinal helminths in perch (*Perca fluviatilis*) from the Macha lake fishpond system, Czechoslovakia. *Vest. cs. Spolec. zool.*, **50**, 300–320.

Schrenkenbach, K, 1975, Research problems in the control of fish diseases to ensure and improve fish production. *Mh. VetMed.*, **30**, 699–703.

Schroeder, D. J., Johnson, A. D. and Mohammad, K. H, 1981, *In vitro* excystment of the black spot trematode *Neascus pyriformis* Chandler, 1951 (Trematoda: Diplostomatidae). *Proc. Helminth. Soc. Wash.*, **48**, 184–189.

Schuurmans Stekhoven, J. H., Jr. and Botman, T. P. J, 1932, Zür Ernauhrungsbiologie von *Proleptus obtusus* Duj. und die von diesen Parasiten hervorgerufenen reaktiven Anderungen des Witsgewebes. *Z. ParasitKde*, **4**, 220–239.

Scott, A. L. and Grizzle, J. M, 1979, Pathology of cyprinid fishes caused by *Bothriocephalus gowkongensis* Yeh, 1955 (Cestoda: Pseudophyllidea). *J. Fish Dis.*, **2**, 69–73.

Scott, J. S, 1969, Trematode populations in the Atlantic argentine, *Argentina silus*, and their use as biological indicators. *J. Fish. Res. Bd Can.*, **26**, 879–891.

Scott, J. S, 1975a, Incidence of trematode parasites of American plaice (*Hippoglossoides platessoides*) of the Scotian Shelf and Gulf of St. Lawrence in relation to fish length and food. *J. Fish. Res. Bd Can.*, **32**, 479–483.

Scott, J. S, 1975b, Geographic variation in incidence of trematode parasites of American plaice (*Hippoglossoides platessoides*) in the Northwest Atlantic. *J. Fish. Res. Bd Can.*, **32**, 547–550.

Scott, J. S, 1981, Alimentary tract parasites of haddock (*Melanogrammus aeglefinus* L.) on the Scotian Shelf. *Can. J. Zool.*, **59**, 2244–2252.

Scott, J. S, 1982, Digenean parasite communities in flatfishes of the Scotian Shelf and southern Gulf of St. Lawrence. *Can. J. Zool.*, **60**, 2804–2811.

Scott, J. S, 1985a, Occurrence of alimentary tract helminth parasites of pollock (*Pollachius virens* L.) on the Scotian Shelf. *Can. J. Zool.*, **63**, 1695–1698.

Scott, J. S, 1985b, Digenean (Trematoda) populations in winter flounder (*Pseudopleuronectes americanus*) from Passamaquoddy Bay, New Brunswick, Canada. *Can. J. Zool.*, **63**, 1699–1705.

Scott, J. S, 1987, Helminth parasites of the alimentary tract of the hakes (*Merluccius*, *Urophycis*, *Phycis*: Teleostei) of the Scotian Shelf. *Can. J. Zool.*, **65**, 304–311.

Scott, J. S, 1988, Helminth parasites of redfish (*Sebastes fasciatus*) from the Scotian Shelf, Bay of Fundy, and eastern Gulf of Maine. *Can. J. Zool.*, **66**, 617–621.

Scott, M. E, 1982, Reproductive potential of *Gyrodactylus bullatarudis* (Monogenea) on guppies (*Poecilia reticulata*). *Parasitology*, **85**, 217–236.

Scott, M. E, 1985, Experimental epidemiology of *Gyrodactylus bullatarudis* (Monogenea) on guppies (*Poecilia reticulata*): short- and long-term studies. In: *Ecology and Genetics of Host–Parasite interactions.* (Eds Rollinson, D. and Anderson, R. M.). London: Academic Press, pp. 22–36.

Scott, M. E, 1987, Temporal changes in aggregation: a laboratory study. *Parasitology*, **94**, 583–595.

Scott, M. E. and Anderson, 1984, The population dynamics of *Gyrodactylus bullatarudis* (Monogenea) within laboratory populations of the fish host *Poecilia reticulata*. *Parasitology*, **89**, 159–194.

Scott, M. E. and Nokes, D. J, 1984, Temperature-dependent reproduction and survival of *Gyrodactylus bullatarudis* (Monogenea) on guppies (*Poecilia reticulata*). *Parasitology*, **89**, 221–227.

Scott, M. E. and Robinson, M. A, 1984, Challenge infections of *Gyrodactylus bullatarudis* (Monogenea) on guppies, *Poecilia reticulata* (Peters) following treatment. *J. Fish Biol.*, **24**, 581–586.

Sedgwick, S. D, 1978, *Trout farming handbook*. London: Seeley, Service and Co. 169 pp.

Sekerak, A. D. and Arai, H. P, 1973, Helminths of *Sebastes alutus* (Pisces: Teleostei) from the north eastern Pacific. *Can. J. Zool.*, **51**, 475–477.

Sekerak, A. D. and Arai, H. P, 1977, Some metazoan parasites of rockfishes of the genus *Sebastes* from the northeastern Pacific Ocean. *Syesis*, **10**, 139–144.

Sekhar, C. S. and Threlfall, S, 1970a, Helminth parasites of the cunner, *Tautogolabrus adspersus* (Walbaum) in Newfoundland. *J. Helminth.*, 44, 169–188.

Sekhar, C. S. and Threlfall, W, 1970b, Infection of the cunner, *Taugogolabrus adspersus* (Walbaum), with metacercariae of *Cryptocotyle lingua* (Creplin, 1825). *J. Helminth.*, 44, 189–198.

Sekretaryuk, K. V, 1980, [Study of transaminase and aldolase activity in *Philometroides* infection in carp.] *Dokl. Vses. Akad. Sel'skokhoz. Nauk V. I. Lenina, No. 11*, 37–38. [In Russian.]

Sekretaryuk, K. V, 1982, [Interrelationships between helminths in bothriocephaliasis in carp.] *Veterinariya, Mosk., No. 10*, 35–37. [In Russian.]

Sekretaryuk, K. V, 1983a, [Morphological and histochemical changes during *Philometroides* infection in carp.] *Veterinariya, Mosk., No. 9*, 45–47. [In Russian.]

Sekretaryuk, K. V, 1983b, [Genetic polymorphism of the proteins of carp muscle and of a pathogenic nematode.] *Sel'skokhoz. Biol., No. 5*, 109–111. [In Russian.]

Sekretaryuk, K. V, 1983c, [Histological and histochemical investigation of the intestine of carp with bothriocephaliasis.] *Parazitologiya*, 17, 203–206. [In Russian.]

Sekretaryuk, K. V, 1984, [Ultrastructure of the zones of contact between the helminth and the intestine during bothriocephaliasis in carp.] *Dokl. Vses. Akad. Sel'skokhoz. Nauk V. I. Lenina, No. 4*, 35–37. [In Russian.]

Self, J. T. and Timmons, H. F, 1955, The parasites of the river carpsucker (*Carpiodes carpio* Raf.) in Lake Texoma. *Trans. Am. Microsc. Soc.*, 74, 350–352.

Self, J. T., Peters, L. E. and Davis, C. E, 1963, The egg, miracidium and adult of *Nematobothrium texomensis* (Trematoda: Digenea). *J. Parasit.*, 49, 731–736.

Seng, L. T., Huat, K. K., Lee, S. F., Tan, E., S. P. and Yong, W. S, 1987, Parasites of fishes from Tasik Temengor in Perak, Malaysia. *Malayan Nature J.*, 41, 75–82.

Seo, B. S., Chai, J. Y., Lee, S. H. and Hong, S. T, 1984, A human case infected by the larva of *Terranova* type A in Korea. *Korean J. Parasit.*, 22, 248–252.

Seo, B. S., Cho, S. Y., Chai, J. Y. and Hong, S. T, 1980, Studies on the intestinal trematodes in Korea. II. Identification of metacercariae *Heterophyes heterophyes* in mullets of three southern coastal areas. *Seol. J. Med.*, 21, 30–38.

Seo, B. S., Lee, S. H., Chai, J. Y. and Hong, S. J, 1984a, Studies on intestinal trematodes in Korea. XII. Two cases of human infection by *Stellantchasmus falcatus. Korean J. Parasit.*, 22, 43–50.

Seo, B. S., Lee, S. H., Chai, J. Y. and Hong, S. J, 1984b, Studies on intestinal trematodes in Korea. XIII. Two cases of natural human infection by *Heterophyopsis continua* and the status of metacercarial infection in brackish water fishes. *Korean J. Parasit.*, 22, 51–60.

Serdyukov, A. M, 1979, [*Difillobotriidy zapadnoi Sibiri.*] Novosibirsk, USSR: 'Nauka', Sibirskoe Otdelenie, 120 pp. [In Russian.]

Serov, V. G, 1984, [Fecundity of *Acanthocephalus lucii* (Echinorhynchidae).] *Parazitologiya*, 18, 280–285. [In Russian.]

Seyda, M, 1973, Parasites of eel *Anguilla anguilla* (L.) from the Szczecin Firth and adjacent waters. *Acta Ichthyol. Pisc.*, 3, 67–76.

Shaharom, F. M. and Lester, R. J. G, 1982, Description of and observations on *Grillotia branchi* n. sp. a larval trypanorynch from the branchial arches of the Spanish mackerel *Scomberomorus commersoni. Syst. Parasit.*, 4, 1–6.

Shamsuddin, M., Nader, I. A. and Al-Azzawi, M. J, 1971, Parasites of common fishes from Iraq with special reference to larval form of *Contracaecum* (Nematoda: Heterocheilidae). *Bull. Biol. Res. Centre, Baghdad*, 5, 66–78.

Shariff, M., Richards, R. H. and Sommerville, C, 1980, The histopathology of acute and chronic infections of rainbow trout *Salmo gairdneri* Richardson with eye flukes, *Diplostomum* spp. *J. Fish Dis.*, 3, 455–465.

Shaw, M. K, 1979a, The ultrastructure of the clamp wall of the monogenean gill parasite *Gastrocotyle trachuri. Z. ParasitKde*, 58, 243–258.

Shaw, M. K, 1979b, The development of the clamp attachment organs of the monogenean *Gastrocotyle trachuri. Z. ParasitKde*, 59, 277–294.

Shaw, M. K, 1980, The ultrastructure of the epidermis of *Diplectanum aequans* (Monogenea). *Parasitology*, 80, 9–21.

Shaw, M. K, 1981, The ultrastructure of synapses in the brain of *Gastrocotyle trachuri* (Monogenea: Platyhelminthes). *Cell Tissue Res.*, 220, 181–189.

Shaw, M. K, 1982, The fine structure of the brain of *Gastrocotyle trachuri* (Monogenea: Platyhelminthes). *Cell Tissue Res.*, 226, 449–460.

Shchepkina, A. M, 1978, [The effect of *Contracaecum aduncum* larvae on the lipid composition of *Engraulis encrasicholus ponticus.*] *Biologiya Morya, Kiev, No. 45*, 109–112. [In Russian.]

Shchepkina, A. M, 1980, [Lipid composition of the tissues of *Engraulis encrasicholus* during its annual cycle and in infection with larvae of the nematode *Contracaecum aduncum*.] *Ekologiya Morya, No. 3*, 33–39. [In Russian.]

Shchepkina, A. M, 1981, [The effect of *Cryptocotyle concavum* metacercariae on the lipid content of the tissue of *Gobius melanostomus*.] *Parazitologiya*, 15, 185–187. [In Russian.]

Shcherban, M. P, 1965, *Cestode infections of carp*. Izdatelstvo Urozhai, Kiev, pp. 1–79.

Shchupakov, I. G, 1954, [New data on the parasite fauna of gwyniad fish acclimatized in the Ural.] *Trudi Problemnikh i Tematicheskikh Soveschchani, Akademiya Nauk SSSR*, 4, 24–28. [In Russian.]

Shendge, S. R. and Deshmukh, P. G, 1977, *Thwaitia macronesi* n. sp. (Nematoda: Philometridae) from the fish *Macrones seenghala*. *Riv. Parassit.*, 38, 157–160.

Shepherd, C. J, 1978, Husbandry and management in relation to disease. In: *Fish pathology* (Ed. Roberts, R. J.). London, UK: Baillière Tindall, pp. 276–282.

Shigin, A. A, 1980, [Trematodes of the genus *Diplostomum* in biocoenoses of the trout farm Skhodnya.] *Trudy Gel'm. Lab. (Gel'm. vod. naz. biotsen.)*, 30, 140–202. [In Russian.]

Shigin, A. A, 1981, [Components of freshwater biocoenoses seen in their role as 'eliminators' of helminths.] *Itogi Nauki i Techniki, Zooparazitologiya, Vol. 7, Popyulyatsionnye i biotsenoticheskie aspekty izucheniya gel'mintov*, 89–133. [In Russian.]

Shigin, A. A, 1986, *{Trematodes of the USSR. Genus Diplostomum. Metacercariae.}* Moscow, USSR: 'Nauka', 256 pp. [In Russian.]

Shigin, A. A. and Gorovaya, T. V, 1974, [The participation of *Cladocera* in the elimination of *Diplostomum* (Diplostomatidae) cercariae.] *Trudy Gel'm. Lab. (Ekol. geogr. gel'm.)*, 24, 232–240. [In Russian.]

Shimazu, T, 1975, [On the parasites of krill, *Euphausia similis*, from Sugura Bay. VI. Larval nematode.] *Jap. J. Parasit.*, 24, 362–264. [In Japanese.]

Shimazu, T, 1979, Developmental stages of *Azygia gotoi* (Digenea: Azygiidae). *Bull. Nat. Sci. Mus. A (Zool.)*, 5, 225–234.

Shimazu, T, 1980, *Dimerosaccus* gen. nov. (Digenea: Opecoelidae) with a redescription of its type species *Dimerosaccus oncorhynchi* (Eguchi, 1931) comb. nov. *Jap. J. Parasit.*, 29, 163–168.

Shimazu, T, 1990a, Trematodes of the genus *Crepidostomum* (Digenea: Allocreadiidae: Crepidostominae) from freshwater fishes of Japan. *Japan J. Nagano Pref. Coll.*, 45, 1–14.

Shimazu, T, 1990b, Trematodes of the genus *Urorchis* (Digenea: Opecoelidae: Urorchiinae) from freshwater fishes of Japan. *Jap. J. Parasit.*, 39, 204–212.

Shimazu, T, 1990c, Trematodes of a new genus, *Neoplagioporus* gen. n. (Digenea: Opecoelidae: Plagioporinae), and an unidentified opecoelid from freshwater fishes of Japan. *Jap. J. Parasit.*, 39, 384–396.

Shimazu, T, 1990d, Trematodes of the genus *Orientocreadium* (Digenea: Orientocreadiidae) from freshwater fishes of Japan. *Zool. Sci.*, 7, 933–938.

Shinde, G. B. and Chincholikar, L. N, 1977, *Mastacembellophyllaeus nandedensis* (Cestoda: Cestodaria) Monticelli, 1892 n.g. et n. sp. from freshwater fish at Nanded Marine Station, India. *Riv. Parassit.*, 38, 171–175.

Shiraki, T, 1974, Larval nematodes of family Anisakidae (Nematoda) in the northern Sea of Japan—as a causative agent of eosinophilic phlegmone or granuloma in the human gastro-intestinal tract. *Acta Med. Biol.*, 22, 57–98.

Shishova-Kasatochkina, O. A. and Dubovskaya, A. Ya, 1977, [Some aspects of protein metabolism in cestodes, parasites of vertebrates of different classes.] *Trudy Gel'm. Lab. (Tsestod. tremat. morf. sist. ekol.)*, 27, 211–221. [In Russian.]

Shishkova-Kasatochkina, O. A., Koloskova, T. G. and Sokhina, L. I, 1976, [Action of the enzymes arginase and urease in a possible regulatory mechanism at tissue level of osmotic and temperature homeostasis in fish nematodes (Abstract).] In: *Kratkie tezisy dokladov II Vsesoyuznogo simpoziuma po parazitam i boleznyam morskikh zhivotnykh*. Kaliningrad, USSR: Ministerstvo Rybnogo Khozyaistva SSSR, AtlantNIRO, [pp. 71–72. [In Russian.]

Shoop, W. L, 1988, Trematode transmission patterns. *J. Parasit.*, 74, 46–59.

Shoshkov, D. and Georgiev, S, 1983, [Problems in the prophylaxis and control of diseases in freshwater fish.] *Vet. Sbirka*, 81, 14–16. [In Bulgarian.]

Shostak, A. W. and Dick, T. A, 1986, Intestinal pathology in northern pike, *Esox lucius* L., infected with *Triaenophorus crassus* Forel, 1868 (Cestoda: Pseudophyllidea). *J. Fish Dis.*, 9, 35–45.

Shotter, R. A, 1976, The distribution of some helminth and copepod parasites in the tissues of whiting, *Merlangus merlangus* L. from Manx waters. *J. Fish Biol.*, 8, 101–117.

Shotter, R. A, 1980, Aspects of the parasitology of the catfish *Clarias anguillaris* (L.) from a river and a lake at Zaria, Kaduna State, Nigeria. *Bull. Inst. Fond. Afr. Noire, Série A*, 42, 836–859.

Shotter, R. A. and Medaiyedu, J. A, 1977, The parasites of *Polypterus endlicheri* Heckel (Pisces: Polypteridae) from the river Galma at Zaria, Nigeria, with a note on its food. *Bull. Inst. Fond. Afr. Noire, Série A*, 39, 177–189.

Shul'man, B. S, 1977, [Seasonal dynamics of monogeneans from the genus *Gyrodactylus* in *Phoxinus phoxinus* from the river Pecha (Kol'skii peninsula).] In: *Isseldovaniya monogenei v SSSR. (Materialy vsesoyuznogo Simpoziuma po monogeneyma, 16–18 Noyabrya, 1976, Leningrad.* Leningrad, USSR: Akademiya Nauk SSR, Zoologicheskii Institut, pp. 71–77. [In Russian.]

Shul'man, R. E. and Shul'man, S. S, 1983, [History of the development of the ideas of V. A. Dogiel in the field of ecological parasitology. Ecology of fish parasites.] In: *Svobodnozhivushchie i paraziticheskie bespozvonochnye (morfologiya biologiya, evolutsiya). (Trudy Biologicheskogo Nauchno-Issledovatel'skogo Instituta No. 34,* (Ed. Polyanskii, Yu. I.). Leningrad, USSR: Izdatel'stvo Leningradskogo Universiteta, pp. 157–174. [In Russian.]

Shul'man, S. S, 1954, [Specificity of fish parasites.] *Zool. Zh.*, 33, 14–35. [In Russian.]

Shul'man, S. S, 1961a, Zoogeography of parasites of USSR freshwater fishes. In: *Parasitology of fishes.* (Eds Dogiel, V. A., Petrushevski, G. K. and Polyanski, Yu. I.) Edinburgh: Oliver and Boyd. pp. 180–245.

Shul'man, S. S, 1961b, Specificity of fish parasites. In: *Parasitology of fishes.* (Eds Dogiel, V. A., Petrushevski, G. K. and Polyanski, Yu. I.). Edinburgh: Oliver and Boyd, pp. 104–117.

Siddiqui, A. A. and Kulkarni, T, 1985, Localization of phosphomonoesterases in *Hamatopeduncularia indicus* Siddiqui and Kulkarni, 1983. *Indian J. Parasit.*, 9, 47–48.

Sidorov, V. S. and Gur'yanova, S. D, 1981, [Content of free amino-acids in the livers of burbot and stickleback infected with *Diphyllobothrium* plerocercoids.] *Parazitologiya*, 15, 126–131. [In Russian.]

Sidorov, V. S. and Smirnov, L. P, 1980, [Fatty acid composition of some helminths of cold-blooded and warm-blooded vertebrates.] *Zh. Evol. Biol. Fiziol.*, 16, 551–555. [In Russian.]

Sidorov, V. S., Vysotskaya, R. U. and Bykhovskaya-Pavlovskaya, I. E, 1972, [Amino-acid composition of some fish helminths.] In: *Losoevye (Salmonidae) Karelii. Vypusk l. Ekologiya, parazitofauna, bio-khimiya.* Petrozavodsk, USSR: Karel'skii filial Akademii Nauk SSR, pp. 144–151. [In Russian.]

Siegel, V, 1980, Parasite tags for some Antarctic channichthyid fish. *Arch. Fisch.*, 31, 97–103.

Silan, P. and Maillard, C, 1986, Modalités de l'infestation par *Diplectanum aequens*, monogène ectoparasite de *Dicentrarchus labrax* en aquaculture. Eléments d'épidémiologie et de prophylaxie. In: *Pathology in Marine Aquaculture (Pathologie en Aquaculture Marine)* (Eds Vivarès, C. P., Bonami, J. R. and Jaspers, E.). Bredene, Belgium: European Aquaculture Society, pp. 139–152.

Silan, P., Cabral, P. and Maillard, C, 1985, Enlargement of the host range of *Polylabris tubicirrus* (Monogenea, Polyopisthocotylea) under fish-farming conditions. *Aquaculture*, 47, 267–270.

Silan, P., Euzet, L. and Maillard, C, 1983, La réproduction chez *Diplectanum* aequans (Monogenea: Monopisthocotylea). Nouvelles données sur l'anatomie du complexe genital et son fonctionnement. *Bull. Soc. Fr. Parasit.*, 1, 31–36.

Silan, P., Euzet, L. and Maillard, C, 1987, Le biotope des ectoparasites branchiaux de poissons: facteurs de variations dans le modèle bar-monogènes. *Bull. Ecol.*, 18, 383–391.

Simmons, J. E, 1961, Urease activity in trypanorhynch cestodes. *Biol. Bull.*, 121, 535–546.

Simmons, J. E, 1969, Composition of the amino acid pools of some cestodes of elasmobranch fishes of the Woods Hole area. *Expl Parasit.*, 26, 264–271.

Simmons, J. E, 1970, Nitrogen metabolism in Platyhelminthes. In: *Comparative biochemistry of nitrogen metabolism I. The invertebrates.* (Ed. Campbell, J. W.). London: Academic Press, pp. 67–90.

Simmons, J. E, 1974, *Gyrocotyle*: a century-old enigma. In: *Symbiosis in the Sea* (Ed. Vernberg, W. B.). Charleston: University of South Carolina Press, pp. 195–218.

Simmons, J. E. and Laurie, J. S, 1972, A study of *Gyrocotyle* in the San Juan Archipelago, Puget Sound, USA, with observations on the host, *Hydrolagus colliei* (Lay and Bennett). *Int. J. Parasit.*, 2, 59–77.

Simmons, J. E., Read, C. P. and Rothman, A. H, 1960, Permeation and membrane transport in animal parasites: permeation of urea into cestodes from elasmobranchs. *J. Parasit.*, 46, 43–50.

Simmons, J. E., Buteau, G. H., MacInnis, A. J. and Kilejian, A, 1972, Characterization and hybridization of DNAs of gyrocotylidean parasites of chimaeroid fishes. *Int. J. Parasit.*, 2, 273–278.

Simon-Martin, F. and Rojo-Vazquez, F, 1984, Scanning electron microscope study of mature and immature worms of a *Sanguinicola* sp. (Digenea: Sanguinicolidae). *Annls Parasit. Hum. Comp.*, 59, 353–360.

Sindermann, C. J, 1961, Parasite tags for marine fish. *J. Wildl. Mgmt*, 25, 41–47.

Sindermann, C. J, 1966, Diseases of marine fishes. *Adv. Mar. Biol.*, 4, 1–89.

Sindermann, C. J, 1970a, *Principal diseases of marine fish and shellfish*. New York: Academic Press, x + 369 pp.

Sindermann, C. J, 1970b, Bibliography of diseases and parasites of marine fish and shellfish with emphasis on commercially important species. *Tropical Atlantic Biological Laboratory Informal Report No. 11*, 440 pp.

Sindermann, C. J, 1974, Diagnosis and control of mariculture diseases in the United States. *Middle Atlantic Coastal Fisheries Center National Marine Fisheries Service; National Oceanic and Atmospheric Administration; U.S. Department of Commerce, Technical Series No. 2*, 8 pp.

Sindermann, C. J, 1983, Parasites as natural tags for marine fish: a review. Presented at Special Session, Bedford Inst. of Oceanography, Dartmouth, Nova Scotia, Canada, 8–10 Sept. 1982. *Northwest Atlantic Fisheries Organization, Scientific Council Studies, No. 6*, 63–71.

Sindermann, C. J, 1986, Effects of parasites on fish populations: practical considerations. In: *Parasitology—Quo Vadit? Proceedings of the sixth International Congress of Parasitology, Brisbane, Australia, 24–29 Aug., 1986* (Ed. Howell, M. J.). Canberra, Australia: Australian Academy of Science, pp. 371–382.

Sindermann, C. J. and Lightner, D. V. (Ed.), 1988, *Disease diagnosis and control in North American marine aquaculture*. Amsterdam, Netherlands: Elsevier Science Publishers, Second Edition, 431 pp.

Sindermann, C. J. and Rosenfield, A, 1954, Diseases of fishes of the western North Atlantic. 3. Mortalities of sea herring (*Clupea harengus*) caused by larval trematode invasion. *Marine Dept. Sea Shore Fish, Res. Bull., No. 21*, 16 pp.

Sinha, D. P. and Hopkins, C. A, 1967, Studies on *Schistocephalus solidus*. 4. The effect of temperature on growth and maturation *in vitro*. *Parasitology*, 57, 555–556.

Sinitsin, D. F, 1901, Einige Beobachtungen über die Entwicklungsgeschichte von *Distomum folium* Olf. Vorläufige Mitteilung. *Zool. Anz.*, 24, 689–694.

Sinitsin, D. F, 1931, Studien über die Phylogenie der Trematoden. IV. Der Entwicklungszyklus von *Plagioporus siliculus* und *Plagioporus virens* mit besondere Berücksichtigung des Ursprungs der Digenea. *Z. Wiss. Zool.*, **138**, 409–456.

Sircar, M. and Sinha, D. P, 1974, Haematological investigations on pigeons and *Clarias batrachus* carrying cestode infections. *Annls Zool., Agra.*, 10, 1–11.

Sircar, M. and Sinha, D. P, 1980, Histopathology of *Lytocestus indicus* infection in the fish, *Clarias batrachus*. *Indian J. Anim. Res.*, 14, 53–56.

Skachkov, D. P, 1978, [Tests with okside, sulphene and halosphene in *Bothriocephalus* infection in carp.] *Byull. Vses. Inst. Gel'm. K. I. Skryabina, No. 22*, 72–74. [In Russian.]

Skachkov, D. P, 1981, [Comparative therapeutic efficacy of halosphene, bithionol, BMC and Lopatol against *Bothriocephalus* infections in carp.] *Byull. Vses. Inst. Gel'm. K. I. Skryabina, No. 29*, 52–54. [In Russian.]

Skachkov, D. P., Kozachenko, N. G. and Skvortsova, F. K, 1981, [Effect of anthelmintic treatment on the haematological picture in carp fry infected with *Bothriocephalus*.] *Byull. Vses. Inst. Gel'm. K. I. Skryabina, No. 30*, 99–103. [In Russian.]

Skachkov, D. P., Moskvin, A. S., Nazarova, N. S., Kozachenko, N. G. and Romashko, V, 1982, [Efficacy of halosphene-medicated granulated mixed feed against *Bothriocephalus* infections in carp.] *Byull. Vses. Inst. Gel'm. K. I. Skryabina, No. 32*, 69–71. [In Russian.]

Skinner, R, 1975, Parasites of the striped mullet, *Mugil cephalus*, from Biscayne Bay, Florida, with descriptions of a new genus and three new species of trematodes. *Bull. Mar. Sci.* 25, 318–345.

Skinner, R. H, 1978, Some external parasites of Florida fishes. *Bull. Mar. Sci.*, **28**, 590–595.

Skinner, R. H, 1982, The interrelation of water quality, gill parasites and gill pathology of some fishes from south Biscayne Bay, Florida. *Fishery Bull.*, **80**, 269–280.

Skomorokhova, N. K. and Kashkovskii, V. V, 1979, [Parasites and diseases of fish at the Bileisk fish farm (USSR).] In: *Bolezni i parazity ryb Ledovitomorskoi provintsii (v predelakh SSSR)*. (Eds Gundrizer, A. N. and Bauer, O. N.). Tomsk, USSR: Izdatel'stvo Tomskogo Universiteta, pp. 110–115. [In Russian.]

Skörping, A, 1980, Population biology of the nematode *Camallanus lacustris* in perch, *Perca fluviatilis* L., from an oligotrophic lake in Norway. *J. Fish Biol.*, 16, 483–492.

Skörping, A, 1981, Seasonal dynamics in abundance, development and pattern of infection of *Bunodera luciopercae* (Müller) in perch, *Perca fluviatilis* L. from an oligotrophic lake in Norway. *J. Fish Biol.*, **18**, 401–410.

Skripchenko, Z. G, 1973, [Changes over several years in the parasite fauna of carp on pond farms in West Siberia.] In: *Vodoemy Sibiri i perspektivy ikh rybokhozyaistvennogo ispol'zovaniya*. Tomsk, USSR: Izdatel'stvo Tomskogo Universiteta, pp. 238–239. [In Russian.]

Skryabina, E. S, 1975, [Helminths of fish in the lower reaches of the river Chaun.] In: *Paraziticheskie organizmy severo-vostoka Azii.* Vladivostok, USSR: Akademiya Nauk SSSR, Dal'nevostochnyi Nauchnyi Tsentr. pp. 180–186. [In Russian.]

Skryabina, E. S, 1978, [A systematic review of Acanthocephala of fish in the USSR.] *Trudy Gel'mintologicheskoi Laboratorii (Nematody i akantotsefaly, morfologiya, sistematika, ekologiya i fiziologiya*, **28**, 166–190. [In Russian.]

Skvortsova, F. K, 1979, [*Valipora* infection in carp.] *Veterinariya, Mosk., No. 3*, 54–56. [In Russian.]

Skvortsova, F. K., Kozachenko, N. T. and Nikulina, A. N, 1985, [Ultrastructural changes in the tegument of fish cestodes caused by anthelmintics.] *Veterinariya, Mosk., No. 7*, 40–42. [In Russian.]

Smija, D, 1982, [Quantitative investigations on the parasitization of fishes from different excavation ponds, with special reference to their succession.] *Gesundheitsprobleme des Menschen im Zusammenhangmit Fischen. Beiträge zur Fischereibiologie.* Stuttgart, GFR: Gustav Fischer Verlag *Fisch und Umwelt, No. 11* 57–75. [In German.]

Smirnov, L. P, 1983, [Comparative study of the protein composition of the membranes of lysosomes of *Eubothrium crassum* and of the liver of the host—*Salvelinus lepechini*—by means of polyacrylamide gel electrophoresis.] In: *Sravnitel'naya biokhimiya vodnykh zhivotnykh.* (Eds Sidorov, V. S. and Vysotskaya, R. U.). Petrozavodsk, USSR: Karel'skii Filial AN SSSR, pp. 150–154. [In Russian.]

Smirnov, L. P. and Sidorov, V. S, 1978, [Comparative study of the protein composition of the cestode *Eubothrium crassum* and some organs of its host, *Salvelinus lepechini* Gmelin, by gel chromatography.] In: *Ekologicheskaya biokhimiya zhivotnykh.* (Ed. Sidorov, V. S.). Petrozavodsk, USSR: Institut Biologii Karel'skii Filial AN SSSR, pp. 53–60. [In Russian.]

Smirnov, L. P. and Sidorov, V. S, 1979, [Fatty acid composition of the cestodes *Eubothrium crassum* and *Diphyllobothrium dendriticum*.] *Parazitologiya*, **13**, 522–520. [In Russian.]

Smirnov, L. P. and Sidorov, V. S, 1981, [Study of the protein composition of the cestodes *Eubothrium crassum* and *Diphyllobothrium dendriticum* by gel chromatography and disc electrophoresis.] In: *Sravnitel'nye aspekty biokhimiii ryb i hekotorykh drugikh zhivotnykh.* (Ed. Sidorov, V. S.). Petrozavodsk, USSR: Institut Biologii, Karel'skii Filial AN SSSR, pp. 80–87. [In Russian.]

Smirnov, L. P. and Sidorov, V. S, 1984, [Comparative study of protein spectra in cestodes and their hosts by means of gel chromatography and disc electrophoresis.] *Parazitologyia*, **18**, 430–435. [In Russian.]

Smith, H. D, 1973, Observations on the cestode *Eubothrium salvelini* in juvenile sockeye salmon (*Oncorhynchus nerka*) at Babine Lake, British Columbia. *J. Fish. Res. Bd Can.*, **30**, 947–964.

Smith, H. D. and Margolis, L, 1970, Some effects of *Eubothrium salvelini* (Schrank, 1790) on sockeye salmon, *Oncorhynchus nerka* (Walbaum), in Babine Lake, British Columbia. *J. Parasit.*, **56**, 321–322.

Smith, J. D, 1984, Development of *Raphidascaris acus* (Nematoda, Anisakidae) in paratenic, intermediate, and definitive hosts. *Can. J. Zool.*, **62**, 1378–1386.

Smith, J. D, 1986, Seasonal transmission of *Raphidascaris acus* (Nematoda), a parasite of freshwater fishes, in definitive and intermediate hosts. *Environ. Biol. Fish.*, **16**, 295–308.

Smith, J. D. and Lankester, M. W, 1979, Development of swim bladder nematodes (*Cystidicola* spp.) in their intermediate hosts. *Can. J. Zool.*, **57**, 1736–1744.

Smith, J. and Tyler, S, 1985, The acoel turbellarians: kingpins of metazoan evolution or a specialised offshoot? In: *The origin and relationships of lower invertebrates. Systematic Association, Special Volume 28,* (Eds Conway Morris, S. *et al.*). Oxford, UK: Clarendon Press, pp. 123–142.

Smith, J. W, 1972, The blood flukes (Digenea: Sanguinicolidae and Spirorchidae) of cold-blooded vertebrates and some comparisons with schistosomes. *Helminth. Abstr.*, **41**, 161–204.

Smith, J. W, 1983a, Larval *Anisakis simplex* (Rudolphi, 1809, det. Krabbe, 1878) and larval *Hysterothylacium* sp. (Nematoda: Ascaridoidea) in euphausiids (Crustacea: Malacostraca) in the North-East Atlantic and northern North Sea. *J. Helminth.*, **57**, 167–177.

Smith, J. W, 1983b, *Anisakis simplex* (Rudolphi, 1809, det. Krabbe, 1878) (Nematoda: Ascaridoidea): morphology and morphometry of larvae from euphausiids and fish, and a review of the life-history and ecology. *J. Helminth.*, **57**, 205–224.

Smith, J. W, 1984a, The abundance of *Anisakis simplex* L3 in the body-cavity and flesh of marine teleosts. *Int. J. Parasit.*, **14**, 491–495.

Smith, J. W, 1984b, *Anisakis simplex* (Rudolphi, 1809, det. Krabbe, 1878): length distribution and viability of L3 of known minimum age from herring *Clupea harengus* L. *J. Helminth.*, **58**, 337–340.

Smith, J. W. and Wootten, R, 1975, Experimental studies on the migration of *Anisakis* sp. larvae (Nematoda: Ascaridida) into the flesh of herring, *Clupea harengus* L. *Int. J. Parasit.*, **5**, 133–136.

Smith, J. W. and Wootten, R, 1978a, *Anisakis* and anisakiasis. *Adv. Parasit.*, 16, 92–163.

Smith, J. W. and Wootten, R, 1978b, Further studies on the occurrence of larval *Anisakis* in blue whiting. *ICESCM1978/H*, 53, 3pp.

Smith, J. W. and Wootten, R, 1984a, Parasitose des poissons par les larves du nematode *Pseudoterranova. Fiches d'Identification des Maladies et Parasites des Poissons, Crustaces et Mollusques.* (Ed. Sinderman, C. J.). Conseil International Pour l'Exploration de la Mer, Palaegrade 2–4, DK-1261 Copenhague K, Danemark.

Smith, J. W. and Wootten, R, 1984b, Parasitose de poissons par les larves du nematode *Anisakis. Fiches d'Identification des Maladies et Parasites des Poissons, Crustaces et Mollusques.* (Ed. Sinderman, C. J.). Conseil International Pour l'Exploration de la Mer, Palaegrade 2–4, DK-1261 Copenhague K, Danemark.

Smith, J. W. and Wootten, R, 1984c, Parasitose des poissons par les larves des nematodes *Phocascaris/ Contracaecum. Fiches d'Identification des Maladies et Parasites des Poissons, Crustaces et Mollusques.* (Ed. Sinderman, C. J.). Counseil International Pour l'Exploration de la Mer, Palaegrade 2–4, DK-1261 Copenhague K, Danemark.

Smith, R. S. and Kramer, D. L, 1987, Effects of a cestode (*Schistocephalus* sp.) on the response of nine-spine sticklebacks (*Pungitius pungitius*) to aquatic hypoxia. *Can. J. Zool.*, 65, 1862–1865.

Smith Trail, D. R, 1980, Behavioural interactions between parasites and hosts: host suicide and the evolution of complex life cycles. *Amer. Naturalist*, 116, 77–91.

Smitherman, R. O, 1968, Effect of the strigeid trematode, *Posthodiplostomum minimum*, upon growth and mortality of bluegill, *Lepomis macrochirus. {Proc. of the FAO World Symp. on warm-water pond fish culture, Rome, Italy, 18–25 May, 1966 Volume 5. FAO Fisheries Reports 1966.}, No. 44*, 380–388.

Smithers, S. R. and Worms, M. J, 1976, Blood fluids—helminths. In: *Ecological aspects of parasitology* (Ed. Kennedy, C. R.) Oxford: North Holland Publishing Company, pp. 349–369.

Smyth, J. D, 1966, *The physiology of trematodes*. Edinburgh: Oliver and Boyd, 256pp.

Smyth, J. D, 1969, *The physiology of cestodes*. Edinburgh: Oliver and Boyd, 279pp.

Smyth, J. D, 1976, *An introduction to animal parasitology*, London, Hodder and Stoughton, 2nd edition, xiv + 466pp.

Smyth, J. D, 1990, Cestoda. In: *In vitro cultivation of parasitic helminths* (Ed. Smyth, J. D.) Boca Raton: CRC Press, pp. 77–154.

Smyth, J. D. and Halton, D. W, 1983, *The physiology of trematodes*. Cambridge: Cambridge University Press, xiii + 446 pp.

Smyth, J. D. and Haslewood, G. A. D, 1963, The biochemistry of bile as a factor in determining host specificity in intestinal parasites, with particular reference to *Echinococcus granulosus. Ann. NY Acad. Sci.*, 113, 234–260.

Smyth, J. D. and McManus, D. P, 1989, *The physiology and biochemistry of cestodes*. Cambridge: Cambridge University Press, xi + 398 pp.

Snieszko, S. F, 1969, Cold-blooded vertebrate immunity to Metazoa. In: *Immunity to parasitic animals.* (Eds Jackson, G. J., Herman, R. and Singer, I.). New York: Appleton-Century-Crofts, pp. 267–275.

Snieszko, S. F, 1970, *A symposium on diseases of fishes and shellfishes*. Washington, D. C.: American Fisheries Society, 526 pp.

Snieszko, S. F, 1973, Diseases of fishes and their control in the U.S.A. In: *The Two Lakes fifth fishery management training course report, Two Lakes, Romsey, Hampshire, 5–7 October, 1973*, Janssen Services, London, pp. 55–66.

Snieszko, S. F., Hoffman, G. L. and McAllister, P. E, 1979, Fish diseases. In: *Wildlife conservation, principles and practice.* (Eds Teague, R. D. and Decker, E.). Washington, DC, USA: Wildlife Society, pp. 269–274.

Sobenin, A. P, 1975, [Feeding competition between acclimatized *Coregonus* spp. and local fish and the parasitological factor.] *Vop. Zool., No. 4*, 104–106. [In Russian.]

Sogandares, L. A, 1961, Nine digenetic trematodes of marine fishes from the Atlantic coast of Panama. *Tulane Stud. Zool.*, 8, 141–153.

Sogandares-Bernal, F, 1959a, *Cleptodiscus kyphosi*, a new trematode (Paramphistomatidae) in *Kyphosus sectatrix* (Linn.) from Bimini, B. W. I. *J. Parasit.*, 45, 148–149.

Sogandares-Bernal, F, 1959b, Digenetic trematodes of marine fishes from the Gulf of Panama and Bimini, British West Indies. *Tulane Stud. Zool.*, 7, 71–117.

Sogandares-Bernal, F. and Bridgman, J. F, 1960, Three *Ascocotyle* complex trematodes (Heterophyidae) encysted in fishes from Louisiana, including the description of a new genus. *Tulane Stud. Zool.*, 8, 31–39.

Sogandares-Bernal, F. and Lumsden, R. D, 1963, The generic status of the heterophyid trematodes of the *Ascocotyle* complex, including notes on the systematics and biology of *Ascocotyle angrense* Travassos, 1916. *J. Parasit.*, 49, 264–274.

Sogandares-Bernal, F., Hietala, H. J. and Gunst, R. F, 1979, Metacercariae of *Ornithodiplostomum ptychocheilus* (Faust, 1917) Dubois, 1936 encysted in the brains and viscera of red-sided shiners from the Clark-Fork and bitterroot Rivers of Montana: an analysis of the infected hosts. *J. Parasit.*, 65, 616–623.

Soleim, Ø, 1984, A synopsis of the genera *Thynnascaris* and *Contracaecum* (Nematoda: Ascaridoidea) with an emendation of the generic definitions. *Acta Parasit. Pol.*, 29, 85–96.

Soleim, Ø. and Berland, B, 1981, The morphology of *Thynnascaris adunca* (Rudolphi) (Nematoda: Ascaridoidea). *Zool. Scr.*, 10, 167–182.

Solonchenko, A. I, 1979, [Life-cycle of the cestode *Bothriocephalus scorpii* (Müller, 1776) [Abstract].] In: *VII Vsesoyuznoe Soveshchanie po parazitam i boleznyam ryb, Leningrad, Sentyabr', 1979 g. (Tezisy dokladov).* Leningrad, USSR: 'Nauka', pp. 104–105. [In Russian.]

Solonchenko, A. I, 1982, [Helminth fauna of fish in the Azov Sea.] *Gel'mintofauna ryb Azovskogo Morya.* Kiev, USSR: Naukova Dumka, 150pp. [In Russian.]

Solonchenko, A. I. and Radchenko, C. A, 1987, [Relationship of *Artemia salina* with some cestode species.] In: *Actual Problems in Fish Parasitology. Proceedings of the Second International Symposium of Ichthyoparasitology, Tihany, Hungary*, pp. 89.

Solonchenko, A. I. and Tkachuk, L. P, 1985, [[Prevalence of helminths in mullets in the Azov-Black Sea basin.] *Ekologiya Morya, Kiev, No. 20*, 39–43. [In Russian.]

Sommerville, C, 1981a, A comparative study of the tissue response to invasion and encystment by *Stephanochasmus baccatus* (Nicoll, 1907) (Digenea: Acanthocolpidae) in four species of flatfish. *J. Fish Dis.*, 4, 53–68.

Sommerville, C, 1981b, Parasites of ornamental fish. [*Waltham Symp. No. 3. 'The diseases of ornamental fishes.' Proc. Symp. 1–2 March 1980, Pedigree Petfoods Anim. Studies Centre, Waltham-on-the Wolds, Leics., UK* (Ed. Ford, D. M.).] *J. Small Anim. Pract.*, 22, 367–376.

Sommerville, C, 1982a, The life history of *Haplorchis pumilio* (Looss, 1896) from cultured tilapias. *J. Fish Dis.*, 5, 233–241.

Sommerville, C, 1982b, The pathology of *Haplorchis pumilio* (Looss, 1896) infections in cultured tilapias. *J. Fish Dis.*, 5, 243–250.

Sommerville, R. I. and Davey, K. G, 1976, Stimuli for cuticle formation and ecdysis *in vitro* of the infective larva of *Anisakis* sp. (Nematoda: Ascaridoidea). *Int. J. Parasit.*, 6, 433–439.

Sonin, M. D. (Ed.), 1985, [*Key to the trematodes of fish-eating birds in the Palaearctic (Brachylaimida, Clinostomida, Cyclocoelida, Fasciolida, Notocotylida, Plagiorchida, Schistosomatida).*] Moscow, USSR: Nauka, 256 pp. [In Russian.]

Soota, T. D, 1983, Studies on nematode parasites of Indian vertebrates. 1. Fishes. *Records of the Zoological Survey of India, Occasional Paper No. 54*, xii + 352 pp.

Soprunov, F. F, 1984, [Achievements in the studies of carbohydrate metabolism in helminths.] *Trudy Gel'm. Lab. (Biokhim. fiziol. gel'm. immun. gel'm.)*, 32, 121–154. [In Russian.]

Sous', S. M, 1979, [The effect of water levels of lakes on the long-term dynamics of the prevalence of *Parasymphylodora markewitschi* (Trematoda: Monorchidae) infection in fish.] *Izv. Sibirsk. Otdel. Akad. Nauk SSSR, Biol. 3, No. 15*, 121–124. [In Russian.]

Southwell, T, 1925, *A monograph on the Tetraphyllidea with notse on related cestodes*, Liverpool: Liverpool University Press, 386 pp.

Soutter, A. M., Walkey, M. and Arme, C, 1980, Amino acids in the plerocercoid of *Ligula intestinalis* (Cestoda: Pseudophyllidea) and its fish host, *Rutilus rutilus. Z. ParasitKde*, 63, 151–158.

Spall, R. D. and Summerfelt, R. C, 1970, Life cycle of the white grub, *Posthodiplostomum minimum* (MacCullum 1921) (Trematoda: Diplostomatidae) and observations on host-parasite relationships of the metacercaria in fish. In: *Symposium on diseases of fishes and shellfishes.* (Ed. Snieszko, S. F.). *Amer. Fish Soc. Spec. Publ. No. 5, Washington D. C.*, p. 218.

Spannhof, L., Reimer, L. and Jürss, K, 1989, Physiology, Biology and Parasitology of Farmed Fish. *VI. Scientific Conference on the Physiology, Biology and Parasitology of Farmed Fish, 29 July to 1 October 1986, Gustrow. Agriculture*, 80, 378–379.

Sparks, A. K, 1954, A new species of *Multitestis* (Trematoda: Allocreadiidae) from the sheeps-head (*Archosargus probatocephalus*) in the Gulf of Mexico. *Trans. Am. Microsc. Soc.*, 73, 36–38.

Sparks, A. K. and Thatcher, V. E, 1958, A new species of *Stephanostomum* (Trematoda: Acanthocolpidae) from marine fishes of the Northern Gulf of Mexico. *Trans. Am. Microsc. Soc.*, 72, 287–290.

Specian, R. D., Ubelaker, J. E. and Dailey, M. D, 1975, *Neoleptus* gen. n. and a revision of the genus *Proleptus* Dujardin, 1845. *Proc. Helminth. Soc. Wash.*, 42, 14–21.

Speed, P. and Pauley, G. B, 1984, The susceptibility of four salmonid species to the eyefluke, *Diplostomum spathaceum. Northwest Sci.*, **58**, 312–316.

Speed, P. and Pauley, G. B, 1985, Feasibility of protecting rainbow trout, *Salmo gairdneri* Richardson, by immunizing against the eye fluke, *Diplostomum spathaceum. J. Fish Biol.*, **26**, 739–744.

Sprent, J. F. A, 1969, Helminth 'zoonoses': an analysis. *Helminth. Abstr.*, **38**, 333–351.

Sprent, J. F. A, 1982, Host–parasite relationships of ascaridoid nematodes and their vertebrate hosts in time and space. *{Deux. Symp. spéc. parasit. parasites de vertébrés, 13–17 Avril, 1987.} Méms Mus. Nat. Hist. Nat., Sér. A, Zool.*, **123**, 255–262.

Sprent, J. F. A, 1983, Observations on the systematics of ascaridoid nematodes. In: *Concepts in nematode systematics. (Proc. Internat. Symp., Cambridge Univ., 2–4 Sept. 1981.) Systematics Association Special Volume, 22.* (Eds Stone, A. R., Platt, H. M. and Khalil, L. F.). London: Academic Press, pp. 303–319.

Sproston, N. G, 1946, A synopsis of the monogenetic trematodes. *Trans. Zool. Soc., Lond.*, **25**, 185–600.

Srivastava, A. B. and Gupta, S. P, 1975, On two new species of the genus *Amplicaecum* Baylis, 1920 from marine fishes of Puri, Orissa. *Indian J. Helminth. (Published 1977)*, **27**, 119–123.

Srivastava, C. B. and Mukherjee, G. D, 1976, Studies on the incidence of infestation of *Isoparorchis hypselobagri* (Billet, 1898) metacercaria in two species of fishes of the genus *Mystus. J. Zool. Soc. India, (1974, publ. 1976)*, **26**, 131–137.

Srivastava, L. P, 1966a, The morphology of *Lepidapedon cambrensis* sp. nov. (Digenea: Lepocreadiidae) from the large intestine of *Onos mustelus* (L.) with a historical review of the genus. *Annls Mag. Nat. Hist., Ser. 13*, **9**, 111–122.

Srivastava, L. P, 1966b, The morphology of *Lecithaster musteli* sp. nov. (Digenea: Hemiuridae) from the intestine of *Onos mustelus* (L.) and a review of the genus *Lecithaster* Lühe, 1901. *Parasitology*, **56**, 543–554.

Srivastava, L. P, 1966c, The helminth parasites of the five-bearded rockling, *Onos mustelus* (L.) from the shore at Mumbles Head, Swansea. *Annls Mag. Nat. Hist., Ser. 13*, **9**, 469–480.

Srivastava, L. P, 1966d, A re-description of *Stephanostomum caducum* (Looss, 1901) (Digenea: Acanthocolpidae) from the intestine of *Onos mustelus* (L.). *Annls Mag. Nat. Hist., Ser. 13*, **9**, 399–403.

Stables, J. and Chappell, L. H, 1986a, Putative immune response of rainbow trout, *Salmo gairdneri* to *Diplostomum spathaceum* infections. *J. Fish Biol.*, **29**, 115–122.

Stables, J. N. and Chappell, L. H, 1986b, The epidemiology of diplostomiasis in farmed rainbow trout from north-east Scotland. *Parasitology*, **92**, 699–710.

Stables, J. N. and Chappell, L. H, 1986c, *Diplostomum spathaceum* (Red. 1819): effects of physical factors on the infection of rainbow trout (*Salmo gairdneri*) by cercariae. *Parasitology*, **93**, 71–79.

Stark, G. T. C, 1965, *Diplocotyle* (Eucestoda), a parasite of *Gammarus zaddachi* in the estuary of the Yorkshire Esk, Britain. *Parasitology*, **55**, 415–420.

Starling, J. A, 1985, Feeding and metabolism. In: *Biology of the Acanthocephala.* (Eds Crompton, D. W. T. and Nickel, B. B.). Cambridge: Cambridge University Press, pp. 125–212.

Starovoitov, V. K, 1986, [Peculiarity of localization of *Ancyrocephalus paradoxus* (Monogenea) on *Stizostedion lucioperca.*] *Parazitologiya*, **20**, 491–492. [In Russian.]

Steele, J. H, 1966, Ecological problems of marine fish farming. *Proc. Nutr. Soc.*, **25**, 126.

Steen, J. B, 1970, The swimbladder as a hydrostatic organ. In: *Fish Physiology* Vol. IV (Eds Hoar, W. S. and Randall, D. J.) London: Academic Press, pp. 413–443.

Sten'ko, R. P, 1976, [The life-cycle of *Crowcrocaecum skrjabini* (Iwanitzky, 1928) (Allocreadiata, Opecoelidae).] *Parazitologiya*, **10**, 9–16. [In Russian.]

Sterry, P. R. and McManus, D. P, 1982, *Ligula intestinalis*: biochemical composition, carbohydrate utilisation and oxygen consumption of plerocercoids and adults. *Z. ParasitKde*, **67**, 87–98.

Stockell, G, 1936, The nematode parasites of Lake Ellesmere trout. *Trans. Proc. R. Soc. N. Z.*, **66**, 80–96.

Stone, A. R., Platt, H. M. and Khalil, L. F. (Ed.), 1983, Concepts in nematode systematics. *Proceedings of an International Symposium held jointly by the Systematics Association and the Association of Applied Biologists, held at Cambridge University, 2–4 September 1981.* (*Systematics Association Special Volume No. 22*). London: Academic Press, x + 390 pp.

Stranack, F, 1972, The fine structure of the acanthor shell of *Pomphorynchus laevis* (Acanthocephala). *Parasitology*, **64**, 187–190.

Strazhnik, L. V, 1980, [The carbohydrate composition of fish cestodes.] *Gidrobiol. Zh.*, **16**, 87–91. [In Russian.]

Strazhnik, L. V, 1981, [The effect of increased temperatures on the carbohydrate metabolism and egg production of *Bothriocephalus gowkongensis* Yeh, 1955—a parasite of carp.] In: *Ekologo-morfologicheskie osobennosti zhivotnykh i sreda ikh obitaniya*. Kiev, USSR: 'Naukova Dumka', pp. 146–148. [In Russian.]

Strazhnik, L. V. and Davydov, O. N, 1971, [Comparison of thiamin content in the tissues of some fish cestodes.] *Gidrobiol. Zh.*, 7, 98–101. [In Russian.]

Strelkov, Yu. A, 1983, [Regulation of parasite abundance in lacustrine ecosystems by different groups of parasitic animals.] *Sbornik Nauchnykh Trudov Gosudarstvennogo Nauchno-Issledovatel'skogo Instituta Ozernogo i Rechnogo Khozyaistva (Problemy ekologii parazitov ryb), No. 197*, 3–16. [In Russian.]

Strelkov, Yu. A. and Shul'man, S. S, 1971, [An ecological and faunistic analysis of the parasites of fish in the Amur.] *Parazit. Sb.*, 25, 196–292. [In Russian.]

Strelkov, Yu. A., Chernysheva, N. B. and Yunchis, O. N, 1981, [Rules regulating the formation of the parasite fauna in freshwater fish fry.] *Trudy Zool. Inst. Leningr. (Ekol. aspek. parazit.)*, **108**, 23–30. [In Russian.]

Stromberg, P. C. and Crites, J. L, 1974a, The life-cycle and development of *Camallanus oxycephalus* Ward and Magath, 1916 (Nematoda: Camallanidae). *J. Parasit.*, **60**, 117–124.

Stromberg, P. C. and Crites, J. L, 1974b, Survival, activity and penetration of the first-stage larvae of *Camallanus oxycephalus* Ward and Magath, 1916. *Int. J. Parasit.*, 4, 417–421.

Stromberg, P. C. and Crites, J. L, 1974c, Triaenophoriasis in Lake Erie white bass, *Morone chrysops*. *J. Wildl. Dis.*, **10**, 352–358.

Stromberg, P. C. and Crites, J. L, 1975a, Population biology of *Camallanus oxycephalus* Ward and Magath, 1916 (Nematoda: Camallanidae) in white bass in western Lake Erie. *J. Parasit.*, 61, 123–132.

Stromberg, P. C. and Crites, J. L, 1975b, An analysis of the changes in the prevalence of *Camallanus oxycephalus* (Nematoda: Camallanidae) in western Lake Erie. *Ohio J. Sci.*, 75, 1–6.

Stumpp, M, 1975, Investigations on the morphology and biology of *Camallanus cotti* (Fujita, 1927). *Z. ParasitKde*, 46, 277–290.

Stunkard, H. W, 1956, The morphology and life-history of the digenetic trematode, *Azygia sebago* Ward, 1910. *Biol. Bull.*, **111**, 248–268.

Stunkard, H. W, 1962a. *Caballerocotyla klawei* sp. n. a monogenetic trematode from the nasal capsule of *Neothunnus macropterus*. *J. Parasit.*, **48**, 883–890.

Stunkard, H. W, 1962b, The organization, ontogeny, and orientation of the Cestoda. *Quart. Rev. Biol.*, 37, 23–34.

Stunkard, H. W, 1963, Systematics, taxonomy and nomenclature of the Trematoda. *Quart. Rev. Biol.*, **38**, 221–233.

Stunkard, H. W, 1964, The morphology, life-history and systematics of the digenetic trematode, *Homalometron pallidum* Stafford, 1904. *Biol. Bull.*, **126**, 163–173.

Stunkard, H. W, 1967, Platyhelminthic parasites of invertebrates. *J. Parasit.*, 53, 673–682.

Stunkard, H. W, 1974a, The trematode family Bucephalidae—problems of morphology, development and systematics. Description of *Rudolphinus* gen. nov. *Trans. N.Y. Acad. Sci., Series II*, **36**, 143–170.

Stunkard, H. W, 1974b, *Rhipidocotyle heptathelata* n. sp., a bucephalid trematode from *Thynnus thunnina* taken in the Red Sea. *Trans. Am. Microsc. Soc.*, **93**, 260–261.

Stunkard, H. W, 1974c, The life-cycle of the gasterostome trematodes, *Rhipidocotyle transversale*, Chandler, 1935 and *Rhipidocotyle lintoni* Hopkins, 1954. *Biol. Bull.*, 147, 500–501.

Stunkard, H. W, 1975, Life histories and systematics of parasitic flatworms. *Syst. Zool.*, 24, 378–385.

Stunkard, H. W, 1976, The life cycles, intermediate hosts and larval stages of *Rhipidocotyle transversale* Chandler, 1935 and *Rhipidocotyle lintoni* Hopkins, 1954: life-cycles and systematics of bucephalid trematodes. *Biol. Bull.*, **150**, 294–317.

Stunkard, H. W, 1978, Metacercariae of digenetic trematodes from ctenophores and medusae in the Woods Hole, Massachusetts area. *Biol. Bull.*, 155, 467–468.

Stunkard, H. W, 1980a, Successive hosts and developmental stages in the life history of *Neopechona cablei* sp. n. (Trematoda: Lepocreadiidae). *J. Parasit.*, **66**, 636–641.

Stunkard, H. W, 1980b, The morphology, life-history, and taxonomic relations of *Lepocreadium areolatum* (Linton, 1900) Stunkard, 1969 (Trematoda: Digenea). *Biol. Bull.*, **158**, 154–163.

Stunkard, H. W, 1980c, The morphology, life-history and systematic relations of *Tubulovesicula pinguis* (Linton, 1940) Manter, 1947 (Trematoda: Hemiuridae). *Biol. Bull.*, **159**, 737–751.

Stunkard, H. W, 1983, Evolution and systematics. In: *Biology of the Eucestoda. Vol. I.* (Eds Arme, C. and Pappas, P. W.). London, UK: Academic Press, pp. 1–25.

Sudarikov, V. E. and Shigin, A. A, 1975, The significance of the components of aquatic biocoenoses in the elimination of trematodes. *Trudy Gel'm Lab.*, 25, 168–180. [In Russian.]

Sulgostowska, T. and Styczynska-Jurewicz, E, 1987, Helminth parasites of flounder *Platichthys flesus* L. in highly eutrophicated gulf of Gdansk, Southern Baltic sea. *2nd International Symposium of Ichthyoparasitology. Actual problems in fish parasitology Sept 27–October 3, 1987*. Tihany, Hungary, p. 93.

Sulgostowska, T., Banaczyk, G. and Grabda-Kazubska, B, 1987, Helminth fauna of flatfish (Pleuronectiformes) from Gdansk Bay and adjacent areas (south-east Baltic). *Acta Parasit. Pol.*, 31, 231–240.

Sullivan, M. X, 1908, The physiology of the digestive tract of elasmobranchs. *Bull, Bur, Fish., Wash.*, 27, 1–27.

Suriano, D. M, 1977, Parasites of elasmobranchs from the coastal region of the Mar del Plata, Argentina (Monogenea: Monopisthocotylea). *Neotropica*, 23, 161–172.

Suzuki, T. and Ishida, K, 1979, *Anisakis simplex* and *Anisakis physeteris*: physicochemical properties of larval and adult haemoglobins. *Expl Parasit.*, 48, 225–234.

Suzuki, T., Ishida, K., Asaishi, K. and Nishino, C, 1976, [Studies on the immunodiagnosis of anisakiasis. 6. Analysis of criteria on intradermal and indirect haemagglutination tests by means of radioimmunoassay.] *Jap. J. Parasit.*, 25, 17–23. [In Japanese.]

Svobodova, Z, 1978, [Values of some external features, condition and physiological indices in two-year-old carp infected by the cestode *Bothriocephalus gowkongensis*.] *Buletin VURH (Vyzkumny Ustav Rybarsky a Hydrobiologicky) Vodnany, CSSR*, 14, 21–25 [In Czech.]

Sweeting, R. A, 1976, Studies on *Ligula intestinalis* (L.) effects on a roach population in a gravel pit. *J. Fish Biol.*, 9, 515–522.

Sweeting, R. A, 1977, Studies on *Ligula intestinalis*. Some aspects of the pathology in the second intermediate host. *J. Fish Biol.*, 10, 43–50.

Swennen, C., Heessen, H. J. L. and Hocker, A. W. M, 1979, Occurrence and biology of the trematodes *Cotylurus (Ichthyocotylurus) erraticus C. (I.) variegatus* and *C. (I.) platycephalus* (Digenea: Strigeidae) in the Netherlands. *Netherl. J. Sea Res.*, 13, 161–191.

Swiderski, Z, 1976, Fine structure of the spermatozoon of *Lacistorhynchus tenuis* (Cestoda: Trypanorhyncha). *Proceedings of the Sixth European Congress on electron microscopy, Jerusalem, 1976*, pp. 309–310.

Swiderski, Z, 1985, Spermiogenesis in the proteocephalid cestode *Proteocephalus longicollis. Electron microscopy of Southern Africa—Proceedings*, 15, 181–182.

Swiderski, Z, 1986, Three types of spermiogenesis in cestodes. *Proc. XIth Int. Cong. on Electron Microscopy, Kyoto*, 2959–2960.

Swiderski, Z. and Eklu-Natey, R. D, 1978, Fine structure of the spermatozoon of *Proteocephalus longicollis* (Cestoda: Proteocephalidea). *Proceedings of the 9th International Congress on electron microscopy, Toronto, 1978, Vol. 2*, 572–573.

Swiderski, Z. and Mackiewicz, J. S, 1976, Fine structure of the spermatozoon of *Glaridacris catostomi* (Cestoidea: Caryophyllidea). *Proceedings of the sixth European Congress on electron microscopy, Jerusalem, 1976*, 307–308.

Swiderski, Z. and Mokhtar-Maamouri, F, 1974, Vitellogenesis in *Bothriocephalus clavibothrium*, Ariola, 1899 (Cestoda: Pseudophyllidea). *Z. ParasitKde*, 43, 135–149.

Swiderski, Z. and Mokhtar-Maamouri, F, 1980, Etude de la spermatogénèse de *Bothriocephalus clavibothrium* Ariola, 1899 (Cestoda: Pseudophyllidea). *Archs Inst. Pasteur, Tunis*, 57, 323–347.

Swiderski, Z., Eklu-Natey, R. D, Subilia, L. and Huggel, H, 1978, Fine structure of the vitelline cells in the cestode *Proteocephalus longicollis* (Proteocephalidea). *Proceedings of the 9th International Congress on electron microscopy, Toronto, 1978, Vol. 2,* pp. 442–443.

Syamsunder, R. P. and Aruna, K, 1984, Succinate dehydrogenase activity in the fish parasite *Isoparorchis hypselobagri* (Trematoda: Digenea). *Proc. Indian Acad. Parasit.*, 5, 39–41.

Sysoev, A. V, 1982, Composition and dynamics of invasion of the first intermediate hosts of *Triaenophorus nodulosus* (Pallas) (Cestoda: Triaenophoridae) under conditions in Karelia. *Helminthologia*, 19, 249–255.

Sysoev, A. V, 1983, The composition of intermediate hosts of *Proteocephalus torulosus* (Batsch) (Cestoda: Proteocephalidae) and the dynamics of invasion of Copepoda with this parasite under conditions in Karelia. *Helminthologia*, 20, 97–102.

Szalai, A. J., Yang, X. and Dick, T. A, 1989, Changes in numbers and growth of *Ligula intestinalis* in the spottall shiner (*Notropis hudsonius*) and their roles in transmission. *J. Parasit.*, 75, 571–576.

Székely, Cs. and Molnár, K, 1987, Mebendazole is an efficacious drug against Pseudodactylogyrosis in the European eel (*Anguilla anguilla*). *2nd International Symposium of Ichthyoparasitology. Actual problems in fish parasitology Sept 27–October 3, 1987. Tihany, Hungary*, p. 94

Szidat, L, 1956, Über den Entwicklungszyklus mit progenetischen Larvenstadien (Cercariaeen) von

Genarchella genarchella Travassos 1928 (Trematoda: Hemiuridae) und die Möglichkeit einer hormonalen Beeinflussung der Parasiten durch ihre Wirtsiere, *Z. Tropenmed. Parasit.*, 7, 132–153.

Szidat, L, 1959, Hormonale Beinflussung von Parasiten durch ihren Wirt. *Z. ParasitKde*, 19, 503–524.

Szidat, L, 1960, La parasitología como ciencia auxiliar para develar problemas hidrobiológicos, zoogeográficos y geofisicos del Atlántico Sud. *Libro Homenaje al Dr. Eduardo Caballero y Caballero, Jubileo, 1930–1960*, 577–594.

Szidat, L, 1969, Parasites of the palometa *Parona signata* (Jenyns, 1842) Berg, 1895, and their application to the zoogeographical problems of the South Atlantic. *Neotropica*, 15, 125–131.

Szuks, H, 1980, Die Verwendbarkeit von Parasiten zur Gruppentrennung beim Grenadierfisch *Macrourus rupestris. Angew. Parasit.*, 21, 211–214.

Tada, I., Otsuji, Y., Kamiya, H., Mimori, T., Sakaguchi, Y. and Makizumi, S, 1983, The first case of a human infected with an acanthocephalan parasite, *Bolbosoma* sp. *J. Parasit.*, 69, 205–208.

Tandon, R. S. and Misra, K. C, 1980, Ascorbic acid levels of trematode parasites of fish and mammalian hosts. *Z. ParasitKde*, 62, 191–193.

Tandon, R. S. and Misra, K. C, 1985, Cholesterol levels of trematode parasites of fish and mammalian hosts. *Indian J. Parasit.*, 9, 67–68.

Tang, C. C. and Ling, S. M, 1975, {*Sanguinicolosis lungensis* n. sp. and outbreaks of sanguinicolosis in Lien-Yüe nursery ponds in South China.] *Xiamen Daxue Xuebao, No. 2.* 139–160. [In Chinese.]

Tang, Z. and Tang, C, 1980, [Life histories of two species of aspidogastrids and the phylogeny of the group.] *Acta Hydrobiol. Sin.*, 7, 153–174. [In Chinese.]

Taranenko, V. M., Davydov, O. N. and Sinits'kii, S. V, 1976, [Effect of potassium ions on electrical and contractile activity of the muscle cells of the cestode *Ligula intestinalis} Dopov Akad Nauk Ukraïns'koi RSR., B, Geol Khim Biol Nauki* No. 7, 656–658. [In Russian.]

Taraschewski, H, 1986, *Heterophyes* sp., sea borne parasites of animals and man. In: *Parasitology-Quo Vadit? Handbook Proceedings of the Sixth International Congress of Parasitology, Canberra, Australia* (Ed. Howell, M. J.). Canberra: Australian Academy of Science; p. 243

Taraschewski, H. and Nicolaidou, A, 1987, *Heterophyes* species in Greece: record of *H. heterophyes, H. aequalis* and *H. dispar* from the first intermediate host, *Pirenella conica. J. Helminth.*, 61, 28–32.

Taraschewski, H., Moravec, F., Lamah, T. and Anders, K, 1987, Distribution and morphology of two helminths recently introduced into European eel populations: *Anguillicola crassus* (Nematoda: Dracunculoidea) and *Paratenuisentis ambiguus* (Acanthocephala: Tenuisentidae). *Dis. Aquat. Org.*, 3, 167–176.

Tarmakhanov, G. D, 1987, [Age variations in the parasite fauna of perch in the Chivyrkuiskii Bay on Lake Baikal.] *Sibirsk Ordel. Akad. Nauk SSSR, Bio. Nauk, No. 3*, 93–96. [In Russian.]

Taylor, A. E. R. and Baker, J. R. (Eds), 1987, *Methods of cultivating parasites in vitro.* London: Academic Press, vii + 465 pp.

Taylor, M. and Hoole, D, 1987, *Ligula intestinalis* (L.) (Cestoda: Pseudophyllidea): plerocercoid-induced changes in the spleen and pronephros of roach, *Rutilus rutilus* (L.) and gudgeon, *Gobio gobio* (L.). *J. Fish Biol.*, 34, 583–596.

Tedesco, J. L, and Coggins, J. R, 1980, Electron microscopy of the tumulus and origin of associated structures within the tegument of *Eubothrium salvelini* Schrank, 1790 (Cestoidea: Pseudophyllidea). *Int. J. Parasit.*, 10, 275–280.

Tedla, S. and Fernando, C. H, 1969, Observations on the seasonal changes of the parasite fauna of yellow perch (*Perca flavescens*) from the Bay of Quinte, Lake Ontario. *J. Fish. Res. Bd Can.*, 26, 833–843.

Tedla, S. and Fernando, G. H, 1970/71, On the characterization of the parasite fauna of yellow perch (*Perca fluviatilis* L.) in five lakes in Southern Ontario, Canada. *Helminthologia*, 11, 23–33.

Terekhov, P. A, 1976, [The parasites of *Rutilus rutilus heckeli* and *Lucioperca lucioperca* fry.] *Veterinariya, Mosk.*, 5, 72–73. [In Russian.]

Terenina, N. B, 1983, [Serotonin and dopamine in the tissue of pseudophyllidean cestodes.] *Zh. Evol. Biokhim. Fiziol.*, 19, 302–304. [In Russian.]

Tesarčik, J, 1971, The present chemoprophylactic and chemotherapeutic methods employed in the treatment of the most important fresh-water fish diseases in the Czech Socialist Republic. *Proceedings of the Symposium held at Ceské Budejovice, 22–24 September 1971, Fisheries Research Institute, Vodnany, Czechoslovakia 1971*, 181–200.

Tesarčik J, 1972, Dehelminthization of carp fry infected with intestinal worm *Neoechinorhynchus rutili* by using Tetrafinol, *Acta Veterinaria Brno*, 41, 207–210.

Thiel, P. H, van 1966a, The final host of the herringworm *Anisakis marina*. *Trop. Geogr. Med.*, 18, 310–328.

Thiel, P. H, van 1966b, The herringworm *Anisakis* spec. and its final host in Corradetti, A. (Ed.) *Proceedings of the first International Congress of Parasitology*, Vol. 1, Oxford: Pergamon, pp. 800–801.

Thiel, P. H, van 1976, The present state of anisakiasis and its causative worms. *Trop. Geogr. Med.*, 28, 75–85.

Thiel, P. H. van and Bakker, P. M, 1981, Wormgranulomen in de maag in Nederland en in Japan. *Nederl. Tijdschr. Geneesk.*, 125, 1365–1370.

Thiel, P. H, van and van Houten, H, 1967, The localization of the herringworm *Anisakis marina* in and outside the human gastro-intestinal wall. *Trop. Geogr. Med.*, 19, 56–62.

Thiel, P. H. van, Kuipers, F. C. and Roskam R. Th., 1960, A nematode parasitic to herring causing acute abdominal syndromes in man. *Trop. Geogr. Med.*, 2, 97–113.

Thomas, J. D, 1958, Three new digenetic trematodes, *Emoleptalea proteropora* n. sp. (Cephalogonimidae: Cephalogoniminae), *Phyllodistomum symmetrorchis* n. sp., and *Phyllodistomum ghanense* n. sp. (Gorgoderidae) from West African freshwater fishes. *Proc. Helminth. Soc. Wash.*, 25, 1–8.

Thomas, J. D, 1959, Trematodes of Ghanaian sub-littoral fishes. I. The family Monorchiidae. *J. Parasit.*, 45, 95–113.

Thomas, J. D, 1964, Studies on the growth of trout, *Salmo trutta* from four contrasting habitats. *Proc. Zool. Soc. Lond.*, 142, 459–509.

Thomas, L. J, 1930, Notes on the life history of *Haplobothrium globuliforme* Cooper, a tapeworm of *Amia calva* L. *J. Parasit.*, 16, 140–145.

Thomas, L. J, 1953, *Bothriocephalus abyssmus* n. sp. a tapeworm from the deep-sea fish, *Echiostoma tanneri* (Gill) with notes on its development. In: *Thapar Commemorative Volume*, (Eds Dayal, J., and Singh, K. S.) pp. 269–276.

Thompson, P. A. and Threlfall, W, 1978, The metazoan parasites of two species of fish from the Port-Cartier-Sept-Iles Park, Québec. *Naturaliste Canadien*, 105, 429–431.

Thompson, R. C. A., Hayton, A. R. and Jue Sue, L. P, 1980, An ultrastructural study of the microtriches of adult *Proteocephalus tidswelli* (Cestoda: Proteocephalidae). *Z. ParasitKde*, 64, 95–111.

Thoney, D. A. and Burreson, E. M, 1986, Ecological aspects of *Multicalyx cristata* (Aspidocotylea) infections in North West Atlantic elasmobranchs. *Proc. Helminth. Soc. Wash.*, 53, 162–165.

Thoney, D. A. and Burreson, E. M, 1987, Morphology and development of the adult and cotylocidium of *Mulicalyx cristata* (Aspidocotylea), a gall bladder parasite of elasmobranchs. *Proc. Helminth. Soc. Wash.*, 45, 96–104.

Threlfall, W, 1970, Some helminth parasites from *Illex argentinus* (de Castellanos, 1960) (Cephalopoda: Ommastrephidae). *Can. J. Zool.*, 48, 195–198.

Thulin, J, 1980a, A redescription of the fish blood-fluke *Aporocotyle simplex* Odhner, 1900 (Digenea: Sanguinicolidae) with comments on its biology. *Sarsia*, 65, 35–48.

Thulin, J, 1980b, Scanning electron microscope observations of *Aporocotyle simplex* Odhner, 1900 (Digenea: Sanguinicolidae). *Z. ParasitKde*, 63, 27–32.

Thulin, J, 1980c, Redescription of *Nematobibothrioides histoidii* Noble, 1974 (Digenea: Didymozooidea). *Z. ParasitKde*, 63, 213–219.

Thulin, J, 1982, Structure and function of the female reproductive ducts of the fish blood-fluke *Aporocotyle simplex* Odhner, 1900 (Digenea: Sanguinicolidae). *Sarsia*, 67, 227–248.

Thune, R. L. and Rogers, W. A, 1981, Gill lesions in bluegill, *Lepomis macrochirus* Rafinesque, infested with *Cleidodiscus robustus* Mueller, 1934 (Monogenea: Dactylogyridae). *J. Fish Dis.*, 4, 277–280.

Timofeeva, S. V. and Marasaeva, E. F, 1984, [The parasite fauna of two forms of cod in the Kandalaksh Bay, White Sea.] In: *Ekologo-parasitologicheskie issledovaniya severnykh morei*. Apatity, USSR: Kol'skii Filial Akademii Nauk SSSR, pp. 62–76. [In Russian.]

Timofeeva, T. A, 1985, [Morphological and ecological aspects of evolution in Monocotylidae (Monogenea).] *Parazit. Sb.*, 33, 44–76. [In Russian.]

Timoshechkina, L. G, 1978, [Dynamics of the parasite fauna of *Abramis brama* in the Gor'kovskii reservoir.] *Ekologiya Gel'mintov, Yaroslavl, USSR, No. 2*, 72–80. [In Russian.]

Timoshechkina, L. G, 1984, [Aspects of the ultrastructure of the tegument and glandular apparatus of *Caryophyllaeus laticeps.*] *Biol. Vnutr. Vod. Inf. Byull., No. 62*, 30–33. [In Russian.]

Tinsley, R. C, 1983, Ovoviviparity in platyhelminth life-cycles. In: *The reproductive biology of parasites. Symposia of the Brit. Soc. for Parasit., Vol. 20*, (Ed. Whitfield, P. J.). *Parasitology*, 86, 161–196.

Tkachuk, L. P, 1985, Special features of the helminth fauna of *Helicolenus maculatus* (Cuvier). In: *Parasitology and pathology of marine organisms of the World Ocean* (Ed. Hargis, W. J. Jr.). Seattle, USA: Department of Commerce, National Oceanic and Atmospheric Administration (NOAA Technical Report NMFS 29), pp. 45–46.

Todorov, I, 1982, The occurrence of nematode larvae from the genus *Anisakis* in some commercial fish of the Atlantic Ocean. *Ribno Stopanstro*, 28, 14–16. [In Bulgarian.]

Toft, C. A, 1986, Communities of species with parasitic life-styles. In: *Community Ecology* (Eds Diamond, J. and Case, T. J.) New York: Harper and Row, pp. 445–463.

Tokobaev, M. M. and Chibichenko, N. T, 1978, [Diplostomatids of fish in the lakes Issyk-kul' and Son-kul'.] In: *Biologicheskie osnovy rybnogo kozyaistva vodoemov srednei Azii i Kazakstana. Materialy konferentsii*. Frunze, USSR: 'Ilim', pp. 494–496. [In Russian.]

Tomasovicova, O, 1981, The role of fresh water fish in transfer and maintenance of trichinellae under natural conditions. *Biológia Bratislava, Sér. B. Zoológia*, **36**, 115–125.

Torisu, M., Iwasaki, K., Tanaka, J., Iino, H. and Yoshida, T, 1983, *Anisakis* and eosinophils: pathogenesis and biologic significance of eosinophilic phlegmon in human anisakiasis. In: *Immunobiology of the eosinophil. (Proceedings of the First International Symposium, 27–28 Nov. 1981, Fukuoka, Japan.)* (Eds Takeshi, Y. and Torisu, M.). New York, USA: Elsevier Biomedical, pp. 343–366.

Torres, P, 1983, Estado actual de la investigacion sobre cestodos del genero *Diphyllobothrium* Cobbold en Chile. *Rev. Méd. Chile*, **110**, 463–470.

Torres, P, 1990, Primeros registras de endohelmintos parasitos en el salmon coho, *Oncorhynchus kisutch* (Walbaum), introducido en Chile. *Arch. Med. Vet.*, 22, 105–107.

Torres, P, Arenas, J., Neira, A., Cabezas, C., Covarrubias, C., Jara, C., Gallardo, C. and Campos, M, 1988, [Nematodos anisákidos en peces autóctonos de la cuenca del rio Valdivia, Chile.] *Bol. Chil. Parasit.*, 43, 37–41. [In Spanish.]

Torres, P., Cabezas, X., Arenas, J., Miranda, J. C., Jara, C. and Gallardo, C, 1991, Ecological aspects of nematode parasites of introduced salmonids from Valdivia river basin, Chile. *Mem Inst. Oswaldo Cruz, R. J.* **86**, 115–122.

Torres, P., Franjola, R., Figueroa, L., Schlatter, R., Gonzalez, H., Contreras, B. and Martin, R, 1981, Researches on Pseudophyllidea (Carus, 1813) in the south of Chile. IV. Occurrence of *Diphyllobothrium dendriticum* (Nitzch). *J. Helminth.*, 55, 173–187.

Torres, P., Pequeno, G., Jeria, M. E. and San Martin, L, 1981, Larvas de Anisakidae (Railliet y Henry, 1912) Skrjabin y Karokhin, 1945 en peces de la costa sur de Chile. *Bol. Chil. Parasit.*, 36, 39–41.

Torres, P., Teuber, S. and Miranda, J. C, 1990, Parasitismo en ecosistemas de agua dulce de Chile. 2. Nematodos parasitos de *percichthys trucha* (Pisces: Serranidae) con la descripción de una nueva especie de *Camallanus* (Nematoda: Spiruroidea). *Studies on Neotropical Fauna and Environment*, 25, 111–119.

Toteja, G. S., Sood, M. L. and Saxena, P. K, 1982, Occurrence of a trematode in the cranial cavity of a cat fish, *Heteropneustes fossilis*. *J. Res. Punjab Agric. Univ. Ludhiana*, **19**, 105–106.

Totterman, G, 1976, On the pathogenesis of pernicious tapeworm anaemia. *Annls Clin. Res.*, **8**, Suppl. 18, 48pp.

Tregubova, E. F, 1972, [A study of the parasites of *Perca fluviatilis* in the Volgograd reservoir.] In: *Voprosy morfologii, ekologii i gematologii pozvonochnykh*. Saratov, USSR, pp. 95–98. [In Russian.]

Troncy, P. M, 1969, Description de deux nouvelles espèces de nématodes parasites de poissons. *Bull. Mus. Natn. Hist. Nat., Paris*, 598–605.

Tsimabyuk, E. M, 1978, [The helminth fauna of some species of Pleuronectidae in Zaliv Petra Velikogo.] *Izv. Tikhook. Nauchno-Issled. Inst. Ryb. Khozyaist. Okeano. (TINRO)*, **102**, 123–128. [In Russian.]

Tsutsumi, Y. and Fujimoto, Y, 1983, Early gastric cancer superimposed on infestation of an *Anisakis*-like larva: a case report. *Tokai J. Exp. Clin. Med.*, **8**, 265–273.

Tuffery, G, 1978, Récherches sur la bucéphalose à *Bucephalus polymorphus* Baer 1827. *Bull. Acad. Vét. Fr.*, 51, 143–145.

Turnpenny, A. W. H., Bamber, R. N. and Henderson, P. A, 1981, Biology of the sand-smelt (*Atherina presbyter* Valenciennes) around Fawley Power Station. *J. Fish Biol.*, **18**, 417–427.

Turovskij, A, 1985, Parasitofauna of fish in the southern gulf of Finland. *Finn. Fish. Res.*, 6, 106–111.

Ubelaker, J. E, 1983, The morphology, development and evolution of tapeworm larvae. In: *Biology of the Eucestoda Volume 1* (Eds Arme, C. and Pappas, P. W.) London: Academic Press. pp. 235–296.

Uglem, G. L, 1972a, The life cycle of *Neoechinorhynchus cristatus* Lynch, 1936 (Acanthocephala) with notes on the hatching of eggs. *J. Parasit.*, **58**, 1071–1074.

Uglem, G. L, 1972b, A physiological basis for habitat specificity in the acanthocephalan parasites *Neoechinorhynchus cristatus* Lynch, 1938, and *N. crassus* Van Cleave, 1919. *Diss. Abstracts Internat.*, **33B**, 2417.

Uglem, G. L, 1980, Sugar transport by larval and adult *Proterometra macrostoma* (Digenea) in relation to environmental factors. *J. Parasit.*, 66, 748–758.

Uglem, G. L. and Beck, S. M, 1972, Habitat specificity and correlated aminopeptidase activity in acanthocephalan *Neoechinorhynchus cristatus* and *Neoechinorhynchus crassus*. *J. Parasit.*, **58**, 911–920.

Uglem, G. L. and Larson, O. R, 1969, The life history and larval development of *Neoechinorhynchus saginatus* Van Cleave and Bangham, 1945 (Acanthocephala: Neoechinorhynchidae). *J. Parasit.*, 55, 1212–1217.

Uglem, G. L. and Prior, D. J, 1983, Control of swimming in cercariae of *Proterometra macrostoma* (Digenea). *J. Parasit.*, 69, 866–870.

Uhazy, L. S, 1976, *Philometroides huronensis* n. sp. (Nematoda: Dracunculoidea) of the common white sucker (*Catostomus commersoni*) from Lake Huron, Ontario, *Can. J. Zool.*, 54, 369–376.

Uhazy, L. S, 1977a, Development of *Philometroides huronensis* (Nematoda: Dracunculoidea) in the intermediate and definitive hosts. *Can. J. Zool.*, 55, 265–273.

Uhazy, L. S, 1977b, Biology of *Philometroides huronensis* (Nematoda: Dracunculoidea) in the white sucker (*Catostomus commersoni*). *Can. J. Zool.*, 55, 1430–1441.

Uhazy, L. S, 1978, Lesions associated with *Philometroides huronensis* (Nematoda: Philometridae) in the white sucker (*Catostomus commersoni*). *J. Wildl. Dis.*, 14, 401–408.

Ulmer, M. J, 1971, Site-finding behaviour in helminths in intermediate and definitive hosts. In: *Ecology and physiology of parasites*. (Ed. Fallis, A. M.). Toronto, Canada: University of Toronto Press, pp. 123–159.

Ulmer, M. J. and James, H. A, 1981, Monogeneans of marine fishes from the Bay of Naples. *Trans. Am. Microsc. Soc.*, **100**, 392–409.

Ulrich, J, 1983, The biology of *Atrispinum labracis* n. comb. (Monogenea) on the gills of the bass, *Decentrarchus labrax*. *J. Mar. Biol. Ass. UK*, 63, 915–927.

Underwood, H. T. and Dronen, N. O. Jr, 1984, Endohelminths of fishes from the Upper San Marcos River, Texas. *Southwest. Nat.*, 29, 377–385.

Uraev, A. F, 1981, [The ecology of fish parasites in the canal Yu. G. K. im. Sarkisova.] *Uzbek. Biol. Zh., No. 1*, 53–56. [In Russian.]

Urazbaev, A. and Allaniyazova, T, 1977, [Biology and life-cycle of *Bothriocephalus gowkongensis* Yeh, 1955 (Cestoda: Pseudophyllidea) in the Amu Darya river.] In: *Diseases of Fish and their control in Kazakhstan and the Central Asian Republics. Proc. 2nd Reg. Sci. Conf.* USSR: Akad. Nauk. Kaz. SSR, pp. 147–150. [In Russian.]

Uspenskaya, A. V, 1953, [Life-cycle of the nematodes belonging to the genus *Ascarophis* van Beneden.] *Zool. Zh.*, 40, 7–12. [In Russian.]

Uspenskaya, A. V, 1960, Parasitofaune des crustaces benthiques de la mer de Barents (Exposé préliminaire), *Annls Parsit. Hum. Comp.*, 35, 221–242.

Uspenskaya, A. V, 1961, [The effect of *Dactylogyrus vastator* Nybelin, 1924 (Monogenoidea: Dactylogyridae) on carp.] *Zool. Zh.*, 40, 7–12. [In Russian.]

Uznanski, R. L. and Nickol, B. B, 1976, Structure and function of the fibrillar coat of *Leptorhynchoides thecatus* eggs. *J. Parasit.*, 62, 569–575.

Uznanski, R. L. and Nickol, B. B, 1980a, Parasite population regulation: lethal and sublethal effects of *Leptorhynchoides thecatus* (Acanthocephala: Rhadinorhynchidae) on *Hyalella azteca* (Amphipoda). *J. Parasit.*, 66, 121–126.

Uznanski, R. L. and Nickol, B. B, 1980b, A sequential ranking system for developmental stages of an acanthocephalan, *Leptorhynchoides thecatus*, in its intermediate host, *Hyalella azteca*. *J. Parasit.*, 66, 506–512.

Uznanski, R. L. and Nickol, B. B, 1982, Site selection, growth, and survival of *Leptorhynchoides thecatus* (Acanthocephala) during the prepatent period in *Lepomis cyanellus*. *J. Parasit.*, 68, 686–690.

Vaes, M, 1978, Infection of the common goby, *Pomatoschistus microps* with *Aphalloides coelomicola* (Trematoda: Digenea). *Vlaans. Diergeneeskundig Tifdschrift*. 47, 274–278.

Vahida, I, 1984, [Parasites and parasitoses of fish in salmonid fish ponds in Bosnia-Hercegovina. (Excerpt from degree theses)]. *Veterinaria*, 33, 305–322. [In Croat.]

Vala, J. C. and Euzet, L, 1977, *Ktariella polyorchis* n.g., n.sp., parasite du teléostéen *Argyrosomus regius* (Asso, 1801) en Mediterranée. *Vie Milieu*, 27, 1–9.

Valdimarsson, G., Einarsson, H. and King, F. J, 1985, Detection of parasites in fish muscle by candling technique. *J. Ass. Off. Analyt. Chem.*, **68**, 549–551.

Valdiserri, R. O, 1981, Intestinal anisakiasis. Report of a case and recovery of larvae from market fish. *Am. J. Clin. Path.*, 76, 329–333.

Valovaya, M. A, 1979, [The biology of *Cucullanus cirratus* Muller, 1777 (Nematoda, Cucullanata). *Parazitologiya*, 13, 540–544. [In Russian.]

Valter, E. D, 1977, [Study of the biology of trematodes of fish in the White Sea.] *Mater. Nauch. Konf. Vses. Obshch. Gel'm. (Trematody i trematodozy)*, 29, 8–11. [In Russian.]

Valter, E. D, 1980, [Observations on the development of *Contracaecum aduncum* (Ascaridata) in *Jaera albifrons* (Crustacea). *Trudy Belomorsk. Biol. Stant. Mosk. Gosud. Univ.*, 5, 155–164. [In Russian.]

Valtonen, E. T, 1979, *Neoechinorhynchus rutili* (Muller, 1780) (Acanthocephala) in the whitefish *Coregonus masus* (Pallas) *sensu* Svardson from the Bay of Bothnia. *J. Fish. Dis.*, 2, 99–103.

Valtonen, E. T, 1980a, *Metechinorhynchus salmonis* infection and diet in the river-spawning whitefish of the Bothnian Bay. *J. Fish Biol.*, 17, 1–8.

Valtonen, E. T, 1980b, *Metechinorhynchus salmonis* (Muller, 1780) (Acanthocephala) as a parasite of the whitefish in the Bothnian Bay. 1. Seasonal relationships between infection and fish size. *Acta. Parasit. Pol.*, 27, 293–300.

Valtonen, E. T, 1980c, *Metechinorhynchus salmonis* (Muller, 1780) (Acanthocephala) as a parasite of the whitefish in the Bothnian Bay. II. Sex ratio, body length and embryo development in relation to season and site in intestine. *Acta Parasit. Pol.*, 27, 301–307.

Valtonen, E. T, 1983a, Relationships between *Corynosoma semerme* and *C. strumosum* (Acanthocephala) and their paratenic fish hosts in the Bothnian Bay, Baltic Sea. *Acta Univ. Oul.*, No.155, Biol. 21, 32pp.

Valtonen, E. T, 1983b, On the ecology of *Echinorhynchus salmonis* and two *Corynosoma* species (Acanthocephala) in the fish and seals of the northern Gulf of Bothnia. *Acta Univ. Oul.*, No. 156, Biol. 22, 1–49.

Valtonen, E. T, and Crompton, D. W. T, 1990, Acanthocephala in fish from the Bothnian Bay, Finland. *J. Zool., Lond.*, 220, 619–639.

Valtonen, E. T, and Helle, E, 1986, Intestinal metazoan parasites of seals in the Baltic sea. In *Proceedings of the Sixth I.C.O.P.A. Canberra Australia. Parasitology—Quo Vadit?* (Ed. Howell, M. J.), Canberra: Australian Academy of Science. Abstract No. 678. p. 248

Valtonen, E. T. and Helle, E, 1988, Host–parasite relationships between two seal populations and two species of *Corynosoma* (Acanthocephala) in Finland. *J. Zool., Lond.*, 214, 361–371.

Valtonen, E. T, and Koskivaara, M, 1987, The effect of environmental stress on trematodes of perch and roach in central Finland. *Second International Symposium on Icthtbyoparasitology. Actual Problems in Fish Parasitology. Tihany, Hungary*. p. 103

Valtonen, E. T. and Niinimaa, A, 1983, Dispersion and frequency distribution of *Corynosoma* spp. (Acanthocephala) in the fish of the Bothnian Bay. *Aquilo, Ser. Zool.*, 22, 1–13.

Valtonen, E. T. and Valtonen, T, 1978, *Cystidicola farionis* as a swimbladder parasite of the whitefish in the Bothnian Bay. *J. Fish Biol.*, 13, 557–561.

Valtonen, E. T. and Valtonen, T, 1980, Comparison of *Metechinorhynchus salmonis* (Muller 1780) (Acanthocephala) infection in the sea-spawning whitefish in the northeastern and central Bothnian Bay. *Bothnian Bay Reports*, 2, 61–66.

Valtonen, E. T., Fagerholm, H. and Helle, E, 1988, *Contracaecum osculatum* (Nematoda: Anisakidae) in fish and seals in Bothnian bay (Northeastern Baltic Sea). *Int. J. Parasit.*, 18, 365–370.

Valtonen, E. T., Gibson, D. I, and Kurttila, M, 1984, Trematodes in Northern Finland. I. Species maturing in fish in the Northeastern Bothnian Bay and in a local lake. *Bothnian Bay Reports*, 3, 31–44.

Valtonen, E. T., Koskivaara, M. Brummer-Korvednkontio, H, 1987, Parasites of fishes in central Finland in relation to environmental stress. *Biological Research Report, University of Jyvaeskylae*, 10, 129–130.

Valtonen, E. T., Maren, M. J. van and Timola, O, 1983, A note on the intermediate hosts of *Echinorhynchus gadi* Zoega, in Müller (Acanthocephala) in the Baltic Sea. *Aquilo, Ser. Zool.*, 22, 93–97; 678.

Van As, J. G. and Basson, L, 1984, Checklist of freshwater fish parasites from southern Africa. *S. Afr. J. Wildl. Res.*, 14, 49–61.

Van Banning, P. and Becker, H. B, 1978, Long-term survey data (1965–1972) on the occurrence of *Anisakis* larvae (Nematoda: Ascaridida) in herring, *Clupea harengus* L. from the North Sea. *J. Fish Biol.*, 12, 25–33.

Van Cleave, H. J, 1952, Speciation and formation of genera in Acanthocephala. *Syst. Zool.*, 1, 72–83.

Varela, M. C, 1975, [General view of ichthyological helminthology.] *Bolm pecuar.*, [In Portuguese.] 43, 35–83.

Varley, R. L, 1977, Economics of fish farming in the United Kingdom. *Fish Farm. Internat.*, 17–18.

Vasil'Kov, G. V, 1968, [The life-cycle of *Philometra lusiana* (Nematoda: Dracunculidae).] *Trudy Vses. Inst. Gel'mint.*, 14, 156–169. [In German.]

Vasil'Kov, G. V, 1980, [Control of Philometroides infection in carp.] *Veterinariya, Mosk.*, 6, 46–48. [In Russian.]

Vasil'Kov, G. V, 1981, [Biological control of *Philometroides* infection in carp.] *Byull. Vses. Inst. Eksp. Vet.*, No. 41, 18–21. [In Russian.]

Vasil'Kov, G. V. and Engashev, V. G, 1981, [Measures for the eradication of fish helminthiases.] *Veterinariya, Mosk.*, No. 8, 49–50. [In Russian.]

Vasquez-Colet, A. and Africa, C. M, 1938, Determination of the piscine intermediate host of Philippine heterophyid trematodes by feeding experiments. *Philip. J. Sci.*, 65, 293–302.

Vaz, Z, 1932, Contribuicao ao conhecimento dos trematoides de peixes fluviaes do Brasil. *Escolas Profissionaes do Lyceu Coracao de Jesus Alameda Barao de Picacicaba*, 48pp.

Velasquez, C. C, 1959, Studies on the family Bucephalidae Poche 1907 (Trematoda) from Philippine food fishes. *J. Parasit.*, 45, 135–147.

Velasquez, C. C, 1961, Further studies on *Transversotrema laruei* Velasquez with observations on the life cycle (Digenea: Transversotrematidae). *J. Parasit.*, 47, 65–70.

Velasquez, C. C, 1964, Life history of *Acanthoparyphium paracharadrii* sp. n. (Trematoda: Echinostomatidae). *J. Parasit.*, 50, 261–265.

Velasquez, C. C, 1969a, Life history of *Paramonostomum philippinensis* sp. n. (Trematoda: Digenea: Notocotylidae). *J. Parasit.*, 55, 289–292.

Velasquez, C. C, 1969b, Life cycle of *Cloacitrema philippinum* sp. n. (Trematoda: Digenea: Philophthalmidae). *J. Parasit.*, 55, 540–543.

Velasquez, C. C, 1973, Observations on some Heterophyidae (Trematoda: Digenea) encysted in Philippine fishes. *J. Parasit.*, 59, 77–84.

Velasquez, C. C, 1975, *Digenetic trematodes of Philippine fishes.* Quezon City: University of the Philippines Press. 140pp.

Velasquez, C. C, 1982, Heterophyidiasis, In: *CRC handbook series in zoonoses. Section C: parasitic zoonoses. Volume III.* (Eds Hillyer, G. V. and Hopla, C. E.) Boca Raton, Florida, U.S.A; CRC Press, Inc. pp 99–107.

Venard, C. E. and Warfel, J. H, 1947, Some effects of Acanthocephala on the largemouthed black bass. *J. Parasit.*, 33, Suppl., p.17.

Venkatanarsaiah, J, 1981, Detection of cholinesterase in the nervous system of the oncomiracidium of a monogenean, *Pricea multae* Chauhan, 1945. *Parasitology*, 82, 241–244.

Venkatanarsaiah, J. and Kulkarni, T, 1980, Studies on the occurrence of cholinesterase and nervous system of monogenetic trematodes (Polyopisthocotylea). *Proc. Indian Acad. Parasit.*, 1, 1–7.

Verbitskaya, I. N., Guseva, N. V., Laptev, V. I. and Musselius, V. A, 1972, *[Principal diseases of pond fish]*. Izdatel'stvo 'Kolos' Moscow, USSR, 1–70. [In Russian.]

Vernberg, W. B, 1968, Platyhelminthes: respiratory metabolism. In: *Chemical Zoology. Vol. II. Porifera, Coelenterata and Platyhelminths.* (Eds Florkin, M. and Scheer, B. T.) New York and London: Academic Press Inc., pp. 359–393.

Vicente, J. J., Rodrigues, H. de O. and Gomes, D. C, 1985, [Nematodes of Brazil, Part One: Nematodes of fish.] *Atas da Sociedade de Biologia do Rio de Janeiro.*, 25, 1–79.

Vik, R, 1954, Investigations on the pseudophyllidean cestodes of fish, birds and mammals in the Anoya water system in Trondelag. Part I. *Cyathocephalus truncatus* and *Schistocephalus solidus*. *Nytt Mag. Zool.*, 2, 5–51.

Vik, R, 1958, Studies of the helminth fauna of Norway. II. Distribution and life cycle of *Cyathocephalus truncatus* (Pallas, 1781) (Cestoda). *Nytt Mag. Zool.*, 6, 97–110.

Vik, R, 1959, Studies of the helminth fauna of Norway. III. Occurrence and distribution of *Triaenophorus robustus* Olsson, 1892 and *T. nodulosus* (Pallas, 1760) (Cestoda) in Norway. *Nytt Mag. Zool.*, 8, 64–73.

Vik, R, 1963, Studies of the helminth fauna of Norway. IV. Occurrence and distribution of *Eubothrium crassum* (Bloch, 1779) and *E. salvelini* (Schrank, 1790) (Cestoda) in Norway, with notes on their life cycles. *Nytt Mag. Zool.*, 11, 47–73.

Vik, R, 1964, Notes on the life history of *Philonema agubernaculum* Simon et Simon, 1935 (Nematoda). *Can. J. Zool.*, 42, 511–512.

Vik, R, 1981, Evolution in cestodes. In: *Evolution of helminths (Procendings EMOP III, Workshop 13). Parasitology*, 82, 163–164.

Vik, R., Halvorsen, O. and Andersen, K, 1969, Observations on *Diphyllobothrium* plerocercoids in three-spined sticklebacks, *Gasterosteus aculeatus* L., from the river Elbe. *Nytt Mag. Zool.*, 17, 75–80.

Vinikour, W. S, 1977, Incidence of *Neascus rhinichthysi* (Trematoda: Diplostomatidae) on longnose dace, *Rhinichthys cataractae* (Pisces: Cyprinidae) related to fish size and capture location. *Trans. Am. Fish. Soc.*, 106, 83–88.

Vinogradov, G. A., Davydov., V. G. and Kuperman, B. I, 1980, [Comparative study of osmoregulation in some fish cestodes] In: *IX Konferentsia Ukrainskogo Parazitologicheskogo Obschestva Tezisy dokladov. Chast' 1.* Kiev, USSR; 'Naukova Duymka', pp. 123–124. [In Russian.]

Vinogradov, G. A., Davydov, V. G. and Kuperman, I, 1982a, [Morphological-physiological study of the mechanisms of adaptation in pseudophyllidean cestodes to different salinities.] *Parazitologiya*, 16, 377–383. [In Russian.]

Vinogradov, G. A., Davydov, V. G. and Kuperman, B. I, 1982b, [Morphological characteristics of water-saline metabolism in some pseudophyllidean cestodes.] *Parazitologiya*, 16, 188–193. [In Russian.]

Vismanis, K. O., Lullu, A. V., Iygis, V. A, 1984, Turovskii, A. M. and Yun, A. I, 1984, [Diseases of Salmonidae in the marine fish farms on the Baltic coast and their prophylaxis.] In: *Biologicheskie osnovy rybovodstva: parazity i bolezni ryb. (Biologicheskie resursy gidrosfery i ikh ispol'zovanie.)* (Eds Bauer, O. N., Musselius, V. A., Skryabina, E. S.) Moscow, USSR, 'Nauka', pp. 56–63. [In Russian.]

Vladimirov, V. L, 1971, [Immunity of fish against *Dactylogyrus*.] *Parazitologiya*, 5, 51–58. [In Russian.]

Vogt, K, 1938, Experimentelle Untersuchungen über die Gründe von Masseninfektionen mit Plerocercoiden des Fischbandwurms *Triaenophorus nodulosus*. (Pall.). *Z. Fisch. Hilf.*, 36, 193–224.

Voth, D. R., Anderson, L. F. and Kleinschuster, S. J, 1974, The influence of waterflow on brown trout parasites. *Progve Fish-Cult.*, 36, 212.

Vykhrestyuk, N. P. and Klochkova, V. I, 1984, [Hexokinase in the trematode *Calicophoron ijimaii*, the cestode *Bothriocephalus scorpii* and the turbellarian *Penecurva sibirica*.] In: *Parazity zhivotnykh i rastenii.* Vladivostok, USSR, Akademiya Nauk SSSR. Dal'nevostochnyi Nauchnyi Tsentr, Biologo-Pochvennyi Institut, pp. 82–86. [In Russian.]

Vykhrestyuk, N. P., Yarygina, G. V. and Nikitenko, T. B, 1977, [The origin of lipids in parasitic platyhelminths.] *Trudy Biologo-Pochovennogo Instituta (Paraziticheskie i svobodnozhivushchie chervi fauny Dal'nego Vostoka), Novaya Seriya*, 47, 117–123. [In Russian.]

Vysotskaya, R. U. and Sidorov, V. S, 1973, [Lipid content of some helminths from freshwater fishes.] *Parazitologiya*, 7, 51–57. [In Russian.]

Vysotskaya, R. U., Sidorov, V. S. and Bykhovskaya-Pavlovskaya, I. E, 1972, [Carbohydrate composition of fish helminths.] In: *Lososevye (Salmonidae) Karelii. Vypusk l. Ekologiya, parazitofauna, biokhimiya.* Petrozavodsk, USSR, pp. 138–143. [In Russian].

Wabuke-Bunoti, M. A. N, 1980, The prevalence and pathology of the cestode *Polyonchobothrium clarias* (Woodland, 1925) in the teleost, *Clarias mossambicus (Peters)*. *J. Fish Dis.*, 3, 223–230.

Wagner, E. D, 1954, The life history of *Proteocephalus tumidocollis* Wagner, 1953 (Cestoda), in rainbow trout. *J. Parasit.*, 40, 489–498.

Wakelin, D, 1985, Genetic control of immunity to helminth infection. *Parasitology Today*, 1, 17–23.

Wales, J. H, 1958a, Two new blood fluke parasites of trout. *California Fish and Game*, 44, 125–136.

Wales, J. H, 1958b. Intestinal flukes as a possible cause of mortality in wild trout. *California Fish and Game*, 44, 350–352.

Walker, R. W. and Barrett, J, 1983, Mitochondrial adenosine triphosphatase activity and temperature adaptation in *Schistocephalus solidus* (Cestoda: Pseudophyllidea). *Parasitology*, 87, 307–326.

Walkey, M, 1967, The ecology of *Neoechinorhynchus rutili* (Muller). *J. Parasit.*, 53, 795–804.

Walkey, M. and Körting, W, 1985, Thermal alterations of pyruvate kinases in the fish tapeworm *Bothriocephalus acheilognathi*, Yamaguti, 1934. *Z. ParasitKde*, 71, 527–532.

Wallace, H. E, 1941, Life history and embryology of *Triganodistomum mutabile* (Cort) (Lissorchiidae, Trematoda). *Trans. Am. Microsc. Soc.*, 60, 309–326.

Wallett, M. and Kohn, A, 1987, [Trematode parasites of marine fish from coastal waters of Rio de Janeiro, Brazil.] *Mem. Inst. Oswaldo Cruz*, 82, 21–27.

Walter, U, 1988, [Parasite fauna of *Stizostedion lucioperca* in the Bodden waters on the Baltic coast of the German Democratic Republic.] *Angew. Parasit.*, 29, 215–219. [In German]

Wang, P. Q, 1980a, [Notes on the Acanthocephala from Fugian. II.] *Acta Zootaxonomica Sinica*, 5, 116–123. [In Chinese.]

Wang, P. Q, 1980b, [Studies on some species of gasterostome trematodes from Fujian, China.] *Acta Zootaxonomica Sinica*, 5, 330–336. [In Chinese.]

Wang, P. Q, 1981a, [Notes on some trematodes from freshwater fishes in Fujian Province.] *Fujian Shida Xuebao (Journal of the Fujian Normal University)*, No. 2, 81–90. [In Chinese.]

Wang, P. Q, 1981b, [Notes on some species of acanthocephalans from fishes of China.] *Acta Zootaxonomica Sinica*, 6, 121–130. [In Chinese.]

Wang, P. Q, 1982a, [Some digenetic trematodes of marine fishes from Fujian Province, China.] *Oceanologia et Limnologia Sinica*, 13, 179–194. [In Chinese.]

Wang, P. Q, 1982b, [Some digenetic trematodes of marine fishes from Fujian Province, China.] *Wuyi Sci. J.* 2, 65–74. [In Chinese.]

Wang, P. Q, 1984, [Notes on some cestodes of fishes in Fujian Province with a list of fish cestodes recorded from China.] *Wuyi Sci. J.* 4, 71–83. [In Chinese.]

Wang, P. Q., Zhao, Y. R. and Chen, C. C, 1978, [On some nematodes from vertebrates in South China.] *Fujian Shida Xuebao,* No. 2, 75–90. [In Chinese.]

Wanson, W. W. and Larson, O. R, 1972, Studies on helminths of North Dakota. V. Life history of *Phyllodistomum nocomis* Fischthal, 1942 (Trematoda: Gorgoderidae). *J. Parasit.,* 58, 1106–1109.

Wanson, W. W. and Nickol, B. B, 1973, Origin of the envelope surrounding larval acanthocephalans. *J. Parasit.,* 59, 1147.

Wanstall, S. T., Robotham, P. W. J. and Thomas, J. S, 1986, Pathological changes induced by *Pomphorhynchus laevis* Müller (Acanthocephala) in the gut of rainbow trout, *Salmo gairdneri* Richardson. *Z. ParasitKde,* 72, 105–114.

Wanstall, S. T., Thomas, J. S. and Robotham, P. W. J, 1988, The pathology caused by *Pomphorhynchus laevis* Müller in the alimentary tract of the stone loach, *Noemacheilus barbatulus* (L.). *J. Fish Diseases,* 11, 511–523.

Ward, H. B, 1911, The distribution and frequency of animal parasites and parasitic diseases in North American freshwater fish. *Trans. Am. Fish. Soc.,* 41, 207–241.

Ward, H. B, 1933, Parasitism and disease among oceanic fishes: economic aspects and epidemics due to animal parasites. *5th Proc. Pacific Cong. Sci.,* 4177–4182.

Ward, H. L, 1940, Studies on the life history of *Neoechinorhynchus cylindratus* (Van Cleave, 1913) (Acanthocephala). *Trans. Am. Microsc. Soc.,* 59, 327–347.

Ward, P. F. V, 1982, Aspects of helminth metabolism. *Parasitology,* 84, 177–194.

Ward, S. M., Allen, J. M. and McKerr, G, 1986a, Neuromuscular physiology of *Grillotia erinaceus* metacestodes (Cestoda: Trypanorhyncha) *in vitro. Parasitology,* 93, 121–132.

Ward, S. M., Allen, J. M. and McKerr, G, 1986b, Action of praziquantel on *Grillotia erinaceus* metacestodes (Cestoda: Trypanorhyncha) *in vitro. Parasitology,* 93, 133–142.

Wardle, R. A. and McLeod, J. A, 1952, *The Zoology of Tapeworms.* New York: Hafner Publishing Co. Inc., 780pp.

Wardle, R. A., McLeod, J. A. and Radinovsky, S, 1974, *Advances in the Zoology of Tapeworms, 1950–1970.* Minneapolis: University of Minnesota Press, 274pp.

Watson, D. E. and Thorson, T. B, 1976, Helminths from elasmobranchs in Central American fresh waters. In: *Investigations of the ichthyofauna of Nicaraguan lakes.* (Ed. Thorson, T. B.). Nebraska, U.S.A; University of Nebraska., pp. 629–640.

Watson, R. A, 1984, The life cycle and morphology of *Tetracerasta blepta*, gen. et sp. nov., and *Stegodexamene callista*, sp. nov. (Trematoda: Lepocreadiidae) from the long-finned eel, *Anguilla reinhardtii* Steindachner. *Aust. J. Zool.,* 32, 177–204.

Watson, R. A. and Dick, T. A, 1979, Metazoan parasites of whitefish *Coregonus clupeaformis* (Mitchill) and cisco, *C. artedii* Lesueur from Southern Indian Lake, Manitoba. *J. Fish Biol.,* 15, 579–587.

Watson, R. A. and Dick, T. A, 1980, Metazoan parasites of pike, *Esox lucius* Linnaeus, from Southern Indian Lake, Manitoba, Canada. *J. Fish Biol.,* 17, 255–261.

Watson, T. G, 1981, *Metorchis conjunctus* (Cobbold, 1860) Looss, 1899 (Trematoda: Opisthorchiidae): isolation of metacercariae from fish hosts. *Can. J. Zool.,* 59, 2010–2013.

Weinmann, C. J, 1966, Immunity mechanisms in cestode infection. In: *Biology of parasites. Emphasis on veterinary parasites.* (Ed. Soulsby, E. J.), New York and London: Academic Press., pp. 301–320.

Weirowski, F, 1979, [The economic importance and distribution of *Khawia sinensis* in carp production in the GDR.] *Z. Binnenfisch. DDR,* 26, 373–376. [In German.]

Weirowski, F, 1984, Occurrence, spread and control of *Bothriocephalus acheilognathi* in the carp ponds of the German Democratic Republic. In: *Fish, pathogens and environment in European polyculture. (Proc. Internat. Seminar. 23–27 June 1981. Szarvas, Hungary. Symposia Biologica Hungarica Vol. 24.* pp. 149–155.

Weisberg, S. B, 1986, *Eustrongylides* (Nematoda) infection in mummichogs and other fishes of the Chesapeake Bay region. *Trans. Am. Fish. Soc.,* 115, 776–783.

Wharton, D. A, 1983, The production and functional morphology of helminth egg-shells. In: *The reproductive biology of parasites.* Symposia of the British Society for Parasitology, Vol. 20, (Ed. Whitfield, P. J.), *Parasitology,* 86, 85–97.

Whitaker, D. J, 1985, A parasite survey of juvenile chum salmon (*Oncorhynchus keta*) from the Nanaimo River. *Can. J. Zool.,* 63, 2875–2877.

Whitfield, A. K. and Heeg, J, 1977, On the life cycles of the cestode *Ptychobothrium belones* and nematodes of the genus *Contracaecum* from Lake St. Lucia, Zululand. *S. Afr. J. Sci.*, 73, 121–122.

Whitfield, P. J, 1971, Phylogenetic affinities of Acanthocephala: an assessment of ultrastructural evidence. *Parasitology*, 63, 49–58.

Whitfield, P. J, 1973, The egg envelopes of *Polymorphus minutus* (Aconthocephala). *Parasitology*, 66, 387–403.

Whitfield, P. J, 1979, The biology of parasitism: an introduction to the study of associating organisms. *Contemporary Biology Series*. (Eds Barrington, E. J. W., Willis, A. J. and Sleigh, M. A.). London: Edward Arnold, x + 277pp.

Whitfield, P. J, (Ed.) 1983, The reproductive biology of parasites. [*Symposia of the British Society for Parasitology, Vol. 20.*] *Parasitology*, 86, 207 pp.

Whitfield, P. J, 1984, Acanthocephala. In: *Biology of the integument. I. Invertebrates*. (Eds Bereiter-Hahn, J., Matolsky, A. G. and Richards, K. S.). Berlin: Springer-Verlag. pp. 234–241.

Whitfield, P. J., Anderson, R. M. and Bundy, D. A. P, 1977, Experimental investigations on the behaviour of the cercariae of an ectoparasitic digenean *Transversotrema patialense*: general activity patterns. *Parasitology*, 75, 9–30.

Whitfield, P. J., Anderson, R. M. and Bundy, D. A. P, 1986, Host-specific components of the reproductive success of *Transversotrema patialense* (Digenea: Transversotrematidae). *Parasitology*, 92, 683–698.

Whitfield, P. J., Anderson, R. M. and Moloney, N. A, 1975, The attachment of cercariae of an ectoparasitic digenean, *Transversotrema patialense*, to the fish host; behavioural and ultrastructural aspects. *Parasitology*, 70, 311–329.

Whitfield, P. J. and Evans, N. A, 1983, Parthenogenesis and asexual multiplication among parasitic platyhelminths. In: *The reproductive biology of parasites. Symposia of the British Society for Parasitology Vol. 20*. (Ed. Whitfield, P. J.). *Parasitology*, 86, 121–160.

Whittaker, F. H., Apkarian, R. P., Curless, B. and Carvajal, G. J, 1985, Scanning electron microscopy of the scolices of the cestodes *Parachristianella monomegacantha* Kruse 1959 (Trypanorhyncha) and *Phyllobothrium* sp. Beneden 1849 (Tetraphyllidea). *J. Parasit.*, 71, 376–381.

Whittington, I. D, 1987, Hatching in two monogenean parasites from the common dogfish (*Scyliorhinus canicula*): the polyopisthocotylean gill parasite, *Hexabothrium appendiculatum* and the microbothriid skin parasite, *Leptocotyle minor*. *J. Mar. Biol. Ass. UK*, 67, 729–756.

Whittington, I. D. and Kearn, G. C, 1986, Rhythmical hatching and oncomiracidial behaviour in the hexabothriid monogenean *Rajonchocotyle emarginata* from the gills of *Raja* spp. *J. Mar. Biol. Ass. UK*, 6, 93–111.

Whittington, I. D. and Kearn, G. C, 1988, Rapid hatching of mechanically-disturbed eggs of the monogenean gill parasite *Diclidophora luscae*, with observations on sedimentation of egg bundles. *Int. J. Parasit.*, 18, 847–852.

Whittington, I. D. and Kearn, G. C, 1989, Rapid hatching induced by light intensity reduction in the polyopisthocotylean monogenean *Plectanocotyle gurnardi* from the gills of gurnards (Triglidae), with observations on the anatomy and behaviour of the oncomiracidium. *J. Mar. Biol. Ass. UK*, 69, 609–624.

Whittington, I. D. and Pichelin, S, 1991, Attachment of eggs by *Concinnocotyla australensis* (Monogenea: Polystomatidae) to the tooth plates of the Australian lungfish, *Neoceratodus forsteri* (Dipnoi). *Int. J. Parasit.*, 21, 341–346.

Whyte, S. K., Chappell, L. H. and Secombes, C. J, 1988, *In vitro* transformation of *Diplostomum spathaceum* (Digenea) cercariae and short term maintenance of post-penetration larvae *in vitro*. *J. Helminth.*, 62, 293–302.

Whyte, S. K., Allan, J. C., Secombes, C. J. and Chappell, L. H, 1987, Cercariae and diplostomules of *Diplostomum spathaceum* (Digenea) elicit an immune response in rainbow trout, *Salmo gairdneri* Richardson. *J. Fish. Biol.*, 31, 185–190.

Wickins, J. F. and MacFarlane, I. S, 1973, Some differences in the parasitic fauna of three samples of plaice (*Pleuronectes platessa* L.) from the southern North Sea. *J. Fish. Biol.*, 5, 9–19.

Wierzbicka, J, 1970, Zmiany spowodowane unawazja Asymphylodora tincae (Modeer, 1790) w jelicie lina (*Tinca tinca* (L.) *Wiad. Parazyt.*, 16, 169–173.

Wierzbicka, J, 1977, An attempt to explain affinities between *Blicca bjoerkna* (L.). *Abramis brama* (L.) and *A. ballerus* (L.), on the grounds of their parasitic fauna. *Acta Ichthyol. Pisc.*, 7, 3–13.

Wierzbicka, J, 1978, Cestoda, Nematoda, Acanthocephala, Hirudinea and Crustacea from *Abramis brama, A. ballerus* and *Blicca bjoercna* [*bjoerkna*] of the Dabie lake, Poland. *Acta Parasit. Pol.*, 25, 293–305.

Wierzbicka, J, 1980, The Didymozoidae trematodes in the north-east Atlantic *Lampris guttatus* (Brunnich, 1788). *Acta Ichthyol. Pisc.*, **10**, 21–34.

Wierzbicka, J. and Einszporn-Orecka, T, 1985, Histopathologic changes in urinary system of *Blicca bjoerkna* (L.) infected with *Phyllodistomum folium* (Olfers 1816). *Acta Parasit. Pol.*, **30**, 57–62.

Wierzbicka, K, 1970, The parasite fauna of the perch, *Perca fluviatilis* L. of Lake Dargin. *Acta Parasit. Pol.*, **18**, 45–55.

Wikerhauser, T, 1986, [Who is who in the parasitic fauna of marine fishes: *Anisakis* and *Kudoa*.] *Praxis Veterinaria,* **33**, 365–372.

Wikgren, B. J, 1956, Studies on Finnish larval flukes with a list of known Finnish adult flukes (Trematoda: Malacocotylea). *Acta Zool. Fenn.*, No. 91, 1–106,

Williams, D. D, 1978, Larval development of *Glaridacris vogei* (Cestoda: Caryophyllaeidae). *Proc. Helminth. Soc. Wash.*, 45, 142–143.

Williams, D. D, 1979a, Seasonal incidence of *Glaridacris laruei* and *G. catostomi* in Red Cedar River, Wisconsin *Catastomus commersoni. Iowa St. J. Res.*, 53, 311–316.

Williams, D. D, 1979b, Seasonal incidence of *Isoglaridacris wisconsinensis* (Cestoda: Caryophyllaeidae) in its fish host. *Iowa St. J. Res.*, 53, 305–310.

Williams, D. D, 1980, Procercoid development of *Isoglaridacris wisconsinensis* Cestoda: Caryophyllaeidae). *Proc. Helminth. Soc. Wash.*, 47, 138–139.

Williams, E. H., Jr. and Phelps, R. P, 1974, Parasites of some mariculture fishes before and after cage culture. In: *Food-drugs from the sea. Proceedings of the 4th Conference. Marine Technology Society, University of Puerto Rico, Mayaguez. 17–21 November 1974.* (Eds Webber, H. H. and Ruggieri, G. D.). Washington, DC. USA: Marine Technology Society, pp. 216–230.

Williams, H. H, 1958a, The anatomy of *Aporocotyle spinosicanalis* (Trematoda Digenea) from *Merluccius merluccius* (L.). *Ann. Mag. Nat. Hist.*, 1, 291–297.

Williams, H. H, 1958b, The anatomy of the trematode *Dictyocotyle coeliaca* Nybelin, 1941 with a discussion of its relationships with species of the genus *Calicotyle* Diesing, 1850. *Ann. Mag. Nat. Hist.*, 1, 465–478.

Williams, H. H, 1958c, Some Phyllobothriidae (Cestoda: Tetraphyllidea) of elasmobranchs from the Western seaboard of the British Isles. *Ann. Mag. Nat. Hist.*, 1, 113–136.

Williams, H. H, 1958d, Some Tetraphyllidea (Cestoda) from the Liverpool School of Tropical Medicine. *Rev. Suisse Zool.*, 65, 867–878.

Williams, H. H, 1959a, The anatomy of *Phyllobothrium sinuosiceps* sp. nov. (Cestoda: Tetraphyllidea) from *Hexanchus griseus* (Gmelin) the six gilled shark. *Parasitology,* 49, 54–69.

Williams, H. H, 1959b, The anatomy of *Kollikeria filicollis* (Rudolphi, 1819) Cobbold, 1860 (Trematoda: Digenea) showing that the sexes are not entirely separate as hitherto believed. *Parasitology,* 49, 39–53.

Williams, H. H, 1960a, *Winkenthughesia bramae* (Parona and Perugia, 1896) a rare monogenetic trematode and a new record for the British Isles. *Ann. Mag. Nat. Hist.*, 2, 551–559.

Williams, H. H, 1960b, Some observations on *Parabothrium gadi-pollachii* (Rudolphi, 1810) and *Abothrium gadi*, van Beneden, 1870 (Cestoda: Pseudophyllidea) including an account of their mode of attachment and of variation in the two species. *Parasitology,* **50**, 303–322.

Williams, H. H, 1960c, The intestine in members of the genus *Raja* and host-specificity in the Tetraphyllidea. *Nature, Lond.*, **188**, 514–516.

Williams, H. H, 1961a, Observations on *Echeneibothrium maculatum* (Cestoda: Tetraphyllidea). *J. Mar. Biol. Ass. UK.*, 41, 631–652.

Williams, H. H, 1961b, Parasitic worms in marine fishes: a neglected study. *New Scientist,* 12, 156–159.

Williams, H. H, 1962b, *Acanthobothrium* sp. nov. (Cestoda: Tetraphyllidea) and a comment on the order Biporophyllaeidea. *Parasitology,* 52, 67–76.

Williams, H. H, 1964, Some new and little known cestodes from Australian elasmobranchs with a brief discussion on their possible use in problems of host taxonomy. *Parasitology,* 54, 737–748.

Williams, H. H, 1965, Observations on the occurrence of *Dictyocotyle coeliaca* and *Calicotyle kroyeri* (Trematoda: Monogenea). *Parasitology,* 55, 201–207.

Williams, H. H, 1966, The ecology, functional morphology and taxonomy of *Echeneibothrium* Beneden, 1849 (Cestoda: (Tetraphyllidea), a revision of the genus and comments on *Discobothrium* Beneden, 1870 *Pseudanthobothrium* Baer, 1956 and *Phormobothrium* Alexander, 1963. *Parasitology,* 56, 227–285.

Williams, H. H, 1967, Helminth diseases of fish. *Helminth. Abstr.*, 36, 261–295.

Williams, H. H, 1968a, *Acanthobothrium quadripartitum* sp. nov. (Cestoda: Tetraphyllidea) from *Raja naevus* in the North Sea and English Channel. *Parasitology,* **58**, 105–110.

Williams, H. H, 1968b, *Phyllobothrium piriei* sp. nov. (Cestoda: Tetraphyllidea) from *Raja naevus* with a comment on its habitat and mode of attachment. *Parasitology*, **58**, 929–937.

Williams, H. H, 1968c, The taxonomy, ecology and host-specificity of some Phyllobothriidae (Cestoda: Tetraphyllidea) a critical revision of *Phyllobothrium* Beneden, 849, and comments on some allied genera. *Trans. Roy. Soc. Lond. B.*, **253**, 231–307.

Williams, H. H, 1969, The genus *Acanthobothrium* (Beneden, 1849) (Cestoda: Tetraphyllidea). *Nytt Mag. Zool.*, 17, 1–56.

Williams, H. H, 1970, Host-specificity of fish parasites. *J. Parasit.*, 56, 482–483.

Williams, H. H. and Bray, R, 1984, *Chimaerocestos prudhoei* gen. et sp. nov., representing a new family of tetraphyllideans and the first record of strobilate tapeworms from a holocephalan. *Parasitology*, **88**, 105–116.

Williams, H. H. and Halvorsen, O, 1971, The incidence and degree of infection of *Gadus morhua* (L.) 1758 with *Abothrium gadi* Beneden, 1871 (Cestoda, Pseudophyllidea). *Norw. J. Zool.*, 19, 193–199.

Williams, H. H. and Jones, A, 1976, Marine helminths and Human Health. *Commonwealth Institute of Helminthology Miscellaneous publication No.3*. Commonwealth Agricultural Bureaux, Farnham Royal, Bucks. pp. 1–47.

Williams, H. H. and Jones, A, 1977, Marine helminths and human health. *Proceedings of the International Symposium on Medicine in a Tropical Environment. Cape Town, South Africa, Medical Research Council*, pp. 98–107.

Williams, H. H. and McVicar, A, 1968a, Aspects of the biology of tetraphyllidean cestodes. *Parasitology*, **58**, 20.

Williams, H. H. and McVicar, A, 1968b, Sperm transfer in Tetraphyllidea (Platyhelminthes: Cestoda). *Nytt Mag. Zool.*, 16, 61–71.

Williams, H. H. and Richards, D. H. H, 1968, Observations on *Pseudanisakis rotundata* (Rudolphi, 1819) Mozgovoi, 1950, a common but little known nematode of *Raja radiata*, Donovan in the northern North Sea. *J. Helminth.*, 42, 199–200.

Williams, H. H. and Richards, D. H. H, 1978, *Proleptus mackenziei* sp. nov. (Nematoda: Spirurida) from *Raia fyllae* in the Barents Sea. *Zool. Scripta*, 7, 85–91.

Williams, H. H., Colin, J. A. and Halvorsen, O, 1987, Biology of gyrocotylideans with emphasis on reproduction, population ecology and phylogeny. *Parasitology*, **95**, 173–207.

Williams, H. H., Mackenzie, K. and McCarthy, A. M, 1992, Parasites as biological indicators of the population biology, migrations, diet, and phylogenetics of fish. *Reviews in Fish Biology and Fisheries*, 2, 144–176.

Williams, H. H., McVicar, A. and Ralph, R, 1970, The alimentary canal of fish as an environment for helminth parasites. In (Eds Taylor, A. E. R. and Muller, R.) *Aspects of Fish Parasitology. Symposia of the British Society for Parasitology, vol. 8*. Oxford, UK: Blackwell Scientific Publications, pp. 43–47.

Williams, I. C. and Bolton, P. A, 1985, The helminth parasites of the European eel *Anguilla anguilla* (L.) from the Driffield Canal, North Humberside. *Naturalist*, **110**, 31–36.

Williams, J. B, 1981, Classification of the Temnocephaloidea (Platyhelminthes). *J. Nat. Hist.*, 15, 277–299.

Williams, J. B, 1988, Ultrastructural studies on *Kronborgia* (Platyhelminthes; Neoophora): the spermatozoon. *Int. J. Parasit.*, 18, 477–483.

Williams, M. O, 1966, Studies on the morphology and life-cycle of *Diplostomum (Diplostomum) gasterostei* (Strigeida: Trematoda). *Parasitology*, **56**, 693–706.

Willomitzer, J, 1980a, Seasonal dynamics of parasitoses in grasscarp (*Ctenopharyngodon idella*) fry and fingerlings. *Acta vet. Brno*, 49, 269–277.

Willomitzer, J, 1980b, Therapy of major ectoparasitoses in grasscarp (*Ctenopharyngodon idella*) fry and fingerlings. *Acta vet. Brno*, **49**, 279–282.

Wilson, K. A. and Ronald, K, 1967, Parasite fauna of the sea lamprey (*Petromyzon marinus* von Linne) in the Great Lakes region. *Can. J. Zool.*, 45, 1083–1092.

Winch, J, 1983, The biology of *Atrispinum labracis* n.comb. (Monogenea) on the gills of the bass, *Dicentrarchus labrax. J. Mar. Biol. Ass. UK*, 63, 915–927.

Wirtz, K. P. and Schreiber, W, 1980, Untersuchungen zum Vordkommen von Nematodenlarven der Gattung *Anisakis* in nordlichen blauen Wittling (*Micromesistius poutassou*). *Fleischwirtschaft*, 60, 282–286.

Wisniewski, L. W, 1932a, *Cyathocephalus truncatus*, Pallas, ein Fischparasit aus dem Vrelo Bosne. *Ribarski List*, 7, 1–4.

Wisniewski, L. W, 1932b, Zur postembryonalen Entwicklung von *Cyathocephalus truncatus* Pallas. *Zool. Anz.*, **98**, 213–218.

Wisniewski, L. W, 1932c, *Cyathocephalus truncatus* Pallas II. Ogolna morphologja. *Bull. Int. Acad. Pol. Sci. Lett. C1, Sci. Math. Nat. Ser. B: Sci. Nat. (II), Annee 1932*, 311–327.

Wisniewski, W. L, 1958, The development cycle of *Bunodera luciopercae* (O. F. Müller). *Acta Parasit. Pol.*, 6, 289–307.

Witenberg, G, 1932, Fish as a source of worm disease in man. *Harefuah*, 6, 127–139.

Witenberg, G, 1964, Zooparasitic diseases. A. Helminthozoonoses. In: *Zoonoses*. (Ed. Hoeden, J. van). Amsterdam: Elsevier Publishing Company, pp. 529–719.

Wobeser, G., Kratt, L. F., Smith, R. J. F. and Acompanado, G, 1976, Proliferative branchitis due to *Tetraonchus rauschi* (Trematoda: Monogenea) in captive Arctic grayling (*Thymallus arcticus*). *J. Fish. Res. Bd Can.*, 33, 1817–1821.

Wolfgang, R. W, 1954, Studies of the trematode *Stephanostomum baccatum* (Nicoll, 1907): I. The distribution of the metacercaria in Eastern Canadian flounders. *J. Fish. Res. Bd Can.*, 11, 954–962.

Wood, B. P. and Matthews, R. A, 1987, The immune response of the thick-lipped grey mullet, *Chelon labrosus* (Risso, 1826), to metacercarial infections of *Cryptocotyle lingua* (Creplin, 1825). *J. Fish Biol.*, 31, 175–183.

Woodhead, A. E, 1929, Life history studies on the trematode family, Bucephalidae. *Trans. Am. Microsc. Soc.*, 48, 256–275.

Woodhead, A. E, 1930, Life history on the trematode family Bucephalidae. No. II. *Trans. Amer. Microsc. Soc.*, 49, 1–17.

Wootten, R, 1973, The metazoan parasite-fauna of fish from Hanningfield Reservoir, Essex in relation to features of the habitat and host populations. *J. Zool.*, 171, 323–331,

Wootten, R, 1974, Studies on the life history and development of *Proteocephalus percae* (Muller) (Cestoda: Proteocephalidea). *J. Helminth.*, 48, 269–281.

Wootten, R. and Smith, J. W, 1976, Observational and experimental studies of larval nematodes in blue whiting from water to the west of Scotland. *ICES CM 1976/H*, 35, 3pp.

Wootten, R. and Smith, J. W, 1979, The occurrence of plerocercoids of *Diphyllobothrium* spp. in wild and cultured salmonids from the Loch Awe area. *Scott. Fish. Res. Rep., DAFS Scotland*, No. 13, 8 pp.

Wootten, R. and Smith, J. W, 1980, Studies on the parasite fauna of juvenile Atlantic salmon, *Salmo salar* L., cultured in fresh water in eastern Scotland. *Z. ParasitKde.*, 63, 221–231.

Wootten, R. and Waddell, I. F, 1977, Studies on the biology of larval nematodes from the musculature of cod and whiting in Scottish waters. *J. Con. Inst. Exp. Mer.*, 37, 266–273.

Wootton, D. M, 1957a, Studies on the life-history of *Allocreadium alloneotenicum* sp. nov. (Allocreadiidae—Trematoda). *Biol. Bull.*, 116, 302–315.

Wootton, D. M, 1957b, The life history of *Cryptocotyle concavum* (Creplin, 1825) Fischoeder, 1903 (Trematoda: Heterophyidae). *J. Parasit.*, 43, 271–279.

Wright, S. A. and Boyce, K. W, 1986, A survey of mosquitofish *Gambusia affinis* parasites in Sacramento County. *Proceedings and Papers of the Annual conference of the California Mosquito and Vector Control Association*. 54, 79–85.

Wu, K, 1938, Progenesis of *Phyllodistomum lesteri* sp. nov. (Trematoda: Gorgoderidae) in fresh-water shrimps. *Parasitology*, 30, 4–19.

Wulker, G, 1929, Der Wirtswechsel der parasitischen Nematoden von Meeresfischen. *Verh. Dt. Zool. Gesell.*, 33, 7–157.

Wurmbach, H, 1937, Zur Krankheitserregender wirkung der Acanthocephalen. Die Kratzerkrankung der Barben der Mosel. *Z. Fisch.*, 35, 217–312.

Xylander, W. E. R, 1984, A presumptive ciliary photoreceptor in larval *Gyrocotyle urna* Grube and Wagener (Cestoda). *Zoomorphology*, 104, 21–25.

Xylander, W. E. R, 1986a, Zur ultrastruktur und Biologie der Gyrocotylida und Amphilinida und ihre Stellung im System der Plathelminthen. *Doctoral thesis, University of Göttingen*. 307pp.

Xylander, W. E. R, 1986b, Ultrastrukturelle Befunde gut Stellung von *Gyrocotyle* im System der parasitischen Plathelminthen. *Verh. Dt. Zool. Ges.*, 79, 193.

Xylander, W. E. R, 1987a, Ultrastructure of the lycophora larva of *Gyrocotyle urna* (Cestoda, Gyrocotylidea). I. Epidermis, neodermis anlage and body musculature. *Zoomorphology*, 106, 352–360.

Xylander, W. E. R, 1987b, Ultrastructure of the lycophora larva of *Gyrocotyle urna* (Cestoda, Gyrocotylidea). II. Receptors and nervous system. *Zool. Anz.*, 219, 239–255.

Xylander, W. E. R, 1987c, Ultrastructure of the lycophora larva of *Gyrocotyle urna* (Cestoda, Gyrocotylidea). III. The protonephridial system. *Zoomorphology*, 107, 88–95.

Xylander, W. E. R, 1987d, Ultrastructural studies on the reproductive system of Gyrocotylidea and

Amphilinidea (Cestoda). II. Vitellarium, vitellocyte development and vitelloduct of *Gyrocotyle urna. Zoomorphology*, 107, 293–297.

Xylander, W. E. R, 1988a, Ultrastructural studies on the reproductive system of Gyrocotylidea and Amphilinidea (Cestoda). I. Vitellarium, vitellocyte development and vitelloduct in *Amphilina foliacea* (Rudolphi, 1819). *Parasit. Res.*, 74, 363–370.

Xylander, W. E. R, 1988b, Early evolution in tapeworms. *Proc. V Europ. Multicoll. Parasit., Budapest, 1988.* 136.

Xylander, W. E. R, 1989, Ultrastructural studies on the reproductive system of Gyrocotylidea and Amphilinidea (Cestoda): spermatogenesis, spermatozoa, testes and vas deferens of *Gyrocotyle. Int. J. Parasitol.*, 19, 897–905.

Yablokov, A. V., Bel'kovich, V. M. and Borisov, V. I, 1972, [Whales and dolphins. (Monographic outline).] Moscow, USSR: Izdatel'stvo 'Nauka', pp. 380–349. [In Russian.]

Yakhod, D. B., Mukhina, V. N. and Gracheva, O. K, 1979, [Intensification of a focus of diphyllobothriasis in the zone of the Kuibyshev reservoir and measures for its sanitation]. *Medskaya Parazit.*, 48, 26–29. [In Russian.]

Yakushev, V. YU, 1984, Seasonal dynamics of the incidence of whitefish (*Coregonus albula* L.) invasion with a cestode *Proteocephalus exiguus* (Cestoda: Proteocephalidae) in Karelia. *Helminthologia*, 21, 123–130.

Yakushev, V. YU, 1985, [Seasonal dynamics in the prevalence of *Proteocephalus exiguus* (Cestoda, Proteocephalidae) in *Coregonus albula* in Karelia.] *Parazitologiya*, 19, 95–100. [In Russian.]

Yamaguti, S, 1958, *Systema helminthum. Volume I. parts 1 and 2. The digenetic-trematodes of vertebrates.* New York: Interscience Publishers, Inc., 1575pp.

Yamaguti, S, 1959, *Systema helminthum. Volume II. The cestodes of vertebrates.* New York: Interscience Publishers, Inc., vii + 860pp.

Yamaguti, S, 1961, *Systema Helminthum. Volume III. The nematodes of vertebrates.* New York: Interscience Publishers, Inc., 1261pp.

Yamaguti, S, 1963a, *Systema Helminthum. Volume IV. Monogenea and Aspidocotylea.* New York: Interscience Publishers (John Wiley and Sons, Ltd.), vii + 699pp

Yamaguti, S, 1963b, *Systema Helminthum. Volume V. Acanthocephala.* New York: Interscience Publishers (John Wiley and Sons, Ltd.), vii + 423pp.

Yamaguti, S, 1969, Special modes of nutrition in some digenetic trematodes. *J. Fish. Res. Bd Can.*, 26, 845–848.

Yamaguti, S, 1971, *Synopsis of digenetic trematodes of vertebrates. Vols. I and II.* Tokyo, Japan: Keigaku Publishing Co., 1074pp.

Yamaguti, S, 1975, *A synoptical review of digenetic trematodes of vertebrates with special reference to the morphology of their larval forms.* Tokyo, Japan: Keigaku Publishing Co., xiii + 575pp., 219 plates.

Yamamoto, K., Takagi, S. and Matsuoka, S, 1984, [Mass mortality of the Japanese anchovy (*Engraulis japonica*) caused by a gill monogenean *Pseudanthocotyloides* sp. (Mazocraeidae) in the Sea of Ivo ('Iyo-nada'), Ehime Prefecture.] *Fish Path.*, 19, 119–123.

Yarygina, G., Vykhrestyuk, N. P. and Klochkova, V. I, 1982, [The amino acid composition of collagenous proteins of the trematodes *Calicophoron erschowi, Eurytrema pancreaticum*, and the cestodes *Bothriocephalus scorpii* and *Nybelinia* sp. larvae.].. *Zh. Evol. Biokhim. Fiziol.*, 18, 564–567. [In Russian.]

Yashchuk, V. D, 1974, [Dynamics of *Philometroides sanguinea* infections in their intermediate hosts.] *Veterinariya Mosk.*, No. 6, 69–71. [In Russian.]

Yashchuk, V. D, 1975, [Ecological links between the final hosts of *Philometroides sanguinea*.] In: *VIII Nauchnaya Konferentsiya Parazitologov Ukrainy. 9Tezisy dokladov). Donetsk, Sentyabr' 1975.* Kiev, USSR: Ukrainskii Nauchno-Issledovatel'skii Institut, Nauchno-Teknicheskoi Informatsii. pp. 202–204. [In Russian.]

Yashchuk, V. D, 1979, [Experiments in the control of bothriocephaliasis.] *Veterinariya, Mosk.*, No. 9, 46–48. [In Russian.]

Yashchuk, V. D., Sventsitskii, M. S. and Rudoi, G. I, 1978, [Determination of the biological bases for the economic loss due to bothriocephaliasis.] *Rybnoe Khozyaistvo*, No. 26, 80–86. [In Russian.]

Yazaki, S., Fukumoto, S., Maeda, T., Maejima, J. and Kamo, H, 1985, [Observations on secretory glands in pleroceroids and strobilae of *Diphyllobothrium* sp. ind.] *Jap. J. Parasit.*, 34, 105–114. [In Japanese.]

Yeh, L. S, 1960, On a collection of camallanid nematodes from freshwater fishes in Ceylon. *J. Helminth.*, 34, 107–116.

Yorke, W. and Maplestone, P. A, 1962, *The nematode parasites of vertebrates*, New York: Hafner Publishing Company i–x + 536 pp.

Young, P. C, 1967, A taxonomic revision of the subfamilies Monocotylinae Gamble, 1896 and Dendromonocotylinae Hargis, 1955 (Monogenoidea: Monocotylidae). *J. Zool., Lond.,* 153, 381–422.

Young, P. C, 1968, Ten new species of *Haliotrema* (Monogenoidea: Dactylogyridae) from Australian fish and a revision of the genus. *J. Zool., Lond.,* 154, 41–75.

Young, P. C, 1972, The relationship between the presence of larval anisakine nematodes in cod and marine mammals in British home waters. *J. Appl. Ecol.,* 9, 459–485.

Young, R. T, 1954, A note on the life cycle of *Lacistorhynchus tenuis* (Van Beneden, 1858), a cestode of the leopard shark. *Proc. Helminth. Soc. Wash.,* 21, 112.

Yousef, M., Mansour, N. S., Hammouda, N. A., Awdalla, H. N. and Boulos, L. M, 1981, Effect of freezing and grilling on *Pygidiopsis genata* metacercariae in *Tilapia. J. Egypt. Soc. Parasit.,* 11, 425–428.

Yu Yi and Wu Huisheng 1989, [Studies on the fauna of Acanthocephala of fishes from middle reaches of the Changjiang (Yangtze) river.] *Acta Hydrobiol. Sin.,* 13, 38–50. [In Chinese.]

Yunchis, O. N, 1977, [Some ecological factors determining the possibility of *Leuciscus rutilus* becoming infected with monogeneans.] *Akademiya Nauk SSSR, Zoologicheskii Institut.* 65–71. [In Russian.]

Yunchis, O. N., Nesterenko, V. N., Kononov, A. A. and Khokhlova, A. N, 1983, [Influence of some higher aquatic plants on the parasite fauna of roach fry.] *Sbornik Nauchnykh Trudov Gosudarstvennogo Nauchno-Issledovatel'skogo Instituta Ozernogo i Rechnogo Khozyaistva (Problemy ekologii parazitov ryb)* No. 197, 39–49. [In Russian.]

Yusufi, A. N. K. and Siddiqi, A. H, 1978, Some aspects of carbohydrate metabolism of digenetic trematodes from Indian water buffalo and catfish. *Z. ParasitKde.,* 56, 47–53.

Zaman, Z. and Leong, T. S, 1987, Seasonal occurrence of *Lytocestus lativitellarium* Furtado and Tan 1973 in *Clarias macrocephalus* Gunther in Kedah and Perak, Malaysia. *Aquaculture,* 63, 319–327.

Zaman, Z. and Leong, T. S, 1988, Some aspects of the biology of the caryophyllid cestode *Djombangia penetrans* (Bovien, 1926) in catfish *Clarias batrachus* (L.) from Kedah, Malaysia. *Malay. Nat. J.,* 41, 473–478.

Zdzitowiecki, K, 1979, Digenetic trematodes in alimentary tracts of fishes of South Georgia and South Shetlands (Antarctica). *Acta Ichthyol. Pisc.,* 9, 15–31.

Zdzitowiecki, K, 1987a, Digenetic trematodes from the alimentary tract of fishes off South Shetlands (Antarctic). *Acta Parasit. Pol.,* 32, 219–232.

Zdzitowiecki, K, 1987b, Acanthocephalans of marine fishes in the regions of South Georgia and South Orkneys (Antarctic). *Acta Parasit. Pol.,* 31, 211–217.

Zdzitowiecki, K, 1988, Occurrence of digenetic trematodes in fishes off South Shetlands (Antarctic). *Acta Parasit. Pol.,* 33, 155–167.

Zelazny, J, 1979, Influence of some physical agents and chemical compounds on hatchability and vitality of coracidia of *Bothriocephalus gowkongensis* Yeh, 1955. *Bull. Vet. Inst. Pulawy,* 23, 20–24.

Zelazny, J, 1980, [Therapeutic value of chosen preparations in the treatment of *Bothriocephalus gowkongensis* Yeh, 1955 in carp.] *Medycyna Wet.,* 36, 295–298. [In Polish.]

Zharikova, T. I, 1979, [Use of ammonium baths for the removal of dactylogyrids from fish.] *Biologiya Vnutrennikh Vod. Informatsionnyi Byulleten',* No. 44, 69–71. [In Russian.]

Zharikova, T. I, 1980, [Preliminary data on the horizontal and vertical movement of free-living larvae of *Dactylogyrus anchoratus* (Monogenea, Dactylogyridae).] *Parazitologiya,* 14, 514–515. [In Russian.]

Zharikova, T. I, 1981, [On the reaction of parasites from the genus *Dactylogyrus* Diesing, 1950 (Monogenoidea) to low temperatures.] In: *Ekologiya gel'mintov.* Yaroslavl', USSR: Yaroslavskii Gosudarstvennyi Universitet, pp. 24–28. [In Russian.]

Zharikova, T. I, 1984a, [Distribution of dactylogyrids in *Abramis brama* populations in relation to the host's sex.] *Ekologiya Gel'mintov (Voprosy ekologicheskoi gel'mintologii)* No. 4, 27–30. [In Russian.]

Zharikova, T. I, 1984b, [Infection of *Abramis brama* with monogeneans from the genus *Dactylogyrus* related to host sex.] *Zool Zh.,* 63, 1779–1784. [In Russian.]

Zharikova, T. J. and Flerov, B. A, 1981, [The effect of chlorophos on the susceptibility of fish to infection with species of *Dactylogyrus* Diesing, 1850 (Monogenoidea).] *Biol. Vnutr. Vod. Informatsi. Byull.,* 49, 43–46. [In Russian.]

Zharikova, T. I., Silkina, N. I. and Stepanova, M. A, 1980, [Dependence of the numbers of *Dactylogyrus* spp. (Dactylogyridae, Monogenea) on the immunophysiological state of the host *Carassius auratus* (L.).] *Dokl. Akad. Nauk SSSR,* 253, 510–512. [In Russian.]

Zhatkanbayeva, D, 1987, Cercaria caused diplostomosis of herbivorous fishes and experiences on the prevention methods in pond fish farms. *Actual Problems in Fish Parasitology; 2nd International symposium of Icthyoparasilology, Tihany, Hungary September 27–October 3, 1987*, p. 111.

Zhukov, E. V, 1976, [Specific features of the trematode fauna of coastal fish in Cuba (Gulf of Mexico).] In: *Probl. Zool. (Zool. Inst. A. N. SSSR). Leningrad. Leningrad, ussr: Izdatel. 'Nauka'*, pp. 37–39. [In Russian.]

Zhukov, E. V, 1985, The flatworm fauna of fishes of the Gulf of Mexico and its genetic relations. In: *Parasitology and pathology of marine organisms of the World Ocean.* (Ed. Hargis, W. J. Jr.) NOAA TECH. REP. NMFS 25 pp. 47–48.

Zietse, M. A., Van Den Broek, E. and Erwteman-Ooms, E. E. A, 1981, Studies on the life-cycle of *Asymphylodora tincae* Modeer, 1790 (Trematoda: Monorchiidae) in a small lake near Amsterdam. Part 2: the relations between *Asymphylodora tincae* and its definitive host, *Tinca tinca. J. Helminth.*, 55, 239–246.

Zitnan, R. and Cankovic, M, 1970/1971 Comparison of the epizootiological importance of the parasites of *Salmo gairdneri irideus* in the two coast areas of Bosna and Herzegovina. *Helminthologia*, 11, 161–166.

Zitnan, R. and Hanzelova, V, 1981, The seasonal dynamics of the invasion cycle of *Dactylogyrus extensus* Mueller et van Cleave 1932 (Monogenea). *Helminthologia*, **18**, 159–167.

Zitnan, R. and Hanzelova, V, 1987, Epizootiological importance of *Proteocephalus meglectus* La Rue 1911 (Cestoda). *Actual Problems in Fish Parasitology, Proceedings of the 2nd International symposium of Ichthyoparasitology. Tihany, Hungary.* p. 112.

Zitnan, R. and Hanzelova, V, 1984a, Negative effects of bothriocephalosis on weight gains in carp. (1982 publ.) *Folia Vet.*, 26, 173–181.

Zitnan, R. and Hanzelova, V, 1984b, Seasonal dynamics of the invasion cycle of *Gyrodactylus shulmani* Ling Mo-En, 1962 (Monogenea). (1982 publ.) *Folia Veterinaria*, 26, 183–194.

Zitnan, R., Hanzelova, V., Prihoda, J. and Kostan, B, 1981, [Evaluation of the efficacy of Taenifugin carp in the treatment of *Bothriocephalus* infections in carp at low water temperature.] *Biologizace e Chemizace Zivocisne vyroby, Veterinaria*, 17, 471–477.

Zmoray, I, 1980, A contribution to make helminthological terminology more precise. *Helminthologia*, 17, 57–62.

Zobel, H, 1975, [Mollusc-eating fish control fish diseases.] *Z. Binnenfisch.* DDR, 22, 150–151.

Zubchenko, A. V, 1980, Parasitic fauna of Anarhichadidae and Pleuronectidae families of fish in the Northwest Atlantic. *International Commission for Northwest Atlantic Fisheries. Selected papers.* 6, 41–46.

Zubchenko, A. V, 1981, [Use of parasitological data for the study of local groupings of *Coryphaenoides rupestris*.] *Tezisy dokladov sovetskikh uchastnikov. Leningrad*, USSR; Nauka, pp. 25–23. [In Russian.]

Zubchenko, A. V, 1984, [Ecological characteristics of the parasite fauna of some Alepocephalidae species.] In: *Ekologoparazitologicheskie issledovaniya severnykh morei.* Apatity, USSR: Kol'skii filial Akademii Nauk SSSR. pp. 77–81. [In Russian.]

Zubchenko, A. V, 1985, Use of parasitological data in studies of the local groupings of roch grenadier, *Coryphaenoides rupestris* Gunner. In: *Parasitology and pathology of marine organisms of the world ocean.* (Ed. J. Hargis, Jr.) NOAA Tech. Rep. NMFS, 25, 19–23.

Index